Handbook of Spine Technology

Boyle C. Cheng
Editor

Handbook of Spine Technology

Volume 1

With 425 Figures and 94 Tables

Editor
Boyle C. Cheng
Neuroscience Institute, Allegheny Health Network
Drexel University, Allegheny General Hospital Campus
Pittsburgh, PA, USA

ISBN 978-3-319-44423-9   ISBN 978-3-319-44424-6 (eBook)
ISBN 978-3-319-44425-3 (print and electronic bundle)
https://doi.org/10.1007/978-3-319-44424-6

© Springer Nature Switzerland AG 2021
All rights are reserved by the Publisher, whether the whole or part of the material is concerned, specifically the rights of translation, reprinting, reuse of illustrations, recitation, broadcasting, reproduction on microfilms or in any other physical way, and transmission or information storage and retrieval, electronic adaptation, computer software, or by similar or dissimilar methodology now known or hereafter developed.
The use of general descriptive names, registered names, trademarks, service marks, etc. in this publication does not imply, even in the absence of a specific statement, that such names are exempt from the relevant protective laws and regulations and therefore free for general use.
The publisher, the authors, and the editors are safe to assume that the advice and information in this book are believed to be true and accurate at the date of publication. Neither the publisher nor the authors or the editors give a warranty, expressed or implied, with respect to the material contained herein or for any errors or omissions that may have been made. The publisher remains neutral with regard to jurisdictional claims in published maps and institutional affiliations.

This Springer imprint is published by the registered company Springer Nature Switzerland AG.
The registered company address is: Gewerbestrasse 11, 6330 Cham, Switzerland

This book is dedicated to my parents, Samuel and Ruth, who were inspiring from an early age and, moreover, instilled within me the power of harnessing personal talent combined with a strong work ethic to achieve the best of all possible outcomes. In turn, I hope this to be a legacy for my wife, Judy and my two sons, Cooper and Jonathan who have in their own ways, motivated, encouraged, and supported this effort.

Preface

Historically, the excitement generated by a new technology for spine or an innovation in spinal interventions has been followed in relatively short order by sobering patient complications or broad-spectrum failures. Such catastrophes often necessitate salvage procedures and ultimately a dramatic decline in interest that ends in a failed pile of debris with a ruinous perception for the categorical technology that may last a generation or more. The disastrous scenarios have been repeated to the point of becoming frequent in spine technologies with very little evidence of slowing. Accordingly, the spine landscape is littered with the burned-out wreckages of abandoned technologies.

This book documents the fundamentals of spinal treatments, original design intent for spinal devices and the clinical outcomes of spine technologies. Often there is little more than tribal knowledge and even less information documenting the history of surgical approaches and supporting hardware. This is evident by the repeated failure modes for similar devices in databases and registries. The goal of this handbook is to provide a repository of information for both successful spine technologies as well as those with poor clinical outcomes and, moreover, the root cause of the failed spinal implants that contributed to the unsatisfactory results. If nothing else, this will serve the healthcare community by memorializing the history of spine technologies and prevent a repeat of the technological cycle that contributed to problematic patient outcomes. The specific aim of this book is to record from both a clinical and a scientific point of view what we have learned about the human spine and the influence of spine technologies that have contributed to the attempted treatment.

Important devices include those that failed catastrophically, for example, nucleus augmentation devices, to those that were ahead of the times but not necessarily a commercial success including motion preservation devices. Beyond the necessary surgical skills, patient selection has often been cited as essential to the commercialization of a device. Patient reported outcomes are a good proxy to the success of the device as well as a relative metric for regulatory purposes. The systematic diagnosis for patients presenting with back pain requires the most appropriate technology for the patient's symptoms. In one patient, immediate fixation and stabilization is the best solution and, in combination with the appropriate adjuncts to fusion, affords the patient the best opportunity for success. In other segments of the population with different

sets of symptoms, the solution may be a motion-preserving technology. The guidance and design rationale will help the audience understand the premise of each technology and, ultimately, how best the technology may be applied and in which patient population.

It is without fail that the lessons of the past can help mitigate potential disasters and even prevent another timely consuming and financially draining iteration. From engineers to clinical scientists, publications frequently discuss the most successful technologies or the most popular techniques. However, the most valuable lessons may be in failure. Failure may be attributable to the design of the device. It may also be a materials limitation. One failure often not discussed is surgeon error. Regardless, understanding the origin of the failure or root cause analysis is essential to future refinements.

Technologies that cover new materials, design failures, and even technologies developed for the sole purpose of finding an indication are presented. Often, there is no prior art, no reported case studies nor hints of potential complications attributable to a technology. Discovery of this information requires the courage to reflect on such mistakes and the willingness to share them. Documentation can lead to helpful preventative warnings resulting in the potential for reduced iterations and, ideally, failed products resulting in voluntary, or worse, mandated product market withdrawals. It is the goal of this handbook to put on display such failures so that we may advance technology for the patients benefit.

Boyle C. Cheng, Ph.D.

Acknowledgments

I want to acknowledge my professors and students that have planted the seeds throughout my career and, in particular, recognize those that have nurtured the growth through the fertile bed of their own experience and wisdom. This would include my colleagues, mentors, co-authors, and Michele Birgelen who worked tirelessly alongside me in completing the handbook and who epitomizes the definition of dedication.

Contents

Volume 1

Part I Low Back Pain Is a Point of View 1

1. **Low Back Patterns of Pain: Classification Based on Clinical Presentation** 3
   Hamilton Hall

2. **Back Pain: The Classic Surgeon's View** 27
   Neil Berrington

3. **Back Pain: Chiropractor's View** 37
   I. D. Coulter, M. J. Schneider, J. Egan, D. R. Murphy, Silvano A. Mior, and G. Jacob

4. **Chiropractors See It Differently: A Surgeon's Observations** 67
   John Street

5. **Medical Causes of Back Pain: The Rheumatologist's Perspective** 93
   Stephanie Gottheil, Kimberly Lam, David Salonen, and Lori Albert

6. **Psychosocial Impact of Chronic Back Pain: Patient and Societal Perspectives** 109
   Y. Raja Rampersaud

Part II Biomaterials and Biomechanics 125

7. **Implant Material Bio-compatibility, Sensitivity, and Allergic Reactions** 127
   Nadim James Hallab, Lauryn Samelko, and Marco Caicedo

8. **Mechanical Implant Material Selection, Durability, Strength, and Stiffness** 151
   Robert Sommerich, Melissa (Kuhn) DeCelle, and William J. Frasier

| 9 | **Material Selection Impact on Intraoperative Spine Manipulation and Post-op Correction Maintenance** | 163 |

Hesham Mostafa Zakaria and Frank La Marca

| 10 | **Biological Treatment Approaches for Degenerative Disc Disease: Injectable Biomaterials and Bioartificial Disc Replacement** | 171 |

Christoph Wipplinger, Yu Moriguchi, Rodrigo Navarro-Ramirez, Eliana Kim, Farah Maryam, and Roger Härtl

| 11 | **Bone Grafts and Bone Graft Substitutes** | 197 |

Jae Hyuk Yang, Juliane D. Glaeser, Linda E. A. Kanim, Carmen Y. Battles, Shrikar Bondre, and Hyun W. Bae

| 12 | **Mechanobiology of the Intervertebral Disc and Treatments Working in Conjunction with the Human Anatomy** | 275 |

Stephen Jaffee, Isaac R. Swink, Brett Phillips, Michele Birgelen, Alexander K. Yu, Nick Giannoukakis, Boyle C. Cheng, Scott Webb, Reginald Davis, William C. Welch, and Antonio Castellvi

| 13 | **Design Rationale for Posterior Dynamic Stabilization Relevant for Spine Surgery** | 293 |

Ashutosh Khandha, Jasmine Serhan, and Vijay K. Goel

| 14 | **Lessons Learned from Positive Biomechanics and Poor Clinical Outcomes** | 315 |

Deniz U. Erbulut, Koji Matsumoto, Anoli Shah, Anand Agarwal, Boyle C. Cheng, Ali Kiapour, Joseph Zavatsky, and Vijay K. Goel

| 15 | **Lessons Learned from Positive Biomechanics and Positive Clinical Outcomes** | 331 |

Isaac R. Swink, Stephen Jaffee, Jake Carbone, Hannah Rusinko, Daniel Diehl, Parul Chauhan, Kaitlyn DeMeo, and Thomas Muzzonigro

| 16 | **The Sacroiliac Joint: A Review of Anatomy, Biomechanics, Diagnosis, and Treatment Including Clinical and Biomechanical Studies (In Vitro and In Silico)** | 349 |

Amin Joukar, Hossein Elgafy, Anand K. Agarwal, Bradley Duhon, and Vijay K. Goel

Part III Considerations and Guidelines for New Technologies ... **375**

| 17 | **Cyclical Loading to Evaluate the Bone Implant Interface** | 377 |

Isaac R. Swink, Stephen Jaffee, Daniel Diehl, Chen Xu, Jake Carbone, Alexander K. Yu, and Boyle C. Cheng

18	**FDA Premarket Review of Orthopedic Spinal Devices** Katherine Kavlock, Srinidhi Nagaraja, and Jonathan Peck	401
19	**Recent Advances in PolyArylEtherKetones and Their In Vitro Evaluation for Hard Tissue Applications** Boyle C. Cheng, Alexander K. Yu, Isaac R. Swink, Donald M. Whiting, and Saadyah Averick	423
20	**Selection of Implant Material Effect on MRI Interpretation in Patients** Ashok Biyani, Deniz U. Erbulut, Vijay K. Goel, Jasmine Tannoury, John Pracyk, and Hassan Serhan	439
21	**Metal Ion Sensitivity** William M. Mihalko and Catherine R. Olinger	459
22	**Spinal Cord Stimulation: Effect on Motor Function in Parkinson's Disease** Nestor D. Tomycz, Timothy Leichliter, Saadyah Averick, Boyle C. Cheng, and Donald M. Whiting	473
23	**Intraoperative Monitoring in Spine Surgery** Julian Michael Moore	483
24	**Oncological Principles** A. Karim Ahmed, Zach Pennington, Camilo A. Molina, and Daniel M. Sciubba	505
25	**Bone Metabolism** Paul A. Anderson	523

Part IV Technology: Fusion 539

26	**Pedicle Screw Fixation** Nickul S. Jain and Raymond J. Hah	541
27	**Interspinous Devices** Douglas G. Orndorff, Anneliese D. Heiner, and Jim A. Youssef	561
28	**Kyphoplasty Techniques** Scott A. Vincent, Emmett J. Gannon, and Don K. Moore	573
29	**Anterior Spinal Plates: Cervical** A. Karim Ahmed, Zach Pennington, Camilo A. Molina, C. Rory Goodwin, and Daniel M. Sciubba	593
30	**Spinal Plates and the Anterior Lumbar Interbody Arthrodesis** Zach Pennington, A. Karim Ahmed, and Daniel M. Sciubba	603
31	**Interbody Cages: Cervical** John Richards, Donald R. Fredericks Jr., Sean E. Slaven, and Scott C. Wagner	633

32	**Anterior Lumbar Interbody Fusion and Transforaminal Lumbar Interbody Fusion** 645
	Tristan B. Fried, Tyler M. Kreitz, and I. David Kaye

33	**Scoliosis Instrumentation Systems** 657
	Rajbir Singh Hundal, Mark Oppenlander, Ilyas Aleem, and Rakesh Patel

34	**SI Joint Fixation** 675
	J. Loewenstein, W. Northam, D. Bhowmick, and E. Hadar

35	**Lateral Lumbar Interbody Fusion** 689
	Paul Page, Mark Kraemer, and Nathaniel P. Brooks

36	**Minimally Invasive Spine Surgery** 701
	Bilal B. Butt, Rakesh Patel, and Ilyas Aleem

37	**Cervical Spine Anatomy** 717
	Bobby G. Yow, Andres S. Piscoya, and Scott C. Wagner

38	**Thoracic and Lumbar Spinal Anatomy** 737
	Patricia Zadnik Sullivan, Michael Spadola, Ali K. Ozturk, and William C. Welch

Volume 2

Part V Technology: Motion Preservation 747

39	**Cervical Total Disc Replacement: FDA-Approved Devices** 749
	Catherine Miller, Deepak Bandlish, Puneet Gulati, Santan Thottempudi, Domagoj Coric, and Praveen Mummaneni

40	**Cervical Total Disc Replacement: Next-Generation Devices** 761
	Tyler M. Kreitz, James McKenzie, Safdar Khan, and Frank M. Phillips

41	**Cervical Total Disc Replacement: Evidence Basis** 771
	Kris E. Radcliff, Daniel A. Tarazona, Michael Markowitz, and Edwin Theosmy

42	**Cervical Total Disc Replacement: Biomechanics** 789
	Joseph D. Smucker and Rick C. Sasso

43	**Cervical Total Disc Replacement: Technique – Pitfalls and Pearls** 807
	Miroslav Vukic and Sergej Mihailovic Marasanov

44	**Cervical Total Disc Replacement: Expanded Indications** ... 823
	Pierce D. Nunley

45	**Cervical Total Disc Replacement: Heterotopic Ossification and Complications**	829
	Michael Paci and Michael Y. Wang	
46	**Lumbar TDR Revision Strategies**	837
	Paul C. McAfee and Mark Gonz	
47	**Posterior Lumbar Facet Replacement and Interspinous Spacers**	845
	Taylor Beatty, Michael Venezia, and Scott Webb	
48	**Cervical Arthroplasty: Long-Term Outcomes**	857
	Thomas J. Buell and Mark E. Shaffrey	
49	**Adjacent-Level Disease: Fact and Fiction**	885
	Jonathan Parish and Domagoj Coric	
50	**Posterior Dynamic Stabilization**	893
	Dorian Kusyk, Chen Xu, and Donald M. Whiting	
51	**Total Disc Arthroplasty**	899
	Benjamin Ebben and Miranda Bice	

Part VI International Experience: Surgery 923

52	**The Diagnostic and the Therapeutic Utility of Radiology in Spinal Care**	925
	Matthew Lee and Mario G. T. Zotti	
53	**Surgical Site Infections in Spine Surgery: Prevention, Diagnosis, and Treatment Using a Multidisciplinary Approach**	949
	Matthew N. Scott-Young, Mario G. T. Zotti, and Robert G. Fassett	
54	**Lumbar Interbody Fusion Devices and Approaches: When to Use What**	961
	Laurence P. McEntee and Mario G. T. Zotti	
55	**Stand-Alone Interbody Devices: Static Versus Dynamic**	997
	Ata G. Kasis	
56	**Allograft Use in Modern Spinal Surgery**	1009
	Matthew N. Scott-Young and Mario G. T. Zotti	
57	**Posterior Approaches to the Thoracolumbar Spine: Open Versus MISS**	1029
	Yingda Li and Andrew Kam	
58	**Lateral Approach to the Thoracolumbar Junction: Open and MIS Techniques**	1051
	Mario G. T. Zotti, Laurence P. McEntee, John Ferguson, and Matthew N. Scott-Young	

59	**Surgical Approaches to the Cervical Spine: Principles and Practicalities** .... Cyrus D. Jensen	1067
60	**Intradiscal Therapeutics for Degenerative Disc Disease** .... Justin Mowbray, Bojiang Shen, and Ashish D. Diwan	1091
61	**Replacing the Nucleus Pulposus for Degenerative Disc Disease and Disc Herniation: Disc Preservation Following Discectomy** .... Uphar Chamoli, Maurice Lam, and Ashish D. Diwan	1111
62	**Spinal Fusion Evaluation in Various Settings: A Summary of Human-Only Studies** .... Jose Umali, Ali Ghahreman, and Ashish D. Diwan	1131
63	**Effects of Reimbursement and Regulation on the Delivery of Spinal Device Innovation and Technology: An Industry Perspective** .... Emma Young	1149
64	**Anterior Lumbar Spinal Reconstruction** .... Matthew N. Scott-Young, David M. Grosser, and Mario G. T. Zotti	1165

Part VII Challenges and Lessons from Commercializing Products 1209

65	**Approved Products in the USA: AxiaLIF** .... Franziska Anna Schmidt, Raj Nangunoori, Taylor Wong, Sertac Kirnaz, and Roger Härtl	1211
66	**Spine Products in Use Both Outside and Inside the United States** .... Tejas Karnati, Kee D. Kim, and Julius O. Ebinu	1217
67	**Trauma Products: Spinal Cord Injury Implants** .... Gilbert Cadena Jr., Jordan Xu, and Angie Zhang	1229
68	**Biologics: Inherent Challenges** .... Charles C. Lee and Kee D. Kim	1251
69	**Robotic Technology** .... Kyle J. Holmberg, Daniel T. Altman, Boyle C. Cheng, and Timothy J. Sauber	1269

Index 1283

About the Section Editors

Part I: Low Back Pain Is a Point of View
Hamilton Hall Department of Surgery, University of Toronto, Toronto, ON, Canada

Part II: Biomaterials and Biomechanics
Hassan Serhan I.M.S. Society, Easten, MA, USA

Tony Tannoury Department of Orthopedics, Boston University Medical Center, Boston, MA, USA

Part III: Considerations and Guidelines for New Technologies
Boyle C. Cheng Neuroscience Institute, Allegheny Health Network, Drexel University, Allegheny General Hospital Campus, Pittsburgh, PA, USA

Vijay K. Goel University of Toledo, Engineering Center for Orthopaedic Research Excellence (E-CORE), Toledo, OH, USA

Departments of Bioengineering and Orthopaedic Surgery, Colleges of Engineering and Medicine, University of Toledo, Toledo, OH, USA

Part IV: Technology: Fusion
Don K. Moore Department of Orthopaedic Surgery, University of Missouri Health Care, Columbia, OH, USA

William C. Welch Department of Neurosurgery, University of Pennsylvania, Philadelphia, PA, USA

Part V: Technology: Motion Preservation
Domagoj Coric Department of Neurological Surgery, Carolinas Medical Center and Carolina Neurosurgery and Spine Associates, Charlotte, NC, USA

Part VI: International Experience: Surgery
Matthew N. Scott-Young Gold Coast Spine, Southport, QLD, Australia

Faculty of Health Sciences and Medicine, Bond University, Varsity Lakes, QLD, Australia

Part VII: Challenges and Lessons from Commercializing Products
R. Douglas Orr S40 Cleveland Clinic, Center for Spine Health, Cleveland, OH, USA

Michael Y. Oh Department of Neurosurgery + Academic Services, Allegheny Health Network, Pittsburgh, PA, USA

Contributors

Anand K. Agarwal Engineering Center for Orthopaedic Research Excellence (E-CORE), University of Toledo, Toledo, OH, USA

Anand Agarwal Department of Orthopaedic Surgery and Bioengineering, School of Engineering and Medicine, University of Toledo, Toledo, OH, USA

A. Karim Ahmed Department of Neurosurgery, The Johns Hopkins School of Medicine, Baltimore, MD, USA

Lori Albert Rheumatology Faculty, University of Toronto, Toronto, ON, Canada

Ilyas Aleem Department of Orthopaedic Surgery, University of Michigan, Ann Arbor, MI, USA

Daniel T. Altman Department of Orthopaedic Surgery, Allegheny Health Network, Pittsburgh, PA, USA

Paul A. Anderson Department of Orthopedic Surgery and Rehabilitation, University of Wisconsin, Madison, WI, USA

Saadyah Averick Department of Neurosurgery, Neuroscience Institute, Allegheny Health Network, Pittsburgh, PA, USA

Hyun W. Bae Surgery, Department of Orthopaedics, Cedars-Sinai Medical Center, Los Angeles, CA, USA

Board of Governors Regenerative Medicine Institute, Cedars-Sinai Medical Center, Los Angeles, CA, USA

Department of Surgery, Cedars-Sinai Spine Center, Los Angeles, CA, USA

Deepak Bandlish Department of Neurological Surgery, SBKS Medical College, Vadodara, India

Carmen Y. Battles Surgery, Department of Orthopaedics, Cedars-Sinai Medical Center, Los Angeles, CA, USA

Department of Surgery, Cedars-Sinai Spine Center, Los Angeles, CA, USA

Taylor Beatty Orthopaedic Surgery Resident PGY5, Largo Medical Center, Largo, FL, USA

Neil Berrington Section of Neurosurgery, University of Manitoba, Winnipeg, MB, Canada

D. Bhowmick Department of Neurosurgery, University of North Carolina, Chapel Hill, NC, USA

Miranda Bice University of Wisconsin School of Medicine and Public Health, Madison, WI, USA

Michele Birgelen Department of Neurosurgery, Neuroscience Institute, Allegheny Health Network, Pittsburgh, PA, USA

Ashok Biyani ProMedica Physicians Biyani Orthopaedics, Toledo, OH, USA

Shrikar Bondre Chemical Engineering, Prosidyan, Warren, NJ, USA

Nathaniel P. Brooks Department of Neurological Surgery, University of Wisconsin, Madison, WI, USA

Thomas J. Buell Department of Neurological Surgery, University of Virginia Health System, Charlottesville, VA, USA

Bilal B. Butt Department of Orthopaedic Surgery, University of Michigan, Ann Arbor, MI, USA

Gilbert Cadena Jr. Department of Neurological Surgery, University of California Irvine, Orange, CA, USA

Marco Caicedo Department of Orthopedic Surgery, Rush University Medical Center, Chicago, IL, USA

Jake Carbone Louis Katz School of Medicine, Temple University, Philadelphia, PA, USA

Department of Neurosurgery, Allegheny Health Network, Pittsburgh, PA, USA

Antonio Castellvi Orthopaedic Research and Education, Florida Orthopaedic Institute, Tampa, FL, USA

Uphar Chamoli Spine Service, Department of Orthopaedic Surgery, St. George & Sutherland Clinical School, University of New South Wales, Kogarah, NSW, Australia

School of Biomedical Engineering, Faculty of Engineering and Information Technology, University of Technology Sydney, Sydney, NSW, Australia

Parul Chauhan Department of Neurosurgery, Neuroscience Institute, Allegheny Health Network, Pittsburgh, PA, USA

Boyle C. Cheng Neuroscience Institute, Allegheny Health Network, Drexel University, Allegheny General Hospital Campus, Pittsburgh, PA, USA

Domagoj Coric Department of Neurological Surgery, Carolinas Medical Center and Carolina Neurosurgery and Spine Associates, Charlotte, NC, USA

I. D. Coulter RAND Corporation, Santa Monica, CA, USA

Reginald Davis BioSpine, Tampa, FL, USA

Melissa (Kuhn) DeCelle Research and Development, DePuy Synthes Spine, Raynham, MA, USA

Kaitlyn DeMeo Department of Neurosurgery, Allegheny Health Network, Pittsburgh, PA, USA

Daniel Diehl Department of Neurosurgery, Neuroscience Institute, Allegheny Health Network, Pittsburgh, PA, USA

Ashish D. Diwan Spine Service, Department of Orthopaedic Surgery, St. George & Sutherland Clinical School, University of New South Wales, Kogarah, NSW, Australia

Bradley Duhon School of Medicine, University of Colorado, Denver, CO, USA

Benjamin Ebben University of Wisconsin, Madison, WI, USA

Julius O. Ebinu Department of Neurological Surgery, University of California, Davis, Sacramento, CA, USA

J. Egan Southern California University of Health Sciences, Whittier, CA, USA

Hossein Elgafy Engineering Center for Orthopaedic Research Excellence (E-CORE), University of Toledo, Toledo, OH, USA

Deniz U. Erbulut Departments of Bioengineering and Orthopaedic Surgery, Colleges of Engineering and Medicine, University of Toledo, Toledo, OH, USA

Robert G. Fassett Faculty of Health Sciences and Medicine, Bond University, Gold Coast, QLD, Australia

Schools of Medicine and Human Movement and Nutrition Sciences, The University of Queensland, St Lucia, QLD, Australia

John Ferguson Ascot Hospital, Remuera, Auckland, New Zealand

William J. Frasier Research and Development, DePuy Synthes Spine, Raynham, MA, USA

Donald R. Fredericks Jr. Department of Orthopaedics, Walter Reed National Military Medical Center, Bethesda, MD, USA

Tristan B. Fried Sidney Kimmel Medical College, Thomas Jefferson University, Philadelphia, PA, USA

Emmett J. Gannon Department of Orthopaedic Surgery and Rehabilitation, University of Nebraska Medical Center, Omaha, NE, USA

Ali Ghahreman Department of Neurosurgery, St. George Hospital and Clinical School, Kogarah, NSW, Australia

Nick Giannoukakis Institute of Cellular Therapeutics, Allegheny Health Network, Pittsburgh, PA, USA

Juliane D. Glaeser Surgery, Department of Orthopaedics, Cedars-Sinai Medical Center, Los Angeles, CA, USA

Board of Governors Regenerative Medicine Institute, Cedars-Sinai Medical Center, Los Angeles, CA, USA

Department of Surgery, Cedars-Sinai Spine Center, Los Angeles, CA, USA

Vijay K. Goel Engineering Center for Orthopaedic Research Excellence (E-CORE), University of Toledo, Toledo, OH, USA

Departments of Bioengineering and Orthopaedic Surgery, Colleges of Engineering and Medicine, University of Toledo, Toledo, OH, USA

Mark Gonz Vascular Surgery Associates, Towson, MD, USA

C. Rory Goodwin Department of Neurosurgery, Duke University Medical Center, Durham, NC, USA

Stephanie Gottheil University of Toronto, Toronto, ON, Canada

David M. Grosser Southern Queensland Cardiovascular Centre, Southport, QLD, Australia

Puneet Gulati Department of Neurological Surgery, Maulana Azad Medical College and Lok Nayak Hospital, New Delhi, India

E. Hadar Department of Neurosurgery, University of North Carolina, Chapel Hill, NC, USA

Raymond J. Hah Department of Orthopaedic Surgery, University of Southern California, Los Angeles, CA, USA

Hamilton Hall Department of Surgery, University of Toronto, Toronto, ON, Canada

Nadim James Hallab Department of Orthopedic Surgery, Rush University Medical Center, Chicago, IL, USA

Roger Härtl Department of Neurological Surgery, Weill Cornell Brain and Spine Center, New York–Presbyterian Hospital, Weill Cornell Medicine, New York, NY, USA

Anneliese D. Heiner Penumbra, Inc., Alameda, CA, USA

Kyle J. Holmberg Department of Orthopaedic Surgery, Allegheny Health Network, Pittsburgh, PA, USA

G. Jacob Southern California University of Health Sciences, Whittier, CA, USA

Stephen Jaffee College of Medicine, Drexel University, Philadelphia, PA, USA

Allegheny Health Network, Department of Neurosurgery, Allegheny General Hospital, Pittsburgh, PA, USA

Nickul S. Jain Department of Orthopaedic Surgery, University of Southern California, Los Angeles, CA, USA

Cyrus D. Jensen Department of Trauma and Orthopaedic Spine Surgery, Northumbria Healthcare NHS Foundation Trust, Newcastle upon Tyne, UK

Amin Joukar Engineering Center for Orthopaedic Research Excellence (E-CORE), University of Toledo, Toledo, OH, USA

Andrew Kam Department of Neurosurgery, Westmead Hospital, Sydney, NSW, Australia

Linda E. A. Kanim Surgery, Department of Orthopaedics, Cedars-Sinai Medical Center, Los Angeles, CA, USA

Board of Governors Regenerative Medicine Institute, Cedars-Sinai Medical Center, Los Angeles, CA, USA

Department of Surgery, Cedars-Sinai Spine Center, Los Angeles, CA, USA

Tejas Karnati Department of Neurological Surgery, University of California, Davis, Sacramento, CA, USA

Ata G. Kasis Northumbria NHS Trust, UK and Nuffield Hospital, Newcastle-upon-Tyne, UK

Katherine Kavlock Center for Devices and Radiological Health, Food and Drug Administration, Silver Spring, MD, USA

I. David Kaye Rothman Institute, Thomas Jefferson University Hospital, Philadelphia, PA, USA

Safdar Khan Division of Spine Surgery, Department of Orthopedic Surgery, Ohio State University, Columbus, OH, USA

Ashutosh Khandha Department of Biomedical Engineering, College of Engineering, University of Delaware, Newark, DE, USA

Ali Kiapour Departments of Bioengineering and Orthopaedic Surgery, Colleges of Engineering and Medicine, University of Toledo, Toledo, OH, USA

Eliana Kim Department of Neurological Surgery, Weill Cornell Brain and Spine Center, New York–Presbyterian Hospital, New York, NY, USA

Kee D. Kim Department of Neurological Surgery, UC Davis School of Medicine, Sacramento, CA, USA

Sertac Kirnaz Department of Neurological Surgery, Weill Cornell Brain and Spine Center, Weill Cornell Medicine, New York, NY, USA

Mark Kraemer Department of Neurological Surgery, University of Wisconsin, Madison, WI, USA

Tyler M. Kreitz Thomas Jefferson University Hospital, Philadelphia, PA, USA

Dorian Kusyk Department of Neurosurgery, Neuroscience Institute, Allegheny Health Network, Pittsburgh, PA, USA

Frank La Marca Department of Neurosurgery, Henry Ford Hospital, Detroit, MI, USA

Department of Neurosurgery, Henry Ford Allegiance Hospital, Jackson, MI, USA

Kimberly Lam University of Toronto, Toronto, ON, Canada

Maurice Lam Spine Service, Department of Orthopaedic Surgery, St. George & Sutherland Clinical School, University of New South Wales, Kogarah, NSW, Australia

Charles C. Lee Department of Cell Biology and Human Anatomy, School of Medicine, University of California, Davis, Davis, CA, USA

Matthew Lee Western Imaging Group, Blacktown, NSW, Australia

Timothy Leichliter Department of Neurology, Neuroscience Institute, Allegheny Health Network, Pittsburgh, PA, USA

Yingda Li Department of Neurosurgery, Westmead Hospital, Sydney, NSW, Australia

Department of Neurological Surgery, University of Miami, Miami, FL, USA

J. Loewenstein Department of Neurosurgery, University of North Carolina, Chapel Hill, NC, USA

Sergej Mihailovic Marasanov Department of Neurosurgery, University Hospital Rebro, Zagreb, Croatia

Michael Markowitz Department of Orthopaedics, Rowan University School of Osteopathic Medicine, Stratford, NJ, USA

Farah Maryam Department of Neurological Surgery, Weill Cornell Brain and Spine Center, New York–Presbyterian Hospital, New York, NY, USA

Koji Matsumoto Department of Orthopedic Surgery, Nihon University School of Medicine, Tokyo, Japan

Paul C. McAfee Spine and Scoliosis Center, University of Maryland St Joseph Medical Center (UMSJMC), Towson, MD, USA

Laurence P. McEntee Gold Coast Spine, Southport, QLD, Australia

Bond University, Varsity Lakes, QLD, Australia

James McKenzie Department of Orthopaedic Surgery, Thomas Jefferson University, Philadelphia, PA, USA

William M. Mihalko Campbell Clinic Department of Orthopaedic Surgery and Biomedical Engineering, University of Tennessee Health Science Center, Memphis, TN, USA

Catherine Miller Department of Neurological Surgery, University of California-San Francisco, San Francisco, CA, USA

Silvano A. Mior Department of Research, Canadian Memorial Chiropractic College, Toronto, ON, Canada

Centre for Disability Prevention and Rehabilitation, Ontario Tech University and Canadian Memorial Chiropractic College, Toronto, ON, USA

Camilo A. Molina Department of Neurosurgery, The Johns Hopkins School of Medicine, Baltimore, MD, USA

Don K. Moore Department of Orthopaedic Surgery, University of Missouri, Columbia, OH, USA

Julian Michael Moore School of Kinesiology, University of Michigan, Ann Arbor, MI, USA

Department of Neurology, University of Michigan, Ann Arbor, MI, USA

Yu Moriguchi Department of Neurological Surgery, Weill Cornell Brain and Spine Center, New York–Presbyterian Hospital, New York, NY, USA

Hesham Mostafa Zakaria Department of Neurosurgery, Henry Ford Hospital, Detroit, MI, USA

Department of Neurosurgery, Henry Ford Allegiance Hospital, Jackson, MI, USA

Justin Mowbray Spine Service, Department of Orthopaedic Surgery, St George and Sutherland Clinical School, The University of New South Wales, Kogarah, NSW, Australia

Praveen Mummaneni Department of Neurological Surgery, University of California-San Francisco, San Francisco, CA, USA

D. R. Murphy Department of Family Medicine, Alpert Medical School of Brown University, Providence, RI, USA

Department of Physical Therapy, University of Pittsburgh, Cranston, RI, USA

Thomas Muzzonigro Department of Neurosurgery, Neuroscience Institute, Allegheny Health Network, Pittsburgh, PA, USA

Srinidhi Nagaraja Center for Devices and Radiological Health, Food and Drug Administration, Silver Spring, MD, USA

Raj Nangunoori Department of Neurological Surgery, Weill Cornell Brain and Spine Center, Weill Cornell Medicine, New York, NY, USA

Rodrigo Navarro-Ramirez Department of Neurological Surgery, Weill Cornell Brain and Spine Center, New York–Presbyterian Hospital, New York, NY, USA

W. Northam Department of Neurosurgery, University of North Carolina, Chapel Hill, NC, USA

Pierce D. Nunley Spine Institute of Louisiana, Shreveport, LA, USA

Catherine R. Olinger Campbell Clinic Department of Orthopaedic Surgery and Biomedical Engineering, University of Tennessee Health Science Center, Memphis, TN, USA

Mark Oppenlander Department of Orthopaedic Surgery, University of Michigan, Ann Arbor, MI, USA

Douglas G. Orndorff Spine Colorado, Durango, CO, USA

Ali K. Ozturk Department of Neurosurgery, University of Pennsylvania, Philadelphia, PA, USA

Michael Paci Department of Neurological Surgery, University of Miami Miller School of Medicine, Miami, FL, USA

Paul Page Department of Neurological Surgery, University of Wisconsin, Madison, WI, USA

Jonathan Parish Department of Neurological Surgery, Carolinas Medical Center, Charlotte, NC, USA

Carolina Neurosurgery and Spine Associates, Charlotte, NC, USA

Rakesh Patel Department of Orthopaedic Surgery, University of Michigan, Ann Arbor, MI, USA

Jonathan Peck Center for Devices and Radiological Health, Food and Drug Administration, Silver Spring, MD, USA

Zach Pennington Department of Neurosurgery, Johns Hopkins Hospital, The Johns Hopkins School of Medicine, Baltimore, MD, USA

Brett Phillips Institute of Cellular Therapeutics, Allegheny Health Network, Pittsburgh, PA, USA

Frank M. Phillips Division of Spine Surgery, Rush University Medical Center, Chicago, IL, USA

Andres S. Piscoya Department of Orthopaedic Surgery, Walter Reed National Military Medical Center, Bethesda, MD, USA

John Pracyk DePuy Synthes Spine, Raynham, MA, USA

Kris E. Radcliff Department of Orthopaedic Surgery, The Rothman Institute, Thomas Jefferson University, Philadelphia, PA, USA

Y. Raja Rampersaud Arthritis Program, Toronto Western Hospital, University Health Network (UHN), Toronto, ON, Canada

Department of Surgery, Division of Orthopaedic Surgery, University of Toronto, Toronto, ON, Canada

John Richards Department of Orthopaedics, Walter Reed National Military Medical Center, Bethesda, MD, USA

Hannah Rusinko Neuroscience Institute, Allegheny Health Network, Pittsburgh, PA, USA

David Salonen University of Toronto, Toronto, ON, Canada

Lauryn Samelko Department of Orthopedic Surgery, Rush University Medical Center, Chicago, IL, USA

Rick C. Sasso Indiana Spine Group, Carmel, IN, USA

Timothy J. Sauber Department of Orthopaedic Surgery, Allegheny Health Network, Pittsburgh, PA, USA

Franziska Anna Schmidt Department of Neurological Surgery, Weill Cornell Brain and Spine Center, Weill Cornell Medicine, New York, NY, USA

M. J. Schneider School of Health and Rehabilitation Sciences, Clinical and Translational Science Institute, University of Pittsburgh, Pittsburgh, PA, USA

Daniel M. Sciubba Department of Neurosurgery, Johns Hopkins Hospital, The Johns Hopkins School of Medicine, Baltimore, MD, USA

Matthew N. Scott-Young Faculty of Health Sciences and Medicine, Bond University, Gold Coast, QLD, Australia

Gold Coast Spine, Southport, QLD, Australia

Jasmine Serhan Department of Biological Sciences, Bridgewater State University, Bridgewater, MA, USA

Hassan Serhan I.M.S. Society, Easten, MA, USA

Mark E. Shaffrey Department of Neurological Surgery, University of Virginia Health System, Charlottesville, VA, USA

Anoli Shah Engineering Center for Orthopaedic Research Excellence (E-CORE), Departments of Bioengineering and Orthopaedic Surgery, Colleges of Engineering and Medicine, University of Toledo, Toledo, OH, USA

Bojiang Shen Spine Service, Department of Orthopaedic Surgery, St George and Sutherland Clinical School, The University of New South Wales, Kogarah, NSW, Australia

Rajbir Singh Hundal Department of Orthopaedic Surgery, University of Michigan, Ann Arbor, MI, USA

Sean E. Slaven Department of Orthopaedics, Walter Reed National Military Medical Center, Bethesda, MD, USA

Joseph D. Smucker Indiana Spine Group, Carmel, IN, USA

Robert Sommerich Research and Development, DePuy Synthes Spine, Raynham, MA, USA

Michael Spadola Department of Neurosurgery, University of Pennsylvania, Philadelphia, PA, USA

John Street Division of Spine, Department of Orthopedics, University of British Columbia, Vancouver, BC, Canada

Patricia Zadnik Sullivan Department of Neurosurgery, University of Pennsylvania, Philadelphia, PA, USA

Isaac R. Swink Department of Neurosurgery, Neuroscience Institute, Allegheny Health Network, Pittsburgh, PA, USA

Jasmine Tannoury Boston University, Boston, MA, USA

Daniel A. Tarazona Department of Orthopaedic Surgery, The Rothman Institute, Thomas Jefferson University, Philadelphia, PA, USA

Edwin Theosmy Department of Orthopaedics, Rowan University School of Osteopathic Medicine, Stratford, NJ, USA

Santan Thottempudi University of California – Santa Cruz, Santa Cruz, CA, USA

Nestor D. Tomycz Department of Neurosurgery, Neuroscience Institute, Allegheny Health Network, Pittsburgh, PA, USA

Jose Umali Spine Service, Department of Orthopaedic Surgery, St. George Hospital and Clinical School, Kogarah, NSW, Australia

Michael Venezia Orthopaedic Specialists of Tampa Bay, Clearwater, FL, USA

Scott A. Vincent Department of Orthopaedic Surgery and Rehabilitation, University of Nebraska Medical Center, Omaha, NE, USA

Miroslav Vukic Department of Neurosurgery, University Hospital Rebro, Zagreb, Croatia

Scott C. Wagner Department of Orthopaedic Surgery, Walter Reed National Military Medical Center, Bethesda, MD, USA

Michael Y. Wang Department of Neurological Surgery, University of Miami Miller School of Medicine, Miami, FL, USA

Scott Webb Florida Spine Institute, Clearwater, Tampa, FL, USA

William C. Welch Department of Neurosurgery, University of Pennsylvania, Philadelphia, PA, USA

Donald M. Whiting Neuroscience Institute, Allegheny Health Network, Pittsburgh, PA, USA

Christoph Wipplinger Department of Neurological Surgery, Weill Cornell Brain and Spine Center, New York–Presbyterian Hospital, New York, NY, USA

Department of Neurosurgery, Medical University of Innsbruck, Innsbruck, Austria

Taylor Wong Department of Neurological Surgery, Weill Cornell Brain and Spine Center, Weill Cornell Medicine, New York, NY, USA

Chen Xu Department of Neurosurgery, Neuroscience Institute, Allegheny Health Network, Pittsburgh, PA, USA

Jordan Xu Department of Neurological Surgery, University of California Irvine, Orange, CA, USA

Jae Hyuk Yang Surgery, Department of Orthopaedics, Cedars-Sinai Medical Center, Los Angeles, CA, USA

Korea University Guro Hospital, Seoul, South Korea

Emma Young Prism Surgical Designs Pty Ltd, Brisbane, QLD, Australia

Jim A. Youssef Spine Colorado, Durango, CO, USA

Bobby G. Yow Department of Orthopaedic Surgery, Walter Reed National Military Medical Center, Bethesda, MD, USA

Alexander K. Yu Department of Neurosurgery, Neuroscience Institute, Allegheny Health Network, Pittsburgh, PA, USA

Joseph Zavatsky Spine & Scoliosis Specialists, Tampa, FL, USA

Angie Zhang Department of Neurological Surgery, University of California Irvine, Orange, CA, USA

Mario G. T. Zotti Orthopaedic Clinics Gold Coast, Robina, QLD, Australia

Gold Coast Spine, Southport, QLD, Australia

Part I
Low Back Pain Is a Point of View

Low Back Patterns of Pain: Classification Based on Clinical Presentation

Hamilton Hall

Contents

Introduction	4
Classification Options	5
Treatment Response	5
Time Based	5
Administrative	5
Risk of Chronicity	6
Anatomic	6
Nonspecific	6
A Syndrome Approach	7
Four Patterns of Pain	7
History	9
Question One	9
Question Two	10
Question Three	10
Question Four	10
Question Five	11
Question Six	11
Questions Seven and Eight	11
Question Nine	12
Physical Examination	12
Observation	13
Movement	13
Prone Passive Extension	13
Nerve Root Irritation	14
Conduction Deficit	15
Upper Motor Neuron Involvement	15
Saddle Sensation	16
Additional Tests	16

H. Hall (✉)
Department of Surgery, University of Toronto, Toronto, ON, Canada
e-mail: hhall@cbi.ca

© Springer Nature Switzerland AG 2021
B. C. Cheng (ed.), *Handbook of Spine Technology*,
https://doi.org/10.1007/978-3-319-44424-6_136

Pattern Identification	16
Pattern 1	16
Pattern 1 PEP	17
Pattern 1 PEN	17
Pattern 2	18
Pattern 3	18
Pattern 4	19
Pattern 4 FA	19
Pattern 4 FR	19
Pattern Directed Care	20
Principles of Nonsurgical Management	21
Pattern 1 PEP	21
Pattern 1 PEN	22
Pattern 2	23
Pattern 3	23
Pattern 4 FA	24
Pattern 4 FR	24
References	24

Abstract

The ubiquitous presence of low back pain with its multiple natural histories makes classification difficult. Any categorization begins by defining the essential elements of the problem to build a structure that reflects the values of the organizer, values determined by experience, personal concerns, and a point of view. Although a grouping of back pain patients based on responses to a particular treatment may be as valid as one based upon the varying degrees of socioeconomic impact produced by the pain, any classification's ultimate value depends on the interests of the user. Patterns of pain focuses on the initial presentation delineated by a specific set of questions in the history and confirmed by selected features of the physical examination. History divides mechanical low back pain into four distinct syndromes while the physical examination further delineates two of these patterns. A pattern of mechanical low back pain can be defined by the location of the dominant pain (back or leg), the consistency (constant or truly intermittent), and the effect of flexion on the symptoms. Response to flexion separates two cohorts of intermittent leg dominant pain patients with very different clinical scenarios and treatment demands. The physical examination divides back dominant pain patients needing only a straightforward treatment strategy from those who require more complex supervision. Additional questions and tests highlight or eliminate sinister, nonmechanical pathologies. The classification both directs initial management and provides a reasonable prognosis for speed and completeness of recovery.

Keywords

Mechanical low back pain · History · Physical examination · Classification · Clinical presentation · Patterns of pain · Referred pain · Radicular pain

Introduction

Low back pain is a human condition. Virtually everyone will, at some time in their lives, suffer pain in the lower back. Those that remain permanently pain free are the exception. Numerous studies have reported a lifetime incidence over 80% (Balagué et al. 2012). Nearly all will suffer from symptoms arising from minor mechanical spinal malfunctions associated with aging and natural degeneration. The pain can be intense but the pathology is overwhelmingly benign (Deyo and Weinstein 2001). Emphasizing the nonthreatening nature of the problem, however, belies its massive impact. A study on the global burden of disease in 2013 found back pain to be the most frequent cause of disability for over half the world's population (Global Burden of Disease Study 2013 Collaborators 2015).

Medicalizing the condition has led to unfortunate consequences, shifting attention from the ubiquitous mechanical causes of pain to rare, albeit more sinister, pathologies. This misdirection is reflected in numerous attempts at a pathology-based classification. How clinicians organize a problem establishes their diagnostic probabilities and assigns priorities for investigations and treatment. To concentrate on the possibility of serious pathology, like malignancy, means screening every back pain sufferer for something present in less than 1% of cases (Henschke et al. 2009). Individual clinical features implying an ominous condition, called Red Flags, have poor accuracy. The Red Flag of night pain, frequent among back pain sufferers, can be raised as a source of concern. In one study of nearly 500 patients, 40% had night pain but not one had serious pathology (Harding et al. 2005). Algorithms beginning with a check for Red Flags are popular and no one can deny the value of a thorough history but that route can lead to unnecessary testing and unwarranted patient anxiety, and may not provide the anticipated certainty (Downie et al. 2013). From a wider perspective, identifying a pathophysiological pain source is possible in only 10–15% of cases, ultimately leading in most cases to the counter-productive diagnosis of "nonspecific" back pain (Krismer et al. 2007).

Classification Options

Treatment Response

Attempts to improve specificity and thereby offer therapeutic guidance have classified low back pain on the patient's reaction to specified mechanical treatments. Results show slight improvement as might be expected from the circular nature of the cohort construction; patients who did well following a particular maneuver were classified, after treatment, as suitable for that category (Fritz et al. 2003). One problem in using this sort of classification in primary care is the requirement that the clinician to be able to properly perform the classifying techniques, direction-specific or trunk-stabilizing exercises or spinal manipulation (Brennan et al. 2006). Even when the determination was made by trained physical therapists, over 30% of the subjects could not be clearly classified (Stanton et al. 2011).

Time Based

Classifications can be time based but even here there remains considerable variability and disagreement. There is no consensus on the length of time back pain must be present before it shifts from acute to chronic and no defined duration for the pain-free interval that distinguishes a new attack from a continuing chronic condition. To address this lack of consistency, the classic designations of acute and chronic have been replaced with more broadly inclusive terms such as "persistent" or "recurrent" (Norton et al. 2016). However, neither set of definitions offers immediate clinical guidance for treating the patients in pain.

Administrative

Administrative classifications typically use the tenth revision of the International Statistical Classification of Diseases and Related Health Problems, Clinical Modification (ICD – 10 – CM) codes to either identify relevant pathologies or support a diagnosis of "nonspecific" back pain. The nonspecific categories include such divergent entities as kissing spines and lumbago, the former a description of putatively abnormal anatomy and the latter an antiquated name for low back pain first used in 1684 to describe "pain in the muscles of the loins" (Oxford English Dictionary 2019). In a comprehensive review of administrative data on health-care utilization, Norton identified and validated four distinct groups of patients: one cohort with immediate total recovery, one with frequent relapses but with little ongoing healthcare utilization, and two groups with high continuing demand – one for therapeutic interventions and the other for medication (Norton et al. 2016). Significantly the groupings were unrelated to either the patients' demographic or clinical characteristics; the classification had no prognostic value. Retrospectively identifying the amount of resource consumption offers no prediction of that outcome nor identifies those patients at risk.

Risk of Chronicity

STarT Back categorizes patients by predicting their risk of chronicity (Hill et al. 2008). This classification has shown promise in directing primary care by identifying those people most likely to develop persistent problems. It was developed in England and uses a short simple questionnaire that takes into account pertinent physical findings while emphasizing a psychosocial subscale gauging bothersomeness, catastrophizing, fear, anxiety, and depression. It characterizes patients as a low, medium, or high risk of chronicity and recommends appropriate intensive therapy. A study by Foster assessing the results found modest overall improvements in patients' outcomes with a more targeted use of health care resources and without increased costs (Foster et al. 2014). The authors noted that the mean difference in patient disability in their study was less than that in the original trial, a fact they attributed to the higher proportion of low-risk patients and the variability in physician engagement. The magnitude of the second problem, the variability of physician engagement, and its negative impact on generalizability was highlighted by the MATCH study at Kaiser Permanente in Washington State. In spite of extensive training for both the participating primary care physicians and physical therapists the trial showed no statistically significant differences in patient outcomes or health care use between the intervention and control groups (Cherkin et al. 2018). Several factors may account for this lack of success including limited access to suitable treatment for the high-risk patients but, regardless of the reason, using the classification did not alter practice patterns. Further, it was never designed nor intended to direct immediate management.

Anatomic

Many, if not most, clinicians believe initial treatment must be determined by and directed toward the source of pain. A pathoanatomic classification seems obligatory. But, unless that treatment is an invasive procedure, management involves the entire patient not just a local painful structure. In cases where the cause of the pain is obscure and the clinical symptoms raise no concerns for an urgent or serious condition, seeking a structural diagnosis simply to fill a physical or pathological category is heading down the wrong track. It promotes needless investigations and excessive imaging. With current technology it is almost always possible to find an aberration in the spine. Whether the identified pathology is the reason for the patient's back pain is an entirely different question. The false positive rate for MRIs of the spine in middle aged patients approaches 90% (Wnuk et al. 2018). Employing MRI as a screening tool to locate abnormalities without a clinical indication has a strong iatrogenic effect, offers no benefits and degrades the outcome (Webster et al. 2013).

Nonspecific

The prevailing medical paradigm dictates that we must establish a cause before we can treat. In the overwhelming majority of back pain patients, however, no pathoanatomic diagnosis is possible (Koes et al. 2006). This pain, designated as "nonspecific," is neither the product of recognizable structural defects or deformities in the spine nor the result of identifiable pathologies including trauma, tumor, systemic disease, or local infection. It denotes pain arising from spinal structures, not pain referred to the back but arising from known causes in other parts of the body or within a sensitized central nervous system. While there is no agreement on the particular pain generator within the spine, there is widespread consensus among clinicians that "nonspecific" back pain is mechanical back pain produced by nothing more sinister than minor mechanical malfunctions, the inevitable consequence of normal wear and tear (Maher et al. 2017). The potential severity of the pain does not reflect the benign reality of the underlying problem but the intensity of the problem can justify immediate treatment. Deferring therapy to conduct unnecessary and predictably futile investigations to isolate the site of the pain is ill-advised.

A Syndrome Approach

From the patient's perspective back pain is never nonspecific; the symptoms are never vague and the mechanical characteristics are obvious. Mechanical pain is pain produced by movement or position and relieved by rest or a change in posture. The pain fluctuates with activity. Again from the patient's perspective, the primary reason for seeking professional help is to relieve that pain. With a definitive diagnosis out of reach the clinician's decisions must be rely on something else. There is another option. In 1987 the Quebec Task Force noted, "Distinct patterns of reliable clinical findings are the only logical basis for back pain categorization and subsequent treatment." (Spitzer et al. 1987). The therapy can be built on the patient's clinical presentation, on a mechanical syndrome. A syndrome is a constellation of signs and symptoms that consistently appear together and respond predictably to treatment. Reluctance to base treatment solely on the clinical picture is understandable but, in the case of mechanical back pain, unjustified. A syndrome has an undetermined but definite etiology; its invariable presentation is not random chance. The only difference between a syndrome and a disease is, in fact, the former's lack of an agreed etiology. Once the cause of a condition becomes known the syndrome becomes a disease. For "nonspecific" mechanical back pain, discovering the exact source of the symptoms would obviously not alter the clinical picture nor diminish the value of already proven effective non-surgical treatment.

A classification that can offer clinicians immediate guidance in the initial management of back pain rests on typical mechanical syndromes or patterns drawn from the history and physical examination without additional imaging or investigations. It should identify unusual presentations and highlight potentially serious features. By emphasizing the regular mechanical patterns, which comprise about 90% of the low back pain presentations in a primary care setting, the classification renders the few sinister presentations plainly visible (Chien and Bajwa 2008). Detecting Red Flags becomes a by-product, not the purpose, of the assessment.

This Patterns of Pain classification has been validated and proven successful. For nearly 50 years, it has been the basis of back pain treatment at the CBI Health Group in its more than 170 rehabilitation clinics across Canada (Hall et al. 2009). It is the foundation of the Saskatchewan Spine Pathway. Instituted in 2011 the Pathway has produced substantial cost savings and improved patient satisfaction with spine care across the province (Wilgenbusch et al. 2014). In 2012, the Ontario Ministry of Health and Long-Term Care launched a pilot project, again using this pattern classification of clinical presentation, to develop the Inter-professional Spine Assessment and Education Clinics (ISAEC), a network of spine triage clinics. The program proved so successful that in 2018 the Ministry expanded it across the entire province. The Ontario Ministry also funded an online aid for primary care practitioners, the CORE (**C**linically **O**rganized **R**elevant **E**xam) Back Tool. It offers a concise method of separating patients with back or leg pain into those who require further investigation or referral and those whose straightforward mechanical picture encourages management by the primary care provider. This differentiation is made using the same mechanical syndrome classification (Alleyne et al. 2016).

Four Patterns of Pain

Mechanical back pain can be divided into four, clearly delineated patterns of pain identified on history (see Table 1) and confirmed or refuted with the physical examination (see Table 2). Each pattern suggests an initial course of treatment, the outcome of which either supports or rejects the

Table 1 The essential components in a "patterns of pain" history

Number	Category	Question	Objective
1	Pattern	Where is your pain the worst?	Discriminate between back dominant (referred) pain and leg dominant (radicular) pain
2		Is your pain constant or intermittent?	Obtain a precise account of the pain's consistency and whether or not it ever completely disappears
3		Does bending forward make your typical pain worse?	Determine the effect of flexion on the pain given in answer to Question One
4	Mandatory	Since the start of your pain, has there been a change in your bladder or bowel function?	Consider possibility of an acute cauda equina syndrome
5	Function	What can't you do now that you could be before you were in pain and why?	Estimate the required treatment intensity; the reason for the impairment should be the pain given in answer to Question One
6	Additional	What positons or movements relieve your typical pain?	Identify features that may assist with management
7		Have you had this pain before?	Establish context for the current episode and the likelihood of further recurrences
8		What treatment have you had and did it work?	Previous successful treatments for the same pattern should be effective again
9	Inflammatory	If you are under 45 years old do you have periods of morning back stiffness lasting longer than 30 minutes?	Screen for spondyloarthritis

Table 2 The essential components in a "patterns of pain" physical examination

Procedure	Optimum Position	Objective/Technique
Observation		Assess general activity both before and during the examination. Back specific elements: gait, contour, color, surgical scars
Movement	Standing	Observe flexion and extension for rhythm of movement and reproduction of the typical pain
	Lying prone	If the patient reports pain on standing flexion evaluate the response to ten prone passive extensions
Nerve root irritation tests	Lying supine	Examiner lifts the patient's straight leg. Nerve root irritation reproduces or exacerbates the typical leg dominant pain. May be performed with patient sitting
	Lying prone	Femoral stretch reproduces the anterior thigh dominant pain. Examiner extends the patient's straight leg. Perform when indicated by history
Nerve root function tests	Detailed in Table 3	Check muscle power or tendon reflexes involving L3, 4, 5, and S1
Upper motor neuron tests	Sitting	Identify spinal cord involvement by plantar response or sustained ankle/ patellar clonus
Saddle sensation	Lying prone	Screen for cauda equina syndrome with light touch to the S2 dermatome, midline between the upper buttocks

Positive findings may prompt further, more comprehensive testing. When suggested by the history additional investigations can include the hips, abdomen, peripheral pulses, or sensation

pattern diagnosis. The classification is constructed to be integrated into early patient management; inconsistencies within the history, between the history and the physical examination or in the anticipated course of treatment for the selected pattern will alert the clinician to potential problems.

History

Question One

The history begins with three pattern questions: questions designed to define the characteristics of the four patterns. The first question is "Where is your pain the worst?" This is not the same as asking the patient "Where do you hurt?" The latter question encourages vague and rambling answers that may not only divert focus from the major symptoms but shift attention to irrelevant details. The important distinction is between back or leg dominant pain. In this classification, back dominant pain is pain felt most intensely in one or more of the following locations: low back, upper buttocks, coccyx, over the greater trochanters (Tortolani et al. 2002). Back dominant pain can occasionally extend to the groin and genitals. This pain is referred pain, pain arising within the musculoskeletal structures of the spine but felt some distance from the source. The concept of referred pain has been recognized for over 100 years but there is still no consensus as to the mechanism by which the pain spreads other than agreement that it does not involve direct irritation of the peripheral nerves (Bogduk 2009).

Back dominant referred pain can radiate into the legs and may extend well below the knee to include the foot (Hill et al. (2011). The clinically important issue is establishing where the patient's pain is most excruciating. Although referred pain can involve the leg, the site of the most severe pain is always somewhere in a band around the lower back, upper buttocks, hips, and groin.

Complicating the recognition of back dominant pain is the fact that areas of referred pain can become locally tender (Smythe 1986). Palpating the trochanteric region may elicit local discomfort misdiagnosed as bursitis. Tenderness over the upper buttock can be falsely attributed to pushing on a painful piriformis muscle. Palpable "trigger points" over the posterior iliac crest are another example of local findings without local pathology. Occasional temporary symptomatic relief following injection of a local anesthetic further compounds the diagnostic confusion.

Leg dominant pain represents radicular pain originating from direct irritation of one or more of the roots of the sciatic or femoral nerves and carried along the nerves into the legs. Radicular pain is pain most intense anywhere at or below the gluteal fold. Pain in the lowest three centimeters of the buttock is considered leg dominant as is pain felt most strongly in the thigh, calf, ankle of foot. Referred back dominant pain can extend to the foot and leg dominant radicular pain may not go below the knee. The demarcation point is the lower buttock, not the knee joint.

Getting a patient to choose the site of the dominant pain can be challenging. Back dominant pain frequently involves the leg and leg dominant pain can be accompanied by pain in the back. Asking the patient to pick only one area when they both hurt may give the erroneous impression that the examiner is not interested in the whole problem and the patient may be unwilling to relinquish any part of the complaint. They refuse to choose or describe them as equally painful. But the pattern of pain classification demands identification of the predominant pain location. The best solution is simply to change the question. Instead of asking, "Where is the pain the worst?" say, "If I could stop only one of your pains, in the back or in the leg, which one would you want me to stop?" The natural reply to this question might be, "I want you to stop them both," but this is a very different conversation from one that tries to determine only which one hurts more. Now the clinician can acknowledge that the patient does indeed have two significant painful areas and that both deserve attention. It is no longer a matter of which pain to treat but merely a decision of which pain to treat first.

In the infrequent situation where the patient still cannot choose between the back and the leg pain, the correct option is to pick the "worst case" scenario. Since leg dominant pain reflects nerve involvement and therefore has the potential, no matter how slight, to be associated with significant neurological impairment leg pain takes precedence. It is prudent throughout the history and physical examination to consider the more serious alternative, bearing in mind that no matter how excruciating the pain, 95% of back pain patients suffer from a benign mechanical condition.

Question Two

The second pattern question in the history addresses consistency. Is the dominant pain constant or intermittent? For fear of minimizing the problem to the examiner, many patients are reluctant to admit the pain ever stops. When asked directly, "Is your pain constant" they respond, "Yes" and once committed patients may be unwilling to change the answer. A correct report of the pain's consistency is essential to assign the pattern. To obtain an accurate report the clinician must give the patient "permission" to relate the all the details, including moments of spontaneous improvement, without appearing to diminish the seriousness of the complaint. The clinician needs to frame the question in a way that does not minimize the patient's concerns. The question is best asked in two parts. The first part lays out the conditions under which the pain might stop: the best time of day or the best situation. These suggestions must be accompanied by a statement that the clinician is fully aware of the severity of the pain and the fact that, even though it may briefly disappear, it will always return. If the patient, however reluctantly, admits the pain does disappear there must be a second follow-up question. "Does that pain disappear completely? Is it totally gone?" There is only one correct answer to describe intermittent pain, "Yes." The patient must state unequivocally that the pain entirely disappears. Any other answer such as "nearly," "almost," "mostly," or "feels much better" is considered as constant pain. The decision to accept the pain as intermittent must take into account the level of analgesic medication; regular narcotic use means the pain must be considered constant. When there is any doubt, the general principle when using this classification is for the clinician to select the more serious option, in this instance constant pain.

This practice is critically important when assessing consistency. Truly intermittent back dominant pain is never the result of spinal malignancy or active spinal infection. Both of these sinister pathologies can produce pain that fluctuates with position or movement but even in the best circumstances, the pain never disappears completely. Whether the pain is constant or intermittent is such an influential factor that the clinician should repeat the patient's words exactly then ask the patient to verify that was what was said. The power of these questions, properly asked and answered, is enormous. At first contact and without any additional investigations, they can eliminate the possibility of two devastating pathologies. Constant pain clearly doesn't confirm malignancy but it does leave the slight possibility of a more serious condition. In this case, it would be appropriate to ask about a history of cancer in the preceding 5 years. Recognizing the fact that the overwhelming majority of back pain whether constant or intermittent is nonthreatening, constant pain still requires further questioning. Truly intermittent back dominant pain permits reassurance that the problem is almost certainly a benign mechanical condition.

Question Three

The third and final pattern question is deliberately direct: "Does bending forward make your typical pain worse?" This is the critical part of the broader open-ended question, "What makes your pain worse?" Understanding the aggravating factors aids planning treatment but knowing the effect of flexion on the typical pain does more. It completes the identification (along with location and consistency) of the principal pain pattern, a pattern which provides direction for the entire therapeutic regimen. The pain under consideration, the typical pain, is the dominant pain given as the answer to the first question. Bending forward may produce discomfort in other areas, like behind the knees from tight hamstrings, but these observations should not distract the examiner from the primary complaint.

Question Four

The fourth question is mandatory since it addresses the only true emergency in low back pain: the acute cauda equina syndrome. Interference with the second, third, and fourth sacral nerve roots, typically from an acute large central lumbar disc rupture, can lead to denervation of the urinary bladder and the rectal sphincter producing the classic triad of a period of urinary retention with eventual overflow, fecal incontinence, and altered perineal sensation (Fraser et al. 2009). Failure to surgically decompress the sacral nerves within the first 48 hours can lead to permanent loss of normal bowel and bladder function, so early recognition is a crucial part of the back examination and, therefore, of the patterns of pain classification.

To avoid confusion with preexisting genitourinary problems and to retain focus on recent onset back and/or leg pain, the fourth question is framed, "Since the start of your current pain has there been a change in your bowel or bladder function?" The temporal limitation keeps the history centered on recent events and avoids a lengthy discussion about prior problems. Another key is the emphasis on change rather than on symptomatic details. A multiparous woman may have longstanding urinary incontinence but that is not a change and therefore not relevant to the current painful episode. Cauda equina syndrome is an extremely rare condition. Most practitioners will spend their entire careers without seeing one so recalling the clinical picture and remaining vigilant for a cauda equina syndrome with every back patient may be unrealistic. Missing the diagnosis is not a matter of negligence as much as it is a matter of extreme improbability. But routinely asking every back pain sufferer if there has been a change in bowel or bladder activity should become a habit and all the clinician needs to remember is "no change…no problem." Any change triggers concern and the opportunity to review the relevant information. Constipation is prevalent and, while distressing, not a sign of ominous pathology. It is, however, a recent change from normal function and so worthy of mention and consideration.

Question Five

Question Five concerns the level of impairment. "What can't you do now that you could do before you were in pain and why?" The degree to which the pain interferes with the patient's daily routine dictates treatment intensity. A pain that occasionally limits a recreational activity does not merit the same degree of medical involvement as one that prevents regular employment. Asking about the reason for the impairment, "…and why?," is a check on the validity of the patient's reports. The cause of the functional limitations in the answer to Question Five should be related to the same pain that the patient reported answering Question One. If the reason for the patient's restrictions is not the dominant pain then treatment is likely to be misdirected. If the patient says back pain is the problem but reports that it is the leg pain which prevents activity or a return to work, this inconsistency must be resolved before proceeding. The patient may have misunderstood the question, the clinician may have misinterpreted the answer or the problem may not be a straightforward mechanical complaint.

Question Six

The next question enquires about relieving factors, what the patient does to reduce or stop the typical pain. The options should be compatible with, that is opposite to, those things which make the pain worse. Mechanical pain is predictable and consistent in its reaction to physical stress. A constant level of pain, unaffected by changes in posture or activity, strongly suggests a nonmechanical etiology. Only the effect of flexion is necessary for pattern determination but the response to other movements or positions is always considered in the selecting the appropriate mechanical therapy.

Questions Seven and Eight

The next two questions involve prior episodes of pain. The first asks if there have been any previous attacks of the same pain, the second deals with any earlier treatment. Both relate to the patient's existing pain as identified by Question One. Back pain is a recurrent complaint and the pattern of pain can change over time (Donelson et al. 2012). Someone suffering a first attack should be cautioned that further episodes are likely. For those with a history of back or leg pain, knowing the outcome of past treatment should influence the current management. If the pattern of pain of a former attack was the same pattern as the present one then treatment that worked in the past will presumably work this time. Conversely if a treatment failed before there is little reason to try it again.

Question Nine

A final question, about unusually prolonged morning back stiffness, addresses the possibility of inflammatory spondyloarthropathies such as ankylosing spondylitis or psoriatic arthritis. This symptom is relevant in young and middle-aged patients but of little significance in the elderly. If the patient is under 40, ask "When you get up in the morning do you have stiffness in your back lasting more than half an hour?" At about 5%, this group of illnesses is the second most frequent cause of back pain after mechanical malfunctions (Weisman et al. 2013). Including a screening question for inflammatory spinal conditions along with the mechanical classification questions encompasses over 95% of patients presenting with back or leg pain.

These nine questions, particularly the first five, are the core of the back assessment. It is not the purpose of this classification to limit the scope of the inquiry but rather to sharpen the evaluation so that the essential elements are not overlooked or obscured by irrelevant detail. The clinician can and will ask for additional information. There are, for example, no questions about potential mechanisms of injury. In a study of over 11,000 patients presenting with nontraumatic, nonspecific back pain two thirds of those without a need to know (claiming worker's compensation or initiating a lawsuit) could not identify any cause for their pain. Spontaneous onset accounted for over 60% of cases (Hall et al. 1998). Moreover, regardless of the purported mechanism, all those with a mechanical presentation could be assigned a pain pattern and it was the pattern, not the precipitating event, that directed treatment. Obviously discovering the mechanism is relevant in situations where liability must be established or where there is a history of significant impact. Supplementary questions should be included whenever the pain is constant and nonmechanical or when there is suspicion of a serious underlying pathology. Progressive neurological deficits, unexplained weight loss, recent infection, disproportionate night pain, or unexplained constitutional symptoms are all reason for concern.

History determines the pattern. The physical examination confirms or refutes the choice. The examination is not an independent activity but rather an integral part of the assessment. It is directed by the information obtained from the history and any inconsistencies between the patient's story and the observed findings, just like inconsistencies within the history, must be resolved in order to clearly establish which pattern will direct treatment. Like the nine points comprising the history, the limited number of tests in the physical examination does not constitute a comprehensive evaluation but are the minimum required to corroborate the selected mechanical pattern while eliminating sinister pathologies. The final examination may incorporate additional steps but must include these components.

Physical Examination

To minimize discomfort and speed the examination, the patient should be assessed in a progression of positions selecting the optimum position for each test. Someone with back pain may take several minutes to lie down. Asking them to get up again for another test prolongs the examination and aggravates the pain. Start with the patient standing then sitting then lying down. Some procedures may be done best with the patient kneeling or sitting on a chair with feet on the floor. Using the chair before sitting on the edge of the examining table is both more efficient and more comfortable.

Observation

The physical examination starts with observation and observation starts before the actual examination. How the patient sits or moves or interacts before the formal assessment starts provides information about normal levels of activity and discomfort. Observe the patient's gait. Inspect the back for deformities, discoloration, and scars from previous surgery. Subtle changes in alignment are generally irrelevant. It is the overall contour or obvious areas of redness and swelling that matter.

Palpating along the spine for areas of tenderness is helpful to elicit sites of acute inflammation but plotting the areas of painful muscle tension is of little diagnostic value. Back dominant pain is referred pain and the location of the muscle tenderness is not necessarily the same as the location of the pathology; a painful L4-5 disc may not hurt at the L4-5 level.

Movement

Assessing spinal movement involves recording the rhythm and the reproduction of the typical pain. The physical examination confirms the history and patients who say that bending forward causes their usual pain should report the same pain when they bend forward for the examiner. The one important exception, which can cause confusion, is the patient who reports back pain only after sitting for a prolonged period. Patients whose pain is produced exclusively by a flexed posture and never by flexion movement should be identified on history and a proper interpretation of the lack of pain with movement on physical examination will support, not contradict, the patient's story.

Normal flexion of the lumbar spine follows a smooth progression cephalad from the pelvis without a catch or hitch. The actual range of movement is less important and, unless the measurement is one of a series, of minimal diagnostic significance. The ability to touch the fingers to the floor has more to do with the length of the arms and the flexibility of the hips than it does with the range of movement in the spine. The range of lumbar extension is similarly inconsequential. The important finding is the exacerbation or relief of the typical pain. To avoid apparent spinal extension produced by bending the knees and to better isolate movement to the low back have the patient stand with the front of the legs against the back of a chair or the examining table. Place the hands on the buttocks and not in the small of the back.

One of the three elements of Pattern recognition is the effect of flexion on the typical pain so the physical examination focuses on sagittal movement. Noting pain on rotation or side-bending (pain which may be present in all four patterns) can be useful in choosing treatment strategies but because these movements do not distinguish between the four mechanical presentations they are not used to establish a pattern.

Prone Passive Extension

When the patient reports feeling the typical pain on bending forward, the physical examination includes prone passive extensions. This maneuver, popularized by physiotherapist Robin McKenzie and referred to as a "sloppy push-up," can have a rapid beneficial effect on flexion-aggravated pain and may ultimately become part of a pain control strategy (Donelson and McKenzie 1992). It has no role in evaluating or treating pain that is not made worse with flexion movement. If used, prone passive extensions are ordinarily carried out at the end of the examination while the patient is lying prone on the examining table. With the hands and palms down and slightly above the head, the patient uses the arms to raise the upper body. The action is passive for the back since all the muscular exertion is in the arms; the paraspinal muscles remain relaxed. At the same time, as the torso is pushed up the hips must remain down on the table. The key to a proper sloppy push-up is to have the elbows fully extended and locked at the same time as the front of the pelvis is in contact with the table. The first error is to allow the hips to stay down by keeping the arms bent. Raising the head and shoulders but not fully extending the elbows engage the back muscles and negate the passive nature of the technique. The second mistake it to allow the hips to rise as the elbows fully extend, as with the conventional push-up exercise. Keeping the spine straight prevents the necessary low back extension. Modifying the patient's hand placement achieves both objectives simultaneously. The more the hands are advanced above the head, the more the arms can be extended without elevating the trunk to the point where the hips are lifted. The quality of the prone passive extension is gauged by the impact on the level of pain not by the distance the sternum is lifted above the bed. The stiffer the spine, the more the hands must be moved above the head. Although the final location of the hands and the amount of lordosis in lumbar spine of a supple young woman are very different from the hand placement and sag in the rigid spine of an old man, the amount of pain relief may be the same. Once the patient has found the proper starting point, suitable for the degree of spinal mobility, he or she slowly repeats the passive extension, pausing briefly between repetitions but without holding the fully elevated positon. Compare the level of typical pain (usually using an 11 point scale of 0 to 10) before the first sloppy push-up to the level of pain at the end of five repetitions. Depending upon the clinical response, another set of five may be required.

Nerve Root Irritation

Straight leg raising (SLR) is a classic test for sciatic nerve root irritation. Lifting the leg with the knee extended puts tension on the nerve and causes the roots to slide though the intervertebral foramen. SLR is widely employed and surprisingly poorly understood. The test is positive only with the reproduction or exacerbation of the patient's typical leg dominant pain – not any leg pain just the patient's preexisting leg dominant pain. The patterns of pain classification rests on distinguishing back dominant referred pain felt in the leg from leg dominant radicular pain that may have associated but secondary pain in the back.

The straight leg raise is a test for radicular pain. A proper interpretation is vital to choosing a correct pattern. If the patient has never had leg dominant pain, the patient cannot have a positive test. You cannot reproduce or exacerbate a pain the patient never had. It is impossible to have a positive straight leg test in a patient with back dominant pain.

Much of the confusion and misapplication of the straight leg raising test arises because the test is interpreted without regard for the history. Because any leg pain is incorrectly taken as a positive finding, posterior leg discomfort from hamstring tightness is misinterpreted as a positive test. To avoid this mistake some physicians consider the SLR to be positive only if pain is produced below 60°, an elevation that does not tense the hamstring muscles. Interpreted correctly the test is positive at any elevation if it reproduces the leg dominant pain identified on the history. The level at which the typical leg pain is produced is a measure of neural irritation. Pain felt at a few degrees of elevation (or even when the knee is extended without lifting the leg) indicates acute inflammation while typical pain that occurs only at 80° or 90° degree, though still a positive test, suggests that the nerve root is well on the way to recovery. SLR is passive; the examiner lifts the patient's extended leg. To minimize confusion with hamstring pain, the contralateral leg can be fully flexed, rotating the pelvis and relaxing the posterior thigh muscles.

A positive SLR indicates radicular, leg dominant pain so the reproduction of back pain cannot be a positive result. Considering both back and leg pain to be a positive test is incompatible with the very definition of the maneuver – a test for nerve root irritation not the presence of mechanical back dominant pain.

The femoral stretch test is designed to assess irritation of the roots of the femoral nerve. It is the reverse of the straight leg raise; the patient lies prone and the examiner lifts the straight leg extending the hip and putting tension on the femoral nerve in the anterior thigh. For most patients, this causes back pain, which is not a positive test. Whether or not to do a femoral stretch depends on the patient's history. Femoral nerve radicular pain is constant in the lower anterolateral thigh and only when this is the chief complaint is the test necessary.

Conduction Deficit

Patients with purely back dominant pain should not have nerve conduction deficits, but since the purpose of the physical examination is to disprove as well as to support the pattern provided by the history, every patient should have a screen of nerve function (see Table 3). This is not intended as a complete neurological examination but simply a quick check on the roots that supply the lower limbs: L3, L4, L5, and S1. The examiner should select one test for each root. A more comprehensive investigation may be necessary if there is an abnormality in the screening exam or when dictated by the history as in cases of leg dominant pain.

Typical choices include the knee reflex for L3 and L4, strength of great toe extension for L5, and the power of great toe flexion for S1. If these are normal bilaterally, no further tests may be necessary. Additional investigations include quadriceps power for L3 and L4; ankle dorsiflexion strength, hip abduction power, and heel walking for L5; ankle reflex, plantar flexion strength, gluteus maximus muscle tone, and toe walking for S1.

Upper Motor Neuron Involvement

Any evidence of spinal cord involvement negates a mechanical pattern diagnosis. Upper motor neuron tests must be part of every examination. Conditions as diverse as a spinal cord meningioma or multiple sclerosis can present as apparently mechanical patterns in the low back, distinguished only by the findings of upper motor pathology: the upgoing toe of a positive plantar reflex, sustained knee, or ankle clonus. One of the goals of the history in this presentation-based classification is to immediately rule out more ominous causes of back pain. A concordant, properly performed physical examination is an indispensable second step to establish the safety and validity of this approach.

Table 3 Nerve root function tests

Optimum position	Procedure	Roots tested	Technique
Gait	Heel walking	L4, L5	Minimum five steps with maximum forefoot elevation
	Toe walking	S1	Minimum five steps with maximum heel elevation
Standing	Trendelenburg test	L5	Examiner's hands on the patient's iliac crests. Assess hip abductor power for the leg on which the patient stands. Contralateral pelvic elevation is the marker. A normal examination is symmetrical elevation
	Toe raises	S1	Ten times bilaterally, then ten times on each leg. Balance by holding the examiner's hands
Kneeling	Ankle tendon reflex	S1	Patient kneels on the chair seat. Tap ankle tendon. Reinforce by squeezing the chair back
Sitting Feet on floor	Ankle dorsiflexion	L4, L5	Elevate forefoot against resistance from the examiner's hand on the midfoot
	Great toe elevation	L5	Elevate great toe against resistance from the examiner's thumb
	Great toe flexion	S1	Keep the great toe flexed and resist pull from the examiner
Sitting Legs free	Patellar tendon reflex	L3, L4	Tap patellar tendon. Reinforce with the Jendrassik maneuver
	Quadriceps power	L3, L4	Patient extends the knee against resistance
Lying prone	Gluteus maximus tone	S1	Palpate buttocks as patient alternately tenses and relaxes. Repeat ten times

Saddle Sensation

This is particularly true of the test for saddle sensation. Cauda equina syndrome is the only diagnosis associated with low back pain where failed recognition on the initial assessment leading to even a short treatment delay can have devastating consequences. Hence Question Four, "Since the start of your current pain has there been a change in your bowel or bladder function?," is mandatory. Testing light touch in the S2 area, midline between the upper buttocks, not only adds an important physical finding, but, when routinely incorporated into the standard back examination, the test itself becomes a prompt to ask the question. Using a tissue or a cotton swab to judge light touch in one outlying area of the perineum is clearly not definitive, and genuine concern will lead to further investigations including a digital rectal examination. But the test is quickly and easily done, nonintrusive and, perhaps most importantly, focuses on cauda equina syndrome, a rare diagnosis that otherwise might not be considered.

Additional Tests

Beyond the six core components of observation, movement, root irritation, nerve function, upper motor neuron involvement, and saddle sensation, the history may suggest further examinations. Ruling out hip pain, a confounding complaint, or checking peripheral pulses in patients with claudication are familiar examples.

Pattern Identification

Combining the history and the physical examination allows classification into one of four mechanical patterns of pain, two of which are subdivided (see Table 4). The patterns are derived from signs and symptoms arising from the underlying mechanical malfunctions but a pattern diagnosis does not require establishing a specific pathoanatomic diagnosis. In some cases, shifting attention from the clinical syndrome to a putative pain generator misleads treatment. Just recognizing a pattern allows valid predictions about symptom duration and the patient's response to selected mechanical therapy. Failure to follow the anticipated course mandates early reassessment and this rapid appreciation of a negative outcome is one of the merits of the system. Back dominant patients constitute the overwhelming majority of the patient population and Pattern 1 is the most frequent presentation.

Pattern 1

Pattern 1 is back dominant pain with pain felt most intensely in the low back, upper buttocks, coccyx, over the flanks, or in the groin; the exact location of the pain should agree with the history. The pain is increased in flexion and may be constant or intermittent. Pattern 1 is the only pattern where the consistency of the pain can vary. The physical examination should support the history so the patient reports the described dominant pain to be increased pain in flexion. The classification defines Pattern 1 as pain worse **in** flexion not **with** flexion. A few Pattern 1 patients have no pain with flexion movement but have pain only after periods of sustained flexion posture. They experience back dominant pain after prolonged sitting; sitting is a flexed posture. In this situation, unless the clinician is prepared to let the patient sit in the examining room for an hour or two, the physical assessment will be negative. A few forward bends will have no effect. For most Pattern 1 patients, however, the typical back pain will be present with both movement and position.

Because Pattern 1 is referred pain without direct involvement of the peripheral nerves, the physical examination will show no signs of nerve root irritation or a loss of normal nerve function associated with the current pattern. An independent defect, such as an absent Achilles tendon reflex from a previous tendon rupture or a long resolved episode of S1 radiculopathy, should not confuse the pattern designation. Single findings – a change in bladder function, for example – should be noted and may significantly change management but it is the combined results of the

Table 4 Patterns of pain

Pattern number	Dominant site	History	Physical examination	Additional features	Subclassification
1	Back	Pain in flexion Constant/ Intermittent	Pain in flexion Neurologically normal	May have pain with extension May have unrelated neurological findings	**PEP** Decrease pain within ten properly performed prone passive extensions
					PEN No change or increase pain within ten properly performed prone passive extensions
2	Back	No pain in flexion Intermittent	No pain in flexion Neurologically normal	Pain with extension May have pain relief with flexion May have unrelated neurological findings	
3	Leg	Constant	Positive irritative and/or conduction findings	Pain with flexion and other movements or positions	
4	Leg	Intermittent	May have positive irritative and/ or conduction findings		**FA** Flexion aggravated
			Negative irritative findings May have conduction loss	Pain with activity in extension Conduction loss may be transient	**FR** Flexion relieved

entire history and physical, not the individual components, that decide the pattern.

Pattern 1 PEP

The change produced by repeated prone passive extensions (the technique is described in detail as part of the physical examination) separates Pattern 1 into two groups. Patients who experience pain reduction within ten repetitions are considered **P**rone **E**xtension **P**ositive or **PEP** patients. For these patients, prone extension is a positive experience. They demonstrate a clear directional preference for unloaded extension and therefore are an easy population to treat. The maneuver used to assess their pain becomes the mainstay of their self-treatment. A few positively responding PEP patients encounter a phenomenon called "centralization" (Aina et al. 2004). As they repetitively extend the lumbar spine, the site of their dominant pain changes in character and sifts toward the midline of the low back, frequently becoming more intense. The change in location toward the center of the back, in spite of the increased pain, is a positive sign and indicates the sloppy push-ups will shortly begin to reduce the typical symptoms. The new central discomfort is always transient. To properly employ this classification a clinician must recognize the favorable significance of centralization.

Pattern 1 PEN

Patients who fail to improve within ten repetitions of the sloppy push-up or whose increased pain prevents any further attempts are labeled Pattern 1 **PEN**, **P**rone **E**xtension **N**egative. For this cohort, the prone passive extension is a negative event and they have neither an obvious directional preference nor a straight path to pain control. Ten repetitions were picked as the demarcation between PEP and PEN because doing ten sloppy push-ups or less should be relatively easy physically and the immediate pain relief highly motivating. That number therefore separates those who should have little difficulty maintaining the routine from those who may not be able to engage and would benefit from alternate strategies, supervision and continued encouragement.

History determines the pattern but the distinction between PEP and PEN is made by the physical examination and specifically by the pain response to repeated prone passive extensions. Having pain on standing extension is not diagnostic. Patients with discomfort in both standing flexion (Pattern 1) and standing extension may still readily respond to unloaded passive movements. During the first few attempts, a sloppy push-up can be uncomfortable and questioning patients about their level of pain as they are performing the maneuver may give the wrong answer; arching a stiff spine can be unpleasant. Estimating pain relief should wait until after the first five push-ups. PEP patients may report initial discomfort but experience relief once the first set is completed.

Pattern 2

Pattern 2 is also back dominant pain. The pain is always intermittent and is never worse in flexion. Constant pain or any pain in flexion marks the patient as Pattern 1. It is not a question of the amount of pain but simply whether there is any pain at all. Pattern 1 patients may have more discomfort on standing extension than they do when they bend forward but flexing also causes recognizable typical discomfort. In contrast, Pattern 2 patients like to bend forward since flexing can reduce or even abolish the back pain; in no circumstance does flexion make their typical pain worse. Although extension aggravates the pain, it is the effect of flexion, the fact that bending forward never increases the symptoms, which, along with the pain location and consistency, define Pattern 2.

Physical examination of the Pattern 2 patient shows back dominant pain aggravated on extension and never increased, at least unchanged and sometimes abolished, in flexion. The site of the pain matches that described in the history. As with Pattern 1, the neurological examination is either normal or any findings are unrelated to the current episode of pain.

Pattern 3

Pattern 3 is leg dominant and therefore represents radicular pain. In the patterns of pain classification, leg dominant pain begins in the lower buttock about 3 cm above the gluteal fold and can be worse anywhere from that point downwards in the thigh, calf, ankle, or foot. The pain is constant and even though it may fluctuate it never disappears completely. This pattern covers "sciatica," a label used so indiscriminately that it has lost much of its diagnostic value. True sciatica describes only radicular pain arising from compression/inflammation of the roots of the sciatic nerve: L4, L5, S1. In practice, however, the term is incorrectly used any time a patient complains of leg pain, as when the more frequent mechanical back dominant pain briefly spreads into the lower limbs. One of the advantages of using this classification is the precision of the definitions. Because of the inflammatory etiology, Pattern 3 pain must be constant. Because the pathology lies within the lumbar spine, the leg pain is altered by spinal movement or posture. This pattern also covers the femoral nerve roots (L2, L3, L4) since the resulting constant anterior thigh pain is also radicular.

To support the history and confirm Pattern 3, the physical examination must show evidence of either nerve root irritation or conduction loss (diminished power, reflexes, sensation) or both. A majority of cases will show evidence of irritation – a positive straight leg raise – without localizing signs. Some will have both irritation and a focal loss of nerve function, locating the involved spinal level. Rarely there will be a significant conduction loss without irritative findings. A totally normal physical examination is inconsistent with a diagnosis of Pattern 3.

Pattern 4

The format of Pattern 4 differs slightly from the other three. It uses the same three basic questions but the designation depends on only the first two. Any patient with leg dominant intermittent pain is Pattern 4. The third item, the effect of flexion on the typical pain, subdivides the pattern; it uses the same three features in a different way.

Pattern 4 FA

If the intermittent leg dominant pain is increased by bending forward the patient is classified as Pattern 4 Flexion Aggravated, Pattern 4 **FA**. This is an unusual clinical picture seen occasionally with a resolving Pattern 3: constant radicular leg pain. Typically as the leg symptoms subside the pain becomes back dominant and the patient reverts to Pattern 1. Presumably, if there has been continued interference with normal nerve function, the leg pain remains the major complaint. Since acute inflammation is no longer the primary cause of pain, the complaints become intermittent. Since a flexed posture raises tension on the exiting roots, bending forward heightens the discomfort. Pattern 4 FA has been attributed to post-inflammatory scarring, an "adherent" nerve root, or to intrinsic damage within the nerve itself but the classification does not demand detailed identification of the pathology. Treatment is chosen according to the clinical presentation and confirmed by the patient's successful achievement of the predicted outcomes.

Findings on the physical examination can vary but, obviously, must include reproduction of the patient's typical leg dominant pain in flexion. There may be indications of residual inflammation or a focal conduction deficit. Since extension minimizes root compression, arching backwards or a gentle sloppy push-up should decease the pain.

Pattern 4 FR

When flexion diminishes the intermittent leg pain, the patient is a Pattern 4 **FR**, Flexion Relieved. This is the clinical picture of neurogenic claudication, a common diagnosis in the older population. Because the symptoms result from vascular compromise of the nerve roots, they are radicular, that is, leg dominant. Because the impact of ischemia varies with activity and posture, the pain is intermittent. Because flexion increases the available space within the intervertebral foramina allowing improved blood supply, sitting or bending forward can eliminate the symptoms. Again it is location, consistency, and the effect of flexion that dictate the Pattern classification. The symptoms are brought on by walking, which is exercise with the back extended, so the differential diagnosis includes intermittent claudication secondary to impaired peripheral circulation. The conditions can coexist making a definitive diagnosis difficult (Nadeau et al. 2013) There are several distinguishing signs, such as the location of the dominant pain above or below the knee (neurogenic claudication/Pattern 4 FR is usually worse in the thigh) however the most reliable differentiating factor is the neurogenic claudicant's need to flex for symptom relief. This is the reason for the "shopping cart sign," the patient's ability to shop comfortably in a supermarket while being unable to walk any distance outside, because the shopping cart permits ambulation in sustained flexion. The history may include what patients describe as a temporary "loss of balance," which is actually a transient motor weakness disrupting normal gait caused by an ischemic nerve root.

The signs and symptoms of Pattern 4 FR, neurogenic claudication, normally disappear at rest so the physical examination can be normal. This is not an inflammatory pathology so the root irritation tests, like the straight leg raise, will be negative. Infrequently long standing cases with substantial vascular impairment may have a permanent focal motor loss.

Using a syndrome-based classification, Pattern 4 FR, reflecting the patient's clinical findings avoids several diagnostic pitfalls and mistakes. One of the most common is misusing spinal stenosis, a description of spinal anatomy, as a diagnosis. A small spinal canal may be asymptomatic and the measurements of canal diameter on a CT scan are not indications for surgery. Concentrating on the clinical picture, the signs and symptoms that drive treatment rather than focusing on an anatomical variant that may or may not be problematic, keeps the clinician on the right path. This is especially important when treating the elderly patient where both back pain and spinal stenosis are prevalent and both are the result of progressive facet joint degeneration with boney encroachment into the canal. A history of walking limited by pain that disappears with bending forward suggests neurogenic claudication. This supposition will be reinforced by the inevitable identification of spinal stenosis on imaging. But without knowing the location of the dominant pain that assumption may be incorrect. If leg pain is the reason for the impairment then Pattern 4 FR is a reasonable diagnosis. If, however, the pain is back dominant the problem is not nerve root ischemia from canal stenosis but rather mechanical Pattern 2 pain possibly arising from the facet joints. The former might benefit from surgical decompression. The latter will only be made worse.

Pattern Directed Care

Patterns of Pain is a robust, comprehensive classification. Its permutations cover every possible presentation of mechanical low back pain including those with predominantly neurological symptoms (see Fig. 1). A patient's patterns can change and some patterns may coexist. It is not possible for a patient to have both patterns of back dominant pain at the same time, but someone with Pattern 1 back pain can certainly develop constant leg dominant symptoms from nerve root inflammation following a sudden disc rupture or intermittent leg pain from recurrent root ischemia. The

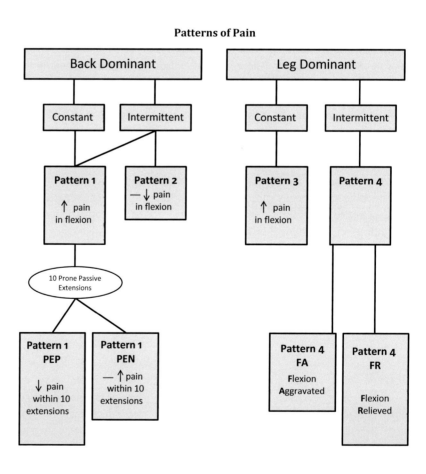

Fig. 1 Patterns of pain

clinical syndromes arise from the underlying pathology but use of the patterns is not tied to a physical diagnosis. Determining a pattern or patterns offers a course of action and usually removes the need for further investigation. Familiarity with the patterns renders the outliers immediately visible and instigates appropriate additional measures. Most mechanical patterns can be managed without recourse to surgery. Non-invasive approaches address the whole patient, not just the pain generator. The anticipated positive response to a therapy chosen by the pattern validates the choice. The goals are pain control and recovery of function not cure. In contrast to nonoperative care, surgery obliges an unequivocally identified, well-defined anatomical target. The aim of an operative intervention is to resolve a local problem, which may be producing widespread symptoms. But in either case, it is the patient's clinical situation, the pattern of the back or leg pain, which shapes treatment.

Principles of Nonsurgical Management

The general principles of nonsurgical management begin with education: advice to the patient about the benign nature of mechanical back pain and the many simple things that can be done every day to reduce the impact of the pain. Patients want to understand the reasons for the pain and need to be reassured that the situation can be controlled. They want assurance that the clinician is capable of successfully managing care and sensing uncertainty in the care provider can make patients less willing to follow sensible recommendations to increase activity in spite of the pain. Offering concrete suggestions related to a recognized pattern rather than resorting to banal platitudes instills confidence in both the patient and the health care provider.

Since the intent is primarily to stop the pain there is a role for purely symptom-relieving procedures. Counter-irritation with heat or cold is hardly a new idea but remains useful. Either modality can be administered professionally as ultrasound or interferential current or self-applied using a hot pack or bag of frozen vegetables. Their application is totally empirical and may be helpful in any pattern.

Correcting posture will change the way the spine is loaded and alter the amount of pain. The correction should be guided by the pattern of pain. Back dominant pain aggravated by flexion will be eased by increasing the lumbar lordosis. Each pattern offers a selection.

Direction-specific movements are at the heart of mechanical therapy. Many are uncomplicated and easily performed, their value determined by their beneficial effect on the pain. Except in Pattern 1 PEN or cases of severe, radicular, leg dominant Pattern 3 pain, pattern-directed repetitive movement should be the first treatment option. Clinicians must prescribe these maneuvers with the same precision and emphasis as they do medicinal remedies. Suggesting they can be done whenever it is convenient or when the patient can find the time belittles their importance and excuses noncompliance.

Analgesic medication should follow not precede mechanical therapy. The pain reduction achieved by changing position or repetitive direction-specific movement usually exceeds that produced by taking a pain pill. Using an analgesic as an adjunct to mechanical treatment is often efficacious but medication should be the second tier. There is no place for opioid medication in the management of Pattern 1 or Pattern 2 pain. No matter how severe, successful control of uncomplicated mechanical back pain can be achieved by physical methods and nonnarcotic analgesia.

Pattern 1 PEP

Patients classified as Pattern 1 PEP (prone extension positive) quickly gain pain control through a variety of activities. Putting one foot up on a footstool arches the low back and reduces the pain. Sitting with a firm foam lumbar roll at about waist level between the spine and the chair back maintains lumbar lordosis. Locating the roll at the correct height places it where it has the most positive influence on the typical pain. The patient is encouraged to make the final adjustments. The roll must be large enough to change sitting posture and that makes it uncomfortable. The question is not, "Does that feel comfortable?" but "What does that do to your typical pain?" It is not uncommon for patients, at the same time, to complain of the discomfort and report that the typical pain has disappeared. Once the pain has been controlled, comfort follows rapidly. A night roll, a firm foam roll longer than the one used in sitting, can be prescribed to treat morning back pain resulting from sleeping on a poor mattress. Side lying provides no support for the spine between the ribcage and the pelvis. The resultant lateral sag can be painful. Placing the roll at waist level across the line of the body reduces the stress and diminishes the discomfort. Once again patients may initially find the lump uncomfortable so focusing on the typical back pain and setting expectations are important. Placing a pillow between the knees reduces tension on the low back and may ease pain but to do that the pillow must be large enough to influence spinal posture. It requires something thick enough to raise the upper leg to the point that the knee is higher than the hip. A couch cushion can be more effective than a pillow off the bed. If the patient finds that standing extension relieves the back pain, then standing extension can be added to the regimen but the decision to use standing extension must be based on the patient's history and confirmed on the physical examination. Allowing the patient to use ineffective standing extensions rather than the demonstrably helpful prone extensions simply because it is too inconvenient to lie down at work misses the point of mechanical therapy.

The key to treating Pattern 1 PEP is the prone passive extension. By definition, a PEP patient experiences pain relief within ten repetitions. Self-treatment is repeating the same exercise in the same way for the same number of times that produced improvement during the assessment. Sessions are scheduled frequently throughout the day, hourly at first. Putting the activity on a timed basis and recording each result allows the patient to appreciate that it is the prone extensions and not something else producing progress. As pain control is established the number of session decreases. Prone passive extensions are treatment not prophylaxis; when the patient is pain free, there is no reason to continue. A return of the symptoms should trigger a return to the exercise.

Pattern 1 PEN

Pattern 1 PEN (prone extension negative) patients have no direct route to pain control so treatment can be challenging. Because this is Pattern 1, back pain aggravated in flexion, the ultimate goal is pain control through repetitive prone extensions. Because these patients initially have too much pain on extension to do sloppy push-ups their management must begin somewhere else. The same things that work for Patten 1 PEPs, a footstool, lumbar or night rolls, a large pillow between the knees should help here as well but with less benefit.

The best ways to start may be to prescribe periods of scheduled rest. Similar to scheduled movement, the duration, frequency, and positions are clearly stated and based on their effect on the patient's pain in the examining room. The length of the rest period is determined by the amount of time the pain remains reduced and selecting the position simply depends on which one works best. The Z-lie is usually the most effective but that choice and the duration of rest are governed by the patient's reports on the pain. For the Z-lie the patient is supine with the lower legs and feet on the seat of a chair or bench and the buttocks underneath. Both the hips and knees are flexed more than 90° so that the thighs are drawn up over the abdomen; generally the greater the tuck the greater the pain relief. Adding a pillow under the head and/or the buttocks may further improve the result. The clinician should experiment with all the factors – the feet on the chair, the distance the buttocks are under the sear, the height of the pillows – to find the best combination. At each modification, the patient is quizzed about the level of pain.

Another useful maneuver is having the patient lie prone over three or four pillows. They are placed in front of the pelvis and adjusted up or down to the most efficacious location. This is, obviously, very different for the Z-lie but can be equally advantageous. The optimum posture depends on the amount of pain reduction but pain control usually improves as the number of pillows increases. As the pain subsides the pillows are sequentially removed.

Managing Pattern 1 PEN is a continuum from rest, typically in flexion, to movement in extension, to a Pattern 1 PEP routine. For the patient with constant back dominant pain where all movement hurts, rest in the way that affords the greatest amount of pain relief is the sensible place to begin. This is frequently the Z-lie. As the symptoms subside movement can be introduced. This can be an unloaded flexion such as knees-to-chest stretches. Paradoxically, although Pattern 1 is aggravated in flexion, most Pattern 1 PEN patients find when starting treatment that unloaded flexion is more comfortable than bending backwards. With increased mobility, treatment progresses to extension: first unmoving, like prone over pillows, then with movement, then the sloppy push-up. The art of managing these patients is choosing how far back along this continuum to begin and how quickly to move forward from static flexion to active extension.

Two other groups qualify as Pattern 1 PEN. The prone passive extension is purely sagittal and involves a full range of movement. Some patients respond only to asymmetrical activity and therefore don't improve with straight line extensions. Others gain relief only with midrange movement and are unable to reach end range. In both instances, ten repetitions of the standard prone passive extension fail to provide pain relief and the patients require modified treatment plans.

Pattern 2

Patients classified as Pattern 2 are never worse in flexion and the back dominant pain is always intermittent. Mechanical therapy is flexion. Except for using a large pillow between the knees when the patient lies down to relax the paraspinal muscles, everything else promotes bending forward. This is easily accomplished in sitting. The patient sits with the knees more that shoulder width apart and bends forward lowering the upper body between the legs. Flexion can be increased by grabbing the ankles and pulling down. To return to an upright posture the patient places the hands on the knees and pushes, using the arms, not the back muscles, to raise the torso. For standing flexion, the patient places one foot up on a bench or chair seat, puts the hands on the elevated thigh, and bends forward to rest the chest on the hands. To straighten up the patient pushes with the arms keeping the back relaxed. The mechanical prescription describes the technique, gives the number of repetitions and specifies the frequency during the day. Pattern 2 responds rapidly and the pain relief is sustained.

Pattern 3

Constant leg dominant pain is managed without movement. Pattern 3 radicular pain results from nerve root inflammation so in the acute phase scheduled rest is most appropriate. Similar to Pattern 1 PEN, the other pattern where scheduled rest is the logical first step, the duration and the spacing of the rest periods are dictated by the patient's pain. Unlike Pattern 1 PEN, the patient with severe radicular pain may need to spend much of time, 30 minutes out of each hour, resting. Several positions can decrease the pain. The Z-lie is the best choice. The setup is the same as described for Pattern 1 PEN, but the deciding factor is now the level of leg pain. The constant leg pain cannot be abolished but the amount can be substantially reduced by slight changes in alignment. Providing precise instructions for a method to achieve some relief also gives the patient a sense of control over a frightening situation, control that can be as beneficial as the mechanical changes. Lying prone over pillows may ease the leg pain. The amount of pain reduction governs the number of pillows; there is no progression to lying flat. The decision to use a Z-lie or to rest prone over pillows is purely pragmatic; the patient is encouraged to experiment. Other options include lying prone on the elbows or even on the hands and knees. Whatever works best is the preferred selection.

As the inflammation and the leg dominant pain subside patients can begin a movement-based routine either as a Pattern 1 PEP or PEN or as a Pattern 4 FA, flexion aggravated.

Pattern 4 FA

The two Pattern 4 categories represent two very different pathologies. The intermittent leg dominant pain of Pattern 4 FA, possibly from residual root impairment or scarring, responds to mechanical treatment. Because the pain increases with flexion, treatment resembles that for Pattern 1, but since the source of the pain is neurogenic rather than purely mechanical, the approach is gentler. The footstool, lumbar roll, and large pillow between the knees can all reduce the leg symptoms. Unloaded back extension, prone over pillows, or extension movements like the sloppy push-ups, may offer relief. The aim of treatment is to diminish the intensity and/or the frequency of the recurrent leg pain so whichever combination works is the best one to use. As with all mechanical therapy, sessions should be specific and repeated frequently during the day.

Pattern 4 FR

The key to managing the symptoms of neurogenic claudication, Pattern 4 FR (flexion relieved), is posture. Flexion increases access to the exiting nerve roots, improving circulation to limit or prevent the symptoms. Maintaining spinal flexion requires strong abdominal muscles so therapy is directed at improving core strength and function. A pelvic tilt is the foundational exercise. Tightening the abdomen rotates the pelvis forward, flattens the lumbar spine, and increases the size of the intervertebral foramina. Performing a pelvic tilt lying down with the knees bent and the feet planted firmly on the floor is relatively easy; maintaining the tilt while walking takes endurance. Core strengthening programs often incorporate using equipment like the Swiss exercise ball or techniques like the one-arm dumbbell bench press, things that are well beyond the ability of the average octogenarian. Pattern 4 FR is most commonly an affliction of the elderly, and although core fitness is a valid principle, in practice it may be impossible to achieve. The affected patient population may not be able to make the long-term commitment to exercise necessary to gain improvement. It is for this reason, not because of a lack of understanding of what mechanical therapy is required to improve function in neurogenic claudication that surgical decompression may be the preferred treatment option.

References

Aina A, May S, Clare H (2004) The centralization phenomenon of spinal symptoms – a systematic review. Man Ther 9:134–143

Alleyne J, Hall H, Rampersaud R (2016) Clinically organized relevant exam (CORE) tool for the low back pain toolkit for primary care providers. J Curr Clin Care 6: 24–31. http://effectivepractice.org/resources/lowback-pain-core-back-tool

Balagué F, Mannion AF, Pellisé F, Cedraschi C (2012) Non-specific low back pain. Lancet 379:482–491

Bogduk N (2009) On the definitions and physiology of back pain, referred pain, and radicular pain. Pain 147:17–19

Brennan GP, Fritz JM, Hunter SJ, Thackeray A, Delitto A, Erhard RE (2006) Identifying subgroups of patients with acute/subacute "nonspecific" low back pain: results of a randomized clinical trial. Spine 31:623–631

Cherkin D, Balderson B, Wellman R, Hsu C, Sherman K, Evers S, Hawkes R, Cook A, Levine M, Piekara D, Rock P Estlin K, Brewer G, Jensen M, La Porte A, Yeoman J, Sowden G, Hill J, Foster N (2018) Effect of low back pain risk-stratification strategy on patient outcomes and core processes: the MATCH randomized trial in primary care. J Gen Intern Med 33:1324–1336

Chien JJ, Bajwa ZH (2008) What is mechanical back pain and how best to treat it? Curr Pain Headache Rep 12:406–411

Deyo RA, Weinstein JN (2001) Low back pain. N Engl J Med 344:363–370

Donelson R, McKenzie R (1992) Effects of spinal flexion and extension exercises on low-back pain and spinal mobility in chronic mechanical low-back pain patients. Spine 17:1267–1268

Donelson R, McIntosh G, Hall H (2012) Is it time to rethink the typical course of low back pain? PM R 4:394–401

Downie A, Williams CM, Henschke N, Hancock MJ, Ostelo RW, de Vet HC, Macaskill P, Irwig L, van Tulder MW, Koes BW, Maher CG (2013) Red flags to screen for malignancy and fracture in patients with low back pain: systematic review. BMJ 347:f7095. https://doi.org/10.1136/bmj.f7095. https://www.bmj.com/content/bmj/347/bmj.f7095.full.pdf

Foster NE, Mullis R, Hill JC, Lewis M, Whitehurst DG, Doyle C, Konstantinou K, Main C, Somerville S, Sowden G, Wathall S, Young J, Hay EM, IMPaCT Back Study Team (2014) Effect of stratified care for low back pain in family practice (IMPaCT Back): a prospective population-based sequential comparison. Ann Fam Med 12:102–111

Fraser S, Roberts L, Murphy E (2009) Cauda equina syndrome: a literature review of its definition and clinical presentation. Arch Phys Med Rehabil 90:1964–1968

Fritz JM, Delitto A, Erhard RE (2003) Comparison of classification-based physical therapy with therapy based on clinical practice guidelines for patients with acute low back pain: a randomized clinical trial. Spine 28:1363–1371

Global Burden of Disease Study 2013 Collaborators (2015) Global, regional, and national incidence, prevalence, and years lived with disability for 301 acute and chronic diseases and injuries in 188 countries, 1990–2013: a systematic analysis for the Global Burden of Disease Study 2013. Lancet 386:743–800

Hall H, McIntosh G, Wilson L, Melles T (1998) The spontaneous onset of back pain. Clin J Pain 14:129–133

Hall H, McIntosh G, Boyle C (2009) Effectiveness of a low back pain classification system. Spine J 9:648–657

Harding IJ, Davies E, Buchanan E, Fairbank JT (2005) The symptom of night pain in a back pain triage clinic. Spine 30:1985–1988

Henschke N, Maher CG, Refshauge KM, Herbert RD, Cumming RG, Bleasel J, York J, Das A, McAuley JH (2009) Prevalence of and screening for serious spinal pathology in patients presenting to primary care settings with acute low back pain. Arthritis Rheum 60:3072–3080

Hill JC, Dunn KM, Lewis M, Mullis R, Main CJ, Foster NE, Hay EM (2008) A primary care back pain screening tool: identifying patient subgroups for initial treatment. Arthritis Rheum 59:632–641

Hill JC, Konstantinou K, Egbewate BE, Dunn KM, Lewis M, van der Windt D (2011) Clinical outcomes among low back pain consulters with referred leg pain in primary care. Spine 36:2168–2175

Koes BW, van Tulder MW, Thomas S (2006) Diagnosis and treatment of low back pain. BMJ 332:1430–1434

Krismer M, van Tulder M, Low Back Pain Group of the Bone and Joint Health Strategies for Europe Project (2007) Strategies for prevention and management of musculoskeletal conditions. Low back pain (non-specific). Best Pract Res Clin Rheumatol 21:77–91

Maher C, Underwood M, Buchbinder R (2017) Non-specific back pain. Lancet 389:736–747

Nadeau M, Rosas-Arellano MP, Gurr KR, Bailey SI, Taylor DC, Grewal R, Lawlor DK, Bailey CS (2013) The reliability of differentiating neurogenic claudication from vascular claudication based on symptomatic presentation. Can J Surg 56:372–377

Norton G, McDonough CM, Cabral HJ, Shwartz M, Burgess JF Jr (2016) Classification of patients with incident non-specific low back pain: implications for research. Spine J 16:567–576

Oxford English Dictionary (2019). www.oed.com

Smythe H (1986) Referred pain and tender points. Am J Med 81:S90–S92

Spitzer WO, LeBlanc FE, Dupuis M (1987) Scientific approach to the assessment and management of activity-related spinal disorders. A monograph for clinicians. Report of the Quebec Task Force on Spinal Disorders. Spine 12:S1–S59

Stanton TR, Fritz JM, Hancock MJ, Latimer J, Maher CG, Wand BM, Parent EC (2011) Evaluation of a treatment-based classification algorithm for low back pain: a cross-sectional study. Phys Ther 91:496–509

Tortolani PJ, Carbone JJ, Quartararo LG (2002) Greater trochanteric pain syndrome in patients referred to orthopedic spine specialists. Spine J 2:251–254

Webster BS, Bauer AZ, Choi Y, Cifuentes M, Pransky GS (2013) Iatrogenic consequences of early magnetic resonance imaging in acute, work-related, disabling low back pain. Spine 38:1939–1946

Weisman MH, Witter JP, Reveille JD (2013) The prevalence of inflammatory back pain: population-based estimates from the US National Health and Nutrition Examination Survey, 2009–10. Ann Rheum Dis 72:369–373

Wilgenbusch C, Wu A, Fourney D (2014) Triage of spine surgery referrals through a multidisciplinary care pathway: a value-based comparison with conventional referral processes. Spine 39:S129–S135

Wnuk N, Alkasab T, Rosenthal D (2018) Magnetic resonance imaging of the lumbar spine: determining clinical impact and potential harm from overuse. Spine 18:1653–1658

Back Pain: The Classic Surgeon's View

2

Neil Berrington

Contents

Introduction	28
The Traditional Approach	28
Specific Spinal Pathology	29
Radicular Syndromes	29
Non-specific Low Back Pain	29
Surgical Treatment of Specific Spinal Pathology	29
Surgical Treatment of Radicular Syndromes	29
Lumbar Disc Prolapse	29
Spinal Stenosis	30
Surgical Treatment of Non-specific Back Pain	31
Degenerative Disc Disease	31
Prolapsed Lumbar Disc	32
Spondylolisthesis	32
Conclusions	33
Cross-References	34
References	34

Abstract

Low back pain is a major cause of disability worldwide and is a major burden on healthcare systems. Treatment strategies are varied and the role of surgery is under constant scrutiny. Many patients benefit from spinal surgery aimed at relieving back pain, radicular symptoms, and neurogenic claudication.

The initial evaluation of patients with categorization into clinical groups may help in appropriately assigning patients to consideration for surgery. Patients presenting with radicular syndromes (radicular pain, radiculopathy, and neurogenic claudication) are widely regarded as potential surgical candidates. Aside from certain distinct groups, non-specific back pain is, as a rule, not regarded as benefiting greatly from surgery. Specific disease entities (such as central disc

N. Berrington (✉)
Section of Neurosurgery, University of Manitoba, Winnipeg, MB, Canada
e-mail: nberrington@hsc.mb.ca

© Springer Nature Switzerland AG 2021
B. C. Cheng (ed.), *Handbook of Spine Technology*,
https://doi.org/10.1007/978-3-319-44424-6_131

prolapse and spondylolysis) might be the exception and should be considered potential surgical candidates.

Patients presenting with radicular syndromes frequently have disc prolapse or spinal stenosis (with or without spondylolisthesis.) After a period of conservative treatment (cognitive behavioral therapy and exercise), surgery may indeed play a major role in treating these patients. Surgical strategies might vary from simple discectomy to complex lumbar decompression and fusions. Outcomes comparable to major joint arthroplasty are sustained for prolonged periods postoperatively.

Careful patient selection and the adoption of less invasive techniques and enhanced recovery after surgery protocols may reduce morbidity and opiate usage in the long run. Surgery remains a valuable and viable option for selected patients presenting with low back pain and associated syndromes.

Keywords

Non-specific back pain · Disc arthroplasty · Spinal fusion · Spondylolisthesis · Guidelines · Opiates · ERAS

Introduction

The almost ubiquitous occurrence of low back pain in the general population makes the symptom a major burden on healthcare systems worldwide. The rising cost of provision of care, the current opiate crisis in North America, and the vast array of less than scientific opinions offered in the popular media have made the condition a major political issue.

In countries with a significant state contribution to healthcare, governments are placing increased scrutiny of healthcare providers to provide cost-effective treatments for spinal conditions. Universal healthcare systems have become bogged down with long wait times for both assessments and treatment of spinal pathology.

It is estimated that the global prevalence of activity limiting low back pain is 7.2% of the population (540 million people). It is now regarded as the single biggest cause of disability worldwide, with the greatest occurrence of disability in lower-income countries (Hartvigsen et al. 2018).

Aside from the cost of care provision, the problem is further compounded by the emergence of a virtual epidemic of opioid use. In Canada alone, the volume of opioid prescriptions has increased by a factor of 3000% since the 1980s (Belzak and Halverson 2018) with a prevalence of opioid-related deaths exceeding accident mortalities (CIHI 2018). As many as 2% of US adult report the regular prescription for opioids with about half of these prescribed for low back pain (Deyo et al. 2011). A recent meta-analysis has shown that prescribing opioids for low back pain is significantly associated with ongoing long-term use (Sanger et al. 2019).

Surgical strategies to alleviate low back pain have become increasingly adopted in North America, with the rates of spinal surgery in the USA now being the highest in the world (Chou et al. 2009). The evidence for surgical efficacy has been inconsistent, and recent critiques have labelled surgery as costly and not providing outcomes any better than intensive multidisciplinary rehabilitation (Foster et al. 2018).

Most spinal surgeons would however contend that the role for surgery in the management of many of the causes of spinal disabilities is clear, and confounding statistics and public opinion with undifferentiated data do a large number of patients a huge disservice.

Successful spinal surgery, like much of medicine, is reliant on the thorough clinical evaluation of a patient, a thoughtful analysis of special investigations and an understanding of the best evidence available in modern literature.

The Traditional Approach

Classical teaching has to date held firm that surgery is key in the successful management of many radicular syndromes, cauda equina syndrome and neurogenic claudication. The role for surgery in treating "undifferentiated back pain" and degenerative disc disease is much less clear. As a consequence of this understanding, many referral pathways are designed such as to direct patients with undifferentiated back pain away from interventional trajectories of care.

The general approach to assessing low back pain is well laid out by a group of Australian primary care providers (Bardin et al. 2017). The initial assessment of a patient with low back pain is contingent on an accurate and thorough history and clinical examination. In this manner, non-spinal pathology can be, for the most part, excluded. Hip pathology, kidney stones, urinary tract infections, and abdominal pathology can then be appropriately directed. The remaining patients are then triaged into three categories:

Specific Spinal Pathology

Fractures, infections, tumors, and cauda equina syndrome should be, for the most part, apparent with a clinical evaluation. This category of patient frequently falls into the realm of spinal surgery as a means to address the spine pathology, relieve neural compression, alleviate pain, and prevent progressive deformity.

Radicular Syndromes

Patients manifesting *radicular pain* account for about 60% of the low back pain population (Hill et al. 2011). Patients with *radiculopathy* on the other hand present with a syndrome characterized by dermatomal sensory loss, myotomal muscle weakness, or absent deep tendon reflexes. Patients with classic spinal stenosis in turn might present with *neurogenic claudication*. Common practice in the treatment of radicular syndromes is a trial of conservative treatment, and surgery is only considered if symptoms extend beyond 6 weeks. The surgical treatment of radicular syndromes remains highly controversial.

Non-specific Low Back Pain

This undifferentiated catch all includes a number of diverse pathologies, mostly regarded as "musculoskeletal." This would include disc prolapse, facet arthropathy, degenerative disc disease, annulus tears, spondylolisthesis, and muscle injuries. The diverse nature of this group makes a universal recommendation (or condemnation) of surgery inappropriate.

Categorizing patients in this manner does help identify a large portion of surgical candidates. Unfortunately a large degree of overlap in symptoms occurs, and patients presenting with a radicular syndrome might have coexisting non-specific low back pain. Attempting to ascribe some degree of dominance to a particular symptom might assist a surgeon in navigating the next course of action.

In healthcare systems with constrained resources, this can aid in identifying patients who might most readily benefit from a surgical consultation or special investigations. A triage pathway has been implemented in Saskatchewan, Canada. The Canadian group was able to demonstrate considerable reductions in wait times and MRI utilization within a universal healthcare system (Fourney et al. 2011). The group regards patients as having syndromes of back pain that are either *leg-dominant* (i.e., radicular syndromes) or *back-dominant*, with an implicit understanding that back-dominant pain represents the non-specific low back pain group and is diverted along a pathway of conservative management.

Surgical Treatment of Specific Spinal Pathology

Much of this tome is devoted to the management of what would seem to be clear-cut indications for spinal surgery. The patient selection and technique adopted will vary widely according to the associated pathology. These conditions are beyond the scope of the discussion in this chapter.

Surgical Treatment of Radicular Syndromes

Lumbar Disc Prolapse

Lumbar disc prolapse is the commonest cause of radicular syndromes, and microdiscectomy is, as a result, the most common neurosurgical procedure in North America. Despite this, the optimal treatment of lumbar disc prolapse remains highly controversial.

Lumbar disc prolapse is the displacement of intervertebral disc material beyond the normal margins of the disc space and might include disc nucleus and or annulus. The prolapse might be contained within the limits of a bulging annulus and posterior longitudinal ligament or be sequestrated from the ligamentous confines above. On rare occasion, sequestrated disc material can even extrude into the dural tube.

Laterally placed prolapse might account for a radicular syndrome, while large central prolapses can cause profound low back pain. Severe compression of neural elements can manifest as *cauda equina syndrome*.

The decision regarding treatment rests with a well-executed physical examination, as special investigations might not be required, and MRI findings can be misleading. Disc prolapses are frequently noted on MRI scans in asymptomatic patients (Brinjikji et al. 2015). The accuracy of findings on clinical evaluation is somewhat variable but taken as a whole can guide initial treatment options.

A history of leg-dominant pain with a typical dermatomal distribution of pain and positive straight leg raising test is sensitive for diagnosing lumbar disc prolapse, while crossed straight leg raising, paresis, muscle atrophy, and loss of deep tendon reflexes are highly specific for disc prolapse (Deyo and Mirza 2016). The presence of psychological distress, depression, and somatization should be sought, as these symptoms are associated with less than ideal surgical outcomes (Kreiner et al. 2014).

It is critical to exclude any suggestion of a cauda equina syndrome – a history and physical examination reflecting saddle anesthesia, incontinence or urine retention, and poor anal sphincter tone establish this syndrome clinically. The duration of symptoms prior to surgical treatment is a determinant of outcome, and the rates of permanent urinary incontinence increase dramatically over time (Qureshi and Sell 2007).

The natural history of lumbar disc prolapse is generally favorable, with up to 87% of patients reporting a reduction in analgesic use at 3 months and 81% of patients with motor deficit reporting improvement at 1 year (Deyo and Mirza 2016).

Attempts at randomized trials comparing surgery to conservative treatment are confounded by a lack of standardization of treatment arms and considerable crossover. The most robust and frequently sited study on the subject, the multicenter Spine Outcomes Research Trial (SPORT), had such a high crossover of subjects that it is difficult to deduce much at all other than patients will make appropriate choices as their symptoms dictate. As many as 60% of patients left the surgical arm as their symptoms were improving spontaneously (Birkmeyer et al. 2002.) Systematic review of the literature does however support the notion of discectomy favoring nonsurgical treatment for the short-term resolution of symptoms (Gibson 2007).

The ideal approach to preforming a discectomy has not been clearly defined, although a tendency toward less invasive strategies has been associated with shorter hospital stay and shorter periods of absence from work. A recent study on endoscopic discectomy demonstrated this effect, although outcomes at 12 months post-surgery are identical to a standard microdiscectomy (Ruetten et al. 2008).

Discectomy procedures have a relatively low complication rate, and surgery is associated with a quicker return to work than conservative treatment (Deyo and Mirza 2016). Surgery is thus an effective treatment for disc prolapse where symptoms have persisted beyond 6 weeks. Considering the benefits of reducing the use of analgesics (particularly opioids), discectomy remains a highly relevant treatment strategy for well-selected cases.

Spinal Stenosis

A radiological or pathological finding of *spinal stenosis* is universal in older patients, although the relationship between the anatomical diagnosis of stenosis is a tenuous one. Many clinicians prefer to regard symptomatic stenosis as an entity defined by the symptom complex, i.e., radiculopathy or neurogenic claudication (Backstrom et al. 2011). An entity of claudicant back pain is seen from time to time. These are patients who seem to have a back-dominant pattern of pain, which is aggravated by walking and relieved by sitting down. Selecting this group of patients for surgery does require considerable deliberation. Further complicating the heterogeneity of this group are the entities of stenosis associated degenerative spondylolisthesis and degenerative kyphoscoliosis.

Due to the diversity of pathology accounting for the symptoms and the lack of uniform terminology in describing what is being treated, studies comparing surgery with conservative treatment are clouded with interfering variables and a heterogeneity of patients.

The commonly cited large, multicenter Spine Patient Outcomes Research Trial (SPORT trial) attempted to analyze stenosis in isolation, excluding patients with degenerative spondylolisthesis. In this study, surgery for stenosis had statistically better outcomes for the surgical group for radicular pain, back pain, function, and patient satisfaction (Weinstein et al. 2010). The favorable outcomes were sustained 4 years post-surgery. The outcome of the SPORT trial seems to be validated by data generated by the Canadian Spine Society "Canadian Spine Surgery Outcomes Network" (CSORN) database. Interestingly these data seem to support a contention that both radicular and back pain improvement is sustained postoperatively (Srinivas et al. 2019).

Although the benefit of surgical treatment would appear to be clear, the choice of intervention is less so. Several trials have attempted to compare decompression alone versus decompression and fusion. Although outcomes appear to be comparable in terms of long-term relief of symptoms, the complication rate of decompression with fusion is somewhat higher (Chou et al. 2009). Most recently it has been proposed that instrumented fusion without decompression might even be appropriate (Goel et al. 2019).

What has become evident of late is that less invasive surgical techniques are associated with a shorter duration of absence from work. In a systematic review, the authors found the predictors of delayed return to work which included age, comorbidities, duration and severity of symptoms, depression, mental stress, lateral disc prolapse, and more invasive surgical techniques (Huysmans et al. 2018). Minimally invasive techniques augmented by Enhanced Recovery After Surgery (ERAS) protocols have been shown to reduce perioperative opiate use (Brusko et al. 2019), which in turn might reduce the burden of opiate abuse in this patient population (Berrington 2019).

Surgical Treatment of Non-specific Back Pain

Of all surgically treated conditions, non-radicular back pain remains the most controversial. The Lancet medical journal featured a series of articles focused on the problem of low back pain. Much of the debate generated by the series related to the efficacy (or lack of) in treating low back pain as group. Arguably one of the most influential journals, the publication found surgery to have "insufficient evidence" of efficacy and the role of spinal fusion to be uncertain (Foster et al. 2018). Spinal surgeons however continue to maintain there are certain groups of patients who benefit from surgical intervention.

Degenerative Disc Disease

Maintaining an active lifestyle, exercise therapy, and cognitive behavioral therapy is likely the most appropriate first-line treatment for low back pain (Foster et al. 2018). Studies on the surgical management of degenerative disc disease have yielded somewhat unimpressive results. Naturally not all procedures are equally as effective, and nor are all patients burdened with equivalent pathology.

Spinal fusion is the most commonly performed surgical procedure for this condition, and whether executed via the use of pedicle screws or augmented with some form of anterior stabilization, surgical strategies yield results not much better than the natural history of the condition.

A commonly quoted trial that proponents of surgery would site is the Swedish Lumbar Spine

Study. The study won the Volvo Award in 2001 and is used as evidence of the efficacy of the surgery (Fritzell et al. 2001). In their study the outcomes were favorable in 46% surgical of cases and 18% of nonsurgical cases. Although statistically different, the net overall benefit of surgery is thus only felt in less than 30% of cases. The conclusion of the article (recommending spinal fusion in this condition) is at odds with the results.

Most trials studying spinal fusion outcomes in degenerative disc disease are inconsistent in outcomes, and the role is indeed uncertain (Chou et al. 2009).

Lumbar disc arthroplasty procedures have been touted as the solution to the marginal gains observed in fusion patients. The two major North American trials studying the efficacy of arthroplasty used fusion patients as their control groups. The Charite artificial disc which was compared to a stand-alone interbody cage failed to demonstrate any significant difference in patient outcomes (Blumenthal et al. 2005), while the Pro-Disc-L trial showed equivalence with 360 degree fusion (Zigler et al. 2007).

The Charite trial in particular is problematic to interpret as it attempted to demonstrate non-inferiority to what was, even at that time, not regarded as the standard surgical treatment for degenerative disc disease (a stand-alone interbody cage). Complications associated with the anterior approach have also led to a slow uptake of the procedure. To what extent lumbar arthroplasty will contribute to treating this group thus remains uncertain.

Emerging technologies attempting to provide motion preservation with a degree of stabilization continue however to intrigue investigators (De Muelenaere and Berrington 2015), but to date the role of any surgery to alter the natural course of degenerative disc disease remains uncertain.

Prolapsed Lumbar Disc

Anecdotal reports of patients with large central herniation of a lumbar disc have indicated a simple discectomy that might improve back-dominant symptoms in selected cases. The mechanism proposed is that dural tension from the disc mass is the meteor for local pain and muscle spasm in the absence of radicular symptoms (Adams 1998; Fig. 1).

Spondylolisthesis

Spondylolisthesis refers to anterior displacement of the vertebral body in reference to the bordering vertebral bodies (Gagnet et al. 2018). Spondylolysis in turn refers to a defect or fracture in the pars interarticularis. Spondylolysis can occur with or without spondylolisthesis.

Five forms of spondylolisthesis are identified (Wiltse et al. 1976):

Type I – Dysplastic: congenital dysplasia and malformation of the first sacral vertebra, resulting in slippage of the L5 vertebra anteriorly

Type II – Isthmic – a defect in the pars interarticularis (IIA) or lengthening of the pars interarticularis due to repetitive fracturing and healing (IIB)

Type III – Degenerative – degenerative failure of the facet joints and ligamentum flavum allowing slippage of the vertebrae anteriorly

Type IV – Traumatic – secondary to trauma to the spine

Type V – Pathologic – lytic tumors, osteopetrosis and osteoporosis resulting in pars defects and subsequent slippage

Spondylolytic spondylolisthesis is the commonest cause of back pain in adolescents and young adults, and conservative management is usually advocated. This takes the form of rest, bracing, and physiotherapy (Blanda et al. 1993). Healing of the pars defect is expected in most early stage cases (Gagnet et al. 2018). When the defect is progressive, success of conservative treatment drops off drastically and generally does not occur once sclerotic change is noted (terminal stage).

Surgical repair of pars defects has taken a number of forms, including screw hook constructs and direct grafting. Where conservative treatments have failed, pars repair remains a viable option. Once a listhesis occurs, most surgeons opt for more substantial implants.

Although a number of procedures have been used to address spondylolytic spondylolisthesis,

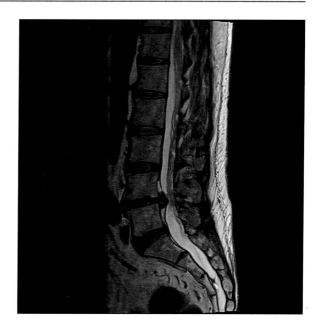

Fig. 1 T2-weighted sagittal MRI scan of a patient with non-specific low back pain and a large central disc herniation. This patient experienced resolution of pain with a simple microdiscectomy

pedicle screw fixation with decompression is generally regarded as the surgical treatment of choice (Violas and Lucas 2016; Gagnet et al. 2018). Lumbar decompression and fusion for degenerative listhesis has been shown to demonstrate sustained improvement comparable to hip and knee arthroplasty (Rampersaud et al. 2014).

Degenerative spondylolisthesis, on the other hand, remains a contested topic with conflicting outcomes in several well-constructed trials. A Scandinavian group showed equivalent outcomes of decompression only when compared to decompression and fusion surgery for stenosis with and without listhesis (Försth et al. 2016), while in the same journal, the frequently cited *Spinal Laminectomy* versus *Instrumented Pedicle Screw* (SLIP) trial demonstrated superiority of outcome at 4 years for the fusion group (Ghogawala et al. 2016).

Both the above studies do show significant and sustained improvement of surgery in patients with spinal stenosis with degenerative listhesis. While it would appear an instrumented fusion that might provide longer term relief, uninstrumented decompression is still clearly a viable option. Careful patient selection might dictate the nature of the surgery, particularly since decompression alone seems to provide meaningful relief of quality of life data.

Conclusions

When confronted with an array of pathology and multiple treatment options, maintaining a clear perspective is difficult. Attempts to offer guidance have been forthcoming from both the North American Spine Society and the Canadian Spine Society.

Choosing Wisely Canada has attempted to provide a consensus opinion of the spine community to allow physicians to make appropriate decisions in this complicated and controversial field. The guidelines are based on both a thorough literature review and expert opinion of the Society members. Their recommendations (1) and (7) are the most relevant to this particular topic. The recommendations follow in Fig. 2.

Patients with low back pain represent a diverse group of pathology, and treatment modalities remain controversial. Identifying which patients are best suited to which type of surgery becomes an art, as it represents the surgeons understanding of the clinical presentation and one's knowledge

1. Don't perform fusion surgery to treat patients with mechanical axial low back pain from multilevel spine degeneration in the absence of: (a) leg pain with or without neurologic symptoms and/or signs of concordant neurologic compression, and (b) structural pathology such as spondylolisthesis or deformity.
2. Don't routinely image patients with low back pain regardless of the duration of symptoms unless: (a) there are clinical reasons to suspect serious underlying pathology (i.e., red flags), or (b) imaging is necessary for the planning and/or execution of a particular evidenced-based therapeutic intervention on a specific spinal condition.
3. Don't use epidural steroid injections (ESI) for patients with axial low back pain who do not have leg dominant symptoms originating in the nerve roots.
4. Don't miss the opportunity to brace the skeletally immature patient with adolescent idiopathic scoliosis (AIS) who has more than one year of growth remaining and a curve magnitude greater than 20 degrees.
5. Don't order peri-operative antibiotics beyond a 24-hour post-operative period for non-complicated instrumented cases in patients who are not at high risk for infection or wound contamination. Administration of a single pre-operative dose for spine cases without instrumentation is adequate.
6. Don't use an opioid analgesic medication as first-line treatment for acute, uncomplicated, mechanical, back-dominant pain.
7. Don't treat post-operative back pain with opioid analgesic medication unless it is functionally directed and strictly time limited.
8. Don't use opioid analgesic medication in the ongoing treatment of chronic, non-malignant back pain.

Fig. 2 Recommendations of *Chooseing Wisely Canada* for patients with spinal symptoms (Choosing Wisely Canada 2019)

of the literature as a whole. As will be seen in this handbook, there are numerous options available to the surgeon.

Certain patients will be identifiable as patients with specific spinal pathology that would potentially require surgical correction ("red flag" patients). Identifying patients with persistent radicular syndromes or claudication will aid in the selection of patients who might benefit from surgery. At that point treatment is individualized, based on which symptoms dominate, the nature of the underlying pathology, and the suitability of the patient for surgery based on comorbid or psychological factors.

Although patients with non-specific low back are usually diverted away from a surgical stream, there might be patients who would benefit from surgery in certain instances, and a blanket condemnation of this form of surgery would seem inappropriate. As technology advances and out understanding of spinal biomechanics changes, hopefully greater numbers of patients will derive benefit form appropriately tailored interventions.

Overall, the modification of surgical stress through the adoption of ERAS and other quality improvement protocols may also in the long run improve outcomes and reduce the use of opiate drugs.

Cross-References

▶ Biological Treatment Approaches for Degenerative Disc Disease: Injectable Biomaterials and Bioartificial Disc Replacement
▶ Posterior Dynamic Stabilization

References

Adams C (1998) A neurosurgeon's notebook – one man's way of trying to avoid trouble. Blackwell Scientific, Oxford

Backstrom KM, Whitman JM, Flynn TW (2011) Lumbar spinal stenosis-diagnosis and management of the aging spine. Man Ther 16:308–317

Bardin L, King P, Maher C (2017) Diagnostic triage for low back pain: a practical approach for primary care. MJA 206(6)

Belzak L, Halverson J (2018) The opioid crisis in Canada: a national perspective. Health Promotion and Chronic Disease Prevention in Canada 38(6):224–232

Berrington NR (2019) Editorial. The opioid crisis: an opportunity to alter morbidity through the implementation of enhanced recovery after surgery protocols during spinal surgery? Neurosurg Focus 46(4):E13

Birkmeyer NJ, Weinstein J, Tosteson A et al (2002) Design of the Spine Patient Outcomes Research Trial (SPORT). Spine 27:1361–1372

Blanda J, Bethem D, Moats W et al (1993) Defects of the pars interarticularis in athletes: a protocol for non-operative treatment. J Spinal Disord 6(5):406–411

Blumenthal S, McAfee PC, Guyer RD et al (2005) A prospective, randomized, multicenter Food and Drug Administration investigational device exemptions study of lumbar total disc replacement with the CHARITE artificial disc versus lumbar fusion: part I: evaluation of clinical outcomes. Spine 30:1565–1575

Brinjikji W, Luetmer PH, Comstock B et al (2015) Systematic literature review of imaging features of spinal degeneration in asymptomatic populations. AJNR Am J Neuroradiol 36:811–816

Brusko GD, Kolcun JPG, Heger JA et al (2019) Reductions in length of stay, narcotics use, and pain following implementation of an enhanced recovery after surgery program for 1- to 3-level lumbar fusion surgery. Neurosurg Focus 46(4):E4

Choosing Wisely Canada – Spine. https://choosingwiselycanada.org/spine/. Accessed 15 Jan 2019

Chou R, Baisden J, Carragee E et al (2009) Surgery for low back pain: a review of the evidence for an American pain society practice guideline. Spine 34(10):1094–1109

CIHI – Canadian Institute for Healthcare Information (2018) Opioid harms. https://www.cihi.ca/sites/default/files/document/opioid-harms-chart-book-en.pdf. Accessed 15 Jan 2019

De Muelenaere P, Berrington N (2015) Initial outcomes of two year follow up of a novel elastomeric disc nucleus prosthesis. Society for Minimally Invasive Spine Surgery Global Forum, Las Vegas

Deyo R, Mirza S (2016) Herniated lumbar intervertebral disk. N Engl J Med 374:1763–1772

Deyo RA, Smith DHM, Johnson ES et al (2011) Opioids for back pain patients: primary care prescribing patterns and use of services. J Am Board Fam Med 24:717–727

Försth P, Ólafsson G, Carlsson T et al (2016) A randomized, controlled trial of fusion surgery for lumbar spinal stenosis. N Engl J Med 374:1413–1423

Foster N, Anema J, Cherkin D et al (2018) Prevention and treatment of low back pain: evidence, challenges, and promising directions. Lancet 391:2368–2383

Fourney DR, Dettori JR, Hall H et al (2011) A systematic review of clinical pathways for lower back pain and introduction of the Saskatchewan Spine Pathway. Spine 36(Suppl):S164–S171

Fritzell P, Hagg O, Wessberg P, Swedish Lumbar Spine Study Group et al (2001) 2001 Volvo Award Winner in Clinical Studies: lumbar fusion versus nonsurgical treatment for chronic low back pain: a multicenter randomized controlled trial from the Swedish Lumbar Spine Study Group. Spine 26:2521–2532

Gagnet P, Kern K, Andrews K et al (2018) Spondylolysis and spondylolisthesis: a review of the literature. J Orthop 15:404–407

Ghogawala Z, Dziura J, Butler W et al (2016) Laminectomy plus fusion versus laminectomy alone for lumbar spondylolisthesis. N Engl J Med 374:1424–1434

Gibson JN (2007) Surgical interventions for lumbar disc prolapse: updated Cochrane review. Spine 32:1735–1747

Goel A, Ranjan S, Patil A et al (2019) Lumbar canal stenosis: analyzing the role of stabilization and the futility of decompression as treatment. Neurosurg Focus 46(5):E7

Hartvigsen J, Hancock MJ, Kongsted A et al (2018) What low back pain is and why we need to pay attention. Lancet 391:2356–2367

Hill JC, Konstantinou K, Egbewale BE et al (2011) Clinical outcomes among low back pain consulters with referred leg pain in primary care. Spine 36:2168–2175

Huysmans E, Goudman L, Van Belleghem G, et al. (2018) Return to work following surgery for lumbar radiculopathy: a ststematic review. The Spine Journal 18:1694–1714

Kreiner DS, Hwang SW, Easa JE et al (2014) An evidence-based clinical guideline for the diagnosis and treatment of lumbar disc herniation with radiculopathy. Spine J 14:180–191

Qureshi A, Sell P (2007) Cauda equina syndrome treated by surgical decompression: the influence of timing on surgical outcome. Eur Spine J 16(12):2143–2151

Rampersaud YR, Lewis SJ, Davey JR et al (2014) Comparative outcomes and cost-utility after surgical treatment of focal lumbar spinal stenosis compared with osteoarthritis of the hip or knee – part 1: long-term change in health-related quality of life. Spine J 14(2):234–243

Ruetten S, Komp M, Merk H et al (2008) Full-endoscopic interlaminar and transforaminal lumbar discectomy versus conventional microsurgical technique. Spine 33:931–939

Sanger N, Bhatt M, Singhal N et al (2019) Adverse outcomes associated with prescription opioids for acute low back pain: a systematic review and meta-analysis. Pain Physician 22:119–138

Srinivas S, Paquet J, Bailey C et al (2019) Effect of spinal decompression on back pain in lumbar spinal stenosis: a Canadian Spine Outcomes Research Network (CSORN) study. Spine J. In Press. https://doi.org/10.1016/j.spinee.2019.01.003

Violas P, Lucas G (2016) L5-S1 spondylolisthesis in children and adolescents. OrthopTraumatol: Surg Res 102(1):S141–S147

Weinstein JN, Tosteson TD, Lurie JD et al (2010) Surgical versus non-operative treatment for lumbar spinal stenosis four-year results of the Spine Patient Outcomes Research Trial (SPORT). Spine (Phila Pa 1976) 35(14):1329–1338

Wiltse LL, Newman PH, Macnab I (1976) Classification of spondylolysis and spondylolisthesis. Clin Orthop Relat Res 117:23–29

Zigler J, Delamarter R, Spivak JM et al (2007) Results of the prospective, randomized, multicenter Food and Drug Administration investigational device exemption study of the ProDisc-L total disc replacement versus circumferential fusion for the treatment of 1-level degenerative disc disease. Spine 32:1155–1162, discussion 1163

Back Pain: Chiropractor's View

I. D. Coulter, M. J. Schneider, J. Egan, D. R. Murphy, Silvano A. Mior, and G. Jacob

Contents

Introduction	38
The Opioid Crisis	38
Widespread Use of Chiropractic for Spinal Care	40
Mounting Evidence About the Efficacy and Safety of Chiropractic Care	40
Evidence for Outcomes for Medical Therapies for Spinal Problems	42
Removal of Legal and Ethical Barriers	42

I. D. Coulter (✉)
RAND Corporation, Santa Monica, CA, USA
e-mail: coulter@rand.org

M. J. Schneider
School of Health and Rehabilitation Sciences, Clinical and Translational Science Institute, University of Pittsburgh, Pittsburgh, PA, USA
e-mail: mjs5@pitt.edu

J. Egan · G. Jacob
Southern California University of Health Sciences, Whittier, CA, USA
e-mail: jonathonegan@scuhs.edu; dclacmph@gmail.com

D. R. Murphy
Department of Family Medicine, Alpert Medical School of Brown University, Providence, RI, USA

Department of Physical Therapy, University of Pittsburgh, Cranston, RI, USA
e-mail: CRISP4PSP@gmail.com

S. A. Mior
Department of Research, Canadian Memorial Chiropractic College, Toronto, ON, Canada

Centre for Disability Prevention and Rehabilitation, Ontario Tech University and Canadian Memorial Chiropractic College, Toronto, ON, Canada
e-mail: smior@cmcc.ca

© Springer Nature Switzerland AG 2021
B. C. Cheng (ed.), *Handbook of Spine Technology*,
https://doi.org/10.1007/978-3-319-44424-6_141

Insurance Coverage	44
What You Need to Know Before Talking to a Chiropractor	45
Education and Training of Chiropractic	46
The Chiropractic Treatment of the Spine	48
The Perspective	49
The Clinical Encounter	51
Chiropractic Manual Therapy	52
What Do Chiropractors Treat?	54
Who Do They Treat	54
Outcomes of Chiropractic Care	56
Can the Elderly Benefit from Chiropractic Care?	57
How Does One Find a Good Chiropractor?	58
Conclusion	59
References	60

Abstract

In educating engineering students about geology, the advice given to the geology tutors is your job is to convince the engineers that when they need a geologist they should send for one. That might also be a good advice now to those in the medical professions treating patients with non-specific back pain: when you need a chiropractor, send for one or at least talk to one. In the face of the escalating opioid deaths from prescription drugs, there is a compelling, and ethical, obligation for health providers to consider non-pharmacological therapies for treating pain. These therapies have evidence for efficacy and safety, are not addictive, and are associated with a very low rate of adverse events. Chiropractic falls squarely within the framework of these therapies. In most jurisdictions chiropractic is defined as the treatment of musculoskeletal conditions *without the use of drugs or surgery*. As with all the complementary and alternative medicine (CAM) professions, pain is the number one condition that drives patients to CAM providers and specifically low back pain in the case of chiropractors. It is estimated that one in five of all adults in the United States and Canada will use chiropractic care at some time in their life. The purpose of this chapter is to provide information that can help medical practitioners in deciding when to refer to chiropractors. This requires some knowledge about chiropractic education and training, the thinking and practice behind chiropractic care, the different views about back pain and health generally, and the different types of management of back pain arising from that point of view.

Keywords

Chiropractic · Non-pharmacological · Complementary and alternative medicine · Spinal manipulation · Primary spine care

Introduction

There are currently several compelling reasons supporting the need for medicine to change its historical stance toward chiropractic. To use the cliché, "times have changed." Among the compelling reasons are the following.

The Opioid Crisis

A total of 70,237 Americans died from drug overdose in 2017, and synthetic opioids are the main contributor, accounting for 47,600 of those deaths. From 1991 until 2017, almost 400,000 people died from an overdose involving any opioid. Of these, from 1991 to 2017, 218,000 people died from overdoses of prescription opioids. Opioids are now the third leading cause of death after heart disease and cancer (https://www.cdc.gov/drugoverdose/data/prescribing/overview.html).

The figures are similar for most western countries. While much of the blame can and should be laid at the feet of drug companies, this crisis is also a medically induced iatrogenic crisis. During an age of evidence-based medicine, it is difficult to understand how new drugs with so little evidence about their addictive nature or associated adverse effects could be so widely prescribed for a common condition like low back pain (Coulter 2018). Systematic reviews of the literature have shown that opioids are actually not very useful in controlling low back pain and are associated with high rates of adverse events (Tucker et al. 2019; Sanger et al. 2019).

There are multiple components to this tragedy, and death is only one part of the picture. Other negative consequences include addiction, lives ruined, crime, and the economic cost to society. The human suffering and social costs are almost incomprehensible. The fact that the crisis came from a desire in medicine to help patients deal effectively with pain adds an element of tragic irony to the crisis. This is somewhat akin to Oedipus declaring he will find and punish the man who caused the plague when Oedipus himself is the cause. But one of the saddest parts of the opioid crisis is that there exists a whole range of nonpharmacological therapies for treating pain. These include such therapies as the chiropractic profession's manipulation, acupuncture, massage therapy, and a whole range of complementary and alternative medicine (CAM) treatments that are available and already treating pain. Estimates from the National Health Survey show that around 38% of the population is using some type of CAM therapy (Clarke et al. 2015).

Of all the available alternative professions, the chiropractic profession is the one most utilized by Americans and Canadians. Low back pain is by far the largest condition category for which CAM therapies are most frequently used, followed by neck pain as the second most common condition (Beliveau et al. 2017). In addition, chiropractic care is rated higher in terms of patient satisfaction (Hertzman-Miller et al. 2002; Yu et al. 2002; Beattie et al. 2011; Herman et al. 2018) as compared to medical care, physical therapy, and osteopathy for low back pain. There is also evidence from one study that initial visits to chiropractors or physical therapists are associated with substantially decreased early and long-term use of opioids. Patients who received initial treatment from chiropractors or physical therapists had decreased odds of short-term and long-term opioid use compared with those who received initial treatment from primary care physicians (Kazis et al. 2019). In the Herman et al. 2019a, observational study, 1,835 chronic back pain patients rated their chiropractic provider at the top of the patient satisfaction scale, and 90% reported that they were extremely confident that their chiropractor would be very or extremely successful in reducing their pain, and over 90% would recommend chiropractic to a friend (Herman et al. 2019a). In the same study (Herman et al. 2018), using 2,024 patients at baseline, over 90% reported high satisfaction with their care, and very few used narcotics. Patients have also stated that avoiding surgery and medications were the most important reasons they chose chiropractic. They also reported high levels of belief in the success of chiropractic in reducing their pain (Hays et al. 2019).

In 2019, UnitedHealthcare was the largest provider of health insurance in the United States. They just announced an innovative new benefit plan for patients in employer-sponsored plans that cover physical therapy and chiropractic services. There will be $0 out-of-pocket cost to patients with low back pain if they choose to see a chiropractor or physical therapist as the first-contact provider, instead of seeing a primary care physician or specialist. To quote from this new benefit plan: "Based on a UnitedHealthcare analysis, by 2021 this benefit design has the potential to reduce the number of spinal imaging tests by 22%, spinal surgeries by 21%, opioid use by 19%, and lower the total cost of care for eligible plan participants and employers." They also note that opioids were still being prescribed to 9% of patients with low back pain and that this condition is the most common reason for giving opioids (https://www.fiercehealthcare.com/payer/unitedhealth-introduces-new-benefit-for-treating-low-back-pain-for-employer-plans? Nov 1st 2019).

Widespread Use of Chiropractic for Spinal Care

The second compelling reason to talk to a chiropractor is the high probability that your patients are already utilizing their services. There are more than 103,000 chiropractors practicing in 90 countries, with the largest number of chiropractors per capita found in the United States (Stochkendahl et al. 2019). Chiropractic care is one of the most commonly used CAM therapies in Europe, Canada, and the United States (Beliveau et al. 2017; Canizares et al. 2017). As noted above (Clarke et al. 2015), chiropractic is the fourth most used CAM therapy in the United States, but if we exclude natural products, deep breathing, and meditation (all non-provider-based therapies), chiropractic is the most used therapy. Globally, the median 12-month utilization of chiropractic services is 9.1% (IQR, 6.7–13.1%) and lifetime utilization of 22.2% (IQR, 12.8–40.0%) (Beliveau et al. 2017). At least 8–14% of the population in the United States seeks chiropractic care each year, there are 190 million patient visits annually, and there are 70,000 actively licensed chiropractors. A Gallup survey in 2015 showed that more than 50% of US adults have previously sought care from a Doctor of Chiropractic (DC) at some point in their lives (Weeks et al. 2015), while 14% had done so within the previous year. Chiropractors can now be found in private practice, multidisciplinary health treatment facilities, professional athletic teams including olympic teams, military health facilities, and the Veterans Affairs (VA) health facilities (Lisi et al. 2009; Green et al. 2009). More than 100 VA healthcare facilities in the United States currently have chiropractic clinics staffed by DCs. In fiscal year 2018, there were 50,000 veterans receiving in-house chiropractic care and another 80,000 veterans referred to community care programs for chiropractic services. The number of veterans receiving care has more than doubled since 2015 with similar future growth expected due to veteran demand, expansion and success of VA chiropractic services, and a shift in healthcare resources to evidenced-based non-pharmacological options for spine-related conditions and chronic pain (Lisi and Brandt 2016; Dunn et al. 2009).

While chiropractors and medical doctors were always linked through their patients, if surreptitiously, the lack of communication between medical and chiropractic providers has not been in the patient's best interest. It behooves any provider to be informed about the remedies patients take for their health, in addition to those provided by the physician. Clearly a physician cannot advise a patient about the interactive effects of combining therapies if the physician does not know about them. In a groundbreaking study on the use of CAM therapies, one of the most striking results was the finding that patients did not inform their medical doctors about their use of CAM therapies. The chief reasons given were they did not need to know, they never asked, and that they would not understand. In addition, patients expressed concerns that their doctor would disapprove and discourage them from using CAM therapies and/or stop being their provider (Eisenberg et al. 2001).

Mounting Evidence About the Efficacy and Safety of Chiropractic Care

A third compelling reason to work with chiropractors is that there is now a strong evidence base for the safety and effectiveness of chiropractic manipulation, which should allay any concerns about vicarious liability (Nahin et al. 2016). As the level of evidence of safety and effectiveness increases and referrals are more common, the threat of direct liability also decreases (Gilmour et al. 2011a). There are some 100 published randomized clinical trials (RCTs) on the effectiveness of spinal manipulation for acute and chronic low back and neck pain (Shekelle and Coulter 1997; Coulter et al. 2018, 2019a; Qaseem et al. 2017). In a recently published systematic review (SR) and meta-analysis of RCTs and other SRs involving spinal manipulative therapy (SMT), the conclusion was that SMT is associated with modest improvements in pain and function and only transient minor musculoskeletal side effects (Paige et al. 2017). Spinal manipulation is now recommended as a frontline treatment in the most current American College of Physicians (ACP) clinical practice guideline as an evidence-based treatment for both acute and chronic back pain (Qaseem et al. 2017). In addition to the ACP guideline, the Veterans Affairs (VA) and Department of Defense (DoD) have created a clinical practice guideline (Provider Summary. Department of Veterans Affairs/Department of Defense 2017; Version 2.0) for the diagnosis and treatment of low back pain, with a major concern about reducing the routine use of opioid medications. This joint VA/DoD guideline also recommends SMT as an important non-pharmacological treatment for low back pain.

The totality of the evidence, although there is some variation, is that for low back and neck pain, chiropractic manipulation has clinical efficacy and a very low rate of adverse events (Shekelle and Coulter 1997; Coulter 1998; Swait and Finch 2017). Serious adverse events were almost unheard in the RAND studies of RCTs for manipulation. The RAND Corporation's groundbreaking study of the appropriateness of chiropractic manipulation for acute low back pain (Shekelle and Coulter 1997), used both RCTs and 135 reported case reports for acute low back pain to estimate there was one serious adverse event (such as cauda equina syndrome) per 100 million manipulations. For their study on cervical manipulation where 110 case studies were deemed acceptable, they estimated the rate of serious complications at 6.39 per 10 million manipulations (Coulter 1998). The difficulty in calculating the risk for an event this rare is that it requires a very large database and huge sample sizes that are simply not provided by clinical trials.

It is no longer legitimate to claim that there is no evidentiary base for the safety of chiropractic manipulation. A systematic review of adverse events reported in spinal manipulation RCTs (Gorrell et al. 2016) that reviewed 368 articles found that adverse events were reported in only 38% of the articles and that there were only 2 major adverse events reported in all of those studies. It is interesting to note that only 22 articles reported adverse events in the abstract. Chiropractors accounted for 55% of the SMT provided and physiotherapists 30%. Rubinstein et al. (2019) reviewed the benefits and harms of SMT based on 47 RCTs for chronic LBP and found that found only 1 serious adverse event possibly attributed to the SMT. Another SR of RCTS involving SMT found that most of the observed adverse events were mild to moderate, transient musculoskeletal symptoms. Rubinstein et al. (2019) in one study, the Data Safety Monitoring Board judged only one serious adverse event to be possibly related to SMT. In two recently published clinical trials of older patients with lumbar spinal stenosis treated with spinal manipulation, no serious adverse events were reported (Schneider et al. 2019). Similar conclusions were recently reached by an independent report commissioned by the Victorian State Government in Australia related to the safety of chiropractic manipulation in children under the age of 12 years. After an extensive review of the literature, regulatory complaints, and stakeholder feedback, the report concluded there was little evidence of harm in Australia (Safe Care Victoria 2019). A scoping review of the risks of manipulation by Swait and Finch (2017) of 250 articles that included RCTs, observational studies, and SRs found that estimates of serious adverse events ranged from 1 per 2 million manipulations to 13 per 10,000. Benign and transient minor adverse events following manipulation were common, but serious adverse events were rare.

To put this in context, in the NIH consensus conference for the diagnosis, treatment, and prevention of dental caries (Coulter 2001), only seven RCTs were found in which it was possible to prove that the patient actually had dental caries. This resulted in an inability of the panel of experts to make any recommendations based on the trial literature for the diagnosis, treatment, or prevention of caries. Even the Cochrane Collaboration was unable to provide substantial evidence for most dental procedures, exemplified by this quote:

> Many standard dental treatments—to say nothing of all the recent innovations and cosmetic extravagances—are likewise not well substantiated by research. Many have never been tested in meticulous clinical trials. Most of the Cochrane reviews reach one of two disheartening conclusions: Either the available evidence fails to confirm the purported benefits of a given dental intervention, or there is simply not enough research to say anything substantive one way or another. (Jabr 2019)

Their conclusion was that dentistry was much less scientific and more prone to gratuitous procedures than the public thinks. Under the standards that are set for evidenced-based practice, this would mean chiropractic is much more evidenced-based than dentistry.

Seeing a chiropractor as first contact has also been shown to decrease duration of episodes (Blanchette et al. 2017) and decrease the likelihood of undergoing surgery, even controlling for severity (Keeney et al. 2013). In addition, an injured worker seeing a chiropractor for low back pain is less likely to experience recurrence of disability (Cifuentes et al. 2011). A recent clinical trial conducted within a military population showed that chiropractic plus medical care for low back pain produced better outcomes than medical care alone (Goertz et al. 2018). Therefore it is clear that if physicians have no qualms about referring patients to dentists despite its lack of evidence about efficacy and safety, the reluctant stance taken toward chiropractic is inconsistent, at the very least. However, such reluctance is diminishing as more patients are asking and seeking referrals for CAM therapies and governments are encouraging collaborative or shared care among healthcare professions (Gilmour et al. 2011a). In Canada, a national survey of family physicians reported that about 12% offer CAM services; however, significant regional variations were noted with higher use in western provinces (Hirschkorn et al. 2009). Others have reported referral rates to chiropractors of about 40% for chronic pain and back problems (Austin et al. 1998). In addition to the availability of a much greater body of evidence on chiropractic, there is a cadre of chiropractic researchers (those with dual DC and PhD degrees) conducting research within prestigious universities both in North America and internationally.

Evidence for Outcomes for Medical Therapies for Spinal Problems

At the same time as this body of positive research on chiropractic is increasing, there is an associated increasing body of literature on the questionable efficacy, effectiveness, and safety of many of the medical procedures for low back pain such as surgery, epidural injections, and even NSAIDs (Bally et al. 2017). This has led to recognition of the over treatment of back pain (Deyo et al. 2009). The questionable results and complications from back surgery have been well documented (Fineberg et al. 2014; Marquez-Lara et al. 2014; Martin et al. 2013), as well as for epidural injections (Manchikanti et al. 2016) and drugs (Machado et al. 2017). Not the least is the evidence for the use of opioids following surgery (Brummett et al. 2017). Increasingly we see in the literature (Chou et al. 2016) calls for noninvasive treatments for low back pain and for non-pharmacological therapies (Chou et al. 2017).

Removal of Legal and Ethical Barriers

Last but not least, in North America there are no longer any legal or ethical barriers for a physician to collaborate with a chiropractor. Until the Wilk et al. versus AMA antitrust trial (Agocs 2011), the AMA stated that it was considered unethical for a physician or hospital to associate in any way with chiropractors, who were considered to be "quacks." While this policy was portrayed as acting in the public interest and protecting the patient from unfounded claims for unscrupulous health practices, the Wilk trial established that the AMA Committee on Quackery was actually a self-serving front for attacking the chiropractic profession. The underlying and stated purpose of this Committee was to first contain and then to eventually eliminate the entire chiropractic profession.

The result of this landmark decision was that CAM providers in general (who were also considered quacks) and chiropractors in particular could now form professional relationships with MDs. This can be seen in the emergence of complementary and integrative medicine clinics in which these inter-professional partnerships were being established (Coulter 2012; Coulter et al. 2008, 2010; Baer and Coulter 2008; Hsiao et al. 2005). It also opened the door for hospitals and the Veterans Administration to include chiropractic services. Prior to this court decision, hospitals could lose their accreditation for accepting referrals from chiropractors. It is ironic that The Joint Commission (2018) (accrediting organization for US hospitals) has published an Advisory Policy on non-pharmacological options for pain management. This Advisory Policy requires that all accredited hospitals include evidence-based, non-opioid treatment options including spinal manipulation, acupuncture, and massage therapy.

One issue at the heart of the Wilk trial (Agocs 2011) was the medical profession's understanding of the legal status and rights of chiropractors, their scope of practice, and what they were licensed to perform. Historically, medicine has frequently questioned the legitimacy of chiropractic, despite the fact that those statements had no basis in law. The state does not give any one profession the legal power to decide who is, or who is not, a legitimate health profession. That power belongs only to the state and once conferred should be recognized. But the health professions as a group and individually have often acted to limit the legal rights of other professions. While this can be seen in conflicts such as optometry and ophthalmology, physical therapy and chiropractic, midwives and nurses, midwives and obstetricians, nurse practitioners and physician assistants, and dentist and denturists, the most extreme case can be seen in medicine and chiropractic. But in all cases, recognizing a scope of practice invariably means confronting other groups' claims for the same scope, either sharing the scope or trying to win exclusivity. Since by definition the scope of medicine is any act carried out by a medical physician, for any other profession to gain a scope of practice, it will be in confrontation with medicine. In some very rare cases, medicine will give over the scope to another profession as in dentistry and the oral cavity, but in most cases, it will be contested as in the case of midwifery, optometry, chiropractic, and, outside of America, osteopathy. As we noted earlier in the Wilk trial, the extent to which organized medicine acted to limit the rights of the chiropractic profession was extreme. The AMA conspired to keep chiropractors out of the military, veterans organizations, hospitals, universities, and the NIH, from access to such things as laboratories, X-rays, and MRI scans. The extent to which this was done can seem staggering and petty to independent observers and is often totally unknown to individual medical doctors.

In 1980, the AMA revised its Principles of Medical Ethics to reflect this new position, allowing medical doctors to be free to choose the patients they served, the environment they served in, and the other types of practitioners they associated with (Agocs 2011). In 1987, US District Judge Susan Getzendanner found the AMA and its co-defendants guilty of violating the Sherman Antitrust Act. In her decision, Getzendanner asserted that "the AMA decided to contain and eliminate chiropractic as a profession" and that it was the AMA's intent "to destroy a competitor" (Getzendanner 1988).

So what is the current legal status of chiropractic in North America? Chiropractic care is licensed and regulated in every state (Lamm et al. 1995; Mootz and Coulter 2002; Sandefur and Coulter 1997) and province and the Yukon Territory, except the Northwest Territories and Nunavut (Boucher et al. 2016) in North America. The legislation for chiropractic covers six dimensions: licensure, the scope of practice, titles, clinical authority (e.g., prescribing authority), self-regulating authority, and reimbursement. In particular, legislation covering chiropractors may include a definition of the scope of practice, specific license to practice as a first-contact provider, title exclusivity, a section on limitations to chiropractic practice, and the specific agency that regulates chiropractors. Legislation is also likely to specify the range of clinical authority for chiropractors, along with reimbursement policies (especially for government schemes such as

Medicare and Medicaid and provincial and federal public funding). Each province or state can be classified according to the nature of the six dimensions mentioned above, so there is a continuum of legal environments under which chiropractors practice ranging from restrictive to expansive. Therefore, the legislation may state: (1) license to practice as a primary provider (primary contact, portal-of-entry, etc.); (2) scope of practice (can be hands only spine only); (3) clinical authority (right to diagnose); (4) reimbursement for services rendered; (5) self-regulatory authority (right to discipline its own members); and (6) exclusive use of the title "Doctor of Chiropractor."

In most jurisdictions, the scope of chiropractic practice will also be influenced by policies or guidelines issued by the regulatory agency responsible for licensing or by court decisions (Sandefur and Coulter 1997). That is, either the licensing agency or the courts may have interpreted the applicable state legislation in ways that affect chiropractic behavior. In particular, most legislation authorizing the licensing of chiropractors provides considerable discretion to the applicable regulatory agency to define the scope of practice. In addition, courts may interpret the standard of care for primary care in a way that would increase or decrease the potential liability exposure for chiropractors who practice primary care. A court may hold a chiropractor to the same standard of care or similar terms that apply to a medical physician, including the principles applied in determining liability (Gilmour et al. 2011b). In Canada, mandated by provincial legislation through enactment of specific chiropractic acts or general health professions/treatment acts, chiropractors must obtain informed consent, written or verbal depending on the jurisdiction, prior to providing care, especially manipulation therapy (Boucher et al. 2016). The courts might also hold chiropractors responsible for performing tasks, such as laboratory tests, traditionally thought to be exclusively a part of primary medical care.

In summary, the legal status may be a restrictive scope of practice, an expansive scope, or a scope that is somewhat ambiguous. For example, California law appears to define the practice of chiropractic very narrowly, seemingly prohibiting the practice of primary care (Deering's California Code Annotated, Business and Professions, Appendix I, Section 7, Chiropractic Act). In Oregon, however, chiropractors have a broad scope, where some forms of primary care (Oregon Revised Statutes Annotated, Vol. 45, Chapter 684.010 and 684.015) such as the practice of obstetrics are permissible. Similarly in Canada, statutes may vary from province to territory because health regulation is a matter of provincial/territorial jurisdiction. So, while all chiropractors in North America have legal status, they vary as to how broad their scope of practice has been defined. However, in all jurisdictions they have the legal right to manipulate the spine and perform diagnosis and to be a primary contact provider. In no jurisdictions is chiropractic restricted by a requirement of referral from the medical professions, as was until recently the case historically with the physical therapy and nursing professions. Both of these professions may now practice as independent, first-contact providers in many jurisdictions.

Insurance Coverage

One of the compelling reasons to recommend chiropractic care in the United States and Canada is the wide coverage of chiropractic care by various types of health insurance. In the recent RAND study in the United States, 68.8% of the patients had some type of health insurance coverage (Herman et al. 2019a). In the United States, chiropractic care is covered by almost all private insurance plans, and chiropractors utilize a majority of the standard diagnostic (ICD-9 and ICD-10) and procedural (CPT) billing codes as other healthcare professionals. Medicare and Medicaid programs as well as most state worker's compensation systems also provide coverage for chiropractic treatment.

Doctors of chiropractic are fully integrated within both the Military Health System and the Veterans Health Administration, caring for patients in healthcare teams, participating in research, training students, and serving in leadership roles (Green et al. 2009). Both active duty and veterans clinics are staffed by doctors of chiropractic who are hired as federal employees or as contractors, depending on each site's needs and structure.

In Canada, coverage for chiropractic services is provided through provincial and federal public funding, extended healthcare (EHC) plans, or out-of-pocket payments. The amount of and access to provincial public funding has varied by province over the years. Such funding has ranged from complete unlimited funding to limited payment for specific subgroups to no coverage at all. Today, only select provinces provide partial coverage for chiropractic services ranging in British Columbia of a limit of ten visits each calendar year for any allied health treatments, including chiropractic, for those eligible for Medical Services Plan Premium Assistance (seniors and low-income citizens) to Alberta where only seniors have access to a government-sponsored health benefit plan with a maximum yearly amount to Manitoba where all have access to a limited number of visits and costs per calendar year (www.chiropractic.ca/about-chiropractic/chiropractic-coverage/). Accident benefits are also provided to those injured at work and in a motor vehicle collision, but coverage may vary by province and nature of injury. In addition, federal government workers (Royal Canadian Mounted Police, Veterans Affairs Canada, Canadian Forces) have access to funded chiropractic services that is limited by either total annual amount or a set number of visits. First Nations have access to the First Nations Non-Insured Health Benefits which may cover chiropractic services that vary from between regional office and by year (www.chiropractic.ca/about-chiropractic/chiropractic-coverage/).

However, for most chiropractic patients, insurance coverage is most likely provided by an EHC. EHC is a supplementary health and medical plan used to complement provincial health coverage and paid by the patient and/or employer. It is estimated that more than 70% of Canadians have EHC coverage. In Canada, national health expenditures are paid by either public (70%) or private sector spending. Of the 30% private sector spending, out-of-pocket spending accounts for about 15%, EHC plans about 12%, and others about 3% (CIHI 2019). In a recent study of Ontario chiropractors, they reported most patient encounters (68%) were paid out of pocket, with about 31% and 1% paid by EHC or work injury plans (Mior et al. 2019).

What You Need to Know Before Talking to a Chiropractor

Given these compelling reasons for talking to and/or referring your patients to a chiropractor, what is it that a physician might need to know to determine if a given patient is an appropriate candidate for seeing a chiropractor. The first is gaining an appreciation for educational background of chiropractors. What do they study, how much do they study it, and how does it compare to the education of a medical physician and other health professions? Do they study pathology, can they determine if a problem is outside their scope of practice, and do they know how to refer such a patient to the relevant healthcare provider or service? Can they perform differential diagnosis? Do they know what is contraindicated for their care? Can they function as primary contact back specialists? Can they function as gatekeepers for back pain patients to enter the healthcare system?

Secondly what does the research show about the outcomes for chiropractic care, and where can I access it? Are there guidelines that are readily available to assist in making a referral decision? How would I know if something is not appropriate for manipulation? Do chiropractors themselves follow guidelines for practice? What are the quality controls for professional practice in chiropractic? How can I advise my patients about the risk and benefits of chiropractic care to meet my obligations for informing them?

Thirdly, if I send a patient to a chiropractor, how will the case be managed? What are the protocols? How do chiropractors view back pain? What do they offer to patients with spine-related conditions? For acute problems is there some rule of thumb about response rates, the number of visits? How do they decide when to terminate care? For chronic patients is this likely to be lifetime care? Is there some way of identifying overutilization by the chiropractor? How will this care differ from medical care of back pain? How might surgeons and chiropractors work together? What would be value-added care from using chiropractic?

Education and Training of Chiropractic

In order to be licensed in most jurisdictions, chiropractors must graduate from an accredited teaching institution. Institutions with university status that grant degrees in chiropractic must be accredited by at least two accrediting bodies: a professional or programmatic accreditor and a regional accreditor.

In the United States, programmatic accreditation is awarded through the Council on Chiropractic Education (CCE) (Council on Chiropractic Education 2018). CCE is itself recognized as an accrediting body by the Council for Higher Education Accreditation (CHEA). CHEA recognizes the accrediting bodies for medical education (Liaison Committee for Medical Education) and education of other health professions that have programmatic accreditation (https://www.chea.org/chea-and-usde-recognized-accrediting-organizations). CCE sets the educational standards and outcomes that each college must meet within their respective chiropractic educational curriculum. In chiropractic, these accreditation standards have been rather prescriptive, though there has been a pronounced shift toward demonstration of student competence in recent chiropractic accreditation standards. All American chiropractic programs that desire programmatic accreditation must be reviewed by CCE. Presently, there are 16 doctors of chiropractic programs in 19 locations in the United States accredited by CCE.

(Note: CCE is also a member of the Association of Specialized and Professional Accreditors (ASPA), whose membership also includes LCME, the accrediting body for medical education.) (https://www.aspa-usa.org/our-members/)

In the United States, universities must also undergo regional accreditation by one of six regional accrediting bodies. In California, for example, universities with chiropractic programs are regionally accredited by the Western Association of Schools and Colleges (WASC) and the Western Senior College and University Commission (WSCUC). In Canada, chiropractic programs are accredited by CCE Canada, and in Ontario the Doctor of Chiropractic degree, a second entry baccalaureate honors degree, is offered under the written consent of the Minister of Training, Colleges and Universities (https://www.cmcc.ca/about-cmcc/accreditation).

The net effect of these dual processes is that though the accrediting bodies vary, universities with chiropractic programs in the United States undergo both programmatic and regional accreditation, just as medical programs do for medical education. In countries like Canada, the same process is followed.

To obtain a chiropractic license in the United States, all states require that graduates of accredited chiropractic programs must pass Parts I, II, III, and IV of the National Board of Chiropractic Examiners (NBCE) examinations (http://directory.fclb.org/Statistics/EducationTesting-US.aspx). One state does not require Part III, and most states require a fifth examination in physiotherapy. Part I is taken during the second year of education and focuses on the basic sciences. Part II is taken in the third year of chiropractic education and focuses on diagnosis and chiropractic practice. Part III is taken in the fourth year of chiropractic education and focuses on clinical practice, diagnosis, and management. Part IV is taken in the fifth year of education and is a practical examination. Examinees work with standardized patients and demonstrate examination, diagnosis, and treatment skills and also interpret imaging as part of a radiology examination (https://mynbce.org).

In addition to these NBCE requirements, each state may have additional requirements for licensure, often including successful completion of a state-specific jurisprudence examination. In Canada, candidates seeking registration in individual provinces must first pass examinations offered by the Canadian Chiropractic Education Board (CCEB) (https://www.cceb.ca/home/).

While undergoing training, chiropractic students receive a biomedical education similar to medical education in many respects. There are some key differences between chiropractic and medical education, particularly in the amount and location of clinical training. A rigorous study comparing the topical content and hours allocated in medical and chiropractic education was published in 1998 (Coulter et al. 1998). That study compared national data for chiropractic and medical curricula and involved site visits and interviews at three chiropractic colleges and three medical schools.

At that time, the basic science programs of medicine and chiropractic were found to be similar, averaging 1,200 h for medical education and 1,420 h for chiropractic education. Chiropractic programs had significantly more anatomy (perhaps not unexpected for practitioners with a neuromuscular and musculoskeletal focus) and physiology instruction. Chiropractors even had more pathology instructional hours than medical education, but this was presumed to be because chiropractic did not have a postgraduate residency program. That is, chiropractors had "lecture learning" in a wide variety of pathologies, where medical education exposes physicians in training to a wider variety of patients and pathologies *clinically*. Medical education also included significantly more training in public health. The type of clinical education varied between the two programs, though the total hours of training were similar when the chiropractic clerkship period (before completing the chiropractic program) is included.

In total, chiropractic and medical education each had curricula of approximately 5,000 h (5,200 for medicine, 4,860 for chiropractic). The most significant difference in the educational programs of the two healthcare disciplines was the medical postgraduate residency. Here, as Coulter et al. (1998) noted, "the difference is drastic, resulting in medical students receiving much more practical clinical education" (p. 73). The other difference was *where* the clinical training occurred – chiropractic's year of clinical training occurred in ambulatory care settings.

Therefore, prior to the postgraduate residency, the education of the two healthcare professions had surprising similarities in 1998 – with the key difference being the presence of the postgraduate residency in medical education. Chiropractic education still resembles dental education (and that of most health professions) in that it does not include a required postgraduate residency. This reflects a funding issue as much as anything else in that residency stipends are rarely available outside of medicine. When osteopathy became recognized in the United States on a par with medicine, it also obtained access to the residencies in medicine. In the United States, the osteopathic academic program mirrors that of medicine. In other countries it more closely mirrors that of chiropractic (Baer 2006).

Since the publication of the 1998 study (Coulter et al. 1998) of chiropractic and medical education, how have things changed? How do chiropractic and medical education compare at present? The Association of American Medical Colleges (AAMC) reported (Association of American Medical Colleges 2019) that the year 1 and year 2 curricula of medical school totaled 1,448.9 h in 2013 (AAMC 2019). The corresponding total was 1,815 in the 1998 study. CCE currently requires that chiropractic programs include a minimum of 4,200 h of education, including the clinical clerkship (CCE 2018). Some states require lengthier programs; for example, California requires 4,400 h (Board of Chiropractic Examiners 2018). In 1998, the total program length of chiropractic programs averaged 4,860 h. Both medical and chiropractic curricula appear to have shortened somewhat since the 1998 study.

The other differences in 1998 between chiropractic and medical education related to the location and amount of postgraduate clinical training. Since 1998, a few chiropractic academic programs have offered limited postgraduate paid residencies in fields such as radiology, sports medicine, primary spine care, and geriatrics. To the authors' knowledge and per recent communication with CCE, none of the above postgraduate residencies have CCE accreditation. However, CCE-accredited residencies have opened on a limited basis within the US Department of Veterans Affairs (VA) (CCE 2017; VA/DoD 2017). Taken together, all of these residencies still impact a small number of chiropractors. On the other hand, since 1998, residency training in medicine has continued to be a central and distinguishing feature of medical education.

In summary, chiropractic education offers comparable training to medical education *in total hours* prior to graduation. Medical education continues to offer much more extensive postgraduate residency hours with a wider variety of patients and patient care settings. Chiropractic clinical training, which occurs over 1 year, is almost exclusively in ambulatory settings, with limited postgraduate residency positions available.

Medical doctors can be reasonably assured that chiropractic education in the basic sciences (anatomy, physiology, etc.) has been similar in total hours and that chiropractors have had a year of clinical training to prepare them for the historic case mix and case complexity which chiropractors typically see – largely, spine-related disorders and musculoskeletal complaints. Chiropractic programs at universities maintain regional and programmatic accreditation. Chiropractors obtain licensure after passing a series of licensure examinations and meeting other state requirements. Medical doctors and chiropractors still have little opportunity to train together outside of the VA residency programs. These VA residencies offer an excellent opportunity to show what may be possible in further improving chiropractic education and building bridges between chiropractic and medicine.

Overall, the objective of chiropractic education is to prepare the student to become a primary contact health professional, a Doctor of Chiropractic (DC). DCs are capable of diagnosing and deciding what is indicated or contraindicated for chiropractic care, who can manage musculoskeletal conditions within a broad-based wellness paradigm that focuses on the whole person and who knows (and is legally obligated to distinguish) when and how to refer patients to other healthcare providers when necessary.

DCs are licensed healthcare professionals who provide non-pharmacological, conservative care focused on the diagnosis, treatment, co-management, or referral for musculoskeletal conditions (most frequently), including back pain, neck pain, headache, and muscle strains or sprains. Some states and provinces have different scopes of practice, and some chiropractors focus on conditions beyond the musculoskeletal or peripheral nervous system.

The modern DC may or may not manipulate or mobilize joints and soft tissues, employ modalities, supervise and prescribe exercise, and counsel lifestyle changes (sleep hygiene, nutrition, etc.). The primary therapeutic procedure used by DCs is spinal manipulation, and chiropractors perform most of the spinal manipulations rendered annually in the United States. However, chiropractors may use a wide range of therapies and may also contribute to treatment of health problems outside musculoskeletal conditions, though this may often be as adjunctive care.

It is important to note that chiropractors are educated to be doctors and members of a profession. Chiropractic is the name of a profession, not a procedure (Herman and Coulter 2015). While manipulation is the skill for which chiropractors are often best known, it is erroneous to equate the word "chiropractic" with "manipulation." Further, joint manipulation is not a skill that is unique to chiropractors, because at least the professions of physical therapy and osteopathy also include manipulation in their treatment tool boxes. Manipulation and manual therapy skills are also taught in some acupuncture, massage therapy, naturopathy, and other programs. Manipulation is not what makes chiropractic unique.

What does make chiropractic unique? How do chiropractors see it differently?

To answer that question, we need to look at the constellation of elements that make up the chiropractic paradigm, and it is the totality of these that distinguishes chiropractic as a profession.

The Chiropractic Treatment of the Spine

Chiropractic is a system of diagnosis and non-pharmacological therapy focused on the neuromusculoskeletal structures of the human body, particularly the spine and nervous system. The mainstay of chiropractic care is spinal manipulation and other manual therapies to improve joint motion, in order to relieve pain, improve function, and help the body heal itself. In addition to manual treatment, chiropractic care may also include other treatments such as postural education and therapeutic exercise, as well as the use of adjunctive modalities such as traction, ultrasound, electrical stimulation, and hot/cold packs.

In the early days of chiropractic, subluxations – defined at the time as a misalignment of one or more vertebrae (Wardwell 1992) – were seen as a cause of "disease" as a result of interruption of afferent and efferent neurological signals. This naturally evolved into chiropractors viewing spine-related disorders primarily in a biomechanical and neurological context, with neurological processes being central to the development and perpetuation of these disorders. Emerging evidence largely supports this viewpoint (Seaman and Winterstein 1998; Reichling and Levine 2009; Ischebeck et al. 2017; Panjabi 2003; Garcia-Larrea and Peyron 2013; Henry et al. 2011; Wenngren et al. 2002).

Increasingly, the biopsychosocial model of back pain is being emphasized by chiropractors (Murphy and Hurwitz 2011a, b; Stilwell and Harman 2017) and incorporated into chiropractic training (Murphy 2013, 2016). In this model, chiropractors recognize the important and interrelated roles that biological (somatic, neurophysiological), psychological (fear, catastrophizing, perceived injustice, etc.), and social (socioeconomic status, home life, work disability, etc.) phenomena play in back pain. Given the chiropractic traditions that focus on holism versus reductionism (see below), the biopsychosocial model is a natural fit in the chiropractic approach.

The Perspective

The best way to understand the chiropractic perspective is through the biopsychosocial paradigm prosed by Engel (Engel 1989). Since this is a paradigm widely used in medical education, it provides a perspective for understanding chiropractic in terms that are common to both groups. In this paradigm, health is a complex mix of body, mind, and society. Clearly it is a paradigm that gives attention to cultural, social, and psychological aspects of health and the health encounter. Therefore, the focus is always on the whole patient (holism) rather than a focus on the individual components. Reductionism, which is a hallmark of modern medicine, focuses on the individual components of health: biological systems (such as the cardiovascular system), disease states (such as cancer), and disrupted function (such as a collapsed disk) from either a disordered pathology or trauma. This reductionistic paradigm, which has been very powerful when we are dealing with disease and trauma, differs considerably from how chiropractors view back pain within a holistic paradigm.

Chiropractors do not view diseases as simply disordered pathology (which in the case of back pain there is frequently no identifiable etiology of disease, i.e., non-specific pain) but as dis-ease, a body at lack of ease. The object is to promote normal physiology as opposed to fighting abnormal physiology. For this reason, although chiropractors disagree among themselves about the philosophical basis of their approach, they do tend to subscribe to the belief that the body largely heals itself (*vis mediatrix naturae*), and the role of the chiropractor is to help the body do that. The chiropractor is a facilitator of health, not a giver of health. To that extent they share two maxims of Andrew Still, the founder of osteopathy: "health comes from within or not at all" and "I can no more give you health than I can give you honesty" (Coulter 1999).

This "vitalistic" concept was historically part of medicine but got lost with the emergence of scientific medicine in the nineteenth and twentieth centuries (Reiser 1979). It was present in the ancient Hygieian philosophy, which focused on the person and on the inherent health-maintaining and health-restoring abilities within the person as the source of recovery and health. Many medical commentators have bemoaned the loss of philosophy in modern medicine (Cassell and Siegler 1979; McWhinney 1986; Gordon 1980; Capra 1986; Pellegrino 1979; Cluff 1987).

It is important to understand, therefore, that the chiropractic paradigm sees health somewhat differently from medicine (not just the absence of disease), sees healthcare somewhat differently (helping the body to restore itself), and sees the health provider somewhat differently (as a facilitator and educator not as a curer) (Coulter 2005). Because of these paradigm differences, MDs and DCs practice somewhat differently. No patient mistakes a chiropractic practice as a medical practice. In numerous studies, patients have commented on the differences in experience between MDs and DCs. This is somewhat surprising because they bring the same health problem to both providers and the therapy may not be radically different (there are MDs who manipulate) but the patients report that the health encounter is very different (Coulter 2018).

To see chiropractic as simply a modality – manipulation or mobilization of the spine – is to do a dis-service to chiropractic and may result in those who suffer spinal complaints being underserved. It is manipulation given within a broad-based paradigm. While other professions also perform manipulation, that does not automatically mean they are practicing chiropractic. Chiropractors provide manipulation within what we might term a wellness paradigm (Coulter 1990, 1996a). That is, while the focus might be on back pain, the chiropractor will be exploring a holistic approach and also focus on the lifestyle of their patients. This might include nutrition, diet, weight, exercise, stress, posture, sleeping habits, alcohol consumption, use of drugs, supplements, and therapeutic/rehabilitative exercise. In addition to manipulation, chiropractic care may also involve a range of adjunctive therapy which may include:

> cryotherapy, trigger point therapy, nutritional counseling, and bracing. The majority of practitioners also use massage, heat, traction and electrical muscle stimulation modalities. Acupressure and meridian therapy are used by about 66% of practitioners with less than 10% reporting that they use acupuncture. (Christensen et al. 2015)

An observational RAND study was conducted to determine treatment utilization patterns based on the records of patient in chiropractic offices with low back pain. This study found the following utilization patterns: 84% of the patients received spinal manipulation, 79% received non-thrust manual therapies such as mobilization, massage, and heat packs, 31% received education, and 5% received other forms of therapy such as acupuncture (Coulter et al. 2002). Similar patterns were reported in a recent scoping review, where median and IQR of treatments provided by chiropractors included spinal manipulation (79.3% (55.4–91.3)), soft tissue therapy (35.1% (16.5–52.0)), formal patient education (31.3% (22.6–65.6)), exercise instruction (26.0% (9.0–68.1)), mobilization/traction (17.2% (12.4–32.0)), and to a less extent physical modalities such as supports, electrical stimulation, ultrasound, and acupuncture (Beliveau et al. 2017). The use of these and other modalities is included in the chiropractic scopes of practice in most jurisdictions.

Coulter (2004) has noted that the story one gets about chiropractic from ethnographic studies of the actual encounter shows a quite different picture than what is obtained from health services research using clinical records. Chiropractic is unique in that there are several ethnographic observation studies of the health encounter (Coulter 2004). As he notes, the view from health services research would seem to depict chiropractic care as a narrow-based, sub-speciality dealing overwhelmingly with back pain and chiefly using spinal manipulation. But those who have done the ethnographic observation studies come to slightly different view and conclusion. In this literature (Kelner et al. 1980; Jamison 1994; Coulehan (1995; Coulter et al. 2019), chiropractic is seen as a broad-based, distinct alternative health paradigm, with its own metaphysic, philosophy, language, therapies, and health practices, and as one providing a unique health encounter. Numerous names have been used to describe this paradigm (patient centered, holistic, a wellness paradigm), but it suggests that chiropractic cannot be reduced simply to the manipulation of the spine and other joints. Coulter suggests it is as though the studies are describing two completely different animals.

One of the more interesting studies was published by a medical physician Coulehan

(1995). He concluded that "Chiropractic care, as opposed to spinal adjustment as an isolated treatment, must be viewed as a process or interaction" (Coulehan 1995). He characterized the chiropractor's view as "the faith that heals." That is, chiropractors use explanations that are understandable, that are both mechanical and holistic, that appeal to the patient in that the person is not subtracted from the encounter, and that are positive and drug free. "The net effect is a logical set of beliefs which appeal to common sense, use scientific terminology, yet promote a holistic approach rather than a biomedical approach" (Coulehan 1995). In addition to the laying on of hands, chiropractic care often includes a program of exercise, nutritional counseling, stress management techniques, and behavioral change. Jamison (1993) writes that chiropractic care involves manual, emotional, and psychosocial contact. Chiropractic care is cooperative and focused on the well-being of the patient; uses a low level of technology; is focused on objective, subjective, and effective data; is directed at understanding the whole person; and is personalized (Jamison 1993).

There is an increasing interest in various components of the entire health encounter, with broad consensus that the health encounter is a social encounter that occurs within cultural, social, and individual history (Coulter et al. 2019c). This includes the content of the doctor-patient communication in the encounter (Van Dulmen and Bensing 2002) and interpersonal elements of the encounter (affective communication and instrumental communication), investigating how the patient and the provider perceive the communication (Adams et al. 2012) by using self-reports or analyzing recorded narratives between the provider and the patient (Tarn et al. 2013). Others have studied the cognitive, psychological, and emotional element of the health encounter (Di Blasi et al. 2001).

Last but not least, studies have focused on the belief and expectations of patients and the impact these have on clinical outcomes (Wirth 1995). It is by focusing on the totality of the elements in the health encounter that we can start to distinguish what is different and unique about the chiropractic approach to healthcare. It is not that any one of these elements is unique but the constellation of all the elements as a whole. As Coulter et al. (2019) have shown in their observation of the health encounter in chiropractic, there is nothing non-specific because it is deliberately constructed. The chiropractors and their staff create a style of practice within which the encounters are quite structured and consistent. They found that not only was the nature of the encounter important to the patients; they can delineate and distinguish chiropractic care as distinct from other health encounters, particularly medical encounters.

In summary, there are two views of chiropractic, even within the profession: one view sees chiropractic as back pain specialists, i.e., spine doctors, while the view sees chiropractors as broad-based "wellness" primary care practitioners (Coulter 1983).

The Clinical Encounter

The clinical diagnosis in chiropractic is similar to that used in all health professions, beginning with a patient history and physical examination. Central to the latter will be a neuromusculoskeletal examination (Haldeman et al. 1993). A study of 4,000 randomly chosen chiropractors in the United States, Christensen et al. (2015) found that a case history and physical examination are routinely performed by most chiropractors. One key objective of the diagnostic procedures is to determine whether the problem is contraindicated for chiropractic – requiring a medical referral – or whether it is within the scope of chiropractic and what type of treatment is indicated. Therefore, taking a case history and performing a physical examination are basic elements of chiropractic practice (Cherkin and Mootz 1997). But as noted earlier, chiropractors also pay attention to the psychosocial aspects of their patients. While the patient history will resemble that performed by a medical physician, the chiropractic musculoskeletal examination will be more extensive and comprehensive than that performed in a general medical practice. The reason for this difference is that historically chiropractors were excluded from the elaborate diagnostic facilities available to MDs; hence they learned to rely more on the history and physical exam findings and less on diagnostic imaging. In this respect, a chiropractor resembles the old-time general practitioner who made house calls where they had to depend on their knowledge of the biological systems, neurological deficits associated with particular diseases, and their palpatory skills, which were all they had to make the diagnosis. Palpation remains a key clinical and low-tech diagnostic skill for chiropractors, where median proportion of use during the assessment is about 90% (Beliveau et al. 2017). While most contemporary chiropractors have access to X-rays, MRIs, CT scans, electromyography, lab tests, etc., they retain the emphasis on physical diagnosis in practice. The sort of physical tests the chiropractor will use may include pain provocation, static palpation, motion palpation, range of motion measurements, observation of postural symmetry, dynamic spinal loading, tissue compliance, gait analysis, muscle strength, and functional capacity (Cherkin and Mootz 1997; Mootz and Coulter 2002; Beliveau et al. 2017).

Another feature of chiropractic care frequently mentioned by patients is their approach to dealing with pain. Kelner et al. (1980) conducted a study of 770 randomly chosen patients, including an ethnographic study of 70 clinics in Canada and interviews of 350 randomly chosen chiropractors. They noted that the majority of patients had tried other types of healthcare for their back problem – usually medicine – before going to the chiropractor. Since their problem was non-specific back pain, they were left without a definitive diagnosis, often with a feeling of rejection with the implied accusation that their problem is all in their head. As noted by Coulehan (1995), the chiropractor not only legitimizes them as a patient; they welcome the opportunity to treat back pain patients. In contrast, the patients reported that in medical encounters their pain was treated almost as a secondary consequence of their being ill.

This has been part of a general cultural belief in western society about being stoic about hardships and pain. There is an assumption that patients are in pain because they are sick. By finding out what is causing the pain and by removing the cause, the assumption is the pain will go away. Alternatively, use drugs to relieve the pain. Patients report that in chiropractic encounters, the pain is not seen as secondary but primary, and the objective of the care immediately, even when the cause is not known definitively, is to target the pain. Having their pain seen as a legitimate focus has a very powerful psychological impact on the patient. This combined with a belief in a positive outcome might be one of the most important predictors of outcomes for chiropractic patients.

The chiropractic clinical encounter may vary for any of the following reasons:

- State or provincial chiropractic statutes
- The philosophical predilection of the individual provider (whether broad-based or focused only on the spine)
- Advanced training the provider may have had in specialized topics such as rehabilitation or sports therapy
- The education the chiropractor had both in chiropractic college and postgraduate education
- The adjunctive therapies/modalities that they might use
- The manipulation system the chiropractor may follow

Manipulation systems are probably the least understood by those outside the chiropractic profession and will be expanded on in the following section. Different systems often have associated with them, specific diagnostic approaches, as in motion palpation or McKenzie technique. The variation in system approach will also include variation in the type of equipment and specialized treatment tables used for the manipulation.

Chiropractic Manual Therapy

Overwhelmingly the main therapy used by chiropractors will be spinal manipulation. In the RAND study (Hurwitz et al. 1998), 84% of patients received manipulation, 79% received non-thrust manual therapies such as joint/soft tissue mobilization and massage, 31% received education, and 5% received other forms of therapy. In addition, most chiropractors incorporate some type of therapeutic or rehabilitative exercise with spinal manipulation (Christensen et al. 2015; Beliveau et al. 2017; Mior et al. 2019). Manipulation is defined as the use of a manual thrust procedure to move joints into the paraphysiological range, without exceeding the anatomical range of motion. Mobilization involves various grades of manual non-thrust oscillatory movements within the physiological range of motion of a joint. This range of motion is the range a joint can be moved into with the application of external force but not exceeding the anatomical limitation of the joints intrinsic connective tissue (e.g., ligaments, joint capsule, tendons, musculature). In chiropractic, manipulation is generally referred to as an "adjustment." However, for the purpose of this chapter, we prefer to use the term "manipulation," as it is more commonly used in medicine and the other health sciences.

The term "manipulation" does not refer to a single procedure. In fact, there are many different types of chiropractic manipulation techniques that have been developed over the years, although Coulter and Shekelle (2005) in their study of chiropractors in North American identified 14 technique systems used routinely or daily in practice. The National Board of Chiropractic Examiners has conducted practice surveys of the chiropractic profession and found that a similar variety of manipulation techniques were regularly used by chiropractors. The most common types of manipulation utilized routinely by chiropractors include standard thrust spinal manipulation procedures ("diversified"); manipulation/mobilization of extremity joints; the use of specialized treatment tables that introduce axial traction or a drop mechanism during the mobilization ("Cox" or "Thompson"); and handheld devices that deliver a mechanical impulse (aka "activator").

The complexity in applying manipulation techniques can best be delineated by examining the variations technique used with respect to cervical manipulation (Bergmann and Peterson 2011). First, the position of the patient during the manipulation can vary. They can be standing, sitting, or lying on a treatment table, while the chiropractor delivers the manipulation procedure. If lying on a treatment table, the patient can be placed on their back (supine), on their front (prone), or on either side. The position of the chiropractor during the manipulation will also vary. The position used will be determined by the size of the patient, the size of the chiropractor, the location of the area to be manipulated, and the direction/vector of the intended manipulation. Secondly, the contact point of the chiropractor's hand will vary. The manipulation can be performed with the side of the hand, with the heal of the hand, or with crossed hands. Thirdly, the speed, angle, and depth of the manipulation will vary. Chiropractors are experts in controlling the thrust, and what is delivered is a highly specific thrust to a specific part of the joint (Triano et al. 2015). Fourthly, the table being used if the patient is prone may have features that are used in the manipulation such as a drop table. The point is there is a wide variation for any type of manipulation used by individual practitioners (Cherkin and Mootz 1997). The chiropractors can also manipulate other joints such as extremities and use soft tissue therapies. As noted previously, in addition to the manipulation or mobilization, chiropractors may use a variety of ancillary therapies such as mechanotherapy, ultrasound, hydrotherapy, electrical therapies, trigger point therapy, acupressure, acupuncture, massage, heat, ice, traction, muscle stimulation therapy, and vibrators. For example, Christensen et al. (2015) reported that 66% of chiropractic practitioners used acupressure and meridian therapy.

The Coulter-Shekelle study (2005) also documented the daily use of some 23 non-manipulative techniques. The most commonly ranked in terms of use were patient education, exercise (both used by over 80% of the chiropractors on a daily basis), physical therapy, ice therapy, ultrasound, massage therapy, electrical therapy (all over 60%), and traction, orthopedic appliances, nutrition supplements, therapeutic supports (all over 40%) and athletic supports, occupational health, orthotics, and vibratory therapy (all over 20%). Acupuncture was used by 10% of the chiropractors along with homeopathy. Therefore, chiropractic care should not be considered consisting exclusively of spinal manipulation.

While most chiropractic patients will receive spinal manipulation, the chiropractor will also promote wellness and lifestyle management by counseling. In Christensen's et al. (2015) study, two-thirds of the chiropractors report using nutritional counseling in practice. He reports that the conditions seen by chiropractors mostly fall into the neuromusculoskeletal category but chiropractors also reported such things as obesity, hypertension, and osteoporosis.

In the Kelner et al. (1980) study, the patients reported lifestyle advice that appears simple, but because a lot of it is achievable, the patients find it very useful. This may be as simple as not sleeping on your stomach with a pillow under you face, not driving with a wallet in your rear pocket, and use the correct chair to sit at a computer. Coulter et al. (1996) examined provider and patient reports for 18 preventive behaviors the chiropractor could counsel patients about. Most of the recommendations that the chiropractors report as giving to at least 25% of their patients can be conceived as recommendations closely related to neuromusculoskeletal complaints and involve active remedies (as opposed to bed rest which is recommended for less than 25% of the patients). As might be expected, given that chiropractic is defined as a drugless therapy, the lowest rated recommendation is for medications. Relaxation techniques are also recommended for relatively few patients (37% of the doctors recommend it for more than 25% of the patients), while reducing stress is recommended by 65% of the chiropractors for less than 50% of the patients. Last, but not least, the results for therapeutic supports indicate that the largest category of the chiropractors (43%) recommend them for less than 25% of the patients. Christensen et al. (2015) summarized several early surveys from 2003, 2009, and 2014. The results show 76% used full spine and extremity manipulations and 96% used diversified technique. Virtually all provided health promotion and wellness care. Three quarters used adjunctive procedures such as ice packs, trigger point therapy, braces, electrical stimulation, and two-thirds of hot packs, massage, and heel lifts. Almost all used corrective and spinal rehabilitation exercises, and over 80% included extremity rehabilitative exercises and advice and training for daily living.

What Do Chiropractors Treat?

There is considerable variation in what chiropractors will sometimes claim to treat. But if we confine the claims for those things for which there is either evidence or a reasonable amount of clinical experience to substantiate treating the condition, it would include the following: acute, subacute, and chronic low back pain (Shekelle and Coulter 1997) as well acute neck pain (Coulter et al. 1996; Shekelle and Coulter 1997), chronic low back pain, and neck pain (Coulter et al. 2019a, b). Christensen et al. (2015) reported that joint dysfunction, headaches, degenerative joint disease, muscular strains, spinal disk problems, myofascitis, radiculopathies, spinal curvatures, tendonitis/tenosynovitis, and peripheral neuralgias are often diagnosed and managed in their practices. They also report that patients with tumors, infectious disease, hereditary disease, and other systemic disorders are virtually never or only rarely evaluated and managed in their practices. About two-thirds of diagnoses recorded in chiropractors practices are for musculoskeletal problems (Hurwitz et al. 1998).

A recent large scoping review of the chiropractic profession reported that almost 50% of patients attending for chiropractic did so for low back/back pain, 22.5% for neck pain, 10% for extremity problems, 7.5% for wellness/maintenance, and 5.5% for headaches. Only 3.1% of reported reasons for seeking chiropractic care was for visceral/non-musculoskeletal problems (Beliveau et al. 2017). Therefore, it appears that chiropractors are treating predominantly spine and other musculoskeletal conditions (Christensen et al. 2015).

Who Do They Treat

Much of the medical concern about chiropractic focuses not on what they are treating but on what they *might be* treating outside their scope of practice. Seldom is this concern based on any data about who constitutes the patient population (Coulter et al. 2002).

Studies of patients using chiropractic care show a prevalence of women (about 60%), whites, those with mid-high levels of income and education with a median age of about 44 years (18–64 years of age primarily (Christensen et al. 2015; Beliveau et al. 2017). Those under 17 years represent 17% of the patients (Christensen et al. 2015) and the over 64, 15%, and with at least partial insurance coverage for care (Coulter and Shekelle 2005; Hurwitz et al. 1998; Mootz et al. 2005). The Coulter-Shekelle study (2005) reported data on 1,275 patients from across the United States and Canada on data collected from three major sources, patient files, practitioner interviews, and patient interviews. The patients were largely white (83%), with an average age of 42 years, predominantly female (61%) and married (57%). The Canadian and US samples were either identical or very similar, but they differed in terms of education. In the American sample, 54% had a degree compared to 38% in Canada, and a further 33% had some college education in the United States compared to 15% in Canada. The patients mostly reported being treated for a back-related problem (76%). When asked to specify their illness or injury, 27% reported it as a neck/cervical problem, 22% as a low back problem, and 21% as a back/spine problem. Extremities accounted for 13% of the health problems. Most had had the symptoms for <3 weeks (45%), but one sizeable group had had them for >6 months (21%). Just over half of the patients (53%) reported having an injury, and the most common reported source for an injury was for nonwork-related events (43%) with work-related accounting for only 16%.

The patients with a back problem (Coulter and Shekelle 2005) were asked to complete a functional self-report questionnaire, the Roland Morris Disability Questionnaire (RMDQ) (Roland and Fairbank 2000). The RMDQ consists of 23 items, and the average score of this sample was 9.7, where a higher value indicates more disability. This sample compares to acute low back pain patients presenting to MDs during the same period, with average RMDQ scores of 10.3 for urban primary care, 12.7 for rural primary care, 11.7 for urban chiropractic, and 9.9 for rural chiropractic (Carey et al. 1995). The patients in this study were asked to complete the Short Form 36 health survey questionnaire (SF 36) to assess their general health status (Jenkinson et al. 1993). The overall results from the SF 36 were compared to age/sex matched norms and to sciatica patients seeing surgeons. Chiropractic patients had values midway between normals and patients with sciatica on physical functioning, role-physical, social functioning, and pain. However for role-emotional, emotional health, and vitality, the chiropractic patients report worse health status than those with sciatica seeking surgery. Compared to the matched norms, the greatest relative difference is for role-limitations physical and pain. The majority of patients (61%) reported that during the last 30 days their pain had been moderate to severe with 33% reporting that the worst bodily pain in the last 30 days was severe to very severe. However, only 8% reported that the pain interfered extremely with their normal work. The majority of patients (58%) reported having no care for the current injury/illness prior to chiropractic. For the majority of these patients, therefore, chiropractors are the primary point of entry for care of these conditions. However, 3% reported having had surgery prior to chiropractic care, 20% reported having medical care other than surgery, and 18% reported having physical therapy. Few patients reported using other forms of "alternative" therapy. They also reported having the current problem for less than 3 months (30%), 6 months to a year (18%), or greater than 1 year (49%).

The mean level of patient satisfaction was quite high (87.4 out of 100). On a scale from 1–10 where 1 represents not confident at all about the treatment and 10 represents very confident, 42% rated the treatment as a 10, and 78% rated it as an 8 or better. Ninety percent would definitely recommend it for their family and friends, and 93% were sure they would return for care.

The results of surveys of acute back pain patients (Herman et al. 2018) show that chiropractic draws the majority of its patients from mainstream healthcare, mostly from medical care. For a majority of patients, the chiropractor is the primary contact provider for the condition being treated by the chiropractor. The patient clientele is largely white, and based on their education and income, middle class. While the patients report considerable pain from their problems, and some limitations, most appear to remain ambulatory and working. These results suggest that, on average, chiropractic patients with acute back problems are similar to those attending other providers.

With regard to **chronic back problems** in a recent observational study (Herman et al. 2018) of 2,024 current chiropractic patients, the mean age was 48 years, 72% were female, 92% were white, 56% had a professional or bachelor's degree or better, 60% were working full time, 16% were retired, and 68% had some form of insurance coverage. Of those who did the screening questionnaire, 23% had chronic low back pain, 15% had chronic neck pain, and 47% had both conditions. The average amount of time for having pain was 14 years and for seeing a chiropractor was 11 years. They have been seeing their current chiropractor for 5 years on average. The sample is composed chiefly of highly educated, white females, with at least partial insurance coverage for chiropractic and who have been in pain and using chiropractic care for many years. The patients who had both chronic neck and low back pain report more pain and disability for more years and have been seeing a chiropractor for more visits.

The patients reported seeing another health provider before seeing their chiropractor (76%) – usually a primary medical care provider (56%) – followed by massage therapist (41%) and physical therapist (28%). But only 32% see another provider concurrently for their back problem. Only 10% had taken prescription drugs in the last 6 months, but 45% had taken over the counter pain medication, and 24% had taken supplements or herbs. Sixty six percent had used exercise in the last 6 months, but only 5% used psychological counseling. Five percent had taken a narcotic, and 2% had injections in the last 6 months. In this sample of chronic patients, the pain and disability scores were low. The average for the chronic low back pain (CLBP) only group was a numeric pain score of 2.8 (0–10 scale) and an Oswestry score of 19.1 points and for the chronic neck pain (CNP) only group was a pain score of 2.8 and a Neck Disability Index (Vernon and Mior 1991) score of 21.4 points, which were all, as would be expected, closer to previous studies' posttreatment values than baseline values. However this may reflect the fact they have been in continuous chiropractic care for an average of 11 years. Again the satisfaction level for these patients was very high. When asked "how confident are you in recommending chiropractic to a friend," 93% answered extremely or very confident. With regard to the question "how successful do you think chiropractic will be in reducing your pain," 72% said very or extremely successful. This is partly because those with chronic back pain have come to see it as a lifelong condition. They reported that their initial pain scores were very high (8 out of 10) and their motivation for continuing chiropractic is pain avoidance, i.e., to make sure that the pain initial level does not return and they credited regular chiropractic care with making sure it does not occur. Only one-third of the patients endorsed a treatment goal of having their pain go away permanently that is of being cured. The rest had goals of preventing their pain from coming back (22% CLBP, 16% CNP), preventing their pain from getting worse (14% CLBP, 12% CNP), or temporarily relieving their pain (31% CLBP, 41% CNP) (Herman et al. 2019b).

Outcomes of Chiropractic Care

The evidence basis for chiropractic care now includes a wide range of studies including long-term clinical experience, observational studies, randomized clinical trials, meta-analyses, systematic literature reviews, formal expert consensus panels, and government reports and guidelines. The type of outcome measures used also covers the gamut and resembles those used in medicine. Among outcomes assessed in manipulation studies, pain level, physical function, and patient satisfaction have all rated highly. A systematic review by Khorsan et al. (2008) reviewed 629 studies on chiropractic. The most common patient-reported outcomes and instruments identified were the Oswestry Disability Index, visual analog scale, and Short Form 36. The most common clinician-reported measures were range of motion (i.e., goniometer, lumbar flexion, and inclinometer), motion palpation, and pain threshold (i.e., total tenderness score, tender joint count, current perception threshold, and algometer). Health service measures used included healthcare consumption (i.e., resource utilization and hospitalization), as well as direct and indirect costs. Therefore, while pain is a major outcome measured, it is not the only outcome that is important to either chiropractors or patients.

In clinical trials, chiropractic has been compared to placebo, exercise and advice, no treatment (natural progression), back school, analgesics and NSAIDs, infrared, shortwave diathermy, ultrasound, flexion exercises, massage, electrical stimulation, and various combinations of these comparators, as well as to usual medical care and physical therapy (Mootz and Coulter 2002; Goertz et al. 2018). In a systematic review of manipulation trials published from 2011 to 2017 for adults with low back pain treated in ambulatory settings, Paige et al. (2017) reported that spinal manipulative therapy was associated with modest improvements in pain and function. These studies included measurements of pain (measured by either the 100-mm visual analog scale, 11-point numeric rating scale, or other numeric pain scale), function (measured by the 24-point Roland Morris Disability Questionnaire or Oswestry Disability Index (range, 0–100)), or any harms measured within 6 weeks. No RCT reported any serious adverse event. Minor transient adverse events such as increased pain, muscle stiffness, and headache were reported 50–67% of the time in large case series of patients treated with SMT.

A systematic review was published by Bronfort et al. (2010) which summarized the scientific evidence regarding the effectiveness of manual treatment for the management of a variety of musculoskeletal and non-musculoskeletal conditions. They found 26 categories of conditions containing randomized controlled trial (RCT) evidence for the use of manual therapy: 13 musculoskeletal conditions, 4 types of chronic headache, and 9 non-musculoskeletal conditions. They also identified 49 recent relevant systematic reviews and 16 evidence-based clinical guidelines, plus an additional 46 RCTs not included in the systematic reviews and guidelines.

They concluded that spinal manipulation/mobilization is effective in adults for acute, subacute, and chronic low back pain, migraine and cervicogenic headache, and cervicogenic dizziness; manipulation/mobilization is effective for several extremity joint conditions; and thoracic manipulation/mobilization is effective for acute/subacute neck pain. They found that the evidence was inconclusive for the use of spinal manipulation for the treatment of various non-musculoskeletal conditions in adults and children.

Several other recent systematic reviews of spinal manipulation have all concluded that manipulation produces modest clinical effects that are similar in effectiveness to other recommended therapies for low back pain and neck pain and that serious adverse events are extremely rare. Also, as noted previously, the most current American College of Physicians guideline for the non-pharmacological management of low back pain recommends spinal manipulation as one of the frontline treatments for low back pain (Rubenstein et al. 2019; Masaracchio et al. 2019).

Can the Elderly Benefit from Chiropractic Care?

Older patients make up approximately 15% of chiropractic patient populations (Coulter 1996a). Of patients between 65 and 75 years of age, 14% report using chiropractic services, but that drops off to 6% among those over the age of 75. However, access issues may account for these numbers. There is great regional variation, and one study examining two rural Midwestern communities found that two-thirds of individuals over the age of 65 used chiropractic care (Lavsky-Shulan et al. 1985). That number is increasing substantially among men over 70 years old. Overall, chiropractic utilization by the elderly mirrors that of the general population. Those that do use chiropractic services tend to be in good health, less likely to use nursing home or hospital services and use fewer prescription drugs, but more over-the-counter medications (Coulter 1996b).

Reliance on clinical experience in the care of elderly patients is the norm within chiropractic, and a great deal of qualitative attention to geriatric care issues can be found in chiropractic training and clinical literature (Killinger 2004; McCarthy 1996). Manipulation techniques are frequently modified to suit the exigencies and tolerances of patients, and specific considerations have been reported in chiropractic literature (Bergmann 1993). Age-appropriate modifications to chiropractic evaluation protocols may also be warranted and have been described as well. Chiropractors also report providing an eclectic host of interventions beyond manipulation for elderly patients including exercise, nutrition, relaxation, and physical therapy (Rupert 2000).

The number of older adults in the United States is increasing yearly, with projections that 20% of the US population will be 65 years of age or older by 2030. Considering the high prevalence of spinal pain and other degenerative musculoskeletal conditions in older adults, chiropractic care should be considered as an option for older adults. Offering nonsurgical, non-pharmacological treatment options to older adults is important, due to the risks associated with opioid medications in this population. It is also important to recognize that chiropractors may incorporate multiple types of manual techniques with older patients. These techniques will be tailored to the individual needs of the patient, with the application of varying levels of biomechanical force and amplitude. Chiropractors also incorporate posture education, health promotion, and therapeutic exercises into multimodal treatment strategies with older adults.

Recently, two large clinical trials were published that involved chiropractic spinal manipulation and exercises provided to older adults with lumbar spinal stenosis (Schneider et al. 2019; Ammendolia et al. 2018). In both of these trials, patients had significant improvements in their walking performance (neurogenic claudication symptoms). Also, it is important to note that no serious adverse events were reported in either trial, which provides evidence that these procedures are relatively safe for use in the older adult population. Dougherty et al. have reported on the safety of SMT in the older adult population, specifically in osteoporosis, anticoagulation therapy, and spinal stenosis. These data are from two randomized controlled trials and also from retrospective data from a chiropractic clinic in a long-term care facility (Dougherty and Killinger 2005; Dougherty et al. 2009).

Therefore, spine surgeons should consider a trial of chiropractic care for patients with mild to moderate levels lumbar spinal stenosis as a reasonable treatment and "screening strategy" before considering decompressive surgery. Failure to respond favorably to chiropractic care would suggest that the patient was more likely to be a proper candidate for surgical decompression.

Hawk et al. (2017) published a systematic review and best practices guideline for chiropractic care of older adults. This document provides a summary of evidence-informed best practices for doctors of chiropractic for the evaluation, management, and manual treatment of older adult patients. This document also provides additional guidance on the importance of tailored approaches to the evaluation of the older adult, specifically in the areas of cognitive impairment and preventive screening. This best practice guideline is an excellent resource for spine surgeons who would like to review an "executive summary" of the literature on the topic of chiropractic care for older adults.

How Does One Find a Good Chiropractor?

The simplest answer is "the same way you find any good doctor." Like any other healthcare practitioner, the expertise, personality, practice style, and availability can all factor into deciding how to find a chiropractor. Different patients may have their own needs and preferences that impact how effective and worthwhile one practitioner might be compared to another. For patients, recommendations of friends or family members are often the most ready source of information. Internal medicine specialists make up one of the more common interdisciplinary referral sources reported by chiropractors (Christensen et al. 2015). Therefore, asking internists or family practitioners for recommendations may be a good starting place (Curtis and Bove 1992).

Obtaining a list of practitioners in your community from a state licensing board is also a starting point. One can inquire if any chiropractors on the list have had complaints or disciplinary actions upheld against them. In general, when looking to establish an inter-referral relationship, it may be worthwhile to meet with and interview a number of chiropractors to get a sense of their educational background and practice style. Asking questions like "how do you determine how much care someone needs?" and "how do you work and communicate with other providers?" can give insight into clinical styles and preferences that can be compared with your own. Will the chiropractor provide written reports and updates of findings, recommendations, and progress? Additionally, asking a chiropractor about what they do when patient progress is slower than expected and what they do to cultivate a patient's own self-reliance may be important to know.

Last, but not least, the extent to which the chiropractor takes care of his/her own health might be an important consideration. Since much of chiropractic care is about increasing the patient's knowledge and behavior modification for self-care and prevention, it seems reasonable to expect the chiropractor to live by the same standards. If the chiropractor's knowledge does not lead to appropriate healthy behaviors for the chiropractor, it seems difficult to believe it will do so for the patient.

Conclusion

We have attempted to show that there is considerable benefit to back pain patients by visiting chiropractors. For back surgeons, and medical physicians generally, chiropractors offer an alternative form of noninvasive, conservative care that does not involve drugs nor opioids. Increasingly this is the recommendation as a first-line option in spine care guidelines. It would also constitute an evidence-based informed choice with regard to spinal care. One overwhelming reason for referring to chiropractors is the fact the patients are not going to be prescribed drugs. This removes the danger of interactive effects (especially important for the elderly), drug addiction, and overdosing (especially important for pain patients).

For non-musculoskeletal problems, the jury is still very much out for two possible reasons. The first reason is that manipulation may have little or no efficacy/effectiveness with such conditions or that the body of research in this area is currently too sparse to come to any conclusion. While impotence and abstinence may have the same clinical outcome – the failure to reproduce – they have very different causes. The best we can say here is that for non-musculoskeletal problems, there are anecdotal reports from patients about the effectiveness of chiropractic care and a body of anecdotal claims made by chiropractors. Since chiropractic patients tend to be well educated and since people do not usually pay for services that are of no value to them, retaining an open mind here may be the best option. However, it is unacceptable for chiropractors to make health claims about conditions for which they have no evidence.

In conclusion, we have discussed in this chapter the limitations of viewing chiropractic care as simply a modality – i.e., manipulation – which misses the contribution that chiropractors make to the overall wellness of patients. In the case of chiropractors, we now have a substantial body of evidence from Health Services Research on the effectiveness of chiropractic care and a substantial body of trial evidence about its efficacy and safety. Chiropractors are also rather unique among the CAM professions by being extensively studied by the use of ethnographic observation so that the health encounter has been well documented. We can close by quoting three such studies: one from the Faculty of Medicine, University of Toronto, by three sociologists, the second by a medical epidemiologist, and the third by an anthropologist.

Kelner, Hall, and Coulter (1980) state:

It offers intelligible care; the chiropractors try to provide their patients with an understanding of their injury or illness, using a language which patients can comprehend. They explain the plan of treatment, the progress of the case, and the relation of their illness to environmental conditions. Finally, they try to make patients aware of their personal role and responsibility in the maintenance of their health. Chiropractic is co-operative care-patients participate as partners in the treatment and enhancing of their own health. (p. 260)

Coulehan (1995) concludes:

Physicians can learn from the success of the clinical art in chiropractic. This art begins with "the faith that heals", and it involves an interaction that may well function as a positive feedback system to promote healing. By healing, I mean a satisfactory outcome for the patient: relief of pain, diminished anxiety, acceptance of one's lot in life, less disability, a positive mental attitude. (p. 389)

Finally Oths and Hinojosa (2004) makes the following conclusion:

Given chiropractic's unified theory of disease etiology, which provides a rational interpretation of a patient's problem and an unambiguous method for treating it, the practitioner and the patient can reach a common level of understanding. The end result is most often a patient highly satisfied with the care received. From the observations made in this study, one might be inclined to agree with Kleinman et al. that the chiropractor is "more interested and skilled in handling illness problems than the M.D."

References

Adams R, Price K, Tucker G et al (2012) The doctor and the patient – how is a clinical encounter perceived? Patient Educ Couns 86:127–133

Agocs S (2011) Chiropractic's fight for survival. Am Med Assoc J Ethics 13(6):384–388

Ammendolia C, Cote P, Southerst D, Schneider M, Budgell B, Bombardier C et al (2018) Comprehensive non-surgical treatment versus self-directed care to improve walking ability in lumbar spinal stenosis: a randomized trial. Arch Phys Med Rehabil 99(12):2408–19.e2

Association of American Medical Colleges. Weeks of instruction and contact hours required at US Medical Schools. https://www.aamc.org/data-reports/curriculum-reports/interactive-data/weeks-instruction-and-contact-hours-required-us-medical-schools. Accessed 20 Oct 2019

Austin JA, Marie A, Pelletier KR, Hansen E, Haskell WL (1998) A review of theincorporation of complementary and alternative medicine by mainstream physicians. Arch Intern Med 158, 2303–2310.

Baer HA (2006) The drive for legitimation by osteopathy and chiropractic in Australia: between heterodoxy and orthodoxy. Complement Health Pract Rev 11(2):77–94

Baer H, Coulter ID (2008) Taking stock of integrative medicine: broadening biomedicine or co-option of complementary and alternative medicine? Health Sociol Rev 17(4):331–342

Bally M, Dendukuri N, Rich B, Nadeau L, Helin-Salmivaara A, Garbe E et al (2017) Risk of acute myocardial infarction with NSAIDs in real world use: Bayesian meta-analysis of individual patient data. BMJ 357:j1909

Beattie PF, Nelson R, Murphy DR (2011) Development and preliminary validation of the MedRisk instrument to measure patient satisfaction with chiropractic care. J Manipulative Physiol Ther 34(1):23–29

Beliveau PJH, Wong JJ, Sutton DA, Simon NB, Bussieres AE, Mior SA, French SD (2017) The chiropractic profession: a scoping review of utilization rates, reasons for seeking care, patient profiles, and care provided. Chiropr Man Therap 25:35. https://doi.org/10.1186/s12998-017-0165-8

Bergmann T (1993) Chiropractic technique principles and procedures. Elsevier Mosby, St Louis, Miss.

Bergmann TF, Peterson DH (2011) Chiropractic technique: principles and procedures, 3rd edn. Elsevier, St. Louis

Blanchette MA, Rivard M, Dionne CE, Hogg-Johnson S, Steenstra I (2017) Association between the type of first healthcare provider and the duration of financial compensation for occupational back pain. J Occup Rehabil 27(3):382–392

Board of Chiropractic Examiners (2018) Rules and regulations. https://www.chiro.ca.gov/laws_regs/regulations.pdf. Accessed 20 Oct 2019

Boucher PB, Brousseau D, Chahine S (2016) A jurisdictional review of the legislation governing informed consent by chiropractors across Canada. J Can Chiropr Assoc 60(1):73–80

Bronfort G, Haas M, Evans R, Leininger B, Triano J (2010) Effectiveness of manual therapies: the UK evidence report. Chiropr Osteopat 18:3. https://doi.org/10.1186/1746-1340-18-3

Brummett CM, Waljee JF, Goesling J, Moser S, Lin P, Englesbe MJ et al (2017) New persistent opioid use after minor and major surgical procedures in US adults. JAMA Surg 152:e170504

Canadian Chiropractic Association website. Accessed 30 Dec 2019

Canadian Institute for Health Information (2019) National health expenditure trends, 1975 to 2019. CIHI, Ottawa

Canizares M, Hogg-Johnson S, Gignac MAM, Glazier RH, Badley EM (2017) Changes in the use practitioner-based complementary and alternative medicine over time in Canada: cohort and period effects. PLoS One

12(5):e0177307. https://doi.org/10.1371/journal.pone.0177307

Capra F (1986) Wholeness and health. Holist Med 1:145–159

Carey TS, Garrett J, Jackman A et al (1995) The outcomes and costs of care for acute low back pain among patients seen by primary care practitioners, chiropractors, and orthopaedic surgeons. N Engl J Med 333:913–917

Cassell EJ, Siegler M (eds) (1979) Changing values in medicine. N.Y. University Publications of America, Washington, DC

Cherkin DC, Mootz RD. (Eds) (1997) Chiropractic in the United States: Training, Practice, and Research. AHCPR Publication No. 98–N002, Washington, D.C.

Chou R, Deyo R, Friedly J, Skelly A, Hashimoto R, Weimer M et al (2016) Noninvasive treatments for low back pain. Agency for Healthcare Research and Quality, Rockville

Chou R, Deyo R, Friedly J, Skelly A, Hashimoto R, Weimer M et al (2017) Nonpharmacologic therapies for low back pain: a systematic review for an American College of Physicians clinical practice guideline. Ann Intern Med 166:493

Christensen MG, Hyland JK, Goertz C, Kollash MW, Shotts B, Blumlein N et al (eds) (2015) Practice analysis of chiropractic 2015. National Board of Chiropractic Examiners, Greeley

Cifuentes M, Willetts J, Wasiak R (2011) Health maintenance care in work-related low back pain and its association with disability recurrence. J Occup Environ Med 53(4):396–404

Clarke TC, Black L, Stussman B, Barnes P, Nahin R (2015) Trends in the use of complementary health approaches among adults: United States, 2002–2012. Natl Health Stat Report 79:1–16

Cluff LE (1987) New agenda for medicine. Am J Med 62:803–810

Coulehan JL (1995) Chiropractic and the clinical art. Social Science & Medicine 21 (4):383–390

Coulter ID (1983) Chiropractic observed: thirty years of changing sociological perspective. Chiropr Hist 3:43–47

Coulter ID (1990) The patient, the practitioner, and wellness: paradigm lost, paradigm gained. J Manipulative Physiol Ther 13:107–110

Coulter ID (1996a) Manipulation and mobilization of the cervical spine: the results of a literature survey and consensus panel. J Musc Med 4(4):113–123

Coulter ID (1996b) Chiropractic approaches to wellness and healing. In: Lawrence D (ed) Advances in chiropractic. Mosby Publications, Chicago

Coulter ID (1998) Efficacy and risks of chiropractic manipulation: what does the evidence suggest? Integr Med 1 (2):61–66

Coulter ID (1999) Chiropractic: a philosophy for alternative health care. Butterworth-Heinemann, Oxford, UK; reprinted in 2001, 2005

Coulter ID (2001) The NIH consensus conference on diagnosis, treatment and management of dental caries throughout life: process and outcome. J Evid-Based Dent Pract 1(1):58–63

Coulter ID (2004) Competing views of chiropractic: health services research versus ethnographic observation. Chapter 3. In: Oths KS, Hinojosa SZ (eds) Healing by hand. Manual medicine and bonesetting in global perspective. AltaMira Press, Walnut Creek

Coulter ID (2005) Communication in the chiropractic health encounter: sociological and anthropological approaches. Chapter 5. In: Haldeman S (ed) Principles and practices of chiropractic, 3rd edn. McGraw Hill, New York, pp 99–109

Coulter I (2012) The future of integrative medicine. A commentary on CAM and IM. In: Adams J, Andrews GJ, Barnes J, Broom A, Margin P (eds) Traditional complementary and integrative medicine. Palgrave Macmillan, Sydney

Coulter I (2018) The intersection of ethics and translational science. In: Chiappelli F, Balenton N (eds) Translational research: recent progress and future directions. NovaScience, Hauppauge

Coulter ID, Shekelle PG (2005) Chiropractic in North America: a descriptive analysis. J Manipulative Physiol Ther 28(2):83–89

Coulter ID, Hays RD, Danielson CD (1996) The role of the chiropractor in the changing health care system: from marginal to mainstream. Res Sociol Health Care 13A:95–117

Coulter I, Adams A, Coggan P, Wilkes M, Gonyea M (1998) A comparative study of chiropractic and medical education. Altern Ther Health Med 4(5):64–75

Coulter ID, Hurwitz EL, Adams AH, Genovese BJ, Hays R, Shekelle PG (2002) Patients using chiropractors in North America: who are they and why are they in chiropractic care? Spine 27(3):291–296; discussion 297–8

Coulter ID, Hilton L, Ryan G, Ellison M, Rhodes H (2008) Trials and tribulations on the road to implementing integrative medicine in a hospital setting. Health Sociol Rev 17(4):368–385

Coulter ID, Khorsan R, Crawford C, Hsiao AF (2010) Integrative health care under review: an emerging field. J Manipulative Physiol Ther 33(9):690–710

Coulter ID, Crawford C, Hurwitz EL, Vernon H, Khorsan R, Booth M, Herman PM (2018) Manipulation and mobilization for treating chronic low back pain: a systematic review and meta-analysis. Spine J 18(5):866–879

Coulter ID, Crawford C, Vernon H, Hurwitz EL, Khorsan R, Suttorp Booth M, Herman PM (2019a) Manipulation and mobilization for treating chronic non-specific neck pain: a systematic review and meta-analysis for an appropriateness panel. Pain Physician 22(2):E55–E70

Coulter ID, Ryan GW, Kraus L, Xenakis L, Hilton L (2019b) A method for deconstructing the health encounter in CAM: the social context. J Complement Med Res 10(2):81–88

Coulter I, Ryan G, Kraus L, Xenakis L, Hilton L (2019c) A Method for Deconstructing the Health Encounter in Complementary and Alternative Medicine: The Social

Context. Journal of Complementary Medicine Research 10(2):81

Council on Chiropractic Education (2017) Residency program accreditation standards: principles, processes, & requirements for accreditation. http://www.cce-usa.org/uploads/1/0/6/5/106500339/2017_cce_residency_accreditation_standards.pdf. Accessed 20 Oct 2019

Council on Chiropractic Education (2018) CCE accreditation standards: principles, processes, & requirements for accreditation. http://www.cce-usa.org/uploads/1/0/6/5/106500339/2018_cce_accreditation_standards.pdf. Accessed 20 Oct 2019b

Curtis P, Bove G (1992) Family physicians, chiropractors, and back pain. J Fam Practice 35(5):551–555

Deerings California Code Annotated, Business and Professions 6090-7079, 1993 Appendix I, Section 7, Chiropractic Act

Deyo RA, Mirza SK, Turner JA, Martin BI (2009) Overtreating chronic back pain: time to back off? J Am Board Fam Med 22(1):62–68

Di Blasi Z, Harkness E, Ernst E et al (2001) Influence of context effects on health outcomes: a systematic review. Lancet 357:757–762

Dougherty P, Killinger L (2005) Role of chiropractic in a long-term care setting. Long-Term Care Interface 6:33–38

Dougherty P, Karuza J, Savino D (2009) Adverse event reporting in a federally funded randomized clinical trial. American Public Health Association annual meeting; November 7–11, 2009. American Public Health Association, Philadelphia

Dunn AS, Green BN, Gilford S (2009) An analysis of the integration of chiropractic services within the United States military and veterans' health care systems. J Manipulative Physiol Ther 32:749–757

Eisenberg DM, Kessler RC, Van Rompay MI, Kaptchuk T, Wilkey S, Appel S, Davis RB (2001) Perceptions about complementary therapies relative to conventional therapies among adults who use both: results from a national survey. Ann Intern Med 135:344–351

Engel JM (1989) The need for a new medical model: a challenge for biomedicine. Holist Med 4:37–53

Fineberg SJ, Nandyala SV, Kurd MF et al (2014) Incidence and risk factors for postoperative ileus following anterior, posterior, and circumferential lumbar fusion. Spine J 14(8):1680–1685

Garcia-Larrea L, Peyron R (2013) Pain matrices and neuropathic pain matrices: a review. Pain 154(Suppl 1):S29–S43

Getzendanner S (1988) Permanent injunction order against AMA. JAMA 259(1):81–82

Gilmour J, Harrison C, Asadi L, Cohen MH, Vohra S (2011a) Referrals and shared or collaborative care: managing relationships with complementary and alternative medicine practitioners. Pediatrics 128:S181–S186

Gilmour J, Harrison C, Asadi L, Cohen MH, Vohra S (2011b) Complementary and alternative medicine practitioners' standard of care: responsibilities to patients and parents. Pediatrics 128 (Supplement 4):S200–S205

Goertz CM, Long CR, Vining RD, Pohlman K, Walter J, Coulter I (2018) Effect of usual medical care plus chiropractic care vs usual medical care alone on pain and disability among US service members with low back pain a comparative effectiveness clinical trial. JAMA Netw Open 1(1):e180105

Gordon JS (1980) The Paradigm of Holistic Medicine. In Hastings AC, Fadiman J, Gordon JS (Eds). Health for the Whole Person. Boulder, Colo, Westview Press, 3–35

Gorrell LM, Engel RM, Brown B, Lystad R (2016) The reporting of adverse events following spinal manipulation in randomized clinical trials -a systematic review. Spine J 16(9):1143–1151

Green BN, Johnson CD, Lisi AJ, Tucker J (2009) Chiropractic practice in military and veterans health care: the state of the literature. J Can Chiropr Assoc 53:194–204

Haldeman S, Chapman-Smith D, Petersen D (eds) (1993) Guidelines for chiropractic quality assurance and practice parameters. Aspen Publishers, Gaithersburg

Hawk C, Schneider MJ, Haas M, Katz P, Dougherty P, Gleberzon B, Killinger LZ, Weeks J (2017) Best practices for chiropractic care for older adults: a systematic review and consensus update. J Manipulative Physiol Ther 40(4):217–229

Hays RD, Sherbourne C, Spritzer K, Hilton LG, Ryan GW, Coulter ID, Herman PM (2019) Experiences with chiropractic care for patients with low back or neck pain. J Patient Exp 1–8. https://doi.org/10.1177/2374373519846022

Henry DE, Chiodo AE, Yang W (2011) Central nervous system reorganization in a variety of chronic pain states: a review. PM & R 3(12):1116–1125

Herman PM, Coulter ID (2015) Complementary and alternative medicine: professions or modalities? Policy implications for coverage, licensure, scope of practice, institutional privileges and research. RAND Corporation, Santa Monica

Herman PM, Kommareddi M, Sorbero ME, Rutter CM, Hays RD, Hilton LG, Ryan GW, Coulter ID (2018) Characteristics of chiropractic patients being treated for chronic low back and chronic neck pain. J Manipulative Physiol Ther 41(6):445–455

Herman PM, Edgington SE, Ryan GW, Coulter ID (2019a) Prevalence and characteristics of chronic spinal pain patients with different hopes (treatment goals) for ongoing chiropractic care. J Altern Complement Med 25(10):1015–1025

Herman PM, Broten N, Lavelle TA, Sorbero ME, Coulter ID (2019b) Exploring the prevalence and construct validity of high-impact chronic pain across chronic low-back pain study samples. The Spine Journal 19(8):1369–1377

Hertzman-Miller RP, Morgenstern H et al (2002) Comparing the satisfaction of low back pain patients randomized to receive medical or chiropractic care: results from the UCLA low-back pain study. Am J Public Health 92(10):1628–1633

Hirschkorn KA, Andersen R, Bourgeault IL (2009) Canadian family physicians and complementary/alternative medicine: the role of practice setting, medical training, and province of practice. Can Rev Sociol 46(2):143–159. PMID: 19831238

Hsiao AF, Hays RD, Ryan GW, Coulter ID, Andersen RM, Hardy ML, Diehl DL, Hui K-K, Wenger NS (2005) A self-report measure of clinicians' orientation toward integrative medicine. Health Serv Res 40(5 Pt 1): 1553–1569

Hurwitz E, Coulter ID, Adams A, Genovese B, Shekelle P (1998) Use of chiropractic services from 1985 through 1991 in the United States and Canada. Am J Public Health 88(5):771–776

Ischebeck BK, de Vries J, Janssen M, van Wingerden JP, Kleinrensink GJ, van der Geest JN et al (2017) Eye stabilization reflexes in traumatic and non-traumatic chronic neck pain patients. Musculoskelet Sci Pract 29:72–77

Jabr F (2019) Is dentistry a science. Atlantic Day, May Issue.

Jamison JR (1993) Chiropractic holism: interactively becoming in a reductionist health care system. Chiro J Aus 23(3):98–105

Jamison JR (1994) Clinical communication: the essence of chiropractic. J Chiropr Human 4:26–35

Jenkinson C, Coulter A, Wright L (1993) Short form 36 (SF 36) health survey questionnaire: normative data for adults of working age. BMJ 306:1437–1440

Kazis LE, Ameli O, Rothendler J, Garrity B, Cabral H, McDonough C et al (2019) Observational retrospective study of the association of initial healthcare provider for new-onset low back pain with early and long-term opioid use. BMJ Open 9(9):e028633

Keeney BJ, Fulton-Kehoe D, Turner JA, Wickizer TM, Chan KC, Franklin GM (2013) Early predictors of lumbar spine surgery after occupational back injury: results from a prospective study of workers in Washington State. Spine 38(11):953–964

Kelner M, Hall O, Coulter ID (1980) Chiropractors Do They Help. Toronto, Fitzhenry Whitesides.

Khorsan R, Coulter ID, Hawk C, Choate CG (2008) Measures in chiropractic research: choosing patient-based outcome assessments. J Manipulative Physiol Ther 31(5):355–375

Killinger LZ (2004) Chiropractic and geriatrics: a review of the training, role, and scope of chiropractic in caring for aging patients. Clinics in Geriatric Medicine 20(2):223–235

Lamm LC, Wegner E, Collard D (1995) Chiropractic scope of practice: what the law allows. J Manipulative Physiol Ther 18:16–20

Lavsky-Shulan M, Wallace RB, Kohout FJ, Lemke JH, Morris MC, MacLean Smith I, (1985) Prevalence and functional correlates of low back pain in the elderly: The Iowa 65+ Rural Health Study. J Am Geriatr Soc 33(1):23–28

Lisi AJ, Brandt CA (2016) Trends in the use and characteristics of chiropractic services in the Department of Veterans Affairs. J Manipulative Physiol Ther 39:381–386

Lisi AJ, Goertz C, Lawrence DJ, Satyanarayana P (2009) Characteristics of Veterans Health Administration chiropractors and chiropractic clinics. J Rehabil Res Dev 46:997–1002

Machado GC, Maher CG, Ferreira PH, Day RO, Pinheiro MB, Ferreira ML (2017) Non-steroidal anti-inflammatory drugs for spinal pain: a systematic review and meta-analysis. Ann Rheum Dis 76(7):1269–1278

Manchikanti L, Knezevic NN, Boswell MV, Kaye AD, Hirsch JA (2016) Epidural injections for lumbar radiculopathy and spinal stenosis: a comparative systematic review and meta-analysis. Pain Physician 19(3):E365–E410

Marquez-Lara A, Nandyala SV, Fineberg SJ, Singh K (2014) Cerebral vascular accidents after lumbar spine fusion. Spine 39(8):673–677

Martin BI, Mirza SK, Franklin GM, Lurie JD, MacKenzie TA, Deyo RA (2013) Hospital and surgeon variation in complications and repeat surgery following incident lumbar fusion for common degenerative diagnoses. Health Serv Res 48(1):1–25

Masaraccio M, Kirker K, States R, Hanney WJ, Liu X, Kolber M (2019) Thoracic spine manipulation for the management of mechanical neck pain: a systematic review and meta-analysis. PLoS One 14(2):e0211877

McCarthy KA (1996) Management considerations in the geriatric patient. Top Clin Chiropr 3(2):66–75

McWhinney IR (1986) Are we on the brink of a major transformation of clinical method. Can Med Assoc J 135:873–878

Mior S, Wong J, Sutton D et al (2019) Understanding patient profiles and characteristics of current chiropractic practice: a cross-sectional Ontario Chiropractic Observation and Analysis STudy (O-COAST). BMJ Open 9:e029851. https://doi.org/10.1136/bmjopen-2019-029851

Mootz RD, Coulter ID (2002) Chiropractic. Chapter 16. In: Kohatsu W (ed) Complementary and alternative medicine secrets. Hanley & Belfus, Philadelphia

Mootz RD, Cherkin DC, Odegard CE, Eisenberg DM, Barassi JP, Deyo RA (2005) Characteristics of chiropractic practitioners, patients, and encounters in Massachusetts and Arizona. J Manipulative Physiol Ther 28(9):645–653

Murphy DR (2013) Clinical reasoning in spine pain volume I: primary management of low back disorders. CRISP Education and Research, Pawtucket

Murphy DR (2016) Clinical reasoning in spine pain volume II: primary management of cervical disorders and case studies in primary spine care. CRISP Education and Research, Pawtucket

Murphy DR, Hurwitz EL (2011a) Application of a diagnosis-based clinical decision guide in patients with neck pain. Chiropr Man Therap 19(1):19

Murphy DR, Hurwitz EL (2011b) Application of a diagnosis-based clinical decision guide in patients with low back pain. Chiropr Man Therap 19(1):26

Nahin RL, Boineau R, Khalsa PS, Stussman BJ, Weber WJ (2016) Evidence-based evaluation of complementary health approaches for pain management in the United States. Mayo Clin Proc 91(9):1292–1306

Oths KS, Hinojosa SZ (eds) (2004) Healing by hand. manual medicine and bonesetting in global perspective. AltaMira Press: Walnut Creek, CA

Paige NM, Miake-Lye I, Boot Mh, . Beroes J, Mardian S, Dougherty P, Branson R, Tang B, Morton S, Shekelle P. Association of spinal manipulative therapy with clinical benefit and harm for acute low back pain: systematic review and meta-analysis. JAMA 2017; 317(14);1451–1460

Panjabi MM (2003) Clinical spinal instability and low back pain. J Electromyogr Kinesiol 13(4):371–376

Pellegrino ED (1979) Medicine, science, art: an old controversy revisited. Man Med 4:43–52

Qaseem A, Wilt TJ, McLean RM, Forciea MA (2017) Noninvasive treatments for acute, subacute, and chronic low back pain: a clinical practice guideline from the American College of Physicians. Ann Internal Med 166(7):514–530

Reichling DB, Levine JD (2009) Critical role of nociceptor plasticity in chronic pain. Trends Neurosci 32(12):611–618

Reiser SJ (1979) Environmental versus biological causation in medicine. In: Cassell EJ, Siegler M (eds) Changing values in medicine. N.Y. University Publications of America, Washington, DC

Roland M, Fairbank J (2000) The Roland–Morris disability questionnaire and the Oswestry disability questionnaire. Spine J 25(24):3115–3124

Rubinstein SM, de Zoete A, van Middelkoop M, Assendelft WJJ, de Boer MR, van Tulder MW (2019) Benefits and harms of spinal manipulative therapy for the treatment of chronic low back pain: systematic review and meta-analysis of randomised controlled trials. BMJ 364:689. https://doi.org/10.1136/bmj.l689

Rupert RL, Manello D, Sandefur R (2000) Maintenance care: Health promotion services administered to US chiropractic patients aged 65 and older, Part II. Journal of Manipulative and Physiological Therapeutics 23(1):10–19

Safe Care Victoria (2019) Chiropractic spinal manipulation of children under 12. Independent Review. Victorian Government, Melbourne. ISBN 978-1-76069-066-3

Sandefur R, Coulter ID (1997) Licensure and legal scope of practice. In: Cherkin DC, Mootz RD (eds) Chiropractic in the United States: training, practice, and research. AHCPR Publication no. 98-N002. Washington, DC

Sanger N, Bhatt M, Singhal N, Ramsden K, Baptist-Mohseni N, Panesar B et al (2019) Adverse outcomes associated with prescription opioids for acute low back pain: a systematic review and meta-analysis. Pain Physician 22(2):119–138

Schneider MJ, Ammendolia C, Murphy DR, Glick RM, Hile E, Tudorascu DL et al (2019) Comparative clinical effectiveness of nonsurgical treatment methods in patients with lumbar spinal stenosis: a randomized clinical trial. JAMA Netw Open 2(1):e186828

Seaman DR, Winterstein JF (1998) Dysafferentation: a novel term to describe the neuropathophysiological effects of joint complex dysfunction a look at likely mechanisms of symptom generation. J Manipulative Physiol Ther 21(4):267–280

Shekelle PG, Coulter ID (1997) Cervical spine manipulation: summary report of a systematic review of the literature and a multidisciplinary expert panel. J Spinal Disord 10(3):223–228

Stilwell P, Harman K (2017) Contemporary biopsychosocial exercise prescription for chronic low back pain: questioning core stability programs and considering context. J Can Chiropr Assoc 61(1):6–17

Stochkendahl MJ, Reza M, Torres P, Sutton D, Tuchin P, Brown R, Cote P (2019) The chiropractic workforce: a global review. Chiropr Man Therap 27:36. https://doi.org/10.1186/s12998-019-0255-x

Swait G, Finch R (2017) What are the risks of manual treatment of the spine? A scoping review for clinicians. Chiropr Man Therap 25:37. https://doi.org/10.1186/s12998-017-0168-5

Tarn DM, Paterniti DA, Good JS et al (2013) Physician–patient communication about dietary supplements. Patient Educ Couns 91:287–294

The Joint Commission. Non-pharmacologic and non-opioid solutions for pain management Issue 44, August 2018. https://www.jointcommission.org/assets/1/23/QS_Nonopioid_pain_mgmt_8_15_18_FINAL1.PDF

Triano JJ, Giuliano D, Kange I et al (2015) Consistency and malleability of manipulation performance in experienced clinicians: a pre post experimental design. J Manipulative Physiol Ther 38:407–415

Tucker HR, Scaff K, McCloud T, Carlomagno K, Daly K, Garcia A et al (2019) Harms and benefits of opioids for management of non-surgical acute and chronic low back pain: a systematic review. Br J Sports Med 1–13. https://doi.org/10.1136/bjsports-2018-099805

VA/DoD Clinical Practice Guideline for Diagnosis and Treatment of Low Back Pain - Provider Summary (2017) Department of Veterans Affairs/Department of Defense. Version 2.0

Van Dulmen AM, Bensing JM (2002) Health promoting effects of the physician-patient encounter, Psychology, Health & Medicine 7(3):289–300. https://doi.org/10.1080/13548500220139421

Vernon H, Mior S (1991) The neck disability index: a study of reliability and validity. J Manipulative Physiol Ther 14(7):409–415

Wardwell WI (1992) Chiropractic – history and evolution of a new profession. Mosby Year-Book, St. Louis

Weeks WB, Goertz CM, Meeker WC, Marchiori DM (2015) Public perceptions of doctors of chiropractic: results of a national survey and examination of variation according to respondents' likelihood to use chiropractic, experience with chiropractic, and chiropractic supply in local health care markets. J Manipulative Physiol Ther 38(8):533–44

Wenngren BI, Pettersson K, Lownhielm G, Hildingsson (2002) Eye mobility and auditory brainstem response dysfunction after whiplash injury. Acta Otolaryngol 122:276–283

Wirth DP (1995) The significance of belief and expectancy within the spiritual healing encounter. Social science & medicine 1995; 41:249–260. 1995/07/01

Yu F, Adams AH, Harber P, Kominski GF (2002) Comparing the satisfaction of low back pain patients randomized to receive medical or chiropractic care: results from the UCLA low-back pain study. Am J Public Health 92(10):1628–1635

Chiropractors See It Differently: A Surgeon's Observations

4

John Street

Contents

Introduction	68
Background of Chiropractic	70
Chiropractic Education	72
Chiropractic Profession at a Crossroad with Mainstream Medicine	74
Characteristics of Chiropractic Health Care and Practice	75
Spinal Manipulation Aspect of Chiropractic	75
Diagnosis and Assessment Methods in Chiropractic	76
Scientific Rationale of the Chiropractic Profession	77
Classification of Low Back Pain	78
Treatment of Lower Back Pain in Chiropractic	79
Examination Procedures in Chiropractic	80
Treatment Frequency and Duration in Chiropractic	80
Reevaluation and Reexamination in Chiropractic	81
Measurement of Health Outcomes in Chiropractic	81
Benefits and Risk of Chiropractic Care	81
Chiropractic Management of Chronic Lower Back Pain	82
Biological Rationale for Using Chiropractic	82
The Medical Approach to Low Back Pain	82
Use of Imaging Technology by Physicians	84
Recognizing Chronicity	84

J. Street (✉)
Division of Spine, Department of Orthopedics, University of British Columbia, Vancouver, BC, Canada
e-mail: John.street@vch.ca

Treatment Methods for Low Back Pain by Physicians or Chiropractors	84
The Place for Surgery	85
Spinal Injections	86
Opioids	87
Adverse Effects of Surgery and Other Invasive Treatment Options	87
Rationale for Including Chiropractic Care into Mainstream Medicine	88
References	88

Abstract

The practice of physical manipulation and manual therapies as treatment options for low back pain and other spinal problems has been prevalent for thousands of years. Manual therapies in the Western world are often offered by chiropractors as part of alternative medicine. Modern and mainstream health care community has greatly benefited from chiropractic practice, especially in alleviating spinal pain and related injuries. Although chiropractic practice has remained consistent over the years, it has gradually transitioned from craft to profession. However, the profession continues to encounter numerous internal and external challenges that have threatened to curtail its aim of becoming a fully-accepted practice in the mainstream health care industry.

This chapter, researched and written by a spine surgeon, is aimed at providing insight into the role of chiropractors in promoting population health by offering interventions for low back pain. The chapter is based on a wide range of literature on the positive contribution of chiropractors in the promotion of patients' well-being. The author discusses the similarities and differences between chiropractors and other health care professionals in terms of education, training, work philosophy, and treatment mechanisms. Much of the focus on this chapter will be on the extent to which chiropractors view their profession as complementary to rather than contrary to mainstream health care community.

A clear understanding about chiropractic is important because it will see the profession gain full legitimacy in the allied health field. Most importantly, the chapter will help in bridging the conceptual gap that exists between chiropractors and other health care providers. Most importantly, policy makers and the general public will change their skepticism with regards to the crucial profession of chiropractic and begin embracing the concepts of traditional and alternative medicine to improve public health.

Keywords

Chiropractic · Healthcare · Legitimacy · Profession · Mainstream · Medical practice

Introduction

The title of this chapter "Chiropractors see it differently," a priori creates limits and expectations on the perspective of the writer, and a belief that traditional medicine and chiropractic are fundamentally different, with different origins, different beliefs, and different treatment philosophies.

To "see it differently" implies having a different perspective on "it." In order to further elaborate on this statement, we must first define what "it" is, and secondly we must consider what shapes one's views or perspectives. "It" may simply refer to axial low back pain alone, or "It" can encompass as broad a view as illness and disease in general and would thus require that we consider the chiropractic "vitalistic theory" in comparison to "traditional" medicine.

This chapter is probably unlike most others in this book. It has been researched and written by a spine surgeon, attempting to provide an unbiased view of a somewhat unfamiliar topic, in a form that can be digested by surgeon and nonsurgeon readers alike.

The practice of chiropractic is concerned with the diagnosis, treatment, and prevention of mechanical disorders of the musculoskeletal system. It primarily focuses on the effects of spinal disorders on the normal functioning of the nervous system and the general health of humans. The primary emphasis of chiropractors is on manual treatment, which includes manipulation and adjustment of the spinal column. By restoring the normal function of the musculoskeletal system, chiropractor professionals play a significant role in relieving pain and discomfort arising from accidents, stress, illness, or daily wear and tear that humans are bound to experience. Essentially, chiropractors adopt a holistic approach to health because they tend to evaluate the human body in terms of aspects such as medical history, experience, mental state, hobbies, occupation, and sporting activities. Various forces and pressure that affect the human body are taken into consideration when restoring the normal function of the musculoskeletal system. In broad terms, the goal of chiropractic treatment is to restore the function of the body and assist humans in eradicating the cycle of noxious stimulus, thus promoting self-healing mechanism.

Over the years, chiropractic has sharply divided opinion among health care practitioners from the mainstream medical community as well as the general public. While chiropractors have made positive contribution in the field of medicine, there has been a general reluctance by mainstream medical practitioners to perceive them as bona fide healthcare professionals. Many people are of the opinion that chiropractors do not qualify to be doctors because they do not hold medical degrees. However, they do possess extensive training in chiropractic care, making them licensed practitioners in administering crucial healthcare services.

Numerous intersections do exist between chiropractors and other medical professionals. First, chiropractors, like medical doctors, primarily rely on the medical history of the patient to determine the best approach to therapy. Second, they follow a systematic treatment plan, which is also a common approach among mainstream medical practitioners. Third, they hold practicing licenses, making them bona fide providers of health care services. However, a notable distinction between chiropractors and doctors from the mainstream medical community is the fact that they do not prescribe drugs, perform surgeries, or cure the underlying pathology.

To put the central premise of this chapter in context, and particularly when comparing to the outcomes of "traditional" medical management of low back pain, the following facts must be considered carefully:

1. Musculoskeletal pain, led by spinal disorders, costs the US health care system $874 billion per year and is the most common cause of severe long-term pain and disability (United States Bone and Joint Initiative 2014).
2. Research has found that prescription opioid pain medications are ineffective in the treatment of chronic low back (spinal) pain (Bone and Joint Decade 1998)
3. Chiropractic care offers a nondrug approach to spinal pain and other musculoskeletal conditions that is effective, saves money, and may help some patients avoid the risks of addiction associated with opioid use (Elton 2014).
4. In 2015, two million Americans had a substance use disorder involving prescription pain relievers with 20,101 overdose deaths related to prescription pain relievers (Abdel Shaheed et al. 2016).
5. From 1999 to 2008, overdose death rates and substance use rates quadrupled in parallel to sales of prescription pain relievers (The Center for Behavioral Health Statistics and Quality 2016).
6. The American College of Physicians Clinical Practice Guideline on Low Back Pain recommends the use of nondrug, noninvasive treatments – including spinal manipulation – before moving on to over-the-counter and prescription pain medications (Paulozzi et al. 2011).
7. Among patients with acute low back pain, spinal manipulative therapy was associated

with modest improvements in pain and function at up to 6 weeks with transient minor musculoskeletal harms (Qaseem et al. 2017).
8. Evidence suggests that therapies involving manual therapy and exercise are more effective than alternative strategies for patients with neck pain (Paige et al. 2017).
9. Patients with chronic low back pain treated by chiropractors showed greater improvement and satisfaction at one month than patients treated by family physicians. Satisfaction scores were higher for chiropractic patients. A higher proportion of chiropractic patients (56% vs. 13%) reported that their low back pain was better or much better, whereas nearly one-third of medical patients reported their low back pain was worse or much worse (Hurwitz et al. 2008).
10. It is unlikely that chiropractic care is a significant cause of injury in older adults. Among Medicare beneficiaries aged 66–99 with an office visit risk for a neuromusculoskeletal problem, risk of injury to the head, neck, or trunk within 7 days was 76% lower among subjects with a chiropractic office visit as compared to those who saw a primary care physician (Nyiendo et al. 2000).
11. In one study, the rate of opioid use was lower for recipients of chiropractic services (19%) as compared to nonrecipients (35%). The likelihood of filling a prescription for opioids was also 55% lower in the chiropractic recipient cohort. The average annual per-person charges for opioid prescription fills were 78% lower for recipients of chiropractic services as compared to nonrecipients (Whedon et al. 2015).
12. In addition, average per person charges for clinical services for low back pain were significantly lower for recipients of chiropractic services, $1,513 for chiropractic management vs. $6,766 for medical management (Whedon 2017).
13. Healthcare plans that formally incorporate chiropractic typically realize a 2 to 1 return for every dollar spent (Feldman 2014).
14. Following work-related low back injury, patients who visited a chiropractor were nearly 30 times less likely (1.5 vs. 42.7%) to require surgery as compared to those who chose a surgeon as their first provider (Keeney et al. 2013).
15. Paid costs for episodes of care initiated with a DC were almost 40% less than episodes initiated with an MD. Even after risk adjusting each patient's costs, we found that episodes of care initiated with a DC are 20% less expensive than episodes initiated with an MD (Liliedahl et al. 2010).
16. For Medicare patients with back and/or neck pain, availability of chiropractic care reduces the number of primary care physician visits, resulting in an annual savings of $83.5 million (Davis et al. 2015).

Background of Chiropractic

Manual therapies have been in existence for many centuries. Many cultural communities had practitioners whose primary role was to administer manual therapies to ease musculoskeletal discomforts and pains (Coulter et al. 1998). For example, the "bone setters" of England and Kung Fu masters in Asia are some of the examples of early chiropractic profession in the twentieth century. In the late twentieth century, the Western world saw a rapid emergence of osteopaths, chiropractors, and physiotherapists, thus changing the entire complexion of the manual therapies. From a traditional perspective, manual therapies were transferred from one generation to the other by parents. Fathers and mothers transferred their knowledge to their children, a practice that is still prevalent in some cultures.

The profession of formal chiropractic is around 120 years old, and it has transitioned from a full alternative medicine concept to become part of complementary health care. In fact, chiropractic is considered an integral part of primary care in some jurisdictions. Some scholars have argued that chiropractic includes elements that are consistent with religion and faith healing (Young 2014). Such views make the practice to assume a broader perspective than conventional health care systems. Regardless, it can be argued that the history of chiropractic and its overall contribution to the health well-being of humans has been checkered by the "good" and by the "bad." From the "good" perspective, the practice has directly contributed to over a century of improvement of public health. Consequently, many people have benefitted from the practice in terms of reduction of suffering caused by low back pain and related disability.

Currently, many countries have streamlined registration and licensing of chiropractors by introducing educational programs aimed at promoting the physical health of populations. The emergence of private colleges and universities is a clear indication that policymakers consider chiropractic as a veritable healthcare alternative. However, there still exists limited private health funding for patient consultations, minimal funding for patients, and little hospital access. The profession graduates competent manual therapists who have excelled in their respective fields and demonstrated their ability to be responsible citizens. However, there have been cases of aberrance, especially when chiropractors have extended their therapies to nonmusculoskeletal areas like ear infections or strabismus. This unwarranted expansion has caused damage not only to the health profession but also to the community at large.

Various scholars have questioned the justification to consider chiropractors as equal and worthy partners in the mainstream health care society. One of the overriding questions has been the ability of chiropractors to command a respect from other practitioners in the health sector, policy makers, and patients. Moreover, there is also the issue of accepting chiropractic professionals as legitimate partners in the national health care industry. To achieve legitimacy in the mainstream health sector, the chiropractors face two critical choices. Practitioners can maintain the status quo and uphold the current practice of "being different" from their counterparts in the mainstream medical profession. Alternatively, they can create a vision in which they endow chiropractors with attributes that make them fit in the mainstream health care community. To advance the chiropract profession globally and achieve the vision of integrating itself fully with the mainstream health industry, it is crucial for stakeholders to consider common grounds between chiropractors and their counterparts, especially in the medical field.

Unlike practitioners in the mainstream medical community, Chiropractic medicine believe that most health problems arise from disturbances in the body's nervous systems. These disturbances, in turn, originate from misalignments or subluxations of the spine. Chiropractic manipulations are aimed at realigning the spinal system and restoring normal function to the nervous system in a self-healing mechanism. The manipulation commonly encompasses a quick thrust of the specific subluxated vertebrae or application of physical pressure, traction, and stretching of the soft tissue attached to the body skeleton (Coulter 1997). Professionals who typically administer the traditional chiropractic medicine and focus entirely on spinal adjustment are referred to as "straights." These professionals have stoked a great deal of controversy, especially among the mainstream medical community because of their unorthodox approach to restoring human health. The main contention is that chiropractors do not belong to the conventional medical realm and thus, should not be considered as bona fide medical practitioners.

Most doctors of chiropractic typically follow traditional approaches when treating their patients. Some of the traditional elements of chiropractic intervention include nutrition and vitamin therapy. These practitioners are referred to as "mixers" because they are inconsistent to the modern medicine practices such as surgery and prescription of drugs. Still, another branch of chiropractic considers the entire musculoskeletal system as a critical determinant of human health. This branch focuses on treating joints and manipulating spinal components. Traditionally, chiropractors have defended their practices against criticism from the medical community and argued that their work is grounded on structural problems. Other than treating lower back pain, some chiropractors intervene in a wide variety of medical problems such as arthritis, asthma, chronic fatigue syndrome, bursitis, headaches, carpal syndrome, menstrual problems, traumatic injuries, and chronic pain syndrome. Chiropractic remains best known for treating lower back pain.

The distinguishing features of chiropractic's professional identity (WFC 2005) have been described as follows:

(a) Ability to improve function in the neuromusculoskeletal system and overall health and quality of life
(b) Specialized approach to examination, diagnosis, and treatment based on the best available research and clinical evidence and with particular evidence on the relationship between the spine and the nervous system
(c) Tradition of effectiveness and patient satisfaction
(d) Without use of drugs and surgery, enabling patients to avoid these where possible
(e) Expertly qualified providers of spinal adjustment, manipulation and other manual treatments, exercise instruction, and patient education
(f) Collaboration with other health professionals
(g) A patient-centered and biopsychosocial approach, emphasizing the mind-body relationship in health, the self-healing powers of the individual, individual responsibility for health, and encouraging patient independence

In proposing a model for chiropractic as a profession of spine care, Nelson offered a coherent and comprehensive model of professional identity (Nelson et al. 2005). He and his co-authors argue that chiropractic's identity is as a provider of spine care. They argued that such a model is consistent with the best available scientific evidence, is consistent with the current public perception, provides benefit to both the profession and the public, and is capable of gaining for the profession the cultural authority it now lacks. In developing the model, they established a set of criteria that the model must meet:

1. It must be consistent with accepted modes of scientific reasoning and knowledge.
2. It must accommodate future changes in scientific understanding.
3. It must represent a set of clinical competencies within the reach of practicing chiropractors.
4. It must be consistent, credible, and communicable to external constituencies on whom the profession relies.
5. It must represent the evidence of practice experience.
6. It must find a substantial presence within the healthcare marketplace.
7. It must be compatible with the training, licensure, history, and heritage of chiropractic.

Chiropractic Education

Many people are often surprised to discover that the education that chiropractors receive in college is quite similar to that of other medical students. Students attending chiropractic college are required to complete a minimum of 3 years of college-level courses before enrolling in the professional program. In addition, they are expected to complete a doctor of chiropractic degree program, which requires between 4 and 5 years of professional coursework. Researchers have further established that the education of a chiropractor is like that of a medical student to the extent of total classroom hours (Coulter 1997). The following table represents a comparison of the overall curriculum structure of chiropractic schools and medical schools in Kansas City (Table 1).

Typically, chiropractors receive more training in anatomy and physiology than physicians, with many chiropractic colleges focusing on therapeutic principles, diagnosis, orthopedics, and nutrition (Coulter et al. 1998). Three key areas, namely, manipulative/spinal analysis, physical diagnosis, and diagnosis imaging account for more than 50% of the education in clinical sciences (Coulter et al. 1998). Chiropractic interns need to complete 2 years of hands-on clinical experience primarily focusing on manipulation/adjustment as the primary treatment procedure. Researchers have established that chiropractic professionals receive more training in the fields of anatomy, physiology, bacteriology, diagnosis, X-ray, and orthopedics than their medical

Table 1 Comparison of class hours between chiropractic and medical students. (Source: Coulter et al. 1998)

Characteristic	Chiropractic schools Average	Percentage	Medical schools Average	Percentage
Contact hours	4,826	100	4,667	100
Basic science hours	1,420	29	1,200	26
Clinical science hours	3,406	71	3,467	76
Chiropractic science hours	1,975	41	N/A	N/A
Clerkshp	1,405	29	3,467	76

counterparts (Coulter 1997). Chiropractic institutions typically devote more time teaching students the basics of clinical sciences.

As a healthcare service, chiropractic offers a conservative management approach and does not necessarily require auxiliary staff. It represents a low-cost way of providing important health care services to patients. The World Health Organization encourages and supports countries in the proper use of safe medication and as a result the need to develop guidelines on chiropractic education and safe practice has steadily gained prominence around the world. Regulations for chiropractic practices vary considerably from country to country. In the USA, Canada, and some European countries, chiropractic has been given legal recognition and is being offered as university programs. In these countries, the chiropractic profession is regulated and the prescribed educational qualifications follow the requirements of the respective accrediting agencies. The emphasis on formalizing chiropractic practice in many Western countries is indicative of the fact that the profession possesses the same medical significance as other health care providers such as nurses, doctors, and surgeons.

At the same time, many countries have not yet established a chiropractic educational frameworks or laws to regulate the qualified presence of the profession. Some countries allow qualified health professionals and untrained medical practitioners to use the same techniques of spinal manipulation and claim to provide chiropractic services to patients. However, it is important to note that such physicians rarely receive chiropractic training in accredited programs. It is evident that many countries still regard chiropractic practice as a viable alternative to mainstream medicine for treating spinal injuries.

With the rapid growth in the demand for chiropractic services, health care practitioners may wish to acquire additional chiropractic qualifications. This has led n governments and policy makers to focus on the role of chiropractic practitioners in promoting health care. Many countries have developed conversion programs to enable persons with background in medical training to acquire supplementary education and skills to become chiropractors. Such conversion programs must be flexible. In countries that lack legislation, there may be no educational, professional, or legal framework governing chiropractic practice. The implementation of educational programs relating chiropractic profession depends on the situation in each country.

Although spinal manipulation dates back several thousand years ago, researchers have attributed the foundation of chiropractic to D.D. Palmer in 1895 (Palmer 1967). In 1897, the first school for the training of chiropractors became operation in Davenport, Iowa. Palmer developed the chiropractic theory from several concepts, including medical manipulation, bone setting, and osteopathy. The term "chiropractic" from the Greek and means "done by hand." It was coined by a patient named Reverend Samuel H. Weed and subsequently adopted by Palmer (1967). Chiropractic education initially developed in the USA during a period of significant reformation in the medical field. During this time, people had multiple treatment options not only within conventional medicine but also traditional and alternative health care approaches.

The principles that distinguish and differentiate chiropractic from conventional medical practice have been studied by scholars in the health care field. The different approach to education between the two fields has influenced chiropractic attitudes. Many professionals from within the profession have maintained principles that include but are not limited to holism, vitalism, naturalism, conservatism, critical rationalism, ethics, and humanism (Coulter 1997). The relationship between structure, especially the spine and musculoskeletal system, and the restoration and preservation of human health is central to the chiropractic approach. Unlike the mainstream medical field, chiropractic focuses on the conservative management of the neuromusculoskeletal system without the need to perform surgery or injections. Biopsychosocial causes and consequences of poor health are also significant factors in the management of patients. When in the best interest of the patient, chiropractic education system emphasizes on the importance of referring to the mainstream health care providers. Medical professionals rarely find the need to refer their patients to chiropractors.

Chiropractic education must involve administrative and academic considerations, including who should be trained, the role and responsibilities of practitioners, the required level of education, the accredited institutions, and the availability of qualified educators. Most countries use national, regional, state, or provincial standards while some authorities delegate to national professional.

The government may wish to evaluate both positive and negative consequences of integrating chiropractic into the mainstream. Many countries have recognized the need to establish "limited" programs as interim measures to establishing full chiropractic educational courses. This approach is deigned to supplement exiting health care education systems rather than replace them. An increased focus on chiropractic indicates that policy makers are recognizing the profession as a viable alternative medicine for promoting the physical wellbeing of citizens. In many countries, chiropractic practitioners who lack formal training are often encouraged to upgrade their education to match the level of their medical counterparts allowing easier integration into the professional workforce.

Chiropractic Profession at a Crossroad with Mainstream Medicine

Chiropractic is an expansive and well-established health care profession in many Western countries, including the United States. The profession is the largest, most regulated, and best recognized practice outside of the mainstream medicine. Research has established that patients seeking alternative health care show great satisfaction with the practice (Coulter et al. 1998). During the past two decades, there has been a drastic change in how medical professionals and learning institution view chiropractic. This change in attitude towards the profession has been partly due to chiropractic's change in its approach to education and treatment.

One point of contention between chiropractic and mainstream medicine is that the former follows a drastically different scientific approach. Chiropractic is an evolving health profession. As envisioned by its originators, the profession is a revolutionary system of healing based on the notion that most of human's suffering and physical discomforts arise from the central nervous system disrupting the healthy expression of life.

From the many schools that were established during the early twentieth century, a stable number of chiropractic training institutions have emerged in many countries around the world. Chiropractic education in the United States, South Africa, Denmark, Canada, and Great Britain is provided in both government-sponsored and private learning institutions and many chiropractic colleges are now accredited by the relevant authorities. For the example, Council of Chiropractic Education is the regulatory agency charged with the responsibility of overseeing all training in the United States (Coulter et al. 1998). In contrast, the mainstream medical practice is regulated by the numerous federal and state laws throughout the country. Each chiropractic college currently requires a minimum of 60 units of prescribed college-level courses – (Coulter 1997). A specialization in the sciences is one prerequisite for enrolling in a chiropractic school.

By the early twenty-first century, chiropractic curricula had an average of 4,820 classroom and clinical hours, with students spending about 30% of these hours in basic science. The rest of the time was spent on clinical work and internships (Coulter et al. 1998). Medical students averaged 4,670 h, with a similar breakdown of subjects. Compared with medical students, chiropractic learners spend more hours focusing on human anatomy and physiology but fewer time on public health (Coulter et al. 1998). They do, however, spend about the same amount of time on important fields such as bio-chemistry, microbiology, and pathology. According to Coulter et al. (1998), chiropractic curricula entails fewer instructions than medical courses in terms of pharmacology, critical/emergency care, and surgery. However, there is a greater emphasis in biomechanics, musculoskeletal function, and manual methods of treatment.

A review of the American medical curriculum indicates that medical students spent more than twice as many hours in clinical experience, but 1,000 fewer hours in didactic and practical clinical courses (Coulter et al. 1998). All chiropractic educational institutions run busy practical clinics to ensure that learners receive training in a chiropractic environment. Specialty programs are available and include 2–3 years' postgraduate residency programs in areas such as radiology, orthopedics, neurology, sports, rehabilitation, and pediatrics. Most states recognize or require students to pass examinations administered by the National Board of Chiropractic Examiners in basic and clinical sciences.

Characteristics of Chiropractic Health Care and Practice

Chiropractic practice has evolved over the years but retains a distinct set of values, traditions, and curricula composition. Modern chiropractic theory has moved away from the original practices envisioned by DD Palmer. Chiropractic practitioners use a novel system of healing based on the premise that neurologic dysfunctions caused by "impinged" nerves at the spinal level is the cause of many human diseases (Coulter et al. 1998). A key belief is that spinal manipulation and adjustment removes the blockage to restore health. Much modern chiropractic theory and practice has moved away from the original mono-causal theory in which practitioners focused on the causes rather than consequences of musculoskeletal injuries. Research is gradually redefining the nature and discipline of chiropractic and its education.

Many observers still associate the practice of chiropractic solely with the practice of spinal manipulation. This is only partly accurate. The modern concept of "complementary and alternative medicine" (CAM) has provided a new meaning and scope to chiropractic practice and many practitioners do not want to be defined by spinal manipulation. Chiropractors perform many of the duties of other primary care providers and often described themselves that way. However, many professionals in health care perceive chiropractic as a profession with limited medical competence, akin to dentistry or podiatry. This is an ongoing debate.

Spinal Manipulation Aspect of Chiropractic

Most chiropractors accept that spinal manipulation is a key element of their practice. Many prefer the term "spinal adjustment" to reflect their belief in the therapeutic and health-enhancing effects of adjusting abnormalities associated with spinal column. There are numerous adjustment techniques but no agreement on the most appropriate or efficacious. Spinal manipulation is the application of a physical force to specific body tissues with therapeutic intent. Traditionally chiropractors applied the external force using their hands, with varying velocity, duration, frequency, and amplitude. Traditional medical practitioners relied on medication, advice, and occasionally surgery to treat spinal problems.

Spinal manipulation is associated with chiropractors, who perform over 90% of the procedures

(Johnson et al. 2012). Additional treatments include heat, cold, TNS, interferential therapy, and active rehabilitation (Johnson et al. 2012). Most chiropractors suggest therapeutic exercises and general physical fitness and combine this with nutritional advice and counseling on weight loss, cessation of smoking, and relaxation techniques (Johnson et al. 2012). Chiropractors also employ massage, acupuncture (and its variants) along with mineral and herb supplements to ease back pain.

Research suggests that for musculoskeletal problems, many patients prefer going to chiropractors than to their family physician with 60% seeking treatment for lower back pain and the remainder for head, neck, and extremity symptoms (Johnson et al. 2012). About 30% of all patients seeking professional intervention for pain in their back consult chiropractors in a primary health care setting. About half of these patients have a chronic musculoskeletal complaint. Studies further indicate that only a very small number of patients – between 2% and 5% – seek care for other conditions (Johnson et al. 2012). A few people visit chiropractors for general health problems, disease prevention, and health advice.

Diagnosis and Assessment Methods in Chiropractic

Chiropractors use a patient-focused diagnostic approach to determine the cause of a problem similar to other health care disciplines. The history, physical examination, and specialty-specific assessments are used by both chiropractors and physicians (Johnson et al. 2012). The Council on Chiropractic Education has specified that institutions must teach basic clinical competencies and that chiropractors are expected to differentiate mechanical musculoskeletal problems from visceral abnormalities. Chiropractic practice guidelines deem obtaining a history, performing a physical examination, and carrying out periodic reassessments are necessary attributes of good practice.

The National Board of Chiropractic Examiners have proposed a framework describing chiropractic practice (Johnson et al. 2012). The emphasis is on arriving at a diagnosis based on information gathered from the patient's history and physical, neurologic, and orthopedic examinations. According to Johnson et al. (2012), chiropractors have the obligation to render "legal and customary" medical diagnoses within the chiropractor's scope of practice. Patients with suspected nonchiropractic diagnoses are referred elsewhere. As distinct from the usual medical primary care physician, the chiropractor frequently does not have the benefit of opinions for practitioners in other specialties.

Unlike medical specialists, chiropractors do not routinely use advanced diagnostic tests. This may reflect their predominantly benign musculoskeletal practice. They do make extensive use of plain-film radiography. Chiropractors spend a considerable amount in their training learning the techniques and interpretation of musculoskeletal radiographs. They typically obtain radiographs of all new patients to diagnose musculoskeletal problems. According to Johnson et al. (2012), chiropractors consider knowledge of normal radiographic procedures to be "extremely important" to an accurate diagnosis.

Clinicians have hotly debated the indications for radiography in chiropractic practice. Its use varies in different regions. A practice-based study comparing chiropractic and physician practices for patients in Oregon hospitals revealed that 26% of patients in both practices had x-rays (Nyiendo et al. 2011). In another study, 67% of chiropractic patients in North Carolina had plain X-rays while 72% of those who saw an orthopedic surgeon were imaged (Johnson et al. 2012). The arrival of Medicare precipitated an increase in ordering radiographs by chiropractic practitioners while subsequent legislation decreased usage.

Chiropractors rely on the patient's medical history to establish the possible etiology of musculoskeletal problems, which can be supplemented by additional studies. Chiropractors are trained to focus on joints, muscle, and soft tissue to determine the potential utility of spinal manipulation. The primary assessment includes palpation, assessing the range and quality of joint motion, and probing for tenderness and inflammations. Based on the findings, chiropractors choose a treatment plan and establish a prognosis. Patients may receive a trial of chiropractic care, be "co-managed" with a physician or be referred to an appropriate specialist. The chiropractic profession has developed detailed guidelines that govern most aspects of management. These guidelines are part of chiropractic training (Smith 2016).

Chiropractors are trained in a high-touch low-tech health model. The primary concern is the person rather than the disease. Chiropractors believe in the inherent self-healing ability of the body and they are taught to communicate that hope of healing to their patients. Using spinal manipulation combined with other "high-touch" techniques requires a high level of trust between the chiropractor and the patient. Repeated visits allow the relationship between the chiropractor and the patient to thrive. Frequent physical contact between the care giver and the patients communicates confidence on both a social and psychological level (Kent 2018). Chiropractic training emphasizes the physical interaction between the patient and the care giver. The positive outcomes of the relationship demonstrate a more humanistic aspect than is generally seen in mainstream health care. Based on this connection, chiropractors may be better able to offer accepted information and advice than the primary care physician.

Anthropologist and sociologists have suggested that chiropractic treatment for chronic back pain can generate a sense of understanding and trust that mainstream medical professionals cannot match (Lisi et al. 2018). The hands-on and personalized "can-do" approach of the chiropractor seems concrete and reassuring compared to the more scientific, overly cautious, distant, and apparently indifferent approach of the physician. Observational studies have revealed that chiropractic patients are more satisfied with the service they receive compared to those treated by doctors. There is a case for including chiropractic in mainstream health care.

Scientific Rationale of the Chiropractic Profession

Over the last two decades, the chiropractic profession has evoked scientific arguments to justify its treatment of low back pain. Critics in the mainstream health care system have suggested that chiropractors rarely use acceptable and scientifically-proven approaches to treatment. A 2006 Gallup Poll rated chiropractic last among health professionals with regards to using ethical principles (Dynamic Chiropractic 2007). Based on the poll, 84% of patients considered nurses' ethics to be "very high" or "high," while only 36% felt that way about chiropractors (Dynamic Chiropractic 2007, par. 1). Although there may be some scientific support for the chiropractic approach to musculoskeletal dysfunctions, the profession faces difficulties when trying to justify a treatment that is partially rooted in quasi-mythical concepts. Confounding the problem is the fact that the sources of chronic musculoskeletal pain remain controversial so proving any treatment has a scientific basis is difficult.

Numerous studies on the rational for chiropractic care have looked at patients experiencing low back and neck, and headache. They chiefly involved placebo-controlled comparisons with other treatment options. Findings suggest much of the benefit of chiropractic comes from their more holistic approach. The same can be said, however, for many practitioners of alternative medicine.

Bussieres et al. using the Arksey and O'Malley framework reviewed the existing literature to establish current state of knowledge on evidence based practice (EBP), research utilization (RU), and knowledge translation (KT) in chiropractic care (Bussières et al. 2016). Nearly 85% (56/67) of the studies were conducted in Canada, USA, UK, or Australia. EBP included the attitudes and beliefs of chiropractors and the implementation of evidence based treatments. RU involved guideline adherence; frequency and sources of information accessed; and the perceived value of websites and search engines. KT looked at knowledge practice gaps; barriers and facilitators to knowledge use; and selection, tailoring, and implementation of interventions. While most practitioners professed a belief in all three areas, their use varied widely. Gaps existed in areas of assessment of activity limitation, determination of psychosocial factors influencing pain, general health indicators, establishing a prognosis, and exercise prescription. The authors' findings suggested that the majority of chiropractors hold favorable attitudes and beliefs toward EBP, RU, and KT but rarely put them into practice. They proposed educational strategies aimed at practicing chiropractors to improve patient care. They concluded that the chiropractic profession requires more robust dissemination and implementation research to improve guideline adherence.

Blanchette et al. examined the clinical effectiveness and economic impact of chiropractic care compared to other commonly used approaches to adult patients with nonspecific low back pain (LBP) (Blanchette et al. 2016). They identified randomized controlled trials (RCTs) and/or full economic evaluations of chiropractic care for low back pain compared to standard care by other healthcare providers. Primary outcomes included pain, functional status, and global improvement. Five RCTs compared chiropractic care to exercise therapy (1), physical therapy (3), and medical care (1). The authors found similar effects for all treatments. Three economic evaluations studies (one cost-effectiveness, one cost-minimization, and one cost-benefit) compared chiropractic to medical care. There were divergent conclusions (one favored chiropractic, one favored medical care, one showed equivalent results). Moderate evidence suggests that chiropractic care for LBP appears to be equally effective to physical therapy. Limited evidence suggests the same conclusion when chiropractic care is compared to exercise therapy and medical care.

Classification of Low Back Pain

Hartvigsen and colleagues provide an excellent review of low back pain, its causes, manifestations, and treatments (Hartvigsen et al. 2018). They point out that low back pain is a very common symptom, occurring in all age, socioeconomic, and geographic groups. Globally, years lived with disability caused by low back pain increased by 54% between 1990 and 2015 due largely to population increase and ageing. The biggest increase was in low- and middle-income countries. Low back pain is now the leading cause of disability worldwide. The authors note that it is usually impossible to identify a specific pain generator and that only a small proportion of people have a well understood pathological cause. The data suggest that people with physically demanding jobs, physical and mental comorbidities, smokers, and obese individuals are at greatest risk of low back pain and that it is overrepresented in people with low socioeconomic status. Most people with new episodes of low back pain recover quickly; however, recurrence is common and in a small proportion of people low back pain becomes persistent and disabling. Initial high pain intensity, psychological distress, and accompanying pain at multiple body sites increase the risk of chronicity. Recent research demonstrates that increasing evidence of central pain-modulating mechanisms and pain cognitions have important roles in the development of persistent disabling low back pain. Cost, health-care use, and disability from low back pain vary substantially between countries influenced by local culture and social systems, as well as by beliefs about cause and effect. Disability and costs attributed to low back pain are projected to increase in coming decades, in particular in countries where healthcare systems are fragile and ill-equipped to cope. The authors conclude that intensified research efforts and global initiatives are needed to address the burden of low back pain as a public health problem.

The key messages of their excellent report are:

- Low back pain was responsible for 60.1 million disability-adjusted life-years in 2015, an increase of 54% since 1990, with the biggest increase seen in low and middle-income countries.
- Disability from low back pain is highest in working age groups worldwide, which is especially concerning in countries where informal employment is common and possibilities for job modification are limited.
- Most episodes of low back pain are short-lasting with little or no consequence, but recurrent episodes are common and low back pain is increasingly understood as a long-lasting condition with a variable course rather than episodes of unrelated occurrences

- Low back pain is a complex condition with multiple contributors to both the pain and associated disability, including psychological factors, social factors, biophysical factors, comorbidities, and pain-processing mechanisms.
- For the vast majority of people with low back pain, it is currently not possible to accurately identify the specific nociceptive source.
- Lifestyle factors, such as smoking, obesity, and low levels of physical activity, that relate to poorer general health, are also associated with occurrence of low back pain episodes
- Costs associated with health care and work disability attributed to low back pain vary. considerably between countries, and are influenced by social norms, health-care approaches, and legislation.
- The global burden of low back pain is projected to increase.

Low back pain may occur as a result of a variety of reasons and pathological conditions. Frequently physicians and chiropractors find it difficult to definitively diagnose low back pain and often rely on a description of the symptoms. Chiropractors and physicians should have a classification of low back pain and be aware of those conditions and disease that can produce symptoms. Table 2 highlights the various etiologies of low back pain and the associated diseases.

Treatment of Lower Back Pain in Chiropractic

A plethora of treatment options for lower back pain has not reduced and may have increased the healthcare burden. The traditional chiropractic approach for treating lower back pain, using noninvasive and natural methods, has gained credibility over the years. The practice has gained wider acceptance among patients suffering from lower back problem, and in many instances chiropractors are the first point of contact for individuals suffering from lower back problems.

The lack of consensus of the etiology of mechanical back pain makes it impossible to prove the scientific rigor of most noninvasive treatments. Yet healthcare practitioners are increasingly obliged to claim scientific and evidence-based reasons for treatment. One reason for the acceptability of chiropractors is their focus on demonstrating a scientific rational (LeFebvre et al. 2013). Striving for a scientific approach helps to ensure that health care practitioners are engaged in the best practices (Slaughter et al. 2015). Evidence-based therapy has been strongly promoted yet estimates based on the 2002 National Health Interview Survey revealed that 62.1% of Americans used complementary and alternative medicine (CAM) therapies in 2001 (LeFebvre et al. 2013). Recent data show chiropractic care is the largest CAM in the

Table 2 Etiologies of low back pain and related diseases

Etiology	Associated disease
Psychological	Psychogenic low back pain, in hysteria, and chronic depression/dementia
Trauma	Low back pain associated with fractures
Inflammation	Purulent spondylitis Tuberculous spondylitis Ankylosing spondylitis
Tumors	Spinal metastasis by malignant tumors Multiple myeloma Spinal cord tumors
Degeneration	Spondylosis deformans Intervertebral disc degeneration Intervertebral articular low back pain Lumbar nonspondylolysis spondylolisthesis Ankylosing spinal hyperostosis Lumbar spinal canal stenosis
Abdominal organs	Liver, gallbladder, and pancreatic diseases

United States with approximately 70,000 members (Chapman-Smith 2010).

With regards to lower back disorder, the availability of CAM presents patients with opportunities to explore treatment methods through alternative and personalized approaches. These are typically more attuned to individual preferences. Mainstream medical practitioners tend to employ more use generalized treatment, denying the patience the opportunity to choose.

The personalized approach is consistent with the three major components of evidence-based practice, namely clinician experience and judgment, patient preference and values, and the best available scientific evidence (LeFebvre et al. 2013). In many cases, doctors and spinal specialists do not have enough time or willingness to focus on the nonmedical aspects of the problem.

Doctors of chiropractic use methods that assist patients in self-management including physical exercise, diet, and modification of lifestyle, which improve outcomes and reduce reliance on health care system resources (Johnson et al. 2012). Chiropractors also recognize that a variety of health care providers play a role at various stages. When it is in the patient's best interest, chiropractor consult practitioners in the mainstream health care community for advice (LeFebvre et al. 2013). The goal is to improve patients' functional capacity and educate them on the need to accept responsibility for their own health.

Examination Procedures in Chiropractic

Obtaining a thorough history and carrying out a comprehensive physical examination are critical for the chiropractic management of lower back pain problems. These provide a clinical rationale for appropriate diagnosis and subsequent plan for treatment. Chiropractic assessment includes several steps:

- Obtain the health history of the patient, including information on pain characteristics, red flags, review of previous treatment systems, and risk factors for chronicity
- Identity the specific causes of low back pain
- Conduct a physical examination on the patient for reflexes, dermatomes, myotomes, and orthopedic tests
- Perform a diagnostic testing for red flags

Imaging and other diagnostic tests may be indicated in the presence of severe or progressive neurologic deficits. They can be conducted when the history and physical examination suggest suspicious underlying pathology (Triano et al. 2010). Chiropractors evaluate patients with persistent low back pain and signs of radiculopathy or spinal stenosis to allow for accurate chiropractic intervention and reduce errors during treatment (Chou et al. 2007). A failure to respond may indicate an underlying anatomical anomaly such as spondylolisthesis and indicate imaging. Chiropractors may recommend lateral flexion-extension views to detect excessive intervertebral translation.

Chiropractors classify the conditions of illness and injury in accordance to severity and duration. There are various common descriptions of the illness stages, including acute, subacute, chronic, and recurrent. The acute stage includes symptoms that have been persistent for a period of less than 6 weeks. Subacute stage relates to symptoms that have persisted for a period of between 6 and 12 weeks. Chronic indicated the symptoms have persisted for at least 12 weeks and recurrent defines a return of the original symptoms after a period of complete remission. These symptoms are further subdivided into mild, moderate, and severe, depending on the intensity and the risk they pose to the patient.

Treatment Frequency and Duration in Chiropractic

The frequency and duration of chiropractic treatment is usually dependent on individual patient factors or characteristics. Some of the factors impeding recovery include comorbidities and clinical "yellow flags." Additional treatment visits can give for time to observe therapeutic responses and the clinician should evaluate the treatment outcomes. A typical therapeutic trial of chiropractic care consists of between 6 and 12 visits over a span of 2–4 weeks (Globe et al. 2016). During these visits, chiropractors monitor their patients' progress to document acceptable gains. Generally acute conditions require fewer treatments.

According to Globe et al. (2016), the initial stages of chiropractic treatment for lower back pain require consistency and clinical methods that reflect the best available evidence. Chiropractors need clinical judgment, experience, and should be aware of the patient's preference. Currently, most literature recommends high-velocity, low amplitude techniques, and mobilizations such as flexion-distraction (Clar et al. 2014).

The initial course of chiropractic treatment typically includes one or more passive, manual therapeutic procedures which require no patient participation. Pain control is achieved with spinal manipulation or mobilization combined with strong reassurance (Globe et al. 2016). As a rule, the chiropractic assessment does not include risk stratification for potential complicating factors.

Clinical judgment and patient preference are important aspects of chiropractic practice and relies on personal assessment of the patient's situation to determine the intervention. Because of the scarcity of definitive evidence about the efficacy of spinal manipulation methods, some researchers have suggested that bracing, taping, and orthoses not be recommended as part of the treatment strategy (Globe et al. 2016). Because so much of chiropractic decision making is highly personal and based on individual experience, however, these options are frequently employed. Such judgments would be less acceptable among medical professionals but are a hallmark of chiropractic care.

Reevaluation and Reexamination in Chiropractic

After concluding the initial phase of treatment, chiropractors re-evaluate the patient's condition. The need for additional treatment is based on the responsiveness to the initial phase of care and the probability achieving additional gains. Patients may plateau in their recovery process necessitating modification or suspension of treatment (Globe et al. 2016). The chiropractor must establish if recovery has paused or if the patient has gained "maximum therapeutic benefit" (Globe et al. 2016). The final treatment visit should provide the necessary education and instructions on effective self-management.

Re-evaluation may suggest continued chiropractic treatment. Chiropractors adopt a proactive system assessing the efficacy of the treatment rather than patient's putative pathology. The chiropractor encourages the patients to resume normal activities while, as much as possible, avoiding aggravating the condition. This advice is consistent with the notion that the human body has an enormous capacity to recover. In cases where the patient plateaus at an unacceptable level of function or pain control, the chiropractor must decide whether to continue treatment or explore for alternative therapeutic interventions. The patients should share in that decision.

Measurement of Health Outcomes in Chiropractic

For a trial of chiropractic care to be considered beneficial, the outcomes must be clinically relevant, the improvement must clearly improve the patient's functional capacity. Some of the observable metrics include pain scales such as the visual analogue and the numeric rating scales, pain diagrams that enable patients to demonstrate the location and character of their conditions and symptoms, an increase in the amount of home and leisure activities such as performing household chores or engaging in physical exercises, an increase in productivity at work, and improvements in validated functional capacity tests such as lifting, flexibility, and endurance.

Benefits and Risk of Chiropractic Care

In comparison to spine surgery, chiropractic care is certainly safe. A 2010 review concluded that the number of serious adverse outcomes among low back pain patients receiving chiropractic lumbar spine manipulation procedures was less than one per million visits (Bronfort et al. 2010). Other studies indicate that the risk of major adverse outcomes with manual therapy for a variety of conditions is less than with usual medical interventions. The most common complaint after manual treatment was short-lived after-treatment discomforts caused by muscle stiffness (Carnes et al. 2010). Patients judged poor candidates for spinal manipulation are offered alternative treatment choices such as soft-tissue, low-velocity, low-amplitude procedures, and mobilization. There are several clinical situations in which manual manipulation of the spine may be contraindicated and the chiropractor must evaluate the associated risks.

There is a need for flexibility in choosing treatment options. These include co-management with medical doctors. In all cases, full documentation is essential. The chiropractor must remain alert to a change in the patient's symptoms that could indicate a significant deterioration or underlying sinister pathology. These case demand immediate diagnostic workup or referral.

Chiropractic Management of Chronic Lower Back Pain

The management of chronic back pain often includes home-directed self-care and scheduled ongoing care. These are patients for whom self-care measures are insufficient to sustain desirable therapeutic gains. Their condition may progressively deteriorate as demonstrated by previous treatment failures. For these patients, the hands-on therapy or manipulation may produce only short term pain relief that allows compliance with the more important aspects of activity modification and lifestyle adjustment. The role of the chiropractor is to provide direction and reassurance more than mechanical treatment. The connection between the patient and the provider established though physical contact can be used to instill confidence and deal with the overwhelming psychosocial devastation of chronic pain. Any ongoing manual therapy must not interfere with the more essential re-establishment of a normal daily routine and lifestyle.

Biological Rationale for Using Chiropractic

Chiropractors typically direct spinal manipulation at a dysfunctional subluxation. In the current context of chiropractic, a subluxation may connote a functional and not necessarily an anatomic entity. Chiropractic dogma suggests that combination of the initial subluxation and the required manipulation can have important physiological effects on patients including increased motion, changes in facet kinematics, improved tolerance to pain, increased muscle strength, and enhanced proprioception. Much of the chiropractic literature focuses on the physical and long-term well-being of patients. Because the primary aim of chiropractors is to improve the health, the benefits of the professional interaction can override the actual anatomical alterations, if any, produced by the manipulation. The treatment has no significant negative side effects risk and whether any resulting perceived improvement in strength, mobility, or pain relief is the result of resolving a subluxation or providing a strong placebo may be moot.

The Medical Approach to Low Back Pain

Low back pain is a considerable health problem for people in both developed and undeveloped countries. The prevalence of low back pain varies from 49% to 70%, with most of these cases being reported in Western societies (Carnes et al. 2010). Individuals suffering from low back problem typically experience pain, muscle tension, or stiffness localized below the costal margin and just above the gluteal folds, with or without associated leg pain.

The diagnosis process differentiates between patients with specific or nonspecific low back pain. Specific low back pain is defined as symptoms that result from a recognized specific pathophysiological mechanism such as disc herniation, infection, rheumatoid disease, tumor, or fracture (Carnes et al. 2010). In the United States, some studies suggest that of those suffering from low back pain, 4% have a compression fracture, 3% spondylolisthesis, 0.7% a tumor or metastasis, 0.3% ankylosing spondylitis, and 0.01% an infection (Carnes et al. 2010). According to Koes et al. (2006), 90% of low back pain problems are nonspecific, that is, lacking a defined etiology. Consequently, physicians often diagnose the problem based on the exclusion of specific pathology. Compared to the chiropractic approach, this tactic places the emphasis in the wrong place, stressing rare pathologies over common minor mechanical malfunctions.

Health care professionals employ a variety of diagnostic labels. General practitioners use terms such as lumbago, physical therapists speak of hypomobility, manual therapists refer to joint disorders, and orthopedic surgeons treat degenerative discs (Koes et al. 2006). There is no widely accepted, reliable, and valid classification method for diagnosing nonspecific low back pain. In most cases, it is classified based on duration of the complaints and its severity, acute when it persists for less than 6 weeks, subacute between 6 and 3 months, and chronic when it lasts longer (Koes et al. 2006). These definitions offer no guidance and may be irrelevant.

The clinical course for most episodes of acute low back pain is favorable, with much of the discomforts dissipating within a couple of weeks. This is consistent with findings that indicate that 90% of patients with low back pain in primary care will cease visiting their doctors within 3 months of active treatment. Symptoms fluctuate over time, making it difficult to establish a consistent trend. Most individuals suffer recurrent attacks, with some estimates at about 70% within the first year. The severity and duration of these recurrences frequently do not precipitate need for treatment. Only about 5% of people with low back pain develop chronic problems and related disability.

Many physicians take a narrow view of back pain and overlook nonphysical factors exacerbating the pain. They tend to use a rigid approach to diagnosis and treatment when compared with chiropractors, who interact with patients long enough to identify the more global issues involved.

For nonspecific LBP, numerous studies have shown that surgery does not offer clinically relevant benefits over conservative interventions, including chiropractic treatment (Peul et al. 2014). For degenerative disc disease, a systematic review, including two randomized controlled trials (121 patients), compared intradiscal electrothermal therapy with sham surgery (Helm II et al. 2012). There was no significant difference in outcomes for pain, disability, and quality of life.

In a meta-analysis of four randomized controlled trials (767 patients) comparing spinal fusion surgery with conservative interventions, surgery was not superior in improving back pain, disability, and quality of life (Chou et al. 2009).

A Cochrane review of five randomized controlled trials with 1301 patients evaluated disc replacement for degenerative low back pain (Jacobs et al. 2012). It included one trial (173 patients) that showed a statistically significant but clinical irrelevant difference in VAS favoring disc replacement over nonoperative treatments including chiropractic. When assessing surgical intervention, the potential harms or side effects of surgery have been inconsistently underreported.

The National Institute for Health and Care Excellence (NICE) guidelines on nonspecific low back pain included only one cost effectiveness analysis. This concluded that if decision makers are willing to pay $50,000 per quality adjusted life year (QALY), the chance that surgery is cost effective compared with nonoperative care is only 20% (Savigny et al. 2009).

Surgery does however have a clear role for relief of radicular pain and LBP caused by specific diagnoses. For sciatica due to herniated lumbar disc, a systematic review with five randomized controlled trials (1135 patients) concluded that early surgery leads to short term benefits in function and reduced leg pain (recovery after 4 vs. 12 weeks) but with similar long-term results. (Jacobs et al. 2011). Interestingly, surgery did not provide superior relief of the back pain in these studies.

For neurogenic claudication due to lumbar stenosis, systematic reviews show that surgical decompression (five randomized controlled trials with 918 patients) is superior in improving pain, disability, and quality of life to conservative treatment up to at least 4 years (Kovacs et al. 2011; May and Comer 2013). These studies intentionally excluded patients with predominantly low back pain as their major complaint.

The National Institute for Health and Care Excellence has recently reviewed the surgical treatment of patients with nonspecific low back pain and given guidelines that fusion should only be offered as part of a randomized controlled trial and that lumbar disc replacement should no longer be performed. The NICE guidelines were based on a small number of prospective RCTs, six for lumbar fusion, and five for disc replacement. Each of these trials had difficulties: some involved few patients; some randomized patients to up to three different surgical techniques; there was inconsistent use nonoperative treatment. There were the usual problems of crossover and loss to follow-up. In an excellent commentary in 2017, Todd conclude that what is now urgently required is a high-quality RCT to identify the "ideal" patients with nonspecific back pain for fusion (Todd 2017).

Use of Imaging Technology by Physicians

Physicians often mistakenly use x-ray or magnetic resonance imaging (MRI) trying to diagnose nonspecific low back pain problems. Most of the abnormalities found imaging individuals without low back pain are the same at in patients with the problem. According to Roland and Van Tulder (1998), up to 50% of asymptomatic people have degenerative findings and imaging as a screening tool for back pain may lead to long term health problems. Most people suffering nonspecific low back pain have radiological abnormalities that bear no relationship to the pain. Recognizing the limited value and harmful effects of screening x-rays, clinical guidelines recommend imaging only when a specific diagnosis is being considered and the pictures will have an impact on the treatment decision.

According to Deyo et al. (2010), computed tomography and magnetic resonance imaging are useful and accurate in confirming the diagnoses of lumbar disc herniation and spinal stenosis. Both can be separated from nonspecific low back pain on the basis of the history and physical examination (Deyo et al. 2006). MRI is useful in assessing spinal infections or tumors but only when the diagnosis is already being considered.

Recognizing Chronicity

Early identification of patients with low back pain at risk of long-term disability allows physicians to intervene in a timely manner. The time to recovery increases the longer the pain exists. The transition from acute to chronic pain is complicated by factors, such as the workplace environment or domestic situation, that are outside the individual's control. A focus solely on the physical findings can lead to misdiagnosis. Due to their holistic approach and long association with the patient, the chiropractor may be in good position to recognize the problem.

A recent study indicated that physicians do not consider factors such as distress, depressive mood, and somatization when assessing individuals for low back pain. According to Coulter et al. (1998), these factors are key etiologies for chronicity. If physicians focus on diagnostic tests to the exclusion of a detailed history, they may miss the relevant psychosocial elements. Relying more on the patient's story and having a broader view of the patient's situation can give the chiropractor an advantage.

Treatment Methods for Low Back Pain by Physicians or Chiropractors

The current evidence for treatment of acute and chronic low back pain is given in Table 3.

There are numerous similarities in the way chiropractors and medical doctors treat low back pain, for example, the advice to stay active and

Table 3 Treatment options for acute and chronic back pain. (Source: Koes et al. (2006))

Effectiveness	Acute low back pain	Chronic back pain
Beneficial	Physicians offer advice for patients to remain active Prescription of nonsteroidal anti-inflammatory drugs (NSAIDs)	Exercise therapy Intensive multidisciplinary treatment programs
Trade-off	Prescribe muscle relaxants	Prescribe muscle relaxants
Likely to be beneficial	Spinal manipulation Behavioral therapy Adopt multidisciplinary treatment programs, especially for subacute low back pain	Analgesics Acupuncture Prescribe antidepressants Behavioral therapy Spinal manipulation NSAIDs Back schools
Unknown	Analgesics Acupuncture Back schools Lumbar supports Massaging Multidisciplinary treatment options Tractions Temperature treatment Electromyographical (EMG) biofeedback	Inject patient with epidural steroids EMG biofeedback Lumbar supports Massage Transcutaneous electrical nerve stimulation Traction Local injections
Unbeneficial	Specified back exercises	N/A
Harmful	Extended bed rest	Facet joint injections

avoid prolonged bed rest. There is strong evidence that nonsteroidal anti-inflammatory drugs relieve pain better than chiropractic intervention. According to Koes et al. (2006), muscle relaxants relieve pain more effectively than placebos but significant side effects such as drowsiness limit their use. For many patients, the chiropractic alternative is more attractive.

Research suggests that spinal manipulation and multidisciplinary intervention for subacute low back pain are effective in relieving pain but there is little evidence that medically applied lumbar supports, traction, massage, or acupuncture are more effective than chiropractic adjustment (Deyo et al. 2010). The fact that medical interventions have proven largely ineffective in alleviating low back pain is a reasonable justification to incorporate chiropractic into the management. The techniques are safe and regardless of disputes over the mechanism of action, moderately successful.

The range of treatment options is immense. Studies have supported the use of analgesics, antidepressants, nonsteroidal anti-inflammatory drugs, spinal manipulation, cognitive behavior therapy, and back school. Exercise is the most prescribed form of conservative management for low back pain. According to Koes et al. (2006), exercises and intensive multidisciplinary pain treatment programs prescribed by either chiropractors or physicians are effective for both acute and chronic low back pain. In most cases, treatment effect is limited and short-term and in almost all instances there is no solid scientific rational for the use of a particular modality or approach. The array of choices reflects this lack of evidence. Reports of success are anecdotal and many clinicians, both medical and chiropractic, simply employ those techniques with which they have had personal success. Against this background it is hard to argue against the patient seeking a trial of chiropractic care.

The Place for Surgery

The role of surgery in treating back pain has been the subject of a vast amount of research and, with a few clear indications, its place is still not clear. Studies have look at the effectiveness of surgery and other invasive interventions for low back pain and sciatica (Van Tulder et al. 2006).

Surgical discectomy for sciatica due to lumbar disc prolapse unresponsive to nonoperative management is widely accepted as safe and effective (Koes et al. 2006). Surgical decompression for focal lumbar canal stenosis causing neurogenic claudication achieves functional improvement comparable to total joint replacement. Spine fusion surgery to treat back pain, however, remains the source of aggressive debate (Van Tulder et al. 2006). Randomized controlled trials comparing fusion surgery with conservative and noninvasive intervention methods for back pain have produced conflicting outcomes (Fairbank et al. 2005). No clear, validated criteria exist to guide decision making for operative intervention in patients with back pain but without discrete, identifiable, localized pathology.

When the proper criteria are met, surgical intervention can be effective and is clearly superior to noninvasive treatments including chiropractic manipulation (Kovacs et al. 2011). The problem is both establishing and meeting the indications for a spinal operation. The overwhelming majority of patients are not and never will be surgical candidates and for this group nonoperative therapy remains the correct choice.

Because of the effects of aging and the normal demands of daily life the spine, like the rest of the body, naturally degenerates. Surgery is a one-time event intended to deal with an immediate problem. It is not prophylactic and its benefits may decline over time as the spine grows older. A good example is the decompression for spinal canal stenosis. Narrowing of the canal from boney overgrowth is a natural response to aging. Enlarging a foramen to relieve nerve root compression is typically a successful procedure but it obviously cannot prevent future compromise. Misunderstanding the purpose and the limits of an operation is one reason for apparent surgical failure. The increasing number of spine surgeries and particularly spine fusions is a source of concern (Deyo et al. 2010). An operation is not an alternative to proper spine care, whether by the chiropractor or the family doctor.

Mannion et al. carried out a long-term (average 11 years) follow-up, comparing the outcome of chronic low back pain patients treated with fusion to those managed with multidisciplinary cognitive-behavioral and exercise rehabilitation. They combined the results of three randomized controlled trials conducted in the United Kingdom and Norway and found no statistically significant or clinically relevant differences between the cohorts (Mannion et al. 2013).

There are more questions than answers with regard to spinal fusion for lumbar degenerative disk disease (Deyo 2015). Data from the online Health Care Utilization Project, sponsored by the Agency for Healthcare Research and Quality, showed that the annual number of fusion operations (all indications and spinal levels) in the United States had increased from about 61,000 in 1993 to over 450,000 in 2011, more than a 600% increase (HCUPNet 2014). These procedures accounted for largest bill of any hospital based surgery: over $40 billion. This surge has occurred despite randomized trials suggesting little, if any, advantage of fusion over well-structured rehabilitation for degenerative discs (Mirza and Deyo 2007; Brox et al. 2003; Fairbank et al. 2005) and despite high and increasing rates of revision (Martin et al. 2007). About one in every five patients who undergo lumbar fusion will have revision surgery within 10 years (Fritzell et al. 2002). Fusion for back pain does not seem like a good idea.

Spinal Injections

Injections generally introduce anesthetics, steroids, or sclerosing agents to specific locations in the spine to diagnose or alleviate low back pain. Facet joint injections are a common form of treatment as well as a diagnostic test for "lumbar facet joint syndrome" (Van Tulder et al. 2006). As in so many instances of managing spinal problems, however, there is no clear-cut criterion for diagnosing this condition and treatment is purely empirical.

Compared with chiropractic care, facet injections have proven ineffective in relieving chronic back pain. One study showed no significant difference in low back pain 6 weeks and 6 months after anesthetic facet injection (Gibson and

Waddell 2005). The European low back pain guidelines identified trials with a range of different outcomes. One trial reported that an epidural injection of methylprednisolone was efficacious in alleviating sciatica. Another found that steroid perineural injections were superior to saline injections in patients suffering from "lumbar syndromes" (Koes et al. 2006). Overall the positive results, if any, were short term. In a study showing positive results of epidural corticosteroid injections, 65% of patients reported pain relief lasting between 1 day and 6 weeks. Adding morphine caused unfavorable side effects including drowsiness and dizziness (Koes et al. 2006). None of the trials described long term pain control. The lack of well-designed trials makes a conclusive decision on the effectiveness of injections impossible. Statistically chiropractic interventions were more effective in providing lasting relief.

A similar technique using local penetration is radiofrequency denervation of the facet joints. Trials have produced results similar to the injection of anesthetics. One showed that radiofrequency ablation relieved the intensity of low back pain and improved function for two months (Van Tulder et al. 2006).

Opioids

Prescribing opioids for low back pain has also been common practice by the mainstream medical profession. Evidence clearly shows that opioid analgesics are not superior to nonopioids treatment strategies (Krebs et al. 2018). Further research demonstrates opioids offer no long-term benefits (Cherkin et al. 2016). Long-term opioid therapy is associated with poor patient-reported pain control, reduced quality of life outcomes, diminished function, and psychiatrist disorders (Turner et al. 2016). There are multiple adverse effects including constipation, confusion, increased pain, and respiratory depression (Oderda et al. 2013). According to Vowles et al. (2015), approximately 20% of patients on long-term opioids have an adverse event.

There is no place for opioids in the management of low back pain. The medical profession has created an epidemic of narcotic addiction and overdose deaths.

Adverse Effects of Surgery and Other Invasive Treatment Options

Every invasive procedure used to treat back pain has the potential for adverse side effects. Every operation is a risk/benefit decision. Complications can be intraoperative or occur after the surgery is complete. They range from misadventure to mistake, from unavoidable to preventable, from unfortunate to malpractice.

Some choices should be simple. Routine microdiscectomy, approaching the disc with a minimum of soft tissue disruption, has a lower risk of reoperation than percutaneous laser disk decompression, a largely discredited procedure (Brouwer et al. 2015). Some choices are more complicated. The more extensive the operation the higher the risk of an adverse event but the success of the surgery may require a more complex and therefore a more perilous procedure.

Spine surgery can create its own iatrogenic problems. Decompression may remove too much bone and render the spine unstable causing increased pain and the need for further surgery. Adjacent structures, nerves or blood vessels, may be damaged. Every operation is a chance for infection. No surgical assault on the spine should be casually dismissed. A fully informed consent is mandatory.

Surgery is followed by recovery and that time can vary depending on the nature and the extent of the operation. According to Johns Hopkins Spine Service, individuals who undergo traditional spine surgery for low back pain should expect to miss between 8 and 12 weeks of work (Seladi-Schulman 2017).

Compared to chiropractic manipulation, surgery is a risky proposition but that does not make it the wrong choice. The nature of the treatment depends on the nature of the problem. Most patients with low back pain do not need spine surgery but for those for whom an operation is needed and necessary delaying surgery to receive useless manual therapy is clearly poor patient care.

Rationale for Including Chiropractic Care into Mainstream Medicine

Chiropractic treatment is an excellent example of creating a rick /benefit balance for treating back pain. Most chiropractors rely on manipulating the spine and limit their practice to spinal or, at most, musculoskeletal problems. Although there are numerous theories as to the mechanism of pain relief, there is not scientific validation. Nevertheless, chiropractic care has been shown repeatedly to be more effective in managing back pain than conventional medical methods. Chiropractic techniques for reducing back pain have few if any significant side effects. The benefits clearly outweigh the risks tilting the risk/benefit balance in favor of chiropractic care.

Unfortunately, some chiropractors have chosen to extend their approach to a wide range of unrelated, nonmechanical conditions, creating doubt in the minds of the medical community on the validity of their claims. Incorporating chiropractic into mainstream medicine will require more rigorous analysis and a clear scope of practice.

Clinical practice is complicated and vulnerable to abuse. No treatment no matter how seemly benign is totally without concerns. Risk/benefit is always a tradeoff. The incorporation of chiropractors into a medical healthcare system raises issue of financial incentives, resource allocation, vested interest, and professional jealousy. It will never be as simple as giving patient's free choice. Many patients experience short-term, mild to moderate muscular pain after manipulation. This can be viewed as a harmless event or can be used to raise concerns. Death after chiropractic treatment is extremely rare but not unheard of (Edzard 2016). Failure to recognize an underlying pathology as a source of apparently nonspecific low back pain can have disastrous results. Repeating an adjustment when the previous attempt caused serious problems is both foolishly endangers the patient. Chiropractic manipulation is safe but it is not trivial.

Chiropractors must obtain informed consent from their patients before commencing therapy. This is an ethical imperative. Chiropractors need to inform their patients about the inherent risks of undergoing an adjustment or manipulation. Although the risks may be less, in this regard the chiropractor is not different than the surgeon.

From its inception at the end of the nineteenth century, the chiropractic profession has overcome many of the challenges raised by the medical community. Many medical doctors treat chiropractors with skepticism and reject the concept of alternative medicine. But combining the aspects of chiropractic care that emphasize patient contact, the holistic aspects of healing and the body's ability to repair itself with the diagnostic skills and technology of modern medicine would be to everyone's advantage, primarily the patient's.

References

Abdel Shaheed C, Maher CG, Williams KA, Day R, McLachlan AJ (2016) Efficacy, tolerability, and dose-dependent effects of opioid analgesics for low back pain: a systematic review and meta-analysis. JAMA Intern Med 176(7):958–968. https://doi.org/10.1001/jamainternmed.2016.1251

Blanchette MA, Stochkendahl MJ, Borges Da Silva R, Boruff J, Harrison P, Bussiéres A (2016) Effectiveness and Economic Evaluation of Chiropractic Care for the Treatment of Low Back Pain: A Systematic Review of Pragmatic Studies. PLoS One 11(8):e0160037. Published 2016 Aug 3. https://doi.org/10.1371/journal.pone.0160037

Bronfort G, Haas M, Evans R, Leininger B, Triano J (2010) Effectiveness of manual therapies: the UK evidence report. Chiropr Osteopat 18(1):3

Brouwer PA, Brand R, van den Akker-van ME, Jacobs WC, Schenk B, van den Berg-Huijsmans AA, … Peul WC (2015) Percutaneous laser disc decompression versus conventional microdiscectomy in sciatica: a randomized controlled trial. Spine J 15(5):857–865

Brox JI, Sorensen R, Friis A, Nygaard Ø, Indahl A, Keller A et al (2003) Randomized clinical trial of lumbar instrumented fusion and cognitive intervention and exercises in patients with chronic low back pain and disc degeneration. Spine 28:1913–1921

Bussières AE, Al Zoubi F, Stuber K, French S, Boruff J, Corrigan J, Thomas A (2016) Evidence-based practice, research utilization, and knowledge translation in chiropractic: a scoping review. BMC Complement Altern Med 16:216

Carnes D, Mars TS, Mullinger B, Froud R, Underwood M (2010) Adverse events and manual therapy: a systematic review. Man Ther 15(4):355–363

Chapman-Smith D (2010) The chiropractic profession: basic facts, independent evaluations, common questions. Chiropr Rep 24:1–8

Cherkin DC, Sherman KJ, Balderson BH, Cook AJ, Anderson ML, Hawkes RJ, ... Turner JA (2016) Effect of mindfulness-based stress reduction vs cognitive behavioral therapy or usual care on back pain and functional limitations in adults with chronic low back pain: a randomized clinical trial. JAMA 315(12):1240–1249

Chou R, Qaseem A, Snow V, Casey D, Cross JT, Shekelle P, Owens DK (2007) Diagnosis and treatment of low back pain: a joint clinical practice guideline from the American College of Physicians and the American Pain Society. Ann Intern Med 147(7):478–491

Chou R, Baisden J, Carragee EJ, Resnick DK, Shaffer WO, Loeser JD (2009) Surgery for low back pain: a review of the evidence for an American Pain Society Clinical Practice Guideline. Spine (Phila Pa 1976) 34:1094–1109

Clar C, Tsertsvadze A, Hundt GL, Clarke A, Sutcliffe P (2014) Clinical effectiveness of manual therapy for the management of musculoskeletal and non-musculoskeletal conditions: systematic review and update of UK evidence report. Chiropr Man Ther 22(1):12

Coulter ID (1997) What is chiropractic. McNamee KP. The chiropractic college directory, 98

Coulter I, Adams A, Coggan P, Wilkes M, Gonyea M (1998) A comparative study of chiropractic and medical education. Altern Ther Health Med 4(5):64–75

Davis MA, Yakusheva O, Gottlieb DJ, Bynum JPW (2015) Regional supply of chiropractic care and visits to primary care physicians for back and neck pain. J Am Board Fam Med 28(4):481–490

Deyo RA (2015) Fusion surgery for lumbar degenerative disc disease: still more questions than answers. Spine J 15(2):272–274. https://doi.org/10.1016/j.spinee.2014.11.004

Deyo RA, Mirza SK, Martin BI (2006) Back pain prevalence and visit rates: estimates from U.S. national surveys, 2002. Spine (Phila Pa 1976). 31(23):2724–2727. https://doi.org/10.1097/01.brs.0000244618.06877.cd

Deyo RA, Mirza SK, Martin BI, Kreuter W, Goodman DC, Jarvik JG (2010) Trends, major medical complications, and charges associated with surgery for lumbar spinal stenosis in older adults. JAMA 303(13):1259–1265

Dynamic Chiropractic (2007) Gallup poll: Americans have low opinion of chiropractors. Retrieved August 27, 2019, from https://www.dynamicchiropractic.com/mpacms/dc/article.php?id=52038

Edzard E (2016) The evidence shows that chiropractors do more harm than good. Retrieved August 29, 2019, from https://life.spectator.co.uk/articles/the-evidence-shows-that-chiropractors-do-more-harm-than-good/

Elton D (2014) The distribution and analysis of annualized claims data for more than 3.7 million commercial health plan members. Data retrieved from the United Healthcare national commercial claims database, July 1, 2013–June 30, 2014. November 10, 2014

Fairbank J, Frost H, Wilson-MacDonald J, Yu LM, Barker K, Collins R (2005) Randomized controlled trial to compare surgical stabilization of the lumbar spine with an intensive rehabilitation program for patients with chronic low back pain: the MRC spine stabilization trial. BMJ 330(7502):1233

Feldman V (2014) Return on investment analysis of Optum offerings – assumes Network /UM/Claims services; Optum Book of Business Analytics 2013. Analysis as of December 8, 2014

Fritzell P, Hagg O, Wessberg P, Nordwall A, Swedish Lumbar Spine Study Group (2002) Chronic low back pain and fusion: a comparison of three surgical techniques: a prospective multicenter randomized study from the Swedish lumbar spine study group. Spine 27:1131–1141

Gibson JNA, Waddell G (2005) Surgery for degenerative lumbar spondylosis: updated Cochrane Review. Spine 30(20):2312–2320

Globe G, Farabaugh RJ, Hawk C, Morris CE, Baker G, Whalen WM, ... Augat T (2016) Clinical practice guideline: chiropractic care for low back pain. J Manip Physiol Ther 39(1):1–22

Hartvigsen, JanBuchbinder, Rachelle et al (2018) What low back pain is and why we need to pay attention. The Lancet 391(10137):2356–2367

HCUPnet (2014) Agency for Healthcare Research and Quality. http://hcupnet.ahrq.gov/HCUPnet.jsp. Accessed 30 Oct 2014

Helm S II, Deer TR, Manchikanti L, Datta S, Chopra P, Singh V et al (2012) Effectiveness of thermal annular procedures in treating discogenic low back pain. Pain Physician 15:E279–E304

Hurwitz EL, Carragee EJ, van der Velde G, Carroll LJ, Nordin M, Guzman J, Bone and Joint Decade 2000–2010 Task Force on Neck Pain and Its Associated Disorders (2008) Treatment of neck pain: noninvasive interventions: results of the Bone and Joint Decade 2000–2010 Task Force on Neck Pain and its Associated Disorders. Spine 33(4 Suppl):S123–S152. https://doi.org/10.1097/BRS.0b013e3181644b1d

Jacobs WC, van Tulder M, Arts M, Rubinstein SM, van Middelkoop M, Ostelo R et al (2011) Surgery versus conservative management of sciatica due to a lumbar herniated disc: a systematic review. Eur Spine J 20:513–522

Jacobs W, van der Gaag NA, Tuschel A, de Kleuver M, Peul W, Verbout AJ et al (2012) Total disc replacement for chronic back pain in the presence of disc degeneration. Cochrane Database Syst Rev 9:CD008326

Johnson C, Killinger LZ, Christensen MG, Hyland JK, Mrozek JP, Zuker RF, ... Oyelowo T (2012) Multiple views to address diversity issues: an initial dialog to advance the chiropractic profession. J Chiropr Humanit 12;19(1):1–11. https://doi.org/10.1016/j.echu.2012.10.003

Keeney BJ, Fulton-Kehoe D, Turner JA, Wickizer TM, Chan KCG, Franklin GM (2013) Early predictors of lumbar spine surgery after occupational back injury: results from a prospective study of workers in Washington State. Spine 38(11):953–964

Kent C (2018) Chiropractic and mental health: history and review of putative neurobiological mechanisms. J Neuro PsyAn Brain Res: 1–10. JNPB-103

Koes BW, Van Tulder M, Thomas S (2006) Diagnosis and treatment of low back pain. BMJ 332(7555):1430–1434

Kovacs FM, Urrútia G, Alarcón JD (2011) Surgery versus conservative treatment for symptomatic lumbar spinal stenosis: a systematic review of randomized controlled trials. Spine (Phila Pa 1976) 36(20):E1335–E1351

Krebs EE, Gravely A, Nugent S, Jensen AC, DeRonne B, Goldsmith ES, … Noorbaloochi S (2018) Effect of opioid vs nonopioid medications on pain-related function in patients with chronic back pain or hip or knee osteoarthritis pain: the SPACE randomized clinical trial. JAMA 319(9):872–882

LeFebvre R, Peterson D, Haas M (2013) Evidence-based practice and chiropractic care. J Evid-Based Complement Altern Med 18(1):75–79

Liliedahl RL, Finch MD, Axene DV, Goertz CM (2010) Cost of care for common back pain conditions initiated with chiropractic doctor vs medical doctor/doctor of osteopathy as first physician: experience of one Tennessee-based general health insurer. J Manip Physiol Ther 33(9):640–643

Lisi AJ, Salsbury SA, Twist EJ, Goertz CM (2018) Chiropractic integration into private sector medical facilities: a multisite qualitative case study. J Altern Complement Med 24(8):792–800

Mannion AF, Brox JI, Fairbank J (2013) Comparison of spinal fusion and nonoperative treatment in patients with chronic low back pain: long-term follow-up of three randomized controlled trials. Spine 13(2013):1438–1448

Martin BI, Mirza SK, Comstock BA, Gray DT, Kreuter W, Deyo RA (2007) Reoperation rates following lumbar spine surgery and the influence of spinal fusion procedures. Spine 32:382–387

May S, Comer C (2013) Is surgery more effective than non-surgical treatment for spinal stenosis, and which non-surgical treatment is more effective? A systematic review. Physiotherapy 99:12–20

Mirza SK, Deyo RA (2007) Systematic review of randomized trials comparing lumbar fusion surgery to nonoperative care for treatment of chronic back pain. Spine 32:816–823

Nelson CF et al (2005) Chiropractic as spine care: a model for the profession. Chiropr Osteopat 9:1–17

Nyiendo J, Haas M, Goodwin P (2000) Patient characteristics, practice activities, and one-month outcomes for chronic, recurrent low-back pain treated by chiropractors and family medicine physicians: a practice-based feasibility study. J Manipulative Physiol Ther 23(4):239–245

Nyiendo J, Haas M, Goldberg B, Sexton G (2011) Patient characteristics and physicians' practice activities for patients with chronic low back pain: a practice-based study of primary care and chiropractic physicians. J Manip Physiol Ther 24(2):92–100

Oderda GM, Gan TJ, Johnson BH, Robinson SB (2013) Effect of opioid-related adverse events on outcomes in selected surgical patients. J Pain Palliat Care Pharmacother 27(1):62–70

Paige NM, Miake-Lye IM, Booth MS, Beroes JM, Mardian AS, Dougherty P et al (2017) Association of spinal manipulative therapy with clinical benefit and harm for acute low back pain: systematic review and meta-analysis. JAMA 317(14):1451–1460. https://doi.org/10.1001/jama.2017.308612

Palmer DD (1967) Three generations: a history of chiropractic. Palmer College of Chiropractic. Davenport, Iowa

Paulozzi LJ, Jones CM, et al (2011) Vital signs: overdoses of prescription opioid pain relievers – United States, 1999–2008. Division of Unintentional Injury Prevention, National Center for Injury Prevention and Control, Centers for Disease Control and Prevention. 60:5

Peul WC, Bredenoord AL, Jacobs WC (2014) Avoid surgery as first line treatment for non-specific low back pain. BMJ 349:g4214

Qaseem A, Wilt TJ, McLean RM, Forciea MA, Clinical Guidelines Committee of the American College of Physicians (2017) Noninvasive treatments for acute, subacute, and chronic low back pain: a clinical practice guideline from the American College of Physicians. Ann Intern Med 166(7):514–530. https://doi.org/10.7326/M16-23672017;166

Roland M, van Tulder M (1998) Should radiologists change the way they report plain radiography of the spine?. Lancet 352(9123):229–230. https://doi.org/10.1016/S0140-6736(97)11499-4

Savigny P, Kuntze S, Watson P. (2009). Low back pain: early management of persistent non-specific low back pain. National Collaborating Centre for Primary Care and Royal College of General Practitioners

Seladi-Schulman J (2017) Laser back surgery: benefits, drawbacks, efficacy, and more. Retrieved August 29, 2019, from https://www.healthline.com/health/back-pain/laser-back-surgery#recovery-time

Slaughter AL, Frith K, O'Keefe L, Alexander S, Stoll R (2015) Promoting best practices for managing acute low back pain in an occupational environment. Workplace Health Saf 63(9):408–414

Smith KM (2016) The relationship between institutional culture and faculty perceptions of online learning in chiropractic higher education. Doctoral dissertation, Trident University International

The Bone and Joint Decade 2000–2010 for prevention and treatment of musculoskeletal disorders (1998) Conference proceedings. Acta Orthop Scand Suppl 218:1–86

The Center for Behavioral Health Statistics and Quality (2016) Key substance use and mental health indicators in the United States: results from the 2015 National Survey on Drug Use and Health (HHS publication no. SMA 16-4984, NSDUH series H-51). Retrieved from https://www.samhsa.gov/data/sites/default/files/NSDUH-FFR1-2016/NSDUH-FFR1-2016.htm

Todd NV (2017) The surgical treatment of non-specific low back pain. Bone Joint J 99-B:1003–1005

Triano JJ, Goertz C, Weeks J, Murphy DR, Kranz KC, McClelland G, ... Nelson CF (2010) Chiropractic in North America: toward a strategic plan for professional renewal – outcomes from the 2006 Chiropractic Strategic Planning Conference. J Manip Physiol Ther 33(5):395–405

Turner JA, Shortreed SM, Saunders KW, LeResche L, Von Korff M (2016) Association of levels of opioid use with pain and activity interference among patients initiating chronic opioid therapy: a longitudinal study. Pain 157(4):849

United States Bone and Joint Initiative (2014) The burden of musculoskeletal diseases in the United States (BMUS), 3rd edn. Rosemont, IL. Available at http://www.boneandjointburden.org.

Van Tulder MW, Koes B, Seitsalo S, Malmivaara A (2006) Outcome of invasive treatment modalities on back pain and sciatica: an evidence-based review. Eur Spine J 15(1):S82–S92

Vowles KE, McEntee ML, Julnes PS, Frohe T, Ney JP, van der Goes DN (2015) Rates of opioid misuse, abuse, and addiction in chronic pain: a systematic review and data synthesis. Pain 156(4):569–576. https://doi.org/10.1097/01.j.pain.0000460357.01998.f1

Whedon J (2017) Association between utilization of chiropractic services and use of prescription opioids among patients with low back pain. Presented ahead of print at the National Press Club in Washington D.C., on March 14, 2017. Accessed online from http://c.ymcdn.com/sites/www.cocsa.org/resource/resmgr/docs/NH_Opioids_Whedon.pdf

Whedon JM, Mackenzie TA, Phillips RB, Lurie JD (2015) Risk of traumatic injury associated with chiropractic spinal manipulation in medicare part B beneficiaries aged 66–99. Spine 40(4):264–270

World Federation of Chiropractic (2005) Summary minutes of the proceedings of the assembly of members; July 14–15, 2005

Young KJ (2014) Gimme that old time religion: the influence of the healthcare belief system of chiropractic's early leaders on the development of x-ray imaging in the profession. Chiropr Man Ther 22(1):36

Medical Causes of Back Pain: The Rheumatologist's Perspective

5

Stephanie Gottheil, Kimberly Lam, David Salonen, and Lori Albert

Contents

Case 1	94
Case Description	94
Salient Features on History and Physical Exam	94
Initial Treatment Approach	95
Appropriate Referrals	95
Case Resolution	96
Clinical Pearls and Myths	96
Case 2	96
Case Description	96
Differential Diagnosis: Spondyloarthritis	97
Salient Features on History and Physical Exam	98
Initial Treatment Approach	102
Appropriate Referrals	103
Case Resolution	103
Clinical Pearls and Myths	103
Case 3	104
Case Description	104
Differential Diagnosis	104
Salient Features on History and Physical Exam	104
Appropriate Use of Investigations	104
Initial Treatment Approach	105
Appropriate Referrals	106
Case Resolution	106
Clinical Pearls and Myths	106

S. Gottheil · K. Lam · D. Salonen
University of Toronto, Toronto, ON, Canada
e-mail: stephanie.gottheil@gmail.com;
Kimberlywclam@gmail.com; david.salonen@uhn.ca

L. Albert (✉)
Rheumatology Faculty, University of Toronto, Toronto, ON, Canada
e-mail: Lori.Albert@uhn.ca

© Springer Nature Switzerland AG 2021
B. C. Cheng (ed.), *Handbook of Spine Technology*,
https://doi.org/10.1007/978-3-319-44424-6_132

Conclusion 106
Investigations to Avoid 106
When to Refer 107

References 107

Abstract

Medical causes of back pain include *infection, inflammation, or malignancy*. These treatable causes of back pain must be recognized by all health providers so that the diagnosis is not missed and appropriate treatments can be started as soon as possible. A medical cause of back pain should be suspected when patients have features of systemic illness – fevers, rash, unexplained weight loss, or joint swelling – or when their medical history contains comorbidities such as psoriasis, inflammatory bowel disease, immunosuppression, or injection drug use. This chapter will focus on how to recognize the clues that will lead to these diagnoses through a series of case studies.

Keywords

Spinal infection · Spondyloarthritis · Osteoporosis · Malignancy · Rheumatology

Case 1

Case Description

A 26-year-old male presents to the emergency department with a 48-h history of worsening back pain. He describes 24 h of fever and chills and is febrile and tachycardic in the emergency room. He is known to be HIV-positive, secondary to intravenous drug use, but has not been taking anti-retroviral therapy for the past several months. His physical exam reveals clear track marks on the upper limbs. The differential diagnosis for suspected spinal infection is listed in Table 1.

Salient Features on History and Physical Exam

Spinal infections present most commonly with acute back pain, fever, and other constitutional symptoms. Limb weakness or other neurologic symptoms are present in less than 50% of cases but, if present, *suggest a surgical emergency*. Patients often have underlying medical comorbidity (diabetes, coronary artery disease, immunosuppression) or use intravenous drugs. The most common sites of *primary* infection are skin and soft tissue, genitourinary, or bacterial endocarditis. Patients with fungal or tuberculous infections are more likely to present with subacute back pain and more prominent constitutional symptoms. Appropriate investigations for suspected spinal infection are outlined in Table 2.

Epidural Abscess

An epidural abscess occurs when pus accumulates between the dural layer of the spinal cord and the spinal ligaments. Over 50% of cases are

Table 1 Differential diagnosis of suspected spinal infection

Condition	Diagnostic clues
Infectious – bacterial (epidural abscess, osteomyelitis, or diskitis)	Acute onset
	History of intravenous drug use
	Recent spinal instrumentation
Infectious – fungal	Subacute presentation
	Immunosuppressed patient
Infectious – tuberculosis	Subacute presentation
	Immunosuppressed patient (particularly HIV+)
	Constitutional symptoms
	Endemic areas – Asia, South America, sub-Saharan Africa; homeless populations

Table 2 Appropriate use of investigations for suspected spinal infection

	Investigation	Relevance
Bloodwork	CBC, electrolytes, creatinine	Leukocytosis in ~60%
	Blood cultures (two sets)	Positive in ~50%
	ESR and CRP	Elevated in >90% of cases
Imaging (Fig. 1)	Plain radiography	Use to Rule out alternative diagnosis
	MRI spine	High sensitivity/specificity
	Gallium scan	Less sensitive than MRI for abscess
Special tests	TB skin test	If from endemic area
	2D echocardiogram	If *S. aureus* bacteremia
	CT-guided tissue biopsy	If blood cultures negative

due to hematogenous spread of bacteria (most commonly *Staphylococcus aureus*), with at least 40% of cases thought to be due to intravenous drug use. If left untreated, epidural abscesses can lead to permanent neurologic compromise and so should be considered a *medical emergency*.

Diskitis and Osteomyelitis

Osteomyelitis refers to infection of the vertebral body and/or adjacent disk (diskitis). Similar to an epidural abscess, most cases are caused by hematogenous spread. Neurologic complications can occur in up to 40% of patients with vertebral osteomyelitis, and epidural abscesses develop in 15–20%.

MRI is the imaging modality of choice to investigate for spinal infection with high sensitivity and specificity. Although there is concern regarding MRI overuse in the back pain population, a recent study of 167 patients with a history of intravenous drug use and acute back pain found that nearly 40% of patients had evidence of spinal infection on their admission MRI (Colip et al. 2018). Thus, it is reasonable to order an MRI for patients with back pain and a history of intravenous drug use.

Initial Treatment Approach

Epidural Abscess

Any patient with neurologic compromise from an epidural abscess requires urgent surgical decompression. In patients without neurologic symptoms, using medical management alone with intravenous antibiotics is controversial. Patients with leukocytosis, positive blood cultures, or history of diabetes are at a higher risk of failure with medical management, and early surgery should be considered in these patients. Initial antibiotic therapy should be broad-spectrum and then tailored after an organism is identified. Infectious diseases consultation may be warranted.

Diskitis and Osteomyelitis

Antibiotic therapy is the mainstay of treatment for vertebral osteomyelitis; if patients are hemodynamically stable without neurologic symptoms, empiric antibiotics may be delayed until a microbiological specimen can be obtained. Osteomyelitis should be treated with at least 4–6 weeks of intravenous antibiotics, and response to therapy can be monitored by following symptoms (back pain, fever) and acute phase reactants (ESR, CRP). Follow-up imaging is not usually required, but can be considered if patients have ongoing symptoms despite therapy. Surgical treatment should be reserved for those with epidural abscess, spinal implants, or progressive neurological symptoms.

Appropriate Referrals

Any spinal infection should be considered a medical emergency and treated rapidly with broad-spectrum intravenous antibiotics. Patients require admission to hospital and frequent monitoring for sepsis and neurologic deterioration. Consultation from infectious disease specialists may be helpful for making decisions regarding antibiotic choice and duration of therapy. For any patient with ongoing intravenous drug use, referrals to social work and addictions specialists will be vital to prevent recurrence.

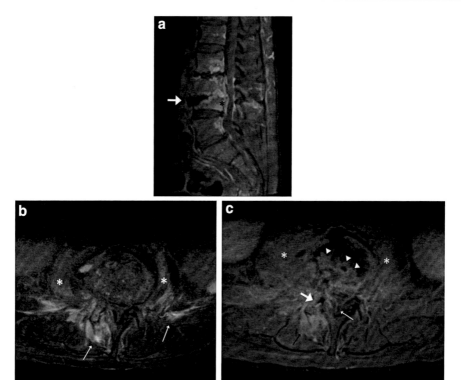

Fig. 1 (**a**) MRI spine showing epidural abscess. Sagittal post-gadolinium T1 FS images of the lumbar spine demonstrate peripherally enhancing intervertebral abscess (white arrow) and significant destruction with loss of height of the L3 and L4 vertebral bodies. Further, endplate destruction is present at L2-3 level. There is intraosseous enhancement in the vertebral bodies from L2 to L5 consistent with osteomyelitis. The enhancing epidural abscess (asterisk) is causing narrowing of the spinal canal. (**b**) MRI spine showing intervertebral discitis. The axial STIR image at the level of L3-4 demonstrates T2-hyperintense intervertebral discitis with significant edema in the posterior elements as well as the paraspinal (white arrows) and psoas muscles (white asterisks). (**c**) MRI spine showing vertebral osteomyelitis. The axial T1 FS post-gadolinium image demonstrates osseous enhancement (thick white arrow) of the posterior elements consistent with vertebral osteomyelitis. There are peripherally enhancing intervertebral (arrowheads) and epidural abscesses (white arrow). There is associated diffuse soft tissue enhancement in the psoas (asterisks) and paraspinal muscles (black arrow)

Case Resolution

This man has a very high pretest probability for spinal infection; he should be admitted to hospital, should have blood cultures drawn, should be treated with broad-spectrum IV antibiotics, and should undergo a spinal MRI to confirm the diagnosis. Careful attention should be paid to his neurological exam, and an infectious diseases consult should also be obtained.

Clinical Pearls and Myths

- Fever with acute back pain is spinal infection until proven otherwise.
- Think of tuberculosis in any patient with fever, weight loss, and subacute back pain, particularly if they are immunosuppressed or from an endemic area.
- Rule out bacterial endocarditis in any patient with S. aureus bacteremia and spinal infection.

Case 2

Case Description

A 35-year-old woman presents with progressive back pain for the past 10 years. The pain localizes to the lower back and radiates down to the buttock, sometimes alternating sides. It is worst during the night as well as in the morning, where it is associated with significant stiffness. She is referred to a surgical spine clinic after an MRI suggested spinal stenosis at L4–L5; however, the radiologist also describes bone marrow edema in the lumbar spine suggestive of spondyloarthritis.

Differential Diagnosis: Spondyloarthritis

Spondyloarthritis (SpA) refers to a group of related but distinct conditions that can cause inflammation of the sacroiliac joints and spine. SpA can be divided into axial- and peripheral- predominant disorders, depending on which joints are most involved. Axial-predominant SpA includes ankylosing spondylitis; peripheral-predominant SpA includes psoriatic arthritis, inflammatory bowel disease-associated arthritis, and reactive arthritis. The major features of each disorder are listed in Table 3.

Clinicians must also be aware of spondyloarthritis mimickers that can also present with similar symptoms. The most common ones are listed in Table 4 below. These conditions can usually be differentiated on the basis of a careful history and physical examination along with imaging. Osteitis condensans ilii (OCI) should be considered in women with postpartum back pain (Fig. 2). Unilateral sacroiliitis without extra-axial features or peripheral arthritis should always raise suspicion for infection and be investigated accordingly (Fig. 3). Diffuse idiopathic skeletal hyperostosis (DISH) can be differentiated from SpA by the lack of SI joint involvement on imaging (Fig. 4).

Table 3 Classification of spondyloarthritis

Condition	Back involvement	Other features
Ankylosing spondylitis	Bilateral, symmetric sacroiliitis	Associated with peripheral large joint arthritis, acute anterior uveitis, enthesitis, dactylitis
	Spondylitis starts in lumbar spine and progresses upward	
Psoriatic arthritis	Unilateral, asymmetric sacroiliitis	Associated with psoriasis
	Spondylitis less common	Peripheral arthritis common
Reactive arthritis	Unilateral, asymmetric sacroiliitis	History of preceding gastrointestinal or genitourinary infection (within 1–3 weeks)
	Spondylitis less common but can occur in up to 20% of patients	Associated with arthritis, urethritis (sterile), and conjunctivitis
IBD-associated arthritis	Can be unilateral or bilateral	Associated with oral ulcers, acute anterior uveitis, and peripheral arthritis
	Spinal inflammation is often independent of IBD activity	

Table 4 Differential diagnosis of SpA

Condition	Diagnostic clues
Osteitis condensans Ilii (OCI) (Fig. 2)	Female preponderance
	Often presents after pregnancy
	X-ray: associated with sclerosis at inferior SI joint
	No erosions or joint space narrowing
Infectious sacroiliitis (Fig. 3)	Acute or subacute onset
	Almost always unilateral
	Presence of constitutional symptoms
	X-ray: early and extensive erosions on the affected side
Diffuse idiopathic skeletal hyperostosis (DISH) (Fig. 4)	Insidious onset
	Usually starts after age 50
	Associated with diabetes and obesity
	X-ray: flowing, anterior osteophytes but no erosions
	Sacroiliac joints should be normal
Spondyloarthritis (Fig. 5)	Insidious onset
	Usually starts before age 45
	Inflammatory-type back pain (see below)
	Presence of extra-axial features
	Imaging described below

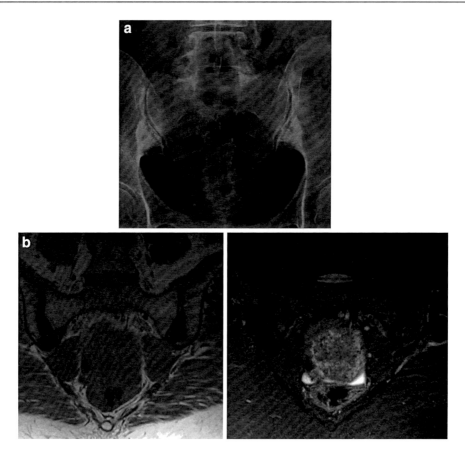

Fig. 2 (**a**) Frontal radiograph showing osteitis condensans ilii. Typical bilateral triangular sclerosis in the ilium adjacent to the sacroiliac joints is depicted. There are no features of erosions or ankylosis. (**b**) MRI showing osteitis condensans ilii. Coronal T1 and STIR images demonstrate signal hypointensity reflecting the sclerosis in the iliac sides of both sacroiliac joints as depicted on the radiographs, in the typical triangular configuration. There are no erosions, ankylosis, or effusion

Salient Features on History and Physical Exam

The history can often be very useful in diagnosing SpA and differentiating it from mimickers. The features of inflammatory back pain can be found in Table 5 and, when present, suggest the need for further investigation. **Extra-axial features**, when present, can help point in the direction of SpA as well: conjunctivitis, uveitis, dactylitis ("sausage" digit), psoriasis, enthesitis (pain at tendon insertion points, commonly the Achilles tendon or plantar fascia), urethritis, peripheral joint arthritis, and inflammatory bowel disease. A family history of psoriasis, Crohn's disease, or ulcerative colitis will also raise the clinical suspicion for SpA.

Physical exam can be helpful in diagnosing SpA and following progression but will often be normal early in the course of disease. The exam should focus on the presence of extra-axial features and measurements of spinal mobility: occiput-to-wall distance, thoracic excursion, lateral spinal flexion, and forward spinal flexion. Abnormalities in spinal mobility may reflect active inflammation and be reversible with treatment, or may reflect chronic damage and be irreversible. Measurement of spinal mobility can be useful to predict disease progression and to follow a patient's response to treatment. The Assessment of SpondyloArthritis International Society (ASAS) handbook is an excellent resource for assessing spinal mobility (Sieper et al.

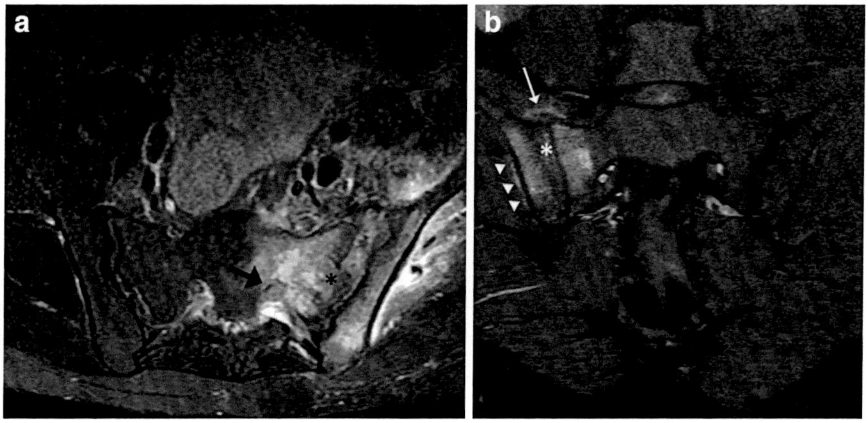

Fig. 3 (**a**) MRI showing infectious sacroiliitis. Axial STIR image shows florid diffuse edema in the sacrum and left ilium as well as in the presacral soft tissues, adjacent gluteal, and iliacus muscles. There is widening of the left sacroiliac joint with an effusion, capsular distension, and cortical destruction (asterisk). The left sacral nerve is thickened (arrow), likely reactive due to the surrounding infective change. (**b**) MRI showing infectious sacroiliitis. The findings on this coronal STIR image are more subtle than the previous example, but the unilateral features of pericapsular edema, effusion with joint distension (arrow) and marrow edema across one sacroiliac joint, are strong indicators of infective sacroiliitis. There is also erosive change with effusion (asterisk) and edema in the adjacent gluteus minimus muscle (arrowheads)

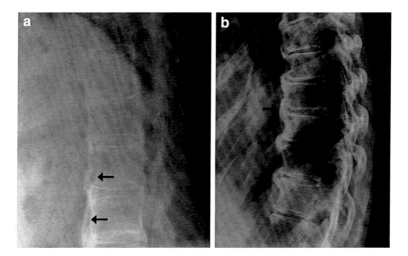

Fig. 4 (**a**) X-ray showing diffuse idiopathic skeletal hyperostosis (DISH). This lateral radiograph of the thoracic spine demonstrates flowing ossification along the anterior longitudinal ligament with radiolucency between the vertebral body and anterior longitudinal ligament (arrow). (**b**) X-ray showing diffuse idiopathic skeletal hyperostosis (DISH). This thoracic radiograph shows exuberant thickening of the anterior longitudinal ligament (black arrows) across contiguous vertebrae. There is only minor endplate degenerative change with preservation of the intervertebral disc heights

2009). Appropriate investigations for suspected SpA are outlined in Table 6.

HLA-B27 positivity is seen frequently in SpA, particularly in ankylosing spondylitis and reactive arthritis. However, it is common enough in the general population (6–8% in North America with significant geographic variation) that its value diagnostically is limited. Only 1 in 50 patients with +HLA-B27 will develop ankylosing spondylitis in their lifetime.

Table 5 Recognizing inflammatory back pain

Insidious onset before age 40
Duration of pain and stiffness at least 3 months
Morning stiffness >30 min
Improvement with exercise but not with rest
Alternating buttock pain
Awakening in second half of the night due to back pain/stiffness
Good response to NSAIDs

Table 6 Appropriate use of investigations for suspected SpA

	Investigation	Relevance
Bloodwork	ESR and CRP	Normal in 30–50% of cases
	HLA-B27	Present in 90% of Caucasians with AS, 50–80% of non-Caucasians
Imaging	Plain radiography	X-ray changes only present in 30% at time of diagnosis
	MRI spine	Highest sensitivity for sacroiliitis

X-ray features of spondyloarthritis include (Figs. 5 and 6):

- Sacroiliac Joints:
  - Joint space narrowing starting at the lower two-thirds of sacroiliac joint
  - Subchondral sclerosis and erosions starting at the iliac side of the sacroiliac joint
  - May see pseudo-widening secondary to erosions
  - Later in disease course will see ankylosis (fusion) of the sacroiliac joint
- Spine:
  - Romanus lesions ("shiny corner sign"): erosions and sclerosis at vertebral body corners
  - Squaring of the vertebral bodies (loss of normal convexity)
  - Syndesmophytes (can be differentiated from osteophytes by the vertical direction of outgrowth and the relative sparing of the disk space)
  - Ossification of the interspinous ligament leading to "dagger spine" appearance

MRI features of spondyloarthritis include (Figs. 5 and 7):

- STIR sequence for active inflammatory lesions:
  - Bone marrow edema
  - Capsulitis
  - Synovitis
  - Enthesitis
- T1-weighted sequence for chronic inflammatory lesions:
  - Subchondral sclerosis
  - Subchondral erosions
  - Syndesmophytes
  - Fat metaplasia
  - Joint space narrowing
  - Ankylosis

Non-radiographic Axial Spondyloarthritis

Axial spondyloarthritis has been divided into "radiographic" (X-ray changes present) and "non-radiographic" (meets diagnostic criteria based on a combination of clinical features, HLAB27 positivity, and/or MRI evidence of sacroiliitis). It is thought that non-radiographic SpA may represent an "early stage" where frank erosions and ankylosis have not yet occurred; however, while some of these patients develop progressive X-ray changes over time, others do not.

Thus, if a patient has a history and physical examination consistent with SpA, or has other suggestive extra-axial manifestations, it is important to consider ordering an MRI, even if plain radiographs are unrevealing. This could also be done in conjunction with a rheumatology consultation.

Subclinical Spondyloarthritis

Recent literature has suggested that MRI changes, particularly bone marrow edema, may be more common in the general population than originally thought. Studies of both elite and recreational athletes without reported back pain have reported rates of sacroiliac joint bone marrow edema up to 30–40% (Lambert et al. 2016). These patients usually do not have axial SpA and do not require treatment. In addition, many patients with

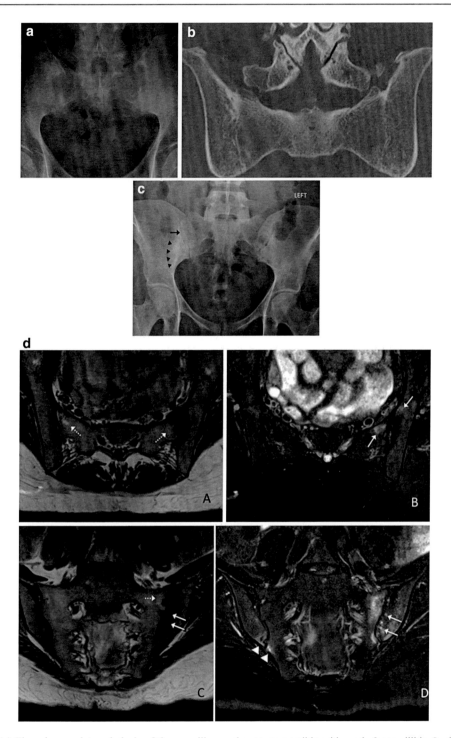

Fig. 5 (**a**) There is complete ankylosis of the sacroiliac joints, in keeping with grade 4 sacroiliitis. (**b**) Coronal image of the CT demonstrates complete ankylosis of the sacroiliac joints, in keeping with grade 4 sacroiliitis. (**c**) Radiograph of the sacroiliac joints demonstrates mild sclerosis in the left sacroiliac joint with minimal erosive change, compatible with grade 2 sacroiliitis. In the right sacroiliac joint, there are erosions (arrow) and marked sclerosis (arrowheads), in keeping with grade 3 sacroiliitis. (**d**) (*A*) Axial T1 and (*B*) STIR images with sclerosis and erosions involving the sacroiliac joints, more prominent on the left. There is multifocal geographic involvement and

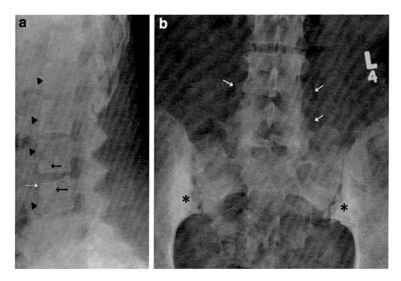

Fig. 6 (**a**) Lateral radiograph of the lumbar spine demonstrates squaring of the vertebral bodies (arrowheads). The irregularity and erosion at the anterior edge of the endplates, known as Romanus lesions (dashed white arrow), are early findings in inflammatory spondyloarthropathies such as ankylosing spondylitis. The "shiny corners" in the anterior aspects of the endplates (black arrows) represent reactive sclerosis and a healing response to the inflammatory erosions. (**b**) Frontal radiograph demonstrates parasyndesmophyte formation (arrows) and features of grade 3 sacroiliitis with erosions and sclerosis (asterisks) involving the sacroiliac joints. Incidental note of spinal dysraphism is present in L5 vertebral body

HLA-B27-related conditions, particularly inflammatory bowel disease, may have radiographic or MRI changes of sacroiliitis without any clinical signs or symptoms (Bandinelli et al. 2016). It is still unclear whether treatment in these patients would lead to improvement in outcomes, but referral to a rheumatologist could be considered for further assessment.

Initial Treatment Approach

All patients with SpA should be counselled on smoking cessation and regular exercise. Appropriate physiotherapy is a cornerstone of management and can help to prevent the restrictions in movement that are common later in the disease course.

Pharmacotherapy has been shown to be effective in symptomatic patients with radiographic and non-radiographic SpA, and the treatment approach is the same. Nonsteroidal anti-inflammatory drugs (NSAIDs) are first-line treatment and can be very effective. There is no evidence that one NSAID is more effective than another. A trial of at least 3 months of daily use is required to determine whether a significant decrease in pain and stiffness has occurred. Switching to a different NSAID is generally indicated if one is unsuccessful. Contraindications to NSAID use include chronic kidney disease, history of gastrointestinal bleeding, and coronary artery disease. Caution

Fig. 5 (continued) subchondral sclerosis with T1-hyperintense regions of sacral fat metaplasia (dashed arrows) on the T1 image. On the STIR image, there is T2-hyperintense edema (solid arrow) around the anterior aspect of the left sacroiliac articulation. (*C*) Coronal T1 and (*D*) STIR images with sclerosis and erosions involving the sacroiliac joints, more prominent on the left. There are erosions (white arrows on both images), more prominent in the iliac side of the left sacroiliac joint where there is also significant edema. In the right sacroiliac joint, there is fat metaplasia (dashed arrow) indicating chronic burnt out inflammatory change on the T1 image. There is a small focus of subchondral edema posteriorly in the right sacroiliac joint

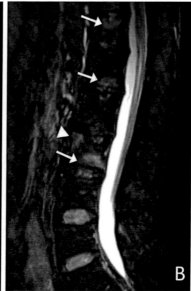

Fig. 7 MRI demonstrating features of spondyloarthritis. (**a**) Sagittal STIR image demonstrates foci of edema at the endplates reflecting inflammatory involvement of the intervertebral discs, described as Andersson lesions (white arrows). The T2-hyperintensities in the anterior corners of the vertebral body (arrowhead) depict "corner inflammatory lesions" in syndesmophyte formation. (**b**) Sagittal T1 image demonstrates T1-hyperintensies surrounding the foci of endplate edema (dashed arrows) depicting fat deposition, reflecting the chronic inflammatory and erosive changes

should be used in the elderly and in those with underlying hypertension or inflammatory bowel disease.

If NSAIDs are unsuccessful, the next treatment option is biological disease-modifying agents. These medications, known as "monoclonal antibodies" or "biologics," are very effective but can be prohibitively expensive for some patients. Common classes of biologics include the anti-TNF antibodies (adalimumab, certolizumab, golimumab, etanercept, and infliximab) as well as an anti-IL-17 antibody, secukinumab. Many other biologic therapies are being developed, and the list of approved medications for SpA is expected to grow. These drugs should be prescribed and monitored by a rheumatologist.

Regular use of NSAIDs and biologics has been shown to improve pain, function, and quality of life and reduce radiographic progression in patients with SpA. Patients are encouraged to remain on therapy long-term, although the optimal treatment duration and approach to tapering or discontinuing medication is still not known.

Appropriate Referrals

All patients with suspected SpA should be referred to rheumatology for further assessment and management. A comprehensive exercise program with involvement of physical therapists should be considered as well.

Case Resolution

This woman's symptoms are classic for inflammatory back pain. She should be investigated for potential spondyloarthritis and treated accordingly; the L4–L5 stenosis is likely incidental.

Clinical Pearls and Myths

- The hallmark of ankylosing spondylitis is a history of inflammatory back pain.
- Myth: ankylosing spondylitis is a male-only disease.

- In fact, newer cohorts suggest that AS is present in a male: female ratio of 2:1.
- Myth: X-ray findings are required to make a diagnosis of SpA.
  - Non-radiographic SpA can be diagnosed based on clinical features, HLA-B27 positivity, and/or consistent MRI findings.
- Think of SpA in patients with back pain and an underlying diagnosis of inflammatory bowel disease, psoriasis, or uveitis.
- Think of OCI as a common SpA mimicker in women with postpartum back pain.

Table 7 Causes of vertebral compression fracture

Osteoporosis
Trauma
Primary bone tumors (hemangioma, giant cell tumor)
Hematologic malignancies (multiple myeloma, lymphoma)
Solid organ metastases (breast, renal, prostate, lung)

Table 8 Red flags in back pain history

Red flags: spinal fracture	Red flags: malignancy
Prolonged corticosteroid use	History of malignancy
Recent trauma (including fall)	Significant weight loss
Older age (men >65, women >75)	No improvement after 1 month

Case 3

Case Description

A 75-year-old woman presents with severe mid-thoracic back pain for the past 4 weeks. She cannot recall any trauma, although it started after carrying some heavy groceries up the stairs. She has a history of breast cancer and underwent curative surgery 10 years ago. On further questioning, she describes 2 inches of height loss since her mid-20s.

Differential Diagnosis

Severe, acute back pain in an older woman should be investigated for vertebral fracture. Osteoporotic fractures usually occur in postmenopausal women and often with little trauma. Risk factors include low body weight, smoking, excessive alcohol use, long-term corticosteroids, and family history of osteoporosis. Other causes of vertebral fractures are listed in Table 7.

Salient Features on History and Physical Exam

Epidemiological studies suggest that 1–4% of patients presenting to primary care with back pain will be found to have a spinal fracture (Downie et al. 2013). Pain from spinal fractures often occurs acutely and eventually diminishes over time. It is often worse with standing and better lying down due to the force of gravity on the vertebrae. Patients may also notice new loss of height or curvature to the spine.

There are multiple "red flags" that have been identified to predict which patients require imaging to rule out fracture or malignancy (Table 8); however, most of these have a high false positive rate according to a recent Cochrane review (Williams et al. 2013). The only physical exam feature suggestive of fracture was a contusion or abrasion at the site of pain consistent with recent trauma. Patients presenting with any of these features and a history of new back pain likely warrant further imaging.

Appropriate Use of Investigations

The suspicion of a vertebral compression fracture should be confirmed radiographically. The gold standard remains lateral thoracolumbar spine X-rays, which have a high sensitivity and specificity for the diagnosis. More recently, Vertebral Fracture Assessment using bone mineral densitometry has been found to provide similar diagnostic information with less radiation.

If an underlying malignancy is suspected, MRI spine is the investigation of choice. Pathologic fractures and underlying lesions can usually be

Table 9 Investigations after identifying osteoporotic fracture

	Investigation	Relevance
Bloodwork	CBC, creatinine, calcium, 25-OH vitamin D level	>Indicated in preparation for starting anti-osteoporotic therapy
	Phosphate, ALP, TSH, protein electrophoresis	>Indicated to workup for secondary causes of osteoporosis
Imaging (Fig. 8)	Thoracolumbar X-ray	>All patients should be assessed for presence of multiple fractures
	Bone mineral density	>Defines baseline so response to treatment can be monitored
	MRI spine	>If malignancy is suspected
	CT spine	>If MRI is unavailable or contraindicated

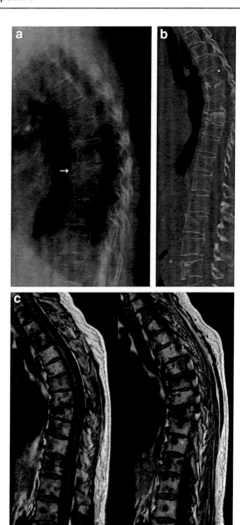

identified and differentiated from osteoporotic fractures with this modality. If an osteoporotic fracture is diagnosed, guidelines suggest screening for secondary causes of osteoporosis as listed in Table 9 (Papaioannou et al. 2010).

Initial Treatment Approach

Most vertebral fractures can be treated conservatively with adequate analgesia, starting with acetaminophen, as well as rehabilitation. There is also low-quality evidence to support use of calcitonin for management of acute pain from vertebral fractures. When the pain from vertebral fractures does not respond to the above treatments, a trial of opioid therapy may be reasonable; these medications are best prescribed and monitored by the family physician. A subset of patients with vertebral fractures will be severely functionally limited and may benefit from surgical intervention. Surgical procedures for compression fractures will be discussed in more detail later in this volume.

Fig. 8 (**a**) Lateral thoracic radiograph demonstrates a vertebral compression fracture of the T8 vertebral body (arrow). There is mild sclerosis within the fractured vertebral body with questionable sclerotic foci in several of the vertebral bodies and degenerative changes present. (**b**) CT performed to evaluate the thoracic compression fracture shows sclerosis in the vertebral body, with small sclerotic foci in the vertebral endplates in the two levels above and another lumbar vertebral body. There are several hypodense foci in the posterior aspect of the vertebral bodies above the compression fracture with posterior bulging, suspicious for metastases. (**c**) MRI was performed for further evaluation of the compression fractures. Two sagittal T1-weighted images demonstrate multiple foci of low T1 signal intensities in the anterior and posterior columns of the thoracolumbar spine, in keeping with diffuse metastases throughout the spine. The metastatic burden is underestimated on radiographs and CT. There is no spinal stenosis or cord compression on this study

Table 10 Evidence-based management for osteoporosis (Papaioannou et al. 2010)

Non-pharmacologic	Regular, weight-bearing exercise
	Education on fall prevention strategies
	Daily calcium intake of 1200 mg (diet or supplementation)
	Daily vitamin D intake of 800–2000 international units
Pharmacologic	Bisphosphonates: alendronate, risedronate, or zoledronate
	Hormonal therapy: estrogen-containing HRT or raloxifene
	Denosumab
	Teriparatide

Additionally, any patient with a fragility fracture at the spine has osteoporosis by definition, regardless of their bone mineral density (BMD). These patients should be strongly encouraged to follow evidence-based treatment guidelines for osteoporosis, including both pharmacologic and non-pharmacologic measures (Table 10). There is no evidence that use of bisphosphonates impairs healing from vertebral fractures; therefore, they should be started as soon as possible to prevent further fracture events.

Appropriate Referrals

Most family physicians will feel comfortable managing osteoporosis and fragility fractures. However, difficult-to-treat cases, especially patients who fracture while on appropriate therapy, may warrant a referral to specialized osteoporosis clinic for assessment.

Case Resolution

Based on her history of breast cancer, this patient underwent an MRI which did not show any evidence of malignancy. A bone mineral density scan was performed which confirmed low bone mass in the lumbar spine, with a T-score of −2.5. Screening bloodwork did not reveal any secondary causes of osteoporosis. This patient was treated with adequate pain control and started on calcium, vitamin D, and an oral bisphosphonate. Ongoing management will be provided by her family physician.

Clinical Pearls and Myths

- Metastatic disease must be ruled out in patients with new back pain and cancer history.
- Any fragility fracture is an indication for treatment with anti-osteoporotic therapy.
- There is no need to delay initiation of osteoporosis therapy after a fracture.

Conclusion

Investigations to Avoid

While it may sometimes seem that back pain has an infinite number of causes, it is important to remember the list of medical conditions that are *not* a cause of back pain so that investigations can be directed appropriately. Too often, a comprehensive "autoimmune workup" is ordered as an attempt to explain many types of chronic pain; however, ordering these tests without concomitant clinical features leaves both patients and healthcare providers with few satisfying answers.

Autoimmune diseases like rheumatoid arthritis and systemic lupus erythematosus are associated with small and large joint pain and swelling, but as a rule do not affect the lower back. The same principle applies to patients with vasculitis, myositis, and autoimmune thyroid disease. For this reason, ordering the following investigations in the workup of chronic back pain is **unlikely to be useful**: antinuclear antibodies (ANA), rheumatoid factor (RF), thyroid-stimulating hormone (TSH), or antithyroid antibodies.

Similarly, when a patient with a known diagnosis of rheumatoid arthritis or lupus presents with low back pain, their underlying condition should not be blamed and an alternate explanation should be sought. These patients have a higher risk of infection and malignancy compared to the general population and are also more likely to suffer from obesity and low levels of physical activity which may also contribute to chronic back pain.

When to Refer

Rheumatologists have expertise in the diagnosis and management of patients with spondyloarthritis, and any patient with this suspected diagnosis can be referred for further assessment. Inflammatory back pain in a young patient, particularly if they have abnormalities on SI joint imaging and extra-articular features, would be a clear reason for referral. The presence of a positive ANA or HLA-B27 *without* concomitant clinical features does not necessarily require evaluation by a rheumatologist.

References

Bandinelli F, Manetti M, Ibba-Manneschi L (2016) Occult spondyloarthritis in inflammatory bowel disease. Clin Rheumatol 35:281–289

Colip CG, Lotfi M, Buch K et al (2018) Emergent spinal MRI in IVDU patients presenting with back pain: do we need an MRI in every case? Emerg Radiol 25:247–256

Downie A, Williams CM, Henschke N et al (2013) Red flags to screen for malignancy and fracture in patients with low back pain: systematic review. BMJ 347:f7095

Lambert RGW, Bakker PAC, van der Heijde D et al (2016) Defining active sacroiliitis on MRI for classification of axial spondyloarthritis: update by the ASAS MRI working group. Ann Rheum Dis 75:1958–1963

Papaioannou A, Morin S, Cheung AM et al (2010) Clinical practice guidelines for the diagnosis and management of osteoporosis in Canada: summary. CMAJ 182:1864–1873

Sieper J, Rudwaleit M, Baraliakos X et al (2009) The assessment of spondyloarthritis international society (ASAS) handbook: a guide to assess spondyloarthritis. Ann Rheum Dis 68:ii1–ii44

Williams CM, Henschke N, Maher CG et al (2013) Red flags to screen for vertebral fracture in patients presenting with low-back pain. Cochrane Database Syst Rev:CD008643

Psychosocial Impact of Chronic Back Pain: Patient and Societal Perspectives

6

Y. Raja Rampersaud

Contents

Epidemiology of Chronic Low Back Pain	110
Burden of Low Back Pain	110
The Need for a Biopsychosocial Model	111
Psychosocial Impact of Chronic Low Back Pain	111
Psychological Factors	112
Social Factors	114
Central Sensitization Syndrome	117
Socioeconomic Impact	117
Overview of Assessment and Management	120
Assessment	120
Management	121
References	122

Abstract

Low back pain (LBP) is a highly prevalent, poorly managed condition that is the number one cause of years lived with disability (YLDs) around the world. It is estimated that one in four prevalent cases of LBP is responsible for 77% of the YLDs. The socioeconomic burden of LBP, particularly in developed countries, is enormous with medical expenditures rivalling that of diabetes or ischemic heart disease. The individual burden of LBP is also tremendous and commonly results in psychosocial distress and dysfunction. The most commonly cited negative prognostic psychological factors are depression, fear-avoidance, and pain catastrophizing. However, pain self-efficacy and patient beliefs have been found to even more strongly associate with actual outcome. Qualitative studies of chronic LBP patients have relieved consistent themes reflecting difficulties in coping with a sense of stigmatization that is associated with an invisible problem, loss of wellness, loss of self, loss of relationships, and loss of the future. For chronic LBP, both exercise therapy and cognitive behavioral therapy (CBT) are now recommended as first-line treatments that should be considered for routine use in addition to providing education regarding the nature of LBP and advice to

Y. R. Rampersaud (✉)
Arthritis Program, Toronto Western Hospital, University Health Network (UHN), Toronto, ON, Canada

Department of Surgery, Division of Orthopaedic Surgery, University of Toronto, Toronto, ON, Canada
e-mail: raja.rampersaud@uhn.ca

© Springer Nature Switzerland AG 2021
B. C. Cheng (ed.), *Handbook of Spine Technology*,
https://doi.org/10.1007/978-3-319-44424-6_135

remain active. These recommendations necessitate timely assessment, regardless of duration of symptoms, in LBP patients for the complex biopsychosocial prognostic factors that may impact patient and societal outcome.

Keywords

Chronic low back pain · Biopsychosocial model · Psychological · Social · Socioeconomic · Impact · Individual · Societal

Epidemiology of Chronic Low Back Pain

Burden of Low Back Pain

The Global Burden of Disease Study (Global Burden of Disease, Injury Incidence, Prevalence Collaborators 2016) has demonstrated that the global prevalence of low back pain (LBP) continues to increase. The global point prevalence of LBP in 2015 was 7.3%, and the estimated median 1-year prevalence in adults was 37% (Hartvigsen et al. 2018). LBP is more common in women, and the peak prevalence is in mid-life. LBP is the number one cause of years lived with disability (YLDs) with 77% of the YLDs accounted for by approximately one in four prevalent cases (Global Burden of Disease, Injury Incidence, Prevalence Collaborators 2016). This suggests that although most people experiencing LBP have low levels of disability, the enormous societal impact of LBP is driven by high prevalence and a subgroup of LBP patients with high levels of persistent disability. In 2013, low back and neck pain were globally ranked the fourth leading cause of disability-adjusted life years (DALYs) just after ischemic heart disease, cerebrovascular disease, and lower respiratory infection (Global Burden of Disease 2015 DALYs and HALE Collaborators 2015). This represents an increase over time from being ranked seventh in 1990 and fifth in 2005. Furthermore, low back and neck pain are ranked as the number one cause of DALYs in most high-income countries.

For decades the overarching public and clinical messaging for LBP (e.g., recommendations from clinical practice guidelines) has been that LBP will get better in the majority of patients. Unfortunately, without providing the full context of this message, many patients perceive the term "get better" to mean resolution. However, multiple studies have demonstrated up to 2/3 of individuals with LBP at both the population and primary care level may have recurrent (i.e., episodic) or persistent LBP at 1 year (Hartvigsen et al. 2018). Most of the personal and societal impact of LBP is in those with chronic LBP (CLBP) which is conventionally considered to be LBP symptoms lasting more the 12 weeks. The National Institute of Health (NIH) Pain Consortium Task Force on research standards for CLBP recently defined CLBP as a "back pain problem that has persisted at least 3 months and/or has resulted in pain on at least half the days in the past 6 months" (Deyo et al. 2014). A call for increase in prognostic research to determine which patients will develop CLBP and system-wide strategies for mitigation of chronicity is at the forefront of the paradigm shift in LBP care (Foster et al. 2018).

The discrepancy between favorable natural history and persistence or recurrence is multifactorial and needs to be considered based on two main perspectives. First is the clinical setting that is being studied: the population, primary allied healthcare (e.g., physiotherapy, chiropractic), primary medical care (i.e., family doctor or nurse practitioner), and secondary or tertiary care (e.g., specialized chronic pain clinic, surgical clinic). Each scenario will provide a progressive increasing prevalence or severity of CLBP (persistent or recurrent). Second is most prevalence studies are cross-sectional or of limited duration and thus do not reflect the longitudinal aspects of LBP. Studies assessing the course of LBP over long periods are limited but can provide valuable insight regarding the true nature of LBP. Recent work from our center has demonstrated a sobering picture of the long-term trajectory of LBP in the Canadian population. Canizares et al. (2019) reported on a representative sample ($n = 12{,}782$) of the Canadian population over a 16-year period from 1994 to 2011. Group-based trajectory analysis was used to group participants based on the nature of their back pain over the 16-year follow-up period.

Overall, 45.6% of participants reported back pain at least once during the study period. Of people with back pain, four distinct trajectories were identified: persistent (18.0%), developing (28.1%), recovery (20.5%), and occasional (33.4%). This is consistent with the findings of the Global Burden of Disease Collaboration study that a subgroup(s) of back pain patient (one in four) generates the majority of back pain impact (Global Burden of Disease, Injury Incidence, Prevalence Collaborators 2016). Specifically, the persistent and developing groups, which made up almost half of the patients reporting back pain, were characterized by having more pain-limited activities, disability, depression, and medical comorbidities. Furthermore, only one in five people with back pain recovered over the 16 years, and one in three continued to report occasional back pain. There is substantial literature that the natural history of LBP is not one of resolution in the vast majority of patients, provides a strong rationale for a change in the basic assumptions and approach to the management of LBP. In short LBP should be viewed as a chronic condition.

The Need for a Biopsychosocial Model

A critical part of changing the approach to LBP assessment and management is identification of risk factors for a less then favorable outcome (Hartvigsen et al. 2018). Identifying who is at risk and what factors are potentially modifiable is of paramount importance in CLBP prevention, assessment, and management. Certainly, prognostic research in LBP is by no means novel. The challenge has been the implementation and practice of prognostic care for LBP. Reviewing various multivariate predictive models, Hartvigsen et al. (2018) note several independent risk factors for individuals that are likely to develop a more disabling course of LBP: high pain intensity, psychological distress, accompanying leg pain, and pain at multiple body sites. A wide variety of well-known risk factors associated with poor outcomes have been published. These are nicely summarized by Hartvigsen et al. (2018) and include but are not limited to *symptom-related factors* (previous episode of LBP, higher pain intensity, presence of leg pain), *lifestyle factors* (smoking, higher body mass index, less physical activity), *psychological factors* (depression, catastrophizing, fear-avoidance behavior), and *social factors* (physical work, lower education, compensation claim, poor work satisfaction). However, as noted by Kent and Keating (2008), these predictive studies only explain a small degree of the variance in the outcome of LBP, with most explaining only 30–40%. Thus, despite great advances in identifying risk factors for CLBP, much of what is known to drive poor outcomes has not been comprehensively studied in the same population or remains unknown. Consequently, what factors should be assessed, when they should be assessed, and best practices for assessment and management of LBP remain a source of ongoing tribal like debate.

Despite broad acknowledgment that a biopsychosocial model is critical to advancing LBP care, the majority of LBP assessment and management remain focused on the biomedical model.

This chapter focuses on the psychosocial aspect of CLBP.

Psychosocial Impact of Chronic Low Back Pain

In the broadest sense, CLBP affects all aspects of an individual's life including day-to-day function, mood, social interactions, recreational activities, and work life. I am not aware of any conclusive studies that assess all of these aspects simultaneously. Thus, we are reliant on quantitative and qualitative systematic reviews to inform us of the broader impact of CLBP. Within the psychosocial realm of CLBP, there is typically greater focus on the psychological aspects. However, the social consequences of CLBP are often the primary drivers of secondary psychological cognition and behavior, and therefore these two dimensions should be considered together. The use of psychosocially oriented screening questions, so-called yellow flags, to identify psychosocial barriers to LBP recovery has been noted in many LBP clinical practice guidelines and clinical tools

(Nicholas et al. 2011). Commonly cited "yellow flags" are an individual's belief that pain and activity are harmful leading to fear-avoidance behavior or sickness behavior (e.g., extended rest), low or negative mood, social withdrawal, overprotective family or lack of support, and treatment expectations or beliefs that are against best evidence or are focused on a passive cure. Increased focus on the biopsychosocial approach to LBP has demonstrated that psychosocial risk factors for developing CLBP are much more prevalent than previously thought. In a recent randomized controlled trial, patients presenting to primary care with LBP demonstrated varying degrees of psychological risk factors with 46% rated as moderate risk and 28% being categorized as high risk of chronicity (i.e., persistent LBP) using the Keele StarT Back screening tool (Hill et al. 2011).

Psychological Factors

A large body of evidence confirms that psychological factors, including emotions, beliefs, or avoidance or other maladaptive behaviors, are linked to poor outcomes in low back pain (Chou and Shekelle 2010). While the majority of studies are in the setting of nonsurgical care, it must be noted that psychological factors are also independently associated with poor outcome in surgical intervention for LBP (Wilhelm et al. 2017). A large number of psychological factors have been assessed over the last few decades and most have been found to have a negative association with LBP patient outcomes. However, there are contradicting studies regarding the impact or dominance of one psychological factor over another, whether factors are individually modifiable or not, and whether these factors are mediated by other psychological or non-psychological (e.g., social) factors. The nature and impact of psychological factors in CLBP is an extremely complex issue and thus is often very difficult for both clinicians and their patients to fully understand.

The most commonly cited and clinically assessed psychological factors are depression, fear-avoidance, and pain catastrophizing (Pincus and McCracken 2013). A detailed description of psychological theory is out of scope for this chapter and is not my area of expertise. Pincus and McCracken (2013) provide an excellent review of this topic. Key components of their review are briefly summarized herein to provide a non-psychologist interpretation of the available LBP psychology literature. Depression or low mood is commonly reported in LBP. Depression tends to occur more in the chronic phase and is a commonly identified poor prognostic factor across different key outcomes (i.e., negatively impacts pain, function, and work status). Fear-avoidance covers a broad spectrum of fearful beliefs or cognitions about pain inducing movement, activities, or re-injury and associated patterns of avoidance behavior. The latter can range from simple (e.g., avoiding any lifting) to very elaborate movements or avoidant activity patterns (e.g., bizarre movement patterns or behavioral responses when attempting to do simple task). Fear-avoidance is often a more significant issue when considering work-related factors but can also significantly impact day-to-day function. Pain catastrophizing is the tendency for an individual to describe a pain experience in a more irrational manner than you would expect from an average person. There are often magnification, rumination, and feelings of helplessness regarding the pain experience. Patients also may have maladaptive cognition regarding future events and a relative inability to inhibit pain focused thoughts in anticipation of, during, or following a painful experience.

Another important factor that is receiving more clinical and academic focus is pain self-efficacy. Described by Bandura (1977), self-efficacy is an individual's belief about how well they can cope with difficult situations or their ability to achieve a desired outcome. Higher levels of pain self-efficacy are typically associated with lower levels of disability despite the presence of pain. Lee et al. (2015) performed a systematic review and meta-analysis of which psychological factors are involved in the process of pain leading to disability. Specifically, they reviewed mediation analysis studies. Mediation analysis seeks to identify and explain the mechanism or process that underlies an observed relationship between an independent variable (e.g., pain) and a dependent variable (e.g., disability) by assessing the influence of the independent variable on a third variable, often termed a mediator variable (e.g., depression), that may influence the dependent variable. Using data from 12 mediation studies, the authors identified self-efficacy, psychological distress (depression/anxiety), and fear were mediators that explained some of the association between LBP pain and developing chronic disability. In a comprehensive study of psychological obstacles to recovery in primary care LBP patients, Foster et al. (2010) assessed the relative strength of 20 different baseline psychological factors to predict patient-reported disability at 6 months. At baseline, most factors were related to degree of disability, with perception of consequences, depression, and pain self-efficacy most strongly associated. However, when considered together, depression, catastrophizing, and fear-avoidance were not independent predictors of outcome. Patients' perceptions regarding the timeline (i.e., belief that their LBP will be chronic), illness identity (i.e., number of symptoms related to their LBP), perception of personal control, and pain self-efficacy were the factors that remained significantly associated with outcome (explaining 56.6% of the variance of outcome). The authors note that patients "who perceive themselves able to exercise control over their back problem, now and in the future, are less likely to develop longer-term disability." I would add that it must be kept in mind that CLBP is a dynamic process; thus psychosocial factors are likely overlapping and/or cumulative to varying degrees over time. Consequently, as much as it is unwise to focus LBP assessment or management on the biomedical model only, focusing on a single psychological factor would likely be equally ineffective.

A patient's acceptance, or lack thereof of chronic pain has been shown to be a significant and independent factor in the outcome of those dealing with CLBP or other types of chronic pain (McCracken and Vowles 2008). Although this may seem intuitive, assessing and managing a patient's ability to accept and deal with chronic pain is not something that is typically addressed in the front-line management of LBP. Similarly, a patient's pain beliefs, and expectations are also independently associated with their recovery from LBP as well as their response to different treatments (Main et al. 2010). In a systematic review by Ramond et al. (2011), the authors found that "depression, psychological distress, passive coping strategies and fear-avoidance beliefs were sometimes found to be independently linked with poor outcome, whereas most social and socio-occupational factors were not." However, a patient's or care provider's perceived beliefs regarding persistent LBP was the factor that was most consistently linked with actual outcome. Although, negative beliefs and expectations are often a part of behavioral/psychological treatment, they are typically not addressed in the early aspects of LBP care and thus may become entrenched (i.e., reinforced) due to persistent or recurrent pain. As noted above, persistence or recurrence is the more likely course of LBP in patients seeking healthcare.

In a Canadian population-based survey of 2400 adults, half of the respondents had pessimistic views of LBP (Gross et al. 2006). Respondents felt that back pain makes "everything in life worse," worsens over time, and eventually will "stop you from working." Interestingly the authors at that time noted: "Contrary to recent evidence-based clinical practice guidelines that advocate that back pain is a benign, self-limiting condition, most subjects in our sample had pessimistic beliefs concerning back pain." The authors concluded: "Public association back pain beliefs in the 2 Canadian provinces sampled are not in harmony with current scientific evidence for this highly prevalent condition. Given the mismatch between public beliefs and current evidence, strategies for re-educating the public are needed." As our knowledge of the natural history of LBP continues to improve, the pateint beliefs reported by Gross et al. (2006) were reflective of the actual experience of many LBP patients. Until recently, the messaging from LBP clinical practice guidelines was in fact misaligned with the reality of LBP patients.

Walker et al. (2004) reported that half of LBP patients do not actually seek healthcare. It is likely that these patients are the ones who have a favorable natural history. In a systematic review of studies assessing the course of non-specific acute LBP patients seeking treatment in primary care, Itz et al. (2013) found that 65% of patients still report pain at 1 year. The authors concluded: "The findings of this review indicate that the assumption that spontaneous recovery occurs in a large majority of patients is not justified." Similar conclusions can be drawn from the cumulative trajectories of back pain (see above) in the population where almost 50% of individuals experience a persistent or developing trajectory of back pain (Canizares et al. 2019). It is the belief of the author that the decades of clinical practice guideline messaging that LBP is a benign and self-limiting condition, while correct from a medical perspective (i.e., no sinister pathology and typically mild in severity), may have (at least in part) led to some of the more strongly held pessimistic beliefs of patients with CLBP.

The optimistic message of "don't worry it gets better in most people" is still reasonable for public health campaigns. This, however, is not the intended audience of clinical practice guidelines. This message has been delivered for decades by health providers to patients who are seeking healthcare. Given the substantial prevalence data on the transition from acute to chronic LBP (Hartvigsen et al. 2018), this messaging was doomed to fail in the majority rather than a minority of patients as intended. For the large number of patients seeking care who do not "resolve" (whether their beliefs and expectation were falsely set by a well-intended provider or put forth by financially incentivized providers promising a "cure" for LBP), persistent or recurrent pain may lead to a negative perception or belief that something more serious or unmanageable is occurring. This also puts the provider in a difficult position to explain why the pain is not getting better as they said it would. It is the belief of this author that this common scenario leads to the perfect storm of overmedicalization of LBP by the both the patient and healthcare provider(s). This is certainly a multifactorial issue driven by persistent/recurrent symptoms, heightened/worsening symptoms, the potential for loss of or lost patient confidence in the provider(s), clinical uncertainty, and the need/desire to do something.

Social Factors

At a population level, CLBP disproportionately affects people with low education and socioeconomic status association (Shmagel et al. 2016). Possible reasons for this may be related to inadequate health literacy, poor access to healthcare, and greater likelihood of being in labor-intensive work (Hartvigsen et al. 2018). At an individual level, the social implications and interplay of LBP are much more complex and are best understood through qualitative methods of inquiry. Three available systematic reviews and meta-syntheses of the qualitative research on patients' lived experience with LBP provide a deeper understanding of the more complex personal and nuanced aspects of LBP. Interestingly all three were published within a year of each other (Froud et al. 2014; Snelgrove and Liossi 2013; Bunzli et al. 2013) and presented different but interrelated perspectives.

In the study by Froud et al. (2014), the authors reviewed 49 articles from 42 studies. They reported on four first-order themes from the qualitative literature:

Theme 1 – Activities: loss of function, particularly regarding domestic chores, important recreational activities (friends and family), and an inability to plan ahead were consistently found across studies.

Theme 2 – Relationships: significant negative impact on personal relationships was a common impact of CLBP. This occurred from two perspectives. First and more common is being worried about how their inability to participate in activities with family or friends was affecting others (e.g., holding others up or ruining it for others), and second is worrying about the pain that would occur if they participated. However, some felt unsupported in these relationships. Regardless of the perspective, the end result was often social withdrawal and isolation.

- **Theme 3 – Work:** the impact of LBP on work was very prevalent and included the need (and difficulty) to modify work, fear of losing their job, and difficulty navigating disbelief from co-workers.
- **Theme 4 – Stigma:** a very prevalent finding in the qualitative literature pertains to worries of not being believed by others (family, friends, employers, healthcare workers, insurance, etc.) and the need to legitimize or validate that their pain was real.

Froud et al. (2014) provide an excellent summary of the pertinent interpretation of the qualitative literature as follows: "People with low back pain seek to regain their pre-pain healthy, and emotionally robust state. They desire not only diagnoses, treatment and cure, but simultaneously reassurance of the absence of pathology. Practically, although sufferers are often chiefly concerned with (re)engagement in meaningful activities, and attenuation of symptoms, the more experientially-focused literature suggests that the impact of back pain is pervasive, with life-changing effects." Consistent with prevalence of persistent or recurrence LBP, the authors also state the following: "Whilst back pain is not itself life-threatening, it does threaten quality of life. In the absence of diagnosis and effective treatment, complex enmeshment and interactions can ensue between chronic LBP, identity, and social roles, having a diverse and pervasive impact of the condition with life-changing psychological and social consequences."

In the review by Snelgrove and Liossi (2013), the authors assess 33 articles from 28 studies. They summarized the qualitative literature in three interrelated themes similar to those put forth by Froud et al. (2014). **Theme 1- Self:** CLBP leads to "loss of a previous lifestyle and changes in personality." Persistent LBP essentially leads to loss of self (former and future) due to "an incremental rise of functional limitation accompanied by feelings of self-loathing, frustration, anger, negativity towards others, self-denigration and even depression." Furthermore stigmatization (see above), perceived or real, threatens personal integrity. **Theme 2 – Relationships:** Snelgrove and Liossi (2013) divided this into two distinct aspects, (1) relationships with family and friends and (2) relationships with health professionals and the organization of care. The impact of CLBP on relationships with family and friends was noted by Froud et al. (2014). The authors highlighted an important issue of the effect of CLBP on relationships over time: "Participants reported being a burden to their families and 'holding people back' with sympathies lessening as time went on with no diagnosis or formal explanation," the latter being another driver for legitimization and over-medicalization of their CLBP. Negative relationships with health professionals and the organization of care were significant issues. As noted by the authors, "Participants described a good consultation as a partnership enabling a sense of security and belonging; promoting feelings of mutual understanding and recognition, and incorporating individualised care, clear explanations, reassurance, discussing psychosocial issues and future options." In many instances, this does not represent real-world health interactions. As noted by the authors "a lack of diagnosis and ongoing unresponsiveness to treatment invoked perceptions of not being believed, leading to a feeling of stigma and distress," "Participants reported being viewed as culpable; accused of imagining their symptoms; seeking secondary gain; symptoms being 'all in the mind' and laziness." Typically, providers and the system (e.g., disability insurer needs a diagnosis to provide benefits) are biased toward the biomedical approach which often will not provide a specific causation or resolution for most CLBP and further drives these negative psychosocial consequences and loss of faith in healthcare provider(s). **Theme 3 – Coping:** The authors reported coping in the context of individuals' attemps to manage their LBP. Snelgrove and Lossi (2013) note "A number of authors identified biomedical beliefs as a determinant of participants' experiences." "Biomedical beliefs were related to less successful rehabilitation to work and perceptions of reduced wellbeing; disappointment with the inefficacy of medical treatments; an overall narrow range of behavioural focused coping strategies, psychological inflexibility and comprehensive enmeshment with pain, with little engagement or acceptance and a loss-orientated focus."

In the third review, Bunzli et al. (2013) reviewed 33 articles representing 28 studies. They also categorized the existing qualitative literature into three interrelated themes similar to those already noted. However, they also provided a provocative conceptualization of CLBP experience as one of "biographical suspension" in which three aspects of suspension were described: "suspended wellness," "suspended self," and "suspended future." **Theme 1 – The Social Construct of CLBP:** This theme emerged based on the highly prevalent biomedical beliefs of back pain patients. This should not be a surprise given the long-standing general biomedical model that is ingrained in both patients and practitioners alike (i.e., "diagnosis-treatment-cure"). As noted by the authors, "A biomedical explanation for the CLBP was critical for an individual to establish their pain as being a legitimate disability, which could then receive the support of the family, workplace, and welfare agencies." "The lack of a satisfactory etiological explanation for their 'invisible' pain meant participants in many studies felt at risk of not being believed." "The participants' experience in the health care system was repeatedly described with feelings of anger and frustration towards professionals who could not fulfill expectations of a diagnosis-treatment-cure pathway." These perceptions were found in most studies to occur with themes of stigmatization as noted above. Even in scenarios where the pain fluctuated, participants reported the need to demonstrate (i.e., sickness/pain behaviors) their pain and its impact all the time as a means of legitimizing their CLBP. The authors put this further into the context of the perceived role of the healthcare provider (HCP), noting that "HCPs were identified as painting an image of the demanding, difficult, and drug-seeking CLBP patient" and "any inference by HCPs of the pain being psychological in origin was felt by participant in several studies to be labeled with the stigma of questionable integrity." **Theme 2 – The psychosocial impact of the nature of CLBP:** The authors noted that "In the studies reviewed, pain was described as omnipresent, salient, and characterized by unpredictable fluctuations in intensity during both waking and sleeping hours"; "studies described participants experiencing disbelief at why they were suffering, prompting feelings of frustration, anger, guilt, and despair"; and "anxiety and distress, in light of an uncertain future, were widely described by study participants." The alterations in mood often resulted in depression. Consistent with the theme of the impact of LBP reported by Snelgrove and Liossi (2013), the authors provide a profound quote that the psychological effects of pain amounted to an "assault on the self." **Theme 3 – Coping with CLBP:** The findings from this theme are consistent with the interpretation of the same theme reported by Snelgrove and Liossi (2013). The authors framed coping strategies that were reflective of a constant fight or struggle to legitimize and control pain and the impact of CLBP in the context of the two other interrelated themes.

It is clear from these qualitative reviews that the biomedical model and practice of medicine are at odds with the lived experience of CLBP patients. The interrelated themes presented by these three reviews certainly resonate with my clinical experience as a spine focused practitioner. For example, the social construct of CLBP noted by Bunzli et al. (2013) is something that I suspect all HCPs dealing with CLBP (or chronic pain of any sort) experience on a regular basis. In my practice, this is a weekly occurrence when attempting to explain to a patient why surgery is not going to fix their pain, a process that takes significantly more time to do than saying "I have a solution that fulfills the biomedical belief and expectation of a given patient." As a strong believer in a holistic approach to CLBP, even under the scenario of a detailed and patient-centric explanation of the nociceptive and centralized mechanisms of pain, explaining the presence of unrelated imaging "abnormalities," etc., patients often simply conclude "so you are saying is this is all in my head" and/or "how is it possible that you cannot fix my – *any given radiographic diagnosis*" (i.e., the highlighted "problem" on their imaging report). As a surgeon in a tertiary-quaternary academic center (i.e., a highly biomedically focused practice), I find the biomedical belief for a solution so ingrained in some patients that it is at times difficult if not impossible to alter. Another scenario that commonly occurs in my practice and is in keeping with the social implications of CLBP such as legitimization of pain is the need for disability or other insurance companies to have a definitive biomedical diagnosis for an individual who in the eye of the insurer (based on decades of messaging regarding resolution of LBP) should be better. I find this to be a profound source of patient frustration and stigmatization. The resultant stress of financial loss added to the common feeling of stigmatization is a tremendous driver for patients to continue to pursue a biomedical approach to their CLBP. This truly represents a vicious negative feedback loop for a significant proportion of the CLBP population.

Central Sensitization Syndrome

For many years the aforementioned "yellow flags" where thought to be predominantly psychosocially driven and in some cases, particularly where the injury or imaging findings where minor or did not remotely match the degree of symptoms, labelled as malingering behavior. In an excellent review by Nijs et al. (2017), the authors provide an update on how contemporary pain neuroscience is providing evidence of pathophysiological changes in pain processing, termed central sensitization that can occur in approximately 25% of patients with CLBP. Woolf (2011) defined central sensitization as "an amplification of neural signaling within the central nervous system (CNS) that elicits pain hypersensitivity." Individuals with central sensitization can have varying degrees of hypersensitivity; however, patients exhibit increased responsiveness to normal or subthreshold afferent input. Clinical features such as allodynia, pressure hyperalgesia, aftersensations, or temporal summation can be objectively detected. This condition is an important consideration in any individual with chronic pain and in fact has a biomedical explanation; however, it can be very difficult to explain to patients and manage, particularly in later stages. A detailed discussion of the pathophysiology, diagnosis, and management of central sensitization is beyond the scope of this chapter.

Although this chapter focused on psychosocial aspects of CLBP, central sensitization is a pivotal advancement in our understanding of the pain experience and must be considered in the context of those with persistent pain. It is not clear whether certain individuals are prone to developing central sensitization or if the psychosocial consequences noted above in some way lead to central sensitization in certain individuals. For example, in a review by Delpech et al. (2015), the authors surmise that "stress-induced microglia dysfunction may underlie neuroplasticity deficits associated to many mood disorders." In a systematic review of the structural and functional brain changes in CLBP, Kregel et al. (2015) noted consistent finding across studies of increased activation not only in somatosensory-discriminative regions of the brain but also in areas of affective and cognitive processing of pain. In a subsequent review by the same group (Kregel et al. 2017), there is limited evidence suggesting that "maladaptive central neuroplastic changes" may not be permanent and can be improved by targeted interventions. For example, behavioral extinction training was shown to shift pain-induced activation back to sensory discriminative regions from affective brain regions.

Socioeconomic Impact

In a widely cited review, Katz (2006) estimated that the cost of LBP in the United States ranged from $100 to $200 billion (2005 dollars) a year with indirect costs (e.g., lost wages) accounting for up to two-thirds of the cost and direct medical expenditures the rest. Around the same period, the cost of medical expenditure on LBP in Canada was estimated to be $6–$12 billion Canadian dollars per year (Brown et al. 2005). These very broad estimates exemplify the challenges of determining the total cost associated with most chronic health conditions. These challenges are due to a variety of factors including, but not limited to the following: region and country specific economic factors (e.g., varying cost of healthcare within and among different countries); assessment of specific subpopulations of LBP patients such as primary care (acute or chronic), workers compensation, surgical or chronic pain patients; reporting of direct (healthcare) cost only; and when indirect cost are reported they are limited to individual productivity losses (e.g., time off work) rather than including caregiver cost or the cost of social support (e.g., food or housing). The latter point is relevant to the growing non-working aging population with CLBP.

Indirect Cost

Determining the indirect cost (often referred to as societal cost) of LBP is a resource intense process, and thus there is a limited amount of studies in this domain. Furthermore, indirect cost can vary widely depending on the specific cost variables assessed and which methods are used to determine productivity losses. The latter of which is the main driver of indirect cost. The two main methods used for determining productivity losses are (1) the human-capital method which takes the patient's perspective and tallies loss based on every hour that a individual does not work over the period that they may be eligible to work and (2) the friction-cost method which takes the payer/employer's perspective and only tallies loss based on those hours not worked until another employee takes over the patient's work (Pike and Grosse 2018). The human-capital approach tends to result in estimating significantly greater losses (i.e., higher indirect cost); thus the friction method has become the preferred method of many health economist and countries. Both methods have their merits and limitation, such that combination of methods may be desirable depending on the perspective taken. For example, for a 40-year-old patient who never returns to any form of work, the impact from the perspective of the employer may be relatively small compared to the impact on the disability insurer that has to pay that patient for the next 25 years until retirement age.

Regardless of method used to determine productivity losses, indirect costs are typically responsible for the majority of cost attributed to LBP. In an international review of national cost of illness studies by Dagenais et al. (2008), eight studies looked at both direct and indirect cost. With the exception of one study, indirect costs were responsible for 55–97% of estimated total national cost associated with back pain. Three of the reviewed studies by Dagenais et al. (2008) used both the human capital and friction methods for determining productivity losses. The friction method yielded estimates that were, on average, 56% lower than the human capital approach. In addition to these methodological differences, it is critical to understand whether the estimated cost per patient is being applied to all LBP patients within a representative sample of the population or a specific subgroup recruited from speciality pain or surgical clinics. In this scenario, the cost per patient (both direct and indirect) will likely be grossly different and is not interchangeable between different subpopulations. Consequently, when interpreting cost of illness studies, it is critical that the reader understand the LBP population being studied and the limitations associated with the methods and costing-data sources being used to determine indirect cost. For example, in a more specific CLBP subgroup (discography confirmed discogenic CLBP) of patients referred to four pain clinics in the Netherlands, Geurts et al. (2018) reported total societal cost of €7911.95 per patient (51% direct and 49% indirect cost) using the friction method and €18,940.58 per patient (22% direct and 78% indirect costs) when using the human capital approach. In this example, the human capital approach attributed more than double the cost per patient. Regardless of methods, the cost per patient is this study would be significantly higher than that of a LBP patient who was being managed only in a primary care setting. Thus, it would be erroneous to assign the cost per patient from this study to a different LBP subpopulation or to all LBP patients.

Direct Cost (Healthcare Expenditures)

Relative to older studies reported in the 2008 review by Dagenais et al. (2008), a more contemporaneous study by Dieleman et al. (2016) reports that the US spending for back and neck pain health services continues to increase annually and was estimated to be $87.6 billion in 2013.

Back and neck pain were ranked third, behind diabetes ($101.4 billion) and ischemic heart disease ($88.1 billion), out of all health conditions. The proportion of spending on ambulatory care, emergency care, and pharmaceuticals for back and neck pain was 60.5%, 4.2%, and 4.1%, respectively. Comparatively, the proportion of spending on ambulatory care, emergency care, and pharmaceuticals for diabetes was 23.5%, 0.4%, and 57.6%, respectively. This clearly demonstrates the differential impact on the healthcare system of a predominantly biomedical condition such as diabetes and predominantly non-biomedical condition such as back pain. To gain insight into some of the specific differences in medical expenditure among CLBP patients, Gore et al. (2012) compared a total of 101,294 patients with CLBP to a 1:1 age-, sex-, and region-matched control cohort CLBP. The authors used settled medical and pharmaceutical claims data from more than 98 commercially managed healthcare plans throughout the United States which represented a broader insured adult population. Relative to controls, CLBP patients had a greater number of medical comorbidities, including higher rates of depression (13.0% vs. 6.1%), anxiety (8.0% vs. 3.4%), and sleep disorders (10.0% vs. 3.4%). As expected, patients with CLBP also were more likely to be on opioids (37.0% vs. 14.8) or other analgesic such as nonsteroidal anti-inflammatory drugs (26.2% vs. 9.6%). The study also reported significantly higher estimated total direct medical costs for CLBP patients ($8386 ± $17,507) compared to those without CLBP ($3607 ± $10,845). One notable driver of cost was the almost three times difference in medical expenditures for outpatient investigations (e.g., imaging) in CLBP compared to the control group ($ 3481.65 vs. $1297.47). Overutilization of diagnostic imaging is perhaps the most targeted area of non-guideline concordant care in LBP, particularly the use of more costly imaging such as magnetic resonance imaging (MRI). Inappropriate MRI utilization is not simply a matter of an unnecessary test. The high likelihood of false-positive findings has been shown to lead to a cascade of further investigations and an increased relative risk of invasive treatments including surgery (Webster et al. 2014). Reduction of imaging for LBP is a very prominent part of the global movement Choosing Wisely. However, as noted above, the biomedical approach to LBP is ingrained into patients and providers, and change of behavior in this area has proven to be very difficult. This is highlighted by a 2017 study by Hong et al. (2017) where the authors only found a 4% relative reduction in low-value imaging for LBP 2.5 years into the Choosing Wisely campaign in the United States. Limited studies assessing the impact of alternative models of care using more active care paths and interprofessional delivery have demonstrated greater potential for reduction of unnecessary imaging and associated cost avoidance (Kim et al. 2011; Rampersaud et al. 2016).

As noted at the beginning of this chapter, LBP is the number one cause of years lived with disability (YLDs) with 77% of the YLDs accounted for by approximately one in four of prevalent cases (Global Burden of Disease 2015 DALYs and HALE Collaborators 2015). Although speculative, the evidence presented in this chapter on the psychological and social impact of CLBP would suggest these one and four prevalent cases are likely those with significant psychosocial (including central sensitization) drivers of persistent pain and disability. From a socioeconomic perspective, this subgroup of patients with a disproportionately greater disability burden is also going to be associated with higher medical expenditures and lost productivity (Hartvigsen et al. 2018). In a study by Luo et al. (2004), the authors performed an analysis of the 1998 Medical Expenditure Panel Survey and reported that 25% of patients with LBP were responsible for at least 75% of the healthcare expenditures in those with LBP. Similarly, Katz (2006) noted that 5% of workers (e.g., mostly those that have been off work for more than 1 year and are very unlikely to return to work) are responsible for the 75% of the loss in productivity cost associated to LBP. There is no current national-level evidence to suggest that these findings are not applicable today. In my opinion, although these individual findings have not been comprehensively assessed in the same study, their overarching context provides a socioeconomic rationale for strategies aimed at prevention or mitigation of CLBP. The early identification of those at risk for persistent pain and disability and implementation of early mitigation strategies to address the psychosocial mediators of persistent pain and disability are clearly the way forward.

Overview of Assessment and Management

A detailed description of the assessment and management of the psychosocial and socioeconomic factors associated with CLBP is not the intended scope of this chapter; however, a brief overview is necessary to provide the clinical implications of the issues outlined in this chapter.

Assessment

A variety of assessment tools as well as integrated model of cares are being developed around the world to address the growing burden of LBP (Foster et al. 2018; Rampersaud et al. 2016). It is clear that the psychosocial impact of CLBP is a principle driver of both patient and societal burden. The paradigm shift in LBP care necessitates primary consideration of psychosocial factors and approaches to care, rather than these being afterthoughts following failure of a primarily biomedical framework. Often, these issues may become ingrained and much more difficult and costly to manage. Consequently, early assessment for psychosocial factors in LBP patients is recommended as part of routine primary care (Foster et al. 2018). Furthermore, the impact of psychosocial factors is not static and thus necessitates prescribed follow-up and reassessment for these barriers to recovery. Incorporating this change enables the move away from a one-size-fits-all approach to a stratified care approach that has been greatly influenced by the literature associated with the use of the Keele STarT Back screening tool in primary care (Hill et al. 2011). The STarT Back screening tool (https://www.keele.ac.uk/sbst/startbacktool/) is a prognostic questionnaire for patients with LBP that aims to identify risk of developing persistent disabling LBP (Hill et al. 2011). It categorizes respondents as low, medium, or high risk of persistent pain/disability (i.e., chronicity) and aims to match treatment to each risk subgroup. Practical tools for primary care are required to enable efficient biopsychosocial assessment and reassessment of LBP in the acute and chronic (persistent or recurrent) scenarios.

In collaboration with interprofessional knowledge experts, front-line primary care providers, and funding from the Ontario Ministry of Health, we have developed the Clinically Organized Relevant Exam (CORE) Tool for the Low Back Pain Toolkit for Primary Care Providers (Alleyne et al. 2016). The CORE Back Tool (https://cep.health/clinical-products/low-back-pain/) provides a primary care toolkit that is evidence informed and interactive and provides a management matrix for early stratified care. It starts with six principle screening questions that allow identification of the biopsychosocial components of a patient presenting with LBP. A response-dependent stepwise progression to more detailed questions and recommended validated tools for more in-depth assessments are also provided. We have had excellent frontline uptake of the CORE Back Tool, and it is now integrated into the medical school curriculum at the University of Toronto. If psychosocial concerns are identified (primary lead question: Is there anything you *can not* do now that you could do before the onset of your low back pain?), then the user is directed to assess for yellow flags and if positive use of a validated prognostic tool such as the STarT Back which has specific questions regarding fear, anxiety, catastrophizing, and low mood (Hill et al. 2011). In addition, I would recommend a brief assessment (Chiarotto et al. 2016) of self-efficacy using the Pain Self-Efficacy Questionnaire – Short Form (PSEQ-4). The PSEQ asks simple questions of patient's confidence in dealing with their LBP such as "I can cope with my pain in most situations" or "I can live a normal lifestyle." Obtaining a basic understanding of a patient's ability to cope or not to cope with a given situation will more broadly help guide the need for psychological as well as social supports as needed. More intensive psychological assessment and treatment is not within the scope of this chapter and is not typically in the scope of practice of many primary care physicians or medical specialists such as rheumatologist or surgeons that deal with spinal conditions. However, identification of psychosocial issues and referral to appropriate assessment and management should be the responsibility of all practitioners that deal with LBP.

Management

The recently published Lancet series on LBP reflects the paradigm change in messaging and first-line management recommendations for both acute (<6 weeks) and CLBP (Foster et al. 2018). The prioritization of the Lancet Low Back Pain Series Working Group recommendations have only recently begun to surface in clinical practice guidelines. For both acute LBP and CLBP, advice to remain active and education are first-line recommendations. For CLBP both exercise therapy and cognitive behavioral therapy (CBT) are also recommended as first line treatments that should be considered for routine use. Additionally exercise therapy and CBT should also be considered for limited use in selected patients with acute LBP. These recommendations are in line with the need for early assessment of the complex biopsychosocial factors that may negatively impact recovery of LBP patients at any time period from the onset of symptoms. All other treatments such as spinal manipulation, massage, acupuncture, medication, injections, and surgery are to be considered second-line or adjunctive and delivered in a limited fashion in highly selected patients. The goal of these secondary or adjunctive treatments should be to enable functional activities (modified as needed) and optimization of non-pharmaceutical treatment options whenever possible.

For those with or identified as at risk for psychosocial barriers to recovery, CBT is the most commonly recommended treatment (Foster et al. 2018). CBT is a structured, time-limited, problem-focused, and goal-oriented form of psychotherapy. As it pertains to LBP, it has been shown that changing an individual's thoughts about pain and the associated negative emotions or beliefs can change not only how that individual's mind responds to pain but also their body. CBT can be effectively delivered in varying degrees of intensity, by a variety of different types of practitioners (not just psychologist), and in a variety of setting such as one-on-one, group, and even virtually (Vitoula et al. 2018; Bostick 2017). Fundamentally, the goals of CBT are to recognize the negative feelings, thoughts, and behaviors that occur as a result of LBP and use of goal-oriented techniques to incrementally transform negative thoughts (cognitive part) and behaviors (behavioral part) to positive ones that improve an individual's ability to manage their pain (i.e., improve pain self-efficacy) and become more active and engage in healthy behaviors that ultimately reduce their pain. Other adjunctive treatments may be required on a case-by-case basis including medical management of more profound psychological dysfunction (e.g., major depression) to address specific issues that exist or may arise. Just as a one-size-fits-all biomedical approach to LBP does not work, a one-size-fits-all psychosocial approach will also fail. For patients who have not responded to recommended first-line treatments, with ongoing significant pain, functional disability, or psychosocial dysfunction, multidisciplinary rehabilitation programs that individualize and coordinate different types of treatment (e.g., pain management, exercise, and CBT) have been shown to be more effective than standard treatments for pain, disability, and return to work (Kamper et al. 2015). However, it must be noted that multidisciplinary rehabilitation programs can be costly, time-consuming, and resource intensive and may not always be accessible to vulnerable populations (Salathé et al. 2018). Unfortunately, significant changes in policy, system-level clinical pathways, available resources (including first-line management of psychosocial issues such as improvement of self-efficacy), and the mindset of frontline clinicians will be required to see meaningful reduction in the increasing individual and socioeconomic impact of LBP (Foster et al. 2014, 2018).

References

Alleyne J, Rampersaud R, Rogers R, Hall H (2016) CORE Back Tool 2016: new and improved. J Curr Clin Care 6(2):21–29

Bandura A (1977) Self-efficacy: toward a unifying theory of behavioral change. Psychol Rev 84:191–215

Bostick GP (2017) Effectiveness of psychological interventions delivered by non-psychologists on low back pain and disability: a qualitative systematic review. Spine J 17(11):1722–1728

Brown A, Angus D, Chen S, Tang Z, Milne S, Pfaff J, Li H, Mensinkai S (2005) Costs and outcomes of chiropractic treatment for low back pain. Technology report no 56. Canadian Coordinating Office for Health Technology Assessment, Ottawa. https://www.cadth.ca/costs-and-outcomes-chiropractic-treatment-low-back-pain-0. Accessed 19 May 2019

Bunzli S, Watkins R, Smith A, Schütze R, O'Sullivan P (2013) Lives on hold: a qualitative synthesis exploring the experience of chronic low-back pain. Clin J Pain 29(10):907–916

Canizares M, Rampersaud YR, Badley EM (2019) The course of back pain in the Canadian population: trajectories, predictors, and outcomes. Arthritis Care Res (Hoboken). 2019 Jan 14. https://doi.org/10.1002/acr.23811. [Epub ahead of print]

Chiarotto A, Vanti C, Cedraschi C, Ferrari S, de Lima E, Sà Resende F, Ostelo RW, Pillastrini P (2016) Responsiveness and minimal important change of the pain self-efficacy questionnaire and short forms in patients with chronic low back pain. J Pain 17(6):707–718

Chou R, Shekelle P (2010) Will this patient develop persistent disabling low back pain? J Am Med Assoc 303(13):1295–1302

Dagenais S, Caro J, Haldeman S (2008) A systematic review of low back pain cost of illness studies in the United States and internationally. Spine J 8(1):8–20

Delpech JC, Madore C, Nadjar A, Joffre C, Wohleb ES, Layé S (2015) Microglia in neuronal plasticity: influence of stress. Neuropharmacology 96(Pt A):19–28

Deyo RA, Dworkin SF, Amtmann D, Andersson G, Borenstein D, Carragee E, Carrino J, Chou R, Cook K, DeLitto A, Goertz C, Khalsa P, Loeser J, Mackey S, Panagis J, Rainville J, Tosteson T, Turk D, Von Korff M, Weiner DK (2014) Report of the NIH Task Force on research standards for chronic low back pain. J Pain 15(6):569–585

Dieleman JL, Baral R, Birger M, Bui AL, Bulchis A, Chapin A, Hamavid H, Horst C, Johnson EK, Joseph J, Lavado R, Lomsadze L, Reynolds A, Squires E, Campbell M, DeCenso B, Dicker D, Flaxman AD, Gabert R, Highfill T, Naghavi M, Nightingale N, Templin T, Tobias MI, Vos T, Murray CJ (2016) US spending on personal health care and public health, 1996–2013. JAMA 316(24):2627–2646

Foster NE, Thomas E, Bishop A, Dunn KM, Main C (2010) Distinctiveness of psychological obstacles to recovery in low back pain patients in primary care. Pain 148(3):398–406

Foster NE, Mullis R, Hill JC, Lewis M, Whitehurst DG, Doyle C, Konstantinou K, Main C, Somerville S, Sowden G, Wathall S, Young J, Hay EM, IMPaCT Back Study Team (2014) Effect of stratified care for low back pain in family practice (IMPaCT Back): a prospective population-based sequential comparison. Ann Fam Med 12(2):102–111

Foster NE, Anema JR, Cherkin D, Chou R, Cohen SP, Gross DP, Ferreira PH, Fritz JM, Koes BW, Peul W, Turner JA, Maher CG, Lancet Low Back Pain Series Working Group (2018) Prevention and treatment of low back pain: evidence, challenges, and promising directions. Lancet 391(10137):2368–2383

Froud R, Patterson S, Eldridge S, Seale C, Pincus T, Rajendran D, Fossum C, Underwood M (2014) A systematic review and meta-synthesis of the impact of low back pain on people's lives. BMC Musculoskelet Disord 15:50

Geurts JW, Willems PC, Kallewaard JW, van Kleef M, Dirksen C (2018) The impact of chronic discogenic low back pain: costs and patients' burden. Pain Res Manag 2018:4696180

Global Burden of Disease 2015 DALYs and HALE Collaborators (2015) Global, regional, and national disability-adjusted life years (DALYs) for 306 diseases and injuries and healthy life expectancy (HALE) for 188 countries, 1990–2013: quantifying the epidemiological transition. Lancet 386:2145–2191

Global Burden of Disease, Injury Incidence, Prevalence Collaborators (2016) Global, regional, and national incidence, prevalence, and years lived with disability for 310 diseases and injuries, 1990–2015: a systematic analysis for the global burden of disease study 2015. Lancet 388:1545–1602

Gore M, Sadosky A, Stacey BR, Tai KS, Leslie D (2012) The burden of chronic low back pain: clinical comorbidities, treatment patterns, and health care costs in usual care settings. Spine (Phila Pa 1976) 37(11):E668–E677

Gross DP, Ferrari R, Russell AS, Battié MC, Schopflocher D, Hu RW, Waddell G, Buchbinder R (2006) A population-based survey of back pain beliefs in Canada. Spine (Phila Pa 1976) 31(18):2142–2145

Hartvigsen J, Hancock MJ, Kongsted A, Louw Q, Ferreira ML, Genevay S, Hoy D, Karppinen J, Pransky G, Sieper J, Smeets RJ, Underwood M, Lancet Low Back Pain Series Working Group (2018) What low back pain is and why we need to pay attention. Lancet 391(10137):2356–2367

Hill JC, Whitehurst DG, Lewis M, Bryan S, Dunn KM, Foster NE, Konstantinou K, Main CJ, Mason E, Somerville S, Sowden G, Vohora K, Hay EM (2011) Comparison of stratified primary care management for low back pain with current best practice (STarT Back): a randomised controlled trial. Lancet 378(9802):1560–1571

Hong AS, Ross-Degnan D, Zhang F, Wharam JF (2017) Small decline in low-value back imaging associated with the 'Choosing Wisely' campaign, 2012-14. Health Aff (Millwood) 36(4):671–679

Itz CJ, Geurts JW, van Kleef M, Nelemans P (2013) Clinical course of non-specific low back pain: a systematic review of prospective cohort studies set in primary care. Eur J Pain 17:5–15

Kamper SJ, Apeldoorn AT, Chiarotto A, Smeets RJ, Ostelo RW, Guzman J, van Tulder MW (2015) Multidisciplinary biopsychosocial rehabilitation for chronic low back pain: Cochrane systematic review and meta-analysis. BMJ 350:h444

Katz JN (2006) Lumbar disc disorders and low-back pain: socioeconomic factors and consequences. J Bone Joint Surg Am 88(Suppl 2):21–24

Kent PM, Keating JL (2008) Can we predict poor recovery from recent-onset nonspecific low back pain? A systematic review. Man Ther 13:12–28

Kim JS, Dong JZ, Brener S, Coyte PC, Rampersaud YR (2011) Cost-effectiveness analysis of a reduction in diagnostic imaging in degenerative spinal disorders. Healthc Policy 7(2):e105–e121

Kregel J, Meeus M, Malfliet A, Dolphens M, Danneels L, Nijs J, Cagnie B (2015) Structural and functional brain abnormalities in chronic low back pain: a systematic review. Semin Arthritis Rheum 45(2):229–237

Kregel J, Coppieters I, DePauw R, Malfliet A, Danneels L, Nijs J, Cagnie B, Meeus M (2017) Does conservative treatment change the brain in patients with chronic musculoskeletal pain? A systematic review. Pain Physician 20(3):139–154

Lee H, Hübscher M, Moseley GL, Kamper SJ, Traeger AC, Mansell G, McAuley JH (2015) How does pain lead to disability? A systematic review and meta-analysis of mediation studies in people with back and neck pain. Pain 156(6):988–997

Luo X, Pietrobon R, Sun SX, Liu GG, Hey L (2004) Estimates and patterns of direct health care expenditures among individuals with back pain in the United States. Spine 29(1):79–86

Main CJ, Foster N, Buchbinder R (2010) How important are back pain beliefs and expectations for satisfactory recovery from back pain? Best Pract Res Clin Rheumatol 24(2):205–217

McCracken LM, Vowles KE (2008) A prospective analysis of acceptance of pain and values-based action in patients with chronic pain. Health Psychol 27(2):215–220

Nicholas MK, Linton SJ, Watson PJ, Main C (2011) Early identification and management of psychological risk factors ("yellow flags") in patients with low back pain: a reappraisal. Phys Ther 91(5):737–753

Nijs J, Clark J, Malfliet A, Ickmans K, Voogt L, Don S, den Bandt H, Goubert D, Kregel J, Coppieters I, Dankaerts W (2017) In the spine or in the brain? Recent advances in pain neuroscience applied in the intervention for low back pain. Clin Exp Rheumatol 35 Suppl 107(5):108–115

Pike J, Grosse SD (2018) Friction cost estimates of productivity costs in cost-of-illness studies in comparison with human capital estimates: a review. Appl Health Econ Health Policy 16(6):765–778

Pincus T, McCracken LM (2013) Psychological factors and treatment opportunities in low back pain. Best Pract Res Clin Rheumatol 27(5):625–635

Ramond A, Bouton C, Richard I, Roquelaure Y, Baufreton C, Legrand E, Huez JF (2011) Psychosocial risk factors for chronic low back pain in primary care – a systematic review. Fam Pract 28(1):12–21

Rampersaud R, Bidos A, Schultz S, Fanti C, Young B, Drew B, Puskas D, Henry D (2016) Ontario's Interprofessional Spine Assessment and Education Clinics (ISAEC): patient, provider and system impact of an integrated model of care for the management of LBP. Can J Surg 59(3 Suppl 2):S39

Salathé CR, Melloh M, Crawford R, Scherrer S, Boos N, Elfering A et al (2018) Treatment efficacy, clinical utility, and cost-effectiveness of multidisciplinary biopsychosocial rehabilitation treatments for persistent low back pain: a systematic review. Glob Spine J 8(8):872–886

Shmagel A, Foley R, Ibrahim H (2016) Epidemiology of chronic low back pain in US adults: national health and nutrition examination survey 2009–2010. Arthritis Care Res 68:1688–1694

Snelgrove S, Liossi (2013) Living with chronic low back pain: a metasynthesis of qualitative research. Chronic Illn 9(4):283–301

Vitoula K, Venneri A, Varrassi G, Paladini A, Sykioti P, Adewusi J, Zis P (2018) Behavioral therapy approaches for the management of low back pain: an up-to-date systematic review. Pain Ther 7(1):1–12

Walker BF, Muller R, Grant WD (2004) Low back pain in Australian adults. Health provider utilization and care seeking. J Manip Physiol Ther 27(5):327–335

Webster BS, Choi Y, Bauer AZ, Cifuentes M, Pransky G (2014) The cascade of medical services and associated longitudinal costs due to nonadherent magnetic resonance imaging for low back pain. Spine (Phila Pa 1976) 39(17):1433–1440

Wilhelm M, Reiman M, Goode A, Richardson W, Brown C, Vaughn D, Cook C (2017) Psychological predictors of outcomes with lumbar spinal fusion: a systematic literature review. Physiother Res Int 22(2):e1648

Woolf CJ (2011) Central sensitization: implications for the diagnosis and treatment of pain. Pain 152(3 Suppl): S2–S15

Implant Material Bio-compatibility, Sensitivity, and Allergic Reactions

7

Nadim James Hallab, Lauryn Samelko, and Marco Caicedo

Contents

Introduction	129
Implant Debris Types: Particles and Ions	129
Particulate Debris	130
Metal Ions (Soluble Debris)	131
Local Tissue Effects of Wear and Corrosion	134
Systemic Effects of Wear and Corrosion	137
Metal Sensitivity Mechanism	140
Testing for Metal Sensitivity	141
Studies of Implant Related Metal Sensitivity Using Diagnostic Testing	143
Conclusions	144
References	144

Abstract

Generally biocompatibility to implant-debris governs long-term clinical performance. The following chapter covers: the kinds of implant-debris and the biologic responses to implant-debris. Implants produce debris from wear and corrosion that take the form of particles and ions. Particulate debris generally ranges from 0.01 to 100 s um. Wear rates of articulating bearing such as total hip arthroplasties generally range from 0.1 to 50mm^3/yr. Metal-on-metal total joint replacement components are well known to produce increases in circulating metal in people (>ten-fold that of people without implant, i.e., 2-5parts per billion-Cobalt and 1-3 ppb-Chromiun). Debris bioreactivity is both local and systemic. Local inflammation is primarily mediated by local immune cells called macrophages, which produce pro-inflammatory mediators/cytokines TNFα, IL-1β, IL-6, and PGE2. Although there are many concerns associated with systemic reactivity to implant-debris, to date well-established systemic reactivity has been limited to developed hypersensitivity/allergy reactions. Elevated amounts of in the remote organs such as the liver, spleen of patients with TJA and high levels of circulating metal have not (yet) been associated with remote toxicological or carcinogenic pathologies. Not all implant debris is similarly biocompatible/

N. J. Hallab (✉) · L. Samelko · M. Caicedo
Department of Orthopedic Surgery, Rush University Medical Center, Chicago, IL, USA
e-mail: nhallab@rush.edu;
nhallab@bioengineeringsolutions.com;
Lauryn_A_Samelko@rush.edu;
mc@orthopedicanalysis.com

© Springer Nature Switzerland AG 2021
B. C. Cheng (ed.), *Handbook of Spine Technology*,
https://doi.org/10.1007/978-3-319-44424-6_29

nonbiocompatible. Additionally, the amount of debris-induced-inflammation depends on both the person and amount/kind/size of implant debris. The inflammation and bone loss associated with debris necessitates continued surveillance by physicians to monitor patients/implants over time using traditional physical exams, x-rays, and when appropriate new biological assays such as the testing of metal content and individual biological response such as hypersensitivity metal-LTT assays.

Keywords

Orthopedic implant · Implant-debris · Biologic responses · Particles · Ions · Inflammation · Macrophages · Innate immune resposne · Adaptive immune response · Cytokines · Hypersensitivity · Allergy · Metal-LTT assays

Abbreviations

Al	Aluminum
ALVAL	Aseptic lymphocyte vasculitis associated lesion
Co	Cobalt
Cr	Chromium
Cr(PO$_4$)4H$_2$O	Chromium orthophosphate
DAMP	Danger associated molecular patterns
DTH	Delayed type hypersensitivity adaptive (lymphocyte mediated) immune response that occurs over days to weeks to years (vs. that of an immediate response).
Hypersensitivity	Adaptive immune responses typically local inflammation mediated by T-cells or B-cells where antigen presenting cells such as macrophages act as gate keepers.
IL-1b	Interleukin 1 almost exclusively produced by inflammasome reaction, such as occurs in a macrophage response to implant debris particles
IL-6	Interleukin 6
IL-18	Interleukin 18
IL-33	Interleukin 33
Inflammasome	Key molecular components of a pro-inflammatory pathway that reacts to danger signals (not pathogens) that are produced when cells are damaged, typically composed of multiprotein oligomers consisting of caspase 1, PYCARD, NALP, and sometimes caspase 5 (also known as caspase 11 or ICH-3).
LALLS	Low angle laser light scattering
metal-LTT	Metal-lymphocyte transformation test (proliferation assay) used as a human diagnostic test for delayed type hypersensitivity responses to implant metals
NALP3/ASC	Inflammasome complex of proteins
PAMP	Pathogen associated molecular pattern
PGE$_2$	Prostaglandin E2
PMMA	Polymethylmethacrylate
ppb	Parts per billion (ng/mL or ug/L)
PTFE	Teflon (polytetraflouroethylene)
RANKL	Receptor activator of nuclear factor Kappa Beta ligand
ROS	Reactive oxygen species
SEM	Scanning electron microscopy
TEM	Transmission electron microscopy
THA	Total hip arthroplasty
Ti	Titanium
TJA	Total joint arthroplasty
TJR	Total joint replacement

TNF-a	Tumor necrosis factor – alpha
UHMWPE	Ultra high molecular weight polyethylene
V	Vanadium

Introduction

Implant debris and not the implant itself causes slow progressive local inflammation that limits the long term performance of over one million total joint arthroplasties implanted each year in the USA (Charnley 1979, 1970). The direct costs of this slow progressive Adverse Reactivity to Implant Debris (ARID) is approximately $20 billion in the USA per year and is expected to double over the next 10 years (Kurtz et al. 2007a, b, 2009). One of the most important human costs of this bio-implant failure is the increased incidence of death during revision orthopedic surgery which is as high as 13% in people older >75–80 years of age while it is <1% in patients <70 years of age. Biocompatibility mediated implant failures also have elevated complication rates associated with re-operation, with a >20% chance of post-operative dislocation (vs <1% in patients <75 years of age) (Radcliffe et al. 1999). Some designs of orthopedic implants release more bioreactive debris (i.e., metal particles and ions) that result in extraordinarily high failure rates, with levels of failure reported as high as 5% at 6 years post-op, such as some past metal-on-metal total hip arthroplasties designs as well as some types of highly modular implants (i.e., several components that press fit together) (Cooper et al. 2013; Jacobs and Hallab 2006; Korovessis et al. 2006; Milosev et al. 2006). The mechanism of implant debris induced inflammation is best known as an activator of local innate immune responses, i.e., monocytes/macrophages activate NFκβ and secretion of potent inflammatory cytokines such as IL-1β, TNFα, IL-6, and IL-8 (Catelas et al. 1999, 2003; Hallab et al. 2003a; Kaufman et al. 2008; Sethi et al. 2003; Trindade et al. 2001) resulting in localized inflammation (Kaufman et al. 2008; Lewis et al. 2003).

Over the long term all accumulating implant debris and the subsequent slow progressive inflammation results in bone loss and loss of implant fixation (Willert and Semlitsch 1977), termed "aseptic osteolysis," and results in pain and premature loosening of orthopedic implants (Archibeck et al. 2001; Arora et al. 2003a; Jacobs et al. 2001). Clinically, aseptic osteolysis (noninfection related bone loss) generally only refers to measureable bone loss as determined on an x-ray (Fig. 1). It is the particulate and soluble degradation products of orthopedic biomaterials (generated by wear and corrosion) that mediate these Adverse Reactivity to Implant Debris (ARID) effects. Debris may be present as particulate material (i.e., as small colloidal nanometer size complexes or larger >0.3um particles), or soluble products such as free metallic ions which can then react with their proteinaceous and cellular environment. Implant particulate debris can have large specific surface areas by virtue of their small size and large number and thus have a large format for interaction with the surroundings. This chapter will focus on orthopedic implant degradation product bio-compatibility, and ensuing local and systemic consequences of this debris including local inflammatory tissue reactivity and sensitivity and allergic reactions, respectively.

Implant Debris Types: Particles and Ions

All orthopedic implants produce debris of two basic types: particles or soluble debris (e.g., metal ions). The biologic consequences of particles and soluble debris blurs as the size of particles decreases into the nanometer range and become "effectively soluble." Particulate debris (metal, ceramic, or polymers) is generally in the range of 40 nm to 1 mm in size, while so-called common forms of "soluble debris" is currently limited to metal and are quickly bound to serum proteins upon release (such as albumin).

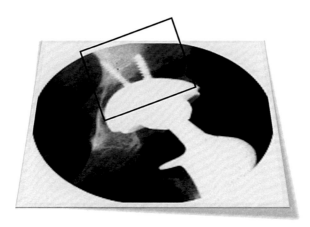

Fig. 1 Peri-implant aseptic osteolysis above the acetabular cup of a metal-on-polymer bearing total hip replacement. Inset shows a granuloma surrounding acetabular fixation screw, which is a common site for bone resorption due to the ease with which particles can migrate and cause inflammatory soft tissue and osteolysis. (Courtesy of Bio-Engineering Solutions Inc.)

Particulate Debris

Different types of orthopedic implants produce different types and amounts of wear debris, with different sizes and shapes of that are generally implant design and material specific. For example, total joint implants with "hard-on-hard" articulating surfaces such as metal-on-metal total hips arthroplasty implants generally produce smaller sized fairly round (submicron), debris. More common metal-on-polymer or ceramic-on-polymer THA bearings produce larger (micron sized) polymeric debris (Fig. 2) that fall into the range of 0.2um to 1um, with little metallic debris. Other sources of metal debris include corrosion and wear at metal-to-metal connections between modular components (Campbell et al. 1995; Jacobs et al. 1994a; Maloney et al. 1993). Highly cross-linked ultrahigh molecular weight polyethylene (X-UMWPE) used in current models of hip replacements provides less war than previous generations of UHMWPE; however, the particle produces are generally smaller (e.g., 0.1microns in size) compared to 0.8–2um of previous generations of UHMWPE (Catelas et al. 2004; Scott et al. 2005). Articulating surfaces comprised of metal and ceramic bearings produce particles that can be an order of magnitude smaller than polymeric particles (at approximately <0.05um in diameter, i.e., in the nanometer range).

Histological analysis of peri-implant tissues has identified different types and sizes of particles (Choma et al. 2009; Jacobs et al. 1998a; Punt et al. 2008, 2009; Urban et al. 1998, 2000; van Ooij et al. 2007). However, the sizes of debris in tissues vary dramatically from that identified using simulators and analysis of synovial fluids and tissues. Metal corrosion based stainless steel debris has been found as closely packed, plate-like particle aggregates mostly at steel screw-plate junctions containing particles of chromium compound ranging in size from 0.5 to 5.0 microns (Urban et al. 1996). Similarly large, cobalt alloy corrosion debris has been shown in tissues to be made of a chromium-phosphate $(Cr(PO_4)4H_2O)$ hydrate rich material termed "orthophosphate" and ranges in size from <1um to >500 micrometers (Urban et al. 1996, 1997).

Particle Characterization: Differently than basic histological analysis, more specific means of characterizing implant debris particles include Scanning Electron Microscopy (SEM) or Transmission Electron Microscopy (TEM) techniques. Both of these characterize particles by counting and sizing particles on a number of high, medium, and low power microscopy fields. These techniques are employed for digested tissues and simulator fluids and synovial fluid analysis, after the particulate debris has been isolated and dried on a membrane/mounting media. Because the particles

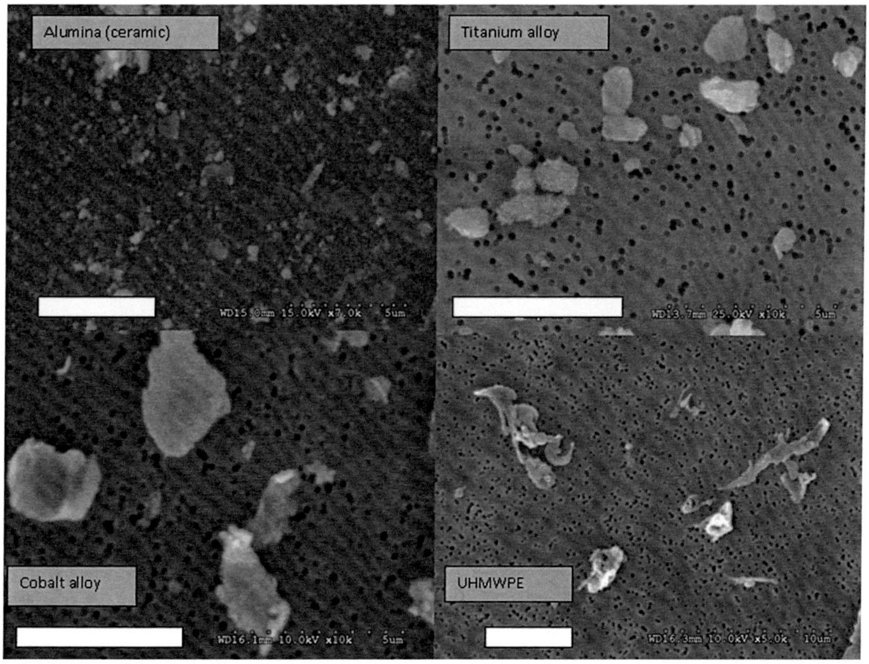

Fig. 2 Implant debris from metal (Cobalt alloy and Titanium) and ceramic (alumina) debris are more rounded in comparison to polymeric (UHMWPE) debris which is more elongated in shape. Note: Bar = 5um. (Courtesy of BioEngineering Solutions Inc.)

observed in the high power fields are over represented when scaled up to the total, these methods have inappropriately biased our understanding that the majority of the wear (mass loss) from an implant is comprised of particles in the nanometer to submicron range. That is while most of the particles identified on a counting (number-based) analysis are in the small ranges (<1um), they do not typically make up the size of debris that is responsible for the majority of the mass loss, i.e., while billions of small particles only add up to 0.01 mg of implant debris it only takes 100's to 1000's of larger particles to equal >10 mg of implant debris. This biased understanding stems from the limited number of particles in tissues and the relatively low numbers of particles (e.g., 100's–1000's) that are counted using image based analysis techniques such as SEM. Other types of analytical techniques, such as low angle laser light scattering (LALLS), have the ability to sample millions to billions of particles, as they pass in front of a laser detection system where the one-in-a-million large particle can be detected and thus provide a more accurate distribution of the total debris.

The ability to comprehensively characterize implant debris is critical to the assessment of consequent biological responses and weigh the effects of new designs and bearing surfaces to older implants. The bias of SEM techniques those of all "number-based" analysis where two very similar number based distributions can look very different when analyzed on a "volume-based" perspective (Fig. 3). Thus for an accurate and comprehensive evaluation of implant debris particulate, both a number and volume based analysis/distribution are required.

Metal Ions (Soluble Debris)

There is continuing clinical concern regarding metal released from orthopedic implant is the form of particles and ions. These ions immediately bind to serum proteins and disseminate into surrounding tissues, bloodstream, and remote organs. Normal

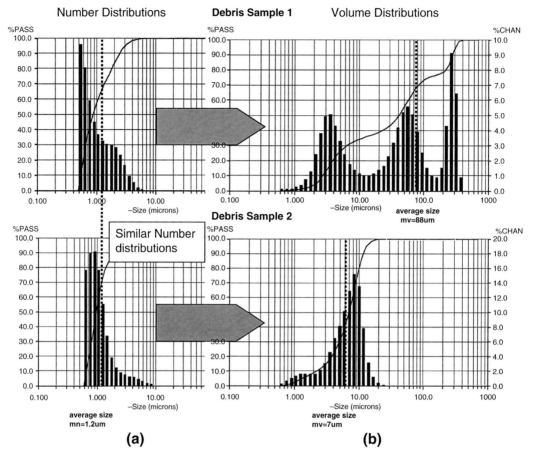

Fig. 3 LALLS analyses of two implant debris samples using a (**a**) volume and (**b**) number distributions demonstrate that similar number distributions and estimates of particle size can result from two very different sizes particles when analyzed using a volume distribution, which shows the size of the particle as a percent of the total volume. Note: The x-axis is particle diameter and the y-axis is (**a**) percentage of total number of particles in each size range and (**b**) the percentage of total mass in each size range. (Courtesy of BioEngineering Solutions Inc)

metal serum levels are generally <1part per billion, ppb (ng/mL): 1–10 ng/ml Al, 0.15 ng/ml Cr, <0.01 ng/ml V, 0.1–0.2 ng/ml Co, and <4.1 ng/ml Ti. Implants do release enough metal to increase these levels systemically following total joint arthroplasty (Table 1). Particles of metal that are released contribute to this increased metal because of the large surface areas available for corrosion (i.e., electrochemical dissolution) (Jacobs et al. 1998a; Urban et al. 1996, 1997, 1998, 2000).

Metal Ion Release: Metal ions released from orthopedic implants have been of concern for over 40 years. Increased levels of systemic circulating Co and Cr are detected following even successful total joint replacements with Co-alloy based components. The same is true of other metal alloy orthopedic implants, e.g., increased serum Ti and Cr concentrations can be found in some individuals with well-functioning Ti and/or Cr containing THR components (Table 1) (Dorr et al. 1990; Jacobs et al. 1994b, 1998a; Michel et al. 1984; Stulberg et al. 1994). Other metals associated with the surgery itself have also been reported where increases in Ni have been noted immediately following surgery, likely related to the use of stainless steel surgical instruments.

Although several factors affect systemic metal ion levels in TJA patients, the most important

7 Implant Material Bio-compatibility, Sensitivity, and Allergic Reactions

Table 1 Approximate concentrations of metal in human body fluids and in human tissue with and without total joint replacements. (Dorr et al. 1990; Jacobs et al. 1994b, 1998a; Michel et al. 1991; Stulberg et al. 1994)

Body fluids (ng/mL or ppb)		Ti	Al	V	Co	Cr	Mo	Ni
Serum	Normal	0.06	0.08	<0.02	0.003	0.001	*	0.007
	TJA	0.09	0.09	0.03	0.007	0.006	*	<0.16
Urine	Normal	<0.04	0.24	0.01	*	0.001	*	*
	TJA	0.07	0.24	<0.01	*	0.009	*	*
Synovial fluid	Normal	0.27	4.0	0.10	0.085	0.058	0.219	0.086
	TJA	11.5	24	1.2	10	7.4	0.604	0.55
Joint capsule	Normal	15.0	35	2.4	0.42	2.6	0.177	69
	TJA-F	399	47	29	14	64	4.65	100
Whole blood	Normal	0.35	0.48	0.12	0.002	0.058	0.009	0.078
	TJA	1.4	8.1	0.45	0.33	2.1	0.104	0.50
Body tissues (µg/g)								
Skeletal muscle	Normal	*	*	*	<12	<12	*	*
	TJA	*	*	*	160	570	*	*
Liver	Normal	100	890	14	120	<14	*	*
	TJA	560	680	22	15,200	1130	*	*
Lung	Normal	710	9830	26	*	*	*	*
	TJA	980	8740	23	*	*	*	*
Spleen	Normal	70	800	<9	30	10	*	*
	TJA	1280	1070	12	1600	180	*	*
Psuedocapsule	Normal	<65	120	<9	50	150	*	*
	TJA	39,400	460	121	5490	3820	*	*
Kidney	Normal	*	*	*	30	<40	*	*
	TJA	*	*	*	60	<40	*	*
Lymphatic Tissue	Normal	*	*	*	10	690	*	*
	TJA	*	*	*	390	690	*	*
Heart	Normal	*	*	*	30	30	*	*
	TJA	*	*	*	280	90	*	*

Normal: Subjects without any metallic prosthesis (not including dental)
TJA: Subjects with total joint arthroplasty
* = Data Not Available

factor is elevated metal implant degradation (wear and/or corrosion). Systemic titanium ion levels up to a hundred times higher than normal have been reported in cases of failed metal-backed patellar components where mechanical implant failures caused high wear such as a wearing through of the polymer liner in a THA and the more wear resistant Co alloy head bores into the titanium alloy acetablular cup. Surprisingly in these cases of excessive Ti-alloy wear and metal release, there was no reported increases in still serum or urine Al, serum or urine V levels, or which are other minor percentages of titanium alloy cups (6% Al and 4% V). Fretting corrosion, of modular implant components has been associated with elevations in serum Co and urine Cr (Jacobs et al. 1998a, b, 1999b). Despite significant increases in Co and Cr concentrations found in the heart, liver, kidney, spleen, and lymphatic tissue from orthopedic implant degradation (Table 1), the majority of metal debris remains local around and in the pseudocapsule that forms around a total joint implant and act much like a joint capsule (Jacobs et al. 1994).

Local Tissue Effects of Wear and Corrosion

The key determining factor of long-term implant performance is implant debris that can trigger a local inflammatory response that causes osteolysis and aseptic implant loosening. Bone homeostasis is dependent upon the intricate balance of bone formation and bone resorption powers which comprises the corresponding function of osteoblasts (bone building cells) vs. osteoclasts (bone resorbing cells) and osteocytes (bone mechanotransduction and signaling network cells). If implant debris induced inflammation causes disruption in bone homeostasis by mitigating osteoblastic bone formation and/or augmenting osteoclastic bone resorption, this will result in a net bone loss (i.e., osteolysis). This osteolysis near the bone-implant interface is the principal pathology associated with the localized effects of TJR degradation. This bone loss happens as a diffuse thinning of the cortical or as focal cyst-like lesions. The first materials to be associated with osteolytic lesions due to massive production amounts of implant debris were particulate polymethylmethacrylate (PMMA) bone cement and old acetabular cups made of PTFE (Teflon). This was based on histological studies showing implant debris associated with macrophages, giant cells, and a vascular granulation tissue. It is now well established that osteolysis in both well-fixed and loose uncemented implants results from the generation of particle debris from any material (Jacobs et al. 2001; Vermes et al. 2001a).

It was first described by Goldring et al. (1983) that the bone-implant interface in patients with loose total hip replacements is comparable to synovial-like membrane and bone resorbing factors such as PGE_2 and collagenase are produced by cells within the membrane. Total hip arthroplasty is more frequently associated with particle induced osteolysis than total knee arthroplasty, and this remains unclear why this is the case. However, it has been postulated that various biomechanical factors such as implant/bone mechanical loading environments, differential mechanisms of hip and knee wear, and differences in interfacial barriers to migration account for this apparent disparity.

All implant debris leads to subtle progressive inflammation that can ultimately result in implant failure. As to exactly how this occurs still remains somewhat contentious, however, increasing evidence continues to indicate that danger signaling by the innate immune system mediates implant debris induced inflammation, which is how the immune system in general detects and reacts to nonpathogen derived biologic stimuli (Caicedo et al. 2008, 2013a; Dostert et al. 2008; Hornung et al. 2008; Naganuma et al. 2016). It has been established over the past 40 years that implant debris induced inflammation is primarily driven by macrophage reactivity to sterile implant debris that results in up-regulation and activation of pro-inflammatory transcription factors (e.g., NFκB) that produce, amplify, and result in the secretion of inflammatory cytokines like IL-1β, TNFα, IL-6, and IL-8 (Jacobs et al. 2001) (Fig. 4). Prostaglandins (e.g., PGE_2) are also involved and mediate implant debris induced inflammation and osteolysis. IL-10 and IL-1Ra are key anti-inflammatory cytokines that act to lessen this inflammatory state induced by implant debris, but it remains less understood the degree which these anti-inflammatory cytokines can decrease the pathology of particle induced osteolysis. Additional factors involved with osteolysis include matrix metalloproteinases collagenase and stromelysin, which are enzymes that mediate the catabolism of the organic component of bone. Also, activated bone and immune cells can generate bone mediators known to play a role in stimulation of osteoclast differentiation and maturation, such as RANKL (also referred to as osteoclast differentiation factor).

Implant debris is sterile and relatively inert and does not have the prototypical molecular characteristics of a pathogen. Therefore, how does implant debris elicit an immune inflammatory response? More specific, how can extra- and intra-cellular mechanisms detect and react to sterile nonbiological material such as implant debris? For the past half century, this question had remained largely unknown. However, new discoveries and advancements in immunology have implicated the NLRP3 inflammasome danger signaling pathway to play a pivotal role in the detection and response to sterile nonbiological stimuli (Fig. 5) (Caicedo et al. 2010).

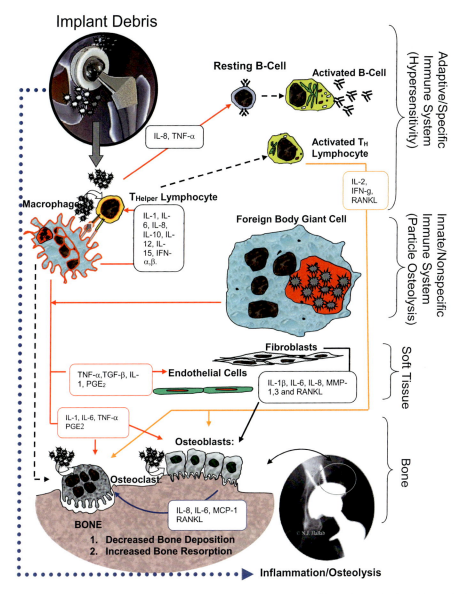

Fig. 4 Numerous cytokines from peri-implant cells reacting to implant debris can negatively affect bone turnover. IL-1, IL-6, and TNF-α are some of the most potent cytokines responsible for increasing bone loss and enhancing pro-inflammatory responses. (Picture courtesy of Bio-Engineering Solutions Inc)

The discovery of the inflammasome danger signaling pathway was pivotal since it was the first biological mechanism to explain how immune cells transduce sterile, nonpathogen derived stimuli (e.g., cell stress and necrosis) into an inflammatory response (Mariathasan et al. 2004; Mariathasan and Monack 2007). Additional nonbiological derived danger signals (e.g., DAMPs) that activate the inflammasome include cell damaging stimuli such as UV light, particulate adjuvants present in modern vaccines (Dostert et al. 2008; Hornung et al. 2008) and, as it turns out, orthopedic implant debris (Caicedo et al. 2008).

When particles activate the inflammasome pathway, immune cells subsequently release

Fig. 5 Metal-induced inflammasome activation occurs when soluble and/or particulate implant debris activate the Nalp3 inflammasome when chemicals inside intracellular compartments used to digest foreign material (such as phagosomal NADPH induced reactive oxygen species and/or Cathepsin B) leaks out of these compartments in an event called phagosomal destabilization. The inflammasome complex Nalp3-ASC then induces the activation of caspase-1, which in turn allows mature IL-1ß to be secreted. IL-1ß is a very potent pro-inflammatory cytokine that exerts an autocrine and paracrine effect inducing a broader more potent inflammatory response (e.g., activation of NFκβ pro-inflammatory responses). (Courtesy of BioEngineering Solutions Inc)

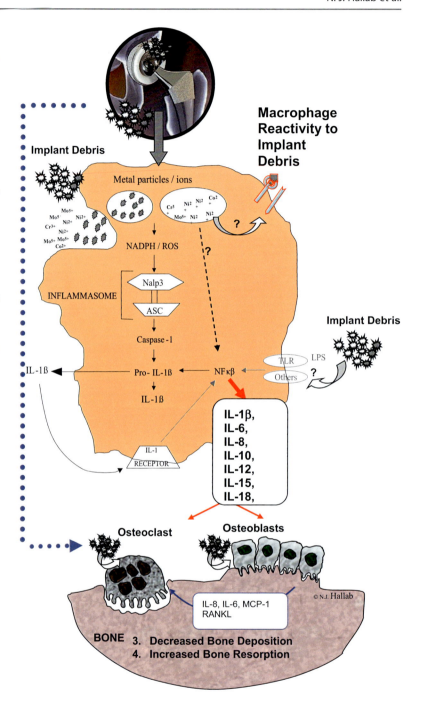

pro-inflammatory cytokines such as IL-1β, IL-18, IL-33, and a multitude of more. The sequence of events is as follows:

Implant Debris → Phagocytosis → Lysosomal damage and rupture of protease enzymes (e.g. Cathepsin-B) → ROS (reactive oxygen species) production → Inflammasome (NALP3/ASC) activation → Caspase-1 → Secretion of mature IL-1β (and other IL-1-family dependent cytokines) (Fig. 20).

More specifically, upon ingestion (phagocytosis) of sterile particles by immune cells (or other DAMPs such as asbestos and implant debris) will cause a degree of lysosomal destabilization. Consequently, lysosomal destabilization will result in the rupture and release of protease enzymes and of the acid rich extreme microenvironment within a lysosome into the cell cytosol, which are used within the lysosome compartment to breakdown ingested DAMPs (e.g., implant debris) and PAMPs. This lysosomal destabilization leads to an increase in NADPH (nicotinamide adenine dinucleotide phosphate-oxidase) and an associated increase in reactive oxygen species (ROS). Subsequently, the release of ROS species leads to the activation of the intracellular multiprotein "inflammasome" complex that is composed of NALP3 (NACHT-, LRR-, and pyrin domain-containing protein 3) in association with ASC (apoptosis-associated speck-like protein containing a CARD domain) (Mariathasan and Monack 2007; Petrilli et al. 2007). Activation of the inflammasome will result in Caspase-1 activation, which then converts cytokines such as pro-IL-1β and pro-IL-18 (and others) into their active mature form. In summary, this illustrates the general numerous steps involved in the activation of the inflammasome danger signaling pathway and the numerous new potential biological points of pharmacologically blocking this response to prevent or mitigate particle induced inflammatory responses and osteolysis.

Systemic Effects of Wear and Corrosion

To some extent, implant surfaces and the implant debris generated are continually releasing chemically active metal ions into the surrounding peri-implant tissues. The released metal ions will bind to serum proteins and may reside in local tissues and also be transported via the bloodstream and the lymphatics to remote organs. This is of concern since it is known the potential toxicity effects of these elements used in modern orthopedic implant alloys: titanium, aluminum, vanadium, cobalt, chromium, and nickel. Metal toxicity can happen by changing: (i) cell/tissue metabolism, (ii) host/parasite interactions, (iii) immunologic interactions, and (iv) by inducing chemical carcinogenesis (Beyersmann 1994; Britton 1996; Goering and Klaasen 1995; Hartwig 1998; Luckey and Venugopal 1979).

Essential trace metals include cobalt and chromium and are necessary for the homeostatic function of various enzyme reactions. However, these elements in excessive quantities can become highly toxic. Accordingly, excessive cobalt can result in heart problems (cardiomyopathy), increased red blood cells (polycythemia), decreased thyroid functions (hypothyroidism), and carcinogenesis, while excessive chromium has been associated to nephropathy, hypersensitivity, and carcinogenesis. Also, metals such as nickel can result in skin rashes (eczematous dermatitis), hypersensitivity reactions, and cancer, and excessive vanadium exposure has been associated to heart and kidney dysfunction, and hypertension and depressive psychosis. Aluminum toxicity can lead to renal failure and blood anemia, bone softening (osteomalacia), and neurological problems. It is important to note, however, that these metal toxicities are generally due to excessively elevated levels of the soluble forms of these elements and most likely do not pertain to the levels of metals released from implant degradation.

Currently, any associated metal toxicity related to metal release from orthopedic implant is conjectural since it has yet to be established the cause and effect of this specific association. It is very difficult, however, to discern any metal toxicity effects related to an implant given the types of health concerns typically associated with the elderly, as well as those expected to occur in any orthopedic patient population (Jacobs et al. 1999a).

Systemic Particle Distribution: It is not well understood as to what determines the amount of implant debris accumulation in remote organs. When the magnitude of particulate debris produced by an implant is augmented, there is a corresponding increase in both the local and systemic burden of implant debris. Mostly, systemic implant debris (located beyond the peri-implant tissue microenvironment) is in the submicron size range. Numerous cases have located metallic, ceramic, or polymeric wear debris from hip and knee prostheses in regional and pelvic lymph nodes along with the findings of gross dark staining by metallic debris, fibrosis (buildup of fibrous tissue), lymph node necrosis, and histiocytosis (abnormal function of tissue macrophages). Moreover, up to 70% of patients with total joint replacement components had metallic wear particles detected in their para-aortic lymph nodes. The consequences of this occurrence are not clear; however, prototypical immune inflammatory responses in lymph nodes to metallic and polymeric debris involve similar responses seen locally, which include activation of macrophages and associated production of cytokines.

Therefore, lymphatic transport is likely the main course for debris dissemination where particles are transported by perivascular channels as independent particles or as phagocytosed particles within macrophages. Disseminated particles within lymph nodes are primarily submicron in size; however, some metallic particles as large as 50 micrometers and polyethylene particles as large as 30 micrometers have also been detected. Additionally, these particles have been located within macrophages in the liver and spleen and in some instances, in nodules of inflammatory tissue granulomas throughout the organs. Typically, metallic particle size is nearly an order of magnitude less in the liver and spleen, than that in lymph nodes, suggesting there is an additional filtration point that occurs prior to particles culminating in those organs. This is not overly concerning since it is a common function of the cells of the liver, spleen, and lymph nodes to accumulate small quantities of a variety of foreign materials without evident clinical significance. However, nodules of inflammatory tissue (granulomas) or granulomatoid lesions in the liver and spleen can be induced by the accumulation of excessive particle debris. The degree of reaction to particles in the liver, spleen, and lymph nodes is probably modulated, as it is in other tissues by: (1) the dose of particles, (2) their rate of accumulation, (3) the period that they are present, and (4) the biologic reactivity of cells to these particles (size and materials composition). It is not unexpected that metallic particles in the liver or spleen are more common in patients with previously failed implants compared to patients with a primary well-functioning TJR.

It would be expected that in diseases which obstruct the continual lymph flow through lymph nodes, such as a metastatic tumor, or those that disrupt the general flow of circulation, such as chronic heart disease or diabetes, would result in reduced particle migration to remote organs, whereas other pathologies, like acute or chronic-active inflammation, likely augment particle migration (Jacobs et al. 1999a, 2001; Vermes et al. 2001b) via the recruitment of more immune cells to transport the debris away.

Hypersensitivity. In general terms, hypersensitivity responses to metal implants can be defined as an adaptive immune response that is mediated by T cells and typically causes a local inflammatory response around the implant. It is imperative to clarify that "hypersensitivity responses" have a wide range of intensity that can span from mild to severe and need not be on the severe end to be termed "hypersensitivity." Early implant failure ($<$7 years) that is caused by an exacerbated immune response to otherwise tolerable amounts of implant debris is likely caused and orchestrated by an adaptive immune response. This response is also often termed "metal-allergy," "implant-allergy," or "implant sensitivity." While soluble metals (i.e., metal ions) released from metal prostheses do not act as sensitizers alone, they are able to combine with self-proteins and form metal-protein complexes (haptens) that have the ability to activate the immune system. On the other hand, polymeric wear debris has not been implicated in allergic type immune responses due to its inability to properly degrade in vivo (Hallab et al. 2000a, b, 2001a, b). The most common metals regarded as sensitizers/allergens (metal haptens) include, but are not limited to beryllium, chromium, cobalt, nickel, tantalum, titanium, and vanadium. Nickel, cobalt, and chromium are the most common metal allergens reported in humans and nickel still constitutes 10–16% of medical grade stainless steel (Table 2). In general, the literature exhibits more case reports of hypersensitivity reactions associated with nickel-containing stainless steel and cobalt-alloy implants compared to Titanium-alloy devices (Burt et al. 1998; Cramers and Lucht 1977; Elves et al. 1975; Gordon et al.

7 Implant Material Bio-compatibility, Sensitivity, and Allergic Reactions

Table 2 Approximate weight percent of different metals within popular orthopedic alloys

Alloy	Ni	N	Co	Cr	Ti	Mo	Al	Fe	Mn	Cu	W	C	Si	V
Stainless steel (ASTM F138)	10–15.5	<0.5	*	17–19	*	2–4	*	61–68	*	<0.5	<2.0	<0.06	<1.0	*
CoCrMo alloys (ASTM F75)	<2.0	*	61–66	27–30	*	4.5–7.0	*	<1.5	<1.0	*	*	<0.35	<1.0	*
(ASTM F90)	9–11	*	46–51	19–20	*	*	*	<3.0	<2.5	*	14–16	<0.15	<1.0	*
Ti alloys														
CPTi (ASTM F67)	*	*	*	*	99		*	0.2–0.5	*	*	*	<0.1	*	*
Ti-6Al-4 V (ASTM F136)	*	*	*	*	89–91	*	5.5–6.5	*	*	*	*	<0.08	*	3.5–4.5
45TiNi	55	*	*	*	45	*	*	*	*	*	*	*	*	*
Zr alloy (97.5% Zr, 2.5% Nb)	*	*	*	*	*	*	*	*	*	*	*	*	*	*

Note: Alloy compositions are standardized by the American Society for Testing and Materials (ASTM vol. 13.01)
* Indicates less than 0.05%

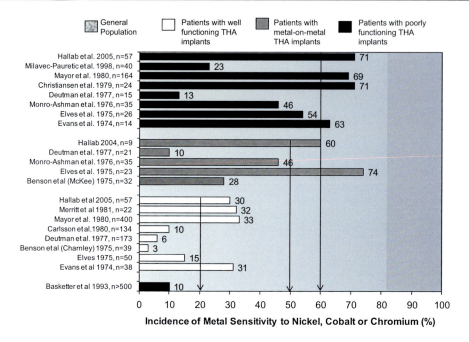

Fig. 6 A compilation of investigations showing the averaged percentage of metal sensitivity among the general population, people with well-functioning implants, people with metal-on-metal implants and people with failing implants (prior to getting them revised). Metal incidence rates include a positive response to allergy testing for nickel, cobalt, and/or chromium. All subjects were tested by means of a patch or metal-LTT (lymphocyte transformation test). (Courtesy of Orthopedic Analysis LLC)

1994; King et al. 1993; Merle et al. 1992; Rostoker et al. 1987; Thomas et al. 1987).

Incidence of Hypersensitivity Responses Among Patients with Metal Implants: People with well-function implants exhibit an incidence of hypersensitivity reactions (25%) twice as high as that of the general population (10%) (Fig. 6). Interestingly, the incidence of metal related hypersensitivity in people with poorly functioning metal prostheses (revision surgery candidates) or well-functioning metal-on-metal hip prostheses is 50–60% (Fig. 6). This higher incidence of metal hypersensitivity in cohorts of patients with metal prostheses has led to speculation that immune reactivity to metal implant components may play a role in implant loosening. Group studies performed over the last three decades have demonstrated a correlation between metal implants and metal sensitization (Hallab et al. 2001a), clearly concluding that metal sensitization can be an important causative factor to implant failure (Merritt and Rodrigo 1996; Rooker and Wilkinson 1980; Rostoker et al. 1987).

Therefore, metal sensitivity testing (metal-LTT) may be beneficial for people with a history of metal allergy before receiving a metal prosthesis. The significance of this line of research cannot be understated, as the use, durability, and performance expectations of metallic spinal implants continue to increase (Black 1996; Jacobs and Goodman 1996).

Metal Sensitivity Mechanism

Generally, metal sensitivity responses can be classified as: 1-Humoral immediate responses that can develop within minutes and are initiated by antibody-antigen complexes (Type I, II, III) and 2-cell-mediated delayed type hypersensitivity responses type IV, which may develop within hours to days (Hensten-Pettersen 1993; Kuby 1991). Immune responses to metal implant degradation products are almost exclusively classified as being delayed type hypersensitivity responses (DTH). This specific type of DTH response has been predominantly classified as a Th1 type of response, where helper T cells are characterized by the release of a unique signature set of cytokines that include interferon-γ (IFN-γ), tumor necrosis factor-α (TNF-α), interleukin-1 (IL-1), and interleukin-2 (IL-2). While this specific subset of cells are intended to detect and eradicate intracellular pathogens, they can also potentially induce autoimmune disorders (i.e., Rheumatoid arthritis, Lups, etc.) when mistakenly activated (Arora et al. 2003b; Hallab et al. 2008).

In this manner, activated and primed antigen presenting cells in combination with metal-activated T helper lymphocytes secrete a variety of pro-inflammatory cytokines that effect the recruitment and activation of innate immune cells (i.e., monocytes, macrophages, neutrophils) (Hallab et al. 2013). Some of these cytokines include, but are not limited to IFN-γ and TNF-β, which in turn induce pro-inflammatory physiological changes on local cells (i.e., endothelial cells) to aid the inflammatory response. The main characteristics of a DTH immune response are recruitment, recognition/activation, and migration inhibition of local immune cells (e.g., macrophages, T lymphocytes). Additionally, the release of potent pro-inflammatory cytokines like IL-1β from activated antigen presenting cells effect further recruitment and activation of T cells, which in turn activate additional macrophages exacerbating the immune response. Therefore, in certain types of DTH responses, including those associated with autoimmune diseases, there is a lack of self-regulation (off-switch) that can result in the perpetuation of the inflammatory response resulting in extensive tissue damage. Immunosuppression has been proposed as a strategy to mitigate the effects of the vicious pro-inflammatory cycle of DTH responses in these individuals in order to aid anti-inflammatory immune mechanisms to operate (Looney et al. 2006; Schwarz et al. 2000).

Testing for Metal Sensitivity

At present, there are two modalities accepted for human diagnostic testing for metal sensitivity: 1-patch testing (dermal testing) and 2-blood testing in vitro using a lymphocyte proliferation test (metal-LTT).

Dermal Skin Testing: Commercially available patch testing kits and protocols for the evaluation of metal induced hypersensitivity reactions have been used for over 40 years for purposes of orthopedic implants (Hensten-Pettersen 1993; Rooker and Wilkinson 1980). While patch testing can be a helpful tool in diagnosing dermal sensitivity to several metals, there are important limitations that must be considered when using this modality to assess DTH responses to orthopedic implant degradation products. (1) Primarily, performing patch testing pre-operatively has the potential to pre-sensitize the patient to one or more implant metals (Merritt and Brown 1980). The process of skin patch testing involves mixing metal ion/salts with an organic vehicle (i.e., petroleum Jelly) and the application of this mixture in direct contact with skin for 48 h. The extent to which dermal patch testing induces metal sensitization in humans is not known, but has been well established as a method to induce metal sensitization in animal models (Bonefeld et al. 2015; Vennegaard et al. 2014); therefore, it can potentially be a hazard for the purposes of diagnosing metal DTH responses in future orthopedic implant patients and a significant concern given how routinely this procedure is performed (Granchi et al. 2012). (2) An additional limitation of patch testing is the simulation of immunological potential of metal haptens in a nonsterile dermal environment compared to a significantly different sterile environment found in the peri-implant tissue (Korenblat 1992; Kuby 1991). For example, Langerhans cells – specialized antigen presenting cells of the skin – possess Birbeck granules which are unique antigen-processing/endosomal-processing organelles not found in macrophages/histiocytes in the peri-implant tissue (Mc et al. 2002; Valladeau et al. 2001). (3) Patch testing results are scored subjectively by a healthcare professional (i.e., Allergist) using a 0 to 3+ system, where results may not be easily compared between providers. (4) Immunological responses to patch testing challenge may be severely diminished due to the nature of the site of challenge and inherent tolerance to environmental factors (i.e., metals) (Benson et al. 1975; Poss et al. 1984; Rooker and Wilkinson 1980; Wang et al. 1997a). The environment of immune challenge during patch testing can be highly variable and is non-standardized as they are usually placed directly on the back of patients (hairless area) for 2 to 3 days and it can be inconsistent from patient to patient. It may also be uncomfortable and the environment under which the test is performed (i.e., cleanliness) cannot controlled or standardized. (5) Lastly, there are no standardized, well-established metal salt concentrations available for patch testing or the availability of all orthopedic implant metals in commercially available patch testing kits (e.g., aluminum, molybdenum, vanadium, and zirconium) (Table 2).

Lymphocyte Transformation Testing (LTT): Also termed lymphocyte proliferation test measures the division/proliferation of peripheral blood T lymphocytes in vitro after exposure to specific antigens during a period of 6 days. Lymphocytes are isolated from a patient's blood sample (simple blood draw) by density gradient separation of mononuclear cells. The proliferation of these lymphocytes is measured 4–6 days (DTH response) after initial antigen exposure using a radiolabeling technique. Radioactive [H^3]-thymidine is incorporated into the DNA of dividing (proliferating) cells and allows for the quantification of actual cell division in response to several metal challenge agents (i.e., $Al+3$, $Co+2$, $Cr+3$, $Mo+5$, $Ni+2$, $V+3$, and $Zr+4$) at different concentrations ranging from 0.001 to 0.1 mM. This specific modality of detection of cell proliferation has the ability to detect the specific subset of cells undergoing cell division in response to the antigen challenge. The final amount of proliferation is measured as a Stimulation Index (SI).

> Proliferation Index or Stimulation Index (SI) = (proliferation with treatment, cpms)/(proliferation of equal amount of starting cells of the same individual without treatment, cpms).

Lymphocyte transformation testing (LTT) has gradually become a more widely used and accepted test modality for the diagnosis of orthopedic implant-related metal sensitivity as well as in cohort and basic science studies of metal-induced DTH responses (Everness et al. 1990; Secher et al. 1977; Svejgaard et al. 1976, 1978; Veien and Svejgaard 1978; Veien et al. 1979). LTT testing is performed by isolating mononuclear cells from a patient's peripheral blood sample (i.e., T-cells, B-cells and other lymphocyte populations) and directly exposing them to metal challenge in order to simulate the local peri-implant environment (not possible with dermal patch testing) (Hallab et al. 1998b, 2000, 2000a, b, 2001b, 2013, 2003b). An advantage of LTT testing is that is highly quantitative and not dependent on subjective assessment of results (vs. patch testing) (Thomas et al. 2009). The stimulation Index (SI) is quantified from multiwell replicates of each challenge agent at each concentration tested that allows for the calculation of an average and standard deviation for each antigen tested. This enables assessment of a dose dependent response where a metal sensitive individual may exhibit lymphocyte proliferation at a lower or higher dose of metal challenge (Fig. 2). LTT has also shown to have a greater sensitivity to detect lymphocyte metal sensitization (>80%) compared to patch testing (Carando et al. 1985; Cederbrant et al. 1997; Federmann et al. 1994; Nyfeler and Pichler 1997; Primeau and Adkinson 2001; Torgersen et al. 1993). A recent study performed by Carossino et al. (2016; Innocenti et al. 2014) where patch testing, LTT, and cytokine analysis were performed concluded that "The lymphocyte transformation is the most suitable method for testing systemic allergies." This testing modality is gaining momentum and is increasingly becoming more relevant to the orthopedic community given the growing numbers of TJA performed each year (Kurtz et al. 2009).

Furthermore, other prospective and longitudinal studies as the one discussed in the next section regarding metal-on-metal devices substantiate the concept that LTT or Patch Testing are necessary in a clinical setting, especially for patients receiving specific types of devices that may be more prone to induce metal sensitization. There are also further case and group studies supporting the clinical utility and routine use of metal sensitivity testing for total joint replacement (TJR) patients that have a history of metal allergy and/or for patients with aseptic/idiopathic implant related pain (Campbell et al. 2010; Hallab et al. 2013; Kwon et al. 2010, 2011; Thomas et al. 2009; Willert et al. 2005; Willert and Semlitsch 1977). Interestingly, while instability and infection are the primary causes of early implant failure, recent reports have put forward algorithms that include metal-induced DTH testing as a possible indication for patients with post-operative pain (Fig. 3) (Park et al. 2016). This specific algorithm suggests that metal-LTT and dermal testing should be performed as a last resort after imaging techniques (MRI, CT) and other infection indications have been ruled out.

Studies of Implant Related Metal Sensitivity Using Diagnostic Testing

Several studies performed over the past four decades have associated metal allergy or metal sensitivity with adverse implant immune responses, where the quantity of implant degradation products has been temporarily linked to symptoms such as severe dermatitis, urticaria, vasculitis (Abdallah et al. 1994; Barranco and Solloman 1972; Halpin 1975; King et al. 1993; Merle et al. 1992; Thomas et al. 1987), and/or nonspecific immune suppression (Bravo et al. 1990; Gillespie et al. 1988; Merritt and Brown 1985; Poss et al. 1984; Wang et al. 1997b). Some case studies have demonstrated cessation of metal sensitivity symptoms after removal of the implant and the reappearance of symptoms once a comparable implant was re-introduced. This agrees with Koch's postulate, an important test for causality in medicine, and demonstrates metal-induced sensitivity responses as causal for early implant failure (Barranco and Solloman 1972). Nevertheless, the majority of the evidence demonstrating the significant clinical utility of metal sensitivity testing can be credited to several retrospective cohort studies that have shown a strong correlation between metal exposure, metal sensitivity, and the performance of metal implants (Benson et al. 1975; Brown et al. 1977; Carlsson et al. 1980; Cramers and Lucht 1977; Deutman et al. 1977; Fischer et al. 1984; Kubba et al. 1981; Mayor et al. 1980; Merritt 1984; Merritt and Brown 1981; Pinkston and Finch 1979; Rooker and Wilkinson 1980). As mentioned previously, these studies demonstrate that people with well performing implants and people with painful/failing implants exhibit rates of metal hypersensitivity two fold or six fold higher compared to the general population, respectively (Caicedo et al. 2013b). It is also clear, based on current and past cohort studies, that specific types of metal implants known to release higher concentrations of ions and/or particles are more likely to induce metal sensitization (Hallab et al. 2013; Kwon et al. 2011).

While metal on metal total hip arthroplasties (MoM THA) provide the advantage of lower implant wear compared to metal-on-polymer (MoP) implants, they are known to release higher concentration of metal ions and particles and thus have a higher incidence of failure attributable to excessive inflammatory responses. Previous studies have shown hypersensitivity-like responses, including histological inflammatory evidence accompanied by severe lymphocyte infiltrates, in as high as 76–100% of patients with poorly performing MoM devices (Korovessis et al. 2006; Milosev et al. 2006). In a prospective study using a cohort of MoM patients, it was shown that in vivo metal sensitivity responses may develop even in well performing (asymptomatic) MoM implants (Hallab et al. 2013) where a significant increase in the rate of diagnosed metal sensitivity increased from 5% preoperatively to 56% within the first 4 years postoperatively (Hallab et al. 2013). In this study increases in serum levels of Co and Cr occurred at early stage, at 3 months postoperatively. However, lymphocyte sensitivity responses only became more evident at 1–4 years post-op. This delay in detection of metal sensitivity responses postoperatively suggests that metal sensitization may develop over-time as exposure to metal ion levels increase. The rates found, while still high compared to conventional implants (25%), are lower than 81% in failing MoM implants previously reported for painful/symptomatic MOM patients by Thomas et al. (2009).

Pain levels have also been shown to correlate with metal sensitivity (Metal-LTT with SI >2) where patients with highly painful implants were significantly higher compared to patients with nonpainful implants (Caicedo et al. 2013b). Furthermore, TJA patients that reported low implant pain levels also exhibited a relatively lower incidence of metal sensitization further supporting a correlation between aseptic implant pain levels and metal sensitivity. Additionally, not only do TJA female patients referred for metal sensitivity testing exhibit a higher average pain level compared to males, but also show a higher incidence and severity of metal sensitization (Caicedo et al. 2017). This supports the utility of metal DTH testing in patients with aseptic implant-related pain, especially for female orthopedic patients.

Conclusions

Implant degradation and debris is unavoidable and results in activation of the immune system resulting in local inflammation that over time causes more bone loss then homeostatic mechanisms can keep up with, and the result is implant loosening, via aseptic osteolysis. This reactivity may activate the adaptive immune systems and result in allergic type responses involving T-cells. Both innate (macrophage) and adaptive (T lymphocyte) immune system reactivity can act to limit the lifetime of current total joint replacement implants. Advances at the molecular and cellular level continue to increase our understanding of immune reactivity based bone loss. There are a new treatment and diagnostic options available for patients and surgeons ranging from diagnosing preexisting or developed conditions of metal allergy (metal-LTT), general management of inflammation (e.g., NSAIDS) to selective blocking of cellular mediators (e.g., anti-IL-6, anti-TNFα, IL-1β-receptor antagonist). These options should be part of the modern arsenal used to help fight the problem of adverse reactivity to implant debris, i.e., induced inflammatory bone loss. There is increasing need for using patient specific diagnosis and treatment to mitigate the role of metal hypersensitivity and genetic susceptibility to implant debris-induced inflammation.

References

Abdallah HI, Balsara RK, O'Riordan AC (1994) Pacemaker contact sensitivity: clinical recognition and management. Ann Thorac Surg 57:1017–1018

Archibeck MJ, Jacobs JJ, Roebuck KA, Glant TT (2001) The basic science of periprosthetic osteolysis. Instr Course Lect 50:185–195

Arora A, Song Y, Chun L, Huie P, Trindade M, Smith RL, Goodman S (2003a) The role of the TH1 and TH2 immune responses in loosening and osteolysis of cemented total hip replacements. J Biomed Mater Res A 64(4):693–697

Arora A, Song Y, Chun L, Huie P, Trindade M, Smith RL, Goodman S (2003b) The role of the TH1 and TH2 immune responses in loosening and osteolysis of cemented total hip replacements. J Biomed Mater Res 64A(4):693–697

Barranco VP, Solloman H (1972) Eczematous dermatitis from nickel. JAMA 220(9):1244

Benson MK, Goodwin PG, Brostoff J (1975) Metal sensitivity in patients with joint replacement arthroplasties. Br Med J 4:374–375

Beyersmann D (1994) Interactions in metal carcinogenicity. Toxicol Lett 72(1–3):333–338

Black J (1996) Prosthetic Materials. VCH Publishers, Inc., New York

Bonefeld CM, Nielsen MM, Vennegaard MT, Johansen JD, Geisler C, Thyssen JP (2015) Nickel acts as an adjuvant during cobalt sensitization. Exp Dermatol 24(3):229–231

Bravo I, Carvalho GS, Barbosa MA, de Sousa M (1990) Differential effects of eight metal ions on lymphocyte differentiation antigens in vitro. J Biomed Mater Res 24(8):1059–1068

Britton RS (1996) Metal-induced Hepatoxicity. Semin Liver Dis 16(1):3–12

Brown GC, Lockshin MD, Salvati EA, Bullough PG (1977) Sensitivity to metal as a possible cause of sterile loosening after cobalt-chromium total hip-replacement arthroplasty. J Bone Joint Surg Am 59-A(2):164–168

Burt CF, Garvin KL, Otterberg ET, Jardon OM (1998) A femoral component inserted without cement in total hip arthroplasty. A study of the tri-lock component with an average ten-year duration of follow-up. J Bone Joint Surg Am 80(7):952–960

Caicedo MS, Desai R, McAllister K, Reddy A, Jacobs JJ, Hallab NJ (2008) Soluble and particulate Co-Cr-Mo alloy implant metals activate the inflammasome danger signaling pathway in human macrophages: a novel mechanism for implant debris reactivity. J Orthop Res 27(7):847–854

Caicedo MS, Pennekamp PH, McAllister K, Jacobs JJ, Hallab NJ (2010) Soluble ions more than particulate cobalt-alloy implant debris induce monocyte costimulatory molecule expression and release of pro-inflammatory cytokines critical to metal-induced lymphocyte reactivity. J Biomed Mater Res A 93(4):1312–1321

Caicedo MS, Samelko L, McAllister K, Jacobs JJ, Hallab NJ (2013a) Increasing both CoCrMo-alloy particle size and surface irregularity induces increased macrophage inflammasome activation in vitro potentially through lysosomal destabilization mechanisms. J Orthop Res 31(10):1633–1642

Caicedo MS, Samelko L, Hallab NJ (2013b) Lymphocyte reactivity to nickel correlates with reported high-pain levels in patients with Total joint arthroplasties: implications for pain-related hypersensitivity responses, Metal-on-metal total hip replacement devices. ASTM STP STP 1560:1–17

Caicedo MS, Solver E, Coleman L, Jacobs JJ, Hallab NJ (2017) Females with unexplained joint pain following total joint arthroplasty exhibit a higher rate and severity of hypersensitivity to implant metals compared with males: implications of sex-based bioreactivity differences. J Bone Joint Surg Am 99(8):621–628

Campbell P, Ma S, Yeom B, McKellop H, Schmalzried TP, Amstutz HC (1995) Isolation of predominantly submicron-sized UHMWPE wear particles from periprosthetic tissues. J Biomed Mater Res 29(1):127–131

Campbell P, Ebramzadeh E, Nelson S, Takamura K, De SK, Amstutz HC (2010) Histological features of pseudotumor-like tissues from metal-on-metal hips. Clin Orthop Relat Res 468(9):2321–2327

Carando S, Cannas M, Rossi P, Portigliatti-Barbos M (1985) The lymphocytic transformation test (L.T.T.) in the evaluation of intolerance in prosthetic implants. Ital J Orthop Traumatol 11(4):475–481

Carlsson AS, Macnusson B, Moller H (1980) Metal sensitivity in patients with metal-to-plastic total hip arthroplasties. Acta Orthop Scand 51:57–62

Carossino AM, Carulli C, Ciuffi S, Carossino R, Zappoli Thyrion GD, Zonefrati R, Innocenti M, Brandi ML (2016) Hypersensitivity reactions to metal implants: laboratory options. BMC Musculoskelet Disord 17(1):486

Catelas I, Petit A, Marchand R, Zukor DJ, Yahia L, Huk OL (1999) Cytotoxicity and macrophage cytokine release induced by ceramic and polyethylene particles in vitro. J Bone Joint Surg Br 81(3):516–521

Catelas I, Petit A, Zukor DJ, Antoniou J, Huk OL (2003) TNF-alpha secretion and macrophage mortality induced by cobalt and chromium ions in vitro-qualitative analysis of apoptosis. Biomaterials 24(3):383–391

Catelas I, Medley JB, Campbell PA, Huk OL, Bobyn JD (2004) Comparison of in vitro with in vivo characteristics of wear particles from metal-metal hip implants. J Biomed Mater Res B Appl Biomater 70(2):167–178

Cederbrant K, Hultman P, Marcusson JA, Tibbling L (1997) In vitro lymphocyte proliferation as compared to patch test using gold, palladium and nickel. Int Arch Allergy Immunol 112(3):212–217

Charnley J (1970) The reaction of bone to self-curing acrylic cement. A long-term histological study in man. J Bone Joint Surg Br 52(2):340–353

Charnley J (1979) Low friction arthroplasty of the hip, theory and practice. Springer, Berlin

Choma TJ, Miranda J, Siskey R, Baxter R, Steinbeck MJ, Kurtz SM (2009) Retrieval analysis of a ProDisc-L total disc replacement. J Spinal Disord Tech 22(4):290–296

Cooper HJ, Urban RM, Wixson RL, Meneghini RM, Jacobs JJ (2013) Adverse local tissue reaction arising from corrosion at the femoral neck-body junction in a dual-taper stem with a cobalt-chromium modular neck. J Bone Joint Surg Am 95(10):865–872

Cramers M, Lucht U (1977) Metal sensitivity in patients treated for tibial fractures with plates of stainless steel. Acta Orthop Scand 48:245–249

Deutman R, Mulder TH, Brian R, Nater JP (1977) Metal sensitivity before and after total hip arthroplasty. J Bone Joint Surg Am 59-A:862–865

Dorr LD, Bloebaum R, Emmanual J, Meldrum R (1990) Histologic, biochemical and ion analysis of tissue and fluids retrieved during total hip arthroplasty. Clin Orthop Relat Res 261:82–95

Dostert C, Petrilli V, Van BR, Steele C, Mossman BT, Tschopp J (2008) Innate immune activation through Nalp3 inflammasome sensing of asbestos and silica. Science 320(5876):674–677

Elves MW, Wilson JN, Scales JT, Kemp HB (1975) Incidence of metal sensitivity in patients with total joint replacements. Br Med J 4:376–378

Everness KM, Gawkrodger DJ, Botham PA, Hunter JA (1990) The discrimination between nickel-sensitive and non-nickel-sensitive subjects by an in vitro lymphocyte transformation test. Br J Dermatol 122(3):293–298

Federmann M, Morell B, Graetz G, Wyss M, Elsner P, von Thiessen R, Wuthrich B, Grob D (1994) Hypersensitivity to molybdenum as a possible trigger of ANA-negative systemic lupus erythematosus. Ann Rheum Dis 53(6):403–405

Fischer T, Rystedt I, Safwenberg J, Egle I (1984) HLA -A, -B, -C and -DR antigens in individuals with sensitivity to cobalt. Acta Derm Venereol (Stockh) 64:121–124

Gillespie WJ, Frampton CM, Henderson RJ, Ryan PM (1988) The incidence of cancer following total hip replacement. J Bone Joint Surg Br 70(4):539–542

Goering PL, Klaasen CD (1995) Hepatoxicity of metals. Academic, New York

Goldring SR, Schiller AL, Roelke M, Rourke CM, O'Neill DA, Harris WH (1983) The synovial-like membrane at the bone-cement interface in loose total hip replacements and its proposed role in bone lysis. J Bone Joint Surg 65A:575–584

Gordon PM, White MI, Scotland TR (1994) Generalized sensitivity from an implanted orthopaedic antibiotic minichain containing nickel. Contact Dermatitis 30:181–182

Granchi D, Cenni E, Giunti A, Baldini N (2012) Metal hypersensitivity testing in patients undergoing joint replacement: a systematic review. J Bone Joint Surg Br 94(8):1126–1134

Hallab NJ, Jacobs JJ, Skipor AK, Black J, Glant T, Mikecz K (1998b) In vitro testing of metal induced luekocyte activation. Trans Soc Biomater 21:76

Hallab NJ, Jacobs JJ, Skipor A, Black J, Mikecz K, Galante JO (2000) Systemic metal-protein binding associated with total joint replacement arthroplasty. J Biomed Mater Res 49(3):353–361

Hallab NJ, Jacobs JJ, Skipor A, Black J, Mikecz K, Galante JO (2000a) Systemic metal-protein binding associated with total joint replacement arthroplasty. J Biomed Mater Res 49(3):353–361

Hallab NJ, Mikecz K, Jacobs JJ (2000b) A triple assay technique for the evaluation of metal-induced, delayed-type hypersensitivity responses in patients with or receiving total joint arthroplasty. J Biomed Mater Res 53(5):480–489

Hallab N, Merritt K, Jacobs JJ (2001a) Metal sensitivity in patients with orthopaedic implants. J Bone Joint Surg Am 83-A(3):428–436

Hallab NJ, Mikecz K, Vermes C, Skipor A, Jacobs JJ (2001b) Differential lymphocyte reactivity to serum-derived metal-protein complexes produced from cobalt-based and titanium-based implant alloy degradation. J Biomed Mater Res 56(3):427–436

Hallab NJ, Cunningham BW, Jacobs JJ (2003a) Spinal implant debris-induced Osteolysis. Spine 28(20):S125–S138

Hallab NJ, Skipor A, Jacobs JJ (2003b) Interfacial kinetics of titanium- and cobalt-based implant alloys in human serum: metal release and biofilm formation. J Biomed Mater Res 65A(3):311–318

Hallab NJ, Caicedo M, Finnegan A, Jacobs JJ (2008) Th1 type lymphocyte reactivity to metals in patients with total hip arthroplasty. J Orthop Surg 3:6

Hallab NJ, Caicedo M, McAllister K, Skipor A, Amstutz H, Jacobs JJ (2013) Asymptomatic prospective and retrospective cohorts with metal-on-metal hip arthroplasty indicate acquired lymphocyte reactivity varies with metal ion levels on a group basis. J Orthop Res 31(2):173–182

Hallab NJ, Caicedo M, McAllister K, Skipor A, Amstutz H, Jacobs JJ (2013) Asymptomatic prospective and retrospective cohorts with metal-on-metal hip arthroplasty indicate acquired lymphocyte reactivity varies with metal ion levels on a group basis. J Orthop Res 31(2):173–182

Halpin DS (1975) An unusual reaction in muscle in association with a vitallium plate: a report of possible metal hypersensitivity. J Bone Joint Surg 57-B(4):451–453

Hartwig A (1998) Carcinogenicity of metal compounds: possible role of DNA repair inhibition. Toxicol Lett 102–103:235–239

Hensten-Pettersen A (1993) Allergy and hypersensitivity. In: Morrey BF (ed) Biological, material, and mechanical considerations of joint replacements. Raven Press, New York, pp 353–360

Hornung V, Bauernfeind F, Halle A, Samstad EO, Kono H, Rock KL, Fitzgerald KA, Latz E (2008) Silica crystals and aluminum salts activate the NALP3 inflammasome through phagosomal destabilization. Nat Immunol 9(8):847–856

Innocenti M, Carulli C, Matassi F, Carossino AM, Brandi ML, Civinini R (2014) Total knee arthroplasty in patients with hypersensitivity to metals. Int Orthop 38(2):329–333

Jacobs JJ, Goodman SL (1996) What in vitro, in vivo and combined approaches can be used to investigate the biologic effects of particles? In: Wright TM, Goodman SB (eds) Implant Wear: the future of total joint replacement. American Academy of Orthopedic Surgeons, Rosemont, pp 41–44

Jacobs JJ, Hallab NJ (2006) Loosening and osteolysis associated with metal-on-metal bearings: a local effect of metal hypersensitivity? J Bone Joint Surg Am 88(6):1171–1172

Jacobs JJ, Gilbert JL, Urban RM (1994) Corrosion of metallic implants. In: Stauffer RN (ed) Advances in Orthopaedic surgery, vol 2. Mosby, St. Louis, pp 279–319

Jacobs JJ, Shanbhag A, Glant TT, Black J, Galante JO (1994a) Wear debris in total joint replacements. J Am Acad Orthop Surg 2(4):212–220

Jacobs JJ, Skipor AK, Urban RM, Black J, Manion LM, Starr A, Talbert LF, Galante JO (1994b) Systemic distribution of metal degradation products from titanium alloy total hip replacements: an autopsy study. Trans Orthop Res Soc New Orleans:838

Jacobs JJ, Gilbert JL, Urban RM (1998a) Corrosion of metal orthopaedic implants. J Bone Joint Surg Am 80(2):268–282

Jacobs JJ, Skipor AK, Patterson LM, Hallab NJ, Paprosky WG, Black J, Galante JO (1998b) Metal release in patients who have had a primary total hip arthroplasty. A prospective, controlled, longitudinal study. J Bone Joint Surg Am 80(10):1447–1458

Jacobs J, Goodman S, Sumner DR, Hallab N (1999a) Biologic response to orthopedic implants. In: Orthopedic basic science. American Academy of Orthopedic Surgeons, Chicago, pp 402–426

Jacobs JJ, Silverton C, Hallab NJ, Skipor AK, Patterson L, Black J, Galante JO (1999b) Metal release and excretion from cementless titanium alloy total knee replacements. Clin Orthop 358:173–180

Jacobs JJ, Roebuck KA, Archibeck M, Hallab NJ, Glant TT (2001) Osteolysis: basic science. Clin Orthop Relat Res 393:71–77

Kaufman AM, Alabre CI, Rubash HE, Shanbhag AS (2008) Human macrophage response to UHMWPE, TiAlV, CoCr, and alumina particles: analysis of multiple cytokines using protein arrays. J Biomed Mater Res A 84(2):464–474

King J, Fransway A, Adkins RB (1993) Chronic urticaria due to surgical clips. N Engl J Med 329(21):1583–1584

Korenblat PE (1992) Contact dermatitis, 2nd edn. W.B. Saunders Company, Philidelphia

Korovessis P, Petsinis G, Repanti M, Repantis T (2006) Metallosis after contemporary metal-on-metal total hip

arthroplasty. Five to nine-year follow-up. J Bone Joint Surg Am 88(6):1183–1191
Kubba R, Taylor JS, Marks KE (1981) Cutaneous complications of orthopedic implants. A two-year prospective study. Arch Dermatol 117:554–560
Kuby J (1991) Immunology, 2nd edn. W.H. Freeman and Company, New York
Kurtz S, Ong K, Lau E, Mowat F, Halpern M (2007a) Projections of primary and revision hip and knee arthroplasty in the United States from 2005 to 2030. J Bone Joint Surg Am 89(4):780–785
Kurtz SM, Ong KL, Schmier J, Mowat F, Saleh K, Dybvik E, Karrholm J, Garellick G, Havelin LI, Furnes O, Malchau H, Lau E (2007b) Future clinical and economic impact of revision total hip and knee arthroplasty. J Bone Joint Surg Am 89(Suppl 3):144–151
Kurtz SM, Ong KL, Schmier J, Zhao K, Mowat F, Lau E (2009) Primary and revision arthroplasty surgery caseloads in the United States from 1990 to 2004. J Arthroplasty 24(2):195–203
Kwon YM, Thomas P, Summer B, Pandit H, Taylor A, Beard D, Murray DW, Gill HS (2010) Lymphocyte proliferation responses in patients with pseudotumors following metal-on-metal hip resurfacing arthroplasty. J Orthop Res 28(4):444–450
Kwon YM, Ostlere SJ, Lardy-Smith P, Athanasou NA, Gill HS, Murray DW (2011) "Asymptomatic" pseudotumors after metal-on-metal hip resurfacing arthroplasty: prevalence and metal ion study. J Arthroplasty 26(4):511–518
Lewis JB, Randol TM, Lockwood PE, Wataha JC (2003) Effect of subtoxic concentrations of metal ions on NFkappaB activation in THP-1 human monocytes. J Biomed Mater Res A 64(2):217–224
Looney RJ, Schwarz EM, Boyd A, O'Keefe RJ (2006) Periprosthetic osteolysis: an immunologist's update. Curr Opin Rheumatol 18(1):80–87
Luckey TD, Venugopal B (1979) Metal toxicity in mammals. Plenum, New York
Maloney WJ, Smith RL, Castro F, Schurman DJ (1993) Fibroblast response to metallic debris in vitro. Enzyme induction cell proliferation, and toxicity. J Bone Joint Surg Am 75(6):835–844
Mariathasan S, Monack DM (2007) Inflammasome adaptors and sensors: intracellular regulators of infection and inflammation. Nat Rev Immunol 7(1):31–40
Mariathasan S, Newton K, Monack DM, Vucic D, French DM, Lee WP, Roose-Girma M, Erickson S, Dixit VM (2004) Differential activation of the inflammasome by caspase-1 adaptors ASC and Ipaf. Nature 430(6996):213–218
Mayor MB, Merritt K, Brown SA (1980) Metal allergy and the surgical patient. Am J Surg 139:477–479
Mc DR, Ziylan U, Spehner D, Bausinger H, Lipsker D, Mommaas M, Cazenave JP, Raposo G, Goud B, de la Salle H, Salamero J, Hanau D (2002) Birbeck granules are subdomains of endosomal recycling compartment in human epidermal Langerhans cells, which form where Langerin accumulates. Mol Biol Cell 13(1):317–335
Merle C, Vigan M, Devred D, Girardin P, Adessi B, Laurent R (1992) Generalized eczema from vitallium osteosynthesis material. Contact Dermatitis 27:257–258
Merritt K (1984) Role of medical materials, both in implant and surface applications, in immune response and in resistance to infection. Biomaterials 5:53–57
Merritt K, Brown S (1980) Tissue reaction and metal sensitivity. Acta Orthop Scand 51:403–4111
Merritt K, Brown S (1981) Metal sensitivity reactions to orthopedic implants. Int J Dermatol 20:89–94
Merritt K, Brown SA (1985) Biological effects of corrosion products from metal. American Society for Testing and Materials, Philadelphia
Merritt K, Rodrigo JJ (1996) Immune response to synthetic materials. Sensitization of patients receiving orthopaedic implants. Clin Orthop Relat Res 326:71–79
Michel R, Hoffman J, Loer F, Zilkens J (1984) Trace element burdening of human tissue due to corrosion of hip-joint prostheses made of cobalt-chromium alloys. Arch Orthop Trama Surg 103:85–95
Michel R, Nolte M, Reich M, Loer F (1991) Systemic effects of implanted prostheses made of cobalt-chromium alloys. Arch Orthop Trauma Surg 110:61–74
Milosev I, Trebse R, Kovac S, Cor A, Pisot V (2006) Survivorship and retrieval analysis of Sikomet metal-on-metal total hip replacements at a mean of seven years. J Bone Joint Surg Am 88(6):1173–1182
Naganuma Y, Takakubo Y, Hirayama T, Tamaki Y, Pajarinen J, Sasaki K, Goodman SB, Takagi M (2016) Lipoteichoic acid modulates inflammatory response in macrophages after phagocytosis of titanium particles through Toll-like receptor 2 cascade and inflammasomes. J Biomed Mater Res A 104(2):435–444
Nyfeler B, Pichler WJ (1997) The lymphocyte transformation test for the diagnosis of drug allergy: sensitivity and specificity. Clin Exp Allergy 27(2):175–181
Park CN, White PB, Meftah M, Ranawat AS, Ranawat CS (2016) Diagnostic algorithm for residual pain after total knee arthroplasty. Orthopedics 39(2):e246–e252
Petrilli V, Dostert C, Muruve DA, Tschopp J (2007) The inflammasome: a danger sensing complex triggering innate immunity. Curr Opin Immunol 19(6):615–622
Pinkston JA, Finch SC (1979) A method for the differentiation of T and B lymphocytes and monocytes migrating under agarose. Stain Technol 54(5):233–239
Poss R, Thornhill TS, Ewald FC, Thomas WH, Batte NJ, Sledge CB (1984) Factors influencing the incidence and outcome of infection following total joint arthoplasty. Clin Orthop 182:117–126
Primeau MN, Adkinson NF Jr (2001) Recent advances in the diagnosis of drug allergy. Curr Opin Allergy Clin Immunol 1(4):337–341
Punt IM, Visser VM, van Rhijn LW, Kurtz SM, Antonis J, Schurink GW, van Ooij A (2008) Complications and

reoperations of the SB Charite lumbar disc prosthesis: experience in 75 patients. Eur Spine J 17(1):36–43

Punt IM, Cleutjens JP, de Bruin T, Willems PC, Kurtz SM, van Rhijn LW, Schurink GW, van Ooij A (2009) Periprosthetic tissue reactions observed at revision of total intervertebral disc arthroplasty. Biomaterials 30(11):2079–2084

Radcliffe GS, Tomichan MC, Andrews M, Stone MH (1999) Revision hip surgery in the elderly: is it worthwhile? J Arthroplasty 14(1):38–44

Rooker GD, Wilkinson JD (1980) Metal sensitivity in patients undergoing hip replacement. A prospective study. J Bone Joint Surg 62-B(4):502–505

Rostoker G, Robin J, Binet O, Blamutier J, Paupe J, Lessana-Liebowitch M, Bedouelle J, Sonneck JM, Garrel JB, Millet P (1987) Dermatitis due to orthopaedic implants. A review of the literature and report of three cases. J Bone Joint Surg 69-A(9):1408–1412

Schwarz EM, Looney RJ, O'Keefe RJ (2000) Anti-TNF-alpha therapy as a clinical intervention for periprosthetic osteolysis. Arthritis Res 2(3):165–168

Scott M, Morrison M, Mishra SR, Jani S (2005) Particle analysis for the determination of UHMWPE wear. J Biomed Mater Res B Appl Biomater 73(2):325–337

Secher L, Svejgaard E, Hansen GS (1977) T and B lymphocytes in contact and atopic dermatitis. Br J Dermatol 97(5):537–541

Sethi RK, Neavyn MJ, Rubash HE, Shanbhag AS (2003) Macrophage response to cross-linked and conventional UHMWPE. Biomaterials 24(15):2561–2573

Stulberg BN, Merritt K, Bauer T (1994) Metallic wear debris in metal-backed patellar failure. J Biomed Mat Res Appl Biomater 5:9–16

Svejgaard E, Thomsen M, Morling N, Hein CA (1976) Lymphocyte transformation in vitro in dermatophytosis. Acta Pathol Microbiol Scand C 84C(6):511–523

Svejgaard E, Morling N, Svejgaard A, Veien NK (1978) Lymphocyte transformation induced by nickel sulphate: an in vitro study of subjects with and without a positive nickel patch test. Acta Derm Venereol 58(3):245–250

Thomas RH, Rademaker M, Goddard NJ, Munro DD (1987) Severe eczema of the hands due to an orthopaedic plate made of Vitallium. Br Med J 294:106–107

Thomas P, Braathen LR, Dorig M, Aubock J, Nestle F, Werfel T, Willert HG (2009) Increased metal allergy in patients with failed metal-on-metal hip arthroplasty and peri-implant T-lymphocytic inflammation. Allergy 64(8):1157–1165

Torgersen S, Gilhuus-Moe OT, Gjerdet NR (1993) Immune response to nickel and some clinical observations after stainless steel miniplate osteosynthesis. Int J Oral Maxillofac Surg 22(4):246–250

Trindade MC, Lind M, Nakashima Y, Sun D, Goodman SB, Schurman DJ, Smith RL (2001) Interleukin-10 inhibits polymethylmethacrylate particle induced interleukin-6 and tumor necrosis factor-alpha release by human monocyte/macrophages in vitro. Biomaterials 22(15):2067–2073

Urban RM, Jacobs JJ, Sumner DR, Peters CL, Voss FR, Galante JO (1996) The bone-implant interface of femoral stems with non- circumferential porous coating: a study of specimens retrieved at autopsy. J Bone Joint Surg Am 78-A(7):1068–1081

Urban RM, Jacobs J, Gilbert JL, Rice SB, Jasty M, Bragdon CR, Galante GO (1997) Characterization of solid products of corrosion generated by modular-head femoral stems of different designs and materials. In: Marlowe DE, Parr JE, Mayor MB (eds) STP 1301 modularity of orthopedic implants. ASTM, Philadelphia, pp 33–44

Urban RM, Hall DJ, Sapienza CI, Jacobs JJ, Sumner DR, Rosenberg AG, Galante JO (1998) A comparative study of interface tissues in cemented vs. cementless total knee replacement tibial components retrieved at autopsy. Trans SFB 21:255

Urban RM, Jacobs JJ, Tomlinson MJ, Gavrilovic J, Black J, Peoc'h M (2000) Dissemination of wear particles to the liver, spleen, and abdominal lymph nodes of patients with hip or knee replacement. J Bone Joint Surg Am 82(4):457–476

Valladeau J, Caux C, Lebecque S, Saeland S (2001) Langerin: a new lectin specific for Langerhans cells induces the formation of Birbeck granules. Pathol Biol (Paris) 49(6):454–455

van Ooij A, Kurtz SM, Stessels F, Noten H, van Rhijn L (2007) Polyethylene wear debris and long-term clinical failure of the Charite disc prosthesis: a study of 4 patients. Spine 32(2):223–229

Veien NK, Svejgaard E (1978) Lymphocyte transformation in patients with cobalt dermatitis. Br J Dermatol 99(2):191–196

Veien NK, Svejgaard E, Menne T (1979) In vitro lymphocyte transformation to nickel: a study of nickel-sensitive patients before and after epicutaneous and oral challenge with nickel. Acta Derm Venereol 59(5):447–451

Vennegaard MT, Dyring-Andersen B, Skov L, Nielsen MM, Schmidt JD, Bzorek M, Poulsen SS, Thomsen AR, Woetmann A, Thyssen JP, Johansen JD, Odum N, Menne T, Geisler C, Bonefeld CM (2014) Epicutaneous exposure to nickel induces nickel allergy in mice via a MyD88-dependent and interleukin-1-dependent pathway. Contact Dermatitis 71(4):224–232

Vermes C, Chandrasekaran R, Jacobs JJ, Galante JO, Roebuck KA, Glant TT (2001a) The effects of particulate wear debris, cytokines, and growth factors on the functions of MG-63 osteoblasts. J Bone Joint Surg Am 83(2):201–211

Vermes C, Glant TT, Hallab NJ, Fritz EA, Roebuck KA, Jacobs JJ (2001b) The potential role of the osteoblast in the development of periprosthetic osteolysis: review of in vitro osteoblast responses to wear debris, corrosion products, and cytokines and growth factors. J Arthroplasty 16(8 Suppl 1):95–100

Wang JY, Wicklund BH, Gustilo RB, Tsukayama DT (1997a) Prosthetic metals impair murine immune response and cytokine release in vivo and in vitro. J Orthop Res 15(5):688–699

Wang JY, Wicklund BH, Gustilo RB, Tsukayama DT (1997b) Prosthetic metals interfere with the functions of human osteoblast cells in vitro. Clin Orthop 339:216–226

Willert HG, Semlitsch M (1977) Reactions of the articular capsule to wear products of artificial joint prostheses. J Biomed Mater Res 11:157–164

Willert HG, Buchhorn GH, Fayyazi A, Flury R, Windler M, Koster G, Lohmann CH (2005) Metal-on-metal bearings and hypersensitivity in patients with artificial hip joints. A clinical and histomorphological study. J Bone Joint Surg Am 87(1):28–36

Mechanical Implant Material Selection, Durability, Strength, and Stiffness

8

Robert Sommerich, Melissa (Kuhn) DeCelle, and William J. Frasier

Contents

Introduction	152
Metallic Implants	152
Porous Metals	154
Polymers	154
Porous PEEK	156
Silicon Nitride	156
Biodegradable Polymers	156
Allograft	159
Hydroxyapatite (HA)	159
Additive Manufacturing	159
Additive Manufactured PEKK	159
Coatings	160
Cross-References	160
References	160

Abstract

Spinal implants are manufactured from a variety of materials to meet user needs as well as the requirements of the physical and environmental demands upon the device. Commonly used materials include titanium, stainless steel, cobalt-chrome, nitinol, carbon fiber reinforced polymer (CFRP), polyetheretherketone (PEEK), silicon nitride, biodegradable polymers, and allograft bone. Material choices can be driven by requirements for strength, biocompatibility, bone ongrowth, flexibility, and radiolucency. Coatings may also be applied to the implants to further enhance physical or biological properties of the implant. These may include

R. Sommerich · M. K. DeCelle · W. J. Frasier (✉)
Research and Development, DePuy Synthes Spine, Raynham, MA, USA
e-mail: BSommeri@its.jnj.com; mdecelle@ITS.JNJ.com; BFrasier@its.jnj.com

hydroxyapatite, titanium plasma, or a combination of these two materials. Additionally, implants may have a porous layer or open structure for improvement of osteointegration. Spinal implants are commonly made using conventional manufacturing methods such as machining and injection molding, but additive manufacturing is becoming more commonly used to produce certain implants.

Keywords

Spinal · Implant · Titanium · PEEK · Cobalt-chrome · Interbody · Pedicle screw · Cage · Rod · Hydroxyapatite

Introduction

Modern spinal surgeries use a variety of implants to decompress neural elements, support spinal segments, and stabilize motion segments. This can be achieved by restricting motion through fusion or preserving the natural motion and kinematics of the spine. Fusion occurs through the interbody space from one end plate to another, and the support of this space is provided by an interbody cage, with stability and compression provided by bone screws or hooks and rods. Spinal plates may also be used to provide stability, restore initial bone mechanics, and speed up the healing process after injury (Caspar et al. 1998; Emery et al. 1997). While the implant must withstand anatomical loading, the implant must not result in stress shielding of the surrounding bone which may result in impeding new bone growth. Additional stability may be provided with the use of bone screws or hooks which are connected to the associated rods using set screws. Multiple materials are used to manufacture these implants. These materials need to provide a balance of strength, stiffness, and biocompatibility, as well as manufacturability. In addition to the base materials, there are often surface treatments and coatings applied which are intended to improve implant performance, usually by increasing the screw's resistance to backing out or pulling out from the bone. This increased resistance to removal is achieved by providing surfaces that have improved ingrowth or adhesion of bone to the implant. When fusion is not the desired outcome, clinicians may opt to use implants such as interspinous process devices (IPDs) or artificial discs for spinal segment stabilization and motion preservation. IPDs, for example, provide indirect decompression of spinal nerve roots and canal. Motion preservation devices aim at allowing for load transfer similar to that of the natural kinematics of the spine (Wilke et al. 2008). When selecting an implant material, multiple factors should be considered such as anatomical location, desired clinical outcome, load sharing capability, desired range of motion (ROM), and degree of biocompatibility. This chapter will focus on implant selection based on material properties.

Metallic Implants

(a) **Titanium** – the most commonly used material to produce bone screws, rods, hooks, and set screws is titanium (Ti). Titanium is a popular choice due to its favorable properties of strength, corrosion resistance, and biocompatibility. Compared to stainless steel, titanium produces a less pronounced imaging artifact during X-ray or computed tomography (CT) scans and is less likely to have bacteria adhere to it (Luca et al. 2013). Titanium has also been shown to have a higher rate of bone ongrowth compared to stainless steel, and when used in pedicle screws, to have an increased resistance to backing out, as measured by removal torque in a mini-pig model (Christensen et al. 2000). Implant grade titanium is available primarily in three varieties: titanium-aluminum-vanadium (Ti-6Al-4V), commercially pure (CP), and titanium-molybdenum (Ti-15Mo). Example mechanical properties of these materials are summarized in Table 1. In general, Ti-6Al-4V is the most commonly used of the three options. Ti-6Al-4V is stronger and stiffer than Commercially Pure Titanium, readily available, and easily to machine. After contouring, such as in the case of spinal rods, Ti-6Al-4V also holds its shape

Table 1 Example implant material properties

Material	Tensile strength, ultimate (MPa)	Tensile strength, yield (MPa)	Modulus of elasticity (GPa)	Elongation at break (%)
Ti-6Al-4V ELI	862	786	110	10
Commercially pure Ti (Grade 4)	550	483	102	15
Ti-15Mo (alpha + beta annealed + aged)	900	800	105	10
316L annealed stainless steel	490	190	193	40
Cobalt- chromium	1290	760	235	25
PEEK		80	4	
CFRP		120	18	

Disegi (2009), Zaman et al. (2017), and Najeeb et al. (2016)
Note: Material properties can vary based on processing and should be verified with the selected supplier

better over temperature changes than commercially pure titanium (Noshchenko et al. 2011). Titanium-molybdenum is more difficult and expensive to obtain, and requires advanced expertise in machining, due to its nature of clogging cutting tools. However, when processed to the alpha+beta phase, Ti-15Mo has superior strength properties and a higher resistance to failure in cyclic loading or crack propagation due to stress risers compared to Ti-6Al-4V.

(b) **Stainless steel** – Stainless steel (SS) has been used for bone screws, rods, and hooks as well. Over the past decade this material has fallen out of favor due to patients with nickel allergies. In the past, stainless steel had historically been the material of choice for spinal rods over titanium when a stronger, stiffer construct was required. The use of a stainless steel rod often drove the use of stainless steel screws, hooks, and set screws. This was intended to prevent galvanic corrosion between dissimilar metals, which was a concern when using titanium bone screws with stainless steel rods. These concerns were proved to be generally unfounded (Serhan et al. 2004). The stainless steel grade used for implants is 316L. This material is available in different treatments, providing multiple strengths and stiffnesses. The material properties of 316L stainless steel are summarized in Table 1.

The material of choice for spinal plates has shifted from stainless steel to titanium alloys such as Ti-6Al-4V. A titanium alloy plate can provide sufficient rigidity and stability to allow for arthrodesis, prevent displacement or collapse of the intervertebral grafts, and maintain cervical lordosis to achieve a better prognosis (Chen et al. 2016). Titanium alloy implants are more ductile than stainless steel implants. It is also a proven biocompatible material.

In general, metal implants produce artifacts that make radiologic interpretation more challenging (Aryan et al. 2007). However, titanium and titanium alloys are more MRI (Magnetic Resonance Imaging) compatible than stainless steel due to its lower X-ray beam attenuation coefficients (Lee et al. 2007; Haramati et al. 1994). Another clinical benefit of titanium implants includes its ability to have a modified surface for improved osseointegration. For example, a rough surface can be induced on titanium implants which results in higher osseointegration compared to the smooth surface present on stainless steel implants.

(c) **Cobalt-chrome** – Cobalt-chrome (Co-Cr) is a relatively new entry into the materials available for implants. It is most commonly used for spinal rods, but not necessarily screws or hooks. The advantages of this material over stainless steel or titanium are numerous.

It provides higher strength and stiffness than titanium given the same rod diameter. This allows for the creation of stiffer constructs with stronger correction, or the use of smaller profile implants. Cobalt-chrome rods that are 5.5 mm in diameter have a greater bending stiffness than 6.35 mm diameter titanium rods. Cobalt-chrome also produces less imaging artifact than stainless steel, and can be combined with titanium screws which have better biocompatibility than stainless steel screws. Although mixing of metals in the body (titanium and cobalt-chrome or titanium and stainless steel) may result in galvanic corrosion, the susceptibility of the Ti-Co-Cr construct to this phenomenon is theorized to be less than in a Ti-SS construct (Piazzolla et al. 2013). Additionally, it has been found that the amount of galvanic corrosion evident with two connected stainless-steel implants is actually greater than the corrosion present between a stainless steel and titanium implant (Serhan et al. 2004). Table 1 summarizes example material properties of cobalt-chromium.

(d) **Nitinol** – Nitinol has been used to manufacture spinal rods with the goal of creating a less stiff construct to help reduce adjacent segment disease and provide a more compliant construct. Nitinol is a nickel–titanium alloy, which can be manufactured to produce unique shape-memory effects. Although it contains nickel, animal studies have found that no measurable amounts of nickel are absorbed into the body after implantation (Kok et al. 2013). Although studied in the literature, Nitinol rods have not proven to be particularly popular in the market. Concerns around fretting or wear and corrosion of the nitinol material where it is connected to conventional titanium or stainless steel screws raise concerns around premature implant failure, and thus would require specially treated screws to be used with nitinol rods. This, along with the processing costs and complexity of nitinol may be factors preventing widespread adoption in the market. An additional potential use of nitinol rods is in sliding growth constructs used in the treatment of early onset scoliosis. The sliding rod component allows for less traumatic adjustment of the construct as the patient grows compared to conventional fixed rod constructs. Nitinol has 100 times greater wear resistance than titanium and similar wear resistance as cobalt-chromium. This increased wear resistance would greatly reduce the amount of wear debris produced by the sliding construct over the implantation period, which spans multiple years, greatly reducing the patient's exposure to metallic particles and the potential irritation these could cause (Lukina et al. 2015)

Porous Metals

Materials having a porous structure have been developed in an attempt to increase the physical integration of bone to the implant structure. For example, with interbody cages, this is intended to result in "enhanced fixation of the device, preventing device migration or movement causing abrasive damage to adjacent tissue" and "may provide a transitional zone between the bone and biomaterial to reduce stress-shielding" (Jarman-Smith et al. 2012). Porous metals such as titanium (PlivioPore, Synthes; Tritanium Stryker), Nitinol (Actipore Biorthex), and Tantalum (Trabecular Metal, Zimmer (Hedrocel, Implex)) have been developed and commercialized to address this issue (Jarman-Smith et al. 2012; Lewis 2013). While these materials may address approximating the modulus of bone and the potential for ingrowth for increased stability, the issue with lack of radiolucency and CT/MRI artifact remains.

Polymers

As more metallic devices were implanted, reported issues with subsidence and stress shielding increased. With interbody cages, for example, metallic implants prevented the assessment of fusion due to lack of radiolucency.

Seaman describes "while Ti (titanium) had favorable fusion rates, a noted shortcoming was subsidence or settling into the adjacent vertebral bodies due to the differences in the modulus of elasticity. As a result, polyetheretherketone (PEEK) cages were introduced in the 1990s as an alternative due to their elastic modulus properties" (Seaman et al. 2017). PEEK allows for improved load sharing within the spine while stabilizing the disease segment and reducing stress on adjacent levels comparing to metallic implants, such as Ti.

Additional materials developed during this time including carbon fiber reinforced polymer (CFRP) cages that consisted of PEEK material with carbon fibers. Both PEEK and CFRP implant materials are biocompatible for safe implantation in the spine.

These polymers have clear advantages of reduced modulus, radiolucency, and reduced CT/MRI artifact in comparison to titanium. The PEEK and CFRP implants have elastic modulus characteristics similar to that of natural bone as compared to titanium. The strength of the CFRP material allows for a reduced implant volume and greater graft volumes as compared to implants manufactured from pure PEEK. Brantigan et al. (1991) reported increased pullout forces and similar compressive strengths for the carbon fiber cage as compared to femoral grafts when placed in cadaveric specimens. The reduced elastic modulus and implant design has been shown to potentially load the interbody graft material to allow for a better load sharing environment (Vadapalli et al. 2006; Kanayama et al. 2000).

In addition to the more commonly used metals, there is some use of PEEK or CFRP for both pedicle screws and spinal rods as well. PEEK has generally been used only for spinal rods, while CFRP has been for screws (Ringel et al. 2017). PEEK/CFRP is obviously much weaker and less stiff than the other metallic choices outlined above. The attraction of PEEK rods was the theory that they would flex as the spine moves and would have a similar modulus of elasticity to a PEEK or CFRP interbody spacer, which would also allow for some compliance. This modulus of elasticity is designed to be between that of cortical and cancellous bone, which allows for improved load sharing while still stabilizing the intended segments, but ultimately reducing the chance of adjacent segment degeneration (Athanasakopoulos et al. 2013). This flexible structure, rather than a completely rigid metal one, would be less likely to result in interbody spacers subsiding into the vertebral end plates and pedicle screws plowing out of or fracturing a pedicle when continually loaded, as in normal activities of daily living. However, PEEK rods have limited application to a smaller number of patients because they are not able to be contoured intraoperatively as compared to titanium or stainless steel rods.

PEEK rods and CFRP screws offer a major advantage over metallic implants when being imaged. They are radiolucent and produce no artifact from magnetic resonance imaging. This is especially useful for patients being treated for spinal tumors, where radiation treatment, planning and execution are negatively impacted by titanium or stainless steel screws (Ringel et al. 2017). PEEK rods have also been used successfully in non-fusion procedures. In these procedures, the flexibility of the rods allows for some motion to be maintained in the segment while still offering support and stabilization to the diseased segments. The results of a multi-patient study were an improvement in pain scores and a reduction in range of motion, with an implant failure rate lower than normally reported in the literature (Huang et al. 2016).

However, there remains a potential concern of direct bone ongrowth onto the implant surfaces of PEEK implants. PEEK is a highly inert, hydrophobic thermoplastic polymer that often results in a lack of direct apposition to bone for proper long-term implant performance. The presence of a fibrous tissue layer between the PEEK implant and the adjacent bone has been documented clinically and in animal studies (Phan and Mobbs 2016; Walsh et al. 2015). Phan has described the resulting radiolucent rim at the bone–implant interface due to the fibrous tissue as a "PEEK-Halo" (Phan et al. 2016).

A number of methods have been used to improve the bioactive surfaces of PEEK implants. Implants have been designed with both PEEK and titanium materials to allow for the titanium surfaces to contact the underlying bone (Rao et al. 2014). Additionally, PEEK implants have been coated with titanium or hydroxyapatite (HA) to improve biocompatibility to increase the resultant direct apposition of bone to the PEEK implant surface (Rao et al. 2014; Robotti and Zappini 2012). However, an early summary of clinical results with the titanium-coated PEEK indicated similar fusion rates as compared to uncoated PEEK (Assem et al. 2015). PEEK is also currently available with HA incorporated into the material (PEEK-OPTIMA HA Enhanced, Invibio) which allows for typical machining of the implant with exposure to HA at the surfaces of the implant. The PEEK HA Enhanced has been shown in animals to result in more direct bone apposition as compared with PEEK bulk material only (Walsh et al. 2016). The addition of bioactive materials to PEEK, surface modification techniques, processing techniques for deposition coating of PEEK implants, and functional and mechanical properties of PEEK are well described (Robotti and Zappini 2012; Roeder and Conrad 2012; Poulsson and Richards 2012). It should be noted that the desire to improve the bone ongrowth onto the PEEK implants must not be at the risk of potential failure of the applied coating during anatomical loading or insertion of the implant. Investigations of coatings have indicated the potential for wear debris or surface damage to occur as a result of procedural impaction to place the implant (Kienle et al. 2016).

Porous PEEK

The solution to the issue of radiopacity that exists with metallic implants may be the development of porous PEEK materials. This may be accomplished through various methods which include particulate leaching, heat sintering, and selective laser sintering. Jarman-Smith describes case studies of porous PEEK that includes mechanical testing and an animal ingrowth in comparison to solid PEEK (Jarman-Smith et al. 2012). In general, bone ingrowth was present in the porous PEEK materials, and more bone ongrowth of the porous PEEK samples which increased at over the 4- to 12-week time periods was demonstrated. Based on the mechanical requirements for a load bearing application a solid–porous PEEK device may be required to meet the functional demands. A solid–porous hybrid has been described using sodium chloride crystals that are leached out to produce a porous surface structure for bone ingrowth. The mechanical properties of the resulting structure have been estimated to support the functional requirements for an interbody device (Torstrick et al. 2016). The mechanical shear properties have been characterized and compared to bulk sintered PEEK in which the surface porous PEEK produced significantly higher results (23.96 MPa vs. 6.81 MPa, for surface porous and bulk porous, respectively). Early clinical results after 1 year with 100 patients have shown no device-related complications (Torstrick et al. 2017) (Fig. 1).

Silicon Nitride

Silicon nitride (Si_3N_4) is a ceramic that has been implanted as an interbody fusion device since 2008 with approximately 25,000 implants up to the year 2015 (McEntire et al. 2015). The materials have also been studied for its characteristics of osteointegration and anti-infection. New bone formation was found to be increased in the absence and presence of a bacterial injection as compared to titanium and PEEK (Webster et al. 2012). However, long-term 10-year clinical history has indicated a potential of adjacent level degeneration that was proposed to be caused by stress shielding due to elastic modulus mismatch (Sorrell et al. 2004). The elastic modulus of silicon nitride is approximately 300 GPa, while that of cortical bone is roughly 10 GPa (Bal and Rahaman 2012).

Biodegradable Polymers

The high stiffness of metallic implants has potential to shield the loading required within the spine to allow for fusion (Chen et al. 2016). This has led

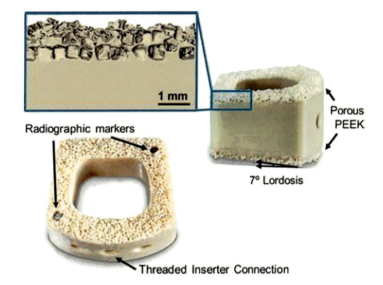

Fig. 1 COHERE implant demonstrating the characteristics of porous PEEK surface (Torstrick et al. 2017)

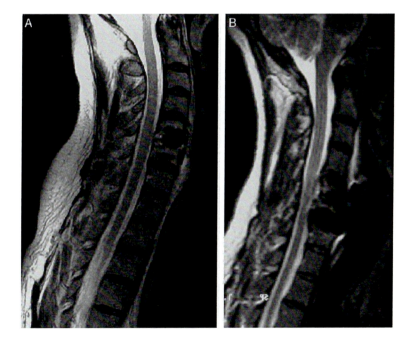

Fig. 2 T2-weighted magnetic resonance imaging of cervical spine showing early postoperative changes after the implantation of bioresorbable plate (**a**) and after implantation of titanium plate (**b**). Notice the obvious imaging artifacts in (**b**) compared with (**a**) (Nabhan et al. 2009)

to the use of biodegradable polymers for use in spine surgery with implants such as cervical plates. The modulus of elasticity of polymers can be altered based on the amount of cross-linking of the polymeric chains present within the material (Cheng et al. 2009). Biodegradable polymers may have lower modulus of elasticity that better represents physiological values when compared to metals (Freeman et al. 2006) which, in turn, can prevent stress shielding. In addition, polymers allow for greater visualization within the interbody space intraoperatively (Aryan et al. 1976) because the material does not produce artifact on MRI or CT scans (Nabhan et al. 2009). This becomes particularly important with specific patient groups, that is, obese patients and patients with shorter necks (Nabhan et al. 2009) (Fig. 2).

One major clinical benefit of biodegradable polymers is the ability of the material to completely hydrolyze within 2 years of initial

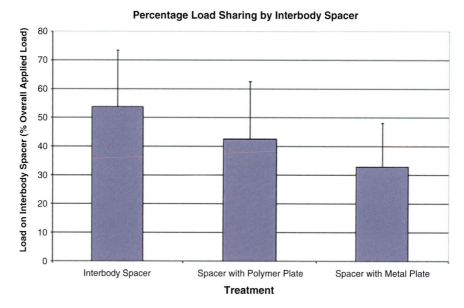

Fig. 3 Graphical representation of percentage load sharing by interbody spacer in the anterior spinal column (Cheng et al. 2009)

surgery. Spinal plates, for example, maintain approximately 90% of its initial strength 6 months post-implantation and approximately 70% of its initial strength 9 months post-implantation. This slow decrease in strength may allow the area of fusion to gradually take more of the load to potentially increase the rate of fusion while reducing stress shielding (Ames et al. 2002; Ciccone et al. 2001). Therefore, there is no need for implant removal in the case of a revision or adjacent segment surgery (Chen et al. 2016). This can reduce the long-term complications that have been historically associated with metallic plating.

In contrast, Boyle et al. compared ROM between an interbody space with a titanium rigid plate and an interbody space fixed with a biodegradable polymer plate (Cheng et al. 2009). They found that the titanium plate in conjunction with the interbody spacer achieved the highest level of motion reduction and also exhibited the lowest mean ROM. In a study by Freeman et al. (2006), the reduction of the ROM for both biodegradable and titanium anterior cervical plates was also compared. The results reported a reduction in the flexion–extension ROM by approximately 50% for a biodegradable plate and approximately 70% with titanium construct.

Boyle et al. also examined the percentage of load sharing with respect to three different conditions:

1. Stand-alone interbody spacer.
2. Spacer with a polymer plate.
3. Spacer with a rigid titanium plate.

The results showed that there was a statistical difference in compressive loading of anterior columns between the stand-alone spacer and the spacer with the Ti plate. However, there was no statistical difference in loading between the stand-alone spacer and the spacer with the polymer plate (see Fig. 3).

Therefore, this study showed that a spacer with a metal plate results in a lower percentage of load shared by the interbody spacer than with a bioresorbable plate. Researchers have reported concerns regarding the reduced rigidity of the biodegradable material and how this will impact its long-term efficacy compared to a rigid metal plate. Brkaric et al. (2007) reported early failure of a bioabsorbable plates, questioning the role of hydrolysis on crack initiation and propagation in polymer plates. In contrast to Boyle's study, there are encouraging clinical results of bioabsorbable plates (Aryan et al. 1976; Franco et al. 2007; Nabhan et al. 2009; Park et al. 2004; Tomasino et al. 2009; Vaccaro et al. 2002). In regard to imaging, Nabhan et al. (2009) confirmed that a single level bioresorbable plate is MRI and tissue compatible, and shows comparable fusion rates to titanium plate.

Allograft

Allograft is the most commonly used non-autogenous grafting material in spinal surgery (Hamer et al. 1996). Mineralized allograft is primarily osteoconductive, with weak osteoinductive capacity. The majority of allografts are primarily composed of cancellous or cortical bone. Cortical bone allografts provide significant mechanical stability and structural support. Cancellous bone allografts have a faster rate of incorporation. Therefore, the clinical application of an allograft should be considered when selecting graft material. Allografts do not have osteogenic potential because graft cells do not survive the processing/transplantation process. Allograft used for orthopedic applications is fresh frozen, freeze-dried, or demineralized.

One concern with the use of human allograft is the potential of disease transmission from donor to recipient. Donor screening, tissue testing and tissue processing have reduced this risk to less than 1 event per million grafts (Stevenson et al. 1996).

Hydroxyapatite (HA)

HA is composed of calcium phosphate mineral, which has both osteointegrative and osteoconductive properties. Osteointegration results from the formation of a layer of HA shortly after implantation. HA has highly osteoconductive properties, which promote bone growth on a surface (Cook et al. 1994). The material is composed of hydroxylated calcium phosphate and is chemically identical to natural HA of bone (Doria and Gallo 2016). It has the ability to bond directly to bone which reproduces the natural bone-cementing mechanism (Eggli et al. 1988). HA is a very brittle ceramic and is prone to fracture with cyclic loading.

Additive Manufacturing

Currently several manufacturers offer a variety of titanium devices that are produced with additive manufacturing for orthopedic implants. These include porous matrices (Zimmer Biomet OsseoTi Porous Metal, Stryker Tritanium, and Smith & Nephew CONCELOC) and designs with open or porous surfaces (4WEB, Joimax, Renovis, K2M, and Spineart). Lewis published a comparison of commercially available porous metals (Lewis 2013).

The manufacturing technique of additive manufacturing by selective laser sintering (SLS) or electron beam (EB) (termed "powder bed fusion" by ASTM) (ASTM F2792) allows for design options not allowed by subtractive manufacturing methods. An example of this is the truss-based designs (4WEB, Camber Spine) for spinal implants. Due to the variability in processes it is difficult to compare mechanical properties of resultant materials. Some of the available devices incorporate the porous–solid hybrid concept. Since these devices are a continuous structure from the solid to porous structure, the issue of coating delamination should be alleviated.

Additive Manufactured PEKK

An alternative implant material to titanium that is currently used in additive manufacturing for implants is polyetherketoneketone (PEKK) (http://oxfordpm.com/cmf-orthopedics). PEKK is from the same family of polyaryletherketone (PAEK) polymer materials as PEEK (Kurtz and Devine 2007; Kurtz 2012). The material properties are very similar to PEEK. PEKK has been used for cranial repair (FDA 510(k) Feb 2013) and interbody fusion devices (FDA 510(k) July 2015). The PEKK material has recently been investigated for antibacterial properties by Wang et al. The authors concluded from the in vitro testing that there was "decreased adhesion and growth of *P. aeruginosa* and *S. epidermidis* on nanorough PEKK surface compared with conventional PEEK surfaces" (Wang et al. 2017).

Coatings

In addition to various material choices for implants, there have been attempts made to improve the strength of the interface between the pedicle bone and screw through the use of surface coatings on the threads of the screw. Examples of coatings used include hydroxyapatite (HA) and titanium plasma spray (TPS). Additionally, these coatings have been combined into a composite coating (HA-TPS). Testing of these coating options on a titanium bone screw in a porcine model has shown improvement in screw back out torque compared to an uncoated titanium screw for all 3 of the coating options (Upasani et al. 2009).

Cross-References

▶ Anterior Lumbar Interbody Fusion and Transforaminal Lumbar Interbody Fusion
▶ Anterior Spinal Plates: Cervical
▶ Implant Material Bio-compatibility, Sensitivity, and Allergic Reactions
▶ Interbody Cages: Cervical
▶ Interspinous Devices
▶ Material Selection Impact on Intraoperative Spine Manipulation and Post-op Correction Maintenance
▶ Pedicle Screw Fixation
▶ Selection of Implant Material Effect on MRI Interpretation in Patients

References

Ames CP, Cornwall GB, Crawford NR, Nottmeier E, Chamberlain RH, Sonntag VK (2002) Feasibility of a resorbably anterior cervical graft containment plate. J Neurosurg 97:440–446
Aryan HE, Lu DC, Acosta FL Jr, Hartl R, McCormick PW, Ames CP (1976) Bioabsorbable anterior cervical plating: initial multicenter clinical and radiographic experience. Spine 32:1084–1088
Aryan HE, Lu DC, Acosta FL Jr, Hartl R, McCormick PW, Ames CP (2007) Bioabsorbable anterior cervical plating: initial multicenter clinical and radiographic experience. Spine 32:1084–1088
Assem Y, Mobbs RJ, Pelletier MH, Phan K, Walsh WR (2015) Radiological and clinical outcomes of novel Ti/PEEK combined spinal fusion cages: a systematic and preclinical evaluation. Eur Spine J 26:593
Athanasakopoulos M, Mavrogenis A, Triantafyllopoulos G, Koufos S, Pneumaticos S (2013) Posterior Spinal Fusion Using Pedicle Screws. Orthopedics 36(7): e951–e957
Bal BS, Rahaman MN (2012) Orthopedic applications of silicon nitride ceramics. Acta Biomater 8:2889–2898
Brantigan JW, Steffee AD, Geiger JM (1991) A carbon fiber implant to aid interbody lumbar fusion. Spine 16: S277–S282
Brkaric M, Baker KC, Israel R, Harding T, Montgomery DM, Herkowitz HN (2007) Early failure of bioabsorbable anterior cervical fusion plates: case report and failure analysis. J Spinal Disord Tech 20:248–254
Caspar W, Geisler FH, Pitzen T, Johnson TA (1998) Anterior cervical plate stabilisation in one and two level degenerative disease: overtreatment or benefit? J Spinal Disord 11:1–11
Chen M, Yang S, Yang C, Xu W, Ye S, Wang J, Feng Y, Yang W, Liu X (2016) Outcomes observed during a 1-year clinical and radiographic follow-up of patients treated for a 1- or 2-level cervical degenerative disease using a biodegradable anterior cervical plate. J Neurosurg Spine 25:205–212
Cheng BC, Burns P, Pirris S, Welch WC (2009) Load sharing and stabilization effects of anterior cervical devices. J Spinal Disord Tech 22:571–577
Christensen FB, Dalstra M, Sejling F, Overgaard S, Bünger C (2000) Titanium-alloy enhances bone-pedicle screw fixation: mechanical and histomorphometrical results of titanium-alloy versus stainless steel. Eur Spine J 9:97–103
Ciccone WJ, Motz C, Bentley C, Tasto JP (2001) Bioabsorbable implants in orthopaedics: new developments and clinical applications. J Am Acad Orthop Surg 9:280–288
Cook SD, Dalton JE, Tan EH, Tejeiro WV, Young MJ, Whitecloud TS 3rd (1994) In vivo evaluation of anterior cervical fusions with hydroxylapatite graft material. Spine 19:1856–1866
Disegi J (2009) Implant materials. Wrought titanium-15% molybdenum, 2nd edn. Synthes: West Chester, PA, USA
Doria C, Gallo M (2016) Roles of materials in cervical spine fusion. In: Cervical spine: minimally invasive and open surgery. Springer International Publishing, Cham, pp 159–171
Eggli PS, Müller W, Schenk RK (1988) Porous hydroxyapatite and tricalcium phosphate cylinders with two different pore size ranges implanted in the cancellous bone of rabbits. A comparative histomorphometric and histologic study of bony ingrowth and implant substitution. Clin Orthop Relat Res 232:127–138

Emery SE, Fisher JR, Bohlman HH (1997) Three level anterior cervical discectomy and fusion: radiographic and clinical results. Spine 15(22):2622–2624

Franco A, Nina P, Arpino L, Torelli G (2007) Use of resorbable implants for symptomatic cervical spondylosis: experience on 16 consecutive patients. J Neurosurg Sci 51:169–175

Freeman AL, Derincek A, Beaubien BP, Buttermann GR, Lew WD, Wood KB (2006) In vitro comparison of bioresorbable and titanium anterior cervical plates in the immediate postoperative condition. J Spinal Disord Tech 19:577–583

Hamer AJ, Strachan JR, Black MM, Ibbotson CJ, Stockley I, Elson RA (1996) Biomechanical properties of cortical allograft bone using a new method of bone strength measurement: a comparison of resh, fresh-frozen and irradiated bone. J Bone Joint Surg Br 78:363–368

Haramati N, Staron RB, Mazel-Sperling K, Freeman K, Nickoloff EL, Barax C, Feldman F (1994) CT scans through metal scanning techniques versus hardware composition. Comput Med Imaging Graph 18:429–434

Huang W, Chang Z, Song R, Zhou K, Yu X (2016) Non-fusion procedure using PEEK rod systems for lumbar degenerative diseases: clinical experience with a 2-year follow up. BMC Musculoskelet Disord 17:53

Jarman-Smith M, Brady M, Kurtz SM, Cordaro NM, Walsh WR (2012) Porosity in polyaryletherketeone. In: Kurtz S (ed) PEEK biomaterials handbook. Elsevier, New York, pp 181–199

Kanayama M, Cunningham BW, Haggerty CJ, Abumi K, Kaneda K, McAfee PC (2000) In vitro biomechanical investigation of the stability and stress-shielding effect of lumbar interbody fusion devices. J Neurosurg Spine 93:259–265

Kienle A, Graf N, Wilke H-J (2016) Does impaction of titanium-coated interbody fusion cages into the disc space cause wear debris or delamination? Spine J 16(2):235–42

Kok D, Firkins PJ, Wapstra FH, Veldhuizen AG (2013) A new lumbar posterior fixation system, the memory metal spinal system: an in-vitro mechanical evaluation. BMC Musculoskelet Disord 14:269

Kurtz SM (2012) An overview of PEEK biomaterials. In: Kurtz SM (ed) PEEK biomaterials handbook. Elsevier, New York, pp 1–7

Kurtz SM, Devine JN (2007) PEEK biomaterials in trauma, orthopedic, and spinal implants. Biomaterials 28:4845–4869

Lee MJ, Kim S, Lee SA, Song HT, Huh YM, Kim DH, Han SH, Suh JS (2007) Overcoming artifacts from metallic orthopedic implants at high-field-strength MR imaging and multi-detector CT. Radiographics 27:791–803

Lewis G (2013) Properties of open-cell porous metals and alloys for orthopedic applications. J Mater Sci 24:2293–2325

Luca A, Lovi A, Bruno B (2013) Titanium vs peek for implants in lumbar surgery. In: European Spine Journal Conference: 36th Italian Spine Society National Congress, Bologna, Italy, p 945

Lukina E, Kollerov M, Meswania J, Wertheim D, Mason P, Wagstaff P, Laka A, Noordeen H, Yoon WW, Blunn G (2015) Analysis of retrieved growth guidance sliding LSZ-4D devices for early onset scoliosis and investigation of the use of nitinol rods for this system. Spine 40:17–24

McEntire BJ, Bal BS, Chevalier J, Pezzotti G (2015) Ceramics and ceramic coatings in orthopedics. J Eur Ceram Soc 35:4327–4369

Nabhan A, Ishak B, Steimer O, Zimmer A, Pitzen T, Steudel WI, Pape D (2009) Comparison of boresorbable and titanium plates in cervical spinal fusion: early radiologic and clinical results. J Spinal Disord Tech 22:155–161

Najeeb S, Zafar M, Jhurshid Z, Siddiqui F (2016) Applications of polyetheretherketone (PEEK) in oral implantology and prosthodontics. J Prosthodont Res 60:12–19

Noshchenko A, Patel VV, Baldini T, Yun L, Lindley EM, Burger EL (2011) Thermomechanical effects of spine surgery rods composed of different metals and alloys. Spine 36:870–878

Park MS, Aryan HE, Ozgur BM, Jandial R, Taylor WR (2004) Stabilization of anterior cervical spine with bioabsorbable polymer in one- and two-level fusions. Neurosurgery 54:631–635

Phan K, Mobbs RJ (2016) Evolution of design of interbody cages for anterior lumbar interbody fusion. Orthop Surg 8:270–277

Phan K, Hogan JA, Assem Y, Mobbs RJ (2016) PEEK-Halo effect in interbody fusion. J Clin Neurosci 24:138–140

Piazzolla A, Solarino G, Gorgoglione F, Mori C, Garofalo N, Carlucci S, Montemurro V, De Giorgio G, Moretti B (2013) The treatment of adolescent idiopathic scoliosis (AIS) with cobalt-chromium-alloy (CoCR-alloy) devices: early results. In: European Spine Journal Conference: 36th Italian Spine Society National Congress, Bologna, Italy, pp 942–943

Poulsson AH, Richards RG (2012) Surface modification techniques of polyetheretherketone, including plasma surface treatment. In: Kurtz S (ed) PEEK biomaterials handbook. Elsevier, New York, pp 145–161

Rao PJ, Pelletier MH, Walsh WR, Mobbs RJ (2014) Spine interbody implants: material selection and modification, functionalization and bioactivation of surfaces to improve osseointegration. Orthop Surg 6:81–89

Ringel F, Ryang YM, Kirschke JS, Müller BS, Wilkens JJ, Brodard J, Combs SE, Meyer B (2017) Radiolucent carbon fiber reinforced pedicle screws for treatment of spinal tumors: advantages for radiation planning and follow-up imaging. World Neurosurg 105:294–301

Robotti P, Zappini G (2012) Thermal plasma spray deposition of titanium and hydroxyapatite on polyaryletherketone implants. In: Kurtz SM (ed) PEEK biomaterials handbook. Elsevier, New York, pp 119–143

Roeder RK, Conrad TL (2012) Bioactive polyaryletherketone composites. In: Kurtz S (ed) PEEK biomaterials handbook. Elsevier, New York, pp 163–179

Seaman S, Kerezoudis P, Bydon M, Torner JC, Hitchon PW (2017) Titanium vs. polyetheretherketone (PEEK) interbody fusion: meta-analysis and review of the literature. J Clin Neurosci 44:23–29

Serhan H, Slivka M, Albert T, Kwak SD (2004) Is galvanic corrosion between titanium alloy and stainless steel spinal implants a clinical concern? Spine J 4:379–387

Sorrell CC, Hardcastle PH, Druitt RK, Howlett CR, McCartney ER (2004) Results of 15-year clinical study of reaction bonded silicon nitride intervertebral spacers. In: 7th World Biomaterials Congress. Australian Society for Biomaterials, Sydney, p 1872

Stevenson S, Emery SE, Goldberg VM (1996) Factors affecting bone graft incorporation. Clin Orthop Relat Res 324:66–74

Tomasino A, Gebhard H, Parikh K, Wess C, Härtl R (2009) Bioabsorbable instrumentation for single-level cervical degenerative disc disease: a radiological and clinical outcome study. J Neurosurg Spine 11:529–537

Torstrick FB, Evans NT, Stevens HY, Gall K, Guldberg RE (2016) Do surface porosity and pore size influence mechanical properties and cellular response to PEEK? Clin Orthop Relat Res 474:2372–2383

Torstrick FB, Safranski DL, Burkus JK, Chappuis JL, Lee CS, Guldberg RE, Smith KE (2017) Getting PEEK to stick to bone: the development of porous PEEK for interbody fusion devices. Tech Orthop 32:158–166

Upasani VV, Farnsworth CL, Tomlinson T, Chambers RC, Tsutsui S, Slivka MA, Mahar AT, Newton PO (2009) Pedicle screw surface coatings improve fixation in nonfusion spinal constructs. Spine 34(4):335–343

Vaccaro AR, Venger BH, Kelleher PM, Singh K, Carrino JA, Albert T, Hilibrand A (2002) Use of a bioabsorbable anterior cervical plate in the treatment of cervical degenerative and traumatic disk disruption. Orthopedics 25:s1191–s1199

Vadapalli S, Sairyo K, Goel VK, Robon M, Biyani A, Khandha A, Ebraheim NA (2006) Biomechanical rationale for using polyetheretherketone (PEEK) spacers for lumbar interbody fusion – a finite element study. Spine 31:E992–E998

Walsh WR, Bertollo N, Christou C, Schaffner D, Mobbs RJ (2015) Plasma-sprayed titanium coating to polyetheretherketone improves the bone-implant interface. Spine J 15:1041–1049

Walsh WR, Pelletier MH, Berollo N, Chrsitou C, Tan C (2016) Does PEEK/HA enhance bone formation compared with PEEK in a sheep cervical fusion model. Clin Orthop Relat Res 474:2364–2372

Wang M, Bhardwaj G, Webster TJ (2017) Antibacterial properties of PEKK for orthopedic applications. Int J Nanomedicine 12:6471–6476

Webster TJ, Patel AA, Rahaman MN, Bal BS (2012) Anti-infective and osteointegration properties of silicon nitride, poly(ether ether ketone), and titanium implants. Acta Biomater 8:4447–4454

Wilke HJ, Drumm J, Häussler K, Mack C, Steudel WI, Kettler A (2008) Biomechanical effect of different lumbar interspinous implants on flexibility and intradiscal pressure. Eur Spine J 17(8):1049–1056

Zaman H, Sharif S, Kim DW, Idris MH, Suhaimi MA, Tumurkhuyag Z (2017) Machinability of cobalt-based and cobalt chromium molybdenum alloys – a review. Procedia Manuf 11:563–570

Material Selection Impact on Intraoperative Spine Manipulation and Post-op Correction Maintenance

Hesham Mostafa Zakaria and Frank La Marca

Contents

Introduction 164
Biomechanics of Spine Correction 164
Commonly Available Materials for Spine Instrumentation 165
Stainless Steel Alloys (SSA) 165
Titanium Alloys (TA) 165
Cobalt Chromium Alloys (CCA) 166
Polyetheretherketone 166
Mixing Metals 167
References 167

Abstract

As spine surgeons, there are a variety of products and technologies available for application within our discipline. The breadth of variety comes from the diverse materials that are available, each with a unique physical, mechanical, and biological property that gives it advantages and disadvantages. It is fundamentally important for a spine surgeon to understand every facet of these materials, because they will ultimately not only have a unique effect on the body's physiology but will also alter the ability to maintain stabilization while the wound heals and arthrodesis is achieved. This chapter will go over the common commercial materials available for spine stabilization and manipulation. It will discuss in depth the advantages and disadvantages of their specific biomechanical properties and biocompatible. Finally, this chapter will discuss how each material can affect spine stabilization and maintenance of correction.

Keywords

Spine surgery · Spine biotechnology · Spine implants · Metal implants · Metal alloys · Stainless steel · Cobalt chromium · Titanium · Polyetheretherketone · PEEK · Spine manipulation · Deformity correction · Metal implants · Metal alloys · Biomaterials · Biotechnology

H. Mostafa Zakaria · F. La Marca (✉)
Department of Neurosurgery, Henry Ford Hospital, Detroit, MI, USA

Department of Neurosurgery, Henry Ford Allegiance Hospital, Jackson, MI, USA
e-mail: hzakari1@hfhs.org; franklamarca@yahoo.com

© Springer Nature Switzerland AG 2021
B. C. Cheng (ed.), *Handbook of Spine Technology*,
https://doi.org/10.1007/978-3-319-44424-6_33

Introduction

It is fundamentally important to understand the mechanical and long term structural properties of spinal implants. Ultimately the goal of spinal instrumentation placement is to provide a transient and load bearing scaffold while wound healing and arthrodesis occurs. Subsequently, implants are required to maintain the biomechanical alignment and correction despite undergoing compressive, torsional, and bending forces. If a selected material is inappropriate for the specific surgery, then it may fail in vivo and thus destabilize the spine and require revision. Failure may be mechanical, or that material itself lacks the tensile strength to maintain correction, or as a result of the material interacting with the local tissue environment, such as corrosion, degradation, or wear. Furthermore, bone remodeling during the healing process as well as over time can affect interaction with the implant material that if not biocompatible could result in inflammatory reaction and/or subsidence of the implant with loss of original corrective alignment results. Tissue reaction to the implant material can also impede bone healing. If bone fusion is not obtained, then all materials may eventually undergo fatigue failure through cyclical force application and loading of the implant. The biomechanical properties of spine alignment correction and how material selection can impact spine manipulation and maintenance of correction will be reviewed.

Biomechanics of Spine Correction

To achieve the required torque for spine manipulation and stabilization, there are a few standard practices: maximum leverage is obtained with fixation of as many possible segments as needed, with rigid rod fixation combined with either pedicle screw fixation, sublaminar hook or wire fixation or anterior vertebral body screw fixation resulting in direct vertebral rotation, distraction and/or compression where needed to restore spinal alignment (Hitchon et al. 2003; Clements et al. 2009; Lee et al. 2004; Bridwell et al. 1993; Bono and Lee 2004; Boos and Webb 1997). The ideal surgical instrumentation will maintain its shape as well as corrective forces at implantation and until arthrodesis occurs. This ability is defined by key mechanical properties, including yield strength (force required to permanently deform the material), stiffness (ability to maintain correction), and fatigue life (how long it can sustain repeated stress).

Young's modulus is the ratio of stress (force per unit area) to strain (deformation of a material), and so it is a measure of the elasticity of a specified material. Ideally, Young's modulus of the implanted material should be similar to bone; differences in elasticity hinder the transfer of forces from the material to the bone tissue, which prevents normal bone remodeling and creates osteopenia, otherwise known as the stress-shielding effect in accordance with Wolf's law (Antunes and de Oliveira 2012; Ebramzadeh et al. 2003; Tahal et al. 2017). Young's modulus of the cortical bone varies depending on bone quality, but it is within the range of 20–30 GPa (El Masri et al. 2012; Dall'Ara et al. 2013). The ultimate and final properties of a construct will depend on the material's diameter, length, and shape. While the minimum parameters required for a given correction is not known, larger and stiffer rods are known to improve deformity correction, and two-rod systems are intuitively stronger than single-rod systems (Fricka et al. 2002; Yoshihara 2013; Abul-Kasim et al. 2011).

An important factor that effects rod fatigue life and a rod's ability to maintain postoperative deformity correction is the intraoperative notching of rods. During normal clinical intraoperative application of rods, straight rods are bent into the required alignment to maintain correction. This is achieved with either a French bender or an in situ bender, which can create a notch in the rod. Notches ultimately weaken the rod by allowing for concentrations of stress at the point of the notch (Shigley 2011; Cook and Young 1985), leading to worsened rod fatigue resistance at that point which can ultimately cause fracture at the point of the notch.

Regardless of the modulus of elasticity of current implant options, there is still a degree of variation between this modulus and that of bone.

Furthermore, the changes occurring during surgery cause for an increased stress riser at the level of transition between fixed spinal segments and those unoperated segments that maintain their physiologic motion. This area of transition and subsequent stress riser is concentrated at the apex of the construct and is known to change the in vivo mechanical properties of the spine, especially fatigue resistance at the transition zone (Yoshihara 2013; Nguyen et al. 2011; Dick and Bourgeault 2001; Lindsey et al. 2006).

Corrosion is material degradation due to reactions with its surrounding environment, which can ultimately lead to hardware failure (Singh and Dahotre 2007; Ratner et al. 2004). As the human body is composed of an aqueous saline solution with numerous cations and anions, it is considered quite corrosive and so corrosion resistance is one of the main features of biocompatibility (Pan et al. 1996; Lin and Bumgardner 2004; Williams and Clark 1982; Hallab et al. 2000; Merritt and Brown 1981, 1985; Baboian 2005). Interaction between different metal implants can also lead to corrosion. Corrosion resistance is impaired passively by the formation of a stable and unreactive surface layer on the material (Gotman 1997); for example, titanium alloys form titanium oxide on their surface which prevents corrosion. For most biomaterials, corrosion progresses by the breakdown of this layer through a variety of means, such as via micro-motions or galvanic processes (Wang et al. 1999; Cunningham et al. 2002, 2013; Hallab et al. 2012). The ability of a material to resist corrosion and failure is one of the main features of biocompatibility, as a corroded material is more likely to fail and corrosive byproducts induce aberrant tissue reactions.

Any foreign material implanted within the human body will produce a reaction in the surrounding tissues. The initial reaction after surgery is always inflammatory, which progresses to fibrosis and scar tissue. At times, this process can become inappropriate for spine surgery. Material selection is important in this regard, as at times, such as in the case where interbody bone tissue formation is required, and not scar tissue. Fibrotic connective tissue growth may supersede bony growth, causing pseudarthrosis. Some materials have been shown to specifically promote bone formation (Matsuno et al. 2001; Blanco et al. 2011). Corrosion and material wear may also produce metallic (nickel, cobalt, and chromium) particulates that form immunogenic metal–protein complexes (Swiontkowski et al. 2001), which can lead to chronic inflammation; chronic inflammation is painful, can affect bone healing (osteolysis), and is more likely to lead to hardware failure by the chronic release of intracellular acids and superoxides (Cook et al. 2000; Gaine et al. 2001; Wang et al. 1997).

Commonly Available Materials for Spine Instrumentation

Stainless Steel Alloys (SSA)

Modern-day surgical grade stainless steel alloys (SSA) are an alloy of iron, nickel, carbon, and molybdenum. SSA are easy to machine, and so they were originally the material of choice for plates and screws. The current practice of surgery seems to be shifting away from the use of SSA, for a variety of reasons; SSA is not corrosive resistant enough for the human body, and so surgical grade alloys require at least 17% chromium to form a stable surface oxide. SSA have the poorest in overall strength of the currently available materials (170 MPa), as well as having a much higher elastic modulus as compared to bone (~200 GPa). It has been shown that SSA have a higher rate of infection than other metals, which may be related to their ability to promote biofilm (Gaine et al. 2001; Soultanis et al. 2008). On MRI, SSA cause a large artifact, which may impair the ability to assess postoperative MRI imaging (Burtscher et al. 1998).

Titanium Alloys (TA)

Titanium naturally forms titanium oxide, making it more corrosive resistant than SSA (Yoshihara 2013; Serhan et al. 2004). Subsequently, titanium alloys are about 90% Ti with the remainder either aluminum or vanadium. However, there is a wide range of titanium alloys, each with a different atomic configuration causing a unique set of strengths and properties. The most common alloy is Ti-6Al-4V, which is has a higher yield strength (869 GPa) and lower Young's modulus than SSA (114 GPa), and may intrinsically resist biofilms and infection as well as promote bone growth and integration (Gaine et al. 2001; Soultanis et al. 2008; Stambough et al. 1997; Pienkowski et al. 1998; Wedemeyer et al. 2007; Yoon et al. 2008; Banerjee et al. 2004; Christensen et al. 2000; Sun et al. 1999). Unfortunately, it is more expensive, is less stiff, and has overall worse fatigue resistance than SSA (Antunes and de Oliveira 2012; Ghonem 2010; Tahal et al. 2017; Chan 2010), and notching of the rod is known to reduce fatigue resistance even further (Dick and Bourgeault 2001; Lindsey et al. 2006). Newer TAs created from only the beta atomic structure have an even lower Young's modulus at 50 GPa, even closer to that of bone (Antunes and de Oliveira 2012; Brailovski et al. 2011). Further developments in titanium manufacturing have recently produced porous titanium, which preserves its biocompatibility and further decreases its Young's modulus to 2–4 GPa (Wu et al. 2013; Fujibayashi et al. 2011). This new material have been promoted as an interbody device, as it has the strength and osteoinductive capacity of TA with a low enough Young's modulus to prevent the stress-shielding effect. Ultimately, due to its superior mechanical and biological properties, TA has become the standard of care in spine surgery.

Cobalt Chromium Alloys (CCA)

The use of cobalt chromium alloys for spine surgery is a recent introduction. Surgical CCA is a composition of cobalt, chromium, molybdenum, and carbon. Similar to TA, the chromium forms a surface oxide layer that imparts inherent corrosion resistance. CCA have a higher yield strength, stiffer, larger fatigue life span, more resistant to notching, and are less likely to fracture in vivo after scoliosis correction than TA (Nguyen et al. 2011; Doulgeris et al. 2013; Marti 2000; Shinohara et al. 2016; Scheer et al. 2011). However, while CCA is highly biocompatible, the milling of CCA is technically more difficult, making it more expensive, and CCA's Young's modulus is 240, much greater than that of bone. It is highly biocompatible. The stiffness of CCA is an important factor when deciding for which material to be used for some deformity correction, as SSA and TA are known to undergo deformations after deformity correction (Lamerain et al. 2014; Cidambi et al. 2012; Cui et al. 2012). It also produces a large artifact on MRI (Tahal et al. 2017).

Polyetheretherketone

Polyetheretherketone (PEEK) is an organic thermoplastic polymer of bisphenolate salts that was initially created in the 1980s (Panayotov et al. 2016). It has many advantageous biomechanical properties; it is strong, has a low fatigue failure rate, is radiolucent and does not cause a large artifact on MRI (Cho et al. 2002). Perhaps its greatest appeal comes from its biological inertness, resistance to corrosion, and its Young modulus being similar to that of bone (3.6 GPa), which prevents the stress-shielding effect when used as an interbody (Hee and Kundnani 2010). As PEEK is considered biologically inert, the PEEK devices used for implantation are composites of PEEK and 30% chopped carbon, which has been shown to promote osteoblast adherence and bone formation (Jockisch et al. 1992). They are commonly filled with autograft to promote for arthrodesis. It has proven efficacy for anterior cervical spine surgery, with no significant differences with titanium cages (Kasliwal and O'Toole 2014; Kersten et al. 2015). PEEK can create a local inflammatory response which eventually results in a biofilm formation surrounding the implant. However, initial reaction can promote local bone remodeling with subsidence of the implant and loss of disc height correction. Further enhancements to PEEK include plasma coating PEEK with titanium, which introduces some of the osteogenic properties of titanium and further enhances osseous integration (Walsh et al. 2015).

Mixing Metals

In the past, orthopedic implant practices cautioned against mixing dissimilar metals in a biologically active environment due to the fear of galvanic corrosion. Surgeons have therefore shied away from mixing different metal implants. However, orthopedic implant designs, materials and passivation processes have evolved considerably since their inception and there is more interest in combining dissimilar metals among orthopedic surgeons today. In the clinical setting, combinations of titanium and stainless steel most frequently occur in spinal fixation constructs. These metals are generally used together in an attempt to form a construct that takes advantage of the mechanical properties of each component.

Due to better biocompatibility and osseointegration, and lower modulus as well as concerns about late onset infection and the potential need for advanced imaging (CT and MRI), most surgeons have migrated to the use of titanium implants worldwide; therefore, most contemporary spinal implant systems are made of titanium. There are clinical scenarios in which a surgeon may want to use stainless steel or cobalt chrome rods, since, for the same rod diameter, the strength and stiffness properties of titanium are not appropriate. For complex, severe, rigid spinal deformities, titanium rods do not provide sufficient correction and long-term durability. Improper sagittal profile, insufficient correction (coronal and sagittal plane), and rod failure due to pseudarthrosis are the consequence.

For junctional degeneration or deformity at the top of a spinal construct known as proximal junctional kyphosis (PJK), the fusion and instrumentation have to be extended up into the cervical spine, across the cervicothoracic junction, sometimes many years later. Cervical systems traditionally have only been available in titanium, and so surgeons were faced with mixing metals to connect up to previous implants, unless they wanted to open up the entire previous incision, remove the old implants, and replace them at considerable morbidity and cost to the patient. Typically, titanium junctional rod-to-rod connectors are attached to stainless steel screws to create a stable revision construct.

Serhan et al. investigated spinal implant constructs of stainless steel rods with mixed stainless steel and titanium alloy components and implants consisting of titanium alloy rods with mixed stainless steel and titanium alloy components (Serhan et al. 2004). Constructs were immersed in saline and subjected to cyclic bending tests. They were then evaluated visually, with electron microscopy and with spectroscopy for evidence of corrosion. The results indicated that the stainless-steel implant components were less resistant to corrosion than the titanium components. This is partly a result of the strong passivating ability of titanium when compared with stainless steel (Serhan et al. 2004). Based on these results and the clinical use of Ti with SS constructs, the FDA has cleared mixing SS and Ti for the first time in history for the EXPEDIUM Spine System and VIPER and VIPER 2 Systems (K160904) in July 1, 2016.

Surgeons have used stainless steel rods with titanium screws in fusion and non-fusion cases for early-onset scoliosis, adolescent idiopathic scoliosis (AIS), neuromuscular scoliosis, and extension of scoliosis constructs to the cervical spine with excellent clinical results and no sequelae (Zartman et al. 2011; Farnsworth et al. 2014). This comfort with mixing metals in spinal instrumentation constructs came out of necessity and has grown based on published literature and discussions in national and international meetings.

References

Abul-Kasim K, Karlsson MK, Ohlin A (2011) Increased rod stiffness improves the degree of deformity correction by segmental pedicle screw fixation in adolescent idiopathic scoliosis. Scoliosis 6:13

Antunes RA, de Oliveira MC (2012) Corrosion fatigue of biomedical metallic alloys: mechanisms and mitigation. Acta Biomater 8(3):937–962

Baboian R ed., (2005) Corrosion Tests and Standards: Application and Interpretation-Second ed. (West Conshohocken, PA: ASTM International) https://doi.org/10.1520/MNL20-2ND-EB

Banerjee R, Nag S, Stechschulte J, Fraser HL (2004) Strengthening mechanisms in Ti–Nb–Zr–Ta and Ti–Mo–Zr–Fe orthopaedic alloys. Biomaterials 25(17):3413–3419

Blanco JF, Sanchez-Guijo FM, Carrancio S, Muntion S, Garcia-Brinon J, del Canizo MC (2011) Titanium and tantalum as mesenchymal stem cell scaffolds for spinal fusion: an in vitro comparative study. Eur Spine J 20(Suppl 3):353–360

Bono CM, Lee CK (2004) Critical analysis of trends in fusion for degenerative disc disease over the past 20 years: influence of technique on fusion rate and clinical outcome. Spine (Phila Pa 1976) 29(4):455–463; discussion Z455

Boos N, Webb JK (1997) Pedicle screw fixation in spinal disorders: a European view. Eur Spine J 6(1):2–18

Brailovski V, Prokoshkin S, Gauthier M et al (2011) Bulk and porous metastable beta Ti–Nb–Zr (Ta) alloys for biomedical applications. Mater Sci Eng C 31(3):643–657

Bridwell KH, Sedgewick TA, O'Brien MF, Lenke LG, Baldus C (1993) The role of fusion and instrumentation in the treatment of degenerative spondylolisthesis with spinal stenosis. J Spinal Disord 6(6):461–472

Burtscher IM, Owman T, Romner B, Stahlberg F, Holtas S (1998) Aneurysm clip MR artifacts. Titanium versus stainless steel and influence of imaging parameters. Acta Radiol 39(1):70–76

Chan K (2010) Changes in fatigue life mechanism due to soft grains and hard particles. Int J Fatigue 32(3):526–534

Cho DY, Liau WR, Lee WY, Liu JT, Chiu CL, Sheu PC (2002) Preliminary experience using a polyetheretherketone (PEEK) cage in the treatment of cervical disc disease. Neurosurgery 51(6):1343–1349; discussion 1349–1350

Christensen FB, Dalstra M, Sejling F, Overgaard S, Bunger C (2000) Titanium-alloy enhances bone-pedicle screw fixation: mechanical and histomorphometrical results of titanium-alloy versus stainless steel. Eur Spine J 9(2):97–103

Cidambi KR, Glaser DA, Bastrom TP, Nunn TN, Ono T, Newton PO (2012) Postoperative changes in spinal rod contour in adolescent idiopathic scoliosis: an in vivo deformation study. Spine (Phila Pa 1976) 37(18):1566–1572

Clements DH, Betz RR, Newton PO, Rohmiller M, Marks MC, Bastrom T (2009) Correlation of scoliosis curve correction with the number and type of fixation anchors. Spine (Phila Pa 1976) 34(20):2147–2150

Cook RD, Young W (1985) Advanced mechanics of materials. Macmillan, New York

Cook S, Asher M, Lai SM, Shobe J (2000) Reoperation after primary posterior instrumentation and fusion for idiopathic scoliosis. Toward defining late operative site pain of unknown cause. Spine (Phila Pa 1976) 25(4):463–468

Cui G, Watanabe K, Nishiwaki Y et al (2012) Loss of apical vertebral derotation in adolescent idiopathic scoliosis: 2-year follow-up using multi-planar reconstruction computed tomography. Eur Spine J 21(6):1111–1120

Cunningham BW, Orbegoso CM, Dmitriev AE, Hallab NJ, Sefter JC, McAfee PC (2002) The effect of titanium particulate on development and maintenance of a posterolateral spinal arthrodesis: an in vivo rabbit model. Spine (Phila Pa 1976) 27(18):1971–1981

Cunningham BW, Hallab NJ, Hu N, McAfee PC (2013) Epidural application of spinal instrumentation particulate wear debris: a comprehensive evaluation of neurotoxicity using an in vivo animal model. J Neurosurg Spine 19(3):336–350

Dall'Ara E, Karl C, Mazza G et al (2013) Tissue properties of the human vertebral body sub-structures evaluated by means of microindentation. J Mech Behav Biomed Mater 25:23–32

Dick JC, Bourgeault CA (2001) Notch sensitivity of titanium alloy, commercially pure titanium, and stainless steel spinal implants. Spine (Phila Pa 1976) 26(15):1668–1672

Doulgeris JJ, Aghayev K, Gonzalez-Blohm SA et al (2013) Comparative analysis of posterior fusion constructs as treatments for middle and posterior column injuries: an in vitro biomechanical investigation. Clin Biomech (Bristol, Avon) 28(5):483–489

Ebramzadeh E, Normand PL, Sangiorgio SN et al (2003) Long-term radiographic changes in cemented total hip arthroplasty with six designs of femoral components. Biomaterials 24(19):3351–3363

El Masri F, Sapin de Brosses E, Rhissassi K, Skalli W, Mitton D (2012) Apparent Young's modulus of vertebral cortico-cancellous bone specimens. Comput Methods Biomech Biomed Engin 15(1):23–28

Farnsworth CL, Newton PO, Breisch E, Rohmiller MT, Kim JR, Akbarnia BA (2014) The biological effects of combining metals in a posterior spinal implant: in vivo model development report of the first two cases. Adv Orthop Surg 2014:1

Fricka KB, Mahar AT, Newton PO (2002) Biomechanical analysis of anterior scoliosis instrumentation: differences between single and dual rod systems with and without interbody structural support. Spine (Phila Pa 1976) 27(7):702–706

Fujibayashi S, Takemoto M, Neo M et al (2011) A novel synthetic material for spinal fusion: a prospective clinical trial of porous bioactive titanium metal for lumbar interbody fusion. Eur Spine J 20(9):1486–1495

Gaine WJ, Andrew SM, Chadwick P, Cooke E, Williamson JB (2001) Late operative site pain with isola posterior instrumentation requiring implant removal: infection or metal reaction? Spine (Phila Pa 1976) 26(5):583–587

Ghonem H (2010) Microstructure and fatigue crack growth mechanisms in high temperature titanium alloys. Int J Fatigue 32(9):1448–1460

Gotman I (1997) Characteristics of metals used in implants. J Endourol 11(6):383–389

Hallab NJ, Jacobs JJ, Skipor A, Black J, Mikecz K, Galante JO (2000) Systemic metal-protein binding associated with total joint replacement arthroplasty. J Biomed Mater Res 49(3):353–361

Hallab NJ, Chan FW, Harper ML (2012) Quantifying subtle but persistent peri-spine inflammation in vivo to submicron cobalt-chromium alloy particles. Eur Spine J 21(12):2649–2658

Hee HT, Kundnani V (2010) Rationale for use of polyetheretherketone polymer interbody cage device in cervical spine surgery. Spine J 10(1):66–69

Hitchon PW, Brenton MD, Black AG et al (2003) In vitro biomechanical comparison of pedicle screws, sublaminar hooks, and sublaminar cables. J Neurosurg 99(1 Suppl):104–109

Jockisch KA, Brown SA, Bauer TW, Merritt K (1992) Biological response to chopped-carbon-fiber-reinforced peek. J Biomed Mater Res 26(2):133–146

Kasliwal MK, O'Toole JE (2014) Clinical experience using polyetheretherketone (PEEK) intervertebral structural cage for anterior cervical corpectomy and fusion. J Clin Neurosci 21(2):217–220

Kersten RF, van Gaalen SM, de Gast A, Oner FC (2015) Polyetheretherketone (PEEK) cages in cervical applications: a systematic review. Spine J 15(6):1446–1460

Lamerain M, Bachy M, Delpont M, Kabbaj R, Mary P, Vialle R (2014) CoCr rods provide better frontal correction of adolescent idiopathic scoliosis treated by all-pedicle screw fixation. Eur Spine J 23(6): 1190–1196

Lee SM, Suk SI, Chung ER (2004) Direct vertebral rotation: a new technique of three-dimensional deformity correction with segmental pedicle screw fixation in adolescent idiopathic scoliosis. Spine (Phila Pa 1976) 29(3):343–349

Lin HY, Bumgardner JD (2004) In vitro biocorrosion of Co-Cr-Mo implant alloy by macrophage cells. J Orthop Res 22(6):1231–1236

Lindsey C, Deviren V, Xu Z, Yeh RF, Puttlitz CM (2006) The effects of rod contouring on spinal construct fatigue strength. Spine (Phila Pa 1976) 31(15): 1680–1687

Marti A (2000) Cobalt-base alloys used in bone surgery. Injury 31(Suppl 4):18–21

Matsuno H, Yokoyama A, Watari F, Uo M, Kawasaki T (2001) Biocompatibility and osteogenesis of refractory metal implants, titanium, hafnium, niobium, tantalum and rhenium. Biomaterials 22(11):1253–1262

Merritt K, Brown SA (1981) Metal sensitivity reactions to orthopedic implants. Int J Dermatol 20(2):89–94

Merritt K, Brown SA (1985) Biological effects of corrosion products from metals. Paper presented at: corrosion and degradation of implant materials: second symposium

Nguyen T, Buckley J, Ames C, Deviren V (2011) The fatigue life of contoured cobalt chrome posterior spinal fusion rods. Proc Inst Mech Eng H J Eng Med 225(2):194–198

Pan J, Thierry D, Leygraf C (1996) Hydrogen peroxide toward enhanced oxide growth on titanium in PBS solution: blue coloration and clinical relevance. J Biomed Mater Res 30(3):393–402

Panayotov IV, Orti V, Cuisinier F, Yachouh J (2016) Polyetheretherketone (PEEK) for medical applications. J Mater Sci Mater Med 27(7):118

Pienkowski D, Stephens GC, Doers TM, Hamilton DM (1998) Multicycle mechanical performance of titanium and stainless steel transpedicular spine implants. Spine (Phila Pa 1976) 23(7):782–788

Ratner BD, Hoffman AS, Schoen FJ, Lemons JE (2004) Biomaterials science: an introduction to materials in medicine. Academic, Boston

Scheer JK, Tang JA, Deviren V et al (2011) Biomechanical analysis of cervicothoracic junction osteotomy in cadaveric model of ankylosing spondylitis: effect of rod material and diameter. J Neurosurg Spine 14(3): 330–335

Serhan H, Slivka M, Albert T, Kwak SD (2004) Is galvanic corrosion between titanium alloy and stainless steel spinal implants a clinical concern? Spine J 4(4): 379–387

Shigley JE (2011) Shigley's mechanical engineering design. Tata McGraw-Hill Education, New York

Shinohara K, Takigawa T, Tanaka M et al (2016) Implant failure of titanium versus cobalt-chromium growing rods in early-onset scoliosis. Spine (Phila Pa 1976) 41(6):502–507

Singh R, Dahotre NB (2007) Corrosion degradation and prevention by surface modification of biometallic materials. J Mater Sci Mater Med 18(5):725–751

Soultanis KC, Pyrovolou N, Zahos KA et al (2008) Late postoperative infection following spinal instrumentation: stainless steel versus titanium implants. J Surg Orthop Adv 17(3):193–199

Stambough JL, Genaidy AM, Huston RL, Serhan H, El-khatib F, Sabri EH (1997) Biomechanical assessment of titanium and stainless steel posterior spinal constructs: effects of absolute/relative loading and frequency on fatigue life and determination of failure modes. J Spinal Disord 10(6):473–481

Sun C, Huang G, Christensen FB, Dalstra M, Overgaard S, Bunger C (1999) Mechanical and histological analysis of bone-pedicle screw interface in vivo: titanium versus stainless steel. Chin Med J (Engl) 112(5):456–460

Swiontkowski MF, Agel J, Schwappach J, McNair P, Welch M (2001) Cutaneous metal sensitivity in patients with orthopaedic injuries. J Orthop Trauma 15(2): 86–89

Tahal D, Madhavan K, Chieng LO, Ghobrial GM, Wang MY (2017) Metals in spine. World Neurosurg 100:619–627

Walsh WR, Bertollo N, Christou C, Schaffner D, Mobbs RJ (2015) Plasma-sprayed titanium coating to polyetheretherketone improves the bone-implant interface. Spine J 15(5):1041–1049

Wang JY, Wicklund BH, Gustilo RB, Tsukayama DT (1997) Prosthetic metals interfere with the functions of human osteoblast cells in vitro. Clin Orthop Relat Res 339:216–226

Wang JC, Yu WD, Sandhu HS, Betts F, Bhuta S, Delamarter RB (1999) Metal debris from titanium spinal implants. Spine (Phila Pa 1976) 24(9):899–903

Wedemeyer M, Parent S, Mahar A, Odell T, Swimmer T, Newton P (2007) Titanium versus stainless steel for anterior spinal fusions: an analysis of rod stress as a predictor of rod breakage during physiologic loading in a bovine model. Spine (Phila Pa 1976) 32(1):42–48

Williams D, Clark G (1982) The corrosion of pure cobalt in physiological media. J Mater Sci 17(6):1675–1682

Wu SH, Li Y, Zhang YQ et al (2013) Porous titanium-6 aluminum-4 vanadium cage has better osseointegration and less micromotion than a poly-ether-ether-ketone cage in sheep vertebral fusion. Artif Organs 37(12):E191–E201

Yoon SH, Ugrinow VL, Upasani VV, Pawelek JB, Newton PO (2008) Comparison between 4.0-mm stainless steel and 4.75-mm titanium alloy single-rod spinal instrumentation for anterior thoracoscopic scoliosis surgery. Spine (Phila Pa 1976) 33(20):2173–2178

Yoshihara H (2013) Rods in spinal surgery: a review of the literature. Spine J 13(10):1350–1358

Zartman KC, Berlet GC, Hyer CF, Woodard JR (2011) Combining dissimilar metals in orthopaedic implants: revisited. Foot Ankle Spec 4(5):318–323

Biological Treatment Approaches for Degenerative Disc Disease: Injectable Biomaterials and Bioartificial Disc Replacement

10

Christoph Wipplinger, Yu Moriguchi, Rodrigo Navarro-Ramirez, Eliana Kim, Farah Maryam, and Roger Härtl

Contents

Pathology, Current Treatments, and Resulting Challenges	172
Biomolecular Treatment (Moriguchi et al. 2016)	174
Recombinant Protein and Growth Factor-Based Therapy	174
Gene Therapy	174
Platelet-Rich Plasma	176
Cell-Based Therapy (Moriguchi et al. 2016)	176
Differentiated Cells	178
Stem Cells	178
Tissue Engineering Strategies	179
Scaffold Development	182
Biological Annulus Fibrosus Repair	182
Bioartificial Total Disc Replacement Therapies	184
Clinical Studies	185

C. Wipplinger (✉)
Department of Neurological Surgery, Weill Cornell Brain and Spine Center, New York–Presbyterian Hospital, New York, NY, USA

Department of Neurosurgery, Medical University of Innsbruck, Innsbruck, Austria
e-mail: christophwipplingermd@gmail.com

Y. Moriguchi · R. Navarro-Ramirez · E. Kim · F. Maryam
Department of Neurological Surgery, Weill Cornell Brain and Spine Center, New York–Presbyterian Hospital, New York, NY, USA
e-mail: mc9087@ommc-hp.jp; ron2006@med.cornell.edu; elianaekim@gmail.com; farahmaryam1@gmail.com

R. Härtl
Department of Neurological Surgery, Weill Cornell Brain and Spine Center, New York–Presbyterian Hospital, Weill Cornell Medicine, New York, NY, USA
e-mail: roger@hartlmd.net; roh9005@med.cornell.edu; rhartl@braintrauma.org

© Springer Nature Switzerland AG 2021
B. C. Cheng (ed.), *Handbook of Spine Technology*,
https://doi.org/10.1007/978-3-319-44424-6_38

Unpublished Clinical Trials 187
Future Perspective 188
References 188

Abstract

Degenerative disc disease (DDD) is a major cause of disability in the western world. Current treatment strategies only address the symptoms of DDD. To meet the clinical need of regenerative treatment strategies, biological treatment approaches have become of increasing interest in the past decade. Currently explored treatment strategies involve biomolecular treatments for early-stage degeneration, cell-based therapies involving differentiated cells as well as stem cells for advanced-stage DDD, as well as tissue engineering strategies for total disc replacement in terminal-stage disc degeneration.

The following chapter will provide a comprehensive overview about recent the recent progress in regenerative treatment strategies. This chapter will elucidate experimental in vivo studies as well as published and ongoing clinical trials.

Keywords

Annulus fibrosus repair · AF repair · Intervertebral disc regeneration · Tissue engineering · MSCs · Mesenchymal stem cells · Growth factors · Gene therapy · TE-IVD · Bioartificial disc · Biological IVD treatment

Pathology, Current Treatments, and Resulting Challenges

Low back pain (LBP) is one of the major causes of morbidity that leads to enormous costs for western healthcare systems (Schmidt et al. 2007; Hoy et al. 2010; McBeth and Jones 2007; CDC 2009; Katz 2006). An association between LBP and degenerative disc disease (DDD) has been established by recent studies, accounting DDD for up to 40% of all LBP cases (Pye et al. 2004; MacGregor et al. 2004; Freemont 2009). The intervertebral disc (IVD) contains the soft and gelatinous nucleus pulposus (NP), the surrounding fibrocartilaginous annulus fibrosus (AF), and the cartilaginous endplate (EP) which connects the IVD to the corpus vertebrae. DDD is characterized by extracellular matrix (ECM) degradation, release of proinflammatory cytokines, altered spine biomechanics, angiogenesis, and nerve ingrowth which is associated with increased pain sensation (Le Maitre et al. 2007; Rannou et al. 2003). Factors including mechanical stress, trauma, genetic predisposition, and inflammation can trigger and exacerbate DDD (Podichetty 2007) (Fig. 1).

Among the most commonly performed spinal procedures to treat disc herniation is lumbar discectomy, with an estimated 300,000 cases per year in the United States (Deyo and Weinstein 2001). However, while the neural tissue is decompressed by the discectomy, it leaves the annular defect untreated. Because of this, the risk of recurrent disc herniation through the open defect is elevated, which occurs in 6–23% of patients following discectomy. It is associated with compromised patient outcomes, the need for revision procedures, and increased healthcare costs (Carragee et al. 2003; Swartz and Trost 2003; Bruske-Hohlfeld et al. 1990; Ambrossi et al. 2009; Frymoyer et al. 1978). Aggressive surgical discectomy can reduce the rate of re-herniation, but is associated with more severe disc degeneration and back pain (Frei et al. 2001; Barth et al. 2008; O'Connell et al. 2011). Since the IVD does not possess a sufficient self-repair capacity, current treatment options for DDD range from conservative treatments to invasive therapies for severe and symptomatic courses of DDD, like spinal fusion or total disc replacement (TDR). However, long-term results do not show significant differences between invasive and conservative therapies, and complications are common (Peul et al. 2007; Lequin et al. 2013; Lurie et al. 2014).

10 Biological Treatment Approaches for Degenerative Disc Disease: Injectable Biomaterials... 173

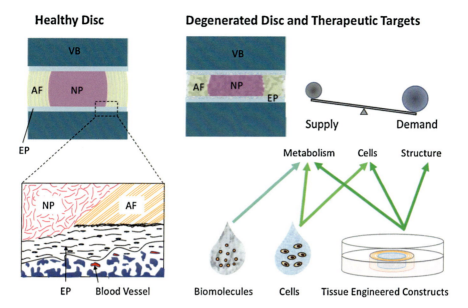

Fig. 1 Schematic pictures of the healthy disc show three components of the disc both macro- and microscopically. In degenerated discs, metabolism, cells, and structure encounter imbalance of supply and demand, one, some, or all of which each strategy will redress. *NP* nucleus pulposus, *AF* annulus fibrosus, *EP* endplate, *VB* vertebral body (Moriguchi et al. 2016)

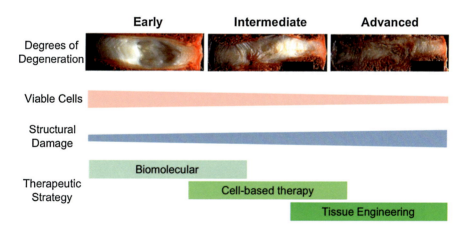

Fig. 2 Treatment strategies for different stages of IVD degeneration (Moriguchi et al. 2016)

To address the limitations of available treatments and enhancing patient outcome, biological approaches to IVD regeneration have become a growing area of interest.

Current strategies for regenerative biological disc treatment can be roughly categorized in three groups: biomolecular therapy, cell therapy, and tissue-engineered IVD construction (An et al. 2011; Zhang et al. 2011a; Maidhof et al. 2012) (Fig. 2).

In the early stage of IVD degeneration, which is defined by beginning structural changes and loss of hydration, a sufficient number of viable cells can still be found.

Thus, these treatment strategies involve recombinant genes, proteins, and stem cell therapies

(Fig. 2). These agents are meant to enhance selective protein expression by stimulating the remaining viable cells in order to promote an intrinsic self-healing within the IVD.

Mid-stage degeneration is characterized by less active and rapidly disappearing viable cells and increasing structural damage. Here, cell transplantations and tissue-engineered biological scaffolds are utilized to recover the damaged IVD.

Finally, the most advanced stage of degeneration is described as severe structural damage to the whole disc and the lack of viable cell activity. For this stage of degeneration, the treatment approaches involve TDR with tissue-engineered constructs.

The following part of this chapter will provide an overview of the current biological treatment approaches for each of the previously described stages, including experimental in vivo studies as well as recent clinical trials.

Biomolecular Treatment (Moriguchi et al. 2016)

A defining compositional change in degenerated discs is the gradual decline of NP water content originating from the loss of proteoglycan. A decrease in swelling pressure within the NP is followed by the reduction of mechanical tension in the AF collagen fibers, resulting in abnormal loading of the spine. The consequence of these alterations often is the instability of segments with subsequent development of neck or back pain and narrowing of the spinal canal, which may induce neurological symptoms. In the early stages of degeneration, the disc undergoes an imbalance of anabolic and catabolic factors that leads to the degradation of extracellular matrix (ECM). Biomolecules such as recombinant proteins and genes can regenerate expression of target molecules through the increase in anabolic or decrease in catabolic factor production, hence facilitating ECM synthesis. The following section will review recent in vivo studies on biomolecules which are used to treat disc degeneration (Table 1).

Recombinant Protein and Growth Factor-Based Therapy

Protein solutions injected directly into discs may have the potential to stimulate cell growth or anabolic response that could reverse disc degeneration. Since the demonstration of the disc's responsiveness to exogenous growth factors in an ex vivo organ culture system (Thompson et al. 1991), various proteins capable of modulating cell growth, differentiation, and ECM synthesis have shown promising for treating DDD. Bone morphogenetic proteins, such as BMP2; BMP7, which is also known as osteogenic protein 1 (OP-1); and BMP14, or growth differentiation factor-5, as well as other transforming growth factor-beta (TGF-β) superfamily such as TGF-beta 1 or 3 have induced bone and cartilage formation. Their usage has been the part of extensive research in cases of spinal arthrodesis and disc regeneration (An et al. 2005, 2011; Imai et al. 2007; Walsh et al. 2004; Masuda et al. 2006; Chujo et al. 2006; Miyamoto et al. 2006; Huang et al. 2007; Chubinskaya et al. 2007; Leckie et al. 2012). In a single in vivo rabbit study by An H. et al., intradiscal OP-1 injection resulted in an increase in proteoglycan content of NP at 2 weeks and disc height at 8 weeks. This treatment has recently been moved on to a clinical trial. Though promising, protein injection is challenged by the short duration of its therapeutic effect. The solution for this may be the development of slow-release carriers or gene-based delivery systems.

Gene Therapy

Gene therapy induces the modification of intradiscal gene expression for a prolonged effect on degenerated discs. Genes that are potentially applicable therefore are delivered through either viral (mostly adenovirus) or non-viral vectors, which are then either directly injected into live tissue (in vivo gene therapy) or transfected into cells cultures in vitro prior to implantation into the IVD (Woods et al. 2011). In one of the pioneering in vivo studies in a rabbit model, NP cells were transfected with TGF-β1 expressing adenovirus vector.

Table 1 Recombinant proteins, growth factors, and gene therapy

Species	Model	Molecules	Dose	Outcome	Refs
Protein injection					
Rat	Compression	IGF-1*	IGF-1 8 ng/8 ul/disc	GDF-5 and TGF-beta aid in expansion of inner annular fibrochondrocytes into the nucleus	Walsh et al. (2004)
		GDF-5	GDF-5 8 ng/8 ul/disc		
		TGF-beta	TGF-beta 1.6 ng/8 ul/disc		
		bFGF	bFGF 8 ng/8 ul/disc		
Rat	Compression	BMP7 (OP-1)	0.2 ug/uL/disc	OP-1 stimulates anabolic response characterized by the restoration of normal disc morphology	Chubinskaya et al. (2007)
Rabbit	Normal	BMP7 (OP-1)	2 ng/10 ul/disc	Increase in disc height	An et al. (2011)
Rabbit	Chemonucleolysis by C-ABC	OP-1	100 ul/10 ul/disc	Increase in disc height and PG content	Imai et al. (2007)
Rabbit	Needle puncture	BMP7 (OP-1)	100 ug/10 ul/disc	Improvement in disc height and MRI findings	Masuda et al. (2006)
Rabbit	Needle puncture	GDF-5	1100 ng,1, 100 ug/10 ul/disc	Increase in disc height	Chujo et al. (2006)
Rabbit	Needle puncture	OP-1	100 ug/10 ul/disc	Increase in disc height and PG content of the NP	Miyamoto et al. (2006)
Rabbit	Annular tear 5 × 7 mm	BMP2	100 ul/10 ul/disc	Exacerbated degeneration	Huang et al. (2007)
Rabbit	Nucleotomy	PRP	20 ul PRP + microsphere/disc	Less degeneration, more PG	Nagae et al. (2007)
Rabbit	Nucleotomy	PRP	20 ul PRP + microsphere/disc	Improvement in disc height and water content	Sawamura et al. (2009)
Rabbit	Annular puncture	PRP-releasate	20 ul/disc	Better X-ray and MRIs	Obata et al. (2012)
Sheep	Annular incision	BMP 13	300 ug/70 ul saline	BMP 13 prevents loss of hydration	Wei et al. (2009)
Gene therapy					
Rat	Degenerative model induced by unbalanced dynamic and static force	Lentiviral CHOP (C/EBP homologous protein) shRNA	1×10^6 PFU/2 ul/disc	Significant decrease of apoptotic incidence in cells treated with CHOP ShRNA at 7 weeks	Zhang et al. (2011b)
Rat	Normal	Plasmid DNA mixed with microbubbles	2 ug/2 ul/disc	Reported genes were expressed up to 24 weeks	Nishida et al. (2006)
Rabbit	Normal	Ad/CMV-hTGFβ1	6×10^6 PFU/15 ul/disc	Leads to double proteoglycan synthesis	Nishida et al. (1999)
Rabbit	Normal	Ad-LMP1	1×10^7 PFU/10 ul/disc	LMP1 overexpression increases PG, BMP2, and BMP7	Yoon et al. (2004)
Rabbit	Annular puncture	ADAMTS5 siRNA oligonucleotide	10 ug/10 ul/disc	Improvement in MRI and histological scores	Seki et al. (2009)

(continued)

Table 1 (continued)

Species	Model	Molecules	Dose	Outcome	Refs
Rabbit	Annulotomy	AAV2-BMP2 or -TIMP1	6×10^6 virus particles/15 ul/disc	AAV-BMP2 and -TIMP1 delayed degeneration	Leckie et al. (2012)
Rabbit	Post-annulotomy	Ad-Sox9	1×10^9 PFU/10 ul/disc	AdSox9 helped retain chondrocytic appearance, cellular morphology, and ECM at 5 weeks	Paul et al. (2003)

Proteoglycan synthesis showed to be increased by 100% in treated tissue (Nishida et al. 2006).

Since, a variety of proteins were discovered as promising targets for gene therapy: upstream proteins such as LMP-1 which regulates BMP2 and BMP7, ECM-degrading enzymes, disintegrin and metalloproteinase with thrombospondin motifs-5, their inhibitors (tissue inhibitor of metalloproteinase-1, TIMP-1), chondrocyte-specific transcription factors (SRY-box 9, Sox9), and apoptosis inducers (C/EBP homologous protein) (Leckie et al. 2012; Nishida et al. 1999, 2006; Yoon et al. 2004; Seki et al. 2009; Zhang et al. 2011b; Paul et al. 2003).Though gene therapy can be advantageous in its sustained effect, inherent risk of viral gene delivery systems becoming infectious or immunogenic has moved the focus toward non-viral gene delivery systems. The development of microbubble-enhanced ultrasound gene therapy and injection of small interfering RNA (siRNA) have proven to achieve long-standing transgene expression in IVD cells in vivo (Nishida et al. 2006; Zhang et al. 2011b). However, non-viral gene delivery systems are limited by low transfection efficiency, which must be overcome to enhance their clinical applicability. The feasibility of ex vivo gene therapy, which reduces the risks of infection and immunogenicity and plays an important role in the future of tissue engineering technology, has been explored in several studies (Xin et al. 2012; Leo et al. 2004).

Platelet-Rich Plasma

Platelet-rich plasma (PRP), an autologous blood product manufactured by the centrifugation of whole blood, offers a variety of proteins for the treatment of degenerative discs due to its high concentration of platelets. Upon activation, these platelets release a variety of multifunctional growth factors such as PDGF (platelet-derived growth factor), IGF-1 (insulin-like growth factor), TGF-β1, (transforming growth factor-beta 1), VEGF (vascular endothelial growth factor), and bFGF (basic fibroblast growth factor). When used in the early stage of disc degeneration, PRP may enhance disc hydration (Gullung et al. 2011). Various PRP technologies have emerged to retard the degenerative cascade, which include a gelatinous hydrogel scaffold, impregnated with PRP (Nagae et al. 2007; Sawamura et al. 2009; Obata et al. 2012) and soluble releasate derived from activated PRP (Obata et al. 2012). The in vivo efficacy of PRP in improving or maintaining disc height and hydration has facilitated its transition to ongoing clinical trials.

Cell-Based Therapy (Moriguchi et al. 2016)

The efficacy of biomolecules is limited when the degeneration of an IVD is more advanced, since there is a correlation between the progress of the degeneration and the decline of the number of cells responsive to injected genes and proteins (Gruber et al. 2002). Mid-stage degeneration is characterized by a decrease in the number of cells within the IVD tissue. Therefore, cell transplantation is a feasible treatment strategy at this stage. A number of in vivo studies report the efficacy of using a vast array of cell sources (Table 2).

Table 2 Cell therapy

Species	Model	Cell type	Dose	Outcome	Refs
Mouse	Post-annular injury	Allogenic bone marrow MSCs	BMSCs 1.0×10^3	ECM augmented in NP via autonomous differentiation and stimulation of endogenous cells at 12 weeks	Yang et al. (2009)
Mouse	Annular puncture	Multipotent stem cells derived from human umbilical cord blood	1.0×10^3 cells intradiscally, 1.0×10^6 cells intravenously	Unlike intradiscal injection, intravenous injection did not preserve the IVD architecture nor disc height at 14 weeks	Tam et al. (2014)
Sand rat	Discectomy	Autologous disc cells	1.0×10^4 cells/ 5 ul/2-mm^3 Gelfoam	Implanted disc engrafted with the host disc for up to 8 months	Gruber et al. (2002)
Rat	Normal	Bone marrow MSCs	5.0×10^5/50 ul hyaluronan gels	MSCs maintained viability and proliferated over 28 days	Crevensten et al. (2004)
Rat	Post-annular puncture	Human bone marrow MSCs	1.0×10^6/15 ul	Human MSCs survived for 2 weeks post transplantation, increasing disc height and MRI intensity	Jeong et al. (2009)
Rat	Post-annular puncture	Adipose-derived MSCs (ADSCs)	1.0×10^6/50 ul	Discs maintained disc height and restored MRI signal intensity	Jeong et al. (2010)
Rat	Nucleotomy	Co-culture of NP cells and MSCs	2.5×10^5 cells (25% NPCs and 75% MSCs)	Bilaminar co-culture pellet of NP cells and MSCs outperformed solely NP cells or MSCs at 5 weeks	Allon et al. (2010, 2012)
Rabbit	Nucleotomy	Allogenic NP cells	5.0×10^4 cells/ 20 ul	Histology indicated delayed degeneration at 16 weeks	Okuma et al. (2000)
Rabbit	Nucleotomy	Autologous articular chondrocytes	2.0×10^6/ 150 ul	Chondrocytes survived and produced hyaline-like cartilage at 6 months	Gorensek et al. (2004)
Rabbit	Normal	Allogenic bone marrow MSCs	1.0×10^5 cells	MSCs survived and enhanced PG synthesis	Zhang et al. (2005)
Rabbit	Post-nucleotomy	Autologous MSCs	4.0×10^4/40 ul atelocollagen	Improved disc height, MRIs, and histology at 48 weeks	Sakai et al. (2003,2005, 2006)
Rabbit	Post-annular Injury	Autologous bone marrow MSCs	1.0×10^5/25 ul	Injection of MSCs significantly increased PG synthesis in severely degenerated discs at 16 weeks	Ho et al. (2008)
Rabbit	Normal	Allogenic MSCs	1.0×105/15 ul	Injected cells engrafted into inner annulus fibrous at 24 weeks	Sobajima et al. (2008)
Rabbit	Post-puncture	Xenogeneic derivatives of embryonic stem cells	1.0×106 cells/ 20 ul	New notochordal cells observed; no immune response elicited	Sheikh et al. (2009)
Rabbit	Nucleotomy	Allogenic synovial MSCs	1.0×107 cells/ 100 ul PBS	Implanted cells labeled with DiI or GFP detected at 24 weeks. Disc height and MRI signal intensity were maintained	Miyamoto et al. (2010)
Rabbit	Compression	Allogenic bone marrow MSCs	0.08 ml of 1.0×106 cells/ ml	Combination of MSC injection and distraction led to better disc height and histology at 8 weeks	Hee et al. (2010)
Rabbit	Post-nucleotomy	Autologous NP cells and allogenic MSCs	1.0×106/20 ul	Both NP cells and MSCSs better maintained disc height and GAG content at 16 weeks	Feng et al. (2011)

(continued)

Table 2 (continued)

Species	Model	Cell type	Dose	Outcome	Refs
Canine	Post-nucleotomy	Disc cells	6.0×10^6 cells/1 ml/disc	Disc remained viable, produced ECM, better maintained disc height	Ganey et al. (2003)
Canine	Post-nucleotomy	Autologous MSCs	1.0×10^6/ml stem cells	MSCs led to better disc height, MRI, and histology grading at 12 weeks	Hiyama et al. (2008)
Canine	Post-nucleotomy	Bone marrow MSCs	$10^5, 10^6, 10^7$ cells	The disc treated with 10^6 MSCs had more viable cells than 10^5 and less apoptotic cells than 10^5 cells at 12 weeks	Serigano et al. (2010)
Porcine	Post-nucleotomy	Human MSCs	0.5×10^6/hydrogel carrier	Implanted cells survived and differentiated into disc-like cells at 6 mos	Henriksson et al. (2009)
Porcine	Nucleotomy	Allogenic juvenile chondrocytes and MSCs	$7–10 \times 10^6$/0.5–75 ml fibrin carrier	JC outperformed MSCs in proteoglycan synthesis at 12 months	Acosta et al. (2011)

Differentiated Cells

Implanted differentiated disc chondrocytes are meant to produce demanded ECM components such as proteoglycan and collagen types II and I under hypoxia and nutrient stress and can meet the increased cellular and metabolic demands of the disc (Rajpurohit et al. 2002).

Accumulated evidence in an array of animal models demonstrate the viability of autologous or allogenic cells in vivo as well as the integration into the host tissue. Thus, a reduction of ECM degradation, recovery of disc height, and MRI signal intensity can be achieved (Table 2). In fact, the pioneering preclinical study in an injured canine model showed that NP disc chondrocyte implantation contributed to ECM regeneration, retarding further disc degeneration (Ganey et al. 2003).

However favorable, disc cell transplantation showed several challenges: (1) donor site morbidity, (2) difficulty in expanding cells in vitro while maintaining cell phenotype, and (3) paucity of allograft donor tissue. Similar to differentiated disc cells, cultured articular chondrocytes (AC) are a well-established non-disc cell source in regenerative medicine (Brittberg et al. 1994). Their effortless extraction from non-weight-bearing parts of the knee and capacity to produce NP-like ECM when transplanted in vivo makes autologous (Gorensek et al. 2004) or allogenic (Acosta Jr et al. 2011) AC a safe and feasible cell source in IVD regeneration. Furthermore, potential immune evasion by juvenile articular chondrocytes supports their applicability in allogenic cell transplantation.

Stem Cells

Multipotent mesenchymal stem cells (MSCs), which are present in adult bone marrow or adipose tissue, can replicate as undifferentiated cells and then differentiate into lineages of mesenchymal tissue: bone, cartilage, fat, tendon, muscle, and marrow stroma (Pittenger et al. 1999). These somatic stem cells are a potentially ideal option for disc repair due to their accessibility and ability to differentiate along a chondrogenic lineage and produce the required proteoglycan and collagen for the disc ECM. The feasibility of MSCs to facilitate disc repair has been substantiated.

Yet it remains controversial whether differentiated cells or stem cells are superior in terms of regenerative capacity of disc morphology.

A porcine study comparing the utility of different cell sources found that committed articular chondrocytes are more suited for the use in disc repair than MSCs due to their aptness for survival in the ischemic disc microenvironment (Acosta Jr et al. 2011). Interestingly, a comparative rabbit study found that MSC transplantation can serve as an ideal substitute for differentiated chondrocytes of disc NP owing to better accessibility with equivalent regenerative potential (Feng et al. 2011). Studies assessing the combination of both cells demonstrated that rather in vitro co-culture (Okuma et al. 2000) or co-implantation (Allon et al. 2010) yields better in vivo performance of the implanted cells. Nonetheless, pluripotent embryonic (Evans and Kaufman 1981; Martin 1981) and induced pluripotent stem cells (iPSCs) (Takahashi and Yamanaka 2006), unlike the lower potent MSCs, have unlimited proliferative and differentiate capacities, which can be strategically exploited in cell-based disc repair.

Sheikh H et al. extracted murine embryonic stem cells (ESCs) and differentiated them into chondro-progenitor cells. Upon implantation into rabbit injured discs, these cells induced notochordal cell formation at site of injury without xenograft-associated immune responses (Sheikh et al. 2009). Unstable in vitro differentiation into desired cell lineages and the potential risks of tumor formation in vivo are still major obstacles in the use of ESCs and iPSCs. However, if these issues are overcome, the use of stem cells may offer abundant potential for intervertebral disc repair.

Tissue Engineering Strategies

The implementation of tissue engineering (TE) pioneered by Langer and Vacanti in 1993 (Langer and Vacanti 1993) has fueled the efforts toward constructing functional biological substitutes for TDR as a novel treatment strategy for DDD. Recently, major efforts have been directed toward developing a replacement for either NP or AF using TE technology.

Tissue engineering originally consists of three, and more recently four components (Langer and Vacanti 1993): scaffolds, cells, growth factors, and physical conditioning using electrical or mechanical stimuli (Fig. 3). Since extensive loss of matrix and structural damages are exhibited in

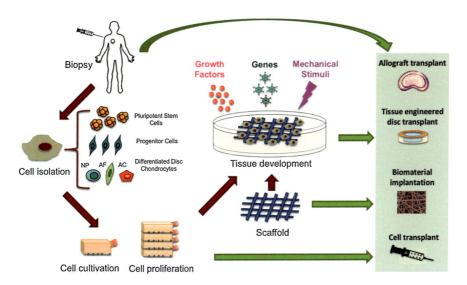

Fig. 3 Cells harvested from different sources can be expanded in vitro and transplanted in vivo in cell transplant for disc regeneration. Scaffolds can be combined with cells, and, if they have bio-mimicking properties, these treatments can be regarded as a part of tissue engineering strategy, which traditionally composes of cells, scaffolds, growth, and factors, but recently including gene treatment and mechanical conditioning. *NP* nucleus pulposus cells, *AF* annulus fibrosus cells, *AC* articular chondrocytes (Moriguchi et al. 2016)

Table 3 Tissue-engineered constructs

Species	Model	Construct	Outcome	Refs
Rat	Subcutaneous implantation	TE-IVD composed of a NP cell-laden alginate surrounded by an AF cell-laden PGL/PLA	Biochemical markers of matrix synthesis, increasing over time, were similar to native tissue at 12 weeks	Mizuno et al. (2004a)
Rat	Subcutaneous implantation	Porous type II collagen/hyaluronate-chondroitin-6-sulfate (CII/HyA-CS)	CII/HyA-CS scaffolds had satisfactory cytocompatibility and histocompatibility, as well as low immunogenicity	Li et al. (2010)
Rat	Subcutaneous implantation	Composite IVD consisting of demineralized bone matrix gelatin and collagen II/hyaluronate/chondroitin-6-sulfate scaffolds seeded AF and NP cells	Implant, similar to native disc in morphology and histology, increased proteoglycan synthesis over 12 weeks	Zhuang et al. (2011)
Rat	Total discectomy	TE-IVD composed of a NP cell-laden alginate surrounded by an AF cell-laden collagen layer	TE-IVD maintained disc space height, produced de novo ECM, and integrated into the spine – yielding intact motion segment with dynamic mechanical properties similar to that of native IVD	Bowles et al. (2011a)
Rat	Subcutaneous implantation	5.0×10^6 cells/ml in pentosan polysulfate-containing polyethylene glycol/hyaluronic acid	MPC/hydrogel composites formed cartilage-like tissue, well tolerated by the host	Frith et al. (2013)
Rabbit	Laser discectomy	2.0×10^6 cells/atelocollagen honeycomb shaped scaffold	AF cells survived and produced hyaline-like cartilage in the disc at 12 weeks	Sato et al. (2003)
Rabbit	Microdiscectomy	Cell-free implant composed of a polyglycolic acid (PGA) felt, hyaluronic acid (HA), and allogenic serum	Implantation of a cell-free PGA-HA implant immersed in serum after discectomy improved disc hydration and preserved disc height 6 months after surgery	Abbushi et al. (2008)
Rabbit	Post-nucleotomy	2.0×10^6 bone marrow MSCs/0.04 ml fibrin glue containing 10-ug/L TGF-β1 (MSC-PFG-TGF-β1)	MSC-PFG-TGF-β1 group had less degeneration and a slower decrease in disc height compared with both degenerative and acellular PFG-TGF-β1 group	Yang et al. (2010)
Rabbit	Nucleotomy	Allogenic NP cell-seeded collagen II/hyaluronan/chondroitin-6-sulfate (CII/HyA/CS) tri-copolymer construct	Viability of allografted NP cells, extracellular matrix deposition, and disc height maintenance; restoration of T2 MRI signal intensity observed at 24 weeks	Huang et al. (2011)
Rabbit	Post-puncture	5.0×10^3 allogenic bone marrow MSCs/10 ul hydrogel	MSCs suppressed collagen I in NP, reduced collagen aggregation, and maintained proper fibrillary properties and function	Leung et al. (2014)
Rabbit	Post-nucleotomy	1.0×10^6 human NP cell line infected with recombinant SV40 adenovirus vector (HNPSV-5) in atelocollagen	Deceleration of disc degeneration was evident after HNPSV-5 transplantation as shown by disc height and histologic examination at 24 weeks	Iwashina et al. (2006)
Canine	Total discectomy	Cell-allograft IVD composites made of allograft and NP cells, with in vitro transduced with recombinant adeno-associated virus (rAAV)-hTERT	The hTERT-loaded NP cells intervention could effectively resist the degeneration of the allogenic transplanted IVD at 12 weeks	Xin et al. (2012)

(continued)

Table 3 (continued)

Species	Model	Construct	Outcome	Refs
Canine	Post-nucleotomy	Autologous adipose tissue-derived stem and regenerative cells in hyaluronic acid carrier (ADRC/HA)	Disc that received ADRC/HA produced matrix and resembled native disc in morphology at 12 months	Ganey et al. (2009)
Canine	Nucleotomy	Cell-scaffold composite made of three-dimensional porous PLGA scaffolds and NP cells	Disc height, segmental stability, and T2-weighted MRI signal intensity were well preserved at 12 weeks	Ruan et al. (2010)
Porcine	Nucleotomy	Cell-scaffold composite made of NP cells and injectable hyaluronan-derived polymeric substitute material HYADDR (1.0×10^5 cells/ml)	Injected discs had a central NP-like region similar to the normal disc biconvex structure and viable chondrocytes forming matrix like that of normal disc at 6 weeks	Revell et al. (2007)
Porcine	Post-annular injury	1.25×10^5 autologous MSCs/ml in either hydrogel PhotoFix or hyaluronic acid	Stem cells in hydrogel treatment had significantly higher T2 MRI intensities and lower degeneration grade at 24 weeks than hydrogel alone treatment	Bendtsen et al. (2011)
Porcine	Partial nucleotomy	5.0×10^5 autologous bone marrow MSCs transduced with retrovirus encoding luciferase in 1 mL hyaluronan-enhanced albumin hydrogel	In vivo 3-day analysis showed persistent metabolically active implanted cells in the disc	Omlor et al. (2014)
Goat	Post-disc injury	2.5×10^5 allogenic bone marrow stromal cells/10 ul PBS + 30 ul chondroitin sulfate-based hydrogel	Significant increase in NP proteoglycan accumulation at 6 months	Zhang et al. (2011c)
Sheep	Total discectomy	Noncrystalline polylactide copolymer interbody cages filled with 1.0×10^6 allogenic mesenchymal progenitor cell (MPC)-laden Gelfoam sponge formulated with the chondrogenic agent pentosan polysulfate (PPS)	Biodegradable cage-contained MPCs in combination with PPS produced cartilaginous tissue at 3 months	Goldschlager et al. (2010)
Sheep	Post-chondroitinase-ABC injection	4.0×10^6 or 0.5×10^6 human mesenchymal precursor cells (MPCs) suspended in hyaluronic acid	High-dose injection improved histopathology scores at 3 months, while low dose at 6 months	Ghosh et al. (2012)
Sheep	Nucleotomy	Allogenic or autologous disc cells ($0.4–2.0 \times 10^6$ cells/0.5–1 ml hydrogel) in hydrogel containing hyaluronic acid and maleolyl-albumin	Biological repair of traumatic damage occurs in sheep discs at 6 months; hydrogel-supported disc cells may be beneficial	Benz et al. (2012)
Canine	Total discectomy	TE-IVD composed of a NP cell-laden alginate surrounded by an AF cell-laden collagen layer	Early displacement in some cases, if stably implanted TE-IVD maintained disc height, produced new ECM, and integrated into host tissue, intact motion segment with dynamic mechanical properties similar to that of native IVD	Moriguchi et al. (2017)

advanced stages of disc degeneration, development of biocompatible and biomimetic scaffolding materials based on engineering innovation can facilitate the recovery of native biological and biomechanical functionality. Numerous studies have assessed tissue-engineered components as well as whole disc constructs of the disc in vivo (Table 3).

Scaffold Development

Numerous scaffold materials, including altinate, silk-fibrin/HA composites, atelocollagen, synthetic polymers, and a collagen 2/hyaluronan/chondroitin-6-sulfate (C2/Hy/CS) composite, which mimic the mechanical and biochemical properties of the native NP, have been part of a study. Extensive research on hyaluronic acid, a native NP extracellular matrix component, has been performed in vivo (Revell et al. 2007; Abbushi et al. 2008; Ganey et al. 2009; Li et al. 2010; Huang et al. 2011). Resorbable cell-free implants consisting of a polyglycolic acid (PGA) felt, hyaluronic acid, and serum were used in a rabbit study. This resulted in improved disc hydration and height 6 months after microdiscectomy (Abbushi et al. 2008). The reason for the frequent use of cells together with bio-mimicking materials is to encourage de novo ECM production. The findings of Ganey T. et al. were that adipose-derived stem cells contribute significantly to the recovery of T2 intensity and disc height in a canine disc injury model. Synthetic polymers such as PGA or poly-L-lactic-co-glycolic acid (PLGA) have also been used either solely or in combination with hydrogels to construct cell-laden TE composites (Abbushi et al. 2008; Ruan et al. 2010).

Biological Annulus Fibrosus Repair

In mid-stage DDD, a commonly occurring pathology is the lumbar disc herniation. Due to the progressive degeneration, the IVD shows reduced hydration. The inadequate hydration of the disc leads to fissure formation, eventually allowing the soft NP to herniate through the defect and thus compress neighboring neural structures (Freemont 2009).

Lumbar discectomy is one of the most commonly performed spinal procedures to treat disc herniation, with an estimated 300,000 cases performed annually in the United States (Deyo and Weinstein 2001). While efficient in relieving acute symptoms by removing the herniated part of the NP and decompressing neural structures, the AF defect typically remains untreated after discectomy. Persistent AF defects increase the risk of re-herniation, which may lead to additional operations including more invasive procedures such as TDR and instrumented fusion (Carragee et al. 2003; Swartz and Trost 2003; Bruske-Hohlfeld et al. 1990; Ambrossi et al. 2009; Frymoyer et al. 1978; Laus et al. 1993).

Previous studies of intervertebral disc repair, which aim to halt, delay, or reverse intervertebral disc degeneration, were primarily focused on NP regeneration (Masuda et al. 2004; Bae and Masuda 2011; Sakai and Grad 2015; Wang et al. 2014; Kepler et al. 2011; Blanquer et al. 2015; Mern et al. 2014). However, the majority of these strategies are delivered through a punctured AF, which can generate a degenerative cascade within the disc affecting IVD biomechanics, cellularity, and biosynthesis even upon modest injury (Elliott et al. 2008; Iatridis et al. 2009; Korecki et al. 2008; Hsieh et al. 2009). Annular defects can emerge not only from needle punctures through the AF to reach the NP but also from the early process of IVD degeneration. Given the sensitivity of the AF, lesions from NP treatment can provoke further degeneration, inducing leakage of the delivered material and eventual failure of the regenerative treatment. In fact, one prospective study with 10-year follow-up found that discography performed with a small needle puncture accelerated disc degeneration rate of same-side disc herniation and changes to the endplate (Carragee et al. 2009). A different study demonstrated that injecting MSCs through the AF into the NP led to cell leakage and augmented osteophyte formation (Vadalà et al. 2012). Combining an injectable NP regenerative strategy with a sealant that repairs annular defects is the optimal strategy that can circumvent leakage of implanted cells or material while enhancing therapeutic outcome. Previous approaches to annular repair have involved mechanical treatments such as suturing and annuloplasty devices, which failed to improve annular healing strength in long-term clinical trials (Ahlgren et al. 2000; Chiang et al. 2011; Bailey et al. 2013). Although several NP regenerative studies and a few in vitro AF studies (Nerurkar et al. 2009) provide critical insight on the

Table 4 Annular repair

Species	Model	Treatment	Outcome	Refs
Rat	Degradation tests with subcutaneous implantation	Fibrin-genipin adhesive hydrogel (fib-gen)	60% of fib-gen remained at 8 weeks and nearly all resorbed at 16 weeks; kinetics show better in vivo longevity compared to fibrin	Likhitpanichkul et al. (2014)
Rat	Needle puncture	Injection of cross-linked high-density collagen (HDC) gels	Cross-liked HDC capable of repairing annular defects most likely due to enhanced stiffness of HDC at 5 weeks	Grunert et al. (2014b)
Porcine	Needle puncture	Injection of Gelfoam, platinum coil, bone cement, and tissue glue	Injection of Gelfoam better improved integrity of punctured disc than the other three to potentially prevent recurrent disc herniation at 2 months	Wang et al. (2007)
Sheep	Box annulotomy	Patch and plug with small intestinal submucosa (SIS) and titanium bone screw	SIS-based treatment led to better maintenance of hydration and intradiscal pressure at 26 weeks after annulotomy	Ledet et al. (2009)
Sheep	Box annulotomy	Triphase AF implant composing two outer phases of absorbable polyglycolic acid (PGA) and a centric phase of a nonabsorbable polyvinylidene fluoride (PVDF) mesh	Implant-treated discs had more reparative tissue. But, contrast media leakage tests under provocative pressure did not show a difference between groups	Hegewald et al. (2015)
Sheep	Microdiscectomy	Allogenic mesenchymal progenitor cells (MPCs) + pentosan polysulfate (PPS) embedded in a gelatin/fibrin scaffold	Discs treated with MPC + PPS showed higher PG content than the untreated or ones treated with solely scaffold at 6 months	Oehme et al. (2014)
Sheep	Box annulotomy	Injection of cross-linked high-density collagen (HDC) gel into annulus defect	IVDs treated with HDC gel showed histologically less degeneration. Imaging difference was not significant	Pennicooke et al. (2017)

reparative process within the AF tissue (Wei et al. 2009; Sakai et al. 2006; Sato et al. 2003; Zhang et al. 2011c), there are a very limited number of in vivo studies focusing primarily on annular repair (Table 4). Current efforts in the biological treatment for in vivo AF repair include either development of injectable material in conjunction with biologics such as biomolecules/cells or construction of rigid implants derived from synthetic polymer or biological tissue.

In order to introduce alternative methods, injectable biomaterials have recently gained further popularity in the field. Injectable genipin cross-linked fibrin collagen gel was suggested to integrate with human AF tissue and presented promising biomechanical and cell-seeding properties in vitro (Schek et al. 2011). Our group successfully tested a high-density collagen gel in vitro and in vivo using a needle puncture rat tail model. Furthermore, we have recently translated this project to a large animal (ovine) model, which demonstrated positive histologic results at 16 weeks following injury (Pennicooke et al. 2017).

Collectively, these studies demonstrate an ability to formulate and deliver injectable biomaterials to the lumbar spine of sheep to seal AF defects, promote sufficient tissue healing, and prevent further disc degeneration.

In another large animal study conducted by Oehme et al., injected mesenchymal progenitor cells combined with chondrogenic agent pentosan polysulfate maintained disc height, disc morphology, and NP proteoglycan content post microdiscectomy in a sheep model (Oehme et al. 2014). Despite the few studies dedicated to annular repair, more attention is now being paid to this field given its enhancement of even NP-targeted therapy.

Bioartificial Total Disc Replacement Therapies

In advanced stages of DDD with significant structural damage and the absence of viable cell activity, the injection of biomolecules or cell transplantation is no longer a feasible option.

A current surgical treatment strategy for advanced DDD is the total removal of the IVD followed by the fusion of the whole segment including the adjacent vertebrae. However, fusion may result in pseudoarthrosis or adjacent segment disease, which may lead to reoperation and long-distance fusion procedures (Maldonado et al. 2011; Sugawara et al. 2009; Bydon et al. 2013). To prevent these complications and to preserve mobility in the treated segment, TDR by synthetic prosthesis has become an alternative treatment strategy. Yet, current mechanical prosthetic TDR devices have not been able to reproduce the biomechanical properties of the natural IVD. Additionally, recent studies have demonstrated that current TDR devices are not without their disadvantages as they also entail the risk of adjacent segment disease (Maldonado et al. 2011; Kelly et al. 2011).

In this case, the total replacement using a tissue-engineered intervertebral disc with the ability to integrate into the host environment is a promising treatment strategy. The current standard in whole IVD implantation involves NP and AF composites that replace the structurally damaged tissues of a severely degenerated disc.

The first tissue-engineered whole IVD, implanted in vitro within the subcutaneous dorsum of athymic mice, comprised of NP cell-laden polyglycolic and polylactic acid (PGA/PLA) and AF cell-laden alginate (Mizuno et al. 2004a, 2006).

More than a decade ago, our group was the first to develop a tissue-engineered disc, composed of NP cells seeded into an alginate hydrogel, surrounded by a polyglycolic acid and polylactic acid scaffold seeded with AF cells (Mizuno et al. 2004b, 2006). This de novo construct was successfully implanted in the subcutaneous space of the dorsum of athymic mice and demonstrated the feasibility of creating a composite IVD including both AF and NP tissues Several other studies have reported the development of composite tissue-engineered IVD constructs, using combinations of materials such as demineralized bone matrix gelatin with type II collagen, hyaluronate and chondroitin-6-sulfate (C2/HyA-CS) (Zhuang et al. 2011), electrospun polycaprolactone and agarose (Martin et al. 2014), and self-assembled NP cells seeded onto calcium polyphosphate (Hamilton et al. 2006).

More recently, we developed a TE-IVD construct composed of an NP cell-laden alginate nucleus encircled by an AF cell-laden collagen annulus (Bowles et al. 2010, 2012). The efficacy of this construct, namely, maintaining disc height and physiological hydration as well as integrating into the host tissue, has been demonstrated through its implantation in a rat tail in vivo model (Bowles et al. 2011a; Gebhard et al. 2010, 2011; Grunert et al. 2014a; James et al. 2011). Although these results are promising, the rat tail has several dissimilarities with the human spine in terms of anatomy and biomechanical properties (O'Connell et al. 2007, 2011; Lotz 2004). Importantly, the rat tail has a significantly different biomechanical loading profile, as the IVDs of the human spine are exposed to higher axial loads. Furthermore, the rat tail lacks a spinal canal containing nervous tissue as well as posterior bone and joint elements. To move our approach closer toward clinical utilization and to mimic the biomechanical loads and anatomy of a human IVD more accurately, we transitioned to a larger animal model.

In a preliminary study, we performed TDR using TE-IVDs in the cervical spine of skeletally mature beagle dogs. Within this, we demonstrated the ability of our TE-IVDs to integrate into the host tissue of a larger animal without any signs of inflammatory response (Moriguchi et al. 2017). Notably, these implants performed quite well when stably implanted in the intervertebral space. However, there was a persistent challenge in ensuring that implants remained firmly implanted in the intervertebral space.

Nonetheless, the addition of growth factors or bioactive molecules can encourage de novo ECM deposition. Goldschlager et al. demonstrated that adult allogenic mesenchymal progenitor cells (MPCs) formulated with a chondrogenic agent pentosan polysulfate (PPS) could synthesize a cartilaginous matrix when implanted into a biodegradable carrier and cage and over time might serve as a bioactive interbody spacer following anterior cervical discectomy (Goldschlager et al. 2010). Furthermore, the integration of tissue engineering and gene therapy has been attempted by a Chinese group that developed a tissue-engineered IVD using an allogenic disc transduced with hTERT gene within its NP cells. When implanted in a canine model, the hTERT-loaded NP cells manifested enhanced antidegenerative effect than unloaded NP cell (Xin et al. 2012). Such constructions of whole disc implants, the most ambitious therapeutic strategy yet, are met with extensive biological and functional challenges in vivo. Yet, the progressing field of TE continues to yield promising modifications to meet the higher demands of implanted discs.

Clinical Studies

Several of the above-described regenerative treatment approaches have already been utilized in a clinical setting. However, to date only a few clinical trials have been published on this topic (Table 5).

In the following section, several representative published clinical studies for the different treatment approaches will be presented.

In 2002, Meisel et al. started a multicenter prospective, randomized, controlled, non-blinded EuroDISC study comparing the safety and efficacy of autologous disc chondrocyte transplant (ADCT) implanted 12 weeks post discectomy. The 2-year interim analysis revealed a significant reduction of low back pain as well as retained disc height in the autologous disc cell transplantation (ADCT) group compared to the discectomy only control group (Meisel et al. 2006, 2007). The ADCT product is currently evaluated in a Phase II clinical trial under the product name NOVOCART® Disc (Meisel 2012; Tschugg et al. 2017).

While to date there is no clinical study using tissue-engineered material, efforts have been made to create functional substitutes for NP (Berlemann and Schwarzenbach 2009; Boyd and Carter 2006). Among many clinical studies focusing on NP replacement, a single-center, non-randomized, prospective feasibility study was undertaken to investigate the use of NuCore Injectable Nucleus hydrogel (Spine Wave, Inc., Shelton, CT, USA) post microdiscectomy prevented early disc collapse to potentially slow the degenerative cascade of the spinal segment over time (Berlemann and Schwarzenbach 2009).

The feasibility of a whole allogenic disc transplantation has first been proven by a group in China. Ruan et al. successfully performed transplantation of fresh frozen disc allografts including endplates in five patients. Implants successfully integrated into the host tissue, over the course of 5 years without any inflammatory reaction, although no immunosuppressive therapy was administered (Ruan et al. 2007). The absence of any immunologic response strongly supports the hypothesis that the intervertebral disc space is immunoprivileged tissue. Although promising, the allogenic transplantation of spinal motion segments has several limitations in terms of availability of healthy donor discs and potential disease transmission.

As mentioned in the section above, a frequently discussed treatment strategy is the intradiscal injection of platelet-rich plasma (PRP) for treating DDD. In 2016, Tuakli-Wosornu et al. published the results of a prospective, double-blind, randomized controlled study. Twenty-nine patients with low back pain, refractory to conservative treatment, received intradiscal PRP injections, while 18 patients who received a

Table 5 Published clinical trials

Trial treatment	No. of patients	Study design	Follow-up (m)	Outcome	Refs
Autologous hematopoietic stem cell injection	10	Case series	12	No patients reported any improvement in their discogenic back pain	Haufe and Mork (2006)
Total disc replacement with allogenic IVD	5	Case series	60	Allograft engrafted disc space without apparent immunoreaction; all minus one disc preserved range of motion	Ruan et al. (2007)
Autologous disc chondrocyte transplantation (EuroDisc)	28	Control study	24	ADCT with discectomy shows more pronounced decrease in OPDQ than discectomy alone	Meisel et al. (2006, 2007)
Injectable biomimetic nucleus hydrogel	14	Case series	24	Significant improvement in leg and back pain after micro-discectomy	Berlemann and Schwarzenbach (2009)
Autologous bone marrow mesenchymal cell injection	2	Case series	24	Both patients showed improvements in the vacuum phenomenon as well as signal intensity of T2-weighed MRIs	Yoshikawa et al. (2010)
Autologous bone marrow mesenchymal cell injection	10	Case series	12	Rapid improvement of pain and disability. Disc height was not recovered, but disc hydration was significantly elevated	Orozco et al. (2011)
Allogenic juvenile chondrocytes injection (NuQu)	15	Case series	12	ODI, NRS, SF-36 improved from baseline. 89% of the patients showed improvement on MRI	Coric et al. (2013)
Injection of autologous bone marrow-concentrated cells	26	Case series	12	Statistically significant improvement in pain scores and impairment was demonstrated. Most dramatic improvement seen in patients with higher CFU-F concentrations. Rehydration of the discs observed in 8 of 20 patients	Pettine et al. (2015)
Intradiscal injection of PRP	47	Prospective double-blinded randomized controlled study	12	Significant improvement in pain scales after 2 months, maintained at the 12-month follow-up	Tuakli-Wosornu et al. (2016)
Intradiscal injection of stromal vascular fraction with PRP	15	Case series	12	Significant improvement in VAS, no worsening, no radiographic changes	Comella et al. (2017)

placebo injection with a contrast agent served as a control group. At the 2-month follow-up, the PRP group showed significant improvement in pain scales. Patients maintained these improvements also in the 12-month follow-up (Tuakli-Wosornu et al. 2016).

Recently the utilization of different stem cell lines has found their way to clinical use. In 2006 Haufe et al. was the first to publish clinical results, reporting about intradiscal autologous hematopoietic stem cell injections. However, in the 12-month follow-up, none of the ten patients reported any improvement in back pain, and 80% of the patients required surgical spinal intervention within a year after injection. Mesenchymal stem cells (MSC) on the other hand showed more promising results in various clinical studies. Due to their relatively easy accessibility and expandability in vivo, the bone marrow has been used as a source for MSCs in several in vitro and in vivo studies. Pettine et al. were the first to utilize bone marrow-concentrated cells (BMCs) as a treatment for discogenic back pain. In 26 patients with chronic low back pain, BMCs harvested from the iliac crest were injected into the IVD. The 1-year follow-up revealed a reduction in pain as well as radiographic improvement in 40% of the patients (Pettine et al. 2015). Yoshikawa et al. reported a case series of two patients who received a collagen sponge soaked with 10^5 cells/mL suspension grafted into a degenerated disc. After 2 years, both patients demonstrated improvement in pain as well as increased hydration on MRI (Yoshikawa et al. 2010). Orozco et al. reported a rapid improvement of pain up to 85% after 3 months in ten patients who underwent intradiscal injection of bone marrow-derived MSCs. Despite the fact that the disc height remained unchanged, an improvement in disc hydration could be observed in the 12-month follow-up MRI (Orozco et al. 2011).

Apart from the bone marrow, the adipose tissue is an abundant source for mesenchymal stem cells (Ganey et al. 2009; Jeong et al. 2010). Due to easier accessibility and less invasive harvest, the utilization of adipose-derived stem cells became more recently of increasing interest. In a recent study, Comella et al. were the first to publish clinical results on the injection of stromal vascular fraction (SVF), containing adipose-derived stem cells as a treatment for low back pain. In this study, SVF was administered along with PRP into lumbar IVDs in 15 patients with discogenic back pain. After a 12-month follow-up, patients showed significant improvement in pain scales. However, this study did not provide any radiographic outcome data (Comella et al. 2017).

Unpublished Clinical Trials

Within the last decade, a clear trend toward regenerative treatment approaches is recognizable. This trend is also represented by the increasing number of clinical studies currently emerging aiming to find new biological treatment approaches for DDD. The following will elucidate several promising ongoing clinical studies that are not published yet.

Due to the similar biological profile as disc chondrocytes and potential immunoprivileged property, allogenic juvenile articular chondrocytes are another promising cell source. In a prospective cohort study, Coric et al. demonstrated that NuQu, an injectable percutaneous fibrin-based delivery of juvenile chondrocytes attenuated otherwise medically refractory low back pain (Coric et al. 2013). A class II study has recently been completed. Despite these study's promising results, further investigation with a prospective, randomized, double-blinded, placebo-controlled study is necessary to make cell transplantation a valid therapeutic option for DDD.

Rathmell et al. are currently the first to evaluate the effects and safety of intradiscal injections with recombinant human growth and differentiation factor 5 (rhGDF5) in a clinical trial. GDF-5 belongs to the transforming growth factor-beta (TGF-β) family which is meant to influence the growth and differentiation of various tissues including the intervertebral disc (Xu et al. 2006). The intradiscal administration has shown to improve the reparative capacity of IVDs in a degenerative rabbit model (Chujo et al. 2006). Within a Phase I/II clinical trial, 32 patients receive a single intradiscal injection of rhGDF5 and will be observed over a 36-month follow-up (J R 2008).

Mesoblast Ltd. developed a commercially available lineage of in vitro differentiated allogenic mesenchymal precursor cells (MPCs). Currently, this product is being evaluated under the name Rexlemestrocel-L in a Phase III prospective, multicenter, randomized, double-blind, placebo-controlled study, comparing Rexlemestrocel-L only vs. Rexlemestrocel-L+ hyaluronic acid (Mesoblast Ltd. 2015).

The recently completed Phase II study included 100 patients with chronic low back pain due to DDD. The outcomes of this study were promising; both treatment groups who received 6 million MPCs and 18 million MPCs, respectively, improved in VAS by 44.4% and 37.9%, whereas the two placebo groups who received saline or hyaluronic acid only improved by 11.8% and 15.8%, respectively. However, no significant improvement in radiographic outcomes could be observed (Mesoblast Ltd. 2019).

The data emerging from these ongoing clinical trials will reinforce findings from published studies and provide new insight for future biological disc repair.

Future Perspective

This present book chapter provides a comprehensive overview on the recent innovations and trends in biological disc repair (Takahashi and Yamanaka 2006). Biomolecular therapies have shown the potential of stimulating the intrinsic healing capacity of the intervertebral discs in early stages (Masuda et al. 2006; Chujo et al. 2006; Huang et al. 2011). In a more advanced setting, cellular therapies are increasingly demonstrating their potential as the understanding of underlying mechanisms of cell differentiation increases (Pittenger et al. 1999; Bernardo et al. 2007; Moroni and Fornasari 2013). A major challenge for cellular therapies remains the determination of the optimal cell type as well as the ideal carrier for application (Acosta Jr et al. 2005).

Another challenge is that all these treatments are inevitably associated with an annular damage caused by the needle puncture, which is necessary for the application of the therapeutic agent. Carragee et al. has shown in a prospective study of notable size that even a small needle puncture may disturb the integrity of the AF enough to accelerate the degeneration of the IVD (Carragee et al. 2009). Therefore, a sufficient annular repair strategy is mandatory in order to seal the defects caused by the necessary needle puncture.

Since the lack of viable cells in advanced DDD makes a stimulating agent, such as growth factors, impossible and the final stages of DDD do not possess enough extracellular matrix to offer an environment for viable cells (Roberts et al. 2006), a replacement will become inevitable. It is known that current mechanical prosthetic devices also involve the risk of adjacent segment disease and thus accelerate further degeneration of the whole spine (Maldonado et al. 2011; Kelly et al. 2011). Therefore, it is inarguable that a biological construct with the ability to integrate into the host tissue will be the better option. Considering the limitations of healthy allogenic transplants (Ruan et al. 2007), tissue engineering will be the best option for end-stage DDD. Although promising, the described in vivo studies for TDR using tissue-engineered constructs (Grunert et al. 2014a; Moriguchi et al. 2017; Bowles et al. 2011b) are still facing challenges that need to be solved before a transition to clinical use will be possible.

Despite all the above-described advances, we still have limited understanding of the physiological concept of a healthy IVD as well as the underlying pathomechanisms of disc degeneration. Also the pathophysiological correlation between back pain and degenerative disc disease is still not entirely explored. Therefore, extensive research about the physiological as well as the pathological processes in intervertebral discs is mandatory before the ideal treatment strategies can be developed.

References

Abbushi A, Endres M, Cabraja M, Kroppenstedt SN, Thomale UW, Sittinger M, Hegewald AA, Morawietz L, Lemke A-J, Bansemer V-G (2008) Regeneration of intervertebral disc tissue by resorbable cell-free polyglycolic acid-based implants in a rabbit model of disc degeneration. Spine (Phila Pa 1976) 33 (14):1527–1532

Acosta FL Jr, Lotz J, Ames CP (2005) The potential role of mesenchymal stem cell therapy for intervertebral disc degeneration: a critical overview. Neurosurg Focus 19 (3):1–6

Acosta FL Jr, Metz L, Adkisson HD IV, Liu J, Carruthers-Liebenberg E, Milliman C, Maloney M, Lotz JC (2011) Porcine intervertebral disc repair using allogeneic juvenile articular chondrocytes or mesenchymal stem cells. Tissue Eng A 17(23–24):3045–3055

Ahlgren BD, Lui W, Herkowitz HN, Panjabi MM, Guiboux JP (2000) Effect of anular repair on the healing strength of the intervertebral disc: a sheep model. Spine (Phila Pa 1976) 25(17):2165–2170

Allon AA, Aurouer N, Yoo BB, Liebenberg EC, Buser Z, Lotz JC (2010) Structured coculture of stem cells and disc cells prevent disc degeneration in a rat model. Spine J 10(12):1089–1097

Allon AA, Butcher K, Schneider RA, Lotz JC (2012) Structured bilaminar co-culture outperforms stem cells and disc cells in a simulated degenerate disc environment. Spine (Phila Pa 1976) 37(10):813

Ambrossi GL, McGirt MJ, Sciubba DM, Witham TF, Wolinsky JP, Gokaslan ZL, Long DM (2009) Recurrent lumbar disc herniation after single-level lumbar discectomy: incidence and health care cost analysis. Neurosurgery 65(3):574–578; discussion 578

An HS, Takegami K, Kamada H, Nguyen CM, Thonar EJA, Singh K, Andersson GB, Masuda K (2005) Intradiscal administration of osteogenic protein-1 increases intervertebral disc height and proteoglycan content in the nucleus pulposus in normal adolescent rabbits. Spine (Phila Pa 1976) 30(1):25–31

An HS, Masuda K, Cs-Szabo G, Zhang Y, Chee A, Andersson GB, Im HJ, Thonar EJ, Kwon YM (2011) Biologic repair and regeneration of the intervertebral disk. J Am Acad Orthop Surg 19(7):450–452

Bae WC, Masuda K (2011) Emerging technologies for molecular therapy for intervertebral disk degeneration. Orthop Clin N Am 42(4):585–601

Bailey A, Araghi A, Blumenthal S, Huffmon GV (2013) Prospective, multicenter, randomized, controlled study of anular repair in lumbar discectomy: two-year follow-up. Spine (Phila Pa 1976) 38(14):1161–1169

Barth M, Diepers M, Weiss C, Thome C (2008) Two-year outcome after lumbar microdiscectomy versus microscopic sequestrectomy: part 2: radiographic evaluation and correlation with clinical outcome. Spine (Phila Pa 1976) 33(3):273–279

Bendtsen M, Bünger CE, Zou X, Foldager C, Jørgensen HS (2011) Autologous stem cell therapy maintains vertebral blood flow and contrast diffusion through the endplate in experimental intervertebral disc degeneration. Spine (Phila Pa 1976) 36(6):E373–E379

Benz K, Stippich C, Fischer L, Möhl K, Weber K, Lang J, Steffen F, Beintner B, Gaissmaier C, Mollenhauer JA (2012) Intervertebral disc cell-and hydrogel-supported and spontaneous intervertebral disc repair in nucleotomized sheep. Eur Spine J 21(9):1758–1768

Berlemann U, Schwarzenbach O (2009) An injectable nucleus replacement as an adjunct to microdiscectomy: 2 year follow-up in a pilot clinical study. Eur Spine J 18 (11):1706

Bernardo ME, Zaffaroni N, Novara F, Cometa AM, Avanzini MA, Moretta A, Montagna D, Maccario R, Villa R, Daidone MG (2007) Human bone marrow–derived mesenchymal stem cells do not undergo transformation after long-term in vitro culture and do not exhibit telomere maintenance mechanisms. Cancer Res 67(19):9142–9149

Blanquer S, Grijpma D, Poot A (2015) Delivery systems for the treatment of degenerated intervertebral discs. Adv Drug Deliv Rev 84:172–187

Bowles RD, Williams RM, Zipfel WR, Bonassar LJ (2010) Self-assembly of aligned tissue-engineered annulus fibrosus and intervertebral disc composite via collagen gel contraction. Tissue Eng Part A 16(4):1339–1348

Bowles RD, Gebhard HH, Hartl R, Bonassar LJ (2011a) Tissue-engineered intervertebral discs produce new matrix, maintain disc height, and restore biomechanical function to the rodent spine. Proc Natl Acad Sci U S A 108(32):13106–13111

Bowles RD, Gebhard HH, Härtl R, Bonassar LJ (2011b) Tissue-engineered intervertebral discs produce new matrix, maintain disc height, and restore biomechanical function to the rodent spine. Proc Natl Acad Sci 108 (32):13106–13111

Bowles RD, Gebhard HH, Dyke JP, Ballon DJ, Tomasino A, Cunningham ME, Hartl R, Bonassar LJ (2012) Image-based tissue engineering of a total intervertebral disc implant for restoration of function to the rat lumbar spine. NMR Biomed 25(3):443–451

Boyd LM, Carter AJ (2006) Injectable biomaterials and vertebral endplate treatment for repair and regeneration of the intervertebral disc. Eur Spine J 15 (3):414–421

Brittberg M, Lindahl A, Nilsson A, Ohlsson C, Isaksson O, Peterson L (1994) Treatment of deep cartilage defects in the knee with autologous chondrocyte transplantation. N Engl J Med 331(14):889–895

Bruske-Hohlfeld I, Merritt JL, Onofrio BM, Stonnington HH, Offord KP, Bergstralh EJ, Beard CM, Melton LJ 3rd, Kurland LT (1990) Incidence of lumbar disc surgery. A population-based study in Olmsted County, Minnesota, 1950–1979. Spine (Phila Pa 1976) 15(1):31–35

Bydon M, Xu R, Macki M, De la Garza-Ramos R, Sciubba DM, Wolinsky J-P, Witham TF, Gokaslan ZL, Bydon A (2013) Adjacent segment disease after anterior cervical discectomy and fusion in a large series. Neurosurgery 74(2):139–146

Carragee EJ, Han MY, Suen PW, Kim D (2003) Clinical outcomes after lumbar discectomy for sciatica: the effects of fragment type and anular competence. J Bone Joint Surg Am 85-A(1):102–108

Carragee EJ, Don AS, Hurwitz EL, Cuellar JM, Carrino J, Herzog R (2009) 2009 ISSLS prize winner: does discography cause accelerated progression of degeneration changes in the lumbar disc: a ten-year matched cohort study. Spine (Phila Pa 1976) 34 (21):2338–2345

CDC (2009) Prevalence and most common causes of disability among adults – United States, 2005. MMWR Morb Mortal Wkly Rep 58:421–426

Chiang C-J, Cheng C-K, Sun J-S, Liao C-J, Wang Y-H, Tsuang Y-H (2011) The effect of a new anular repair after discectomy in intervertebral disc degeneration: an experimental study using a porcine spine model. Spine (Phila Pa 1976) 36(10):761–769

Chubinskaya S, Kawakami M, Rappoport L, Matsumoto T, Migita N, Rueger DC (2007) Anti-catabolic effect of OP-1 in chronically compressed intervertebral discs. J Orthop Res 25(4):517–530

Chujo T, An HS, Akeda K, Miyamoto K, Muehleman C, Attawia M, Andersson G, Masuda K (2006) Effects of growth differentiation factor-5 on the intervertebral disc– in vitro bovine study and in vivo rabbit disc degeneration model study. Spine (Phila Pa 1976) 31(25):2909–2917

Comella K, Silbert R, Parlo M (2017) Effects of the intradiscal implantation of stromal vascular fraction plus platelet rich plasma in patients with degenerative disc disease. J Transl Med 15(1):12

Coric D, Pettine K, Sumich A, Boltes MO (2013) Prospective study of disc repair with allogeneic chondrocytes presented at the 2012 joint spine section meeting. J Neurosurg Spine 18(1):85–95

Crevensten G, Walsh AJ, Ananthakrishnan D, Page P, Wahba GM, Lotz JC, Berven S (2004) Intervertebral disc cell therapy for regeneration: mesenchymal stem cell implantation in rat intervertebral discs. Ann Biomed Eng 32(3):430–434

Deyo RA, Weinstein JN (2001) Low back pain. N Engl J Med 344(5):363–370

Elliott DM, Yerramalli CS, Beckstein JC, Boxberger JI, Johannessen W, Vresilovic EJ (2008) The effect of relative needle diameter in puncture and sham injection animal models of degeneration. Spine (Phila Pa 1976) 33(6):588–596

Evans MJ, Kaufman MH (1981) Establishment in culture of pluripotential cells from mouse embryos. Nature 292(5819):154–156

Feng G, Zhao X, Liu H, Zhang H, Chen X, Shi R, Liu X, Zhao X, Zhang W, Wang B (2011) Transplantation of mesenchymal stem cells and nucleus pulposus cells in a degenerative disc model in rabbits: a comparison of 2 cell types as potential candidates for disc regeneration. J Neurosurg Spine 14(3):322–329

Freemont AJ (2009) The cellular pathobiology of the degenerate intervertebral disc and discogenic back pain. Rheumatology (Oxford) 48(1):5–10

Frei H, Oxland TR, Rathonyi GC, Nolte LP (2001) The effect of nucleotomy on lumbar spine mechanics in compression and shear loading. Spine (Phila Pa 1976) 26(19):2080–2089

Frith JE, Cameron AR, Menzies DJ, Ghosh P, Whitehead DL, Gronthos S, Zannettino AC, Cooper-White JJ (2013) An injectable hydrogel incorporating mesenchymal precursor cells and pentosan polysulphate for intervertebral disc regeneration. Biomaterials 34(37):9430–9440

Frymoyer JW, Hanley E, Howe J, Kuhlmann D, Matteri R (1978) Disc excision and spine fusion in the management of lumbar disc disease. A minimum ten-year followup. Spine (Phila Pa 1976) 3(1):1–6

Ganey T, Libera J, Moos V, Alasevic O, Fritsch K-G, Meisel HJ, Hutton WC (2003) Disc chondrocyte transplantation in a canine model: a treatment for degenerated or damaged intervertebral disc. Spine (Phila Pa 1976) 28(23):2609–2620

Ganey T, Hutton WC, Moseley T, Hedrick M, Meisel H-J (2009) Intervertebral disc repair using adipose tissue-derived stem and regenerative cells: experiments in a canine model. Spine (Phila Pa 1976) 34(21):2297–2304

Gebhard H, Bowles R, Dyke J, Saleh T, Doty S, Bonassar L, Hartl R (2010) Total disc replacement using a tissue-engineered intervertebral disc in vivo: new animal model and initial results. Evid Based Spine Care J 1(2):62–66

Gebhard H, James AR, Bowles RD, Dyke JP, Saleh T, Doty SP, Bonassar LJ, Hartl R (2011) Biological intervertebral disc replacement: an in vivo model and comparison of two surgical techniques to approach the rat caudal disc. Evid Based Spine Care J 2(1):29–35

Ghosh P, Moore R, Vernon-Roberts B, Goldschlager T, Pascoe D, Zannettino A, Gronthos S, Itescu S (2012) Immunoselected STRO-3+ mesenchymal precursor cells and restoration of the extracellular matrix of degenerate intervertebral discs. J Neurosurg Spine 16(5):479–488

Goldschlager T, Ghosh P, Zannettino A, Gronthos S, Rosenfeld JV, Itescu S, Jenkin G (2010) Cervical motion preservation using mesenchymal progenitor cells and pentosan polysulfate, a novel chondrogenic agent: preliminary study in an ovine model. Neurosurg Focus 28(6):E4

Gorensek M, Joksimovic C, Kregar-Velikonja N, Gorensek M, Knezevic M, Jeras M, Pavlovcic V, Cor A (2004) Nucleus pulposus repair with cultured autologous elastic cartilage derived chondrocytes. Cell Mol Biol Lett 9(2):363–374

Gruber HE, Johnson TL, Leslie K, Ingram JA, Martin D, Hoelscher G, Banks D, Phieffer L, Coldham G, Hanley EN Jr (2002) Autologous intervertebral disc cell implantation: a model using *Psammomys obesus*, the sand rat. Spine (Phila Pa 1976) 27(15):1626–1633

Grunert P, Gebhard HH, Bowles RD, James AR, Potter HG, Macielak M, Hudson KD, Alimi M, Ballon DJ, Aronowitz E et al (2014a) Tissue-engineered intervertebral discs: MRI results and histology in the rodent spine. J Neurosurg Spine 20(4):443–451

Grunert P, Borde BH, Hudson KD, Macielak MR, Bonassar LJ, Härtl R (2014b) Annular repair using high-density collagen gel; a rat-tail in vivo model. Spine (Phila Pa 1976) 39(3):198

Gullung GB, Woodall WB, Tucci MA, James J, Black DA, McGuire RA (2011) Platelet-rich plasma effects on degenerative disc disease: analysis of histology and imaging in an animal model. Evid Based Spine Care J 2(04):13–18

Hamilton DJ, Seguin CA, Wang J, Pilliar RM, Kandel RA (2006) Formation of a nucleus pulposus-cartilage endplate construct in vitro. Biomaterials 27(3):397–405

Haufe SM, Mork AR (2006) Intradiscal injection of hematopoietic stem cells in an attempt to rejuvenate the intervertebral discs. Stem Cells Dev 15(1):136–137

Hee H, Ismail H, Lim C, Goh J, Wong H (2010) Effects of implantation of bone marrow mesenchymal stem cells, disc distraction and combined therapy on reversing degeneration of the intervertebral disc. J Bone Joint Surg Br 92(5):726–736

Hegewald AA, Medved F, Feng D, Tsagogiorgas C, Beierfuß A, Schindler GAK, Trunk M, Kaps C, Mern DS, Thomé C (2015) Enhancing tissue repair in annulus fibrosus defects of the intervertebral disc: analysis of a bio-integrative annulus implant in an in-vivo ovine model. J Tissue Eng Regen Med 9(4):405–414

Henriksson HB, Svanvik T, Jonsson M, Hagman M, Horn M, Lindahl A, Brisby H (2009) Transplantation of human mesenchymal stems cells into intervertebral discs in a xenogeneic porcine model. Spine (Phila Pa 1976) 34(2):141–148

Hiyama A, Mochida J, Iwashina T, Omi H, Watanabe T, Serigano K, Tamura F, Sakai D (2008) Transplantation of mesenchymal stem cells in a canine disc degeneration model. J Orthop Res 26(5):589–600

Ho G, Leung VY, Cheung KM, Chan D (2008) Effect of severity of intervertebral disc injury on mesenchymal stem cell-based regeneration. Connect Tissue Res 49(1):15–21

Hoy D, March L, Brooks P, Woolf A, Blyth F, Vos T, Buchbinder R (2010) Measuring the global burden of low back pain. Best Pract Res Clin Rheumatol 24(2):155–165

Hsieh AH, Hwang D, Ryan DA, Freeman AK, Kim H (2009) Degenerative anular changes induced by puncture are associated with insufficiency of disc biomechanical function. Spine (Phila Pa 1976) 34(10):998–1005

Huang K-Y, Yan J-J, Hsieh C-C, Chang M-S, Lin R-M (2007) The in vivo biological effects of intradiscal recombinant human bone morphogenetic protein-2 on the injured intervertebral disc: an animal experiment. Spine (Phila Pa 1976) 32(11):1174–1180

Huang B, Zhuang Y, Li C-Q, Liu L-T, Zhou Y (2011) Regeneration of the intervertebral disc with nucleus pulposus cell-seeded collagen II/hyaluronan/chondroitin-6-sulfate tri-copolymer constructs in a rabbit disc degeneration model. Spine (Phila Pa 1976) 36(26):2252–2259

Iatridis JC, Michalek A, Purmessur D, Korecki C (2009) Localized intervertebral disc injury leads to organ level changes in structure, cellularity, and biosynthesis. Cell Mol Bioeng 2(3):437–447

Imai Y, Okuma M, An HS, Nakagawa K, Yamada M, Muehleman C, Thonar E, Masuda K (2007) Restoration of disc height loss by recombinant human osteogenic protein-1 injection into intervertebral discs undergoing degeneration induced by an intradiscal injection of chondroitinase ABC. Spine (Phila Pa 1976) 32(11):1197–1205

Iwashina T, Mochida J, Sakai D, Yamamoto Y, Miyazaki T, Ando K, Hotta T (2006) Feasibility of using a human nucleus pulposus cell line as a cell source in cell transplantation therapy for intervertebral disc degeneration. Spine (Phila Pa 1976) 31(11):1177–1186

J R (2008) Intradiscal rhGDF-5 Phase I/II Clinical Trial. https://clinicaltrials.gov/show/NCT00813813

James AR, Bowles RD, Gebhard HH, Bonassar LJ, Hartl R (2011) Tissue-engineered total disc replacement: final outcomes of a murine caudal disc in vivo study. Evid Based Spine Care J 2(4):55–56

Jeong JH, Jin ES, Min JK, Jeon SR, Park C-S, Kim HS, Choi KH (2009) Human mesenchymal stem cells implantation into the degenerated coccygeal disc of the rat. Cytotechnology 59(1):55–64

Jeong JH, Lee JH, Jin ES, Min JK, Jeon SR, Choi KH (2010) Regeneration of intervertebral discs in a rat disc degeneration model by implanted adipose-tissue-derived stromal cells. Acta Neurochir 152(10):1771–1777

Katz JN (2006) Lumbar disc disorders and low-back pain: socioeconomic factors and consequences. J Bone Joint Surg Am 2:21–24

Kelly MP, Mok JM, Frisch RF, Tay BK (2011) Adjacent segment motion after anterior cervical discectomy and fusion versus Prodisc-c cervical total disk arthroplasty: analysis from a randomized, controlled trial. Spine (Phila Pa 1976) 36(15):1171–1179

Kepler CK, Anderson GD, Tannoury C, Ponnappan RK (2011) Intervertebral disk degeneration and emerging biologic treatments. J Am Acad Orthop Surg 19(9):543–553

Korecki CL, Costi JJ, Iatridis JC (2008) Needle puncture injury affects intervertebral disc mechanics and biology in an organ culture model. Spine (Phila Pa 1976) 33(3):235

Langer R, Vacanti JP (1993) Tissue engineering. Science 260(5110):920–926

Laus M, Bertoni F, Bacchini P, Alfonso C, Giunti A (1993) Recurrent lumbar disc herniation: what recurs?(a morphological study of recurrent disc herniation). Chir Organi Mov 78(3):147–154

Le Maitre CL, Hoyland JA, Freemont AJ (2007) Interleukin-1 receptor antagonist delivered directly and by gene therapy inhibits matrix degradation in the intact degenerate human intervertebral disc: an in situ zymographic and gene therapy study. Arthritis Res Ther 9(4):R83.

Leckie SK, Bechara BP, Hartman RA, Sowa GA, Woods BI, Coelho JP, Witt WT, Dong QD, Bowman BW, Bell KM (2012) Injection of AAV2-BMP2 and AAV2-TIMP1 into the nucleus pulposus slows the course of intervertebral disc degeneration in an in vivo rabbit model. Spine J 12(1):7–20

Ledet EH, Jeshuran W, Glennon JC, Shaffrey C, De Deyne P, Belden C, Kallakury B, Carl AL (2009) Small intestinal submucosa for anular defect closure: long-term response in an in vivo sheep model. Spine (Phila Pa 1976) 34(14):1457–1463

Leo BM, Li X, Balian G, Anderson DG (2004) In vivo bioluminescent imaging of virus-mediated gene transfer and transduced cell transplantation in the intervertebral disc. Spine (Phila Pa 1976) 29(8):838–844

Lequin MB, Verbaan D, Jacobs WC, Brand R, Bouma GJ, Vandertop WP, Peul WC (2013) Surgery versus prolonged conservative treatment for sciatica: 5-year results of a randomised controlled trial. BMJ Open 3(5):002534

Leung VY, Aladin DM, Lv F, Tam V, Sun Y, Lau RY, Hung SC, Ngan AH, Tang B, Lim CT (2014) Mesenchymal stem cells reduce intervertebral disc fibrosis and facilitate repair. Stem Cells 32(8):2164–2177

Li C-q, Huang B, Luo G, C-z Z, Zhuang Y, Zhou Y (2010) Construction of collagen II/hyaluronate/chondroitin-6-sulfate tri-copolymer scaffold for nucleus pulposus tissue engineering and preliminary analysis of its physicochemical properties and biocompatibility. J Mater Sci Mater Med 21(2):741–751

Likhitpanichkul M, Dreischarf M, Illien-Junger S, Walter B, Nukaga T, Long R, Sakai D, Hecht A, Iatridis J (2014) Fibrin-genipin adhesive hydrogel for annulus fibrosus repair: performance evaluation with large animal organ culture, in situ biomechanics, and in vivo degradation tests. Eur Cell Mater 28:25

Lotz JC (2004) Animal models of intervertebral disc degeneration: lessons learned. Spine (Phila Pa 1976) 29(23):2742–2750

Lurie JD, Tosteson TD, Tosteson AN, Zhao W, Morgan TS, Abdu WA, Herkowitz H, Weinstein JN (2014) Surgical versus nonoperative treatment for lumbar disc herniation: eight-year results for the spine patient outcomes research trial. Spine (Phila Pa 1976) 39(1):3–16

MacGregor AJ, Andrew T, Sambrook PN, Spector TD (2004) Structural, psychological, and genetic influences on low back and neck pain: a study of adult female twins. Arthritis Rheum 51(2):160–167

Maidhof R, Alipui DO, Rafiuddin A, Levine M, Grande DA, Chahine NO (2012) Emerging trends in biological therapy for intervertebral disc degeneration. Discov Med 14(79):401–411

Maldonado CV, Paz RD-R, Martin CB (2011) Adjacent-level degeneration after cervical disc arthroplasty versus fusion. Eur Spine J 20(3):403

Martin GR (1981) Isolation of a pluripotent cell line from early mouse embryos cultured in medium conditioned by teratocarcinoma stem cells. Proc Natl Acad Sci 78(12):7634–7638

Martin JT, Milby AH, Chiaro JA, Kim DH, Hebela NM, Smith LJ, Elliott DM, Mauck RL (2014) Translation of an engineered nanofibrous disc-like angle-ply structure for intervertebral disc replacement in a small animal model. Acta Biomater 10(6):2473–2481

Masuda K, Oegema TR Jr, An HS (2004) Growth factors and treatment of intervertebral disc degeneration. Spine (Phila Pa 1976) 29(23):2757–2769

Masuda K, Imai Y, Okuma M, Muehleman C, Nakagawa K, Akeda K, Thonar E, Andersson G, An HS (2006) Osteogenic protein-1 injection into a degenerated disc induces the restoration of disc height and structural changes in the rabbit anular puncture model. Spine (Phila Pa 1976) 31(7):742–754

McBeth J, Jones K (2007) Epidemiology of chronic musculoskeletal pain. Best Pract Res Clin Rheumatol 21(3):403–425

Meisel HJ (2012). Safety and efficacy with NOVOCART® disc plus (ADCT) for the treatment of degenerative disc disease in lumbar spine. https://clinicaltrials.gov/show/NCT01640457

Meisel HJ, Ganey T, Hutton WC, Libera J, Minkus Y, Alasevic O (2006) Clinical experience in cell-based therapeutics: intervention and outcome. Eur Spine J 15(3):397–405

Meisel HJ, Siodla V, Ganey T, Minkus Y, Hutton WC, Alasevic OJ (2007) Clinical experience in cell-based therapeutics: disc chondrocyte transplantation: a treatment for degenerated or damaged intervertebral disc. Biomol Eng 24(1):5–21

Mern DS, Beierfuß A, Thomé C, Hegewald AA (2014) Enhancing human nucleus pulposus cells for biological treatment approaches of degenerative intervertebral disc diseases: a systematic review. J Tissue Eng Regen Med 8(12):925–936

Mesoblast Ltd. (2015) A prospective, multicenter, randomized, double-blind, placebo-controlled study to evaluate the efficacy and safety of a single injection of rexlemestrocel-L alone or combined with hyaluronic acid (HA) in subjects with chronic low back pain. https://clinicaltrials.gov/show/NCT02412735

Mesoblast Ltd. (2019) Trial results: MPC-06-ID phase chronic low back pain due to Disc degeneration clinical trial. http://www.mesoblast.com/clinical-trial-results/mpc-06-id-phase-2

Miyamoto K, Masuda K, Kim JG, Inoue N, Akeda K, Andersson GB, An HS (2006) Intradiscal injections of osteogenic protein-1 restore the viscoelastic properties of degenerated intervertebral discs. Spine J 6(6):692–703

Miyamoto T, Muneta T, Tabuchi T, Matsumoto K, Saito H, Tsuji K, Sekiya I (2010) Intradiscal transplantation of synovial mesenchymal stem cells prevents intervertebral disc degeneration through suppression of matrix metalloproteinase-related genes in nucleus pulposus cells in rabbits. Arthritis Res Ther 12(6):R206

Mizuno H, Roy AK, Vacanti CA, Kojima K, Ueda M, Bonassar LJ (2004a) Tissue-engineered composites of anulus fibrosus and nucleus pulposus for intervertebral disc replacement. Spine (Phila Pa 1976) 29(12):1290–1297

Mizuno H, Roy AK, Vacanti CA, Kojima K, Ueda M, Bonassar LJ (2004b) Tissue-engineered composites of anulus fibrosus and nucleus pulposus for intervertebral disc replacement. Spine 29(12):1290–1297; discussion 1297–1298

Mizuno H, Roy AK, Zaporojan V, Vacanti CA, Ueda M, Bonassar LJ (2006) Biomechanical and biochemical characterization of composite tissue-engineered intervertebral discs. Biomaterials 27(3):362–370

Moriguchi Y, Alimi M, Khair T, Manolarakis G, Berlin C, Bonassar LJ, Härtl R (2016) Biological treatment approaches for degenerative disk disease: a literature review of in vivo animal and clinical data. Global Spine J 6(05):497–518

Moriguchi Y, Mojica-Santiago J, Grunert P, Pennicooke B, Berlin C, Khair T, Navarro-Ramirez R, Ricart Arbona RJ, Nguyen J, Härtl R, Bonassar LJ (2017) Total disc replacement using tissue-engineered intervertebral discs in the canine cervical spine. PLoS One 12(10):e0185716

Moroni L, Fornasari PM (2013) Human mesenchymal stem cells: a bank perspective on the isolation, characterization and potential of alternative sources for the regeneration of musculoskeletal tissues. J Cell Physiol 228(4):680–687

Nagae M, Ikeda T, Mikami Y, Hase H, Ozawa H, Matsuda K-I, Sakamoto H, Tabata Y, Kawata M, Kubo T (2007) Intervertebral disc regeneration using platelet-rich plasma and biodegradable gelatin hydrogel microspheres. Tissue Eng 13(1):147–158

Nerurkar NL, Baker BM, Sen S, Wible EE, Elliott DM, Mauck RL (2009) Nanofibrous biologic laminates replicate the form and function of the annulus fibrosus. Nat Mater 8(12):986–992

Nishida K, Kang JD, Gilbertson LG, Moon SH, Suh JK, Vogt MT, Robbins PD, Evans CH (1999) Modulation of the biologic activity of the rabbit intervertebral disc by gene therapy: An in vivo study of adenovirus-mediated transfer of the human transforming growth factor beta 1 encoding gene. Spine (Phila Pa 1976) 24(23):2419–2425

Nishida K, Doita M, Takada T, Kakutani K, Miyamoto H, Shimomura T, Maeno K, Kurosaka M (2006) Sustained transgene expression in intervertebral disc cells in vivo mediated by microbubble-enhanced ultrasound gene therapy. Spine (Phila Pa 1976) 31(13):1415–1419

O'Connell GD, Vresilovic EJ, Elliott DM (2007) Comparison of animals used in disc research to human lumbar disc geometry. Spine (Phila Pa 1976) 32(3):328–333

O'Connell GD, Malhotra NR, Vresilovic EJ, Elliott DM (2011) The effect of nucleotomy and the dependence of degeneration of human intervertebral disc strain in axial compression. Spine (Phila Pa 1976) 36(21):1765–1771

Obata S, Akeda K, Imanishi T, Masuda K, Bae W, Morimoto R, Asanuma Y, Kasai Y, Uchida A, Sudo A (2012) Effect of autologous platelet-rich plasma-releasate on intervertebral disc degeneration in the rabbit anular puncture model: a preclinical study. Arthritis Res Ther 14(6):R241

Oehme D, Ghosh P, Shimmon S, Wu J, McDonald C, Troupis JM, Goldschlager T, Rosenfeld JV, Jenkin G (2014) Mesenchymal progenitor cells combined with pentosan polysulfate mediating disc regeneration at the time of microdiscectomy: a preliminary study in an ovine model. J Neurosurg Spine 20(6):657–669

Okuma M, Mochida J, Nishimura K, Sakabe K, Seiki K (2000) Reinsertion of stimulated nucleus pulposus cells retards intervertebral disc degeneration: an in vitro and in vivo experimental study. J Orthop Res 18(6):988–997

Omlor G, Fischer J, Kleinschmitt K, Benz K, Holschbach J, Brohm K, Anton M, Guehring T, Richter W (2014) Short-term follow-up of disc cell therapy in a porcine nucleotomy model with an albumin–hyaluronan hydrogel: in vivo and in vitro results of metabolic disc cell activity and implant distribution. Eur Spine J 23(9):1837–1847

Orozco L, Soler R, Morera C, Alberca M, Sánchez A, García-Sancho J (2011) Intervertebral disc repair by autologous mesenchymal bone marrow cells: a pilot study. Transplantation 92(7):822–828

Paul R, Haydon RC, Cheng H, Ishikawa A, Nenadovich N, Jiang W, Zhou L, Breyer B, Feng T, Gupta P (2003) Potential use of Sox9 gene therapy for intervertebral disc disease. Spine (Phila Pa 1976) 28(8):755

Pennicooke B, Hussain I, Berlin C, Sloan SR, Borde B, Moriguchi Y, Lang G, Navarro-Ramirez R, Cheetham J, Bonassar LJ et al (2017) Annulus fibrosus repair using high-density collagen gel: an in vivo ovine model. Spine (Phila Pa 1976) 43(4):E208–E215

Pettine KA, Murphy MB, Suzuki RK, Sand TT (2015) Percutaneous injection of autologous bone marrow concentrate cells significantly reduces lumbar discogenic pain through 12 months. Stem Cells 33(1):146–156

Peul WC, van Houwelingen HC, van den Hout WB, Brand R, Eekhof JA, Tans JT, Thomeer RT, Koes BW (2007) Surgery versus prolonged conservative treatment for sciatica. N Engl J Med 356(22):2245–2256

Pittenger MF, Mackay AM, Beck SC, Jaiswal RK, Douglas R, Mosca JD, Moorman MA, Simonetti DW, Craig S, Marshak DR (1999) Multilineage potential of adult human mesenchymal stem cells. Science 284(5411):143–147

Podichetty VK (2007) The aging spine: the role of inflammatory mediators in intervertebral disc degeneration. Cell Mol Biol (Noisy-le-Grand) 53(5):4–18

Pye SR, Reid DM, Smith R, Adams JE, Nelson K, Silman AJ, O'Neill TW (2004) Radiographic features of lumbar disc degeneration and self-reported back pain. J Rheumatol 31(4):753–758

Rajpurohit R, Risbud MV, Ducheyne P, Vresilovic EJ, Shapiro IM (2002) Phenotypic characteristics of the nucleus pulposus: expression of hypoxia inducing factor-1, glucose transporter-1 and MMP-2. Cell Tissue Res 308(3):401–407

Rannou F, Revel M, Poiraudeau S (2003) Is degenerative disk disease genetically determined? Joint Bone Spine 70(1):3–5

Revell P, Damien E, Di Silvio L, Gurav N, Longinotti C, Ambrosio L (2007) Tissue engineered intervertebral disc repair in the pig using injectable polymers. J Mater Sci Mater Med 18(2):303–308

Roberts S, Evans H, Trivedi J, Menage J (2006) Histology and pathology of the human intervertebral disc. JBJS 88:10–14

Ruan D, He Q, Ding Y, Hou L, Li J, Luk KD (2007) Intervertebral disc transplantation in the treatment of degenerative spine disease: a preliminary study. Lancet 369(9566):993–999

Ruan D-K, Xin H, Zhang C, Wang C, Xu C, Li C, He Q (2010) Experimental intervertebral disc regeneration with tissue-engineered composite in a canine model. Tissue Eng A 16(7):2381–2389

Sakai D, Grad S (2015) Advancing the cellular and molecular therapy for intervertebral disc disease. Adv Drug Deliv Rev 84:159–171

Sakai D, Mochida J, Yamamoto Y, Nomura T, Okuma M, Nishimura K, Nakai T, Ando K, Hotta T (2003) Transplantation of mesenchymal stem cells embedded in Atelocollagen® gel to the intervertebral disc: a potential therapeutic model for disc degeneration. Biomaterials 24(20):3531–3541

Sakai D, Mochida J, Iwashina T, Watanabe T, Nakai T, Ando K, Hotta T (2005) Differentiation of mesenchymal stem cells transplanted to a rabbit degenerative disc model: potential and limitations for stem cell therapy in disc regeneration. Spine (Phila Pa 1976) 30(21):2379–2387

Sakai D, Mochida J, Iwashina T, Hiyama A, Omi H, Imai M, Nakai T, Ando K, Hotta T (2006) Regenerative effects of transplanting mesenchymal stem cells embedded in atelocollagen to the degenerated intervertebral disc. Biomaterials 27(3):335–345

Sato M, Kikuchi M, Ishihara M, Asazuma T, Kikuchi T, Masuoka K, Hattori H, Fujikawa K (2003) Tissue engineering of the intervertebral disc with cultured annulus fibrosus cells using atelocollagen honeycomb shaped scaffold with a membrane seal (ACHMS scaffold). Med Biol Eng Comput 41(3):365–371

Sawamura K, Ikeda T, Nagae M, Okamoto S-i, Mikami Y, Hase H, Ikoma K, Yamada T, Sakamoto H, Matsuda K-i (2009) Characterization of in vivo effects of platelet-rich plasma and biodegradable gelatin hydrogel microspheres on degenerated intervertebral discs. Tissue Eng A 15(12):3719–3727

Schek RM, Michalek AJ, Iatridis JC (2011) Genipin-crosslinked fibrin hydrogels as a potential adhesive to augment intervertebral disc annulus repair. Eur Cell Mater 21:373–383

Schmidt CO, Raspe H, Pfingsten M, Hasenbring M, Basler HD, Eich W, Kohlmann T (2007) Back pain in the German adult population: prevalence, severity, and sociodemographic correlates in a multiregional survey. Spine (Phila Pa 1976) 32(18):2005–2011

Seki S, Asanuma-Abe Y, Masuda K, Kawaguchi Y, Asanuma K, Muehleman C, Iwai A, Kimura T (2009) Effect of small interference RNA (siRNA) for ADAMTS5 on intervertebral disc degeneration in the rabbit anular needle-puncture model. Arthritis Res Ther 11(6):R166

Serigano K, Sakai D, Hiyama A, Tamura F, Tanaka M, Mochida J (2010) Effect of cell number on mesenchymal stem cell transplantation in a canine disc degeneration model. J Orthop Res 28(10):1267–1275

Sheikh H, Zakharian K, De La Torre RP, Facek C, Vasquez A, Chaudhry GR, Svinarich D, Perez-Cruet MJ (2009) In vivo intervertebral disc regeneration using stem cell–derived chondroprogenitors. J Neurosurg Spine 10(3):265–272

Sobajima S, Vadala G, Shimer A, Kim JS, Gilbertson LG, Kang JD (2008) Feasibility of a stem cell therapy for intervertebral disc degeneration. Spine J 8(6):888–896

Sugawara T, Itoh Y, Hirano Y, Higashiyama N, Mizoi K (2009) Long term outcome and adjacent disc degeneration after anterior cervical discectomy and fusion with titanium cylindrical cages. Acta Neurochir 151(4):303–309

Swartz KR, Trost GR (2003) Recurrent lumbar disc herniation. Neurosurg Focus 15(3):E10

Takahashi K, Yamanaka S (2006) Induction of pluripotent stem cells from mouse embryonic and adult fibroblast cultures by defined factors. Cell 126(4):663–676

Tam V, Rogers I, Chan D, Leung VY, Cheung K (2014) A comparison of intravenous and intradiscal delivery of multipotential stem cells on the healing of injured intervertebral disk. J Orthop Res 32(6):819–825

Thompson JP, Oegema TR, Bradford D (1991) Stimulation of mature canine intervertebral disc by growth factors. Spine (Phila Pa 1976) 16(3):253–260

Tschugg A, Diepers M, Simone S, Michnacs F, Quirbach S, Strowitzki M, Meisel HJ, Thome C (2017) A prospective randomized multicenter phase I/II clinical trial to evaluate safety and efficacy of NOVOCART disk plus autologous disk chondrocyte transplantation in the treatment of nucleotomized and degenerative lumbar disks to avoid secondary disease: safety results of phase I-a short report. Neurosurg Rev 40(1):155–162

Tuakli-Wosornu YA, Terry A, Boachie-Adjei K, Harrison JR, Gribbin CK, LaSalle EE, Nguyen JT, Solomon JL, Lutz GE (2016) Lumbar Intradiskal platelet-rich plasma (PRP) injections: a prospective, double-blind, randomized controlled study. PM & R 8(1):1–10; quiz 10

Vadalà G, Sowa G, Hubert M, Gilbertson LG, Denaro V, Kang JD (2012) Mesenchymal stem cells injection in degenerated intervertebral disc: cell leakage may induce osteophyte formation. J Tissue Eng Regen Med 6(5):348–355

Walsh AJ, Bradford DS, Lotz JC (2004) In vivo growth factor treatment of degenerated intervertebral discs. Spine (Phila Pa 1976) 29(2):156–163

Wang Y-H, Kuo T-F, Wang J-L (2007) The implantation of non-cell-based materials to prevent the recurrent disc herniation: an in vivo porcine model using quantitative discomanometry examination. Eur Spine J 16(7):1021–1027

Wang S, Rui Y, Lu J, Wang C (2014) Cell and molecular biology of intervertebral disc degeneration: current understanding and implications for potential therapeutic strategies. Cell Prolif 47(5):381–390

Wei A, Williams LA, Bhargav D, Shen B, Kishen T, Duffy N, Diwan AD (2009) BMP13 prevents the effects of annular injury in an ovine model. Int J Biol Sci 5(5):388

Woods BI, Vo N, Sowa G, Kang JD (2011) Gene therapy for intervertebral disk degeneration. Orthop Clin N Am 42(4):563–574

Xin H, Zhang C, Wang D, Shi Z, Gu T, Wang C, Wu J, Zhang Y, He Q, Ruan D (2012) Tissue-engineered allograft intervertebral disc transplantation for the treatment of degenerative disc disease: experimental study in a beagle model. Tissue Eng A 19(1–2):143–151

Xu D, Gechtman Z, Hughes A, Collins A, Dodds R, Cui X, Jolliffe L, Higgins L, Murphy A, Farrell F (2006) Potential involvement of BMP receptor type IB activation in a synergistic effect of chondrogenic promotion between rhTGFbeta3 and rhGDF5 or rhBMP7 in human mesenchymal stem cells. Growth Factors 24(4):268–278

Yang F, Leung VY, Luk KD, Chan D, Cheung KM (2009) Mesenchymal stem cells arrest intervertebral disc degeneration through chondrocytic differentiation and stimulation of endogenous cells. Mol Ther 17 (11):1959–1966

Yang H, Wu J, Liu J, Ebraheim M, Castillo S, Liu X, Tang T, Ebraheim NA (2010) Transplanted mesenchymal stem cells with pure fibrinous gelatin-transforming growth factor-β1 decrease rabbit intervertebral disc degeneration. Spine J 10(9):802–810

Yoon ST, Park JS, Kim KS, Li J, Attallah-Wasif ES, Hutton WC, Boden SD (2004) ISSLS prize winner: LMP-1 upregulates intervertebral disc cell production of proteoglycans and BMPs in vitro and in vivo. Spine (Phila Pa 1976) 29(23):2603–2611

Yoshikawa T, Ueda Y, Miyazaki K, Koizumi M, Takakura Y (2010) Disc regeneration therapy using marrow mesenchymal cell transplantation: a report of two case studies. Spine (Phila Pa 1976) 35(11):E475–E480

Zhang Y-G, Guo X, Xu P, Kang L-L, Li J (2005) Bone mesenchymal stem cells transplanted into rabbit intervertebral discs can increase proteoglycans. Clin Orthop Relat Res 430:219–226

Zhang Y, Chee A, Thonar EJ, An HS (2011a) Intervertebral disk repair by protein, gene, or cell injection: a framework for rehabilitation-focused biologics in the spine. PM & R 3(6 Suppl 1):S88–S94

Zhang Y-H, Zhao C-Q, Jiang L-S, Dai L-Y (2011b) Lentiviral shRNA silencing of CHOP inhibits apoptosis induced by cyclic stretch in rat annular cells and attenuates disc degeneration in the rats. Apoptosis 16 (6):594–605

Zhang Y, Drapeau S, An HS, Thonar EJA, Anderson DG (2011c) Transplantation of goat bone marrow stromal cells to the degenerating intervertebral disc in a goat disc-injury model. Spine (Phila Pa 1976) 36 (5):372

Zhuang Y, Huang B, Li CQ, Liu LT, Pan Y, Zheng WJ, Luo G, Zhou Y (2011) Construction of tissue-engineered composite intervertebral disc and preliminary morphological and biochemical evaluation. Biochem Biophys Res Commun 407(2):327–332

Bone Grafts and Bone Graft Substitutes

11

Jae Hyuk Yang, Juliane D. Glaeser, Linda E. A. Kanim, Carmen Y. Battles, Shrikar Bondre, and Hyun W. Bae

Contents

Introduction	201
Design Requirements for Engineered Biomaterials (Table 2)	201
Spinal Fusion	201

The co-first authors Jae Hyuk Yang and Juliane D. Glaeser shares an equal contribution of this chapter.

J. H. Yang
Surgery, Department of Orthopaedics, Cedars-Sinai Medical Center, Los Angeles, CA, USA

Korea University Guro Hospital, Seoul, South Korea
e-mail: kuspine@naver.com

J. D. Glaeser · L. E. A. Kanim · H. W. Bae (✉)
Surgery, Department of Orthopaedics, Cedars-Sinai Medical Center, Los Angeles, CA, USA

Board of Governors Regenerative Medicine Institute, Cedars-Sinai Medical Center, Los Angeles, CA, USA

Department of Surgery, Cedars-Sinai Spine Center, Los Angeles, CA, USA
e-mail: Juliane.Glaeser@cshs.org; Linda.Kanim@cshs.org; Hyun.Bae@cshs.org; baemd@me.com

C. Y. Battles
Surgery, Department of Orthopaedics, Cedars-Sinai Medical Center, Los Angeles, CA, USA

Department of Surgery, Cedars-Sinai Spine Center, Los Angeles, CA, USA
e-mail: carmen.battles@cshs.org

S. Bondre
Chemical Engineering, Prosidyan, Warren, NJ, USA
e-mail: sbondre@prosidyan.com; shrikar.bondre@outlook.com

© Springer Nature Switzerland AG 2021
B. C. Cheng (ed.), *Handbook of Spine Technology*,
https://doi.org/10.1007/978-3-319-44424-6_36

Current Materials for Spinal Fusion 213
Autografts 213
Allografts 214
Exogenous Inductive Differentiation Growth Factors and Other Peptides (Table 5) ... 232
Synthetic Materials and Drafts (Table 6) 240
Mixed Use Graft Materials with Antibacterial Effects (Table 7) 265

Conclusion 266

References 266

Abstract

Bone grafting has been uniquely practiced since the early 1600s. Historically, bone grafting includes autologous bone, a variety of allograft bone, and synthetic based-materials utilized in surgical interventions to treat spinal diseases or fractures. One of the most common uses of bone grafts is in spinal surgery to promote fusion between two functional vertebral segments. During a spinal surgery procedure wherein host bone is prepared, bone grafts are employed to optimize the biological environment to augment healing of the bony tissues for a desired outcome of a solid union – successful spinal fusion. State-of-the art bone graft materials have been effectively used to enhance bone induction and healing, providing more predictable outcomes resulting in spinal fusion.

Autograft has traditionally been recognized as the gold standard for bone grafts. However, differing grafting modalities are currently replacing autograft as the standard of care due to patient donor site morbidity, limitation to autograft, and the cessation of training young surgeons in the technique of autograft harvest. This has led to the research and development of various next-generation osteoconductive, osteoinductive, and osteogenic materials. In this chapter, various options to augment or replace autograft bone have been reviewed. Current options for spinal fusion discussed herein include autografts, allografts, and osteoconductive, osteoinductive, osteogenic, and osteostimulative materials. Further, novel materials such as engineered bioactive glass and peptide-based materials are presented. Choice of graft material with consideration of anatomical location, surgical application, spinal fusion technique, and patient characteristics will optimize bone healing and clinical outcomes.

Keywords

Autograft · Allograft · Synthetic bone grafts · Viable bone grafts · Cell-based bone grafts · Bioactive glass · Growth differentiation factors · Bone graft extenders · Substitutes · Combination products · Osteoconductive · Osteoinductive · Osteogenic · Interbody fusion · Posterolateral fusion · Posterior fusion · BMP · rhBMP-2 · rhBMP-7 · P-15

Abbreviations

AATB	American Association of Tissue Bank
ABM	Anorganic bone matrix
ACDF	Anterior cervical discectomy and fusion
ACS	Absorbable collage sponge, used as a carrier
AIBG	Autologous iliac bone graft
ALIF	Anterior lumbar interbody fusion
Allograft	Graft derived from unrelated human donor and transplanted to another person/patient, cadaver via bone bank; live donor patients undergoing removal, i.e., hip replacement

APC, PRP	Autologous platelet concentrate, serum derived from patient himself is concentrated via centrifugation (contains cytokines, growth factors; theorized to promote fusion)	CHO	Chinese hamster ovary (used to derive rhBMP-2)
		Collagen Carrier	ACS, bovine type I collagen matrix
		CS	Calcium sulfate, synthetic ceramic material composed of calcium-sulfate (1:1)
Autograft, Autologous	Bone harvested from patient self from one site of the body and implant to another site of same patient	DBM powder	Demineralized bone matrix powder (human derived)
		DBM, hDBM	Demineralized bone matrix (allograft), bone powder (allograft), hDBM (human derived)
BAG, BG	Bioactive glass, a ceramic, biologically compatible synthetic material of crystalline components		
		DBM-based product	Demineralized bone matrix-based product, DBM powder (human derived) mixed with other material substances, or carriers
BCG	Biocompatible glass		
BCP	Biphasic calcium phosphate		
BIC	Bone-implant contact	DFBA	Freeze-dried bone allograft
BMA	Bone marrow aspirate		
BMC	Bone marrow concentrate	E.BMP, E.BMP-2	Escherichia coli-derived BMP-2 (used to derive rhBMP-2)
BMP, rhBMP-2, rhBMP-7	Bone morphogenetic protein, recombinant human bone morphogenetic protein	ECM	Extracellular matrix
		Enhancer	Acts to add properties of osteogenicity or osteoinductivity to a graft material
βTCP, β-TCP	Beta-tricalcium phosphate		
Cage	Cages are cylindrical or square-shaped devices usually threaded. Used as instrumentation/fixation and to hold graft material in a surgical site, employed for interbody fusion, i.e., LT-Cage	Extender	Bone graft extender, osteoconductive material, compounds, scaffolds added to other grafting materials (ideally inductive or osteogenic), to increase the volume of graft. May add structural support
Cell-based	Bone grafts with viable cells preserved or substitutes wherein cells are added	FDA CFR	Code of Federal Regulations
		FDA, US-FDA	Food and Drug Administration, US FDA
CGTP	Current Good Tissue Practice		

Growth factors/ growth Differentiation factors	BMP, rhBMP-2, rhBMP-7, bone morphogenetic protein, recombinant human bone morphogenetic protein	Osteoconductive	Provides structural scaffolding upon which matrix-producing cells deposit new bone
GvHD	Graft-versus-host disease	Osteogenic	Presence of osteoblast precursor cells that contribute to new bone growth
h	Human-derived or human-like version	Osteoinductive	Presence of molecular growth factors that stimulate precursors cells to migrate to graft site, mature into osteoid-producing cells, increase production of bone matrix
HA	Hydroxyapatite, calcium-containing porous crystal, accounts for a majority of bone natural mineral component		
HCO	Bicarbonate (Bae to review)	PEEK	Polyetheretherketone synthetic material, hydrophobic material to which cells have a limited ability to bond (polyaryletherketone family colorless organic thermoplastic polymer), used to fabricate spinal devices such as cages
HCT/P	Human cell and tissue product		
ICBG	Iliac crest bone graft, autograft – bone morsels harvested from iliac crest		
ISO	International Organization for Standardization		
LBG, LAG	Local bone graft, local autograft – bone morsels harvested from the surgical dissection site	Peptides/growth factors/growth differentiation factors	BMP, rhBMP-2, rhBMP-7 (OP-1 osteogenic protein), bone morphogenetic protein, recombinant human bone morphogenetic protein
LLIF	Lateral lumbar interbody fusion		
MED	Minimally effective dose		
MIS	Minimally Invasive Surgery	rh	Recombinant human form
MRI	Magnetic resonance imaging	PLF	Posterolateral fusion
MSC	Mesenchymal stem cell	PLIF	Posterior lumbar interbody fusion
ncHA	Nanocrystal hydroxyapatite	PLLF	Posterolateral lumbar fusion
OIF	Osteoinductive factor	PRP	Platelet-rich plasma (PRP) platelets (thought to have target
OLIF	Oblique lumbar interbody fusion		

	growth factors) from patients' own blood
Segment, spinal segment	Spinal segment of the spine includes a superior vertebral body, disc, inferior vertebral body. Upper vertebral body, target vertebral body, lower vertebral body
Substitute	Graft substitute used instead of autologous bone grafting
TCP	Tricalcium phosphate, synthetic ceramic material composed of calcium and phosphate (3:2)
TI	Titanium (Ti)
TLIF	Transforaminal lumbar interbody fusion
TNF	Tumor necrosis factor
UDI	FDA rule that requires medical device manufacturers to update products with a unique device identifier
Xenograft	Grafted from one species to another species (i.e., bovine to human; porcine to human)

Introduction

Bone grafts and graft substitutes are materials that are used to rapidly induce or support biologic bone remodeling after surgical procedures to reconstruct bony structures and correct deformities and/or to provide initial structural support (Wang and Yeung 2017). In the spine, bone grafts are most often used to support biological healing with bony union of vertebral segments after a spinal fusion surgical procedure.

Bone grafts may come from a patient's own bone (autograft), may come from a human cadaver or living donor via bone bank (allograft), or may be fabricated from a synthetic material such as ceramics or bioactive glass. Furthermore, combination materials including composites of allografts, growth factors, osteogenic cells, synthetic materials of ceramic and/or cements, bioactive glass, and peptide-based materials have been developed and are offered for clinical use in spine fusion. See Table 1 for a description of sources of grafting materials and their associated bone-forming properties.

Design Requirements for Engineered Biomaterials (Table 2)

The selection of bone graft alternatives to be used for spinal fusion should be conducted carefully by considering the different healing environments, reviewing the preclinical and clinical data, and also considering the regulatory burden of proof for products not subjected to high levels of regulation (Boden 2002).

The development of products used for bone regeneration has followed the basic criteria of providing a biocompatible three-dimensional scaffold with controlled architecture capable of stimulating or supporting bone growth in the natural in vivo environment. The ability of the material to be amalgamated with cellular and signal (differentiation/growth factors)-based products is a key strategy in maximizing the efficacy and likely success of fusion. The primary characteristics and significance of bone graft substitutes is shown in Table 2.

Spinal Fusion

Spinal fusion is usually performed to provide stability to the spine when its biomechanics have been disturbed or altered. The surgical concept underlying spinal fusion is to reduce clinically important abnormal motion and add immediate and long-term stability, therefore decreasing or

Table 1 Description of specific graft materials and bone forming properties

Grafting material	Grafting material (typical abbreviation)	Grafting material category and description	Variability	Osteogenic	Osteo inductive	Osteo conductive	Immunogenicity/ disease transmission	Strength (immediate)	Donor site morbidity
Autograft									
Autograft	Iliac crest bone graft (ICBG)	More cancellous (mercerized and/or strut form)	Patients' own bone quality	+++	++	+++	–	+++	++
Autograft	Local bone (LB, LAG)	More cortical (mostly mercerized form)		+/–	+	+	–	+/–	+/–
Autograft	Bone dust (Gao et al. 2018; Street et al. 2017)	Generated via high-speed burr on bone surface		+/– (less than local bone)	+/– (less than local bone)	+/– (less than local bone)	–	–	+/–
Autograft	PRP (platelet-rich plasma) (Elder et al. 2015)	Plasma preparation with increased platelet concentration	Patients' own health status	+/–	++ (Elder et al. 2015) (activation of growth factors)	–	–	–	–
Autograft/bone marrow aspirate (Robbins et al. 2017)	Osteogenic cell with growth factors	Most common source of MSC		+ (variable according to donor's condition)	+ (variable according to donor's condition)	–	–	–	+

Allograft								
Allograft	Fresh (Meyers 1985)	1. Living donor (patient to patient transfer) 2. Cadaveric donor (harvested within 12 h and allotransplantation within 72 h) → Femoral head (as osteochondral form)	Lot-to-lot variability donor's bone condition + sterilization processing techniques	? (no data for osteogenic graft for human)	? (no data for osteogenic graft for human)	? (no data for osteogenic graft for human)	+++ (generally causes an unacceptable host immune reaction as osteogenic graft) → Not used commercially, only animal studies	
Allograft	Fresh (osteochondral graft) (Rauck et al. 2019)	1. Living donor (patient to patient transfer) 2. Cadaveric donor (harvested within 12 h and allotransplantation within 24 h) → Femoral head (as osteochondral form) (Torrie et al. 2015)		+/− (only chondrocyte viability remains)	−	++	+ (reduced/mild immune reaction by cartilaginous portion of graft) + (infection risk due to storage media)	++ (grafted at articular portion for weight supporting)
Allograft	Fresh-frozen (Kawaguchi and Hart 2015)	From: 1) Living donor 2) Cadaveric donor		−	−	++ (less than autogenous bone) (Miyazaki et al. 2009)	+ (reported cases)	++ (Kawaguchi and Hart 2015)	−
Allograft	Freeze-dried	From: 1) Living donor 2) Cadaveric donor		−	−	++	+/−	+/− (significantly affected by drying process) (Wheeless 2013)	−
Allograft	Gamma sterilization			−	−	++	+/−	+ (by radiation effect) (Hamer et al. 1999)	

(continued)

Table 1 (continued)

Grafting material	Grafting material (typical abbreviation)	Grafting material category and description	Variability	Osteogenic	Osteo inductive	Osteo conductive	Immunogenicity/ disease transmission	Strength (immediate)	Donor site morbidity
Allograft	Demineralized bone matrix (Morris et al. 2018)	Mostly cadaveric donors	Demineralization processes + particle sizes	−	+/−	++	+/−	−	−
	Selective cell retained allografts		Patient characteristics	+	+/−	−	−	−	−
Differentiation factors									
Differentiation/ growth factor (Burke and Dhall 2017)	rhBMP-2	Differentiation/ growth factor	High manufacturing consistency	−	+++	−	−	−	−
	rhBMP-7	Differentiation/ growth factor		−	+++	−	−	−	−
Peptides	B2A (Glazebrook and Young 2016)	Bioactive synthetic peptide		−	−	++	−	−	−
	P15 (Hsu et al. 2017)	Bioactive synthetic peptide		−	−	++	−	−	−
Synthetic ceramics									
Tricalcium phosphate	TCP	Synthetic ceramics		−	−	++	−	−	−
Hydroxy apatite	HA	Synthetic ceramics		−	−	++	−	+/−	−

Biphasic calcium phosphate	BCP	Synthetic ceramics	–	–	++	–	+/–	–
Calcium sulfate	CS	Synthetic ceramics	–	–	++	–	–	–
Synthetic bioactive glasses								
Synthetic bioactive glass	45S5	Bioactive glass with 45% silicate	–	+	++	–	+/– (Hench and Jones 2015)	–
	S53P4	Bioactive glass with 53% silicate		+	+ (less bioactive than 45S5) (Hench and Jones 2015)			
Others								
Xenograft	Xenograft	From nonhuman species, mainly bovine-based bone graft	Donor variability	+/– (according to sterilization process)	+	+++ (more than allograft) (Shibuya and Jupiter 2015)	++ (in spine, foot, and ankle and trauma part) (Shibuya and Jupiter 2015)	–
Type I collagen	Xenograft carrier (bovine)	Osteoconductive scaffold/ hemostasis	Anatomic source location and species	–	+	–	+/–	–

+++ Characteristic is definitely observed from biologic, clinical, and preclinical studies
++ Characteristic is somewhat observed from biologic, clinical, and preclinical studies
+ Suggested by clinical and preclinical studies. There may be some controversy, or effect is minimal
+/– Debate status
– None/no effect

Table 2 Optimal characteristics of engineered biomaterials (O'Brien 2011)

Characteristic/*sub-characteristic*	Significance
Biocompatibility	The very first criterion of any biomaterials for tissue engineering is biocompatibility. Cells must adhere, function normally, and migrate onto the surface and eventually through the scaffold and begin to proliferate before laying down new matrix. The host's immune reaction to the material must be negligible in order to allow for proper healing
Capacity to bind cells or growth factors	For this purpose, collagen often used as a method to enhance cell and growth factor attachment
Biodegradability	The biomaterials must be biodegradable to allow cells to produce their own extracellular matrix. The by-products of this degradation should also be nontoxic and able to exit the body without interference with other organs
Resorption rate balanced with rate of bone formation	The biomaterials must resorb and allow formation of new bone. Otherwise material may remain and become an inert obstacle to fusion or healing
Mechanical properties	The biomaterials should have mechanical properties consistent with the anatomical site into which it is to be implanted
Intraoperative handling	From a practical perspective, it must be strong enough to allow surgical handling during implantation
Ability to visualize by fluorography intraoperatively	Radio-density important to visualize location and to determine healing with subsequent x-rays
Biomaterials architecture	The biomaterials should have an interconnected pore structure and high porosity to ensure cellular penetration and adequate diffusion of nutrients to cells and of waste products out of the scaffold
Controlled architecture, i.e., porosity, interconnected pores, and pore size that permits cell ingrowth	Cells need to be allowed to interact with each other and have continuity
Promotes revascularization and bone ingrowth	Essential in aiding bone healing
Manufacturing technology	In order for a particular tissue-engineered construct to become clinically and commercially available, it should be safe and cost-effective, manufactured following GMP, GLP, US-FDA, EU-EMA, and WHO International Conference of Harmonization technology standards (ISO) for scale-up from research laboratory-based small batch to large scale production lot reproducibility, maintaining reliability, stability, with optimized production processes meeting manufacturing requirements of country of manufacture and distribution

For the USA, PHS Act (Public Health Service Act), FDA-established regulations for Human Cells, Tissues, and Cellular and Tissue-Based Products (HCT/Ps), set forth in Title 21, Code of Federal Regulations, Part 1271 (21 CFR 1271), Public Health Service Act (42 USC 264), Good Manufacturing Practices (GMP), Good Laboratory Practices (GLP), GTP, Minimal Manipulation of Human Cells, Tissues, and Cellular and Tissue-Based Products etc. HCT/Ps to prevent the introduction, transmission, and spread of communicable diseases. These regulations can be found in 21 CFR Part 1271, section 361 (US FDA 2017, 2018)

eliminating pain thought to be aggravated by the abnormal motion (Herkowitz et al. 2004; Adams 2013). Spinal fusion is performed in patients with degenerative diseases like spinal instability, vertebral fractures, degenerative disc disease, and scoliosis. After a surgical decompression procedure has been performed to relieve pressure on the nerve roots or spinal cord, a fusion procedure may be completed as well to address the instability and provide long-term bony stability and structural reinforcement. The two main types of spinal fusion procedures are posterolateral fusion (PLF) and interbody fusion (IBF) performed from among a large variety of surgical approaches and techniques (Makanji et al. 2018; Morris et al. 2018).

11 Bone Grafts and Bone Graft Substitutes

Fig. 1 Example of posterolateral fusion with consolidated bone mass (BB) approximately 1 year after spinal fusion procedure with instrumentation and autograft placed in the posterolateral bed

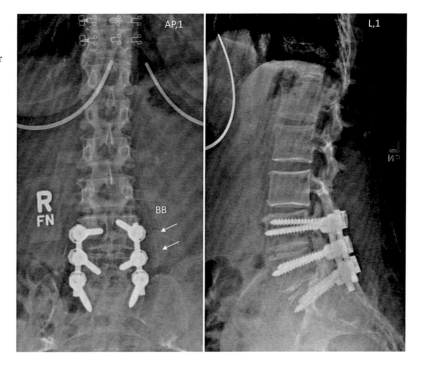

Fig. 2 Example of four-level posterolateral fusion with dense solid bone mass (BB) bilaterally at 1 year after posterior spinal fusion procedure with instrumentation and grafting with Fibergraft (Prosidyan), allografts, and bone marrow aspirate (BMA)

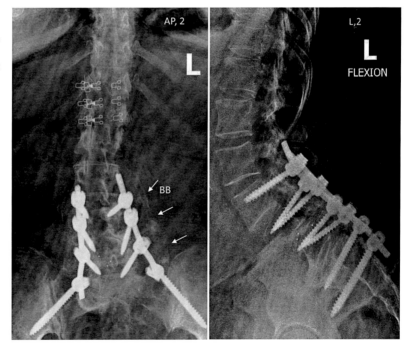

Radiographic images of example patients after surgical fusion procedures in lumbar and cervical spine are provided (Figs. 1, 2, 3, 4, 5, 6, 7, 8, 9, 10, and 11).

In posterolateral fusion (PLF), the bone graft or bone graft substitute is surgically placed between the transverse processes, lateral to the side of the superior vertebral body and inferior vertebral

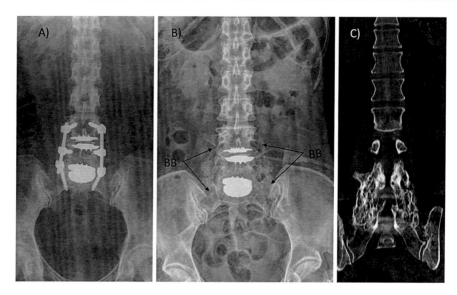

Fig. 3 A 59-year-old female patient was surgically treated for failed artificial disc in the lumbar spine. For treatment, a posterolateral fixation was performed with allograft cancellous chip bone (Medtronic) mixed with autologous local bone and pedicle screw fixation device (Medtronic). A radio-dense bone bridge *was not* observed between transverse process between L3 and S1 on the initial postoperative anterior-posterior radiographs (A). On radiographs taken at 3-year follow-up after removal of pedicle screws, there was a radio-dense bone bridge (BB black arrows) on anterior-posterior (AP) radiographs. On CT image (C), definitive radio-dense bone (BB) was observed with contact between transverse process, facet joint, and grafted bone

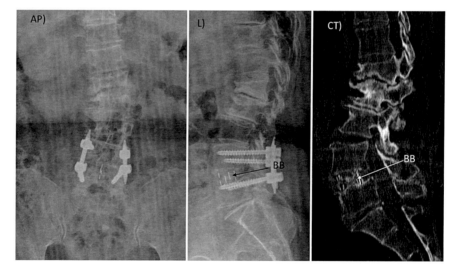

Fig. 4 A 60-year-old female patient was surgically treated with a diagnosis of spinal-stenosis L4-L5. Surgery: an indirect decompression of spinal nerve and fusion via oblique interbody fusion (OLIF) technique with PEEK cage containing DBM-based product (Medtronic). A hemilaminectomy via MIS surgery technique was performed using percutaneous screw fixation. On anterior-posterior (AP), lateral (L) radiographs taken at 1.5-year follow-up, a radio-dense bony line was observed between upper and lower vertebral body through the inserted cage. On sagittal CT view, a dense bridge (bony incorporation) was formed at fusion site. Wedge-shaped vertebral deformation of L1 and L2 compression fracture was observed. L2 fracture was treated with PMMA bone cement

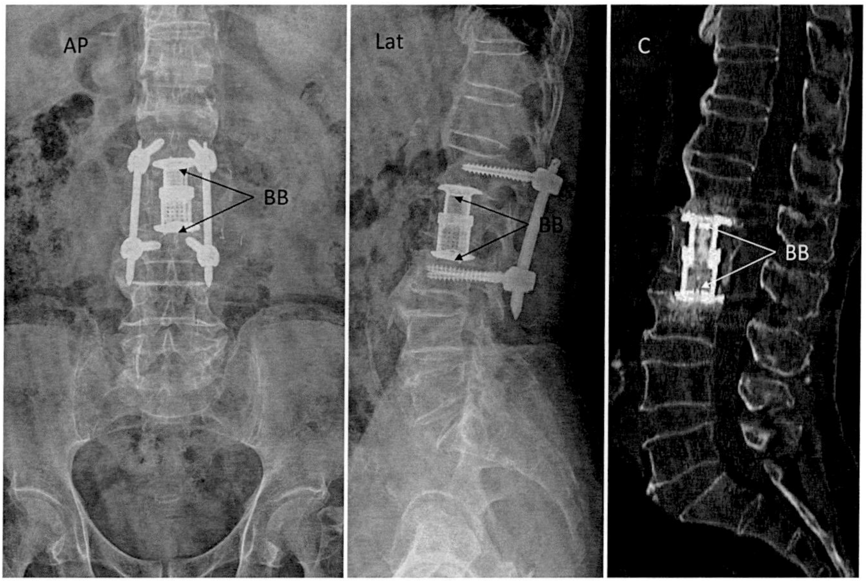

Fig. 5 A 78-year-old male patient was surgically treated for TB spondylitis in L2. The infected vertebrae were removed through corpectomy process. Fusion procedures were performed using a distractible cage (DePuy Synthes) with allograft cancellous chip bone (Medtronic) mixed with autologous local bone and a percutaneous pedicle screw fixation device (Medtronic). Titanium distractible cage with keel on contact surface with vertebral endplate was used for load bearing architecture. On anterior-posterior (AP), lateral (L) radiographs taken at 2-year follow-up, a radio-dense bone bridge (black BB) was observed between upper and lower vertebral body. On CT image (C), definitive radio-dense bone (white BB) was observed connecting through the cage between the two vertebral endplates (white arrows indicate endplates)

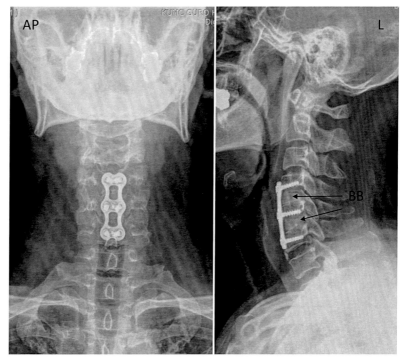

Fig. 6 A 49-year-old male patient was surgically treated for herniated intervertebral disc C4–C5, C5–C6. A total discectomy was performed for decompression through an anterior surgical approach. A machined cortico-cancellous allograft (Medtronic) and cervical plate (Medtronic) were used in the fusion procedure. On anterior-posterior (AP), lateral (L) radiographs taken at 1-year follow-up, a radio-dense bone bridge (BB) was observed between upper and lower vertebral bodies though the inserted machined cortico-cancellous allograft

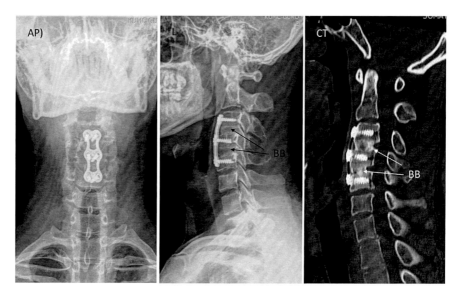

Fig. 7 A 34-year-old male patient was surgically treated for two-level herniated intervertebral disc C3–C4, C4–C5. A total discectomy was performed for decompression through an anterior approach. For the fusion procedure, iliac autogenous bone graft and a cervical plate (Medtronic) were used. On anterior-posterior (AP), lateral (L) radiographs taken at 1-year follow-up, a radio-dense bone bridge (BB) was observed between upper and lower vertebral body through the inserted iliac autogenous bone graft. On sagittal CT view, there is bone formed and complete incorporation of C3–C4–C5 (BB) at the fusion site

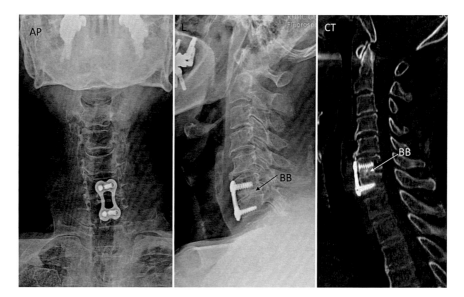

Fig. 8 A 64-year-old male patient was surgically treated for herniated intervertebral disc C6–C7. A total discectomy was performed for decompression through an anterior surgical approach. For fusion procedure, a machined cortico-cancellous allograft (Medtronic) and a cervical plate (Medtronic) were used. On anterior-posterior (AP), lateral (L) radiographs taken at 1.5-year follow-up, a radio-dense bone bridge (BB) was observed between upper and lower vertebral body through the inserted machined cortico-cancellous allograft. On sagittal CT view, a bone bridge (BB, complete bony incorporation) was formed at fusion site

Fig. 9 A 49-year-old female patient was surgically treated for herniated intervertebral disc at C5–C6. A total discectomy was performed for decompression through the anterior approach. For the fusion procedure, a Zero-p system (DePuy Synthes) and DBM-based putty (DBX, DePuy Synthes) were used. On anterior-posterior (AP) and lateral (L) radiographs taken at 2-year follow-up, a radio-dense bone bridge (BB) was observed between upper and lower vertebral body

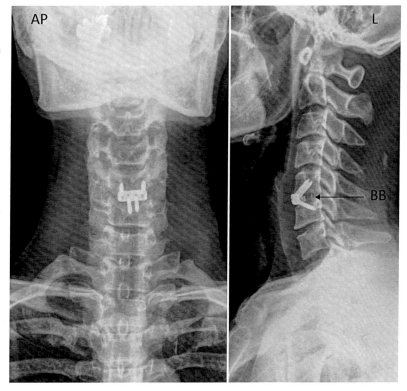

Fig. 10 A 59-year old patient was surgically treated for fracture dislocation injury at C3 vertebrae after fall from height. Treatment of fracture was performed by a decompression through corpectomy and fusion using auto-iliac crest strut bone graft with cervical metal plate fixation spanning C3–4–5). On anterior-posterior (AP), lateral (L) radiographs taken at 3-year follow-up, a radio-dense bone bridge (black BB) was observed between upper and lower vertebral bodies (C3–C5 a solid bone unit). Note no radio-opaque gap between graft material and endplate of adjacent vertebral bodies

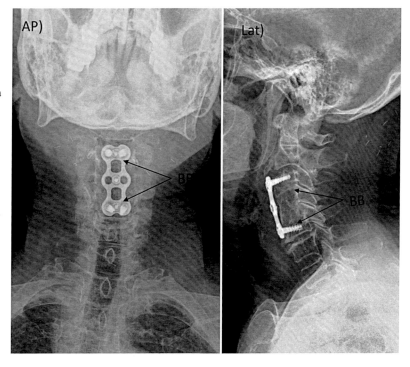

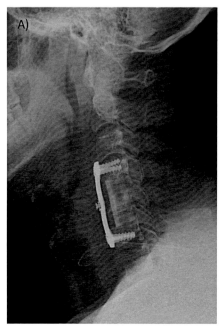 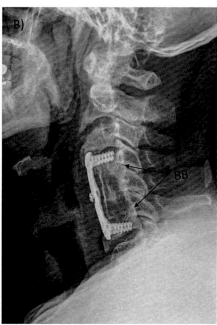

Fig. 11 A 62-year-old male patient was surgically treated for OPLL from C4 to C6 cervical spine. Decompression of the spinal cord was performed through removal of vertebral body of C5 and C6. Fusion was performed using allograft strut fibula bone with a cervical metal plate (Medtronic). After initial postoperative radiography, a radio-opaque gap is seen between grafted material and endplate vertebral body. At 1-year follow-up, a complete incorporation between allograft and host bone is observed as a solid bone unit – no radio-opaque gap between grafted material and vertebral endplates. There is radio-density of grafted material indicating bony consolidation and incorporation with fusion from C4-C7

body. During the healing process, the graft material is remodeled and incorporated into a solid bony "bridge" (BB) between the transverse processes and lamina. Once healed, the spine segment is stabilized, and motion between vertebral functional segments is eliminated or reduced (Fig. 1, example lumbar spine).

In interbody spinal fusion, compared to PLF, the bone grafts or bone graft substitutes are placed between the endplates of two adjacent vertebrae (e.g., Figs. 4 and 5, lumbar spine; Figs. 6, 7, 8, 9, 10, and 11, cervical spine). The bone graft's and/or instrument with graft's contact with the endplates of the adjacent vertebral bodies resists relatively high loading forces.

Due to these biomechanical differences in graft sites, interbody fusion bone grafts are commonly placed with cages that hold the graft in place and are designed to withstand the compressive forces of the vertebrae. When the bone graft or bone graft in a cage is placed between the endplates of the vertebral body, it creates a framework of mechanical support during the early time of graft incorporation. This mechanical fixation and support eventually aids in the biologic bony union connecting one vertebral body to the other. Similarly to posterolateral fusion, once the vertebrae are fused, the spine is stabilized, and movement between operated spine segments should disappear. A systematic recent review of 12 studies (565 IBF-treated patients) by Baker et al. (2017) concluded that interbody fusion was a good surgical option in spondylolisthesis patients with instability. Interbody fusion can be performed by several different surgical approaches and techniques such as anterior (anterior lumbar interbody fusion (ALIF)), posterior (posterior lumbar interbody fusion (PLIF)), transforaminal (transforaminal lumbar interbody fusion (TLIF)), and lateral (lateral lumbar interbody fusion (LLIF)).

After a fusion procedure, the bone healing process occurs in different phases: inflammation, soft callus formation, hard callus formation, and bone remodeling.

This includes hematoma formation, release of native growth factors/cytokines, and recruitment of inflammatory cells (e.g., macrophages and bone-forming cells); cell differentiation to bone-forming cells and mineralization of the extracellular matrix (ECM); bone resorption and remodeling; and formation of lamellar bone and hematopoietic marrow cavities (Rausch et al. 2017). These complex biologic processes of consolidation of grafted materials into new bone and remodeled into mature bone can be negatively affected by various systemic and local factors. Typical host or patient-based negative factors are advanced age (Ajiboye et al. 2017), concomitant use of tobacco or other drugs, poor nutritional status, and metabolic comorbidities (e.g., diabetes or osteoporosis) (Campbell et al. 2012; Ajiboye et al. 2017). Negative local factors are remaining structural instability, poor vascularity around surgical site, revision surgery, previous or current infection, and other local/surgical site considerations including surgical technical factors such as inadequate preparation of host bone, lack of fixation, inadequate bone graft volume and preparation, and improper use of graft materials (Yoo et al. 2015). Critical challenges for both interbody and posterolateral fusion are the excessive distances for the cells to migrate within and between host bone beds in order to attach to targeted neighboring anatomic bony structures; the limited durability of concentrations of growth factors, peptides, exogenous cells, biochemical, and other agents; and the biomechanical stability. These biologic challenges are particularly deterring in geriatric spine patients with severe osteoporosis.

To achieve successful bone fusion in the spine, surgeons vigilantly adhere to the requirements of bone regeneration and fracture healing mentioned above in deciding use of grafting materials. Bone formation requires three critical elements: osteoconduction, osteoinduction, and osteogenesis. *Osteoconduction* relies on a scaffold that supports cell ingrowth, facilitates vascularization, and provides a network for cells to attach. *Osteoinduction* relies on the provision of signals that act on the precursor cells and encourage cell migration, proliferation, and differentiation into bone-forming cells leading to rapid bone formation. *Osteogenesis* relies on the immediate provision of viable cells emanating from the host to the defect site differentiating into bone-forming cells. Autograft or autogenous bone possesses all three properties essential for bone formation and is therefore considered to be the gold standard graft material for inducing bone healing, consolidation, and fusion of the spine.

Current Materials for Spinal Fusion

The graft material used in spinal fusion procedures can be generally categorized into three main types of materials: autogenous bone graft (autograft) from the patient's own body, allograft from human cadavers and/or living donors, and synthetic bone graft or substitutes (Table 1).

The use of autogenous bone graft has been a standard practice in spine surgery for over a century. The first reported use of autogenous bone graft for spine fusion was reported in 1911 when Fred Albee, MD, placed a tibia between spinal lamina in order to fuse and stabilize the spine (Albee 2007). Autograft has been considered the "gold standard" of bone grafting primarily because it contains all the elements required for successful fusion mentioned above: osteoconductive matrix, osteoinductive factors, and pluripotent bone-forming cells (Gupta and Maitra 2002; Whang and Wang 2003).

Autografts

Autograft for spinal fusion can be obtained via different surgical approaches and dissection methods. Firstly, resected lamina, spinous process, facet, and osteophytes during the surgical decompression process yield bone graft which is then morselized – "local bone graft" (LBG). Local bone is commonly limited in amount and quality as mixture consists of mostly cortical bone vs. cancellous bone (Tuchman et al. 2016). Secondly, a bone graft can be obtained from iliac crest using a separate surgical incision and various dissections (White and Hirsch 1971), which then can be used as strut or morselized bone. Iliac crest bone graft (ICBG) is relatively abundant, providing good-quality graft (mainly cancellous bone). However, iliac bone and local bone autografts have similar effectiveness in terms of fusion rates, pain scores, and functional outcomes in view of lumbar spine fusion (Tuchman et al. 2016).

Autograft bone is safe to use due to the low risk of disease transmission and offers the optimal chance of acceptance and effectiveness in the transplant site without immune reaction (Campana et al. 2014). However, the limitations with autogenous iliac bone graft such as relatively limited quantity, increased surgical time, and donor site morbidity are well recognized (Vaccaro et al. 2002). Due to these limitations, the use of autograft has declined.

The reduction in the use of autograft from the iliac crest in the recent practice has led to the increase in the use of local bone graft and has created new demands for the identification of cost-effective biologic materials that will "extend" the bone healing effects of local autograft (Ito et al. 2013). To achieve optimal outcomes, these materials should be biocompatible and biodegradable and have beneficial mechanical properties and microarchitecture that facilitates the biological healing process (Table 2).

Allografts

Allografts are primarily osteoconductive with minimal osteoinductive potential and traditionally not osteogenic because the donor cells are eradicated during processing (Campana et al. 2014; Duarte et al. 2017). Allografts have the advantage for a surgeon of easy procurement (off-the-shelf), availability (commercially available), and many varieties of structural and non-structural form. However, allografts consist of nonviable tissue and cannot stimulate bone formation without the addition of bone-stimulating factors and cells (Goldberg and Stevenson 1993; Garbuz et al. 1998; Stevenson 1999). These limitations lead to slower and less complete incorporation with native bone. Additionally, allografts have potential risk of disease transmission even if the incidence is very low and the risk can be controlled during procurement and sterilization process (Campana et al. 2014).

Allo-bone Graft: Cortico-cancellous Allograft (Table 3)

Allograft bone obtained from cadaver sources is added to the most widely used substitute or extender for autogenous bone graft. In the 1980s, femoral head from living donors (after total hip replacement surgery) was also introduced as another form of allograft and has demonstrated good clinical results in lumbar spine fusion (Urrutia and Molina 2013).

Allograft bone may be morselized to various sizes of particulate (i.e., chip bone) formed or machined to create structural spacers and then applied to site of desired bone formation. Cortical allograft is most often used as mechanical strut graft and is suited for interbody fusion, while cancellous allograft serves as a useful osteoconductive scaffold for bone formation.

The efficacy of allograft alone has been shown to have more clinical variability and lower fusion rates in challenging animal models and human studies of spinal fusion (Morris et al. 2018). These overall clinical results suggest that allograft be cautiously used in conjunction with either autograft or osteogenic material (e.g., bone marrow aspiration) to achieve good fusion rates and clinical outcomes (Morris et al. 2018). However, while the actual risk of transmission is negligible, issues of immunogenicity are present (Manyalich et al. 2009).

Allo-bone Graft: Demineralized Bone Matrix (DBM)-Based Product (Table 4)

The DBM technology is based on the observation by Urist MR (Urist 1965) that soluble signals contained within the organic phase of bone were capable of promoting bone formation. The processing of transforming ground cortical bone into DBM powder base involves the use of hydrochloric acid to progressively remove mineral while attempting to preserve the organic phase containing type 1 collagen, non-collagenous

Table 3 Commercially available structural and/or nonstructural allografts[a]

Company	Allograft spinal graft products	Formulation product composition	Clinical evidence ClinicalTrials.gov ongoing study	Regulatory clearance/approvals US by FDA registered tissue bank establishments 21CFR1270, CFR 1271 AATB; US; Pharmacopoeia USP standard 71
AlloSource®, Centennial, Co, USA, 1995 Allosource.org	AlloFuse®	Cortical/cancellous spacers Cancellous cervical spacers Cortical cervical spacers	n/a	Regulated human tissue CFR 1270, 1271
	Spinal grafts freeze-dried	Bicortical blocks Dowel Patella wedge Cervical spacers, parallel spacer/textured lordotic Femoral rings Fibular rings, radial rings, ulna rings	n/a	Regulated human tissue CFR 1270, 1271
	Spinal grafts freeze-dried/frozen	Cortical strut TriCortical Ilium Wedges, Strips	n/a	Regulated under CFR 1270, 1271 as a human tissue http://activize.com/wp-content/uploads/2014/05/Allograft-Catalog.pdf
ATEC, Carlsbad, CA	AlphaGRAFT® structural allografts	Vacuum level allograft designed to hydrate	n/a	AATB standards and Good Tissue Practices
Aziyo, Richmond, CA	OsteSpine	Structural cortical (femur, tibia, cortical shafts/struts)/cancellous spacers, (OsteSpine) precision-machined allografts	n/a	Regulated under CFR 1270, 1271 as a human tissue
Bone Bank Allografts, Texas, USA	SteriSorb™	Osteoconductive Sponge Allografts (100% cancellous bone) Characteristics of a sponge by absorbing saline, blood, or bone marrow aspirate[a]	n/a	Bone Bank Allografts Registration – FDA BBA Manufacturing Registration – FDA (previously THB)
	SteriFlex™	Wrappable Bone Allografts (100% cortical bone) Can be bent, contoured, rolled, trimmed, molded, or sewn, making this flexible bone material	n/a	Bone Bank Allografts – Accreditation American Association of Tissue Banks (AATB) CTO Registration Certificate – Bone Bank Allografts (International Registration)

(continued)

Table 3 (continued)

Company	Allograft spinal graft products	Formulation product composition	Clinical evidence ClinicalTrials.gov ongoing study	Regulatory clearance/approvals US by FDA registered tissue bank establishments 21CFR1270, CFR 1271 AATB; US; Pharmacopoeia USP standard 71
	SteriGraft™ – Cervical ACF	Fully machined. Constructed of 100% human cortical bone (femur or tibia)	n/a	
	SteriGraft™ – ACF Cortical-Cancellous Spacer	Fully machined. Constructed of 100% human cortical bone with an internal cancellous plug (femur or tibia)		
	SteriGraft™ – ALIF	Fully machined. Constructed of 100% human cortical bone (femur)		
	SteriGraft™ – PLIF			
	SteriGraft™ – Unicortical Dense Cancellous Block	Fully machined. Constructed of 100% human cortical bone (femur or tibia) Unicortical Dense Cancellous Block (femoral head, patella, distal tibia, talus, or calcaneus)		
	SteriGraft™ – Dense Cancellous Block	Dense Cancellous Block (femoral head, patella, distal tibia, talus, or calcaneus)		
	Traditional bone/cancellous bone allografts Traditional bone/cortical-cancellous bone allografts	Traditional-type cancellous bone chip or tricortical allo-iliac bone	n/a	
DePuy Synthes Spine	Zero-P Natural Plate System	Zero-Profile Plate with Allograft Spacer (cervical spine)	n/a	21 CFR 888.3060, K152239, 2015 Dec FDA 510(k) cleared
Hospital Innovations Ltd. Pontyclum, Wales, UK	Ilium Tricortical strips Bone blocks Whole and hemi shaft	Traditional cortical/cancellous bone graft Available freeze-dried (FD) or frozen (FZ) Sterilized SAL 10^{-6}	n/a	#22512, CF729FG, 2014 Jul [Regulation 7(1) Schedule 2 of Human Tissue Authority (Quality and Safety for Human Application (non-departmental public body, Department of Health and Social Care, UK) Regulations, 20070]

Globus Medical Inc	FORGE® FORGE® Oblique	Fully machined corticocancellous spacer (cervical spine fusion) Fully machined cortical spacer designed to provide a natural option for transforaminal lumbar fusion	n/a	FDA 510(k) cleared K153203, 2015 Dec
LifeLink Tissue Bank	Cortical Cancellous Spacer	Fully machined cortical-cancellous spacer (from femur and tibia) for cervical spine		AATB FDA Florida, California, and New York Holds a permit to provide tissue in Maryland
Mountain States Medical → Merged into Zimmer	OsteoStim®	Fully machined cortical spacer (from femur and tibia) for cervical spine		Processed at an AATB-accredited facility
Medtronic Spinal and Biologics	Allograft structural Cornerstone SR Cornerstone™ ASR Cornerstone-Reserve™	Fully machined cortical block (from femur or tibia) with capital D shape Fully machined cortical lateral wall with a cancellous center with capital D shape Fully machined cortical ring with cancellous plug	ClinicalTrials.gov Identifier: NCT01491399, no results posted	AATB standards, FDA regulations, and applicable Public Health Service Guidelines for donor screening
	Cornerstone™ tricortical Cornerstone™ bicortical Cornerstone™ unicortical Cornerstone™ dense cancellous block	Freeze-dried cortical/cancellous (iliac crest) Freeze-dried cortical/cancellous (iliac crest) Freeze-dried anterior cortical wall with cancellous center Freeze-dried dense cancellous with capital D shape		AATB standards, FDA regulations, and applicable Public Health Service Guidelines for donor screening
	Cornerstone™ selective/cortical wedge	Freeze-dried cortical ring		AATB standards, FDA regulations, and applicable Public Health Service Guidelines for donor screening
Orthofix®	AlloQuent-s, Monolithic Cortical Structural allograft	Structural allograft (cervical fusion, lumbar fusion) Different sizes and shapes (ALIF, PLIF, TLIF)	NCT00637312, has results posted – cervical disc: Trial was stopped. Approval not being pursued for device (clinicaltrial.gov)	unk

(continued)

Table 3 (continued)

Company	Allograft spinal graft products	Formulation product composition	Clinical evidence ClinicalTrials.gov ongoing study	Regulatory clearance/approvals US by FDA registered tissue bank establishments 21CFR1270, CFR 1271 AATB; US; Pharmacopoeia USP standard 71
RTI Surgical®	Elemax® Cortical Spacer Allograft Elemax® Cortical Spacer Allograft Elemax® PLIF Allograft	Precision-machined cortical spacer for anterior cervical discectomy and fusion procedures Fully machined cortical lateral wall with a cancellous center with capital D shape for anterior cervical discectomy and fusion procedures Fully machined cortical spacer designed to provide a natural option for PLIF	n/a	AATB Accreditation Certificate – (Florida) FDA Establishment Registration and Listing for Human Cells (FDA-HCT/Ps) Florida, USA Tutogen Medical, GmbH (Germany) International Organization of Standards (ISO) Tutogen Medical, GmbH (Germany) CMDCAS – RTI Surgical (Florida) CE Certificates Pioneer Surgical Technology (Michigan) International Facility Registrations Health Canada CTO Registration State Tissue Banking Licenses California, Florida, Maryland, New York, Oregon, Illinois, Delaware
	AlloWedge® Bicortical Allograft Bone	Options for approaching opening wedge osteotomies in the foot and ankle Pre-shaped bicortical allografts	n/a	FDA Establishment Registration and Listing for Human Cells (FDA-HCT/Ps)
	Cross-Fuse® Advantage Lateral Allograft	All cortical bone implant designed for a lateral approach to provide maximum potential for fusion Produced from femoral or tibial tissue	n/a	FDA Establishment Registration and Listing for Human Cells (FDA-HCT/Ps)
	Bigfoot® ALIF Allograft	All cortical bone implant designed for use as an intervertebral spacer in anterior lumbar interbody fusion (ALIF) approach	n/a	FDA Establishment Registration and Listing for Human Cells (FDA-HCT/Ps)

	Traditional cortical and/or cancellous strut allobone	Freeze-dried: rehydrate for a minimum of 30 s Frozen: thaw for a minimum of 15 min Femoral head, hemi femoral shaft, humeral head, ilium tricortical block, ilium tricortical strip, proximal and distal femur, proximal and distal humerus, proximal and distal tibia, unicortical block, whole femur, fibula and humerus, and bicortical block	n/a	FDA Establishment Registration and Listing for Human Cells (FDA-HCT/Ps)
SeaSpine, Carlsbad, CA, USA	Capistrano™ System	Cervical allograft spacer system is precision machined from cortical and cancellous allograft bone	n/a	361-HCT/P US FDA 21 CFR 1271 Restricted to homologous use for the repair, replacement, or reconstruction of bony defects by a qualified healthcare professional (e.g., physician)
Stryker	AlloCraft™ CA, CL, CP, CS	Machined from femoral/tibial allograft → ACDF Freeze dried Chamfered edge	n/a	AATB US FDA regulations for tissue management. US FDA 21 CFR 1271
Xtant, USA	Ilium tricortical blocks, unicortical blocks, fibula segments, and femoral struts	Traditional allografts	n/a	Processed by tissue banks that are members of the American Association of Tissue Banks (AATB)
X-spine Systems, Inc./Xtant, USA	Atrix-C™ Cervical Allograft Spacer	Precision-milled cortical bone w/ teeth like keel surfaces	n/a	Processed by tissue banks that are members of the American Association of Tissue Banks (AATB)
Zimmer Biomet	OsteoStim® Cervical Allograft System OsteoStim® PLIF OsteoStim® ALIF/[a]	Fully machined cortical spacer bone for cervical and lumbar w/ teeth like keel surfaces	n/a	Processed by tissue banks that are members of the American Association of Tissue Banks (AATB)

AATB (American Association of Tissue Banks) Policies (2018)
FDBA Freeze-dried bone allograft
n/a not available on ClinicalTrials.org/no clinical data found or clinical trial registered
[a]Indicates cancellous chips "crunch" available

Table 4 Commercially available DBM-based products

Company	DBM-based product (human)	Formulation	Product composition	Peer-reviewed clinical evidence/ongoing study ClinicalTrials.gov identifier	Regulatory clearance/approval FDA 510(k), CFR 1270, CFR 1271
AlloSource®, Centennial, Co, USA, 1995 Allosource.org	AlloFuse® Gel AlloFuse® Putty (identical to StimuBlast Putty and Gel manufactured for Arthrex)	Injectable gel and putty	DBM, reverse phase medium (RPM) carrier Carrier comprised polyethylene oxide-polypropylene oxide block copolymer dissolved in water exhibiting reverse phase characteristics (i.e., an increase in viscosity as temperature increases)	n/a	510(k) cleared K071849, 2008 Dec
	AlloFuse Plus	Paste, putty	DBM, RPM, cancellous chips	n/a	510(k) cleared K103036, 2011 Jan
	AlloFlex	Strips, blocks, fillers	Cancellous bone allograft, DBM, strip form, no carriers added	n/a	Marketed as human tissue
Amend Surgical, Inc	NanoFUSE® Bioactive Matrix	Putty	DBM + 45S5 bioactive glass: bond void filler	n/a	510(k) cleared K161996 2017 Feb
	NanoFUSE® DBM	Putty 2 cc to 10 cc	45S5 bioactive glass + porcine gelatin + demineralized bone matrix (DBM) 45S5 bioactive glass: osteoconductive scaffold, DBM: osteoinductive potential	Kirk et al. 2013 Xynos et al. 2000a Xynos et al. 2000b	Regulated under CFR 1270, 1271 as a human tissue K110976, 2011 May www.accessdata.fda.gov/cdrh_docs/pdf11/K110976.pdf accessdata.fda.gov/cdrh_docs/pdf16/K161996.pdf 510(k)
ATEC, Carlsbad, CA	AlphaGRAFT® DBM AlphaGRAFT ProFuse DBM	Putty or gel	A reverse phase medium sponge-like DBM, superior handling characteristics, and ready-to-use application (thickens at body temp.)	n/a	unk
Aziyo Biologics	OsteoGro		Cancellous bone and partially demineralized bone		

11 Bone Grafts and Bone Graft Substitutes

Bacterin International, Inc. → Changed to Xtant Medical	OsteoSelect DBM	Putty	74% DBM dry weight	n/a	510(k) cleared K091321, 2009 Sept K130498, 2013 May
	OsteoSelect Plus DBM	Putty	74% DBM dry weight + demineralized cortical chips (1–4 mm)	n/a	510(k) cleared K150621, 2015 Aug HCT/P (FEI 3005168462)
	OsteoSponge®	The malleable sponge	DBM (100% human demineralized cancellous bone)	Shehadi and Elzein 2017	510(k) cleared HCT/P (FEI 3005168462), 2017 Nov
	OsteoSponge® SC	The malleable sponge	Demineralized cancellous bone intended to treat the pathology of damaged subchondral bone of the articulating joints	Galli et al. 2015	510(k) cleared HCT/P (FEI 3005168462), 2017 Nov
	OsteoWrap®	Flexible handling characteristics with a scalpel or scissors	100% human demineralized cortical bone	n/a	510(k) cleared HCT/P (FEI 3005168462), 2017 Nov
	3Demin®	Various shapes (fiber, boat shape, strip)	100% human demineralized cortical bone fiber Contain BMPs and other growth factor 3Demin allografts are also available as loose cortical fibers in three volume options	n/a	Compliance with FDA guidelines regarding human cells, tissues, and cellular tissue-based products HCT/P 361 regulated viable allogeneic bone scaffold American Association of Tissue Banks guidelines
Berkeley Advanced Biomaterials, CA, USA	H-GENIN™	Putty matrix sponge powder	100% demineralized bone matrix putty and crush mix	n/a	510(k) cleared (as B-GENIN, R-GENIN) K092046, 2010 Mar
Biomet Osteobiologics → Merged into Zimmer Biomet.	InterGro® DBM	Putty (40% DBM), paste (35% DBM)	DBM, lecithin carrier (resorbable, biocompatible, semi-viscous lipid)	Prospective case series	510(k) cleared K082793, 2009 Apr K031399, 2005 Feb

(continued)

Table 4 (continued)

Company	DBM-based product (human)	Formulation	Product composition	Peer-reviewed clinical evidence/ongoing study ClinicalTrials.gov identifier	Regulatory clearance/approval FDA 510(k), CFR 1270, CFR 1271
Bioventus® Surgical	Exponent™	Putty form	Demineralized bone matrix is composed of human demineralized bone (DBM) mixed with resorbable carrier, carboxymethylcellulose (CMC)	n/a	AATB US FDA 21 CFR 1271. (HCT/P)
	PUREBONE	Sponge shape (available in block or strip format)	100% demineralized cancellous bone (osteoconductive matrix with osteoinductive potential that provides a natural scaffold for cellular ingrowth and revascularization) Sterilized by gamma irradiation	n/a	FDA 510(k) cleared AATB US FDA 21 CFR 1271. (HCT/P)
Bone Bank™ Allografts 2017/ Texas Human Biologics	SteriFuse™ DBM Putty	Flowable, formable putty	100% demineralized bone matrix from human bone	n/a	Regulated under 21 CFR Part 1271 (h FDA requirements for human cellular and tissue-based products (HCT/P))
	SteriFuse™ Crunch	Flowable, formable crunch	SteriFuse™ DBM putty with cortical cancellous bone chips	n/a	Regulated under 21 CFR Part 1271 (h FDA requirements for human cellular and tissue-based products (HCT/P))
DePuy Synthes	DBX®	Putty type	DBM + sodium hyaluronate	ClinicalTrials.gov Identifier: NCT02005081: RCT, results are not reported	510(k) cleared K103795, 2011 Apr

	Synthes® Dento	Powder type Granule type Putty type	Powder type: demineralized cortical powder, mineralized cancellous powder, mineralized cortical powder Granule type: demineralized cortical (80%)/cancellous granules, mineralized cortical (80%)/cancellous granules DBM putty type: 93% DBM	n/a	unk
ETEX (Zimmer Biomet, 2014 October)	CaP Plus	CaP Plus	Synthetic calcium phosphate, an inert carrier, carboxymethyl cellulose (CMC), and DBM	n/a	510(k) cleared K063050, 2007 Nov K080329, 2008 Apr
	EquivaBone Osteoinductive Bone Graft	Powder and hydration solution	Synthetic calcium phosphate, an inert carrier, carboxymethyl cellulose (CMC) and DBM		510(k) cleared K090855, 2009 Sep K090310, 2009 Mar
Exactech	Optecure	Injectable paste	DBM (81% by dry weight), hydrogel carrier	Prospective RCT: ClinicalTrials.gov Identifier: NCT00254852	510(k) cleared K121989, 2012 Nov K061668, 2006 Sept K050806, 2006 Feb
	Optecure® + CCC	Injectable paste	Polymer powder, DBM, cortical cancellous chips (1–3 mm)	Comparative study allograft vs Optecure® + CCC: ClinicalTrials.gov Identifier: NCT02127112	510(k) cleared K061668, 2006 Sep K121989, 2012 Nov
	Optefil (OSTEOFIL® DBM Paste, OSTEOFIL® RT DBM Paste)	DBM paste or dry powder – hydrated to become injectable paste	DBM in gelatin carrier	n/a	510(k) cleared K043420, 2005 Feb
	Opteform	Putty or dry powder – hydrated to become paste	Gelatin, DBM, and cortical-cancellous bone chips	n/a	510(k) cleared K043421, 2005 Feb

(continued)

Table 4 (continued)

Company	DBM-based product (human)	Formulation	Product composition	Peer-reviewed clinical evidence/ongoing study ClinicalTrials.gov identifier	Regulatory clearance/approval FDA 510(k), CFR 1270, CFR 1271
Integra OrthoBiologics (IsoTis OrthoBiologics), Inc., Irvine, CA/SeaSpine 2018	Accell Connexus	Injectable putty	DBM (70% by weight), RPM	Retrospective comparative: Schizas et al. 2008	510(k) cleared K060306, 2006 Mar K061880, 2007 Aug
	Accell Evo3™	Injectable putty	DBM (Accell Bone Matrix), RPM	NCT02018445 ("Efficacy and Safety of Integra Accell Evo3™ Demineralized Bone Matrix in Instrumented Lumbar Spine Fusion") NCT01714804 (Integra Accell Evo3 Demineralized Bone Matrix) NCT01430299	510(k) cleared K103742, 2011 Mar
	Accell TBM	Preformed matrix (strip, square, round)	100% DBM (Accell Bone Matrix)	n/a	510(k) cleared K081817, 2008 Sep
	Dynagraft II	Injectable gel, putty	DBM (Accell Bone Matrix), RPM, cancellous bone chips	n/a	510(k) cleared K040419, 2005 Mar
	Orthoblast II	Injectable paste, putty	DBM (Accell Bone Matrix), RPM, cancellous bone chips from same donor	n/a	510(k) cleared K050642, 2005 Dec
LifeNet Health	IC Graft Chamber	Freeze dried in injectable delivery chamber, can be mixed with whole blood, PRP, or BMA	DBM, cancellous chips	n/a	Regulated under CFR 1270, 1271 as a human tissue
	Optium DBM Putty	Putty	DBM, glycerol carrier	n/a	510(k) cleared K053098, 2005 Nov
	Optium® DBM Gel	Gel	Particulate DBM & Glycerol	n/a	510(k) cleared K053098, 2005 Nov
	Cellect DBM®	Provided in a specialized cartridge	DBM fibers + cancellous chips	Case reports: Lee and Goodman 2009	510(k) cleared Regulated under CFR 1270 and 1271

Medtronic Spinal and Biologics	OSTEOFIL DBM	Injectable paste, moldable strips	DBM (24% by weight) in porcine gelatin	Prospective case series: Epstein and Epstein 2007	510(k) cleared K043420, 2005 Feb
	Progenix TM Plus	Putty with demineralized cortical chips	DBM in type I bovine collagen and sodium alginate	n/a	510(k) cleared K081950, 2008 Jul
	Progenix Putty	Injectable putty	DBM in type 1 bovine collagen and sodium alginate	n/a (human) Blinded observations/assessment of study in rabbit (Smucker and Fredericks 2012)	510(k) cleared K080462, 2008 May
	Magnifuse™ Family 1) Magnifuse Bone Graft substitute/bone void filler 2) Magnifuse II Bone Graft		DBM mixed with autograft in 1:1 ratio packed into polyglycolic acid (PGA) resorbable mesh bag 1) DBM + surface-demineralized chips 2) Combination of surface demineralized cortical chips and allograft fibers that have been processed removing the mineral component leaving only the organic portion	ClinicalTrials.gov Identifier: NCT02684045: retrospective case series study, results are not posted	510(k) cleared K123691, 2013 Jan K082615, 2008 Oct
MTF/Synthes	DBX	Paste, putty mix, strip	DBM (32% by weight), sodium hyaluronate carrier (mix vary for paste, putty, mix)	n/a	510(k) cleared K040262, 2005 Mar (putty, paste, matrix mix) K040501, 2005 Apr – (putty, paste, matrix mix) K053218, 2006 Dec (putty, paste, matrix mix) K063676, 2007 Mar (putty, paste, matrix mix) K080399, 2008 Oct (paste) K091217, 2009 Oct (putty) K091218, 2009 Sep (putty) K103795, 2011 Apr (putty K103784, 2011 Apr (putty) K042829, 2006 Jan (strip)

(continued)

Table 4 (continued)

Company	DBM-based product (human)	Formulation	Product composition	Peer-reviewed clinical evidence/ongoing study ClinicalTrials.gov identifier	Regulatory clearance/ approval FDA 510(k), CFR 1270, CFR 1271
Nanotherapeutics, Inc.	Origen™ DBM with Bioactive Glass (NanoFUSE® DBM)	A malleable, putty-like, bone void filler	Human demineralized bone matrix (DBM) and synthetic calcium-phospho-silicate particulate material particles (45S5 bioactive glass), both coated with gelatin derived from porcine skin	n/a	510(k) cleared K120279, 2012 Apr K110976, 2011 May
NuTech Medical, Inc.	Matrix: Osteoconductive Matrix Plus	Putty type	Allograft cancellous and demineralized cortical mixture Freeze-dried for convenient ambient temperature storage	n/a	
	Matrix: FiberOS	Putty type	Demineralized cortical fibers, mineralized cortical powder, and demineralized cortical powder Gamma sterilized for patient safety Freeze-dried for convenient ambient temperature storage		Nutec
Osteotech/ Medtronic	GRAFTON A-Flex	Round flexible sheet	DBM	n/a	510(k) cleared K051188, 2006 Jan
	GRAFTON Crunch	Packable graft	DBM, demineralized cortical cubes	n/a	510(k) cleared K051188, 2006 Jan
	GRAFTON Flex	Flexible sheets, varying sizes	DBM	Retrospective comparative study	510(k) cleared K051195, 2005 Dec
	GRAFTON Gel	Injectable syringe	DBM	RCT, prospective case series	510(k) cleared K051195, 2005 Dec
	GRAFTON Matrix PLF	Troughs	DBM	RCT	510(k) cleared K051195, 2005 Dec

	GRAFTON Matrix Scoliosis Strips	Strips, various sizes	DBM	Retrospective case series	510(k) cleared (Recalled 10/18/2012)
	GRAFTON Orthoblend Large Defect	Packable graft	DBM, crushed cancellous chips	n/a	510(k) cleared
	GRAFTON Orthoblend Small Defect	Packable, moldable graft	DBM, crushed cancellous chips	n/a	510(k) cleared
	GRAFTON PLUS® DBM Paste	Paste	Human bone allograft demineralized bone matrix (DBM) + inert starch-based carrier has been added	n/a	510(k) cleared K043048, Nov 2005 (Osteotech)-traditional K042707, Nov 2005 (Osteotech)
	Grafton Putty 22076647	Packable, moldable graft	DBM (17% by weight), glycerol	Kang et al. 2012	510(k) cleared K051195, 2005 Dec
Pioneer® Surgical Technology and Regeneration Technologies → All companies merged into RTI Surgical®	BioSet TM	Injectable paste, putty, strips, and blocks with cortical cancellous chips	DBM, inert porcine, gelatin carrier	n/a	510(k) cleared K080418, 2008 Apr Regulated under 21 CFR Part 1271 (h FDA requirements for human cellular and tissue-based products (HCT/P)) 12.07.2016 (validated by FDA)
	BioAdapt® DBM	Powder form	Dried granular powder form (70% of DBM by weight) from 100% human bone matrix	n/a	Regulated under 21 CFR Part 1271 (h FDA requirements for Human Cellular and Tissue-based Products (HCT/P)) 12.07.2016 (validated by FDA)

(continued)

Table 4 (continued)

Company	DBM-based product (human)	Formulation	Product composition	Peer-reviewed clinical evidence/ongoing study ClinicalTrials.gov identifier	Regulatory clearance/ approval FDA 510(k), CFR 1270, CFR 1271
	BioReady® DBM Putty and Putty with Chips	Putty/putty with bone chip	Putty: 56% DBM by weight Putty with chips: 42% DBM by weight + small or large mineralized cortical cancellous chip → 100% allograft DBM	n/a	Regulated under 21 CFR Part 1271 (h FDA requirements for human cellular and tissue-based products (HCT/P)) 12.07.2016 (validated by FDA)
SeaSpine, Carlsbad, CA	OsteoBallast™ Demineralized Bone Matrix	DBM in resorbable mesh	100% DBM	n/a	FDA 510(k) cleared
	OsteoSurge® 300 Demineralized Bone Matrix	The moldable putty form	DBM + Accell® bone matrix (it is an open-structured, dispersed form of DBM) + cancellous bone	NCT01430299	It is same material to Accell Evo3
	OsteoSurge® 300c Demineralized Bone Matrix	The moldable putty including cancellous chips	DBM + Accell® bone matrix (it is an open-structured, dispersed form of DBM) + cancellous bone + bioresorbable, reverse phase medium carrier.	n/a	It is same material to Accell Evo3c SeaSpine new sponsor, same material
	OsteoSparx® Demineralized Bone Matrix	Gel or putty-like consistency	DBM + reverse phase medium carrier	n/a	It is same material to Accell Evo3 (NCT01430299)
	OsteoSparx® C Demineralized Bone Matrix	Gel or putty-like consistency.	DBM + reverse phase medium carrier + cancellous bone	n/a	It is same material to Accell Evo3c
	Accell Total Bone Matrix®	Pre-formed shape (round or rectangular)	DBM + Accell® bone matrix → 100% DBM	NCT01430299	It is same material to Accell Evo3
	Accell Evo3c™	Putty	DBM + Accell® bone matrix (it is an open-structured, dispersed form of DBM) + cancellous bone + bioresorbable, reverse phase medium carrier	n/a	FDA510(k) cleared K103742, 2011 Mar

	Accell Evo3™	Putty	DBM + Accell® bone matrix (it is an open-structured, dispersed form of DBM) + bioresorbable, reverse phase medium carrier	1. Case study on posterolateral fusion (12/2013 ~ 06/2017, ClinicalTrials.gov Identifier: NCT02018445) 2. Prospective study on posterolateral fusion (12/2017 ~ 01/2018, ClinicalTrials.gov Identifier: NCT01714804) 3. RCT on posterolateral fusion Accell Evo3 Demineralized Bone Matrix (DBM) (93.5% fused) vs. rhBMP-2(100% fused). (ClinicalTrials.gov Identifier: NCT01430299)	FDA 510(k) cleared K103742, 2011 Mar
	Capistrano™	DBM + allobone	DBM (demineralized bone matrix) + machined cortical and cancellous allograft bone	n/a	FDA 510(k) cleared
Smith & Nephew	VIAGRAF	Putty, paste, gel, crunch, and flex	DBM, glycerol	n/a	510(k) cleared K043209 – 2005 Dec
Spinal Elements	Hero DBM	Putty, paste, gel	DBM, RPM	n/a	Regulated under CFR 1270, 1271 as human tissue
	Hero DBM Powder	Powder	DBM	n/a	Regulated under CFR 1270, 1271 as human tissue
Wright Medical technology	ALLOMATRIX®	Various volumes, consistency varies depending on proportion of cancellous chips utilized	DBM (86% by volume) +/− cancellous bone matrix (CBM) in surgical-grade calcium sulfate powder	Retrospective comparative study	510(k) cleared K041663, 2004 Sept
	ALLOMATRIX ®RCS	Formable putty	DBM, synthetic resorbable conductive scaffold (RCS), calcium sulfate, and hydroxypropylmethylcellulose (HPMC)	n/a	510(k) cleared K041663, 2004 Sept

(continued)

Table 4 (continued)

Company	DBM-based product (human)	Formulation	Product composition	Peer-reviewed clinical evidence/ongoing study ClinicalTrials.gov identifier	Regulatory clearance/ approval FDA 510(k), CFR 1270, CFR 1271
	ALLOMATRIX™ C	Putty	ALLOMATRIX™ + small cancellous chips	n/a	510(k) cleared K040980, 2004 Jul
	ALLOMATRIX™ CUSTOM	Putty	ALLOMATRIX™ + large cancellous chips	n/a	510(k) cleared K040980, 2004 Jul
	ALLOMATRIX™	Injectable	DBM (86% by volume) + OSTEOSET™ (surgical-grade calcium sulfate)	RCT in trauma treatment ClinicalTrials.gov Identifier: NCT00274378	510(k) cleared K020895, 2004 Mar
	ALLOMATRIX® DR	Putty	Calcium sulfate, DBM, and small cancellous chips	n/a	510(k) cleared K040980, 2004 Jul
	PROSTIM™	Procedure kits, various volumes of injectable paste/formable putty	50% calcium sulfate, 10% calcium phosphate, and 40% DBM by weight	n/a	FDA 510(k) cleared K190283, 2019 Feb
Zimmer → It merged into Zimmer Biomet company	IGNITE	Percutaneous graft for fracture mal/nonunion	DBM in surgical-grade calcium sulfate powder to be mixed with bone marrow aspirate	n/a	510(k) cleared K052913, 2005 Nov
	OSTEOSET DBM Pellets	Packable pellets	3.0 mm or 4.8 mm pellets Surgical-grade calcium sulfate, DBM (53% by volume), stearic acid	n/a	510(k) cleared K022828, 2004 Apr K053642, 2006 Jan
	PROSTIM™ Injectable Inductive Graft	Injectable paste/ formable putty	DBM (40% by weight), calcium sulfate (50% by weight), calcium phosphate (10% by weight)	n/a	510(k) cleared K190283, 2019 Feb
	Puros DBM with RPM Gel and Paste	Gel, paste	DBM, RPM, ground cancellous bone (<500 microns)	n/a	Regulated under CFR 1270, 1271 as human tissue
	Puros DBM with RPM Putty and Putty with chips	Putty	DBM, RPM, +/− cortical bone chips (850 microns–4 mm)	n/a	Regulated under CFR 1270, 1271 as human tissue
	Puros DBM Block and Strip	Blocks, strips in varying sizes	DMB (100%)	n/a	Regulated under CFR 1270, 1271 as human tissue

	Putty type			
Bonus® CC Matrix		Demineralized cortical bone (DBM) + mineralized cancellous chips All-inclusive bone grafting kit	n/a	FDA registration number: FEI 1000160576 (till 06.30.2020) AATB and HTC/P
StaGraft™ DBM Putty and Plus		DBM + natural lecithin carrier + resorbable coralline hydroxyapatite/calcium carbonate granules Available as a 40% DBM Putty, or 35% DBM PLUS	n/a	FDA registration number: FEI 1000160576 (till 06.30.2020)
StaGraft™ Cancellous DBM Sponge and Strips		Cancellous DBM Sponge and Strips are machined from a single piece of cancellous bone Osteoinductive bone, trabecular structure, sponge-like handling	n/a	FDA registration number: FEI 1000160576 (till 06.30.2020)
FiberStack™ Demineralized Bone Matrix (DBM)		Manufactured entirely from cortical bone, which has been demonstrated to maintain higher osteoinductivity than cancellous bone after demineralization 100% DBM (without carrier)	n/a	FDA registration number: FEI 1000160576 (till 06.30.2020)

510(K) is a premarket submission made to FDA to demonstrate that the device to be marketed is at least as safe and effective, that is, substantially equivalent, to a legally marketed device that is not subject to premarket approval. 501(k) documentation for individual products is available via FDA online database (http://www.accessdata.fda.gov)
Code of Federal Regulations (CFR) 1270 (Human tissue intended for transplantation) and 1271 (Human cells, tissues and tissue-based products) are federal regulations relating to the procurement and processing of human-derived tissues
Human Tissues Banks: https://images.magnetmail.net/images/clients/AATB/attach/Bulletin_Links/18_2/AATB_Accreditation_Policies_February_08_2018.pdf (last update 2018 Feb)
TBI/Tissue Banks International National Processing Center (an AATB-accredited tissue bank)
US Human Tissue Bank Lic States: California, Florida, Maryland, and New York

proteins, and inductive growth factors (Gupta et al. 2015). Even after processing, DBM possesses osteoconductivity and osteoinductivity, but as a putty-/paste-like substance, it lacks structural integrity (Gupta et al. 2015). Since DBM base powder is derived from human bone allograft, disease transmission related with implantation is low, yet possible, although still less than structural-type allografts (bacterial infection estimated at 0.7 for non-massive to 11.7% for massive bone) (Zamborsky et al. 2016; Kwong et al. 2005; Lord et al. 1988).

Due to lack of structural integrity and relatively low osteoinduction potential comparing to autograft, DBM mixed with a carrier (DBM-based product, DBMs) is frequently used as a bone graft extender/carrier in interbody fusion. Commonly, DBMs are mixed with morselized autografts and exogenous peptide/differentiation factors along with collagen matrix, bone marrow aspirate, and/or isolated native blood-derived growth factors to stimulate new bone growth. In previous clinical reports on spine surgery, DBMs with autograft, and DBMs with growth factors (bone marrow aspiration), DBMs mixed with peptides (rhBMP-2/ACS) may be substituted for ICBG (Kang et al. 2012; Morris et al. 2018). DBM-based products or DBM powder are rarely used as a stand-alone graft material (Kinney et al. 2010).

There are several limitations to overcome in the clinical use of DBM-based products. The clinical effectiveness of DBM-based products is known to be variable according to manufacturer, form of product, as well as different lot-based batches from the same product form and manufacturer (Bae et al. 2006, 2010). The possible features of DBM-based products that contribute to varied reliability are varying native BMPs, growth/differentiation factors (donor bone), and dosages (Bae et al. 2006, 2010); forms such as putty, gel, flexible sheets, or mixed with cortical chips; compositions of carriers, scaffolds, gels, and other fillers; particle sizes of final bone powder; quality of the donor bone; and manufacturers processing procedures and sterilization method of products (Peterson et al. 2004; Bae et al. 2006, 2010). Amid these limitations, DBM-based products provide a diverse range of DBM-based grafting options that have been commonly employed for specific applications. DBM-based products introduced to the market over the last two decades and currently used are presented in Table 4.

Exogenous Inductive Differentiation Growth Factors and Other Peptides (Table 5)

Bone Morphogenetic Protein

Bone morphogenetic proteins (BMPs) are soluble members of the transforming growth factor-β superfamily that are involved in the differentiation, maturation, and proliferation of mesenchymal stem cells (MSCs) into osteogenic cells (Miyazono et al. 2005). To describe the acting mechanism, BMPs act via serine-threonine kinase receptors found on the surface of target cells and often transduce their signal via the SMAD pathway, leading to nuclear translocation and subsequent expression of target genes involved in osteogenesis (Hoffmann and Gross 2001; Sykaras and Opperman 2003). The reaction mechanism of BMP is mainly osteoinduction and reactively much less osteogenic potential (Campana et al. 2014). The graft material includes rhBMP-2 (exogenous protein) along with absorbable collagen sponge (ACS, rhBMP-2 carrier). The carrier (ACS) has BMP binding competence in order to decrease diffusion away from the desired site for bone formation and increase controlled continual delivery of protein at the site. Although numerous carriers such as metals, collagen, ceramic such as tricalcium phosphate (TCP) and HCO, bioactive glass (BG), and polymers have been described (Agrawal and Sinha 2017), the most commercially available scaffold is an absorbable type 1 collagen sponge (ACS) bovine derived (Kannan et al. 2015).

For several decades, over 20 BMPs have been identified and described. Among them, BMP-1, BMP-2 (BMP-2A), BMP-3 (osteogenin, less osteoinductive) BMP-4 (BMP-2B), BMP-5, BMP-6, BMP-7, and osteoinductive factor (OIF) have been shown to induce bone formation

11 Bone Grafts and Bone Graft Substitutes   233

Table 5 Commercially available bone inductive peptides, proteins-based products, recombinant versions

Company	Peptide, growth factor product	Formulation	Product composition	Peer-reviewed clinical evidence/ongoing study ClinicalTrials.gov identifier	Regulatory clearance/US approval/PMA, FDA 510 (k)
CeraPedics Inc.	P-15L Bone Graft	Bone graft putty	ABM, lecithin carrier (resorbable, biocompatible, semi-viscous lipid) P-15 synthetic peptide, calcium phosphate particles, porcine derived ABM anorganic bone matrix, along with bovine collagen carrier, P-15L	Ongoing clinical trials NCT03438747 (P-15L Bone Graft in an instrumented TLIF)	PMA
	i-FACTOR Bone Graft 'i-FACTOR™ Peptide Enhanced Bone Graft P-15 Putty, or iFACTOR Putty'	Synthetic small peptide (P-15) and peptide bone matrix (PBM) used in an allograft bone ring and with supplemental anterior plate fixation C3–C4 to C6–C7 following single-level discectomy for intractable radiculopathy (arm pain and/or a neurological deficit), with or without neck pain, or myelopathy due to a single-level abnormality localized to the disc space	A composite bone substitute consisting of a synthetic collagen fragment (P-15) bound to calcium phosphate particles. As an engineered product, P-15 quantity and viability remain consistent from lot to lot. P-15 in anorganic bone mineral (ABM). This unique combination replicates the organic (type I human collagen) and inorganic (calcium phosphate) components of autograft bone	Arnold et al. 2016b, 2018 NCT00310440 (P-15 synthetic osteoconductive bone substitute, ACDF) NCT01618435 (i-FACTOR Bone Graft in non-instrumented Posterolateral Spondylodesis in Elderly with LSS) NCT02895555 (5-10 CC i-FACTOR putty mixed with local harvested autograft, fusion lumbar spinal stenosis)	PMA cleared P140019, 2015 Nov www.accessdata.fda.gov/cdrh_docs/pdf14/p140019a.pdf

(continued)

Table 5 (continued)

Company	Peptide, growth factor product	Formulation	Product composition	Peer-reviewed clinical evidence/ongoing study ClinicalTrials.gov identifier	Regulatory clearance/US approval/PMA, FDA 510 (k)
Medtronic Sofamor Danek, Inc., USA (Medtronic BioPharma B.V., Netherlands)	INFUSE™ Bone Graft/LT-CAGE™	Recombinant human bone morphogenetic protein and a carrier/scaffold inserted into a hard LT-CAGE™	rhBMP-2 (derived from a recombinant Chinese Hamster Ovary (CHO) cell line) Bovine absorbable collagen I scaffold (ACS) (filler) Ti metal prosthesis (Ti-6Al-4V)	NCT01491386 NCT01491425	FDA approved, regulated under PMA P000058, 2002 Jul www.accessdata.fda.gov/cdrh_docs/pdf/p000058b.pdf
	INFUSE® Bone Graft	LT-CAGE® Lumbar Tapered Fusion Device and INFUSE® Bone Graft	rhBMP-2/ACS/INTERFIX™ rhBMP-2, collagen scaffold (ACS) (filler) Ti metal prosthesis (Ti-6Al-4V)	NCT01491451 NCT00635843 Litrico et al. 2018	
	INFUSE® Bone Graft	INFUSE® Bone Graft	INFUSE® Bone Graft consists of two components – recombinant human bone morphogenetic protein-2 (rhBMP-2, known as dibotermin alfa) 1.5 mg/mL of rhBMP-2; 5.0 mg sucrose, NF; 25 mg glycine, USP; 3. 7 mg L-glutamic acid, FCC; 0.1 mg sodium chloride, USP; 0.1 mg polysorbate 80, NF; and 1.0 mL of sterile water. Reconstituted rhBMP-2 solution has a pH of 4.5 and is clear, colorless, and essentially free from plainly visible particulate matter Placed on an absorbable collagen sponge (ACS); bovine type I collagen obtained from the deep flexor (Achilles) tendon	ClinicalTrials.gov listings	FDA approved, regulated under PMA P000053, 2007 Mar www.accessdata.fda.gov/cdrh_docs/pdf5/p050053b.pdf

INFUSE/ MASTERGRAFT™	AMPLIFY™	Posterolateral Revision Device The INFUSE/ MASTERGRAFTTM Posterolateral Revision Device is indicated for the repair of symptomatic, posterolateral lumbar spine pseudarthrosis. This device is intended to address a small subset of patients for whom autologous bone and/or bone marrow harvest is not feasible or is not expected to promote fusion. These patients are diabetics and smokers. This device is indicated to treat two or more levels of the lumbar spine Orthopedics Adult Oct. 10, 2008	NCT01491542 https://www.accessdata.fda.gov/cdrh_docs/pdf4/h040004c.pdf	INFUSE/ MASTERGRAFTTM This device has been withdrawn at the request of the sponsor effective Mar 2010
INFUSE® Bone Graft	Polyetheretherketone (PEEK) in oblique lateral interbody fusion (OLIF)	OLIF 51 procedures with Divergence-L Interbody Fusion Device at a single level from L5–S1 OLIF 25 procedures with Pivox Oblique Lateral Spine System at a single level from L2–L5 ALIF procedures with Divergence-L at a single level from L2–S1	NCT01415908, NCT04073563	PMA supplement approval, 2015 Dec

(continued)

Table 5 (continued)

Company	Peptide, growth factor product	Formulation	Product composition	Peer-reviewed clinical evidence/ongoing study ClinicalTrials.gov identifier	Regulatory clearance/US approval/PMA, FDA 510 (k)
	INFUSE® Bone Graft PEEK ACDF			NCT00485173, 2013 Arnold et al 2016a; Zadegan et al. 2017b IDE# G010188/ NCT00642876 and IDE# G000123/NCT00437190 (www.clinicaltrials.gov) Arnold et al. 2016a	Not cleared.
	InductOs® (Medtronic Spinal and Biologics)	Dibotermin alfa	Dibotermin alfa (recombinant human bone morphogenetic protein-2; rhBMP-2) is a human protein derived from a recombinant Chinese Hamster Ovary (CHO) cell line Bovine collagen I carrier	NCT02280187, InductOs® in Real World Spine Surgery; A Retrospective, French, Multi-centric, Study (InductOR)	Cleared in EU 1/02/226/002
Stryker 2004 (Olympus Biotech 2010–2014, then closed facility)	OP-1 Putty	OP-1 Putty	OP-1/rhBMP-7 produced delivered on a purified type I bovine collagen carrier and Carboxymethylcellulose	For revision posterolateral lumbar spinal fusion spinal fusion Vaccaro et al. 2008 PMID: 17588821 Guerado and Fuerstenberg 2011	Center for Devices and Radiological Health (CDRH) of the Food and Drug Administration (FDA) humanitarian device exemption (HDE) (check current 2018, not

Stryker 2004 Pfizer (Wyeth is now a wholly owned subsidiary of Pfizer)	Op-1 Putty	OP-1(rhBMP-7) / purified Type I bovine collagen carrier and carboxymethylcellulose	Delawi et al. 2016 OP-1 group (54% versus 74% in the autograft group, $p = 0.03$)	sold in the USA) OP-1 Implant is approved in 28 additional countries, including Australia, Canada, and the European Union OP-1 no longer marketed in the USA (note, 2019) FDA Advisory Committee: ("against expanded approval") Not recommended
	rhBMP-2/CPM rhBMP-2/CPM matrix	Recombinant human bone morphogenetic protein-2 (rhBMP-2)/calcium phosphate matrix (CPM)	NCT00752557 (bone mineral density) Closed fractures NCT00161629 (radius) NCT00384852 (humerus)	Not yet approved
BioAlpha Inc.	ExcelOS-Inject ExcelOS 14-01 InjectBMP	Injectable Ceramics Bone Graft (beta-TCP) containing rhBMP-2	NCT02714829	Not USA, Korea

Includes US FDA-tracked products (excludes products controlled exclusively under other regulatory agencies outside of USA)

(Wozney 1989, 2002). However, only two commercial forms of recombinant human BMPs currently are available for clinical use (Kannan et al. 2015). Recombinant human forms of BMP-2 (Infuse®; Medtronic) and BMP-7 (OP-1; Stryker) have been developed and approved both in the USA and Europe for commercial purposes by employing mammalian cells transfected with the corresponding human BMP sequence (Campana et al. 2014). Extensive research (over 30 years) has been conducted in support of the US-FDA approval process of rhBMP-2 and rhBMP-7; translational problems include scaling up, as super physiologic concentrations of rhBMP-2 are needed to meet MEDs in widely applicable orthopaedic indications in humans (Vallejo et al. 2002).

Internationally, in Korea, several products were developed employing various production processes for BMP-2 and different carriers. Its approved by Korea Food and Drug Administration (KFDA); to date its not approved by US-FDA. For spine fusion, the product carrier is granular HA and is based on *Escherichia coli*-derived rhBMP-2 (E.BMP-2, CGBio, Korea; E. BMP-2/HA, Novosis®, Korea) designed to improve the protein yield over the production process of using mammalian origin cell lines, such as Chinese hamster ovary (CHO) cells that incur low yield and high cost. There are several animal and clinical studies demonstrating the effectiveness and safety of Novosis® (Lee et al. 2012; Kong et al. 2014; Kim et al. 2015). According to the study of Cho et al. (2017), a fusion rate of 100% for E.BMP-2/HA (Novosis®) was comparable with that of 94.1% for AIBG demonstrating clinical efficacy and safety in PLF. E.BMP-2 production of rhBMP-2-based products and clinically used or in investigation are a rhBMP-2/Beta-TCP putty type (NCT01764906, Novosis® Korea), another Beta-TCP product containing rhBMP-2 (ExcelOS-inject, ExcelOS 14-01, NCT02714829, BioAlpha Inc., Korea), and a collagen gel +DBM containing rhBMP-2 (50 ug/cc) (rhBMP-2 produced from CHO cells, RafugenTM BMP-2, Cellumed Co Ltd., Seoul, Korea) employed as graft for interbody spinal fusion (pivotal RCT completed, 2017; submitted KFDA 2018, approved for dental application KFDA 2013).

US Regulatory approval by the FDA was initially granted for rhBMP-2/ACS (Infuse, Medtronic) in single-level anterior lumbar interbody fusion procedures in 2002 (Burkus et al. 2002). rhBMP-2 was then approved for tibia nonunion as an alternative to autograft in 2004 and for oral maxillofacial reconstructions in 2007 (Rengachary 2002). During last decade, rhBMP-2 has been commonly used off-label in posterolateral lumbar fusion surgery (Morris et al. 2018).

RhBMP-7, an osteogenic growth factor related to BMP-2, was first approved by the FDA in 2001 for use as an alternative to autograft for long bone fracture repair. In 2004, approval was expanded to cover PLLF (Morris et al. 2018). RhBMP-7 or OP-1 was approved for limited use under humanitarian device exemption (HDE) (no longer marketed in the USA (https://www.transparencymarketresearch.com/bone-morphogenetic-protein-market.html). In the last decade, several types of rhBMPs were developed and commoditized to medical market. RhBMP-based products were introduced to the market over the last two decades. Currently used products are presented in Table 5.

The osteogenic/osteoinductive potential of rhBMPs was strongly investigated in both preclinical and clinical studies, with a reported performance that is comparable to autogenous cancellous bone, with fusion rates between 80% and 99% (Campana et al. 2014). There are approximately 80 clinical studies on rhBMP-2 testing various surgical indications. According to a Level I comparison study of ICBG vs. rhBMP-2 with collagen sponge and ceramic granule by Dawson et al. (2009), at 24 months the rhBMP-2-/CS-/CM-treated patients had significantly higher solid fusion rates than those in the iliac crest autograft group (95% vs. 70%). Additionally, patients in the rhBMP-2/CS/CM group reported significantly greater improvement in clinical outcomes than did those in the iliac crest autograft group. According to the studies of Vaccaro et al. (2004, 2005), the use of rhBMP-7 (as OP-1 putty from) in conjunction with bovine collagen and carboxymethylcellulose (carrier) showed similar or slightly superior clinical result in spine fusion (posterolateral non-instrumented fusion) compared with autograft from the iliac crest.

However, limitations for general use of BMPs and complete substitution for autograft remain. First, rhBMPs have marked species-specific concentration requirements for osteogenesis, and thus results from preclinical studies are not considered as valuable background information for human application. Second, the dose-dependent efficacy in humans of rhBMPs has been observed in previous studies, and various clinical trials are aimed toward elucidating the optimal dosage of rhBMP-2/ACS (Govender et al. 2002). However, the optimal dosage/concentration for various off-label applications has rarely been reported or suggested in spine surgery. Third, during clinical trials, several major and minor adverse effects like ectopic bone formation in the neural canal, dysphagia when used in cervical fusion applications, prevertebral swelling, seroma/hematoma formation, radiculitis, osteolysis, heterotopic ossification, retrograde ejaculation, increased rates of new malignancy, and implant subsidence due to end-plate osteolysis are reported (Shields et al. 2006; James et al. 2016). Because of these limitations, numerous ongoing areas of investigation target alterations in dosage for optimal minimal dosage, scaffold to maintain concentration, and the implementation of supplemental proteins or growth factors to regulate the nonspecific action of rhBMP-2 (Agrawal and Sinha 2017; Burke and Dhall 2017; Poorman et al. 2017). Outside the USA, alternative-type protein products are in development (BoneAlbumin™, plasma protein) used to enhance bone allograft (Gmbh, OrthoSera, Austria).

Peptide-Based Materials

Although naturally derived extracellular matrix (ECM) has demonstrated some degree of success in selected studies, it is challenging to modify, characterize, and control the presentation of natural ECM biomaterials (Shekaran and Garcia 2011). The limitations of ECM molecules have spurred the use of ECM-derived peptides or recombinant fragments that incorporate the minimal functional sequence of their parent protein to convey bioactivity to implant materials.

Cerapedics

P-15 is a synthetic 15-amino acid peptide derived from the (766)GTPGPQGIAGQRGVV(780) sequence found in the α1(I) chain of type I collagen. Several preclinical studies have demonstrated that P-15 enhances cell adhesion, osteoblastic gene expression, and mineralization when implanted on anorganic bone matrix (ABM) in vitro and accelerates early bone formation in porcine and rat cranial defects (Shekaran and Garcia 2011). In a head-to-head comparison of DGEA peptide and P-15-coated hydroxyapatite discs implanted into rat tibiae, both peptides improved new bone formation, but P-15 failed to enhance bone implant contact. A recent study in the larger bovine model, ABM/P-15 (ABM an allograft in this application), failed at 4.5 months after uninstrumented posterior lumbar spine surgery; 68% fusion in allograft implanted sheep vs. 0% fusion as determined by bridging between transverse processes was found in ABM/P-15 implanted sheep (Axelsen 2019).

For human applications of a xenograft carrier (a sinterized cancellous bovine bone matrix), the implemented carrier has been employed. P-15 peptide-coated ABM has been used in human periodontal osseous defects resulting in better clinical outcomes than open flap debridement alone and has also been used in a pilot clinical study for long-bone defects (Shekaran and Garcia 2011). In a prospective, randomized, single-blinded trial of single-level ACDF using P-15/CBM in an allograft spacer versus local autograft in an allograft spacer, 89.0% vs. 85.8% fusion rates were reported, respectively, at 1-year follow-up with equivalent clinical outcomes and complications (Hsu et al. 2017).

B2A

B2A is a bioactive synthetic multi-domain peptide that augments osteogenic differentiation via increasing endogenous cellular BMP-2 by pre-osteoblast receptor modulation at spine fusion site (Lin et al. 2012). The empirical formula of B2A is C241H418N66O65S2 containing 42 amino acids and 3 lysine analogue residues of 6-aminohexanoic (Glazebrook and Young 2016). This peptide has osteoinductive potential; it is used with a scaffold. The osteoconductive scaffold is a ceramic granule from which B2A elutes in vivo. Two commercialized products PREFIX® (Ferring Pharmaceuticals, Saint-Prex, Switzerland) and AMPLEX® (Ferring Pharmaceuticals, Saint-Prex, Switzerland) are based on this converged technology. After grafting of B2A with ceramic granules, complete absorption of B2A occurs within approximately 6–8 weeks (Glazebrook and Young 2016).

There is great interest in the benefits of conjugation technology for modulating release kinetics in grafting materials. However, there are limited preclinical and clinical studies on the safety and effectiveness of B2A/ceramics. B2A/ceramic granule was tested in two animal studies (rabbit and sheep). B2A/ceramic significantly improved the fusion rate in PLLF and PLIF over simple autograft bone graft (Smucker et al. 2008; Cunningham et al. 2009). In a clinical study, higher fusion rates were observed in B2A-coated ceramic granule (formulated as PREFIX®)-grafted patients than in ICBG-grafted patients after an interbody fusion procedure (Sardar et al. 2015). Studies are limited; Clinicaltrials.gov indicates two registered multicenter studies with "unknown" status. Validating the safety and efficacy of this bone graft material necessitates high-quality clinical studies and/or multicenter studies enrolling large number of patients. To date, there is only one published pilot study (Sardar et al. 2015, Canada) with a small/insufficient sample size.

Synthetic Materials and Drafts (Table 6)

Synthetic graft materials are typically employed during fusion surgery as bone graft extenders and sometimes substitutes. Traditionally, these materials provide an osteoconductive scaffold with ideally no reactive inflammatory immunogenic response from host tissues. The known advantageous properties of synthetic materials like ceramics include osteoconductive, biodegradable, no risk of infection, no donor site morbidity, unlimited supply, relatively easy sterilization, easiness of molding sized and shape, and lack of immunogenicity and toxicity (Gupta et al. 2015; Kannan et al. 2015). More recently, the emerging novel synthetics involve new technological advances in material science and/or incorporate a menagerie of cross-product materials in order to address the molecular biologic demands for bone induction, consolidation or healing, and fusion mass incorporation. Design innovation may lead to a true potent autograft substitute.

For making an ideal bone graft extender or graft substitute, several characteristics should be considered. The development of products used for bone regeneration has followed the basic criteria of providing a biocompatible three-dimensional scaffold with controlled architecture capable of stimulating or supporting bone growth in the natural in vivo environment (O'Brien 2011). The ability of the material to be used in conjunction with other cellular and signal-based therapies (peptides, growth factors) is a key strategy in maximizing the efficacy and likely success of fusion. The primary characteristics of bone graft substitutes are shown in Table 6.

Calcium Phosphate Materials

Calcium phosphates are a common base for synthetic graft materials. This is primarily because 70–90% of inorganic material in the body is a type of calcium phosphate. Calcium phosphate materials have been cleared in the USA for use as "bone void fillers" (FDA MQV, MBP) that can be used for spine fusion and orthopedic applications. The common types of calcium phosphate materials are beta tricalcium phosphate $Ca_3(PO_4)_2$ (TCP) and hydroxyapatite $Ca_{10}(PO_4)_6(OH)_2$ (HA).

TCP was one of the earliest synthesized forms of calcium phosphate materials that was used as an osteoconductive bone void filler. TCP in the form of granules or blocks is available as a three-dimensional structure with interconnected pores from 1 to 1,000 microns. However, TCP and all

Table 6 Commercially available synthetic bone void fillers, extenders, and substitutes products

Company	Synthetic product	Formulation	Product composition	Peer-reviewed clinical evidence ClinicalTrials.gov/ongoing study	Regulatory clearance/approval FDA 510(k), CFR 1270, CFR 1271 MQV, filler, bone void, calcium compound common name – bone grafting material classification name – bone grafting material, synthetic
Amend Surgical, Inc., FL, USA	0.5 cc NanoFUSE® (posterolateral fusion)	Putty-like, malleable	Amend Surgical, Inc., NanoFUSE® comprised synthetic calcium-phospho-silicate particulate material particles (45S5 bioactive glass) coated with gelatin derived from porcine skin. (there is a DBM version of this material)	n/a	510(k) cleared 21 CFR 888.3045 K161996, 2017 Feb
	45S5 bioactive glass (1–10 cc)	Particulate material	45 wt% SiO2, 24.5 wt% CaO, 24.5 wt% Na2O, and 6.0 wt% P2O5 Synthetic binder		510(k) cleared K110368, 2017 Jan
Aspine USA, Oakland, CA	Osteo-G Bone Void Filler System	Pellets/paste	Bioabsorbable, calcium sulfate dihydrate (prefabricated or kit to form into various shape implants), radiopaque	n/a	FDA 510(k) cleared K031319, 2003 Jul
APATECH LTD, United Kingdom → Merged into Baxter Healthcare	ACTIFUSE synthetic bone graft		Phase-pure silicon-substituted calcium phosphate Osteoconductive bone graft substitutes, comprising a single-phase calcium phosphate scaffold, either granules or granules delivered in a matrix of resorbable polymer	NCT01833962, no results posted	510(k) cleared K090850 {K040082, K071206, K080736, K082073, K081979, K082575}
	ACTIFUSE Shape	Flexible shape	Distinctive moldability and versatility allowing the unique contours of each defect to be addressed Silicon-substituted calcium phosphate	NCT02005081 Anterior cervical corpectomy (ACC), no results posted	FDA 510(k) cleared K082575, 2008 Nov

(continued)

Table 6 (continued)

Company	Synthetic product	Formulation	Product composition	Peer-reviewed clinical evidence ClinicalTrials.gov/ongoing study	Regulatory clearance/approval FDA 510(k), CFR 1270, CFR 1271 MQV, filler, bone void, calcium compound common name – bone grafting material classification name – bone grafting material, synthetic
	ACTIFUSE ABX		A sculptable synthetic bone graft substitute Silicon-substituted calcium phosphate	ClinicalTrials.gov Identifier: NCT01852747 (fusion rate for multilevel fusion in spine) ClinicalTrials.gov Identifier: NCT01013389 (RCT, interbody fusion, comparative study with infuse) ClinicalTrials.gov Identifier: NCT01018771 (RCT, PLF fusion, comparative study with infuse)	FDA 510(k) cleared K082575, 2008 Nov
	ACTIFUSE Microgranules	Granules	Synthetic bone graft substitute designed for smaller defects Silicon-substituted calcium phosphate	n/a	FDA 510(k) cleared K082575, 2008 Nov
	ACTIFUSE Granules	Granules	Synthetic bone graft substitute, designed for larger defects Silicon-substituted calcium phosphate	n/a	FDA 510(k) cleared K082575, 2008 Nov
	ACTIFUSE MIS System	Injectable type	A ready-to-use applicator and cartridge designed for controlled delivery during minimally invasive procedures It contains ACTIFUSE ABX	NCT02845141, revision ACL	FDA 510(k) cleared K082575, 2008 Nov
	ACTIFUSE E-Z-Prep Syringe	Putty Matrix	A preloaded syringe containing ACTIFUSE Microgranules	n/a	FDA 510(k) cleared K082575, 2008 Nov
	Inductigraft		MAUDE adverse event report Active bone graft substitute 45S5 bioactive glass (M-45 granules,	NCT01452022, Inductigraft in posterolateral fusion	Not registered in the USA

Berkeley Advanced Biomaterials, Berkeley, CA, USA, 1996	Cem-Ostetic®, Bi-Ostetic™, GenerOs™	Granule, block Injectable putty	MS-45 microspheres) and bovine type I collagen (hydration with saline or blood). For use in posterolateral fusion	Bi-Ostetic Bioactive Glass Foam	FDA 510(k) cleared K170917, 2017 Oct ISO 13485 certified and CGMP
			Bone void fillers are based on nanocrystalline hydroxyapatite (HAP) and tricalcium phosphate (TCP)		
Biomatlante, France	MBC PTM	Various shapes and sizes	Microporous and macroporous biphasic calcium phosphate ceramic consisting of 60% hydroxyapatite (HA) and 40% beta-Tricalcium Phosphate (β-TCP)	NCT00206791	K043005, 2015 May
BioAlpha Inc., Korea	Novomax	Intervertebral spacer	Bioactive glass-ceramic spacer	NCT03532945	unk
	Bongros®-HA		Hydroxyapatite ($Ca_{10}(PO_4)_6(OH)_2$) highly pure with trabecular structure, 3-dimensional interconnected pores		K090793, 2009 May
Biomet Osteobiologies →Merge into Zimmer-Biomet	Calcigen™-S	Paste granules	Calcium sulfate dihydrate, isothermic	n/a	FDA 510(k) cleared K013790, 2002 Jun
	BonePlast®	Powder	Calcium sulfate with or without HA/CC	n/a	FDA 510(k) cleared K070864, 2007 Jun
	BonePlast® Quick Set	Quick setting paste	Calcium sulfate (mixed setting solution, QS) (limited to be used in posterolateral fusion)	n/a	FDA 510(k) cleared K070864, 2007 Jun
Biomimetic Therapeutics, Inc., Franklin, TN	Augment Bone Graft		β-Tricalcium phosphate-containing matrices + recombinant human platelet-derived growth factor-BB	Solchaga 2012 (ovine) rhPDGF-BB solution (0.3 mg/mL) was mixed with the β-TCP (1:1, v:v), allowed to incubate at room temperature for 10–15 min and then transferred to a 3 mL syringe with the end removed	

(continued)

Table 6 (continued)

Company		Formulation	Product composition	Peer-reviewed clinical evidence ClinicalTrials.gov/ongoing study	Regulatory clearance/approval FDA 510(k), CFR 1270, CFR 1271 MQV, filler, bone void, calcium compound common name – bone grafting material classification name – bone grafting material, synthetic
	Synthetic product				
	Augment Injectable		β-Tricalcium phosphate-containing matrices + recombinant human platelet-derived growth factor-BB	Solchaga 2012 (ovine) rhPDGF-BB solution (0.3 mg/mL) was mixed with the β-TCP/collagen (3:1, v:w), allowed to incubate at room temperature for 5–15 min and then transferred to a 3 mL syringe with the end removed. The syringe was used to dispense 0.4 mL of Augment Injectable to the interior of the PEEK spacer	
Bioventus® Surgical	Signafuse™	Putty type	Microporous and macroporous biphasic calcium phosphate + bioactive glass Multidirectional interconnected porosity structure similar to that of human cancellous bone (20–30% microporous (pore size <10 μm) and 50–55% macroporous)	n/a in clinical study Comparative study in rabbit (Fredericks et al. 2016)	FDA 510(k) cleared K132071, 2014 Jan (permitted as Biostructures, LLC)
	INTERFACE	Powder type	45S5 bioactive glass Particle size of 200–420 microns is designed for a faster speed of bone fill	n/a	FDA 510(k) cleared K112857, 2011 Dec (permitted as Biostructures, LLC)
	OsteoMatrix™	Strip type	60% hydroxyapatite (HA) + 40% beta-tricalcium phosphate (β-TCP) + type 1 collagen A synthetic two-phase calcium phosphate embedded in a cross-linked collagen carrier	n/a	FDA 510(k) cleared K051774, 2006 Jan (as a product name of MBCP™)

	Osteo Plus™	Granules in delivery syringe	Biphasic calcium phosphate (BCP): 60% hydroxyapatite (HA) + 40% beta-tricalcium phosphate (β-TCP) Synthetic two-phase calcium phosphate granules with interconnected macro and 3D micropores	n/a	FDA 510(k) cleared K051774 (01.20.2006) (as a product name of MBCP™)
Bone Bank Allografts, Texas, USA	Confirm™ Bioactive: Confirm™ Gel: Confirm™ Crunch	Gel and crunch type	Gel type Composition: Bioglass + hyaluronic acid + glycerol Sterile-packed in a syringe Available in three sizes: 2, 5, and 10 cc Uniform Bioglass particle sizes Crunch type Composition: Bioglass + hyaluronic acid + glycerol Sterile-packed in a syringe Available in three sizes: 2, 5, and 10 cc Mixture of Bioglass particle sizes	n/a	FDA 510(k) cleared K133678, 2014 Aug
BonAlive Biomaterials, Biolinja 12, 20750 Turku, Finland	BonAlive®	Granule and putty type Various sizes	S53P4 bioactive glass (53% SiO2, 23% Na2O, 20% CaO, 4% P2O)	Long bone defect treatment, Aurégan and Bégué 2015 Osteomyelitis treatment, Malat et al. 2018	BonAlive® products are not sold in the USA. (05.22.2018)
BONESUPPORT AB, Scheelevägen 19, SE-223 70, Lund, Sweden	CERAMENT® BONE VOID FILLER	Injectable type	Combination of two natural materials – hydroxyapatite and calcium sulfate – with a radiopacity enhancing agent 40% hydroxyapatite +60% calcium sulfate + iohexol (as a radio-opacity enhancer)	Ongoing RCT NCT01828905 (active status) BONESUPPORT AB: NCT02820363 Other Study ID Numbers: CLIN001 – FORTIFY US Food and Drug Administration (FDA) ongoing Investigational Device Exemption (IDE) study of its product	FDA 510(k) cleared K073316, 2008 Jun

(continued)

Table 6 (continued)

Company	Synthetic product	Formulation	Product composition	Peer-reviewed clinical evidence ClinicalTrials.gov/ongoing study	Regulatory clearance/approval FDA 510(k), CFR 1270, CFR 1271 MQV, filler, bone void, calcium compound common name – bone grafting material classification name – bone grafting material, synthetic
CAM Bioceramics BV/CAM implants BV (University of Leiden)/ 1993 Osteotech Inc.	Camceram TCP	Granules (1–4 mm) or block	Beta-tricalcium phosphate with 90% porous (Available specialty compound form; β –TCP, Milled β-TCP, α –TCP Cement Powder)	n/a	FDA 510(k) cleared K050357, 2005 Apr
DePuy Synthes, West Chester, Pennsylvania	HEALOS® Bone Graft Substitute	Putty type	Type 1 collagen + HA scaffold	Yousef MAA, 2017 (MSCs) Villa et al. 2015 Kunakornsawat et al. 2013 Ploumis et al. 2010 Carter et al. 2009 Birch and D'Souza 2009 (Implant Removal) Magit et al. 2006 (w/rhBMP2) Neen et al. 2006 Kraiwattanapong et al. 2005 Jahng et al. 2004 Furstenberg et al. 2010	HEALOS® Bone Graft Substitute (K012751, 2001 Nov and K043308, 2005 Feb)
	HEALOS Fx Injectable Bone Graft Replacement	Injectable type	Type I bovine collagen + hydroxyapatite HEALOS Fx is approximately 20–30% mineral by weight	Same material of HEALOS® Bone Graft Substitute	HEALOS® Fx Bone Graft Substitute (K062495) Mixing Device K081758, 2008 Sep
	Conduit	Putty type	Pure β-tricalcium phosphate (β-TCP) + a non-animal-derived sodium hyaluronate	ClinicalTrials.gov Identifier: NCT02056834 (tibia plateau fracture) (observatory case series study) ClinicalTrials.gov Identifier: NCT01615328 (cervical spine fusion)	510(k) cleared K041350, 2004 Jul

	Synthes chronOS™ chronOS™ inject	Granules, blocks, wedges, and cylinders	β-Tricalcium phosphate (β-TCP)	NCT02803177 (vs. cells) NCT02056834 (chronOS Inject, fractures) NCT00943384 (strip +BMA+ local bone) (posterolateral fusion) NCT00841152 (b-TCP vs. bioactive glass) filling defects (tumor)	FDA 510(k) cleared K0430453, 2005 Jan
	CONDUIT® TCP	Granules type	100% β-TCP TCP Granules obtained after high-temperature ceramicization of tribasic calcium phosphate. Interconnected pores 70% of volume (1 and 600 μm)	n/a (MAUD report)	FDA 510(k) cleared K014053, 2002 Mar
ETEX (Zimmer Biomet 2014 October)	CaP Plus	CaP Plus	Synthetic calcium phosphate, an inert carrier, carboxymethyl cellulose (CMC), and DBM	n/a	510(k) cleared K063050, 2007 Nov K080329, 2008 Apr
Globus Medical, Inc.	MicroFuse® Putty and MicroFuse® ST MIS		Resorbable calcium salt bone void filler	Is a bone graft extender	K102392, 2010 Dec MicroFuse® Bone Void Filler (K071187, K082442)
	MicroFuse™ Bone Void Filler MicroFuse™ granules MicroFuse™ blocks	Granules, sheets, and pre-formed blocks	Porous bone graft scaffold composed of bonded poly(lactide-co-glycolide) or poly(lactic acid) microspheres with and without barium sulfate and calcium sulfate form of granules, sheets, and pre-formed blocks. MicroFuse™ implants are available in short-term (ST), mid-term (MT), or long-term (LT) compositions	n/a	K083232, 2008 Dec
Isto Biologics, USA	InQu®	Past Mix, Matrix Granules	Synthetic PLGA (poly(lactide-co-glycolide)) with HyA (hyaluronic acid)	NCT01746212 Bone graft extender	K063359, 2007 Apr
Inion Oy, Lääkärinkatu 2, FIN-33520 Tampere, Finland	Bioactive glass (S53P4)	Variable shape: cylinders, blocks, and morsels	Different size degradable bioactive glass (S53P4)	NCT01304121, no results posted	K070998, 2007 Oct

(continued)

Table 6 (continued)

Company	Synthetic product	Formulation	Product composition	Peer-reviewed clinical evidence ClinicalTrials.gov/ongoing study	Regulatory clearance/approval FDA 510(k), CFR 1270, CFR 1271 MQV, filler, bone void, calcium compound common name – bone grafting material classification name – bone grafting material, synthetic
Medtronic Spinal and Biologics	MASTERGRAFT®	Granule form Putty type (combined with a type I collagen) Strip type	Biphasic, resorbable ceramics composed of hydroxyapatite (HA) and β-TCP	ClinicalTrials.gov Identifier: NCT01491542 (PLF as Pilot study) ClinicalTrials.gov Identifier: NCT00549913 (PLF, clinical study to evaluate the feasibility, safety, and tolerability of 3 different doses of immunoselected, culture-expanded, nucleated, allogeneic MPCs (NeoFuse))	FDA 510(k) cleared, K081784, 2008 Sep: putty form FDA 510(k) cleared, K082166, 2008 Sep: strip form
Molecular Matrix	Osteo-P bone graft substitute		Osteo-P is a non-mineralized, synthetic bone void filler made of a hyper-crosslinked carbohydrate polymer. It is highly porous, biocompatible, and biodegradable	n/a	510(k) cleared K170165, 2017 Dec
NovaBone Jacksonville, FL, USA → Osteogenics Biomedical	NovaBone Putty – Bioactive Synthetic Bone Graft	Soft malleable putty	Bimodal particle distribution of calcium-phospho-silicate (CPS, Bioglass) + polyethylene glycol (PEG) as additive + glycerin as binder Volume of active ingredient is 70%	n/a	510(k) cleared K082672, 2008 Dec CE approval
	NovaBone-AR	Packable graft	Synthetic calcium-phospho-silicate (Bioglass) particulate, fused into a bulk porous form having a multidirectional interconnected porosity	n/a	510(k) cleared K041613

	NovaBone IRM™	Flexible sheets, varying sizes	IRM (irrigation resistant matrix) Bioactive calcium-phospho-silicate particulate and a synthetic, absorbable binder	Retrospective comparative study	510(k) cleared KO4 16 13, 2005 Dec November 19, 2016 (21 CFR 888.3045)
	NovaBone Bioactive Strip	Strip type	Purified fibrillar collagen and resorbable bioactive synthetic granules (Bioglass)	n/a	K141207, 2014 May
	NovaBone MacroFORM	Moldable type	Open porous structure to facilitate the absorption of bone marrow aspirate Purified collagen and resorbable bioactive synthetic granules (Bioglass)	n/a	510(k) cleared K0140946, 2014 Aug
	NovaBone porous	Powder	Synthetic calcium-phospho-silicate (Bioglass)	n/a	510(k) cleared K090731, 2009 Apr
ORTHOReBIRTH Co., Ltd.	ReBOSSIS	Cottony type: glass wool-like physical form	A synthetic, resorbable bone void filler 40% beta-tricalcium phosphate (β-TCP), 30% siloxane-containing vaterite (a form of calcium carbonate, CaCO3), and 30% poly(L-lactide-co-glycolide) The electrospinning process used in manufacturing ReBOSSIS results in a cotton like form, which had a merit like easier-to-handle, good elasticity and resilient capability	n/a	K142090 ReBOSSIS ORTHOReBIRTH CO., LTD. 2014 Oct K172573/K170620 ReBOSSIS85, 2017 Dec Primary Predicate K140375 scaffold is type I bovine collagen scaffold
Orthovita Inc. → Merged into Stryker	Vitoss	Various types (original, foam pack, foam strip, Morsel, and block)	Highly porous beta-tricalcium phosphate (>90% porous) + type I bovine collagen (may be combined with saline, autogenous blood, and/or bone marrow)	ClinicalTrials.gov Identifier: NCT00147823: RCT (comparative study)	510(k) cleared STRIP and PACK – K081439, 2008 Nov K032288 – Vitoss Scaffold Foam Bone Graft Material
	Vitoss BA (Bioactive Bone Graft Substitute)	Various types (original, foam pack, foam strip)	Highly porous β-TCP + bioactive glass	n/a	510(k) cleared K083033, 2008 Nov
	Vitoss® Bone Graft Substitute-Bioactive Foam Strip	Strip type	Highly porous β-TCP + bioactive glass	n/a	510(k) cleared K072184, 2007 Sept

(continued)

Table 6 (continued)

Company	Synthetic product	Formulation	Product composition	Peer-reviewed clinical evidence ClinicalTrials.gov/ongoing study	Regulatory clearance/approval FDA 510(k), CFR 1270, CFR 1271 MQV, filler, bone void, calcium compound common name – bone grafting material classification name – bone grafting material, synthetic
	Vitoss BBTrauma	Putty type (foam pack)	Highly porous β-TCP + bioactive glass. A broader range of bioactive glass particle size distribution and has a unique porosity, structure, and chemistry to help drive 3D regeneration of bone	n/a	510(k) cleared
	Vitoss BA2X (Bioactive Bone Graft Substitute)	Putty type (foam pack)	Highly porous β-TCP + bioactive glass. Increased levels of bioactive glass compared to Vitoss BA and has a unique porosity, structure, and chemistry to help drive 3D regeneration of bone	n/a	510(k) cleared K103173, 2011 Feb; K16321, (2017 Mar (BA Injectable)
	HydroSet HydroSet XT	Injectable type	Tetracalcium phosphate that is formulated to convert to hydroxyapatite, the principal mineral component of bone HydroSet XT is simple and easy form of HydroSet	n/a	510(k) cleared K161447, 2016 Oct
Pioneer Surgical Technology, MI USA → Merged into RTI Surgical	Pioneer FortrOss Bone Void Filler	Putty type	Porous calcium phosphate material mixed with a porcine gelatin carrier	n/a	510(k) cleared K091031, 2009 Nov
	Pioneer E-Matrix Bone Void Filler	Granular gelatin-based	Porous calcium phosphate material mixed with a porcine gelatin carrier	n/a	510(k) cleared K083449, 2009 Jun
Progentix Orthobiology BV, The Netherlands	CuriOS™		Micro-structured calcium phosphate resorbable bone void filler for the repair of bony		K090641, 2009 Oct

11 Bone Grafts and Bone Graft Substitutes

Progentix Orthobiology BV, The Netherlands/ NuVasive	AttraX Putty	AttraX Putty cylinders, strips, and blocks	defects. The product comprises a beta-tricalcium phosphate and hydroxyapatite AttraX Putty is a synthetic ceramic granule premixed with a polymeric binder that provides cohesion between the granules Beta-tricalcium phosphate (β-TCP > 90%) and hydroxyapatite (HA <10%) The granule size range: 500–1000 μm The premixed binder is alkylene oxide copolymer (AOC)	NCT02250248, XLIF, Brazil RCT NCT01982045, Netherlands Sponsors: UMC Utrecht NuVasive	K151584, 2015 Jun
Prosidyan Inc., USA	FIBERGRAFT® BG	Morsels	FIBERGRAFT® BG Morsels is an ultraporous synthetic bone graft substitute made entirely from crystalline 45S5 bioactive glass	(Fortier et al. 2017)	Class II (Special Controls) K151154, K141956, K132805, 2017 May (posterolateral fusion)
	FIBERGRAFT® BG Putty Bone Graft Substitute	Putty	FIBERGRAFT™ 45S5 bioactive glass (M-45 granules, MS-45 microspheres) and bovine type I collagen (hydration with saline or blood). For use in posterolateral fusion		K143533, K170306, K180080, 2018 Mar The technological characteristics of the FIBERGRAFT™ BG Putty are similar to FIBERGRAFT™ BG Morsels Bone Graft Substitute
Regeneration Technologies, Inc., FL, USA → Merges with Tutogen to form RTI Biologics® (2008) → Change to RTI Surgical	nanOss® Loaded Advanced Bone Graft Substitute	Prefilled mixing syringe	Nano-structured hydroxyapatite (HA) has extremely high surface area	NCT02586116, ongoing Cervical Spine Belgium Ahn and Webster 2009	K081558 – NanOss Bone Void Filler, 2008 Aug
	nanOss® 3D Advanced Bone Graft Substitute	Strip type	Nano-structured hydroxyapatite granules suspended in a porous gelatin-based foam matrix	ClinicalTrials.gov Identifier: NCT01829997: case series study (active status)	510(k) cleared K132050, 2013 Jun
	nanOss® Advanced Bone Graft Substitute	Putty type	Nano-structured hydroxyapatite granules and an open structured engineered collagen carrier	ClinicalTrials.gov Identifier: NCT01968993: A Prospective, Nonrandomized Study (PLF)	510(k) cleared K141600, 2014 Oct
Science for Biomaterials, France	BIOSORB RESORBABLE BONE VOID FILLER	Variable shape	Resorbable calcium salt bone void filler 9 shapes: stick, granule, cube, block (6 shapes)	n/a	K130953, 2013 Jul K(021963 and K(071155)

(continued)

Table 6 (continued)

Company	Synthetic product	Formulation	Product composition	Peer-reviewed clinical evidence ClinicalTrials.gov/ongoing study	Regulatory clearance/approval FDA 510(k), CFR 1270, CFR 1271 MQV, filler, bone void, calcium compound common name – bone grafting material classification name – bone grafting material, synthetic
SeaSpine, Carlsbad, CA	Accell Connexus®			NCT01873586 1: Schizas et al. 2008	
SeaSpine, Carlsbad, CA	OsteoStrux® Putty		Moldable, osteoconductive scaffold composed of purified collagen and β-TCP 20% type I bovine collagen and 80% highly purified β-TCP The matrix was developed to resemble the composition and pore structure of natural human bone	NCT01873586: Case series study; results are not posted.	K073316 2008 Jun
	OsteoStrux® Strip		Strip is a compression-resistant, osteoconductive scaffold composed of purified collagen and β-TCP	NCT01873586	K073316 2008 Jun
Synergy Biomedical, LLC	BioSphere® Putty	Putty type	80% bioactive glass spheres; 20% phospholipid carrier	n/a	K122868, 2013 Apr
	BioSphere MIS Putty (BioSphere MIS)	Prefilled type	Medical-grade 45S5 bioactive glass particles + carrier (same composition of BioSphere® Putty)	n/a	K173301, 2018 Jan
THERICS, LLC	TheriGraft™ TCP Putty Bone Void Filler	Putty type	Synthetic ß-tricalcium phosphate granules (0.1–0.4 mm diameter) in a poloxamer based carrier Approximately 0.1–0.4 mm granules	n/a	510(k) cleared K053228, 2006 Jan

Vivoxid Ltd.: Turku, Finland	Bioactive Glass (S53P4) as granules (BonAlive™)	Granules Plates	By weight, SiO$_2$ 53%, Na$_2$O 23%, CaO 20%, and P$_2$O$_5$ 4% (synthetic, osteoconductive, and bacterial growth-inhibiting material)	*NCT00935870* Sponsor Turku, Finland NCT00841152 Bonalive (Vivoxid Ltd., Turku, Finland)	K071937 2007 Oct
Wright Medical Technology	OSTEOSET®	Pellet type → Moldable in operation room	Engineered calcium sulfate hemihydrate	n/a	K053642, 2006 Jan
	PRO-DENSE™	Injectable type (4, 10, 15, 20 cc)	75% calcium sulfate and 25% tricalcium phosphate	n/a	K113871, 2013 Mar
Zimmer Biomet/ BONESUPPORT AB, Sweden	CERAMENT™	Injectable type	Biphasic ceramic bone substitute: 60% synthetic calcium sulfate hemihydrate and sintered hydroxyapatite) + 40% hydroxyapatite (HA) pellet + radio-contrast agent iohexol (180 mg/ml)	NCT01828905 ongoing trial	K073316, 2008 Jun
	Pro Osteon® 200R	Powder type	Hydroxyapatite and calcium carbonate diameter of 190–230 microns Calcium carbonate matrix covered by a very thin outer layer of calcium phosphate, approximately 2–10 microns in thickness. The calcium phosphate is located on the outer surface of the porosity throughout the entire structure of the implant	Walsh et al. 2003: A resorbable porous ceramic composite bone graft substitute in a rabbit metaphyseal defect model. Journal of Orthopaedic Research, 2003	510(k) cleared K000515, 2000 Sep
Zimmer Biomet Spine, Inc. (Interpore Cross International, Irvine, CA)	Pro Osteon® 500R (Chapman and Madison 1993)	Powder type	Thin, 2–10 micron layer of hydroxyapatite over a calcium carbonate core Provides a natural scaffold for new bone growth when placed in contact with viable bone	NCT00858598 Thalgott et al. 2001 Harris et al. 1995	510(k) cleared K031336, 2002 Jul

Materials in this class are most always used with BMA, whole blood, serum, and physiological saline and/or also mixed with autograft
BMA, bone marrow aspirate. BMA is typically added at the time of surgery to these synthetic materials for grafting into the spinal fusion surgical bed/site
Whole blood or serum (patient's own) may be used
https://510k.directory/clearances/MQV/1. Accessed April–June 2018

calcium phosphate materials are brittle, as they do not possess the tensile properties of bone. Therefore, TCPs and calcium phosphates have been used in areas of relatively low tensile stress or non-load-bearing applications. Thus, the calcium phosphate-based materials are not recommended alone for use in load-bearing applications (Park et al. 2013). It is important to recognize that most osteoconductive products have been approved for use only in posterolateral spine fusion applications and not in interbody fusion applications. Since TCP has only osteoconductive effects, these TCP-type products may be used in conjunction with biologic osteoinductive or osteogenic supplements of autograft, BMPs, growth factors, mesenchymal stem cell (MSC) derivatives, etc. (see combined products, Table 7) (Gupta et al. 2015; Duarte et al. 2017).

The most widely recognized TCP product is Vitoss® Bone Graft Substitute (Stryker, Allendale, NJ). This material was first commercialized in 2004, and its application in different formats has established it as the preferred TCP material. Another TCP-based material that has been reported is the Augment® Bone Graft from Wright medical. Augment® Bone Graft combines recombinant human platelet-derived growth factor B homodimer (rhPDGF-BB) with a bio-resorbable synthetic bone matrix (β-TCP). This product has been developed for use in bone repair. It is reported that the use of this product eliminates the need for using autograft, proposed as a "substitute." However, Augment® Bone Graft is only indicated for use as an alternative to autograft in the ankle or hindfoot (Augment® bone graft – FDA. https://www.accessdata.fda.gov/cdrh_docs/pdf10/P100006d.pdf). There are several TCP-based products combined with different carriers to provide improved handling characteristics (see combined products, Table 7).

HA is another calcium phosphate material of significance, since x-ray diffraction and chemical studies have demonstrated that the primary mineral phase in bone is HA. HA is a biomaterial for medical devices and is available in the form of nanocrystalline powders, porous granules, and dense blocks. It can be manufactured from natural coral, bovine cortical bone, or synthesized by chemical reactions. HA is stronger (less brittle) than TCP providing high compression strength but is still somewhat brittle. Due to its brittle quality, HA use is limited in load-bearing applications (Zdeblick et al. 1994; Park et al. 2013). Unlike autograft, allograft, and TCP, the absorption rate of HA is very slow (with incomplete absorption/resorption), and HA remains at the site of implantation for years (Zadegan et al. 2017a). In most circumstances, this prolonged resorption may not be advantageous. Grafting materials are ideally completely resorbed and replaced by new bone eventually. If the material does not resorb, it can act as an obstacle or inhibit new bone formation. Historically, coralline HA has been used effectively as a bone graft extender in patients as an adjunct to autologous bone for PLLF (Morris et al. 2018). The critical amount of graft volume per area of functional level (spine) has not been reported. Yoo et al. suggest that an amount of at least 12 mL of bone graft is needed to achieve a satisfactory bone fusion in minimal invasive TLIF surgery regardless of mixture ratio of HA with autograft bone (Yoo et al. 2015). There are several HA-based products combined with different carriers to provide improved handling characteristics (Tables 6 and 7).

According to study of Nickoli MS et al., ceramic-based bone grafts (TCP) with an osteoinductive stimulus represent a promising bone graft extender in lumbar spine fusion (Nickoli and Hsu 2014). In a meta-analysis review of 1,332 patients in 30 studies, from 1980 to 2013, ceramics used in combination with local autograft resulted in significantly higher fusion rates compared with all other adjuncts and bone marrow aspirate and platelet concentrates (Nickoli and Hsu 2014). Previous clinical studies on HA-based bone graft such as HA when used alone, or in combination with BAG (bioactive glass), BMA (bone marrow aspirate), or rhBMP-2 have been shown to improve function to the and reduce preoperative pain same extent as ICBG, yet have been associated with suboptimal radiographic fusion rates in lumbar spine (Singh et al. 2006; Acharya et al. 2008; Ploumis et al. 2010).

Table 7 Commercially available combination grafting products, naturally occurring peptides, growth differentiating factors, cellularized grafts, cellular bone matrices (CBMs)

Company	Combination product	Formulation	Product composition	Peer-reviewed clinical evidence/ongoing study	Regulatory clearance/approval FDA 510(k), FDA 361, 21 CFR Part 1271 CFR 1270, CFR 1271 21 CFR 3.2(e) HCT/P 361, Human allografts (No Clinical Studies) Biologic Drugs and Devices 351 (Clinical Trials)
Advanced Biologics, Carlsbad, CA., 2009 (marketed OsteoAMP in the USA since 2009)	OsteoAMP	Granules or sponge	OsteoAMP, an allogeneic growth factor implant, exploits the angiogenic, mitogenic, and osteoinductive growth factors that are within marrow cells	Field et al. 2014 Cervical Spine-Fusion	Bioventus manages orders and sales of HCT/Ps (not a distributor, FDA)
Bioventus' Surgical, Durham, NC, USA (original developer)			Growth factor-rich naturally occurring growth factors including BMP-2, BMP-7, aFGF, and TGF-β1 bone graft substitute: intended for homologous use repair, replacement, or reconstruction of musculoskeletal defects	ClinicalTrials.gov Identifier: NCT02225444 Lumbar Spine-PLF Roh et al. 2013	Regulated under CFR 1270, 1271 as a human tissue, registration held by Tissue Bank Permit: Millstone Medical Outsourcing, LLC, Olive Branch, MS (Bone, Demineralized Bone Matrix, Ligament, Musculoskeletal Tissues, Tendon.) Maryland, New York State Tissue Bank Permit: Advanced Biologics, LLC (Bone Demineralized Bone Matrix)
AlloSource®, Centennial, Co, USA, 1995 Allosource.org	AlloStem Cellular Bone Autograft	Strips, blocks, cubes, morselized	Partially demineralized allograft bone combined with adipose-derived mesenchymal stem cells (MSC)	n/a	Regulated under CFR 1270, 1271 as a human tissue

(continued)

Table 7 (continued)

Company	Combination product	Formulation	Product composition	Peer-reviewed clinical evidence/ongoing study	Regulatory clearance/approval
Aziyo Biologics, Inc. MD, Ga, CA	OsteoGro™	Cancellous Bone	Partially demineralized cortical bone	NCT03425682	FDA 510(k), FDA 361, 21 CFR Part 1271 CFR 1270, CFR 1271 21 CFR 3.2(e) HCT/P 361, Human allografts (No Clinical Studies) Biologic Drugs and Devices 351 (Clinical Trials)
	VBone	Structural allografts Package Bone matrix	Preserve natural components of the matrix Viable bone matrix		Regulated under CFR 1270, 1271 as a human tissue
BBS-Bioactive Bone Substitutes Oyj (Finland)	ARTEBONE®		Tricalcium phosphate (TCP) + natural cocktail of bone proteins (growth factors)	ClinicalTrials.gov Identifier: NCT02480868: case series study for ankle fusion	FDA 510(k) ~2020
Bioventus® Surgical	OsteoAMP	Granule, putty, and sponge form	Cervical and lumbar spine fusion procedures Allograft with growth factors (such as BMP-2, BMP-7, TGF-β1, aFGF, VEGF, and ANG1, within bone marrow cells)	Active but no results posted: Posterolateral Lumbar Fusions (PLF) With OsteoAMP® (ClinicalTrials.gov Identifier: NCT02225444) Comparative study with rhBMP-2 with OsteoAMP (Roh et al. 2013)	AATB US FDA 21 CFR 1271 (HCT/P)
BONESUPPORT AB, Scheelevägen 19 SE-223 70, Lund, Sweden	CERAMENT® G	Injectable type	Injectable antibiotic-eluting bone graft substitute that provides local sustained bactericidal effect and scaffold for fusion CERAMENT (40% hydroxyapatite +60% calcium sulfate) + 17.5 mg gentamicin/mL paste	ClinicalTrials.gov Identifier: NCT02820363: Clinical Trial (RCT) for open tibial fracture (recruiting status) ClinicalTrials.gov Identifier: NCT02128256: Case series study (unknown status)	Combination product, HCT/P GS1, HIBCC, ICCBBA FDA-PMA approval underway (communication from BoneSupport 2020 Feb

	CERAMENT® V	Injectable type	Injectable antibiotic-eluting bone graft substitute that provides local sustained bactericidal effect and scaffold for fusion CERAMENT (40% hydroxyapatite +60% calcium sulfate) + iohexol (as a radio-opacity enhancer) + 66 mg vancomycin/mL paste	US-NCT03389646 trial of Cerament TM V, G for hip or knee prosthesis infection	CE Mark approval FDA-PMA approval underway (communication from BoneSupport 2020 Feb
DePuy Synthes	ViviGen Cellular Bone Matrix Vertigraft®	Cryo Cortical Cortical cancellous bone matrix and demineralized bone	ViviGen® Cellular Bone Matrix comprised cryopreserved viable cortical cancellous bone matrix and demineralized bone. ViviGen Cellular Bone Matrix is a human cells, tissues, and cellular and tissue-based product (HCT/P). ViviGen Cellular Bone Matrix is processed from donated human tissue, resulting from the generous gift of an individual or his/her family	NCT02814825 HCT/P (Divi and Mikhael 2017)	(HCT/P) as defined by the US Food and Drug Administration in 21 CFR 1271.3(d). 21CFR 1271
	CONFORM CUBE®	Cube shape	Demineralized cancellous bone, organic matrix (osteoinductive, promotes cellular ingrowth and vascularization) General bone void filler and use with lumens of allograft spinal spacers	n/a	(HCT/P) as defined by the US Food and Drug Administration in 21 CFR 1271.3(d). 21CFR 1271
	CONFORM SHEET®	Sheet shape	Demineralized cancellous bone, organic matrix (osteoinductive, promotes cellular ingrowth and vascularization) For PLF (posterolateral gutters of the spine)	n/a	(HCT/P) as defined by the US Food and Drug Administration in 21 CFR 1271.3(d). 21CFR 1271

(continued)

Table 7 (continued)

Company	Combination product	Formulation	Product composition	Peer-reviewed clinical evidence/ongoing study	Regulatory clearance/approval
Mesoblast Ltd., Australia	NeoFuse(TM)	Cells + Granules	Allogenic mesenchymal precursor cells (MPCs) combined w/ MasterGraft in PEEK cage	NCT00549913 (Lumbar PLF)	FDA 510(k), FDA 361, 21 CFR Part 1271 CFR 1270, CFR 1271 21 CFR 3.2(e) HCT/P 361, Human allografts (No Clinical Studies) Biologic Drugs and Devices 351 (Clinical Trials) FDA 510(k) cleared, K153615 2016 Jun (HA Enhanced PLIF/TLIF)
Angioblast Systems Inc., USA	NeoFuse(TM)	Cells + Granules	Allogenic mesenchymal precursor cells (MPCs) combined w/ MasterGraft in anterior cervical discectomy and fusion (ACDF) Anterior cervical plate fixation	NCT01106417 (Cervical fusion)	FDA 510(k) cleared, K170318 2017 Jun
MTF Orthofix	Trinity Evolution TM	Moldable allograft fibers, varying sizes	Allogenic DBM, osteoprogenitor cells (OPC), MSC (minimum of 500,000 cells/cc; 100,000 of which are MSC and/or OPC)	NCT00951938 (Anterior cervical) Peppers et al. 2017 Vanichkachorn et al. 2016	Regulated under CFR 1270, 1271 as a human tissue
	Trinity Elite	Moldable allograft fibers, varying sizes	DBM, osteoprogenitor cells, MSC (minimum of 500,000 cells/cc; 100,000 of which are MSC and/or OPC) Trinity Elite and/or local bone	NCT02969616 (PLF, TLIF,ALIF, XLIF, etc., lumbar fusion) NCT00965380	Regulated under CRF 1270, 1271 as a human tissue

11 Bone Grafts and Bone Graft Substitutes

NuVasive			with supplemental pedicle screw fixation allogeneic cancellous bone matrix containing viable osteoprogenitor cells, mesenchymal stem cells, and a demineralized cortical bone (DCB)	
	Osteocel	Moldable bone matrix	DBM, OPC, MSC (<50,000 cells/cc, >70% viability)	Retrospective case series	Regulated under CFR 1270, 1271 as a human tissue
	Osteocel Plus	Moldable bone matrix	DBM, OPC, MSC (<50,000 cells/cc, >70% viability)	McAnany et al. 2016, Retrospective comparative study: NCT00948532 (Osteocel® Plus in extreme lateral interbody fusion (XLIF®). Kerr et al. 2011; Tohmeh et al. 2012 Extreme lateral interbody fusion (XLIF) Ammerman et al. 2013 Eastlack et al. 2014: (Osteocel Plus in a polyetheretherketone cage and anterior plating at 1 or 2 consecutive levels) Prospective case series Retrospective case series, clinical trial: ClinicalTrials.gov Identifier: Evaluation of Radiographic and Patient Outcomes Hollawell 2012	Regulated under CFR 1270, 1271 as a human tissue
Organogenesis 2017 Mar/NuTech Medical, Inc.	NuCel®	Putty type	Cryopreserved, bioactive amniotic suspension allograft Cellular, growth factor, and	ClinicalTrials.gov Identifier: NCT02023372: A Prospective, Efficacy Study (RCT) for PLF	unk

(continued)

Table 7 (continued)

Company	Combination product	Formulation	Product composition	Peer-reviewed clinical evidence/ongoing study	Regulatory clearance/approval
			extracellular matrix components	NCT02070484: NuCel vs. DBM	FDA 510(k), FDA 361, 21 CFR Part 1271 CFR 1270, CFR 1271 21 CFR 3.2(e) HCT/P 361, Human allografts (No Clinical Studies) Biologic Drugs and Devices 351 (Clinical Trials)
Osteotech's → Merged into Medtronic	Plexur M(TM)	Moldable type (putty like)	Human allograft bone tissue + resorbable polymer Processed human bone particles that are mixed with resorbable/ biodegradable non-tissue components	MAUDE Adverse Event Report	FDA 510(k) cleared K073405 2008 Mar
RTI Surgical Inc. Allendale, NJ, USA	map3® Cellular Allogeneic Bone Graft	Putty type Strip type	Cortical cancellous bone chip (or strip shape bone) + DBM + cryogenically preserved, viable multipotent adult progenitor (MAPC®)-class cells	ClinicalTrials.gov Identifier: NCT02161016: case series study in foot and ankle. Results posted ClinicalTrials.gov Identifier: NCT02628210: A Prospective, Multi-Center, Non-Randomized Study for lumbar interbody fusion (active status)	Unk status (Regulated under 361 PHS Act 42 U.S.C. 264 and reg. 21CFR Part 1271.1O(s)(4)(ii) (b) + FDA Act {21 U.S. C. 321 (g)++)

Stryker	BIO (Hollawell 2012)	Putty type (1, 2.5, 5, 10 cc)	Allograft bone (cortical and cancellous) + periosteum A viable bone matrix containing endogenous bone forming cells (including mesenchymal stem cells, osteoprogenitor cells, and osteoblasts) as well as osteoinductive and angiogenic growth factors	ClinicalTrials.gov Identifier: NCT03077204: Clinical case series study (cervical spine), recruiting status	AATB US FDA regulations for tissue management. US FDA 21 CFR 1271
Vericel Corporation		Bone repair cells	Bone repair cells (BRCs) with allogeneic, demineralized bone matrix	NCT00797550 (posterolateral spinal fusion) terminated no results NCT00424567 repair pseudarthrosis atrophic nonunion	Biologics License Application (BLA) w/ post-marketing commitments (~2017 June), status unknown
Xtant	OsteoVive™	Putty type	A cell population that includes marrow-isolated adult multilineage-inducible (MIAMI) cells Blend of microparticulate cortical, cancellous, and demineralized cortical allograft bone (particle size range of 100–300 microns)	n/a	FDA 510(k) cleared Compliance with FDA guidelines regarding human cells, tissues, and cellular tissue-based products HCT/P 361 regulated viable allogeneic bone scaffold American Association of Tissue Banks guidelines

510(K) is a premarket submission made to FDA to demonstrate that the device to be marketed is at least as safe and effective, that is, substantially equivalent, to a legally marketed device that is not subject to premarket approval. 501(k) documentation for individual products is available via FDA online database (http://www.accessdata.fda.gov) Code of Federal Regulations (CFR) 1270 (Human tissue intended for transplantation) and 1271 (Human cells, tissues, and tissue-based products) are federal regulations relating to the procurement and processing of human-derived tissues

Claims: grafting with component to provide the required osteoconduction, osteogenesis, and osteoinduction necessary for successful bone grafting

GSI: it is an international, not-for-profit association that creates and implements standards to bring efficiency and visibility to supply chains across industries

HIBCC Health Industry Business Communications Council

ICCBBA International Council for Commonality in Blood Banking Automation

PEEK, a polyetheretherketone material used for cage devices employed as instrumentation in anterior interbody spinal fusion procedures

Silicate-Substituted Calcium Phosphate

Silicate-substituted calcium phosphate (Si-CaP) constitutes a newer generation of ceramics produced by adding silicate which has been found to play role in bone metabolism to previously developed calcium phosphate ceramics (Gao et al. 2001). This combination provides superior biocompatibility and osteoconductivity. In addition combining Si-CaP with a graft provides negative surface charge that results in enhanced osteoblast activity and neovascularization of the bone which lead to more ideal spine fusion as a substitute of ICBG (Campion et al. 2011; Alimi et al. 2017).

Silicated hydroxyapatite has been prepared by the addition of a small amount of silicon (0.4% to 0.8% by wt.) into the structure of HA. The role of silicate-based materials in improving tissue implant interactions has been reported (Zhou et al. 2017). Silica-substituted HA, such as Actifuse™ from Baxter, is available in the form of granules, pastes, and blocks. The performance of these products has been investigated in preclinical models and clinical study. According to study of Jenis and Banco (2010), a silica-substituted hydroxyapatite (Actifuse™) with BMA has been shown to be effective as a graft substitute as ICBG with significant pain improvement in PLLF. According to study of Licina P et al. (Licina et al. 2015), silicate-substituted calcium phosphate (Actifuse™) and rhBMP-2 with ceramic granule were comparable in view of achieving PLLF.

Clinical data are limited for various types of lumbar surgery and the numbers of enrolled patients in trials. For confirming the efficacy and safety of Si-CaP and/or silicated hydroxyapatite as a bone-grafting substitute, further investigations using greater numbers of subjects will be necessary. And the radio-opaque nature of Si-CaP allows for intra- and post-operative localization, but this radio-dense characteristic immediately after surgery resembling bone and the long residence time exceeding a year has decreased the accurate assessment of the process of bone formation.

Bioactive Glass (Table 8)

Bioactive glass (BAG) is a class of glass-based graft substitute or extender products having a compositional range that allows the formation of nanocrystalline hydroxyapatite (ncHA) as a surface layer when exposed to an aqueous phosphate-containing solution, such as simulated body fluid. The ncHA layer that forms within an aqueous phosphate-containing solution plays a significant role in forming a strong bond with natural bone.

BAG has an established history of bone bonding that occurs as a result of a rapid sequence of reactions on its surface when implanted into living tissues (Hench and Jones 2015). There are two mechanisms of bioactivity for bioactive glass products. Bone bonding is attributed to the (1) formation of an HA layer, which interacts with collagen fibrils of damaged bone to form a bond (Hench and Jones 2015), while the action of the (2) dissolution products from the bioactive glass is reported to simulate osteogenesis (Hench and Polak 2002). When hydrated, a layer of silica gel forms on the surface of the bioactive glass. The adhesion of amorphous calcium, phosphate, and carbonate ions to the silica surface leads to an eventual crystallization of a bone-like HA as early as 24 hours. Bone-forming cells migrate and colonize the surface of the bioactive glass and promote the production of a new bone-like matrix (Beckham et al. 1971). Gao et al. (2001) observed increased expressed detectable mRNA levels of BMP-2 from Saos-2 osteoblastic cells when cultured on two types of BAG (BAG containing 6% Na_2O, 12% K_2O, 20% CaO, 4% P_2O_5, 5% MgO and *53% SiO_2* and biocompatible glass (BCG) containing 6% Na_2O, 12% K_2O, 15% CaO, 4% P_2O_5, 5% MgO and *58% SiO_2* (wt.%)) than on control inert glass (Gao et al. 2001).

The mechanism for the formation of the ncHA layer is now quite well understood and well characterized, but the biological interactions at the ncHA–host bone interface are still under intense investigation in view of potential employment with stem cells (Tsigkou et al. 2014).

In addition, the high pH and the subsequent osmotic effect caused by dissolution of the bioactive glass have been suggested as an antibacterial material quality (Stoor et al. 1998; Allan et al. 2001). Recently, Sanchez-Salcedo et al. (2017) introduce the design and synthesis of a new

Table 8 Composition and properties of bioactive glasses and glass-ceramics used clinically for ontological, musculoskeletal, and dental grafting applications (Baino et al. 2018; Hench and Jones 2015)

Product	Composition wt %								
	Na_2O	CaO	CaF_2	MgO	P_2O_5	SiO_2	B_2O_3	K_2O	CuO	ZnO
45S5 Bioglass Otology: MEP® a, Douek-MED™, Ceravital® a, Bioglass-EPI® a Dental graft: EMRI® a, Biogran®, PerioGlas®, NovaMin® Orthopedics: NovaBone®, GlassBone™, FIBERGRAFT®, BioSphere® Putty	24.5	24.5	0	0	6	45	0	0	0	0
S53P4 Dental graft: AdminDent1 Orthopedics: BonAlive®	23.0	20.0	0	0	4	53	0	0	0	0
A-W glass-ceramic Dental graft: Cerabone®	0	44.7	0.5	4.6	16.2	34	0	0	0	0
Strontium substituted bioactive glasses: StronBone®	4	17.8	0	7.5	4.5	44.5	0	0	0	0
13-93	6	20	0	5	4	53	0	0	0	0
Bioactive glass by the sol-gel process TheraGlass® a	0	30	0	0	0	70	0	0	0	0
Boron bioactive	6	20	0	5	4	0	51.6	12	0.4	1

[a]This product is not commercially available due to side effects, structural problems, lack of clinical effect, etc.

nano-structured zwitterionic mesoporous bioactive glasses (MBGs) with incorporation with amino acid for antibio-fouling capability that inhibits bacterial adhesion (formation of biofilm) wherefrom they report successful results in vitro.

BAG has been used for a variety of clinical applications since it was first created in 1969 (Hench and Jones 2015). There are many types of BAG (Table 6) and glass-based products used (Hench and Jones 2015) in periodontal repair and orthopaedic applications (Table 8).

The originally developed composition was bioactive glass 45S5 (Food and Drug Administration (FDA) approved in 1993 (Jones 2015). 45S5 bioactive glass consists of 45 wt.% SiO_2, 24.5 wt.% CaO, 24.5 wt.% Na_2O, and 6.0 wt.% P_2O_5 which demonstrated effective biological properties. NovaBone®, a product based on this 45S5 technology, has been approved as a bone graft substitute in 1999 (Jones 2013; Hench and Jones 2015). The NovaBone® material is considered an early generation of bioactive glass. This is due to the lack of inherent porosity of the NovaBone® granules or granules in which porosity has been manufactured by the fusion of smaller granules. NovaBone® was compared to autograft in posterior spinal fusion procedures for treatment of adolescent idiopathic scoliosis in 88 patients (Ilharreborde et al. 2008). NovaBone® showed improved clinical results in terms of reduced infection, donor site complication, and fewer mechanical failures in a 4-year follow-up. However, its clinical use for spine fusion applications has not been reported widely.

A commercially available bioactive glass product is BonAlive® (BonAlive Biomaterials, Turku, Finland), which was programmed in Finland based on S53P4 bioactive glass. BonAlive® received European approval for orthopedic use as a bone graft substitute in 2006 (Jones 2015). The S53P4 bioactive glass contains 53 wt.% SiO2, 23 wt.% Na_2O, 20 wt.% CaO, and 4 wt.% P_2O_5. According to Frantzen et al. (2011) a prospective long-term study (11 years) of Frantzen et al., the fusion rate of all fusion sites for BAG-S53P4 with autograft as a bone substitute was 88% at the L4/L5 level and 88% at the L5/S1 level compared to 100% for autograft in degenerative spondylolisthesis patients. Similar results were seen after surgical treatment of a spondylitis patient (Lindfors et al. 2010). BonAlive® was also compared to autograft in the same patients in PLF procedures for treatment of spine burst fractures. At the 10-year follow-up, 5 out of 10 implants had full fusion compared to all 10 autografts (Rantakokko et al. 2012).

Fibergraft® BG Morsels (Prosidyan Inc., USA) is a 100% BAG material (no additives) specifically FDA cleared for orthopedic and spine grafting applications. Traditional bioactive glass does not allow for ease of handling and has slow resorption due to low porosity. Fibergraft® BG Morsels is the first osteostimulative (or bioactive) material engineered to take advantage of the unique properties of bioactive glass. The morsels are engineered with overlapping and interlocking bioactive glass fibers with pores dispersed throughout. The material structure and ultra-porous, nano-, micro-, and macro-porosity provides direct connectivity for cell in-growth and material resorption, enabling new bone formation.

A 95% radiographic success rate was reported in a retrospective study of Fibergraft® BG Morsels use when mixed with local autograft and bone marrow aspirate in 63 patients at 1 year after 1-, 2-, and 3-level posterolateral fusions (Barcohana et al. 2017). Additionally, a high rate of 88.5% (46/52 levels with complete fusion) together with a 5.8% (3/52, levels partial fusion) in anterior cervical fusion was demonstrated after use of Fibergraft® BG Morsels mixed with BMA, bone dust, and or local bone in 27 patients (51 levels of fusion) at approximately 6 months after anterior cervical discectomy and fusion study (Fortier et al. 2017).

Fibergraft® BG Morsels (Prosidyan Inc., USA) is also provided in a putty form as Fibergraft® BG Putty and in a Matrix form as Fibergraft® BG Matrix. All Fibergraft® products are specifically FDA cleared for orthopedic and spine grafting applications. The BG Putty can be used for Minimally Invasive Surgery (MIS) applications, while the BG Matrix can be combined with bone marrow aspirate and used as a compression-resistant strip that can be molded to the shape of the defect.

Clinical and in vivo studies on commercially available bioactive glass particulates show that BAG can perform better than other bio-ceramic particles and have performed similarly to autograft in multiple in vivo studies (Walsh et al. 2017; Bedi 2017).

Unmet Challenges for Engineered Bioactive Glass Matrices

The major scientific and technical challenges exist with previously developed bioactive glass. Glass based materials lack osteogenesis, are difficult in clinical handling, not load bearing due to brittleness, and have slow resorption due to low porosity (Hench and Jones 2015; Jones 2015). To overcome these limitations and use BAG as effective substitute for autograft, several experiments were attempted to combat these limitations.

First, to enhance osteogenesis, tissue regeneration through gene activation by controlled release of inorganic ions from BAG is required. However, the role of the dissolution products from implanted BAG on bone marrow-derived mesenchymal stem cells (MSC) is not yet controllable. In some studies dissolution products induced osteogenic differentiation into osteoblast-like cells, and in others, it did not (Reilly et al. 2007; Karpov et al. 2008; Brauer et al. 2010). To control this problem, the fundamental mechanisms involved in ionic stimulation in the stem cell nucleus and the exact mechanism of "how the bioactive glass particles/dissolution products" should be explained (Hench and Jones 2015).

Second, particles and putties containing a variety of BAG particulates are in widespread clinical use, but large interconnected macroporous scaffolds for regeneration of large bone defects were not developed. To overcome and address this, the bottom-up sol–gel process, where gelation of nanoparticles in a sol (polycondensation) forms a glass network by avoiding sintering of crystalized Bioglass 45S5, was initially developed (Li et al. 1991). After, a room temperature gelation process was employed, allowing pores interconnection with a compression strength equivalent to porous bone (Jones et al. 2006). Melt-derived glass scaffolds were introduced to make macroporous scaffolds (Wu et al. 2011). According to a review by

Hench and Jones (2015), none of described techniques are being further developed for use by medical device companies even though sol–gel and melt-derived scaffolds still exist.

Third, tissue-engineered constructs for replacement of large bone defects have been investigated for many years but are still not available as routine clinical products. To achieve this, a stable vasculature is necessary during initial grafting. Tsigkou et al. (2010) demonstrated that it is possible in mice models (Tsigkou et al. 2010). More research is needed to test the possible enhancement of angiogenesis optimal activity duration in humans (Azevedo et al. 2015).

Fourth, load-bearing devices that can be used in orthopedics over the long term, which also regenerate living bone, are still not available clinically. Therefore, the 3D printing technology was adapted to bioactive glass scaffolds to generate interconnected pores similar in diameter to the porous foam scaffolds but with higher compressive strengths (Fu et al. 2011; Kolan et al. 2011). However, BAG scaffolds are still brittle and therefore not suitable for all grafting applications, such as sites that are under cyclic loads.

Mixed Use Graft Materials with Antibacterial Effects (Table 7)

Infection Prevention and Treatment of Previous Surgical Site Infection

For improvement of bone graft materials including substitutes, dual-functional graft materials have been designed. Among several possible additional options, prevention or treatment of surgical site infection with/without bone destruction is needed for clinical application (Turner et al. 2005; Anderson et al. 2014). Risk factors associated with surgical conditions (relatively wide soft tissue dissection, muscular damage, long operation time, and limited control of bleeding during operation) and patient characteristics and health status (old age, comorbidities like diabetes mellitus, renal failure and vasculopathy, and smoking, etc.) in spine fusion operations.

For prevention or control of the post-operative infection, systemic and localized bactericide are necessary. However systemic delivery of antibiotics to infected site or vulnerable to infection is limited by abnormal blood supply in operated site, drug toxicity to organs, antimicrobial-resistant form of bacteria, etc. (Shiels et al. 2017). Due to mentioned causes, newly designed graft materials have been developed for local bactericidal carrier, which may increase the safety and satisfaction after treatment (Lentino 2003; Radcliff et al. 2015).

A variety of materials including calcium-based substitutes, synthetic polymers, DBM, and protein-based materials have been proposed as alternative delivery vehicles with bone fusion function (McLaren 2004; Nelson 2004). Because the most common pathogen responsible for spinal infections after surgery is the gram-positive bacteria *Staphylococcus aureus*, the antibiotic candidates for biomaterials for infection-targeted delivery (or prevention) may be limited to vancomycin, aminoglycoside series like tobramycin, gentamicin, amikacin, and quinolone series like ciprofloxacin (Turner et al. 2005; Logoluso et al. 2016; Shiels et al. 2017; Boles et al. 2018; Wells et al. 2018).

Several animal studies have shown that calcium sulfate pellets are substantially resorbed and replaced with new bone formation by 6 weeks and a similar rate of pellet resorption has been reported clinically (Turner et al. 2001; McKee et al. 2002). According to study by Shiels SM.et al., vancomycin continued to be released from the DBM over the course of 6 days while maintaining sufficient eluate concentrations to maintain a zone of inhibition similar or larger than a vancomycin control in spine fusion in rabbit (Shiels et al. 2017).

There are several obstacles to overcome in order to use this newly designed bone graft material in clinical spine fusion. First, the ideal shape, desired materials of bone graft, and release concentrations are not established. McLaren et al. questioned the effect of laboratory sampling methods on characterizing the elution of tobramycin from calcium sulfate and the reliability of in vitro elution data in predicting the in vivo release of antibiotics (McLaren et al. 2002). Second, local site effects by eluted antibiotics are of concern. Since neither the optimal level of antibiotic nor the duration of its release has been established, the effect of high local levels of antibiotics on the ability of grafted material to enhance bone healing is largely unknown. In a rabbit study, the use of vancomycin-loaded DBM showed a decrease in the fusion rate compared to DBM when used in a sterile wound (Shiels et al. 2017). Furthermore, an in vitro study suggests that vancomycin has toxic effects on hMSCs, a cell population particularly important for bone formation (Chu et al. 2017). Finally, clinical studies on the use of antibiotic-impregnated graft materials for spine fusion in humans are few. Pilot studies focused on the use of antibiotic-impregnated graft material in total joint arthroplasty and osteomyelitis (Logoluso et al. 2016) (Table 7).

Conclusion

A wide variety of bone graft materials are used in spinal surgery applications. Increasingly, over the past decade, diverse materials and composites are being developed as grafting options for use in spinal surgery. Consideration of the ideal properties of a grafting material and the material's mechanism of action, structural and handling characteristics, FDA classification and related approval or registration, and available clinical and preclinical data will optimize appropriate grafting choice for a certain surgical application for spinal fusion. Moreover, bone grafts do not fuse immediately; instead, they provide a foundation or scaffold for the patient's body to grow new bone in anatomical sites wherein bone did not previously exist such as in a spinal fusion site.

The development of products used for bone regeneration has followed the basic criteria of providing a biocompatible three-dimensional scaffold with controlled architecture capable of stimulating or supporting bone growth in the natural in vivo environment. The ability of the material to be used in conjunction with other cellular and signal (growth factors)-based therapies is a key strategy in maximizing the efficacy and likely success of fusion. However, while many bone graft substitutes perform well as bone graft extenders, only autogenous bone grafts are osteogenic and BMPs are osteoinductive.

Variations in anatomical location, surgical application (meticulous surgical preparation including adequate decortication), instrumentation type, and the patient's risk factors (metabolic and nutritional status, vitamin D, diabetes, smoking, drug and alcohol abuse) are critically important factors to consider in choosing an ideal grafting agent or bone graft to achieve a successful biologic bone union.

Acknowledgments Authors thank Samantha Thordarson, BS, for her editorial comments.

References

AAoT Banks (2018) Accreditation policies. https://images.magnetmail.net/images/clients/AATB/attach/Bulletin_Links/18_2/AATB_Accreditation_Policies_February_08_2018.pdf

Acharya NK, Kumar RJ, Varma HK, Menon VK (2008) Hydroxyapatite-bioactive glass ceramic composite as stand-alone graft substitute for posterolateral fusion of lumbar spine: a prospective, matched, and controlled study. J Spinal Disord Tech 21(2):106–111

Adams MA (2013) The biomechanics of back pain. Churchill Livingstone Elsevier, Edinburgh

Agrawal V, Sinha M (2017) A review on carrier systems for bone morphogenetic protein-2. J Biomed Mater Res B Appl Biomater 105(4):904–925

Ahn E, Webster T (2009) Enhanced osteoblast & osteoclast function on nanOss a calcium phosphate nanotechnology. nanOss® Bioactive Overview, 2016. http://www.lifehealthcare.com.au/wp-content/uploads/2017/06/nanOss_Bioactive_OUS_System_Overview.pdf

Ajiboye RM, Alas H, Mosich GM, Sharma A, Pourtaheri S (2017) Radiographic and clinical outcomes of anterior and transforaminal lumbar interbody fusions: a systematic review and meta-analysis of comparative studies. Clin Spine Surg 31:E230–E238

Albee FH (2007) Transplantation of a portion of the tibia into the spine for Pott's disease: a preliminary report 1911. Clin Orthop Relat Res 460:14–16

Alimi M, Navarro-Ramirez R, Parikh K, Njoku I, Hofstetter CP, Tsiouris AJ, Hartl R (2017) Radiographic and clinical outcome of silicate-substituted calcium phosphate (Si-CaP) ceramic bone graft in spinal fusion procedures. Clin Spine Surg 30(6):E845–E852

Allan I, Newman H, Wilson M (2001) Antibacterial activity of particulate bioglass against supra- and subgingival bacteria. Biomaterials 22(12):1683–1687

Ammerman JM, Libricz J, Ammerman MD (2013) The role of Osteocel Plus as a fusion substrate in minimally invasive instrumented transforaminal lumbar interbody fusion. Clin Neurol Neurosurg 115(7):991–994

Anderson DJ, Podgorny K, Berrios-Torres SI, Bratzler DW, Dellinger EP, Greene L, Nyquist AC, Saiman L, Yokoe DS, Maragakis LL, Kaye KS (2014) Strategies to prevent surgical site infections in acute care hospitals: 2014 update. Infect Control Hosp Epidemiol 35 (Suppl 2):S66–S88

Arnold PM, Anderson KK, Selim A, Dryer RF, Kenneth Burkus J (2016a) Heterotopic ossification following single-level anterior cervical discectomy and fusion: results from the prospective, multicenter, historically controlled trial comparing allograft to an optimized dose of rhBMP-2. J Neurosurg Spine 25(3):292–302

Arnold PM, Sasso RC, Janssen ME, Fehlings MG, Smucker JD, Vaccaro AR, Heary RF, Patel AI, Goulet B, Kalfas IH, Kopjar B (2016b) Efficacy of i-Factor bone graft versus autograft in anterior cervical discectomy and fusion: results of the prospective, randomized, single-blinded Food and Drug Administration Investigational Device Exemption Study. Spine (Phila Pa 1976) 41(13):1075–1083

Arnold PM, Sasso RC, Janssen ME, et al. (2018) i-Factor™ Bone Graft vs Autograft in Anterior Cervical Discectomy and Fusion: 2-Year Follow-up of the Randomized Single-Blinded Food and Drug Administration Investigational Device Exemption Study. Neurosurgery 83(3):377–384. https://doi.org/10.1093/neuros/nyx432

Aurégan JC, Bégué T (2015) Bioactive glass for long bone infection: a systematic review. Injury 46:S3–S7

Axelsen MG, Overgaard S, Jespersen SM, Ding M (2019) Comparison of synthetic bone graft ABM/P-15 and allograft on uninstrumented posterior lumbar spine fusion in sheep. J Orthop Surg 14(1):2. Res

Azevedo MM, Tsigkou O, Nair R, Jones JR, Jell G, Stevens MM (2015) Hypoxia inducible factor-stabilizing bioactive glasses for directing mesenchymal stem cell behavior. Tissue Eng Part A 21(1–2):382–389

Bae HW, Zhao L, Kanim LE, Wong P, Delamarter RB, Dawson EG (2006) Intervariability and intravariability of bone morphogenetic proteins in commercially available demineralized bone matrix products. Spine (Phila Pa 1976) 31(12):1299–1306; discussion 1307–1308

Bae H, Zhao L, Zhu D, Kanim LE, Wang JC, Delamarter RB (2010) Variability across ten production lots of a single demineralized bone matrix product. J Bone Joint Surg Am 92(2):427–435

Baino F, Hamzehlou S, Kargozar S (2018) Bioactive glasses: where are we and where are we going? J Funct Biomater 9(1):25

Baker JF, Errico TJ, Kim Y, Razi A (2017) Degenerative spondylolisthesis: contemporary review of the role of interbody fusion. Eur J Orthop Surg Traumatol 27 (2):169–180

Barcohana B, Gravori TT, Kasimian S, Feldman L (2017) Use of FIBERGRAFT® BG Morsels mixed with local bone autograft and BMA in posterolateral lumbar spine fusion: a retrospective analysis of fusion results. Podium presentation, SMIS 2017 annual meeting, Las Vegas, 14–16 September 2017

Beckham CA, Greenlee TK Jr, Crebo AR (1971) Bone formation at a ceramic implant interface. Calcif Tissue Res 8(2):165–171

Bedi H (2017) Use of FIBERGRAFT BG morsels mixed with BMA in anterior cervical discectomy and fusion at 1, 2, 3 and 4 levels: A retrospective analysis of fusion results. Podium presentation, the 14th annual cabo meeting: state of spine surgery think tank, Los Cabos, 15–17 June 2017

Birch N, D'Souza WL (2009) Macroscopic and histologic analyses of de novo bone in the posterior spine at time of spinal implant removal. J Spinal Disord Tech 22 (6):434–438

Boden SD (2002) Overview of the biology of lumbar spine fusion and principles for selecting a bone graft substitute. Spine (Phila Pa 1976) 27(16 Suppl 1):S26–S31

Boles LR, Awais R, Beenken KE, Smeltzer MS, Haggard WO, Jessica AJ (2018) Local delivery of amikacin and vancomycin from chitosan sponges prevent polymicrobial implant-associated biofilm. Mil Med 183 (Suppl 1):459–465

Brauer DS, Karpukhina N, O'Donnell MD, Law RV, Hill RG (2010) Fluoride-containing bioactive glasses: effect of glass design and structure on degradation, pH and apatite formation in simulated body fluid. Acta Biomater 6(8):3275–3282

Burke JF, Dhall SS (2017) Bone morphogenic protein use in spinal surgery. Neurosurg Clin N Am 28(3):331–334

Burkus JK, Transfeldt EE, Kitchel SH, Watkins RG, Balderston RA (2002) Clinical and radiographic outcomes of anterior lumbar interbody fusion using recombinant human bone morphogenetic protein-2. Spine (Phila Pa 1976) 27(21):2396–2408

Campana V, Milano G, Pagano E, Barba M, Cicione C, Salonna G, Lattanzi W, Logroscino G (2014) Bone substitutes in orthopaedic surgery: from basic science to clinical practice. J Mater Sci Mater Med 25 (10):2445–2461

Campbell PG, Yadla S, Nasser R, Malone J, Maltenfort MG, Ratliff JK (2012) Patient comorbidity score predicting the incidence of perioperative complications: assessing the impact of comorbidities on complications in spine surgery. J Neurosurg Spine 16(1):37–43

Campion CR, Chander C, Buckland T, Hing K (2011) Increasing strut porosity in silicate-substituted calcium-phosphate bone graft substitutes enhances osteogenesis. J Biomed Mater Res B Appl Biomater 97 (2):245–254

Carter JD, Swearingen AB, Chaput CD, Rahm MD (2009) Clinical and radiographic assessment of transforaminal lumbar interbody fusion using HEALOS collagen-hydroxyapatite sponge with autologous bone marrow aspirate. Spine J 9(6):434–438

Chapman MW, Madison M (1993) Operative orthopaedics, 2nd edn. J.B. Lippincott, Philadelphia

Cho JH, Lee JH, Yeom JS, Chang BS, Yang JJ, Koo KH, Hwang CJ, Lee KB, Kim HJ, Lee CK, Kim H, Suk KS, Nam WD, Han J (2017) Efficacy of Escherichia coli-derived recombinant human bone morphogenetic protein-2 in posterolateral lumbar fusion: an open, active-controlled, randomized, multicenter trial. Spine J 17(12):1866–1874

Chu S, Chen N, Dang ABC, Kuo AC, Dang ABC (2017) The effects of topical vancomycin on mesenchymal stem cells: more may not be better. Int J Spine Surg 11:12

Cunningham BW, Atkinson BL, Hu N, Kikkawa J, Jenis L, Bryant J, Zamora PO, McAfee PC (2009) Ceramic granules enhanced with B2A peptide for lumbar interbody spine fusion: an experimental study using an instrumented model in sheep. J Neurosurg Spine 10(4):300–307

Dawson E, Bae HW, Burkus JK, Stambough JL, Glassman SD (2009) Recombinant human bone morphogenetic protein-2 on an absorbable collagen sponge with an osteoconductive bulking agent in posterolateral arthrodesis with instrumentation. A prospective randomized trial. J Bone Joint Surg Am 91(7):1604–1613

Delawi D, Jacobs W, van Susante JL, Rillardon L, Prestamburgo D, Specchia N, Gay E, Verschoor N, Garcia-Fernandez C, Guerado E, Quarles van Ufford H, Kruyt MC, Dhert WJ, Oner FC (2016) OP-1 compared with iliac crest autograft in instrumented posterolateral fusion: a randomized, multicenter non-inferiority trial. J Bone Joint Surg Am 98(6):441–448

Divi SN, Mikhael MM (2017) Use of allogenic mesenchymal cellular bone matrix in anterior and posterior cervical spinal fusion: a case series of 21 patients. Asian Spine J 11(3):454–462

Duarte RM, Varanda P, Reis RL, Duarte ARC, Correia-Pinto J (2017) Biomaterials and bioactive agents in spinal fusion. Tissue Eng Part B Rev 23(6):540–551

Eastlack RK, Garfin SR, Brown CR, Meyer SC (2014) Osteocel Plus cellular allograft in anterior cervical discectomy and fusion: evaluation of clinical and radiographic outcomes from a prospective multicenter study. Spine (Phila Pa 1976) 39(22):E1331–E1337

Elder BD, Holmes C, Goodwin CR, Lo SF, Puvanesarajah V, Kosztowski TA, Locke JE, Witham TF (2015) A systematic assessment of the use of platelet-rich plasma in spinal fusion. Ann Biomed Eng 43(5):1057–1070

Epstein NE, Epstein JA (2007) SF-36 outcomes and fusion rates after multilevel laminectomies and 1 and 2-level instrumented posterolateral fusions using lamina autograft and demineralized bone matrix. J Spinal Disord Tech 20(2):139–145

Field J, Yeung C, Roh J (2014) Clinical Evaluation of Allogeneic Growth Factor in Cervical Spine Fusion. J Spine 3:158

Fortier L, Bauer L, Chung E (2017) Use of FIBERGRAFT BG Morsels mixed with BMA in anterior cervical discectomy and fusion at 1, 2, 3 and 4 levels: a retrospective analysis of fusion results. J Spine Neurosurg 6(3):200–205

Frantzen J, Rantakokko J, Aro HT, Heinanen J, Kajander S, Gullichsen E, Kotilainen E, Lindfors NC (2011) Instrumented spondylodesis in degenerative spondylolisthesis with bioactive glass and autologous bone: a prospective 11-year follow-up. J Spinal Disord Tech 24(7):455–461

Fredericks D, Petersen EB, Watson N, Grosland N, Gibson-Corley K, Smucker J (2016) Comparison of two synthetic bone graft products in a rabbit posterolateral fusion model. Iowa Orthop J 36:167–173

Fu Q, Saiz E, Tomsia AP (2011) Direct ink writing of highly porous and strong glass scaffolds for load-bearing bone defects repair and regeneration. Acta Biomater 7(10):3547–3554

Furstenberg CH, Wiedenhofer B, Putz C, Burckhardt I, Gantz S, Kleinschmidt K, Schroder K (2010) Collagen hydroxyapatite (Healos) saturated with gentamicin or levofloxacin. In vitro antimicrobial effectiveness – a pilot study. Orthopade 39(4):437–443

Galli MM, Protzman NM, Bleazey ST, Brigido SA (2015) Role of demineralized allograft subchondral bone in the treatment of shoulder lesions of the talus: clinical results with two-year follow-up. J Foot Ankle Surg 54(4):717–722

Gao T, Aro HT, Ylanen H, Vuorio E (2001) Silica-based bioactive glasses modulate expression of bone morphogenetic protein-2 mRNA in Saos-2 osteoblasts in vitro. Biomaterials 22(12):1475–1483

Gao R, Street M, Tay ML, Callon KE, Naot D, Lock A, Munro JT, Cornish J, Ferguson J, Musson D (2018) Human spinal bone dust as a potential local autograft: in vitro potent anabolic effect on human osteoblasts. Spine (Phila Pa 1976) 43(4):E193–E199

Garbuz DS, Masri BA, Czitrom AA (1998) Biology of allografting. Orthop Clin North Am 29(2):199–204

Glazebrook M, Young DS (2016) B2A polypeptide in foot and ankle fusion. Foot Ankle Clin 21(4):803–807

Goldberg VM, Stevenson S (1993) The biology of bone grafts. Semin Arthroplast 4(2):58–63

Govender S, Csimma C, Genant HK, Valentin-Opran A, Amit Y, Arbel R, Aro H, Atar D, Bishay M, Borner MG, Chiron P, Choong P, Cinats J, Courtenay B, Feibel R, Geulette B, Gravel C, Haas N, Raschke M, Hammacher E, van der Velde D, Hardy P, Holt M, Josten C, Ketterl RL, Lindeque B, Lob G, Mathevon H, McCoy G, Marsh D, Miller R, Munting E, Oevre S, Nordsletten L, Patel A, Pohl A, Rennie W, Reynders P, Rommens PM, Rondia J, Rossouw WC, Daneel PJ, Ruff S, Ruter A, Santavirta S, Schildhauer TA, Gekle C, Schnettler R, Segal D, Seiler H, Snowdowne RB, Stapert J, Taglang G, Verdonk R, Vogels L, Weckbach A, Wentzensen A, Wisniewski T, BMP-2 Evaluation in Surgery for Tibial Trauma (BESTT) Study Group (2002) Recombinant human bone morphogenetic protein-2 for treatment of open tibial fractures: a prospective, controlled, randomized study of four hundred and fifty patients. J Bone Joint Surg Am 84-A (12):2123–2134

Guerado E, Fuerstenberg CH (2011) What bone graft substitutes should we use in post-traumatic spinal fusion? Injury 42(Suppl 2):S64–S71

Gupta MC, Maitra S (2002) Bone grafts and bone morphogenetic proteins in spine fusion. Cell Tissue Bank 3(4):255–267

Gupta A, Kukkar N, Sharif K, Main BJ, Albers CE, El-Amin Iii SF (2015) Bone graft substitutes for spine fusion: a brief review. World J Orthop 6(6):449–456

Hamer AJ, Stockley I, Elson RA (1999) Changes in allograft bone irradiated at different temperatures. J Bone Joint Surg Br 81(2):342–344

Harris AI, Poddar S, Gitelis S, Sheinkop MB, Rosenberg AG (1995) Arthroplasty with a composite of an allograft and a prosthesis for knees with severe deficiency of bone. J Bone Joint Surg Am 77(3):373–386

Hench LL, Jones JR (2015) Bioactive glasses: frontiers and challenges. Front Bioeng Biotechnol 3:194

Hench LL, Polak JM (2002) Third-generation biomedical materials. Science 295(5557):1014–1017

Herkowitz HN, Dvorak J, Bell GR, Nordin M, Grob D, International Society for the Study of the Lumbar Spine (2004) The lumbar spine. Lippincott Williams & Wilkins, Philadelphia

Hoffmann A, Gross G (2001) BMP signaling pathways in cartilage and bone formation. Crit Rev Eukaryot Gene Expr 11(1–3):23–45

Hollawell SM (2012) Allograft cellular bone matrix as an alternative to autograft in hindfoot and ankle fusion procedures. J Foot Ankle Surg 51(2):222–225

Hsu WK, Goldstein CL, Shamji MF, Cho SK, Arnold PM, Fehlings MG, Mroz TE (2017) Novel osteobiologics and biomaterials in the treatment of spinal disorders. Neurosurgery 80(3S):S100–S107

Ilharreborde B, Morel E, Fitoussi F, Presedo A, Souchet P, Pennecot GF, Mazda K (2008) Bioactive glass as a bone substitute for spinal fusion in adolescent idiopathic scoliosis: a comparative study with iliac crest autograft. J Pediatr Orthop 28(3):347–351

Ito Z, Imagama S, Kanemura T, Hachiya Y, Miura Y, Kamiya M, Yukawa Y, Sakai Y, Katayama Y, Wakao N, Matsuyama Y, Ishiguro N (2013) Bone union rate with autologous iliac bone versus local bone graft in posterior lumbar interbody fusion (PLIF): a multicenter study. Eur Spine J 22(5):1158–1163

Jahng TA, Fu TS, Cunningham BW, Dmitriev AE, Kim DH (2004) Endoscopic instrumented posterolateral lumbar fusion with Healos and recombinant human growth/differentiation factor-5. Neurosurgery 54(1):171–180; discussion 180-171

James AW, LaChaud G, Shen J, Asatrian G, Nguyen V, Zhang X, Ting K, Soo C (2016) A review of the clinical side effects of bone morphogenetic protein-2. Tissue Eng Part B Rev 22(4):284–297

Jenis LG, Banco RJ (2010) Efficacy of silicate-substituted calcium phosphate ceramic in posterolateral instrumented lumbar fusion. Spine (Phila Pa 1976) 35(20):E1058–E1063

Jones JR (2013) Review of bioactive glass: from Hench to hybrids. Acta Biomater 9(1):4457–4486

Jones JR (2015) Reprint of: review of bioactive glass: from Hench to hybrids. Acta Biomater 23(Suppl):S53–S82

Jones JR, Ehrenfried LM, Hench LL (2006) Optimising bioactive glass scaffolds for bone tissue engineering. Biomaterials 27(7):964–973

Kang J, An H, Hilibrand A, Yoon ST, Kavanagh E, Boden S (2012) Grafton and local bone have comparable outcomes to iliac crest bone in instrumented single-level lumbar fusions. Spine (Phila Pa 1976) 37(12):1083–1091

Kannan A, Dodwad SN, Hsu WK (2015) Biologics in spine arthrodesis. J Spinal Disord Tech 28(5):163–170

Karpov M, Laczka M, Leboy PS, Osyczka AM (2008) Sol-gel bioactive glasses support both osteoblast and osteoclast formation from human bone marrow cells. J Biomed Mater Res A 84(3):718–726

Kawaguchi S, Hart RA (2015) The need for structural allograft biomechanical guidelines. Instr Course Lect 64:87–93

Kerr EJ 3rd, Jawahar A, Wooten T, Kay S, Cavanaugh DA, Nunley PD (2011) The use of osteo-conductive stem-cells allograft in lumbar interbody fusion procedures: an alternative to recombinant human bone morphogenetic protein. J Surg Orthop Adv 20(3):193–197

Kim HJ, Chung JH, Shin SY, Shin SI, Kye SB, Kim NK, Kwon TG, Paeng JY, Kim JW, Oh OH, Kook MS, Yang HJ, Hwang SJ (2015) Efficacy of rhBMP-2/hydroxyapatite on sinus floor augmentation: a multicenter, randomized controlled clinical trial. J Dent Res 94(9 Suppl):158S–165S

Kinney RC, Ziran BH, Hirshorn K, Schlatterer D, Ganey T (2010) Demineralized bone matrix for fracture healing: fact or fiction? J Orthop Trauma 24(Suppl 1):S52–S55

Kirk JF, Ritter G, Waters C, Narisawa S, Millan JL, Talton JD (2013) Osteoconductivity and osteoinductivity of NanoFUSE((R)) DBM. Cell Tissue Bank 14(1):33–44

Kolan KC, Leu MC, Hilmas GE, Brown RF, Velez M (2011) Fabrication of 13-93 bioactive glass scaffolds for bone tissue engineering using indirect selective laser sintering. Biofabrication 3(2):025004

Kong CB, Lee JH, Baek HR, Lee CK, Chang BS (2014) Posterolateral lumbar fusion using Escherichia coli-derived rhBMP-2/hydroxyapatite in the mini pig. Spine J 14(12):2959–2967

Kraiwattanapong C, Boden SD, Louis-Ugbo J, Attallah E, Barnes B, Hutton WC (2005) Comparison of Healos/bone marrow to INFUSE(rhBMP-2/ACS) with a collagen-ceramic sponge bulking agent as graft substitutes for lumbar spine fusion. Spine (Phila Pa 1976) 30(9):1001–1007; discussion 1007

Kunakornsawat S, Kirinpanu A, Piyaskulkaew C, Sathira-Angkura V (2013) A comparative study of radiographic results using HEALOS collagen-hydroxyapatite sponge with bone marrow aspiration versus local bone graft in the same patients undergoing posterolateral lumbar fusion. J Med Assoc Thail 96 (8):929–935

Kwong FN, Ibrahim T, Power RA (2005) Incidence of infection with the use of non-irradiated morcellised allograft bone washed at the time of revision arthroplasty of the hip. J Bone Joint Surg Br 87(11):1524–1526. https://doi.org/10.1302/0301-620X.87B11.16354

Lee K, Goodman SB (2009) Cell therapy for secondary osteonecrosis of the femoral condyles using the Cellect DBM System: a preliminary report. J Arthroplast 24(1):43–48

Lee JH, Yu CH, Yang JJ, Baek HR, Lee KM, Koo TY, Chang BS, Lee CK (2012) Comparative study of fusion rate induced by different dosages of Escherichia coli-derived recombinant human bone morphogenetic protein-2 using hydroxyapatite carrier. Spine J 12(3):239–248

Lentino JR (2003) Prosthetic joint infections: bane of orthopedists, challenge for infectious disease specialists. Clin Infect Dis 36(9):1157–1161

Li R, Clark AE, Hench LL (1991) An investigation of bioactive glass powders by sol-gel processing. J Appl Biomater 2(4):231–239

Licina P, Coughlan M, Johnston E, Pearcy M (2015) Comparison of silicate-substituted calcium phosphate (actifuse) with recombinant human bone morphogenetic protein-2 (infuse) in posterolateral instrumented lumbar fusion. Global Spine J 5(6):471–478

Lin X, Guo H, Takahashi K, Liu Y, Zamora PO (2012) B2A as a positive BMP receptor modulator. Growth Factors 30(3):149–157

Lindfors NC, Hyvonen P, Nyyssonen M, Kirjavainen M, Kankare J, Gullichsen E, Salo J (2010) Bioactive glass S53P4 as bone graft substitute in treatment of osteomyelitis. Bone 47(2):212–218

Litrico S, Langlais T, Pennes F, Gennari A, Paquis P (2018) Lumbar interbody fusion with utilization of recombinant human bone morphogenetic protein: a retrospective real-life study about 277 patients. Neurosurg Rev 41(1):189–196

Logoluso N, Drago L, Gallazzi E, George DA, Morelli I, Romano CL (2016) Calcium-based, antibiotic-loaded bone substitute as an implant coating: a pilot clinical study. J Bone Jt Infect 1:59–64

Lord CF, Gebhardt MC, Tomford WW, Mankin HJ (1988) Infection in boneallografts. Incidence, nature, and treatment. J Bone Joint Surg Am 70(3):369–376

Magit DP, Maak T, Trioano N, Raphael B, Hamouria Q, Polzhofer G, Drespe I, Albert TJ, Grauer JN (2006) Healos/recombinant human growth and differentiation factor-5 induces posterolateral lumbar fusion in a New Zealand white rabbit model. Spine (Phila Pa 1976) 31(19):2180–2188

Makanji H, Schoenfeld AJ, Bhalla A, Bono CM (2018) Critical analysis of trends in lumbar fusion for degenerative disorders revisited: influence of technique on fusion rate and clinical outcomes. Eur Spine J 27:1868–1876

Malat TA, Glombitza M, Dahmen J, Hax PM, Steinhausen E (2018) The use of bioactive glass S53P4 as bone graft substitute in the treatment of chronic osteomyelitis and infected non-unions – a retrospective study of 50 patients. Z Orthop Unfall 156(2):152–159

Manyalich M, Navarro A, Koller J, Loty B, de Guerra A, Cornu O, Vabels G, Fornasari PM, Costa AN, Siska I, Hirn M, Franz N, Miranda B, Kaminski A, Uhrynowska I, Van Baare J, Trias E, Fernandez C, de By T, Poniatowski S, Carbonell R (2009) European quality system for tissue banking. Transplant Proc 41(6):2035–2043

McAnany SJ, Ahn J, Elboghdady IM, Marquez-Lara A, Ashraf N, Svovrlj B, Overley SC, Singh K, Qureshi SA (2016) Mesenchymal stem cell allograft as a fusion adjunct in one- and two-level anterior cervical discectomy and fusion: a matched cohort analysis. Spine J 16(2):163–167

McKee MD, Wild LM, Schemitsch EH, Waddell JP (2002) The use of an antibiotic-impregnated, osteoconductive, bioabsorbable bone substitute in the treatment of infected long bone defects: early results of a prospective trial. J Orthop Trauma 16(9):622–627

McLaren AC (2004) Alternative materials to acrylic bone cement for delivery of depot antibiotics in orthopaedic infections. Clin Orthop Relat Res (427):101–106

McLaren AC, McLaren SG, Nelson CL, Wassell DL, Olsen KM (2002) The effect of sampling method on the elution of tobramycin from calcium sulfate. Clin Orthop Relat Res (403):54–57

Meyers MH (1985) Resurfacing of the femoral head with fresh osteochondral allografts. Long-term results. Clin Orthop Relat Res (197):111–114

Miyazaki M, Tsumura H, Wang JC, Alanay A (2009) An update on bone substitutes for spinal fusion. Eur Spine J 18(6):783–799

Miyazono K, Maeda S, Imamura T (2005) BMP receptor signaling: transcriptional targets, regulation of signals, and signaling cross-talk. Cytokine Growth Factor Rev 16(3):251–263

Morris MT, Tarpada SP, Cho W (2018) Bone graft materials for posterolateral fusion made simple: a systematic review. Eur Spine J 27:1856–1867

Neen D, Noyes D, Shaw M, Gwilym S, Fairlie N, Birch N (2006) Healos and bone marrow aspirate used for lumbar spine fusion: a case controlled study comparing healos with autograft. Spine (Phila Pa 1976) 31(18):E636–E640

Nelson CL (2004) The current status of material used for depot delivery of drugs. Clin Orthop Relat Res (427):72–78

Nickoli MS, Hsu WK (2014) Ceramic-based bone grafts as a bone grafts extender for lumbar spine arthrodesis: a systematic review. Global Spine J 4(3):211–216

O'Brien FJ (2011) Biomaterials & scaffolds for tissue engineering. Mater Today 14(3):88–95

Park JJ, Hershman SH, Kim YH (2013) Updates in the use of bone grafts in the lumbar spine. Bull Hosp Jt Dis (2013) 71(1):39–48

Peppers TA, Bullard DE, Vanichkachorn JS, Stanley SK, Arnold PM, Waldorff EI, Hahn R, Atkinson BL, Ryaby

JT, Linovitz RJ (2017) Prospective clinical and radiographic evaluation of an allogeneic bone matrix containing stem cells (Trinity Evolution(R) Viable Cellular Bone Matrix) in patients undergoing two-level anterior cervical discectomy and fusion. J Orthop Surg Res 12(1):67

Peterson B, Whang PG, Iglesias R, Wang JC, Lieberman JR (2004) Osteoinductivity of commercially available demineralized bone matrix. Preparations in a spine fusion model. J Bone Joint Surg Am 86-A(10):2243–2250

Ploumis A, Albert TJ, Brown Z, Mehbod AA, Transfeldt EE (2010) Healos graft carrier with bone marrow aspirate instead of allograft as adjunct to local autograft for posterolateral fusion in degenerative lumbar scoliosis: a minimum 2-year follow-up study. J Neurosurg Spine 13(2):211–215

Poorman GW, Jalai CM, Boniello A, Worley N, McClelland S 3rd, Passias PG (2017) Bone morphogenetic protein in adult spinal deformity surgery: a meta-analysis. Eur Spine J 26(8):2094–2102

Radcliff KE, Neusner AD, Millhouse PW, Harrop JD, Kepler CK, Rasouli MR, Albert TJ, Vaccaro AR (2015) What is new in the diagnosis and prevention of spine surgical site infections. Spine J 15(2):336–347

Rantakokko J, Frantzen JP, Heinanen J, Kajander S, Kotilainen E, Gullichsen E, Lindfors NC (2012) Posterolateral spondylodesis using bioactive glass S53P4 and autogenous bone in instrumented unstable lumbar spine burst fractures. A prospective 10-year follow-up study. Scand J Surg 101(1):66–71

Rauck RC, Wang D, Tao M, Williams RJ (2019) Chondral delamination of fresh osteochondral allografts after implantation in the knee: a matched cohort analysis. Cartilage 10:402–407

Rausch V, Seybold D, Konigshausen M, Koller M, Schildhauer TA, Gessmann J (2017) [Basic principles of fracture healing]. Orthopade 46(8):640–647

Reilly GC, Radin S, Chen AT, Ducheyne P (2007) Differential alkaline phosphatase responses of rat and human bone marrow derived mesenchymal stem cells to 45S5 bioactive glass. Biomaterials 28(28):4091–4097

Rengachary SS (2002) Bone morphogenetic proteins: basic concepts. Neurosurg Focus 13(6):e2

Robbins MA, Haudenschild DR, Wegner AM, Klineberg EO (2017) Stem cells in spinal fusion. Global Spine J 7(8):801–810

Roh JS, Yeung CA, Field JS, McClellan RT (2013) Allogeneic morphogenetic protein vs. recombinant human bone morphogenetic protein-2 in lumbar interbody fusion procedures: a radiographic and economic analysis. J Orthop Surg Res 8:49

Sanchez-Salcedo S, Garcia A, Vallet-Regi M (2017) Prevention of bacterial adhesion to zwitterionic biocompatible mesoporous glasses. Acta Biomater 57:472–486

Sardar Z, Alexander D, Oxner W, du Plessis S, Yee A, Wai EK, Anderson DG, Jarzem P (2015) Twelve-month results of a multicenter, blinded, pilot study of a novel peptide (B2A) in promoting lumbar spine fusion. J Neurosurg Spine 22(4):358–366

Schizas C, Triantafyllopoulos D, Kosmopoulos V, Tzinieris N, Stafylas K (2008) Posterolateral lumbar spine fusion using a novel demineralized bone matrix: a controlled case pilot study. Arch Orthop Trauma Surg 128(6):621–625

Shehadi JA, Elzein SM (2017) Review of commercially available demineralized bone matrix products for spinal fusions: a selection paradigm. Surg Neurol Int 8:203

Shekaran A, Garcia AJ (2011) Extracellular matrix-mimetic adhesive biomaterials for bone repair. J Biomed Mater Res A 96(1):261–272

Shibuya N, Jupiter DC (2015) Bone graft substitute: allograft and xenograft. Clin Podiatr Med Surg 32(1):21–34

Shields LB, Raque GH, Glassman SD, Campbell M, Vitaz T, Harpring J, Shields CB (2006) Adverse effects associated with high-dose recombinant human bone morphogenetic protein-2 use in anterior cervical spine fusion. Spine (Phila Pa 1976) 31(5):542–547

Shiels SM, Raut VP, Patterson PB, Barnes BR, Wenke JC (2017) Antibiotic-loaded bone graft for reduction of surgical site infection in spinal fusion. Spine J 17(12):1917–1925

Singh K, Smucker JD, Gill S, Boden SD (2006) Use of recombinant human bone morphogenetic protein-2 as an adjunct in posterolateral lumbar spine fusion: a prospective CT-scan analysis at one and two years. J Spinal Disord Tech 19(6):416–423

Smucker JD, Fredericks DC (2012) Assessment of Progenix((R)) DBM putty bone substitute in a rabbit posterolateral fusion model. Iowa Orthop J 32:54–60

Smucker JD, Bobst JA, Petersen EB, Nepola JV, Fredericks DC (2008) B2A peptide on ceramic granules enhance posterolateral spinal fusion in rabbits compared with autograft. Spine (Phila Pa 1976) 33(12):1324–1329

Stevenson S (1999) Biology of bone grafts. Orthop Clin North Am 30(4):543–552

Stoor P, Soderling E, Salonen JI (1998) Antibacterial effects of a bioactive glass paste on oral microorganisms. Acta Odontol Scand 56(3):161–165

Street M, Gao R, Martis W, Munro J, Musson D, Cornish J, Ferguson J (2017) The efficacy of local autologous bone dust: a systematic review. Spine Deform 5(4):231–237

Sykaras N, Opperman LA (2003) Bone morphogenetic proteins (BMPs): how do they function and what can they offer the clinician? J Oral Sci 45(2):57–73

Thalgott JS, Giuffre JM, Fritts K, Timlin M, Klezl Z (2001) Instrumented posterolateral lumbar fusion using coralline hydroxyapatite with or without demineralized bone matrix, as an adjunct to autologous bone. Spine J 1(2):131–137

Tohmeh AG, Watson B, Tohmeh M, Zielinski XJ (2012) Allograft cellular bone matrix in extreme lateral interbody fusion: preliminary radiographic and clinical outcomes. ScientificWorldJournal 2012:263637

Torrie AM, Kesler WW, Elkin J, Gallo RA (2015) Osteochondral allograft. Curr Rev Musculoskelet Med 8(4):413–422

Tsigkou O, Pomerantseva I, Spencer JA, Redondo PA, Hart AR, O'Doherty E, Lin Y, Friedrich CC, Daheron L, Lin CP, Sundback CA, Vacanti JP, Neville C (2010) Engineered vascularized bone grafts. Proc Natl Acad Sci U S A 107(8):3311–3316

Tsigkou O, Labbaf S, Stevens MM, Porter AE, Jones JR (2014) Monodispersed bioactive glass submicron particles and their effect on bone marrow and adipose tissue-derived stem cells. Adv Healthc Mater 3 (1):115–125

Tuchman A, Brodke DS, Youssef JA, Meisel HJ, Dettori JR, Park JB, Yoon ST, Wang JC (2016) Iliac crest bone graft versus local autograft or allograft for lumbar spinal fusion: a systematic review. Global Spine J 6 (6):592–606

Turner TM, Urban RM, Gitelis S, Kuo KN, Andersson GB (2001) Radiographic and histologic assessment of calcium sulfate in experimental animal models and clinical use as a resorbable bone-graft substitute, a bone-graft expander, and a method for local antibiotic delivery. One institution's experience. J Bone Joint Surg Am 83-A Suppl 2(Pt 1):8–18

Turner TM, Urban RM, Hall DJ, Chye PC, Segreti J, Gitelis S (2005) Local and systemic levels of tobramycin delivered from calcium sulfate bone graft substitute pellets. Clin Orthop Relat Res (437):97–104

Urist MR (1965) Bone: formation by autoinduction. Science 150(3698):893–899

Urrutia J, Molina M (2013) Fresh-frozen femoral head allograft as lumbar interbody graft material allows high fusion rate without subsidence. Orthop Traumatol Surg Res 99(4):413–418

US FDA (2017) CFR – Code of Federal Regulations title 21. www.accessdata.fda.gov/scripts/cdrh/cfdocs/cfCFR/CFRSearch.cfm?CFRPart=1271

US FDA (2018) Jurisdictional update: human demineralized bone matrix. www.fda.gov/combinationproducts/jurisdictionalinformation/jurisdictionalupdates/ucm106586.ht

Vaccaro AR, Chiba K, Heller JG, Patel T, Thalgott JS, Truumees E, Fischgrund JS, Craig MR, Berta SC, Wang JC, North American Spine Society for Contemporary Concepts in Spine Care (2002) Bone grafting alternatives in spinal surgery. Spine J 2(3):206–215

Vaccaro AR, Patel T, Fischgrund J, Anderson DG, Truumees E, Herkowitz HN, Phillips F, Hilibrand A, Albert TJ, Wetzel T, McCulloch JA (2004) A pilot study evaluating the safety and efficacy of OP-1 Putty (rhBMP-7) as a replacement for iliac crest autograft in posterolateral lumbar arthrodesis for degenerative spondylolisthesis. Spine (Phila Pa 1976) 29 (17):1885–1892

Vaccaro AR, Anderson DG, Patel T, Fischgrund J, Truumees E, Herkowitz HN, Phillips F, Hilibrand A, Albert TJ, Wetzel T, McCulloch JA (2005) Comparison of OP-1 Putty (rhBMP-7) to iliac crest autograft for posterolateral lumbar arthrodesis: a minimum 2-year follow-up pilot study. Spine (Phila Pa 1976) 30 (24):2709–2716

Vaccaro AR, Whang PG, Patel T, Phillips FM, Anderson DG, Albert TJ, Hilibrand AS, Brower RS, Kurd MF, Appannagari A, Patel M, Fischgrund JS (2008) The safety and efficacy of OP-1 (rhBMP-7) as a replacement for iliac crest autograft for posterolateral lumbar arthrodesis: minimum 4-year follow-up of a pilot study. Spine J 8(3):457–465

Vallejo LF, Brokelmann M, Marten S, Trappe S, Cabrera-Crespo J, Hoffmann A, Gross G, Weich HA, Rinas U (2002) Renaturation and purification of bone morphogenetic protein-2 produced as inclusion bodies in high-cell-density cultures of recombinant Escherichia coli. J Biotechnol 94(2):185–194

Vanichkachorn J, Peppers T, Bullard D, Stanley SK, Linovitz RJ, Ryaby JT (2016) A prospective clinical and radiographic 12-month outcome study of patients undergoing single-level anterior cervical discectomy and fusion for symptomatic cervical degenerative disc disease utilizing a novel viable allogeneic, cancellous, bone matrix (trinity evolution) with a comparison to historical controls. Eur Spine J 25(7):2233–2238

Villa MM, Wang L, Huang J, Rowe DW, Wei M (2015) Bone tissue engineering with a collagen-hydroxyapatite scaffold and culture expanded bone marrow stromal cells. J Biomed Mater Res B Appl Biomater 103(2):243–253

Walsh WR, Chapman-Sheath PJ, Cain S, Debes J, Bruce WJ, Svehla MJ, Gillies RM (2003) A resorbable porous ceramic composite bone graft substitute in a rabbit metaphyseal defect model. J Orthop Res 21 (4):655–661

Walsh WR, Oliver R, Chistou C, Wang T, Walsh ER, Lovric V, Pelletier M, Bondre S (2017) Posterolateral fusion in a NZ white rabbit model using a novel bioglass fiber combined with autograft. Poster#390, ORS 2017 annual meeting, San Diego, 19–22 March

Wang W, Yeung KWK (2017) Bone grafts and biomaterials substitutes for bone defect repair: A review. Bioact Mater 2(4):224–247

Wells CM, Beenken KE, Smeltzer MS, Courtney HS, Jennings JA, Haggard WO (2018) Ciprofloxacin and rifampin dual antibiotic-loaded biopolymer chitosan sponge for bacterial inhibition. Mil Med 183 (Suppl_1):433–444

Whang PG, Wang JC (2003) Bone graft substitutes for spinal fusion. Spine J 3(2):155–165

Wheeless CR III (2013) Allografts, February 19. http://www.wheelessonline.com/ortho/allografts

White AA 3rd, Hirsch C (1971) An experimental study of the immediate load bearing capacity of some commonly used iliac bone grafts. Acta Orthop Scand 42 (6):482–490

Wozney JM (1989) Bone morphogenetic proteins. Prog Growth Factor Res 1(4):267–280

Wozney JM (2002) Overview of bone morphogenetic proteins. Spine (Phila Pa 1976) 27(16 Suppl 1):S2–S8

Wu ZY, Hill RG, Yue S, Nightingale D, Lee PD, Jones JR (2011) Melt-derived bioactive glass scaffolds produced by a gel-cast foaming technique. Acta Biomater 7(4):1807–1816

Xynos ID, Edgar AJ, Buttery LD, Hench LL, Polak JM (2000a) Ionic products of bioactive glass dissolution increase proliferation of human osteoblasts and induce insulin-like growth factor II mRNA expression and protein synthesis. Biochem Biophys Res Commun 276(2):461–465

Xynos ID, Hukkanen MV, Batten JJ, Buttery LD, Hench LL, Polak JM (2000b) Bioglass 45S5 stimulates osteoblast turnover and enhances bone formation In vitro: implications and applications for bone tissue engineering. Calcif Tissue Int 67(4):321–329

Yoo JS, Min SH, Yoon SH (2015) Fusion rate according to mixture ratio and volumes of bone graft in minimally invasive transforaminal lumbar interbody fusion: minimum 2-year follow-up. Eur J Orthop Surg Traumatol 25(Suppl 1):S183–S189

Zadegan SA, Abedi A, Jazayeri SB, Bonaki HN, Vaccaro AR, Rahimi-Movaghar V (2017a) Clinical application of ceramics in anterior cervical discectomy and fusion: a review and update. Global Spine J 7(4):343–349

Zadegan SA, Abedi A, Jazayeri SB, Nasiri Bonaki H, Jazayeri SB, Vaccaro AR, Rahimi-Movaghar V (2017b) Bone morphogenetic proteins in anterior cervical fusion: a systematic review and meta-analysis. World Neurosurg 104:752–787

Zamborsky R, Svec A, Bohac M, Kilian M, Kokavec M (2016) Infection in bone allograft transplants. Exp Clin Transplant 14(5):484–490

Zdeblick TA, Cooke ME, Kunz DN, Wilson D, McCabe RP (1994) Anterior cervical discectomy and fusion using a porous hydroxyapatite bone graft substitute. Spine (Phila Pa 1976) 19(20):2348–2357

Zhou X, Zhang N, Mankoci S, Sahai N (2017) Silicates in orthopedics and bone tissue engineering materials. J Biomed Mater Res A 105(7):2090–2102

Mechanobiology of the Intervertebral Disc and Treatments Working in Conjunction with the Human Anatomy

Stephen Jaffee, Isaac R. Swink, Brett Phillips, Michele Birgelen, Alexander K. Yu, Nick Giannoukakis, Boyle C. Cheng, Scott Webb, Reginald Davis, William C. Welch, and Antonio Castellvi

Contents

Introduction	276
Mechanobiology	277
Degenerative Matrix and Utility	277
Common Device Categories	279
Interspinous Spacer Devices	279
Pedicle Screw and Rod-Based Devices	281
Total Facet Replacement Systems	287
Spinal Fusion	288
Merits and Downfalls of the PDS System	288
Conclusions	289
Dedication	289
References	290

Abstract

Degenerative conditions of the spine benefit from a methodical approach for the management of patients with chronic low back pain when offered surgery. Surgical solutions should consider the severity of the disease along with the approach in order to provide the patient with the best potential long-term outcomes. Posterior dynamic stabilization is considered to be an alternative therapy to rigid spinal fusion and is intended to produce equal stability within the affected vertebral space, while promoting additional mobility. Through its use in treating conditions such as spondylolisthesis, disc degeneration, and disc herniation, posterior dynamic stabilization has emerged as a potential solution to unintended consequences of more conventional therapeutic modalities, like rigid spinal fusion. Complications, such as adjacent disc disease, may be

---

Antonio Castellvi is deceased

S. Jaffee (✉)
College of Medicine, Drexel University, Philadelphia, PA, USA

Allegheny Health Network, Department of Neurosurgery, Allegheny General Hospital, Pittsburgh, PA, USA
e-mail: Stephen.Jaffee@AHN.ORG

mitigated through an approach that permits additional mobility, returning the pathological segments to their intact range of movement and functionality. This chapter will review the history and development of posterior dynamic stabilization devices from their early inception to the current state of the art, as well as analyze the current pros and cons (garnered through both biomechanical and clinical testing) of each. Specifically, it will focus on the following device categories: interspinous spacers, pedicle screw and rod-based devices, and total facet replacement systems. Finally, there will be a discussion regarding the shortcomings of current metrics used to test such devices, along with an analysis on the cooperation between industry leaders and surgeons in designing said devices.

I. R. Swink · M. Birgelen · A. K. Yu
Department of Neurosurgery, Neuroscience Institute, Allegheny Health Network, Pittsburgh, PA, USA
e-mail: Isaac.Swink@ahn.org; Michele.BIRGELEN@ahn.org; Alexander.YU@ahn.org

B. Phillips · N. Giannoukakis
Institute of Cellular Therapeutics, Allegheny Health Network, Pittsburgh, PA, USA
e-mail: Brett.Phillips@AHN.ORG; Nick.GIANNOUKAKIS@ahn.org

B. C. Cheng
Neuroscience Institute, Allegheny Health Network, Drexel University, Allegheny General Hospital Campus, Pittsburgh, PA, USA
e-mail: bcheng@wpahs.org; Boyle.CHENG@ahn.org

S. Webb
Florida Spine Institute, Clearwater, Tampa, FL, USA

R. Davis
BioSpine, Tampa, FL, USA
e-mail: rjdavismd@aol.com

W. C. Welch
Department of Neurosurgery, University of Pennsylvania, Philadelphia, PA, USA
e-mail: William.welch@pennmedicine.upenn.edu

A. Castellvi
Orthopaedic Research and Education, Florida Orthopaedic Institute, Tampa, FL, USA

Keywords

Mechanobiology · Posterior dynamic stabilization · Interspinous spacers · Pedicle rods and screws · Total facet arthroplasty · Fusion · Rigid · VAS · ODI

Introduction

The motion of the spine can be studied in the most basic form by investigating a single index level or functional spinal unit (FSU). The FSU is a three-joint complex comprised of two vertebral bodies with three articulations, including the intervertebral, disc as well as the two posterior facet joints. The intervertebral disc forms an integral part of the FSU and has a propensity for degeneration with increasing age. Anatomically, the disc consists of highly oriented unidirectional layers arranged concentrically in alternating lamellar structures in conjunction with a gelatinous inner core, referred to as the nucleus pulposus. The nucleus has the ability to absorb transient forces, of which shock loads may have highest magnitudes, and to subsequently distribute loads to the end plates of the vertebrae. The other important articulations within the FSU are the facet joints which are also susceptible to disease. Facets, in the normal condition, play a role in controlling the motion of the FSU. This three-joint complex within each FSU controls the kinematic response to load. The primary modes of loading taken into consideration when evaluating the kinematic response to physiologic loads include axial compression, flexion extension bending, lateral bending, and axial torsion.

As degeneration occurs the disc may become fibrotic, compromising its ability to dissipate and distribute loads. Consequently, non-physiologic loads are then distributed to the vertebral end plates and the annulus of the disc which may lead to morphologic end plate changes and annular fissuring. With the onset of the degenerative cascade, both the intervertebral disc along with facets becomes compromised. The degeneration within the FSU may lead to the inability to withstand even physiological loads and eventually, depending on the severity, instability may develop. Both clinically and biomechanically, instability can be defined by the inability of the FSU to control physiological displacement. With instability, the neurological structures are prone to impingement and injury. Instability of the intervertebral disc changes the kinematic loading profile of the spine with increased load transfer through the facet joints and ligamentum flavum. With time, these structures all undergo hypertrophy with narrowing of the central neural canal as well as the lateral recesses and neural foramina.

Mechanobiology

The intervertebral disc is comprised of at least two distinct cellular populations. Within the nucleus pulposus resides a chondrocyte like cellular population, while the cells of the annulus and cartilaginous endplate are primarily fibroblast like with an elongated shape. In a healthy state these cells work to continuously remodel the ECM, maintaining a balance of catabolic and anabolic remodeling. The cellular populations which reside in the soft tissue structures of the intervertebral disc (IVD) respond to applied mechanical stimuli, a phenomenon known as mechanotransduction (Johnson and Roberts 2003). The loads transmitted to the FSU are applied from various vector orientations, with axial compressive forces being converted to a hydrostatic pressure by the nucleus pulposus and then shear stress on the collagen fibers within the annulus (Vergroesen et al. 2015).

The local tissue environment is a key factor in the outcome when treating spinal pathologies. In a diseased state the accumulation of inflammatory cytokines, such as IL-6, IL-1β, and TNF-α, disrupt the balance between anabolic and catabolic remodeling leading to increased matrix degradation. Cytokine accumulation within the IVD may be the product of native cellular activity or the result of immune cells infiltrating the region and disrupting the microenvironment. Recent research has identified the presence of immune cells within degenerated or injured disc tissue. In the case of disc herniation, both neutrophils and macrophage have been identified in pathological tissues removed during microdiscectomy procedures. These cells are biologically active, producing inflammatory factors such as TNF-α. Even if the pathologic disc tissue is removed, pain may persist due to the continued presence of inflammation. Furthermore, inflammation of the disc may accelerate the degenerative process. For example, the presence of TNF-α has been associated with loss in disc height due to matrix destruction (Kang et al. 2015; Wang et al. 2017). Early intervention may be crucial in halting the degenerative cascade, with evidence suggesting the local tissue environment of the disc can be modulated with conservative therapies such as steroid injection or physical therapy (Fig. 1).

Additionally, the classical surgical treatments of degenerative disc disease also alter the kinematic response of an FSU. This abnormal motion may be caused by decompressive-type destabilizing procedures or by increasing the range of motion of the level adjacent to a fusion, also referred to as a "neo hinge." Finite element analysis of the von Misses stresses at the level above a fused level may also exhibit abnormal increases in the loads, and over time, hyper mobility may become evident. The von Misses stresses applied to that segment are altered both in distribution as well as in magnitude (Castellvi et al. 2007). Thus, the need for additional treatments which aim to restore the appropriate kinematic signature has led to serious consideration for the exploration of motion preservation technology as an alternative to fusion in the treatment of lumbar DDD. The goal of these motion preservation systems is to restore the mechanics of the intervertebral disc, thus disrupting the positive feedback loop which results in continued degeneration.

Degenerative Matrix and Utility

With the introduction of motion preservation technology, the matrix shown in Table 1 is proposed as a means to discretize the severity of the pathology by providing three distinct categories: mild, moderate and severe. Similarly, the targeted FSU for treatment can be further broken down into three distinct

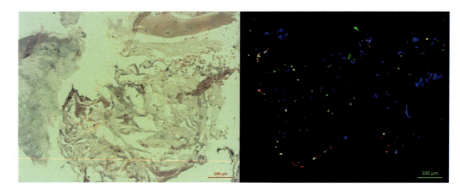

Fig. 1 Inflammatory response of a moderately degenerated human intervertebral disc. (**a**) Tissue sections were stained with hematoyxlin and eosin. Hematoxylin stains cell nuclei blue, while eosin stains extracellular matrix and cytoplasm pink (**b**) Immunofluorescence microscopy techniques were used to identify specific cellular markers of infiltrating immune cells for the same section. Red coloration depicts macrophages identified by the presence of surface marker CD68. Green coloration depicts neutrophils and granulocytes identified by the presence of the surface maker CD66b. Nuclear staining shown is shown in blue

Table 1 Matrix of degenerative condition versus the region of the spine within an FSU

		Region		
		Anterior	Middle	Posterior
Severity	Mild	Nucleus replacement Nucleus augmentation Biologics	Annuloplasty	Ligament replacement Interspinous spacers
	Moderate	Nucleus replacement Nucleus augmentation Biologics Total disc replacement	Porsterior pedicle-based systems Facet replacement	Interspinous Interlaminar
	Severe	Fusion	Fusion	Fusion

regions: the anterior, middle and posterior columns. The resulting intersections of these two variables (level of degeneration and region within FSU) provide potential treatment solutions with appropriate implant class descriptions at the junctions shown. This matrix is intended to methodically classify the severity of the pathology, origin or source of pain, and identify a potential implant or procedural solution in a systematic fashion. As technology increases, ideally the design and application of these technologies may be more precise and further refinement of technology classification may result.

Posterior Region

Degenerative conditions that affect the posterior regions may compromise anatomical structures including the facets joints and osteoligamentous tissue. Pathologically, these conditions may occur in combination with a degenerated anterior column. The potential consequences of facet degeneration are clinically well recognized and contribute to conditions such as stenosis. Other consequences include ligamentum flavum infolding into the spinal canal, osteophyte production with subsequent neuroforaminal stenosis, generation of inflammatory proteins with subsequent pain, reduced range of motion from hypertrophic facets and degenerative spondylolisthesis. The posterior degeneration of the lumbar spine is part of the overall degenerative cascade, but interventions through a posterior approach can stabilize or reverse this degenerative cascade, potentially obviating the need for intervention of the middle or posterior columns. The classification in terms of degeneration has also been aided by the prevalence of imaging modalities and other diagnostic tools, such as diffusion weighted imaging.

A posterior approach provides bone anchoring locations and access to the anterior column via the pedicles. Also, the anatomical layout allows for bone and bone graft substitutes to be placed in the lateral gutters, and between spinous and along transverse processes from the same posterior approach. The posterior column, in combination with the anterior column, absorbs stresses and loads placed onto the spinal column. The articulating cartilaginous surfaces of the two facets within the FSU provide guided motion as a kinematic response to load. Furthermore, facets articulate in combination with the third joint, the intervertebral disc in order to offload, to some degree, a portion of the high loads from the anterior column.

The bony structures of the posterior region, including the lamina and pedicles, present excellent bone anchoring for fixation hardware. The cortical strength and designs that have taken advantage of cortical implants have been well documented. Ease of access is a main benefit of posterior approached. Implants intended to facilitate arthrodesis or preserve motion can be anchored in the posterior column with relatively simple access procedures. In particular, the cortical bone that comprises the pedicle provides a competent bone implant interface for the attachment of fusion constructs and motion preservation devices alike. Both implant designs require osteointegration at the bone implant interface for immediate and long-term stability.

Posterior approaches to the spine are well understood and described, provide direct and extensive access to locations with good bone anchoring and a portal to the vertebral bodies, and allow the surgeon to perform extensive corrective procedures indirectly to the anterior column. Despite the strength of cortical bone fixation in the posterior column, pedicle-based fixation devices often require a strong bone-implant interface access the anterior column via the middle region through a posterior approach. The large surface area of the vertebral endplates allow for the development of a wide variety of mechanical and potentially biological corrective forces to be applied so as to improve FSU mechanics. Direct replacement or removal of degenerative encroaching tissue or material may reduce pain and inflammation, promoting further healing of the diseased FSU while allowing easy access to the lateral gutters and other important structures.

Common Device Categories

Within the PDS space, three major technological approaches have emerged: (1) Interspinous Spacer Devices, (2) Pedicle Screw/Rod-based devices and (3) Total Facet Replacement. Each category has its own counter and normal indication for use in patients. This chapter summarizes all the major modern modalities of treatment for each and will present a detailed list of differing technologies, their claims, and an analysis of them. Notably, this is not all encompassing, as this industry is bristling with new improvements and technologies, some of which are not made public and are in various stages of preliminary research and development (Khoueir et al. 2007).

Interspinous Spacer Devices

Interspinous Spacer Devices are widely recognized as a means to address lumbar spinal stenosis via decompression. In fact, they have a reputation as devices with few significant negative effects. The general premise of this technology involves the placement of a device between the vertebral spinous processes to stabilize the structure and inhibit the compression of the spinal cord. The four major kinds of interspinous devices that are used heavily in the market are as follows: Wallis, XSTOP, DIAM, and Coflex. Each has been assessed in patients and an analysis of their efficacy is as follows.

Wallis implants are comprised of a Polyetheretherketone (PEEK) block, a common material used in both orthopedic and neurosurgical implants. This implant classifies as a floating system, and it adheres to the spinous process via two Dacron ribbons which warp around them (the spinous processes), creating a tight fit (Sobottke et al. 2009). It has been consistently demonstrated that the Wallis implant can prevent further disc degeneration and pain in patients with spinal stenosis. Floman et al. showed this in their 2007 study where they analyzed whether the Wallis interspinous implant may reduce the number of recurrent lumbar disc herniation in patients with primary disc excision (Floman et al. 2007). The research concluded that while the device did not impact the rate of recurrent herniation, there was a marked decrease in the Visual Analog Scale (VAS) of pain, in both the back and the legs. A study conducted by Senegas et al., performed in 1988, showed a similar point (Senegas et al. 1988). They demonstrated that widening the lumbar vertebral canal served as an effective treatment for patients suffering from spinal stenosis and postoperative spinal stability. The researchers mentioned the method's virtues: it did not need the whole lumbar laminectomy, which usually causes spinal instability. Sobottke et al. further proves the point. After the study analyzed the various interspinous implants, Wallis, X-STOP and DIAM, they found that all devices created significant and long-lasting symptom control (Sobottke et al. 2009). Despite no statistically significant difference in device performance, between the three brands, it should be noted that all produced favorable results in terms of patient satisfaction and treatment of Lumbar Spinal Stenosis (LSS) pain.

In a series of similar technologies within the interspinous Spacer Device market, one example, X-STOP, is an implant crafted out of titanium and coated with a PEEK composite. The spacer is oval in shape and carries two wings on its lateral sides which are intended to prevent lateral migration (Sobottke et al. 2009). Sobottke et al. found that X-STOP displayed a positive ability to combat LSS pain and served as a means to surgically decompress the spine. Puzzilli et al. highlighted this conclusion in their study on the efficacy of X-STOP as a treatment for LSS. This study involved a 3-year patient follow up with 542 patients in total. Of these 542, 422 underwent surgical implantation of X-STOP, while just 120 patients served as the control and were managed conservatively. Results showed a substantial 83.5% of X-STOP treated patients reported positive results in the later follow-up appointments, while 50% of the control group reported these same results. Notably, 38 out of the 120 control cases selected to receive another surgery to decompress the spine, as they found the control (conservative therapy) unsatisfactory. The authors concluded that interspinous process decompression via an interspinous spacer device offered an effective and less invasive alternative to classical microsurgical posterior decompression. This was specifically true in selected patients with spinal stenosis and lumbar degenerative disk diseases (Puzzilli et al. 2014). Furthermore, less than 6% of the patients that did receive the X-STOP intervention had the device removed because of worsening neurological complications.

The Device for Intervertebral Assisted Motion, or (DIAM), is comprised of a sleeve of polyester that surrounds a core of silicon. This device is situated between two adjacent spinous processes. It is bound by three mesh bands which tether it to the spinous process and to the supraspinous ligament for extra support (Sobottke et al. 2009). In a study done by Fabrizi et al., the DIAM device and the Aperius PercLID system were compared in patients, DIAM: 1,315; Aperius PercLID: 260. The patient population was comprised of patients with a spectrum of spinal pathologies including: degenerative disc disease (478), foraminal stenosis (347), disc herniation (283), black disc and facet syndrome (143), and topping-off (64). The study also differentiated between a single level (1,100) and a multilevel (475) intervention and resulted in an overwhelming majority of patients displaying symptom resolution and improvement. They, therefore, declared that both technologies showed clinical benefits, displaying the merits of the system (Fabrizi et al. 2011).

Sharing a similar role to the above devices, Coflex is a titanium-based implant that exists in a characteristically "U" shape. It adheres to the spinous processes by means of wings that are crimped to the bone. It is believed that the compliant "U" shape of the implant allows for additional load to be transferred through the disc as well (Kettler et al. 2008). Xu et al., in their publication, "Complication in degenerative lumbar disease treated with a dynamic interspinous spacer (Coflex)," resolved that the technology employed in this device was relatively safe, with only 11 patient complications in a sample size of 131. These complications involved three device-related issues (spinal process fracture, Coflex loosening, and fixed wing breakage), two tissue injuries (dura mater tear), and one superficial wound infection. The low complication and reoperation rate of the Coflex technology demonstrates its clinical utility (Xu et al. 2013). The authors of this study mentioned that care should be taken to prevent non-device-related complications emphasizing the importance of surgical proficiency and technique.

Pintauro et al. comprehensively reviewed the different interspinous spacer devices (Pintauro et al. 2017). There, the authors systematically analyzed each of the above technologies and sought to determine if the preliminary generation of implants is preferable to the second generation in terms of outcomes and complications. This review used 37 studies conducted from 2011 to 2016 to gain an up-to-date depiction on the current measures of success. This analysis generated an impressive finding, in that second-generation devices had a significantly lower rate of reoperation as compared to first generation devices (3.7% vs. 11%), which was not influenced by the type of Interspinous process device. This claim argued that older technologies were marginally obsolete, noting that the long-term functionality of first generation is questionable, and that newer devices did not suffer from the same degree of reemergence of symptoms in patients. The authors hypothesized that the differences in outcomes between first and second generation devices was due to two key factors: (1) they do not require additional decompression surgery with their utilization and (2) they are more frequently comprised of PEEK, which may be a more robust and nondegenerative material. The study acknowledged that there was insufficient randomized control trial data to emphatically make the claim that newer generation implants are superior. No statistically significant difference between the symptom relief of patients when their treatment with older versus newer devices was analyzed. The paper also acknowledged the influence of patient selection on the success rate of the surgery, emphasizing the importance of the proposed degeneration matrix and consideration of the stage of degeneration in selecting the appropriate treatment strategy (Pintauro et al. 2017).

In a study conducted by Richter et al., 60 patients were isolated, 30 were treated with decompression surgery for lumbar spinal stenosis and 30 with both decompression surgery and Coflex (a second-generation device). The study found, "... no significant difference between both groups in all parameters, including patient satisfaction and subjective operation decision." (Richter et al. 2009). The implementation of interspinous spacer devices on the whole has shown positive outcomes for patients in a myriad of different ways (pain and re-operation rate); however this study demonstrates that more research into this issue must be conducted to gain better insight into the significance of this treatment modality (implants) compared to spinal decompression surgery.

Pedicle Screw and Rod-Based Devices

One of the more versatile modalities of treatment within the posterior dynamic stabilization device space is that of pedicle screws and rods. These implants differ in terms of materials, design and efficacy in patients. Moreover, this section will be divided into two parts. The first subsection discusses the role of rigid rod-based systems, while the second subsection discusses one of pedicle screw-based systems.

The use of the Isobar TTL (Fig. 2) is considered as a means of mitigating lumbar degenerative disease. This technology is one of the preliminary semirigid rods that was used for dynamic fusion. The physical makeup of this device is a rod comprised of titanium alloy and a dampener which is made out of titanium O-rings that are stacked

Fig. 2 Isobar TTL construct on an anatomical model

upon one another vertically (Gomleksiz et al. 2012). The Isobar TTL device was utilized in a study conducted by Zhang et al., in which 38 cases of lumbar degenerative disease were analyzed in a retrospective study done between June 2007 and May 2011 (Zhang et al. 2012). The cases broke down into the following categories of pathology: 4 cases of grade I spondylolisthesis, 11 cases of lumbar instability and lumbar disc protrusion, 21 cases of lumbar spinal stenosis and lumbar disc protrusion, and 2 cases of postoperative recurrence of lumbar disc protrusion. Of the cases presented in the study, 22 of them displayed adjacent segment disc degeneration. The cases all shared a similar procedure of posterior decompression and the implantation of the Isobar TTL device. The evidence conferred in this study demonstrated near unanimous success in treatment of patients' symptoms. In fact, in the study's 38 cases, 32 were considered "excellent," 3 cases "good," 2 cases "fair," and only 1 case displayed a poor result. The final conclusion showed that the Isobar TTL stabilization system was a more than adequate means of treating lumbar degenerative disease characterized by a lower VAS score.

In a separate study, Gao et al. suggested that using Isobar TTL in a posterior approach provided a fixation system that had shown to "delay degeneration of intervertebral discs" (Gao et al. 2014). They appeared interested in Isobar TTL's unique features that "allowed for mobility of the fixation segments, maintained intervertebral space height, reduced the bearing load in both facet joints and discs and could prevent intervertebral disc degeneration." This study utilized MRI imaging to retrospectively assess 54 patients that had undergone dynamic lumbar fixation using the Isobar TTL. There was a heavy emphasis on both pre- and postoperative imaging to determine how this technology affected spinal health. It was found that after 24 months postoperatively, the associated diffusion coefficient (ADC) values increased significantly. The ADC is an indicator of the health of the nucleus pulposus, the central component of the intervertebral disc. Thus, an increase in the ADC showed an increase in health and hydration (concentration of water) of the disc. It should also be noted that DWI (Diffusion weighted imaging) was used to measure in vivo water molecule diffusion. Thus, this DWI value can demonstrate the structural characteristics of tissue. In effect, the DWI and ADC score are correlated, as a DWI may demonstrate disc health through the ADC value.

Barrey et al. commented on the use of Isobar TTL as a dynamic fusion system without the supposed effects of pseudoarthrosis, bone refraction and mechanical failure that other rigid apparatus suffered from (Barrey et al. 2013). This study is unique because of its long-term patient follow-up, totaling 10.2 years, of 18 patients with degenerative lumbar disc disease. The most important conclusion of this study is that within the 18-patient sample size, there were no adverse reactions to the treatment, and all patients showed positive signs of a successful treatment. Thus, there were no observed complications or revision surgeries in their sample. Notably this observation did not match those previously reported by other dynamic systems such as Dynesys (27.5% of patients) in a study done by Bothmann et al. (2008). Stoll et al. found that 10% of 73 patients with a Dynesys system displayed complications following the implantation of the device (Stoll et al. 2002). However, it is difficult to compare the efficacy of the two devices (Isobar TTL system vs. the Dynesys system) due to the limited sample size. Therefore, the author indicated that more work needs to be done to assess both systems and measure their respective outcomes.

Cook et al. analyzed the properties of the implant when placed into a cadaveric model. By performing a comprehensive analysis of each FSU's kinematic response to load they found the Isobar TTL rod is uniquely suited as a dynamic fusion system and it provided the same immediate stabilization as that of a rigid fixture, but with a greater potential to handle greater compressive loads, the evidence of which was proven statistically significant. The author, therefore, believed that the Isobar TTL system could mitigate the common problems facing more rigid implantable systems, specifically greater load sharing between the anterior and posterior columns and axial distraction, the latter of which they found could lead to pedicle travel during bending. This data was garnered through the biomechanical analysis of ten human lumbar cadaveric specimens measured upon various indices such as range of motion, anterior column load sharing, facet engagement via vertex distance map (VDM), interpedicular distance excursion, and finite helical screw axis (HSA). Analysis of which showcased the robust mobility of the device and its ability to assist in sharing of loads as previously mentioned (Cook et al. 2015). This evidence may aid in understanding why physicians see clinical benefits in patients, as this potentially sheds light on some of the factors that attribute to the success of an implant under physiological conditions.

Another rod technology to emerge was BalanC, a dynamic rod-based system. The device itself was comprised of two portions marked "dynamic" and "fusion." The dynamic portion of the device contained a complex of PEEK and Silicone, and the more rigid fusion portion was made entirely out of PEEK. Under testing performed by Cheng et al., the device did not display a statistically significant difference in biomechanical performance when compared to titanium and pure PEEK rods (Cheng et al. 2010).

One of the first posterior dynamic stabilization devices to gain wide usage was the Graf Ligamentoplasty system. This technology was notable for its braided polypropylene to connect two titanium pedicle screws (one on the superior and one on the inferior vertebra- on the symptomatic level) to create an apparatus that would provide structural integrity but still maintain a robust mobile characteristic. The intention of the device was to permit load sharing, primarily to the posterior annulus, and to allow micro-tears in the anterior annulus fibrosus to heal (Gomleksiz et al. 2012). Rigby et al. conducted a mid- and long-term follow up study on 51 patients that received the Graf ligament stabilization surgery. There was a very high rate of complication (12 out of the 51 suffered complications), and of those that had complications, four patients required additional follow-up surgeries due to their unresolved condition. A poll conducted during the study showed that 41% of patients indicated that they would choose to not have the operation again. Seven of the patients in the group later went on to have full bony fusion procedures due to unresolved issues. This study's conclusion indicated that the device should be used with caution (Rigby et al. 2001). Hadlow et al. criticized the Graf ligament technology, as they found that this modality of treatment was associated with a worse outcome at 1 year and a significant higher revision rate at 2 years (Hadlow et al. 1998). Sengupta mentions that the Graf ligament has a propensity for producing lateral canal stenosis in patients, particularly in cases where the patient suffers from degeneration of the facet joints or in-folding of the ligamentum flavum, demonstrating early clinical failure (Sengupta 2004). The author further mentions that evidence has elucidated the exact mechanism in which the device may treat symptoms, as clinical success may be from restriction of movement or from shifting loads to the posterior annulus.

In contrast, Madan et al. showed that the Graf system demonstrated superior results to that of a more conventional rigid fixation and fusion device. The author assessed the outcomes of two groups of 27–28 patients, the first of which was treated with the Graf ligamentoplasty and the second was an anterior lumbar interbody fusion device (ALIF) known as a Hartshill Horseshoe (Madan and Boeree 2003). After a follow-up period of 2.1 years, it was found that the Graf system and ALIF system had successful outcomes in 93% and 77.8%, respectively. The authors attributed this result to the increased lumbar segment mobility and better stabilization results. Likewise, a study performed by Grevitt et al., followed 50 patients postoperatively after Graf stabilization had been performed. A marked decrease was observed in the mean disability score (59% preoperatively compared to 31% postoperatively) and noted that 72% of patients stated the procedure produced good or excellent results (Grevitt et al. 1995). Markwalder and Wenger stated the same, although with the caveat that patient selection was primarily "young patients with painful mechanical disease who are resistant to conservative treatment and yield favorable long-term results" (Markwalder and Wenger 2003). This study demonstrated that while the patient population may be narrow, the device still had potential to combat the demonstrated symptoms. It is evident that more work in this space must be done to gain more knowledge regarding the benefits and potential complications of this technology.

The Dynamic Neutralization System (Dynesys) sought to stabilize the spine without bone grafting (Molinari 2007). The exact specifications of this device apparatus involve a titanium-alloy pedicle screw system connected by an elastic compound. Welch et al. stated that the device showed the ability to mitigate symptoms in patients (back and leg pain) and seemed to avoid any major surgical or device-related complications, some of which are more common in fusion approaches. A group of 101 patients were analyzed using the Oswestry Disability Index (ODI), and postoperative treatment groups displayed a near 30% reduction in disability (55.6% to 26.3%). Additionally, the pain data was conveyed by use of a 12-month follow-up questionnaire in which leg pain and back pain saw substantial reductions in mean values (80.3 to 25.5 and 54 to 29.4, respectively). While they did share a positive outlook on the device's ability to confer strong clinical results, they admitted that more research was needed (Welch et al. 2007). But Schwarzenbach et al. stated "Dynesys technology suggested it had limitations in elderly patients with osteopathic bone or those with severe segmental macro-instability with degenerative olisthesis and advanced disc degeneration," denoting an extra risk of failure. The study highlighted that no complications were found in their analysis and stated that more studies were needed to show that this technology definitively demonstrated a decrease in postsurgical complications (Schwarzenbach et al. 2005) (Fig. 3).

The use of PEEK rods has been a known method of posterior dynamic stabilization for some time. The material properties of such a technology are tremendously advantageous to this type of intervention due to its nonrigid physical nature, its radiolucent quality, and its versatility. Ormond et al., in their retrospective case series, showed that PEEK rods demonstrated similar fusion to Titanium rods. They argued initially that the semirigidity of PEEK rods would provide a reduction in stress-shielding and increased anterior load-sharing properties. This clinical evaluation of the technology showed that these assertions were well founded (Ormond et al. 2016). Additionally, a study in which the PEEK rods were retrieved from 12 patients conducted by Kurtz et al., demonstrated that the rods were comparable to their Titanium counterparts and displayed no cases of PEEK rod or pedicle screw fracture. This study shows that this modality of treatment (PEEK rods) is effective in not producing any major material-specific complications (Kurtz et al. 2013). While the study is limited in its sample size, it seems evident that the semirigid nature of PEEK serves as a comparable material for future device innovation within this space.

Fig. 3 Dynesys construct on an anatomical model

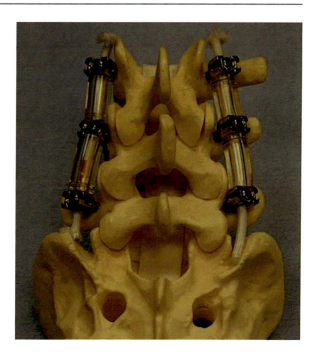

Notably, however, the same study mentioned that "seven out of eight periprosthetic tissue samples taken from the PEEK rods displayed signs of extensive degeneration, four of which had areas of tissue calcification." Also, PEEK wear shedding and PEEK debris were found in two out of the eight patients and was minimal, producing no significant inflammation.

The Bioflex Spring Rod Pedicle Screw System is comprised of a special Nitinol coil spring made of a small 4 mm diameter wire. The wire is set between the screws for the purpose of generating increased flexibility (Sengupta and Herkowitz 2012). An example of this technology being implemented in patients is shown in a study conducted by Heo et al. The study found that this approach was not significantly beneficial in preventing adjacent level degeneration completely. Based on MRI scans, only 2 of the 13 discs in the implantation segment showed any improvement in their disc degeneration, while 3 of the cranial adjacent discs (out of 25) and 4 of the caudal (out of 25) demonstrated a progression of disc degeneration (Heo et al. 2012). The biomechanics of this system were evaluated by Zhang et al., in which they found that the Bioflex system did not preserve ROM at implantation segments to that of any preoperative values but did preserve functional motion to these same levels (Zhang et al. 2009). This demonstrates that the biomechanical properties are indicative of a stable and effective PDS system; however, more clinical trials are needed to determine if the biomechanical advantages can translate into clinical utility.

Sengupta and Mulholland discussed the Fulcrum Assisted Soft Stabilization (FASS) in their publication assessing whether or not the aforementioned system could be a new means of treatment for degenerative lower back pain. The biomechanical properties of the technology displayed an ability to unload the affected disc and maintain a controlled range of motion. This was achieved by stabilizing the lumbar spine using pedicle screws, ligament, and a fulcrum to permit unloading. The thought process involved the transition of force from the disc to the ligament and fulcrum to achieve this characteristic unloading. Although done in cadaveric models, this study conveyed a new innovation to the PDS field (Sengupta and Mulholland 2005). While little clinical information has been produced as of late, the idea of circumventing any load on the affected disc by means of a mechanical transfer poses an interesting means of combating the problems that consistently affect PDS systems.

The AccuFlex Rod system was composed of a metal rod with a distinct double helical cut inside of it to permit increased flexibility, primarily in the flexion and extension direction. Because this implant is quite similar to that of conventional metal rod constructs, it may be easily adapted to a surgeon's repertoire of procedures, according to Mandigo et al. (2007). In their study, they compared patients treated with Accuflex rod and with conventional rigid fusion devices. They resolved that the Accuflex technology displayed extremely similar characteristics to rigid fusion devices, demonstrating no significant differences in rate of fusion and highlighting the device's ability to serve as an alternative to other rod-based therapies. Reyes-Sanchez et al. conducted a study with a 2-year follow-up and found that 83% of patients showed a benefit in clinical symptoms after lumbar stabilization with the Accuflex system. They also showcased that the device had a 22% hardware failure rate, which is relatively high compared to other technologies. These competing claims show that the Accuflex system, like others mentioned before, have demonstrated clinical efficacy, in terms of relieving problems associated with lumbar destabilization, but may also show signs of common device complications (Reyes-Sánchez et al. 2010).

Cosmic Posterior Dynamic System is another variant of the pedicle screw and rod system. It employs a 6.25 mm rod which is attached in a non-rigid fashion by pedicle screws with a distinctive hinged screw head, which according to Kim et al., causes load sharing between the anterior vertebral column and the implant. The device is used for conditions such as symptomatic lumbar stenosis, chronically recurring lumbagly in the case of discogenic pain and facet syndrome, recurrent disk herniation, and spondylodesis (Maleci et al. 2011). Moreover, Stoffel et al. analyzed these claims and reviewed 103 patients that were implanted with the Cosmic system and found that 91% of the patients in their study were satisfied with their treatment. Some of the problems displayed in cases involved screw loosening (two patients), disk protrusion in an instrumented segment (three patients), symptomatic degeneration of an adjacent segment (six patients) and osteoporotic fracture of an adjacent vertebra (one patient). Importantly, pain scores were significantly reduced (VAS pre-op 65% +/−1; post op 21% +/−2) and disability scores also decreased showing a marked reduction in ODI by approximately 30% (Stoffel et al. 2010). Safinas, a system similar to the cosmic rod and screw system mentioned above, allows limited motion due to the hinged screw design. Ozer et al. demonstrated that the implementation of this technology resulted in "comparable relief of pain and maintenance of sagittal balance to that of a standard rigid screw-rod fixation" (Ozer et al. 2010). It is evident that the dynamic screw design shows promise in its ability to assist in PDS. There has been wide recognition of the positive outcomes with the use of this technology. While the clinical results are not significantly different than the current rigid fixation techniques, it demonstrates an opportunity for further investigation and research. Additionally, better design clinical studies may highlight the quality-of-life improvements that are currently demonstrated from clinical trials.

An ideological culmination of these technologies presents itself as a dynamic rod and dynamic screw apparatus. This set up entails the utilization of pedicle screws with hinges for increased load sharing and rods that are capable of moving to accommodate for stabilization. Bozkus et al. demonstrated in their biomechanical study that dynamic hinged pedicle screws had a unique ability to increase ROM (flexion extension and lateral bending, and axial rotation). It was noted that this improvement showed a much closer range of motion compared to normal than that of a rigid pedicle screw (30% less than normal ROM, but 160% greater than standard rigid screws) (Bozkus et al. 2010). Kaner et al. reinforces this conclusion. In that study, they assess the use of both dynamic screws and dynamic rods. They observed a significant improvement in the ODI and VAS values of their patients. They also observed "that using dynamic rods with dynamic screws prevented deformity in the rods due to the lower load transfer because of a decrease in the stress shield." This provides an exciting example of a synergistic effect of current technologies with the potential of providing more mobility for patients (Kaner et al. 2009).

Total Facet Replacement Systems

Total Facet Replacement Systems serve the purpose of fully replacing the facet joints of the spine with a mechanical fixture. This surgery has the potential to be "an alternative treatment to lumbar fusion and instrumentation after laminectomy for spinal stenosis" (Serhan et al. 2011). One of the emerging technologies within this space is known as the Total Facet Arthroplasty System (TFAS). The TFAS is "a sliding ball-in-bowl type joint with a pedicle anchor to treat spinal stenosis," according to Serhan et al. The technology was tested biomechanically using cadaveric spines to assess the loading of this type of implant compared to a more conventional rigid posterior instrumentation system (Sjovold et al. 2012). Sjovold et al. found that TFAS implementation produced near intact anterior column load sharing, which was measured by a disc pressure gauge. It was also found that the rigid system displayed larger implant loads than the TFAS system, potentially demonstrating a successful finding that the TFAS system has loading characteristics preferable to those of more rigid systems. However, the study claimed that more testing was needed to understand the physiological implications of such data (Fig. 4).

The total posterior arthroplasty system (TOPS) is a pedicle screw-based device containing an elastic core. This elastic core serves as a flexible apparatus, permitting additional movement in the treated segment. A study evaluating TOPS was conducted by Anekstein et al., in which they sought to measure the clinical outcomes of patients with the TOPS system implanted to relieve their degenerative spondylolisthesis and spinal stenosis. It was found that there was a substantial decrease in VAS scores (88 to 8.8) in a 7-year follow-up. The results from the long-term follow-up permits the discussion that the device is a solid means of mitigating symptoms associated with Spondylolisthesis and Spinal Stenosis. ODI also dropped dramatically (from 49.1 to 7.8) during the 7-year follow-up (Anekstein et al. 2015).

StabilimaxNZ is built upon the neutral zone hypothesis of back pain, according to Panjabi and Timm (2007). The neutral zone was defined as the region of intervertebral laxity around a neutral position. This assumption is contingent on the relationship between spinal instability, movement, and pain. Thus, they hypothesized that an increase in the neutral zone, due to instability or injury, results in accelerated degeneration of discs and the manifestation of back pain. The device was designed with these biomechanical principles in mind and incorporated a pedicle

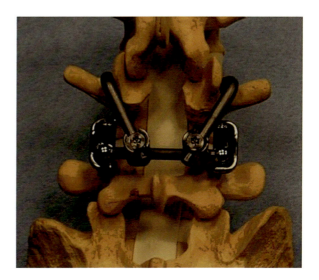

Fig. 4 TFAS construct on an anatomical model

Fig. 5 Stabilimax on an anatomical model

screw-based dynamic stabilization system, dual concentric springs combined with a ball and socket joint at the end. Therefore, according to their hypothesis, the device intends not only to maintain and maximized the range of motion but also add resistance to the passive spinal system to retain a normalized neutral zone, and thus mitigating symptoms. While there has not been any major clinical data published on this device, early biomechanical studies found the device shows promise for single level procedures (Fig. 5).

Posterior stabilization devices that provide immediate postoperative stability and improve chances of arthrodesis in the spinal column have also evolved in parallel with anterior stabilization devices. Cripton et al. investigated the load-sharing properties of lumbar spine segments after being stabilized with a rigid posterior implant (Cripton et al. 2000). Uniaxial strain gauges were used to create six-axis load cells to measure loads and forces through these implants, and pressure transducers measured the IDP. The authors concluded that these implants were not suitable for severe anterior column injuries in the absence of anterior stabilization systems.

These studies showed that PDS devices allow load sharing, but they may not be more efficacious than rigid rod posterior constructs. The rigid systems may also lead to excess load-transfer through the anterior column which can't be handled without anterior plates. Nevertheless, clinical validation through long-term investigations can improve our understanding of these systems.

Spinal Fusion

Posterior Dynamic Stabilization (PDS) and the technology that accompanies it, have remained a vital instrument for surgical implementation. Likewise, there has been tremendous innovation within this space considering the various technologies and approaches to combating common conditions, such as spondylolisthesis, disc degeneration, and other spinal movement disorders. The history of motion preservation requires an examination of the predated rigid body devices. Spinal fusion is a procedure where vertebrae are conjoined thereby creating a greater stabilized structure. While the current gold standard of care remains as rigid spinal fusion, many have argued that the consequences and unintended complications of this system call for a new method of treatment. Thus, the posterior dynamic stabilization (PDS) system emerged as a potential solution. The PDS of vertebrae claimed to yield a beneficial characteristic: it can allow a kinematic signature not found in rigid rod constructs (Gomleksiz et al. 2012; Cheng et al. 2010).

Merits and Downfalls of the PDS System

Understanding the market pressure to adapt to a more dynamic system is contingent upon recognizing the specific pathologies that are resultant of the rigid fusion system. These systems had a propensity of causing disc degeneration at both the upper and lower margins of the therapeutic window, which often manifested as significant osteoporosis. Rigid systems also had an anterior loading preference, and thus resulted in an imbalance of load sharing between the posterior and anterior elements of the vertebra. PDS was intended to ameliorate these specific concerns and, by virtue, engender a new wave of medical device innovation.

Conclusions

The efficacy of a tool is a function heavily influenced by its effectiveness and ease of use. Technologies that need expansive series of training may dissuade surgeons from adapting such a device. The devices covered within this chapter have showed not only innovation within the posterior dynamic stabilization space, but also a conservation of treatment modality in terms of tools and methods used to treat relevant conditions. A surgeon may have a propensity to retain tools and techniques that have been proven rather than explore new alternative forms of treatment, and so it is evident that the devices mentioned above all display characteristics that are similar to the current state of the art (pedicle screw, rod, drill usage). This observation is reinforced by the findings in the World Health Organization report titled, "Increasing complexity of medical technology and consequences for training and outcome of care," (World Health Organization 2010). In conjunction with its analysis of the use of complicated technology, and the burdens that they may cause, the report emphasizes the importance of training and procedural practice to combat any nondevice-related complications. It is imperative that both surgeons and device innovators work in a synergistic manner to achieve a robust and long-standing educational and co-operational relationship to permit the smooth transition of new technologies into the operating space.

An important observation regarding the value of already existing metrics for rigid fixation technology was found during this review: the merit of applying existing metrics rigid fixation technologies to more motion preserving technologies is debatable. Moreover, there may come a time in which new methods of scoring and characterizing PDS technology compared to rigid instrumentation may be necessary to permit the observation of the novel properties of PDS, which may not be easily elucidated through conventional metrics. An example of this principle is the use of the interpedicular travel characteristic (IPT), which was implemented by Cheng et al. in their biomechanical evaluation of the StabilimaxNZ and ScientX technologies. The author found that most biomechanical testing catered to the specifications and characteristics of rigid systems. Therefore, a new metric was needed to characterize more motion preservation devices. The use of IPT was advantageous because it was a novel property that was more founded in motion preserving technology than its more rigid counterpart. While it remains to be seen if there is a direct correlation between this measurement and positive clinical outcome, it provides an example of researchers recognizing that the novel properties of motion preserving technology may not be best tested through the same procedures as more rigid technology. Moreover, this consideration would need to be a joint effort among both biomechanical and clinical scientists to trace the correlation between these new characteristics and their clinical outcome. It has been shown in the literature throughout this chapter that PDS technology may result in similar, preferable results compared to more conventional rigid implants. Thus, this difference must be studied in more detail if there is generally no statistically significant difference between existing rigid technologies and more dynamic ones with current metrics.

Dedication

Dr. Antonio Castellvi was an early adopter and a pioneer in the field of motion preservation technology. Through his clinical work in motion preservation and design contributions to posterior dynamic stabilization constructs, posterior dynamic stabilization gained special consideration and credibility including the development of technologies such as the Scient'X Isobar TTL, Archus Total Facet Arthroplasty System (TFAS), while also challenging the status quo in that rigid fixation is preferable to a motion preservation technology. Graduating with honors from the University of Zaragoza Medical School in Spain and training in orthopedic surgery at the University of South Florida with a fellowship in spine at the University of Rochester, Dr. Castellvi's career spanned continents and brought the leading minds in spine surgery together. A surgeon, a prolific researcher, a mentor, and a friend, Dr. Castellvi's curiosity was only second to his compassion for others. He continues to be missed and remembered; this chapter is in his honor.

References

Anekstein Y, Floman Y, Smorgick Y, Rand N, Millgram M, Mirovsky Y (2015) Seven years follow-up for total lumbar facet joint replacement (TOPS) in the management of lumbar spinal stenosis and degenerative spondylolisthesis. Eur Spine J 24:2306–2314

Barrey C, Perrin G, Champain S (2013) Pedicle-screw-based dynamic systems and degenerative lumbar diseases: biomechanical and clinical experiences of dynamic fusion with isobar TTL. ISRN Orthop 2013:1–10

Bothmann M, Kast E, Boldt GJ, Oberle J (2008) Dynesys fixation for lumbar spine degeneration. Neurosurg Rev 31:189–196

Bozkus H, Senoglu M, Baek S, Sawa AG, Ozer AF, Sonntag VK, Crawford NR (2010) Dynamic lumbar pedicle screw-rod stabilization: in vitro biomechanical comparison with standard rigid pedicle screw-rod stabilization. J Neurosurg Spine 12:183–189

Castellvi AE, Huang H, Vestgaarden T, Saigal S, Clabeaux DH, Pienkowski D (2007) Stress reduction in adjacent level discs via dynamic instrumentation: a finite element analysis. SAS J 1(2):74–81. https://doi.org/10.1016/SASJ-2007-0004-RR. PMID: 25802582; PMCID: PMC4365575

Cheng BC, Bellotte JB, Yu A, Swidarski K, Whiting DM (2010) Historical overview and rationale for dynamic fusion. ArgoSpine News J 22:53–56

Cook DJ, Yeager MS, Thampi SS, Whiting DM, Cheng BC (2015) Stability and load sharing characteristics of a posterior dynamic stabilization device. Int J Spine Surg 9:1–10

Cripton PA, Jain GM, Rwittenberg RH, Nolte LP (2000) Load-sharing characteristics of stabilized lumbar spine segments. Spine 25:170–179

Fabrizi AP, Maina R, Schiabello L (2011) Interspinous spacers in the treatment of degenerative lumbar spinal disease: our experience with DIAM and Aperius devices. Eur Spine J 20:20–26

Floman Y, Millgram MA, Smorgick Y, Rand N, Ashkenazi E (2007) Failure of the Wallis interspinous implant to lower the incidence of recurrent lumbar disc herniations in patients undergoing primary disc excision. J Spinal Disord Tech 20:337–341

Gao J, Zhao W, Zhang X, Nong L, Zhou D, Lv Z, Sheng Y, Wu X (2014) MRI analysis of the ISOBAR TTL internal fixation system for the dynamic fixation of intervertebral discs: a comparison with rigid internal fixation. J Orthop Surg Res 9:1–6

Gomleksiz C, Sasani M, Oktenoglu T, Ozer AF (2012) A short history of posterior dynamic stabilization. Adv Orthop 2012:1–12

Grevitt MP, GArdner ADH, Spilsbury J, Shackleford IM, Baskerville R, Pursell LM, Hassaan A, Mulholland RC (1995) The Graf stabilisation system: early results in 50 patients. Eur Spine J 4:169–175

Hadlow SV, Fagan AB, Hillier TM, Fraser RD (1998) The Graf ligamentoplasty procedure. Spine 23:1172–1179

Heo DH, Cho YJ, Cho SM, Choi HC, Kang SH (2012) Adjacent segment degeneration after lumbar dynamic stabilization using pedicle screws and a nitinol spring rod system with 2-year minimum follow-up. J Spinal Disord Tech 25:409–414

Johnson WEB, Roberts S (2003) Human intervertebral disc cell morphology and cytoskeletal composition: a preliminary study of regional variations in health and disease. J Anat 203:605–612

Kaner T, Sasani M, Oktenoglu T, Cosar M, Ozer AF (2009) Utilizing dynamic rods with dynamic screws in the surgical treatment of chronic instability: a prospective clinical study. Turk Neurosurg 19:319–326

Kang R, Li H, Rickers K, Ringgaard S, Xie L, Bünger C (2015) Intervertebral disc degenerative changes after intradiscal injection of TNF-α in a porcine model. Eur Spine J 24(9):2010–2016. https://doi.org/10.1007/s00586-015-3926-x. Epub 2015 Apr 8. PMID: 25850392

Kettler A, Drumm J, Heuer F, Haeussler K, Mack C, Claes L, Wilke HJ (2008) Can a modified interspinous spacer prevent instability in axial rotation and lateral bending? A biomechanical in vitro study resulting in a new idea. Clin Biomech 23:242–247

Khoueir P, Kim KA, Wang MY (2007) Classification of posterior dynamic stabilization devices. Neurosurg Focus 22:1–8

Kurtz SM, Lanman TH, Higgs G, Macdonald DW, Berven SH, Isaza JE, Phillips E, Steinbeck MJ (2013) Retrieval analysis of PEEK rods for posterior fusion and motion preservation. Eur Spine J 22:2752–2759

Madan S, Boeree NR (2003) Outcome of the Graf ligamentoplasty procedure compared with anterior lumbar interbody fusion with the Hartshill horseshoe cage. Eur Spine J 12:361–368

Maleci A, Sambale RD, Schiavone M, Lamp F, Ozer F, Von Strempel A (2011) Nonfusion stabilization of the degenerative lumbar spine. J Neurosurg Spine 15:151–158

Mandigo CE, Sampath P, Kaiser MG (2007) Posterior dynamic stabilization of the lumbar spine: pedicle-based stabilization with the AccuFlex rod system. Neurosurg Focus 22:1–4

Markwalder TM, Wenger M (2003) Dynamic stabilization of lumbar motion segments by use of Graf's ligaments: results with an average follow-up of 7.4 years in 39 highly selected, consecutive patients. Acta Neurochir 145:209–214

Molinari RW (2007) Dynamic stabilization of the lumbar spine. Curr Opin Orthop 18:215–220

Ormond DR, Albert L, Das K (2016) Polyetheretherketone (PEEK) rods in lumbar spine degenerative disease. Clin Spine Surg 29:E371–E375

Ozer AF, Crawford NR, Sasani M, Oktenoglu T, Bozkus H, Kaner T, Aydin S (2010) Dynamic lumbar pedicle screw-rod stabilization: two-year follow-up and comparison with fusion~!2009-12-30~!2010-02-04~! 2010-03-04~! Open Orthop J 4:137–141

Panjabi MM, Timm JP (2007) Development of stabilimax NZ from biomechanical principles. Int J Spine Surg 1:2–7

Pintauro M, Duffy A, Vahedi P, Rymarczuk G, Heller J (2017) Interspinous implants: are the new implants better than the last generation? A review. Curr Rev Musculoskelet Med 10:189–198

Puzzilli F, Gazzeri R, Galarza M, Neroni M, Panagiotopoulos K, Bolognini A, Callovini G, Agrillo U, Alfieri A (2014) Interspinous spacer decompression (X-STOP) for lumbar spinal stenosis and degenerative disk disease: a multicenter study with a minimum 3-year follow-up. Clin Neurol Neurosurg 124:166–174

Reyes-Sánchez A, Zárate-Kalfópulos B, Ramírez-Mora I, Rosales-Olivarez LM, Alpizar-Aguirre A, Sánchez-Bringas G (2010) Posterior dynamic stabilization of the lumbar spine with the Accuflex rod system as a stand-alone device: experience in 20 patients with 2-year follow-up. Eur Spine J 19:2164–2170

Richter A, Schütz C, Hauck M, Halm H (2009) Does an interspinous device (Coflex™) improve the outcome of decompressive surgery in lumbar spinal stenosis? One-year follow up of a prospective case control study of 60 patients. Eur Spine J 19:283–289

Rigby M, Selmon G, Foy M, Fogg A (2001) Graf ligament stabilisation: mid- to long-term follow-up. Eur Spine J 10:234–236

Schwarzenbach O, Berlemann U, Stoll TM, Dubois G (2005) Posterior dynamic stabilization systems: DYNESYS. Orthop Clin N Am 36:363–372

Senegas J, Etchevers JP, Vital JM, Baulny D, Grenier F (1988) Recalibration of the lumbar canal, an alternative to laminectomy in the treatment of lumbar canal stenosis. Rev Chir Orthop Reparatrice Appar Mot 74(1):15–22. French. PMID: 2967989

Sengupta DK (2004) Dynamic stabilization devices in the treatment of low back pain. Orthop Clin North Am 35:43–56

Sengupta DK, Herkowitz HN (2012) Pedicle screw-based posterior dynamic stabilization: literature review. Adv Orthop 2012:1–7

Sengupta DK, Mulholland RC (2005) Fulcrum assisted soft stabilization system. Spine 30:1019–1029

Serhan H, Mhatre D, Defossez H, Bono CM (2011) Motion-preserving technologies for degenerative lumbar spine: the past, present, and future horizons. SAS J 5:75–89

Sjovold SG, Zhu Q, Bowden A, Larson CR, De Bakker PM, Villarraga ML, Ochoa JA, Rosler DM, Cripton PA (2012) Biomechanical evaluation of the Total Facet Arthroplasty System® (TFAS®): loading as compared to a rigid posterior instrumentation system. Eur Spine J 21:1660–1673

Sobottke R, Schlüter-Brust K, Kaulhausen T, Röllinghoff M, Joswig B, Stützer H, Eysel P, Simons P, Kuchta J (2009) Interspinous implants (X Stop®, Wallis®, Diam®) for the treatment of LSS: is there a correlation between radiological parameters and clinical outcome? Eur Spine J 18:1494–1503

Stoffel M, Behr M, Reinke A, Stüer C, Ringel F, Meyer B (2010) Pedicle screw-based dynamic stabilization of the thoracolumbar spine with the Cosmic®-system: a prospective observation. Acta Neurochir 152:835–843

Stoll TM, Dubois G, Schwarzenbach O (2002) The dynamic neutralization system for the spine: a multi-center study of a novel non-fusion system. Eur Spine J 11(Suppl 2):S170–S178

Vergroesen PP, Kingma I, Emanuel KS, Hoogendoorn RJ, Welting TJ, Van Royen BJ, Van Dieen JH, Smit TH (2015) Mechanics and biology in intervertebral disc degeneration: a vicious circle. Osteoarthritis Cartilage 23:1057–1070

Wang C, Yu X, Yan Y, Yang W, Zhang S, Xiang Y, Zhang J, Wang W (2017) Tumor necrosis factor-alpha: a key contributor to intervertebral disc degeneration. Acta Biochim Biophys Sin (Shanghai) 49:1–13

Welch WC, Cheng BC, Awad TE, Davis R, Maxwell JH, Delamarter R, Wingate JK, Sherman J, Macenski MM (2007) Clinical outcomes of the Dynesys dynamic neutralization system: 1-year preliminary results. Neurosurg Focus 22:1–8

World Health Organization (2010) Increasing complexity of medical technology and consequences for training and outcome of care: background paper 4, August 2010. World Health Organization

Xu C, Ni W-F, Tian N-F, Hu X-Q, Li F, Xu H-Z (2013) Complications in degenerative lumbar disease treated with a dynamic interspinous spacer (Coflex). Int Orthop 37:2199–2204

Zhang HY, Park JY, Cho BY (2009) The BioFlex system as a dynamic stabilization device: does it preserve lumbar motion? J Korean Neurosurg Soc 46:1066–1070

Zhang L, Shu X, Duan Y, Ye G, Jin A (2012) Effectiveness of ISOBAR TTL semi-rigid dynamic stabilization system in treatment of lumbar degenerative disease. Zhongguo Xiu Fu Chong Jian Wai Ke Za Zhi 26:1066–1070

Design Rationale for Posterior Dynamic Stabilization Relevant for Spine Surgery

13

Ashutosh Khandha, Jasmine Serhan, and Vijay K. Goel

Contents

Introduction	294
Spinal Fusion and Structural Integrity	294
Spinal Fusion and Related Complications	295
Rationale for Dynamic Stabilization and Device Classification	296
Posterior Dynamic Stabilization: Methods for Testing and Performance Evaluation	298
Pedicle-Based PDS Devices: Preclinical In Vitro Mechanical Testing	298
Preclinical In Vitro Biomechanical Testing and Simulation	298
Evaluating In Vivo Performance	299
Pedicle Screw-Based PDS Devices: In Vivo Performance and Failure Modes	301
Accuflex System (Globus Medical Inc.)	301
BioFlex System (Bio-Spine)	301
CD Horizon Agile (Medtronic Sofamor Danek)	302
Cosmic Posterior Dynamic System (Ulrich Medical)	302
Dynesys (Zimmer Biomet)	302
Graf Ligament (SEM Co.)	303
Isobar TTL (Scient'x)	303
NFlex (Synthes Spine)	304
Stabilimax NZ (Rachiotek LLC)	304

A. Khandha (✉)
Department of Biomedical Engineering, College of Engineering, University of Delaware, Newark, DE, USA
e-mail: ashutosh@udel.edu

J. Serhan
Department of Biological Sciences, Bridgewater State University, Bridgewater, MA, USA

V. K. Goel
Engineering Center for Orthopaedic Research Excellence (E-CORE), University of Toledo, Toledo, OH, USA

Departments of Bioengineering and Orthopaedic Surgery, Colleges of Engineering and Medicine, University of Toledo, Toledo, OH, USA
e-mail: Vijay.Goel@utoledo.edu

© Springer Nature Switzerland AG 2021
B. C. Cheng (ed.), *Handbook of Spine Technology*,
https://doi.org/10.1007/978-3-319-44424-6_24

Percudyn (Interventional Spine) 306
Posterior Dynamic Interspinous Devices: In Vivo Performance and Failure Modes ... 307

Discussion 311

References 312

Abstract

Motion sparing posterior dynamic stabilization (PDS) devices have been introduced as an alternative to spinal fusion. A majority of these devices are based on instrumentation and techniques that surgeons are most familiar with, due to their experience with posterior fixation for spinal fusion. The goal of this new generation of devices is to allow controlled motion of the treated spinal segment that closely mimics physiologic spinal kinetics and kinematics, with the most common indication for use being spinal stenosis. The rationale for dynamic stabilization as an alternative to spinal fusion is to restore spinal stability, while avoiding (or delaying) degeneration of adjacent segments. Most commonly used PDS devices are either pedicle screw-based or interspinous process-based. The pedicle screw-based devices are commonly approved for use in spinal fusion, or as an adjunct to fusion, but not as stand-alone devices in the absence of fusion. Despite familiar surgical techniques and extensive preclinical testing, most pedicle screw-based PDS devices are still considered investigational for the treatment of disorders of the spine. One of the main reasons is that it is not yet clear whether PDS truly offer advantages over conventional spinal fusion or decompression alone, in terms of patient reported outcome scores. Other technical factors that pose a challenge for PDS devices are long-term fixation to the spine via pedicle screws or interspinous fixation, and variations in device stiffness, level of stabilization offered, and the range of motion allowed by PDS devices over time. This chapter presents an overview of in vitro testing methodologies used to evaluate PDS devices, followed by a summary of clinical performance of stand-alone dynamic stabilization devices with or without direct decompression.

Keywords

Spine · Dynamic stabilization · Biomechanics · Posterior Stabilization · Design rationale · Metrics · Spine surgery · Interspinous devices

Introduction

Spinal Fusion and Structural Integrity

Spinal surgery may be performed to address biomechanical instability introduced in the spinal column due to trauma (Puttlitz et al. 2000; Benzel 2001c), infection (Weiss et al. 1997), or tumors (Bakar et al. 2016). Besides addressing instability, the most common objective for performing surgery is treating pain by achieving neural decompression, correcting deformity, and addressing aberrant spinal kinematics (Schlenk et al. 2003; Panjabi and Timm 2007).

Surgery disrupts either the passive load sharing elements (ligaments and bone) or active musculature, or both. Hence, the surgical procedure itself can destabilize the spine (Hasegawa et al. 2013; Vadapalli et al. 2006; Benzel 2001b). To address biomechanical instability and to compensate for the destabilization introduced by surgery, fusion devices are considered the "gold standard" for treatment (Serhan et al. 2011). Over 400,000 fusion discharges occur annually in the United States (Rajaee et al. 2012).

An intervertebral fusion device contains bone graft (or substitute) that promotes bone healing and osteogenesis, and this process is enhanced during weight-bearing activities (Egger et al. 1993). However, to avoid excessive loading and motion, particularly during the bone healing process, the spinal segment is immediately immobilized by additional hardware commonly implanted in the posterior region. This allows for early overall mobility for patients, while also needing less external support (Shono et al. 1998). Over time, as the structural integrity of the bone fusion increases, the integrity of posterior fixation device component can decrease (Benzel 2001a). As shown in Fig. 1, the theoretical net structural integrity (combination of bone fusion and posterior fixation device) stays the same over time. In the absence of adequate bone fusion, late failure of posterior fixation can occur (Bellato et al. 2015; Agarwal et al. 2009).

This highlights an important functional requirement for posterior dynamic stabilization (PDS) devices that will be discussed further in this chapter: a PDS device, which is commonly used "without" a bone graft, needs to maintain its structural integrity over a longer period of time. Hence, fatigue-strength-enhancement is crucial for a PDS device (Bhamare et al. 2013).

Spinal Fusion and Related Complications

When the goal is spinal segment immobilization to address gross instability, whether due to spine deformation-related issues, trauma, or tumors, spinal fusion surgery may be the only viable alternative. However, irreversible bone fusion can have a negative impact when addressing a smaller amount of instability, as in the case of spinal decompression surgery for stenosis. When a spinal segment is irreversibly fused, and overall patient mobility is desirable, the vertebral levels adjacent to the fused segment are subjected to additional loading and stress during activities of daily living (Lee and Langrana 1984). This phenomenon is termed as adjacent segment disease (Fig. 2), or ASD (Saavedra-Pozo et al. 2014; Panjabi and Timm 2007; Lindsey et al. 2015). ASD is defined as the presence of new degenerative changes at adjacent spinal levels, accompanied by radiculopathy, myelopathy, or instability (Saavedra-Pozo et al. 2014). The incidence of ASD is approximately 3% in the cervical spine and approximately 8% in the lumbar spine (Saavedra-Pozo et al. 2014). When considering the occurrence of ASD, it is important to differentiate between radiographic and symptomatic ASD (Virk et al. 2014). Also, given the average age of the population being treated, ASD, at least in part, is also related to the natural history of disc degeneration and not just altered biomechanics due to surgical treatment (Saavedra-Pozo et al. 2014). Hence, determining a cause-and-effect relationship in vivo is challenging.

In accordance with Wolff law, some level of compressive forces borne by the bone fusion mass is necessary for fusion and healing to occur (Kowalski et al. 2001). Excessively rigid posterior spinal fixation devices can also lead to stress shielding of the fusion mass (Saphier et al. 2007; Kanayama et al. 2000). Stress shielding refers to a

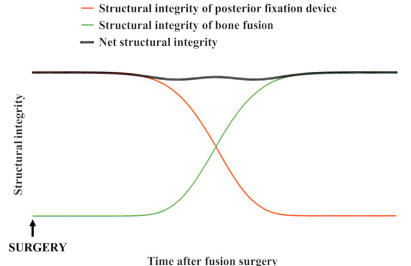

Fig. 1 Structural integrity after fusion surgery. (Source: Created in Microsoft Excel, adapted from "Benzel, E.C., 2001a. Spinal Fusion. In Biomechanics of spine stabilization. American Association of Neurological Surgeons, pp. 121–133")

reduction of load and stress seen by bone fusion mass (< ~70% of the total load), as a disproportionately large amount of the total load may be borne by the posterior fixation device (Fig. 3). This occurrence can further be complicated due to low bone-mineral density and osteoporosis (Bhamare et al. 2013; Park et al. 2013).

Failed bony fusion, or pseudarthrosis, is also an iatrogenic complication, with incidence rates ranging from 5% to 35% in the lumbar spine (Chun et al. 2015). While controversial, it is important to note that according to the United States Food and Drug Administration (FDA) guidelines, greater than 3 mm of translation motion and greater than 5 of angular motion on flexion-extension radiographs should be considered as a failed bony fusion (Gruskay et al. 2014; Chun et al. 2015).

Donor-site morbidity (due to bone grafting for fusion mass) is also a complication reported after spinal fusion (Vaz et al. 2010), which may be addressed by using alternatives such as recombinant human bone morphogenetic proteins (rhBMPs). Prolonged recuperation time also remains a concern (Serhan et al. 2011). Overall, patient satisfaction rate for lumbar spinal fusion averages around 60–70% (Turner et al. 1992; Slosar et al. 2000).

Rationale for Dynamic Stabilization and Device Classification

To address some of the limitations posed by fusion surgery, there has been a growing interest in the field of dynamic spine stabilization (Bhamare

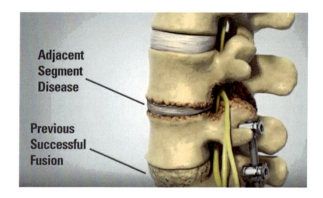

Fig. 2 Adjacent segment disease (ASD) after fusion surgery. (Source: https://www.youtube.com/watch?v=yQwYISvBkzo)

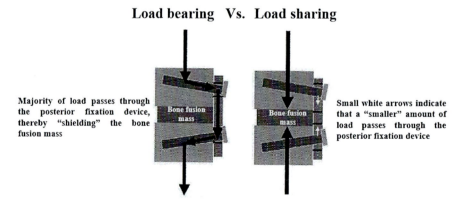

Fig. 3 Load bearing vs. load sharing after fusion surgery. (Source: http://www.bioline.org.br/showimage?ni/photo/ni05146f1.jpg, adapted from "Benzel, E.C., 2005. Spine Surgery: Techniques, Complication Avoidance, and Management")

et al. 2013). These devices may be viable alternatives addressing a range of spinal disorders, including stenosis and discogenic low back pain (Serhan et al. 2011). The rationale for dynamic stabilization is that by preserving functional range of spinal motion, one can alleviate at least some of the complications related to spinal fusion listed above. It should be noted that up to 5° of angular motion may be present on flexion-extension radiographs in the case of a successful fusion (Gruskay et al. 2014; Chun et al. 2015). Hence, if the ROM allowed under a similar radiographic evaluation for a dynamic stabilization device is less than 5°, justifying the use of the device as a truly non-fusion dynamic stabilization device is controversial. To the best of our knowledge, no pedicle screw-based PDS device has been approved by the FDA for use other than an adjunct to spinal fusion (Fig. 4).

While a dynamic stabilization device may not increase the range of motion (ROM) of the segment being treated, the objective is to preserve normal motion as much as possible, while at the same time limiting abnormal motion (Sengupta and Herkowitz 2012). In the case of a PDS device, some loss of ROM (compared to ROM before surgery) may be unavoidable (Sengupta and Herkowitz 2012). Another important consideration for a dynamic stabilization device is to ensure the adequate level of load transfer through the joint. In the case of a PDS device, it has to sustain loads for a longer amount of time, compared to posterior fixation devices used for fusion, since there is no bone fusion mass (Fig. 1). Hence, to avoid fatigue failure and implant loosening, which are often seen in a PDS device (Bhamare et al. 2013), the PDS device should be load-sharing, and not load-bearing (Sengupta and Herkowitz 2012). While there is no fusion mass to share load with (Fig. 3), the PDS device should be able to share load with other load-bearing spinal components. It should be noted that ROM and loading can be interdependent (Grob et al. 2005; Mulholland and Sengupta 2002; Kirkaldy-Willis and Farfan 1982; Doria et al. 2014), and hence, alteration (or restoration) of one may also impact the other.

One way to classify dynamic stabilization devices is by defining whether the device replaces an existing joint or a mobile anatomical region, or whether it augments it. Thus, preservation of motion after surgery can be achieved by either replacing the entire intervertebral disc (disc replacement), just the nucleus (nucleus replacement), or the facet joints (facet replacement). Alternately, preservation of motion after surgery can be achieved by augmenting the posterior spinal elements. The indications for use of each of these devices can be very different. However, from a biomechanical perspective, each device aims to address the instability introduced by surgery by allowing "some" motion at the joint

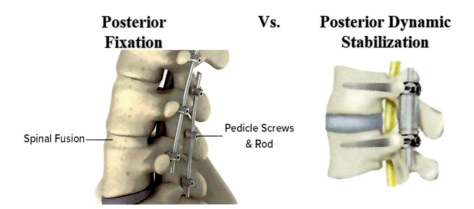

Fig. 4 Posterior fixation vs. Dynesys posterior dynamic stabilization device with flexible components. (Sources: SpinalFusion.jpg and https://www.hindawi.com/journals/aorth/2013/753470.fig.0012.jpg, and "What is Spinal fusion?." *Atlantic Brain and Spine*, www.brainspinesurgery.com/spinal-fusion/)

(vs. fusing the joint) and sharing load within the joint.

This chapter will focus on posterior dynamic stabilization: devices that either allow some motion or control motion at a spinal joint, by augmenting the posterior spinal elements, that is, PDS devices, with a focus pedicle screw and interspinous PDS devices used in the lumbar spine.

Posterior Dynamic Stabilization: Methods for Testing and Performance Evaluation

Pedicle-Based PDS Devices: Preclinical In Vitro Mechanical Testing

Static and dynamic reliability testing of PDS devices is based on standards developed by the American Society for Testing and Materials (ASTM) and/or the International Standardization Organization (ISO). For pedicle screw-based PDS devices, the ASTM F1717 and/or ISO 12189 standards are used for assembly level testing (Fig. 5) (La Barbera et al. 2015), wherein the complete instrumentation system is subjected to bending loads and stresses (Bhamare et al. 2013). These standards describe implant assembly with simulated vertebral body test blocks in either a vertebrectomy model (ASTM F1717) or a model with anterior support (ISO12189 – calibrated springs – Fig. 5) (La Barbera and Villa 2017). While the F1717 standard reflects the worst-case load-bearing scenario, the ISO12189 standard reflects a load-sharing scenario (Fig. 3). In the context of PDS devices, an important distinction between the two standards is the ASTM F1717 may not be directly usable, due to the combination of the allowable degree of freedom in the simulated vertebral body test blocks and the allowable motion of PDS device itself.

Component and Interface Level Static and Dynamic Testing

Component and interface (bone-implant as well as inter-component) testing is also performed both statically and dynamically. In the case of pedicle screw-based PDS devices, component level performance is commonly performed for pedicle screw pullout (ASTM F543) and bending loads (ASTM F1798) as well as for flexible rod component bending strength (ASTM F2193).

For both component and interface level testing, dynamic cyclic testing for pedicle screw-based PDS systems is performed to a runout of 10 million cycles. With 125 significant bends performed annually, 10 million cycles represents 80 years of wear (Vermesan et al. 2014; Schwarzenbach et al. 2005). This testing characterizes the asymptotic endurance level for load/stress, that is, the level below which the implant/ component/material does not fail and can be cycled infinitely.

Preclinical In Vitro Biomechanical Testing and Simulation

PDS devices are commonly evaluated for biomechanical performance characterization using

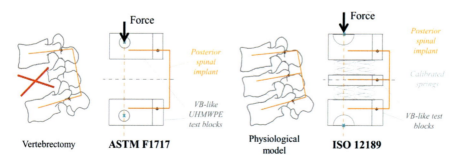

Fig. 5 Testing setups for posterior spinal implants per ASTM F1717 and ISO12189 standards. (Source: https://ars.els-cdn.com/content/image/1-s2.0-S1529943015012024-spinee56502-fig-0001_lrg.jpg)

cadaveric experiments. The primary modes of loading tested in these experiments are shown in Fig. 6 below.

Physiologic loading and range of motion are applied to cadaveric specimens by applying pure moments and a compressive follower load (Patwardhan et al. 1999). Specimens are tested intact, after destabilization surgery, and finally after device implantation under load control or by using a hybrid testing protocol (Goel et al. 2005; Bennett et al. 2015). Testing can also be simulated using finite element (FE) modeling, provided the FE model is validated against experimental results. Figure 7 shows an FE model of the lumbar spine and the corresponding cadaveric experimental setup for testing a dynamic stabilization system.

In addition to characterizing range of motion (ROM), biomechanical testing and simulation also allow for quantification of interpedicular travel (IPT) and displacement (Fig. 8), which is particularly useful for design, development and optimization of dynamic stabilization devices (Cook et al. 2012; Yeager et al. 2015). Limiting interpedicular motion in PDS implants may lead to implant loosening over time (Lima et al. 2017). Using these testing and simulation methods, it has been determined that an axial stiffness of 45 N/mm and bending stiffness of 30 N/mm can reduce spinal ROM by 30% (compared to intact specimen ROM), and this is thought to be an optimal level of motion reduction after surgery (Erbulut et al. 2013; Schmidt et al. 2009). When pedicle screw-based PDS have stiffness characteristics that are greater than optimum, there can be a larger reduction in ROM, thereby rendering their performance almost similar to fusion devices.

Evaluating In Vivo Performance

In addition to ROM measurements from in vivo flexion extension radiographs, IPT measurements can also be characterized in vivo. More recently, translation per degree of rotation (TPDR – Fig. 9) and qualitative stability index (QSI) have been used to characterize instability in vivo (Hipp et al. 2015). A QSI score of 2 indicates a TPDR value 2 standard deviations compared to values observed in healthy controls, and this in turn may indicate instability and poor quality of motion. Similar measurements may also be performed using fluoroscopy (Davis et al. 2015), and these instability measurements can be adapted for evaluating in vivo ROM quality and characterizing in vivo performance of PDS devices. Finally, patient reported outcome measures (PROMs) (Nayak et al. 2015) that quantify quality of life, pain, and disease-specific disability after surgery

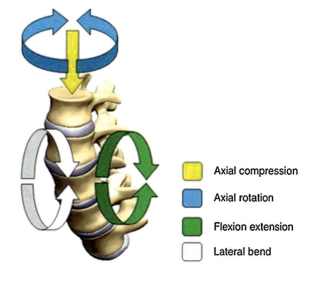

Fig. 6 Primary modes of loading tested in a cadaveric experimental setup. (Source: https://clinicalgate.com/dynamic-stabilization-of-the-lumbar-spine-indications-and-techniques/)

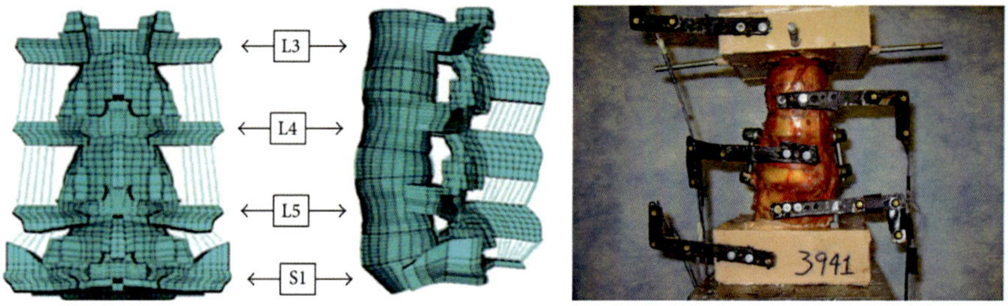

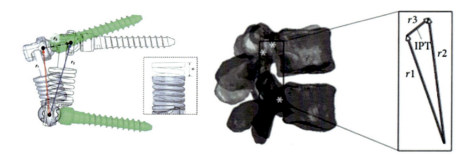

Fig. 7 Finite element modeling of the spine and the corresponding experimental setup. (Source: https://www.ncbi.nlm.nih.gov/pmc/articles/PMC3626386/)

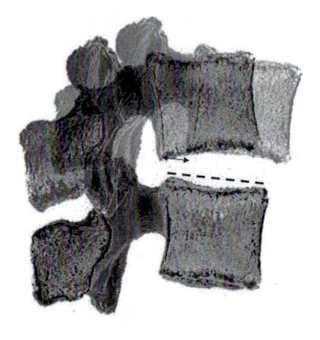

Fig. 8 Interpedicular travel (r3) and displacement measurement for a PDS device during in vitro biomechanical testing. (Sources: http://www.isass.org/pdf/sas10/4-Friday/Abstract_301.pdf and https://www.hindawi.com/journals/aorth/2015/895931/)

Fig. 9 Measurement of TPDR (translation per degree of rotation) from radiographs. (Source: https://www.ncbi.nlm.nih.gov/pmc/articles/PMC4528437/)

are critical for evaluating the long-term performance of PDS devices.

Pedicle Screw-Based PDS Devices: In Vivo Performance and Failure Modes

Pedicle screw-based PDS devices are based on instrumentation and techniques that surgeons are most familiar with, due to their experience with posterior fixation for spinal fusion (Barrey et al. 2008). These devices are commonly approved for use in spinal fusion, or as an adjunct to fusion, but not as stand-alone devices in the absence of fusion. Despite familiar surgical techniques and extensive preclinical testing, pedicle screw-based PDS devices are still considered investigational for the treatment of disorders of the spine. One of the main reasons is that it is not yet clear from randomized clinical trials (RCTs) whether pedicle screw-based PDS truly offer advantages over conventional spinal fusion, in terms of health outcomes. Other reasons range from some PDS devices not being truly dynamic (in vivo range of motion is similar to fusion) to device failure and screw loosening (Kaner et al. 2010b; Stoffel et al. 2010; Kocak et al. 2010; Grob et al. 2005; Chen et al. 2011).

Below is a summary of some of the pedicle screw-based PDS devices that have been studied in vivo as stand-alone devices, that is, without fusion and bone graft. A discussion of failure modes, where applicable, is also included.

Accuflex System (Globus Medical Inc.)

The Accuflex system (Fig. 10) consists of a flexible rod anchored by pedicle screws made of titanium alloy. Flexibility in the rod is achieved by helical cuts along the length of the rod. The flexible rod system has undergone extensive in vitro static and dynamic biomechanical testing (Reyes-Sánchez et al. 2010). In a 20-patient study with 2-year follow-up, improvements in all clinical measurements and PROMs were observed (Reyes-Sánchez et al. 2010). However, hardware fatigue failure was also observed in ~22% of the subjects. Failure included rod breakage as well as pedicle screw breakage in the bone. Both these failure mechanisms were caused due a combination of a large bending moment and stress concentration in the failure regions.

BioFlex System (Bio-Spine)

The BioFlex system (Fig. 11) consists of a flexible spring made out of Nitinol (a shape memory alloy) anchored by pedicle screws made out of titanium alloy. In a 12-patient study with 2-year follow-up, reduced ROM was observed at the treated level (compared to ROM before surgery), with minimal

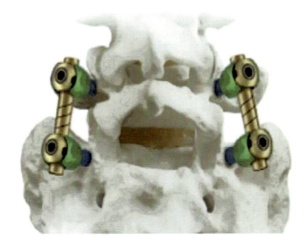

Fig. 10 Accuflex system with a flexible rod. (Source: https://www.ncbi.nlm.nih.gov/pmc/articles/PMC4365627/)

Fig. 11 BioFlex system with flexible springs. (Source: https://pubmed.ncbi.nlm.nih.gov/20401848/)

changes at adjacent levels (Zhang et al. 2009). In another study with short-term follow-up (less than 1 year), 28 patients treated solely with the BioFlex (Kim et al. 2007), a similar reduced ROM was observed at the treated level. Limited long-term data is available for this device. It should also be noted that Nitinol is a notch-sensitive material which can reduce fatigue strength (Yoshihara 2013). Notch sensitivity describes the sensitivity of a material to geometric discontinuities and can have a significant negative effect on fatigue strength.

CD Horizon Agile (Medtronic Sofamor Danek)

In the CD Horizon Agile system (Fig. 12), the rod component between the pedicle screws is available in different sizes to offer a less stiff (longer spacer) or a more stiff (shorter spacer) option for dynamic stabilization. The spacer, made out of a thermoplastic polymer (polycarbonate urethane or PCU), encloses a titanium alloy cable. While allowing a greater ROM that most other PDS devices, the implant was noted to break due to shear-related failure of the cable component, particularly in cases of advanced instability (Doria et al. 2014). Shear-related failure occurred due to kinking of the cable component during anterior-posterior translation of the spinal segment (Hoff et al. 2012).

Cosmic Posterior Dynamic System (Ulrich Medical)

The Cosmic posterior dynamic system (Fig. 13) includes a hinged pedicle screw which can reduce stresses at the bone screw interface while allowing segmental motion (Gomleksiz et al. 2012). The pedicle screw (threads) includes a calcium phosphate coating to promote osteointegration. The rod in this system is rigid. In a study with 30 patients and over 3 years of follow-up (Kaner et al. 2010a), significant improvement in PROMs were observed, and no screw breakage was observed. One instance of screw loosening was reported.

Dynesys (Zimmer Biomet)

Dynesys (Fig. 4, right) has the largest amount of clinical follow-up data, compared to other pedicle screw-based PDS. Between the pedicle screws, the system consists of a thermoplastic spacer (PCU) that encloses a cord (made out of polyethylene terephthalate or PET). A comprehensive literature review (Pham et al. 2016) spanning 21 studies and a total of 1166 patients with mean follow-up of almost 3 years has shown that the pedicle screw loosening rate is ~12% (higher than the rate commonly observed after fusion) and ASD rate is ~7%, (slightly lower than the rate commonly observed after fusion). The pedicle screw fracture rate for Dynesys was less than 2%. In another study with 46 patients and mean follow-up of over 4 years (Zhang et al. 2016), significant improvements in PROMs were observed for patients treated with Dynesys,

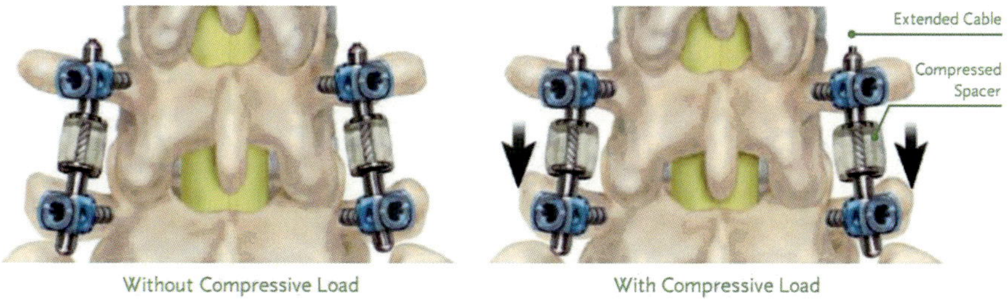

Fig. 12 CD Horizon Agile. (Source https://pubmed.ncbi.nlm.nih.gov/20401848/)

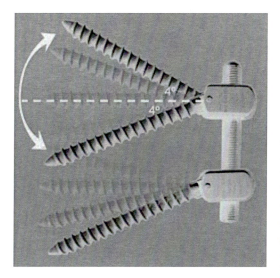

Fig. 13 Cosmic PDS with a hinged pedicle screw. (Source: https://pubmed.ncbi.nlm.nih.gov/20401848/)

as well as for patients treated with fusion. While the mean ROM (flexion-extension radiographs) was lower than 5° for both groups (patients treated with Dynesys or fusion), the Dynesys system did allow slight greater ROM and lower ASD rate, compared to patients treated with fusion.

Graf Ligament (SEM Co.)

The Graf ligament (Fig. 14) represents the earliest attempts in using a flexible PDS. The device includes a braided polyester (polypropylene) tension band between titanium pedicle screws.

The hypothesis for this device was that abnormal rotational motion was responsible for pain generation, and this device was designed to control the same by locking the lumbar facets in an extended position (Doria et al. 2014; Erbulut et al. 2013). The Graf ligament transfers load from the anterior disc to the posterior annulus, increasing disc pressure, which can accelerate disc degeneration (Gomleksiz et al. 2012) and even cause lateral recess stenosis. In a review of 43 patients with a minimum of 8 years follow-up (Choi et al. 2009), angular instability was observed in 28% of the segments, while translational instability was observed in 5% of the segments. Additionally, adjacent segment instability was observed in 42% and 30% of the subjects at the upper and lower segments, respectively. No instrumentation failures were reported. In another study with 31 patients and 7-year follow-up, significant improvements in PROMs have been reported, despite an established degenerative process (Gardner and Pande 2002).

Isobar TTL (Scient'x)

The Isobar TTL system (Fig. 15) is composed of a semirigid titanium alloy rod with a dampener stacked with titanium alloy rings. This rod is inserted between titanium alloy pedicle screws and the system allows some axial and angular motion. In a review of 37 patients with a mean follow-up of 2 years, excellent improvement PROMs have been reported (Li et al. 2013). However, ROM after surgery was significantly lower (compared to ROM before surgery) and new signs

Fig. 14 Graf ligament inserted between pedicle screws. (Source: https://www.ncbi.nlm.nih.gov/pmc/articles/PMC4365627/)

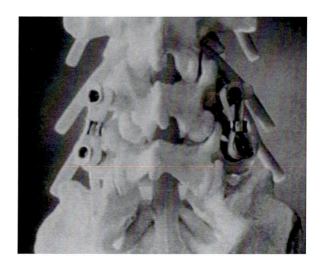

Fig. 15 Isobar semi-rigid rod. (Source https://www.ncbi.nlm.nih.gov/pmc/articles/PMC4365627/)

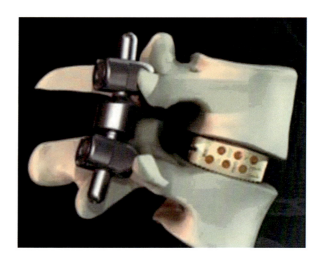

of degeneration were observed at adjacent levels in 39% of the patients, with 8% of the patients requiring revision due to ASD.

NFlex (Synthes Spine)

In the NFlex device (Fig. 16), a polyaxial titanium alloy pedicle screw is affixed to a central titanium core which is integrated with a PCU spacer. This design allows for a physiologic change in interpedicular distance (Fig. 8). In a study reporting 2-year clinical outcomes in 65 patients (Coe et al. 2012), 25 patients received non-fusion dynamic stabilization solely with Isobar TTL. Significant improvements in PROMs were observed in these patients, with one instance each of rod fracture and pedicle screw loosening.

Stabilimax NZ (Rachiotek LLC)

The Stabilimax NZ device (Fig. 17) aims to provide maximum support in the neutral zone (NZ – the initial portion of the total range of

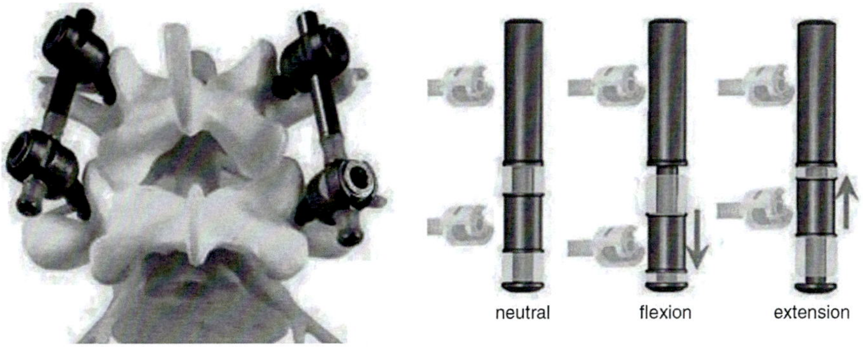

Fig. 16 NFlex device in neutral, flexion and extension positions. (Source: https://www.ncbi.nlm.nih.gov/pmc/articles/PMC3424174/)

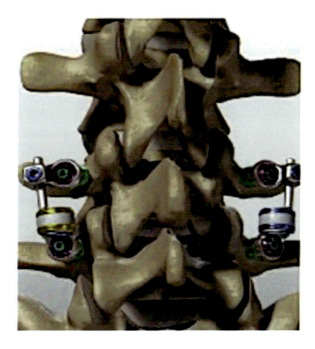

Fig. 17 Stabilimax NZ device dual springs and ball-socket joints. (Source: https://www.ncbi.nlm.nih.gov/pmc/articles/PMC4365627/)

motion, where minimal resistance to motion is offered by passive spinal structures) while maintaining maximum possible total range of motion (reduced support in the final portion of total range of motion, where maximal resistance to motion is offered by active and passive spinal structures) (Panjabi and Timm 2007). This is achieved through the use of dual concentric springs that permit physiologic interpedicular travel and the use of ball and socket joints to reduce bending moment at the bone screw interface and permitting axial rotation. In a preliminary report on 60 patients with 2-year follow-up (Neel Anand et al. 2012), significant improvements in PROMs were observed. IPT travel (Fig. 8) was also physiologic. However, pedicle screw breakage was also seen in 10% of the cases. Grit blasted surface of the pedicle screws was found to be the root cause of failure (grit blasting of titanium alloy screws can promote osteointegration, but it can also make the surface notch sensitive, thereby reducing fatigue life). The surface treatment was later changed using laser shock peening (LSP). LSP improves fatigue life by impacting residual stresses (Bhamare et al. 2013).

Percudyn (Interventional Spine)

In the Percudyn device (Fig. 18), a PCU stabilizer is installed onto an anchor. This is a pedicle screw-based device without an interpedicular connection. Biomechanically, the Percudyn device serves to augment the posterior elements of the functional facet by serving as a mechanical stop between the inferior and superior articular facets (Smith et al. 2011). In a study reporting on 96 patients at a 2-year follow-up period (Canero and Carbone 2015), significant improvements were observed in PROMs, with more than 70% of the patients satisfied with the procedure, while 10% of the patients required revision surgery at longer follow-up.

Interspinous Devices: Preclinical In Vitro Mechanical Testing

The motion preserving interspinous devices could be divided into devices that oppose motion in a rigid manner and devices that oppose it in a flexible manner. Rigid, or static, devices consist of relatively noncompressible solid materials like titanium or PEEK; their main function is to ensure a consistent level of posterior distraction during extension. The flexible interspinous devices allow for compression during extension and could be classified as flexible/dynamic devices. They offer a higher level of elasticity that allows their deformation during extension. This is achieved by the material and/or their shape.

Parchi et al. 2014 have characterized the biomechanical effects of interspinous devices by:

1. Modifying/Stabilizing the motion segment and altering the range of motion (ROM)
2. Decompression of the spinal canal and foramina via posterior distraction
3. Reduction of intradiscal pressure and facet load
4. Impact on sagittal alignment and instantaneous axis of rotation (IAR) of the treated segment

Human cadaveric studies to investigate the range of motion, instantaneous access or rotation, or measuring the intradiscal pressures of intact condition and post-decompression and/or interspinous device insertion are commonly used to evaluate the in vitro performance of these devices. Several biomechanical studies on interspinous device are reported in literature (Lindsey et al. 2003, Phillips et al. 2006, Tsai et al. 2006, Lafage et al. 2007). In cadaveric studies, interspinous devices improve the stability of the treated motion segment in flexion-extension but do not stabilize the spine in axial rotation or lateral bending. Zheng et al. (2010) found also that size of the interspinous device affect their performance, smaller interspinous device did not provide the stabilization of larger devices. He found that using a spacer with height equal to the distance of the interspinous process was associated with a slight flexion of the segment and less effects on the dimension of the spinal canal and foramen. An oversized device, on the other hand, could induce a kyphotic position and may increase disc loading. Selecting the appropriate device design, size, and material while taking in consideration the treatment goal, patients' pathology, bone quality, and

Fig. 18 Percudyn PCU spacer inserted onto the anchor. (Source: https://www.ncbi.nlm.nih.gov/pmc/articles/PMC4365627/)

symptoms should be carefully considered to achieve the best biomechanical and clinical outcome.

Posterior Dynamic Interspinous Devices: In Vivo Performance and Failure Modes

The interspinous devices were designed as an alternative treatment for neurogenic claudication and pain which is attributed to facet joint disease. The spine is kept in a flexed position by which the interspinous devices increase the total canal and foraminal size, which decompresses the cauda equina, which is responsible for neurogenic claudication. This device allows for neural decompression with minimal tissue resection; thus, the device is less invasive and can be implanted without a laminectomy. It avoids the risk of epidural scaring and cerebrospinal fluid leakage by functioning through indirect decompression. In some cases, interspinous dynamic stabilization is used to prevent the instability that occurs after decompression.

These devices limit extension of the spine, allow for the unloading of the facet joint, and allow for the relief of pain attributed to facet disease as well (Khoueir et al. 2007). The notion of interspinous device to produce segmental posterior distraction was first introduced in the 1960s by Dr. Fred Knowles. He is better known for his hip pin design; however, he reported limited success with the spinal device due to subsidence and displacement. His ideas were latter improved upon, in the form of the Xstop device (Kyphon, Sunnyvale, California). There have been multiple interspinous devices which have been developed, such as the X-stop, DIAM, Wallis system, and the CoFlex system. All these devices work to limit spinal extension. The interspinous spacers may be helpful when more conservative (nonoperative) care does not improve symptoms. All of these devices allow the spine to be held in a position of slight flexion, in order to decompress the spinal cord or nerve roots. The spine, however, may still rotate axially or bend laterally when the device is in place.

The Wallis System (Zimmer)

The Wallis system was the first interspinous device introduced in Europe around 1986 and was developed by Sénégas (Fig. 19). The design originated with a titanium block inserted between adjacent processes, which is then held in place with a flat Dacron cord or ribbon wrapped around the spinous process above and below the block. This first-generation device provided positive results and so the second generation of Wallis implants was developed. The main change was seen in the material used for the interspinous block, which was changed to PEEK, which is a

Fig. 19 Wallis® posterior dynamic stabilization system (https://www.ncbi.nlm.nih.gov/pmc/articles/PMC4365627/)

plastic like polymer that has more flexibility than titanium. The design and material allow for the minimization of the need for bone resection. In a controlled study which was done between 1998 and 1993, more than 300 patients were treated for degenerative lesions, in which positive results were found. Trials of the first-generation implant provided evidence that the interspinous system of nonrigid stabilization is effective against lower back pain caused by degenerative instability (Anderson et al. 2006). More recently Song et al. (2019)) provided information on 33 patients treated for degenerative lumbar spine diseases with the Wallis system. ROM of surgical segments was significantly lower than those before operation ($P < 0.05$), while ROM of the upper and lower adjacent segments and disc height did not change significantly ($P > 0.05$).

X-STOP (Medtronic)

The X-stop is made of titanium and PEEK components, with side wings encapsulating the lateral sides of the spinous processes to reduce the risk of implant migration (Fig. 20). FDA approval was obtained in 2005 after a 2-year clinical study. The device is approved for use in patients aged 50 years or older with lower-extremity neurogenic pain from lumbar spinal stenosis and can be implanted under local anesthesia. In the pilot study, inclusion criteria were mild or moderate symptoms that were relieved by flexion and the ability to walk at least 50 ft. Exclusion criteria were a fixed motor deficit or prior treatment with X-stop (Anderson et al. 2006).

DIAM (Medtronic)

The Device for Intervertebral Assisted Motion (DIAM) is made of a silicon H-shaped spacer encased within a Polyethylene terephthalate (Polyester) jacket that is secured (after removal of the interspinous ligament) with two associated tethers, around the supra-adjacent and sub-adjacent spinous processes (Fig. 21). In the past, DIAM has been successful in long-term treatment of lower back pain caused by degenerative disc disease. The first clinical case was performed in 1997 in France, and 25,000 patients have been treated outside the United States since then. In 2010 study, Buric et al. found that over two-thirds of patients achieved and maintained significant, clinically apparent differences in both VAS scores and Roland-Morris Disability Questionnaire (RMDQ) scores over a 48-month period (Buric and Pulidori 2011). FDA

Fig. 20 X-Stop device interspinous spacer Medtronic (https://www.ncbi.nlm.nih.gov/pmc/articles/PMC4365627/)

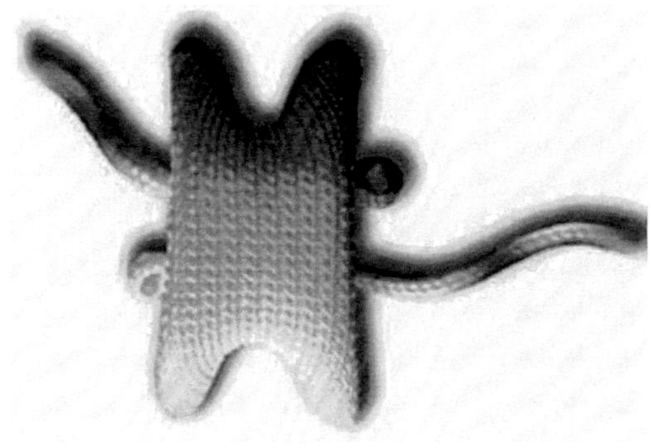

Fig. 21 Device for intervertebral assisted motion (DIAM) (https://www.ncbi.nlm.nih.gov/pmc/articles/PMC4365627/)

Fig. 22 Coflex® interlaminar stabilization (https://www.ncbi.nlm.nih.gov/pmc/articles/PMC4365627/)

randomized clinical trials to evaluate the effectiveness of DIAM versus decompression versus posterolateral fusion were completed in December 2010. However, in 2016 the FDA's Orthopedic and Rehabilitation Devices Panel of the Medical Devices Advisory Committee recommended against approval for the DIAM spinal stabilization system.

Recent study by Krappel et al. (2017) reported on a multicenter prospective randomized clinical study of 146 patients with a single level disc herniation (L2 to L5): 75 investigational (herniectomy and DIAM) and 71 control (herniectomy alone) treated and followed up for 24 months. Leg pain, back pain, and the level of disability were not significantly different between groups; however, the number of patients reaching the minimum clinically important difference (MCID) improvement for back pain was significantly higher in the investigational group at 6 through 24 months.

Coflex Interlaminar Stabilization Device (RTI Surgical)

The CoFlex is based on the interspinous-U design from Fixano (Péronnas, France) that was clinically used from 1995 onward (Fig. 22). It is made in its classic form as a titanium U-shaped metal design that is maintained between spinous processes with side wings, so as to control movement while allowing motion, being marketed as a non-fusion device. In 2012 the FDA approved the Coflex device after an IDE study.

Schmidt et al. (2018) performed a prospective, randomized, multicenter study with 2-year follow-up to compare the performance of decompression with and without Coflex interlaminar stabilization. This study reports a multicenter, randomized controlled trial in which decompression with interlaminar stabilization (D + ILS) was compared with decompression alone decompression alone (DA) for treatment of moderate to severe lumbar spinal stenosis. 230 patients (1:1 ratio) randomized to either DA or D + ILS (Coflex) were treated at seven sites in Germany. There was no significant difference in the individual patient-reported outcomes (e.g., ODI, VAS, ZCQ) between the treatments. However, microsurgical D + ILS increases walking distance, decreases compensatory pain management, and maintains radiographic foraminal height, extending the durability and sustainability of a decompression procedure. To date, Coflex has been implanted in more than 163,000 patients in over 60 countries worldwide.

In recent years multiple companies have offered various devices, such as NuVasive with ExtendSure; Biomech's (Taipei, Taiwan) Promise and Rocker designs, made of PEEK and mobile core and articulated design, respectively; Cousin Biotech (Wervicq-Sud, France) with Biolig silicon encapsulated in woven synthetics; Alphatec (Carlsbad, California) with the HeliFix screwtype PEEK space design; Vertiflex (San Clemente, California) with the Superion implant whose deployable wings aim at less invasive insertion (FDA cleared after completing PMA clinical studies in 2016); Orthofix (Bussolengo, Italy) with InSWing; Pioneer with BacJac; Maxx Spine (Bad Schwalbach, Germany) with I-MAXX; Sintea Plustek (Assago, Italy) with Viking; Globus Medical with Flexus; and Privelop (Neunkirchen-Seelscheid, Germany) (Serhan et al. 2011) (Fig. 23).

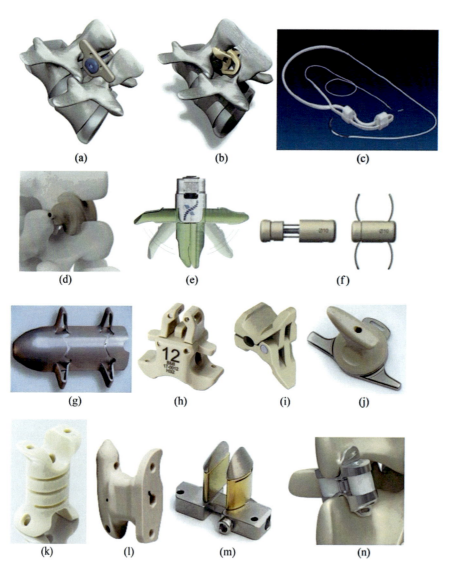

Fig. 23 Other interspinous spacer alternatives: (a) Promise; (b) Rocker; (c) Biolig; (d) HeliFix; (e) Superion; (f) InSpace; (g) Aperius; (h) InSWing; (i) BacJac; (j) I-MAXX; (k) Viking; (l) Flexus; (m) Spinos; and (n) Wellex (Eden Spine) source (https://www.ncbi.nlm.nih.gov/pmc/articles/PMC4365627/)

Discussion

Traditional fusion continues to be the gold standard for treating degenerative spinal disorders. Dynamic spinal stabilization is based on the concept of restricting movement of spinal segments rather than preventing the movement, that is, it restricts movements in the directions that may cause pain or instability, but permits motion in other directions. Dynamic spinal stabilization can achieve spinal stability and prevent diseases of adjacent segments without requiring fusion. Clinical indications for the use of PDS devices are still very broad and lack sufficient evidence. Scientific reviews have indicated that use of PDS pedicle-based systems as an adjunct to fusion may be acceptable. In fact, a majority of the devices described above as well as other devices (Transition: Globus Medical, and CD Horizon Legacy PEEK rod: Medtronic Sofamor Danek, to name a few) are successfully used as an adjunct to fusion across one or multiple spinal levels. However, fatigue failure is a concern when pedicle screw-based PDS systems are used as stand-alone stabilization devices. Failures have been reported at both the implant component interfaces as well as the bone implant interface. In terms of patient reported scores, PDS systems have produced clinical outcomes comparable to that of fusion, and the incidence of ASD is lower when compared to fusion, at least during short-term follow-up. RCTs with long-term follow-up are required to confirm whether the incidence of symptomatic ASD (and not just radiographic ASD) continues to stay lower when compared to fusion, as well as to prove the safety and efficacy of PDS devices. In summary, improvements in in vitro testing modalities, fatigue behavior, long-term follow-up, and a clear definition of clinical indications for using PDS as stand-alone stabilization devices are required to verify the benefits of this technology.

Similar to pedicle-based dynamic stabilization, interspinous devices are indicated to treat skeletally mature patients suffering from pain, numbness, and/or cramping in the legs (neurogenic intermittent claudication) secondary to a diagnosis of moderate degenerative lumbar spinal stenosis, with or without grade 1 spondylolisthesis, confirmed by x-ray, MRI, and/or CT evidence of thickened ligamentum flavum, narrowed lateral recess, and/or central canal or foraminal narrowing. Interspinous devices are also indicated for patients with impaired physical function who experience relief in flexion from symptoms of leg/buttock/groin pain, numbness, and/or cramping, with or without back pain, and who have undergone at least 6 months of nonoperative treatment. Interspinous devices may be implanted at one or two adjacent lumbar levels in patients in whom treatment is indicated at no more than two levels, from L1 to L5 (Khoueir et al. 2007; Senegas 2002).

Interspinous dynamic stabilization has theoretical advantages over conventional fusion, as it maintains stability by restricting mobility, whereas fusion simply prevents motion. Relatively good clinical results have been reported in the literature. However, despite the increasing use of this technology, few long-term review studies have been conducted to assess its safety and efficacy. Interspinous dynamic stabilization produced slightly better clinical outcomes than conservative treatments for spinal stenosis. The complication rate of interspinous dynamic stabilization has been reported to be 0–32.3% in 3- to 41-month follow-up studies. The complication rate of combined interspinous dynamic stabilization and decompression treatment (32.3%) was greater than that of decompression alone (6.5%), but no complication that significantly affected treatment results was found (Anderson et al. 2006; Zucherman et al. 2005). The typical complications of interspinous devices include spinous process fracture, especially with stiff design; novel radiculopathy, especially with devices with limited motion-constraining ability; and returning or increased pain around the implant area. Implant dislodgement is also a potential complication, particularly in those designs with limited fixation means. Compared to stiff and rigid interspinous designs, dynamic designs such as the Wallis or Coflex have relatively lower device complications.

References

Agarwal R et al (2009) Osteoinductive bone graft substitutes for lumbar fusion: a systematic review. J Neurosurg Spine 11(6):729–740

Anderson PA, Tribus CB, Kitchel SH (2006) Treatment of neurogenic claudication by interspinous decompression: application of the X STOP device in patients with lumbar degenerative spondylolisthesis. J Neurosurg Spine 4(6):463–471. https://doi.org/10.3171/spi.2006.4.6.463

Bakar D et al (2016) Decompression surgery for spinal metastases: a systematic review. Neurosurg Focus 41(2):E2

Barrey CY et al (2008) Biomechanical evaluation of pedicle screw-based dynamic stabilization devices for the lumbar spine: a systematic review. SAS J 2(4):159–170

Bellato RT et al (2015) Late failure of posterior fixation without bone fusion for vertebral metastases. Acta Ortopédica Brasileira 23(6):303–306

Bennett CR, DiAngelo DJ, Kelly BP (2015) Biomechanical comparison of robotically applied pure moment, ideal follower load, and novel trunk weight loading protocols on L4-L5 cadaveric segments during flexion-extension. Int J Spine Surg 9:33

Benzel EC (2001a) Spinal fusion. In: Biomechanics of spine stabilization. American Association of Neurological Surgeons, Rolling Meadows, pp 121–133

Benzel EC (2001b) The destabilizing effects of spinal surgery. In: Biomechanics of spine stabilization. American Association of Neurological Surgeons, Rolling Meadows, pp 111–120

Benzel EC (2001c) Trauma, tumor and infection. In: Biomechanics of spine stabilization. American Association of Neurological Surgeons, Rolling Meadows, pp 61–82

Bhamare S et al (2013) Design of dynamic and fatiguestrength-enhanced orthopedic implants. In: Multiscale simulations and mechanics of biological materials. Wiley, Oxford, UK, pp 333–350

Buric J, Pulidori M (2011) Long-term reduction in pain and disability after surgery with the interspinous device for intervertebral assisted motion (DIAM) spinal stabilization system in patients with low back pain: 4-year follow-up from a longitudinal prospective case series. Eur Spine J 20(8):1304–1311

Canero G, Carbone S (2015) The results of a consecutive series of dynamic posterior stabilizations using the PercuDyn device. Eur Spine J 24(S7):865–871

Chen H et al (2011) Influence of 2 different dynamic stabilization systems on sagittal spinopelvic alignment. J Spinal Disord Tech 24(1):37–43

Choi Y, Kim K, So K (2009) Adjacent segment instability after treatment with a Graf ligament at minimum 8 years' followup. Clin Orthop Relat Res 467(7):1740–1746

Chun DS, Baker KC, Hsu WK (2015) Lumbar pseudarthrosis: a review of current diagnosis and treatment. Neurosurg Focus 39(4):E10

Coe JD et al (2012) NFlex dynamic stabilization system: two-year clinical outcomes of multi-center study. J Korean Neurosurg Soc 51(6):343–349

Cook DJ, Yeager MS, Cheng BC (2012) Interpedicular travel in the evaluation of spinal implants. Spine 37(11):923–931

Davis R et al (2015) Measurement performance of a computer assisted vertebral motion analysis system. Int J Spine Surg 9:1–13

Doria C, Muresu F, Leali PT (2014) Dynamic stabilization of the lumbar spine: current status of minimally invasive and open treatments. In: Minimally invasive surgery of the lumbar spine. Springer, London, pp 209–227

Egger EL et al (1993) Effects of axial dynamization on bone healing. J Trauma 34(2):185–192

Erbulut DU et al (2013) Biomechanics of posterior dynamic stabilization systems. Adv Orthop 2013:451956

Gardner A, Pande KC (2002) Graf ligamentoplasty: a 7 year follow-up. Eur Spine J 11(Suppl 2):S157–S163

Goel VK et al (2005) Effects of charité artificial disc on the implanted and adjacent spinal segments mechanics using a hybrid testing protocol. Spine 30(24):2755–2764

Gomleksiz C et al (2012) A short history of posterior dynamic stabilization. Adv Orthop 2012:629698

Grob D et al (2005) Clinical experience with the Dynesys semirigid fixation system for the lumbar spine: surgical and patient-oriented outcome in 50 cases after an average of 2 years. Spine 30(3):324–331

Gruskay JA, Webb ML, Grauer JN (2014) Methods of evaluating lumbar and cervical fusion. Spine J 14(3):531–539

Hasegawa K et al (2013) Biomechanical evaluation of destabilization following minimally invasive decompression for lumbar spinal canal stenosis. J Neurosurg Spine 18(5):504–510

Hipp JA et al (2015) Development of a novel radiographic measure of lumbar instability and validation using the facet fluid sign. Int J Spine Surg 9:37

Hoff E et al (2012) Which radiographic parameters are linked to failure of a dynamic spinal implant? Clin Orthop Relat Res 470(7):1834–1846

Kanayama M et al (2000) In vitro biomechanical investigation of the stability and stress-shielding effect of lumbar interbody fusion devices. J Neurosurg 93(2 Suppl):259–265

Kaner T, Sasani M et al (2010a) Clinical outcomes of degenerative lumbar spinal stenosis treated with lumbar decompression and the cosmic "semi-rigid" posterior system. SAS J 4(4):99–106

Kaner T, Dalbayrak S, Oktenoglu T, Sasani M, Aydin AL, Ozer AF (2010b) Comparison of posterior dynamic and posterior rigid transpedicular stabilization with fusion to treat degenerative spondylolisthesis. Orthopedics 33(5). Published 2010 May 12. https://doi.org/10.3928/01477447-20100329-09

Khoueir P, Kim KA, Wang MY (2007) Classification of posterior dynamic stabilization devices. Neurosurg Focus 22(1):E3

Kim Y-S et al (2007) Nitinol spring rod dynamic stabilization system and Nitinol memory loops in surgical treatment for lumbar disc disorders: short-term follow up. Neurosurg Focus 22(1):E10

Kirkaldy-Willis WH, Farfan HF (1982) Instability of the lumbar spine. Clin Orthop Relat Res 165:110–123

Kocak T et al (2010) Screw loosening after posterior dynamic stabilization – review of the literature. Acta Chir Orthop Traumatol Cechoslov 77(2):134–139

Kowalski RJ, Ferrara LA, Benzel EC (2001) Biomechanics of bone fusion. Neurosurg Focus 10(4):E2

Krappel F, Brayda-Bruno M, Alessi G, Remacle JM, Lopez LA, Fernández JJ, Maestretti G, Pfirrmann CWA (2017) Herniectomy versus Herniectomy with the Diam spinal stabilization system in patients with sciatica and concomitant low back pain: results of a prospective randomized controlled multicenter trial. Eur Spine J 26(3):865–876. https://doi.org/10.1007/s00586-016-4796-6. Epub 2016 Oct 4

La Barbera L, Villa T (2017) Toward the definition of a new worst-case paradigm for the preclinical evaluation of posterior spine stabilization devices. Proc Inst Mech Eng H J Eng Med 231(2):176–185

La Barbera L, Ottardi C, Villa T (2015) Comparative analysis of international standards for the fatigue testing of posterior spinal fixation systems: the importance of preload in ISO 12189. Spine J 15(10):2290–2296

Lafage V, Gangnet N, Sénégas J et al (2007) New interspinous implant evaluation using an in vitro biomechanical study combined with a finite-element analysis. Spine (Phila Pa 1976) 32:1706–1713

Lee CK, Langrana NA (1984) Lumbosacral spinal fusion. A biomechanical study. Spine 9(6):574–581

Li Z et al (2013) Two-year follow-up results of the isobar TTL semi-rigid rod system for the treatment of lumbar degenerative disease. J Clin Neurosci 20(3):394–399

Lima LVPC et al (2017) Limiting interpedicular screw displacement increases shear forces in screws: a finite element study. Orthop Traumatol Surg Res 103(5):721–726

Lindsey DP, Swanson KE, Fuchs P, Hsu KY, Zucherman JF, Yerby SA (2003) The effects of an interspinous implant on the kinematics of the instrumented and adjacent levels in the lumbar spine. Spine 28(19):2192–2197

Lindsey DP et al (2015) Sacroiliac joint fusion minimally affects adjacent lumbar segment motion: a finite element study. Int J Spine Surg 9:64

Mulholland RC, Sengupta DK (2002) Rationale, principles and experimental evaluation of the concept of soft stabilization. Eur Spine J 11(Suppl 2):S198–S205

Nayak NR et al (2015) Tracking patient-reported outcomes in spinal disorders. Surg Neurol Int 6(Suppl 19):S490–S499

Neel Anand M, et al (2012) 24 Month functional outcomes from the US IDE Trial Evaluating The Stabilimax®, A lumbar Posterior Dynamic Stabilization (PDS) system with Interpedicular Travel (IPT). In: Meeting of the Lumbar Spine Research Society

Panjabi MM, Timm JP (2007) Development of stabilimax NZ from biomechanical principles. SAS J 1(1):2–7

Parchi PD, Evangelisti G, Vertuccio A, Piolanti N, Andreani L, Cervi V, Giannetti C, Calvosa G, Lisanti M (2014) Biomechanics of interspinous devices. Biomed Res Int 2014:839325

Park Y-S et al (2013) The effect of zoledronic acid on the volume of the fusion-mass in lumbar spinal fusion. Clin Orthop Surg 5(4):292–297

Patwardhan AG et al (1999) A follower load increases the load-carrying capacity of the lumbar spine in compression. Spine 24(10):1003–1009

Pham MH et al (2016) Complications associated with the Dynesys dynamic stabilization system: a comprehensive review of the literature. Neurosurg Focus 40(1):1–8

Phillips FM, Voronov LI, Gaitanis IN, Carandang G, Havey RM, Patwardhan AG (2006) Biomechanics of posterior dynamic stabilizing device (DIAM) after facetectomy and discectomy. Spine J 6(6):714–722

Puttlitz CM et al (2000) Pathomechanisms of failures of the odontoid. Spine 25(22):2868–2876

Rajaee SS et al (2012) Spinal fusion in the United States. Spine 37(1):67–76

Reyes-Sánchez A et al (2010) Posterior dynamic stabilization of the lumbar spine with the Accuflex rod system as a stand-alone device: experience in 20 patients with 2-year follow-up. Eur Spine J 19(12):2164–2170

Saavedra-Pozo FM, Deusdara RAM, Benzel EC (2014) Adjacent segment disease perspective and review of the literature. Ochsner J 14(1):78–83

Saphier PS et al (2007) Stress-shielding compared with load-sharing anterior cervical plate fixation: a clinical and radiographic prospective analysis of 50 patients. J Neurosurg Spine 6(5):391–397

Schlenk RP, Stewart T, Benzel EC (2003) The biomechanics of iatrogenic spinal destabilization and implant failure. Neurosurg Focus 15(3):E2

Schmidt H, Heuer F, Wilke H-J (2009) Which axial and bending stiffnesses of posterior implants are required to design a flexible lumbar stabilization system? J Biomech 42(1):48–54

Schmidt S, Franke J, Rauschmann M, Adelt D, Bonsanto MM, Sola S (2018) Prospective, randomized, multicenter study with 2-year follow-up to compare the performance of decompression with and without interlaminar stabilization. J Neurosurg Spine 28(4):406–415. https://doi.org/10.3171/2017.11.SPINE17643. Epub 2018 Jan 26

Schwarzenbach O et al (2005) Posterior dynamic stabilization systems: DYNESYS. Orthop Clin N Am 36(3 Spec Iss):363–372

Senegas J (2002) Mechanical supplementation by nonrigid fixation in degenerative intervertebral lumbar segments: the Wallis system. Eur Spine J 11(Suppl 2):S164–S169

Sengupta DK, Herkowitz HN (2012) Pedicle screw-based posterior dynamic stabilization: literature review. Adv Orthop 2012(424268):1–7

Serhan H, Mhatre D, Defossez H, Bono CM (2011) Motion-preserving technologies for degenerative lumbar spine: the past, present, and future horizons. SAS J 5(3):75–89. https://doi.org/10.1016/j.esas.2011.05.001. Published 2011 Sep 1

Shono Y et al (1998) Stability of posterior spinal instrumentation and its effects on adjacent motion segments in the lumbosacral spine. Spine 23(14):1550–1558

Slosar PJ et al (2000) Patient satisfaction after circumferential lumbar fusion. Spine 25(6):722–726

Smith ZA et al (2011) A minimally invasive technique for percutaneous lumbar facet augmentation: technical description of a novel device. Surg Neurol Int 2:165

Song KP, Zhang B, Ma JL, Wang B, Chen B (2019) Midterm follow-up efficacy of interspinous dynamic stabilization system for lumbar degenerative diseases. Zhongguo Gu Shang 32(11):991–996. https://doi.org/10.3969/j.issn.1003-0034.2019.11.004

Stoffel M et al (2010) Pedicle screw-based dynamic stabilization of the thoracolumbar spine with the cosmic®-- system: a prospective observation. Acta Neurochir 152(5):835–843

Tsai K, Murakami H, Lowery GL, Hutton WC (2006) A biomechanical evaluation of an interspinous device (Coflex) used to stabilize the lumbar spine. J Surg Orthop Adv 15(3):167–172

Turner JA et al (1992) Surgery for lumbar spinal stenosis. Attempted meta-analysis of the literature. Spine 17(1):1–8

Vadapalli S, Sairyo K, Goel VK et al (2006) Biomechanical rationale for using polyetheretherketone (PEEK) spacers for lumbar interbody fusion-a finite element study. Spine (Phila Pa 1976) 31(26):E992–E998. https://doi.org/10.1097/01.brs.0000250177.84168.ba. [published correction appears in Spine. 2007 Mar 15;32(6):710]

Vaz K et al (2010) Bone grafting options for lumbar spine surgery: a review examining clinical efficacy and complications. SAS J 4(3):75–86

Vermesan D et al (2014) A new device used in the restoration of kinematics after total facet arthroplasty. Med Devices (Auckland, NZ) 7:157–163

Virk SS et al (2014) Adjacent segment disease. Orthopedics 37(8):547–555

Weiss LE et al (1997) Pseudarthrosis after postoperative wound infection in the lumbar spine. J Spinal Disord 10(6):482–487

Yeager MS, Cook DJ, Cheng BC (2015) In vitro comparison of Dynesys, PEEK, and titanium constructs in the lumbar spine. Adv Orthop 2015:1–8

Yoshihara H (2013) Rods in spinal surgery: a review of the literature. Spine J 13(10):1350–1358

Zhang HY, Park JY, Cho BY (2009) The BioFlex system as a dynamic stabilization device: does it preserve lumbar motion? J Korean Neurosurg Soc 46(5):431–436

Zhang Y et al (2016) Comparison of the Dynesys dynamic stabilization system and posterior lumbar interbody fusion for lumbar degenerative disease. PLoS One 11(1):e0148071

Zheng S, Yao Q, Cheng L et al (2010) The effects of a new shape- memory alloy interspinous process device on the distribution of intervertebral disc pressures in vitro. J Biomed Res 24(2):115–123

Zucherman JF, Hsu KY, Hartjen CA et al (2005) A multicenter, prospective, randomized trial evaluating the X STOP interspinous process decompression system for the treatment of neurogenic intermittent claudication: two-year follow-up results. Spine (Phila Pa 1976) 30(12):1351–1358

Lessons Learned from Positive Biomechanics and Poor Clinical Outcomes

14

Deniz U. Erbulut, Koji Matsumoto, Anoli Shah, Anand Agarwal, Boyle C. Cheng, Ali Kiapour, Joseph Zavatsky, and Vijay K. Goel

Contents

Spinal Fusion	317
Dynamic Stabilization	322
References	328

D. U. Erbulut (✉) · A. Kiapour
Departments of Bioengineering and Orthopaedic Surgery, Colleges of Engineering and Medicine, University of Toledo, Toledo, OH, USA
e-mail: deniz.erbulut@utoledo.edu; ali.kiapour@utoledo.edu

K. Matsumoto
Department of Orthopedic Surgery, Nihon University School of Medicine, Tokyo, Japan

A. Shah
Engineering Center for Orthopaedic Research Excellence (E-CORE), Departments of Bioengineering and Orthopaedic Surgery, Colleges of Engineering and Medicine, University of Toledo, Toledo, OH, USA
e-mail: anoli.shah@rockets.utoledo.edu

A. Agarwal
Department of Orthopaedic Surgery and Bioengineering, School of Engineering and Medicine, University of Toledo, Toledo, OH, USA
e-mail: anand.agarwal@utoledo.edu

B. C. Cheng
Neuroscience Institute, Allegheny Health Network, Drexel University, Allegheny General Hospital Campus, Pittsburgh, PA, USA
e-mail: bcheng@wpahs.org

J. Zavatsky
Spine & Scoliosis Specialists, Tampa, FL, USA

© Springer Nature Switzerland AG 2021
B. C. Cheng (ed.), *Handbook of Spine Technology*,
https://doi.org/10.1007/978-3-319-44424-6_27

Abstract

Biomechanical testings are essential to the research process. However, understanding the assumptions, inevitable part of engineering solution as it translates to clinical outcomes is paramount to the iterative process in implant design. Therefore, it is critical to consider that beneficial biomechanical data may not actually yield good clinical outcomes.

Keywords

Spine biomechanics · Clinical outcome · Spinal fusion · Dynamic Stabilization · Adjacent segment degeneration

Design excellence tools have been widely used for many years, historically mainly in the automotive industry, and more recently in the medical device industry. A plethora of tools are available; however, only their appropriate selection and deployment during the various stages of the new medical device development can optimize the overall process.

In every lab-based biomechanical study (i.e., finite element (FE) analyses, in vitro, in vivo, etc.), there is a "limitations" section stating the assumptions that the reader should keep in mind for proper interpretation of the findings. Such assumptions for real-life scenarios are inevitable and are the part of engineering solutions. Similarly, long- and short-term clinical outcomes of a study will highlight the discrepancies between actual biomechanical data and surgical outcomes, for example. Thus, beneficial biomechanical data may not actually yield good clinical outcomes. The design and development of a medical device is an iterative process, and decreasing the number of iterations can be helpful. This chapter addresses these issues for the devices specifically used to provide stability to the spine with the ultimate goal of improving patient satisfaction.

Preclinical tests are necessary for evaluating the safety and efficacy of new techniques and devices including: biocompatibility, structural integrity, and biomechanical performance before clinical trials. Biomechanical testing for a spinal device involves multiple steps, including benchtop testing as per ASTM and ISO standards, cadaver studies, etc. Safety can be evaluated with biocompatibility and mechanical tests, and efficacy can be investigated with finite element analysis (FEA), in vitro, and animal tests. Although finite element modeling (FEM) (Fig. 1a) technique has limitations, nowadays, this technique is part of the first step in biomechanical investigations due to its many advantages (Agarwal et al. 2017). The advantages include relatively shorter completion time, ability to analyze complex engineering problems, and undertaking parametric studies, to name a few. Additionally, many engineering parameters important for the design and development of a medical device, such as reaction force, stress, strain, and load transmission, are among the various components that can be obtained. It is not practical to understand the roles of these parameters using any other testing protocol. FEM's ability to analyze failures and to modify the parameters accordingly with minimal effort and time makes FEM a strong engineering tool in the biomedical device industry.

In mechanical testing, a spinal implant must stay functional during both static and dynamic tests (predicting lifetime) to determine the worst-case scenarios for failure. Commonly, test protocols are developed as per the international standards, i.e., ISO and ASTM (Fig. 1b). In vitro cadaver or animal studies assess the performance of a device in response to clinically relevant loads or displacements. These studies allow for the measurement of stiffness or flexibility of the construct (Fig. 1c). In flexibility protocols, range of motion in three main planes in response to the applied loads can be compared to the intact spine. The data can be further processed to determine

---

V. K. Goel
Engineering Center for Orthopaedic Research Excellence (E-CORE), University of Toledo, Toledo, OH, USA

Departments of Bioengineering and Orthopaedic Surgery, Colleges of Engineering and Medicine, University of Toledo, Toledo, OH, USA
e-mail: Vijay.Goel@utoledo.edu

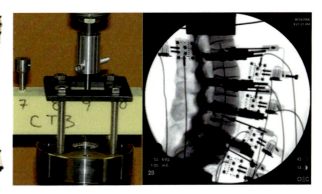

Fig. 1 (**a**) Intact and instrumented scoliotic FEM (Agarwal et al. 2017), (**b**) Mechanical setup for pull-out testing of a pedicle screw, and (**c**) Radiograph of a cadaveric lumbar spine testing setup. Showing follower preload path to simulate muscle action in an in vitro protocol. The LED markers attached to the vertebral bodies allow determination of the construct kinematics (Goel et al. 2006)

instantaneous center of rotation, neutral zone, disc height, disc bulge, spinal alignment, creep, and viscoelastic behavior of the system, etc. Also, transducers can be placed at relevant locations on the constructs to investigate other parameters such as intradiscal pressure and elongation (Goel et al. 2006).

Positive data from such studies can allow for the manufacturing of devices that are better targeted for clinical applications in patients. Nonetheless, there are also inherent limitations as the cadaveric specimens do not represent the variabilities in patient populations and differences in age, gender, and pathologies encountered in real life. Therefore, the final safety and efficacy of a new technique or system can only be revealed following clinical trials – an iterative process as stated above.

Spinal fusion technique may be the best example to start with to raise the issue of "positive biomechanics and poor clinical outcomes." For instance, although more than 90% of the fusion treatments deliver the expected biomechanical fusion outcome immediately after the surgery, reported successful clinical outcomes have only been reported up to 70%.

Spinal Fusion

Adjacent segment degeneration: Spinal fusion is widely accepted for the treatment of discogenic pain, segmental spinal instability or spondylolisthesis, and spinal deformity. It is well-known that spinal fusion provides stability to the index segment/s and can relieve back pain after the surgery. Targeted area/s within the spinal functional unit for bony fusion are the spinous processes, the transverse processes, the laminae, the facet joints, and the intervertebral disc spaces (Evans 1985). Lumbar fusion shows positive biomechanical research outcomes in terms of immediate segmental stability and does not represent the fusion that matures over time. Furthermore, it is almost impossible to predict devices functionality in the body, as the device interacts with spinal boney tissues during daily activities that could lead to device loosening, subsidence, and/or migration.

Spinal fusion treatment adjusts the motion and load transmission at the index levels. This segmental adjustment causes higher motion and stress distribution on the adjacent segments, which can lead to subsequent degeneration called adjacent segment degeneration (ASD). Clinical reports of ASD following spinal fusion in patients were sparse in the 1950s (Unander-Scharin 1950; Anderson 1956). It became evident once long-term clinical outcomes were analyzed. We now know that spinal fusion techniques can be one of the contributing factors to ASD. The prevalence of radiographic ASD postoperatively after spinal fusion is high. Potential risk factors associated with adjacent segment disease after spinal fusions include injuries to the facet joint of the adjacent segment, fusion length, sagittal alignment, pre-existing degenerated disc disease at the adjacent levels, lumbar stenosis, age, osteoporosis, female gender, and postmenopausal state (Park et al. 2004; Moreau et al. 2016). Due to many risk factors not predicted by earlier biomechanical studies, the importance of future research with expended objectives is paramount (Eck et al. 2002; Chin et al. 2016; Voronov et al. 2016; Lafage et al. 2017; Natarajan and Andersson 2017).

Spinal instability can cause back pain. In the case of instability, spinal fusion has been advocated as the gold standard treatment to relieve associated back pain. Spinal instability can be described as abnormal motion at the joint and/or load transmission. For example, changes in the mechanical properties of the intervertebral disc (due to degeneration, replacement with artificial discs, and interbody augmentation) can alter the normal load transmission. Studies suggest that the altered load transmission pattern causes the pain, rather than changes in load magnitude. Obviously, fusion techniques will lead to altered load transmission at the index and adjacent segments. Over time, at the adjacent levels, the increased force and hypermobility may lead not only to disc degeneration, but also to hypertrophic degenerative arthritis of the facet joints, spinal stenosis, degenerative spondylolisthesis, or herniated nucleus pulposes (Brodsky 1976; Lee 1988; Wimmer et al. 1997; Etebar and Cahill 1999; Kumar et al. 2001). Biomechanical studies, at least in the past, were not able to address all of these variables.

Many studies have reported the effects of fusion techniques and related incidence of ASD as fusion can lead to additional surgeries (Whitecloud et al. 1994; Schlegel et al. 1996; Phillips et al. 2000; Chen et al. 2001a, b). Lee's biomechanical study published in 1984 (Lee and Langrana 1984) reported the altered kinematics and biomechanics of three different fusion procedures (posterior, bilateral, and anterior) at the index and adjacent segments. The study demonstrated the desirable stability effects of all the fusion techniques, but increased stress on the adjacent segments specifically at the facet joints was observed. The same authors published a clinical study reporting 18 patients who developed adjacent segment symptoms after the first 5 years of fusion treatment (Lee 1988).

The study of Evans (1985) indicated lists of essential biomechanical criteria of a good fusion device. For example, the intensity of force or stress on the graft should not be so great as to damage the fusion device. Optimal grafts should bear loads without graft migration immediately following surgery, and it should resist shear force to prevent sliding on the host bone. The study also showed the potential restoration of normal anatomic functionality of the treated segment. This raises the following question: If the segment is fused and the segmental motion eliminated, how can the force be transferred comparable to a normal intact spinal segment? Brodke et al. (1997) investigated initial stiffness of posterior lumbar interbody fusions with rigid posterior instrumentation. The additional posterior instrumentation along with interbody graft led to increased stiffness and prevented graft reposition. Eck et al. (2002) performed a cadaver study to investigate biomechanical effect of cervical spinal fusion on adjacent level intradiscal pressure. They found that the intradiscal pressure increased by up to 73.2% and 45.3% at superior and inferior adjacent level, respectively. These results suggest a biomechanical cause of disc degeneration at the adjacent levels. Other studies have reported that up to 40% of postfusion low back pain involves the sacroiliac joint (SIJ) (Katz et al. 2003; Maigne and Planchon 2005). FE studies have predicted that posterior lumbar spinal fusion causes increases in motion and stress across the SIJ (Fig. 2) (Ivanov et al. 2009).

One of the risk factors for adjacent segment degeneration is postoperative lumbar sagittal misalignment caused by rigid fixation, as well as the number of treated level (Umehara et al. 2000). At the end, postoperative lumbar sagittal malalignment and loss of motion at the fused level change the structural response to external loads. Therefore, one can expect accelerated degenerative changes at the levels adjacent to the fusion site secondary to hypo-lordosis in the instrumented segment. Additionally,

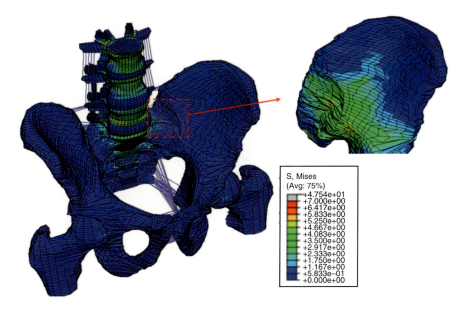

Fig. 2 Finite element model (FEM) of the lumbar spine and pelvis. Stress distribution at the sacroiliac area after simulation is shown. Sacroiliac joint reaction was observed using the model with L4-S1 fusion under 25 Nm bending moment and 400 N compressive follower load (Ivanov et al. 2009)

the load across the posterior transpediclular devices increases due to hyperlordosis. Similarly, Akamaru et al. (2003) reported a very large increase in flexion/extension motion at proximal and distal adjacent levels with hypo-lordotic alignment, compared to the intact spine.

Reports have demonstrated good result with posterior lumbar fusion with pedicle screw fixation. In spite of the procedure at achieving successful spinal fusion, long-term follow-up studies have revealed postoperative segmental instability, spinal stenosis, intervertebral disc herniation, retrospondylolisthesis, and fracture at the adjacent segments (Brunet and Wiley 1984; Whitecloud et al. 1994; Kerr et al. 2015). More recently, significant correlation between thoracic kyphosis, lumbar lordosis, pelvic tilt (PT) and pelvic incidence (PI) parameters, and the occurrence of ASD after lumbar fusion has been reported (Rothenfluh et al. 2015; Nakashima et al. 2015; Yamasaki et al. 2017; Matsumoto et al. 2017). Patients having a pelvic incidence-lumbar lordosis mismatch (greater than +/− 10°) exhibit a ten times higher risk for undergoing revision surgery compared to nonmismatched patients (Rothenfluh et al. 2015). The risk of ASD incidence was 5.1 times greater in subjects with preoperative PT greater than 22.5° (Yamasaki et al. 2017). Matsumoto et al., through finite element model simulations of various parameters, found that the preoperative global sagittal imbalance (SVA >50 mm and higher PT), lower pre- and postoperative LL, and a PI-LL mismatch were significantly associated with ASD (Matsumoto et al. 2017; Shah et al. 2019). This study has revealed the biomechanical relationship between ASD (proximal and distal junctions) and different spinal and pelvic parameters following lumbar arthrodesis. A validated FE model from T1 to femur without rib cage was used. The sagittal vertical axis (SVA), lumbar lordosis (LL), thoracic kyphosis (TK), pelvic incidence (PI), pelvic tilt (PT), and sacral slope (SS) were modified to develop three different sagittally balanced models, simulating different compensate-mechanisms. As shown in Fig. 3, these are (a) Normal (Balanced: SVA = 0 mm, LL = 50°, TK = 25°, PI = 45°, PT = 10°, and SS = 35°); (b) Flat back (Balanced with

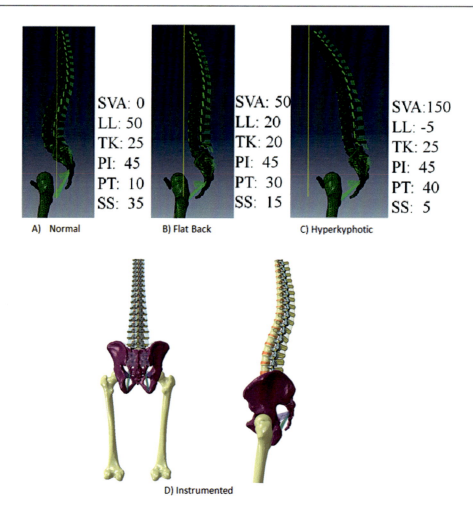

Fig. 3 Three different finite element models (FEM) used for the investigations. Yellow line indicates the C7 Plum line. (**a**) Normal spine, (**b**) flat back, (**c**) hyperkyphotic, and (**d**) posterior and sagittal view of instrumented models from L2-L5. *SVA* sagittal vertical axis, *LL* lumbar lordosis, *TK* thoracic kyphosis, *PI* pelvic incidence, *PT* pelvic tilt, and *SS* sacral slope (Matsumoto et al. 2019)

compensatory mechanism: SVA = 50 mm, LL = 20°, TK = 20°, PI = 45°, PT = 30°, and SS = 15°); and (c) Hyperkyphotic (Imbalance: SVA = 150 mm, LL = -5°, TK = 25°, PI = 45°, PT = 40°, and SS = 5°). A posterior rigid pedicle screw fixation system was simulated across L2-L5 (Fig. 3). The model was fixed at the distal femurs, and 2 Nm moments were applied at T1 to simulate flexion (FLEX), extension (EXT), right bending (RB), left bending (LB), right rotation (RR), and left rotation (LR) in intact and instrumented models. The von Mises stress on the proximal vertebra (L1) and distal vertebra (S1) as an indicator of proximal junctional kyphosis (PJK) and distal junctional kyphosis (DJK) was calculated and compared.

The maximum von Mises stress at the proximal vertebra increased by up to 143% (average of all motions: 74.9%) in the flat back model, and 18% (6.0%) in kyphotic model, compared to the normal balanced model (Fig. 4a). The maximum von Mises stress at the distal vertebra increased by up to 196% (average of all motions: 49.5%) in the flat back model, and 527% (141.8%) in kyphotic model, compared to normal (Fig. 4b). In the instrumented flat back, the maximum von Mises

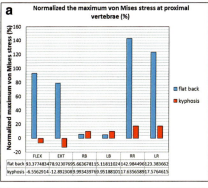
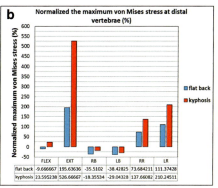
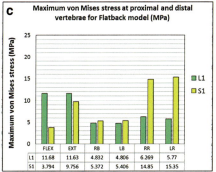
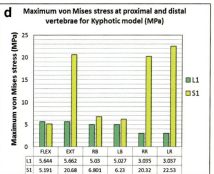

Fig. 4 (**a**) The normalized maximum von Mises stress at the proximal vertebrae (%), (**b**) the normalized maximum von Mises stress at the distal vertebrae (%), (**c**) the maximum von Mises stress at the proximal and distal vertebrae for the instrumented flat back model (%), and (**d**) the maximum von Mises stress at the proximal and distal vertebrae for the instrumented kyphotic model (%) (Matsumoto et al. 2018)

stresses at the proximal vertebra and distal vertebra were up to 11.7 MPa (average of all motions: 7.5 MPa), and 15.4 MPa (9.1 MPa) (Fig. 4c). In an instrumented kyphotic model, the maximum von Mises stresses at the proximal vertebra and distal vertebra were up to 5.6 MPa (average of all motions: 4.6 MPa) and up to 22.5 MPa, respectively (Fig. 4d) (Matsumoto et al. 2019).

The results show that the von Mises stress on adjacent vertebra increased by up to 196% in the flat back model, and 527% in the kyphotic model, compared to the normal model. The data suggests that when considering L2-5 fixation in flat back and kyphotic models, care should be taken to restore the normal lumbar alignment. Our data tends to suggest that the kyphotic model may contribute to higher incidences of DJK than PJK. Surgeons may consider using dynamic stabilization devices in the distal region for kyphotic patients (Matsumoto et al. 2018; Shah et al. 2019).

Subsequent to poor clinical outcomes, biomechanical studies have emerged to address ASD. Agarwal et al. (2016) conducted an in vitro study using 12 L2-S1 specimens instrumented with either titanium rods or PEEK (polyetheretherketone) rods and compared the effects of these materials on the kinematics of the adjacent level. They found that lower rigidity (PEEK rods) did not make a significant difference in terms of superior adjacent level motion in flexion and extension, compared to titanium. However, other biomechanical studies have shown that dynamic posterior stabilization constructs provide posterior band stiffness closer to the normal spine and protect adjacent levels, from excessive motion compare to more rigid constructs (Erbulut et al. 2014). The finite element models provide a good platform to study the biomechanical effects of new implant designs and surgical techniques. For example, the comparison of TLIF versus PLIF using a validated L4-L5 model showed that the TLIF cages showed the higher stresses compared to ALIF and PLIF on the endplate stresses. In this FE model, the footprints of the cages were modified showing that the increase in footprint showed to lower the stresses compared to smaller footprints. Another modification of simulating different material properties of the cages showed that PEEK cages produced lower stresses compared to the titanium cages (Sudershan 2017). The change in parameters led to the simulation of parametric analyses which was difficult and expensive for in vitro studies. Thus, FEA is a very useful tool in studying in-depth biomechanics.

Dynamic Stabilization

Center of rotation (COR) of artificial disc designs: Initially, biomechanical studies of total disc replacement (TDR) devices reported that adjacent segment degeneration can be eliminated or reduced by implanting a dynamic device, instead of after anterior cervical discectomy and fusion (ACDF) (DiAngelo et al. 2003; Dmitriev et al. 2005). Therefore, TDR for the cervical spine instead of ACDF is potentially favorable by decreasing the incidence of adjacent segment degeneration. A recent study (Laxer et al. 2017) investigated the potential bias in the reports of outcomes from patients with TDR, because many of the studies' investigators reported financial conflicts of interest related to industry support, stock ownership, or consulting income from the TDR device company. They did not identify such bias. Therefore, the biomechanical data correlated with the clinical outcomes, regarding adjacent segment degeneration.

However, the relevance of the sliding articulation feature of a cervical artificial disc has emerged as studies have revealed the relationship between segmental sagittal translation and COR of the segment. The COR is a function of rotation and translation within the cervical spinal functional unit (Bogduk and Mercer 2000). Protraction is defined as flexion of the lower cervical spine and extension of the upper cervical spine, and the opposite is true for retraction (Ordway et al. 1999). Amevo et al. showed that 77% of patients with neck pain displayed abnormal instantaneous COR (Amevo et al. 1991). Similarly, Hanten et al. reported that patients with neck pain have little to no cervical sagittal plane translation mobility compared to healthy subjects (Hanten et al. 2000).

Consequently, the COR of an artificial disc affects the biomechanics of the spine. Fundamentally, the facet joints of the cervical spine play a major role in controlling the segmental axes of rotation (Penning 1988). Mo et al. (2015) postulated that facet joint stress was eliminated when an artificial disc had a sliding articulation feature to allow translation in the sagittal plane. This is supported by an early study of Nowitzke et al. (1994) that showed the relationship between cervical zygapophyseal joints geometry and the patterns of movement of the cervical vertebrae in the sagittal plane. The study further exhibits the location of the instantaneous center of location of the segment as a function of the facet joints. Similarly, a biomechanical study showed that an artificial disc with a fixed center of location may increase facet loading at the index segment (Faizan et al. 2012).

Our recent FE study investigated an elastomer total disc replacement (Fig. 5c) in the lumbar spine. The objective was to compare the biomechanical effects of the TDR in the lumbar spine as compared to natural spinal kinematics. An experimentally validated FE model of a ligamentous L1-S1 lumbar spinal model was used (Fig. 5a). To simulate the surgical procedure for disc replacement, the CAD model for the elastomer TDR was imported into the FEA software, and inserted into the L4-L5 segment following removal of the anterior annulus fibrosus, ALL ligament, and entire nucleus pulposus. A pure moment of 10 Nm combined with an applied follower load of 400 N was used to simulate the model in flexion-extension, lateral bending, and axial rotation. Range of motion, intradiscal

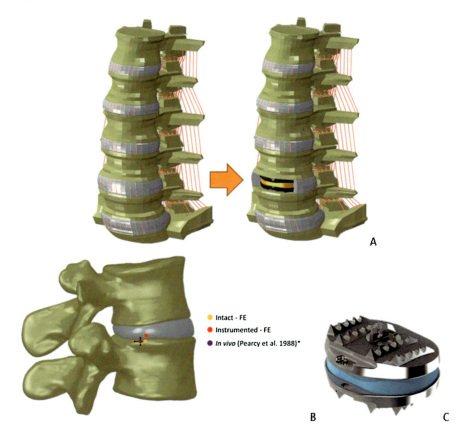

Fig. 5 (**a**) FE models of the intact and instrumented spine, (**b**) location of extension-to-flexion range of motion for the intact and instrumented spine, at index level, compared to in vitro data, and (**c**) total disc replacement device with titanium alloy endplates and CarboSil 20 80A silicone rubber flexible component

pressure, and facet loads across the segments were calculated. COR for full extension to full flexion was also calculated for the intact and instrumented models (Fig. 5b).

Figure 6 shows the kinematic data for index and superior adjacent levels. Range of motion of the instrumented model was 88% and 80% of the intact model in extension and flexion, respectively. COR was in good agreement with in vitro study reported by Pearcy and Bogduk (1988). Following instrumentation, the motion at the superior adjacent level was increased slightly in flexion (8%) and extension (20%). When comparing the intact model and the in vitro study, there was up to a 12% increase in intradiscal pressure at the superior adjacent segment after instrumentation. The loads at the facet increased by 1% in flexion and decreased by 5% in extension, compared to the intact model.

Our study showed that the elastomer lumbar TDR device not only preserved the range of motion, but also maintained the loading condition at the facet joints. In addition, the device may minimize the risk of adjacent segment disc and facet degeneration.

Further, poor clinical outcomes such as heterotopic ossification, wear debris, or metal hypersensitivity, of total disc replacements have yet to be biomechanically addressed (Sengupta 2015).

Posterior dynamic stabilization systems: Dynesys (Zimmer Spine, Minneapolis MN) is the most widely implanted posterior stabilization system (PDS) in the world. In addition, numerous

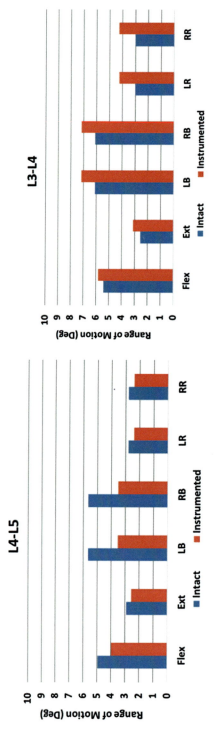

Fig. 6 Range of motion for intact and instrumented spines at index level and superior adjacent levels under anatomical loading

clinical and biomechanical articles have been published about the system since 1999. Dynesys is a second generation PDS system and was introduced in 1994. The first set of results was presented by the inventor Gill Dubois (Freudiger et al. 1999). In this study, Dynesys was tested on four cadaveric lumbar spines, three cases at L4-L5 and one case at L3-L4 level under bending, compressive, and shear loads. The loads were different for flexion and extension, average of 18.4 Nm flexion moment with 2296 N compressive and 458 N anterior shear load, and average of 12.5 Nm extension moments, with 667 N compressive and 74 N shear loads. The study reported motion reduction in flexion and extension at the index segment with Dynesys, which was described as an efficient supporting structure. The first clinical experience of the device was also published by the inventor in 2002 (Stoll et al. 2002). The study investigated the safety and efficacy of Dynesys for lumbar instability conditions. A consecutive series of 73 patients underwent Dynesys implantation and were evaluated pre- and postoperatively. The follow-up time range was between 11.2 and 79.1 months. No screw breakage was observed and loosening was suspected in only seven screws, based on radiological appearances, and only one of the patients had to be reoperated on due to screw loosening. Clinically, the authors claimed that Dynesys is a safe and efficient device to treat degenerative lumbar disease compared to spinal fusion techniques. They hypothesized that Dynesys could address the incidence of adjacent segment degeneration long-term. However, Sengupta and Herkowitz (2012) analyzed their data, and it was discovered that 60.2% of the patients had spinal stenosis, and questioned the etiology of the good clinical outcome reports. Was the success primarily due to the Dynesys instrumentation or the decompression? Successful clinical reports decreased more than half of the Dynesys cases without accompanying decompression (Grob et al. 2005; Würgler-Hauri et al. 2008).

Additionally, studies have revealed that Dynesys acts like a rigid system and does not address the adjacent-segment degeneration problem (Schmoelz et al. 2003; Schnake et al. 2006).

Schnake et al. (2006) reported the clinical outcomes of 26 patients with spondylolisthesis, treated with decompression and dynamic stabilization with Dynesys. Twenty-nine percent of patients had indications of adjacent segment degeneration with regard to osteochondritis and arthritis of the facet joints. Similarly, Kumar et al. (2008) reported that 22% of 32 patients treated with Dynesys developed adjacent segment degeneration. A cadaver study (Schmoelz et al. 2003) reported similar range of motion in flexion and limited range of motion in extension with Dynesys compared to intact case. Clinical outcomes agreed with the cadaver study that the system limits extension motion. Motion limitation in extension may also be the reason for the poor clinical outcomes in terms of screw loosening or breakages, which is a relatively common complication with Dynesys (Stoll et al. 2002; Grob et al. 2005; Ko et al. 2010; Pham et al. 2016). Ultimately, Dynesys can only be used at the adjunct level of the fusion as approved by the FDA in the United States, and stand-alone use of Dynesys without fusion is considered an off-label use.

Similarly, most of the posterior dynamic stabilization systems (PDSS) have been reported as a possible alternative technique to spinal fusion, as a way to reduce the incidence of ASD. For example, Graft system was designed and proposed in 1991 for degenerative spinal diseases. This stabilization technique was aimed to prevent the abnormal flexion and restore segmental lordosis. Initial clinical results of the system were promising. Grevitt et al. (1995) reported on 50 consecutive patients who underwent Graft stabilization for intractable symptomatic degenerative disc disease. After a follow-up period of 24 months, the clinical results were good or excellent in 36 patients, and fair in 5 patients. Only one patient was reported as worse, and the system was suggested as a reasonable alternative to fusion. Other reports have demonstrated similar positive clinical results on the Graft system (Markwalder and Wenger 2003). However, one report demonstrated superior clinical outcomes for posterolateral fusion as compared to the Graft system (Hadlow et al.

1998). Additionally, there was a higher rate of revision surgery in Graft patients as compared to fusion patients, 18% and 29% at 1 and 2 years postoperatively, respectively. Late failure of the system was discussed by Mulholland and Sengupta (2002), and failure was related to transferring the load to the posterior annulus.

There are many advantages of using dynamic stabilization systems over fusion systems. It is designed to provide desirable environments for spinal movement and can limit abnormal motion, along with restoring load transmission patterns. However, careful consideration of the design must take place since a dynamic system must provide stability for its life time, unlike rigid fusion systems. Rigid spinal fusion systems need to provide stability only until the segment has fused. Dynamic systems have to withstand constant loading, which usually leads to fatigue failure or screw loosening (Schnake et al. 2006; Ko et al. 2010; Wu et al. 2011). Due to this reason, most of the dynamic stabilization devices, particularly pedicle screw-based systems, are designed to be stiffer to bear constant loading (Erbulut et al. 2014). However, if an implant is too stiff, instead of load-shearing it becomes load-bearing, which causes implant failure or screw loosening (Welch et al. 2007; Wu et al. 2011). Rohlmann et al. suggested that implant stiffness greater than 1000 N/mm may not be considered as dynamic because the mechanical effect would be similar to a fusion system (Rohlmann et al. 2007). Schmidt et al. (2009) determined the desirable stiffness parameter ranges for posterior dynamic stabilization systems. They found that only axial stiffness of 45 N/mm and bending stiffness of 30 N/mm are enough to reduce the segmental flexibility by 30%. Wilke et al. suggested that a dynamic device should allow 70% of the intact motion to achieve the desirable segmental stabilization.

For example, interspinous process (ISP) devices are considered to be dynamic stabilization systems and known to be useful for spinal pathologies such as spinal stenosis or facet arthritis. However, lack of design considerations of the device has been seen after the clinical and biomechanical tests. Clinical cases of ISP device failure have been reported. Not only has the gradual erosion at the spinous process due to consistent dynamic interaction at bone-device interface been observed (Miller et al. 2010), but also spinous process fracture due to stress concentration (Bowers et al. 2010; Kim et al. 2012). In our FE study, the biomechanical effects of an implanted interspinous process (ISP) device on load shearing at the index and adjacent segments were evaluated by using a hybrid protocol (Erbulut et al. 2015). Our stress results at the spinous process were in agreement with the literature that the maximum von Miss stress increased with implanted models up to 53 MPa. In addition, although the facet loads were decreased at the implanted level, FE model predicted that it was increased up to 60% at both the superior and inferior adjacent levels with extension (Table 1). Similarly, ISP devices decreased intradiscal pressure (IDP) at the index level, but not at the adjacent segments (Table 2).

Biomechanical data is essential in research process. Understanding the limitations of biomechanical data as it translates to clinical outcomes is paramount to the iterative process in implant design. Knowing that beneficial biomechanical data may not actually yield good clinical outcomes is important. Biomechanical data can also be utilized to analyze poor clinical outcomes to understand the reason for failure. Optimizing this process may help to address the poor clinical outcome by engaging biomechanical work again.

Although, the finite element models provide a good platform to study the biomechanical effects of a new implant designs and surgical techniques, care must be taken to consider the limitations. Some of the assumptions and simplifications include the homogeneity of tissue materials, lack of accurate muscle representations, linear material behavior, and loading and boundary conditions. For example, the various components of spine models are still to be considered as homogeneous and isotropic for stress distribution analysis in most published studies. Most of these assumptions are clearly inevitable, but a researcher should be knowledgeable to avoid misleading results from a FE simulation.

Table 1 Calculated facet loads of intact and instrumented models at index and adjacent levels (*Rt* Right, *Lt* Left)

Condition	Flexion			Extension			Lateral Bending			Axial Rotation		
	L2–3	L3–4	L4–5	L2–3	L3–4	L4–5	L2–3	L3–4	L4–5	L2–3	L3–4	L4–5
Intact												
Rt	45.63	76.68	25.89	127.13	147.36	120.37	29.40	54.67	76.17	0	0	0
Lt	54.40	77.44	19.46	151.72	156.99	194.33	37.51	59.35	21.93	172.11	175.72	214.00
Implanted												
Rt	45.65	76.68	25.89	184.02	0.06	192.95	29.84	66.32	73.95	0	0	0
Lt	54.40	77.44	19.46	230.38	11.06	299.92	37.53	46.12	23.55	172.16	150.12	215.17

Values expressed in N (Newton)

Table 2 Calculated intradiscal pressures of intact and instrumented models at index and adjacent levels

	Flexion			Extension			Lateral Bending			Axial Rotation		
Condition	L2–3	L3–4	L4–5	L2–3	L3–4	L4–5	L2–3	L3–4	L4–5	L2–3	L3–4	L4–5
Intact	1.44	1.34	1.46	0.87	0.98	0.90	2.04	1.80	2.11	1.34	1.33	1.22
Implanted	1.44	1.34	1.46	1.21	0.47	0.96	2.04	1.7	2.11	1.34	1.33	1.22

Values expressed in MPa (Megapascal)

Acknowledgments The work was supported in part by NSF Industry/University Cooperative Research Center at The University of California at San Francisco, CA, and The University of Toledo, Toledo, OH. A sincere thank you to Ronit Shah for his diligent proofreading of this book chapter.

References

Agarwal A, Ingels M, Kodigudla M et al (2016) Adjacent-level hypermobility and instrumented-level fatigue loosening with titanium and PEEK rods for a pedicle screw system: an in vitro study. J Biomech Eng 138:51004–51008. https://doi.org/10.1115/1.4032965

Agarwal A, Agarwal AK, Jayaswal A, Goel VK (2017) Outcomes of optimal distraction forces and frequencies in growth rod surgery for different types of scoliotic curves: an in silico and in vitro study. Spine Deform 5:18–26

Akamaru T, Kawahara N, Tim Yoon S et al (2003) Adjacent segment motion after a simulated lumbar fusion in different sagittal alignments: a biomechanical analysis. Spine (Phila Pa 1976) 28:1560–1566

Amevo B, Worth D, Bogduk N (1991) Instantaneous axes of rotation of the typical cervical motion segments: a study in normal volunteers. Clin Biomech (Bristol, Avon) 6(2):111–117. https://doi.org/10.1016/0268-0033(91)90008-E

Anderson CE (1956) Spondyloschisis following spine fusion. JBJS 38:1142–1146

Bogduk N, Mercer S (2000) Biomechanics of the cervical spine. I: normal kinematics. Clin Biomech (Bristol, Avon) 15:633–648

Bowers C, Amini A, Dailey AT, Schmidt MH (2010) Dynamic interspinous process stabilization: review of complications associated with the X-stop device. Neurosurg Focus 28:E8. https://doi.org/10.3171/2010.3.FOCUS1047

Brodke DS, Dick JC, Kunz DN et al (1997) Posterior lumbar interbody fusion A biomechanical comparison, including a new threaded cage. Spine (Phila Pa 1976) 22:26–31

Brodsky AE (1976) Post-laminectomy and post-fusion stenosis of the lumbar spine. Clin Orthop Relat Res 115:130–139

Brunet JA, Wiley JJ (1984) Acquired spondylolysis after spinal fusion. J Bone Joint Surg (Br) 66:720–724

Chen CS, Cheng CK, Liu CL, Lo WH (2001a) Stress analysis of the disc adjacent to interbody fusion in lumbar spine. Med Eng Phys 23:483–491

Chen W-J, Lai P-L, Niu C-C et al (2001b) Surgical treatment of adjacent instability after lumbar spine fusion. Spine (Phila Pa 1976) 26:E519–E524

Chin KR, Newcomb AGU, Reis MT et al (2016) Biomechanics of posterior instrumentation in L1-L3 lateral interbody fusion: pedicle screw rod constructvs. transfacet pedicle screws. Clin Biomech 31:59–64

DiAngelo DJ, Roberston JT, Metcalf NH et al (2003) Biomechanical testing of an artificial cervical joint and an anterior cervical plate. J Spinal Disord Tech 16:314–323

Dmitriev AE, Cunningham BW, Hu N et al (2005) Adjacent level intradiscal pressure and segmental kinematics following a cervical total disc arthroplasty: an in vitro human cadaveric model. Spine (Phila Pa 1976) 30:1165–1172

Eck JC, Humphreys SC, Lim T-H et al (2002) Biomechanical study on the effect of cervical spine fusion on adjacent-level intradiscal pressure and segmental motion. Spine (Phila Pa 1976) 27:2431–2434

Erbulut DU, Kiapour A, Oktenoglu T et al (2014) A computational biomechanical investigation of posterior dynamic instrumentation: combination of dynamic rod and hinged (dynamic) screw. J Biomech Eng 136:51007

Erbulut DU, Zafarparandeh I, Hassan CR et al (2015) Determination of the biomechanical effect of an interspinous process device on implanted and adjacent lumbar spinal segments using a hybrid testing protocol: a finite-element study. J Neurosurg Spine 23:200–208

Etebar S, Cahill DW (1999) Risk factors for adjacent-segment failure following lumbar fixation with rigid instrumentation for degenerative instability. J Neurosurg Spine 90:163–169

Evans JH (1985) Biomechanics of lumbar fusion. Clin Orthop Relat Res 193:38–46

Faizan A, Goel VK, Biyani A et al (2012) Adjacent level effects of bi level disc replacement, bi level fusion and disc replacement plus fusion in cervical spine – a finite element based study. Clin Biomech (Bristol, Avon) 27:226–233. https://doi.org/10.1016/j.clinbiomech.2011.09.014

Freudiger S, Dubois G, Lorrain M (1999) Dynamic neutralisation of the lumbar spine confirmed on a new lumbar spine simulator in vitro. Arch Orthop Trauma Surg 119:127–132

Goel VK, Panjabi MM, Patwardhan AG et al (2006) Test protocols for evaluation of spinal implants. J Bone Joint Surg Am 88(Suppl 2):103–109

Grevitt MP, Gardner ADH, Spilsbury J et al (1995) The Graf stabilisation system: early results in 50 patients. Eur Spine J 4:169–175

Grob D, Benini A, Junge A, Mannion AF (2005) Clinical experience with the Dynesys semirigid fixation system for the lumbar spine: surgical and patient-oriented outcome in 50 cases after an average of 2 years. Spine (Phila Pa 1976) 30:324–331

Hadlow SV, Fagan AB, Hillier TM, RDF (1998) The graft ligamentoplasty procedure: comparison with posterolat- eral fusion in the management of low back pain. Spine (Phila Pa 1976) 23:1172–1179

Hanten WP, Olson SL, Russell JL et al (2000) Total head excursion and resting head posture: normal and patient comparisons. Arch Phys Med Rehabil 81:62–66

Ivanov AA, Kiapour A, Ebraheim NA, Goel V (2009) Lumbar fusion leads to increases in angular motion and stress across sacroiliac joint: a finite element study. Spine (Phila Pa 1976) 34:E162–E169

Katz V, Schofferman J, Reynolds J (2003) The sacroiliac joint: a potential cause of pain after lumbar fusion to the sacrum. Clin Spine Surg 16:96–99

Kerr D, Zhao W, Lurie JD (2015) What are long-term predictors of outcomes for lumbar disc herniation? A randomized and observational study. Clin Orthop Relat Res 473:1920–1930

Kim DH, Shanti N, Tantorski ME et al (2012) Association between degenerative spondylolisthesis and spinous process fracture after interspinous process spacer surgery. Spine J 12:466–472

Ko C-C, Tsai H-W, Huang W-C et al (2010) Screw loosening in the Dynesys stabilization system: radiographic evidence and effect on outcomes. Neurosurg Focus 28:E10

Kumar MN, Jacquot F, Hall H (2001) Long-term follow-up of functional outcomes and radiographic changes at adjacent levels following lumbar spine fusion for degenerative disc disease. Eur Spine J 10:309–313

Kumar A, Beastall J, Hughes J et al (2008) Disc changes in the bridged and adjacent segments after Dynesys dynamic stabilization system after two years. Spine (Phila Pa 1976) 33:2909–2914

Lafage R, Schwab F, Glassman S et al (2017) Age-adjusted alignment goals have the potential to reduce PJK. Spine (Phila Pa 1976) 42:1275–1282

Laxer EB, Brigham CD, Darden BV et al (2017) Adjacent segment degeneration following ProDisc-C total disc replacement (TDR) and anterior cervical discectomy and fusion (ACDF): does surgeon bias effect radiographic interpretation?. Eur Spine J 26(4):1199–1204. https://doi.org/10.1007/s00586-016-4780-1

Lee CK (1988) Accelerated degeneration of the segment adjacent to a lumbar fusion. Spine (Phila Pa 1976) 13:375–377

Lee CK, Langrana NA (1984) Lumbosacral spine fusion, A biomechanical study. Spine (Phila Pa 1976) 9:574–581

Maigne JY, Planchon CA (2005) Sacroiliac joint pain after lumbar fusion. A study with anesthetic blocks. Eur Spine J 14:654–658

Markwalder TM, Wenger M (2003) Dynamic stabilization of lumbar motion segments by use of Graf's ligaments: results with an average follow-up of 7.4 years in 39 highly selected, consecutive patients. Acta Neurochir 145:209–214

Matsumoto T, Okuda S, Maeno T et al (2017) Spinopelvic sagittal imbalance as a risk factor for adjacent-segment disease after single-segment posterior lumbar interbody fusion. J Neurosurg Spine 26:435–440

Matsumoto K, Shah A, Agarwal A, Goel V (2018) Biomechanics of the relationship between adjacent segment disease (ASD) after lumbar arthrodesis and sagittal imbalance: a finite element study. Glob Spine J 8(1S):174S–374S

Matsumoto K, Shah A, Agarwal A, Goel V (2019) Biomechanics of vertebral stress and sagittal imbalance in an adult spine deformity: a finite element study. Orthop Res Soc 44(PS1-025):0809

Miller JD, Miller MC, Lucas MG (2010) Erosion of the spinous process: a potential cause of interspinous process spacer failure. J Neurosurg Spine 12:210–213

Mo Z, Zhao Y, Du C et al (2015) Does location of rotation center in artificial disc affect cervical biomechanics? Spine (Phila Pa 1976) 40:E469–E475

Moreau PE, Ferrero E, Riouallon G et al (2016) Radiologic adjacent segment degeneration 2 years after lumbar fusion for degenerative spondylolisthesis. Orthop Traumatol Surg Res 102:759–763

Mulholland RC, Sengupta DK (2002) Rationale, principles and experimental evaluation of the concept of soft stabilization. Eur Spine J 11(Suppl 2):S198–S205

Nakashima H, Kawakami N, Tsuji T et al (2015) Adjacent segment disease after posterior lumbar interbody fusion: based on cases with a minimum of 10 years of follow-up. Spine (Phila Pa 1976) 40:E831–E841

Natarajan RN, Andersson GBJ (2017) Lumbar disc degeneration is an equally important risk factor as lumbar fusion for causing adjacent segment disc disease. J Orthop Res 35:123–130

Nowitzke A, Westaway M, Bogduk N (1994) Cervical zygapophyseal joints: geometrical parameters and relationship to cervical kinematics. Clin Biomech 9:342–348

Ordway N, Seymour R, Donelson R (1999) Cervical flexion, extension, protrusion, and retraction: a radiographic segmental analysis. Spine (Phila Pa 1976) 24:240–247

Park P, Garton HJ, Gala VC et al (2004) Adjacent segment disease after lumbar or lumbosacral fusion: review of the literature. Spine (Phila Pa 1976) 29:1938–1944

Pearcy MJ, Bogduk N (1988) Instantaneous axes of rotation of the lumbar intervertebral joints. Spine (Phila Pa 1976) 13:1033–1041

Penning L (1988) Differences in anatomy, motion, development and aging of the upper and lower cervical disk segments. Clin Biomech 3:37–47

Pham MH, Mehta VA, Patel NN et al (2016) Complications associated with the Dynesys dynamic stabilization system: a comprehensive review of the literature. Neurosurg Focus 40:1–8

Phillips FM, Carlson GD, Bohlman HH, Hughes SS (2000) Results of surgery for spinal stenosis adjacent to previous lumbar fusion. Clin Spine Surg 13:432–437

Rohlmann A, Burra NK, Zander T, Bergmann G (2007) Comparison of the effects of bilateral posterior dynamic and rigid fixation devices on the loads in the lumbar spine: a finite element analysis. Eur Spine J 16(8):1223–1231. https://doi.org/10.1007/s00586-006-0292-8

Rothenfluh DA, Mueller DA, Rothenfluh E, Min K (2015) Pelvic incidence-lumbar lordosis mismatch predisposes to adjacent segment disease after lumbar spinal fusion. Eur Spine J 24:1251–1258

Schlegel JD, Smith JA, Schleusener RL (1996) Lumbar motion segment pathology adjacent to thoracolumbar, lumbar, and lumbosacral fusions. Spine (Phila Pa 1976) 21:970–981

Schmidt H, Heuer F, Wilke H-J (2009) Which axial and bending stiffnesses of posterior implants are required to design a flexible lumbar stabilization system? J Biomech 42:48–54

Schmoelz W, Huber JF, Nydegger T et al (2003) Dynamic stabilization of the lumbar spine and its effects on adjacent segments: an in vitro experiment. J Spinal Disord Tech 16:418–423

Schnake KJ, Schaeren S, Jeanneret B (2006) Dynamic stabilization in addition to decompression for lumbar spinal stenosis with degenerative spondylolisthesis. Spine (Phila Pa 1976) 31:442–449

Sengupta DK (2015) Cervical and lumbar disc replacement. In: Kim DH, Sangupta DK, FPC J et al (eds) Dynamic reconstruction of the spine, 2nd edn. Thieme Medical Publisher, New York, pp 7–19

Sengupta DK, Herkowitz HN (2012) Pedicle screw-based posterior dynamic stabilization: literature review. Adv Orthop 2012:1–7

Shah A, Lemans JVC, Zavatsky J, Agarwal A, Kruyt MC, Matsumoto K et al (2019) Spinal balance/alignment-clinical relevance and biomechanics. J Biomech Eng 141(7):1–14

Stoll TM, Dubois G, Schwarzenbach O (2002) The dynamic neutralization system for the spine: a multicenter study of a novel non-fusion system. Eur Spine J 11(Suppl 2):S170–S178

Sudershan S (2017) Biomechanical evaluation of lumbar interbody fusion surgeries with varying interbody device shapes, material properties, and supplemental fixation. Electronic thesis or dissertation. Retrieved from https://etd.ohiolink.edu

Umehara S, Zindrick MR, Patwardhan a G et al (2000) The biomechanical effect of postoperative hypolordosis in instrumented lumbar fusion on instrumented and adjacent spinal segments. Spine (Phila Pa 1976) 25:1617–1624

Unander-Scharin L (1950) A case of spondylolisthesis lumbalis acquisita. Acta Orthop Scand 19:536–544

Voronov LI, Mica MRC, Carandang G et al (2016) Biomechanics of transforaminally deployed expandable lumbar interbody fusion cage. Spine J 16:S255

Welch WC, Cheng BC, Awad TE et al (2007) Clinical outcomes of the Dynesys dynamic neutralization system: 1-year preliminary results. Neurosurg Focus 22:E8

Whitecloud TSIII, Davis JM, Olive PM (1994) Operative treatment of the degenerated segment adjacent to a lumbar fusion. Spine (Phila Pa 1976) 19:531–536

Wimmer C, Gluch H, Krismer M et al (1997) AP-translation in the proximal disc adjacent to lumbar spine fusion: a retrospective comparison of mono- and polysegmental fusion in 120 patients. Acta Orthop Scand 68:269–272

Wu J-C, Huang W-C, Tsai H-W et al (2011) Pedicle screw loosening in dynamic stabilization: incidence, risk, and outcome in 126 patients. Neurosurg Focus 31:E9

Würgler-Hauri CC, Kalbarczyk A, Wiesli M et al (2008) Dynamic neutralization of the lumbar spine after microsurgical decompression in acquired lumbar spinal stenosis and segmental instability. Spine (Phila Pa 1976) 33:E66–E72

Yamasaki K, Hoshino M, Omori K et al (2017) Risk factors of adjacent segment disease after transforaminal interbody fusion for degenerative lumbar disease. Spine (Phila Pa 1976) 42:E86–E92

Lessons Learned from Positive Biomechanics and Positive Clinical Outcomes

15

Isaac R. Swink, Stephen Jaffee, Jake Carbone, Hannah Rusinko, Daniel Diehl, Parul Chauhan, Kaitlyn DeMeo, and Thomas Muzzonigro

Contents

Introduction	332
Anterior Cervical Plating	333
Plating vs. Non-Plating	334
Dynamic vs. Fixed	336
Pedicle Screws and Rods	338

Isaac R. Swink and Stephen Jaffee contributed equally with all other contributors.

I. R. Swink (✉) · D. Diehl · P. Chauhan · T. Muzzonigro
Department of Neurosurgery, Neuroscience Institute, Allegheny Health Network, Pittsburgh, PA, USA
e-mail: Isaac.Swink@ahn.org

S. Jaffee
College of Medicine, Drexel University, Philadelphia, PA, USA

Allegheny Health Network, Department of Neurosurgery, Allegheny General Hospital, Pittsburgh, PA, USA
e-mail: Stephen.Jaffee@AHN.ORG

J. Carbone
Louis Katz School of Medicine, Temple University, Philadelphia, PA, USA

Department of Neurosurgery, Allegheny Health Network, Pittsburgh, PA, USA

H. Rusinko
Neuroscience Institute, Allegheny Health Network, Pittsburgh, PA, USA
e-mail: rusink_h1@denison.edu

K. DeMeo
Department of Neurosurgery, Allegheny Health Network, Pittsburgh, PA, USA

© Springer Nature Switzerland AG 2021
B. C. Cheng (ed.), *Handbook of Spine Technology*,
https://doi.org/10.1007/978-3-319-44424-6_28

Unilateral v. Bilateral Rod Constructs 338
Trajectory 340

Interbody Devices 341
Device Design 342
Interbody Fusion Approaches 343
Perioperative Factors 345

Conclusion 346

References 346

Abstract

This chapter seeks to explore how biomechanical studies positively influence clinical procedures by reviewing the literature relevant to three of the largest modalities in spinal surgery: anterior cervical plating, pedicle screws and rods, and interbody fusion devices. The area of focus within anterior cervical plating includes the introduction of plating systems to increase stability and the recent shift towards more dynamic systems. Furthermore, the differences between various pedicle screw and rod constructs as well as lumbar interbody fusion device configurations and approaches will be examined in detail to demonstrate the correlation between biomechanical results and clinical outcomes. The lessons garnered throughout this review will demonstrate how biomechanics can be best utilized to evaluate the efficacy of new devices, provide possible explanations to device complications, and refine design interactions to improve care and compare various device designs. Therefore, researchers and surgeons should be able to distill the important elements of using biomechanical and clinical data synergistically to prove both device and procedural success.

Keywords

Pullout strength · Bone implant interface · Cervical plating · Interbody fusion · Stability · Pedicle screw · VAS score · ODI score · Screw trajectory

Introduction

This chapter presents an introduction to the relationship between biomechanical and clinical testing of spine implant technologies, with the intended goal of highlighting specific areas of interest in which biomechanical studies have had a positive influence on clinical results rather than providing a comprehensive review of the literature within the field. It is important for both device innovators and physicians to understand the role of biomechanics testing to evaluate not only the physical characteristics of the devices, but predict their real-world performance under physiologic conditions. Thus, the integration of the breakthroughs garnered through this research aids in creating optimized designs and procedural tactics. In fact, the standard procedure for introducing a new technology to the market place involves a series of checkpoints in which the device is tested in a multitude of mediums: biomaterials, biomechanics, surgical safety, and both short- and long-term surgical benefits and complications. Ideally, biomechanical testing paves the way for further clinical research if the conclusions distilled from such work demonstrate an effective solution to a clinical problem that has yet to be solved.

This chapter will highlight this pipeline through its discussion of both entities of development: biomechanical and clinical efficacy. Furthermore, the successful integration of these research areas is paramount in creating an effective and useful device. Biomechanical success criteria include metrics like range of motion (ROM) and neutral zone (NZ) measured in response to axial compression (AC), flexion-extension (FE), lateral bending (LB), or axial torsion (AT) loading conditions. All of these testing conditions are intended to produce physiological loads by placing the device in a cadaveric model simulating real-world conditions. Clinically relevant metrics include visual analogue scale (VAS) scores, Oswestry disability index ODI scores, fusion rates, non-union rates, pseudoarthrosis rates, infection rates, and device complication rates. The chapter will utilize three major categories of devices to denote how biomechanical testing has proven an effective means of predicting clinical success, and the respective lessons learned from such discoveries.

Anterior Cervical Plating

Cervical plating, a fixation system that is primarily utilized for spine segment stabilization, has indications for usage in pathologies such as spondylosis due to its hypothesized ability to enhance the rate of arthrodesis, Fig. 1. Moftakhar et al., in their paper providing a comprehensive overview of the anterior cervical plating technology, mention that plating has been shown to aid with earlier patient mobilization, a reduction in the need for postoperative collars, increased in graft loading, a reduction in graft dislodgement, and an increase in the ability to treat spinal deformities (Moftakhar and Trost 2004).

A bounty of technologies exist within the anterior cervical plate (ACP) space, as surgeons and device manufacturers aim to prevent and mitigate common complications associated with anterior cervical discectomy fusion (ACDF), primarily non-unions and pseudoarthrosis. While the first ACDF was performed in the mid-twentieth century by Bailey and Badgley, the market is still full of different ACP technologies, which necessitate further investigation of this surgical approach (Moftakhar and Trost 2004). ACP systems can be categorized based on their design characteristics, including plate-screw interaction and system rigidity, Fig. 2. This portion of the chapter will first highlight the topics of plating versus non-plating and rigid versus dynamic plating systems. It will include a discussion on the various biomechanical studies performed on ACP systems to evaluate their robustness and practicality, with respect to classic biomechanical models and metrics. The chapter will then focus on the clinical outcomes and the efficacy of these devices. Lastly, there will be an integration of these conclusions garnered from each of the aforementioned sections in order to provide the reader with a concise means of

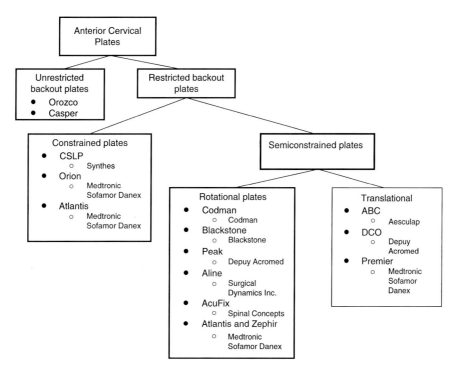

Fig. 1 Categories of Anterior Cervical Plates (Moftakhar and Trost 2004)

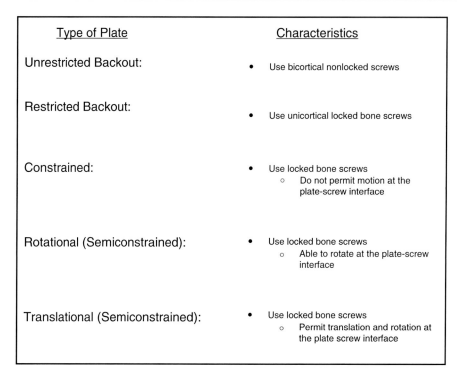

Fig. 2 Characteristics of Plate Designs (Moftakhar and Trost 2004)

reconciling whether the biomechanical models reflect the efficacy of this technology.

Plating vs. Non-Plating

Biomechanics

Ideally, a biomechanically successful ACP system would be able to significantly reduce the motion of a destabilized spinal segment in order to reduce pain associated with cervical trauma or degeneration, thus producing a stable spine segment (Rubin and Lanyon 1984). Hakalo et al. (2008) investigated the benefits associated with cervical plating in an in vivo biomechanical study that compared the stability and subsidence of different instrumentation systems in a porcine model.

Using a C3-C4 porcine model, ACDF procedures were performed in order to compare stabilization with either a cage alone or with a cage and plate. The specimens were instrumented and then dissected in order to evaluate the stability of the instrumented levels' devices using the MTS 858 Mini Bionix testing machine. A 2.5 Nm moment was applied to the cervical segments in flexion-extension and LB at a speed of 40 cm/min and the displacement of C3 with respect to C4 was measured. In order to keep the results relative, the stability of the vertebrae after discectomy and instrumentation was calculated and normalized to the intact condition. Then, a subsidence test was performed in the specimens that received instrumentation; in which, each segment underwent 21,000 cycles of axial loading ranging from 20 to 200 N at a frequency of 2.5 Hz (subsidence was measured by subtracting displacements before and after the cyclical test while the specimen was subjected to a 200 N preload).

The cage and plate systems resulted in significantly increased segmental stability when compared to the cage alone cohort. Stability is often the metric used to demonstrate the ability of a spinal segment ability to withstand anatomical loading. It may also permit proper load distribution at the cage-bone interface, impacting the rate of arthrodesis and fusion time (Rubin and Lanyon 1984). Subsidence was also shown to be largest in the cage-alone group and was reduced by 50% in the cage and plate group. It is a relatively common phenomenon in spinal instrumentation caused by the difference between the mechanical properties of bone and implant materials, as well as the patterns of trabecular structure within vertebrae. Thus, the cage and the graft-bone bed interface may lead to subsidence, potentially manifesting in poor clinical results (Gercek et al. 2003). Biomechanical studies may be limited in their clinical relevance; however, the increase in stability and decrease in subsidence demonstrated with the implementation of cervical plates and interbody cages may indicate improved clinical outcomes when compared to ACDF procedures using interbody cages alone.

Clinical Efficacy

The clinical efficacy of single-level ACDF was assessed in the study conducted by Wang et al. They intended to determine whether or not cervical plating contributed to preferable clinical outcomes and reduced rates of pseudoarthrosis, which were measured by factors such as patient reported outcome success (based on Odom's criteria), graft collapse, and kyphotic deformity angle increase. Utilizing a patient population of 80 (36 with fusions without plating and 44 with it) and implementing a follow-up period of 6 years, they were able to demonstrate the benefits cervical plating. While they determined that the resultant rates of pseudoarthrosis between the plated and non-plated patients were not statistically significant, it was shown that cervical plating was safe and not associated with a significant increase in complication rates. This is a noteworthy conclusion because while the study may not have been sufficiently powered to detect a statistically significant difference, the authors conclude that the cervical plating method was as effective as the current standard practice. Additionally, the use of anterior cervical plates showed a marked decrease in the kyphotic deformity angle when compared to the non-plated population (1.2 degrees to 1.9 degrees respectfully). Moreover, rates of pseudoarthrosis were lower in the plated group than those in the non-plated group (4.5–8.3%, respectively). Although this finding was also not statistically significant, it is a promising conclusion regarding the efficacy of this technology and may have paved the way for future inquiry into the cervical plating (Wang and McDonough 1999). The finding that single-level anterior cervical plating is inconsequential and does not substantially aid in fusion rates was reinforced by the research published by Connolly et al. Similar to the study performed by Wang and McDonough (1999), researchers assessed the efficacy of using plating for the treatment of spondylosis using a total of 43 patients, 25 of which were treated with anterior cervical discectomy, autograft fusion, and plate fixation, while 18 were treated in the same manner without plating. The study found that plating did not significantly affect the fusion rate for single-level interventions (Connolly and Esses 1996). However, Connolly did foreshadow to the potential for using plating technology for reducing multilevel fusion complications.

Compared to single-level cervical fusion, three-level fusions are known to carry a higher risk of non-union and pseudoarthrosis. Wang et al. also conducted a study highlighting the benefits of using anterior cervical plates as a method of stabilization. By tracking 59 patients over a 7-year period, with a 3.2-year follow-up, they were able to compare the success of anterior plates on multi-level cervical discectomy to the success of standalone cages. The researchers found that 14 of the 59 patients monitored developed the pseudoarthrosis. However, a breakdown of this ratio engendered an interesting conclusion, of the 40 total patients in the plating group 7 (18%) had some degree of pseudoarthrosis compared to 7 out of 19 (37%) of the patients in the non-plate group. While this result was not statistically significant, the anterior cervical plate provided increased stability and higher rates of fusion when compared to a more conventional treatment. The authors noted that, while the incorporation of anterior cervical plates produced favorable results when compared to the standard treatment, it still had a fairly high percentage of failure of fusion. This conclusion provides justification for further study into the efficacy of cervical plating in order to distinguish how best to stabilize and support fusion within the spine (Wang and McDonough

2001). Through these two respective studies, ACP primarily demonstrated clinical efficacy in multilevel discectomy.

The dichotomy between the effect of cervical plating between single and multilevel discectomies is further demonstrated through the study conducted by Kaiser et al. Through a retrospective review of 540 patients who underwent either one- or two-level ACDF procedures, 251 with and 289 without plating, the researchers concluded that plating marginally increased the rate of fusion in the single level but resulted in a substantial increase in the rate of fusion in the case of two-level ACDF procedures. The fusion rates of plating versus non-plating two-level ACDF procedures were 91% and 72%, respectively, and the fusion rates of plating versus non-plating for one-level ACDF were much closer, 96% and 90%, respectively. These results along with the conclusions from the aforementioned studies provide substantial evidence that cervical plating has benefits in multilevel ACDF procedures (Kaiser and Haid 2002).

Integration

As demonstrated by the biomechanical test conducted by Hakalo et al., cervical plating showed promise for clinical use because of additional increase in stability to degenerated specimens and its theoretical ability to reduce subsidence of the implant. This conclusion is reinforced by the clinical data which also shows that plating can be advantageous, more so in multilevel cases than in single level ones. Plating contributes to reduced pseudoarthrosis, although not statistically significant, and comparable amounts of complications to that of the standard treatment. Therefore, it is evident that the biomechanical testing rightfully and accurately forecasted the technology's clinical success. It should be noted that the ability to utilize knowledge, such as the evidence that plating permits a more stable environment in a porcine model, is paramount to refining clinical practice and availability of different modalities of therapy. Thus, the anterior cervical plating model is a positive result for the use of both biomechanical and clinical testing in synchrony.

Dynamic vs. Fixed

Biomechanical Efficacy

While the use of plates is supported throughout the literature, their design has seen some fluctuation in recent years with evidence suggesting that the use of rigid fixation may reduce mechanical loading of the graft. A lack of mechanical stimulation may cause a negative effect on new bone formation as well as bone remodeling (Churches et al. 1979). This idea is supported by Wolff's Law, in that stress shielding and reduced load sharing lead to a decrease in bone (Frost 1994). Dynamic cervical plate systems have been developed to provide less rigid fixation, thus allowing for graft loading to accelerate the time to union.

Brodke et al. investigated the benefits associated with the use of dynamic cervical plates by performing an in vitro biomechanical study using a simulated cervical corpectomy model to compare the load-sharing properties of four cervical plate systems (Brodke and Gollogly 2001). The study consisted of two static plate systems – Synthes CSLP (Synthes Spine, Paoli, PA) and Sofamor-Danek Orion (Medtronic Sofamor Danek, Memphis, TN), and two dynamic systems – Depuy Acromed DOC (Depuy Acromed, Raynham, MA) and Aesculap ABC (Aesculap, Tuttlingen, Germany). Six specimens of each of the four plate types were mounted on ultra-high-molecular-weight polyethylene blocks intended to simulate vertebral bodies. Load-sharing between the graft and plate was measured under two conditions (30-mm and a 27-mm graft) while subjected to a linearly increased AC load. The 27-mm graft condition was intended to simulate a 10% loss in graft height or subsidence. Load transmission through the graft was measured at the inferior graft-endplate interface using a thin film force transducer. Measurements of load sharing were expressed as a percentage of the load applied to the vertebral segment.

Under the 30-mm graft condition, all four instrumentation systems transmitted more than 60% of the axial load through the graft, indicating that both locked and dynamic cervical plates effectively shared loads under parameters where subsidence or collapse did not occur. However, under the 10% graft loss condition, only the dynamic plates shared significant portions of the load to the graft, with values of 88% for ABC and 96% for DOC. The locked cervical plates prevented graft from sharing any portion of the axial load until a minimum of 90 N was applied, even at the maximum load of 120 N, only reported 11% and 17% of the applied load was transferred through the graft for the Orion and CSLP, respectively.

This preliminary biomechanical study demonstrated the benefits associated with dynamic cervical plate systems when compared to traditional static systems. Physiological strain levels with an appropriate distribution can produce an osteogenic stimulus that is capable of increasing bone mass (Rubin and Lanyon 1984). Therefore, load-sharing is desirable in a cervical plate system because the transfer of load through the graft would lead to increased rate of bone formation and increased fusion rate. Static systems were able to effectively share loads under the 30-mm graft condition; however, their performance in the simulated loss condition (27 mm) indicated that there could be a significant reduction in their performance, should graft collapse or subsidence occur. On the contrary, the dynamic plate systems demonstrated their ability to share load through the graft over a range of conditions and could lead to consistently improved patient outcomes.

ACP has also been shown to have a significant effect on the loading characteristics of the posterior cervical spine. Petterson et al. performed a study in which the instantaneous axis of rotation (IAR), anterior column load sharing, and posterior element strain of cadaveric specimens were compared between rigid and dynamic plating systems (Peterson et al. 2018a, b). This study found that rigid plates cause a shift in the IAR of the spinal segment from the posterior third to the anterior periphery of the disc space. As a result of this anterior shift, the distance between the posterior elements and center of rotation is larger, therefore increasing the forces acting on the posterior elements. They also showed that the magnitude of the anterior IAR shift is directly correlated to the stiffness of the plating system used. Thus, a dynamic plate with lower material stiffness could minimize the anterior shift in IAR and reduce posterior element strain, which is a significant finding given that mechanical loading of the posterior elements has been identified as a possible contributor to low back pain (Cohen and Raja 2007).

Clinical Efficacy

Building upon the conclusion derived the biomechanical literature, Saphier et al. conducted a prospective study to determine if dynamic plates yielded any additional clinical benefit compared to static plates. The study focused on two Medtronic ACP systems, the ORION ACP system (rigid plating) and the PREMIER ACP system (dynamic plating) (Saphier and Arginteanu 2007). Twenty-five of the 50 patients underwent one- or two-level ACDF procedures with the rigid system and the other 25 with the dynamic system between 1998 and 2002. The procedure's success was determined radiographically by measuring vertical translation and in terms of patient reported outcomes using a 6-month follow-up questionnaire focused on pain, disability, and overall satisfaction.

Seventeen of the patients in the ORION group underwent one-level fusion and the remaining eight underwent two-level fusion. In the PREMIER cohort, 18 patients received one-level fusion and the other seven had two-level fusion performed. Patients treated with the dynamic system reported significantly lower pain scores and increased functionality on average when compared to the patients with rigid plates. While the difference between the rigid and dynamic groups was not statistically significant the overall satisfaction of patients in the dynamic plating group was on average higher.

Other clinical studies designed to compare static and dynamic plating systems have come to similar conclusions. In a 2013 meta-analysis of clinical studies comparing dynamic and static plating systems Li et al. found static plating systems to have higher complications rates and slower fusion rates compared to dynamic systems. This literature review included 315 patients across five studies, including 172 patients implanted with dynamic cervical plated and 143 patients with fixed plates. Complications were reported in two of the five studies and were only in the static plate group. These complications included plate break, screw dislocation, and screw back-out. Also included in the meta-analysis was a blinded study evaluating the success of 66 patients divided evenly between dynamic and screw plating systems (Li et al. 2013). Based on VAS, neck disability index, and radiologic evaluation it was determined that plating systems produce improved clinical outcomes in multilevel fusions, but are not significantly improved with respect to single-level fusions (Nunley et al. 2013).

Integration

Cervical plating has been shown to serve as an effective means of providing stability, specifically in multilevel pathology. This conclusion was garnered via both biomechanical and clinical testing, thus demonstrating the benefit of utilizing both research approaches. Through well-planned biomechanical studies researchers may determine how both healthy and pathological spine segments function. Furthermore, from this understanding the spine research community can better evaluate the effect of instrumentation and other therapeutic modalities on these specimens. The increased number of levels in a diseased spinal segment coincides with an increase in instability, pain, and degeneration. Moreover, as the number of diseased levels increases, the effectiveness of traditional plating therapies decreases, while that of the dynamic systems is maintained. This conclusion may not have been reached without an understanding of the biomechanics of both single and multilevel spinal segments. The development of cervical plating systems from rigid to dynamic constructs should serve as an example of how biomechanics can guide developing technologies.

Pedicle Screws and Rods

Pedicle screw fixation is intended to provide spinal stabilization for the treatment of traumatic injuries, deformity, and degenerative diseases.

Roy-Camille introduced early pedicle screw instrumentation to the US healthcare market in 1979 as rigid systems with thick rods or plates designed to provide maximum stability. However, due to a series of complications, the Food and Drug Administration (FDA) was forced to prohibit manufacturers from promoting the use of screws in the spine in 1993 and required patients be warned regarding the pedicle screws' experimental nature (Mulholland 1994). Both clinical and biomechanical studies have since been performed to evaluate various aspects of pedicle screw and rod systems and prove their efficacy in fixating the spine. As a result, their design has been refined as evidenced by their 95% success rate with respect to facilitating fusion (Choi and Park 2013). Pedicle screw development relied on both biomechanical and clinical success; thus, it is important to analyze how this development progressed.

Unilateral v. Bilateral Rod Constructs

Pedicle screws and rods are widely popular for single- and multilevel spinal fusions and have been used in various lumbar disorders. Several posterior fixation techniques are currently available to assist spinal fusion, with bilateral fixation being considered as the "gold standard" (Liu et al. 2016). Unilateral fixation has been introduced as an alternative to bilateral fixation because it is a less invasive procedure which still provides necessary stability. However, unilateral versus bilateral fixation has been widely debated as each method has both positive and negative indications. Thus, there may not be a clearly superior system. Traditionally, most surgeons would perform a bilateral screw fixation, but recently it was discovered that internal fixation of this type can result in a decrease in bone mineral content caused by excessive rigidity. A brief overview of the biomechanical and clinical comparison of the two constructs will be presented in an effort to determine how a mechanical evaluation of the construct may guide clinical practice.

Biomechanics

Godzik et al. conducted a biomechanical study assessing the stability of unilateral (UPS) versus bilateral (BPS) pedicle screw fixation with and without interbody support, using lateral lumbar interbody fusion approach (Godzik et al. 2018). The researchers determined that when an interbody cage was used, there was a negligible difference in terms of the stability generated between a unilateral versus a bilateral pedicle screw approach. They were able to come to this conclusion by using 13 cadaveric specimens divided into two groups. Specimens in group 1 were tested in three stages: intact, UPS alone, and BPS. Group two specimens were tested in four stages: intact, interbody alone, interbody cage with UPS, and interbody cage with BPS. Conventional biomechanical metrics were used for the assessment of segmental stability: FE, LB, and AT. ROM was calculated for each stage of testing. Results showed the bilateral construct appeared to have an increase in immediate stability, but the difference in stability between unilateral and bilateral screw constructs was statistically insignificant when an interbody cage was included in the construct. This conclusion may indicate that in a clinical setting, when both an interbody cage and pedicle screw fixation are implemented, the clinical outcomes between patients instrumented with unilateral and bilateral pedicle screw fixation would be comparable.

A similar conclusion was drawn from a study conducted by Liu et al., in which the researchers also compared three different posterior fixation techniques for two-level lumbar spine disorders in cadaveric specimens (Liu et al. 2016). The three instrumentation systems included UPS, UPS with contralateral translaminar facet screw (UPSFS), and BPS. Polyaxial pedicle screws (6 mm diameter, 45 mm length) were used in the study. Eight intact cadaveric lumbar spines (four from L1-L5, four from L1-S1) were tested by applying pure moments of ±8 Nm followed by testing left facetectomized L3-L4 and L4-L5 segments (to simulate unstable conditions). ROM and NZ of L3-L5 were recorded, and the results of the study showed that all fixation types could significantly reduce the ROM of L3-L5 in all loading conditions when compared to the intact state, with the exception of AT. Only BPS significantly reduced the AT ROM in comparison to intact condition. With respect to NZ, there was significant reduction in all the three conditions under flexion extension when compared to intact stage of testing; however, no significant difference was found in LB and AT. Overall, BPS offered the highest stability, with UPSFS being the least invasive with good fixation strength – which could possibly be used to replace BPS. These results further support the above assertion that unilateral and bilateral fixation techniques are not statistically significantly different, while bilateral fixation has the highest stabilization effect.

Clinical Efficacy

According to studies conducted by Ding et al., unilateral and bilateral pedicle screw fixation systems produce similar clinical results (Ding and Chen 2014). Unilateral and bilateral screw systems were compared in a meta-analysis conducted by Ding et al., which included information from over 400 patients across five studies. The results revealed that there was no significant difference between the fusion rates of unilateral and bilateral pedicle screw fixation; however, UPS patients benefited from improved perioperative results (significantly shorter operative time and significantly less blood loss for unilateral pedicle screw fixation). The conclusions of this study suggest there is little difference in the clinical outcomes of unilateral and bilateral pedicle screw fixation. While bilateral screw fixation may produce marginally improved fusion rates, unilateral screw fixation presents shorter operative time and less blood loss.

Moreover, a comprehensive overview on the outcome differentials between unilateral and bilateral rod constructs was performed by Molinari et al. to address the controversy between the two constructs in terms of efficacy and safety. Through a series of analysis on studies ranging from the topics such as complications, non-union rates, infection ODI and VAS scores, cage migration, and screw failure, it was deemed that there were no statistically significant differences between the two treatment modalities. The authors did encourage further investigation into this conclusion because most reports only involve single-level lumbar unilateral instrumentation (Molinari et al. 2015). A study conducted by Cheriyan et al. mirrored this conclusion by emphasizing that there was no statistically significant difference in terms of fusion rates and complications regarding unilateral or bilateral instrumentation (Cheriyan et al. 2015). Although the study did indicate that there was an increased likelihood of cage migration in the unilateral construct, the results were statistically insignificant and thus would require additional follow-up research to make a definitive conclusion. Thus, it is apparent that the unilateral instrumentation approach is a functional alternative to a bilateral approach, and may be indicated or contraindicated depending on the circumstance of the procedure.

Integration

Clinical success of spinal instrumentation is not solely dependent on fusion rates, and perioperative measures are also an important consideration. As the success of pedicle screw and rod systems surpasses 95% surgeons and device manufacturers must begin to investigate means to improve other success metrics, such as blood loss and operation time. Replacing BPS with UPS would allow for significant reductions in both blood loss and operation time. Biomechanical evaluation of unilateral and bilateral constructs indicated that both systems were able to provide the necessary stability to facilitate fusion. Clinical evaluation of both systems validates this biomechanical conclusion. This is a noteworthy correlation as it indicates that biomechanical comparison of two systems with respect to segmental stability is a successful method of predicting clinical results. While the unilateral construct requires more research and scrutiny, results indicate that it may serve as a promising alternative to the bilateral approach.

Trajectory

Achieving solid implant fixation when using pedicle screws is a problem which may be intensified in patients with osteoporosis and other disorders that lead to the diminution of bone density. Innovations aimed at addressing this issue have been seen in recent years with the hopes that better implant fixation will lead to improved patient outcomes. Cortical bone trajectory (CBT) is an alternative approach to traditional pedicle screw placement intended to increase screw-bone purchase in the lumbar spine by placing the screws in environments with higher bone mineral density than the traditional approach (Matsukawa et al. 2017). Common pedicle screw complications include screw loosening – which has been estimated to range from 1% to 15% in non-osteoporotic patients and exceed 60% in osteoporotic bones (El Saman et al. 2013). A more reliable avenue is necessary due to the complications of traditional transpedicular screw trajectory and surgical approach. The CBT trajectory differs from the transpedicular approach by maximizing cortical bone contact via a caudocephalad and mediolateral trajectory, leading to a reduction in soft tissue dissection, blood loss, and postoperative complications.

Biomechanics

Santoni et al. performed a human cadaveric biomechanical study to evaluate the efficacy of the CBT approach compared to the traditional pedicle screw placement approach (Santoni et al. 2009). Five fresh human lumbar spines were utilized from L1-L5 and the vertebral bodies were stripped of all muscular and ligamentous tissue, and then they were implanted with a set of pedicle screws. Each specimen received a pedicle screw using the traditional approach on one side and the CBT approach on the other. The vertebral body was then coupled to a six degree of freedom load cell, via a custom designed fixture and a screw pullout test was performed. Screws were withdrawn uni-axially at a rate of 10 mm/min until a sharp drop in the force profile was observed or there was observation of bone failure, and contralateral pedicle screws were then evaluated in the same fashion and yield force for each screw was calculated.

The study showed that the mean resistance against pullout for the new CBT was 30% greater than the traditional approach, 367 N and 287 N, respectively. This is a significant finding given that CBT screws were smaller in both length and diameter compared to traditional ones. Other biomechanical studies have demonstrated improved performance associated with the use of CBT as well. A study conducted by Wray et al. suggests this improvement in pullout strength is due to an alignment of screws with denser bone, regardless of DXA or qCT evaluation of bone quality (Wray et al. 2015). While screw pullout tests should not be the deciding factor in care pathways, the improvement in performance and decrease in size seen with the use of CBT pedicle screws could indicate that they may be preferable to the traditional screw and approach.

Clinical Efficacy

Two clinical studies conducted by Takenaka et al. and Kasukawa et al. found a significant reduction in blood loss using the CBT technique versus the traditional transpedicular screw placements during PLIF and TLIF procedures, respectively (Takenaka et al. 2017; Kasukawa et al. 2015).

Kasukawa's study evaluated clinical and radiological results of TLIF performed with CBT pedicle screw insertion versus traditional screw placement. Twenty-six patients were separated into three groups: minimally invasive pedicle screw insertion (M-TLIF), percutaneous pedicle screw insertion (P-TLIF), or pedicle screw insertion following the CBT (CBT-TLIF). Blood loss was significantly less with patients in the CBT-TLIF cohort compared to M-TLIF and P-TLIF groups. Operation time, postoperative bone union, lordotic angle maintenance, and screw placement accuracy were similar between the three groups. Takenaka et al. compared the effectiveness of PLIF using CBT and transpedicular screw techniques and found no significant difference between groups in operative time or fusion rates (Takenaka et al. 2017). However, the CBT group experienced a significant reduction in blood loss, postoperative creatine kinase levels, and pain scoring when compared to the traditional pedicle screw approach. CBT provided additional benefits of reduced perioperative pain and earlier return to normal activity levels. An additional study conducted by Mizuno et al. found that the CBT proved to be less invasive and equally effective as the traditional approach (Mizuno et al. 2014). The authors concluded that midline lumbar fusion procedures should follow the CBT when treating single-level degenerative pathologies in combination with midline insertion of an interbody graft. The CBT technique has been shown to be a viable alternative to the transpedicular approach in the treatment of spinal instability, degenerative disease, trauma, and spinal deformities because of its improved screw pull-out strength and reduction in intra and postoperative complications.

Integration

CBT development demonstrated how biomechanical analysis can be used to improve clinical outcomes. Osteoporotic patients have traditionally been a challenge for surgeons and device manufacturers due to their reduced trabecular structures. Augmenting purchase using techniques such as vertebroplasty has proven difficult, making the optimization of inherent bony purchase essential for improvement of pedicle screw–based treatments. The pattern of more dense trabecular bone structures in the most cephalad region of the pedicle and vertebral body has allowed surgeons and manufacturers to best adapt their approach to screw placement and design. As a result, the CBT is able to provide sufficient fixation to facilitate fusion while also significantly reducing blood loss, decreasing perioperative pain, and allowing patients to recover faster.

Interbody Devices

According to a historical review conducted by de Kunder et al., anterior lumbar interbody fusion (ALIF) was first introduced in 1933 by Burns and Capener and paved the way for future innovation in the treatment of degenerative disc disease, spondylolisthesis, and other spine destabilizing conditions (Kunder et al. 2018). By 1944, Briggs and Milligan began utilizing the now conventional procedure known as posterior lumbar interbody fusion (PILF) to treat disc herniations that impinged on spinal nerves and aid in spine stabilization. Through the years, spinal fusion using an interbody device has become the "gold standard" in the treatment for spinal destabilization pathology. Moreover, because of its wide use, spinal fusion procedures have evolved to include new approaches and mechanisms with which devices are able to stabilize the respective treatment levels.

Examples of innovations within this treatment modality include the incorporation of expandable cages to increase the tension within the spinal column and provide increased structural integrity, additional screws or blades to provide increased fixation into the vertebral bone, use of different materials such as PEEK and Titanium composites to permit preferable bone growth and stability of fusion, incorporation of biomaterial surface coatings to increase osteoconduction, and the addition of large bone graft windows to increase the rate of bone growth and osteoblastic differentiation. Initial adjustments to interbody cage design lead to substantial improvements in clinical success (fusion and patient reported measures have both shown to increase over time with the integration of new technologies). However, with the wide range of recently introduced innovative device designs, fusion and clinical success rates have been relatively consistent, thus creating the need for analysis on the current state of the art and a better understanding of what other metrics could be amended to provide clinical improvements.

Device Design

Biomechanics

Current ALIF cages vary widely in their designs, which may suggest differences in initial stability and long-term performance. Tstantrizos et al. investigated the biomechanical stability of five stand-alone ALIF constructs by utilizing an in vitro cadaveric model (Tstantrizos et al. 2000). The five cages of interest included the paired BAK cage (a threaded design), the Anterior Lumbar I/F cage (an oval fenestrated carbon implant with saw teeth), the Titanium Interbody Spacer or TIS (a round titanium implant with long serrated teeth), the SynCage (an oval titanium implant with short serrated teeth), and the ScrewCage (a rectangular titanium body with saw teeth housing a conical threading component). Forty-two lumbar spines L1-S1 were tested in the intact condition and then again after the insertion of an ALIF cage into the L3-L4 disc space. Each spinal segment was mounted on a custom six degrees of freedom testing machine and an electromagnetic tracking system was used to measure relative segmental motion via rigid sensors attached to the vertebral bodies. The loading protocol consisted of AR, FE, and LB conditions. NZ and ROM were extrapolated from the load-displacement curves of each loading condition. Pull-out force was also determined using a strain-gauge force transducer. With regard to anterior column stability, the five stand-alone cages were shown to be effective in reducing ROM but increased in NZ under all loading conditions. ROM represents an absolute measure of the total joint compliance under an applied moment. NZ represents an absolute measure of joint laxity and can be understood as the region of physiologic motion where the osteoligamentous structure of a functional spine unit does not provide resistance to motion (Panjabi 1992). The increase in NZ demonstrated in these devices could be from the absence of muscle contraction seen in a cadaveric model. Clinically, an increase in NZ may depict segmental instability.

Due to their shared ability to reduce ROM, a wide variety of cage designs are utilized in interbody fusion procedures today; a majority of which are capable of achieving arthrodesis. Significant differences were demonstrated between the devices above, with regard to specific directional loading and pullout force. However, each device managed to show that it could provide substantial stabilization to an FSU after implementation. Such results suggest that alternative metrics need to be created in order for biomechanical analysis of interbody devices to provide more valuable information. Specific attention should be focused on subsidence and migration associated with the interbody devices in addition to the traditional biomechanical measures of ROM and NZ. While biomechanics are effective in demonstrating that a device has the ability to provide stability to an unstable FSU, new metrics of quantifying clinical success associated with various cage designs need to be generated in order to further differentiate between the variety of interbody cages designs.

Clinical Efficacy

A series of recent clinical studies evaluating new interbody cage designs did not demonstrate significantly different outcomes compared to more traditional designs. This is evidenced by a study conducted by Sasso et al. which compared the clinical efficacy of a cylindrical threaded titanium cage (INTER FIX device) to a femoral ring allograft (control group) in patients with degenerative disc disease undergoing an ALIF surgery (Sasso et al. 2004). Patients who received the INTER FIX device had a reduction of ODI scores from 51.1 (preoperative) to 33.7 (postoperative), whereas the control group also had reduced ODI scores from 52.7 (preoperative) to 38.4 (postoperative) demonstrating no statistical difference between the mean scores of the two groups. In a separate clinical study, researchers compared the clinical outcomes of the Stablilis Stand Alone Cage (SAC) to the Bagby and Kuslich (BAK) device in an ALIF procedure for patients with degenerative disc disease (Lavelle et al. 2014). They found that there was no significant difference between the SAC and BAK devices in terms of mean operative time or blood loss or ODI scores. The ODI scores for patients in the BAK and SAC groups improved significantly (53.6 to 38.6 and 50.5 to 35.8, respectively).

Integration

Despite design differences, various interbody fusion devices possess similar biomechanical and clinical outcomes. The shared ability to reduce the ROM in cadaveric spine models and the reduction in ODI confer the impression that additional design features may not significantly affect device performance characteristics. This evidence reinforces the conclusion that interbody fusion implant biomechanical models correlate with their clinical results. Moreover, this information permits surgeons and device manufacturers an opportunity to develop new means of assessing performance because the current metrics do not generate significantly different results based on attempted design changes. This is not to say that these devices are performing exactly the same, rather it is more accurate to deduce that scale of these differences is not being recognized by standard measures.

Interbody Fusion Approaches

Spine interbody fusion is an effective treatment option for the stabilization of painful motion segments to relieve nerve compression, restore lordosis, and correct deformities (Mobbs et al. 2015). Depending on the direction of approach, there are several surgical techniques utilized during lumbar interbody fusion: posterior lumbar interbody fusion (PLIF), transforaminal lumbar interbody fusion (TLIF), minimally invasive transforaminal lumbar interbody fusion (MI-TLIF), oblique lumbar interbody fusion/anterior to psoas (OLIF/ATP), lateral lumbar interbody fusion (LLIF), extreme lateral interbody fusion (XLIF), and anterior lumbar interbody fusion (ALIF). Each technique has its own pros and cons; thus, comparative biomechanical analysis coupled with clinical studies could provide valuable insight in choosing the right treatment strategy. In order to understand the differentiation in clinical outcome of the various surgical approaches, it is important to recognize the indications and contraindications for each. Once the nuances of each surgical approach are appreciated, one may then apply scrutiny regarding the patient data resultant in their comparison.

PLIF is suitable for patients with segmental instability, disc herniation, spinal stenosis, and pseudoarthrosis. Advantages include increased visualization of nerve roots, adequate interbody height restoration, and the potential for 360-degree fusion through a single incision. However, disadvantages entail iatrogenic injury due to prolonged muscle retraction, retraction of nerve roots causing chronic radiculopathy, and difficulty restoring lordosis.

TLIF was developed to address the limitations of PLIF like the length of neural retraction, dural tears, and nerve root injury. TLIF involves direct access to the intervertebral foraminal space, which is advantageous due to its decreased muscle retraction time and bypassing the nerve roots, dura, and ligamentum flavum. This technique also permits a minimally invasive approach while preserving ligamentous structures allowing for increased biomechanical stability. A disadvantage of the TLIF approach, much like the PLIF, is the risk of paraspinal iatrogenic injury due to retraction.

ALIF has become the predominant surgical procedure in patients with discogenic low back pain especially in the areas of L4-L5 and L5-S1. This approach permits adequate access to the entire ventral surface of the disc and completed surgical dissection and allows for proper lateral exposure to the vertebral bodies permitting dynamic disc space clearance and endplate preparation. The reduction in posterior paraspinal muscle retraction decreases postoperative pain and impairment. One of the major limitations of this approach is the risk of vascular injury to the superior mesenteric artery which may cause thrombosis.

LLIF involves accessing the pathological disc space by a lateral retroperitoneal transpsoas entrance and is more suitable for disc spaces from T12-L1 to L4-L5, but is contraindicated below L5-S1 due to the iliac crest obstructing access and potential damage to the lumbar plexus. There is also risk of psoas muscle injury, bowel perforation, and vascular injury; however, it remains an option for sagittal and coronal deformities, especially lumbar scoliosis with laterolisthesis.

Lastly, OLIF, a minimally invasive approach, allows access through a small corridor between the peritoneum and the psoas muscle. Like the LLIF, the OLIF does not require laminectomy, facetectomy, posterior surgery, or stripping of paraspinal musculature. Neuromonitoring is not needed in the OLIF approach, as compared to the LLIF, because of the lack of psoas muscle dissection; therefore, it is most suitable for levels L1-S1. Comparable to LLIF, OLIF is also reasonable approach for sagittal and coronal deformities, especially in lumbar degeneration. Advantages include aggressive deformity correction, high fusion rates with complete disc space clearance, decreased psoas muscle and lumbar plexus injury, but the potential risks for sympathetic dysfunction and vascular injury are still apparent (Mobbs et al. 2015).

Biomechanics

Mica et al. evaluated the biomechanical stability of an expandable interbody cage (Luna 360) deployed in situ using a TLIF approach and compared it to a traditional lumbar interbody cage using an ALIF approach (control) (Mica et al. 2017). Twelve cadaveric spine specimens (L1-L5) were tested in the intact condition and after implantation of both the control and test device in the L2-L3 and L3-L4 index levels of each specimen. Additionally, the effect of supplemental pedicle screw-rod stabilization was assessed. Moments were applied to the segments under three loading conditions: FE, LB, and AR and segmental motions were recorded using an optoelectronic motion measurement system in order to calculate ROM. It was determined that the expandable TLIF cage and ALIF control device significantly reduced FE, LB, and AR motion with and without compressive preload when compared to the intact condition. Under all loading conditions (FE under 400 N preload, LB, and AR), the postoperative motions of the two constructs did not differ statistically. Adding bilateral pedicle screws resulted in further reduction of ROM for all the test modes compared to intact condition, with no statistical significance between the test and control device. The two approaches were found to be equivalent biomechanically demonstrating the consistency of performance among the different approaches and technologies within this modality of treatment.

Niemeyer et al. performed a similar study using titanium cages deployed as either TLIF or ALIF with and without posterior pedicle fixation and showed that the different cage design and approach resulted in only minor differences in segmental stability when combined with posterior pedicle screw fixation (Niemeyer et al. 2006). However, with pedicle screw fixation, the ALIF cage provided more stability than the TLIF cage in flexion-extension and axial rotation, but the absolute biomechanical differences were minor. While it is noteworthy that there was an observable difference in the ROM of the ALIF and TLIF groups with posterior pedicle screw fixation, the size of such a difference may be clinically negligible.

Ames et al. compared PLIF to TLIF at one and two levels with and without posterior fixation (Ames et al. 2005). Fourteen cadaveric specimens were subjected to either PLIF or TLIF at L2-L3 (single-level) and L3-L4 (two-level). ProSpace Interbody allograft was inserted into disc spaces in both cases. Pure moments (max. 4 Nm) were applied to the specimens. No significant differences were found in ROM between the approaches. Results also showed that posterior fixation with a pedicle-screw-rod construct was beneficial and could be used to achieve stability after fusion across one or two levels using either technique. These biomechanical studies conclude that the various interbody fusion approaches do not differ significantly, suggesting that perioperative parameters should be assessed to decide the safest treatment as per a patient's needs. However, more research may be required to understand the stabilizing effect of augmentation techniques in conjunction with the fusion approaches to clearly understand their effects on spine stability.

Clinical Efficacy

A number of publications have sought to reconcile whether or not various interbody fusion approaches posed significant advantages over one another in terms of clinical outcome. Zhang et al. compared PILF to TLIF outcomes in a meta-analysis containing seven comparative studies. They determined that while the respective approaches possessed different complication profiles, they had no statistically significant difference between important metrics such as clinical satisfaction or radiographic fusion (Zhang et al. 2014). Likewise, Phan et al. compared fusion characteristics between ALIF and TLIF and found a distinctly similar outcome (Phan et al. 2015). They noted that the complication profiles may vary, but that this might be a result of the different techniques and directions of approach. The fusion rates were 88.6% and 91.9%, respectively, demonstrating the similarity between the end results of the fusion surgery despite the noticeably different surgical approach.

Moreover, Watkins et al. compared three approaches against one another: ALIF, LLIF, and TLIF. Utilizing a patient population of 220 (309 operative levels) and radiographic analysis to measure variables such as lordosis restoration, disc height, and spondylolisthesis reduction, they determined that all groups showed a reduction in spondylolisthesis (Watkins et al. 2014). There was some degree of differentiation in regards to the lordosis improvement (ALIF, 4.5 degrees; ranked superiorly to TLIF, 0.8 degrees; and LLIF, 2.2 degrees) and disc height (ALIF, 2.2 mm; LLIF, 2.0 mm; and ranked superiorly to TLIF, 0.5). However, even when one approach appears to be superior the benefits are marginal when compared to the other techniques, demonstrating that the different surgical approaches could in some ways still be ubiquitous to one another. These studies show that while different approaches may have some degree of variability in their outcomes, their fusion results are generally similar and lead to reasonably comparable results.

Integration

Both biomechanical and clinical data suggest that despite the change in approach direction, stabilization and fusion (two common measures of implant success) are comparable amongst all. Thus, as argued in the aforementioned section, it would behoove both industry and surgeons alike to create additional means of distinction between the different device approaches in order to better understand how they affect the body under physiological loads.

Perioperative Factors

As the modalities of interbody fusion have reached a performance plateau, factors outside of pure fusion and stabilization are of increasing importance in the comparison of different methods. Perioperative measures such as infection rates, blood loss, device complications, surgical complications, and invasiveness are noteworthy metrics.

Tormenti et al., in their retrospective analysis of 531 patients, found that open-TLIF was a successful means of achieving arthrodesis in the lumbar spine (Tormenti et al. 2012). However, the researchers noted that there was a propensity for complications (durotomy and infection) in revision or multilevel fusion cases. Wong et al. compared intraoperative and perioperative complications between minimally invasive and conventional open-TLIF using a retrospective analysis of 513 patients (Wong et al. 2015). They were able to conclude that the minimally invasive approach was as successful or better than the open approach as mentioned by Tormenti et al. This argument is further echoed by Sulaiman et al., in which they concluded that minimally invasive TLIF actually performed superiorly to an open-TLIF approach in the metrics of average length of surgical time, estimated blood loss, ODI score, VAS rating, and direct cost of treatment (Sulaiman and Singh 2014). Due to the similar fusion results in both surgeries, surgeons should be concerned with perioperative measure as much or more so than classical metrics of success, like fusion and non-union. Furthermore, they may progress from questioning the fundamental specifications of device design to bettering the procedural variables. Choy et al. analyzed the perioperative results and complications in 1474 patients who had undergone ALIF and found the overall rate of surgical and medical complication was 14.5%. It was noted that complications were often associated with longer operation times (Choy et al. 2017). The anterior approach exposes more of the body (especially blood vessels and organs) increasing the risk of vascular damage.

These studies show how different approaches may correspond to different intra and postoperative complications. As devices grow more complex and achieve marginally fewer positive outcome differentials, perioperative measures will still be a means of increasing positive outcomes at a faster rate. Surgeons can still affect a large difference that may manifest itself in fewer complications and more successful procedures via the analysis and attention to these perioperative measures.

Conclusion

It has been demonstrated throughout this chapter that both biomechanical and clinical testing have proven beneficial in driving industry leaders and surgeons to adopt successful devices and practices. While the methods of such discoveries are far from perfect, and could benefit from additional inquiry and research, the correlation of both testing modalities serves as a vehicle for advancing the state of the art within the spinal surgery space. It is imperative that future research continues to challenge the therapeutic status quo and promotes the use of innovative designs and methods of intervention to further the quality standards of the industry. Biomechanical testing has been shown to pave the way for preliminary discoveries that may have robust clinical ramifications. It also serves to determine when new technologies do not lead to their intended outcome before they are moved into clinical models. Thus, the synergistic relationship among these two testing modalities should be enhanced to ensure the future prosperity of spinal surgery.

References

Ames C, Acosta F, Chi J, Iyengar J, Muiru W, Acaroglu E, Puttlitz C (2005) Biomechanical comparison of posterior lumbar interbody fusion and transforaminal lumbar interbody fusion performed at 1 and 2 levels. J Biomech 30(19):562–566

Brodke D, Gollogly S (2001) Dynamic cervical plates biomechanical evaluation of load sharing and stiffness. Spine 26(12):1324–1329

Cheriyan T, Lafage V, Bendo J, Spivak J, Goldstein J, Errico T (2015) Complications of unilateral versus bilateral instrumentation in Transforaminal lumbar Interbody fusion: a meta-analysis. Spine J 15(10). https://doi.org/10.1016/j.spinee.2015.07.251

Choi U, Park J (2013) Unilateral versus bilateral percutaneous pedicle screw fixation in minimally invasive transforaminal lumbar interbody fusion. Neurosurg Focus 35(2):1–8

Choy W, Barrington N, Garcia R, Kim R, Rodriguez H, Lam S, Dahdaleh N, Smith Z (2017) Risk factors for medical and surgical complications following single-level ALIF. Global Spine J 7(2):141–147

Churches A, Howlett C, Waldron K, Ward G (1979) The response of living bone to controlled time-varying loading: method and preliminary results. J Biomech 12:35–45

Cohen S, Raja S (2007) Pathogenesis, diagnosis, and treatment of lumbar Zygapophysial (facet) joint pain. Anesthesiology 106:591–614

Connolly P, Esses S (1996) Anterior cervical fusion: outcome analysis of patients fused with and without anterior cervical plates. J Spinal Discord 9(3):202–206

Ding W, Chen Y (2014) Comparison of unilateral versus bilateral pedicle screw fixation in lumbar interbody fusion: a meta-analysis. Eur Spine J 23:395–403

El Saman A, Meier S, Sander A, Kelm A, Marzi I, Laurer H (2013) Reduced loosening rate and loss of correction following posterior stabilization with or without PMMA augmentation of pedicle screws in vertebral fractures in the elderly. Eur J Trauma Emerg Surg 39:455–460

Frost H (1994) Wolff's law and Bone's structural adaptations to mechanical usage: an overview for clinicians. Angle Orthod 64(3):175–188

Gercek E, Arlet V, Delisle J, Marchesi D (2003) Subsidence of stand-alone cervical cages in anterior interbody fusion: warning. Eur Spine J 12:513–516

Godzik J, Martinez-Del-Campo E, Newcomb A, Reis M, Perez-Orribo L, Whiting A, Singh V, Kelly B, Crawford N (2018) Biomechanical stability afforded by unilateral versus bilateral pedicle screw fixation with and without Interbody support using lateral lumbar Interbody fusion. World Neurosurg 113. https://doi.org/10.1016/j.wneu.2018.02.053

Hakalo J, Pezowicz C, Wronski J, Bedzinski R, Kasprowicz M (2008) Comparative biomechanical study of cervical spine stabilization by cage alone, cage with plate, or plate-cage: a porcine model. J Orthop Surg 16(1):9–13

Kaiser M, Haid R (2002) Anterior cervical plating enhances arthrodesis after discectomy and fusion with cortical allograft. J Neurosurg 2(1):229–238

Kasukawa Y, Miyakoshi N, Hongo M, Ishikawa Y, Kudo D, Shimada Y (2015) Short-term results of Transforaminal lumbar Interbody fusion using pedicle screw with cortical bone trajectory compared with conventional trajectory. Asian Spine J 9(3):440–448. https://doi.org/10.4184/asj.2015.9.3.440

Kunder S, Rijkers K, Caelers I, Bie R, Koehler P, Santbrink H (2018) Lumbar Interbody fusion, a historical overview and a future perspective. Spine 1. https://doi.org/10.1097/brs.0000000000002534

Lavelle M, Mclain R, Rufo-Smith C, Gurd D (2014) Prospective randomized controlled trial of the Stabilis stand alone cage (SAC) versus Bagby and Kuslich (BAK) implants for anterior lumbar interbody fusion. Int J Spine Surg 8(8). https://doi.org/10.14444/1008

Li H, Min J, Zhang Q, Yuan Y, Wang D (2013) Dynamic cervical plate versus static cervical plate in the anterior cervical discectomy and fusion: a systematic review. Eur J Orthop Surg Traumatol 23:41–46

Liu F, Feng Z, Liu T, Fei Q, Jiang C, Li Y, ... Dong J (2016) A biomechanical comparison of 3 different posterior fixation techniques for 2-level lumbar spinal disorders. J Neurosurg:375–380. https://doi.org/10.3171/2015.7.SPINE1534

Matsukawa K, Yato Y, Hynes R, Imabayashi H, Hosogane N, Asazuma T, Matsui T, Kobayashi Y, Nemoto K (2017) Cortical bone trajectory for thoracic pedicle screws: a technical note. Clin Spine Surg 30(5):497–504. https://doi.org/10.1097/BSD.0000000000000130

Mica M, Voronov L, Carandang G, Havey R, Wojewnik B, Patwardhan A (2017) Biomechanics of an expandable lumbar Interbody fusion cage deployed through Transforaminal approach biomechanics of an expandable lumbar Interbody fusion cage deployed through Transforaminal approach. Int J Spine Surg 11(4):193–200. https://doi.org/10.14444/4024

Mizuno M, Kuraishi K, Umeda Y, Sano T, Tsuji M, Suzuki H (2014) Midline lumbar fusion with cortical bone trajectory screw. Neurol Med Chir 54:716–721. https://doi.org/10.2176/nmc.st.2013-0395

Mobbs R, Phan K, Malham G, Seex K, Rao P (2015) Lumbar interbody fusion: techniques, indications and comparison of interbody fusion options including PLIF, TLIF, MI-TLIF, OLIF/ATP, LLIF and ALIF. J Spine Surg 1(1). https://doi.org/10.3978/j.issn.2414-469X.2015.10.05

Moftakhar R, Trost G (2004) Anterior cervical plates: a historical perspective. J Neurosurg 16(1):1–5

Molinari RW, Saleh A, Molinari R, Hermsmeyer J, Dettori JR (2015) Unilateral versus bilateral instrumentation in spinal surgery: a systematic review. Glob Spine J 5(3):185–194. https://doi.org/10.1055/s-0035-1552986

Mulholland R (1994) Pedicle screw fixation in the spine. J Bone Joint Surg 76(4):517–519

Niemeyer T, Koriller M, Claes L, Kettler A, Werner K, Wilke H (2006) In vitro study of biomechanical behavior of anterior and transforaminal lumbar interbody instrumentation techniques. Neurosurgery 59(6):1271–1277. https://doi.org/10.1227/01.neu.0000245609.01732.e4

Nunley P, Jawahar A, Kerr E, Cavanaugh D, Howard C, Brandao S (2013) Choice of plate may affect outcomes for single versus multilevel ACDF: results of a prospective randomized single-blind trial. Spine J 9(2):121–127

Panjabi MM (1992) The stabilizing system of the spine. Part II. Neutral zone and instability hypothesis. J Spinal Disord 5(4):390–396

Peterson J, Chlebek C, Clough A, Wells A, Batzinger K, Houston J, Kradinova K, Glennon J, DiRisio D, Ledet E (2018a) Stiffness matters: part II – the effects of plate stiffness on load-sharing and the progression of fusion following ACDF in vivo. Spine. https://doi.org/10.1097/BRS.0000000000002644

Peterson J, Chlebek C, Clough A, Wells A, Ledet E (2018b) Stiffness matters: part I – the effects of plate stiffness on the biomechanics of ACDF in vitro. Spine. https://doi.org/10.1097/BRS.0000000000002643

Phan K, Thayaparan G, Mobbs R (2015) Anterior lumbar interbody fusion versus transforaminal lumbar interbody fusion – systematic review and meta-analysis. Br J Neurosurg 29(5):705–711. https://doi.org/10.3109/02688697.2015.1036838

Rubin C, Lanyon L (1984) Regulation of bone formation by applied dynamic loads. J Bone Joint Surg Am 66:397–402

Santoni B, Hynes R, McGilvray K, Rodriguez-Canessa G, Lyons A, Henson M, Womack W, Puttlitz C (2009) Cortical bone trajectory for lumbar pedicle screws. Spine J 9(5):366–373

Saphier P, Arginteanu M (2007) Stress-shielding compared with load-sharing anterior cervical plate fixation: a clinical and radiographic prospective analysis of 50 patients. J Neurosurg Spine 6(5):391–397

Sasso R, Kitchel S, Dawson E (2004) A prospective, randomized controlled clinical trial of anterior lumbar Interbody fusion using a titanium cylindrical threaded fusion device. Spine 29(2):113–122

Sulaiman W, Singh M (2014) Minimally invasive versus open transforaminal lumbar interbody fusion for degenerative spondylolisthesis grades 1–2: patient-reported clinical outcomes and cost-utility analysis. Ochsner J 14(1):32–37

Takenaka S, Mukai Y, Tateishi K, Hosono N, Fuji T, Kaito T (2017) Clinical outcomes after posterior lumbar Interbody fusion: comparison of cortical bone trajectory and conventional pedicle screw insertion. Clin Spine Surg 30(10):E1411–E1418. https://doi.org/10.1097/BSD.0000000000000514

Tormenti M, Maserati M, Bonfield C, Gerszten P, Moossy J, Kanter A, Sprio R, Okonkwo D (2012) Perioperative surgical complications of transforaminal lumbar interbody fusion: a single-center experience. J Neurosurg Spine 16(1):44–50. https://doi.org/10.3171/2011.9.spine11373

Tstantrizos A, Andreou A, Aebi M, Steffen T (2000) Biomechanical stability of five stand-alone anterior lumbar Interbody fusion constructs. Eur Spine J 9:14–22

Wang J, McDonough P (1999) The effect of cervical plating on single-level anterior cervical discectomy and fusion. J Spinal Discord 12(6):467–471

Wang J, McDonough P (2001) Increased fusion rates with cervical plating for three-level anterior cervical discectomy and fusion. J Spine 26(6):643–646

Watkins RG, Hanna R, Chang D, Watkins R (2014) Sagittal alignment after lumbar Interbody fusion. J Spinal Disord Tech 27(5):253–256. https://doi.org/10.1097/bsd.0b013e31828a8447

Wong A, Smith Z, Nixon A, Lawton C, Dahdaleh N, Wong R, Auffinger B, Lam S, Song J, Liu J, Koski T, Fessler R (2015) Intraoperative and perioperative complications in minimally invasive transforaminal lumbar interbody fusion: a review of 513 patients. J Neurosurg Spine 22(5):487–495. https://doi.org/10.3171/2014.10.spine14129

Wray S, Mimran R, Vadapalli S, Shetye SS, McGilvray KC, Puttlitz CM (2015) Pedicle screw placement in the lumbar spine: effect of trajectory and screw design on acute biomechanical purchase. J Neurosurg Spine 22:503–510

Zhang Q, Yuan Z, Zhou M, Liu H, Xu Y, Ren Y (2014) A comparison of posterior lumbar interbody fusion and transforaminal lumbar interbody fusion: a literature review and meta-analysis. BMC Musculoskelet Disord 15(1). https://doi.org/10.1186/1471-2474-15-367

The Sacroiliac Joint: A Review of Anatomy, Biomechanics, Diagnosis, and Treatment Including Clinical and Biomechanical Studies (In Vitro and In Silico)

16

Amin Joukar, Hossein Elgafy, Anand K. Agarwal, Bradley Duhon, and Vijay K. Goel

Contents

Background	350
Anatomy	350
Ligaments	351
Muscles	351
Function and Biomechanics	351
Range of Motion	354
Sexual Dimorphism	355
Causes of SIJ Pain	357
Diagnosis of SIJ Dysfunction	358
Nonsurgical Management	358
Open SIJ Fusion	360
Minimally Invasive SIJ Fusion	360
Clinical Studies	361
In Vitro and In Silico Studies	365
Summary	369

A. Joukar · H. Elgafy · A. K. Agarwal
Engineering Center for Orthopaedic Research Excellence (E-CORE), University of Toledo, Toledo, OH, USA

B. Duhon
School of Medicine, University of Colorado, Denver, CO, USA

V. K. Goel (✉)
Engineering Center for Orthopaedic Research Excellence (E-CORE), University of Toledo, Toledo, OH, USA

Departments of Bioengineering and Orthopaedic Surgery, Colleges of Engineering and Medicine, University of Toledo, Toledo, OH, USA
e-mail: Vijay.Goel@utoledo.edu

© Springer Nature Switzerland AG 2021
B. C. Cheng (ed.), *Handbook of Spine Technology*,
https://doi.org/10.1007/978-3-319-44424-6_130

Conclusion 370

References 370

Abstract

Sacroiliac joint (SIJ) is one of the most overlooked sources of LBP. The joint is responsible for the pain in 15–30% of people suffering from LBP. Fixation is increasingly recognized as a common surgical intervention for the treatment of chronic pain originating from sacroiliac joint (SIJ). Many studies have investigated the clinical outcomes and biomechanics of various SIJ surgical procedures. However, there is currently no agreement on the surgical indications for SIJ fusion or the best and most successful surgical technique for sacroiliac joint fixation and SIJ pain treatment.

Biomechanics of normal, and injured SIJs and biomechanical differences due to sex are well documented. Various studies have investigated the clinical outcomes of different surgical techniques and devices intended for treatment of the SIJ pain, and they have shown that these techniques are effective indeed. Several questions related to clinical and biomechanical effects of surgical parameters such as number, design/shape and positioning of implants, and unilateral versus bilateral placement remain unanswered. Biomechanical studies using in vitro and in silico techniques are crucial in addressing such unanswered questions. These are synthesized in the review.

Keywords

Sacroiliac joint · Fusion · Biomechanics · Surgery · Anatomy · Diagnosis · Treatment · Clinical · In vitro · In silico

Background

Low back pain (LBP) is the most common reason for primary care visits after common cold, with approximately 90% of adults being impacted by this condition at some point in their lives (Weksler et al. 2007; Frymoyer 1988). Apart from hindering the quality of life of those affected by LBP, if left untreated or improperly diagnosed, LBP may also profoundly impact affected patients' work productivity and therefore economic success. LBP accounts for annual cost up to 60 billion dollars due to decreased productivity and income as well as medical expenses (Koenig et al. 2016; Rudolf 2012; Murray 2011).

The majority of LBP cases originate from the lumbar spine. One of the most overlooked sources of LBP is the sacroiliac joint (SIJ) due to its complex nature and the fact that the pain emanating from this region can mimic other hip and spine conditions (Weksler et al. 2007; Smith 1999). However, recent studies have reported a higher prevalence of the SIJ as a source for LBP, with some reports estimated that the SIJ is the actual source of pain in 15–30% of cases of LBP (Sachs and Capobianco 2012; Lingutla et al. 2016; Schwarzer et al. 1995). Increased physicians' awareness of the prevalence of the SIJ as a source of LBP has given rise to an increased clinical suspicion of SIJ dysfunction as a pain generator and planning treatment accordingly.

Lumbar spine fusion, particularly L5–S1 segment, directly impacts the biomechanics of the SIJ by increasing both the motion and stress across the articular surface of the joint (Ivanov et al. 2009). As a significant source of LBP, focus on the SIJ is presently quite high. Current nonsurgical treatment and pain management strategies include physical therapy, SI joint injections, and radiofrequency (RF) ablation. When patients continue to present chronic LBP characteristic with the SIJ, surgical procedures become a final resort.

Anatomy

The SIJ, the largest axial joint in the body, is the articulation of the spine with the pelvis that allows for the transfer of loads to the pelvis and lower extremities (Dietrichs 1991; Cohen 2005). The

SIJ lies between the sacrum and the ilium, spanning about 1–2 mm in width and held together by fibrous capsule (Fig. 1). The sacral side of the joint is covered with hyaline cartilage thicker than iliac cartilage, which appears more fibrocartilaginous (Foley and Buschbacher 2007).

Ligaments

Several ligaments support and limit the movement and mobility of the SIJ. These ligaments include the interosseous sacroiliac ligament, the posterior and anterior ligaments, and sacrotuberous, sacrospinous, and iliolumbar ligaments. The interosseous ligament, also known as the axial ligament, connects the sacrum and ilium at S1 and S2 levels. The posterior sacroiliac ligament is quite strong and consists of multiple bundles which pass from the lateral crest of the sacrum to the posterior superior iliac spine and the posterior end of the iliac crest. The anterior sacroiliac ligament is a thin ligament that is weaker than the posterior ligament and runs over the joint obliquely from sacrum to ilium. The sacrotuberous ligament is located at the inferior-posterior part of the pelvis and runs from the sacrum to the ischial tuberosity. The sacrospinous ligament's attachment is behind of the sacrotuberous ligament, and it connects the outer edge of the sacrum and coccyx to the Ischia of the ilium. The iliolumbar originates from the tip of the fifth lumbar vertebral body to the iliac crest (Fig. 2) (Ombregt 2013). The long dorsal sacroiliac ligament can stretch in periods of reduced lumbar lordosis, such as during pregnancy, which will be discussed further. Table 1 summarizes sacroiliac joint ligaments' locations and their functions.

Muscles

While no muscles are designed to act on the SIJ to produce active movements, the joint is still surrounded by some of the largest and most powerful muscles of the body. These muscles include the erector spinae, psoas, quadratus lumborum, piriformis, abdominal obliques, gluteal, and hamstrings. While they do not act directly on the SIJ, the muscles that cross the joint act on the hip or the lumbar spine (Miller et al. 1987; Solonen 1957; Albee 1909). Movements of the SIJ are indirectly produced by gravity and muscles acting on the trunk and lower limbs rather than active movements of the sacrum (Ombregt 2013). Table 2 summarizes sacroiliac joint muscles' actions and their effect on SIJ.

Function and Biomechanics

The flat shape of SIJ along with its ligaments helps it to transfer large bending moments and compression loads. However, it is weak against shear loads; it is counteracted by compression of SIJ which is generated by a self-bracing mechanism. The self-bracing mechanism consists of a loading mode of pelvis and forces produced by muscles and ligaments which are normal to the joint surface. The loading mode of the pelvis due to gravity and the free body diagram of the self-bracing mechanism which involves normal and tangential forces of the joint surface, hip joint force, and muscle or ligament force are shown in Fig. 3a, b, respectively. The friction coefficient of SIJ surfaces without grooves and ridges was measured as 0.4. This resistance can be increased by grooves and ridges and wedge angle β to prevent sliding of SIJ surfaces due to shear (Snijders et al. 1993). It was shown that M. transversus abdominis and the pelvic floor muscles are playing a major rule in SIJ stability by enlarging the SIJ compression load to resist shear loads (Pel et al. 2008).

Pool-Goudzwaard et al. (2003) conducted a study on 12 human cadavers to assess the effect of the iliolumbar ligament (IL) on SIJ stability. Four cases were tested: (1) Intact IL, (2) random dissection of IL, (3) further dissection of IL, and (4) cut IL. The moment-rotation relationships were assessed by applying various moments to SIJ and measuring the rotation in the sagittal plane. The sacrum and iliac bones were fixed, and the moment was applied by a traction device to generate a tension in the string. Eight light-reflecting markers were utilized to calculate the rotation. Dissection of the ventral side of the iliolumbar ligament is causing less SIJ stability

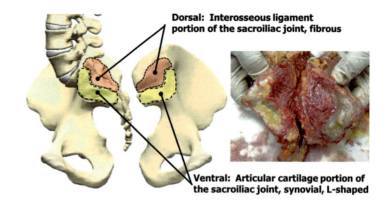

Fig. 1 Articular surfaces of the sacroiliac joint (Dall et al. 2015)

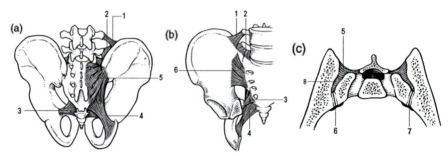

Fig. 2 (**a**) Posterior view; (**b**) anterior view; and (**c**) sacroiliac joint cut in transverse plane. (1, 2) Superior and inferior iliolumbar ligaments; (3) sacrospinous ligament; (4) sacrotuberous ligament; (5) posterior sacroiliac ligaments; (6) anterior sacroiliac ligaments; (7) sacroiliac joint; (8) interosseous ligament (Ombregt 2013)

Table 1 Sacroiliac joint ligaments' locations and their functions (Dall et al. 2015)

Ligament	Location	Primary restraint

Dorsal ligaments
   Long ligament
   Short ligament
	PSIS to sacral tubercles	Sacral extension

Sacrotuberous	PSIS and sacrum to ischial tuberosity	Sacral flexion
Sacrospinous	Apex of sacrum to ischial spine	Sacral flexion
Ventral ligament	Crosses ventral and caudal aspect of SIJ	Sacral flexion Axial rotation
Interosseous	Between sacrum and ilium dorsal to SIJ	Sacral flexion Axial rotation

Iliolumbar
   Ventral band
   Dorsal band
   Sacroiliac part
	Transverse process of L5 to iliac tuberosity and crest	Lateral side bending
		Ventral band
		Forward flexion
		Dorsal band

in the sagittal plane. Dorsal side and sacroiliac part of the IL does not have a significant role in providing SIJ stability (Pool-Goudzwaard et al. 2003). It is also stabilizing the lumbar vertebra on the sacrum (Yamamoto et al. 1990).

The posterior sacroiliac ligaments are contributed most to the SIJ mobility, while the anterior sacroiliac ligament has little influence (Vrahas et al. 1995). The motion of ilium respect to sacrum is called nutation which is anterior sacral tilt and counternutation which is posterior sacral tilt. Resisting the nutation and counternutation of the joint is done by the sacrotuberous ligament (STL), the sacrospinous ligament (SSL), and the long

Table 2 Sacroiliac joint muscles' actions and their effect on SIJ (Dall et al. 2015)

Muscle	Primary action	Effect on SIJ

Erector spinae
  Iliocostalis lumborum
  Longissimus thoracis
	Bilateral: back extension
	Unilateral: side bending
		Hydraulic amplifier effect

Multifidus	Back extension, side bending, and rotation	Imparts sacral flexion, force closure of SIJ with deep abdominals
Gluteus maximus	Hip extension, hip lateral rotation	Stabilizes SIJ
Piriformis	Hip lateral rotation	May alter SIJ motion via direct attachment to ventral aspect of sacrum
Biceps femoris	Hip extension, knee flexion	Long head: Imparts sacral extension via attachment to sacrotuberous ligament
Deep abdominals *Transversus abdominis*	Compression of abdominal cavity	Force closure of SIJ
Iliacus	Hip flexion (open chain) and tilts pelvis/sacrum ventrally (closed chain)	Synchronous tilting of the pelvis/sacrum ventrally (closed chain)
Pelvic floor	Support pelvic viscera	Imparts sacral extension

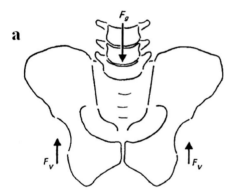 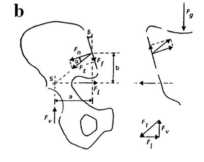

Fig. 3 (a) Pelvis free-body diagram due to gravity. Trunk weight (F_g) and hip joint forces (F_v). (b) Free-body diagram of self-bracing effect of the sacroiliac joint. SIJ reaction force: normal and tangential (F_n and F_f), ligament or muscle force (F_l), and hip joint force (F_v) (Snijders et al. 1993)

dorsal ligament (LDL), respectively (Vleeming et al. 1992a; Sashin 1930). During pregnancy by increased laxity of SIJ ligaments, the pain is mostly experienced in LDL due to its counteraction to the counternutation (Eichenseer et al. 2011). Pain in this region is also common in men due to its location which is superficial and will put asymmetric stress on the SIJ. Flattening of lumbar lordosis brings about a decrease in SIJ nutation (Vleeming et al. 2012).

A cadaveric study was done by Wang et al. (Wang and Dumas 1998) to calculate the SIJ motion and influence of anterior and posterior ligaments on the SIJ stability. Four female cadaver specimens were tested by applying five different eccentric compressive loads (combination of compression, bending moment, and forward shear due to inclination angle) to the sacrum. The main motions of the sacrum were lateral rotation and nutation rotation which were less than 1.2°. The lateral rotation is restricted by transverse portions of anterior and posterior ligaments. Also, the nutation rotation is prevented by the top portion of anterior and lower portion of posterior ligaments (i.e., Shear resisting couple), and dissection of these two ligaments has a significant influence on the joint stability. It was shown that interosseous ligaments are the strongest ligaments which provide less motion in the joint's translation.

Dujardin et al. (2002) assessed the SIJ micromotion under compression load applied to the ischial tuberosity. By sectioning SSL and STL, SIJ stability will decrease. Buyruk et al. (1995) using Doppler imaging of vibrations showed that left and right SIJ stiffnesses are different in various conditions, which means there is asymmetry in the SIJ stiffness resulting in low back pain and pelvic pain. Rothkotter et al. (Rothkotter and Berner 1988) indicated that the SIJ ligamentous structure failed at 3368 N under transverse loading with displacement range from 5.5 to 6.6 mm. They found that under dorsocranial loading, the self-bracing mechanism of the SIJ between the sacrum and ilium is working better than other loading directions.

Range of Motion

The sacrum can move with respect to the ilium in six degrees of freedom which is shown in Fig. 4. The intersection of the middle osteoligamentous column and the lumbosacral intervertebral disc is defined as the lumbosacral pivot point. Placing constructs posterior to this pivot point extending to the anterior of the point would provide rotational stability (McCord et al. 1992).

While the primary function of the SIJ is to absorb and transmit forces from the spine to the pelvis, it is also responsible for facilitating parturition and limiting x-axis rotation (Dietrichs 1991; Cohen 2005). The SIJ is unique in that it is rather stable, and motion of the joint is quite minimal (Foley and Buschbacher 2007). The exact range of motion (ROM) of the SIJ has been debated and studied extensively, with varying results. There are different methods to measure the SIJ motion such as roentgen stereophotogrammetric, radiostereometric, ultrasound, and Doppler (Vlaanderen et al. 2005; Jacob and Kissling 1995; Sturesson et al. 1989, 2000a); they indicated that the SIJ rotation and translation in different planes do not exceed 2–3° and 2 mm, respectively (Foley and Buschbacher 2007; Zheng et al. 1997). The joint's ROM is greatest in flexion-extension with a value of approximately 3°. Axial rotation of the SIJ is about 1.5°, and lateral bending provides the least ROM with approximately 0.8° (Miller et al. 1987). As the characteristics of the SIJ change with aging, these values can increase or decrease depending on the circumstance.

Many studies have been conducted concerning the biomechanics of the SIJ, and the results can be summarized quite simply: the SIJ rotates about all three axes, and these incredibly small movements are very difficult to measure (Walker 1992; White and Panjabi 1990). In an attempt to understand the load-displacement behavior of single and paired SI joints, a study involving eight elderly cadavers was conducted by Miller et al. (1987). In this

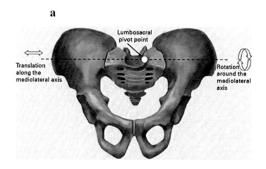 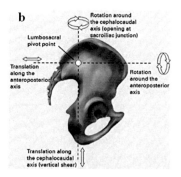

Fig. 4 Pelvis six degrees of movement and lumbosacral pivot point: (**a**) coronal plane, (**b**) sagittal plane (Berber et al. 2011)

study, rotations about all three axes were measured for one and both iliac fixed, with static test loads applied in superior, lateral, anterior, and posterior directions. According to their results, movements in all planes with one leg fixed ranged from 2 to 7.8 times greater than those measured with both legs fixed.

Another series of cadaveric studies by Vleeming et al. (1992a, b) was conducted to investigate the biomechanics of the SIJ, reporting that the ROM for flexion and extension rarely exceeded 2°, with an upper limit of 4° during sagittal rotation. To compare male and female SIJ ROM, a cadaver study by Brunner et al. (1991) found that the maximum ROM for men and women was 1.2° and 2.8°, respectively. Another study by Sturesson et al. (1989) involved measuring SIJ movements in 25 patients diagnosed with SIJ pain. According to their results, all movements were incredibly small, with translations never exceeding 1.6 mm and an upper rotational limit of 3°. This study also found that no differences in ROM existed between symptomatic and asymptomatic SI joints, which led the authors to conclude that three-dimensional motion analysis is not a useful tool for identifying painful SI joints in most patients (Sturesson et al. 1989). Jacob et al. (Jacob and Kissling 1995) reported mobility of SIJ of 15 healthy people using a three-dimensional stereophotogrammetric method. The average total rotation and translation were 1.7 and 0.7 mm, respectively.

Sexual Dimorphism

Sexual dimorphism exists in the pelvis with the male pelvis being larger, a distinction that decreases in the later years of childhood. While the sacral base articular facet for the fifth lumbar vertebra occupies more than a third of the width of the sacral base, it occupies less than a third in females. Compared to the male sacrum, the female sacrum is wider, more uneven, less curved, and more backward tilted. Males tend to have a relatively long and narrow pelvis, with a longer and more conical pelvic cavity than those of females (Figs. 5 and 6). In the second decade of life, women develop a groove in the iliac bone, the paraglenoidal sulcus, which usually does not occur for men. Such gender-related differences in the development of the SIJ can lead to a higher rate of SIJ misalignment in young women (Vleeming et al. 2012).

According to a study by Ebraheim and Biyani (2003), the SIJ surface area is relatively greater in adult males than females, which consequentially allows males to withstand greater biomechanical loading. While the average auricular surface area for females has been reported to range from 10.7 to 14.2 cm^2 (Miller et al. 1987; Ebraheim and Biyani 2003) with an upper limit of 18 cm^2 (Sashin 1930), this ligamentous area for males is approximately 22.3 cm^2 (Miller et al. 1987). Another reason that males can withstand greater biomechanical loading can be attributed to the fact that males possess significantly higher lumbar isometric strength, almost twice as strong as those of females, thus requiring more significant load transfers through the SI joints (Graves et al. 1990; Masi 1992).

Another significant influence on the development of particular SIJ form is the center of gravity, which has been reported to exist in different positions for males and females. Compared to men, who have a more ventral center of gravity, the center of gravity in females commonly passes in front of or through the SIJ (Tischauer et al. 1973; Bellamy et al. 1983). This difference implies that men would have a greater lever arm than women, accounting for the higher loads on the joints and stronger SI joints in males (Vleeming et al. 2012). This characteristic also may explain why males have more restricted mobility, as the average movement for men is approximately 40% less than that of women (Vleeming et al. 2012; Sturesson et al. 2000a, b).

The increased mobility of the SIJ in women can be attributed to individual anatomical correlations. Two features that allow for higher mobility in women are the less pronounced curvature of the SIJ surfaces and a greater pubic angle compared to those of males (Vleeming et al. 2012). While males typically have an average pubic angle of 50–82°, females have an average pubic angle of 90° (Bertino 2000). A possible reason for these

Structural Aspects	Female	Male
General structure	Light and thin	Heavy and thick
Pelvic brim (inlet)	Wide and more oval	Narrow and heart-shaped
Pubic arch	Greater than 90 angle	Less than 90 angle

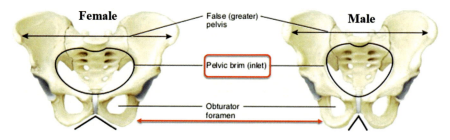

Fig. 5 Comparison of the female and male pelvic brim (inlet) (Tortora and Derrickson 2010)

Structural Aspect	Female	Male
Pelvic Outlet	Wider	Narrower

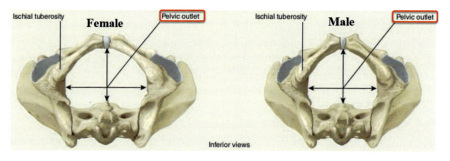

Fig. 6 Comparison of the female and male pelvic outlet (Tortora and Derrickson 2010)

differences can be attributed to the facilitation of parturition in females, which involves the influence of hormones such as relaxin (Dietrichs 1991; Cohen 2005; Ross 2000). Under the effect of relaxin, relative symphysiolysis appears to occur, and both of these factors loosen the SIJ fibrous apparatus, thus increasing mobility (Vleeming et al. 2012). While these unique aspects of the SIJ provide females with the necessary ability to give birth, they also may predispose females to a greater risk of experiencing pelvic pain (Brooke 1924; Hisaw 1925; Chamberlain 1930; Borell and Fernstrom 1957). One factor that plays a major role in determining the severity of this predisposition involves the laxity of the female SI joints during pregnancy. According to a study by Damen et al. (2001), females who experience asymmetric laxity of the SI joints during pregnancy are three times more likely to develop moderate to severe pelvic girdle pain (PGP) than females who experience symmetric laxity. As the particular form of the SIJ differs immensely between males and females, it becomes rather clear that women are more likely to develop PGP and are therefore at greater risk of experiencing LBP. Figures 5 and 6 and Table 3 show the anatomical and biomechanical differences between male and female pelvis.

Table 3 A biomechanical comparison of the female and male SIJ

Biomechanical aspects	Female	Male
SIJ motions	More rotational	More translational
SIJ surface area	Lesser	Greater
Interosseous sacroiliac ligament	Larger	Smaller
Anterior sacroiliac ligaments	Smaller	Larger
Posterior sacroiliac ligaments	Smaller	Larger

Causes of SIJ Pain

The mechanism of SIJ injury has been viewed as a combination of axial loading and abrupt rotation (Dreyfuss et al. 1995). From an anatomical perspective, pathologic changes specific to different SI joint structures can result in SIJ pain. These changes include, but are not limited to, capsular and ligamentous tension, hypomobility or hypermobility, extraneous compression or shearing forces, microfractures or macrofractures, soft tissue injury, and inflammation (Cohen 2005). Also, numerous other factors can predispose a person to a gradual development of SIJ pain.

As the primary function of the SIJ is to transfer loads between the spine and lower extremities effectively, simple daily activities such as walking and lifting objects can also cause stress and wear on the joint over time. However, dysfunction and pain of the joint often are not solely due to these activities. Many other causes of SIJ pain exist and impact the joint in combination with daily load bearing and aging. Some of the most common sources of SIJ pain include injuries sustained from falling directly on the buttocks, and collisions during sports and car accidents. Abnormal loading due to lumbar spinal fusions, limb length discrepancy, or prior medical procedures may also play a role in SIJ pain and dysfunction.

As mentioned, many studies have reported that prior lumbar fusion can directly increase angular motion and stress across the patient's SIJ, and the magnitude of both of these parameters is strongly correlated to the specific lumbar levels fused as well as the number of segments fused (Ivanov et al. 2009). When surgical arthrodesis causes degeneration of an adjacent segment, such as the SIJ, this profound adverse effect is known as adjacent segment disease (ASD) (Ivanov et al. 2009; Park et al. 2004; Ha et al. 2008; Hilibrand and Robbins 2004).

Other causes of SIJ pain and dysfunction have also been studied extensively – one of which involves limb length discrepancy (LLD). While it has commonly been accepted that LLD is related to LBP, the exact mechanism of this relation is unknown. However, several authors have reported the correlation between LLD and LBP to be strongly related to SIJ dysfunction (Cohen 2005; Schuit et al. 1989; Winter and Pinto 1986; Golightly et al. 2007). Due to the length discrepancy, the mechanical alignment of the SI joints becomes increasingly imbalanced, resulting in an increased load distribution across both SI joints (Cohen 2005; Winter and Pinto 1986; Golightly et al. 2007).

Apart from injuries, prior lumbar fusion, and LLD, several other factors can also cause the gradual development of SIJ pain. Additional sources of increased stress and pain across the SI joints include joint infection, spondyloarthropathies such as ankylosing spondylitis, inflammatory bowel disease (Cohen 2005), gait abnormalities (Herzog and Conway 1994), scoliosis (Schoenberger and Hellmich 1964), and excessive exercise (Marymount et al. 1986). Regardless of the cause, the association of pain with SIJ dysfunction is rather consistent.

Symptoms of SIJ dysfunction include pain in the lower back that sometimes radiates to the back of the thigh, and knee. Patients with LBP often experience pain when sitting, leaning forward, and with an increase in intra-abdominal pressure (DonTigny 1985). While these pain characteristics are associated with SIJ dysfunction, they also are consistent with other hip and spine conditions, making accurate diagnosis and confirmation of the SIJ as the pain source a rather difficult task. Table 4 summarizes the causes of intra-articular and extra-articular SIJ pain.

During pregnancy, many hormonal and biomechanical changes are occurring which contribute to ligaments laxity. One of the leading

Table 4 Causes of intra-articular and extra-articular SIJ pain (Holmes et al. 2015)

Intra-articular pain	Extra-articular pain
• Arthritis	• Ligamentous injury
• Spondyloarthropathy	• Bone fractures
• Malignancies	• Malignancies
• Trauma	• Myofascial pain
• Infection	• Enthesopathy
	• Trauma
	• Pregnancy

musculoskeletal changes is increasing the mass of uterus and breast which causes anterior displacement of the center of gravity. This effect heightens joint loads (e.g., increased hip-joint anterior torque by eight times) and is aggravated by the laxity of other ligaments and other joints which may contribute to pain and risk of injury (Fitzgerald and Segal 2015).

Diagnosis of SIJ Dysfunction

Symptoms of SIJ dysfunction include pain in the lower back, buttock, back of the thigh, and knee. Patients with LBP often experience pain when sitting, leaning forward, and with an increase in intra-abdominal pressure (DonTigny 1985). While these pain characteristics are associated with SIJ dysfunction, they also are consistent with other hip or spine conditions, making accurate diagnosis and confirmation of the SIJ as the pain source a rather difficult task.

Due to the complexity of diagnosing the SIJ as the pain source, numerous physical examination tests have been utilized, many of which incorporate distraction of the sacroiliac joints. Two of the most commonly performed tests are the Gaenslen's test and Patrick's test, also known as the FABER test (Cohen 2005). Other provocation tests for assessing SIJ pain include distraction/compression tests, the thigh thrust test, and the sacral thrust test (Table 5) (Laslett et al. 2005). It is commonly accepted that if three or more of these tests are deemed positive, then they can be considered reliable for diagnosing the SIJ as the source of pain (Laslett 2006). Despite the various physical diagnostic tests available, many clinical studies have shown rather inconsistent findings in the success of identifying the pain source to be SIJ dysfunction (Schwarzer et al. 1995; Cohen 2005). For this reason, other techniques have been suggested in conjunction with physical diagnostic tests to improve reliability.

Two techniques that are implemented in addition to physical examinations include radiological studies and diagnostic blocks, or intra-articular injections. Radiological imaging tests, however, have proven to be rather insufficient, yielding reports of low sensitivities and poor correlations with diagnostic injections and symptoms (Cohen 2005). However, an exception is the high specificity of MRI in the setting of the seronegative spondyloarthropathies (90–100%) (Battafarano et al. 1993; Docherty et al. 1992; Murphey et al. 1991). Diagnostic blocks, on the other hand, are often considered to be one of the most reliable methods for diagnosing SIJ pain. These blocks, which are typically fluoroscopically guided, are used to determine if the patient experiences a significant reduction in pain while the anesthetic is active (Foley and Buschbacher 2007). A controversial aspect of diagnostic blocks is that no actual "gold standard" exists for this technique, though it is commonly accepted that a successful injection helps the diagnosis of SIJ dysfunction (Cohen 2005; Foley and Buschbacher 2007; Broadhurst and Bond 1998). After determining that the sacroiliac joint is the pain generator in patients with LBP, there are several treatment strategies for relieving SIJ pain.

Nonsurgical Management

The first step in the treatment of SIJ dysfunction involves nonsurgical management (NSM). Nonsurgical treatment options include physical therapy, steroid injections, radiofrequency (RF) ablation, and prolotherapy. For patients with leg length discrepancy (LLD), only utilizing shoe inserts can help eliminate LLD, consequentially equalizing and decreasing the load distribution across the joints over time (Cohen 2005; Kiapour et al. 2012). This conservative management strategy, however, is not a valid treatment option for

Table 5 A comparison of provocation tests

Provocation test	Patient position	Technique description
Gaenslen's test	Supine	With a symptomatic leg resting on the edge of a table and the nonsymptomatic hip and knee flexed, a force is applied to the symptomatic leg while a counterforce is simultaneously applied to the flexed leg, producing pelvic torque (Kokmeyer et al. 2002; Dreyfuss et al. 1996)
Distraction test	Supine	A vertical, posteriorly directed force is applied to both anterior superior iliac spines (ASIS) (Sashin 1930; Cook and Hegedus 2013; Laslett 2008; Laslett et al. 2003)
Compression test	On side	Pressure is applied to the upper part of the iliac crest, producing forward pressure on the sacrum (Magee 2008)
Thigh thrust test	Supine	The hip is flexed to 90° to stretch posterior structures. With one hand fixated below the sacrum, the other applies downward axial pressure along the femur, which is used as a lever to push the ilium posteriorly (Vercellini 2011; Broadhurst and Bond 1998; Laslett 1997; Laslett and Williams 1998)
Sacral thrust test	Prone	With one hand placed directly on the sacrum and the other hand reinforcing it, an anteriorly directed pressure is applied over the sacrum (Vercellini 2011; Broadhurst and Bond 1998)

patients with causes of SIJ pain irrelevant to LLD. For such patients, other measures must be taken.

For patients with SIJ pain not related to LLD, physical therapy and chiropractic manipulation are typically advocated for NSM strategies. Several studies of physical therapy and chiropractic manipulation programs have reported promising long-term results, achieving reductions in pain and disability, as well as enhanced mobility (Sasso et al. 2001; Cibulka and Delitta 1993; Osterbauer et al. 1993); however, there is currently a lack of prospectively controlled studies to back up these treatment strategies (Cohen 2005). Other stabilization plans have also been introduced, such as pelvic belts. These belts have shown to decrease sagittal rotation and consequentially enhance pelvic stability, especially in pregnant women (Vleeming et al. 1992c; Damen et al. 2002). In addition to therapeutic measures, intra-articular injections have also been advocated for SIJ pain relief.

Studies regarding the effectiveness of corticosteroid injections have been conducted to quantify the magnitude of pain reduction in patients with varying reported results. A controlled study by Maugars et al. (1996) reported that after a 6-month follow-up, the subjects experienced a mean pain reduction of 33%. While this is one of the lowest pain reduction rates that have been reported, it should be noted that the sample size was rather small with ten subjects. In contrast, another study conducted by Bollow et al. (1996) consisted of a mean follow-up duration of 10 months and reported a statistically significant pain reduction in 92.5% of the subjects. With a larger sample size of 66 subjects, such a high-pain reduction rate in the majority of subjects indicates that there is effectiveness in administering intra-articular corticosteroid injections for many patients despite the different reported results. For those who do not find significant reductions in pain from intra-articular injections, alternative treatment measures must be considered.

Radiofrequency (RF) denervation procedures are utilized as another treatment strategy with a goal of providing intermediate-term pain relief. Several studies have proven that lateral branch RF denervation strategies may improve the pain, disability, and quality of life for patients suffering from chronic SIJ pain (Cohen et al. 2008; Patel et al. 2012). However, similar to intra-articular injections, the reported success rates of RF denervation vary immensely. A retrospective study conducted by Ferrante et al. (2001) involved the targeting of the intra-articular nerves via a bipolar leapfrog RF technique, and a success rate of 36.4% was reported at follow-up of 6 months. In contrast, a prospective, observational study conducted by Burnham and Yasui (2007) focusing on the targeting of the L5–S3 nerves via the same RF procedure reported a success rate of 89% after 12 months. With such inconsistent reported

Open SIJ Fusion

When NSM strategies fail to reduce the pain and discomfort of patients with suspected SIJ dysfunction, surgical measures become an option, beginning with open arthrodesis, or fusion of the SIJ. A study of open fusion of the SIJ was conducted by Smith-Petersen and Rogers to determine the success of arthrodesis. According to their results, in approximately 96% of cases, the patients were able to return to their previous work, though it should also be noted that the average time required to go back to regular activities was approximately four and a half months (Smith-Peterson and Rogers 1926).

While the success of open arthrodesis of the SIJ has been reported in numerous studies (Smith-Peterson and Rogers 1926; Wheeler 1912; Harris 1933; Ledonio et al. 2014a; Alaranta et al. 1990), several aspects of this procedure have also been deemed worthy of improvement. Smith et al. conducted a multicenter comparison between open and minimally invasive SIJ fusion procedures using triangular titanium implants to compare the clinical outcomes. According to their results, open surgical fusion required longer operative time, greater blood loss, and longer hospital stays. Apart from having less advantageous operative measures, open arthrodesis of the SIJ also showed less superior SIJ pain rating changes over the duration of 12 and 24 months (Smith et al. 2013). According to their study, the mean change in VAS pain score at 24 months was approximately -2.0 and -5.6 for open surgical fusion and minimally invasive fusion, respectively, demonstrating the advantage of minimally invasive surgery in regard to pain-recovery ratings. Results of the study also further confirm the superiority of minimally invasive approaches compared to open surgical fusion, as minimally invasive techniques are accompanied by less tissue damage, blood loss, and duration of hospitalization (Ledonio et al. 2014a; Smith et al. 2013).

Minimally Invasive SIJ Fusion

To date, numerous studies have been conducted to investigate the effectiveness of minimally invasive SIJ fusion techniques. Among the various studies, several of the parameters measured included pain scores, disability indices, quality of life, patient satisfaction, and economical outcomes.

One of the most commonly used outcome instruments for assessing variations in pain is the visual analog scale (VAS) (Damen et al. 2002). The VAS is obtained by marking on the patient a 100-mm line along which the patient indicates the intensity of the pain they are experiencing (Wise and Dall 2008). The scoring of the VAS typically ranges from 0 to 100, though it can also be expressed between 0 and 10. Due to its high degree of reliability, validity, and responsiveness, the VAS is a widely utilized instrument for gauging pre- to posttreatment outcomes (Gatchel 2006; Alaranta et al. 1990; Million et al. 1982).

Another commonly used measure of pain and disability is the Oswestry Low Back Pain Disability Index (ODI), which is a self-rating questionnaire that measures a patient's degree of functional impairment. Advantageous aspects that make the ODI a popular outcome instrument include the ease of administration and the short amount of time needed to complete and evaluate. Another commonly used questionnaire that measures health-related quality of life is the Medical Outcomes Short Form-36 Health-Status Survey (SF-36), which is comprised of eight separate scales, along with a standardized mental component scale (MCS) and physical component scale (PCS) (Gatchel 2006). While the SF-36 consists of 36 questions, a shorter, yet still valid version known as the SF-12 has been adapted to have only 12 questions (Ware et al. 2002). The short form surveys allow for assessment of a patient's quality of life from the health care recipient's point of view (Gatchel 2006).

In conclusion, there is a wide range of treatment options for sacroiliitis, and most do improve with conservative, nonsurgical interventions. For those with refractory SI joint-mediated pain, minimally invasive SI Joint fusion has been found to be a safe and effective alternative.

Clinical Studies

Wise et al. (Wise and Dall 2008) performed percutaneous posterior minimally invasive SIJ fusion for 13 consecutive patients to assess the outcome of this technique within 24–35 months follow-up. It was shown that the total fusion rate was 89% and there was a significant improvement in pain scores. After Wise, a new percutaneous lateral SIJ arthrodesis technique using a hollow modular anchorage screw was introduced by Al-Khayer et al. (2008). No one had combined MIS method and bone grafting for SIJ fusion before Al-khayar. Nine patients underwent surgery with 2 years follow-up, and it was shown that the VAS score fell from 8.1 Preoperation to 4.6 postoperation. This new technique provided a safe and successful fusion for SIJ pains. Hollow modular anchorage screw was also utilized by Khurana et al. (2009) for 15 patients during 9–39 months follow-up. They observed good results regarding pain score improvement and concluded that this method is a suitable surgery process for SIJ fusion. Mason et al. (2013) did a study using this fixation system for 55 patients within 12–84 months follow-up. This fusion resulted in reduced VAS score from 8.1 to 4.5 and reduced pain.

As one key focus of the medical field is the improvement of surgical procedures and the discovery of novel treatment approaches, various studies have been performed to further confirm the important trend toward less invasive arthrodesis procedures.

Among the different techniques for minimally invasive SIJ fusion, perhaps the most popular fusion system involves triangular titanium implants with a porous titanium plasma spray coating. The shape, coating, and interference fit of these implants allow for initial stabilization or mechanical fixation, and then effective stabilization of the joint is eventually achieved from long-term biological fixation (Rudolf 2012; Smith et al. 2013; Lindsey et al. 2014). They have various unique features which make them different from traditional cages and screws. Due to their design, an interference fit was provided to allow them the proper fixation. Their triangular profile reduces implant rotation significantly, and their porous surface minimizes the implant micromotion and enhances bone ingrowth resulting in better fusion. Biomechanical studies showed that an 8 mm cannulated screw is three times weaker in shear and bending than a triangular implant (Fig. 7). In this fusion system, no grafts are placed in the sacroiliac joint, therefore all fusions are obtained by their porous coating (Wang et al. 2014).

During a minimally invasive SIJ fusion, the patient is administered general anesthesia and is placed in the prone position to use intraoperative fluoroscopy (Rudolf 2012; Sachs and Capobianco 2012; Smith et al. 2013). A 3 cm lateral incision is then made in the buttock region, and the gluteal fascia is penetrated and dissected to reach the outer table of the ilium. A Steinmann pin is then passed through the ilium across the SI joint to the middle of the sacrum and lateral to the neural foramen (Cher et al. 2013). A soft tissue protector is inserted over the pin, and a drill is utilized to create a pathway and decorticate the bone. Upon removal of the drill, a triangular broach is malleated across the joint to prepare the triangular channel for the first implant. Finally, using a pin guidance system, the implants can be placed, which is followed by irrigation of the incision and closure of the tissue layers (Rudolf 2012; Sachs and Capobianco 2012, 2013; Smith et al. 2013; Cher et al. 2013).

A prospective study by Duhon et al. (Cher et al. 2013) was conducted to determine the safety and effectiveness of MIS fusion with a follow-up duration of 6 months. In this study, the safety cohort consisted of 94 subjects while the effectiveness cohort consisted of 32 subjects, 26 of which were available for postoperative follow-up at 6 months. According to the results, mean SI joint pain at baseline was about 76, while the 6-month follow-up pain score was approximately

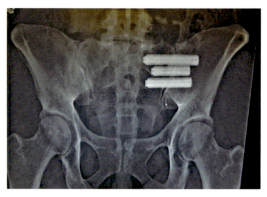 

Fig. 7 Triangular titanium implant with porous-coating – lateral approach (Wang et al. 2014)

29.3, indicating an improvement of about 49 points. Furthermore, the mean ODI at baseline was about 55.3 and decreased to approximately 38.9 points, showing an improvement of about 15.8 points. To determine the 6-month outcome of quality of life, this study incorporated Short Form-36 (SF-36) PCS and MCS questionnaires. The results from this study revealed that the SF-36 PCS and SF-36 MCS improved by about 6.7 and 5.8 points, respectively. Finally, patient satisfaction was assessed and recorded to be approximately 85%, a rather high rate of satisfaction.

A similar study was conducted by Cummings and Capobianco (2013), except with a longer follow-up duration of 1 year involving 18 subjects. Similarly, the parameters measured were pain score, disability index for back functionality, quality of life via Short Form-12 questionnaires, and patient satisfaction. Upon a 12-month follow-up, the results of this study revealed an improvement in VAS pain score of about 6.6 points, ODI improvement of −37.5 points, and SF-12 PCS and SF-12 MCS improvements of 11.19 and 20.37 points, respectively. Similar to the study by Duhon et al. (Cher et al. 2013), patient satisfaction was again rather high with a value of 95% satisfaction and 89% of patients claiming that they would undergo the same surgery again.

A study by Sachs and Capobianco (2012) was performed to investigate the successful outcomes for minimally invasive arthrodesis after a 1-year follow-up duration for the first 11 consecutive patients who underwent MIS SIJ fusion using triangular porous plasma coated titanium implants by a single surgeon. At baseline, the mean pain score was approximately 7.9, which decreased to about 2.3 after 12 months. This improvement in mean pain score of about 6.2 points from baseline was considered clinically and statistically significant, and patient satisfaction was immensely high with 100% of subjects claiming that they would again undergo the same surgery.

Sachs and Capobianco (2013) also conducted a retrospective 1-year outcome analysis of MIS–SIJ fusion in 40 patients. The parameters measured in this study primarily involved pain score changes and patient satisfaction; postoperative complications were also taken into consideration. The pain scores in this study were measured on a numerical rating scale (NRS) from 0 to 10, with 10 indicating the highest amount of pain. At baseline, the mean pain score was approximately 8.7, while at follow-up of 12 months, the average pain score decreased to about 0.9, indicating an improvement of approximately 7.8 points. According to the results, patient satisfaction was highest in this study with a value of 100% of the subjects declaring that they would undergo the same surgery again.

It is shown that lumbosacral fusion is contributed to 75% of SIJ degeneration (Ha et al. 2008). Schroeder et al. (2013) performed a clinical study on six patients who had SIJ fusion besides long fusions ending in sacrum with the 10.25 months average follow-up. SIJ fixation improved the results of all scores like Leg VAS score, Back VAS score, SRS 22, and also ODI score from 22.2 to 10.5. They indicated that the SIJ fixation in patients with long fusions results in back pain reduction. The SIJ fusion was achieved by using titanium triangular implants within the follow-up which led to minimized rotation and micromotion due to osteogenic interference fit used in this study and not having implant loosening and breakage. Long fusions to the sacrum are providing increased motion and force at the SIJ resulting in an increase in SIJ pain (Rudolf 2012; Ha et al. 2008). Unoki et al. (2015) reported a retrospective study to determine the effect of multiple segment fusion on the incidence of SIJ pain for 262 patients. It was indicated that multiple segment fusion (at least 3) could enhance the incidence of SIJ pain. Another clinical study conducted by Shin et al. (2013) indicated that greater pelvic tilt and insufficient restored lumbar lordosis by far play a role in generating SIJ pain after PLIF surgery.

While the effectiveness and safety of minimally invasive fusion of the SIJ have been reported to be significant over the duration of 6 and 12 months, studies of longer follow-up durations have been conducted to confirm the long-term success of these implants. A study by Duhon et al. (2016) was carried out to determine the long-term results over a 2-year follow-up duration from a prospective multicenter clinical trial. Similar to the 6-month study by Duhon et al. (Cher et al. 2013), this analysis also measured parameters of SIJ pain rating, ODI, Short Form-36 PCS and MCS, and patient satisfaction. According to their results, SIJ pain decreased from a baseline value of 79.8–26.0 after 2 years, and the ODI decreased from 55.2 at baseline to 30.9 at 2 years. Furthermore, SF-36 PCS and MCS improved by approximately 8.9 and 10.1 points, respectively, and 88.5% of subjects reported decreased pain at follow-up of 2 years (Duhon et al. 2016). A similar 2-year retrospective follow-up study of 45 subjects was conducted by Rudolf (2012), which reported a mean pain score improvement of approximately 5.9 points and an 82% patient satisfaction rate.

To further investigate and confirm the previous findings of the effectiveness and safety of minimally invasive fusion procedures, Rudolf and Capobianco (2014) conducted a 5-year clinical and radiographic outcome study of 17 patients treated with MIS–SIJ fusion for degenerative sacroiliitis and sacroiliac joint disruptions. The parameters measured in this study include pain on a visual analog scale (VAS) from 0 to 10, mean ODI score, and patient satisfaction. The results of this study revealed an improvement in VAS pain score from 8.3 at baseline to 2.4 after 5 years, with a patient satisfaction rate of 82% after 1 year. While a preoperative mean ODI score was not reported, the reported mean ODI score at the 5-year follow-up was approximately 21.5.

Regardless of the duration of follow-up time and the parameters measured, the numerous studies of the outcomes of MIS SI joint fusion reveal that fusion of the SIJ via minimally invasive approaches with triangular titanium implants can be considered a safe and efficient option for treatment of SIJ pain (Rudolf 2012; Sachs and Capobianco 2012, 2013; Wang et al. 2014; Cher et al. 2013; Cummings and Capobianco 2013; Duhon et al. 2016; Rudolf and Capobianco 2014). A comparison of the studies performed and the outcomes of MIS SIJ fusion is shown in Table 6.

While pain scores, disability indices, and quality of life questionnaires have served as important measures for determining the long-term effects of SI joint-fusion procedures, other studies have been conducted to investigate the success of such operations from a unique perspective involving work productivity and economic concerns.

One study conducted by Saavoss et al. (Koenig et al. 2016) analyzed the productivity benefits for patients with chronic SIJ dysfunction to compare worker function and economic outcomes between nonsurgical management and MIS SIJ fusion. The importance of this study was to determine the impact of arthrodesis on worker productivity, a relationship which has not been previously examined. According to their results, patients who underwent MIS–SIJ fusion were expected to have an increase in the probability of working for 16% compared to patients who received nonsurgical management, and the expected difference in earnings among the groups was deemed to be not statistically significant with a value of approximately $3128. When the metrics of working probability and expected change in earnings were combined, the annual increase in worker productivity between patients receiving MIS SIJ

Table 6 SIJ fusion with triangular implants outcome reports

Study	Patients included	Prior lumbar fusion	Follow-up duration	Pain score improvement	Patient satisfaction
Sachs and Capobianco (2012)	11 (10F/1M)	18%	12 months	70%	100%
Rudolf (2012)	50 (34F/16M)	44%	12 months	56%	82%
Rudolf (2013)	18 (12F/6M)	No prior fusion	24 months	80%	89%
	15 (11F/4M)	Prior lumbar fusion	24 months	73%	92%
	7 (3F/4M)	Prior lumbar pathology treated conservatively	24 months	63%	63%
Schroeder et al. (2013)	6 (6F/0M)	100%	10.25 months (4–15)	61%	100%
Gaetani et al. (2013)	12 (12F/0M)	8.3%	10 months (8–18)	4	100%
Cummings and Capobianco (2013)	18 (12F/6M)	61%	12 months	74%	95%
Sachs and Capobianco (2013)	40 (30F/10M)	30%	12 months	90%	100%
Duhon et al. (2013)	32 (21F/11M)	69%	6 months	67%	85%
Smith et al. (2013)	114 (82F/32M)	47.4%	24 months	79%	82%
Kim et al. (2013)	31 (24F/7M)	48%	12 months	N/A	87%
Ledonio et al. (2014a)	17 (11F/6M)	82%	12 months	78%	94%
Ledonio et al. (2014b)	22 (17F/5M)	64%	15 months (12–26)	54% (17%)	73%
Smith et al. (2013)	144 (102F/42M)	62%	12 months	68%	80%
Rudolf and Capobianco (2014)	17 (13F/4M)	47%	60 months	71%	82%
Vanaclocha-Vanaclocha et al. (2014)	24 (15F/9M)	8%	23 months (1–4.5 years)	43%	89%
Whang et al. (2015)	102 (75F/27M)	38%	6 months	63%	79%
Duhon et al. (2015)	172 (120F/52M)	44.2%	24 months	67%	78%
Polly et al. (2015)	102 (75F/27M)	38%	24 months	83%	73%
Sturesson et al. (2016)	52 (38F/14M)	N/A	6 months	55%	55%

fusion and those receiving nonsurgical management was estimated to be approximately $6924.

SI-LOK is another MI SIJ fixation system which locates three hydroxyapatite-coated screws across the sacroiliac joints laterally (Fig. 8). There are optional bone graft slots inside the screw which can be used to enhance fusion. Also, the optional lag screw thread allows applying

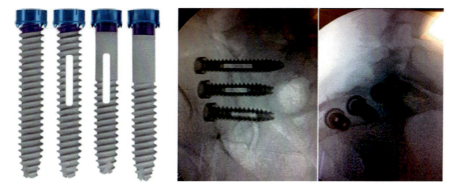

Fig. 8 SI-LOK sacroiliac joint fixation system – lateral approach (Wang et al. 2014)

compression force during placement (Wang et al. 2014). There is no biomechanical study on this screw yet, however, prospective 1-year outcomes of 32 patients were reported. VAS back pain improved from 55.8 ± 26.7 to 28.5 ± 21.6 ($P < 0.01$) and ODI improved from 55.6 ± 16.1 to 34.6 ± 19.4 at 1 year (Rappoport et al. 2017).

SImmetry is another cannulated titanium screw type SIJ fixation system which usually is used with two screws (one is antirotation screw) laterally across the SIJ (Fig. 9). There is no bone graft slot in this system, and the bone graft is placed across the articular part of the joint (Wang et al. 2014). This surgery technique is defined comprehensively in (Beaubien et al. 2015). One-year outcomes of 18 patients were reported as follows: VAS reduced from 81.7 (15.2) to 44.1 (22.9), and radiographic arthrodesis was identified on CT scan in 15 of 17 patients (88%) (Kube and Muir 2016).

SIFix is one of the posterior MI SIJ fixation systems and uses two-threaded cancellous bone to stabilize the joint. This method can be done bilaterally with a single midline incision (Fig. 10).

Beck et al. (2015) conducted posterior fusion surgery utilizing RI-ALTO implants for 20 patients during 17–45 months follow-up. The fusion rate and satisfaction ratings were 97% and 76%, respectively. It was shown that this method is safe and effective in SIJ fusion and reduces surgical morbidity due to posterior approach (Fig. 11).

From significantly successful reports of surgical outcomes, patient satisfaction, recovery rate, and implant survivorship, minimally invasive procedures have now become the predominant focus for treating patients with chronic SIJ pain.

In conclusion, the results of clinical studies showed that the minimally invasive approaches, compared to open surgical fusion, as minimally invasive techniques are accompanied by less tissue damage, blood loss, and duration of hospitalization. Furthermore, there are various techniques and different types of SIJ fusion implants for minimally invasive approaches. Since some clinical questions could not be answered through clinical studies, in vitro and in silico studies have been used to address these questions.

In Vitro and In Silico Studies

Soriano-Baron et al. (2015) conducted a cadaver study to investigate the effect of placement of sacroiliac joint fusion implants which were triangular implants. Nine human cadaveric specimens from L4-pelvis were used to perform the range of motion testing for one leg stance under three conditions: intact, cut pubic symphysis to allow the right and left SI joints to move freely, and treated. The treated condition was performed using two different approaches for SIJ fusion implant placement which were posterior and transarticular techniques. In the posterior procedure, the three implants were placed inline in the inlet view, and parallel in the outlet and lateral views. In the transarticular approach, the superior and inferior implants were placed similar to

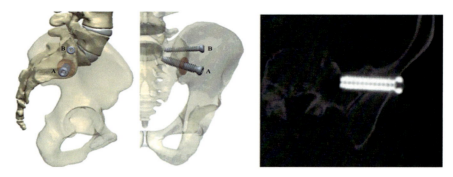

Fig. 9 SImmetry sacroiliac joint fusion system – lateral approach (Wang et al. 2014)

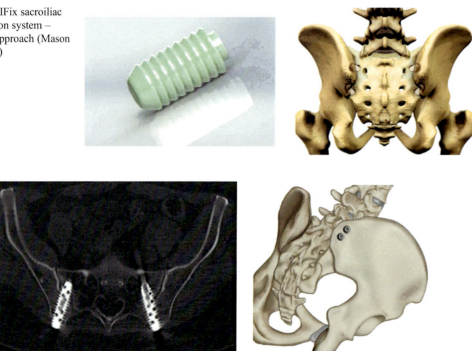

Fig. 10 SIFix sacroiliac joint fixation system – posterior approach (Mason et al. 2013)

Fig. 11 RI-ALTO sacroiliac joint fusion system – posterior approach (Beck et al. 2015)

the posterior technique, and the middle implant was positioned toward the anterior third of the sacrum across the cartilaginous portion of the SI joint. The 7.5 Nm pure moment was applied to simulate the flexion, extension, lateral bendings, and axial rotations under one-leg stance condition. They showed that placement of three implants in both approaches significantly reduced the ROM in all motions. Interestingly, there was no significant difference between these two techniques regarding motion reduction (Soriano-Baron et al. 2015).

Hammer et al. (2013) using finite element analysis showed that SIJ cartilage and ligaments are playing a significant role in pelvic stability. By increasing in SIJ cartilage and ISL, IL, ASL, and PSL stiffness would decrease the pelvic motion with highest strains at ISL, and pubic ligaments have the least effect on the pelvic motion. These ligaments are contributed to transferring loads horizontally at the acetabulum and ilium. In contrast, increasing stiffness of SS and ST has opposite effect and causes an increase in the pelvic motion, and both are doing vertical load transfer followed by sacrum translation. Moreover, in standing position, the ligaments strain is higher than in sitting position.

Eichenseer et al. (2011) also evaluated the correlation between ligaments stiffness and SIJ stress and motion. They showed that decreasing ligaments stiffness results in an increase in stress and motion at SIJ. Moreover, ISL has the highest strains under different spine motions which confirmed the finding of Hammer's study.

Mao et al. (2014) investigated the effect of lumbar lordosis alteration on sacrum angular displacement after lumbosacral fusion. Decreasing and increasing lumbar lordosis result in increased sacrum angular motion. In addition, fusion at L4–S1 level is providing higher sacrum angular displacement compared to L3–L5 level. Therefore, it can be the reason why SIJ degeneration incidence is higher in fusions at S1 rather than L5.

Lindsey et al. (2015) assessed the range of motion of SIJ and the adjacent lumbar spinal motion segments after SIJ fusion using triangular implants via finite element analysis. They evaluated the ROM of their model which was L3-Pelvis under 10 Nm moment to simulate flexion, extension, lateral bendings, and axial rotation. They showed that SIJ fusion using three triangular implants provided a significant reduction in SIJ motion in all six motions. Moreover, SIJ motion reduction by fusion resulted in least increase in adjacent lumbar segment motion.

Bruna-Rosso et al. (2016) used finite element method to analyze SIJ biomechanics under RI-ALTO fusion implant which is a new sacroiliac fusion device. Thousand newton compression load was applied to the pelvis to simulate the experimental test. They evaluated the effect of number of implants (one and two implants) and their placement at SIJ. Proximal insertion of the implant which was farther from the SIJ center of rotation was more efficient than distal insertion of the implant. Proximal insertion of one implant even had better performance than using two implants in terms of motion reduction. There is no significant difference in providing stability between two trajectories of placement which were medial and oblique for using one-implant instrumentation, although medial placement provided higher stability compared to oblique in two-implant instrumentation. Overall, the more parallel and farther the implant was inserted from the SIJ center of rotation, the more stability is provided.

Lindsey et al. (2018) performed another finite element study on SIJ fusion with triangular implants to assess the biomechanical effects of length, orientation, and number of implants under all six spine motions. The variables were one, two, and three implants; superior implant lengths of 55 and 75 mm; midline implant length of 45 mm; and inferior implant length of 45 mm for inline orientation and 50 mm for transarticular orientation. They showed that the transarticular orientation provided better fixation compared to inline orientation due to crossing more the cartilaginous portion of SIJ, although Soriano-Barron revealed that there was no significant difference between these two approaches. Using longer superior implant led to more reduced SIJ motion under different spine motions. In addition, placing two implants close together is less stable than two implants far from each other. Overall, placing implants in the thicker cortical bone areas and a more dense bone region is providing more stability.

A finite element analysis was conducted by Kiapour et al. (2012) to quantify the changes in load distribution through the SIJ as a result of LLD. In this study, the peak stresses and contact loads across the SIJ were measured for leg-length discrepancies of 1, 2, and 3 cm. The results showed that the peak loads and stresses of both legs were always higher than that of the intact model, with a greater magnitude consistently occurring on the longer leg side. Furthermore, as the length discrepancies increased from 1 to 3 cm, the stresses increased accordingly.

Zhang et al. (2014) studied the biomechanical stability of four different SI screw fixations under two types of SI dislocation using finite element method. They placed implants at SIJ in four different configurations: Single screw in S1, single screw in S2, two screws in S1, and one screw in

S1 and another one in S2. Then biomechanical analysis of implanted pelvis was done under inferior translation, flexion, and lateral bending. In type B dislocation, except LPS and SPS ligaments, all ligaments are damaged, and in type C, all ligaments are damaged. The weakest placement configuration was the single screw in S2 in both injury types due to placement farther from S1 end plate which confirmed the study of Bruna-Rosso. Two screws at S1 and S2 were the strongest placement compared to placing two screws closely in S1 in both dislocation types which is in contrast to the finding by Bruna-Rosso.

Ivanov et al. (2009) evaluated sacrum angular motion and stress across SIJ after lumbar fusion. Fusion was performed at different levels of L4–L5, L5–S1, and L4–S1. They showed that lumbar fusion would result in an increase in SIJ motion and stress across SIJ. L4–S1 level fusion provided the greatest SIJ motion and stress across SIJ compared to fusions at other levels.

Another study conducted by Lindsey et al. (2014) investigated the outcomes of minimally invasive SIJ fusion from an in vitro biomechanical approach, comparing the initial and cycled properties. Because the goal of fusion is a reduction in joint motion, the effectiveness of the implants was measured by joint-motion properties in flexion-extension, lateral bending, and axial rotation. The results of this study revealed a significant decrease in flexion-extension range of motion (ROM), and an insignificantly altered lateral bending and axial rotation in the treated specimen compared to the intact condition. Although deemed statistically insignificant, lateral bending and axial rotation were decreased in the majority of subjects, indicating that the implants effectively reduced joint motion in most of the specimens.

A recent study performed by Lindsey et al. (2017) evaluated and compared the biomechanical impact of unilateral and bilateral triangular implant placement across the SI joint. They found that the unilateral and bilateral SIJ fusion lead significant motion reduction across SIJ.

Lee et al. (2017) investigated the biomechanics of intact and treated pelvis via FE and experimental analysis. The spine-pelvis-femur FE model included ligaments and muscles as truss elements. It was demonstrated that posterior iliosacral screw fixation provided higher stability and lower risk of implant failure compared to sacral bar fixation and a locking compression plate fixation.

Joukar et al. (2017) studied the biomechanical differences between male and female SIJs using finite element analysis. They found out that female SIJ had higher mobility, stresses, loads, and pelvis ligament strains compared to the male SIJ which led to higher stress across the joint, especially on the sacrum under identical loading conditions. This could be a possible reason for higher incidence of SIJ pain and pelvic-stress fracture in females.

Joukar et al. (Joukar 2017) investigated the effect of unilateral and bilateral SIJ fusion and different placements of fully threaded screw and half threaded screw during standing upright (similar to RI-ALTO and SI-LOK implant systems), respectively, on the SIJ male and female models' range of motion and stresses. The fully-threaded and half-threaded screws were located posterior and lateral into the SI joint, respectively. Unilateral stabilization significantly reduced the fused SIJ range of motion along with reduction in contralateral (nonfused) SIJ motion during standing upright. Moreover, regardless of sex, lateral and posterior placements of the implants had similar performance on the SIJ stability. Both male and female models showed high reduction in stress and range of motion after treatment compared to the intact model, however, female model showed more stress and motion reductions after SIJ fusion due to higher stress and range of motion values in prior fusion compared to the male model. SIJ implants are more effective in females in terms of stability but may be more prone to higher rate of loosening/failure compared to males. The motion reduction at the SI joint after unilateral and bilateral fusions resulted in minimal changes at the adjacent lumbar levels for both male and female models. Although, the implant shape effects were minimal, the implant placements played a major role in stresses on the bone and implant. In both unilateral and bilateral fusions, SIJ stabilization was primarily due to the inferior and superior implants.

Joukar et al. (2019) developed a validated finite element (FE) model of lumbo-pelvic segment to investigate the biomechanical effects of fixation of the sacroiliac joint using triangular implants on the hip joint. Their model included the most critical anatomical features including connective tissue and articular cartilage across the hip joint. They performed an analysis with femurs fixed in double-leg-stance configuration and application of a 400 N compressive follower preload applied across the lumbo-sacral segment followed by a 10 Nm bending moment applied to the topmost level of the spine segment. Intact model was modified to include SIJ fixation and unilateral and bilateral joint instrumentations. The analyses demonstrated a decrease in range of motion of the SI joint in the instrumented model, compared to the intact. The bilateral fixation resulted in a greater reduction in motion compared to unilateral fixation. The contact stresses and load sharing did not significantly change in contralateral SI joint, following unilateral fixation.

The average hip contact stress and contact area changed less than 5% and 10% respectively in instrumented models relative to intact in most of anatomical motions. The data suggested a low risk of developing adjacent segment disease across the hip joint due to minimal changes in contact area and load sharing at the hip joint following instrumentation with the triangular implant compared to the intact. The changes in the lumbar spine segment were minimal as well.

In conclusion, in vitro studies were performed to address different unanswered questions in clinical studies such as implant failure, range of motion, and bone failure. Since in vitro studies were unable to record some biomechanical data like stresses across bones and implants, and ligament strains, in silico studies were used to overcome these limits of experimental tests.

Summary

SIJ is a complex joint sitting in between the sacrum and iliac bone on either side. The joint plays a vital role in transmitting upper body loads to lower extremities via the hip joints. The wedging of the sacrum in between pelvic bones, irregular and rough surface of the joint itself, and tight banding due to ligaments and pelvic floor muscles (levator ani and coccygeus muscles) make the SIJ extremely stable. SIJ pain can be due to, but are not limited to, capsular and ligamentous tension, hypo- or hypermobility, extraneous compression or shearing forces, and a host of other factors. Other sources of pain are the surgical arthrodesis at one level causing degeneration of an adjacent segment, leg length discrepancy, and spondylo-arthropathies. There are anatomical differences between male and female pelvis, including SIJ characteristics. In females, ligaments become lax during pregnancy. These factors may make females more prone to low back pain. To restore quality of life and alleviate LBP due to SIJs, conservative and surgical treatments are available.

The first step in the treatment of SIJ dysfunction involves a thorough diagnostic workup followed by nonsurgical management. When nonsurgical management strategies fail, surgical management (open or minimal fusion) is considered. Several studies have investigated the clinical outcomes of surgical techniques for the sacroiliac joint. The studies have shown that minimally invasive techniques involve less tissue damage, blood loss, and duration of hospitalization, thus leading to superior clinical outcomes.

Despite the satisfactory data on clinical outcomes of SIJ fixation surgery, the data on biomechanics of SIJ in general and fixation techniques in particular are sparse. The existing literature suggests that at least two fixation devices spaced apart in their locations on either side of the pivot point of SIJ facilitate "solid" fixation/stabilization across the joint. Both unilateral and bilateral SIJ fusions reduce motion. However, if bilateral SIJ fusion is considered, it is essential to ensure that implant design and SIJ morphology permit such a procedure.

Both males and females showed high performance after SIJ fusion treatment, however, females showed more stress and motion reductions after SIJ fusion. Regardless of sex, lateral and posterior placements of the implants had similar performance on the SIJ stability. SIJ implants are more effective in females in terms of stability but may be more prone to higher rate of loosening/failure compared to males. The optimum number of implants and implant placement location is two or three implants (depending on the bone quality and implant type) across S1 and S2 levels of the sacrum. Having more parallel and farther from SIJ-pivot-point implant placement results in higher stability of the joint. Using longer superior implant placed in S1 level (proximally) closer to the sacral midline leads to higher reduction in SIJ motion. It is better to place the implant in thicker cortical bone areas and a more dense bone region leading to better stability. Most importantly, SIJ fusion has no effect on the adjacent segments on either sides, spine or hip.

Finally, regarding the shapes of the implants, currently, there are two popular designs on the market: circular sections such as SImmetry, SI-LOK, and RI-ALTO; and triangular design such as iFuse. Further biomechanical studies and long-term clinical follow-ups are required to delineate the optimum design (e.g., implant shape) since the existing literature on biomechanics of circular SIJ devices (SImmetry and SI-LOK implant systems) is limited.

In conclusion, despite the existing literature, there are several unanswered questions related to the effect of surgical parameters on the clinical outcome of the SIJ fixation procedures. For example, the effects of different implant shapes on the biomechanical and long-term clinical outcomes of the sacroiliac joint are not fully understood. It is particularly crucial to understand the relationship between bone quality/density and effectiveness of the surgical technique from a biomechanics perspective and the long-term clinical outcomes. Such questions can be answered by looking at parameters such as load-sharing at the bone-implant interface, distribution of the load across the implant, failure mechanism of the bone/implant, and bone remodeling. The clinical studies, due to their inherent limitations, are unable to address such issues. Such knowledge will be crucial for improvement of existing techniques or development of more efficient instrumentation that would yield superior clinical outcomes for SIJ fixation.

Conclusion

The sacroiliac joint (SIJ) is one of the most overlooked sources of LBP. The joint is responsible for the pain in 15–30% of people suffering from LBP. Various studies have investigated the clinical outcomes of different surgical settings intended for treatment of the pain, and they have shown that these techniques are effective indeed. Several questions related to clinical and biomechanical effects of surgical parameters such as number and positioning of implants, unilateral versus bilateral placement, etc., remain unanswered. Biomechanical studies using in vitro and in silico techniques are crucial in addressing such unanswered questions. These were synthesized in the review.

Acknowledgments The work was supported in part by NSF Industry/University Cooperative Research Center at The University of California at San Francisco, CA, and The University of Toledo, Toledo, OH.

References

Alaranta H, Soukis A, Harjula et al (1990) Developing techniques used for diagnosing musculoskeletal diseases (Finnish with English summary). Publications of the Finnish Work Environmental Fund, Helsinki

Albee FH (1909) A study of the anatomy and the clinical importance of the sacroiliac joint. JAMA 53(16):1273–1276

Al-Khayer A, Hegarty J, Hahn D, Grevitt MP (2008) Percutaneous sacroiliac joint arthrodesis: a novel technique. J Spinal Disord Tech 21:359–363

Battafarano DF, West SG, Rak KM, Fortenbery EJ, Chantelois AE (1993) Comparison of bone scan, computed tomography, and magnetic resonance imaging in the diagnosis of active sacroiliitis. Semin Arthritis Rheum 23(3):161–176

Beaubien B, Salib RM, Fielding LC, Block JE (2015) SImmetry sacroiliac joint fusion system with SImmetry decorticator. Surg Sci 6:282–291

Beck CE, Jacobson S, Thomasson E (2015) A retrospective outcomes study of 20 sacroiliac joint fusion patients. Cureus 7(4):e260

Bellamy N, Parl W, Rooney PJ (1983) What do we know about the sacroiliac joint? Semin Arthritis Rheum 12:282–313

Berber O, Amis AA, Day AC (2011) Biomechanical testing of a concept of posterior pelvic reconstruction in rotationally and vertically unstable fractures. J Bone Joint Surg Br 93:237–244

Bertino AJ (2000) Forensic science – fundamentals and investigations. South-Western Cengage Learning, Mason, p 386

Bollow M, Braun J, Taupitz M et al (1996) CT-guided intraarticular corticosteroid injection into the sacroiliac joints in patients with spondyloarthropathy: indication and follow-up with contrast-enhanced MRI. J Comput Assist Tomogr 20:512–521

Borell U, Fernstrom I (1957) The movements at the sacroiliac joints and their importance to changes in the pelvic dimensions during parturition. Acta Obstet Gynecol Scand 36:42–57

Broadhurst NA, Bond MJ (1998) Pain provocation tests for the assessment of sacroiliac joint dysfunction. J Spinal Disord 11(4):341–345

Brooke R (1924) The sacro-iliac joint. J Anat 58:299–305

Bruna-Rosso C, Arnoux PJ, Bianco RJ et al (2016) Finite element analysis of sacroiliac joint fixation under compression loads. Int J Spine Surg 10:16

Brunner C, Kissling R, Jacob HA (1991) The effects of morphology and histopathologic findings on the mobility of the sacroiliac joint. Spine 16:1111–1117

Burnham RS, Yasui Y (2007) An alternate method of radiofrequency neurotomy of the sacroiliac joint: a pilot study of the effect on pain, function, and satisfaction. Reg Anesth Pain Med 32(1):12–19

Buyruk HM, Stam HJ, Snijders CJ, Vleeming A, Laméris JS, Holland WP (1995) The use of color Doppler imaging for the assessment of sacroiliac joint stiffness: a study on embalmed human pelvises. Eur J Radiol 21:112–116

Chamberlain WE (1930) The symphysis pubis in the roentgen examination of the sacroiliac joint. Am J Roentgenol 24:621–625

Cher D, Duhon B, Wine K, Lockstadt H, Kovalsky D, Soo C (2013) Safety and 6-month effectiveness of minimally invasive sacroiliac joint fusion: a prospective study. Med Devices Evid Res 6:219–229

Cibulka MT, Delitta A (1993) A comparison of two different methods to treat hip pain in runners. J Orthop Sports Phys Ther 17:172–176

Cohen SP (2005) Sacroiliac joint pain: a comprehensive review of anatomy, diagnosis, and treatment. Anesth Analg 101(5):1440–1453

Cohen SP, Hurley RW, Buckenmaier CC, Kurihara C, Morlando B, Dragovich A (2008) Randomized placebo-controlled study evaluating lateral branch radiofrequency denervation for sacroiliac joint pain. Anesthesiology 109:279–288

Cook C, Hegedus E (2013) Orthopedic physical examination test: an evidence based approach. Prentice Hall, Upper Saddle River

Cummings J, Capobianco RA (2013) Minimally invasive sacroiliac joint fusion: one-year outcomes in 18 patients. Ann Surg Innov Res 7(1):12

Dall BE, Eden SV, Rahl MD (eds) (2015) Surgery for the painful, dysfunctional sacroiliac joint. Springer, Cham

Damen L, Buyruk HM, Guler-Uysal F et al (2001) Pelvic pain during pregnancy is associated with asymmetric laxity of the sacroiliac joints. Acta Obstet Gynecol Scand 80:1019–1024

Damen L, Spoor CW, Snijders CJ, Stam HJ (2002) Does a pelvic belt influence sacroiliac joint laxity? Clin Biomech 17:495–498

Dietrichs E (1991) Anatomy of the pelvic joints – a review. Scand J Rheumatol 88(Suppl):4–6

Docherty P, Mitchell MJ, MacMillan L, Mosher D, Barnes DC, Hanly JG (1992) Magnetic resonance imaging in the detection of sacroiliitis. J Rheumatol 19(3):393–401

DonTigny RL (1985) Function and pathomechanics of the sacroiliac joint. Phys Ther 65(1):35–44

Dreyfuss P, Cole AJ, Pauza K (1995) Sacroiliac joint injection techniques. Phys Med Rehabil Clin N Am 6:785–813

Dreyfuss P, Michaelsen M, Pauza K et al (1996) The value of medical history and physical examination in diagnosing sacroiliac joint pain. Spine 21:2594–2602

Duhon B, Cher D, Wine K, Lockstadt H, Kovalsky D, Soo C-L (2013) Safety and 6-month effectiveness of minimally invasive sacroiliac joint fusion: a prospective study. Med Devices Evid Res 6:219–229

Duhon BS, Cher DJ, Wine KD et al (2015) Triangular titanium implants for minimally invasive sacroiliac joint fusion: a prospective study. Glob Spine J 6:257–269

Duhon B, Bitan F, Lockstadt H, Kovalsky D, Cher D, Hillen T (2016) Triangular titanium implants for minimally invasive sacroiliac joint fusion: 2-year follow-up from a prospective multicenter trial. Int J Spine Surg 10:13

Dujardin FH, Roussignol X, Hossenbaccus M, Thomine JM (2002) Experimental study of the sacroiliac joint micromotion in pelvic disruption. J Orthop Trauma 16(2):99–103

Ebraheim NA, Biyani A (2003) Percutaneous computed tomographic stabilization of the pathologic sacroiliac joint. Clin Orthop Relat Res 18:60–69

Eichenseer PH, Sybert DR, Cotton JR (2011) A finite element analysis of sacroiliac joint ligaments in response to different loading conditions. Spine 36:1446–1452

Ferrante FM, King LF, Roche EA et al (2001) Radiofrequency sacroiliac joint denervation for sacroiliac syndrome. Reg Anesth Pain Med 26(2):137–142

Fitzgerald CM, Segal NA (eds) (2015) Musculoskeletal health in pregnancy and postpartum. Springer, Cham

Foley BS, Buschbacher RM (2007) Re: sacroiliac joint pain: anatomy, biomechanics, diagnosis, and treatment. Am J Phys Med Rehabil 86(12):1033

Frymoyer JW (1988) Back pain and sciatica. N Engl J Med 318:291–300

Gaetani P, Miotti D, Risso A et al (2013) Percutaneous arthrodesis of sacro-iliac joint: a pilot study. J Neurosurg Sci 57(4):297–301

Gatchel RJ (ed) (2006) Compendium of outcome instruments for assessment & research of spinal disorders. North American Spine Society, LaGrange

Golightly YM, Tate JJ, Burns CB et al (2007) Changes in pain and disability secondary to shoe lift intervention in subjects with limb length inequality and chronic low back pain: a preliminary report. J Orthop Sports Phys Ther 37:380–388

Graves JE, Pollock ML, Carpenter DM et al (1990) Quantitative assessment of full range-of-motion isometric lumbar extension strength. Spine 15:289–294

Ha KY, Lee JS, Kim KW (2008) Degeneration of sacroiliac joint after instrumented lumbar or lumbosacral fusion: a prospective cohort study over five-year follow-up. Spine 33(11):1192–1198

Hammer N, Steinke H, Lingslebe U, Bechmann I, Josten C, Slowik V et al (2013) Ligamentous influence in pelvic load distribution. Spine J 13:1321–1330

Harris CT (1933) Operative treatment of sacroiliac disease: analysis of cases and end results. J Bone Joint Surg 15:651–660

Herzog W, Conway PJ (1994) Gait analysis of sacroiliac joint patients. J Manip Physiol Ther 17:124–127

Hilibrand AS, Robbins M (2004) Adjacent segment degeneration and adjacent segment disease: the consequences of spinal fusion? Spine J 4(6):190S–194S

Hisaw FL (1925) The influence of the ovary on the resorption of the pubic bones. J Exp Zool 23:661

Holmes SL, Cohen SP, Cullen ML et al (2015) Sacroiliac joint pain. In: Pain medicine: an interdisciplinary case-based approach. Oxford University Press, New York, pp 160–182

Ivanov AA, Kiapour A, Ebrahim NA, Goel VK (2009) Lumbar fusion leads to increases in angular motion and stress across sacroiliac joint. Spine 34(5):E162–E169

Jacob HA, Kissling RO (1995) The mobility of the sacroiliac joints in healthy volunteers between 20 and 50 years of age. Clin Biomech 10(7):352–361

Joukar A (2017) Gender specific sacroiliac joint biomechanics: a finite element study. University of Toledo, Toledo

Joukar A, Chande RD, Carpenter RD, Lindsey DP, Erbulut DU, Yerby SA, Duhon B, Goel VK (2019) Effects on hip stress following sacroiliac joint fixation: a finite element study. JOR Spine 2:e1067

Khurana A, Guha AR, Mohanty K, Ahuja S (2009) Percutaneous fusion of the sacroiliac joint with hollow modular anchorage screws: clinical and radiological outcome. J Bone Joint Surg Br 91:627–631

Kiapour A, Abdelgawad AA, Goel VK, Souccar A, Terai T, Ebrahim NA (2012) Relationship between limb length discrepancy and load distribution across the sacroiliac joint-a finite element study. J Orthop Res 30 (10):1577–1580

Kim JT, Rudolf LM, Glaser JA (2013) Outcome of percutaneous sacroiliac joint fixation with porous plasma-coated triangular titanium implants: an independent review. Open Orthop J 7:51–56

Koenig L, Saavoss J, Cher D (2016) Productivity benefits of minimally invasive surgery in patients with chronic sacroiliac joint dysfunction. Clinicoecon Outcomes Res 8:77–85

Kokmeyer DJ, van der Wurff P, Aufdemkampe G et al (2002) The reliability of multitest regimens with sacroiliac pain provocation tests. J Manip Physiol Ther 25:42–48

Kube RA, Muir JM (2016) Sacroiliac joint fusion: one year clinical and radiographic results following minimally invasive sacroiliac joint fusion surgery. Open Orthop J 10:679

Laslett M (1997) Pain provocation sacroiliac joint tests: reliability and prevalence. In: Vleeming A, Mooney V, Snijders CJ, Dormann TA, Stoeckart R (eds) Movement, stability and low back pain: the essential role of the pelvis, 1st edn. Churchill Livingstone, New York

Laslett M (2006) Pain provocation tests for diagnosis of sacroiliac joint pain. Aust J Physiother 52(3):229

Laslett M (2008) Evidence-based diagnosis and treatment of the painful sacroiliac joint. J Man Manip Ther 16 (3):142–152

Laslett M, Williams M (1998) The reliability of selected pain provocation test for sacroiliac joint pathology. Spine 19(11):1243–1249

Laslett M, Young SB, Aprill CN, McDonald B (2003) Diagnosing painful sacroiliac joints: a validity study of a McKenzie evaluation and sacroiliac joint provocation tests. Aust J Physiother 49:89–97

Laslett M, Aprill CN, Mcdonald B, Young SB (2005) Diagnosis of sacroiliac joint pain: validity of individual provocation tests and composites of tests. Man Ther 10 (3):207–218

Ledonio CG, Polly DW Jr, Swiontkowski MF, Cummings JT Jr (2014a) Comparative effectiveness of open versus minimally invasive sacroiliac joint fusion. Med Devices Evid Res 7:187–193

Ledonio CGT, Polly DW, Swiontkowski MF (2014b) Minimally invasive versus open sacroiliac joint fusion: are they similarly safe and effective? Clin Orthop 472 (6):1831–1838

Lee CH, Hsu CC, Huang PY (2017) Biomechanical study of different fixation techniques for the treatment of sacroiliac joint injuries using finite element analyses and biomechanical tests. Comput Biol Med 87:250–257

Lindsey D, Perez-Orribo L, Rodriguez-Martinez N, Reyes PM, Cable A, Hickam G, ... Newcomb A (2014) Evaluation of a minimally invasive procedure for sacroiliac joint fusion – an in vitro biomechanical analysis of initial and cycled properties. Med Devices Evid Res 7:131–137

Lindsey D, Kiapour A, Yerby S, Goel V (2015) Sacroiliac joint fusion minimally affects adjacent lumbar segment motion: a finite element study. Int J Spine Surg 9:64

Lindsey DP, Kiapour A, Yerby SA, Goel VK (2018) Sacroiliac joint stability: finite element analysis of implant number, orientation, and superior implant length. World J Orthop 9(3):14

Lindsey D, Parrish R, Gundanna M, Leasure J, Yerby S, Kondrashov D (2017) Unilateral sacroiliac joint implant placement does not reduce contralateral sacroiliac joint range of motion – a biomechanical study. Spine J

Lingutla KK, Pollock R, Ahuja S (2016) Sacroiliac joint fusion for low back pain: a systematic review and meta-analysis. Eur Spine J 25(6):1924–1931

Magee DJ (2008) Orthopedic physical assessment, 5th edn. W. B. Saunders, Philadelphia

Mao N, Shi J, He D, Xie Y, Bai Y, Wei X, Shi Z, Li M (2014) Effect of lordosis angle change after lumbar/

lumbosacral fusion on sacrum angular displacement: a finite element study. Eur Spine J 23(11):2369–2374

Marymount JV, Lynch MA, Henning CE (1986) Exercise-related stress reaction of the sacroiliac joint: an unusual cause of low back pain in athletes. Am J Sports Med 14:320–323

Masi AT (1992) Do sex hormones play a role in ankylosing spondylitis? Rheum Dis Clin N Am 18:153–176

Mason LW, Chopra I, Mohanty K (2013) The percutaneous stabilisation of the sacroiliac joint with hollow modular anchorage screws: a prospective outcome study. Eur Spine J 22:2325–2331

Maugars Y, Mathis C, Berthelot JM et al (1996) Assessment of the efficacy of sacroiliac corticosteroid injections in spondylarthropathies: a double-blind study. Br J Rheumatol 35:767–770

McCord DH, Cunningham BW, Shono Y, Myers JJ, McAfee PC (1992) Biomechanical analysis of lumbosacral fixation. Spine 17:S235–S243

Miller JA, Schultz AB, Andersson GB (1987) Load-displacement behavior of sacroiliac joints. J Orthop Res 5:92–101

Million S, Hall W, Haavik NK et al (1982) Assessment of the progress of the back-pain patient. 1981 Volvo Award in Clinical Science. Spine 7:204–212

Murphey MD, Wetzel LH, Bramble JM, Levine E, Simpson KM, Lindsley HB (1991) Sacroiliitis: MR imaging findings. Radiology 180(1):239–244

Murray W (2011) Sacroiliac joint dysfunction: a case study. Orthop Nurs 30(2):126–131

Ombregt L (2013) Applied anatomy of the sacroiliac joint. In: A system of orthopaedic medicine. Elsevier, Amsterdam, pp e233–e238

Osterbauer PJ, De Boer KF, Widmaier R et al (1993) Treatment and biomechanical assessment of patients with chronic sacroiliac joint syndrome. J Manip Physiol Ther 16:82–90

Park P, Garton HJ, Gala VC, Hoff JT, Mcgillicuddy JE (2004) Adjacent segment disease after lumbar or lumbosacral fusion: review of the literature. Spine 29(17):1938–1944

Patel N, Gross A, Brown L, Gekht G (2012) A randomized, placebo-controlled study to assess the efficacy of lateral branch neurotomy for chronic sacroiliac joint pain. Pain Med 13:383–398

Pel JJM, Spoor CW, Pool-Goudzwaard AL, Hoek van Dijke GA, Snijders CJ (2008) Biomechanical analysis of reducing sacroiliac joint shear load by optimization of pelvic muscle and ligament forces. Ann Biomed Eng 36:415–424

Polly DW, Cher DJ, Wine KD et al (2015) Randomized controlled trial of minimally invasive sacroiliac joint fusion using triangular titanium implants vs nonsurgical management for sacroiliac joint dysfunction: 12-month outcomes. Neurosurgery 77:674–691

Pool-Goudzwaard AL, Hoekvan Dijke G, Mulder P, Spoor C, Snijders CJ, Stoeckart R (2003) The iliolumbar ligament: its influence on stability of the sacroiliac joint. Clin Biomech 18:99–105

Rappoport LH, Luna IY, Joshua G (2017) Minimally invasive sacroiliac joint fusion using a novel hydroxyapatite-coated screw: preliminary 1-year clinical and radiographic results of a 2-year prospective study. World Neurosurg 101:493–497

Ross J (2000) Is the sacroiliac joint mobile and how should it be treated? Br J Sports Med 34:226

Rothkotter HJ, Berner W (1988) Failure load and displacement of the human sacroiliac joint under in vitro loading. Arch Orthop Trauma Surg 5(107):283–287

Rudolf L (2012) Sacroiliac joint arthrodesis-MIS technique with titanium implants: report of the first 50 patients and outcomes. Open Orthop J 6(1):495–502

Rudolf L (2013) MIS fusion of the SI joint: does prior lumbar spinal fusion affect patient outcomes? Open Orthop J 7:163

Rudolf L, Capobianco R (2014) Five-year clinical and radiographic outcomes after minimally invasive sacroiliac joint fusion using triangular implants. Open Orthop J 8(1):375–383

Sachs D, Capobianco R (2012) One year successful outcomes for novel sacroiliac joint arthrodesis system. Ann Surg Innov Res 6(1):13

Sachs D, Capobianco R (2013) Minimally invasive sacroiliac joint fusion: one-year outcomes in 40 patients. Adv Orthop 2013:1–5

Sashin D (1930) A critical analysis of the anatomy and pathological changes of the sacroiliac joints. J Bone Joint Surg 12:891–910

Sasso RC, Ahmad RI, Butler JE, Reimers DL (2001) Sacroiliac joint dysfunction: a long-term follow-up study. Orthopedics 24:457–460

Schoenberger M, Hellmich K (1964) Sacroiliac dislocation and scoliosis. Hippokrates 35:476–479

Schroeder JE, Cunningham ME, Ross T, Boachie-Adjei O (2013) Early results of sacro-iliac joint fixation following long fusion to the sacrum in adult spine deformity. Hosp Spec Surg J 10(1):30–35

Schuit D, McPoil TG, Mulesa P (1989) Incidence of sacroiliac joint malalignment in leg length discrepancies. J Am Podiatr Med Assoc 79:380–383

Schwarzer AC, Aprill CN, Bogduck M (1995) The sacroiliac joint in chronic low back pain. Spine 20:31–37

Shin MH, Ryu KS, Hur JW et al (2013) Comparative study of lumbopelvic sagittal alignment between patients with and without sacroiliac joint pain after lumbar interbody fusion. Spine 38(21):E1334–E1341

Smith AG (1999) The diagnosis and treatment of the sacroiliac joints as a cause of low back pain. The management of pain in the butt. Jacksonv Med 50:152–154

Smith A, Capobianco R, Cher D, Rudolf L, Sachs D, Gundanna M, ... Shamie A (2013) Open versus minimally invasive sacroiliac joint fusion: a multi-center comparison of perioperative measures and clinical outcomes. Ann Surg Innov Res 7(1):14

Smith-Peterson MN, Rogers WA (1926) End-result study of arthrodesis of the sacroiliac joint for arthritis – traumatic and non-traumatic. J Bone Joint Surg 8:118–136

Snijders CJ, Vleeming A, Stoeckart R (1993) Transfer of lumbosacral load to iliac bones and legs. 1: biomechanics of self-bracing of the sacroiliac joints and its significance for treatment and exercise. Clin Biomech 8:285–294

Solonen KA (1957) The sacroiliac joint in the light of anatomical, roentenological and clinical studies. Acta Orthop Scand 27(Suppl):1–127

Soriano-Baron H, Lindsey DP, Rodriguez-Martinez N et al (2015) The effect of implant placement on sacroiliac joint range of motion: posterior vs trans-articular. Spine 40(9):E525–E530

Sturesson B, Selvik G, Uden A (1989) Movements of the sacroiliac joints: a roentgen stereophotogrammetric analysis. Spine 14:162–165

Sturesson B, Uden A, Vleeming A (2000a) A radio-stereometric analysis of movements of the sacroiliac joints during the standing hip flexion test. Spine 25(3):364–368

Sturesson B, Uden A, Vleeming A (2000b) A radio-stereometric analysis of the movements of the sacroiliac joints in the reciprocal straddle position. Spine 25:214–217

Sturesson B, Kools D, Pflugmacher R, Gasbarrini A, Prestamburgo D, Dengler J (2016) Six-month outcomes from a randomized controlled trial of minimally invasive SI joint fusion with triangular titanium implants vs conservative management. Eur Spine J 26(3):708–719

Tischauer ER, Miller M, Nathan IM (1973) Lordosimetry: a new technique for the measurement of postural response to materials handling. Am Ind Hyg Assoc J 1:1–12

Tortora GJ, Derrickson B (2010) Introduction to the human body: the essentials of anatomy and physiology. Wiley, New York

Unoki E, Abe E, Murai H, Kobayashi T, Abe T (2015) Fusion of multiple segments can increase the incidence of sacroiliac joint pain after lumbar or lumbosacral fusion. Spine 41(12):999–1005

Vanaclocha-Vanaclocha V, Verdú-López F, Sánchez-Pardo M et al (2014) Minimally invasive sacroiliac joint arthrodesis: experience in a prospective series with 24 patients. J Spine 3:185

Vercellini P (2011) Chronic pelvic pain. Wiley-Blackwell, Oxford, pp 118–119

Vlaanderen E, Conza NE, Snijders CJ et al (2005) Low back pain, the stiffness of the sacroiliac joint: a new method using ultrasound. Ultrasound Med Biol 31:39–44

Vleeming A, van Wingerden JP, Snijders CJ et al (1992a) Load application to the sacrotuberous ligament: influence on sacroiliac joint mechanics. Clin Biomech 4:204–209

Vleeming A, van Wingerden JP, Dijkstra PF et al (1992b) Mobility in the sacroiliac joints in the elderly: a kinematic and radiological study. Clin Biomech 7:170–176

Vleeming A, Buyruk HM, Stoeckart R et al (1992c) An integrated therapy for peripartum pelvic instability: a study of the biomechanical effects of pelvic belts. Am J Obstet Gynecol 166:1243–1247

Vleeming A, Schuenke MD, Masi AT, Carreiro JE, Danneels L, Willard FH (2012) The sacroiliac joint: an overview of its anatomy, function and potential clinical implications. J Anat 221(6):537–567

Vrahas M, Hern TC, Diangelo D, Kellam J, Toile M (1995) Ligamentous contributions to pelvic stability. Orthopedics 18:271–274

Walker JM (1992) The sacroiliac joint: a critical review. Phys Ther 72:903–916

Wang M, Dumas GA (1998) Mechanical behavior of the female sacroiliac joint and influence of the anterior and posterior sacroiliac ligaments under sagittal loads. Clin Biomech 13:293–299

Wang MY, Lu Y, Anderson DG, Mummaneni PV (2014) Minimally invasive sacroiliac joint fusion. Springer, Cham

Ware JE, Kosinski M, Turner-Bowker D et al (2002) How to score version 2 of the SF-12 Health Status Survey. Quality Metric Incorporated, Lincoln

Weksler N, Velan GJ, Semionov M, Gurevitch B, Klein M, Rozentsveig V et al (2007) The role of sacroiliac joint dysfunction in the genesis of low back pain: the obvious is not always right. Arch Orthop Trauma Surg 127(10):885–888

Whang PG, Cher D, Polly D, Frank C, Lockstadt H, Glaser J et al (2015) Sacroiliac joint fusion using triangular titanium implants vs. non-surgical management: six-month outcomes from a prospective randomized controlled trial. Int J Spine Surg 9:6

Wheeler W (1912) Surgery of the sacroiliac joint. Br Med J 22:877–880

White AA, Panjabi MM (1990) Clinical biomechanics of the spine, 2nd edn. J. B. Lippincott, Philadelphia

Winter RB, Pinto WC (1986) Pelvic obliquity. Its causes and its treatment. Spine 11:225–234

Wise CL, Dall BE (2008) Minimally invasive sacroiliac arthrodesis: outcomes of a new technique. J Spinal Disord Tech 21:579–584

Yamamoto I, Panjabi MM, Oxland TR, Crisco JJ (1990) The role of the iliolumbar ligament in the lumbosacral junction. Spine 15:1138–1141

Zhang L, Peng Y, Du C, Tang P (2014) Biomechanical study of four kinds of percutaneous screw fixation in two types of unilateral sacroiliac joint dislocation: a finite element analysis. Injury 45(12):2055–2059

Zheng N, Watson LG, Yong-Hing K (1997) Biomechanical modelling of the human sacroiliac joint. Med Biol Eng Comput 35:77–82

Part III

Considerations and Guidelines for New Technologies

Cyclical Loading to Evaluate the Bone Implant Interface

17

Isaac R. Swink, Stephen Jaffee, Daniel Diehl, Chen Xu, Jake Carbone, Alexander K. Yu, and Boyle C. Cheng

Contents

Introduction	378
Fatigue Testing Review	380
Fatigue Testing Terminology	380
Standardized Cyclical Loading Protocols	380
In Vitro Cyclical Loading	381
In Vitro Cyclical Loading Methodology	381
Universal Methods	381
Specimen Preparation	381

I. R. Swink (✉) · D. Diehl · C. Xu · A. K. Yu
Department of Neurosurgery, Neuroscience Institute, Allegheny Health Network, Pittsburgh, PA, USA
e-mail: Isaac.Swink@ahn.org; diehlda@gmail.com; Chen.XU@ahn.org; Alexander.YU@ahn.org

S. Jaffee
College of Medicine, Drexel University, Philadelphia, PA, USA

Allegheny Health Network, Department of Neurosurgery, Allegheny General Hospital, Pittsburgh, PA, USA
e-mail: Stephen.Jaffee@AHN.ORG; sejaffee8@gmail.com

J. Carbone
Louis Katz School of Medicine, Temple University, Philadelphia, PA, USA

Department of Neurosurgery, Allegheny Health Network, Pittsburgh, PA, USA
e-mail: Jjcarbone20@gmail.com

B. C. Cheng
Neuroscience Institute, Allegheny Health Network, Drexel University, Allegheny General Hospital Campus, Pittsburgh, PA, USA
e-mail: bcheng@wpahs.org; Boyle.CHENG@ahn.org; boylecheng@yahoo.com; boylecheng@gmail.com

© Springer Nature Switzerland AG 2021
B. C. Cheng (ed.), *Handbook of Spine Technology*,
https://doi.org/10.1007/978-3-319-44424-6_121

Specimen Selection and Handling	382
Recording Frequency	382
Loading Rate	382
Outcome Measures	383
Pedicle Screws	383
Pedicle Screw Constructs	383
Interbody Devices	384
Loading Modality	384
Pedicle Screws	385
Pedicle Screw Constructs	385
Interbody Devices	386
Magnitude of Loading	386
Pedicle Screws	387
Pedicle Screw Constructs	387
Interbody Devices	388
Failure Criteria	388
Pedicle Screws	388
Pedicle Screw Constructs	389
Interbody Devices	389
Cyclical Loading of Spinal Constructs Literature Review	390
Basic Science	390
Pedicle Screws	391
Pedicle Screw Constructs	395
Interbody Devices	396
Conclusion	398
References	398

Abstract

The goal of in vitro cyclical loading studies in spine biomechanics is to provide empirical data related to the long-term efficacy of spinal implants. Ultimately, these studies are used to determine if an implant has the ability to provide biomechanical stability over extended periods of time until arthrodesis achieves. In these studies, the bone implant interface should be gradually stressed according to physiological loading patterns to determine the rate at which the interfacial strength between the bone and implant degrades. When designed properly, these studies may able be used to determine the ultimate failure load of the bone implant interface. While study design is always an important aspect in benchtop research, the repetitive nature of in vitro cyclical loading studies exacerbates the effects of study design and emphasizes the importance of rigorous planning based on a strong understanding of the boundary conditions. This chapter is therefore focused on the most important aspects of study design surrounding in vitro spine biomechanics including specimen preparation, loading rate, loading magnitude, loading modality, outcome measures, and failure criteria and how they influence the results of these studies.

Keywords

Cyclical loading · Bone implant interface · Biomechanics · Pure Moment · Fatigue · Stair step loading · Functional spinal unit

Introduction

Knowledge about the spine has been fortified by insight gained through biomechanics research. New technologies and treatment modalities are often a result of in vitro biomechanical studies using human cadaveric spine segments. Comparative biomechanics studies are utilized in the engineering design process to determine if an instrumentation system or stand-alone device meets design requirements – a conclusion often measured by the ability of the system to provide mechanical stabilization. Moreover, biomechanical studies provide a critical tool for researchers and clinicians to increase basic understanding of the spine and to facilitate evidence-based clinical decision-making.

Biomechanics studies have largely been conducted according to the pure moment protocols pioneered by Panjabi's conceptual framework in 1988 (Panjabi 1988). Adherence to this basic framework across biomechanics literature has served to prevent confusion due to unique and individualized testing protocols, allowing for broader comparisons. As illustrated in Fig. 1 below, pure moment application creates a consistent force throughout the construct without introducing shear forces. A moment is defined as a force that causes rotation of a rigid body about a specific point or axis, while a pure moment is defined as a pair of parallel forces, applied in equal measure but opposite direction to create rotation while mitigating translation. Shear forces, forces acting in parallel to the cross-sectional plane of the construct, cannot be accurately measured and are therefore detrimental to reproducibility. Biomechanics studies conducted according to Panjabi's framework are often referred to as flexibility, stability, or static tests and provide measures of initial stability through metrics such as range of motion (ROM), neutral zone, and construct stiffness.

While Panjabi's work delineated an integral scheme, which has been invaluable to our understanding of the spine and various treatment modalities thereof, we acknowledge limitations of these studies. Standard in vitro biomechanics studies provide information relative to the immediate postoperative time period, and while initial stability measures are important, instrumentation intended to provide stabilization necessary to facilitate fusion must exist for extended periods until solid arthrodesis has formed. Additionally, standard flexibility tests cannot identify common potential risk and failure patterns. As modern spine testers advance, so does our ability to test these constructs using advanced cyclical loading protocols.

The goal of this chapter will be to review testing protocols and provide specific guidance on considerations for in vitro biomechanical studies focused on cyclical loading. As the prevalence and complexity of these cyclical studies continues to advance, it is imperative that the research

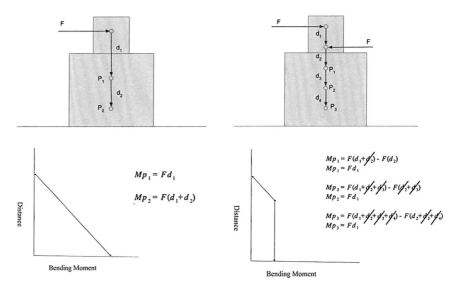

Fig. 1 Illustration of a pure moment and the resulting bending moment

community follow common boundary conditions and testing methodologies to more effectively compare results between research studies and prevent the misinterpretation due to testing variability. As such, we will review standardized cyclical testing protocols, considerations for in vitro cyclical testing protocols, and highlight results of current notable in vitro fatigue studies in the literature.

Fatigue Testing Review

Fatigue Testing Terminology

Long-term performance of implanted systems is of utmost importance given the longitudinal nature of their use in the spine and the severe consequences of construct failure. Proper evaluation and understanding of the literature surrounding cyclical fatigue tests is contingent upon understanding some of the basic testing terminology.

Stress is the force per unit area of a given construct. Compressive stresses are conventionally represented as positive forces while tensile stresses are negative. The amount of deformation (δ) caused by an applied stress can be expressed as a fraction of construct length (L) to calculate strain ($\varepsilon = \delta/L$). Plots of stress versus strain are used to calculate the Young's modulus – determined using the linear portion of a stress–strain curve and may be interpreted as a measure of construct stiffness. Stress–strain curves can also be used to determine the ultimate failure strength of a construct, representing the maximum sustained stress that can be applied before failure. However, repetitive loading can cause materials to fail at loads significantly lower than their ultimate failure strength. This is known as fatigue failure. Unfortunately, it is not possible to generate a proper stress–strain curve to determine the interfacial fatigue strength of spinal constructs due to the limitations of working with cadaveric tissue. However, that does not mean that fatigue factors, such as creep, do not play a role in the failure resulting from in vitro and in vivo repetitive loading of spinal constructs.

In materials science, a material's fatigue strength represents the maximum stress which can be sustained for more than 10^7 loading cycles without failure. Fatigue failure is driven by the accumulation of microcracks which occur below the ultimate strength of a material and is influenced by material properties including strength and toughness and environmental factors like creep and corrosion. A material's strength refers to its resistance to plastic flow while toughness refers to its ability to resist crack propagation. Creep is the slow deformation of a material in response to consistent loading.

Standardized Cyclical Loading Protocols

Standardized testing protocols for cyclical loading of spinal constructs, governed by ASTM 1717, were developed to evaluate the strength of these systems in response to repetitive loading conditions (ASTM 2014). The goal of these test protocols is not to predict clinical performance but to provide a controlled environment in which past, present, and future implant systems may be evaluated comparatively. Notably, the tests intended to evaluate the strength of the instrumentation itself and not the stability provided to a lumbar spine segment.

Cadaveric tissue is supplemented with ultra-high-molecular-weight polyethylene (UHMWPE) blocks, which are machined and instrumented to create a vertebrectomy model. Precise dimensional control ensures that the appropriate forces are transmitted through the testing construct as they are eccentrically loaded. Replacing cadaveric tissue with polymer blocks eliminates variability attributable to differences in bone density and geometric anatomy of the system. This test setup produces an ideal scenario in terms of the bone implant interface (BII) and worst-case scenario for the construct, as all of the applied load must be transferred from the superior to inferior levels through the test system.

To conduct these tests, a compressive bending moment is generated via a sinusoidal cyclic compressive load, which is applied in load control to the superior block at a rate of up to 5 Hz. Prior to cyclic fatigue testing, the magnitude of the compressive force applied is determined by the static bending strength of the construct. Testing is conducted until either failure occurs or the system reaches predefined runoff criteria of 5,000,000 cycles, a benchmark intended to represent approximately 2 years of loading at an average of 7,000 loading cycles per day.

In Vitro Cyclical Loading

While the information related to construct failure gained from experiments conducted according to ASTM 1717 are valuable, they are not well suited to predict clinical failure. In a clinical scenario, instrumentation failure is most often attributable to a failure of the BII, either by aseptic loosening or abrupt fracture. A testing paradigm, which addresses questions related to the physiologic performance of spinal constructs outside of the postoperative period, is therefore necessary to evaluate the long-term strength of the BII. This interfacial strength is often measured indirectly through measures of construct stability or determination of failure conditions. The goal of these tests is to provide information as to how failure occurs, when this failure may occur, the failure modality, and any patient characteristics which may increase the risk of failure.

Common forms of fatigue failure include implant migration, subsidence, and aseptic screw loosening – the latter of which is characterized by an initial phase of micromotion followed by a slower continuous phase of migration (Xie et al. 2019). Cyclical loading studies are therefore designed to force the failure of instrumentation systems with respect to criteria based on these aforementioned patterns. Typical outcome measures in cyclical in vitro biomechanics studies include range of motion (ROM) as an indicator of aseptic loosening, axial displacement as an indicator of subsidence, and construct stiffness as an indicator of interfacial stiffness. Active measurement of these parameters during cyclic loading is not always possible; therefore, cyclical loading is often conducted in series with nondestructive static biomechanics experiments to assess stability at various time points.

In Vitro Cyclical Loading Methodology

There is no standardized in vitro fatigue protocol to evaluate spinal implants. Rather, experiments are designed and conducted to answer specific research questions because the destructive nature of these tests limits their scope. While testing constructs and experimental conditions may change between experiments, the underlying goal remains the same: to gradually stress the BII through the application of physiological loads to demonstrate longitudinal efficacy. The methods of each publication should be critically reviewed to understand the boundary conditions and the study conclusions. Tests for pedicle screws, pedicle screw constructs, and interbody devices are often different, and testing protocols and setups are often tailored to a specific construct. In this section, we will review the important parameters in the cyclical loading of cadaveric tissue and how they may vary based on the test construct. Constructs were categorized as follows: (1) pedicle screws (individual screw cyclical loading), (2) pedicle screw constructs (pedicle screw and rod constructs), and (3) interbody devices.

Universal Methods

Regardless of the test article, there are certain aspects of cyclical loading studies which should remain relatively standard: specimen preparation, hydration, measurement time points, and measurement frequency.

Specimen Preparation

As with standard static biomechanics tests, specimens are often mounted in a test fixture with a resin-based potting compound, wherein all soft tissue is removed from the bony elements.

Failure to properly remove soft tissues will result in increased micromotion between the specimen and potting compound, eventually leading to construct failure. Due to the destructive nature of cyclical loading, construct failure often leads to specimen disqualification, deficits in statistical power, and costly repetition of testing procedures.

Specimen Selection and Handling

The results of in vitro cyclical loading studies are highly dependent on the characteristics of the specimens used for testing due to large variations in tissue quality between donors, thereby necessitating that tissue used for these protocols is representative of the intended population. Measurement techniques like DEXA and calibrated CT scans should be used to quantify bone density and morphological characteristics when possible (McCubbrey et al. 1995; Weiser et al. 2017a). In doing so, statistical tests can be used to ensure that the outcomes of the given paper were not influenced by factors such as bone mineral density or pedicle diameter.

A single cyclical loading experiment may take up to 14 h to complete depending on the loading frequency and failure criteria and as such active measures must be taken to maintain the hydration and physiologic performance of the tissue. This may include using a saline spray to hydrate the specimen periodically (at least every 20 min) or wrapping the specimen in saline soaked gauze, which is periodically replenished. Alternatively, testing may be performed in a controlled environment with 100% humidity to prevent dehydration over the course of testing procedures.

Proper techniques should also be followed for the storage and handling of cadaveric tissue including freezing of specimens at $-20\,^\circ$C and wrapping them in saline-soaked gauze while not in use. It is also recommended to avoid any freeze thaw cycles after potting procedures have been performed to prevent weakening of the specimen and test fixture interface. Specimens are generally thawed overnight at 4 $^\circ$C before testing. Testing at room temperature is the de facto standard, and testing before a specimen is allowed to reach room temperature could result in artificially high interfacial strength or measured stability due to the increased stiffness of soft tissues. Specimens may be warmed by leaving them at room temperature for at least 2 h or with the use of a warm saline bath.

Recording Frequency

In order to capture the extent of induced motion, measurements must be done at sufficient rates which vary from 10 to 200 Hz throughout the literature. While a higher sampling frequency leads to more accurate data collection, it also translates to larger file sizes requiring more computational power for analysis. Generally speaking, higher sampling rates should be used with higher loading frequencies, and the sampling rate should be maximized within the technical constraints of the researcher.

Loading Rate

Loading rate refers to the speed with which loads are applied to a spinal construct. This metric is usually expressed in Hertz (Hz), which indicates the number of loading cycles applied during each second of testing. ASTM 1717 dictates that 5 Hz is the absolute maximum loading rate for spinal constructs; however, these tests were conducted with polyethylene blocks allowing for a more aggressive loading rate. When designing studies with cadaveric tissue, the loading rate should be reduced in order to provide a more physiologic test scenario and prevent unwarranted damage to the specimen.

Given the viscoelastic nature of bony tissues, loading at a rate which exceeds the rate of physiologic loading may artificially increase the interfacial stiffness or cause the formation of microcracks which lead to reduced fatigue capacity. On the other hand, slower loading rates may lead to increased displacement and lead to issues with the rate of specimen degradation exceeding the rate of fatigue damage of the BII (Xie et al. 2019). The majority of in vitro biomechanics studies report loading rates between 0.5 and 2 Hz.

Outcome Measures

The metrics used to evaluate the efficacy of spinal instrumentation vary based on the type of construct being tested. Some are evaluated during cyclical fatigue loading, while most outcomes are based on static in vitro biomechanics tests performed before and after cyclic loading. Both active and periodic assessments are intended to show how the strength of the BII changes in response to repetitive loads. In vitro cyclical loading protocols are often designed as comparative studies in which a new technology or treatment strategy is compared to the current clinical standard. In this case, nondestructive measurement techniques are used to assess the stability of a construct before and after cyclical loading. These repeatable measurements allow for each specimen to serve as its own control. Therefore, by normalizing data collected after cyclical loading to a baseline measurement, the damage caused at the BII can be accurately evaluated. Destructive tests (i.e., screw pullout tests) are also used; however, studies designed to include destructive outcomes require the use of multiple cohorts thereby increasing biologic variability and cost.

Pedicle Screws

The fatigue strength of pedicle screw fixation is often assessed by comparing the number of cycles to failure and peak pullout-strength after cyclic loading. The number of cycles required to induce fatigue failure assesses the fatigue strength of the BII of the screw. Pedicle screw failure may be determined by a displacement criterion, relative screw motion, bony fracture, or other study-specific means. Failure based on displacement criteria generally refers to the amount of screw head displacement measured in response to an applied load. Screw head displacement may be measured directly by attaching a tracking body to the screw or indirectly by monitoring displacement of the loading frame. Relative screw motion is assessed using optoelectronic measurement systems with a tracking body attached to both the pedicle screw and tissue sample, allowing for calculation of relative motion between the bone and pedicle screw. Comparison of the axial pullout strength after cyclical loading and fatigue failure to that of a control group without cyclical loading determines the damage at the BII.

Other more advanced means of evaluating the strength of the bone screw interfaced have been described but have yet to gain popularity in the literature. Lai et al. describe the determination of a microfracture event in which a sharp increase in displacement is observed during cyclical loading, helping to identify the method of fatigue failure (Lai et al. 2018). Advanced imaging techniques like micro computed tomography (micro-CT) allow for precise and accurate identification of screw breach, quantification of trabecular damage or pedicle fractures.

Pedicle Screw Constructs

Unlike the evaluation of individual screws, pedicle screw constructs are evaluated using spinal segments or functional spinal units. Therefore, the strength of the BII is measured indirectly through measures of construct stability like range of motion, which should be periodically evaluated in each specimen to evaluate the efficacy at various time points (Cheng et al. 2011; Agarwal et al. 2016; Wilke et al. 2006, 2016; Duff et al. 2018). At minimum, the stability of each specimen should be evaluated at intact, baseline, and post-cyclical loading conditions. Inclusion of additional stability measurements throughout cyclical loading allows for the determination of further outcome measures such as number of cycles to failure and failure load (Wilke et al. 2016). Measurements of stability should be conducted in a nondestructive fashion to ensure that damage is not accumulated outside of the cyclical loading schemes. At the very least, flexibility measurements should be conducted according to the loading magnitudes published in the literature, with loads applied to the cervical spine topping out at ±2.5 Nm and loads applied to the lumbar spine topping out at ±10 Nm. Some research groups have opted to reduce the loading magnitude used to measure stability in cyclical loading protocols to further reduce the risk of introducing additional damage to the BII, which is especially important if multiple stability measurements are performed throughout the study.

Interbody Devices

The evaluation of interbody device performance in response to cyclical loading is similar to the process described for pedicle screw constructs; metrics of construct stability are measured at various time points to determine fatigue performance (Freeman et al. 2016; Palepu et al. 2017). Other common interbody failure methods like implant subsidence (evaluated radiographically by periodic measurements of disc height performed in series with cyclical loading or with the use of motion tracking bodies attached to the vertebral body) and migration (movement of the interbody device with respect to the cadaveric specimen quantified by fixing a motion tracking system directly to the interbody device) should also be considered (Freeman et al. 2016; Pekmezci et al. 2016; Alkalay et al. 2018). Subsidence may also be evaluated by measurement of the interfacial stiffness at the BII, as described by Alkalay et al. This stiffness may be extracted from the hysteresis plot generated during compressive loading of interbody constructs (Alkalay et al. 2018).

Stand-alone interbody devices often rely on spikes, fins, or anchoring blades to achieve fixation. The damage caused to bony structures, specifically the vertebral endplate and trabeculae, by these means of fixation are not well understood. Therefore, advanced imaging techniques such as micro computed tomography (micro-CT) may be used to identify the presence of endplate fractures or to quantify the damage by measuring the volume of bone displaced (Palepu et al. 2018).

Loading Modality

Loading modality refers to the method in which a specimen is loaded. This is done in isolation of a particular mode of loading such as screw toggling, axial compression, or flexion-extension bending thereby mitigating the application of shear forces and controlling to match physiologic loads. Ideally, cyclical loads should be applied based on the typical failure patterns observed clinically.

The loading modality is also highly dependent on the type of testing machine used, with standard uniaxial MTS machines limiting researchers to the application of uniaxial compressive forces. These machines are therefore combined with test fixtures to produce bending moments through eccentric loading patterns as seen in Fig. 2. When a column or spinal construct is loaded asymmetrically (with respect to the central axis), a bending moment is produced through the column, the magnitude of which is driven by the moment arm formula. The magnitude of the moment is calculated as the product of the applied force by the distance to the center axis. Offset loads are often applied through a hinge joint, connecting the test article to the MTS machine via a rigid connection that serves as the lever. While these MTS machines can be used in conjunction with lever arms to offset forces and generate bending moments, readers should be aware of the uneven load distribution these constructs create. Per Panjabi's conceptual framework, the application of eccentric bending moments creates an uneven bending moment through the column, or in this case spinal construct, which peaks towards the middle of the construct.

More advanced spine testers allow for the application of pure moment loads in a dynamic fashion, allowing for active control of all six degrees of freedom. Use of a dynamic spine tester for cyclical loading of spinal constructs allows for the application of pure moment loads in a given loading pattern, while all other modes of loading are allowed to move freely. In doing so, researchers can apply flexion-extension, lateral bending, and axial rotation moments and evaluate how fatigue affects the performance of instrumentation with respect to each individual

Fig. 2 Illustration of an FSU mounted to facilitate eccentric moment loading

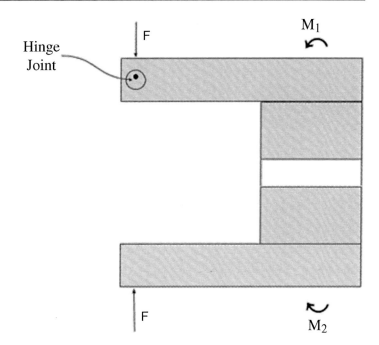

loading modality. While separate cohorts must be used to test the effects of each aforementioned loading modality, these machines do not require a change in testing setup to change the type of fatigue loading being evaluated.

Pedicle Screws

Pedicle screws are loaded perpendicular to the long axis of the screw (in the cranio-caudal direction), accomplished directly by applying a compressive force to the screw head or indirectly by applying the force through a connecting rod (Weiser et al. 2017a; Lai et al. 2018; Akpolat et al. 2016; Baluch et al. 2014; Bostelmann et al. 2017; Kueny et al. 2014; Lindtner et al. 2018; Qian et al. 2018). Screw loading with a compressive force perpendicular to the screw's long axis is intended to reproduce the toggle failure commonly observed during aseptic screw loosening. Trends in recent literature demonstrate the application of compressive forces through a connecting rod as this provides a more physiological loading modality. By attaching the pedicle screw to an MTS machine via a standard rod in combination with the use of a lever arm, a bending moment is created about the head of the screw, approximating flexion-extension and compressive loading (Lindtner et al. 2018). To accomplish this, both the superior and inferior vertebral bodies are rigidly fixed. Use of an x-y slide or bearing table allows for translation in the transverse plane and may help reduce the effects of shear forces during cyclic loading. Failure to reduce shear forces, which act as axial pullout forces in cranio-caudal loading, may lead to early fatigue failure (Akpolat et al. 2016).

Pedicle Screw Constructs

Cyclical loading of pedicle screw constructs is intended to produce physiologic loading patterns that are often subjected to combined compressive and flexion bending moments. Bending moments are applied eccentrically using custom test fixtures where the magnitude of the applied moment is given by the magnitude of the compressive force multiplied by the distance from the central axis of the specimen to the axis of loading (Agarwal et al. 2016; Duff et al. 2018; Shimamoto et al. 2001). Typically, the center of the vertebral body or test fixture is used to approximate the central axis of the specimen (Shimamoto et al. 2001). The central axis of the loading machine or test fixture represents the loading axis. All specimens should be prepared and mounted in a reproducible fashion as geometric variations may have an effect on the study results.

These experiments are performed with the cranial aspect of the test construct fixed to the test machine and the caudal aspect attached to an x-y slide or similar fixture. Alternatively, the specimen may be mounted in a spine tester with the ability to apply pure moments in a cyclical manner as described by Cheng et al. (2011). Programmable spine testers allow for active control (with respect to one individual mode of loading), while all other motors are passively controlled to allow for coupled motion and thereby reduce the effects of shear loading. This is commonly referred to as load control because the motors are set to maintain zero force in any given direction, functioning like an x-y slide allowing motion in the transverse plane.

While most studies focus on compressive forces and flexion-extension bending, it is important that other loading modalities such as lateral bending and axial rotation are considered. Wilke et al. demonstrated a technique in which specimens were cyclically loaded with eccentrically applied compressive forces while the specimen was rotated 360°, producing a cycle of flexion, left lateral bending, right lateral bending, and extension bending moments. This more complex loading scheme allows for the application of compression, shear, and bonding forces in a reproducible fashion (Wilke et al. 2006, 2016). As a result, the applied forces are more physiological and may improve the quality of the data. Other research groups have suggested breaking protocols into blocks of flexion-extension, lateral bending, and axial rotation loading patterns; however, these techniques have yet to be applied to pedicle screw constructs.

Interbody Devices

Cyclical loading protocols conducted to evaluate interbody devices most often include the application of flexion-extension bending moments as flexion-extension is considered the main mode of loading in the lumbar spine. Unlike tests designed to evaluate pedicle screws and pedicle screw constructs, interbody constructs are most often subject to pure moment loads, with or without the use of a follower load. Inclusion of a follower load is intended to provide a more physiologic loading schema, with compressive forces transmitted throughout the spine via a system of pulleys attached to each index level. Compressive preloads may also be employed as a simplified alternative to the use of a follower load.

Although most studies focus on flexion-extension loading, lateral bending and axial torsion loading modalities should also be evaluated if possible. Cheng et al. advocate for independent cohorts for the evaluation of flexion-extension, lateral bending, and axial rotation fatigue performance of interbody devices. Other authors have applied multiple loading schemes to each specimen, alternating between flexion-extension and lateral bending paradigms until failure has occurred (Freeman et al. 2016). Each of these strategies has their own merit; separation into multiple cohorts allows for the researcher to determine if the construct is vulnerable to failure with respect to a given loading modality, while combined loading techniques may provide a more physiologic result. When evaluating the effect of compressive loading on interbody performance, the anatomical differences between cadaveric specimens should be considered in order to reduce the effects of coupled moments. Alkalay et al. describe the application of compressive forces through each specimen's instantaneous center of rotation (ICR) to mitigate these effects (Alkalay et al. 2018).

Magnitude of Loading

The magnitude of loading should be adjusted to achieve physiological loading if possible, reflecting values of the forces acting on the spine during daily activities, i.e., walking (Rohlmann et al. 1997). Stair-step loading protocols (initial

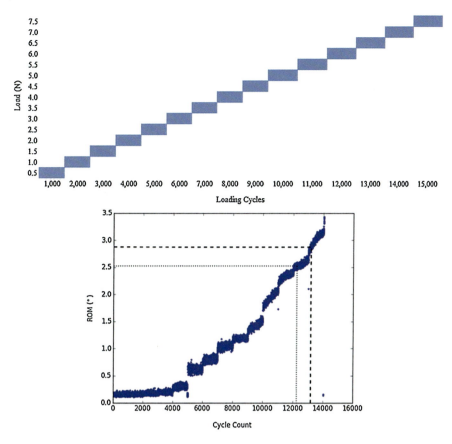

Fig. 3 Example of a stair-step loading scheme (top) and angular corresponding angular measurements produced by cyclically loading an FSU according to this loading protocol (bottom)

cyclical loading at a low magnitude with steady increases as the test progresses) may be employed to ensure that fatigue failure is achieved. This enables the gradual loading of the BII and allows for the inclusion of failure load as an additional outcome measure to compare the fatigue performance of the given constructs. Figure 3 illustrates a stair-step loading protocol and the angular measurements corresponding to the cyclical loading of an instrumented FSU according to the given protocol.

Pedicle Screws

The loads applied to pedicle screw constructs are generally based on the peak forces measured on a spinal fixator during walking (Rohlmann et al. 1997). When forces are applied at a constant magnitude, they typically range from a 50–200 N of force. Stair-step loading protocols for cyclical loading of pedicle screws employ a minimum of 20–50 N force historically, which is increased by small increments ranging anywhere from 0.1 to 20 N until failure is achieved. Maximum applied forces in stair-step loading protocols have been reported as high as 550 N (Bostelmann et al. 2017).

Pedicle Screw Constructs

Protocols designed to evaluate pedicle screw constructs tend to parallel the biomechanics literature for standard flexibility protocols. In the cervical spine, bending moments between 1.5 and 2.5 Nm are achieved through the application of up to 200 N of eccentric loads or in a pure moment fashion using a spine tester (Cheng et al. 2011; Duff et al. 2018; Koller et al. 2015). For tests centered around the lumbar spine, larger bending moments of up to 10 Nm are induced through the application of eccentric loads ranging from 100 to 600 N (Agarwal et al. 2016; Wilke et al. 2006, 2016; Shimamoto et al. 2001). Stair-step loading protocols are especially used in the lumbar spine to force fatigue failure before the specimen degrades. These protocols generally start with loading magnitudes ranging from 100–300 N or 3–9 Nm and bending moments of up to 600 N or 18 Nm of force.

Interbody Devices

Protocols designed to evaluate the performance of interbody devices are intended to gradually stress the bone implant interface. As such, the magnitude of applied loads is often below those applied in traditional in vitro biomechanics tests, which range from 7.5 to 10 Nm in all modes of loading. When applying a stair-step loading protocol, the initial magnitude may be as low as 0.5 Nm which is gradually increased until failure has occurred (Cheng et al. 2018). Alternatively, authors have described the consistent application of up to 5 Nm loads until failure has occurred (Freeman et al. 2016). To evaluate the effects of combined loading, follower loads of up to 400 N and compressive forces of up to 1200 N have been described.

Failure Criteria

A cyclical loading experiment may be stopped based on a set number of loading cycles or predetermined failure criteria, which are metrics used to determine when fatigue failure occurs and can be unique to a given experimental protocol. Notably, the metric in question must be continuously monitored throughout the experiment in order for an experimental protocol to have a stopping condition based on a predefined measure of fatigue failure. Examples include actuator displacement or optoelectronic motion data. Failure may also be assessed after cyclical loading has been conducted to the desired number of loading cycles.

These criteria are used to evaluate the strength of the BII and quantify the number of loading cycles applied to induce failure, allowing for comparison of the fatigue strength across cohorts. When possible, failure criteria should be derived from clinical assessments of construct failure, such as radiographic assessments or motion measured using dynamic flexion-extension radiographs. Failure criteria may also be based on the determination of key events in the life cycle of a testing construct. Common significant events include the identification of a bony fracture, hardware deformation, or sharp change in construct stiffness.

Pedicle Screws

Screw loosening has been associated with other complications including hardware failure, pseudoarthrosis, non-union or progressive kyphosis. Currently, there is no standardized classification system for clinicians to evaluate screw loosening; however, according to a recent literature review conducted by Galbusera et al., the most common assessment of screw loosening is a radiographic assessment of the BII (marked by the presence of a radiopaque zone extending 1 mm or more from the screw in all directions) (Galbusera et al. 2015).

Screw loosening may occur due to exceedingly high loads, microfractures at the BII, or bony remodeling as a result of stress shielding. In vitro cyclical loading of pedicle screws may be used to assess failure due to the application of loads exceeding the bone screw interface (potentially resulting in a sharp increase in screw motion) or the accumulation of microfractures within it (identified by a slow and gradual increase in motion) (Bostelmann et al. 2017).

In vitro fatigue failure of pedicle screws is defined by the amount of displacement, measured directly by attaching tracking bodies to the screw head to measure angular displacement or indirectly by tracking the displacement of the test system. Measurement of screw angulation is preferred as this provides more physiologically relevant data, while stopping criteria based on the displacement of the test fixture does not directly translate to screw displacement. This is especially true in testing constructs where pedicle screws are loaded eccentrically and displacement of the axial ram accounts not only for screw displacement but rotation about the axis of the hinge joint and deformation of the lever arm as well (Bostelmann et al. 2017). If the stopping criteria cannot be based on a continuous measurement of angular screw displacement or other fatigue metric, it is recommended that the data be post processed to determine when fatigue failure occurred.

Recent literature has defined fatigue failure of a pedicle screw during cyclical loading to be associated with anywhere from 6 to 8° of angular motion or an increase in the angular displacement of greater than 1° over a short period of time (Akpolat et al. 2016; Baluch et al. 2014; Lindtner et al. 2018). A limit of 10–15 mm of displacement may be applied when stopping criteria are based on the axial displacement of the test frame used to apply a cyclical compressive force (Bostelmann et al. 2017).

Pedicle Screw Constructs

Fatigue failure of pedicle screw constructs is largely related to screw loosening, bending, pullout, or rod failure (Wittenberg et al. 1992). From an in vitro perspective, success of a pedicle screw construct is given by the stability of the spinal construct as a whole, often measured in degrees of relative motion between vertebral bodies. This is largely due to the fact that the forces generated during in vitro testing are not sufficient to damage the instrumentation in question. Wilke et al. have developed a series of success criteria to evaluate when fatigue failure occurs in vitro; they are facilitated by repeated stability measurements performed in series with blocks of cyclical loading. Here, loosening or failure is defined as either a visible fracture, motion which exceeds baseline instrumented stability by a factor of three, or motion which exceeds the value corresponding to the intact state (Wilke et al. 2006, 2016).

Numerous studies have been conducted to evaluate pedicle screw constructs without the inclusion of a fatigue failure criteria. In this case, success of a construct is determined by how well it performs in comparison to a control cohort based on a clinical standard of care (Cheng et al. 2011; Agarwal et al. 2016; Duff et al. 2018; Shimamoto et al. 2001).

Interbody Devices

Failure of the BII for an interbody device may lead to decreased stability, device subsidence, or device migration, and as such failure criteria should be established for all three possible options. Stability is most often measured through range of motion, and as such increases in motion beyond a specific threshold are often used to determine when failure has occurred. Various research groups have their own definition as to what constitutes an increase in motion large enough to be considered failure; however, all of these researchers agree that failure should be based on a baseline stability measurement. For example, Cheng et al. define failure as a 110% increase in ROM, based on the idea that a 10% increase in segmental ROM could have a substantially negative impact on arthrodesis (Cheng et al. 2018). As mentioned in the previous section, Wilke et al. argue that a return of ROM to the values measured at the "intact" stage of testing, before instrumentation, is performed should also be considered failure.

Cage migration may be considered clinically relevant when it exceeds 3 mm in any direction (Abbushi et al. 2009). Failure with respect to subsidence may be considered clinically relevant when subsidence exceeds 3 mm as well; however, it is often difficult to actively monitor cage subsidence during cyclic loading. As such, subsidence failure criteria are often based on the displacement of the test machine, as described by Alkalay et al. In those instances where subsidence is based on machine displacement failure, criteria are often much higher, on the order of 10 mm (Alkalay et al. 2018). This relatively large criteria compared to the 3 mm threshold for clinical relevance is due to the inherent flexibility of the construct, as not all 10 mm of displacement may be attributed to cage subsidence, but rather plastic deformation of the specimen in addition to any subsidence.

Cyclical Loading of Spinal Constructs Literature Review

Basic Science

Cyclical loading studies help to further our basic understanding of the spine in addition to providing evidence on the in vitro performance of spinal instrumentation. Numerous studies have been conducted on cadaveric tissues evaluating their response to repeated loads and damage accrued from low magnitude repetitive loads to further understand spinal injuries and determine ideal boundary conditions for in vitro evaluation of instrumentation. The number of loading cycles a specimen may tolerate and the appropriate magnitude and rate of loading are integral to this process. In doing so, we can ensure that any damage created during experimental procedures can be attributed to the test article rather than the testing conditions.

Schmidt et al. used compressive cyclical loading data from 6 previously published works to generate a lumbar spine injury risk model based on fatigue failure data collected from 105 cadaveric samples subjected to varying loading magnitudes and lengths (Schmidt et al. 2012). All but one of these specimens consisted of a single FSU, subjected to compressive loads until failure or runout was achieved. As expected, age and sex were both significant predictors for the risk of failure. Notably, Schmidt found that the applied stress has a stronger influence than the number of loading cycles on the failure probability of a vertebral segment – a finding that gives credence to the use of stair-step loading protocols and underscores the importance of reporting failure loads when conducting cyclical loading experiments.

Liu et al. set a tremendous foundation for modern cyclical loading studies through their novel methodology, notably replicated in 2018 by Alkalay et al. in their investigation of performance of interbody cage designs (Alkalay et al. 2018; Liu et al. 1983). Through their compressive study, Liu hypothesized that clinical instability is the result of fatigue failure of the soft tissue structures of a spinal segment. Liu first identified the instantaneous center of rotation for each FSU by identifying the stiffest loading axis and aligning this axis with the loading axis of the MTS machine. Up to 10,000 cycles of compressive forces were then applied to this location at 0.5 Hz (ranging from 37% to 80% of the ultimate failure load for the respective index level). Failure loads used to determine the loading conditions for each specimen were based on those values reported in the literature. Furthermore, they divided the 11 FSU's that were loaded until failure into two groups: abrupt failure ($n = 5$) and gradual failure ($n = 6$). Abrupt failing specimens were classified as unstable and were associated with large changes in axial displacement. Gradually failing specimens were classified as stable with a consistent linear increase in displacement. Liu further characterized the damage caused to each vertebral body through radiographic imaging (abrupt failure specimens suffered endplate fractures which originated peripherally and propagated towards the center of the vertebral body) and morphological characterization facilitated by digestion of the organic matrix with sodium hydroxide to isolate the bony structures (the unstable group lost their structure and essentially disintegrated into small pieces of bone). This is opposed to those specimens from the stable group which maintained their structure after removal of the organic matrix. As such, recorded displacement values from the unstable group were attributed to bony failure while displacement in the stable group was attributed to creeping deformation of the intervertebral disc.

Parkinson et al. has further corroborated these results with their work relative to the cervical spine by evaluating the injury risk to both bony structures and soft tissue in response to dynamic flexion-extension and compressive loading (Parkinson and Callaghan 2009). Specimens were cyclically loaded and divided into five cohorts based on the magnitude of applied load (ranging from 10% of ultimate failure load to 90% of the ultimate failure load). Loading was performed until failure (>9 mm vertical displacement) or runoff (12 h or 21,600 cycles) was achieved. Throughout testing, the compressive load, vertical displacement, and angular motion were collected to facilitate determination of a precise failure event, and the authors also quantified the cumulative moments and compression to achieve failure using trapezoidal integration. They concluded that failure was most often identified by examining the compressive stiffness of the construct and noted that a sharp decline in stiffness was associated with an increase in vertical displacement. Similarly to the work of Liu et al., this study found the magnitude of loading had a significant effect on the fatigue life of the constructs, with an average injury cycle of 14,400, 5,031, 155, 22, and 4 for the 10%, 30%, 50%, 70%, and 90% loading cohorts, respectively. Only 5 of the 46 specimens included in the study survived to the runout criteria of 21,600 loading cycles: 4 from the 10% loading cohort and 1 from the 30% loading cohort. Furthermore, the specimens in the 10% and 30% loading cohorts sustained a significantly larger cumulative compressive load than specimens from the 50%, 70%, or 90% groups. Post fatigue injury analysis revealed 34 instances of fracture, with the large majority of these injuries occurring within the endplate.

Both the Liu and Parkinson studies emphasize the importance of including multiple measurement techniques in studies designed to evaluate the fatigue performance of spinal instrumentation. By showing that an abrupt increase in recorded displacement is associated with radiographic and morphological evidence of bony failure, these studies support the inclusion of failure criteria based on abrupt changes in displacement or stiffness of the tested construct. The disintegration of samples which suffered bony failure after removal of the organic matrix proves the accumulation of microcracks in the trabeculae of vertebrae after cyclic loading. These results also provide insight into the number of cycles a cadaveric specimen can be expected to withstand, allowing for better understanding of construct efficacy through the window of cycles to failure.

Considerable work has also been put into studying the viscoelastic nature of trabecular bone. The Pankaj lab performed a series of experiments to develop a finite element model of the BII in order to evaluate the interface deformation as a function of cycle number and loading frequency (Xie et al. 2019; Manda et al. 2016). An idealized 2D plane-strain construct represented interactions between trabecular bone, a screw, and a 1 mm thick cortical shell in response to 300 N cyclic loads applied at seven different loading frequencies from 0.1 to 10 Hz for a total of 2,000 cycles. The trabecular bone volume was also varied to study the effect of bone density on displacement of the BII. The results of these models show that displacement at the BII is a function of both cycle number and loading frequency. At a low cycle number, slower loading rates produce more deformation; however, at higher cycles (above 500), faster loading rates produce much greater deformation. Unsurprisingly, the study also demonstrated a correlation between bone volume fraction and displacement, with lower volume fractions showing higher displacement compared to models run according to the same parameters but with a higher volume fraction. It is important to understand how the applied loading rate, number of applied cycles, and bone density influence reported outcomes, and the results of these studies as well as other similar reports should be considered when planning or evaluating cyclical loading research.

Pedicle Screws

One of the keystone designs for spinal fixation comes from the polyaxial screw system which was designed to connect the vertebrae to a rigid rod construct for spinal stabilization. Screws were designed with spherical heads enclosed within a housing that allowed for extreme mobilization within multiple axes relative to the housing. The ball joint allowed for use in most spinal surgeries because of its flexibility and ability to adapt to the most severe degenerative diseases. The system was placed above and below the vertebrae to bridge the fusion, giving it strength while arthrodesis occurs. Rods were then connected to the polyaxial screws to prevent movement for ultimate stability. If rigid fixation is not maintained, complications may arise from screw breakage or loosening. Changes to the system's approach, screw diameter, length, pitch attempt to overcome many obstacles the original design imposed.

One area of focus in pedicle screw placement is maximizing fixation strength through pedicle screw trajectory. Using traditional fatigue studies to compare pedicle screw pullout strengths, Blauch et al. sought to determine fixation strength of laterally directed, cortical pedicle screw under physiological cyclical loads compared to a traditional approach. Lateral trajectory was designed to minimize soft-tissue dissection during instrumentation while providing improved spinal fixation (Baluch et al. 2014). Seventeen vertebral levels in total were obtained from three cadaveric spine specimens (T11-L5). Radiographic testing ensured no prior trauma or fracture, and pQCT was obtained for each level to determine bone mineral density (mean 202 mg/cm^3). Alternating sides of each level were instrumented with cortical then laterally and medially directed screws (tercet Triple-lead and Preference 2 Pedicle Screws). Each screwhead was connected to a 5.5 mm rod and secured via setscrews. Once in the testing apparatus, each screw underwent cyclic cranio-caudal toggling under increasing physiological loads until 2 mm head displacement was recorded and uniaxial pullout of each toggled screw was performed. The load (N) and toggle cycles to each pedicle screw movement were recorded and compared between the two techniques. Notably, cortical screws required 184 cycles to reach 2 mm of displacement compared to 102 cycles for traditional screws. Moreover, the force necessary to displace these screws was much greater for cortical than that of traditional pedicle screw trajectory; however, no statistical difference in axial pullout strength between toggled cortical and traditional pedicle screws was found. The results supported the hypothesis that laterally directed pedicle screws have greater resistance to cranio-caudal toggling than traditional trajectories – demonstrating the efficacy of alternative pedicle screw trajectories for spinal fusion surgery.

Akpolat et al. conducted a similar investigation to perform fatigue studies on specimens with poor bone quality to assess efficacy of the cortical bone trajectory (CBT) vs. the standard pedicle screw fixation (Akpolat et al. 2016). They postulated that the use of a hinge joint does not accurately represent physiological loading and therefore designed a new fixture based on ASTM1717 to allow for pure moments to be applied to the construct for fatigue testing purposes to better simulate physiologic screw loading and motion. CBT and standard trajectory pedicle screws were inserted in the same vertebrae and cyclically loaded to failure. A 5.5 × 130 mm rod was used to connect the screws to another screw mounted in a polyethylene block, intended to represent the superior vertebral level. Both the block and vertebral body were loaded into the testing fixture. Two equal and opposite forces were applied to generate pure moment bending through hinge forces designed to limit shear forces which may induce axial pullout forces. Loosening was monitored optoelectronically via reflective markers and infrared cameras, so relative motion between the vertebral body and pedicle screws indicated motion at the bone-screw interface. Six cadaveric lumbar spines were obtained using 12 vertebrae. Bone mineral density was obtained for each vertebral level. Each vertebral body was instrumented with screws from the cortical bone trajectory (4.5 × 25 mm) and standard pedicle screw trajectory (6.5 × 55 mm). A load was then applied under displacement control at 1 Hz in sagittal bending at 4 Nm. Appropriate force was determined by measuring the distance from the rod to the hinge joint. The construct was loaded for 100 cycles or until 6° of loosening was observed. Once fatigue testing was completed, the screws were pulled out axially at 5 mm/min. Standard pedicle screws had (1) significantly longer fatigue life than cortical bone trajectory screw (3592 +/− 4564 and 84 +/− 24, respectively), (2) showed better resistance to motion (6.9 +/− 4.8° of motion at 100 cycles and 15.2 +/− 5.5°, respectively), and (3) had significantly higher pullout strength than CBT screws (776 N +/− 370 vs. 302 N +/− 332, respectively). Damage to the bone along its shaft by rotating around a fulcrum, located at the pars, pedicle isthmus, or the junction of the pedicle and superior endplate was the primary limitation of the CBT screws – laminar anatomy may prevent proper CBT screw insertion trajectory.

While imperfect, pedicle screw fixation has been considered the gold stand in posterior spinal stabilization and fusion for many years, with screw loosening at the bone-screw interface in an aging osteoporotic population (generally the highest source of spinal fixations) as an especially notable complication (Halvorson et al. 1994). Screw designs like fenestrated pedicle screws for bone cement applications, pre-cemented pedicle screws, in situ augmentation and augmentations to pedicle screw diameters have attempted to limit poor outcomes. Pedicle screw augmentation was designed for limited applications in spinal fusion and was focused on specific conditions like osteoporosis (Weiser et al. 2017b). The most widely used techniques are pedicle screw insertion in non-cured cement and in situ-augmentation with cannulated fenestrated screws. Bostelmann et al. focused on assessing the different augmentation techniques to explore pedicle screw loosening under physiological cyclic cranio-caudal loading (Bostelmann et al. 2017). While previous tests focused on axial pullout strength to determine anchorage, it does not simulate the in vivo nature of cranio-caudal loading and failure of pedicle screws. Two test groups were created, each containing 15 vertebral bodies (L1–L5, three of each level per group). High viscosity bone cement was used for all augmentation. Pedicle screws were placed on both the right (instrumented with solid pedicle screws in standard orientation with no bone cement added) and left side (cannulated and augmented with bone cement) of the vertebral bodies for the first test. For the second round of testing, the right pedicle screws were enhanced with the cement first technique, while the left were inserted with cannulated fenestrated screws within situ augmentation. Screws were tested using cranio-caudal cyclical loading started at 20–25 N and loading increased 0.1 N per cycle (1 Hz) for a total of 5,000 cycles or fatigue failure with stress X-rays were taken to determine screw integrity. In the first group, augmented screws showed significantly higher load cycles compared to control pedicle screws. Completed stress X-rays determined much lower screw toggling for the augmented screws than that of the solid pedicle screws. The second group of testing determined high load cycles until failure for cement first augmentation compared to in situ augmentation were not significantly different. Stress X-rays for this group revealed screw toggling for in situ augmentation vs. cement first augmentation was not significantly different. This test demonstrates the strength added for augmented pedicles screws compared to those with standard orientations. The significantly higher load cycles and failure loads add considerable strength to weakened osteoporotic bone. For all augmentation techniques tested (cement first, in situ augmentation, percutaneous application), no effects were exhibited on failure of these pedicle screws. The screws that did fail by cranio-caudal cyclical loading occurred via a "windshield-wiper effect" through the superior endplate. This is typically seen in clinical practice with osteoporotic patients.

Kueny et al. conducted a study to better understand the complications with aging osteoporotic bone on pedicle screw loosening at the bone-screw interface focusing on three different fixation techniques: traditional prefilled augmentation, screw injected augmentation, and unaugmented screws with increased diameters, while also determining whether pullout testing can be translated to physiological fatigue testing (Kueny et al. 2014). Thirty-nine osteoporotic lumbar vertebrae were instrumented with pedicle screws covering four testing groups: (1) screw only, (2) prefilled augmentation, (3) screw injected augmentation, and (4) unaugmented screws with increased diameters. Toggling testing was performed using cranio-caudal cyclical loading (1 Hz). The initial compression forces started at 25–75 N and was increased in a stepwise fashion by 25 N every 250 cycles until 5.4 mm of screw head displacement was noted. Once completed, the contralateral screw was subjected to pure axial pullout (5 mm/min). All instrumentation techniques were compared to control. Screw injected augmentation increased fatigue force by 27% ($p = 0.045$), while prefilled augmentation reduced fatigue force by -7% ($p = 0.73$). Both of these techniques increased pullout force compared to control ($p < 0.04$). Increase in screw diameter (1 mm) increased pullout force by 24% ($p = 0.19$) while inducing the least amount of stiffness loss at -29% from control. They concluded that the highest biomechanical stability lies within the augmentation of the injected pedicle screws. Strong considerations should be noted from these studies indicating utilization of screw injected cement augmentation along with maximal screw diameter for increased biomechanical stability in osteoporotic patients.

Lai et al. further investigated the optimal screw diameter for osteoporotic bone (2018). Larger screws may provide more fixation strength but risk purchase failure during instrumentation, while smaller diameters minimized pedicle fracture but compromise stability. Focus was directed towards screw diameter and pullout of pedicle screws after fatigue loading in osteoporotic vertebrae. Five human cadavers were harvested for testing on 27 osteoporotic vertebrae (T3–T8). Two different size polyaxial pedicle screws were instrumented into each pedicle (5.0 mm × 35 and 4.35 mm × 35). Specimens were then randomly distributed into three groups: (1) control group, (2) fatigue group of 5,000 cycles, and (3) fatigue group of 10,000 cycles. This was accomplished with peak-to-peak loadings of 10–100 N at 1 Hz. After fatigue loading was completed, each specimen was subjected to axial pullout tests at a rate of 5 mm/min and the maximum pullout strength (N) and stiffness (N/mm) were obtained. During fatigue testing, displacement curves were used to determine microfracture analysis. No specimens incurred microfractures during the 5,000 loading cycles, but during the 10,000-cycle group, some specimens experienced discontinued curves and abrupt jumps – they were further categorized into two subgroups with and without microfractures. No statistical difference was noted between pedicle height and width from the 5.0 mm and 4.35 mm screw groups. Micro-CT and X-ray showed lateral breaches in most specimens. Specimens with no abrupt jumps in their loading displacement curve showed no noticeable changes along their screw trajectories. Those with abrupt jumps showed radiolucent lines between bone and screw interfaces after fatigue loading. Pullout strength of the 5.0 mm group (363.3 (138.3) N) was statistically higher than those of the 4.35 mm group (259.0 (159.3) N). The 5,000-cycle group pullout strength was reduced to 33.25 (165.2) N, but the 4.35 mm group increased to 316.1 (106.0) N. The pullout strength between the 5.0 mm group and 4.35 mm group was not statistically significant after 5,000 cycles. The pullout strength between the two screw diameters in the 10,000-cycle group was significantly decreased to 208.9 (90.1) N in the 5.0 mm group and 229.3 (99.1) N in the 4.35 mm group. No statistically significant change in pullout strength between the two groups were found. 5.0 mm screws at 10,000 cycles without microfractures showed pullout strengths that were not statistically significant to those with one microfracture. 4.35 screws at 10,000 cycles without microfractures show statistically higher pullout strengths than those with microfractures. Pullout stiffness in the 5,000-cycle loading group of 5.0 mm screws increased to 679.1 (290.8) N/mm and one of the 4.35 mm groups increased to 583.5 (215.8) N/mm. The pullout stiffness of the 5.0 mm group is higher but not statistically significant. In the 10,000-cycle group, the pullout stiffness decreased to 405.7 (236.2) N/mm and 444.5 (238.0) N/mm between the 5.0 mm and 4.35 mm screws. For the 5.0 mm screws in the 10,000 cycle groups, the pullout stiffness was not statistically significant between fracture and microfracture subgroups, even though slightly higher in the group without fractures. The same goes for the 4.35 mm screws in the 10,000-cycle group. Non-fracture pullout stiffness was slightly higher but not statistically significant. For bone mineral density vs. pullout strength and stiffness, the 5.0 mm screw groups were statistically significantly correlated with BMD in the control group and 5,000 cycle loading group but not in the 10,000-cycle loading group. For the 4.35 mm group in all loading conditions, the pullout strength and stiffness were not statically correlated with BMD. This study provides evidence for the preferential use of smaller over larger diameter screws as the only positive outcome from using a larger diameter screw in osteoporotic patients is immediate strength after implantation. After considerable fatigue loading, this study did not show any added fixation strength for larger diameter screws. The results show that smaller diameter screws may be more beneficial in osteoporotic patients in reducing risk of surgical breaching of the pedicle cortical layer while providing comparable fixation strength to larger diameter screws. Further studies are likely needed in this area, especially with testing in the cranio-caudal cyclic loading fatigue testing of the prior two articles.

Pedicle Screw Constructs

Pedicle screw constructs have been one of the foundational pillars to spinal instrumentation; their presence permits the creation of immediate stability with the additional benefit of long-term fusion. There has been focused investigation into how the specificities of pedicle screw construct design may be tailored to improve the efficacy of these constructs in response to cyclical loading protocols. The works presented below provide an example of these experiments and what we can learn from them as a research community.

Irrespective of concerns regarding bone mineral density, pedicle screw constructs may cause significant unintended damage to adjacent levels due to rigid body constructs, abnormal forces, stress shielding, and hypermobility – a notion widely reported in the literature. This phenomenon was investigated by Agarawal et al. in which they assessed the significance of rod rigidity when with respect to screw loosening in a fatigue model (Agarwal et al. 2016). They postulated that less rigid rods may confer the advantage of minimizing these aforementioned complications. They utilized two groups of cadaveric lumbar spines, one with pedicle screws connected to titanium rods and the other connected to PEEK rods. Specimens were subjected to 10 Nm of pure moment in flexion and extension and later the L4–L5 segments from each were segmented and tested with cyclical loading up to 100,000 cycles followed by post-fatigue kinematic analysis. The titanium rod construct demonstrated a significant increase in flexion and extension ROM in pre/post-fatigue kinematic analysis, whereas the PEEK group did not show any significant difference, thereby potentially demonstrating an indirect association between rod material flexibility and fatigue.

Wilke et al. utilized both a pedicle screw and lamina hook system to evaluate its long-term efficacy (Wilke et al. 2016). As Wilke mentioned in their manuscript, the advantage of a pedicle screw is the primary means of rigid bone fixation, although bone mineral density is a limiting factor, especially in osteoporotic or osteopenic patients. Meanwhile, the lamina hook system provides direct connection to the lamina itself and possibly confers reduced risk of damage to neighboring structures, with the added patient benefit of reduced harm to anatomic structures if revision is indicated. To do this, the authors utilized a small sample of cadaveric thoracolumbar specimens and pedicle or lamina hook systems. The samples were subsequently placed into 100,000 cyclical loading (flexion/extension, lateral bending, and compression) cycles with moments of 3–66 Nm or until failure criteria was met: (1) failure at bony structure, (2) exceeding of the threefold ROM of the primary stability after implantation in flexion plus extension, and (3) reaching of the ROM based on the intact state before implantation both in flexion plus extension. Both implant systems demonstrated a significant reduction in ROM in all motion planes. Moreover, the pedicle screw and the lamina hook system were comparable in terms of loading cycles reached before failure (30,000–32,500, respectively) and corresponding moments (24 Nm and 25.5 Nm, respectively). The conclusion garnered from this exercise was that both pedicle screw and lamina hook constructs provide substantial and comparable characteristics – providing a potential biomechanically comparable alternative for patients with lower bone density.

The authors noted that basing failure criteria off baseline measurements of construct stability may influence the study outcomes, specifically related to the number of cycles to failure. In instances where failure is determined based on a percentage increase relative to the baseline condition, a higher initial ROM would translate to a higher ROM in degrees but may have a higher level of instability and not be relevant clinically. For this reason, it makes sense to return to an intact ROM of motion to account for baseline ROM after fixation.

While biomechanics models have proven effective at delineating basic science characteristics of the spine, it is important to also recognize their importance in establishing and influencing clinical decision-making. Cheng et al. developed a cyclical loading methodology for cadaveric cervical spine models and further demonstrated the potential for pedicle screw constructs and biomechanics testing to influence clinical guidelines for patients in their postoperative course (Cheng et al. 2011). This study was also significant for use of cadaveric cervical spine columns, which may be subjected to significant physiological stressors due to daily movement, that were fused both posteriorly and laterally. A C5 corpectomy was performed and the cervical columns were fused from C3–C7 with posterior instrumentation and screws placed in the lateral mass of the vertebrae. Flexion/extension, lateral bending, axial torsion were each cycled to +/− 2.5 Nm at 1 Hz and axial compression cycled between 0 and 150 N. All testing was performed under continuous load control protocol and a sinusoidal waveform corresponding to 200 s/cycle. Cheng found that there was a significant change in pre- and post-fatigue cycling in axial torsion (pre-cycling mean 22.9 +/− 15.3, post-cycling mean 27.5 +/− 21.3, $p = 0.0030$) but none in flexion/extension, lateral bending, or axial compression. The aforementioned result led to the recommendation of potentially limiting axial torsion in the immediate postoperative course to decrease risk of instrumentation failure.

Interbody Devices

The use of interbody devices to increase anterior column support has improved spinal fusion outcomes. However, the potential for subsidence and cage migration necessitates biomechanics research to determine the risk of these complications based on various aspects of interbody cage design and fixation technique. Alkalay et al. performed a series of cyclic loading experiments to determine the influence of cage design on the long-term stability of TLIF constructs with three different cage designs: (1) a unilateral oblique cage (UOL) placed anteriorly, (2) an anterior conformal shaped interbody (ACS) intended to match the geometry of the endplate, and (3) bilateral linear implants (BLL) (Alkalay et al. 2018). An MTS machine was used to apply compressive loads ranging from 400 to 1200 N through the instantaneous center of rotation of each lumbar FSU at a rate of 2 Hz. Loading was performed until 20,000 loading cycles or 10 mm of displacement was achieved. All of the specimens tested survived for the entirety of the loading regime, with no more than 6 mm of recorded displacement. In addition, all three cohorts showed a continual increase in displacement throughout cyclical loading, with the rate of displacement highest in the first 500 loading cycles. This result indicates that 500 cycles at a constant loading rate may be a sufficient period to generate damage at the BII and could be used to validate stair-step loading protocols. This conclusion correlates well with basic sciences studies related to aseptic loosening which indicate there is an initial period of rapid loosening followed by a slow and steady increase in motion (Xie et al. 2019). The authors found that the compressive stiffness of the BLL cohort decreased at a significantly higher rate compared to the ACS and UOL cohorts and attribute this difference in cohorts to the centralized position of the BLL implants as opposed to the UOL and ACS implants, which are located closer to the endplate periphery where bone quality is superior (Alkalay et al. 2018). These results reinforce the notion that the majority of implant subsidence occurs within the first few weeks postoperation.

As stand-alone interbody devices continue to gain popularity, the diversity of cage fixation techniques has advanced in parallel. New fixation techniques require rigorous characterization to determine their fatigue capacity and elucidate common failure patterns and risk categories based on patient demographics like bone quality. These results are of particular interest to the FDA as they continue to evaluate the safety of these devices. Indeed, the FDA's Office of Science and Engineering Laboratories have conducted a number of studies centered around the ability of these novel fixation techniques to retain fixation for extended periods (Palepu et al. 2017, 2018; Nagaraja and Palepu 2017). Screws placed across the endplate and into the vertebral body are often used to fix stand-alone interbody devices; however, bone quality varies across both the endplate and within the trabeculae of the vertebral body and may have a negative impact on the efficacy of these screws and other fixation strategies that rely on bony purchase within the endplate. Palepu et al. used a combination of high resolution micro-CT imaging and cyclical loading to evaluate the stability of integrated fixation cage screws based on their trajectory and bone quality (Palepu et al. 2018). Three different screw trajectories based on commercially available designs were evaluated: lateral to medial (LM), mid-sagittal (MS), and medial to lateral (ML). After a 3.5 mm pilot hole was drilled and micro-CT scans were performed, 5.5 mm screws were placed and subjected to cranio-caudal cyclic loads of 10–50 N for 10,000 cycles. The maximum load was increased by 25 N every 5,000 cycles until the 25,000 cycle mark was reached. Testing was stopped prior to 25,000 cycles if the screw displacement exceeded 5 mm. The authors found significant differences between the bone quality surrounding screws of the three trajectories, with the ML trajectory having greater total bone volume and bone volume fraction compared to MS and LM trajectories. Screws placed according to a MS trajectory had significantly higher bone volume fraction compared to LM screws. The higher bone quality surrounding ML screws translated to improved fatigue capacity with specimens from the ML cohort sustaining significantly higher counts of cycles to failure compared to both LM and MS screws.

The same authors furthered this work with another study focused on determining how stand-alone interbody devices compare to current clinical standards of constructs that combined interbody support with anterior plating to achieve fusion (Palepu et al. 2017). Lumbar segments L2–3 and L4–5 were instrumented with either an integrated fixation cage with four screws (two superior and two inferior) or a standard interbody cage with anterior plate secured with four screws. Both constructs were subjected to 5,000 cycles of 3 Nm flexion-extension loading at 1 Hz followed by 15,000 cycles of 5 Nm flexion-extension loading. Measurements of construct stability (ROM and lax zone) in response to flexion-extension, lateral bending, and axial rotation were performed in the intact condition, implanted baseline, 5,000, 12,500, and 20,000 loading cycle time points. The authors found both constructs demonstrated comparable stability in response to all modes of loading at all time points. Both constructs were able to significantly reduce ROM from the intact condition and saw significant increases in motion at the 12,500 and 20,000 loading cycle when compared to the instrumented baseline for both flexion-extension and lateral bending motions.

A 2016 study conducted by Freeman et al. addressed a similar research question: is there a difference in the fatigue performance of interbody devices placed through an anterior approach as opposed to an oblique approach (Freeman et al. 2016). The authors instrumented lumbar segments (L2–3 and L4–5) with one of two interbody devices, an ALIF and OLIF. Both interbody devices were of the similar design and employed the same fixation strategy, a curved anchor plate which is impacted across the endplate into the vertebral body. Specimens were subject to 30,000 cycles of moment loading, alternating between flexion-extension and lateral bending loading patterns according to a protocol similar to the hybrid or follower-load methodologies previously described Panjabi and Patwardhan (Panjabi 2007; Patwardhan et al. 1999). A custom follower load system was used to apply a cumulative 400 N compressive load in conjunction with ± 5 Nm pure moments at a 1 Hz loading rate. The authors reported no instances of fatigue failure and minimal subsidence averaging 0.8 mm and 1.4 mm in the anterior and oblique cohorts respectively, with no significant differences between the cohorts. Similar to the work of Alkalay et al., this study found that the majority of implant subsidence occurs early in the cyclical loading protocol, in this instance before the 2500th cycle. Device migration was also minimal with an average of less than 1 mm of anterior migration.

Cheng et al. have evaluated a number of interbody devices using the same protocol in an effort to build a library of data on the long-term efficacy of these systems. To accomplish this goal, standard measures of stability are conducted in series with stair-step cyclic loading protocols intended to gradually load the BII. Measures of flexibility are performed in the intact, instrumented baseline, mid-fatigue (5,000 loading cycles), and post-fatigue time points in addition to continuous tracking of segmental ROM throughout cyclic loading. Pure moment cyclical loading is applied in either flexion-extension, lateral bending, or axial rotation with an initial loading magnitude of 0.5 Nm which is increased by 0.5 Nm every 1,000 loading cycles until failure is achieved. Cheng defined failure as an increase in ROM to a value corresponding to 125% of baseline instrumented ROM. To date, these experiments have been conducted on interbody devices with various methods of integrated fixation including anteriorly impacted fins, curved anchor plates, endplate hooks, vertically driven anchor plates, and screw-based designs. These designs were associated with fatigue lives of 15,600, 13,124, 12,826, 14,052, and 12,500 respectively.

It is important to note that the three previously described experimental protocols used to evaluate the fatigue life of various interbody devices produced varying results. This illustrates how variations in experimental techniques may affect the reported outcomes. For example, although its effects are not well understood from a fatigue perspective, the inclusion of a follower load is known to increase the compressive load carrying capacity of the lumbar spine in static measures of stability (Patwardhan et al. 1999). Unless devices are tested in a head to head comparison, it is difficult to compare their ability to retain fixation for extended periods of time.

Conclusion

Fatigue testing of instrumented spinal constructs is a valuable tool in furthering our understanding of how instrumentation is able to maintain long-term stability. These studies provide evidence as to how long instrumented spinal constructs may be able to maintain stability in the absence of biological healing. The nuances of how these studies are conducted, including everything from the type of machine used to apply loads to how well specimen hydration is maintained, could have significant effects on the results of these studies and readers should therefore consider the differences between various manuscripts when comparing results between various laboratories. Future work is necessary to determine how these constructs respond to more complex loading schemes.

References

Abbushi A, Cabraja M, Thomale UW, Woiciechowsky C, Kroppenstedt SN (2009) The influence of cage positioning and cage type on cage migration and fusion rates in patients with monosegmental posterior lumbar interbody fusion and posterior fixation. Eur Spine J 18(11):1621–1628

Agarwal A, Ingels M, Kodigudla M, Momeni N, Goel V, Agarwal AK (2016) Adjacent-level hypermobility and instrumented-level fatigue loosening with titanium and PEEK rods for a pedicle screw system: an in vitro study. J Biomech Eng 138(5):051004

Akpolat YT, Inceoglu S, Kinne N, Hunt D, Cheng WK (2016) Fatigue performance of cortical bone trajectory screw compared with standard trajectory pedicle screw. Spine (Phila Pa 1976) 41(6):E335–E341

Alkalay RN, Adamson R, Groff MW (2018) The effect of interbody fusion cage design on the stability of the instrumented spine in response to cyclic loading: an experimental study. Spine J 18(10):1867–1876

ASTM (2014) Standard test methods for spinal implant constructs in a vertebrectomy model. Medical devices and services. 2014 annual book of ASTM standards: ASTM International, West Conshohocken, PA, pp 658–679

Baluch DA, Patel AA, Lullo B, Havey RM, Voronov LI, Nguyen NL et al (2014) Effect of physiological loads on cortical and traditional pedicle screw fixation. Spine (Phila Pa 1976) 39(22):E1297–E1302

Bostelmann R, Keiler A, Steiger HJ, Scholz A, Cornelius JF, Schmoelz W (2017) Effect of augmentation techniques on the failure of pedicle screws under

cranio-caudal cyclic loading. Eur Spine J 26(1):181–188
Cheng BC, Cook DJ, Cuchanski M, Pirris SM, Welch WC (2011) Biomechanical cyclical loading on cadaveric cervical spines in a corpectomy model. J ASTM Int 8(7):154–168. https://doi.org/10.1520/JAI103408
Cheng BC, Swink ITM, Carbone JJ, Diehl D (eds) (2018) Biomechanical fatigue evaluation of anterior lumbar interbody fusion devices with respect to flexion-extension loading. NASS, Los Angeles
Duff J, Hussain MM, Klocke N, Harris JA, Yandamuri SS, Bobinski L et al (2018) Does pedicle screw fixation of the subaxial cervical spine provide adequate stabilization in a multilevel vertebral body fracture model? An in vitro biomechanical study. Clin Biomech (Bristol, Avon) 53:72–78
Freeman AL, Camisa WJ, Buttermann GR, Malcolm JR (2016) Flexibility and fatigue evaluation of oblique as compared with anterior lumbar interbody cages with integrated endplate fixation. J Neurosurg Spine 24(1):54–59
Galbusera F, Volkheimer D, Reitmaier S, Berger-Roscher-N, Kienle A, Wilke HJ (2015) Pedicle screw loosening: a clinically relevant complication? Eur Spine J 24(5):1005–1016
Halvorson TL, Kelley LA, Thomas KA, Whitecloud TS III, Cook SD (1994) Effects of bone mineral density on pedicle screw fixation. Spine 19(21):2415–2420. https://doi.org/10.1097/00007632-199411000-00008
Koller H, Schmoelz W, Zenner J, Auffarth A, Resch H, Hitzl W et al (2015) Construct stability of an instrumented 2-level cervical corpectomy model following fatigue testing: biomechanical comparison of circumferential antero-posterior instrumentation versus a novel anterior-only transpedicular screw-plate fixation technique. Eur Spine J 24(12):2848–2856
Kueny RA, Kolb JP, Lehmann W, Puschel K, Morlock MM, Huber G (2014) Influence of the screw augmentation technique and a diameter increase on pedicle screw fixation in the osteoporotic spine: pullout versus fatigue testing. Eur Spine J 23(8):2196–2202
Lai DM, Shih YT, Chen YH, Chien A, Wang JL (2018) Effect of pedicle screw diameter on screw fixation efficacy in human osteoporotic thoracic vertebrae. J Biomech 70:196–203
Lindtner RA, Schmid R, Nydegger T, Konschake M, Schmoelz W (2018) Pedicle screw anchorage of carbon fiber-reinforced PEEK screws under cyclic loading. Eur Spine J 27(8):1775–1784
Liu YK, NJUS G, Buckwalter J, Wankano K (1983) Fatigue Reponse of lumbar intervertebral joints under axial cyclic loading. Spine (Phila Pa 1976) 8(8):857–865
Manda K, Xie S, Wallace RJ, Levrero-Florencio F, Pankaj P (2016) Linear viscoelasticity – bone volume fraction relationships of bovine trabecular bone. Biomech Model Mechanobiol 15(6):1631–1640
McCubbrey DA, Peterson DDC, Kuhn JL, Flynn MJ, Golstein SA (1995) Satic and fatigue failure properties of thoracic and lumbar vertebral bodies and their relation to regional density. J Biomech 28(8):891–899
Nagaraja S, Palepu V (2017) Integrated fixation cage loosening under fatigue loading. Int J Spine Surg 11:20
Palepu V, Peck JH, Simon DD, Helgeson MD, Nagaraja S (2017) Biomechanical evaluation of an integrated fixation cage during fatigue loading: a human cadaver study. J Neurosurg Spine 26(4):524–531
Palepu V, Helgeson MD, Molyneaux-Francis M, Nagaraja S (2018) Impact of bone quality on the performance of integrated fixation cage screws. Spine J 18(2):321–329
Panjabi M (1988) Biomechanical evaluation of spinal fixation devices: I. A conceptual framework. Spine (Phila Pa 1976) 13(10):1129–1134
Panjabi MM (2007) Hybrid multidirectional test method to evaluate spinal adjacent-level effects. Clin Biomech (Bristol, Avon) 22(3):257–265
Parkinson RJ, Callaghan JP (2009) The role of dynamic flexion in spine injury is altered by increasing dynamic load magnitude. Clin Biomech (Bristol, Avon) 24(2):148–154
Patwardhan AG, Havey RM, Meade KP, Lee B, Dunlap B (1999) A follower load increases the load-carrying capacity of the lumbar spine in compression. Spine (Phila Pa 1976) 24(10):1003–1009
Pekmezci M, Tang JA, Cheng L, Modak A, McClellan RT, Buckley JM et al (2016) Comparison of expandable and fixed interbody cages in a human cadaver corpectomy model: fatigue characteristics. Clin Spine Surg 29(9):387–393
Qian L, Jiang C, Sun P, Xu D, Wang Y, Fu M et al (2018) A comparison of the biomechanical stability of pedicle-lengthening screws and traditional pedicle screws. Bone Joint J 100(B):516–521
Rohlmann A, Bergmann G, Graichen F (1997) Loads on an internal spinal fixation device during walking. J Biomech 30(1):41–47
Schmidt AL, Paskoff G, Shender BS, Bass CR (2012) Risk of lumbar spine injury from cyclic compressive loading. Spine (Phila Pa 1976) 37(26):E1614–E1621
Shimamoto N, Kotani Y, Shono Y, Kadoya K, Kaneda K, Minami A (2001) Biomechanical evaluation of anterior spinal instrumentation systems for scoliosis. Spine (Phila Pa 1976) 26(24):2701–2708
Weiser L, Huber G, Sellenschloh K, Viezens L, Puschel K, Morlock MM et al (2017a) Insufficient stability of pedicle screws in osteoporotic vertebrae: biomechanical correlation of bone mineral density and pedicle screw fixation strength. Eur Spine J 26(11):2891–2897
Weiser L, Huber G, Sellenschloh K, Viezens L, Püschel K, Morlock MM, Lehmann W (2017b) Insufficient stability of pedicle screws in osteoporotic vertebrae: biomechanical correlation of bone mineral density and pedicle screw fixation strength. Eur Spine J 26(11):2891–2897. https://doi.org/10.1007/s00586-017-5091-x

Wilke HJ, Mehnert U, Claes LE, Bierschneider MM, Jaksche H, Boszcyk BM (2006) Biomechanical evaluation of vertebroplasty and kyphoplasty with polymethyl methacrylate or calcium phosphate cement under cyclical loading. Spine (Phila Pa 1976) 31(25):2934–2941

Wilke HJ, Kaiser D, Volkheimer D, Hackenbroch C, Puschel K, Rauschmann M (2016) A pedicle screw system and a lamina hook system provide similar primary and long-term stability: a biomechanical in vitro study with quasi-static and dynamic loading conditions. Eur Spine J 25(9):2919–2928

Wittenberg RH, Shea M, Edwards WT, Swartz DE, White AA, Hayes WC (1992) A biomechanical study of the fatigue characteristics of thoracolumbar fixation implants in a calf spine model. Spine (Phila Pa 1976) 17(6):S121–S1S8

Xie S, Manda K, Pankaj P (2019) Effect of loading frequency on deformations at the bone-implant interface. Proc Inst Mech Eng H 233(12):1219–1225

FDA Premarket Review of Orthopedic Spinal Devices

18

Katherine Kavlock, Srinidhi Nagaraja, and Jonathan Peck

Contents

Organizational Structure of the FDA and CDRH	402
Device Classification	402
Premarket Submission Types	403
Premarket Notification (510(k))	403
Premarket Approval (PMA)	404
Humanitarian Device Exemption (HDE)	405
Evaluation of Automatic Class III Designation (De Novo)	405
Investigational Device Exemption (IDE)	406
Premarket Submission Device Evaluations	407
Mechanical Bench Testing	407
Cadaver Testing	410
Computational Modeling (Finite Element Analysis)	414
Animal Testing	416
Clinical Trials	417
Conclusions	420
References	420

Abstract

Spinal implants are regulated by the Food and Drug Administration (FDA) in the Center for Devices and Radiological Health (CDRH). This chapter focuses on the premarket activities at CDRH that help determine the safety and effectiveness of orthopedic spinal devices prior to reaching the market. The specific topics discussed in this chapter include:

- FDA organizational structure and medical device classification
- The main FDA premarket submission types
- The types of evaluations used to assess the performance of spinal devices prior to reaching the market including mechanical testing, cadaver testing, computational modeling, animal testing, and clinical trials

K. Kavlock (✉) · S. Nagaraja (✉) · J. Peck (✉)
Center for Devices and Radiological Health, Food and Drug Administration, Silver Spring, MD, USA
e-mail: katherine.kavlock@fda.hhs.gov; srin78@gmail.com; jonathan.peck@fda.hhs.gov

© This is a U.S. Government work and not under copyright protection in the U.S.; foreign copyright protection may apply 2021
B. C. Cheng (ed.), *Handbook of Spine Technology*, https://doi.org/10.1007/978-3-319-44424-6_97

Keywords

Spinal implant · Premarket evaluation · Medical device regulation · FDA classification · Premarket notification – 510 (k) · Premarket Approval – PMA · Biomechanical testing · Computational modeling · Animal testing · Clinical trials

Organizational Structure of the FDA and CDRH

FDA-regulated products account for approximately 20 cents of every dollar of annual spending by US consumers (Fact Sheet: FDA at a Glance). The FDA traces its founding back to the establishment of a consumer protection agency within the US Department of Agriculture in 1906 with the passage of the Pure Food and Drug Act, which prohibited the manufacture and sale of adulterated food products and poisonous drugs. The Federal Food, Drug, and Cosmetic Act (FD&C Act), enacted in 1938, replaced the 1906 law and further tightened controls over drugs and food and extended control to include medical devices and cosmetics. Importantly, the Medical Device Amendments of 1976 amended the FD&C Act to give the FDA authority to evaluate medical devices prior to the devices being marketed. The FD&C Act has been amended several additional times and today authorizes the FDA to regulate medical devices, human and animal drugs, foods and dietary supplements, and cosmetics. These products are regulated in CDRH, the Center for Drug Evaluation and Research (CDER), the Center for Veterinary Medicine (CVM), and the Center for Food Safety and Applied Nutrition (CFSAN). The Public Health Service Act (PHS Act), first passed in 1944, extends FDA's authority to include regulation of biological products. The Center for Biological Evaluation and Research (CBER) regulates vaccines, blood products including devices such as blood separators, human tissues for transplantation, and cellular and gene therapies. Lastly, the Family Smoking Prevention and Tobacco Control Act, passed in 2009, established the Center for Tobacco Products (CTP) which regulates the manufacturing, distribution, and marketing of tobacco products. The most current set of rules administered by the FDA can be found in Title 21 of the Code of Federal Regulations (CFR).[1]

The mission of CDRH is to protect and promote public health and to ensure that patients and providers have timely and continued access to safe, effective, and high-quality medical devices and safe radiation-emitting products. To accomplish this mission, CDRH plays several roles in the total product life cycle of regulated medical devices and radiation-emitting products. These roles include providing evaluation of the safety and effectiveness of various devices and diagnostic tests prior to their market release, monitoring the safety and effectiveness of devices after they have reached the market, ensuring compliance with medical device laws and regulations, when necessary, taking action against firms that violate these laws. Additionally, CDRH performs research studies that aid in developing appropriate evaluation strategies, testing standards, and communications for healthcare professionals and patients.

Device Classification

Medical devices are defined in Section 201(h) of the FD&C Act. A device is:

"an instrument, apparatus, implement, machine, contrivance, implant, in vitro reagent, or other similar or related article, including a component part, or accessory which is:

1. recognized in the National Formulary, or the United States Pharmacopoeia, or any supplement to them,
2. intended for use in the diagnosis of disease or other conditions, or in the cure, mitigation, treatment, or prevention of disease, in man or other animals, or,

---

[1]The electronic code of federal regulations can be accessed here: www.eCFR.gov

3. intended to affect the structure or any function of the human body of man or other animals, and

which does not achieve its primary intended purposes through chemical action within or on the body of man or other animals and which is not dependent upon being metabolized for the achievement of its primary intended purposes."

As noted above, the Medical Device Amendments in 1976 amended the FD&C Act to give the FDA authority to impose premarket approval requirements for medical devices and called for devices to be divided into classes with varying amounts of control required for each class in order to provide reasonable assurance of safety and effectiveness. CDRH categorizes medical devices into one of three classes based on their level of risk. As device class increases from Class I to Class III, the regulatory controls also increase with Class I devices subject to the least regulatory control and Class III devices subject to the most stringent regulatory control as defined by Part 513 (a)(1) of the Federal Food, Drug, and Cosmetic Act:

- **Class I (low to moderate risk)**: Devices are subject to a comprehensive set of regulatory authorities called general controls[2] that are applicable to all classes of devices.
- **Class II (moderate to high risk)**: Devices for which general controls, by themselves, are insufficient to provide reasonable assurance of the safety and effectiveness of the device and for which there is sufficient information to establish special controls[3] to provide such assurance.
- **Class III (high risk)**: Devices for which general controls, by themselves, are insufficient and for which there is insufficient information to establish special controls to provide reasonable assurance of the safety and effectiveness of the device.[4]

Table 1 contains examples of orthopedic spinal devices in each of the three classes.

Premarket Submission Types

The regulatory class to which a device is assigned guides the type of premarket submission required to obtain FDA marketing authorization. Additionally, CDRH grants approval of and oversees clinical trials of significant risk, investigational devices conducted in the United States, through submissions called investigational device exemptions (IDE), which allow clinical evaluation of devices which have not yet been cleared for marketing in the United States.

Premarket Notification (510(k))

The premarket notification or 510(k) process, named after Section 510(k) of the FD&C Act, is generally the regulatory process by which CDRH evaluates Class II medical devices (FDA 2014a). As previously mentioned, most Class I and a few Class II devices are exempt from 510(k) requirements subject to the limitations on exemptions. If a device falls into a generic category of exempted Class I devices as defined in 21 CFR Parts 862–892, FDA "clearance" before marketing the device in the United States is not required. However, generally before marketing a Class II device, a submitter must receive a letter from the FDA which clears the device for marketing by stating that the FDA finds the device to be substantially equivalent to a similar

---

[2]All medical devices are subjected to general controls which include, for example, registration and listing, medical device reporting, and good manufacturing practices

[3]Special controls can include activities such as special labeling requirements, demonstration that the device components are biocompatible, or non-clinical performance testing such as mechanical testing or electromagnetic compatibility

[4]Certain types of devices classified into Class III that were in commercial distribution in the United States prior to May 28, 1976 (i.e., preamendment devices), may be cleared through the 510(k) process until the FDA issues an order requiring them to go through the premarket approval process or reclassifying them into Class I or Class II

Table 1 Classifications of orthopedic spinal devices

Classification	Examples
Class I: Generally exempt from premarket review	Manual surgical instruments used in orthopedic spinal device procedures such as retractors, scalpels, rongeurs External orthotic braces
Class II: Devices generally requiring 510(k) submission and intended for stabilization until fusion occurs	Pedicle screw systems Intervertebral body fusion devices Vertebral body replacements Spinous process plates Anterior/lateral plating systems Surgical instruments specific to Class II implants such as device inserters or trials
Class III: Devices generally requiring PMA submission and intended for non-fusion use or are drug/device or biologic/device combination products	Total disc replacements Interspinous process spacers Intervertebral body fusion devices used with any therapeutic biologic (e.g., BMP2) Device-specific surgical instruments provided with Class III device

legally marketed device with the same intended use and similar technological characteristics (also referred to as a predicate device).

Within a 510(k) notification, submitters must compare their device to one or more predicate devices to demonstrate that the new device is as safe and as effective as the predicate device. Substantial equivalence of a new device can be claimed to a device that has been previously cleared through the 510(k) process, a device marketed prior to May 28, 1976 (preamendments device), a device which has been reclassified to Class II from Class I or III, or a device which has been granted marketing authorization via the De Novo classification process (discussed below). Examples of Class II orthopedic spinal implants that are subject to the 510(k) process include pedicle screw systems and intervertebral body fusion devices. Figure 1 shows the distribution of various orthopedic spinal device types submitted to the FDA through the 510(k) process between 2008 and 2017.

Modifications to a 510(k) cleared device may or may not require a new 510(k), depending on the significance of the changes (FDA 2017). Examples of modifications that may require a 510(k) include but are not limited to a change to the device geometry or material type, a change in indications for use, changes to the environment of use such as the use in a magnetic resonance (MR) environment, or changes in sterilization or cleaning. A new 510(k) must be submitted if it is determined that a modification to a 510(k) cleared device is significant enough not to be covered under the existing 510(k). It is possible that submission of a Special 510(k), a submission which includes only summary information resulting from the design controls process, may be appropriate if the modification does not affect the intended use of the device or alter the fundamental scientific technology of the device.

Premarket Approval (PMA)

A Premarket Approval (PMA) as described in 21 CFR 814 is generally the process by which the FDA evaluates the safety and effectiveness of Class III medical devices. Contrary to the 510(k) premarket notification process, in which a submitter can leverage existing information on predicate devices including applicable clinical data, a PMA application must provide sufficient valid scientific evidence to independently demonstrate a reasonable assurance of safety and effectiveness of the device. PMAs typically involve data from clinical trials of the specific device that support both safety and effectiveness, as well as detailed manufacturing information for the device. Many orthopedic spinal devices intended for non-fusion applications, such as total disc replacements and interspinous process spacers, are reviewed and approved through the PMA process. Table 2 lists the orthopedic spinal device original PMAs approved by CDRH between 2002 and 2017.

Unlike a 510(k), a PMA holder must report all design, manufacturing, and labeling changes

Fig. 1 Percentage of orthopedic spinal device 510(k)s received by CDRH from 2008 to 2017 (2452 total submission). (Information on orthopedic spinal devices cleared through the 510(k) premarket notification process can be found in a searchable database on the FDA website: https://www.accessdata.fda.gov/scripts/cdrh/cfdocs/cfpmn/pmn.cfm)

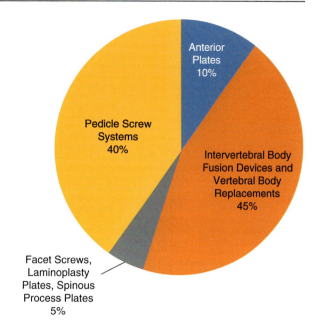

made to the approved device to the FDA via PMA supplements and PMA annual reports once a PMA is approved (FDA 2011; FDA 2014b).

Humanitarian Device Exemption (HDE)

The Humanitarian Device Exemption (HDE), as described in 21 CFR 814 Subpart H, is a marketing application for a humanitarian use device (HUD) and intended to benefit patients in the treatment of diseases or conditions that affect small (rare) populations. A HUD designation is reserved for those devices used in the treatment or diagnosis of a disease or condition that affects or is manifested in no more than 8000 individuals in the United States per year and for which there is no other comparable device that is legally marketed for the same intended use, other than another approved HUD. An HDE is exempt from the effectiveness requirements of Sections 514 and 515 of the FD&C Act and instead requires demonstration of safety and *probable benefit*. Additionally, HDE applicants are barred from selling their devices for a profit unless the device is intended and labeled for the treatment or diagnosis of a disease or condition that either (1) occurs in pediatric patients or in a pediatric subpopulation, and such device is labeled for use in pediatric patients or in a pediatric subpopulation in which the disease or condition occurs, or (2) is intended for the treatment or diagnosis of a disease or condition that does not occur in pediatric patients or that occurs in pediatric patients in such numbers that the development of the device for such patients is impossible, highly impracticable, or unsafe (FDA). Examples of orthopedic devices that received HDE approval include The Tether - Vertebral Body Tethering System (HDE number H190005) and the Minimally Invasive Deformity Correction (MID-C) System (HDE number H170001) which are non-fusion spinal devices intended to treat idiopathic scoliosis.[5]

Evaluation of Automatic Class III Designation (De Novo)

As previously described, all novel devices that have not been previously classified by the FDA are "automatically" determined to be Class III

---

[5]Information on orthopedic HDE approvals can be found in the searchable HDE database on the FDA website: https://www.accessdata.fda.gov/scripts/cdrh/cfdocs/cfHDE/hde.cfm

Table 2 Original Premarket Approval (PMA) applications approved for orthopedic spinal devices (2002–2017)

Device type	PMA number	Applicant	Device
Interspinous process spacer	P040001	Medtronic Sofamor Danek	X-STOP Interspinous Process Decompression System
	P110008	Paradigm Spine	Coflex® Interlaminar Technology
	P140004	Vertiflex	Superion Interspinous Spacer
Cervical total disc replacement	P060018	Medtronic Sofamor Danek	PRESTIGE Cervical Disc System
	P060023	Medtronic Sofamor Danek	BRYAN Cervical Disc Prosthesis
	P070001	Synthes Spine	ProDisc-C Total Disc Replacement
	P090029	Medtronic Sofamor Danek	PRESTIGE-LP Cervical Disc
	P100003	Globus Medical	SECURE-C Cervical Artificial Disc
	P100012	NuVasive	NuVasive PCM Cervical Disc System
	P110002	LDR Spine	Mobi-C Cervical Disc Prosthesis (one-level indication)
	P110009	LDR Spine	Mobi-C Cervical Disc Prosthesis (two-level indication)
Lumbar total disc replacement	P120024	Aesculap Implant Systems	ActivL Artificial Disc
	P040006	DePuy Spine	Charite Artificial Disc
	P050010	Synthes Spine	ProDisc-L Total Disc Replacement device

Additional information on orthopedic spinal device PMA approvals can be found in the searchable PMA database on the FDA website: https://www.accessdata.fda.gov/scripts/cdrh/cfdocs/cfPMA/pma.cfm

regardless of risks associated with the device. However, there may be low to moderate risk devices for which there is no legally marketed predicate device, but for which special or general controls are sufficient to provide a reasonable assurance of safety and effectiveness. For these devices, an applicant may submit a De Novo request for the FDA to evaluate the automatic Class III designation and consider whether a lower level of regulatory controls may be appropriate. Devices that are classified through the De Novo process into Class II may be used as predicate devices for future 510(k) submissions. An example of a granted orthopedic device De Novo request is an intraoperative orthopedic strain sensor (De Novo number DEN180012).

Investigational Device Exemption (IDE)

An investigational device exemption (IDE), as defined in 21 CFR 812, is an exemption that allows for an unapproved device to be shipped lawfully without complying with other requirements of the FD&C Act that would apply to marketed devices (e.g., registration and listing or quality system requirements except for the requirements for design controls). More simply, an IDE allows a significant risk device that has not received marketing authorization for a particular intended use to be investigated in a clinical study to collect safety and effectiveness information that can potentially be used to support a future marketing application. All clinical studies for evaluation of a new device or use of an existing device for a new intended use should include:

(1) An investigational plan approved by an institutional review board (IRB). If the study involves a significant risk device, the IDE must also be approved by the FDA.
(2) Informed consent from all patients.
(3) Labeling stating that the device is for investigational use only.
(4) Monitoring of the study.
(5) Required records and reports (FDA).

FDA's *Guidance Document for the Preparation of IDEs for Spinal Systems* provides information on the design of clinical trials for orthopedic spinal devices (FDA 2000). Note that the FDA often requires non-clinical evaluations (discussed in the remainder of this chapter) to be completed at the time of IDE submission as they may offer preliminary assurance of safety prior to implantation in human subjects.

Premarket Submission Device Evaluations

The success of orthopedic spine surgery depends on many factors including surgeon technique, patient characteristics, and performance of the orthopedic spinal implant. This chapter is primarily focused on the factors related to performance of the orthopedic spinal implant. A well-designed orthopedic spinal implant should have the following characteristics:

- Mechanical strength: The spinal device should be strong enough to endure the forces and moments it will be subjected to over its implanted lifetime.
- Device function: The spinal device should enable or restrict motion as intended. For example, fusion devices are typically designed to restrict motion, while non-fusion devices are often designed to allow motion similar to the healthy spine.
- Tissue-implant integrity: Many spinal devices depend on a strong interface between the spinal implant and bone both initially after surgery and over time through the development of a fusion mass or through bone ingrowth into the device.
- Safe implantation: The spinal device should have a surgical technique that allows safe access to the spine and does not result in unintended damage to bone or other anatomical structures.

In addition to the device characteristics listed above, all premarket submissions for orthopedic spinal devices should include information on the ability to sterilize the device, information demonstrating that the device is biocompatible (i.e., that the final, finished form of the device does not cause an inappropriate local or systemic reaction) and that the proposed labeling provides adequate instructions for use. While inclusion of biocompatibility assessment, sterility/reprocessing information, and draft labeling are generally necessary in all premarket submissions for orthopedic spinal devices, the information required for these elements will not be discussed in this chapter as the requirements are not unique to orthopedic spinal implants.

There are scientific methods available to assess each of the characteristics listed above ranging from simple mechanical bench tests to complex, randomized clinical trials. In this section, we will discuss common spinal device evaluation methods such as mechanical bench testing, cadaver testing, computational modeling, animal testing, and clinical trials. We will discuss the device characteristics each of these methods assess, including advantages and limitations of each method. We will also discuss how the FDA uses this information in the regulatory review process.

Mechanical Bench Testing

Mechanical bench testing involves the attachment of an orthopedic spinal device to a test machine and application of static or cyclic forces, moments, and/or linear or angular displacements. This type of testing is used to evaluate the susceptibility of spinal implants to experience failure modes such as yield, fracture, bone-implant interface failures, and excessive wear. Additionally, mechanical bench testing is used to characterize device function, such as stiffness or mobility, to help ensure that the device will perform as intended once implanted. For example, pedicle screw systems are subjected to static and cyclic mechanical testing to ensure adequate strength to resist the loads and moments they will be subjected to throughout the expected use life of the device. In addition, mechanical testing is utilized to ensure a pedicle screw system will provide adequate stiffness to stabilize the spine in order to promote development of a fusion. The FDA provides recommendations for the mechanical testing of various spinal implants in guidance documents specific to implant types (FDA 2004, 2007, 2008).

Mechanical testing is one of the most common evaluation methods used to assess the performance of spinal devices, as the monetary and time costs are relatively low. With the exception of finite element analysis (FEA), mechanical testing represents one of the lowest cost methods of device evaluation compared to most other types of evaluation including cadaver testing, functional animal models, or clinical trials. Mechanical testing is particularly useful for the comparison of devices, which is the core concept of the 510(k) process. For example, a company seeking clearance of a new intervertebral body fusion device through the 510(k) process may perform axial compression, compression-shear, and torsion testing on the new device to evaluate how the device performs as compared to a legally marketed predicate device. This mechanical comparison helps to establish that the new device will have adequate strength and stiffness to perform its intended use. In essence, the comparison of mechanical testing information in a 510(k) premarket notification allows the FDA to apply the long history of clinical use of one device to another substantially equivalent device, thereby eliminating the need for clinical trials in most cases to show that the new or modified device is as safe and effective as the predicate device.

Standardized test methods are critical to performing the mechanical testing comparisons discussed above. Standards development organizations (SDOs), specifically ASTM International and the International Standards Organization (ISO), develop mechanical testing standards for orthopedic spinal implants by utilizing the expertise of spinal device manufacturers, third-party testing laboratories, physicians, and regulatory agencies from around the world. These standards are written through a consensus process in order to allow for the mechanical characterization and comparison of orthopedic spinal devices across different companies throughout the world. Without standard test methods, there would be substantial variability between device testing performed across different laboratories, and the device evaluation and regulatory process would be far more burdensome for both device manufacturers and regulatory agencies. A well-developed standard test method offers the user confidence that the test results will be repeatable assuming consistent manufacturing of the device and reproducible across different test laboratories. In addition, the user can feel confident that differences in test results between devices are due to differences in device performance and not variability in the test method. In order to ensure test methods are repeatable within a given test laboratory and reproducible between different test laboratories, ASTM International sponsors precision and bias studies of their test standards. These studies typically involve sending identical test specimens to several test laboratories to be tested per a given standard. The results are then compiled and compared, and the variability within individual laboratories and between laboratories is assessed. Subsequently, additional studies may be conducted, and revisions to the standard may be made in order to reduce variability within and between test laboratories. Precision and bias studies have been performed on orthopedic spinal device standards such as *ASTM F1717 – Standard Test Methods for Spinal Implant Constructs in a Vertebrectomy Model*,[6] and the results are published within the standard.

Pedicle screw systems and intervertebral body fusion devices used in the previous examples represent devices used to stabilize the spine while a fusion develops. As previously stated, these devices are evaluated through the 510(k) process by comparison to predicate devices. However, standardized mechanical testing is also useful for non-fusion devices reviewed through the PMA process. In these cases, mechanical testing can reduce the burden associated with conducting clinical studies by providing an assessment of long-term device performance in a relatively short amount of time. While clinical studies offer

---

[6]ISO and ASTM standards mentioned in this chapter are published on an annual basis and are available via the organizations' websites: www.astm.org and www.iso.org

the most realistic environment for device assessment, clinical studies have a finite duration and do not cover the entire intended life of the spinal implant. Accelerated mechanical testing can be performed to simulate long-term use of the implant. For example, ISO 18192-1 – *Implants for Surgery – Wear of Total Intervertebral Spinal Disc Prostheses – Part 1: Loading and Displacement Parameters for Wear Testing and Corresponding Environment Conditions for Test* contains methods for performing wear testing on total disc replacements. The standard test method suggests performing ten million cycles of combined flexion-extension, lateral bending, and axial rotation and states that this approximates 10 years of in vivo function. Therefore, by performing this wear testing to supplement a clinical trial, additional confidence can be drawn that the device will last substantially longer than the length of the clinical trial (e.g., 5 years or less).

One common limitation of mechanical testing is that the test may not accurately replicate the complex and highly variable loading scenarios in the human body. For this reason, the utility of many mechanical tests is not in the precise simulation of a physiologic condition, but rather in its ability to characterize specific aspects of device performance particularly in comparison to similar devices. This is important to consider when determining the acceptance criteria for a mechanical test. The two most common sources for acceptance criteria for mechanical tests are (1) test results from a currently marketed device and (2) expected physiologic loads (Graham and Estes 2009). Due to the high variability in physiologic loading scenarios and the fact that the mechanical test may not be accurately replicating physiologic loading, it is often recommended that test results be compared to another similarly designed device tested in the same manner (ASTM Standard F2077 2014, "Test Methods for Intervertebral Body Fusion Devices" 2014; Graham and Estes 2009; Peck et al. 2017, 2018). Attempting to create test methods that better simulate the physiologic conditions may lead to more accurate loading on a spinal device during testing but also may lead to test results that are more difficult to interpret and compare across devices. As an example, *ASTM F1717 – Standard Test Methods for Spinal Implant Constructs in a Vertebrectomy Model* and *ISO 12189 – Implants for Surgery – Mechanical Testing of Implantable Spinal Devices – Fatigue Test Method for Spinal Implant Assemblies Using an Anterior Support* are mechanical testing standards for the testing of pedicle screw systems (and other devices that primarily reside outside the anterior spinal column). ASTM F1717 involves testing the device in a scenario without anterior column support that would almost always be present in vivo from either the intact anterior spinal column or other spinal implants such as vertebral body replacement devices. The benefit of testing in this manner is that all of the load is transferred through the device and the results are easier to interpret and compare to other devices. A limitation of testing in this manner is that the device loading may not be entirely indicative of clinical use. ISO 12189 attempts to address this by adding anterior support to the testing model in the form of a spring intended to replicate the stiffness of the anterior spinal column structures. This method may result in loading on the device that is more physiologically realistic, as load is now shared between the device being tested and simulated anterior structures. However, load sharing with the simulated anterior column support makes it difficult to determine how much load is actually being applied to the device being tested. There has been some debate in the literature regarding which method is more appropriate (Graham et al. 2014; Villa et al. 2014). In most cases, the FDA prefers the use of ASTM F1717 as the loading on the device during testing can be easily determined and comparisons can be made between devices. There are cases where alternate test methods to ASTM F1717 are necessary, such as during the testing of low stiffness devices that cannot resist the bending loads applied under ASTM F1717. However, in these cases the use of ISO 12189 is still limited by the inability to easily determine the load that is borne by the device itself during testing.

In summary, mechanical testing is one of the most common performance evaluation methods for spinal implants and is present in some form in nearly every premarket application for a spinal implant submitted to the FDA. Mechanical testing is relatively low cost compared to some of the other evaluation methods and is often repeatable and reproducible making it potentially ideal for comparisons of devices to one another. However, mechanical tests often do not fully replicate complex physiologic loading environments; therefore, testing should not be considered to represent a true simulation of in vivo conditions. Despite this limitation, the advantages of mechanical testing and the need to assess device strength, stiffness, and wear resistance prior to implantation in humans will ensure that mechanical testing remains a cornerstone of the performance evaluations of spinal devices. Table 3 lists the current mechanical testing standards for spinal devices that have been developed by the ASTM International and ISO (International Standards Organization) standards development organizations.

Cadaver Testing

Cadaver biomechanical testing is useful for assessing the performance of spinal devices, particularly when other non-clinical models are not suitable to address a specific question. For example, bone-implant interface assessments often use test methods that utilize uniform density polyurethane foam as a surrogate for bone (e.g., subsidence testing per ASTM F2267 or screw pullout testing per ASTM F543). However, the polyurethane foam does not incorporate the regional variations in bone density that exist in human vertebrae. Therefore, cadaver testing may be more advantageous when assessing bone-implant issues since cadaveric spinal segments represent human spinal anatomy and disease/aging conditions. Since cadaver testing can incorporate bone and disc quality for different disease and aging states, cadavers are the most realistic non-clinical model for assessing adverse events related to the tissue-implant interface such as screw loosening, device subsidence, device migration, or bone fracture. Numerous publications have used cadaveric biomechanical testing to measure the risk of pullout of pedicle screws from vertebral bone (Bianco et al. 2017; Cunningham et al. 1993; Frankel et al. 2007; Lehman Jr. et al. 2003; Pishnamaz et al. 2017). For spinal cages and total disc replacement devices specifically, cadaver test methods have been employed to understand adverse events such as subsidence into the vertebral body and migration of devices beyond the intervertebral space (Briski et al. 2017; Cho et al. 2008; Labrom et al. 2005; Pitzen et al. 2000). Furthermore, cadaver testing incorporates load sharing with other tissues (e.g., bone, cartilage, tendons/ligaments) to provide a more accurate representation of spinal device function (e.g., stability, in situ behavior) compared to other non-clinical testing options. For example, range of motion testing is commonly used to assess stability of an implant in physiologically relevant conditions such as flexion-extension, lateral bending, and axial rotation of the spine. Several previous publications have used cadavers to compare range of motion of a spinal device to the intact condition or to another device (Beaubien et al. 2010; Cain et al. 2005; Kornblum et al. 2013; Kuzhupilly et al. 2002; O'Leary et al. 2005; Oxland and Lund 2000; Voronov et al. 2014). For devices with novel or complicated surgical techniques (e.g., minimally invasive placement of screws or intervertebral body fusion devices), ease of implantation can be assessed with cadavers (Dixon et al. 2017; Kim et al. 2004; Luo et al. 2017; Ma et al. 2012; Ryken et al. 1995).

Cadaver biomechanical testing also has several limitations that can influence results and should be carefully considered when developing a cadaver test method and interpreting the results. One important consideration is that cadaver tissue cannot mimic biological response to the implant. For example, bone growth during spinal fusion is not captured in cadavers. Therefore, long-term assessment of the interaction of an intervertebral body fusion device with the surrounding tissue will not be captured with cadaveric testing. In addition, soft tissue and bone degradation occurs in ambient conditions, which limits biomechanical testing duration. This is important as most spinal implants undergo cyclic loading during daily activities such as walking, sitting, or standing. Adverse events such as device loosening, migration, and subsidence may occur gradually over time due to these daily activities and may be better assessed

Table 3 Consensus standards for mechanical testing of spinal devices

Device type	Standard	Methods used to assess
Pedicle screw systems and anterior plating systems	ASTM F1717 – *Standard Test Methods for Spinal Implant Constructs in a Vertebrectomy Model*	Static and fatigue strength of pedicle screw/hook systems and anterior plate constructs consisting of hardware necessary to span two spinal levels (e.g., pedicle screws and rods)
	ASTM F1798 – *Standard Test Method for Evaluating the Static and Fatigue Properties of Interconnection Mechanisms and Subassemblies Used in Spinal Arthrodesis Implants*	Interconnection strength of pedicle screw system components (e.g., pedicle screws, hooks, cross-connectors, etc.) to the rod under various loading modes
	ASTM F2193 – *Standard Specifications and Test Methods for Components Used in the Surgical Fixation of the Spinal Skeletal System*	Strength of individual spinal implant components such as screws, rods, and plates using methods such as cantilever bending and 4-point bend tests
	ASTM F2706 – *Standard Test Method for Occipital-Cervical and Occipital-Cervical-Thoracic Spinal Implant Constructs in a Vertebrectomy Model*	Static and fatigue strength of occipital-cervical-thoracic system constructs. The methods are derived from those in ASTM F1717 but include test setups that allow the attachment of occipital components
	ISO 12189 – *Implants for Surgery – Mechanical Testing of Implantable Spinal Devices – Fatigue Test Method for Spinal Implant Assemblies Using an Anterior Support*	Static and fatigue strength of pedicle screw systems and plating systems in a construct that includes a simulation of the anterior anatomical structures
Intervertebral body fusion devices and vertebral body replacements	ASTM F2077 – *Test Methods for Intervertebral Body Fusion Devices*	Static and fatigue strength of intervertebral body fusion devices in axial compression, compression-shear, and torsion loading modes
	ASTM F2267 – *Standard Test Method for Measuring Load Induced Subsidence of Intervertebral Body Fusion Device Under Static Axial Compression*	Propensity of an intervertebral body fusion device to subside using polyurethane foam as a surrogate for the vertebral body.
Artificial total disc replacements	ISO 18192-1 – *Implants for Surgery – Wear of Total Intervertebral Spinal Disc Prostheses – Part 1: Loading and Displacement Parameters for Wear Testing and Corresponding Environment Conditions for Test*	Wear of cervical and lumbar artificial disc replacements under loads and motions the device is expected to experience in vivo. The ISO and ASTM methods were developed by the separate SDOs independently but share many of the same principles
	ASTM F2423 – *Standard Guide for Functional, Kinematic, and Wear Assessment of Total Disc Prostheses*	
	ISO 18192-3 – *Implants for Surgery – Wear of Total Intervertebral Spinal Disc Prostheses – Part 3: Impingement-Wear Testing and Corresponding Environmental Conditions for Test of Lumbar Prostheses Under Adverse Kinematic Conditions*	Impingement wear and damage of lumbar artificial disc replacements to determine how the device behaves when operating at the limits of its designed range of motion
	ASTM F2346 – *Standard Test Methods for Static and Dynamic Characterization of Spinal Artificial Discs*	Static and fatigue strength of cervical and lumbar artificial disc replacements in axial compression, compression-shear, and torsional loading modes
Motion-sparing pedicle screw-based systems	ASTM F2624 – *Standard Test Method for Static, Dynamic, and Wear Assessment of*	Static and fatigue strength and wear characteristics of devices that reside outside of the disc space implanted across a single

(continued)

Table 3 (continued)

Device type	Standard	Methods used to assess
	Extra-Discal Single Level Spinal Constructs	level. This standard was developed primarily for motion-sparing devices
Motion-sparing facet replacement systems	ASTM F2694 – *Standard Practice for Functional and Wear Evaluation of Motion-Preserving Lumbar Total Facet Prostheses*	Wear of motion-sparing lumbar facet replacement devices under loads and motions that the device is expected to experience in vivo
	ASTM F2790 – *Standard Practice for Static and Dynamic Characterization of Motion-Preserving Lumbar Total Facet Prostheses*	Static and fatigue strength of motion-sparing lumbar facet replacement devices
Disc nuclear replacements	ASTM F2789 – *Standard Guide for Mechanical and Functional Characterization of Nucleus Devices*	Static and fatigue strength of disc nucleus replacements as well as various characterizations of device behavior such as swelling and lifting force
	ISO 18192-2 – *Implants for Surgery – Wear of Total Intervertebral Spinal Disc Prostheses – Part 2: Nucleus Replacements*	Wear of nucleus replacement devices under loads and motions the device is expected to experience in vivo

through long-term fatigue testing. However, cadaver specimens may be unable to endure the number of testing cycles required to simulate these activities. Although fatigue testing in cadavers has been performed to assess longer-term performance of spinal implants, it is a more challenging test to perform due to tissue degradation over time. A previous publication found that cadaver spine range of motion increases linearly in axial rotation, lateral bending, and flexion-extension over 72 h when exposed to ambient testing conditions (Wilke et al. 1998b). Therefore, cadaver fatigue testing is typically performed to a maximum of 24 h (up to 180,000 cycles at 0.5–2 Hz) to minimize degradation effects, particularly range of motion changes due to loss of hydration of the intervertebral discs in cadaver spine segments (Cook et al. 2015; Ferrara et al. 2003; Freeman et al. 2016; Heth et al. 2001; Hitchon et al. 2000; Lu et al. 2000; Palepu et al. 2017; Pfeiffer et al. 1997; Trahan et al. 2014; Vadapalli et al. 2006a; Wang et al. 2005).

Another disadvantage of cadaveric biomechanical testing is heterogeneity in the cadavers used for testing. Variability is observed in donor demographics such as age, gender, and race. In particular, it is difficult to obtain donors that are less than approximately 60 years old with normal bone mineral density. In general, cadavers obtained for biomechanical testing are elderly individuals with low bone quality (i.e., osteopenic or osteoporotic based on T-scores). In addition, various conditions (e.g., osteophytes, Schmorl's nodes) may be present and can affect bone (cortex, endplate, trabecular bone) and/or soft tissue (intervertebral disc, ligaments, tendons) quality. Therefore, cadavers available for biomechanical testing may not be representative of the intended patient population for a specific device. The substantial heterogeneity in cadavers typically leads to high variability in biomechanical testing results. For example, the average coefficient of variation[7] (COV) for screw pullout strength in studies using polyurethane foam was 8.8% (Hsu et al. 2005; Nagaraja and Palepu 2016; Thompson et al. 1997), which is substantially lower than an average COV of 37.5% (Beaubien et al. 2010; Helgeson et al. 2013; Nagaraja et al. 2015) for cadaveric screw pullout studies (Table 4). Furthermore, range of motion testing in cadaveric spine segments (e.g., functional spine units) had an average COV of 42.8%, even greater than cadaveric screw pullout studies (Cain et al. 2005; O'Leary et al. 2005; Palepu et al. 2017).

---

[7]Coefficient of variation for a given test result is the standard deviation normalized to the mean. This parameter allows for comparisons of variability across tests

Table 4 Coefficient of variation in published spine biomechanics studies

Screw pullout testing – foam	6.1% (2.3–10.2%) (Nagaraja and Palepu 2016)	13.3% (7.8–19.6%) (Thompson et al. 1997)	6.9% (3.8–10.0%). (Hsu et al. 2005)
Screw pullout testing – cadaver	29.1% (26.8–31.3%) (Helgeson et al. 2013)	35.8% (28.4–43.8%) (Nagaraja et al. 2015)	47.6% (36.9–64.6%) (Beaubien et al. 2010)
Range of motion testing – cadaver	37.6% (17.2–64.7%) (O'Leary et al. 2005)	39.7% (22.4–73.3%) (Cain et al. 2005)	51.1% (24.4–71.4%) (Palepu et al. 2017)

The large standard deviations observed in cadaveric testing may mask a true difference in outcomes between devices due to the heterogeneous set of cadavers used for testing. High variability is an important consideration in the regulatory framework, particularly when comparing a device to another device in the 510 (k) process. A lack of statistical difference between groups should not be interpreted to imply equivalence between two devices, but rather the variability inherent in the cadaver testing may be large enough to prevent observing differences between groups. In addition, obtaining and testing a large number of specimens to achieve a desired level of statistical confidence and reliability is difficult, particularly when comparing between two devices. Therefore, it is important to reduce this variability through cadaver preparation and screening. Nondestructive characterization such as visual inspection or image analysis (e.g., DEXA, micro-CT, MRI) can provide valuable information regarding the anatomy and quality of the vertebral structures to aid in excluding samples that do not meet pre-specified criteria. For example, BMD or T-scores from DEXA scans can be used to exclude osteoporotic cadavers. For devices that interface with vertebral endplates and/or trabecular bone, three-dimensional micro-CT imaging can be used to exclude samples with endplate sclerosis or Schmorl's nodes. Visual assessments are also particularly useful for identifying macroscopic abnormalities such as scoliosis, osteophytes, or disc degeneration. Degeneration within the nucleus or inner annulus can be assessed with magnetic resonance imaging (MRI). For cadaver tests intended to compare the performance between two or more devices, it is important to make the cadaver groups as similar as possible. Demonstrating similarity in cadavers can be achieved by stratifying cadavers into groups based on key factors (e.g., BMD, anatomical features, age, trabecular bone volume fraction) that influence the performance metric(s) of interest. For example, bias may be reduced by selecting cadaveric vertebrae with similar endplate thicknesses and trabecular bone volume fractions when comparing subsidence resistance between a subject and predicate interbody device. In addition, variability may be reduced by normalizing results to a parameter that can act as an internal control. For example, ROM after device implantation can be normalized by the preimplantation ROM (i.e., intact condition) to more directly understand the stability obtained due to device implantation (Kornblum et al. 2013). Another method of reducing variability is testing devices on the same cadaveric sample, if possible. For example, previous studies have performed matched pair testing (e.g., pullout of screws on the same vertebrae) to reduce variability and increase the statistical power and thus more directly compare two devices (Helgeson et al. 2013; Nagaraja et al. 2015). Although these methods aid in minimizing variability between cadaver groups, the inherent variability within and between cadaver spines cannot be completely eliminated.

Another limitation with cadaver testing is the lack of standardized methods for cadaver spine biomechanical testing. Cadaver studies reported in literature have variable testing protocols that currently prevent comparisons of results between different testing facilities. In order to address interlaboratory variability in cadaver test protocols, previous publications have provided general recommendations for multi-directional cadaver testing such as loading magnitude, control mode (load/torque vs. displacement/angle), preloads (e.g., follower loads), use of a passive/active sliding table for pure moment loading, and reporting of spinal stability parameters (Crawford et al. 1998; Goel et al. 2006; Hanlon et al. 2014; Panjabi 1988; Patwardhan et al. 2003; Wilke et al. 1998a). However, fatigue testing introduces additional considerations such as testing frequency, physiologic loading modes and magnitudes, and duration that are inconsistent between testing facilities.

Overall, cadaveric testing of spinal implants is important not only in research/academic environments, but also can be a powerful tool during product development. In a regulatory setting, biomechanical testing can be used to assess device implantation, device function, and the bone-implant interface. However, heterogeneity between cadavers and variability in testing procedures are important limitations that currently prevent broad utilization of cadaver testing results in regulatory submissions. Careful consideration of these factors is critical when interpreting cadaveric testing in a regulatory framework. To increase the use of cadaveric testing in regulatory submissions, future work should focus on developing standardized cadaver selection and testing procedures where possible for spinal devices.

Computational Modeling (Finite Element Analysis)

There are several advantages to using computational modeling when evaluating spinal devices. Finite element analysis (FEA) can assess changes to a spinal device design quickly and can be accomplished relatively inexpensively compared to mechanical testing of various design iterations. In addition, FEA can assess local mechanics (e.g., stress concentrations), which cannot be determined from experimental mechanical testing. Although mechanical test methods provide a relative comparison between two devices, FEA can incorporate spinal structures to more accurately replicate the clinical scenario and understand interactions at the bone-implant interface. In fact, there have been many publications that have used FEA to understand the biomechanical environment surrounding spinal implants (Dreischarf et al. 2014; Goel et al. 2005; Newcomb et al. 2017; Polikeit et al. 2003; Trautwein et al. 2010). Another important advantage is that FEA can simulate complex physiologic loading (e.g., multi-axial loading) that may be difficult or even impossible with bench testing. However, an adequate understanding of expected in vivo loading conditions is necessary to simulate physiologic loads.

During spinal device development, FEA is used extensively to optimize device design in order to achieve desired performance and assess whether certain failure modes are mitigated. For regulatory submissions, FEA has also been used to assess device strength and function, particularly to determine the worst-case implant size for mechanical bench testing; thus, a least-burdensome approach is utilized so that mechanical testing is not performed on every size implant within a system. Additionally, when failures occur after approval or clearance, a root cause analysis is performed to assess these failures and subsequently determine a corrective action plan. FEA can be used in these situations where the knowledge gained from simulations is useful for reevaluating the design inputs and assessing potential design changes to ultimately improve device performance and safety. The FDA published a guidance document (*Reporting of Computational Modeling Studies in Medical Device Submission – Guidance for Industry and FDA Staff*) on reporting of FEA studies intended to support a regulatory submission (FDA 2016). Table 5 provides a summary of FDA recommendations in an FEA report.

Two important items recommended in this guidance document are verification and validation of the computational model to demonstrate that the simulation adequately represents the intended reality. A subcommittee within the American Society of Mechanical Engineers (ASME) is developing verification and validation (V&V) activities for computational modeling within

Table 5 Recommended format for FDA submissions of computational modeling studies

Section	Contents
Executive summary	Overview of the report including the use of modeling with respect to regulatory submission, scope and quantity(s) of interest (QOI) of the analysis, validation activities, results, and conclusions
Background/introduction	Description of the context of use for the modeling study, device description, and intended use environment
Code verification	Description of the software quality assurance (SQA) and numerical code verification (NCV) performed
System geometry	Description of the device and/or tissue geometry and the method used to create the computational representation of device/tissue
Constitutive laws	Constitutive relationships used to describe the behavior (e.g., linear elastic) of the device material (s) and surrounding anatomy (if applicable)
Material properties	Inputs necessary to fully characterize the behavior (e.g., elastic modulus) of the device or tissue
Boundary and initial conditions	Diagram/schematic of the location and direction of the imposed boundary conditions (e.g., forces, displacements) including constraints
Mesh	Description of the type and number of elements used for the mesh and a convergence analysis used to demonstrate that the QOIs are independent of element size
Solver	Description of the software used in the computational analysis
Validation	Validation study that supports the context of use by comparing QOI (s) between model and comparator (e.g., physical testing, in vivo study, literature)
Results	Presentation of the quantitative results (e.g., von Mises stress/strain, fatigue safety factor)
Limitations	Discussion of major limitations to the computational model
Discussion/conclusions	Overall conclusions of the modeling study and how the objective(s) have been met with respect to the results

different medical device product areas (e.g., orthopedics, stents). Establishment of an adequate level of V&V is necessary for FEA simulations that are intended for regulatory evaluation. This subcommittee has developed a framework called the "risk informed credibility assessment method" which can be used to help determine the appropriate level of V&V necessary to support using the computational model for regulatory decision-making within a specific intended use of the model (e.g., determine worst-case implant size). For example, if the risk is determined to be low based on this method, credibility of the FEA may be established by comparing the force-displacement behavior of experimental testing per a standard test method (e.g., ASTM F1717) to the simulation results of that same test. This validation activity provides objective evidence whether the FEA results are representative of the physical test method. However, validation activities will depend on the level of credibility needed based on the risk and intended use of the model. Although there have been recent efforts to include validation activities in spinal device publications using FEA, many of these investigations model a single spinal anatomical structure with one set of tissue properties and are typically validated with limited experimental data (Campbell et al. 2016; Grauer et al. 2006; Kallemeyn et al. 2010; Vadapalli et al. 2006b; Wagnac et al. 2012). A recent publication reported the results of a multi-center study to compare eight previously validated FE models of the lumbar spine (Dreischarf et al. 2014). This study found that interlaboratory variability in mechanical parameters was low and fell within published in vitro ranges. These findings demonstrate that validated FE models can accurately estimate the response of the lumbar spine. Another challenge of FEA is accurately modeling interconnections between different components in a spinal construct. For example, polyaxial pedicle screws pose challenges to modeling the polyaxial mechanism (e.g., degrees of freedom) and the interaction (e.g., frictional contact) between the screw tulip, rod, and set screw components of this construct. Assumptions and simplifications to the computational model (e.g., geometry, material model, boundary conditions) are typically used but may limit the applicability of the results and conclusions.

Overall, computational modeling, such as FEA, is a powerful tool that is used throughout the total product life cycle of medical devices. Spinal device manufacturers have used computational modeling in implant design, some pre-market submissions, and making iterative changes to device design based on post-market monitoring. However, additional computational modeling V&V efforts are needed in the spinal device area to serve as a source of valid scientific evidence and ultimately play a greater role in regulatory decision-making. Further collaboration within the spinal device medical community is needed to establish a framework for model credibility and develop best practices for computational modeling of spinal devices.

Animal Testing

Animal testing is routinely used to address questions related to biologic response. Many of the standard biocompatibility evaluations referenced in the FDA Guidance and the associated ISO 10993 series of standards utilize animal models to evaluate potential hazards. When considering new materials for use in spinal implants, the biological response to particulate for novel materials is compared to the response generated by common orthopedic device materials using a similar animal model. With specific relevance to spinal devices, Cunningham et al. developed a rabbit model to examine the biological responses to epidural application of particles released from bearing surfaces or resulting from implant interconnection loosening (Cunningham et al. 2003). To date, more than 10 types of biomaterial particles, representing the mostly commonly utilized orthopedic implant materials including stainless steel, titanium alloy, cobalt chromium alloy, ultra-high-molecular-weight polyethylene (UHMWPE), and polyether ether ketone (PEEK), have been evaluated using this same rabbit model (Cunningham et al. 2013; Rivard et al. 2002). Based on the results of the aforementioned studies, it is known that particles of a phagocytosable size (between 1 and 10 µm) may be expected to be the most reactive particle types. Particles of smaller size (< 0.5 µm) were less inflammatory and resulted in less cellular injury as they were able to be taken up by the cells with the extracellular fluid through pinocytosis. Therefore, it is important to consider not only the device materials, but also the size and morphology of the particles expected to be generated by a device when evaluating the biologic response to wear debris, which may be determined based on the results of device fatigue or wear testing.

While animal testing may be used to address questions related to biocompatibility of the device materials or particulate that may be generated by the device as discussed above, animal testing is also used to address questions related to the functionality of a device. It should be noted that some animal tests may include assessments of both device biocompatibility and function. Functional animal testing can be used to assess development of an intervertebral fusion, bone integration into a porous coating, wear and durability of a device, or propensity of a device to migrate or expulse. An ideal animal model is one that adequately mimics the anatomical structures, physiologic loads, and relevant biological processes (e.g., bone remodeling) of a human. Common examples for large animal models used to evaluate spinal devices are sheep, goat, pig, and baboon (Abbah et al. 2009; Anderson et al. 2004; Cunningham et al. 2004; Di Martino et al. 2005; Drespe et al. 2005; Kotani et al. 2002). Drespe et al. summarized the various animal models used for spinal fusion studies stating that goats are suitable for cervical fusion studies and sheep for lumbar studies because their vertebrae are similar in size to the cervical and lumbar vertebrae in humans, respectively. In addition, Drespe et al. stated that primates have the greatest genetic similarities to humans and also approximate the human upright posture (Drespe et al. 2005).

Functional animal testing offers the advantage of being able to study a spinal implant in vivo without risks to humans. For this reason, animal testing results are often submitted to the FDA to demonstrate initial safety and/or effectiveness of a novel spinal implant prior to initiation of a clinical trial. Additionally, because animals can be sacrificed, extensive local and systemic evaluations can be performed at various time points after implantation. For example, if intervertebral fusion is being assessed by the animal model, the spine of the animal can be removed and mechanically tested at various time points after device implantation with verification of bony fusion through histology at different time points.

Overall, selection of the most appropriate animal model for a given situation is important for scientific and ethical reasons. In fact, the FDA recommends that spinal device manufacturers discuss animal study protocols with the FDA prior to initiation of the study to ensure that the most appropriate animal model is chosen and that the protocol developed is suitable to address the relevant scientific question. Alternatively, it may be determined that no animal study should be performed depending on the scientific questions that need to be addressed. No animal test is fully representative of a human, and models that have characteristics most representative of humans (e.g., the baboon) tend to be very expensive and have the most serious ethical considerations. It is therefore important to balance many factors when determining whether an animal study can help provide safety and effectiveness evidence that would be translatable to humans, and if so which animal model is most appropriate.

Clinical Trials

In general, spinal device clinical trials involve the implantation of an investigational device into humans to assess the safety and/or effectiveness of the device in treating a particular spinal pathology. Clinical data are typically provided in original PMAs for Class III devices as these devices are considered to have higher risk and/or incorporate new intended uses or new technology. De Novo submissions often include clinical data to help demonstrate that a novel device or intended use that is automatically designated as Class III is actually low to moderate risk and that special and general controls are adequate to assure safety and effectiveness. Clinical data are also sometimes necessary in 510(k) submissions for Class II devices to help demonstrate substantial equivalence when other assessment methods (e.g., mechanical bench testing, cadaver testing, animal testing) are unable to adequately answer the question at hand. Unlike the other evaluations discussed in this chapter which rely on assumptions/simulations of the clinical setting, the advantage of clinical data is that experience is gained about an orthopedic spinal implant in the environment in which the device is used. However, collection of clinical data for orthopedic spinal devices is often very expensive and time-consuming.

Clinical data concerning spinal implants may be derived from a variety of sources and study designs ranging from case reports to randomized clinical trials. However, in order for the clinical data to be utilized in a regulatory submission to make a determination of device safety and effectiveness, the data must constitute valid scientific evidence. Valid scientific evidence is defined in 21 CFR 860.7(c)(2) as "evidence from well-controlled investigations, partially controlled studies, studies and objective trials without matched controls, well-documented case histories conducted by qualified experts, and reports of significant human experience with a marketed device, from which it can fairly and responsibly be concluded by qualified experts that there is reasonable assurance of the safety and effectiveness of the device under its conditions of use." This section of the CFR also states, "Isolated case reports, random experience, reports lacking sufficient details to permit scientific evaluation, and unsubstantiated opinions are not regarded as valid scientific evidence...." The amount and characteristics of the clinical data which are necessary depend on the questions that need to be answered. For example, if the safety and effectiveness profiles of a new technology are not well understood, a multicenter, prospective, randomized clinical trial may be necessary. Alternatively, there are instances where more is known about the performance of a particular technology and only limited clinical data specifically tailored to answer a narrower set of questions are required.

Study Evaluations

During a clinical trial for an orthopedic spinal device, study subjects are evaluated prior to the surgical procedure, intraoperatively, and at pre-defined time points postoperatively. Study assessments are primarily focused on the safety and effectiveness of the investigational device and surgical implantation procedure and include evaluations from the patients' perspective through the use of patient-reported outcome measures. Common assessments for orthopedic spinal device studies include:

- Pain and function evaluations, such as the Oswestry Disability Index (Fairbank and Pynsent 2000) and the Neck Disability Index (Vernon and Mior 1991), are used to assess changes in relevant pain symptoms experienced by the patient as well as assessments of the patient's ability to perform activities of daily living (e.g., walking, tying shoes). General health and disease-specific patient-reported outcome measures are used to objectively document the response to surgical intervention.
- Neurologic assessments (e.g., reflex, motor, and sensory evaluations) are performed to assess whether the procedure or device caused neurological damage.
- Medical imaging evaluations are utilized to assess the status and function of the device, bone/implant interface, and presence or absence of a solid arthrodesis if fusion is a goal of the surgical procedure.
- Adverse events are collected and categorized by seriousness, severity, and relationship to the device or procedure.
- Patient satisfaction measures are used to assess how satisfied the patient is with their outcome.
- Additional assessments are determined on a case-by-case basis depending on the intended use or technological features of a device.

Individual Patient Success

In clinical trials for orthopedic spinal devices, the success of a spinal procedure for a given patient is most commonly assessed using a composite primary endpoint. The individual components of a composite endpoint for a spinal device trial generally include pain, function, neurological status, subsequent surgical interventions, serious adverse events, and radiographic success for both the investigational and control groups. A predefined success criterion is identified for each component of the composite endpoint prior to initiating the clinical trial. Examples of success criteria include the following: pain and function scores must improve by a certain absolute amount or percentage, neurologic status must maintain or improve from preoperative status, there should be no serious device or procedure-related adverse events, there should be no significant subsidence or migration (based on pre-specified quantitative criteria) of the device seen on medical imaging, and depending on the device type and intended use, there should be no unplanned subsequent surgical interventions. If all of these criteria are met at the agreed-upon assessment time point (typically 2 years post-operation for a spinal device study), the patient is considered a success in the study. However, if one of these criteria is not met, the patient is considered a failure in the study.

Overall Study Success

Determination of overall study success in a clinical trial for a spinal device most commonly involves comparison of the proportion of patients that achieve individual success between the investigational and control groups. It is important to select an active control intervention with proven efficacy compared to a placebo (Mirza et al. 2011). There are two common types of pre-defined, statistical evaluations performed: non-inferiority and superiority. If non-inferiority is demonstrated, this means that the proportion of patients determined to be a success in the investigational group was statistically shown to be not inferior to the control group by a pre-specified margin. This type of study is appropriate when studying two comparable procedures such as anterior cervical discectomy and fusion (ACDF) with an intervertebral body fusion device filled with autograft bone graft compared to ACDF with an intervertebral body fusion device filled with a bone void filler or a therapeutic biologic. Once non-inferiority to the control is established, superiority compared to the control can be tested. If superiority is demonstrated, this means that the proportion of patients experiencing success was statistically shown to be higher than the control group. Superiority evaluations are important if the investigational device treatment has additional inherent risks associated with it as compared to an active control treatment. For example, if the investigational treatment involves the use of a spinal device and the control treatment does not require the use of a spinal device (e.g., the control treatment is a spinal injection procedure), superiority is expected to be shown in the investigational group to offset the inherent risks associated with implantation of a spinal device. Superiority evaluations are also performed if a company wants to show their device to be better than an alternative treatment regardless of risk level.

Medical Imaging

One of the biggest advantages of clinical trials is their ability to assess the device in the actual environment in which it is used. While most of the clinical assessments collected in a clinical trial are related to the primary purpose for the surgical intervention (e.g., relief of pain, increase in function), it is also important to assess the status and function of the device and its relation to surrounding tissues. However, once the device is implanted, inspection of the device in vivo raises challenges. Medical imaging is necessary to view the device and surrounding bone and soft tissues and potentially correlate device status to adverse events such as new or increased pain. In clinical trials, medical imaging is generally performed according to a pre-specified imaging protocol, to assess various factors such as:

- Loosening of a device at the bone-implant interface.
- Changes in device position – device migration, expulsion from the disc space (if applicable), and subsidence of the device into bone can be visualized.
- Changes in device condition – imaging can be used to assess whether the device has fractured, excessively worn, or otherwise experienced damage.
- Fusion status – assessment of fusion status can be performed by evaluating motion across the spinal segment (or lack thereof) and the presence of bridging bone.
- Motion-sparing device function – assessment of device function for motion-sparing spinal devices such as a total disc replacement can be performed by assessing whether the device appears to be functioning appropriately during spinal motion (e.g., flexion-extension motion).

Imaging modalities utilized postoperatively can include plain film radiographs (X-rays), computed tomography (CT) scans, and magnetic resonance imaging (MRI). Each of these imaging modalities offers unique advantages and limitations, and a combination of imaging modalities may be used in clinical studies for orthopedic spinal devices. For example, X-rays are the most common method used in clinical trials because X-ray machines are widely available and images can be quickly obtained on the patient in various positions. However, X-rays cannot adequately visualize certain radiolucent polymeric materials commonly used to manufacture spinal devices. To address this issue, medical device manufacturers often add metallic radiographic markers to the polymeric components in order to visualize the components on X-ray. Additionally, X-rays provide a 2-dimensional representation of a 3-dimensional environment which can make it challenging to assess the true position and status. CT scans can be utilized to create a 3-dimensional reconstruction of the implant and surrounding region. Therefore, CT scans are useful to inspect the implant and surrounding bone in more detail, but the frequency of scans must be minimized due to the amount of radiation exposure associated with the use of this imaging modality. As MRI does not use ionizing radiation, it is often the initial imaging study obtained to evaluate spinal anatomy. However, imaging artifacts present from metal components may limit the utility of this imaging modality for many orthopedic spinal implants. Overall, medical imaging is often the primary method used to assess the condition of a spinal device while implanted and is therefore a critical assessment during clinical trials.

Conclusions

In order to market a new Class II or Class III spinal device in the United States, spinal device manufacturers are generally required to submit a pre-market application to the FDA to receive marketing authorization. These submissions contain many types of data including mechanical testing, animal testing, cadaver testing, and clinical data intended to demonstrate the safety and effectiveness of a new spinal implant prior to marketing. Each type of evaluation has limitations that should be well understood prior to selecting a method to answer specific safety and/or effectiveness questions. For patients to have timely access to safe, effective, and high-quality medical devices, a balance between scientifically sound review processes and minimizing the burden on the industry to obtain regulatory approval is needed.

References

Abbah SA, Lam CX, Hutmacher DW, Goh JC, Wong H-K (2009) Biological performance of a polycaprolactone-based scaffold used as fusion cage device in a large animal model of spinal reconstructive surgery. Biomaterials 30:5086–5093

Anderson PA, Sasso RC, Rouleau JP, Carlson CS, Goffin J (2004) The Bryan Cervical Disc: wear properties and early clinical results. Spine J 4:S303–S309

ASTM Standard F2077 2014 Test methods for intervertebral body fusion devices (2014). ASTM International, West Conshohocken. https://doi.org/10.1520/F2077-14

Beaubien BP, Freeman AL, Turner JL, Castro CA, Armstrong WD, Waugh LG, Dryer RF (2010) Evaluation of a lumbar intervertebral spacer with integrated screws as a stand-alone fixation device. J Spinal Disord Tech 23:351–358

Bianco R-J, Arnoux P-J, Wagnac E, Mac-Thiong J-M, Aubin C-É (2017) Minimizing pedicle screw pullout risks: a detailed biomechanical analysis of screw design and placement. Clinic Spine Surg 30:E226–E232

Briski DC et al (2017) Does spanning a lateral lumbar interbody cage across the vertebral ring apophysis increase loads required for failure and mitigate endplate violation. Spine 42:E1158–E1164

Cain CM, Schleicher P, Gerlach R, Pflugmacher R, Scholz M, Kandziora F (2005) A new stand-alone anterior lumbar interbody fusion device: biomechanical comparison with established fixation techniques. Spine 30:2631–2636

Campbell J, Coombs D, Rao M, Rullkoetter P, Petrella A (2016) Automated finite element meshing of the lumbar spine: verification and validation with 18 specimen-specific models. J Biomech 49:2669–2676

Cho W, Wu C, Mehbod AA, Transfeldt EE (2008) Comparison of cage designs for transforaminal lumbar interbody fusion: a biomechanical study. Clin Biomech 23:979–985

Cook DJ, Yeager MS, Oh MY, Cheng BC (2015) Lumbar intrafacet bone dowel fixation. Neurosurgery 76:470–478

Crawford NR, Peles JD, Dickman CA (1998) The spinal lax zone and neutral zone: measurement techniques and parameter comparisons. J Spinal Disord 11:416–429

Cunningham BW, Sefter JC, Shono Y, PC MA (1993) Static and cyclical biomechanical analysis of pedicle screw spinal constructs. Spine 18:1677–1688

Cunningham BW, Orbegoso CM, Dmitriev AE, Hallab NJ, Sefter JC, Asdourian P, PC MA (2003) The effect of spinal instrumentation particulate wear debris: an in vivo rabbit model and applied clinical study of retrieved instrumentation cases. Spine J 3:19–32

Cunningham BW et al (2004) Total disc replacement arthroplasty using the AcroFlex lumbar disc: a non-human primate model. In: Arthroplasty of the Spine. Springer, Berlin, pp 59–67

Cunningham BW, Hallab NJ, Hu N, PC MA (2013) Epidural application of spinal instrumentation particulate wear debris: a comprehensive evaluation of neurotoxicity using an in vivo animal model. J Neurosurg Spine 19:336–350

Di Martino A, Vaccaro AR, Lee JY, Denaro V, Lim MR (2005) Nucleus pulposus replacement: basic science and indications for clinical use. Spine 30:S16–S22

Dixon D, Darden B, Casamitjana J, Weissmann KA, Cristobal S, Powell D, Baluch D (2017) Accuracy of a dynamic surgical guidance probe for screw insertion in the cervical spine: a cadaveric study. Eur Spine J 26:1149–1153

Dreischarf M et al (2014) Comparison of eight published static finite element models of the intact lumbar spine: predictive power of models improves when combined together. J Biomech 47:1757–1766

Drespe IH, Polzhofer GK, Turner AS, Grauer JN (2005) Animal models for spinal fusion. Spine J 5:S209–S216

Fact Sheet: FDA at a Glance. https://www.fda.gov/AboutFDA/Transparency/Basics/ucm553038.htm. Accessed 27 Feb 2018

Fairbank JC, Pynsent PB (2000) The Oswestry disability index. Spine 25:2940–2953

FDA (2000) Guidance Document for the Preparation of IDEs for Spinal Systems. https://www.fda.gov/downloads/MedicalDevices/DeviceRegulationandGuidance/GuidanceDocuments/ucm073772.pdf

FDA (2004) Guidance for Industry and FDA Staff: Spinal System 510(k)s. https://www.fda.gov/MedicalDevices/DeviceRegulationandGuidance/GuidanceDocuments/ucm072459.htm

FDA (2007) FDA Guidance for Industry and FDA Staff – Class II Special Controls Guidance Document: Intervertebral Body Fusion Device. http://www.fda.gov/MedicalDevices/DeviceRegulationandGuidance/GuidanceDocuments/ucm071408.htm

FDA (2008) Guidance for Industry and FDA Staff: Preparation and Review of Investigational Device Exemption Applications (IDEs) for Total Artificial Discs. https://www.fda.gov/MedicalDevices/DeviceRegulationandGuidance/GuidanceDocuments/ucm071154.htm

FDA (2011) Guidance for Industry and FDA Staff: 30-Day Notices, 135-Day Premarket Approval (PMA) Supplements and 75 Day Humanitarian Device Exemption (HDE) Supplements for Manufacturing Method or Process Changes. https://www.fda.gov/downloads/MedicalDevices/.../ucm080194.pdf

FDA (2014a) The 510(k) Program: Evaluating Substantial Equivalence in Premarket Notifications [510(k)] – Guidance for Industry and Food and Drug Administration Staff. http://www.fda.gov/downloads/MedicalDevices/.../UCM284443.pdf

FDA (2014b) Guidance for Industry and Food and Drug Administration Staff: Annual Reports for Approved Premarket Approval Applications (PMA). https://www.fda.gov/downloads/MedicalDevices/DeviceRegulationandGuidance/GuidanceDocuments/ucm089398.pdf

FDA (2016) Reporting of Computational Modeling Studies in Medical Device Submissions – Guidance for Industry and Food and Drug Administration Staff

FDA (2017) Deciding When to Submit a 510(k) for a Change to an Existing Device: Guidance for Industry and Food and Drug Administration Staff. https://www.fda.gov/downloads/medicaldevices/deviceregulationandguidance/guidancedocuments/ucm514771.pdf

Ferrara LA, Secor JL, Jin B-H, Wakefield A, Inceoglu S, Benzel EC (2003) A biomechanical comparison of facet screw fixation and pedicle screw fixation: effects of short-term and long-term repetitive cycling. Spine 28:1226–1234

Frankel BM, D'Agostino S, Wang C (2007) A biomechanical cadaveric analysis of polymethylmethacrylate-augmented pedicle screw fixation. J Neurosurg 7:47–53

Freeman AL, Camisa WJ, Buttermann GR, Malcolm JR (2016) Flexibility and fatigue evaluation of oblique as compared with anterior lumbar interbody cages with integrated endplate fixation. J Neurosurg Spine 24:54–59

Goel VK et al (2005) Effects of Charite artificial disc on the implanted and adjacent spinal segments mechanics using a hybrid testing protocol. Spine 30:2755–2764

Goel VK, Panjabi MM, Patwardhan AG, Dooris AP, Serhan H (2006) Test protocols for evaluation of spinal implants. JBJS 88:103–109

Graham J, Estes BT (2009) What standards can (and can't) tell us about a spinal device. SAS J 3:178–183

Graham JH, Anderson PA, Spenciner DB (2014) Letter to the editor in response to Villa T, La Barbera L, Galbusera F, "Comparative analysis of international standards for the fatigue testing of posterior spinal fixation systems". Spine J 14:3067–3068

Grauer JN et al (2006) Biomechanics of two-level Charite artificial disc placement in comparison to fusion plus single-level disc placement combination. Spine J 6:659–666

Hanlon AD, Cook DJ, Yeager MS, Cheng BC (2014) Quantitative analysis of the nonlinear displacement–load behavior of the lumbar spine. J Biomech Eng 136:081009

Helgeson MD, Kang DG, Lehman RA, Dmitriev AE, Luhmann SJ (2013) Tapping insertional torque allows prediction for better pedicle screw fixation and optimal screw size selection. Spine J 13:957–965

Heth JA, Hitchon PW, Goel VK, Rogge TN, Drake JS, Torner JC (2001) A biomechanical comparison between anterior and transverse interbody fusion cages. Spine 26:e261–e267

Hitchon PW et al (2000) In vitro biomechanical analysis of three anterior thoracolumbar implants. J Neurosurg Spine 93:252–258

Hsu C-C, Chao C-K, Wang J-L, Hou S-M, Tsai Y-T, Lin J (2005) Increase of pullout strength of spinal pedicle screws with conical core: biomechanical tests and finite element analyses. J Orthop Res 23:788–794

Kallemeyn N, Gandhi A, Kode S, Shivanna K, Smucker J, Grosland N (2010) Validation of a C2–C7 cervical spine finite element model using specimen-specific flexibility data. Med Eng Phys 32:482–489

Kim YJ, Lenke LG, Bridwell KH, Cho YS, Riew KD (2004) Free hand pedicle screw placement in the thoracic spine: is it safe? Spine 29:333–342

Kornblum MB, Turner AW, Cornwall GB, Zatushevsky MA, Phillips FM (2013) Biomechanical evaluation of stand-alone lumbar polyether-ether-ketone interbody cage with integrated screws. Spine J 13:77–84

Kotani Y et al (2002) Artificial intervertebral disc replacement using bioactive three-dimensional fabric: design, development, and preliminary animal study. Spine 27:929–935

Kuzhupilly RR, Lieberman IH, McLain RF, Valdevit A, Kambic H, Richmond BJ (2002) In vitro stability of FRA spacers with integrated crossed screws for anterior lumbar interbody fusion. Spine 27:923–928

Labrom RD, Tan J-S, Reilly CW, Tredwell SJ, Fisher CG, Oxland TR (2005) The effect of interbody cage positioning on lumbosacral vertebral endplate failure in compression. Spine 30:E556–E561

Lehman RA Jr, Polly DW Jr, Kuklo TR, Cunningham B, Kirk KL, Belmont PJ Jr (2003) Straight-forward versus anatomic trajectory technique of thoracic pedicle screw fixation: a biomechanical analysis. Spine 28:2058–2065

Lu WW, Zhu Q, Holmes AD, Luk K, Zhong S, Leong C (2000) Loosening of sacral screw fixation under in vitro fatigue loading. J Orthop Res 18:808–814

Luo J et al (2017) The accuracy of the lateral vertebral notch-referred pedicle screw insertion technique in subaxial cervical spine: a human cadaver study. Arch Orthop Trauma Surg 137:517–522

Ma T, Xu Y-Q, Cheng Y-B, Jiang M-Y, Xu X-M, Xie L, Lu S (2012) A novel computer-assisted drill guide template for thoracic pedicle screw placement: a cadaveric study. Arch Orthop Trauma Surg 132:65–72

Mirza S, Konodi M, Martin B, Spratt K (2011) Safety and functional outcome assessment in spine surgery. Orthop Knowl Update Spine 4:589–606

Nagaraja S, Palepu V (2016) Comparisons of anterior plate screw pullout strength between Polyurethane Foams and Thoracolumbar Cadaveric Vertebrae. J Biomech Eng 138:104505

Nagaraja S, Palepu V, Peck JH, Helgeson MD (2015) Impact of screw location and endplate preparation on pullout strength for anterior plates and integrated fixation cages the. Spine Journal 15(11):2425–2432

Newcomb AG, Baek S, Kelly BP, Crawford NR (2017) Effect of screw position on load transfer in lumbar pedicle screws: a non-idealized finite element analysis. Comput Meth Biomech Biomed Eng 20:182–192

O'Leary P et al (2005) Response of Charite total disc replacement under physiologic loads: prosthesis component motion patterns. Spine J 5:590–599

Oxland TR, Lund T (2000) Biomechanics of stand-alone cages and cages in combination with posterior fixation: a literature review. Eur Spine J 9:S095–S101

Palepu V, Peck JH, Simon DD, Helgeson MD, Nagaraja S (2017) Biomechanical evaluation of an integrated fixation cage during fatigue loading: a human cadaver study. J Neurosurg Spine 26:524–531

Panjabi MM (1988) Biomechanical evaluation of spinal fixation devices: I. Concept Framew Spine 13:1129–1134

Patwardhan AG et al (2003) Effect of compressive follower preload on the flexion–extension response of the human lumbar spine. J Orthop Res 21:540–546

Peck JH, Sing DC, Nagaraja S, Peck DG, Lotz JC, Dmitriev AE (2017) Mechanical performance of cervical intervertebral body fusion devices: A systematic analysis of data submitted to the Food and Drug Administration. J Biomech 54:26–32

Peck JH, Kavlock KD, Showalter BL, Ferrell BM, Peck DG, Dmitriev AE (2018) Mechanical performance of lumbar intervertebral body fusion devices: an analysis of data submitted to the Food and Drug Administration. J Biomech 78:87–93

Pfeiffer M, Hoffman H, Goel V, Weinstein J, Griss P (1997) In vitro testing of a new transpedicular stabilization technique. Eur Spine J 6:249–255

Pishnamaz M et al (2017) The quantity of bone cement influences the anchorage of augmented pedicle screws in the osteoporotic spine: a biomechanical human cadaveric study. Clin Biomech 52:14–19

Pitzen T, Geisler FH, Matthis D, Müller-Storz H, Steudel W-I (2000) Motion of threaded cages in posterior lumbar interbody fusion. Eur Spine J 9:571–576

Polikeit A, Ferguson SJ, Nolte LP, Orr TE (2003) Factors influencing stresses in the lumbar spine after the insertion of intervertebral cages: finite element analysis. Eur Spine J 12:413–420

Rivard CH, Rhalmi S, Coillard C (2002) In vivo biocompatibility testing of peek polymer for a spinal implant system: a study in rabbits. J Biomed Mater Res A 62:488–498

Ryken TC, Clausen JD, Traynelis VC, Goel VK (1995) Biomechanical analysis of bone mineral density, insertion technique, screw torque, and holding strength of anterior cervical plate screws. J Neurosurg 83:324–329

Thompson JD, Benjamin JB, Szivek JA (1997) Pullout strengths of cannulated and noncannulated cancellous bone screws. Clin Orthop Relat Res 341:241–249

Trahan J, Morales E, Richter EO, Tender GC (2014) The effects of lumbar facet dowels on joint stiffness: a biomechanical study. Ochsner J 14:44–50

Trautwein FT, Lowery GL, Wharton ND, Hipp JA, Chomiak RJ (2010) Determination of the in vivo posterior loading environment of the Coflex interlaminar-interspinous implant. Spine J 10:244–251

Vadapalli S, Robon M, Biyani A, Sairyo K, Khandha A, Goel VK (2006a) Effect of lumbar interbody cage geometry on construct stability: a cadaveric study. Spine 31:2189–2194

Vadapalli S, Sairyo K, Goel VK, Robon M, Biyani A, Khandha A, Ebraheim NA (2006b) Biomechanical rationale for using polyetheretherketone (PEEK) spacers for lumbar interbody fusion–a finite element study. Spine 31:E992–E998

Vernon H, Mior S (1991) The Neck Disability Index: a study of reliability and validity. J Manip Physiol Ther 14:409–415

Villa T, La Barbera L, Galbusera F (2014) Comparative analysis of international standards for the fatigue testing of posterior spinal fixation systems. Spine J 14:695–704

Voronov LI et al (2014) Biomechanical characteristics of an integrated lumbar interbody fusion device. Int J Spine Surg 8. https://doi.org/10.14444/1001

Wagnac E, Arnoux P-J, Garo A, Aubin C-E (2012) Finite element analysis of the influence of loading rate on a model of the full lumbar spine under dynamic loading conditions. Med Biol Eng Comput 50:903–915

Wang S-T, Goel VK, Fu C-Y, Kubo S, Choi W, Liu C-L, Chen T-H (2005) Posterior instrumentation reduces differences in spine stability as a result of different cage orientations: an in vitro study. Spine 30:62–67

Wilke H-J, Wenger K, Claes L (1998a) Testing criteria for spinal implants: recommendations for the standardization of in vitro stability testing of spinal implants. Eur Spine J 7:148–154

Wilke HJ, Jungkunz B, Wenger K, Claes LE (1998b) Spinal segment range of motion as a function of in vitro test conditions: effects of exposure period, accumulated cycles, angular-deformation rate, and moisture condition. Anat Rec 251:15–19

Recent Advances in PolyArylEtherKetones and Their In Vitro Evaluation for Hard Tissue Applications

19

Boyle C. Cheng, Alexander K. Yu, Isaac R. Swink, Donald M. Whiting, and Saadyah Averick

Contents

Conclusions	430
Cross-References	430
References	430

Abstract

The advent of thermoplastic semicrystalline polymeric materials in the design of medical devices has allowed for the widespread use of polymeric interbody spacers for spinal arthrodesis to treat spinal degeneration. These polymers come from the PolyArylEtherKetone class of materials which are inert, readily machined, and serializable and have mechanical modules closely matching bone. Unfortunately, the inert nature of this class of materials may prevent osseointegration and can potentially generate a negative immune response. To overcome the inert character of PolyArylEtherKetones, researches have investigated several approaches to improving the biological properties of this important class of material. This review summarizes the history of PolyArylEtherKetones within the context of spinal arthrodesis and the recent approaches to improving the osseointegrative properties of this polymer.

Keywords

Polymer · Biomaterial · PEEK · PolyArylEtherKetone · Osseointegration

Degenerate issues of the spine affect nearly 20% of the American population (Brinjikji et al. 2015; Buser et al. 2018; Ravindra et al. 2018). Of that group, nearly 5% require surgical intervention to restore stability, reduce pain, and reestablish health function (Weinstein et al. 2006; Friedly et al. 2010; Martin et al. 2019). This means that nearly 400,000 spinal arthrodesis procedures take place annually.

B. C. Cheng
Neuroscience Institute, Allegheny Health Network, Drexel University, Allegheny General Hospital Campus, Pittsburgh, PA, USA
e-mail: bcheng@wpahs.org; Boyle.CHENG@ahn.org; boylecheng@yahoo.com; boylecheng@gmail.com

A. K. Yu · I. R. Swink · S. Averick (✉)
Department of Neurosurgery, Neuroscience Institute, Allegheny Health Network, Pittsburgh, PA, USA
e-mail: Alexander.Yu@ahn.org; Isaac.Swink@AHN.ORG; Saadyah.Averick@ahn.org

D. M. Whiting
Neuroscience Institute, Allegheny Health Network, Pittsburgh, PA, USA
e-mail: donald.whiting@ahn.org

© Springer Nature Switzerland AG 2021
B. C. Cheng (ed.), *Handbook of Spine Technology*,
https://doi.org/10.1007/978-3-319-44424-6_99

Spinal degeneration is caused by a series of factors including genetic predisposition, lifestyle and injury, weight management, as well as chronic disorders including diabetes and inflammation (Elfering et al. 2002; Steelman et al. 2018). With better management of many health conditions, coupled with longer expected life spans, the probability of degenerative disc disease occurring in the United States has risen and the number of surgical interventions to relieve this medical challenge has followed suit (Mihailidis et al. 2017; Kurucan et al. 2018).

Degenerative issues of the spine cause pain due to a loss of vertebral tertiary structural integrity due to the degeneration and potentially, instability (Mobbs et al. 2013; Battie et al. 2019). This loss of structure induces non-native conformation of vertebra translating into pressure on the spinal cord with symptoms ranging from pain to immobility (Donnally et al. 2020). The degeneration may occur at any point in the spinal column including the cervical, lumbar, thoracic, and sacroiliac joint.

While surgical intervention to stabilize degenerative issues has benefited from advances in materials, alternative pain management approaches are initially considered before surgical intervention is offered (Vaishnav and McAnany 2019; Winebrake et al. 2020). Hence, surgical intervention to stabilize and fix a motion segment may be the only viable medical approach available to patients to allow for redress to their health challenges. The goal of surgical intervention is to stabilize the degenerated tertiary vertebral structure by providing a means of fusion through the damaged disc space (Vaishnav and McAnany 2019; Lykissas and Aichmair 2013; Yavin et al. 2017). By allowing spinal arthrodesis to occur between two adjacent vertebrae through the intra-disc space, relief from pain and return of motor function can occur, thus making spinal arthrodesis a critical medical intervention.

The goal of spinal arthrodesis is to fix two adjacent vertebrae together by forming a de novo bone mass at the interface of the end plates (Spoor and Oner 2013; Baliga et al. 2015). Generally speaking, the approach requires exposure of the region of interest on the spine. Surgeons remove damaged disc and implant a temporary adjunct to fixation and may add additional bone growth supporting matrixes to aid in stability until arthrodesis occurs. Typically, fusion of two vertebra takes place over a period of 6–12 months with initial detectable bone deposition taking place as early as 2–3 months (Lee et al. 2011). The design and use of the temporary adjuncts to fixative devices play an important role in induction of fusion, radiological diagnosis of arthrodesis progression, and support of vertebra faceplate. (Danison et al. 2017) These adjuncts are hard metals, plastics, or ceramics that are meant to support the vertebra till new de novo bone is formed (Nouh 2012; Patel et al. 2019).

This chapter focuses on methods to improve the spinal arthrodesis outcomes through innovation in polymer materials used as temporary adjuncts to fusion. Included in this review are approaches for next-generation composites, changes to surface structure, and changes to surface chemistry. While advances in metallic implant devices have played an important role in recent clinical results, this review will not cover the use of metal implants. Detailed reviews on these implants can be found here (Ni et al. 2019). Furthermore, impact of spinal arthrodesis due to use of various bone growth extenders including allografts and synthetic bone void fillers has been well documented in the literature. Finally, the surgical approach can play an important role in device design as well as recovery post-spinal surgery. While important, this is the focus of other reviews and will not be covered in this review.

It was recognized in the late 1950s and early 1960s that spinal arthrodesis was a legitimate and effective tool to treat degenerative disc conditions and was first used in lumbar spinal fusion with metallic implants pioneered by Cushing (Prolo 1969; Bydon et al. 2011), Dommisse (Dommisse 1959), Boucher, (Boucher 1959) and others (Winter and Lonstein 1999). Initial surgical approaches for fixative devices to create space between degraded discs utilized sterilized autograft bone. While autograft bone alone is somewhat effective, major challenges to device fracture, sizing, placement, and device migration made the initial approach a suboptimal solution and the quest for new materials (Gupta et al. 2015; Buser et al.

2016; Fernandez de Grado et al. 2018; Sohn and Oh 2019).

Metallic implants were a breakthrough, as these implants could be machined and precision shapes could be readily prepared on commercial scales (Chong et al. 2015; Phan and Mobbs 2016; Tarpada et al. 2017). Initial medical devices custom made for spinal arthrodesis were comprised of stainless steel (DeBowes et al. 1984; Bagby 1988). The use of stainless steel allowed for some degree of customization in terms of device size and shape, but due to x-ray and MRI reactivity imaging, vertebral fusion and potential for challenges in future imaging studies limited the appeal of this material (Rupp et al. 1993; Kumar et al. 2006; Knott et al. 2010). Furthermore, an additional issue arose due to the use of stainless steel implants. It was found that the interface between the vertebra plates adjacent to the implant had a propensity to crack/shatter due to a mismatch between the tensile strength of metal and the bone (Ordway et al. 2007; Herrera Herrera et al. 2013; Choy et al. 2019).

While the promise of spinal arthrodesis as a tool to treat degenerative spinal conditions had begun to be achieved through the use of metal implants, the quest for new materials that were radiolucent and had strength indexes better matched to native cortical bone had begun to correct defects identified in metallic implants. At the time, circa 1980, a new polymeric material had begun making headway in advanced engineering applications due to the polymer's strength and inert nature (Parker et al. 2012; Manoukian et al. 2019). This polymer, PolyEtherEtherKetone (PEEK), held several advantages over traditional metals including radiolucency, easily machined from various sized rodstock, and had a strength profile closely matching that of bone (Kurtz and Devine 2007; Selim et al. 2018). Once identified and approved for use as an adjuvant to fusion by the FDA, PEEK has remained dominate in the spinal arthrodesis pace with only recent challenges from innovations in titanium-based devices (Li et al. 2016; Pelletier et al. 2016; Seaman et al. 2017).

The current space of devices serving as temporary adjuncts to fixation in spinal arthrodesis has rapidly expanded with devices having a litany of shapes, compositions, and surface treatments (Jain et al. 2016; Phan and Mobbs 2016). The goal of these new devices is to create a material that encourages rapid and effective bone growth to minimize wound healing time through the promotion of spinal arthrodesis. Recently, PEEK use has suffered from the perception that the material has inherent limitations that are not present in metal-based systems (Toth et al. 2006; Torstrick et al. 2017a). These limitations relate to the polymer's inert surface chemistry that prevents cell adhesion and growth.

While each material has inherent limitations, the goal of scientists and engineers has been the creation of an ideal implant. An ideal implant is one that would promote arthrodesis, while matching mechanical properties of bone, and have radiolucent properties to allow for visualization of wound healing and bone ongrowth (Martz et al. 1997; Kadam et al. 2016). These properties have thus far proved elusive in a single material, leading to cost-benefit analysis choices to be made in device selection by the surgeon.

When comparing radiolucency, electron-dense materials, such as metal like stainless steel and titanium, prevent the penetration of x-ray beams and interfere with MRI imaging. This radiopaque property leads to errors in quantification of bone ongrowth, an imaging of arthrodesis limiting outcome measurement post-surgery (Hayashi et al. 2012; Thakkar et al. 2012). Polymers such as PEEK do not suffer from this limitation. To compensate for metal's limitation, additive manufacturing of metallic implant with void space and low density has been proposed (Wilcox et al. 2017).

The implant used as an adjuvant to fusion must be capable of supporting the vertebra until the formation of a bony mass occurs (Cole et al. 2009). If the material has a mechanical strength greatly exceeding that of bone, the potential for the faceplate of the vertebra to crack or for the implant to sink into the vertebra is possible (Proietti et al. 2013; Tang et al. 2014), thus defeating the goal of surgical intervention, as the failed implant will not provide sufficient relief of pain and pressure removal from the spinal cord. If the implant is too weak, it may shatter, causing pain and potential further damage to an already degenerated wound (Chou et al. 2016). The mechanical strength of an implant can play an important role in spinal arthrodesis (Hoshijima et al. 1997; Steffen et al. 2000). PEEK implants have a mechanical strength closely matching that of native bone compared to metallic implants that have a known risk of bone fracture due to mechanical strength mismatch. (Heary et al. 2017)

The ultimate goal of spinal surgery is to fix a degenerative spine. This is accomplished by the fusion of two adjacent vertebrae, (Gittens et al. 2014) making osteogenic promotion a crucial property of the implant material. Implants used as adjuncts to spinal arthrodesis must provide a matrix that allows for mineralization, bone ongrowth, and support for cell deposition. (Agarwal and Garcia 2015; Lewallen et al. 2015) Materials that have good surfaces for cell adhesion include those with surface charge or surface functional groups that allow for protein and cellular interaction (Mastrogiacomo et al. 2005; Stevens 2008; Amini et al. 2012; Qu et al. 2019). Materials that are neutral and without cell binding motifs generally are poor biomaterials due to their inability to promote cell ongrowth and adhesion. PEEK suffers in this category, as it is a neutral hydrophobic inert polymer with no native motifs for cell binding and growth. (Toth 2012) This is in contrast to typical metal-based implants that have a native surface charge that allows for cell and protein deposition and growth (Shayesteh Moghaddam et al. 2016; Gao et al. 2017).

A devices materials properties and device design must therefore be balanced between (a) radiolucency, (b) mechanical strength, and (c) device osseointegration as no currently available device is capable of excelling in all three categories (Sohn and Oh 2019; Warburton et al. 2019). Therefore, research into metallic implants has focused on improving the radiolucency and better matching mechanical strength to that of bone. Research to improve PEEK's properties has focused on changing the bone-PEEK interface to increase osseointegration while still retaining PEEK's positive mechanical and imaging attributes (Torstrick et al. 2016; Walsh et al. 2016; Honigmann et al. 2018).

Industrially, PEEK is synthesized with a step-growth polymerization with AA BB type monomers with a phenol or phenolic salt and a halogenated benzophenone (Scheme 1). The reaction requires high temperatures, organic solvent, and generated alkyl halide salts. These polymers must be precipitated to remove unreacted monomers, solvent, and generated salts. Unfortunately, the residual starting material and generated salts are toxic and may induce biological effects. Once isolated, the polymer pellets are dried and the rods are extruded and can be machined into medical devices.

Despite PEEK's poor cell growth characteristics, this material has found widespread adoption and use and has been implanted in millions of patients globally. The adoption and use of PEEK was heavily influenced by the shortcoming of the initial medical devices used as temporary adjuncts to spinal fixation, as these materials were radiopaque and were not matched with bone's mechanical properties, resulting in a higher probability of poor medical outcomes (Kurtz and Devine 2007).

Historically, PEEK's commercial development by Vitrex (in the mid-1980s) occurred around the same time as issues with metal implants were becoming apparent. It was realized that good spinal arthrodesis required a strong material that supported vertebra without inducing stress fractures while still allowing for direct imaging to monitor wound healing. Furthermore, the use of a material readily manufactured from rodstock available in a variety of dimensions was a boon to device designers. These advantages lead to the first PEEK spinal devices developed by AcroMed in the 1990s (Wenz et al. 1990; Brantigan and Steffee 1993). PEEK was accepted as a material with desirable properties and found widespread adoption by all major medical device manufactures.

Since PEEK's initial use as a medical device, manufacturers have studied approaches to increase bone ongrowth, custom composites, and novel structural elements (Reid et al. 2019; Zhang et al. 2019; Buck et al. 2020; Enders et al. 2020). Recent innovations that are currently in clinical use have focused on PEEK's unique features, described herein. This has led to the creation of

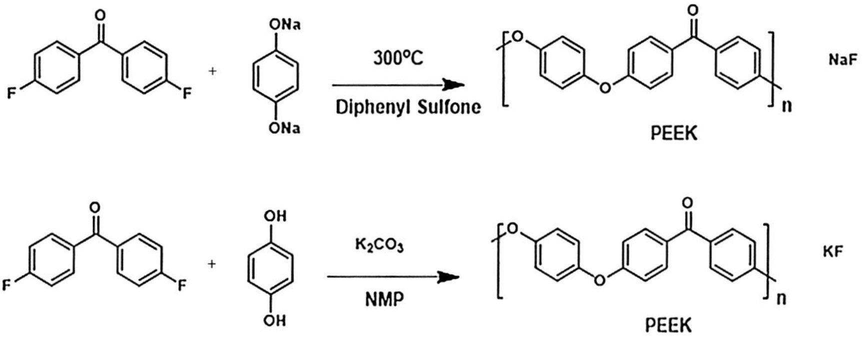

Scheme 1 Synthesis of PEEK with: (**a**) Hydroquinone sodium salt or (**b**) Hydroquinone with potassium carbonate

spinal cages capable of expansion on implant to facilitate minimally invasive surgery (Alimi et al. 2015a; Kale et al. 2017; Zhang et al. 2018). Expandable cages are not possible using metal-based devices due to the rigid nature of metal. PEEK rodstock is formed from extruded PEEK pellets. Therefore, the opportunity to incorporate additives to form a composite that has increased bone ongrowth potential is afforded (Evans et al. 2015; Zhong et al. 2019; Petersmann et al. 2020). Finally, coating of PEEK with a thin metallic layer in order to increase the cell ongrowth potential of the implant has recently been introduced into the market (Gardon et al. 2014; Yang et al. 2015; Hasegawa et al. 2020). These innovations point to a strong appetite for next-generation evolved PEEK-based devices that retain the polymer's advantages but incorporate new functionalities for improved cell ongrowth and eventual fusion.

Although PEEK implants have found widespread adoption and use as temporary adjuncts to fusion as a material, PEEK has significant limitations not found with metallic implants. PEEK's major limitation relates to its inert hydrophobic character and lack of surface functional groups that promote cell binding. Typical polymer-based scaffolds optimized to support cell growth have an overall surface charge, or they may possess binding motifs for cell surface interactions (Polo-Corrales et al. 2014; Nikolova and Chavali 2019; Richbourg et al. 2019). For example, in simple in vitro assays, it is clear that PEEK does not support cell proliferation and has relatively low early- and late-stage osteogenic proliferation markers compared to titanium or positive control tissue culture polystyrene (Olivares-Navarrete et al. 2015; Cheng et al. 2018). Clinically, the formation of a fibrous layer around PEEK indicates a foreign body response (Anderson et al. 2008; Torstrick et al. 2016; Walsh et al. 2016). Due to this lack of surface charge and cell interacting groups, it is theorized that spinal arthrodesis may be delayed.

There is an emerging body of preclinical and clinical evidence related to heretofore unknown liabilities associated with PEEK (Walsh et al. 2016; Kao et al. 2014). This evidence is represented by both *preclinical* and clinical studies demonstrating PEEK's inability to support the proliferation of osteoblasts, leading to longer times required to successful vertebral fusion. A recent series of evidence has emerged that points to a potential immunogenic effect caused by PEEK's implantation. (Boyan et al. 2014; Krause et al. 2018)The immunogenic effect of PEEK, while only recently realized, may be due to a foreign body response or leaching of impurities from synthesis in PEEK pellets (Kurtz 2012). The immunogenic effects of PEEK are still being explored and have not been widely confirmed in the peer-reviewed literature. Clinical evidence has emerged that rates of pseudarthrosis in medical procedures that utilized PEEK intervertebral bodies as spacers are greater than in cases that used metallic spacers (Sardana et al. 2019; Teton et al. 2020). While this recent evidence does not preclude the use of PEEK, the questions raised have led to a surge of innovations in PEEK design to overcome the emerging liabilities.

Due to PEEK's importance as a material in spinal arthrodesis, extensive research in several solutions are currently employed to overcome PEEK's hydrophobic inert character to improve cell ongrowth and remove any potential immunological limitations. In terms of risk and reward, these approaches include the physical modification of PEEK, preparation of PEEK composites, and modification of PEEK's surface with new chemical functionalities. Physical modification of native PEEK is readily translated to a clinical setting without any risks to moving outside of the FDA 505(B)2 regulatory pathway, but this approach also limits the device to retaining PEEK's inherent cell interface weakness. Generation of PEEK composites using two FDA cleared materials has the potential to introduce a cell ongrowth component to the polymer implant but introduces liabilities for machining due to changes to device mechanical properties and a potential for regulatory oversight under non-505 (B)2 pathways. Finally, introduction of novel surface chemistry to modify the cell-PEEK interface has the greatest potential for changing how the body "sees" the implant but embodies the greatest risk for clinical translation.

Physical modification of PEEK through new PEEK cage design is considered the lowest risk approach and aims to accomplish surgical convenience and increase PEEK's surface. Upon implantation via in situ expansion with devices like the StaXx XD™ (Alimi et al. 2015b), the FlareHawk devices allow for more surface contact and areas for bone growth to occur using a relatively more minimally invasive approach. These devices accomplish their expansion via incorporation of an expandable metallic core that upon implantation is opened. NuVasive has recently demonstrated the clinical viability of a novel surface, porous PEEK, created via the etching of salt crystals emended into PEEK's surface yielding a material with the highest clinically available surface area for cell growth and proliferation (Torstrick et al. 2017b; Carpenter et al. 2018).

New materials are being designed to overcome the current clinical challenges known to PEEK. The goal of this research is the creation of a material that promotes cell growth while retaining bone strength matching and radiolucent properties. Critical to these efforts is to design a material that retains the best of PEEK, remains straightforward to machine and manufacturer, yet promotes cell growth and adhesion. There are two major categories of next-generation PEEK materials covered in this review: (1) filled PEEK composites, wherein PEEK remains chemically unmodified but is filled with inorganic minerals that promote cell growth, (2) surface-modified PEEK, PEEK's surface is modified with surface relative motifs that introduce a functional charge that allows for inherently better cell growth.

Of the innovative PEEK materials, filled PEEK composites have been present longest and have seen some recent clinical success. This class of material takes advantage of the extensive amounts of inorganic materials that may have known positive cell adhesion properties and combines them with PEEK to make a hybrid material that can support cell growth (Walsh et al. 2016). These materials are made through co-extrusion of PEEK pellets mixed with inorganic filler to a form hybrid rodstock that can be machined into desired spacer design (Ma and Guo 2019). Only inorganic filler materials are possible due to the relatively high temperatures required for PEEK extrusion. There are several key parameters that impact the properties of the extruded material and the eventual cell ongrowth properties. These include the inorganic crystal size, composition, and loading (Zhong et al. 2019; Ma et al. 2014; Kutikov et al. 2015; Zhu et al. 2019). All have influence over final device stability, biocompatibility, machinability, and material properties.

Examples of two recently developed and FDA-approved PEEK composites are PEEK OptimaHA® and ZFuse®. PEEK OptimaHA® is a PEEK composite filled with ~5 micron hydroxyapatite crystals at an ~20% loading. This material has shown clinical evidence of superiority vs PEEK with faster fusion times while retaining radiotransparency. ZFuse® is a PEEK zeolite (aluminosilicate) composite with an ~10% zeolite loading. Zeolites are inorganic sorbent materials capable of sequestering inorganic and organic small molecules. The introduction of zeolites into PEEK may aid in the sequestering of the impurities found in PEEK. In addition, the introduction of zeolite into PEEK may provide surface charges that enable cell binding and growth to occur. While ZFuse® devices have not undergone clinical evaluation, this material has recently gained clinical approval for use in spinal applications. In a large animal study, the use of ZFuse® was found to lower local tissue inflammatory markers. These two newly approved PEEK-based biomaterials' impact on the medical device space has yet to be determined as implant makers begin the process of testing and potentially fielding devices based on these composites.

When considering how a biomaterial implant imparts its effect, both the bulk and surface properties must be considered (Ikada 1994; Angelova and Hunkeler 1999; Lucke et al. 2000). The bulk properties will influence device strength and stability, while the device surface interfaces directly with the body and influences cell attachment and biological response properties. Therefore, the direct chemical derivation of PEEK's surface to introduce cell binding groups can impart the desired biological response utilizing currently available PEEK manufacturing technology without impacting material mechanical or physical properties (other than hydrophilic and surface charge) (Poulsson et al. 2019; Wang et al. 2019). Changing PEEK's properties through direct surface modification is a new approach to creating next-generation materials with inherently improved cell growth properties (Kassick et al. 2018). Scheme 2 demonstrates possible reactions of PEEK found in the literature.

Unfortunately, the synthetic toolbox available to modify PEEK's backbone has been limited due to this polymer's lack of readily modified functional groups on its backbone (Franchina and McCarthy 1991; Díez-Pascual et al. 2009; Shukla et al. 2012). Two recent chemical-based approaches have broken through this barrier to create surface-modified PEEK. These approaches are the introduction of sulfonic acid moieties onto the aryl ring via etching with concentrated sulfuric acid. (Wang et al. 2019; Shibuya and Porter 1992; Chaijareenont et al. 2018) The introduction of the anionic surface charge promotes protein binding and deposition, creating an environment conducive to cell growth. In vivo evidence has emerged demonstrating the superiority of sulfonic acid-modified PEEK compared to native PEEK for support of bone ongrowth. (Zhao et al. 2013; Ouyang et al. 2016) A limitation to the sulfuric acid modification of PEEK is the potential for over-modification of the PEEK, (Daoust et al. 1994) leading to a rapid deterioration of the polymer's mechanical properties. Additional surface modification approach is the borohydride reduction of the ketone group of PEEK to an alcohol introducing a slight increase in hydrophilicity (Erik et al. 2016); additionally this approach allows for covalent medication of the reactive alcohol group (Fukuda et al. 2018). For example, the cell binding motif GRGD was attached to PEEK in a three-step procedure wherein the ketone was reduced to an alcohol PEEK was silanized and then reacted with the peptide GRGD to improve cell binding (Zheng et al. 2014). A recent breakthrough is the realization that the ketone groups of PEEK could be directly modified with compounds bearing aminooxy or hydrazine moieties forming a stable link via formation of a Schiff base. This approach was used to introduce peptides, amino acid mimetics such as the P15 peptide, and charged functional groups onto the surface of PEEK (Kassick et al. 2018). While only tested in vitro, the materials show great promise with significantly greater signs of both late-stage and early-stage mineralization and cell binding and growth. These new biomaterials have the most complex regulatory path forward but have demonstrated the greatest ability to support cell ongrowth and proliferation, potentially heralding the future of polymer-based medical devices.

The rapid assessment of a biomaterial's osteoconductive properties in vitro allows for straightforward go/no-go points. Therefore, in vitro assays are an excellent tool to establish if a surface modification or composite strategy imparts positive bone growth. Although advances in real-time PCR allow for assessment of osteogenic mRNA expression levels (Jadlowiec et al. 2004; Koch et al. 2005; Tuzmen and Campbell 2018), direct phenotypic expression of bone tissue biomarkers remains the gold standard for biomaterial assessment (Lian and Stein 1995; Yang et al. 2018; de Wildt et al. 2019). While the literature is

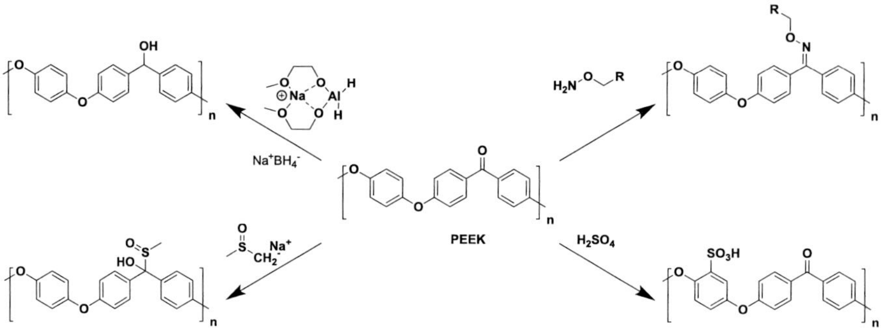

Scheme 2 Chemical modification of PEEK

replete with a host of variables in experimental design including a broad range of cell types used in cell line (Chin et al. 2012; Burmester et al. 2014; Yoo et al. 2016; Hwang and Horton 2019), primary (Noori et al. 2017; Lopes et al. 2018; Fu et al. 2019), and induced stem cells (Pirraco et al. 2010; Mattioli-Belmonte et al. 2015) and experimental conditions (Black et al. 2015; Dang et al. 2018; Levin et al. 2018), the major goal of these experiments is the measure of early phase bone biomarkers such as alkaline phosphatase and late phase presence of calcium deposits and collagen I (Lopes et al. 2018; Zadpoor 2015; Turnbull et al. 2018). Additionally, in vitro experiments allow a materials immunological profile to be screened and characterized (Chen et al. 2016). The lack of standard experimental conditions prohibits clear comparison between differing different literature reports highlighting the importance for the inclusion of positive control experimental cohorts. This will be an important area to standardize and correlate with in vivo data for a standard platform for new bone biomaterial assessment. This standard screen can lead to faster study of new materials, leading to improved patient outcomes.

Conclusions

PolyArylEtherKetones remain a critical implant material in bony tissue applications. Although PolyArylEtherKetone-based materials have found widespread use as adjuncts to spinal fixation, the future for these materials is questioned. The questions are based upon recent preclinical and clinical studies revealing that the inert nature of this material, while once heralded as a feature, can limit bone ongrowth and the potential to induce local inflammation, leading to pseudarthrosis. Fortunately, research into improving PolyArylEtherKetone osseointegrative properties has demonstrated promise for this material and the potential for these PolyArylEtherKetone derivatives' bright future as next-generation hard tissue implants while maintaining the critical design features of radiolucency, mechanical strength matching bone, and pro-cell ongrowth properties.

Cross-References

▶ Biological Treatment Approaches for Degenerative Disc Disease: Injectable Biomaterials and Bioartificial Disc Replacement
▶ Biologics: Inherent Challenges
▶ Bone Grafts and Bone Graft Substitutes
▶ Implant Material Bio-compatibility, Sensitivity, and Allergic Reactions
▶ Mechanical Implant Material Selection, Durability, Strength, and Stiffness
▶ Metal Ion Sensitivity
▶ Selection of Implant Material Effect on MRI Interpretation in Patients

References

Agarwal R, Garcia AJ (2015) Biomaterial strategies for engineering implants for enhanced osseointegration and bone repair. Adv Drug Deliv Rev 94:53–62. https://doi.org/10.1016/j.addr.2015.03.013. Epub 2015/04/12. PubMed PMID: 25861724; PMCID: PMC4598264

Alimi M, Shin B, Macielak M, Hofstetter CP, Njoku I Jr, Tsiouris AJ, Elowitz E, Hartl R (2015a) Expandable polyaryl-ether-ether-ketone spacers for interbody distraction in the lumbar spine. Global Spine J 5(3):169–178. https://doi.org/10.1055/s-0035-1552988. Epub 2015/07/02. PubMed PMID: 26131383; PMCID: PMC4472284

Alimi M, Shin B, Macielak M, Hofstetter CP, Njoku I, Tsiouris AJ, Elowitz E, Härtl R (2015b) Expandable Polyaryl-ether-ether-ketone spacers for Interbody distraction in the lumbar spine. Global Spine J 5(3):169–178. https://doi.org/10.1055/s-0035-1552988

Amini AR, Laurencin CT, Nukavarapu SP (2012) Bone tissue engineering: recent advances and challenges. Crit Rev Biomed Eng 40(5):363–408. https://doi.org/10.1615/critrevbiomedeng.v40.i5.10. Epub 2013/01/24. PubMed PMID: 23339648; PMCID: PMC3766369

Anderson JM, Rodriguez A, Chang DT (2008) Foreign body reaction to biomaterials. Semin Immunol 20(2):86–100. https://doi.org/10.1016/j.smim.2007.11.004. Epub 2007/12/29. PubMed PMID: 18162407; PMCID: PMC2327202

Angelova N, Hunkeler D (1999) Rationalizing the design of polymeric biomaterials. Trends Biotechnol 17(10):409–421. https://doi.org/10.1016/s0167-7799(99)01356-6

Bagby GW (1988) Arthrodesis by the distraction-compression method using a stainless steel implant. Orthopedics 11(6):931–934. Epub 1988/06/01. PubMed PMID: 3387340

Baliga S, Treon K, Craig NJ (2015) Low back pain: current surgical approaches. Asian Spine J 9(4):645–657. https://doi.org/10.4184/asj.2015.9.4.645. Epub 2015/08/05. PubMed PMID: 26240729; PMCID: PMC4522460

Battie MC, Joshi AB, Gibbons LE, Group IDSP (2019) Degenerative disc disease: what is in a name? Spine 44(21):1523–1529. https://doi.org/10.1097/BRS.0000000000003103. Epub 2019/05/29. PubMed PMID: 31135628

Black CRM, Goriainov V, Gibbs D, Kanczler J, Tare RS, Oreffo ROC (2015) Bone tissue engineering. Curr Mol Biol Rep 1(3):132–140. https://doi.org/10.1007/s40610-015-0022-2

Boucher HH (1959) A method of spinal fusion. J Bone Joint Surg Br 41-B(2):248–259. Epub 1959/05/01. PubMed PMID: 13641310

Boyan BD, Olivares-Navarrete R, Hyzy SL, Slosar PJ, Ullrich PF, Schwartz Z (2014) Osteoblasts secrete pro-inflammatory cytokines on PEEK but anti-inflammatory cytokines on microstructured titanium. Spine J 14(11). https://doi.org/10.1016/j.spinee.2014.08.221

Brantigan JW, Steffee AD (1993) A carbon fiber implant to aid interbody lumbar fusion. Spine 18(Suppl):2106–2117. https://doi.org/10.1097/00007632-199310001-00030

Brinjikji W, Luetmer PH, Comstock B, Bresnahan BW, Chen LE, Deyo RA, Halabi S, Turner JA, Avins AL, James K, Wald JT, Kallmes DF, Jarvik JG (2015) Systematic literature review of imaging features of spinal degeneration in asymptomatic populations. AJNR Am J Neuroradiol 36(4):811–816. https://doi.org/10.3174/ajnr.A4173. Epub 2014/11/29. PubMed PMID: 25430861; PMCID: PMC4464797

Buck E, Li H, Cerruti M (2020) Surface modification strategies to improve the osseointegration of poly (etheretherketone) and its composites. Macromol Biosci 20(2):e1900271. https://doi.org/10.1002/mabi.201900271. Epub 2019/11/30. PubMed PMID: 31782906

Burmester A, Luthringer B, Willumeit R, Feyerabend F (2014) Comparison of the reaction of bone-derived cells to enhanced MgCl2-salt concentrations. Biomatter 4(1). https://doi.org/10.4161/21592527.2014.967616

Buser Z, Brodke DS, Youssef JA, Meisel HJ, Myhre SL, Hashimoto R, Park JB, Tim Yoon S, Wang JC (2016) Synthetic bone graft versus autograft or allograft for spinal fusion: a systematic review. J Neurosurg Spine 25(4):509–516. https://doi.org/10.3171/2016.1.SPINE151005. Epub 2016/05/28. PubMed PMID: 27231812

Buser Z, Ortega B, D'Oro A, Pannell W, Cohen JR, Wang J, Golish R, Reed M, Wang JC (2018) Spine Degenerative Conditions and Their Treatments: National Trends in the United States of America. Global Spine J 8(1):57–67. https://doi.org/10.1177/2192568217696688. Epub 2018/02/20. PubMed PMID: 29456916; PMCID: PMC5810888

Bydon A, Dasenbrock HH, Pendleton C, McGirt MJ, Gokaslan ZL, Quinones-Hinojosa A (2011) Harvey cushing, the spine surgeon: the surgical treatment of Pott disease. Spine 36(17):1420–1425. https://doi.org/10.1097/BRS.0b013e3181f2a2c6. Epub 2011/01/13. PubMed PMID: 21224751; PMCID: PMC4612634

Carpenter RD, Klosterhoff BS, Torstrick FB, Foley KT, Burkus JK, Lee CSD, Gall K, Guldberg RE, Safranski DL (2018) Effect of porous orthopaedic implant material and structure on load sharing with simulated bone ingrowth: a finite element analysis comparing titanium and PEEK. J Mech Behav Biomed Mater 80:68–76. https://doi.org/10.1016/j.jmbbm.2018.01.017

Chaijareenont P, Prakhamsai S, Silthampitag P, Takahashi H, Arksornnukit M (2018) Effects of different sulfuric acid etching concentrations on PEEK surface bonding to resin composite. Dent Mater J 37(3):385–392. https://doi.org/10.4012/dmj.2017-141

Chen Z, Klein T, Murray RZ, Crawford R, Chang J, Wu C, Xiao Y (2016) Osteoimmunomodulation for the development of advanced bone biomaterials. Mater Today 19(6):304–321. https://doi.org/10.1016/j.mattod.2015.11.004

Cheng BC, Koduri S, Wing CA, Woolery N, Cook DJ, Spiro RC (2018) Porous titanium-coated polyetheretherketone implants exhibit an improved bone-implant interface: an in vitro and in vivo biochemical, biomechanical, and histological study. Med Devices 11:391–402. https://doi.org/10.2147/MDER.S180482. Epub 2018/11/23. PubMed PMID: 30464653; PMCID: PMC6211303

Chin W-C, Tsai S-W, Liou H-M, Lin C-J, Kuo K-L, Hung Y-S, Weng R-C, Hsu F-Y (2012) MG63 osteoblast-like cells exhibit different behavior when grown on electrospun collagen matrix versus electrospun gelatin matrix. PLoS One 7(2). https://doi.org/10.1371/journal.pone.0031200

Chong E, Pelletier MH, Mobbs RJ, Walsh WR (2015) The design evolution of interbody cages in anterior cervical discectomy and fusion: a systematic review. BMC Musculoskelet Disord 16(1). https://doi.org/10.1186/s12891-015-0546-x

Chou PH, Ma HL, Liu CL, Wang ST, Lee OK, Chang MC, Yu WK (2016) Is removal of the implants needed after fixation of burst fractures of the thoracolumbar and lumbar spine without fusion? A retrospective evaluation of radiological and functional outcomes. Bone Joint J 98-B(1):109–116. https://doi.org/10.1302/0301-620X.98B1.35832. Epub 2016/01/07. PubMed PMID: 26733523

Choy WJ, Abi-Hanna D, Cassar LP, Hardcastle P, Phan K, Mobbs RJ (2019) History of integral fixation for anterior lumbar Interbody fusion (ALIF): the Hartshill Horseshoe. World Neurosurg 129:394–400. https://doi.org/10.1016/j.wneu.2019.06.134. Epub 2019/06/30. PubMed PMID: 31254709

Cole CD, McCall TD, Schmidt MH, Dailey AT (2009) Comparison of low back fusion techniques: transforaminal lumbar interbody fusion (TLIF) or posterior lumbar interbody fusion (PLIF) approaches. Curr Rev Musculoskelet Med 2(2):118–126. https://doi.org/10.1007/s12178-009-9053-8. Epub 2009/05/27. PubMed PMID: 19468868; PMCID: PMC2697340

Dang M, Saunders L, Niu X, Fan Y, Ma PX (2018) Biomimetic delivery of signals for bone tissue engineering. Bone Res 6(1). https://doi.org/10.1038/s41413-018-0025-8

Danison AP, Lee DJ, Panchal RR (2017) Temporary stabilization of unstable spine fractures. Curr Rev Musculoskelet Med 10(2):199–206. https://doi.org/10.1007/s12178-017-9402-y. Epub 2017/03/21. PubMed PMID: 28316056; PMCID: PMC5435633

Daoust D, Godard P, Devaux J, Legras R, Strazielle C (1994) Chemical modification of poly(ether ether ketone) for size exclusion chromatography at room temperature: 1. Absolute molecular-mass determination for sulfonated PEEK. Polymer 35(25):5491–5497. https://doi.org/10.1016/s0032-3861(05)80013-8

de Wildt BWM, Ansari S, Sommerdijk NAJM, Ito K, Akiva A, Hofmann S (2019) From bone regeneration to three-dimensional in vitro models: tissue engineering of organized bone extracellular matrix. Curr Opin Biomed Eng 10:107–115. https://doi.org/10.1016/j.cobme.2019.05.005

DeBowes RM, Grant BD, Bagby GW, Gallina AM, Sande RD, Ratzlaff MH (1984) Cervical vertebral interbody fusion in the horse: a comparative study of bovine xenografts and autografts supported by stainless steel baskets. Am J Vet Res 45(1):191–199. Epub 1984/01/01. PubMed PMID: 6367560

Díez-Pascual AM, Martínez G, Gómez MA (2009) Synthesis and characterization of poly(ether ether ketone) derivatives obtained by carbonyl reduction. Macromolecules 42(18):6885–6892. https://doi.org/10.1021/ma901208e

Dommisse GF (1959) Lumbo-sacral interbody spinal fusion. J Bone Joint Surg Br 41-B(1):87–95. Epub 1959/02/01. PubMed PMID: 13620711

Donnally IC, Hanna A, Varacallo M (2020) Lumbar degenerative disk disease. StatPearls, Treasure Island

Elfering A, Semmer N, Birkhofer D, Zanetti M, Hodler J, Boos N (2002) Risk factors for lumbar disc degeneration: a 5-year prospective MRI study in asymptomatic individuals. Spine 27(2):125–134. https://doi.org/10.1097/00007632-200201150-00002. Epub 2002/01/24. PubMed PMID: 11805656

Enders JJ, Coughlin D, Mroz TE, Vira S (2020) Surface Technologies in Spinal Fusion. Neurosurg Clin N Am 31(1):57–64. https://doi.org/10.1016/j.nec.2019.08.007. Epub 2019/11/20. PubMed PMID: 31739930

Erik H, Roger W, Richard W, Liam G (2016) Chemically linked PEEK/HA composite - mechanical properties and in vitro cell response. Front Bioeng Biotechnol 4. https://doi.org/10.3389/conf.FBIOE.2016.01.00490

Evans NT, Torstrick FB, Lee CS, Dupont KM, Safranski DL, Chang WA, Macedo AE, Lin AS, Boothby JM, Whittingslow DC, Carson RA, Guldberg RE, Gall K (2015) High-strength, surface-porous polyether-ether-ketone for load-bearing orthopedic implants. Acta Biomater 13:159–167. https://doi.org/10.1016/j.actbio.2014.11.030. Epub 2014/12/03. PubMed PMID: 25463499; PMCID: PMC4294703

Fernandez de Grado G, Keller L, Idoux-Gillet Y, Wagner Q, Musset AM, Benkirane-Jessel N, Bornert F, Offner D (2018) Bone substitutes: a review of their characteristics, clinical use, and perspectives for large bone defects management. J Tissue Eng 9:2041731418776819. https://doi.org/10.1177/2041731418776819. Epub 2018/06/15. PubMed PMID: 29899969; PMCID: PMC5990883

Franchina NL, McCarthy TJ (1991) Surface modifications of poly(ether ether ketone). Macromolecules 24(11):3045–3049. https://doi.org/10.1021/ma00011a003

Friedly J, Standaert C, Chan L (2010) Epidemiology of spine care: the back pain dilemma. Phys Med Rehabil Clin N Am 21(4):659–677. https://doi.org/10.1016/j.pmr.2010.08.002. Epub 2010/10/28. PubMed PMID: 20977955; PMCID: PMC3404131

Fu L, Omi M, Sun M, Cheng B, Mao G, Liu T, Mendonça G, Averick SE, Mishina Y, Matyjaszewski K (2019) Covalent attachment of P15 peptide to Ti alloy surface modified with polymer to enhance Osseointegration of implants. ACS Appl Mater Interfaces 11(42):38531–38536. https://doi.org/10.1021/acsami.9b14651

Fukuda N, Kanazawa M, Tsuru K, Tsuchiya A, Sunarso, Toita R, Mori Y, Nakashima Y, Ishikawa K (2018) Synergistic effect of surface phosphorylation and micro-roughness on enhanced osseointegration ability of poly(ether ether ketone) in the rabbit tibia. Sci Rep 8(1). https://doi.org/10.1038/s41598-018-35313-7

Gao C, Peng S, Feng P, Shuai C (2017) Bone biomaterials and interactions with stem cells. Bone Res 5(1). https://doi.org/10.1038/boneres.2017.59

Gardon M, Melero H, Garcia-Giralt N, Dosta S, Cano IG, Guilemany JM (2014) Enhancing the bioactivity of polymeric implants by means of cold gas spray coatings.

J Biomed Mater Res B Appl Biomater 102(7):1537–1543. https://doi.org/10.1002/jbm.b.33134. Epub 2014/03/07. PubMed PMID: 24599842

Gittens RA, Olivares-Navarrete R, Schwartz Z, Boyan BD (2014) Implant osseointegration and the role of microroughness and nanostructures: lessons for spine implants. Acta Biomater 10(8):3363–3371. https://doi.org/10.1016/j.actbio.2014.03.037. Epub 2014/04/12. PubMed PMID: 24721613; PMCID: PMC4103432

Gupta A, Kukkar N, Sharif K, Main BJ, Albers CE, El-Amin Iii SF (2015) Bone graft substitutes for spine fusion: a brief review. World J Orthop 6(6):449–456. https://doi.org/10.5312/wjo.v6.i6.449. Epub 2015/07/21. PubMed PMID: 26191491; PMCID: PMC4501930

Hasegawa T, Ushirozako H, Shigeto E, Ohba T, Oba H, Mukaiyama K, Shimizu S, Yamato Y, Ide K, Shibata Y, Ojima T, Takahashi J, Haro H, Matsuyama Y (2020) The titanium-coated PEEK cage maintains better bone fusion with the endplate than the PEEK cage 6 months after PLIF surgery-a multicenter, prospective, randomized study. Spine. https://doi.org/10.1097/BRS.0000000000003464. Epub 2020/03/10. PubMed PMID: 32150128

Hayashi D, Roemer FW, Mian A, Gharaibeh M, Muller B, Guermazi A (2012) Imaging features of postoperative complications after spinal surgery and instrumentation. AJR Am J Roentgenol 199(1):W123–W129. https://doi.org/10.2214/AJR.11.6497. Epub 2012/06/27. PubMed PMID: 22733920

Heary RF, Parvathreddy N, Sampath S, Agarwal N (2017) Elastic modulus in the selection of interbody implants. J Spine Surg 3(2):163–167. https://doi.org/10.21037/jss.2017.05.01. Epub 2017/07/27. PubMed PMID: 28744496; PMCID: PMC5506312

Herrera Herrera I, Moreno de la Presa R, Gonzalez Gutierrez R, Barcena Ruiz E, Garcia Benassi JM (2013) Evaluation of the postoperative lumbar spine. Radiologia 55(1):12–23. https://doi.org/10.1016/j.rx.2011.12.004. Epub 2012/04/24. PubMed PMID: 22520556

Honigmann P, Sharma N, Okolo B, Popp U, Msallem B, Thieringer FM (2018) Patient-specific surgical implants made of 3D printed PEEK: material, technology, and scope of surgical application. Biomed Res Int 2018:1–8. https://doi.org/10.1155/2018/4520636

Hoshijima K, Nightingale RW, Yu JR, Richardson WJ, Harper KD, Yamamoto H, Myers BS (1997) Strength and stability of posterior lumbar interbody fusion. Comparison of titanium fiber mesh implant and tricortical bone graft. Spine 22(11):1181–1188. https://doi.org/10.1097/00007632-199706010-00002. Epub 1997/06/01. PubMed PMID: 9201853

Hwang PW, Horton JA (2019) Variable osteogenic performance of MC3T3-E1 subclones impacts their utility as models of osteoblast biology. Sci Rep 9(1). https://doi.org/10.1038/s41598-019-44575-8

Ikada Y (1994) Surface modification of polymers for medical applications. Biomaterials 15(10):725–736. https://doi.org/10.1016/0142-9612(94)90025-6

Jadlowiec J, Koch H, Zhang X, Campbell PG, Seyedain M, Sfeir C (2004) Phosphophoryn regulates the gene expression and differentiation of NIH3T3, MC3T3-E1, and human mesenchymal stem cells via the integrin/MAPK signaling pathway. J Biol Chem 279(51):53323–53330. https://doi.org/10.1074/jbc.M404934200. PubMed PMID: 15371433

Jain S, Eltorai AE, Ruttiman R, Daniels AH (2016) Advances in spinal interbody cages. Orthop Surg 8(3):278–284. https://doi.org/10.1111/os.12264. Epub 2016/09/15. PubMed PMID: 27627709; PMCID: PMC6584167

Kadam A, Millhouse PW, Kepler CK, Radcliff KE, Fehlings MG, Janssen ME, Sasso RC, Benedict JJ, Vaccaro AR (2016) Bone substitutes and expanders in spine surgery: a review of their fusion efficacies. Int J Spine Surg 10:33. https://doi.org/10.14444/3033. Epub 2016/12/03. PubMed PMID: 27909654; PMCID: PMC5130324

Kale A, Oz II, Onk A, Kalayci M, Buyukuysal C (2017) Unilaterally posterior lumbar interbody fusion with double expandable peek cages without pedicle screw support for lumbar disc herniation. Neurol Neurochir Pol 51(1):53–59. https://doi.org/10.1016/j.pjnns.2016.11.001. Epub 2016/12/03. PubMed PMID: 27908615

Kao TH, Wu CH, Chou YC, Chen HT, Chen WH, Tsou HK (2014) Risk factors for subsidence in anterior cervical fusion with stand-alone polyetheretherketone (PEEK) cages: a review of 82 cases and 182 levels. Arch Orthop Trauma Surg 134(10):1343–1351. https://doi.org/10.1007/s00402-014-2047-z. Epub 2014/08/08. PubMed PMID: 25099076; PMCID: PMC4168225

Kassick AJ, Yerneni SS, Gottlieb E, Cartieri F, Peng Y, Mao G, Kharlamov A, Miller MC, Xu C, Oh M, Kowalewski T, Cheng B, Campbell PG, Averick S (2018) Osteoconductive enhancement of polyether ether ketone: a mild covalent surface modification approach. ACS Appl Bio Mater 1(4):1047–1055. https://doi.org/10.1021/acsabm.8b00274

Knott PT, Mardjetko SM, Kim RH, Cotter TM, Dunn MM, Patel ST, Spencer MJ, Wilson AS, Tager DS (2010) A comparison of magnetic and radiographic imaging artifact after using three types of metal rods: stainless steel, titanium, and vitallium. Spine J 10(9):789–794. https://doi.org/10.1016/j.spinee.2010.06.006. Epub 2010/07/14. PubMed PMID: 20619749

Koch H, Jadlowiec JA, Campbell PG (2005) Insulin-like growth factor-I induces early osteoblast gene expression in human mesenchymal stem cells. Stem Cells Dev 14(6):621–631. https://doi.org/10.1089/scd.2005.14.621. PubMed PMID: 16433617

Krause KL, Obayashi JT, Bridges KJ, Raslan AM, Than KD (2018) Fivefold higher rate of pseudarthrosis with polyetheretherketone interbody device than with structural allograft used for 1-level anterior cervical discectomy and fusion. J Neurosurg Spine 30(1):46–51. https://doi.org/10.3171/2018.7.SPINE18531. Epub 2018/11/30. PubMed PMID: 30485200

Kumar R, Lerski RA, Gandy S, Clift BA, Abboud RJ (2006) Safety of orthopedic implants in magnetic resonance imaging: an experimental verification. J Orthop Res 24(9):1799–1802. https://doi.org/10.1002/jor.20213. Epub 2006/07/14. PubMed PMID: 16838376

Kurtz SM (2012) Synthesis and processing of PEEK for surgical implants. In: PEEK biomaterials handbook. Kurtz: William Andrew, pp 9–22

Kurtz SM, Devine JN (2007) PEEK biomaterials in trauma, orthopedic, and spinal implants. Biomaterials 28(32):4845–4869. https://doi.org/10.1016/j.biomaterials.2007.07.013. Epub 2007/08/10. PubMed PMID: 17686513; PMCID: PMC2040108

Kurucan E, Bernstein DN, Thirukumaran C, Jain A, Menga EN, Rubery PT, Mesfin A (2018) National trends in spinal fusion surgery for neurofibromatosis. Spine Deform 6(6):712–718. https://doi.org/10.1016/j.jspd.2018.03.012. Epub 2018/10/24. PubMed PMID: 30348349

Kutikov AB, Skelly JD, Ayers DC, Song J (2015) Templated repair of long bone defects in rats with bioactive spiral-wrapped electrospun Amphiphilic polymer/hydroxyapatite scaffolds. ACS Appl Mater Interfaces 7(8):4890–4901. https://doi.org/10.1021/am508984y

Lee CS, Hwang CJ, Lee DH, Kim YT, Lee HS (2011) Fusion rates of instrumented lumbar spinal arthrodesis according to surgical approach: a systematic review of randomized trials. Clin Orthop Surg 3(1):39–47. https://doi.org/10.4055/cios.2011.3.1.39. Epub 2011/03/04. PubMed PMID: 21369477; PMCID: PMC3042168

Levin A, Sharma V, Hook L, García-Gareta E (2018) The importance of factorial design in tissue engineering and biomaterials science: optimisation of cell seeding efficiency on dermal scaffolds as a case study. J Tissue Eng 9. https://doi.org/10.1177/2041731418781696

Lewallen EA, Riester SM, Bonin CA, Kremers HM, Dudakovic A, Kakar S, Cohen RC, Westendorf JJ, Lewallen DG, van Wijnen AJ (2015) Biological strategies for improved osseointegration and osteoinduction of porous metal orthopedic implants. Tissue Eng Part B Rev 21(2):218–230. https://doi.org/10.1089/ten.TEB.2014.0333. Epub 2014/10/29. PubMed PMID: 25348836; PMCID: PMC4390115

Li ZJ, Wang Y, Xu GJ, Tian P (2016) Is PEEK cage better than titanium cage in anterior cervical discectomy and fusion surgery? A meta-analysis. BMC Musculoskelet Disord 17:379. https://doi.org/10.1186/s12891-016-1234-1. Epub 2016/09/03. PubMed PMID: 27585553; PMCID: PMC5009677

Lian JB, Stein GS (1995) Development of the osteoblast phenotype: molecular mechanisms mediating osteoblast growth and differentiation. Iowa Orthop J 15:118–140. Epub 1995/01/01. PubMed PMID: 7634023; PMCID: PMC2329080

Lopes D, Martins-Cruz C, Oliveira MB, Mano JF (2018) Bone physiology as inspiration for tissue regenerative therapies. Biomaterials 185:240–275. https://doi.org/10.1016/j.biomaterials.2018.09.028

Lucke A, Teßmar J, Schnell E, Schmeer G, Göpferich A (2000) Biodegradable poly(d,l-lactic acid)-poly (ethylene glycol)-monomethyl ether diblock copolymers: structures and surface properties relevant to their use as biomaterials. Biomaterials 21(23):2361–2370. https://doi.org/10.1016/s0142-9612(00)00103-4

Lykissas MG, Aichmair A (2013) Current concepts on spinal arthrodesis in degenerative disorders of the lumbar spine. World J Clin Cases 1(1):4–12. https://doi.org/10.12998/wjcc.v1.i1.4. Epub 2013/12/05. PubMed PMID: 24303453; PMCID: PMC3845930

Ma R, Guo D (2019) Evaluating the bioactivity of a hydroxyapatite-incorporated polyetheretherketone biocomposite. J Orthop Surg Res 14(1). https://doi.org/10.1186/s13018-019-1069-1

Ma R, Tang S, Tan H, Qian J, Lin W, Wang Y, Liu C, Wei J, Tang T (2014) Preparation, characterization, in vitro bioactivity, and cellular responses to a Polyetheretherketone bioactive composite containing Nanocalcium silicate for bone repair. ACS Appl Mater Interfaces 6(15):12214–12225. https://doi.org/10.1021/am504409q

Manoukian OS, Sardashti N, Stedman T, Gailiunas K, Ojha A, Penalosa A, Mancuso C, Hobert M, Kumbar SG (2019) Biomaterials for tissue engineering and regenerative medicine. In: Encyclopedia of biomedical engineering. Manoukian: Academic Press, pp 462–482

Martin BI, Mirza SK, Spina N, Spiker WR, Lawrence B, Brodke DS (2019) Trends in lumbar fusion procedure rates and associated hospital costs for degenerative spinal diseases in the United States, 2004 to 2015. Spine 44(5):369–376. https://doi.org/10.1097/BRS.0000000000002822. Epub 2018/08/04. PubMed PMID: 30074971

Martz EO, Goel VK, Pope MH, Park JB (1997) Materials and design of spinal implants – a review. J Biomed Mater Res 38(3):267–288. https://doi.org/10.1002/(sici)1097-4636(199723)38:3<267::aid-jbm12>3.0.co;2-8. Epub 1997/10/01. PubMed PMID: 9283973

Mastrogiacomo M, Muraglia A, Komlev V, Peyrin F, Rustichelli F, Crovace A, Cancedda R (2005) Tissue engineering of bone: search for a better scaffold. Orthod Craniofacial Res 8(4):277–284. https://doi.org/10.1111/j.1601-6343.2005.00350.x

Mattioli-Belmonte M, Teti G, Salvatore V, Focaroli S, Orciani M, Dicarlo M, Fini M, Orsini G, Di Primio R, Falconi M (2015) Stem cell origin differently affects bone tissue engineering strategies. Front Physiol 6. https://doi.org/10.3389/fphys.2015.00266

Mihailidis HG, Manners S, Churilov L, Quan GMY (2017) Is spinal surgery safe in octogenarians? ANZ J Surg 87(7-8):605–609. https://doi.org/10.1111/ans.13885. Epub 2017/01/27. PubMed PMID: 28124479

Mobbs RJ, Loganathan A, Yeung V, Rao PJ (2013) Indications for anterior lumbar interbody fusion. Orthop Surg 5(3):153–163. https://doi.org/10.1111/os.12048. Epub 2013/09/05. PubMed PMID: 24002831; PMCID: PMC6583544

Ni J, Ling H, Zhang S, Wang Z, Peng Z, Benyshek C, Zan R, Miri AK, Li Z, Zhang X, Lee J, Lee KJ, Kim HJ, Tebon P, Hoffman T, Dokmeci MR, Ashammakhi N,

Li X, Khademhosseini A (2019) Three-dimensional printing of metals for biomedical applications. Mater Today Bio 3:100024. https://doi.org/10.1016/j.mtbio.2019.100024. Epub 2020/03/12. PubMed PMID: 32159151; PMCID: PMC7061633

Nikolova MP, Chavali MS (2019) Recent advances in biomaterials for 3D scaffolds: a review. Bioact Mater 4:271–292. https://doi.org/10.1016/j.bioactmat.2019.10.005. Epub 2019/11/12. PubMed PMID: 31709311; PMCID: PMC6829098

Noori A, Ashrafi SJ, Vaez-Ghaemi R, Hatamian-Zaremi A, Webster TJ (2017) A review of fibrin and fibrin composites for bone tissue engineering. Int J Nanomedicine 12:4937–4961. https://doi.org/10.2147/ijn.S124671

Nouh MR (2012) Spinal fusion-hardware construct: Basic concepts and imaging review. World J Radiol 4(5):193–207. https://doi.org/10.4329/wjr.v4.i5.193. Epub 2012/07/05. PubMed PMID: 22761979; PMCID: PMC3386531

Olivares-Navarrete R, Hyzy SL, Slosar PJ, Schneider JM, Schwartz Z, Boyan BD (2015) Implant materials generate different peri-implant inflammatory factors: polyether-ether-ketone promotes fibrosis and microtextured titanium promotes osteogenic factors. Spine 40(6):399–404. https://doi.org/10.1097/BRS.0000000000000778. Epub 2015/01/15. PubMed PMID: 25584952; PMCID: PMC4363266

Ordway NR, Lu YM, Zhang X, Cheng CC, Fang H, Fayyazi AH (2007) Correlation of cervical endplate strength with CT measured subchondral bone density. Eur Spine J 16(12):2104–2109. https://doi.org/10.1007/s00586-007-0482-z. Epub 2007/08/23. PubMed PMID: 17712574; PMCID: PMC2140123

Ouyang L, Zhao Y, Jin G, Lu T, Li J, Qiao Y, Ning C, Zhang X, Chu PK, Liu X (2016) Influence of sulfur content on bone formation and antibacterial ability of sulfonated PEEK. Biomaterials 83:115–126. https://doi.org/10.1016/j.biomaterials.2016.01.017. PubMed PMID: 26773668

Parker D, Bussink J, van de Grampel HT, Wheatley GW, Dorf E-U, Ostlinning E, Reinking K, Schubert F, Jünger O (2012) Polymers, high-temperature. In: Ullmann's encyclopedia of industrial chemistry. Parker: Wiley

Patel DV, Yoo JS, Karmarkar SS, Lamoutte EH, Singh K (2019) Interbody options in lumbar fusion. J Spine Surg 5(Suppl 1):S19–S24. https://doi.org/10.21037/jss.2019.04.04. Epub 2019/08/06. PubMed PMID: 31380489; PMCID: PMC6626748

Pelletier MH, Cordaro N, Punjabi VM, Waites M, Lau A, Walsh WR (2016) PEEK versus Ti interbody fusion devices: resultant fusion, bone apposition, initial and 26-week biomechanics. Clin Spine Surg 29(4):E208–E214. https://doi.org/10.1097/BSD.0b013e31826851a4. Epub 2012/07/18. PubMed PMID: 22801456

Petersmann S, Spoerk M, Van De Steene W, Ucal M, Wiener J, Pinter G, Arbeiter F (2020) Mechanical properties of polymeric implant materials produced by extrusion-based additive manufacturing. J Mech Behav Biomed Mater 104:103611. https://doi.org/10.1016/j.jmbbm.2019.103611. Epub 2020/01/14. PubMed PMID: 31929095

Phan K, Mobbs RJ (2016) Evolution of design of interbody cages for anterior lumbar interbody fusion. Orthop Surg 8(3):270–277. https://doi.org/10.1111/os.12259

Pirraco RP, Marques AP, Reis RL (2010) Cell interactions in bone tissue engineering. J Cell Mol Med 14(1–2):93–102. https://doi.org/10.1111/j.1582-4934.2009.01005.x

Polo-Corrales L, Latorre-Esteves M, Ramirez-Vick JE (2014) Scaffold design for bone regeneration. J Nanosci Nanotechnol 14(1):15–56. https://doi.org/10.1166/jnn.2014.9127. Epub 2014/04/16. PubMed PMID: 24730250; PMCID: PMC3997175

Poulsson AHC, Eglin D, Richards RG (2019) Surface modification techniques of PEEK, including plasma surface treatment. In: PEEK biomaterials handbook. Poulsson: William Andrew, pp 179–201

Proietti L, Scaramuzzo L, Schiro GR, Sessa S, Logroscino CA (2013) Complications in lumbar spine surgery: a retrospective analysis. Indian J Orthop 47(4):340–345. https://doi.org/10.4103/0019-5413.114909. Epub 2013/08/21. PubMed PMID: 23960276; PMCID: PMC3745686

Prolo D (1969) Harvey cushing selected papers on neurosurgery. Arch Surg 99(5). https://doi.org/10.1001/archsurg.1969.01340170128031

Qu H, Fu H, Han Z, Sun Y (2019) Biomaterials for bone tissue engineering scaffolds: a review. RSC Adv 9(45):26252–26262. https://doi.org/10.1039/c9ra05214c

Ravindra VM, Senglaub SS, Rattani A, Dewan MC, Hartl R, Bisson E, Park KB, Shrime MG (2018) Degenerative lumbar spine disease: estimating global incidence and worldwide volume. Global Spine J 8(8):784–794. https://doi.org/10.1177/2192568218770769. Epub 2018/12/19. PubMed PMID: 30560029; PMCID: PMC6293435

Reid PC, Morr S, Kaiser MG (2019) State of the union: a review of lumbar fusion indications and techniques for degenerative spine disease. J Neurosurg Spine 31(1):1–14. https://doi.org/10.3171/2019.4.SPINE18915. Epub 2019/07/02. PubMed PMID: 31261133

Richbourg NR, Peppas NA, Sikavitsas VI (2019) Tuning the biomimetic behavior of scaffolds for regenerative medicine through surface modifications. J Tissue Eng Regen Med 13(8):1275–1293. https://doi.org/10.1002/term.2859. Epub 2019/04/05. PubMed PMID: 30946537; PMCID: PMC6715496

Rupp R, Ebrahim NA, Savolaine ER, Jackson WT (1993) Magnetic resonance imaging evaluation of the spine with metal implants. General safety and superior imaging with titanium. Spine 18(3):379–385. Epub 1993/03/01. PubMed PMID: 8475442

Sardana H, Sahu R, Kedia S (2019) Letter to the editor. PEEK interbody device and pseudarthrosis. J Neurosurg Spine 31(4):617–618. https://doi.org/10.3171/2019.3.Spine19244

Seaman S, Kerezoudis P, Bydon M, Torner JC, Hitchon PW (2017) Titanium vs. polyetheretherketone (PEEK) interbody fusion: meta-analysis and review of the literature. J Clin Neurosci 44:23–29. https://doi.org/10.1016/j.jocn.2017.06.062. Epub 2017/07/25. PubMed PMID: 28736113

Selim A, Mercer S, Tang F (2018) Polyetheretherketone (PEEK) rods for lumbar fusion: a systematic review and meta-analysis. Int J Spine Surg 12(2):190–200. https://doi.org/10.14444/5027. Epub 2018/10/03. PubMed PMID: 30276079; PMCID: PMC6159719

Shayesteh Moghaddam N, Taheri Andani M, Amerinatanzi A, Haberland C, Huff S, Miller M, Elahinia M, Dean D (2016) Metals for bone implants: safety, design, and efficacy. Biomanuf Rev 1(1). https://doi.org/10.1007/s40898-016-0001-2

Shibuya N, Porter RS (1992) Kinetics of PEEK sulfonation in concentrated sulfuric acid. Macromolecules 25(24):6495–6499. https://doi.org/10.1021/ma00050a017

Shukla D, Negi YS, Uppadhyaya JS, Kumar V (2012) Synthesis and modification of poly(ether ether ketone) and their properties: a review. Polym Rev 52(2):189–228. https://doi.org/10.1080/15583724.2012.668151

Sohn HS, Oh JK (2019) Review of bone graft and bone substitutes with an emphasis on fracture surgeries. Biomater Res 23:9. https://doi.org/10.1186/s40824-019-0157-y. Epub 2019/03/28. PubMed PMID: 30915231; PMCID: PMC6417250

Spoor AB, Oner FC (2013) Minimally invasive spine surgery in chronic low back pain patients. J Neurosurg Sci 57(3):203–218. Epub 2013/07/24. PubMed PMID: 23877267

Steelman T, Lewandowski L, Helgeson M, Wilson K, Olsen C, Gwinn D (2018) Population-based risk factors for the development of degenerative disk disease. Clin Spine Surg 31(8):E409–EE12. https://doi.org/10.1097/BSD.0000000000000682. Epub 2018/07/10. PubMed PMID: 29985801

Steffen T, Tsantrizos A, Aebi M (2000) Effect of implant design and endplate preparation on the compressive strength of interbody fusion constructs. Spine 25(9):1077–1084. https://doi.org/10.1097/00007632-200005010-00007. Epub 2000/05/02. PubMed PMID: 10788851

Stevens MM (2008) Biomaterials for bone tissue engineering. Mater Today 11(5):18–25. https://doi.org/10.1016/s1369-7021(08)70086-5

Tang H, Zhu J, Ji F, Wang S, Xie Y, Fei H (2014) Risk factors for postoperative complication after spinal fusion and instrumentation in degenerative lumbar scoliosis patients. J Orthop Surg Res 9(1):15. https://doi.org/10.1186/1749-799X-9-15. Epub 2014/03/13. PubMed PMID: 24606963; PMCID: PMC3995926

Tarpada SP, Morris MT, Burton DA (2017) Spinal fusion surgery: a historical perspective. J Orthop 14(1):134–136. https://doi.org/10.1016/j.jor.2016.10.029. Epub 2016/11/23. PubMed PMID: 27872518; PMCID: PMC5107724

Teton ZE, Cheaney B, Obayashi JT, Than KD (2020) PEEK interbody devices for multilevel anterior cervical discectomy and fusion: association with more than 6-fold higher rates of pseudarthrosis compared to structural allograft. J Neurosurg Spine:1–7. https://doi.org/10.3171/2019.11.Spine19788

Thakkar RS, Malloy JP, Thakkar SC, Carrino JA, Khanna AJ (2012) Imaging the postoperative spine. Radiol Clin N Am 50(4):731–747. https://doi.org/10.1016/j.rcl.2012.04.006. Epub 2012/05/31. PubMed PMID: 22643393

Torstrick FB, Evans NT, Stevens HY, Gall K, Guldberg RE (2016) Do surface porosity and pore size influence mechanical properties and cellular response to PEEK? Clin Orthop Relat Res 474(11):2373–2383. https://doi.org/10.1007/s11999-016-4833-0

Torstrick FB, Safranski DL, Burkus JK, Chappuis JL, Lee CSD, Guldberg RE, Gall K, Smith KE (2017a) Getting PEEK to stick to bone: the development of porous PEEK for interbody fusion devices. Tech Orthop 32(3):158–166. https://doi.org/10.1097/BTO.0000000000000242. Epub 2017/12/12. PubMed PMID: 29225416; PMCID: PMC5720158

Torstrick FB, Safranski DL, Burkus JK, Chappuis JL, Lee CSD, Guldberg RE, Gall K, Smith KE (2017b) Getting PEEK to stick to bone. Tech Orthop 32(3):158–166. https://doi.org/10.1097/bto.0000000000000242

Toth JM (2012) Biocompatibility of polyaryletheretherketone polymers. In: PEEK biomaterials handbook. Toth: William Andrew, pp 81–92

Toth JM, Wang M, Estes BT, Scifert JL, Seim HB 3rd, Turner AS (2006) Polyetheretherketone as a biomaterial for spinal applications. Biomaterials 27(3):324–334. https://doi.org/10.1016/j.biomaterials.2005.07.011. Epub 2005/08/24. PubMed PMID: 16115677

Turnbull G, Clarke J, Picard F, Riches P, Jia L, Han F, Li B, Shu W (2018) 3D bioactive composite scaffolds for bone tissue engineering. Bioactive Mater 3(3):278–314. https://doi.org/10.1016/j.bioactmat.2017.10.001

Tuzmen C, Campbell PG (2018) Crosstalk between neuropeptides SP and CGRP in regulation of BMP2-induced bone differentiation. Connect Tissue Res 59(Suppl 1):81–90. https://doi.org/10.1080/03008207.2017.1408604. PubMed PMID: 29745819

Vaishnav AS, McAnany SJ (2019) Future endeavors in ambulatory spine surgery. J Spine Surg 5(Suppl 2):S139–SS46. https://doi.org/10.21037/jss.2019.09.20. Epub 2019/10/28. PubMed PMID: 31656867; PMCID: PMC6790808

Walsh WR, Pelletier MH, Bertollo N, Christou C, Tan C (2016) Does PEEK/HA enhance bone formation compared with PEEK in a sheep cervical fusion model? Clin Orthop Relat Res 474(11):2364–2372. https://doi.org/10.1007/s11999-016-4994-x

Wang W, Luo CJ, Huang J, Edirisinghe M (2019) PEEK surface modification by fast ambient-temperature sulfonation for bone implant applications. J R Soc Interface 16(152). https://doi.org/10.1098/rsif.2018.0955

Warburton A, Girdler SJ, Mikhail CM, Ahn A, Cho SK (2019) Biomaterials in spinal implants: a review. Neurospine. https://doi.org/10.14245/ns.1938296.148

Weinstein JN, Lurie JD, Olson PR, Bronner KK, Fisher ES (2006) United States' trends and regional variations in lumbar spine surgery: 1992–2003. Spine 31(23):2707–2714. https://doi.org/10.1097/01.brs.0000248132.15231.fe. Epub 2006/11/02. PubMed PMID: 17077740; PMCID: PMC2913862

Wenz LM, Merritt K, Brown SA, Moet A, Steffee AD (1990) In vitro biocompatibility of polyetheretherketone and polysulfone composites. J Biomed Mater Res 24(2):207–215. https://doi.org/10.1002/jbm.820240207

Wilcox B, Mobbs RJ, Wu AM, Phan K (2017) Systematic review of 3D printing in spinal surgery: the current state of play. J Spine Surg 3(3):433–443. https://doi.org/10.21037/jss.2017.09.01. Epub 2017/10/24. PubMed PMID: 29057355; PMCID: PMC5637198

Winebrake JP, Lovecchio F, Steinhaus M, Farmer J, Sama A (2020) Wide variability in patient-reported outcomes measures after fusion for lumbar spinal stenosis: a systematic review. Global Spine J 10(2):209–215. https://doi.org/10.1177/2192568219832853. Epub 2020/03/25. PubMed PMID: 32206520; PMCID: PMC7076598

Winter RB, Lonstein JE (1999) Congenital scoliosis with posterior spinal arthrodesis T2-L3 at age 3 years with 41-year follow-up. A case report. Spine (Phila Pa 1976) 24(2):194–197. https://doi.org/10.1097/00007632-199901150-00023. Epub 1999/02/02. PubMed PMID: 9926393

Yang YJ, Tsou HK, Chen YH, Chung CJ, He JL (2015) Enhancement of bioactivity on medical polymer surface using high power impulse magnetron sputtered titanium dioxide film. Mater Sci Eng C Mater Biol Appl 57:58–66. https://doi.org/10.1016/j.msec.2015.07.039. Epub 2015/09/12. PubMed PMID: 26354240

Yang YK, Ogando CR, Wang See C, Chang TY, Barabino GA (2018) Changes in phenotype and differentiation potential of human mesenchymal stem cells aging in vitro. Stem Cell Res Ther 9(1):131. https://doi.org/10.1186/s13287-018-0876-3. Epub 2018/05/13. PubMed PMID: 29751774; PMCID: PMC5948736

Yavin D, Casha S, Wiebe S, Feasby TE, Clark C, Isaacs A, Holroyd-Leduc J, Hurlbert RJ, Quan H, Nataraj A, Sutherland GR, Jette N (2017) Lumbar fusion for degenerative disease: a systematic review and meta-analysis. Neurosurgery 80(5):701–715. https://doi.org/10.1093/neuros/nyw162. Epub 2017/03/23. PubMed PMID: 28327997

Yoo C-K, Jeon J-Y, Kim Y-J, Kim S-G, Hwang K-G (2016) Cell attachment and proliferation of osteoblast-like MG63 cells on silk fibroin membrane for guided bone regeneration. Maxillofac Plast Reconstr Surg 38(1). https://doi.org/10.1186/s40902-016-0062-4

Zadpoor AA (2015) Bone tissue regeneration: the role of scaffold geometry. Biomater Sci 3(2):231–245. https://doi.org/10.1039/c4bm00291a

Zhang J, Pan A, Zhou L, Yu J, Zhang X (2018) Comparison of unilateral pedicle screw fixation and interbody fusion with PEEK cage vs. standalone expandable fusion cage for the treatment of unilateral lumbar disc herniation. Arch Med Sci 14(6):1432–1438. https://doi.org/10.5114/aoms.2018.74890. Epub 2018/11/06. PubMed PMID: 30393499; PMCID: PMC6209698

Zhang J, Tian W, Chen J, Yu J, Zhang J, Chen J (2019) The application of polyetheretherketone (PEEK) implants in cranioplasty. Brain Res Bull 153:143–149. https://doi.org/10.1016/j.brainresbull.2019.08.010. Epub 2019/08/20. PubMed PMID: 31425730

Zhao Y, Wong HM, Wang W, Li P, Xu Z, Chong EY, Yan CH, Yeung KW, Chu PK (2013) Cytocompatibility, osseointegration, and bioactivity of three-dimensional porous and nanostructured network on polyetheretherketone. Biomaterials 34(37):9264–9277. https://doi.org/10.1016/j.biomaterials.2013.08.071. PubMed PMID: 24041423

Zheng Y, Xiong C, Li X, Zhang L (2014) Covalent attachment of cell-adhesive peptide Gly-Arg-Gly-Asp (GRGD) to poly(etheretherketone) surface by tailored silanization layers technique. Appl Surf Sci 320:93–101. https://doi.org/10.1016/j.apsusc.2014.09.091

Zhong G, Vaezi M, Mei X, Liu P, Yang S (2019) Strategy for controlling the properties of bioactive poly-ether-ether-ketone/hydroxyapatite composites for bone tissue engineering scaffolds. ACS Omega 4(21):19238–19245. https://doi.org/10.1021/acsomega.9b02572

Zhu Y, Cao Z, Peng Y, Hu L, Guney T, Tang B (2019) Facile surface modification method for synergistically enhancing the biocompatibility and bioactivity of poly(ether ether ketone) that induced Osteodifferentiation. ACS Appl Mater Interfaces 11(31):27503–27511. https://doi.org/10.1021/acsami.9b03030

Selection of Implant Material Effect on MRI Interpretation in Patients

20

Ashok Biyani, Deniz U. Erbulut, Vijay K. Goel, Jasmine Tannoury, John Pracyk, and Hassan Serhan

Contents

Pedicle Screw System 440
Study I 441
Study II 441
MRI Imaging 442
Results 444

Cervical Disc Devices 445
Study I 447

A. Biyani
ProMedica Physicians Biyani Orthopaedics, Toledo, OH, USA
e-mail: spineless1025@gmail.com

D. U. Erbulut
Departments of Bioengineering and Orthopaedic Surgery, Colleges of Engineering and Medicine, University of Toledo, Toledo, OH, USA
e-mail: erbulut.deniz@gmail.com; deniz.erbulut@utoledo.edu

V. K. Goel
Engineering Center for Orthopaedic Research Excellence (E-CORE), University of Toledo, Toledo, OH, USA

Departments of Bioengineering and Orthopaedic Surgery, Colleges of Engineering and Medicine, University of Toledo, Toledo, OH, USA
e-mail: Vijay.Goel@utoledo.edu

J. Tannoury
Boston University, Boston, MA, USA
e-mail: cbcbforever1@gmail.com

J. Pracyk
DePuy Synthes Spine, Raynham, MA, USA
e-mail: jpracyk@its.jnj.com

H. Serhan (✉)
I.M.S. Society, Easten, MA, USA
e-mail: hassan.serhan@IMSSociety.org

© Springer Nature Switzerland AG 2021
B. C. Cheng (ed.), *Handbook of Spine Technology*,
https://doi.org/10.1007/978-3-319-44424-6_35

Study II 447
MR Imaging 448
Results 449

Discussion 453

Conclusion 456

References 457

Abstract

MRI is an important modality in the evaluation of the posttraumatic and postoperative spine. The use of MRI and its advantages over other modalities in evaluation of the spine with metal implants have been well documented in the literature (Sekhon et al., Spine 32(6):673–680, 2007; Malik et al., Acta Radiologica 42:291–293, 2001; Rupp et al., Spine, 1993; Rupp et al., J Spinal Disord 9(4):342–346, 1996; Tominaga et al., Neurosurgery 36(5):951–955, 1995). Although MRI is an established radiological method in spinal diagnostics, the clarity of images produced through magnetic resonance can be sensitive to the size and magnetic susceptibility of the materials used. Incompatible materials produce an artifact in the image, which can make assessments of adjacent osseous and neural structures difficult, if not impossible. Also, obtaining a high-quality diagnostic MR image in a patient can become challenging especially in certain postoperative patients as the image quality is affected by so many factors like the location of the device, implant materials, trajectory angle, etc (Sekhon et al., Spine 32(6):673–680, 2007). Therefore, at a minimum, implant materials not only must be MRI conditional (nonferrous) but also optimally allow meaningful diagnostic information to be obtained. Furthermore, implants should possess certain biomechanical properties such as weight-bearing strength, stiffness, biocompatibility, and resistance to corrosion and fatigue.

To that end, this chapter covers case studies that characterize the artifacts seen in magnetic resonance imaging (MRI) of the axial spine created by the constituent properties of the implanted material. These investigational studies are designed to characterize magnetic resonance image distortion associated with:

- Posterior pedicle screw systems
- Total cervical disc replacement intravertebral implants comprised of various materials

Keywords

MRI compatibility · Metallic implants · MRI artifacts · MRI image distortion · Geometric distortion · MRI magnetic compatibility

Pedicle Screw System

Metallic spinal implants like pedicle screws and rods are commonly used to provide stability and maintenance of spinal correction during a time in which bone fusion occurs. Combinations of different materials, like titanium (Ti) screws and cobalt-chromium (CoCr) rods, were recently adopted into clinical practice by spine surgeons. CoCr rods have the advantages of high stiffness and strengths required to correct rigid scoliotic deformities. While stainless steel (SS) has almost the same desirable mechanical characteristics, it produces high levels of artifacts in magnetic resonance imaging. In this study, implant volume and imaging parameters were kept constant to assess the artifacts produced by implants made of stainless steel, titanium, and cobalt-chromium. In this case study, postoperative MRI quality among three different constructs was compared: (1) Ti pedicle screws w/ Ti rods, (2) Ti pedicle screws w/ CoCr rods, and (3) SS pedicle screws w/ SS rods. This study was performed in two groups. The first study consisted of two human torsos to qualify the images, and the second study was performed in a phantom setup to quantify the artifact produced.

Study I

Two fresh-frozen human torsos (cervical-pelvis) were used for bilateral implantation at two levels as shown in Table 1. Torso A (81, male) and torso B (84, male) did not show any structural damage or bony deformity on radiographs.

Implantation in Torso: The EXPEDIUM® Spine System (DePuy Spine, Raynham, MA) with 5.5 mm rods was used for implantation. 6-mm-diameter and 45-mm-long pedicle screws made from stainless steel, titanium, or carbon fiber were used along with 5.5-mm-diameter rods made from titanium, stainless steel, cobalt-chromium, or PEEK.

Midline incision was made with muscle retraction, and pedicle preparation was performed utilizing a selection of awls, pedicle probes, ball tip feelers, and bone taps. Polyaxial crews were inserted into the pedicles. Appropriate length rods with the desired lordosis were selected and placed into the polyaxial screw heads. Different material constructs were implanted at levels based on a randomized protocol. Surgical incisions were sutured closed and the torso maintained in a supine position for MRI scanning. Each specimen was scanned with one of the four different implant groups using 1.5 T and 3 T scanners. A torso on the 3 T scanner is shown in Fig. 1. The six different pedicle screw and rod combinations used at the vertebral levels are shown in Table 1.

Study II

Phantom grids were used for precise quantitative measurement of the artifacts produced by different material as shown in Fig. 2. This method is similar to the ASTM F 2119-01 "Standard Test Method for Evaluation of MR Image Artifacts from Passive Implants" (ASTM F 2119-01).

Procedure: A plastic bin was assembled with a grid fixed at the bottom as a reference for quantitative measurement of artifacts. Different screw-rod construct combinations were assembled and then fixed in the plastic sheets and then suspended in a phantom bath normal to the grid. Constructs

Table 1 Specimen implantation levels

Specimen/torso	Upper level	Lower level
A	L1–L2	L3–L4
B	L2–L3	L4–L5

Table 2 Screw and rod combinations

Material	
Pedicle screw	Rod
Stainless steel	Stainless steel
Carbon fiber reinforced PEEK	PEEK
Titanium	Titanium
Titanium	Cobalt-chromium
Titanium	PEEK

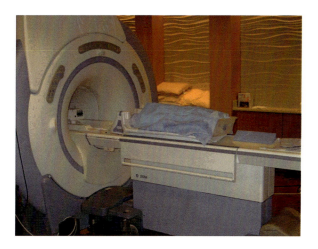

Fig. 1 3 T MRI scanner with implanted torso on coil

used for the phantom study are shown in Table 2. The phantom bath consisted of CuSo₄ solution (1–2 g/L), shown in Fig. 3, which reduces T1 and keeps TR at a reasonable level (Rupp et al. 1996). The bin was then placed on the same spine coil as the torsos and placed in the scanner as shown in Fig. 4.

MRI Imaging

MRI is a very useful diagnostic tool for spinal disorders due to its excellent soft tissue contrast. In this study, we compare MR artifact production by identical pedicle screws and rods with different materials: stainless steel (SS), titanium (Ti), cobalt-chromium (CoCr), carbon fiber reinforced PEEK (CFRP), and PEEK. Scans of instrumented constructs implanted in both the fresh cadavers and a phantom were then obtained.

The torsos and phantom were tested both on 1.5 T and 3 T MRI scanners. Parameters were varied according to standard clinical protocols, including echo time (TE), repetition time (TR), echo train length (ETL), number of excitations (NEX), bandwidth (BW), number of cycles (N Freq), phase (N Ph), phase direction (Ph dir), field of view (FOV), (Th), and slice gap (gap).

MR images were obtained using a clinical MRI system. A metallic object that displayed "weak" ferromagnetic qualities in association with a 1.5 T MRI system may exhibit substantial magnetic field interactions during exposure to a 3 T MRI system (Shellock et al. 2002). Problems presented

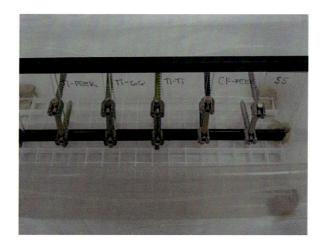

Fig. 2 Phantom construct, from left: Ti-PEEK, Ti-CoCr, Ti-Ti, CF-PEEK, SS-SS

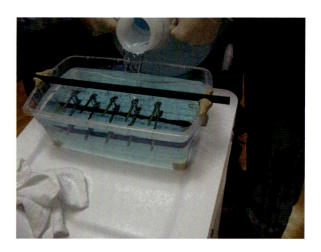

Fig. 3 Phantom setup with CuSO₄ solution

by 3 T systems for metallic implants include translational attraction and torque. Transitional attraction is what has been referred to as the *projectile effect* and results when an object moves toward the magnet at a high rate of speed. Torque, as it relates to MRI, is the shifting or twisting of the medical device or implant inside the patient's body. This movement is caused by the static magnetic field and can cause discomfort, injury, or death if the implant is displaced (Jerrolds and Keene 2009). Using muscle as a calibration standard at 3 T showed the highest accuracy value of 89 percent versus 80.5% at 1.5 T, whereas accuracies of the disc were 78.1% at both 3 T and 1.5 T (Zhao et al. 2009).

Images were acquired using a common 1.5 T MRI (Siemens Magnetom Espree 1.5 T MRI, Malvern, PA) and using 3 T MRI (General Electric Signa HDX 3.0 T MRI scanner, Milwaukee, WI) systems. T1 and T2 fast spin-echo (FSE) protocols typical for clinical images were acquired in the axial and sagittal plane.

IDEAL, a special GE software package, was used for artifact reduction. First, cursory analysis didn't find significant improvement with this package, but it is believed that there would have been a significant difference if we compared with scans with the fat saturation correction modality had been enabled. This software can reduce the fat signal by other mechanisms that are not magnetic field homogeneity dependent as well as providing a separate fat-sensitive image.

The hydrogen in fat has a slightly reduced resonant frequency due to its molecular environment. Normal clinical scans, especially with contrast, would be performed with fat saturation enabled to reduce the bright fat signal. Fat saturation options precede the pulse sequence with a saturation pulse tuned to the fat resonant frequency (which is 224 Hz below the 63 MHz frequency for 1.5 T systems or 448 Hz below the 127 MHz frequency for 3 T systems) in order to reduce the fat signal. Fat saturation, however, requires good magnetic field homogeneity, which is perturbed by the presence of metal implants. Hence, fat saturation correction modality is usually turned off when implants are present.

A list of the scan sequence and the specific materials used is shown in Table 3. The 3 T scans were performed at 41.67 Hz (clinically used) with change in bandwidth to 83.33 Hz with fat saturation turned off for better visualization. Field of view was maintained at 192–224 mm. Slice thickness was varied between 2 and 4 mm with slice separation of 1–1.5 mm. Abbreviations used for the description of the various parameters are shown in Table 4.

Four 1.5 T scans were performed with the distortion correction option enabled. The Siemens Magnetom Espree 1.5 T MRI, Malvern, PA, has a short and wide magnet bore, which adds to patient comfort, but reduces the area of magnetic field homogeneity which can distort images if not corrected.

Fig. 4 3 T MRI scanner with phantom setup

Table 3 MRI scan sequence

MRI	Scan	Specimen/torso	Implanted level			
			Upper level		Lower level	
			Screw	Rod	Screw	Rod
3 T	Scan 1	A	SS	SS	Ti	Ti
	Scan 2		CF	PEEK	Ti	PEEK
	Scan 3	B	Ti	CoCr	SS	SS
	Scan 4		Ti	Ti	Ti	CoCr
1.5 T	Scan 5	A	SS	SS	Ti	Ti
	Scan 6		Ti	Ti	Ti	CoCr
	Scan 7	B	Ti	Ti	Ti	CoCr
	Scan 8		Ti	Ti	Ti	PEEK

	Configurations
SS – Stainless steel	SS/SS
Ti – Titanium	CF/PEEK
PEEK – Polyether ether ketone	Ti/Ti
CF – Carbon fiber	Ti/CoCr
CoCr – Cobalt-chromium	Ti/PEEK
Total scan = 10 (4 for each specimen and 2 for the phantom)	

Table 4 MRI parameter abbreviations

FRFSE	Fast-recovery fast spin-echo
ETL	Echo train length
NEX	Number of excitations
FOV	Field of view
Th	Slice thickness
Gap	Slice spacing
FR	Fast recovery
FC	Flow compensation
NPW	No phase wrap
TRF	Tuned R/F
SEQ	Sequential acquisition
ETL/sl	Number of echo trains/scanned slice
TF	Echo train length (turbo factor)
BW/pix	Bandwidth per pixel
Th	Slice thickness
FOV	Field of view
Gap	Slice spacing

BW/pix were used in the range of 117–170 Hz (clinically used) with increase of bandwidth for 290–651 Hz for implantation. ETL was kept in the range of 5–23 and 96 for few scans.

Axial scans were performed aligned with the disc spaces (separate series for each space). The "nonstandard" scans were performed with increased bandwidth.

Results

Phantom and torso scans showed that stainless steel causes the highest distortion of the MR images. Titanium does not contribute much for artifacts in the image. Titanium screws with cobalt-chromium rods produces larger artifacts than titanium screw with titanium rods. Nonmetallic implants like carbon fiber screw and PEEK rods do not cause image artifacts. Cobalt-chromium rods can cause artifacts at the junction where it meets the titanium screw head.

Study 1 was performed on torsos, where 1.5 T MRI examination revealed no visible metal artifact prior to the implantation of the spinal implants. After implantation, the images showed that specific neural structures (foramina and spinal canal) were unreadable in cases where stainless steel implant was used as shown in Fig. 5.

Study II was performed on a phantom grid that accurately measured the artifact sizes. The artifact sizes for 1.5 T were on an average greater in area for the stainless steel implants as opposed to Ti64 implants and the CCM rods. The artifact sizes for carbon fiber reinforced polymer (CFRP) screws were on an average less than titanium (Fig. 6).

For the 3 T scanners, the artifact sizes were on an average greater in area for the stainless steel

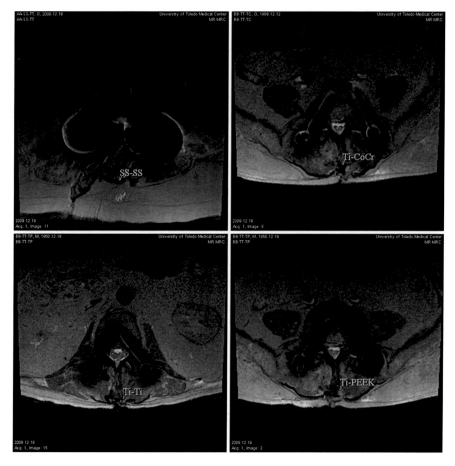

Fig. 5 1.5 T MRI of pedicle screw instrumented and rods

implants as opposed to Ti64 implants and the CCM. The artifact sizes for carbon fiber reinforced polymer (CFRP) screws were less than titanium.

There was a statistical difference in the artifacts and the quality of the images between 3 T and 1.5 T MRI. However, this difference was negligible for the CFRP implants.

Cervical Disc Devices

Materials used in the design of total disc replacement (TDR) devices are of utmost importance as they determine not only the mechanical stability and biocompatibility but also the amount of artifacts produced in magnetic resonance (MR) imaging of patients following surgery.

MRI is considered the diagnostic imaging procedure of choice for intervertebral disc herniation and disc degeneration (Practice parameters 1994), as it can provide exquisite morphologic detail of the disc abnormality (Herzog et al. 1995, Modic and Ross 1991). However, diagnosing a patient using MRI can become challenging as the quality of MRI is affected by previously mentioned factors. Artifacts may be far more important in the cervical spine than in the lumbar spine because of the relatively small size of the cervical vertebrae, which may cause image distortion into the adjacent level discs.

A variety of TDR designs today use a combination of metals and polymers. Metals used in cervical prostheses provide a base of support for the polymer surfaces as well as a surface for fixation to bone. The commonly used metals include stainless steel, titanium carbide alloy,

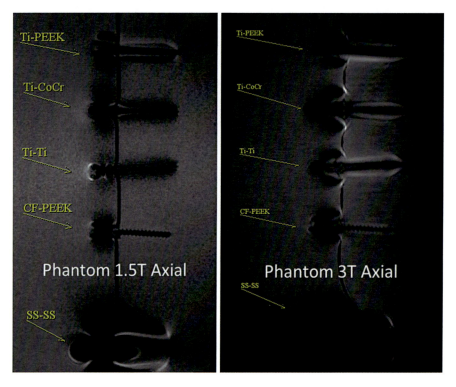

Fig. 6 Phantom 1.5 and 3 T axial views illustrating the significant artifacts associated with SS screws and rods

cobalt-chromium-molybdenum alloy (CoCr), and traditional titanium alloys. Polymers include polyurethane and ultrahigh molecular weight polyethylene. Polymers provide a low-friction surface for articulation as well as some degree of "shock absorption" (Oskouian et al. 2004).

The usage of titanium-based spinal implants is increasing. One of the many reasons is that titanium produces less MR susceptibility artifacts when subjected to MR imaging (Malik et al. 2001). Other advantages of using titanium include improved biocompatibility, increased resistance to corrosion and fatigue, and decreased hypersensitivity response by the body when compared to other nickel- or chromium-containing alloys like stainless steels and cobalt-chromium (Malik et al. 2001). Titanium implants also satisfactorily fulfill the requisite biomechanical demands. The usage of cobalt-chromium alloys has also been reported in literature for several other reasons of which one is its minimal wear (Tranelis 2002; Link et al. 2004; Ernstberger et al. 2007). In another study, H.D. Link et al. concluded that metal/polyethylene material combination is the best and the most suitable for an artificial cervical disc and preferred the proven combination of a highly polished cobalt-chromium alloy component articulating against an UHMWPE sliding partner (Link et al. 2004). Stainless steel on the other hand has met the biomechanical requirements but is a barrier to optimal imaging evaluation. In addition, stainless steel alloys corrode the most of all the alloys used in arthroplasty.

To date, titanium, cobalt-chromium, and stainless steel all have been used in the manufacturing of cervical arthroplasty devices. Previous studies have highlighted the advantages of using Ti alloys (Sekhon et al. 2007; Malik et al. 2001; Tominaga et al. 1995) from an MRI perspective, but there are very few studies comparing titanium and cobalt-chromium alloys (Sekhon et al. 2007; Ernstberger et al. 2007). These studies have examined the extent of cobalt-chromium (CCM) and titanium (Ti64) artifacts in cadaver spines and attested to the importance of artifact effects. However, the present case study not only emphasizes the important impact of material selection in MRI interpretation of patients with cervical disc devices but also compares the amount of distortion produced by these devices on the adjacent segments.

Therefore, we conducted two studies to assess the extent of MRI artifacts produced by artificial cervical disc replacement devices on adjacent neural structures and to evaluate the influence of scanning parameters with respect to artifact size. In the first study, we used fresh human torsos to evaluate the extent of obscurement of peri-prosthetic tissues. The second study employed phantom grids to make precise quantitative measurements of the artifacts produced by CCM and Ti64 devices.

The ASTM F 2119-01 testing protocol provides a more controlled environment for precise study of artifacts from different materials and scanning parameters (ASTM F 2119-01).

For both studies, the DISCOVER™ artificial cervical discs provided by DePuy Spine Inc. (Fig. 7) were used. The CCM and Ti64 materials are compliant with the ASTM standards: ASTM F75-07 Standard Specification for Cobalt-28 Chromium-6 Molybdenum Alloy Castings and Casting Alloy for Surgical Implants (UNS R30075) and ASTM F136-02a Standard Specification for Wrought Titanium-6 Aluminum-4 Vanadium ELI (Extra Low Interstitial) Alloy for Surgical Implant Applications (UNS R56401). Disc implants were stored at room temperature in plastic bags and all tests were performed at room temperature.

Study I

In Study 1, three fresh-frozen human torsos (head-pelvis), two females and one male with an average age of 60 years, were used. CT and MRI scans of the torsos prior to implantation were taken at room temperature to evaluate visible metal artifact prior to spinal implantation. Images were obtained for a single-level implantation at C5–C6 or C6–C7 and bi-level implantation at C5–C6–C7 with identical devices made of either Ti64 or CCM. An experienced arthroplasty surgeon performed all the implantations on the specimen. Apart from making artifact measurements using a display software, hard copy images were also presented to an independent radiologist to evaluate the distortion of MR implant image itself and distortion of the MR image of adjacent neural structures (foramina and spinal canal). Approximate artifact measurements were performed in Study I to show that CCM resulted in larger artifacts and to see the effect of increasing bandwidth on the artifact size.

Study II

Study II was performed to make precise quantitative comparisons between artifacts produced by Ti64 and CCM which was relatively difficult due to variability in torso anatomy and void spaces. A phantom was used to evaluate the metal artifacts produced by the disc devices. Two artificial cervical disc replacements (Ti64 and CCM) were suspended in a phantom bath, which consisted of

Fig. 7 Different views of DISCOVER artificial cervical disc, featuring titanium alloy (Ti64) and polyethylene components. (**a**) Oblique view; (**b**) Side view

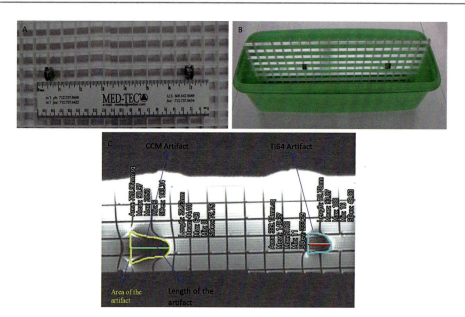

Fig. 8 Image showing disc placement and artifact measurement on the phantom grid. (**a**) The two discs (Ti64 and CCM) were placed sufficiently far apart (distance >4 cm) so that there is field homogeneity and no overlap of artifacts produced by the two discs. (**b**) Discs were aligned the same way as with the human discs in a container that was filled with CuSo$_4$ (1–2 g/L) solution. (**c**) Length and area of the artifact produced by CCM are greater than Ti64 alloy disc

a plastic container filled with CuSO$_4$ solution (1–2 g/L) (Fig. 8). This method was similar to the ASTM F 2119-01 "Standard Test Method for Evaluation of MR Image Artifacts from Passive Implants" (ASTM F 2119-01). It quantitatively assessed the extent of magnetic resonance imaging artifacts of cervical disc replacement devices (made of titanium alloy (Ti64) or cobalt-chromium (CCM)) and evaluated the influence of scanning parameters with respect to artifact size.

CuSo4 (1–2 g/L) solution following ASTM F 2119-01 (ASTM F 2119-01) was used as it reduces T1 and keeps TR at a reasonable level. First, the two discs were placed over 4 cm apart, simulating a non-contiguous two-level instrumentation, and the artifacts were measured. The clearance between the two discs and each side of the container was kept at more than 4 cm to achieve adequate field homogeneity. Then the discs were kept at a distance of 1.5 cm (Fig. 7), simulating two contiguous functional spinal units instrumented with artificial disc devices, and MR images were obtained and artifacts were evaluated.

To study the influence of bandwidth on artifact size, images were obtained at increased bandwidths (2.02–62.5 kHz) with ETL kept constant. Conversely, images were obtained with the ETL increased (4–60) and BW kept constant. As a determinant of image quality, the signal-to-noise (S/N) ratio was measured as the ratio of image signal value at a fixed location within the phantom grid to the standard deviation of the signal outside of the liquid phantom.

MR Imaging

MR images were obtained using a common MRI system (GE Signa 1.5 T, General Electric Company, Milwaukee, WI). Fast spin-echo protocols typical for clinical imaging were acquired using a repetition time (TR) of 4000 mil sec and a minimum echo time (TE). Scan parameters used: a 256 × 192 matrix for sagittal, sagittal STIR, and axial sequences and an FOV of 24-cm and 3-mm-thick sections for all sequences. The following MRI sequence and parameters were included: T1-weighted sagittal (TR, 435 ms; TE, 14 ms; NEX, 4), T1-weighted axial (TR, 600 ms; TE, 14; NEX, 4), T2-weighted sagittal (TR,

4000 ms; TE, 102 ms; NEX, 4), and T2-weighted axial (TR, 485 ms; TE, 9 ms; NEX, 4).

Images were displayed and artifacts were measured using Aquarius-NET (TeraRecon Inc., Tokyo, Japan) software. Susceptibility artifacts were seen as a bright-displaced signal and a void of signal loss. Artifact size (area and length) was measured in the frequency encode direction from the hyper-signal region to the edge of the signal void. For all sequences, artifact size for each artificial disc was measured on two images that contained the most well-defined artifact, and the average value was taken. We measured the artifact along the short axis of the metal implant to be consistent in measurements. Artifacts measured on the sagittal images were used for quantitative comparison and those on the axial images for comparison of the extent of artifacts on the adjacent neural structures.

Results

In general, the results showed that artifact sizes were greater with CCM than those seen with Ti64. CCM produced greater amount of distortion at the index level as well as at the adjacent segments when compared to Ti64 (Fig. 8). Artifact size was reduced at increased BW but degraded the image quality. Increasing ETL did not seem to significantly vary the artifact size.

In Study I performed on torsos, CT and MRI examination revealed no visible metal artifact prior to the implantation of the spinal implants (Fig. 9). After implantation, the images showed that specific neural structures (foramina and spinal canal; the spinal cord was visible in both cases) but were less readable in cases where a CCM implant was used (Fig. 8). Alternatively, the Ti64 implant allowed uncompromised imaging of the spine with only

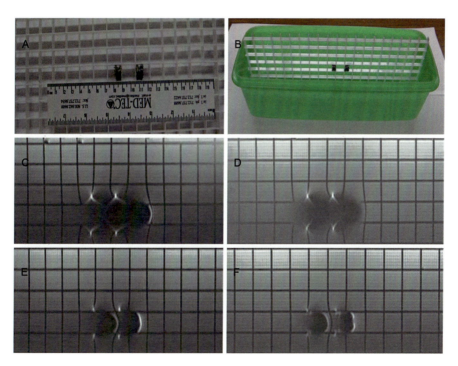

Fig. 9 Image showing disc placement and artifacts on the phantom grid with two similar types of discs separated by a distance equal to 2.5 cm which is approximately the distance between the center of the C5–C6 disc and center of the C6–C7 disc. (**a**) Placement of two Ti64 or CCM discs on the phantom grid with a separation of 2.5 cm. (**b**) Disc alignment in the container that was filled with CuSo$_4$ (1–2 g/L) solution. (**c**) Sagittal T2-weighted FSE image of bi-level CCM alloy disc obtained at 15 KHz BW. (**d**) When a similar image was obtained at 31.2 KHz BW using bi-level CCM alloy disc, the artifact size was comparatively lesser. (**e**) Image obtained with same scan parameters (15 KHz) but this time using Ti64 resulted in smaller size artifacts when compared to CCM (**c**). (**f**) The BW was increased to 31.2 KHz and artifact size was reduced when compared to (**e**)

minimal artifact production. The artifacts, in both cases, seem to skew more in the right direction (distorting the right foramen) as compared to the left in the axial images and downward when compared to upward direction in sagittal images (Fig. 8). However, the distortion caused in any direction was greater with CCM compared to Ti64. In case of dual CCM, the artifacts appeared to encompass over the entire index level, the adjacent level vertebra, and the adjacent disc while using Ti64 implant barely distorted the index level (Fig. 10).

Study II performed on a phantom grid accurately measured the artifact sizes (Fig. 11). The artifact sizes were on an average 58% greater in length and 44% greater in area for the CCM implants as opposed to Ti64 implants. Increasing bandwidth alone (Fig. 12 and Table 5) from 2.02 kHz to 62.5 kHz decreased the artifact size (lengths by 67% and areas by 78% for CCM and 70% and 81% for Ti64, respectively).

However, this increase in BW can degrade the image quality as seen (Table 6) by decrease in S/N (Figs. 13 and 14). Increasing the ETL did not seem to significantly vary the artifact size (Fig. 12 and Table 7).

Recent advancement in biomaterials has allowed the creation of the next-generation cervical total disc replacement with PEEK on ceramic articulated surfaces and plasma-sprayed titanium coating of the endplates to achieve better osteointegration and significantly reduce the artifacts. The 80 micron titanium plasma spray coating on the prosthesis endplates casts a minimal amount of artifact (Fig. 15).

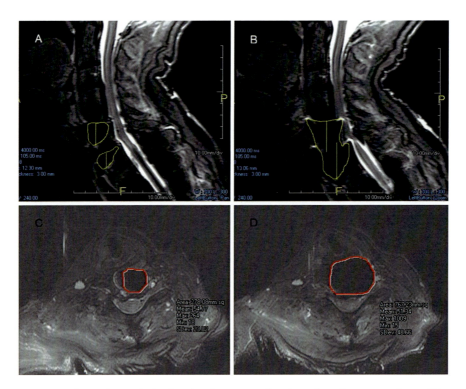

Fig. 10 MRI scans of one of a cervical spine showing image distortion produced by two adjacent (bi-level) artificial discs. (**a**) Sagittal T2 FSE image showing artifacts (length and area) produced by two adjacent levels (C5–C6–C7) Ti64 alloy discs. (**b**) A similar scan for two adjacent levels, CCM alloy discs clearly shows that the artifacts (circled) produced by CCM are greater in size and extending into the adjacent vertebrae. C6 vertebra is completely invisible in case of CCM. (**c**) An axial image with bi-level Ti64 alloy discs. (**d**) A similar scan obtained with bi-level CCM alloy discs clearly shows the artifact produced extends into the adjacent structures by a greater amount when compared to Ti64

Fig. 11 MRI scan of one of a cervical spine prior to implantation of artificial cervical disc showing no visible metal artifacts at the intervertebral disc spaces of C2–T2

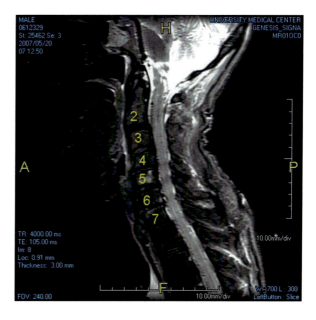

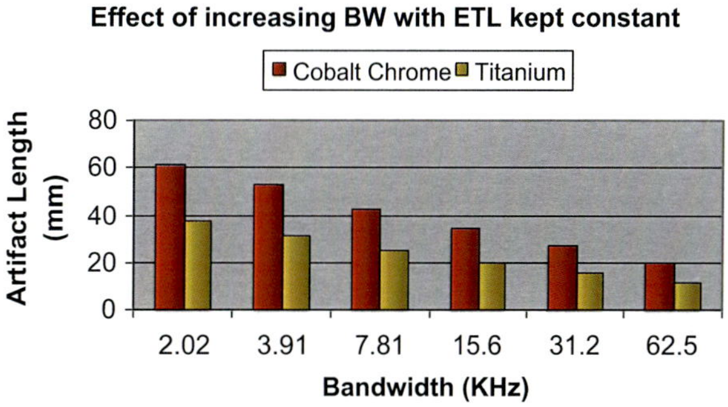

Fig. 12 Increasing bandwidth alone from 2.02 kHz to 62.5 kHz decreased the artifact size, lengths by 67% and areas by 78% for CCM and 70% and 81% for Ti64, respectively

Table 5 Artifact measurements with ETL kept constant. The use of CCM resulted in an average increase in artifact lengths by 58% and areas by 44% compared with Ti64 for identical scanning parameters. Increasing bandwidth alone, from 2.02 to 62.5 KHz, decreased the artifact size (lengths by 67% and areas by 78% for CCM and 70% and 81% for Ti64, respectively)

Artifact measurements with ETL kept constant					
ETL	BW	CC length	CC area	Ti length	Ti area
20	2.02	61.045	1788.495	37.545	888.9
20	3.91	52.7	1466.035	30.73	587.715
20	7.81	42.645	1078.375	24.875	439.49
20	15.6	34.145	703.045	19.78	319.99
20	31.2	27.505	485.72	15.605	221.28
20	62.5	20.185	387.635	11.43	166.705

Table 6 Variation of signal-to-noise ratio with bandwidth. The signal-to-noise (S/N) ratio was measured as the ratio of image signal value at a fixed location within the phantom grid to the standard deviation of the signal outside of the liquid phantom

BW	ETL	S/N	Normalized value
2.02	20	167	3.707977
3.91	20	131.3083	2.915498
15.6	20	76.7	1.703005
31.2	20	62.86	1.395709
62.5	20	45.03803	1

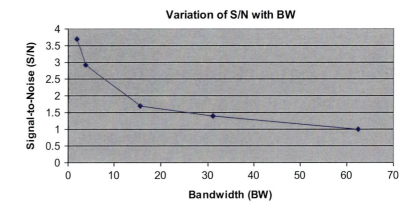

Fig. 13 Increasing bandwidth (BW) degraded the image quality as seen by decrease in signal-to-noise ratio (S/N)

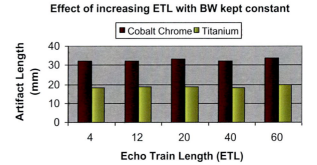

Fig. 14 Increasing the ETL did not seem to significantly vary the artifact size for both Ti64 and CCM

Table 7 Artifact measurements with BW kept constant. Increasing ETL alone did not significantly change the artifact size. Note: All lengths are in mm and areas in sq mm

Artifact measurements with BW kept constant					
BW	ETL	Length		Area	
		CC	Ti	CC	Ti
15.6	4	33.53	19.315	709.485	313.41
15.6	12	34.145	19.93	718.83	321.585
15.6	20	34.145	19.78	703.045	319.99
15.6	40	33.525	19.47	735.4	317.25
15.6	60	32.955	19.61	727.72	293.405

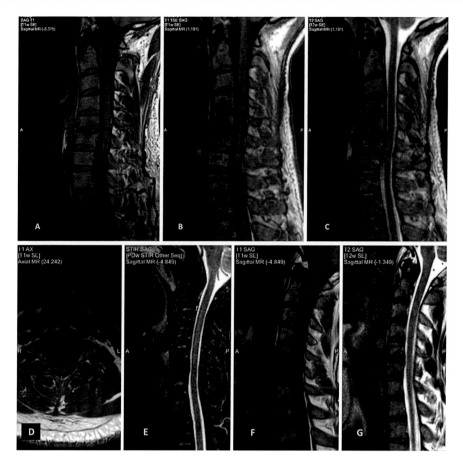

Fig. 15 Post-implantation MRI scans of double levels (**a, b, c**) or single-level (**d, e, f, g**) PEEK/zirconia-toughened alumina ceramic cervical artificial discs

Discussion

In the first section, we examined imaging artifacts in the lumbar spine. This pedicle screw study evaluated the artifacts produced by four materials Ti64, SS, CCM, and CFRP using 1.5 T and 3 T MRI scans. The ease of interpretation of MR scans with devices in place was evaluated. Approximate artifact measurements were performed in Study I and demonstrated that stainless steel produced in larger artifacts than all other materials yet CFRP produced the least artifacts. Precise quantitative measurements were performed in Study 2.

In the second section, we performed a cervical disc device study to evaluate the artifacts produced by two materials Ti64 and CCM commonly used in cervical arthroplasty and examined the ease of interpretation of MR scans with these TDR devices in place. In addition, we reported the effect of scan parameters on the artifact size.

In this section, we shall synthesize how the characteristics of the materials of implants and devices affect the quality of MR spine examination and how to optimize the images.

Pulse sequences appear to have some effect on the quality of the scans. Spin-echo sequences are less sensitive to field inhomogeneities and are preferred to gradient-echo sequences (Young et al. 1988). The protons located in the zone of field inhomogeneity diphase more rapidly than the others. In the conventional spin-echo sequence, the 180° pulse re-phases the magnetization, causing a spin-echo signal to appear, and compensates for the de-phasing effect of the inhomogeneities of the main field and of most of the effects of patient-related susceptibility.

Using the gradient-echo techniques, there is no 180° re-focalization pulse. Thus, there is no correction for the de-phasing effect of field inhomogeneities. The loss of signal will increase, even with less intense magnetic fields. We did not measure the artifact on gradient-echo image, as it was very severe.

In one of the studies conducted by A.S. Malik et al., it was found that FSE sequences led to a decrease in perceptible MR artifacts. In this study, fast spin-echo (FSE) protocols typical for clinical imaging were used because FSE imaging, especially when performed with shorter echo spacing, increases the amount of T2-weighted information in the presence of metallic artifact because it decreases magnetic susceptibility effects (Rudisch et al. 1998). FSE pulse sequences seem to be more diagnostically useful than conventional spin-echo images, especially on T2-weighted images. Artifacts are not eliminated, but they are reduced enough to provide useful information. Moreover, these benefits are achieved in a shorter period of time. Spinal imaging with the fat suppression technique can also be degraded by the presence of metallic fixation (Leclet 1994).

The plane of the scan is a significant parameter in achieving clinically useful imaging studies. Theoretically, the artifact is greater along the long axis of the implant. For postoperative spine (instrumented) MRI, the sagittal plane is the best orientation because preference should be given to the slice perpendicular to the axis of the implant (Leclet 1994). For the current study, artifacts measured on the sagittal images were used for quantitative comparison, and those on the axial images were used for comparison of the extent of artifacts on the adjacent neural structures. Joined sagittal slices often make possible an interpretable median slice between the spinal instrumentation devices, which are often posterolateral (Leclet 1994). Acquisitions at the ends of the devices should be avoided. Slice thickness is not a determining factor.

Errors and misinterpretations of MRI images are inevitable (Leclet 1994). Artifacts impede interpretation by deteriorating the quality of the image and its informational content by masking the anatomical and pathological structures (Leclet 1994). Measurement of artifacts especially in human torsos can be difficult because of void spaces contained within it. To overcome this problem and predict the extent of artifacts into the adjacent structures, we modeled the in vitro study by separating the discs by about 1.5 cm, which is approximately the midsagittal height of the C6 vertebra (Fig. 13). Efforts were also made to minimize the error in measurement by firstly carefully selecting the void region and secondly by measuring artifacts in two images that contained the most well-defined artifact and then averaging them (Fig. 16).

Artifacts are signal intensities that have no relation to the spatial distribution of the tissues being imaged. There are four types of artifacts (Leclet 1994) (based on appearance): (a) edge artifacts (ghosting, chemical shifts, and ringing), (b) distortions, (c) aliasing (wraparound) artifacts, and (d) flow artifacts.

Motion artifacts (ghosting and smearing) often result from involuntary movements (e.g., respiration, cardiac motion and blood flow, eye movements, and swallowing) and minor voluntary subject movements. Motion artifacts appear only in the phase-encoding direction and appear as ghosts or smears. Motion artifacts can be flipped 90° by swapping the phase-/frequency-encoding directions. Flow effects can be reduced by using gradient motion rephasing (GMR) or synchronization of acquisition with motion rhythms or increasing the number of acquisitions (NEX) (Rudisch et al. 1998).

Metallic implant artifacts may be reduced by a number of imaging factors including shorter echo times, lower field strengths, higher readout bandwidths (BWs), and smaller voxel sizes. Techniques such as view angle tilt (VAT), orienting the long axis of the metal implants along the frequency-encoding direction, seem to reduce such artifacts (Kolind et al. 2004). Several studies examined the effect of various scanning parameters in order to reduce these resultant artifacts (Kolind et al. 2004; Tartaglino et al. 1994; Olsrud et al. 2004; Czervionke et al. 1988; Ludeke et al. 1985; Laakman et al. 1985; Orlando et al.

Fig. 16 MR image of one of the cadaveric specimens showing the midsagittal height of C6 vertebra measurement and the distance between the discs. C6 vertebra was about 13.14 mm and distance between the discs was about 24.16 mm

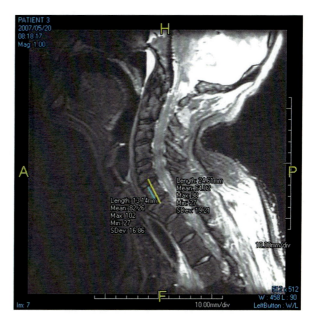

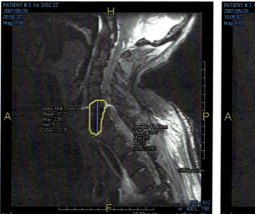 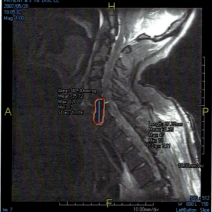

Fig. 17 Comparison of MR images at different bandwidths (BWs). (**a**) T2 FSE image – single level. (**b**) T2 FSE image – single level and higher bandwidth

1996; Farahani et al. 1990; Young et al. 1988; Vinitski et al. 1987, 1993). In one such study conducted by Shannon H. Kolind et al. (2004), it was found that artifact size could be reduced on an average of 60% by increasing BW from ±16 kHz to ±64 kHz. Although the scan parameters used in our study (Study II) were different, we found that increasing bandwidth alone (Fig. 10, 17, and Table 1) from 2.02 kHz to 62.5 kHz decreased the artifact size (lengths by 67% and areas by 78% for CCM and 70% and 81% for Ti64, respectively).

The bandwidth is defined as the sampling rate of the received RF signal at the processional resonant frequency of the hydrogen. The overall bandwidth as presented here can also be expressed as a frequency per pixel length by dividing the bandwidth by the matrix size, which is precisely 256 in these images. The 2.02–62.5 KHz therefore is equivalent to

7.9–244 Hz/pixel. As the sampling bandwidth is increased, the errors in the appropriate magnetic field strength result in a smaller spatial displacement of the signal and a resultant smaller artifact. In a related phenomenon, higher sampling bandwidths are also known to reduce chemical shift artifacts derived from fat. The primary cost with the increased bandwidth is an increase in the image noise.

The ETL, or the number of echoes obtained per excitation pulse, does not appear to have a significant effect on the size of the artifact, although very long ETLs may affect other scan parameter options and the presence of other artifacts.

When using CCM, a BW of 15.6 KHz or 31.2 KHz may be used for ease of readability of spinal MR images. Alternatively, the Ti64 allowed uncompromised imaging of the spine with only minimal artifact production for BWs 15.6 KHz and 31.2 KHz. The decision of selecting 15.6 KHz or 31.2 KHz is left to the discretion of the radiologist.

In a study conducted by Lali H. S. Sekhon et al. (2007), it was shown that titanium devices with or without polyethylene allow for satisfactory monitoring of the adjacent and operated levels. The results in our study were similar. Titanium devices induced more artifacts than CoCr alloys. The reason for this kind of behavior by the titanium materials may be explained as discussed below.

MR imaging localizes signals from hydrogen by varying the magnetic field across the object, causing a change in the resonant frequency dependent on position. The insertion of materials with significantly different magnet susceptibility properties from the surrounding tissues will distort the local magnetic fields, causing the signal to be improperly located. This susceptibility artifact can result in image distortion, as well as significant signal loss. The high ferromagnetic properties of the SS yielded significantly higher artifact than both CCM and Ti64. The magnetic susceptibility or paramagnetic properties of CCM produced higher artifacts than Ti64 but did not approach the level seen with the SS implants. CFRP and titanium which is a non-ferromagnetic metal did not exhibit as high a degree of deflection forces in a static magnetic field as ferromagnetic metals (e.g., certain stainless steel) (Zhao et al. 2009).

Conclusion

Magnetic resonance imaging artifacts are subject to a variety of influences, most notably implant material, implant volume, measurement criteria, and imaging parameters. In the cervical TDR study, of the two materials considered in our study, CCM resulted in significantly larger artifacts than Ti64. In the pedicle screw study, of the four materials considered in our study, stainless steel resulted in significantly larger artifacts than CCM, Ti64, or CFRP. MRI provides adequate visualization of neural structures at the operated and adjacent levels when CCM is used with Ti64 pedicle screws.

Today, titanium fixation implants and devices are recommended as a substitute for stainless steel in a patient who may need further MR examination. MRI provides adequate visualization of neural structures at the operated and adjacent levels when Ti64 is used which implies that the quality of the spinal MR images depends on the type of materials used to construct the implants or devices. Before beginning the MR study, the radiologist must critically review the X-rays showing the location and orientation of the implant. Only then is he able to adapt the MR acquisition technique.

Increasing the sampling bandwidth reduces the artifact at the expense of increased image noise, while changing ETL has relatively little effect on the susceptibility artifact. Since the extent of the artifact is primarily in the frequency-encoding direction, the extent and orientation of the artifact can be altered by swapping the phase and frequency-encoding directions of the data acquisition or by scanning in an alternate anatomical plane.

Of note, these results are true for cervical arthroplasty and may vary for thoracic and lumbar spine instrumented procedures. For example, Ortiz et al. (Orlando et al. 1996) showed that titanium implants of the thoracic and lumbar spine produce extensive artifacts that make interpretation of postoperative MRI studies extremely difficult, if not impossible. In their study, they found that the MR images of titanium implants are superior than those obtained with stainless steel in some cases but were useless in other cases.

The results of the study show that MRI provides adequate visualization of neural structures at the operated and adjacent levels in devices containing titanium alloys. Neural imaging is required to define the pathologic anatomy when clinical neurologic symptoms are present, and this is typically best performed by MRI. Titanium and ceramic materials are the most MRI-compatible materials in use today and will afford the greatest versatility and visibility in postoperative imaging studies. Operational knowledge about MRI imaging techniques following spine surgery and metallic implant-induced artifacts can improve the quality of MRI postoperative studies now and in the future.

The findings of the study can be used in the selection of implant materials for optimal MRI interpretation of patients with cervical disc devices. The knowledge of artifact size and the amount of distortion caused into the adjacent structures and neural elements helps in the design of the devices that produce minimal artifact. The importance of artifacts depends on the volume and shape of the fixation device and therefore on the total amount of metal. The geometric distortions are more significant close to the extremities of the implant and in any region of sharp contour and shape change. All of these things need to be kept in mind during the design considerations for artificial cervical disc devices.

References

ASTM F 2119-01. Standard test method for evaluation of MR image artifacts from passive implants. https://www.astm.org/DATABASE.CART/HISTORICAL/F2119-01.htm

Czervionke LF, Daniels DL, Werhti FW et al (1988) Magnetic susceptibility artifacts in gradient-recalled echo MR imaging. AJNR 9:1149–1155

Ernstberger T, Heidrich G, Buchhorn G (2007) Post-implantation MRI with cylindric and cubic intervertebral test implants: evaluation of implant shape, material, and volume in MRI artifacting- an in vitro study. Spine J 7(3):353–359

Farahani K, Sinha U, Sinha S, Chiu LC-L, Lufkin RB (1990) Effect of field strength on susceptibility artifacts in magnetic resonance imaging. Comput Med Imaging Graph 14:409–413

Herzog RJ, Guyer RD, Graham-Smith A et al (1995) Magnetic resonance imaging. Use in patients with low back or radicular pain. Spine 20:1834–1838

Jerrolds J, Keene S (2009) MRI safety at 3T versus 1.5T. Int J World Health Soc Politics 6(1). https://eproofing.springer.com/books_v2/mainpage.php?token=PPBMWP81kVjWjKiJreeKCsHeNfYVP9BkOc97jATBTnw

Kolind SH, MacKay AL, Munk PL, Xiang Q-S (2004) Quantitative evaluation of metal artifact reduction techniques. J Magn Reson Imaging 20:487–495

Laakman AW, Kaufman B, Han JS et al (1985) MR imaging in patients with metallic implants. Radiology 157:711–714

Leclet H (1994) Artifacts in magnetic resonance imaging of the spine after surgery with or without implant. Eur Spine J 3:240–245

Link HD, McAfee PC, Pimenta L (2004) Choosing a cervical disc replacement. Spine J 4:294S–302S

Ludeke KM, Röschmann P, Tischler A (1985) Susceptibility artifacts in NMR imaging. Magn Reson Imaging 3:329–343

Malik AS, Boyko O, Aktar N, Young WF (2001) A comparative study of MR imaging profile of titanium pedicle screws. Acta Radiologica 42:291–293

Modic MT, Ross JS (1991) Magnetic resonance imaging in the evaluation of low back pain. Orthop Clin North Am 22:283–301

Olsrud J, Latt J, Brockstedt S, Romner B, Bjorkman-Burtscher IM (2004) Magnetic resonance imaging artifacts caused by aneurysm clips and shunt valves: dependence on field strength (1.5 and 3 T) and imaging parameters. J Magn Reson Imaging 22:433–437

Orlando O, Glenn PT, Phillip MA, Kent S (1996) Postoperative magnetic resonance imaging with titanium implants of the thoracic and lumbar spine. Neurosurgery 38(4):741–745

Oskouian RJ, Whitehill R, Samii A, Shaffrey ME, Johnson JP (2004) The future of spinal arthroplasty: a biomaterial perspective. Neurosurg Focus 17(3):E2

Practice parameters: magnetic resonance imaging in the evaluation of low back syndrome (summary statement). Report of the Quality Standards Subcommittee of the American Academy of Neurology. (1994) Neurology 44:767–770

Rudisch A, Kremser C, Peer S, Kathrein A, Judmaier W, Daniaux H (1998) Metallic artifacts in magnetic resonance imaging of patients with spinal fusion: a comparison of implant materials and imaging sequences. Spine 23(6):692–699

Rupp RE, Ebraheim NA, Wong FF (1996) The value of magnetic resonance imaging of the postoperative spine with titanium implants. J Spinal Disord 9(4):342–346

Sekhon LHS, Duggal N, Lynch JJ, Haid RW, Heller JG, Riew KD, Seex K, Anderson PA (2007) Magnetic resonance imaging clarity of the Bryan®, Prodisc-C®, Prestige LP®, and PCM® cervical arthroplasty devices. Spine 32(6):673–680

Shellock et al (2002) Biomedical implants and devices: assessment of magnetic field interactions with a 3.0-Tesla MR system. Journal of Magnetic Resonance Imaging 16:721–732

Tartaglino LM, Flanders AE, Vinitski S, Friedman DP (1994) Metallic artifacts on MR images of the postoperative spine: reduction with fast spin-echo technique. Radiology 190:565–569

Tominaga T, Shimizu H, Koshu K, Kayama T, Yoshimoto T (1995) Magnetic resonance imaging of titanium anterior cervical spine plating systems. Neurosurgery 36(5):951–955

Tranelis VC (2002) Spine arthroplasty. Neurosurg Focus 13(2):Article 10

Vinitski S, Griffey A, Fuka M, Matwiyoff N, Prost A (1987) Effect of the sampling rate on magnetic resonance imaging. Magn Reson Med 5:278–285

Vinitski S, Mitchell DG, Einstein SG et al (1993) Conventional and fast spin-echo MA imaging: minimizing echo time. J Magn Reson Imaging 3:801–507

Young IA, Cox IJ, Bryant DJ, Bydder GM (1988) The benefits of increasing spatial resolution as a means of reducing artifacts due to field in homogeneities. Magn Reson Imaging 6:585–590

Zhao J et al (2009) MRI of the spine: image quality and normal–neoplastic bone marrow contrast at 3 T versus 1.5 T. Am J Roentgenol 192:873–880

Metal Ion Sensitivity

William M. Mihalko and Catherine R. Olinger

Contents

Introduction	460
Metal Hypersensitivity Physiology	460
Implant Sources of Particulate Debris	462
Implant Debris Physical Attributes and Local Physiological Response	463
Systemic Response to Metal Debris and Prevalence in the General Population	465
Testing for Metal Hypersensitivity	465
Risk Factors for Metal Hypersensitvity	467
Clinical Presentation of Metal Hypersensitivity	467
Metal Sensitivity to Spinal Implants	468
Spinal Implant Composition	468
Treatment	469
Conclusion	469
Summary	470
References	470

Abstract

Metal hypersensitivity to biomaterial alloys have been reported since the 1970s. While most reports have been in the total joint literature, in the last 10 years isolated spinal implant reactions have been reported. Much of this is because spine implants have been developed with bearing surfaces that may be a trigger for sensitizing patient from the local wear debris. Reaction to metal alloys and debris is a type IV hypersensitivity immunologic reaction in that it does not produce anaphylaxis. The adverse local tissue reactions (ALTR) around the implant can be substantial and lead to further surgery. The metal alloys used in spinal implants typically have an oxide passivation

W. M. Mihalko (✉) · C. R. Olinger
Campbell Clinic Department of Orthopaedic Surgery and Biomedical Engineering, University of Tennessee Health Science Center, Memphis, TN, USA
e-mail: wmihalko@campbellclinic.com; colinger@campbellclinic.com

© Springer Nature Switzerland AG 2021
B. C. Cheng (ed.), *Handbook of Spine Technology*,
https://doi.org/10.1007/978-3-319-44424-6_139

layer that can protect the body from these local reactions, but any type of fretting from modular connections of wear from a metal bearing can lead to exposure of the alloy below the passivation layer and be the trigger to the start of a reaction leading to ALTR.

Knowing the frequency of these sensitivities in the general population can help surgeons identify hypersensitive patients and notify them of the possible risk.

Keywords

Biomaterial · Metal alloy · Passivation layer · Hypersensitivity

Introduction

Metal hypersensitivity to biomaterial alloys has been reported since the 1970s. While most reports have been in the total joint literature, there also are case reports of spinal implant reactions in the last 10 years. Reaction to metal alloys and debris is a type IV hypersensitivity immunologic reaction that does not produce anaphylaxis. The adverse local tissue reactions (ALTR) around the implant can be substantial and lead to further surgery or significant morbidity. Knowing how common these sensitivities are in the general population, understanding the physiology of metal hypersensitivity, identifying appropriate testing protocols, and recognizing clinical signs of sensitivity can help surgeons properly diagnose this uncommon complication and notify patients of the possible risk of ALTR.

Metal Hypersensitivity Physiology

The pathophysiology behind metal hypersensitivity is a type IV hypersensitivity or delayed type hypersensitivity reaction. When a hypersensitivity reaction occurs, activated T-lymphocytes react to a foreign antigen presented via co-stimulatory molecules, which play a critical role in sustaining the chronic inflammatory response (Goodman 2007). Through this inflammatory cascade, T-lymphocytes CD4 and CD8 cells are activated and release a multitude of cytokines including IFN-gamma, IL-1, IL-6, and TNF-alpha (Merritt and Brown 1980).

The immune system can mount an adaptive or innate immune response to metal debris. The innate or nonspecific foreign body reaction is composed primarily of macrophages, foreign body giant cells, fibroblasts, and occasional lymphocytes. The aggressive inflammatory granulomatosis found in the monocyte-macrophage mediated clearance of debris is normally followed by the resolution of the reaction via the fibroblast mediated synthesis of remodeling the extracellular matrix. Metal implants may have osteoclasts that line the bone implant interface and in the presence of metallic or bearing debris the tissue may have high levels of proinflammatory cytokines, indicating an immune response that can put the longevity of the implant at risk (Goodman 2007). Animal models of exposed rabbits with implanted nickel demonstrated tissue reaction to screws with inflammatory cells and macrophages in induced sensitivity models (Merritt and Brown 1980). A combination of both innate and acquired immune response has been elucidated in the metal hypersensitivity reaction pathway.

In the case of metal hypersensitivity, the foreign antigen is metal debris from an implanted medical device. Metal wear degradation products combined with serum proteins form haptens. Haptens are then recognized via antigen presenting T-cells and initiate the activated T cell cascade. This activation of T-cells locally produces an inflammatory response and lessens circulating T cells. One study demonstrated that the serum analysis of patients with aseptic loosening showed decreased levels of circulating T-cells indicating an inflammatory consumptive process (Goodman 2007).

Proposed intracellular indigestible particles, via metal implant debris, together with elevated costimulatory molecule expression via antigen presenting cells and macrophages, promote T-cell inflammatory reactions in the surrounding tissues. Cobalt-chromium (Co-Cr) alloys are common metal compounds used in spinal implants. In vitro proliferation of cellular responses to Co-Cr has been found to be significantly higher in patients with revision surgery for aseptic loosening compared to patients with revision for infection. Furthermore, patients demonstrated higher proliferative responses and cytokine production in response to Co-Cr challenge postoperatively after total joint replacement (TJR) than preoperatively (Goodman 2007).

Tissue samples from retrieved failed implants that formed pseudocapsules around metal implants have been analyzed for inflammatory cells. One study examined 123 tissue samples excised from reoperations for loosening, fracture, or mechanical irritation (infections were excluded). The removed pseudocapsule represented a crude joint capsule of scar tissue without defined layers. The inflammatory response to foreign bodies inside the fibrous tissue was characterized pathologically as granulation tissue. This inflammatory tissue was surmised to be from production of continual foreign material from wear particles. Patients with retrieved tissue from loosening had a marked tendency towards fibrosis, which gave rise to numerous lymphoplasmacellular infiltrations surrounding the implants (Willert and Semlitsch 1977). Metal hypersensitivity-induced osteolysis and aseptic loosening have been suggested to represent an underappreciated and ignored subset of failure mechanisms within TJR (Jacobs and Hallab 2006).

Knowing the mechanism behind these reactions in aseptic lymphocytic vasculitis-associated lesion (ALVAL) from metal-on-metal hip prosthesis is important since similar bearings are now being used in spinal implants. Patients deemed to have either a high- or low-wear pattern were identified and during retrieval had tissue analysis regarding the ALVAL score as determined by the histologic scoring of the synovial lining, inflammatory infiltrate, and tissue organization (low, 0–4; moderate, 5–8; high, 9–10). Tissues from patients who had revisions for suspected high wear had a lower ALVAL score, fewer lymphocytes, more macrophages, and more metal particles than tissues from patients who had revisions for pain and suspected metal hypersensitivity (Campbell et al. 2010). The characterization of the type of local tissue response and the patient's presenting symptoms of pain and dermatitis could help differentiate between metal type IV hypersensitivity reaction and metal-on-metal ion release from failed components (Verma et al. 2006).

Total joint literature has demonstrated periprosthetic pseudocapsule tissues harvested from failed TJR implants containing titanium and Co-Cr alloy to have pathologic demonstration of abundant macrophages containing titanium particles, numerous T-cells, but few B-cells. This pathologic evaluation was further used with tissue marker enzyme-linked immunosorbent assay (ELISA) studies in these samples for T-cell markers Cd11c, CD25, IL-2R, HLA-DR, CD35, CD36, CD2, and CD22. IL-2 was used as the main cell marker for activated T-cells in this population (Goodman 2007). The type IV hypersensitivity response surrounding failed arthroplasties has been supported by the presence of activated T-cells in vivo, and both pathologic and ELISA testing confirms their presence in these tissues.

Osteolysis mechanisms surrounding failed aseptic TJR have been investigated on the cellular levels. Receptor activator of nuclear factor-kappa B (RANK) production, determined by ELISA testing of harvested pseudocapsule tissue in failed implants, has been shown to be increased, as have abnormally high levels of RANK. The RANK-RANKL mediation has been shown to contribute substantially to aseptic implant loosening. Activation of this pathway via the type IV hypersensitivity response has been elucidated. Activated T-cells have been shown to express RANKL activating osteoclastogenesis. TNF-a and IL-1 are pro-inflammatory cytokines present in type IV hypersensitivity reactions that also upregulate the expression of RANK/RANKL (Holt et al. 2007).

Other cytokines have been demonstrated to contribute to bone homeostasis surrounding metal implants. IL-18 is a novel cytokine involved in the role of disturbance in bone homeostasis observed in numerous systemic disorders, specifically inflammatory arthritis. It initially was characterized in properties of acquired immune response via activated T1 and T2 helper cells. After inflammatory responses are initiated, IL-18 is widely distributed, even in pseudocapsules from retrieved failed implants, as identified by PCR/ELISA testing (Goodwin et al. 2018).

The method of reaction to metal hypersensitivity in vivo may be a combination of several factors. Local inflammatory responses to the presence of metal alloys are recognized by the innate immune response, leading to T-cell activation, inflammatory cytokine production, osteolysis, and loosening via RANKL/RANK activation. Loosening of the implant-bone surface can lead to further wear debris and propagation of this cascade.

Implant Sources of Particulate Debris

Mechanisms that produce increased metal hypersensitivity require a nidus for metal debris. Metal implant wear can produce local tissue infiltration of metal ions and particles. Wear involves the loss of the material (mass) as a consequence of relative motion between two surfaces. Gravimetric wear is measured by the weight loss of the individual component after simulator or retrieval in vivo use. The amount of wear depends on two factors: the amount of force pressing the two materials together and the type or amount of lubrication between the two surfaces (Hallab 2009).

Wear is a mechanical or physical degradation of materials characterized as either abrasive or adhesive. The primary sources of articulating wear debris from hard-on-hard material couples, such as metal-on-metal articulations, generally produce less wear (volumetric loss) than metal-on-polymers. Corrosion is a chemical or electrochemical form of degradation of metal implants. Implant corrosion reduces structural integrity and causes release of by-products that interact locally and systemically. Stainless steel alloys generally corrode to a greater extent than cobalt or titanium. Fretting corrosion can take place at mechanical connections between implants. This is a common occurrence in spinal reconstructive surgery. With this kind of fretting in spinal instrumentation, chemical degradation is enhanced by mechanical factors such as a crevice and abrasive wear. Corrosion products typically are oxides, metal phosphates, metal salts, metal-ions bound to proteins, or organometallic complexes (Hallab 2009).

Implant debris types can be characterized as particles or ions. Particulate wear debris (metal or ceramic) exists from the submicron size up to thousands of microns in size. Soluble debris is limited to metal ions that are bound to plasma proteins. The most numerous particulate debris to measure is typically less than 1 μ in size. Particles generated in simulator studies of articulating spinal implants match the sizes and types of particles produced from hip and knee arthroplasty. Metal-on-metal articulations generally produce smaller-sized (submicron) fairly round debris, whereas traditional metal-on-polymer bearings produce larger (micron) debris that is more elongated in shape (Hallab 2009). Polymeric particles produced from implants generally fall into the range 0.23–1 μ. During articulating implant studies, 70–90% of recovered particulates were submicron, with the mean size being 0.2–1 μ. Newer polymer implant debris from highly cross-linked polymers have demonstrated the production of smaller, more rounded debris in the submicron range as small as 0.1 μ (Hallab 2009).

Metal-on-metal particles are one to three orders of magnitude in number over those produced by metal-on-polymer articulating surfaces, but with far less volume. Cobalt alloy corrosion mechanisms also produce a chromium phosphate hydrate-rich material termed "orthophosphate," which ranges in size from submicron to aggregates of particles up to 500 μ. Low-angle laser light scattering (LALLS) can perform particulate characterization and increase the number of counted and sized particles from hundreds to millions. It is important to perform a number-based and volume-based analysis. Ability to accurately and comprehensively characterize implant debris is important where weight loss from the implant after a year of use (<0.2 mm^3 volume loss) could be attributed to the loss of a relatively few large particles or hundreds of millions of small particles (Hallab 2009).

For soluble metal ions, metal levels measured in people with disc arthroplasties have comparable levels of circulating metal ions as people with TJR. Normal human serum levels of prominent implants metals are approximately

(a) 1–10 mg/mL AL
(b) 0.15 ng/mL Cr
(c) <0.01 ng/mLV
(d) 0.1–0.2 ng/Ml Co
(e) <4.1 ng/mL Ti

Recent studies of metal-on-metal total disc arthroplasty found serum levels of Co-Cr to concentrations of 3–4 ng/mL Co, 1–2 ng/mL for Cr (Guyer et al. 2011; Hallab et al. 2003; Seo et al. 2016). The concentrations of circulating Co-Cr metal in serum with total disc arthroplasty are similar to levels measured in well-functioning metal-on-metal THA. This has not been demonstrated in nonarticulating implants, where recent studies have failed to detect elevated amounts of Cr or Ni from stainless steel scoliosis rod fixation (Hallab 2009).

In vitro assessment of ion levels from spinal implants with 20% volumetric wear in comparison between serum and saline testing found 1000-fold more particles in saline testing, demonstrating a protective effect of serum proteins and demonstrating a worst case scenario in saline testing (Hallab et al. 2008).

Implant Debris Physical Attributes and Local Physiological Response

Particle-sizing techniques such as scanning electron microscopy (SEM) or transmission electron microscopy can determine the size of the wear particles ranging from nanometer to submicron range. New low-angle laser light scattering (LALLS) techniques sample millions to billions of particles that determine the significant portion of the total mass loss (the total amount of debris). A volume-based analysis that also can characterize implant debris with a number bases is very important, where different samples of particles look demonstrably very different when viewed as a volume-based distribution compared to number-based distribution. Collected metal particles are characterized for size and number by laser diffraction technology and have a mean diameter of less than 10 μ, usually approximately 1–2 μ with a size range of 1–10 [(Garcia et al. 2020)].

General particle characteristics on which local inflammation has been shown to depend are particle load (particle size and volume), aspect ratio, and chemical reactivity. (Bio Reactivity index: particle load x aspect ratio x material type x K unknown). Greater particle load can increase inflammation and is directly correlated to the concentration of phagocytosable particles per tissue volume. The degree to which equal numbers (dose) of large versus small particles (10 μ vs. 1 μ) induce an inflammatory response on a per-particle basis in vivo has not been thoroughly investigated. However, some studies have shown that in equal amounts of debris mass, small particles (0.4 μ) produced a greater inflammatory response than larger (7.5 μ) particles (Hallab 2009).

Elongated fibers are more pro-inflammatory than round particles. Currently, fibers can be categorized as particles with an aspect ratio greater than 3 to be more inflammatory. More chemically reactive particles are more pro-inflammatory. Despite reported differences, there is a growing consensus that metallic particles that are capable of corroding and releasing ions are associated with hypersensitivity responses, cytotoxicity, and DNA damage. Thus, they are more capable of eliciting proinflammatory responses than relatively inert polymers and ceramics (Hallab 2009).

To produce an in vitro inflammatory response, particles need to be less than 10 μ that are within phagocytosable range. Particle mean sizes of 0.2–10 μ are generally the most proinflammatory. The relationship between bacteria and aseptic loosening has been inferred because antibiotic-eluting bone cement and systemically administered antibiotics reportedly reduce the frequency of aseptic loosening (Hallab 2009).

Implant debris from wear causes local inflammation and granulomatous invasion of bone-implant contact that, over time, results in implant loosening and pain, necessitating revision in total joint arthroplasty. Implant debris is known to cause inflammation, osteolysis, and, in some cases, hypersensitivity and concerns persist about implant debris becoming carcinogenic or toxic. Other systemic conditions from implant debris, such as renal failure, have been reported in patients with Co, Cr levels over 100-fold in comparison with individuals with stable prostheses with no aseptic loosening.

Metal debris becomes antigens for T-cell recognition. Once debris is ingested by macrophages and other peri-implant cells, host pro-inflammatory reactions occur, such as activation of metal reactive T-cells. Cobalt-chromium-molybdenum (CoCrMo) alloy debris form metal protein complexes that activate the macrophage inflammasome pathway. CoCrMo alloy debris has been shown to induce macrophage activation, which stimulates secretion of IL-1b TNFa, IL-6, IL-8, and upregulates NFKb and downstream inflammatory cytokines (Mitchelson et al. 2015). Titanium particles induced IL-8, monocyte chemoattractant protein-1 (MCP-1). The study demonstrated that osteoblast chemokine expression with increased NFKB inducing osteoblast activated periprosthetic osteolysis (Fritz et al. 2006).

Biologic reactivity to spinal implant debris has been clinically observed with all the hallmarks of traditional particle-induced osteolysis; granulomatous epithelioid membranes coating the metal implants have been reported, similar to the fibrous membranes associated with loose total hip replacements. Case reports of painful granuloma associated with spinal implant debris demonstrate that spinal implant debris-induced inflammation can result in bone destroying granuloma (Hallab 2009). There are relatively few reports of human retrieval studies of loose spinal implants, but granulomatous epithelioid membranes coating the metal implants, similar to the fibrous membranes associated with loose total hip replacements, have been identified. Metallosis often accompanies metal implant debris-related osteolysis, aseptic fibrosis, local necrosis, or loosening (Hallab 2009).

In a cohort of 12 loosened spinal implants, metallosis of the internal membrane was associated with the outer layer of membrane containing an infiltrate of leukocytes and macrophages and all 12 patients had radiolucency around part of the spinal instrumentation. During the study, 11 of 12 patients demonstrated elevated TNFa levels and an increased osteoclastic response in the vicinity of wear debris caused by dry frictional wear particles of titanium or stainless steel. The focal areas of osteolysis involved loose transverse connectors. Removal of the loose metal implants and tissue surrounding them in the fibro-inflammatory zones resulted in resolution of clinical symptoms in all 12 patients (Hallab 2009).

Particles activate macrophages that secrete TNFa, IL-1b, IL-6, IFNgamma, and PGE2, stimulating differentiation of osteoclast precursors into mature osteoclasts and increasing periprosthetic bone resorption. Wear debris particles also have been shown to compromise mesenchymal stem cell differentiation into functional osteoblasts, and particles can directly inhibit collagen synthesis by mature osteoblasts and induce apoptosis of osteoblasts (Hallab 2009). Protein chip assays of ELISA performed on resected inflammatory tissue surrounding failed arthroplasty demonstrates local increase in IL-6, IL-8 cytokines, driving local osteoclastogenesis and osteolysis (Shanbhag et al. 2007).

The release of IL-1b is a powerful inflammatory cytokine response. Co-Cr-Mo alloy particles were found to activate the inflammatory pathway in part through NADPH-mediated monocyte macrophage production of reactive oxygen species. Activation of the inflammatory pathway leads to cleavage of intracellular pro-IL-1b and pro-IL-18 into their mature forms and ultimately leads to their secretion of pro-inflammatory responses through autocrine and paracrine activation of NFKb, which initiates a powerful pro-inflammatory response. The identification of the inflammatory involvement in particle and metal ion-induced inflammation will likely provide new therapeutic strategies to pharmacologically treat implant debris-induced inflammation and hypersensitivity by specifically interrupting the initiation of the inflammatory response that leads to aseptic osteolysis (Hallab 2009).

Systemic Response to Metal Debris and Prevalence in the General Population

Debris-induced systemic effects with implant metals such as Co, Cr, V, and possibly Ni are rare and typically occur with extremely high serum levels of Co. Distant organ levels of cobalt have been found at necropsy with both total hip and knee implants (Arnholt et al. 2020; Urban et al. 2000, 2004). Isolated cases of cardiomyopathy, optic neuritis, and neuropathies from a failing implant have been reported after metal-on metal total hip replacements (Choi et al. 2019; Devendra and Kumar 2017; Garcia et al. 2020; Goodwin et al. 2018; Mikhael et al. 2009; Mosier et al. 2016; Runner et al. 2017; Sabah et al. 2018; Sanz Pérez et al. 2019). A review of the literature, however, does not produce reports of such high levels or systemic symptoms from spinal implants. Neuropathic effects have been reported around both well-functioning and failing articulating implants, but these were generated from a granulomatous response to implant debris and not directly from the implant debris. Inflammation of unknown etiology associated with spinal implants has been shown to resolve after implant removal (Hallab 2009; Zielinski et al. 2014).

Metal hypersensitivity is well documented in case reports and group studies, though overall it remains a relatively unpredictable and poorly understood phenomenon in the context of orthopedic spinal implants. The specific T-cell subpopulations, the cellular mechanism of recognition and activation, and the antigenic metal-protein determinants created by these metals remain incompletely characterized. Nickel is the most common metal sensitizer in humans, followed by Co and Cr. The prevalence of metal sensitivity among the general population is approximately 10–15%, with nickel sensitivity as the highest. Clinical studies of metal implant-related sensitivity link immunogenic reactions with adverse performance of metallic cardiovascular, orthopedic, plastic surgical, and dental implants (Merritt and Brown 1980). Dermatitis, urticaria, and itching, round red wheals, and/or vasculitis have been linked with the relatively more general phenomena of metallosis, excessive periprosthetic fibrosis, and muscular necrosis. Hypersensitivity reactions associated with stainless steel and cobalt alloy implants are more severe than those associated with titanium alloy components (Hallab 2009).

Specific types of implants with a greater propensity to release metal in vivo may be more prone to induce metal sensitivity, as has been shown in metal-on-metal total joint arthroplasty. Spinal implants have been rarely implicated in case reports or group studies of hypersensitivity; thus, metal lymphocyte transformation testing (LTT) prior to receiving an implant may be warranted for people with a history of metal allergy (Hallab 2009).

Toxicity investigations of implant-related metal toxicity include a variety of cell types, including fibroblasts endothelial cells and non-human osteoblast like cells, but these generally have been limited to in vitro studies and animal studies. Concentrations at which this will occur are not known and the degree to which soluble metals are able to contribute induced toxic effects will likely be difficult to distinguish from well-established pro-inflammatory effects of metal particles (Hallab 2009).

While reports of titanium hypersensitivity are absent in the total joint literature, there is a case report of one patient with titanium metal hypersensitivity following VEPTR rod insertion for congenital scoliosis confirmed with testing; symptoms improved with removal of the rod (Zielinski et al. 2014). Testing of a carbon coated VEPTR rod was undertaken with rod desensitization under the skin in the forearm for a 3-month trial. The patient tolerated the carbon rod and the metal rod was replaced with a VEPTR carbon-coated implant. No documented hypersensitivity was found following reimplantation with the carbon-coated implants (Zielinski et al. 2014).

Testing for Metal Hypersensitivity

In 2012 the dermatology literature published a report stating that all patients should be patch tested for skin sensitivity before any elective surgery using a metal orthopedic device. A rebuttal of this practice was published soon after, pointing out multiple issues with patch testing as a gold standard for diagnosing metal hypersensitivity. The skin reactions are driven by a dendritic cell called the Langerhans cell. These cells are not what drive the deep tissue reactions that are seen around implants. There have been reports in knee replacement patients showing no correlation to skin patch results and outcomes of patients who test positive for the metal in the alloy of the implant (Bravo et al. 2016). There also are multiple reports of patients changing their skin patch test results from negative to positive after undergoing a total hip or knee replacement. The incidence of sensitization to metals in orthopedic implants by patch testing increased by 6.5% following hip and knee arthroplasty (Mihalko et al. 2012). Sensitivity to Ni, Co, Cr was 25% in well-functioning implants; this is more than twice the rate in the normal population. In patients with a failed or failing hip prosthesis, the rate of metal sensitivity rises dramatically to 60% or six times that of the general population (Hallab et al. 2001). Nickel is the metal that most often leads to hypersensitivity reaction and studies place the prevalence of nickel sensitivity in the general population between 8% and 25% (Mitchelson et al. 2015).

While a skin patch test may be helpful in the identification of a patient with a metal hypersensitivity, there remains no proof that routine screening will make a difference and may complicate treatment plans for many patients who otherwise will have no reaction to their implants after surgery. There are other options for identifying patients who may be at risk. Testing for hypersensitivity with lymphocyte transformation testing (LTT) in vitro involves measuring the proliferative response of lymphocytes obtained from peripheral blood by routine blood draw (Hallab 2009). Testing for metal sensitivity with metal-LTT testing generally is preferable since there is no subjectivity to the results as in skin patch testing. LTT testing is better suited for the testing of implant-related sensitivity because there is no risk of inducing metal sensitization using skin exposure, thus metal-LTT is highly quantitative (Hallab 2009).

Cutaneous patch testing is considered by some to be the gold standard for in vivo evaluation of delayed hypersensitivity reactions. It can be argued to be invalid because of the differences in antigen presentation between superficial and deep tissue responses in delayed type hypersensitivity reactions (Mitchelson et al. 2015). Some physicians also suggest that it can be subjective as far as grading dermal reactions from 1 to 3 (Merritt and Brown 1980).

One study has demonstrated that despite six positive skin tests before implantation of metal-on-metal (MOM) hip, five patients subsequently lost their sensitivity with repeat skin testing. All patients had good clinical outcomes with no evidence of loosening (Jacobs and Hallab 2006). Another disadvantage of patch testing is that the process of in vivo patch testing could potentially induce sensitization in a previously nonsensitized patient (Mitchelson et al. 2015). Patients' patch test results will shift from negative to positive after joint replacement surgery, suggesting that in vivo metal exposure can cause sensitization (Merritt and Brown 1980).

Postoperative patch testing has been advocated in patients presenting with suspected metal hypersensitivity implant failure in the absence of infection (Mitchelson et al. 2015). The rate of positive patch test results to metals is highest in patients with MOM implants and in those with failed prosthesis (Ooij et al. 2007). Regular preoperative skin testing is not supported; in patients with 21 positive patch results, hypoallergenic TKA components produced no hypersensitivity reactions (Mitchelson et al. 2015). A correlation has been established between patients who had poor outcomes after TKA and positive skin patch testing that indicated metal sensitivity (Maldonado-Naranjo et al. 2015). Routine screening for metal hypersensitivity prior to TKA is not supported by the literature.

Lymphocyte transformation testing (LTT) involves measuring the proliferative response of lymphocytes, following activation, by using a radioactive marker added to patients spun down lymphocytes along with the desired agent (the metal ions) measured in counts per minute of stimulation (Hallab et al. 2001). LTT can be used as an alternative method to determine metal sensitivity by in vitro testing of sensitivity via venipuncture. It has been found to be more sensitive than patch testing and is highly quantifiable and reproducible. LTT does NOT confer sensitization to the patient as does patch testing, and LTT prior to arthroplasty may be effective as a preoperative screening tool for metal hypersensitivity.

In vitro leukocyte migration testing can be performed by capillary tube testing with leukocyte migration in response to antigen, membrane migration, leukocyte migration agarose technique, and collagen gel electrophoresis (Hallab et al. 2001).

Implantable metal testing for sensitivity has not established guidelines regarding the depth or duration of subcutaneous metal implantation as screening tests for hypersensitivity (Mitchelson et al. 2015). The timing of implantable metal testing is not supported in all TKA/THA patients and has only been found to be indicated in patients with a history of a metal allergy or previous aseptic orthopedic implant failure. Postoperative testing should be limited in patients with allergic contact dermatitis, arthralgia, and radiolucencies surrounding the implant or aseptic loosening without infection (Mitchelson et al. 2015).

Routine use of radiographs is supported for identification of periprosthetic radiolucent lines or aseptic loosening after TKA/THA (Mitchelson et al. 2015). Loosening or fracture of spinal implants has not been routinely documented in metal hypersensitivity reactions in the literature.

Risk Factors for Metal Hypersensitvity

One study of 28 TKA patients determined that those with hypersensitivity were more likely to be female; seven patients had a history of metal hypersensitivity before arthroplasty (Mihalko et al. 2012). Twenty-two patients had self-reported allergies, and skin patch testing was positive in 19 patients. Dermatologic symptoms resolved in patients who had revision with hypoallergenic implants with no further instability. A similar study found positive skin patch results in 68% of patients with reported metal allergy (Mitchelson et al. 2015). Another study found that 32% of patients who had TJR with no known prior history of metal allergies developed a positive leukocyte migration inhibition test of Ti, Co, Cr, or Ni 3 months to 1 year following surgery (Goodman 2007). Implant failure was reported to be up to 4 times greater in patients with a self-reported history of preoperative metal allergy compared with patients who did not have an allergy (Mitchelson et al. 2015).

Age, gender, and occupation are all risk factors for developing nickel hypersensitivity. Exposure to costume jewelry may account for the higher rates in women. Nickel sensitization has been reported to be present in 17–32% of women and 3–10% in men. Cr is more common sensitization in men at 10% compared to 7% in women. Cr is associated with concrete exposure in the construction industry, leatherworking, and occupations involving cleaning. Co sensitization is common in hairdressers and textile industry workers. Nickel sensitization is associated with healthcare, agriculture, mechanics, and metal work.

One study demonstrated acquired hypersensitivity following Ti spinal implants and tattoos. Skin biopsy of reaction and surrounding tissues of TI spinal implants demonstrated high levels of Ni and Cr, Ti, skin testing was negative for Ni and Cr (de Cuyper et al. 2017).

The North American Skin Patch testing group reported 21% sensitivity to Ni 21% and 8% to Co and Cr in 5,000 patients (Merritt and Brown 1980). As more patients are repeatedly exposed to metal variants commonly used in orthopedic implants, the possibility of increased sensitivity reactions to these metals may rise.

Clinical Presentation of Metal Hypersensitivity

Metal hypersensitivity may result in localized or systemic allergic dermatitis, loss of joint function, implant failure, and pain. Pruritic erythematous, eczematous, edematous, and sometimes exudative lesions may present over implant sites (Mitchelson et al. 2015). Symptoms ascribed to metal hypersensitivity include pain, swelling, cutaneous rash, patient dissatisfaction, and loss of function (Merritt and Brown 1980). The degree to which the known condition of metal hypersensitivity induced failure is not well known. No clear association between the prevalence of metal sensitivity and duration of implant in situ has been identified, and no clear objective lines have been found between pain-related failure in metal sensitive and nonsensitive patients undergoing revision (Hallab et al. 2001). This may represent an extreme complication or may be a more subtle contribution to implant failure overall (Hallab et al. 2001).

Metal hypersensitivity-induced allergic dermatitis, pain, and implant failure have clinically relevant laboratory markers associated with the conditions (Mihalko et al. 2012). Elevated levels of IL-6, INFa, and IL-17 are common identifiable markers in implant failure. Increased Ni and Ti have been demonstrated to increase expression of RANKL, macrophage colony stimulating factor, TNFa, and CCR4 receptors. Clinicians should have a high level of suspicion when patients present with arthralgia, periprosthetic radiolucent lines, or aseptic implant loosening. Ordering appropriate inflammatory laboratory markers during a workup for suspected metal hypersenstivity is necessary to determine the extent of the response (Mitchelson et al. 2015).

Metal Sensitivity to Spinal Implants

Spinal Implant Composition

Spinal implant composition is dependent on the implant function and location. Multiple implants ranging from pedicle screw instrumentation, metal on polyethylene disc replacements, and PEEK fusion grafts are implanted in patients undergoing spinal surgery for various reasons. Reports of metal sensitivity in patients with spinal implants are primarily single case reports or very small series (2–4 patients).

Disc Replacements

Disc replacements are metal and polyethylene combination implants used to replace symptomatic disc pathology while preserving the motion segment. In one study of 4 patients who had failed lumbar disc replacements, retrieval showed evidence of wear of the polyethylene cores, but the extent and severity varied among the four patients. Wear and fracture of the core were associated with osteolysis of the underlying sacrum. Histologic examination confirmed the presence of wear debris in inflammatory fibrous tissue. Evidence of failure prior to retrieval included subsidence, migration, undersizing, and reactionary adjacent fusion on radiographic analysis. The mechanism of wear was determined by adhesive wear of the central domed region of the polyethylene core and chronic rim impingement resulting in rim fatigue and fracture (Ooij et al. 2007).

In another report of four patients with TDR who had an uncomplicated initial postoperative course followed by worsening pain months after surgery, retrieval found an avascular soft-tissue mass was found to be causing an epidural mass effect scar and causing symptoms to re-emerge. Laboratory analysis of the tissue found lymphocytic reaction tissue, dominated by a large number of lymphocytes and small number of macrophages (Guyer et al. 2011).

Total Disc Replacement (TDR) Materials

The TDR can be composed of stainless steel alloys which confer greater ductility; Co-Cr alloys which confer increased corrosion resistance, hardest, strongest, and most fatigue resistance; or titanium alloys which good flexural rigidity and toughness and high corrosion resistance compared to stainless steel and Co-Cr (Hallab et al. 2003).

Polyetheretherketone (PEEK)

PEEK cages have a high biocompatibility profile and are radiolucent. One case report regarding chronic allergic response to interbody PEEK material reported diffuse erythema and itching, tongue swelling, and erythema in the throat following PEEK implantation. No significant inflammatory tissue or response was found in the retrieval (Maldonado-Naranjo et al. 2015).

Device for Intervertebral Assisted Motion (DIAM)

DIAM is a silicone disc enveloped in a polyethylene terephthalate fiber sack. One case report described granulation tissue 5 years after DIAM. Histology demonstrated wear debris and chronic inflammation with a hypersensitivity reaction and subsequent bone osteolysis surrounding the implant (Seo et al. 2016).

Metal-on-Metal Facet Replacements

Two patients with MOM facet replacements were reported to develop local tissue reactions with pseudotumor formation, characteristic soft yogurt-like chalky white scars surrounding the implants (Goodwin et al. 2018).

Carbon-Coated Implants

Metal hypersensitivity was described in one patient after VEPTR titanium rod insertion for congenital scoliosis. No hypersensitivity was documented after reimplantation with carbon-coated implants (Zielinski et al. 2014).

Zirconium Rods

Plasma sprayed zirconium interface rods cannot be contoured, and significant implant brittleness precludes their use in deformity correction, limiting the use of zirconium for spinal implants (Zielinski et al. 2014).

ACDF Implants

Reported allergic reaction to PEEK implant and Ti ACDF plate and screws include system rash, congestion, dysphasia, and urticaria. Symptoms resolved once implants were removed, with no visible osteolysis (Urban et al. 2000).

Treatment

Treatment of metal hypersensitivity can range from symptomatic treatment to revision surgery. Reactions around the spine obviously play a different role than those in total joint replacement where symptomatic treatment of the dermatologic symptoms may resolve completely with a use of topical corticosteroid (Mitchelson et al. 2015). In the spine these reactions caused by proximity of vital neurologic structures need a more heightened awareness and investigation of possible deep tissue reactions. Metal artifact reduction sequence (MARS) MRI can help determine if a reaction is occurring and can help grade the reaction if any (Connelly et al. 2018). This can aid in the choice of an approach for treatment, which can be difficult depending on the purpose of the implant in place.

New technologies involving immune modulation have emerged, but are still investigative. The use of Nac, an antioxidant inhibitor of NFkB, can potentially be used to augment the inflammatory response via glutathione (GSH), which inhibits serine phosphorylation of iKB, thereby preventing the dissociation of NFkB induced cellular response to particulate debris. Reduction in the stimulation of NFkB leads to decreased osteolysis surrounding an implant (Willert and Semlitsch 1977).

Further use of disease modifying anti-rheumatic drugs (DMARDs) has expanded the pharmacologic treatment of metal hypersensitivity reactions. Numerous in vivo and in vitro animal model studies suggest that bisphosphonates may be a potential benefit for treatment of particle-induced osteolysis. Antitumor necrosis factor alpha (TNFa) therapy with etanercept has been reported to inhibit osteoclastic bone resorption; however, in an underpowered study it was found to produce no change in volumetric wear osteolysis compared to a placebo (Holt et al. 2007).

Conclusion

Over the last 10 years, metal hypersensitivity has been documented in patients with spinal implants with bearing surfaces that may be a trigger for sensitizing patients due to the local wear debris. Reaction to metal alloys and debris is a type IV hypersensitivity immunologic reaction that does not produce anaphylaxis and has led to increased exposure. Realizing that 10–15% of the population has these sensitivities, identifying who is at risk, and noting clinical and radiographic signs of hypersensitivity can help surgeons notify their patients of the possible risk and determine appropriate treatment if required. Treatment can range from symptomatic treatment of dermatologic conditions to revision surgery and use of hypoallergenic implants.

Summary

While most reports of metal hypersensitivity have been in the total joint literature, in the last 10 years there have been a number of case reports of spinal implant reactions. Much of this is because spine implants have been developed with bearing surfaces that may be a trigger for sensitizing patients from the local wear debris. Recognizing the signs of metal hypersensitivity, becoming familiar with testing procedures, and identifying risk factors for hypersensitivity, in addition to knowing the frequency of these sensitivities in the general population, can help surgeons identify these patients and notify them of the possible risk.

References

Arnholt CM, White JB, Lowell JA, Perkins MR, Mihalko WM, Kurtz SM (2020) Postmortem retrieval analysis of metallosis and periprosthetic tissue metal concentrations in total knee arthroplasty. J. Arthroplasty 35(2):569–578

Bravo D, Wagner ER, Larson DR, Davis MP, Pagnano MW, Sierra RJ (2016) No increased risk of knee arthroplasty failure in patients with positive skin patch testing for metal hypersensitivity: a matched cohort study. J Arthroplast 31:1717–1721

Campbell P, Ebramzadeh E, Nelson S, Takamura K, Smet KD, Amstutz HC (2010) Histological features of pseudotumor-like tissues from metal-on-metal hips. Clin Orthop Relat Res 468:2321–2327

Choi HI, Hong JA, Kim MS, Lee SE, Jung SH, Yoon PW, Song JS, Kim JJ (2019) Severe cardiomyopathy due to arthroprosthetic cobaltism: report of two cases with different outcomes. Cardiovasc Toxicol 19(1):82–89

Connelly JW, Galea VP, Matuszak SJ et al (2018) Indications for MARS-MRI in patients treated with metal-on-metal resurfacing arthroplasty. J Arthroplast 33:9199–1925

de Cuyper CD, Lodewick E, Schreiver I, Hesse B, Seim C, Castillo-Michel H, Luch A (2017) Are metals involved in tattoo-related hypersensitivity reactions? A case report. Contact Dermatitis 77:397–405

Devendra L, Kumar P (2017) Pseudotumour complicated by implant loosening one year after revision ceramic on metal total hip arthroplasty: a case report. J Orthop Case Rep 7:82–86

Fritz EA, Glant TT, Vermes C, Jacobs JJ, Roebuck KA (2006) Chemokine gene activation in human bone marrow-derived osteoblasts following exposure to particulate wear debris. J Biomed Mater Res 77A:192–201

Garcia MD, Hur M, Chen JJ, Bhatti MT (2020) Cobalt toxic optic neuropathy and retinopathy:case report and review of the literature. Am J Ophthalmol Case Rep 17:100606

Goodman SB (2007) Wear particles, periprosthetic osteolysis and the immune system. Biomaterials 28:5044–5048

Goodwin ML, Spiker WR, Brodke DS, Lawrence BD (2018) Failure of facet replacement system with metal-on-metal bearing surface and subsequent discovery of cobalt allergy: report of 2 cases. J Neurosurg Spine 29:1–84

Guyer RD, Shellock J, Maclennan B, Hanscom D, Knight RQ, Mccombe P et al (2011) Early failure of metal-on-metal artificial disc prostheses associated with lymphoytic reaction. Spine (Phila Pa 1976) 36: E492–E497

Hallab NJ (2009) A review of the biologic effects of spine implant debris: fact from fiction. SAS J 3:143–160

Hallab N, Merritt K, Jacobs JJ (2001) Metal sensitivity in patients with orthopaedic implants. J Bone Joint Surg Am 83:428–436

Hallab N, Link HD, Mcafee PC (2003) Biomaterial optimization in total disc arthroplasty. Spine (Phila Pa 1976) 28:S139–S152

Hallab N, Khandha A, Malcolmson G, Timm J (2008) In vitro assessment of serum-saline ratios for fluid simulator testing of highly modular spinal implants with articulating surfaces. SAS J 2:171–183

Holt G, Murnaghan C, Reilly J, Meek RMD (2007) The biology of aseptic osteolysis. Clin Orthop Relat Res 460:240–252

Jacobs JJ, Hallab NJ (2006) Loosening and osteolysis associated with metal-on-metal bearings. J Bone Joint Surg Am 88:1171–1172

Maldonado-Naranjo AL, Healy AT, Kalfas IH (2015) Polyetherether-ketone (PEEK) intervertebral cage as a cause of chronic systemic allergy: a case report. Spine J 15:e1–e3

Merritt K, Brown SA (1980) Tissue reaction and metal sensitivity: an animal study. Acta Orthop Scand 51:403–411

Mihalko WM, Goodman SB, Hallab NJ, Jacobs JJ (2012) Skin patch testing and associated total knee outcomes. AAOS Now September, pp 40–41

Mikhael MM, Hassen AD, Sierra RJ (2009) Failure of metal-on-metal total hip arthroplasty mimicking hip infection: a report of two cases. J Bone Joint Surg Am 91:443–446

Mitchelson AJ, Wilson CJ, Mihalko WM et al (2015) Biomaterial hypersensitivity: is it real? Supportive evidence and approach considerations for metal allergic patients following total knee arthroplasty. Biomed Res Int 2015:137287

Mosier BA, Maynard L, Sotereanos NG, Sewecke JJ (2016) Progressive cardiomyopathy in a patient with elevated cobalt ion levels and bilateral metal-on- metal hip arthroplasties. Am J Orthop (Belle Mead NJ) 45(3): E132–E135

Ooij AV, Kurt SM, Stessels F, Noten H, Rhijn LV (2007) Polyethylene wear debris and long-term clinical failure of the Charité disc prosthesis. Spine (Phila Pa 1976) 32:223–229

Runner RP, Briggs M, Ahearn MD, Guild GN 3rd (2017) Case report. Unusual presentation of failed metal-on-metal total hip arthroplasty with features of neoplastic process. Arthroplasty Today 3:71–76

Sabah SA, Moon JC, Jenkins-Jones S, Morgan CLI, Currie CJ, Wilkinson JM et al (2018) The risk of cardiac failure following metal-on-metal hip arthroplasty. J Bone J 100-B:20–27

Sanz Pérez MI, Rico Villoras AM, Moreno Velasco A, Bartolomé García S, Campo Loarte J (2019) Heart transplant secondary to cobalt toxicity after hip arthroplasty revision. Hip Int 29(4):NP1–NP5

Seo JY, Ha KY, Kim YH, Ahn JH (2016) Foreign body reaction after implantation of a device for intervertebral assisted motion. J Korean Neurosurg Soc 59:647–649

Shanbhag AS, Kaufman AM, Hayata K, Rubash HE (2007) Assessing osteolysis with use of high-throughput protein chips. J Bone Joint Surg Am 89:1081–1089

Urban RM, Jacobs JJ, Tomlinson MJ, Gavrilovic J, Black J, Peoc'h M (2000) Dissemination of wear particles to the liver, spleen, and abdominal lymph nodes of patients with hip or knee replacement. J Bone Joint Surg Am 82:457–476

Urban RM, Tomlinson MJ, Hall DJ, Jacobs JJ (2004) Accumulation in liver and spleen of metal particles generated at nonbearing surfaces in hip arthroplasty. J Arthroplast 19(8 Suppl 3):94–101

Verma SB, Mody B, Gawkrodger DJ (2006) Dermatitis on the knee following knee replacement: allergy to chromate, cobalt or nickel but a causal association is unproven. Contact Dermatitis 54(4):228–229

Willert HG, Semlitsch M (1977) Reactions of the articular capsule to wear products of artificial joint prostheses. J Biomed Mater Res 11:157–164

Zielinski J, Lacy TA, Phillips JH (2014) Carbon coated implants as a new solution for metal allergy in early-onset scoliosis: a case report and review of the literature. Spine Deform 2:76–80

Spinal Cord Stimulation: Effect on Motor Function in Parkinson's Disease

Nestor D. Tomycz, Timothy Leichliter, Saadyah Averick, Boyle C. Cheng, and Donald M. Whiting

Contents

Introduction	474
Animal Studies	474
Human Studies	476
Possible Mechanisms of SCS in PD	478
Future Directions	479
Conclusion	480
References	480

N. D. Tomycz · S. Averick (✉)
Department of Neurosurgery, Neuroscience Institute, Allegheny Health Network, Pittsburgh, PA, USA
e-mail: Nestor.Tomycz@ahn.org; Saadyah.Averick@ahn.org

T. Leichliter
Department of Neurology, Neuroscience Institute, Allegheny Health Network, Pittsburgh, PA, USA
e-mail: Timothy.Leichliter@ahn.org

B. C. Cheng
Neuroscience Institute, Allegheny Health Network, Drexel University, Allegheny General Hospital Campus, Pittsburgh, PA, USA
e-mail: bcheng@wpahs.org; Boyle.CHENG@ahn.org; boylecheng@yahoo.com; boylecheng@gmail.com

D. M. Whiting
Neuroscience Institute, Allegheny Health Network, Pittsburgh, PA, USA
e-mail: donald.whiting@ahn.org

© Springer Nature Switzerland AG 2021
B. C. Cheng (ed.), *Handbook of Spine Technology*,
https://doi.org/10.1007/978-3-319-44424-6_142

Abstract

Invasive high frequency electrical stimulation of the brain, deep brain stimulation (DBS), has become a standard of care intervention for improving motor symptoms in Parkinson's disease (PD). Although DBS has been shown to improve many of the cardinal motor symptoms of PD including dyskinesias, bradykinesia, tremor, and rigidity, DBS has not shown consistent benefit for gait dysfunction in PD. Spinal cord stimulation (SCS) is an older form of electrical neuromodulation and has been used in humans for decades to treat primarily chronic pain disorders. Over the past decade, there has been a growing numbers of animal and human studies suggesting that SCS may improve motor symptoms, especially gait dysfunction problems such as freezing, in patients with PD. SCS has no current regulatory approval for usage in PD motor symptomatology and many of the benefits of SCS in PD patient have been incidentally observed in PD patients who were implanted with SCS for chronic pain. This chapter will review the published evidence for SCS in PD and discuss possible mechanisms for motor improvement in PD in addition to pain alleviation.

Keywords

Spinal cord stimulation · Parkinson's disease · Deep brain stimulation · Gait · Thoracic level · Basal ganglia

Introduction

Parkinson's disease (PD) is a complex neurodegenerative disorder with myriad motor and non-motor symptoms. Medications, primarily in the form of dopamine replacement therapy, and deep brain stimulation (DBS) have proven successful in alleviating many of the most common PD motor symptoms. However, axial motor symptoms such as truncal postural abnormalities and gait problems such as freezing are often resistant to both medications and DBS. In fact, DBS is often withheld from PD patients with significant disability due to gait dysfunction and postural disability since DBS may not only prove ineffective but could potentially worsen such symptoms. Dissatisfaction with current DBS brain targets for axial symptoms in PD has led to exploration of new brain DBS targets such as the pedunculopontine nucleus (PPN) (French and Muthusamy 2018). Spinal cord stimulation (SCS) is an attractive neuromodulation technology for improving motor symptoms in PD for multiple reasons (Cai et al. 2020). First, SCS may be considered less invasive and risky since it does not require cranial surgery. SCS has a longer safety track record in humans due to its use for decades in chronic pain disorders. Moreover, SCS, unlike DBS, is able to be percutaneously trialed in a minimally invasive fashion prior to implant and this may facilitate better candidate selection for surgical implants. Further, chronic pain is a common yet underappreciated symptom in PD and SCS is most established for pain relief as shown in the Fig. 1. Finally, market forces have driven SCS technological innovation faster than that in DBS and consequently have generated a greater option of hardware and software technologies in SCS than in DBS. Although traditional tonic SCS engenders stimulation-induced paresthesias or tingling in parts of a patient's body, a recent innovation in SCS is paresthesia-free programming modes which lend themselves to blinded studies in both pain and movement disorders (De Ridder et al. 2010).

SCS was first used in humans for pain relief in 1967 (Shealy et al. 1967). The first report suggesting that SCS may be able to ameliorate motor aspects of a disease was seen in 1973 when a patient with multiple sclerosis receiving SCS for pain was noted to have improvements in weakness, speech, and swallowing after SCS implantation (Cook and Weinstein 1973). Over the past 30 years there have been multiple, mostly open-label studies of SCS in various movement disorders including dystonia (generalized dystonia, focal dystonia, and spasmodic torticollis), nonparkinsonian tremor, painful leg and moving toes (PLMT), and Parkinson's disease (Thiriez et al. 2014). The focus below will be on the published studies that have explored the gait and movement disorder consequences of SCS in patients with PD as well as in animal models of PD. We will also discuss recommendations for future study design to better investigate the motor function effects of SCS in PD.

Animal Studies

In 2009, rodent models of PD secondary to dopamine depletion with tyrosine hydroxylase inhibitor, alpha-methy-para-tyrosine, and 6-hydroxydopamine showed significant improvement in locomotion with SCS (biphasic square pulses at 300 Hz) applied to the upper thoracic levels. Fuentes et al. furthermore showed that SCS in these rat models altered both cortical and striatal local field potentials, suggesting that SCS, although applied to the epidural space within the spine, may provide motor benefit in PD by directly modulating brain function (Fuentes et al. 2009). Interestingly, SCS was shown to reduce aberrant synchronous

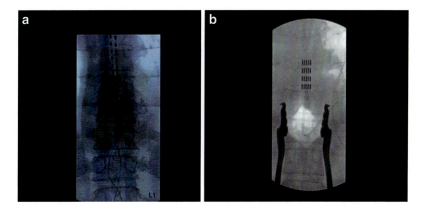

Fig. 1 Spinal cord stimulation leads: (**a**) 2 percutaneous or cylindrical leads (**b**) one plate or paddle lead

low-frequency oscillations in the basal ganglia which is similar to current theories for the therapeutic mechanism of DBS in PD (Beudel et al. 2019).

Moreover, in addition to dopamine depletion rodent models of PD, rats rendered parkinsonian by overexpression of alpha-synuclein (using unilateral injection of adeno-associated virus serotype 6 into the substania nigra) also were shown to improve use of their affected forepaw via SCS (Brys et al. 2017).

Santana et al. applied upper thoracic (T3-T4) SCS to a primate model of PD and observed improvements in freezing, hypokinesia, postural instability, and bradykinesia. Motor improvements were assessed by observers blinded to the SCS "ON" or "OFF" condition and the motor deficit which showed the highest improvement with SCS was freezing. The motor improvements in these primate models of PD treated with SCS showed a similar degree of motor improvement to that observed with UPDRS III motor score reduction with DBS in humans. They did not observe that altering the frequency of SCS led to differences in the motor improvements. Concurrent microelectrode brain recording in these animals showed that when SCS was activated, many neurons significantly decreased their beta rhythmicity (Santana et al. 2014). Thus, this study demonstrated that dopamine agonists, DBS, and SCS may also share the ability to improve parkinsonian motor symptoms by reducing excessive neuronal synchronization in the basal ganglia.

Yadav et al. reported that thoracic (T2) SCS applied just twice a week in bilateral 6-hydroxydopamine striatal lesioned rats significantly improved both posture and locomotion. Interestingly, similar to that observed in humans receiving subthalamic nucleus DBS for PD, SCS treated rats showed more weight gain compared to the control groups. Most significant in this study was the finding that striatal immunostaining for tyrosine hydroxylase, a marker for dopaminergic innervation, and substania nigra pars compacta neuronal cell count was significantly preserved in rats receiving SCS as compared to the controls (Yadov et al. 2014). Although requiring further study, this remarkable finding suggests that SCS may have a neuroprotective effect in PD.

Zhong et al. tried SCS at much lower levels in the spine, L2-S1, in a rat model of Parkinson's disease engendered by unilateral 6-hydroxydopamine lesion of the nigrostriatal pathway. During SCS, the lesioned rats which exhibited severe parkinsonism did show improved step initiation and step quality (Zhong et al. 2019).

Shinko et al. applied SCS to high cervical spinal cord (C1-C2) in unilaterally 6-hydroxydopamine lesioned rats and also demonstrated forepaw mobility improvement with SCS. Further support for neuroprotection with SCS was shown with tyrosine hydroxylase (TH) immunostaining which demonstrated preservation of TH-positive fibers in SCS treated rats. This group tried three different SCS frequencies (2 Hz, 50 Hz, and 200 Hz) and found that 50 Hz SCS engendered the greatest motor improvement and largest neuroprotective effect on dopamine cells within the striatum, particularly the substantia nigra pars compacta. They furthermore investigated the mechanism of TH-cell preservation and showed that 50 Hz SCS significantly increased levels of the growth factor VEGF in the lesioned striatum (Shinko et al. 2014).

Human Studies

Despite heterogeneity in the location of SCS epidural electrode placement and programming parameters, preclinical studies overall with various parkinsonian animal models have shown improvements in motor function with application of SCS. Human studies thus far are comprised of case reports and small case series which can be divided into three groups: (1) patients with PD receiving SCS for motor symptoms, (2) patients with PD implanted with SCS for pain in which SCS was observed for benefits on motor symptoms, and (3) patients with PD implanted with DBS for which SCS was added as adjunctive neuromodulation to improve motor symptoms.

Thevathasan et al. published on 2 PD patients with moderate to severe motor impairments who received high cervical epidural SCS which were implanted surgically. One patient received SCS at 130 Hz and the other at 300 Hz. Ten days after surgery, the patients participated in a double-blind crossover study of the motor effects of SCS. The primary outcome was motor subsection of UPDRS (mean score of 2 blinded neurologists). Despite trying a range of SCS frequencies and intensities, there was no difference detected in the primary outcome measure of motor UPDRS. The authors speculated that perhaps the benefits of SCS seen in animal models of PD were secondary to SCS precipitating movements merely from its startling arousal effect (Thevathasan et al. 2010). This study has been criticized for the different frequencies chosen for each patient and for placing the SCS electrode in the high cervical spine, whereas most of the animal work showed benefit using thoracic spinal level SCS.

In contrast, Hassan et al. reported a single case of a 43-year-old woman with PD who underwent high cervical (C2) SCS for neuropathic upper extremity pain after trauma and was found to have significant improvement in her rigidity and tremor during SCS activation (Hassan et al. 2013).

The largest study of cervical SCS for PD to date was published by Mazzone et al. This study applied both tonic SCS and burstDR mode (burst rate 40 Hz, intraburst rate 500 Hz, pulse width 1000 microseconds) SCS. Tonic cervical SCS was applied to 6 PD patients (Group 1) suffering from low back pain and burst cervical SCS was implanted in 12 PD patients (Group 2) primarily to improve motor symptoms including tremor, rigidity, gait, and posture disturbances. Criteria for inclusion into the burst cervical SCS group included ineligibility for DBS, ineffectiveness of STN DBS, or decay of benefit from pedunculopontine DBS. The electrode tip was located at C2 (16 patients) and C2-C3 (2 patients). Group 1 patients had either a quadripolar electrode implanted (5 patients, Medtronic mod. 3487-A) or an octapolar electrode (1 patient, Medtronic mod. 3898), whereas group 2 patients all had an MRI-compatible octapolar electrode implanted (Abbott). In the Group 1, patients receiving tonic cervical SCS (3 months) were required to shown reduction in the UPDRS and Hoehn and Yahr scale (Mazzone et al. 2019).

The rest of the published human experience has highlighted thoracic SCS in PD. Nishioka and Nakajima described three cases of thoracic (T8-T11) SCS in PD patients in which SCS was implanted for low back and lower extremity pain. One patient had failed back surgery syndrome and the other two patients had lumbar stenosis. Gait was not examined, but SCS led to statistically significant improvements in both rigidity and tremor based on UPDRS. SCS improved pain in all patients but did not improve dementia or activities of daily living (Nishioka and Nakajima 2015).

Fenelon et al. reported a single patient with PD who underwent thoracic SCS for post-laminectomy pain syndrome. After 29 months of follow-up, UPDRS motor scores (off drug, on SCS) were reduced by 50% (Fenelon et al. 2012).

Agari and Date published on 15 patients (5 men, 10 women) with advanced PD (7 of which already had DBS) undergoing thoracic SCS for low back and/or lower extremity pain. The follow-up period was 12 months and motor function was evaluated with UPDRS, Timed Up and Go tests, and Timed 10-Meter Walk test. Percutaneous leads with 4 or 8 electrodes were implanted using local anesthesia after a trial provided paresthesia coverage for more than 80% of the painful region. Posture, postural stability, bradykinesia, and gait showed significant improvement at 3 months, but the improvement decreased at 12 months. Timed 10-Meter Walk times were also improved at both 3 months and 12 months after SCS implant. No changes had been made to DBS settings if DBS or medications were present, so motor improvements in these patients were attributed to SCS (Agari and Date 2012).

More recently, Samotus et al. described five male patients with advanced PD and significant gait disturbances and freezing of gait who underwent thoracic SCS. The group used advanced gait analysis technology (Protokinetics Walkway) to measure gait parameters including timed sit-to-stand and automated freezing-of-gait detection via foot pressures. SCS programming combinations were tested over a 1–4 month period to find optimal programming settings for each patient. SCS led to significant improvements in mean UPDRS and step length, stride velocity, and sit-to-stand and also significantly reduced the number of freezing-of-gait episodes (Samotus et al. 2018).

Much of the remaining thoracic SCS data in PD patients involves patients already implanted with DBS. Pinto de Souza et al. treated four patients with PD and significant postural instability and gait disturbance with high thoracic (T2-T4) SCS. All patients had previously been implanted with bilateral subthalamic nucleus DBS. Timed-Up-GO and 20-meter-walk tests, UPDRS III, freezing of gait questionnaires, and quality of life scores were measured. Blinded assessments of gait were performed with sham stimulation as well as with SCS at 300 Hz and 60 Hz. Overall, it was reported that SCS at 300 Hz was well tolerated and led to significant improvement in gait. SCS at 300 Hz decreased freezing of gait, improved self-reported quality of life, and improved UPDRS motor scores (Pinto de Souza et al. 2017). Lima-Pardini et al. further investigated this group of four patients with more advanced biomechanical gait assessment tools. Testing was performed to determine effects of SCS on freezing of gait, postural reactive responses, and anticipatory postural adjustment (APA). The authors explained that APA is deficient in freezing of gait episodes and APA is crucial for normal gait. Freezing of gait was analyzed using wireless accelerometry and APA was quantified with a force platform. Overall, SCS engendered improvement in APA and freezing of gait but did not change postural reactive responses (de Lima-Pardini et al. 2018).

Lai et al. recently described a 73-year-old man with PD who underwent thoracic SCS for chronic low back pain. The authors describe an interesting accidental blinding that occurred when the SCS percutaneous lead dislocated and was replaced with a paddle lead. Despite SCS being below the threshold of perception for causing paresthesias, the SCS was shown to improve gait, suggesting that this is not a placebo effect (Lai et al. 2020).

Another case report by Akiyama et al. described thoracic SCS placed for painful camptocormia with Pisa syndrome in a 65-year-old woman with PD. Camptocormia refers to the exaggerated thoracolumbar flexion that occurs in PD, and Pisa syndrome is lateral trunk flexion also seen in neurodegenerative disorders such as PD. The patient was previously treated with DBS that did improve these truncal postural abnormalities for some time. However, the patient had reappearance of these truncal abnormalities and progressive resultant pain. SCS lead to immediate pain improvement and gradual improvement (after 10 days) in both her camptocormia and Pisa syndrome (Akiyama et al. 2017). This delay in postural improvement suggests that SCS may improve motor symptoms directly and not just indirectly from pain alleviation.

Additional small case reports and series have supported that thoracic SCS, both tonic and burst mode, can improve motor function in PD patients, particularly their gait and posture (Hubsch et al. 2019; Kobayashi et al. 2018). There is also some early evidence that burst mode SCS may additionally improve emotional symptoms and mental status in PD patients (Kobayashi et al. 2018). This is interesting since studies of brain imaging during SCS for chronic pain suggest that burstDR SCS, unlike tonic SCS, modulates the medial pathway of pain involving the cingulate cortex and therefore may be able to positively influence emotions (Chakravarthy et al. 2019).

Overall, most of the human evidence has shown a beneficial effect of SCS, mostly thoracic level, on motor symptoms of PD and there is a growing body of evidence suggesting that SCS may be particularly beneficial for the levodopa-resistant motor symptoms of PD, such as gait and axial symptoms, which may be difficult to treat with basal ganglia DBS (de Andrade et al. 2016).

Possible Mechanisms of SCS in PD

The evidence that SCS provides benefits in PD, a neurodegenerative brain disease, provides further support that SCS a supraspinal site of action. Nashold et al., by studying the effect of SCS on EEG potentials, were the first to suggest that SCS works by blocking pain processing at the cerebral level rather than at the spinal cord (Nashold et al. 1972). Since SCS has been used for many decades to alleviate pain, it is natural to start with the assumption that SCS may be improving motor function in PD secondary to ameliorating pain in PD. Pain is an under-appreciated nonmotor symptom of PD and occurs through all stages of the disease. Moreover, pain has been found to significantly diminish quality of life in Parkinson's disease and may occur in >60% of PD patients (Skogar and Lokk 2016). Pain in PD is multifactorial and has been characterized in musculoskeletal, radicular/neuropathic, dystonia-related, akathitic, and central pain (Ford 1998). Most pain in PD is musculoskeletal and it is difficult to determine how much motor function in Parkinson's disease is influenced by chronic pain. The studies reviewed herein support that SCS can significantly and durably reduce pain in PD, especially pain involving the low back and lower extremities. However, there is evidence that SCS can produce motor benefits in PD independent of pain relief. Most of the research thus far has employed traditional tonic SCS, which consists of continuous square wave pulses at low frequencies (20–120 Hz) and at amplitudes chosen to create stimulation-induced paresthesias. At least two studies have suggested that there may be a latency between motor benefits and tonic SCS, despite immediate pain relief from the SCS, which supports that the motor function benefits of SCS in PD patients cannot solely be attributed to pain reduction (Mazzone et al. 2019; Akiyama et al. 2017).

Could the placebo effect be a mechanism behind the motor benefits observed in PD patient receiving SCS? The placebo effect is well-described and has been shown to be strong and durable with other implanted medical devices. The SCS literature has been criticized for a lack of blinding and placebo controls. However, it has been generally regarded that the inclusion of good placebo controls with tonic SCS is not possible since patients feel stimulation-induced paresthesias that would unblind them. Sham surgery or inactivated SCS implants could be considered for some reduction of the placebo effect in tonic SCS, yet there are significant ethical problems with adding placebo controls to surgical trials (Kjaer et al. 2020). The recent advent of paresthesia-free SCS programming modes, including burstDR mode, high density mode, and 10 kHz high frequency SCS, has opened the possibility of true double blinded SCS studies (Morales et al. 2019).

Animal models of PD have supported that SCS may be able to reduce abnormal synchronization in the basal ganglia, which has become a leading theory of the pathogenesis of PD. There is also evidence from PD animal models that SCS may be neuroprotective, particularly towards dopamine cells. DBS may also provide benefit by desynchronizing pathological oscillations in the brain of patients with PD; however, there is no strong current evidence that DBS is neuroprotective in PD (McKinnon et al. 2019). Reliable biomarkers of PD will be necessary to determine whether neuromodulation modalities such as DBS or SCS will be able to impart neuroprotection in PD.

It is intriguing to consider that SCS may be able to modulate the brain in the same way as DBS. Obviously SCS as a noncranial neuromodulation approach and would therefore be a potential therapy for other brain diseases that may be considered as emerging from abnormal neuronal oscillations. Yadav and Nicolelis have argued that the spinal cord may be increasingly viewed as a "channel" that has the ability to transmit therapeutic electrical signals to the brain in PD and other neurological disorders (Yadov and Nicolelis 2017). A future strategy to determine whether SCS in humans with PD can also desynchronize abnormal basal ganglia activity might employ SCS and closed-loop DBS in the same patient; the SCS could be used to stimulate and the closed-loop DBS could be used to record from the basal ganglia without stimulating (Parastarfeizabadi and Kouzani 2017).

Finally, in addition to modifying brain circuits, SCS improves motor symptoms such as gait by its direct action on the spinal cord. Gait is a complex behavior which involves interaction between brain circuits and spinal pattern generation centers located within the spinal cord parenchyma. Local spinal circuits help to regulate limb muscle control and some have hypothesized that SCS improves gait by directly facilitating such local circuits within the spinal cord (Fonoff et al. 2019).

Future Directions

There are both animal and human studies supporting that SCS may have a significant therapeutic benefit in PD; however, these studies are limited by heterogeneity in spinal levels targeted by SCS, various hardware and software technology utilized for SCS, and lack of blinding and placebo controls. Another weakness of the human studies thus far has been the various metrics utilized for measuring motor improvement and the failure of routinely more objective measurements of motor outcomes such as gait. Although it has become a standard for assessing degree of PD severity and helping to determine candidacy for DBS, the United Parkinson's Disease Rating Scale (UPDRS) may be a poor indicator of gait and postural problems in PD. The UPDRS has good interrater reliability, but has been criticized for its inability to measure mobility in PD patients (Brusse et al. 2005).

Given the success of DBS in treating many of the motor problems associated with PD, future studies could focus on using SCS in PD patients with gait and postural problems as a chief complaint and to exclude PD patients with severe pain. Further studies of SCS in PD patients already implanted with DBS would be useful to determine whether there is a synergistic effect to these neuromodulation modalities. In addition, there is especially a need to perform high class evidence trials of SCS in PD patients who are not considered candidates for DBS. For example, presumably since SCS does not require cranial surgery, there should be less risk of cognitive decline with SCS, and SCS could be an option for those patients whose neuropsychological scores may preclude DBS implantation. To reduce bias from the placebo effect and to permit double blinding, it would be prudent to employ a paresthesia-free programming mode of SCS such as burstDR for all future studies of SCS in PD.

Based on the work in humans and animals thus far, it would make most sense to perform thoracic and not cervical SCS in patients with PD. Moreover, it would be helpful for SCS to be trialed percutaneously in PD patients similar to how it is currently trialed for chronic pain. A trial of SCS in PD could be used to help identify which candidates will respond the best to implantation and would be useful to set reasonable patient and caretaker expectations. Any trial of SCS in PD should include a "wash-out" period of SCS inactivation after implantation and between any change in programming settings.

Finally, future studies of SCS in PD could employ wearable sensory-based gait analysis systems to objectively assess gait parameters. Wearable sensor systems combining accelerometers and gyroscopes have been successfully tested in PD patients over the past decade (Schlachetzki et al. 2017; Brognara et al. 2019). These technologies are affordable, noninvasive and can provide multiple objective metrics about the complex gait behavior in PD patients including freezing of gait episodes. Furthermore, inertial sensors can be used by PD patients both in the clinic setting and at home and can therefore give clinicians a larger snapshot of PD gait abnormalities without requiring expensive gait analysis laboratories. There is still a need for standardizing the measurement setup and selecting the most valuable gait parameters to evaluate in these patients. However, the quality, comparability, and reproducibility of neuromodulation research in PD could be significantly improved if such wearable motion sensors become a routine way to objectively measure motion. Furthermore, answering the question of whether neuromodulation in the form of DBS or SCS might change the progression of PD may be facilitated by long-term motion sensor monitoring of both nonimplanted and implanted patients.

Conclusion

Despite the well-established beneficial effects of DBS in PD, not all patients are candidates and DBS has known limitations in improving nonmotor symptoms and axial motor symptoms such as gait disturbance. SCS is a promising new therapeutic approach based on animal and human studies. Although most of the current evidence is low level due to lack of control arms, the vast majority of studies have shown a motor benefits of SCS in PD despite different epidural SCS targets, heterogeneous programming parameters, and various SCS lead and battery technology. The recent development of paresthesia-free SCS programming modes and technological advances in wearable motion sensors has well-poised the field of neuromodulation for high level, blinded studies of SCS in PD.

References

Agari T, Date I (2012) Spinal cord stimulation for the treatment of abnormal posture and gait disorder in patients with Parkinson's disease. Neurol Med Chir (Tokyo) 52(7):470–474

Akiyama H, Nukui S, Akamatu M, Hasegawa Y, Nishikido O, Inoue S (2017) Effectiveness of spinal cord stimulation for painful camptocormia with Pisa syndrome in Parkinson's disease: a case report. BMC Neurol 17:148

Beudel M, Sadnicka A, Edwards M, de Jong BM (2019) Linking pathological oscillations with altered temporal processing in Parkinsons disease: neurophysiological mechanisms and implications for neuromodulation. Front Neurol 10:462

Brognara L, Palumbo P, Grimm B, Palmerini L (2019) Assessing gait in Parkinson's disease using wearable motion sensors: a systematic review. Diseases 7:18

Brusse KJ, Zimdars S, Zaleswki KR, Steffen TM (2005) Testing functional performance in people with Parkinson's disease. Phys Ther 85(2):134–141

Brys I, Bobela W, Schneider BL, Aebischer P, Fuentes R (2017) Spinal cord stimulation improves forelimb use in an alpha-synuclein animal model of Parkinson's disease. Int J Neurosci 127(1):28–36

Cai Y, Reddy RD, Varshney V, Chakravarthy KV (2020) Spinal cord stimulation in Parkinson's disease: a review of the preclinical and clinical data and future prospects. Bioelectron Med 6:5

Chakravarthy K, Fishman MA, Zuidema X, Hunter CW, Levy R (2019) Mechanism of action in burst spinal cord stimulation: review and recent advances. Pain Med 20(Suppl 1):S13–S22

Cook AW, Weinstein SP (1973) Chronic dorsal column stimulation in multiple sclerosis. Preliminary report. N Y State J Med 73:2868–2872

de Andrade EM, Ghilardi MG, Cury RG, Barbosa ER, Fuentes R, Teixeira MJ, Fonoff ET (2016) Spinal cord stimulation for Parkinson's disease: a systematic review. Neurosurg Rev 39(1):27–35

de Lima-Pardini AC, Coelho DB, Souza CP, Souza CO, Ghilardi MGDS, Garcia T, Voos M, Milosevic M, Hamani C, Teixeira LA, Fonoff ET (2018) Effects of spinal cord stimulation on postural control in Parkinson's disease patients with freezing of gait. elife 7:e37727

De Ridder D, Vanneste S, Plazier M, van der Loo E, Menovsky T (2010) Burst spinal cord stimulation: toward paresthesia-free pain suppression. Neurosurgery 66(5):986–990

Fenelon G, Goujon C, Gurruchaga JM, Cesaro P, Jarraya B, Palfi S, Lefaucheur JP (2012) Spinal cord stimulation for chronic pain improved motor function in a patient with Parkinson's disease. Parkinsonism Relat Disord 18(2):213–214

Fonoff ET, de Lima-Pardini AC, Coelho DB, Monaco BA, Machado B, de Souza CP, dos Santos Ghilardi MG, Hamani C (2019) Spinal cord stimulation for freezing of gait: from bench to bedside. Front Neurol 10:905

Ford B (1998) Pain in Parkinson's disease. Clin Neurosci 5(2):63–72

French IT, Muthusamy KA (2018) A review of the pedunculopontine nucleus in Parkinson's disease. Front Aging Neurosci 10:99. https://doi.org/10.3389/fnagi.2018.00099

Fuentes R, Petersson P, Siesser WB, Caron MG, Nicolelis MA (2009) Spinal cord stimulation restores locomotion in animal models of Parkinson's disease. Science 323:1578–1582

Hassan S, Amer S, Alwaki A, Elborno A (2013) A patient with Parkinson's disease benefits from spinal cord stimulation. J Clin Neurosci 20(8):1155–1156

Hubsch C, D'Hardemare V, Ben Maccha M, Ziegler M, Patte-Karsenti N, Thiebaut JB, Gout O, Brandel JP (2019) Tonic spinal cord stimulation as therapeutic option in Parkinsons disease with axial symptoms: effects on walking and quality of life. Parkinsonism Relat Disord 63:235–237

Kjaer SW, Rice ASC, Wartolowska K, Vase L (2020) Neuromodulation: more than a placebo effect? Pain 161(3):491–495

Kobayashi R, Kenji S, Taketomi A, Murakami H, Ono K, Otake H (2018) New mode of burst spinal cord stimulation improved mental status as well as motor function in a patient with Parkinson's disease. Parkinsonism Relat Disord 57:82–83

Lai Y, Pan Y, Wang L, Zhang C, Sun B, Li D (2020) Spinal cord stimulation with surgical lead improves pain and gait in Parkinson's disease after a dislocation of percutaneous lead: a case report. Stereotact Funct Neurosurg 25:1–6

Mazzone P, Viselli F, Ferraina S, Giamundo M, Marano M, Paoloni M, Masedu F, Capozzo A, Scarnati E (2019) High cervical spinal cord stimulation: a one year follow-up study on motor and non-motor functions in Parkinson's disease. Brain Sci 9(4):78

McKinnon C, Gros P, Lee DJ, Hamani C, Lozano AM, Kali LV, Kalia SK (2019) Deep brain stimulation: potential for neuroprotection. Ann Clin Transl Neurol 6(1):174–185

Morales A, Yong RJ, Kaye AD, Urman RD (2019) Spinal cord stimulation: comparing traditional low-frequency tonic waveforms to novel high frequency and burst stimulation for the treatment of chronic low back pain. Curr Pain Headache Rep 23(4):25

Nashold B, Somjen G, Friedman H (1972) Paresthesias and EEG potentials evoked by stimulation of the dorsal funiculi in man. Exp Neurol 36:273–287

Nishioka K, Nakajima M (2015) Beneficial therapeutic effects of spinal cord stimulation in advanced cases of Parkinson's disease with intractable chronic pain: a case series. Neuromodulation 18(8):751–753

Parastarfeizabadi M, Kouzani AZ (2017) Advances in closed-loop deep brain stimulation devices. J NeuroEng Rehabil 14:79

Pinto de Souza C, Hamani C, Oliveira Souza C, Lopez Contreras WO, Dos Santos Ghilardi MG, Cury RG, Reis Barbosa E, Jacobsen Teizeira M, Talamoni Fonoff E (2017) Spinal cord stimulation improves gait in patients with Parkinson's disease previously treated with deep brain stimulation. Mov Disord 32(2):278–282

Samotus O, Parrent A, Jog M (2018) Spinal cord stimulation therapy for gait dysfunction in advanced Parkinson's disease patients. Mov Disord 33(5):783–792

Santana MB, Halje P, Simplicio H, Richter U, Freire MAM, Petersson P, Fuentes R, Nicolelis MAL (2014) Spinal cord stimulation alleviates motor deficits in a primate model of Parkinson disease. Neuron 84(4):716–722

Schlachetzki JCM, Barth J, Marxreiter F, Gossler J, Kohl Z, Reinfelder S, Gassner H, Aminian K, Eskofier BM, Winkler J, Klucken J (2017) Wearable sensors objectively measure gait parameters in Parkinson's disease. PLoS One 12(10):e0183989

Shealy CN, Mortimer JT, Reswick JB (1967) Electrical inhibition of pain by stimulation of the dorsal columns: preliminary clinical report. Anesth Analg 45:489–491

Shinko A, Agari T, Kameda M, Yasuhara T, Kondo A, Tayra JT, Sata K, Sasaki T, Sasada S, Takeuchi H, Wakamori T, Borlongan CV, Date I (2014) Spinal cord stimulation exerts neuroprotective effects against experimental Parkinson's disease. PLoS One 9(7):3101468

Skogar O, Lokk J (2016) Pain management in patients with Parkinson's disease: challenges and solutions. J Multidiscip Healthc 9:469–479

Thevathasan W, Mazzone P, Jha A, Djamshidian A, Dileone M, Di Lazzaro V, Brown P (2010) Spinal cord stimulation failed to relieve akinesia or restore locomotion in Parkinson disease. Neurology 74:1325–1327

Thiriez C, Gurruchaga J, Goujon C, Fenelon G, Palfi S (2014) Spinal stimulation for movement disorders. Neurotherapeutics 11:543–552

Yadov AP, Nicolelis MAL (2017) Electrical stimulation of the dorsal columns of the spinal cord for Parkinson's disease. Mov Disord 32(6):820–832

Yadov AP, Fuentes R, Zhang H, Vinholo T, Wang CH, Freire MAM, Nicolelis MAL (2014) Chronic spinal cord electrical stimulation protects against 6-hydroxydopamine lesions. Sci Rep 4:3839

Zhong H, Zhu C, Minegishi Y, Richter F, Zdunowski S, Roy RR, Vissel B, Gad P, Gerasimenko Y, Chesselet MF, Edgerton VR (2019) Epidural spinal cord stimulation improves motor function in rats with chemically induced parkinsonism. Neurorehabil Neural Repair 33(12):1029–1039

Intraoperative Monitoring in Spine Surgery

23

Julian Michael Moore

Contents

History of Intraoperative Neuromonitoring 484

Principles of Neurophysiology 485
Recording of Electrical Current 485
Delivery of Electrical Stimulation 486
Troubleshooting 487

Functioning within the Operating Room 488
IOM Setup 488
Anesthetic Effect on IOM 488
Baselines and Documentation 489
Protocol for Intraoperative Change 489

Electromyography (EMG) 490
Neurophysiology 490
Recording Methods 490
Anesthetic Consideration 494

Somatosensory Evoked Potentials (SSEPs) 494
Neurophysiologic System 494
Recording Method 495
Parameters and Technique 495
Anesthetic Consideration 496
Obligate Waveforms and Warning Criteria 496

Motor Evoked Potentials 497
Neurophysiologic System 497

Recording Method 497
Parameters and Technique 498
Anesthetic Considerations 498

J. M. Moore (✉)
School of Kinesiology, University of Michigan, Ann Arbor, MI, USA

Department of Neurology, University of Michigan, Ann Arbor, MI, USA
e-mail: jumich@umich.edu

© Springer Nature Switzerland AG 2021
B. C. Cheng (ed.), *Handbook of Spine Technology*,
https://doi.org/10.1007/978-3-319-44424-6_128

Obligate Waveforms and Warning Criteria 499
Utility of Neuromonitoring in Spine Procedures 500
Multimodality Approach 501
IOM Services 501

Conclusions 501

Cross-References 502

References 502

Abstract

Intraoperative neuromonitoring (IOM) has been introduced into the field of surgical medicine as a series of diagnostic modalities. IOM may help to prevent perioperative injury to the spinal cord and nerve roots during spine procedures. Neurophysiologists are trained specialists who work together with other members of the surgical team to create and record electrical signals within the nervous system closely related to motor and sensory function. Electromyography (EMG) and evoked potentials (EPs) are monitored throughout the operation, and surgeons are alerted when waveforms change and specific alarm criteria have been breached. A multimodality approach to spinal cord monitoring seems to be the most effective method of using IOM. The use of IOM has not been established as a standard of care, IOM continues to be used more frequently by surgeons across the world and its favorability continues to expand. The future of IOM as it relates to spine surgery will depend largely on the advancements in electrical technology and continued research on the effectiveness of IOM techniques within certain populations criteria.

Keywords

Intraoperative neuromonitoring · SSEP · TcMEP · EMG · Guidelines · Utility

History of Intraoperative Neuromonitoring

The advancement of surgical knowledge in the early 1970s created greater possibilities for the treatment of scoliosis and spinal deformity. However, these new innovations also posed great risks which could be detrimental to patient outcome. In an effort to recognize intraoperative harm to the patient, neuromonitoring was explored as a tool to reduce the instance of unintended injuries during surgical maneuvers. Intraoperative monitoring of the spinal cord for spinal surgery was initially developed by a group of surgeons who began stimulating and recording EPs for their own procedures (Tamaki and Kubota 2007). The field of intraoperative testing for the integrity of the nervous system has been given a number of abbreviations. In this chapter, it will be referred to as IOM.

At the inception, IOM was limited exclusively to somatosensory evoked potentials (SSEPs) which consisted of cortical and spinal recordings. In the following decade, Merton and Morton (1980) experimented on the effects of stimulating the motor and visual regions of the human cortex. Electromyography (EMG) had already been introduced much earlier in 1849 and was officially named in 1890 after Marey had successfully recorded the electrical activity of a voluntary muscle contraction in a human subject for the first time (Reaz et al. 2006). Pairing the progress of Merton and Morton with Marey led to the creation of transcranial motor evoked potentials (TcMEP), which made intraoperative tests for motor function available to surgeons.

Complications during spine surgeries typically result in a deficit to motor and/or sensory function. In most cases, the affected region of the body is at or distal to the sight of an injury to the nervous system. Hence, a severe injury to the cervical region may cause quadriplegia with loss of sensation. Neurophysiologists will generally select the SSEP, TcMEP, and EMG modalities for monitoring surgery around the spinal column, spinal cord, or nerves. The combination of these modalities provide immediate feedback from the cortical, subcortical, spinal, and peripheral neurogenic structures which may be damaged in a procedure of spinal surgery involving large mechanical manipulation and placement of mechanical fixation devices.

The field of IOM has continued to progress over time, and it has allowed for more refined monitoring of various pathways in which neural tissues travel and interact. Even the most subtle and delicate features of the nervous system (e.g. spinal reflex synaptic activity, vision, and hearing) can be monitored under intraoperative settings. The use of IOM has enhanced an array of surgical approaches including: spinal cord mapping for tumor resection, cranial nerve monitoring for parotidectomy and thyroidectomy, nerve grafting procedures, provocative testing for awake brain mapping, musculoskeletal tumor cryoablation, microvascular decompression for hemi-facial spasm, thoracolumbar aortic aneurysm repair, and carotid endarterectomy. In this chapter, only those intraoperative modalities commonly found in spinal surgery will be discussed. A more comprehensive and inclusive understanding of neurophysiologic intraoperative monitoring can be found in other literature (Husain 2015) (Nuwer 2008).

Principles of Neurophysiology

The nervous system is comprised of neurons and other supporting cell types which create a network that communicates information within the brain, spinal cord, and peripheral nerves. Motor commands, sensory feedback, autonomic regulation, and other higher cognitive features are emergent properties of this system.

Charged ions circulating within and around a neuron create the foundational membrane potential that results from an electrochemical gradient between the intra- and extracellular fluid. When the dendritic tree of a neuron receives enough excitatory input to reach a threshold gradient at the axon hillock, the neuron proceeds to fire a single action potential down its axon and release a neurotransmitter onto downstream neurons, glands, or motor end plates. As an action potential propagates down an axon, it does so by a wave of depolarization. Information is coded within these electrical impulses through frequency and rhythm. IOM records these currents as they pass along specific pathways or evokes them by some form of stimulation.

Recording of Electrical Current

A general knowledge in electricity should be applied when understanding the practice of neurophysiology. The mechanisms of capturing EPs and EMG are based on differential amplification. This process requires two input electrodes referred to as active and inverted. Monitoring begins by creating a recording channel between these two electrodes. As charge flows between a pair of electrodes, the voltage within each electrode field varies. The voltage values are differentiated and the signal is magnified. A ground electrode is necessary for optimal signal acquisition along with low recording impedance. The grounded reference is placed remotely with electrical fields isolated from the other electrodes. The versatility of differential amplification allows for various types of electrical activity to be monitored with IOM including central and peripheral synapses, axon impulses, muscle activation, subcortical, and cortical activity.

In the case of all recordings, amplified signals are processed and digitized into a format that graphs upward peaks and downward deflections of voltage on the y-axis and time on the x-axis based on a sampling rate. Recording is sampled with respect to the Nyquist frequency, which is twice the value of the highest frequency of activity recorded. These recordings are put together similar to the way a movie pieces together a collection of consecutive images sampled over time. The result is a waveform that represents positive and negative fluctuations of charge. Obligate waveforms for each modality are correlated to specific generators within that pathway and indicate the overall status of these specific structures. They repeat consistently and are therefore able to be monitored for change.

Neuromonitoring can record continuously to allow for accurate temporal feedback during surgery. This method of IOM is used to identify nerve root irritation during pedicle screw placement. Continuous monitoring such as EMG or electroencephalography (EEG) requires constant attention of the neurophysiologist because symptomatic nerve firing may only be temporary, lasting a couple seconds. This type of monitoring focuses primarily on the presence of particular wave morphologies and general activity, opposed to changes in latency or amplitude of specific repeating waves which are seen in EPs.

Evoked potentials such as SSEPs and TcMEPs are captured within a specific time period. This time period is determined by the distance the current travels before passing into the recording field of an electrode. A trial is a single snapshot of recording channels generated after the application of stimulation. Obligate waveforms in different channels will have variation due to the distance between two input electrodes, the location of these electrodes along the pathway, and the type of activity being monitored. Evoked recordings of electrical potentials allow for monitoring of signal generators within a specific pathway.

Latency

Latency is an important characteristic attributed to the obligate waveforms of IOM. It describes the speed the current moves along a pathway and has units of milliseconds. For EPs, each waveform has a latency associated with a typical healthy adult. This value is based on nerve conduction velocity.

Many factors can influence the latency of responses including nerve thickness and myelination, perfusion to neurons, anesthetic interference with synaptic communication, and surgical injury. Patients with a previous history of myelopathy, diabetes, or other neurological conditions are expected to have greater latency in SSEP and TcMEP recordings in addition to low amplitude responses.

Amplitude

Amplitude is another aspect used to describe a waveform and has units of millivolts. The amplitude of a signal is measured by the difference in voltage between the upward and downward polarity of a wave. This characteristic is mainly determined by the strength of the patient's motor or sensory function prior to surgery.

A number of intraoperative factors can affect the amplitude of waveforms including the number of nerve fibers activated, anesthetic drug interference within synaptic pathways, perfusion to neurons, age, and surgical injury. Edema or large amounts of subcutaneous fat covering parts of the body may decrease the amplitude of recording channels due to the spatial distance between electrodes and the electric current, as seen in needle EMG. Larger needles may be used to access appropriate recording proximity (Daube and Rubin 2009). Patients with somatosensory or motor deficits prior to operation are expected to have lower amplitude responses.

Morphology

Morphology is the general shape of a waveform and can be used to describe complex multipeak waveforms such as those seen in muscle contractions after stimulation of the motor cortex or EEG. A loss of morphological value in a waveform may appear as a shift from polyphasic to biphasic. This is simply the shortening of a wave and the loss of positive and negative variation in the activity. Morphological data is used to subtly understand the underlying mechanisms, but it has not been promoted for the use of alarm criteria due to the variation in morphology that naturally occurs from surgical stress to the nervous system (Langeloo et al. 2007).

Delivery of Electrical Stimulation

The use of electrical stimulation to elicit nerve conduction has become a standard practice in medicine and the field of IOM. Nerve root differentiation, tissue identification, and EPs are all examples of stimulation-based monitoring. Electric charge can be administered through specialized electrode strips, a pair of electrodes, or a single hand-held stimulator.

Ohm's Law states that voltage equals the product of resistance and current. Constant voltage stimulation is used in TcMEPs because resistance is not expected to vary much between individuals at the head region. Constant current is used in most other cases of stimulation because resistance is largely variable throughout the body. Stimulation intensities vary in IOM depending on the application. Bone has much a higher resistivity than soft tissue and fluid. Therefore, more stimulation is needed for TcMEPs than peripheral nerve stimulation. Corkscrew and needle electrodes have lower impedance than disk and adhesive electrodes and are often suggested for both stimulation and recording.

Stimulation can also be classified as anodal or cathodal depending on the polarity of charge that is applied. Unlike a battery, the stimulator's negative end is the cathode and the positive end is the anode. Neurons are preferentially activated in different anatomical regions based on the polarity of the charge introduced to the extracellular fluid. TcMEPs depend on anodal stimulation in which positively charged cations (+) are discharged under the anode and flow back toward the cathode. Pyramidal cell bodies depolarize as negative ions are pulled to the surface where cations (+) are being discharged. Cathodal stimulation, in which anions (−) are discharged from the cathode and flow toward the anode, preferentially activates axons because negative charges build up along the outer membrane and cause greater electric gradients and positive ions to move toward the cell membrane from the inside, thus causing depolarization.

Monopolar stimulation is useful when current can spread to surrounding tissue. These instances will identify the presence or absence of nerve bundles hidden under layers of fascia. Alternatively, bipolar stimulation applies a more localized current that can be specific in nerve activation. However, it is not as effective if used near pooled fluid which may cause current to dissipate.

Troubleshooting

The equipment used for a standard spine surgery consist of a computer, digital amplifier, stimulating and recording pods, electrodes, and possibly a hand-held probe when stimulated EMG is used. The wires connecting all these components together create the potential for technical error. IOM equipment should always be accessible to neurophysiologists in the event that changes in monitoring are noted of technical troubleshooting is required.

Electrical Interference

Electrical recording is not exempt from limitation. Even if all electrodes are placed correctly and the hardware is properly linked, differential amplifiers are subject to electrical interference. Mains electricity, also known as the power grid, emits a 60-Hz electrical frequency from all appliances, wires, or outlets. Electrocautery artifact completely obliterates any IOM recording with large amplitude and high frequency activity. The movement of an electrode tip from external pressure on the surface of the skin results in a low-frequency artifact termed a DC shift. Other interferences monitored in recording channels may be from fluoroscopy equipment, EKG, or other machines such as blood warmer or warming blanket heater.

Parameters for Signal Acquisition

In order to obtain the clearest and most accurate waveforms, neurophysiologists use a number of recording and stimulation parameters. Averaging is a method of processing consecutive trials to create a more precise waveform. The use of averages prevents the transient loss or disruption of a waveform to cause an alarm. Individual trials often appear noisy and vary from one to the other. Averages cancel out extraneous waves as the trials are summated, and the actual electrical impulses from the nervous system are distinguished. Signals may be clarified with digital smoothing and filters. A low-frequency noise filter can be used in all recording channels to quiet the electrical hum produced from the power grid. High-frequency filters help distinguish important waveforms from activity that may be present in other parts of the nervous system. Time-lock simulation is another technique used in IOM that keeps the stimulation and recording window of EPs consistent. It is also a way to eliminate "stimulation artifact" in EPs.

The charge applied during current delivery is the product of stimulation intensity and pulse width and it is given units of coulombs. Stimulation intensity is the quantity of energy being applied in volts or amperes. Raising stimulation intensity increases the depth and range of electrical current between the anode and cathode. Therefore, high stimulation intensities result in the activation of more nerve fibers within a fasciculus. Pulse width is the duration for which stimulation lasts. Most stimulation occurs in increments of microseconds. Supra-maximal stimulation refers to the technique of maximizing the charge delivered to ensure all fibers of a specific area are saturated and that complete nerve pathway activation is achieved. Safety of the patient becomes a risk with greater stimulation intensities and pulse durations, and neurophysiologists should adhere to a recognized protocol. The American Society of Neurophysiological Monitoring (ASNM) is a group of internationally recognized members who published an updated set of guidelines for IOM in 2009 that is used in many standard practices today.

Repetition rate (rep-rate) is the frequency of stimulating and recording an EP. Because the nervous system relies so heavily on the timing of currents and synapses to calculate information, adjusting rep-rate can optimize nerve activation. Pathways of the nervous system relay at different intervals, so their recording parameters differ. Patients with deteriorated nerve fibers require lower frequency simulations in order for the action potentials to summate and neurons to synapse in the correct fashion. Similar to the Nyquist frequency, rep-rate can also be useful to cancel out specific noise frequencies found in operating rooms.

Functioning within the Operating Room

IOM Setup

The preparation for monitoring begins before the patient has entered the room. A proper discussion of modalities and verification should take place with a surgeon present before any needle electrodes are placed. A reading neurologist must immediately confirm any changes or the status of IOM responses. Safety of the patient should come before any accommodations are made for neuromonitoring. Special attention should be brought to adhesive allergies, drug allergies, external shunts, pacemakers, or other metallic implants.

The application of subdermal needle electrodes occurs after induction, intubation, and approval from the other members of the surgical team. IOM setup may delay the positioning process due to limited patient access during this time from placement of other devices on the body (e.g., arterial lines, EKG, Foley catheter). A well-trained neurophysiologist should understand the priority of tasks and execute their setup with efficiency.

Anesthetic Effect on IOM

A continuous dialogue between the neurophysiologist and anesthesiologist is crucial to optimize neuromonitoring and patient safety. Anesthetic drugs have a major effect on synaptic activity in the entire nervous system. In almost all instances of IOM in spinal column surgery, anesthetic concentrations and regimes must be altered in order to allow for the nervous system to maintain specific functions. Temperature drops will cause neurons to decrease their activity in an attempt to protect the cell from energy loss and death. Changes in blood pressure and perfusion of neurovasculature structures can lead to transient signal loss (Sloan and Jäntti 2008).

Neuromuscular blockers prevent the activation of skeletal muscle. Therefore, EMG and TcMEP do not tolerate paralytics. Volatile inhalant agents act against cortical synapses and those within the spinal cord. Nitrous oxide is generally avoided because of the global depression in signals within all modalities (Liem 2016). Other drugs of consideration include benzodiazepines, barbiturates, and Propofol, which can diminish cortical activity at high concentrations. Systemic anticoagulants (e.g., Heparin) can result in delayed clotting at the sight of IOM needle electrodes.

Typically, monitoring continues until after the closure of skin or once the surgeon has verbally ended monitoring. After decompression, derotation, and hardware placement, the neurological structures of the spinal cord and nerve roots are no longer at large risk for further insult. The early dismissal of monitoring during case closure can allow for faster wake-up times and earlier neurologic exams, however prolonged monitoring may reveal a delayed change from a hematoma. The concentration of gases can be titrated more readily to alter anesthetic depth and is less expensive when compared to infusions of Propofol, which is often requested in the case of IOM (Gertler and Joshi 2010).

Baselines and Documentation

Baseline monitoring of EMG, SSEP, and TcMEP traces are recorded and stored before incision. Establishing baselines will depend on anesthetic cooperation and the extent of technical troubleshooting required. Baselines are used for comparison to responses monitored throughout the surgery. Any deviations in obligate waveform latency, amplitude, or morphology are quantified, and alerts are communicated based on a specific alarm criterion for each modality.

In the event that a patient has an unstable neurological condition, whether from trauma or degenerative instability, caution is necessary during the transition from hospital bed to operating table. Pre-positioning baselines can be obtained before the patient is placed into the prone position after induction and intubation. There are instances in which TcMEP and SSEP are lost due to the degree of cervical manipulation (Jameson 2017). Baselines are promptly set under low anesthetic values to ensure optimal neurologic function and response acquisition.

Communication with the surgeon and anesthesia staff is crucial throughout all of the procedure. Documentation by a neurophysiologist takes place throughout the entirety of monitoring. Time-stamped events of surgery, anesthesia, and other aspects of the operation allow for retrospective analysis of the case as it correlates to changes in IOM signals (ASNM 2009a). Thorough documentation of steps in the procedure offer evidence and proof that may be supportive in medicolegal situations. Any and all communication between the surgeon, neurologist, and anesthesiologist should be documented by the neurophysiologist.

Protocol for Intraoperative Change

Checklists are commonly used tools in medicine because they effectively organize priorities and can reduce harm or death in certain medical practices. As an effort to prevent error in the event of intraoperative changes in IOM waveforms, checklists have been proposed specifically for spinal deformity procedures. They can reduce the probability for human error and serve as a memory aid in highly crucial situations in a prompt fashion. Unfortunately, checklists are not a perfect way of addressing intraoperative change because every situation in the operating room will present in a different manner (Ziewacz et al. 2012).

In the event of intraoperative change, neurophysiologists are trained to alert the surgeon immediately. The surgeon then has the opportunity to try to determine if there is a correctable surgical cause of the neurophysiologic change. The neurophysiologists may then address the items outlined in the checklists. Documentation of surgical actions with anesthetic concentrations should regularly take place and IOM responses should be free of any technical errors from malfunctioning equipment or electrical interference (ASNM 2009a). The cause of intraoperative change should be considered by all members of the surgical team. IOM requires that a reading neurologist be available to confirm any observations to the operating physician.

Remote access to the case via digital network allows the machine in the operating room to be in other locations. A neurologist interprets the IOM data with the neurophysiologist. Clear communication between the anesthesia team, surgeon, and the neurophysiologist is vital to the success and ease of using IOM (Fig. 1).

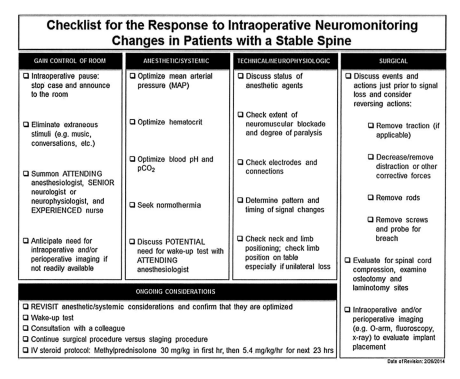

Fig. 1 Standard checklist for spine surgery created by Vitale et al. (2018).

Electromyography (EMG)

Neurophysiology

Electromyography is a standard technique used to observe and record the electrical impulses transmitted through muscle activation. It can be used intraoperatively to detect injury to nerve roots during decompression, hardware insertion, realignment of the spine, and tissue retraction. A clear understanding of muscle contraction physiology is necessary to interpret EMG accurately.

Alpha motor neurons within the gray matter of the corticospinal tract receive excitatory input from upper neurons. Action potentials propagate toward the neuromuscular junction and calcium influx occurs near the axonal terminal. This causes the release of the neurotransmitter acetylcholine which enters the synaptic cleft and binds to receptors on the motor end plate. Ligand-gated channels open and an influx of sodium creates waves of depolarization across the sarcolemma that travel throughout the muscle. This causes the release of calcium and activation of cross-bridge cycling.

As muscle fibers activate, compound muscle action potentials (CMAPs) occur as a result of synchronous co-contractions. CMAP do not always create a visible twitch. CMAP is directly reflective of muscle activation but can also give information on the irritation of specific nerve roots that innervate that muscle. Perturbation of the motor nerves distal to the spinal cord can cause spontaneous firing of action potentials along the axon and activates the release of acetylcholine and muscle contraction. This is often seen in EMG and is the reason why it is a useful tool in detecting any injury or sensitivity of nerve roots during parts of surgery.

Recording Methods

Paired needle electrodes inserted into bilateral muscle groups form channels for recording activity. An optional stimulator device may be used by the surgeon to activate nerve roots or verify distance of the spinal cord from the pedicle space. The muscles chosen by neurophysiologist are easily identified anatomically and should be at least one spinal level above and below the site of surgery. Monitoring multiple levels of nerve roots is crucial because this method reduces the chance of false responses. For example, activity observed in the biceps brachii is more likely to originate from the C5 nerve root than the C6 nerve root, if there is also EMG activity in the deltoid channel. For example, C5 nerve palsy is a major complication for cervical spine operations, hence monitoring must be made as accurate as possible by monitoring adjacent levels (Nichols and Manafov 2012). Surgeons operating at or near the sacral plexus should consider monitoring the anal sphincter. Vodušesk and Deletis (2002) concluded that motor function of the sacral region is reflective of the integrity of the entire sacral plexus because of the anatomical nerve bundle which is comprised of efferent and afferent axons. Bladder, bowel, and sexual function can also be monitored through other IOM techniques, but these are not generally used in degenerative spine surgery (Table 1).

Spontaneous EMG

This modality is used in almost all instances of monitoring spine surgery. Spontaneous EMG is also termed "free-running" because it continuously records activity throughout the entire surgery. This offers excellent temporal resolution by changes in nerve function seen immediately as EMG pattern firing. When nerve roots experience injury from mechanical or thermal events, subsequent firing of high frequency bursts (termed "neurotonic discharges" or "A-trains") are seen in the recording of the same muscle levels (Holland 2002). Spontaneous EMG can also reveal a number of pathologies in nerve and muscle fibers, therefore creating activity which may not be caused from any surgical action. Surgeons may elect to decline intervention based on the alert of any kind during any time, such as rhythmic tibialis anterior activity seen during exposure. Although activity like this seems mistaken, patients with myelopathy and palsies experience fibrillation potentials due to muscle fibers losing contact with innervating nerves and spontaneously producing CMAPs as a result. An EMG baseline is taken after induction before the patient has been fully relaxed. Muscle activity in all channels is documented by the neurophysiologist.

Train of Four

The train of four (TO4) test is used to assess the degree of neuromuscular blockade (NMB). Four successive stimuli are applied through needle electrodes at a peripheral nerve. These stimuli produce a train of action potentials that propagate down the axon to the motor end plate of distal muscles. Recording the following CMAPs produced by these four stimuli notify the depth of patient relaxation. TO4 monitoring should always be done in the extremity from which EMG or EPs are being monitored. In most cases, the peripheral stimulators located at the wrist or ankle are used for a TO4. Often, these are the same electrodes used in evoking SSEPs and deliver cathodal stimulation. Typical parameters for this test are four supra-maximal bursts lasting 0.2–0.3 seconds are applied at a rate of 2 Hz (Nichols and Manafov 2012).

Paralytic is often requested during the exposure of the spine. During this time, EMG and TcMEP responses are not accurate. When

Table 1 Common choices for EMG monitoring based on nerve root level. (Standard knowledge not requiring copyright permission)

Common EMG recording channels and their corresponding nerve root levels	
Nerve root	Muscle group
C3-C4	Trapezius
C5	Deltoid
C5-C6	Biceps brachii
C6-C7	Triceps brachii
C8-T1	Abductor pollicis brevis, abductor digiti minimi
T2-T6	Intercostal muscles
T7-T12	Intercostal muscles
L1-L2	External oblique, rectus abdominus
L2-L4	Illiopsoas
L4-L5	Vastus medialis, vastus lateralis
L5-S1	Tibialis anterior
S1-S2	Biceps femoris, abductor hallucis longus
S3-S5	Gastrocnemius
	Anal sphincter

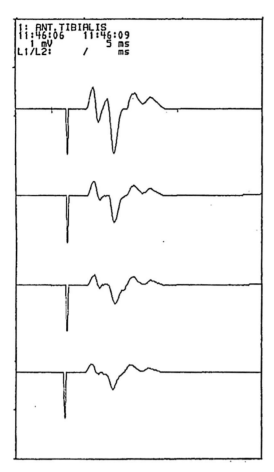

Fig. 2 TO4 test reveals four twitches with anesthetic fade. These results indicate adequate clearance of the NMB. (Holland 2002)

assessing the NMB from nondepolarizing relaxants (e.g., Rocuronium), a TO4 appears with fade. The amplitude of the first CMAP is the largest and they subsequently diminish. Four twitches of equal amplitude may not be achieved for a patient with a severe neurological condition. Baseline monitoring and continuous TO4 recordings provide the neurophysiologist with a good estimation of the extent of NMB. TO4 is interpreted largely by visual analysis (Fig. 2).

Triggered EMG

Stimulus triggered EMG is a surgeon-directed method to help to determine functional connectivity of nerve roots, or identify nerves from the surrounding tissue (e.g., tumors, scar tissue from revision, herniated disc). Nerve root testing, especially in the lumbar region, is not a good indicator of vertebral level; however, considering stimulation typically consists of a bipolar, cathodal stimulator that is held by the surgeon and an anode return is placed near the skin of the incision. Constant current is delivered at 1–3 mA for pulse durations of 0.2 ms at a frequency of 2 Hz. Surgeons direct the neurophysiologist during the stimulation to activate, deactivate, and adjust the intensity of the stimulator. The neurophysiologist provides positive or negative feedback seen in EMG channels and documents any observed activity. Previous studies have shown that nerve thresholds with these stimulation parameters are within 0.2–5.7 mA (Leppanen 2005) (Fig. 3).

Pedicle Screw Threshold Testing

Pedicle screw placement is a critical step in posterior spinal surgery. Stimulation of the pedicle hole, probe, tap, or screw with a ball tip probe can be useful in detecting a breach of the pedicle wall. Surgeons may choose to remove screws with a low threshold and stimulate the pedicle track with a probe, redirect the screw trajectory and replace the screw, or leave it out completely. There may be practicality in stimulating with the pedicle probe as it is driven into the pedicle wall using a real-time testing. In doing so, any alteration in trajectory of the live-probe can be made based on unacceptable low threshold before a violation of the pedicle walls occurs and the probe is fully seated. This is achieved by substituting the handheld probe with an alligator clamp connected pedicle probe. Screws can be stimulated after placement as well in these approaches (Rose et al. 1997).

The stimulation parameters for pedicle testing have been studied by many physicians, and there is a general consensus that cathodal stimulation at constant current offers the best result. Single pulses of stimuli lasting 100–200 microseconds are applied at a frequency of 2–3 Hz. Stimulation intensity is applied at low level and increased until a cutoff "threshold" value is reached. If EMG activity is observed before the threshold is reached, that is considered a positive response and suggests a breach in the

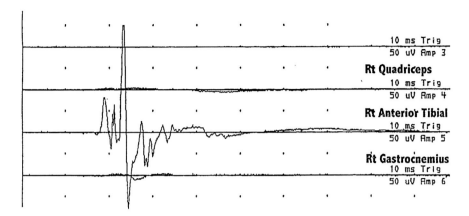

Fig. 3 Anterior tibialis activity recording after direct stimulation of the L5 nerve root. (Holland 2002)

pedicle. Higher thoracic levels are more difficult to monitor based on the level of deformity in the region and the ability to monitor the corresponding musculature (intercostal muscles). Recent publications have utilized the combination of pulse-train stimulation parameters in TcMEPs in addition to EMG monitoring of the lower limbs to test for malposition of thoracic level screws (Donohue et al. 2008).

The most appropriate threshold criteria and cutoff stimulation for pedicle threshold testing is of wide debate. There are a number of reasons why pedicle threshold testing may not be the most accurate assessment of neurological injury. Higher cutoff thresholds lead to a greater occurrence of false positives while lower threshold increase the incidence of false negatives (Skinner and Rippe 2012). For example, a 91% positive predictive value (PPV) with a 43% sensitivity and 99% specificity has been found when using a threshold cutoff of 5 mA in the lumbosacral region. Threshold cutoffs of 8 mA had sensitivity around 87% and specificity of 94% with a PPV of only 26%. Calancie et al. (1994) found that probe stimulation of the lumbosacral region to a threshold ≥7 mA with no EMG activity indicates good placement. A slightly higher threshold of 10 mA confirmed well placed screws. Surgeons may choose to stimulate up to intensities of 30–40 mA to confirm the adequate placement of screws. However, 15 mA of stimulation with no EMG responses is strongly suggestive of no postoperative deficit.

Misplaced pedicle screws may be clinically inconsequential and a safe zone may extend up to 4 mm depending on the anatomy. Small breaches of the pedicle wall could potentially cause current spread into the spinal cord or nerve roots and create an EMG alarm. However, these breaches are not typically clinically significant to deterioration of motor function. The conductivity of metallic screws and bone density can affect the impedance of charge delivery and therefore alter cutoff values as well. These factors contribute to inaccuracies between pedicle screw testing criteria and clinical outcomes. A study by Holland (2012) in cervical and thoracic pedicle screws found that 88% of medial breaches in the cervical spine can be detected by using a threshold of 15 mA. Medial breaches in the thoracic spine were detected 85% of the time while using a threshold of 6 mA. However, none of the screws detected by IOM correlated to postoperative deficits.

Although there is no technique to guarantee pedicle screw placement with 100% accuracy, there are many techniques to avoid symptomatic malpositioned screws. The traditional technique of pedicle screw placement relies on the surgeon's anatomical knowledge and confidence in finding the correct entry point; however, this method has been linked with failure rates of

15–42%. Plain radiographic CT scans also have a chance of being misinterpreted. Djurasovic et al. (2005) conducted a study comparing IOM threshold testing results with those from a postoperative CT scan. They found 7 out of 11 instances where intraoperative screw stimulation detected unacceptable screws that were not identified on plain lateral radiographs.

The utility of IOM in pedicle screw placement is questioned because of the technological advancements in image-based navigation systems. Lee et al. (2014) compared various imaging approaches to pedicle screw placement and found that CT-based navigation, along with 2-d and 3-d fluoroscopy-based navigation, has a high accuracy for pedicle screw insertion. CT and 3-d navigation seem to be more accurate than 2-d navigation; however, a 96.7% accuracy was still found in 2-d fluoroscopy. Unfortunately there are technical limitations with these technologies as well, and a 3.3% failure rate has been observed. Intraoperative CT and fluoroscopy are not always available to surgeons because of a lack in hospital resources. The cost of imaging equipment and the need for lead shielding and a trained staff person are all considerations. This is contrast to the IOM technique which requires very little preparation and minimal interruption to the operating surgeon. Also, images of the bony anatomy have no direct reflection on the physiologic condition. EMG is able to detect the proximity of pedicle insertion to the neurologic tissue through current activation and weighs this information based on a standard threshold criterion.

Anesthetic Consideration

The main anesthetic concern for EMG monitoring is the NMB, which is established through the use of short- or long-acting paralytics. Reversal agents (e.g., Neostigmine, Sugammadex) allow for quicker recovery from paralysis. EMG may be performed with as much as 75% of receptors blocked. This percentage is not correlated to the loss of CMAP amplitude in a TO4 under paralytic. In fact, changes in EMG are only seen after 75% of NMB (Holland 2002). Minahan et al. (2000) studied the effect of NMB on pedicle screw testing and concluded that largest CMAP in a TO4 must not be more than 80% decreased from the baseline in order for optimal pedicle screw threshold testing. Surgeons who request a NMB for exposure should inform anesthesia of the appropriate time for reversal if they wish to monitor EMG. By monitoring the TO4, neurophysiologists communicate when the blockade has been adequately reversed.

Somatosensory Evoked Potentials (SSEPs)

Neurophysiologic System

SSEPs test the dorsal column-medial lemniscal (DCML) pathway, which is responsible for upper and lower extremity somatosensory function. This pathway is a series of large diameter afferent nerve fibers, which carry information from the periphery about fine touch and proprioception to the brain. The origin of these nerves is found in the muscle spindle receptors surrounded by skeletal muscle fibers. As the human body moves through space, a number of internal and external forces act on the musculature. Muscle spindles activate in response to static and dynamic changes in muscle length.

The activation of these afferent nerve fibers can also be achieved through stimulation over the axon. Afferent somatosensory nerves travel in the same common nerve bundle as efferent motor axons that innervate similar muscles. These "mixed nerves" split into a dorsal and ventral tract once they reach the spinal cord. The advantage of stimulating a mixed nerve is that visible twitches from the hand or foot ensure correct stimulation delivery during SSEP monitoring.

Proprioceptive nerve fibers enter into the dorsal root ganglia of correlating dermatome. The fasciculus gracilis and fasciculus cuneatus are white matter tracts of the DCML that carry upper and lower extremity sensation, respectively. A synapse in the medulla decussates the information and sends impulses into the thalamus and onto the contralateral hemisphere. Therefore, the right side of the body is perceived in the left hemisphere and vice versa.

Recording Method

Stimulation electrodes 2 cm apart are placed in the wrist for upper extremity SSEPs, while lower extremity electrodes are placed in the ankle. Stimulation is applied over mixed nerves and both motor and sensory nerve fibers activate. Depending on the degree of NMB and patient condition, a twitch is not always observed. The activation of sensory fibers creates a wave of depolarization that flows proximally. Recording sites for SSEPs are variable depending on the training and technique of the neurophysiologist. However, in most setups, at least three channels of peripheral, subcortical, and cortical EPs are captured (ASNM 2009b).

The first recording along the sensory pathways is over a peripheral nerve or plexus that is just distal to the spinal cord. This type of axonal activity is termed a volley. For upper extremity SSEPs, electrodes are placed bilaterally behind the clavicle (Erb's point) or near the axilla. Two electrodes placed ~5 cm apart over each popliteal fossa are used for monitoring of the right and left lower extremity. Additional recording channels can be added to the lumbar or thoracic midline for lower extremity SSEPs. However, this activity is not frequently monitored.

The International 10–20 system is a method of measuring and describing the placement of electrodes for an EEG headset with a combination of letters and numbers. Capital letters are used to describe the region of the brain over which the electrode is placed. For example, C represents the central sulcus and F represents the frontal lobe. Z indicates midline. Odd numbers are located on the left hemisphere and higher order electrodes are placed more laterally. Therefore C1 and C3 would be placed over the left central sulcus, but would not lie in the same sagittal plane. References to the 10–20 system are used for the placement of cranial electrodes in all IOM applications. In order to monitor subcortical SSEPs of the upper and lower extremity, a channel is created between an electrode placed at the spinous process of a cervical vertebrae and an electrode placed midline at few centimeters rostral from the natural hairline. These locations are termed "Cs" and "Fpz," respectively. If placement of C lead is not possible due to the surgical field, it can be placed at the left or right auricular point (A1 or A2).

Cortical SSEPs are commonly monitored by electrodes placed over the left and right somatosensory cortex, 2 cm back from the central sulcus. Therefore these electrodes are given a prime mark because of their modified location (C3' and C4'). This channel is reversed depending on the stimulation side of the SSEP. This keeps the waveforms of each extremity in the same polarity so that waves are not flipped, which would make comparison more difficult. Upper extremity cortical SSEPs also use an alternative channel of C3' (right SSEP) or C4' (left SSEP) reference with Fpz. Lower extremity cortical SSEPs are also monitored with C3' and C4', but a perpendicular channel between Cz and Fpz is recommended (Nuwer and Packwood 2008).

Parameters and Technique

According to the recommended standards for SSEP monitoring set by ASNM (2009b), cathodal stimulation of rectangular pulses lasting 100–300 microseconds should be kept to 30–40 mA at a frequency of 2–8 stimuli per second. As in the case with all IOM application of stimulation, the charge deliver should not injure the patient in any way. Rep-rates that lie within 60 Hz frequency, such as 5.00 stimuli per second, should be avoided. Each pulse of stimulation creates one EP along the pathway. SSEPs in all four limbs can be interleaved by most IOM equipment. This technique cycles each limb at a rate so that averages are obtained most readily.

For surgeries above the C6 level, median nerves are suggested because this nerve has the largest cortical response from the upper extremity. Surgeries extending to C8 or instances where the patient is prone with the arms above the head, stimulation of the ulnar nerve is selected. Surgery in the thoracic and lumbosacral regions requires stimulation of the posterior tibial nerve, but the peroneal nerve or the popliteal fossa can be used if the patient suffers from neuropathy. In most spine surgeries, both upper and lower SSEPs are monitored in order to control for global changes that may occur.

Peripheral channels assure that the DCML pathway is at supra-maximal stimulation because the amplitude of this response indicates the totality of afferent nerve fiber activation. Once maximum amplitude has been reached in the peripheral response, cortical and subcortical waveforms will not strengthen with additional stimulation.

Anesthetic Consideration

All anesthetic agents have some impact on the synapses between neurons. Therefore cortical SSEP activity diminishes before subcortical or epidural activity because of the large number of cell bodies and synapses in the neocortex. Halogenated gases (e.g., isoflurane, desflurane, sevoflurane) are commonly administered at minimum alveolar concentrations (MAC) of 0.5–1.0, with the addition of an analgesic agent (e.g., fentanyl, sufentanil, remifentanil) (Slimp and Holdefer 2014). MAC is a unitless measurement of vapor in the lungs that is required for 50% of the population to not respond to a surgical stimuli. Less potent gas (e.g., nitrous oxide) has higher MAC value because more is needed to keep a patient sedated. Nitrous oxide is not compatible with neuromonitoring; however, halogenated gases can be used as a part of a balanced anesthetic at low MAC.

Opiates show little effect on EPs at lower dosages, and muscle relaxants may improve SSEPs because there is less artifact from patient movement. Ketamine and Etomidate have been shown to increase the strength of cortical SSEP signals due to their mechanisms of cortical excitability and disinhibition (Sloan and Jäntti 2008). Other small concerns may be noted for SSEPs. Pulse oximetry is not always compatible when the stimulation of the wrist or ankle causes muscle twitches in the fingers and toes. Corkscrew electrodes at the scalp can be substituted for needle electrodes or disk electrode if a patient requires anticoagulation for the surgery and bleeding is expected.

Obligate Waveforms and Warning Criteria

The obligate waveforms of SSEPs are distinguished by the polarity (positive peak or negative dip) and latency. The latencies of these waveforms should be generally similar, but can have variation. Therefore, an individual baseline is recorded before incision, which is used to compare with the SSEPs throughout surgery. The last position of the ASNM (2009b) for alert criteria is an amplitude decrease of 50% or more or a latency increase of 10% or more in any obligate waveform (Fig. 4).

Cortical amplification is the process of thalamic input summating before it projects onto the cortex. Because of this, cortical recordings appear larger in amplitude than the subcortical and peripheral recordings (Slimp and Holdefer 2014). Subcortical and peripheral SSEP changes are considered more significant in spine surgery because insult to the system is likely in the spinal cord or nerve roots. The cortical channels will not be as affected by this damage because of thalamic summation and cortical amplification. As a result, an amplitude criterion is generally used in cortical and subcortical waveforms while latency is assessed in the periphery.

Causes of SSEP intraoperative change has been outlined by Gonzalez and Shilian (2015). Damage to the nerve roots will result in the change of subcortical and cortical waveforms. Peripheral channels distal to the site of injury would not show immediate decrement because the nerve underneath these electrodes is still intact. Therefore, insult to the SSEP pathway causes the loss of signal conduction proximal to the injury site. This is indicated by the loss of subcortical and cortical signals with no changes in peripheral response.

Global changes in SSEPs are usually a result of ischemic situations when the limb is malpositioned. Whenever the patient is positioned prone with the arms overhead, ulnar nerve SSEPs are used. It is the nerve at greatest risk for perioperative neuropathy from external pressure. However, up to one half of male patients may fail to perceive or experience clinical symptoms of ulnar nerve compression sufficient to elicit SSEP changes (Prielipp et al. 1999).

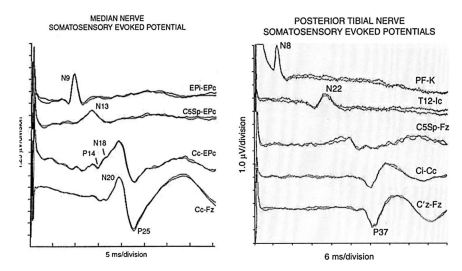

Fig. 4 Peripheral, spinal, subcortical, and cortical SSEPs of median (left) and posterior tibial nerves (right). (Nuwer and Packwood 2008)

Motor Evoked Potentials

Neurophysiologic System

Spinal cord function is not completely covered through the use of SSEPs alone. Motor evoked potentials allow for the monitoring of motor function via stimulation of the motor cortex. The lateral corticospinal tract is system of neurons originating in the motor cortices and terminating at the voluntary muscles throughout our entire body. Similar to somatosensory regions, the motor cortex is partitioned bilaterally on contralateral hemispheres in a homunculus arrangement.

Inputs from the basal ganglia, cerebellum, and the pre- and supplementary motor cortices excite the corticomotor neurons in the primary motor cortex. The axons of these cells outline the corticospinal tract as they travel through the internal capsule and cross hemispheres in the medulla oblongata. Depolarization along white matter continues inferiorly into the corticobulbar tract and the corticospinal tract. Excitation of alpha motor neurons at the ventral horn activates the physiological process of muscle contraction.

Indirect motor pathways (e.g., rubrospinal, reticulospinal, and tectospinal) consist of functions that heavily influence the corticospinal tract. These pathways do not have direct interactions with alpha motor neurons, but instead are connected through interneuronal synapses. These pathways have influence in facilitating a level of excitation in the corticospinal tract that makes TcMEPs possible (MacDonald et al. 2013). They are also responsible for what makes this system difficult to monitor.

Recording Method

TcMEPs deliver charge to the cortex through the skull. Other methods of evoking motor potential include direct stimulation of the cortex or deeper regions of the subcortical pathways. Stimulation electrodes for TcMEPs are placed 2 cm anterior to the central sulcus over the motor cortex and 4 cm anterior to SSEP head electrodes. A charge is delivered between these two electrodes to activate the left or right hemisphere.

Recording channels within the muscles are observed for the presence of CMAPs. Muscles that are more distal have greater latency because of the time it takes for the corticospinal tract activation to reach these motor units. Muscles are chosen in the upper and lower extremity to control for global change and confirm true positive results. Epidural recordings can also be performed to monitor the descending portion of corticospinal tract within the spinal cord. In these instances, electrode pairs or strips are positions distal and proximal to the surgical site (Legatt et al. 2016).

Parameters and Technique

Anodal stimulation is used for TcMEPs most often with constant voltage delivery. The method of mutli-pulse, or pulse train stimulation, refers to a series of short consecutive shocks lasting about 50 microseconds. In awake subjects, 3 pulses with a 2 millisecond pause between each pulse successfully elicited motor activation at 25–50 volts. However, under anesthesia, the stimulation intensity required was as high as 400 volts (Calancie et al. 1998).

Neurophysiologists follow separate guidelines and no official protocol for stimulation parameters have been set forth. Calancie et al.'s (1998) original methods are still used but are often changed slightly. Double trains allow for two separate clusters of stimuli to be applied. Larger pauses between each stimulus can also reach 5 milliseconds. This technique of troubleshooting TcMEPs is often used because the timing of the excitation is crucial to successfully activating corticomotor nuclei. A maximum of 9 pulses can be given.

TcMEPs create patient movement, and surgeons must stop their actions and clear the surgical field every time. Bilateral bite blocks are mandatory in all instances of stimulating the motor cortex. Bite injuries are the most common complication with TcMEPs, appearing in 0.2% of all cases (Macdonald et al. 2013). Once the patient has been positioned prone, gauze padding should be confirmed secure before obtaining baselines by using a mirrored faceplate. Pacemakers and other electronic devices may also jeopardize the safety of collecting TcMEPs and are contraindicated. Patients with a history of seizures or superficial shunts should not undergo TcMEPs if the risk for harm is made greater by the stimulation.

Motor evoked potentials are used in spine surgery to monitor the connection of the corticospinal tract. It has been shown to be more sensitive for spinal cord compromise than SSEPs (Macdonald et al. 2015). Injury to the spinal cord will alter the CMAP in all muscle groups distal to the lesion. Therefore, nerve roots can be intact but the connection to the alpha motor neuron is blocked due to damage in the proximal corticospinal tract. TcMEPs do not provide the nerve root specific feedback that is monitored with spontaneous EMG. A damaged nerve root may not affect the CMAP of a muscle that has innervations from multiple levels of the spinal cord.

Epidural d-wave recordings represent the status of the corticospinal tract within the spinal cord from direct activation of the corticomotor neurons. The D-wave is recorded caudal and/or distal to the site of surgery and lies between the motor cortex and alpha motor neuron cell body. Therefore, no synapse occurs before the recording electrode. This makes D-waves less sensitive to general interference and yet very precise in determining the conductivity of the descending volley down the corticospinal tract (Deletis & Sala 2008).

Anesthetic Considerations

TcMEPs are the most sensitive to anesthesia than EMG or SSEPs. In a recent study by Acharya et al. (2017), over 50% total TcMEP alerts were accredited to anesthetic management. As aforementioned, TcMEPs prohibit the use of neuromuscular blocking agents to allow for CMAP activity. Reversal agents are administered to allow for enough time for the NMB to dissipate. By monitoring the TO4, neurophysiologists communicate when the blockade has been reduced optimally for TcMEPs.

Another concern for TcMEPs includes perfusion. The anterior aspect of the spinal cord, which contains the descending motor tract, is only perfused by one anterior spinal artery opposed to two posterior spinal arteries on the dorsal aspect. This makes TcMEPs more sensitive than SSEPs to decreases in blood pressure caused by anesthesia. Similar to SSEPs, needle or disk electrodes can be used at the head instead of corkscrews if anticoagulant is planned for the surgery.

Total Intravenous Anesthesia (TIVA)

TIVA refers to a mixture of drugs other than inhalants to achieve the four elements of anesthesia: analgesia, amnesia, "sleep," and muscle relaxation. TcMEPs are very incompatible of inhalational agents because of their influence on the indirect motor pathways and the excitation of alpha motor neurons (Legatt et al. 2016). Interneurons do not remain stable with anesthetic gases, and TIVA should be prioritized for the most accurate TcMEP monitoring. Studies also show that TIVA regiments also carry an increased risk for intraoperative awareness (Gertler and Joshi 2010). As a result, anesthesiologists work together with neurophysiologists to create a balanced anesthetic that can include low MAC values of halogenated gas along with infusions of opiate and other amnesic drugs (Fig. 5).

Obligate Waveforms and Warning Criteria

The ideal waveforms for TcMEPs consist of robust CMAPs in all recording channels. Baseline CMAP latency, amplitude, and morphology vary depending on patient's health. TcMEPs are extremely sensitive to change due to the effects of anesthetic agents, perfusion of the spinal cord, and the activation of all the corticomotor neurons using appropriate parameters. Anesthetic fade over the course of a long procedure will often decrease the amplitude of EPs. A looser alarm criterion is used to prevent the instance of false positives, and an attenuation of 80% amplitude from baseline has been suggested for significant change (Langeloo et al. 2007). An all-or-none criterion is used in many instances of monitoring. Alternatively, increasing the stimulation intensity by 100 V or more to elicit the same responses as baseline qualifies as a standard criteria as well (ASNM 2009a) (Fig. 6).

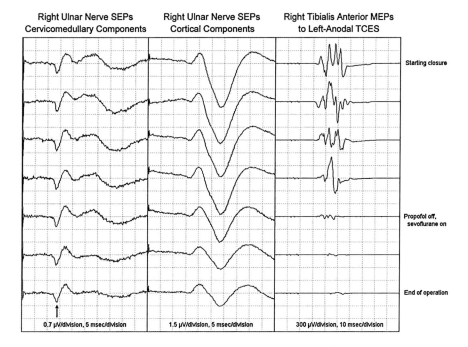

Fig. 5 SSEP and TcMEP amplitude and morphology diminishing as Propofol infusions were ended and anesthesia was switched to sevoflurane at the end of surgery. (Legatt et al. 2016)

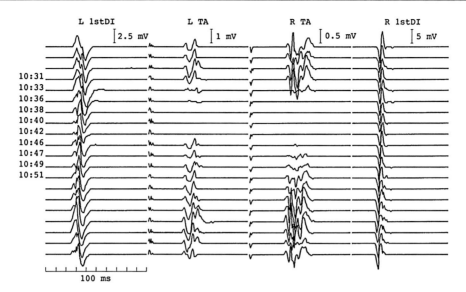

Fig. 6 Transient loss in TcMEPs from acute spinal cord ischemia during thoracoabdominal aneurysm surgery. Responses quickly recover after blood pressure was restored. MacDonald et al. 2013)

Utility of Neuromonitoring in Spine Procedures

Before deciding to use intraoperative neuromonitoring, the limitations of IOM should be known to spine surgeons. As previously discussed, the presence of IOM in the operating room can be cumbersome and result in surgical delay. Anesthesia protocol requires alteration and more closely managed throughout the case. Electrical recording has its own drawbacks as well. The variation in patient neuroanatomy and condition can make it difficult to distinguish the true utility of IOM for every situation. Preoperative condition will largely influence the efficacy of IOM in producing repeating and monitorable waveforms. The strength of baseline signals provide surgeons with an idea of how effective IOM will be during the rest of the surgery.

There is a wide debate on the effectiveness of neuromonitoring preventing or reducing neurological complications during spinal surgery, and that is accredited to a number of inconsistencies. Studies performed in this field are often retrospective and do not share the same independent variables. Differences in electrical recording and stimulation parameters between authors confound different conclusions as a result. The chosen alarm criteria used in different studies also affect the rate of sensitivity, specificity, and predictive value. The outcomes reported by recent authors vary along with their perspectives on the utility of IOM for spine surgery.

Ibrahim et al. (2017) used SSEPs, EMG, and TcMEPs and found three instances of false negatives in a study of 121 with a sensitivity of 57% with a specificity of 98%. In addition, they calculated that IOM has a PPV of 67% and an NPV of 97%. However, the three deficits that went undetected intraoperatively were eventually resolved after several weeks. This finding can be explained by neuroplasticity and the nervous system's ability to reorganize itself. Neira et al. (2016) found that TcMEPs were slightly more successful than SSEPs in spinal column surgery in detecting postoperative deficit, based on the findings in a population of children. This is in accordance with Zuccaro et al. (2017), who conducted a retrospective study on 809 pediatric patients in posterior spinal fusions, vertebral column resection, or other spinal surgeries. Each patient included in the study had one of the following diagnoses: idiopathic, congenital, or syndromic scoliosis, which included various neuromuscular syndromes. TcMEPs showed 100% sensitivity with 93% specificity and

SSEPs had 13% sensitivity with 100 specificity. These results were achieved using an alarm criteria of 50% decrease in the amplitude of TcMEPs or SSEPs with disregard to any latency alarm criteria.

Some physicians believe that IOM is a predictive tool to establish an increased risk for the adverse outcomes of paraparesis, paraplegia, and quadriplegia in spinal surgery. Tamkus et al. (2017) found the overall rate of IOM failing to predict a postoperative deficit was 0.04%. Animal studies reveal that surgical response to intraoperative change greatly reduces the risk of postoperative deficits, when compared to not intervening at all after a change in IOM responses. Neuromonitoring for human subjects has not been directly linked to decreasing the severity or rate of surgical injuries during spine operations because these studies raise ethical issues (Nuwer et al. 2012).

Multimodality Approach

Surgeons may choose to select their own modalities and alarm criteria for the procedure, or the neurophysiologist can suggest recommendations. A recent analysis on the field of IOM revealed that the most frequent unimodality monitoring consists of EMG, but SSEPs are also monitored alone. A multimodality approach of SSEP and EMG, or all three modalities including TcMEPs, appears to be the most preferential among surgeons. Trends in scoliosis surgery show that the use of neuromonitoring has steadily increased within the past decade, although the prevention of neurological injury has not significantly improved with the use of monitoring (Ajiboye et al. 2017).

IOM Services

The demand for neuromonitoring has risen and the field is projected to grow exponentially in the next decade. Neuromonitoring performed in medical centers is done by IOM staff or the use of a contracted neuromonitoring service. Qualifications of the neurophysiologist are beginning to become standardized, as this field continues to expand. A Certificate in Neurophysiologic Intraoperative Monitoring (CNIM) is the current specification preferred for neurophysiologists and it is credentialed by the American Board of Registration of Electroencephalographic and Evoked Potential Technologists (ABRET) (Nuwer 2015). Neurophysiologists should be properly trained and complete a certification process. Educational programs within universities that teach the knowledge and skills of neuromonitoring have become available in the past decade. Mergos et al. (2015) proposed a training curriculum for such programs which include theoretical study and clinical experience.

Conclusions

The historical evolution of surgical technology and the advancements in anesthesiology have supported the development of intraoperative neuromonitoring within the past several decades. The field of IOM has grown quite extensively and can be used in a variety of operative procedures. The diagnostic tests available through EMG can be used to inform the level of NMB, injuries to specific nerve roots, the identification of nerve tissue, and the integrity of pedicle screw placement. Spine surgeries often require the patient to be in a prone position with their arms stretched out over head. Monitoring SSEPs can prevent brachial plexus ischemia and detect injuries along the DCML pathway. SSEPs also give insight into the perfusion of the spinal cord, specifically the posterior aspect. TcMEPs test the connection between the motor cortex, corticospinal tract, and peripheral muscles. They are closely associated with overall motor function and can indicate the perfusion of the anterior spinal cord.

Ajiboye et al. (2017) performed a retrospective analysis between 2004 and 2011 and found a statistical increase of about 20% for the use of IOM. It appears that multimodality combinations are considered necessary for practical and effective neuromonitoring. Laratta et al. (2018) have described the importance of an experienced multimodality approach to neuromonitoring in spine surgery. They emphasized that the efficacy of IOM depends on the familiarity of the surgeon and neurophysiologist with both the surgical procedure and the IOM modalities. The prompt detection of potential neurologic injury is accredited to a firm understanding of the surgical maneuvers and anesthetic concentrations. These authors point out that rapid intervention and successful outcomes can only be achieved through clear communication and interdisciplinary respect between the surgeon, anesthesia staff, and neurophysiologist.

Although neuromonitoring has been successful in predicting postoperative outcomes and eliminated the need for wake-up tests, it has not been proven to reduce the neurological complications associated with spine surgery. The future holds much promise for the field of neurodiagnostics and the application of IOM. Technological innovation coupled with standardization and uniform practice would yield more precise, physiologic feedback. As with all surgical techniques, the primary objective of IOM is to optimize patient safety to maximize postoperative outcomes.

Cross-References

▶ Pedicle Screw Fixation
▶ The Diagnostic and the Therapeutic Utility of Radiology in Spinal Care

References

Acharya S, Palukuri N, Supta P, Kohli M (2017) Transcranial motor evoked potentials during spinal deformity corrections- safety, efficacy, limitations, and the role of a checklist. Front Surg 4(8):1–11

Ajiboye RM, Park HY, Cohen JR, Vellios EE, Lord EL, Ashana AO, Buser Z, Wang JC (2017) Demographic trends in the use of intraoperative neuromonitoring for scoliosis surgery in the United States. Int J Spine Surg 11:271–277

ASNM (2009a) Guideline 11A: recommended standards for neurophysiologic intraoperative monitoring – principles. American Clinical Neurophysiology Society. https://www.acns.org/pdf/guidelines/Guideline-11A.pdf. Accessed 10 May 2018

ASNM (2009b) Guideline 11B: recommended standards for neurophysiologic intraoperative monitoring of somatosensory evoked potentials. American Clinical Neurophysiology Society. https://www.acns.org/pdf/guidelines/Guideline-11B.pdf. Accessed 10 May 2018

Calancie B, Madsen P, Lebwohl N (1994) Stimulus-evoked EMG monitoring during transpedicular lumbosacral spine instrumentation. Spine 19:2780–2786

Calancie B, Harris W, Broton JG, Alexeeva N, Green BA (1998) "Threshold-level" multipulse transcranial electrical stimulation of motor cortex for intraoperative monitoring of spinal motor tracts: description of method and comparison to somatosensory evoked potential monitoring. J Neurosurg 88:457–470

Daube JR, Rubin DI (2009) Needle electromyography. Muscle Nerve 39:244–270

Deletis V, Sala F (2008) Corticospinal tract monitoring with D- and I- waves from the spinal cord and muscle MEPs from limb muscles. In: Nuwer MR (ed) Intraoperative monitoring of neural function: handbook of clinical neurophysiology, vol 8. Elsevier, Amsterdam, pp 235–251

Djurasovic M, Dimar JR, Glassman SD (2005) A prospective analysis of intraoperative electromyographic monitoring of posterior cervical screw fixation. J Spinal Disord Tech 18:515–518

Donohue ML, Murtagh-Schaffer C, Basta J, Moquin RR, Bashir A, Calancie B (2008) Pulse-train stimulation for detecting medial malpositioning of thoracic pedicle screws. Spine 33:E378–E385

Gertler R, Joshi GP (2010) General anesthesia. In: Twersky RS, Philip BK (eds) Handbook of ambulatory anesthesia, 2nd edn. Springer, New York, pp 234–251

Gonzalez AA, Shilian P (2015) Somatosensory evoked potentials in: Barry B (ed) a practical approach to neurophysiologic intraoperative monitoring, 2nd edn. Demos Medical, New York, pp 18–32

Holland NR (2002) Intraoperative electromyography. J Clin Neurophysiol 19:444–453

Holland N (2012) Neurophysiological assessment of thoracic and cervical pedicle screw integrity. J Clin Neurophysiol 39:489–492

Husain AM (2015) A practical approach to neurophysiologic intraoperative monitoring. Demos Medical, New York

Ibrahim T, Mrowczynski O, Zalatimo O, Chinchilli V, Sheehan J, Harbaugh R, Rizk E (2017) The impact of neurophysiological intraoperative monitoring during spinal cord and spine surgery: a critical analysis of 121 cases. Cureus 9:e1861. https://doi.org/10.7759/cureus.1861

Jameson L (2017) Acute loss of intraoperative evoked potential signals. In: McEvoy MD, Furse CM (eds) Advanced perioperative crisis management. New York, Oxford, pp 504–510

Langeloo DD, Journée HL, Kleuver M, Grotenhuis JA (2007) Criteria for transcranial electrical motor evoked potential monitoring during spinal deformity surgery a review and discussion of the literature. Clin Neurophysiol 37:431–439

Laratta JL, Ha A, Shillingford JN, Makhni MC, Lombardi JM, Thuet E, Lehman RA, Lenke LG (2018) Neuromonitoring in spinal deformity surgery: a multimodality approach. Glob Spine J 8:68–77

Lee KD, Lyo U, Kang BS, Sim HB, Kwon SC, Park ES (2014) Accuracy of pedicle screw insertion using fluoroscopy-based navigation-assisted surgery: computed tomography postoperative assessment in 96 consecutive patients. J Korean Neurosurg 56:16–20

Legatt AD, Emerson RG, Epstein CM, MacDonald DB, Deletis V, Bravo RJ, López JR (2016) ACNS guideline: transcranial electrical stimulation motor evoked potential monitoring. J Clin Neurophysiol 33:42–50

Leppanen RE (2005) Intraoperative monitoring of segmental spinal nerve root function with free-run and electrically-triggered electromyography and spinal cord function with reflexes and f-responses. J Clin Monit Comput 19:437–461

Liem L K (2016) Intraoperative neurophysiological monitoring: overview, history, Electromyography. Medscape. https://emedicine.medscape.com/article/1137763-overview#a2. Accessed 21 April 2018

MacDonald DB, Skinner S, Shils J, Yingling C (2013) Intraoperative motor evoked potential monitoring – a position statement by the American Society of Neurophysiological Monitoring. J Clin Neurophysiol 124:2291–2316

Macdonald DB, Al-Enazi M, Zayed A-Z (2015) Vertebral column surgery. In: Barry B (ed) A practical approach to neurophysiologic intraoperative monitoring, 2nd edn. Demos Medical, New York, pp 87–105

Mergos J, Kale EB, Husain AM (2015) Training curriculum for NIOM. In: Barry B (ed) A practical approach to neurophysiologic intraoperative monitoring, 2nd edn. Demos Medical, New York, pp 334–353

Merton PA, Morton HB (1980) Stimulation of the cerebral cortex in the intact human subject. Nature 285:227

Minahan RE, Riley LH, Lukaczyk T, Cohen DB, Kostuik JP (2000) The effect of neuromuscular blockade on pedicle screw stimulation thresholds. Spine 25:2526–2530

Neira VM, Ghaffari K, Bulusu S, Moroz P, Jarvis JG, Barrowman N, Splinter W (2016) Anesth Analg 123:1556–1566

Nichols GS, Manafov E (2012) Utility of electromyography for nerve root monitoring during spinal surgery. J Clin Neurophysiol 29:140–148

Nuwer MR (2008) Intraoperative monitoring of neural function. Elsevier, Amsterdam

Nuwer MR (2015) Coding, billing and ethical issues. In: Barry B (ed) A practical approach to neurophysiologic intraoperative monitoring, 2nd edn. Demos Medical, New York, pp 363–367

Nuwer MR, Packwood JW (2008) Chapter 11: somatosensory evoked potential monitoring with scalp and cervical recording. In: Nuwer MR (ed) Intraoperative monitoring of neural function: handbook of clinical neurophysiology, vol 8. Elsevier, Amsterdam, pp 180–189

Nuwer MR, Emerson RG, Galloway G, Legatt AD, Lopez J, Minahan R, Yamada T, Goodin DS, Armon C, Chaudhry V, Gronseth GS, Harden CL (2012) Evidence-based guideline update: intraoperative spinal monitoring with somatosensory and transcranial electrical motor evoked potentials. J Clin Neurophysiol 29:101–108

Prielipp RC, Morell RC, Walker FO, Santos CC, Bennet J, Butterworth J (1999) Ulnar nerve pressure. Anesthesiology 91:345–354

Reaz MBI, Hussain MS, Mohd-Yasin F (2006) Techniques of EMG signal analysis: detection, processing, classification and applications. Biol Proced Online 8:11–35

Rose RD, Welch WC, Balzer JR, Jacobs GB (1997) Persistently electrified pedicle stimulation instruments in spinal instrumentation. Technique and protocol development. Spine 22:334–343

Skinner SA, Rippe DM (2012) Threshold testing of lumbosacral pedicle screws: a reappraisal. J Clin Neurophysiol 29:493–501

Slimp JC, Holdefer RN (2014) Chapter 36: somatosensory evoked potentials: an electrophysiological tool for intraoperative monitoring. In: Intraoperative neuromonitoring, 1st edn. McGraw-Hill Education, China, pp 405–423

Sloan TB, Jäntti V (2008) Anesthetic effects on evoked potentials. In: Nuwer MR (ed) Intraoperative monitoring of neural function: handbook of clinical neurophysiology, vol 8. Elsevier, Amsterdam, pp 94–126

Tamaki T, Kubota S (2007) History of the development of intraoperative spinal cord monitoring. J Eur Spine 16:S140–S146

Tamkus AA, Rice KS, McCaffrey MT (2017) Perils of intraoperative neurophysiological monitoring: analysis of "false-negative" results in spine surgeries. Spine J 18:276–284

Vitale MG, Skaggs DL, Pace GI, Wright ML, Matsumoto H, Anderson RCE, Brockmeyer DL, Dormans JP, Emans JB, Erickson MA, Flynn JM, GLotzbecker MP, Ibrahim KN, Lewis SJ, Luhmann SJ, Mendiratta A, Richards BS, Sanders JO, Shah SA, Smith JT, Song KM, Sponseller PD, Sucato DJ, Roye DP, Lenke LG (2018) Best practices in intraoperative neuromonitoring in spine deformity surgery: development of and intraoperative checklist to optimize response. Spine Deform 2:333–339

Vodušesk DB, Deletis V (2002) Intraoperative neurophysiological monitoring of the sacral nervous system. In: Deletis V, Shils J (eds) Neurophysiology in neurosurgery: a modern intraoperative approach, 1st edn. Elsevier Science, California, pp 197–217

Ziewacz JE, Berven SH, Mummaneni VP, Tu T, Akinbo OC, Lyon R, Mummaneni PV (2012) The design, development, and implementation of a checklist for intraoperative neuromonitoring changes. Neurosurg Focus 33:1–10

Zuccaro M, Zuccaro J, Samdani AF, Pahys JM, Hwang SW (2017) Intraoperative neuromonitoring alerts in a pediatric deformity center. Neurosurg Focus 43:1–7

Oncological Principles

24

A. Karim Ahmed, Zach Pennington, Camilo A. Molina, and Daniel M. Sciubba

Contents

Introduction	506
Patient Evaluation	506
Multidisciplinary Approach	507
Primary Spinal Column Tumors	507
Enneking Classification System	507
Case Illustration	508
Metastatic Spine Disease	511
Neurologic Oncologic Mechanical Systemic (NOMS) Framework	511
Predictive Analytics	512
Case Illustration	512
Intradural Spinal Tumors	513
Operative Considerations	513
Intradural Extramedullary Tumors	515
Case Illustration	515
Intramedullary Tumors	517
Case Illustration	518
References	519

Abstract

Surgical resection of neoplastic spinal pathology requires a multifactorial understanding and appreciation of the anatomy and biomechanics of the spinal column, pathologic characterization of the lesion, multidisciplinary options available, best guidelines for treatment, patient presentation, and feasibility of resection. When counseling patients with spinal pathologies, in my practice, I distill decision-making into three key questions: "(1) What is the tumor histology? (2) What is it doing to the patient neurologically and mechanically? (3) What are the patient's options?" The first question aims at identifying a conclusive radiographic and pathologic

A. K. Ahmed · C. A. Molina
Department of Neurosurgery, The Johns Hopkins School of Medicine, Baltimore, MD, USA
e-mail: aahmed33@jhmi.edu; cmolina2@jhmi.edu

Z. Pennington · D. M. Sciubba (✉)
Department of Neurosurgery, Johns Hopkins Hospital, The Johns Hopkins School of Medicine, Baltimore, MD, USA
e-mail: zpennin1@jhmi.edu; dsciubb1@jhmi.edu

© Springer Nature Switzerland AG 2021
B. C. Cheng (ed.), *Handbook of Spine Technology*,
https://doi.org/10.1007/978-3-319-44424-6_127

diagnosis. The second assesses the patient's symptomatology, clinical presentation, spinal stability, and systemic burden. These two dictate the final query – describing treatment options and indications/extent of surgical resection. This chapter aims to provide an overview of essential principles in spinal oncology.

Keywords

Spinal tumor · Spinal metastasis · Vertebral column tumor · Oncologic resection · Enneking system · Negative margins · Multidisciplinary care

Introduction

The management of spinal malignancies is one of the most complex clinical tasks encountered by the academic spine surgeon. Coarsely, spinal malignancies can be grouped along two main axes: (1) primary vs. metastatic and (2) intrathecal vs. extrathecal location. This division gives rise to four main lesion types: (1) intradural, intramedullary lesions, (2) intradural, extramedullary lesions, (3) primary vertebral column tumors, and (4) vertebral column metastases. Though metastatic disease of the vertebral column is far and away the most common lesion class, spinal oncologists must be prepared to address all four classes, including the goals of surgery (palliative, curative, etc.), the proper strategy for resection (en bloc vs. piecemeal), and the indications for instrumentation and reconstruction following resection. Here we address each tumor class in turn, focusing on preoperative patient selection, intraoperative surgical strategies, and the need for adjuvant or neoadjuvant therapy to mediate an optimal outcome.

Patient Evaluation

Magnetic resonance imaging (MRI), X-ray, and computed tomography (CT) are the cornerstone of imaging for spinal pathologies. X-ray often serves as the initial form of radiographic evaluation and may be beneficial to identify a lesion, motion, or alignment (Ilaslan et al. 2004; Greenspan 2004; Rodallec et al. 2008). Dynamic X-rays, such as flexion-extension films, are useful to assess sagittal translation, suggesting possible destabilization between adjacent segments. This may result in mechanical pain, exacerbated by motion, as can occur with lytic disease involving the facet joints (Ciftdemir et al. 2016). Disruption in the anterior vertebral line, posterior vertebral line ("George's Line"), spinolaminar line, and posterior spinous line may indicate adjacent segment motion (Horne et al. 2016). Alignment is assessed on standing 36-inch scoliosis films and should be considered in cases of extensive lesions resulting in spinal deformity, or possibility of deformity following iatrogenic destabilization and reconstruction (Mehta et al. 2015).

MRI is the gold standard to assess soft tissue and neural structures, with the ability to identify cystic or lobulated structures. In the presence of neurologic deficit, spinal cord and/or nerve root compression should be carefully identified. One must be mindful to rule out vascular abnormalities, vitamin deficiency (i.e., subacute combined degeneration), or immune conditions (i.e., multiple sclerosis). Generally, dense osteoblastic lesions are hypointense on T1-weighted and T2-weighted MRI, whereas osteoclastic lesions are frequently T1-hypointense and T2-hyperintense. Based on vascularity, most tumors enhance following gadolinium administration, which is especially useful to characterize epidural soft tissue invasion. CT is unparalleled to assess bone mineralization, ideal for the characterization of bony invasion, fracture, vertebral height loss, and the classification of osseous lesions (i.e., blastic, lytic, or mixed) (Rodallec et al. 2008; Kim et al. 2012a).

MRI and CT can be useful to provide diagnostic insight, but pathologic diagnosis should always be attempted in non-emergent cases. CT-guided needle biopsy should be repeated or a core biopsy attempted for nondiagnostic lesions (Sciubba et al. 2010; Aaron 1994; Laufer et al. 2013). For locally aggressive malignancies, such as chordoma, marking the percutaneous trajectory with methylene blue or a tattoo can minimize the risk of tumor seeding to be included in the surgical resection (Mehta et al. 2015). PET-CT characterizes systemic burden, particularly necessary for suspected metastatic disease (Metser et al. 2004). Patients with spinal tumors most often present with pain, and/or neurologic dysfunction, requiring a thorough physical examination. Frailty and medical comorbidities are significant factors for surgical decision-making, with a substantial impact on postoperative adverse events and long-term function, and should be considered as well (Sciubba et al. 2010; Aaron 1994; Laufer et al. 2013).

Multidisciplinary Approach

Optimal treatment of spinal neoplasms relies on a multidisciplinary approach of surgeons, medical oncologists, radiation oncologists, interventional radiologists, pathologists, and neuroradiologists. These specialists play an integral role in not only the treatment of spinal lesions but should be involved early during patient evaluation to achieve the best outcomes (i.e., imaging, biopsy, and staging).

The indications for surgery depend on tumor type, clinical presentation, systemic burden, and response to chemotherapy or radiotherapy. In many centers, such options are presented in multidisciplinary conferences to determine the most appropriate treatment plan (Wallace et al. 2015; Kim et al. 2012b; Ropper et al. 2012; Kaloostian et al. 2014). Prior to surgery, neoadjuvant chemotherapy or radiation can play a critical role in feasibility of resection. Postoperatively, adjuvant therapy may substantially increase prognosis in both primary and metastatic lesions and may even delay recurrence for subtotally resected intramedullary tumors (Uei et al. 2018; Oh et al. 2013; Dea et al. 2017). Other tools for optimizing surgical outcomes include preoperative embolization for vascular lesions (i.e., renal cell carcinoma, pheochromocytoma, hemangioma, follicular thyroid carcinoma).

Access surgeons including general surgeons, vascular surgeons, oral maxillofacial (OMF), and thoracic surgeons may be consulted especially during technically demanding or high-risk exposures. This may involve surgical exposures to the upper cervical spine such as transoral or transmandibular circumglossal; to the thoracic spine such as thoracotomies, or sternotomies; and to the lumbar spine such as anterior transperitoneal or anterior retroperitoneal (Chiriano et al. 2009; Fourney and Gokaslan 2004; Walsh et al. 1997; Samudrala et al. 1999; DeMonte et al. 2001). Additionally, more extensive surgical resection may require complex plastic and reconstructive surgery closure to optimize aesthetic outcome and minimize the risk of postoperative infection (Chieng et al. 2015; Epstein 2013).

Primary Spinal Column Tumors

Primary tumors originate in the spinal column and are far less common compared to metastatic spine disease. Patients frequently present with progressively worsening back pain and/or neurologic deficit. Soft tissue extension to the surrounding subcutaneous tissue or muscle results in paraspinal pain, whereas spinal cord or nerve root compression results in myelopathy and radiculopathy, respectively (Ropper et al. 2012; Kaloostian et al. 2014; Dea et al. 2017).

Unlike metastatic spine disease, the focality of primary spinal lesions allows for the potential to surgically cure these patients of disease. Consequently, the goal of surgery for primary lesions is curative, when feasible, unlike metastatic lesions, which is palliative. Several factors are considered to decide the optimal treatment for these patients.

Enneking Classification System

The Enneking Staging System was proposed in 1986 for the classification of primary spinal neoplasms, with three characterizing factors (Enneking 1986): tumor grade (G), local extent (T), and the presence or absence of metastasis (M). Tumor grade is based on histopathological diagnosis of the lesion; local extent refers to lesions confined to the compartment (i.e., intracompartmental vs. extracompartmental); and metastasis describes systemic burden. Benign lesions are given a grade of G0, with the stage denoted by Arabic numbering (i.e., 1,2,3) (Dea et al. 2017). In 2009, the Spinal Oncology Study Group (SOSG) proposed a modified version of the original staging system, with suggestions for appropriate treatment of benign and malignant primary spinal tumors (From Chan et al. 2009),

benign tumor stage (all grade G0, with no metastasis M0):

- Stage 1: Latent tumor, with well-defined margin, intracompartmental (T0), low biological activity, that often resolves spontaneously (i.e., non-ossifying fibroma). Surgical management not required for oncologic control but may be warranted for decompression or stabilization if symptomatic.
- Stage 2: Active tumor, with lytic bone destruction, intracompartmental (T0), often symptomatic, and can result in pathologic fracture (i.e., aneurysmal bone cyst). Treatment involves intralesional resection with the possibility for local adjuvant therapy.
- Stage 3: Aggressive tumor, without well-defined margins. High rates of recurrence with subtotal resection, frequently symptomatic (i.e., giant cell tumor). Optimal treatment consists of marginal en bloc resection.

Malignant primary spinal lesions, however, have grades from G1 to G2, and stages are denoted with Roman numerals (i.e., I, II, III):

- Stage IA: Grade G1 (low-grade malignant), intracompartmental (T0), without metastases (M0). Best treatment consists of wide en bloc resection.
- Stage IB: Grade G1 (low-grade malignant), extracompartmental (T1), without metastases (M0). Best treatment consists of wide en bloc resection.
- Stage IIA: Grade G2 (high-grade malignant), intracompartmental (T0), without metastases (M0). Best treatment includes wide en bloc resection with postoperative adjuvant therapy.
- Stage IIB: Grade G2 (high-grade malignant), extracompartmental (T1), no metastases (M0). Optimal treatment is wide en bloc resection with postoperative adjuvant therapy.
- Stage III: Presence of metastases (M1) is the key characterizing factor. May be low grade or high grade, intracompartmental, or extracompartmental. Treatment is palliative.

The feasibility of resection is further illustrated by the Weinstein-Boriani-Biagini classification (Dea et al. 2017; From Chan et al. 2009; Boriani et al. 1996, 1997), which defines the extent of vertebral involvement. The axial vertebral body is segmented into 12 sectors with 1 and 12 situated on the left and right side of the spinous process, respectively. In addition, lateral to medial involvement are defined by the following:

A. Extraosseous soft tissue
B. Intraosseous (superficial, cortical)
C. Intraosseous (deep, medullary)
D. Extraosseous within the spinal canal (epidural)
E. Extraosseous within the spinal canal (intradural)
F. Vertebral artery involved (cervical)

Case Illustration

Radiation-Induced Osteosarcoma

A 69-year-old white male with a history of radiation-treated seminoma was referred to our service with diagnosis of osteosarcoma of the lumbosacral spine. The patient initially presented to an outside facility with chief complaint of non-mechanical lower back pain with right radicular leg pain. Workup revealed a right-sided mass involving the right posterior L5 vertebral body, right pedicle, and right L5 and S1 facets. This led to a CT-guided biopsy demonstrating a high-grade (3 out of 3) osteosarcoma. The patient was referred to our center for further management.

During consultation, the patient reported radicular pain in the right L5 distribution, and motor testing demonstrated mild weakness in the right extensor hallucis longus with commensurate decrease in the deep tendon reflex at the right ankle. PET-CT revealed the patient's disease to be localized, and recommendation was made for the patient to undergo en bloc resection of the L5 mass with preoperative embolization to minimize intraoperative morbidity (Fig. 1).

The patient elected to pursue this therapeutic regimen and underwent a three-stage operation to resect the lesion. During stage I, a posterior midline incision was made from L3 to S2 with subperiosteal dissection bilaterally over L3, L4, S1, and S2. The dissection over the sacral level was carried laterally to expose the iliac wings for instrumentation. Over the level of L5, subperiosteal dissection was carried laterally on the left side to expose the pedicle for instrumentation; dissection was restricted on the right side to maintain a muscular cuff around the lesion. An ultrasonic cutting device was then used to perform an L4 laminectomy and resect the inferior articulating process of L4 bilaterally. The pedicle and facet surfaces were also resected from the left L5 and S1 vertebrae to expose the descending S1 and S2 nerve roots. Left-sided hemidiscectomies were then performed at the L4/5 and L5/S1 levels, and the ultrasonic cutting device was used to create a parasagittal osteotomy through the L5 vertebral body. The osteotomy did not involve the anterior cortical surface to avoid injury to the IVC. Pedicle

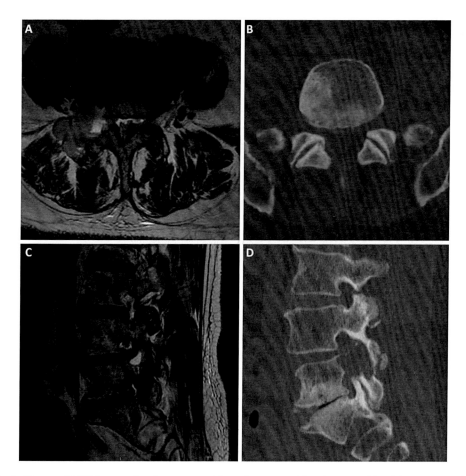

Fig. 1 A 69-year-old WM presenting from an outside hospital with nonmechanical back pain secondary to an aggressive (3 out of 3) radiation-induced osteosarcoma arising from the right L5 vertebral body. MR (**a** and **c**) demonstrated involvement of the right L5 root, and CT (**b** and **d**) demonstrated an osteoblastic lesion with poorly defined margins. PET (not shown) demonstrated no systemic metastases

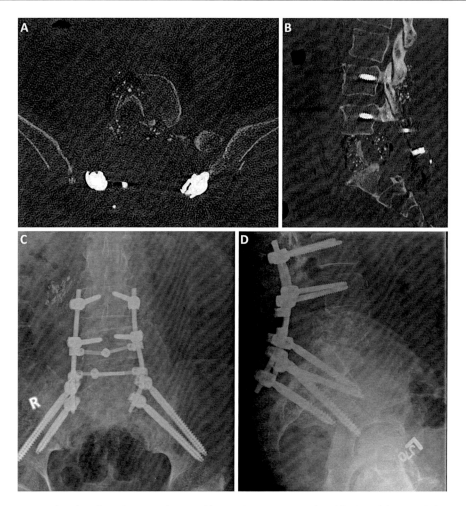

Fig. 2 Postoperative imaging demonstrating en bloc resection of the right L5 hemivertebrae (**a** and **b**). Standing radiographs acquired at 3 months (**c** and **d**) postoperatively demonstrate good positioning of the L3-pelvis fusion with no signs of anterior column insufficiency

screws were then placed bilaterally at L3 and on the left side of L4. Double iliac bolts were placed on the left side, and a 5.5-mm titanium rod was used to connect the left-sided instrumentation. Plastic surgery was consulted for wound closure, which was performed using adipocutaneous flaps in circumferential fashion.

During stage II, a right-sided retroperitoneal approach was employed. The patient was placed in the left lateral decubitus position on a beanbag with an axillary roll. An incision was made in the midaxillary line extending from the 12th rib to the iliac crest. The abdominal wall musculature was incised, and the peritoneal contents were deflected medially. A plane was developed between the abdominal musculature and the peritoneum, allowing identification of the right iliopsoas muscle and the great vessels. Bipolar coagulation and sharp dissection with Metzenbaum scissors were used to elevate the iliopsoas from the underlying spine. The segmental arteries were then ligated and transected to allow mobilization of the iliac vessels, IVC, and aorta. The L4/5 and L5/S1 discectomies that begun in stage I were then completed, as was the parasagittal osteotomy, mobilizing the rostral aspect of the tumor. Hemostasis was obtained, and the wound was closed in layers.

During the final stage of the operation, a posterior midline approach was again adopted. The prior incision was reopened to expose the lumbosacral junction. The right-sided S1 pedicle and superior articulating process were resected to expose the L5/S1 disc space and to allow identification of the L4 and L5 roots. The specimen was rotated away from the midline, with sequential sectioning of adhesions to the surrounding structure. The sympathetic branch of L4 and entire L5 nerve root were found to be involved in the tumor and were sacrificed to preserve negative margins. Further dissection freely mobilized the tumor, which was delivered en bloc, including a cuff of healthy paraspinal soft tissue. Hemostasis was obtained, and instrumentation was performed at the right L3 and L4 pedicles and right ilium, using the holes cannulated in stage I. Intraoperative testing suggested that a cage could not be safely seated in the corpectomy defect, and so no anterior column reconstruction was performed (Fig. 2). Iliac crest autograft was employed for arthrodesis, and the wound was again closed by the plastic surgery service using adipocutaneous flaps.

The patient's postoperative course was uncomplicated, and he was transferred to inpatient rehabilitation for recovery. Postoperative follow-up demonstrated that the wound was healing appropriately, and the patient's pain was well controlled. The patient passed shortly after the 3-month postoperative follow-up.

Metastatic Spine Disease

The vast majority of extradural spinal tumors are metastases, with the spinal column representing the most common site of skeletal metastasis. The greatest incidence of metastatic spine disease occurs in those from 40–65 years old, and symptomatic lesions are most commonly located in the thoracic spine (Sciubba et al. 2010; Aaron 1994; Laufer et al. 2013; Perrin and Laxton 2004; Fisher et al. 2010a; Bilsky et al. 2010; Posner 1987). These most often present with pain, followed by neurologic dysfunction, frequently originating from a primary breast, prostate, or lung cancer (Sciubba et al. 2010). Unlike primary spinal tumors, the aim of surgery of metastatic spine disease is palliative nature, to treat neurologic compromise, pain, and/or mechanical instability.

Neurologic Oncologic Mechanical Systemic (NOMS) Framework

A multidisciplinary algorithm for the management of spinal metastasis, the NOMS Framework, encompasses four clinical aspects to inform decision-making (Laufer et al. 2013):

- **N**eurologic: characterization of symptomatic epidural spinal cord or nerve root compression due to extradural metastasis
- **O**ncologic: pathologic diagnosis of the tumor, aggressiveness, and its response to treatments
- **M**echanical stability: "loss of spinal integrity as a result of a neoplastic process that is associated with movement-related pain, symptomatic or progressive deformity, and/or neural compromise under physiologic loads" (Fisher et al. 2010b)
- **S**ystemic: metastatic disease burden, stage, grade, medical comorbidities, benefits and suitability for surgery, prognosis, and risks

The neurologic factor is an extension of the epidural spinal cord compression scale (ESCC, Table 1), a 6-point grading system from axial MRI images, reliably validated by the Spinal Oncology Study Group (SOSG) (Fisher et al. 2010b). Low-grade compression is classified as Grades 0–1b that are generally suggested to undergo radiation, in the absence of mechanical instability. The oncologic assessment classifies tumors by normal response to conventional external beam radiation therapy (cEBRT), as radiosensitive or radioresistant. In the absence of mechanical instability, the authors advocate for cEBRT to treat radiosensitive tumors (i.e., lymphoma, seminoma, myeloma, breast, prostate, ovarian, neuroendocrine carcinoma), even in the presence of high-grade ESCC. Treatment of radioresistant tumors (i.e., renal, thyroid, hepatocellular, colon, non-small cell lung, sarcoma, melanoma)

Table 1 Epidural spinal cord compression scale (Tokuhashi et al. 2005)

Epidural spinal cord compression scale	
Grade	
0	Osseous only disease
1a	Soft tissue component effaces the dura without deformation of the thecal sac or abutting the spinal cord
1b	Soft tissue component deforms the thecal sac but does not abut the spinal cord
1c	Soft tissue component deforms the thecal sac and abuts the spinal cord, without cord compression
2	Spinal cord compression with CSF visible circumferentially around the cord
3	Spinal cord compression with no CSF around the cord

relies on the degree of ESCC. For low-grade ESCC, radioresistant lesions can be adequately managed by stereotactic radiosurgery (SRS) alone, unlike radioresistant high-grade ESCC, which is treated by surgical decompression and separation surgery.

Mechanical instability is an absolute indication for surgical stabilization or cement augmentation, regardless of radiosensitivity or other factors. Clinically, this commonly presents as mechanical pain exacerbated by movement, distinct from tumorigenic or biologic pain (Sciubba et al. 2010; Laufer et al. 2013; Bauer and Wedin 1995). The Spinal Instability Neoplastic Score (SINS), consisting of seven radiographic and clinical features, was developed by the SOSG to quantify instability due to metastatic spine disease (Tatsui et al. 1996). The greatest contributing factors for mechanical instability include metastatic lesions located at a junctional level (i.e., occiput – C2, C7 – T2, T11 – L1, L5 – S1), mechanical pain on clinical presentation, lytic lesions on CT, vertebral subluxation/translation, >50% vertebral body collapse from pathologic fracture, and bilateral involvement of the posterolateral elements.

Predictive Analytics

Predictive analytic scoring systems may play a role in decision-making for spinal tumors. Several scoring systems have emerged in the literature (Tokuhashi et al. 2005; Bauer and Wedin 1995; Tatsui et al. 1996; Katagiri et al. 2005; Tomita et al. 2001), in addition to machine learning prediction tools (Senders et al. 2018). These scoring systems take various factors into consideration: including performance status, primary tumor diagnosis, age, visceral/brain metastases, neurologic deficit, and number of vertebral metastases to predict survival for patients undergoing surgery. This may be instrumental for surgeons and clinicians to design the most appropriate treatment for patients with metastatic spine disease, in order to maximize quality of life and offer the most robust treatment. Algorithms have been recently proposed that consider tumor-specific and prognosis-specific survival to select the most accurate scoring system for a given patient (Ahmed et al. 2018).

Case Illustration

Mechanically Unstable Breast Metastasis to Lumbar Spine

A 39-year-old Caucasian female presented to our service with a primary complaint of right-sided mechanical back pain. She had a recent history of breast cancer treated with right-sided lumpectomy and axillary lymph node dissection followed by systemic chemotherapy (doxorubicin, cyclophosphamide, and paclitaxel) and hormonotherapy. Her disease was felt to be in remission, but outside imaging demonstrated a large lytic lesion of the L3 vertebral body and pedicle, which was biopsied and found to be consistent with ER-, PR-positive breast cancer (Fig. 3). Systemic imaging identified no other lesions, and a recommendation was made for the patient to undergo resection of the lesion with instrumented fusion.

The patient elected to pursue this intervention, which proceeded as a single-stage operation. The patient was placed prone, and a midline incision was formed over the L2–4 vertebral levels. Subperiosteal dissection was performed over these vertebral levels, extending laterally to the pedicles, which were cannulated and instrumented bilaterally at L2 and L4 and on the left side of

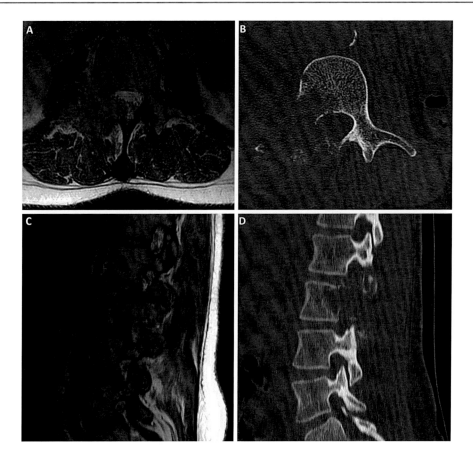

Fig. 3 A 39-year-old WF with a recent history of breast adenocarcinoma in remission presented with mechanical lower back pain. Imaging demonstrated a mass involving the right vertebral body, pedicle, and facets (**a** and **c**). The mass was lytic (**b** and **d**) with a SINS score of 8 (potentially unstable lesion)

L3. A laminectomy was then performed at the level of L3, along with a piecemeal transpedicular corpectomy of the right hemivertebra, resecting more than 50% of the body. During this, the L3 root was skeletonized and decompressed. The wound was washed with vancomycin for infection prophylaxis and solumedrol to reduce irritation of the L3 nerve root. Rods were placed bilaterally at L2–4, and following confirmation of positioning, the wound was closed in typical neurosurgical fashion (Fig. 4).

The patient's recovery was unremarkable, and she was recommended for adjuvant radiotherapy (3000 cGy in 10 fractions to the L3–5 vertebrae). She then underwent six cycles of systemic chemotherapy with docetaxel, trastuzumab, and pertuzumab. By the 3-month follow-up, the patient reported complete relief of her pain, and she remained neurologically intact. However, imaging at the 6-month follow-up demonstrated progression of disease, with new metastases involving the lungs, liver, and axial skeleton. As of last follow-up (10mo post-op), the patient remains neurologically intact, and her surgical site remains asymptomatic, though she continues to show progression of her disease, for which she is receiving trastuzumab emtansine.

Intradural Spinal Tumors

Operative Considerations

Intradural spine tumors represent 4–16% of all central nervous system neoplasms, consisting of ~60% intradural extramedullary (IDEM) lesions,

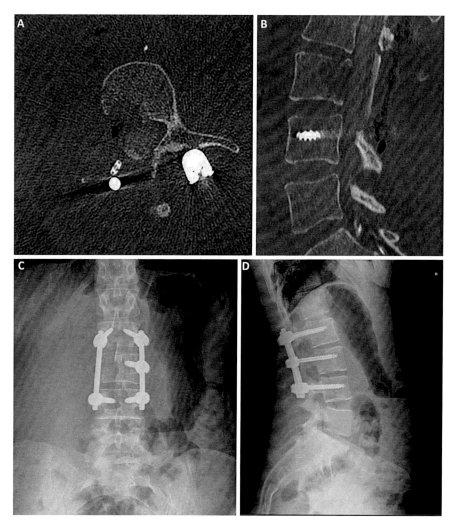

Fig. 4 Postoperative imaging demonstrating partial L3 corpectomy with adequate decompression of the thecal sac (**a** and **b**). Standing films acquired at 4-month follow-up (**c** and **d**) demonstrated good positioning of the hardware without signs of anterior column breakdown

~20% primary extradural lesions, and ~20% intramedullary lesions in adults (Hirano et al. 2012; Fehlings and Rao 2000). In pediatric patients, intramedullary, IDEM, and primary extradural lesions represent 40%, 10%, and 50% of spinal cord neoplasms (Fehlings and Rao 2000). Intradural lesions typically demonstrate indolent growth and slowly progressive neurological symptoms, with pain as the most frequent initial symptom (Samartzis et al. 2015). The majority of intramedullary lesions are benign or have a low-grade histologic diagnosis. These should be carefully distinguished from inflammatory, autoimmune, infections, or vascular conditions (i.e., multiple sclerosis, tuberculosis, granulomatous angiitis, Guillain-Barre, dural AV fistula, AVM) (Tobin et al. 2015).

Intraoperative ultrasound and neuromonitoring are critical for identification and preservation of neurologic function, respectively. Cavitron ultrasonic surgical aspirator (CUSA) is utilized for tumor debulking and resection. Neuromonitoring consists of continuous somatosensory evoked potentials (SSEP), serial transcranial motor evoked potential (MEP), and epidural (D-wave) evoked potentials (Fehlings and Rao 2000;

Samartzis et al. 2015; Tobin et al. 2015; Costa et al. 2013). Surgeons should monitor for the presence or absence of MEPs intraoperatively as well as the proportion of D-wave decrease. Intraoperative loss of MEPs and 50% reduction in D-wave amplitude have been associated wtih permanent postoperative neurologic deficit (Costa et al. 2013). Following tumor resection, laminoplasty has been demonstrated to reduce the risk of postoperative CSF leak (Nagasawa et al. 2011).

Intradural Extramedullary Tumors

Intradural extramedullary (IDEM) lesions include meningioma, schwannoma, neurofibroma, or malignant peripheral nerve sheath tumor (MPNST). Meningiomas are classically broad-based with a dural tail, isointense, homogenously enhancing, and more common in female patients, arising from arachnoidal cells of the meninges. Schwannomas are enhancing nerve sheath tumors, eccentrically located, commonly arising from the sensory nerve root, with T2-hyperintensity and possible cystic components (De Verdelhan et al. 2005; Turel and Rajshekhar 2014; Ahn et al. 2009). Myxopapillary ependymoma (WHO I) is typically located at the cauda equina, with contrast enhancement and T1-hyperintensity. These arise from the central canal, similar to traditional intramedullary ependymomas, but may seed via drop metastasis (Kucia et al. 2011).

Case Illustration

Symptomatic WHO I Meningioma

A 51-year-old Caucasian female presented acutely to our emergency department with an 8-month history of back pain and 2-month history of progressive lower extremity numbness with the subjective sensation of leg tightness and weakness. Neurological examination demonstrated minor weakness in hip flexion for the left lower extremity and clonus in the right lower extremity. Imaging demonstrated a 1.7 × 1.4 × 1.0 cm (craniocaudal, mediolateral, anterior-posterior) intradural extramedullary mass at the level of T7/8, posteriorly displacing and compressing the cord, with concomitant T2/STIR signal change consistent with cord edema. These findings were felt to be most consistent with meningioma, and given the patient's clinical picture, a recommendation was made for the patient to undergo operative management (Fig. 5).

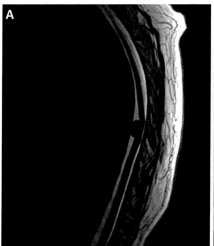 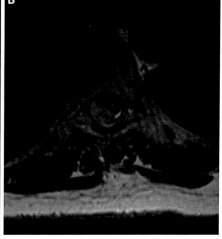

Fig. 5 A 51-year-old WF presented to the emergency department with a 2-month history of progressive BLE weakness. MR imaging (**a** and **b**) demonstrated a large intradural, extramedullary lesion at the level of T7/8 that significantly compressed the cord. Biopsy was consistent with WHO I meningioma

The patient underwent a single-stage operation using a posterior midline approach. Subperiosteal dissection was performed over the T6–8 levels, dissecting laterally to the facets. An ultrasonic bone cutter was then used to perform and en bloc laminectomy of the T6–8 levels; the laminae were delivered off the field and preserved for post-resection reconstruction. An ultrasound was then brought into the field to verify the tumor location. Under microscopic visualization, the dura was sharply incised in the midline using a #15 blade, and the flaps were retracted laterally using 4–0 silk sutures. Dissection of the arachnoid was performed using microscissors, revealing a left-sided lesion. A 5–0 Prolene suture was placed through the left-sided dentate ligament to allow manipulation of the cord, which was rotated to the right to better expose the lesion. The tumor was detached from the overlying arachnoid, followed by division of the dural attachment and separation from the adjacent cord. The lesion was delivered en bloc, and the dura was repaired with 5–0 Prolene. Valsalva demonstrated no CSF leakage, and fibrin glue was applied over the repair. The T6–8 laminae were then reattached with titanium plates, and the wound was closed in typical neurosurgical fashion (Fig. 6). The patient's

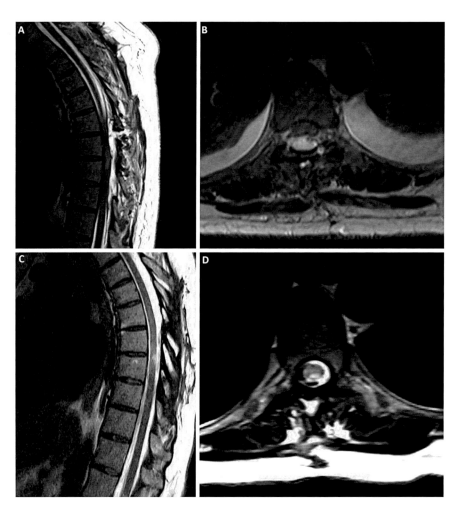

Fig. 6 Postoperative MR (**a** and **b**) demonstrated adequate decompression of the cord with residual cord deformity and T2 signal hyperintensity. MR acquired at the 4-month follow-up (**c** and **d**) showed improvement in cord deformity, though residual T2 signal hyperintensity was noted

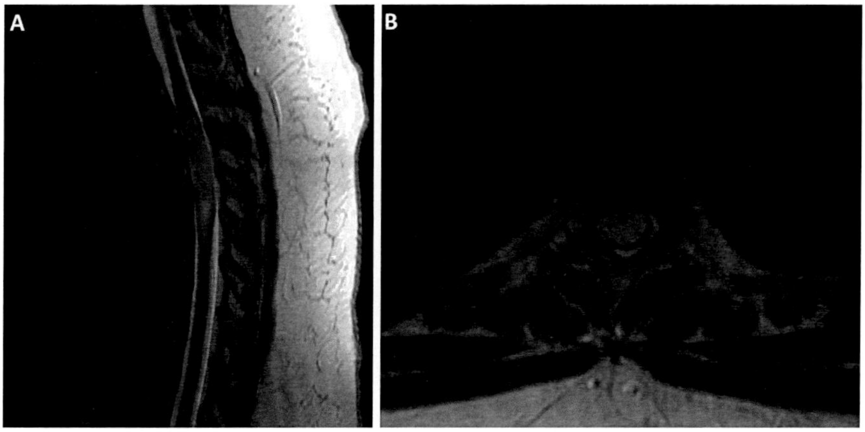

Fig. 7 A 26-year-old female presented with a 2-year history of nonmechanical upper back pain and progressive BLE weakness. MR imaging (**a** and **b**) demonstrated a large intradural, intramedullary lesion extending from T2/3 to T5/6. Histology of the surgical specimen was consistent with WHO III anaplastic astrocytoma

recovery was uncomplicated, and she was discharged on post-op day 3. At 3-month follow-up, the patient reported near-complete relief of her symptoms, with only residual circumferential numbness at the level of the lesion, which had been noted at discharge.

Intramedullary Tumors

Intramedullary lesions frequently present with a long history of symptoms, predominated by sensory dysesthesia (Samartzis et al. 2015; Tobin et al. 2015; Ferrante et al. 1992; Nakamura et al. 2008). Preoperative ambulatory status is graded by the McCormick Scale, from grade I to V (McCormick and Stein 1990). The majority of intramedullary lesions are ependymoma or astrocytoma. Astrocytoma represents ~30% of intramedullary spinal cord tumors and is the most common spinal cord tumor of children. Astrocytomas are generally hypointense on T1-weighted MRI; are hyperintense on T2-weighted MRI, eccentrically located, and cystic; and have variable enhancement (benign often enhances, i.e., WHO I juvenile pilocytic astrocytoma). In contrast, ependymomas are centrally located, originate from the central canal, and are the most common intramedullary tumor in adults, contrast enhancing, T1-hypointense, and T2-hyperintense. The "Rule of C's" can be a useful mnemonic to summarize ependymomas: cervical, contrast-enhancing, cavity (syrinx)-associated, cap of hemosiderin, and centrally located (Tobin et al. 2015; Ferrante et al. 1992; Nakamura et al. 2008).

More rarely, intramedullary hemangioblastoma, subependymoma, or intramedullary metastasis may also occur. Hemangioblastoma are subpial, highly vascular lesions, associated with a large syrinx with possible flow voids (Samartzis et al. 2015; Ferrante et al. 1992; Nakamura et al. 2008). Removal of these lesions is similar to that of an AVM, whereby arterial feeders are primarily coagulated with preservation of the dilated veins. Subependymomas are minimally enhancing lesions, located eccentrically, with T2-hyperintensity, often infiltrating the dorsal or ventral spinocerebellar tracts (Krishnan et al. 2012).

To avoid epidural bleeding and excess CSF obscuring the operative field, the dura is opened at the posterior midline before the arachnoid is incised. Myelotomy is performed at the posterior median sulcus, through either mapping of the dorsal columns or identification of vessels entering the spinal cord at this location. Due to the location of myelotomy, dorsal column dysfunction is a significant risk of the procedure (Samartzis et al. 2015; Tobin et al. 2015).

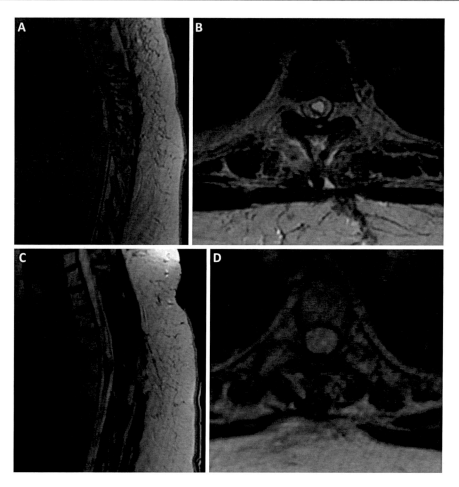

Fig. 8 Postoperative MR (**a** and **b**) demonstrated subtotal resection of the lesion with formation of large syrinx. Sequential follow-up MR demonstrated little change, with the most recent images (**c** and **d**) acquired at 41-month follow-up demonstrating slight decrease in the syrinx size

Case Illustration

Anaplastic Astrocytoma

A 26-year-old Hispanic female presented to our service with a 2-year history of progressive upper back pain accompanied by pain and tinging in her bilateral lower extremities. The patient also complained of urinary retention and progressive lower extremity weakness over the preceding 2 months, forcing her to employ knee braces and crutches for ambulation. Physical examination demonstrated mild weakness of the bilateral lower extremity, most pronounced in dorsiflexion and plantarflexion; she was also noted to be hyperreflexive in the bilateral lower extremities. Imaging demonstrated a 1.3 × 1.3 × 5.0 cm T2-hyperintense intramedullary lesion extending from T2/3 to T5/6, suggestive of an intramedullary neoplasm (Fig. 7). Surgical intervention was recommended, which the patient elected to pursue due to her worsening neurological symptoms.

Surgery was performed in a single stage using a posterior midline approach. The patient was placed prone on the Jackson table, and a #15 blade was used to incise the skin overlying the T2–5 vertebrae. Bovie cautery was used to carry the dissection through the subcutaneous tissues, and a subperiosteal dissection was carried out over the laminae, facets, and spinous processes of the T2–5 levels. An ultrasonic bone cutter was then used to perform an en bloc laminectomy of the T2–5 levels; the specimen was delivered off the field and preserved for post-resection reconstruction. An ultrasound was then used to confirm the lesion position and to confirm that both the rostral and caudal poles were exposed. Epidural electrodes were then placed to demonstrate that stimulation of the area overlying the lesion led to no MEPs in the lower extremities. Under microscopic visualization, the dura was then sharply incised in the midline with a #15 blade, and the dural leaflets were tacked down laterally with 4–0 Nurolon sutures. The arachnoid was sharply incised and dissected laterally to expose the cord. A midline myelotomy was then performed with a #11 blade, extending the incision to the level of the rostral and caudal poles. This revealed a large lesion, which herniated through the myelotomy defect. Dissection proceeded by developing a plane between the tumor and cord parenchyma. Bipolar cautery and gentle suction were used to develop this plane circumferentially, and the tumor was resected en bloc. Hemostasis was obtained, and the dura was reapproximated using 5–0 Prolene sutures. Valsalva was performed to confirm a CSF tight seal. The laminae were then replaced and affixed with titanium microplates, and the wound was closed in typical neurosurgical fashion with interrupted 0 Vicryl in the fascia, 3–0 Vicryl in the dermis, and 4–0 Biosyn in the skin (Fig. 8).

Pathology revealed a WHO Grade III anaplastic astrocytoma with positive margins that was negative for R132H mutation in IDH-1. Postoperatively the patient was paraplegic and was discharged to an inpatient rehabilitation center on postoperative day seven. The decision was made to start the patient on temozolomide and forego radiation in order to increase the potential for spinal cord recovery. At 6-month follow-up, the patient was deemed neurologically stable with minimal recovery of neurological function in her lower extremities. Given her stable neurological status and positive surgical margins, the patient underwent conventional multifractionated radiation to the site of tumor (4500 cGy) at this time. The patient is alive at 53 months postoperatively with no evidence of disease recurrence.

References

Aaron AD (1994) The management of cancer metastatic to bone. JAMA 272:1206–1209

Ahmed AK, Goodwin CR, Heravi A et al (2018) Predicting survival for metastatic spine disease; a comparison of nine scoring systems. Spine J 18:1804. [Epub ahead of print]

Ahn D, Park H, Choi D et al (2009) The surgical treatment for spinal Intradural extramedullary Tumors. Clin Orthop Surg 1(3):165–172

Bauer HC, Wedin R (1995) Survival after surgery for spinal and extremity metastases. Prognostication in 241 patients. Acta Orthop Scand 66:143–146

Bilsky MH, Laufer I, Fourney DR et al (2010) Reliability analysis of the epidural spinal cord compression scale. J Neurosurg Spine 13(3):324–328

Boriani S, Biagini R, DeLure F (1996) En bloc resections of bone tumors of the thoracolumbar spine. A preliminary report on 29 patients. Spine 21:1927–1931

Boriani S, Weinstein JN, Biagini R (1997) Primary bone tumors of the spine. Terminology and surgical staging. Spine 22:1036–1044

Chieng LO, Hubbard Z, Salgado CJ et al (2015) Reconstruction of open wounds as a complication of spinal surgery with flaps: a systematic review. Neurosurg Focus 39(4):e17

Chiriano J, Abou-Zamzam AM, Urayeneza O et al (2009) The role of the vascular surgeon in anterior retroperitoneal spine exposure: preservation of open surgical training. J Vasc Surg 50(1):148–151

Ciftdemir M, Kaya M, Selcuk E, Yalniz E (2016) Tumors of the spine. World J Orthop 7(2):109–116

Costa P, Peretta P, Faccani G (2013) Relevance of intraoperative D wave in spine and spinal cord surgeries. Eur Spine J 22(4):840–848

De Verdelhan O, Haegelen C, Carsin-Nicol B, Riffaud L, Amlashi SF, Brassier G et al (2005) MR imaging features of spinal schwannomas and meningiomas. J Neuroradiol 32(1):42–49

Dea N, Gokaslan ZL, Choi D, Fisher C (2017) Spinal oncology – primary spine tumors. Neurosurgery 80(3):S124–S130

DeMonte F, Diaz EM, Callendar D et al (2001) Transmandibular, circumglossal, retropharyngeal approach for chordomas of the clivus and upper cervical spine. Technical note. Neurosurg Focus 10(3):e10

Enneking WF (1986) A system of staging musculoskeletal neoplasms. Clin Orthop Relat Res 204:9–24

Epstein NE (2013) When does a spinal surgeon need a plastic surgeon? Surg Neurol Int 4(Suppl 5):S299–S300

Fehlings MG, Rao SC (2000) Spinal cord and spinal column tumors. In: Bernstein M, Berger MS (eds)

Neuro-oncology – the essentials. Thieme, New York, pp 445–464

Ferrante L, Matronardi L, Celli P et al (1992) Intramedullary spinal cord ependymomas – a study of 45 cases with long-term follow-up. Acta Neurochir 119(1–4):74–79

Fisher CG, DiPaola CP, Ryken TC et al (2010a) A novel classification system for spinal instability in neoplastic disease: an evidence-based approach and expert consensus from the spine oncology study group. Spine 35: E1221–E1229

Fisher CG, DiPaola CP, Ryken TC et al (2010b) A novel classification system for spinal instability in neoplastic disease: an evidence-based approach and expert consensus from the Spine Oncology Study Group. Spine 35:E1221–E1229

Fourney DR, Gokaslan ZL (2004) Anterior approaches for thoracolumbar metastatic spine tumors. Neurosurg Clin N Am 14:443–451

From Chan P, Boriani S, Fourney DR et al (2009) An assessment of the reliability of the Enneking and Weinstein-Boriani-Biagini classifications for staging of primary spinal tumors by the Spine Oncology Study Group. Spine 34:385

Greenspan A (2004) Radiologic evaluation of tumors and tumor-like lesions. In: Orthopedic imaging: a practical approach, 4th edn. Lippincott Williams & Wilkins, Philadelphia, pp 529–570

Hirano K, Imagama S, Sato K et al (2012) Primary spinal cord tumors: review of 678 surgically treated patients in Japan. A multicenter study. Eur Spine J 21(10): 2019–2026

Horne PH, Lampe P, Nguyen JT et al (2016) A novel radiographic Indicator of developmental cervical stenosis. J Bone Joint Surg Am 98(14):1206–1214

Ilaslan H, Sundaram M, Unni KK, Shives TC (2004) Primary vertebral osteosarcoma: imaging findings. Radiology 230(3):697–702

Kaloostian PE, Zadnik PL, Etame AB et al (2014) Surgical management of primary and metastatic spinal tumors. Cancer Control 21(22):133–139

Katagiri H, Takahashi M, Wakai K et al (2005) Prognostic factors and a scoring system for patients with skeletal metastasis. J Bone Joint Surg Br 87:698–703

Kim YS, Han IH, Lee IS et al (2012a) Imaging findings of solitary spinal bony lesions and the differential diagnosis of benign and malignant lesions. J Korean Neurosurg Soc 52(2):126–132

Kim JM, Losina E, Bono CM et al (2012b) Clinical outcome of metastatic spinal cord compression treated with surgical excision ± radiation versus radiation therapy alone: a systematic review of literature. Spine 37:78–84

Krishnan SS, Panigrahi M, Pendyala S et al (2012) Cervical subependymoma: a rare case report with possible histogenesis. J Neurosci Rural Pract 3(3):366–369

Kucia EJ, Maughan PH, Kakarla UK, Bambakidis NC, Spetzler RF (2011) Surgical technique and outcomes in the treatment of spinal cord ependymomas: Part II: myxopapillary ependymoma. Neurosurgery 68 (1 Suppl Operative):90–94; discussion 4

Laufer I, Rubin DG, Lis E, Cox BW, Stubblefield MD, Yamada Y et al (2013 Jun) The NOMS framework: approach to the treatment of spinal metastatic tumors. Oncologist 18(6):744–751

McCormick PC, Stein BM (1990) Intramedullary tumors in adults. Neurosurg Clin N Am 1:609–630

Mehta VA, Amin A, Omeis I (2015) Implicatinos of spinopelvic alignment for the spine surgeon. Neurosurgery 76(Suppl 1):S42–S56

Metser U, Lehrman H, Blank A et al (2004) Malignant involvement of the spine: assessment by 18F-FDG PET/CT. J Nuc Med 45(2):279–284

Nagasawa DT, Smith ZA, Cremier N et al (2011) Complications associated with the treatment for spinal ependymomas. Neurosurg Focus 31(4):E13

Nakamura M, Ishii K, Watanabe K et al (2008) Surgical treatment of intramedullary spinal cord tumors: prognosis and complications. Spinal Cord 46:282–286

Oh MC, Ivan ME, Sun MZ et al (2013) Adjuvant radiotherapy delays recurrence following subtotal resection of spinal cord ependymomas. Neuro-Oncology 15(2): 208–215

Perrin RG, Laxton AW (2004) Metastatic spine disease: epidemiology, pathophysiology, and evaluation of patients. Neurosurg Clin N Am 15:365–373

Posner JB (1987) Back pain and epidural spinal cord compression. Med Clin North Am 71(2):185–205

Rodallec MH, Feydy A, Larousserie F et al (2008) Diagnostic imaging of solitary tumors of the spine: what to do and say. Radiographics 28(4):1019–1041

Ropper AE, Cahill KS, Hanna JW et al (2012) Primary vertebral tumors: a review of epidemiologic, histological and imaging findings, part II: locally aggressive and malignant tumors. Neurosurgery 70 (1):211–219

Samartzis D, Gills CG, Shih P et al (2015) Intramedullary spinal cord Tumors: Part I – epidemiology, pathophysiology, and diagnosis. Global Spine J 5(5):425–435

Samudrala S, Khoo LT, Rhim SC et al (1999) Complications during anterior surgery of the lumbar spine: an anatomically based study and review. Neurosurg Focus 7(6):e9

Sciubba DM, Petteys RJ, Dekutoski MB et al (2010) Diagnosis and management of metastatic spine disease: a review. J Neurosurg Spine 13(1):94–108

Senders JT, Staples PC, Karhade AV et al (2018) Machine learning and neurosurgical outcome predictin: a systematic review. World Neurosurg 109:476–486

Tatsui H, Onomura T, Morishita S et al (1996) Survival rates of patients with metastatic spinal cancer after scintigraphic detection of abnormal radioactive accumulation. Spine (Phila Pa 1976) 21:2143–2148

Tobin MK, Geraghty JR, Engelhard HH et al (2015) Intramedullary spinal cord tumors: a review of current and future treatment strategies. Neurosurg Focus 39(2): E14

Tokuhashi Y, Matsuzaki H, Oda H et al (2005) A revised scoring system for preoperative evaluation of metastatic spine tumor prognosis. Spine (Phila Pa 1976) 30:2186–2191

Tomita K, Kawahara N, Kobayashi T et al (2001) Surgical strategy for spinal metastases. Spine (Phila Pa 1976) 26:298–306

Turel MK, Rajshekhar V (2014) Magnetic resonance imaging localization with cod liver oil capsules for the minimally invasive approach to small intradural extramedullary tumors of the thoracolumbar spine. J Neurosurg Spine 21(6):882–885

Uei H, Tokuhashi Y, Maseda M et al (2018) Clinical results of multidisciplinary therapy including palliative posterior spinal stabilization surgery and postoperative adjuvant therapy for metastatic spinal tumor. J Orthop Surg Res 13:30

Wallace AN, Robinson CG, Meyer J et al (2015) The metastatic spine disease multidisciplinary working group algorithms. Oncologist 20(10):120501215

Walsh GL, Gokaslan ZL, McCutcheon IE et al (1997) Anterior approaches to the thoracic spine in patients with Cancer: indications and results. Ann Thorac Surg 64(6):1611–1618

Bone Metabolism

25

Paul A. Anderson

Contents

Introduction	524
Epidemiology	524
Morbidity and Mortality of Fragility Fractures	524
Secondary Fracture Risk	525
Secondary Fracture Prevention	525
Outcomes of Secondary Fracture Prevention	526
Own the Bone Quality Improvement Program	527
Diagnosis of Osteoporosis	527
Dual-Energy X-Ray Absorptiometry (DXA)	528
Advances in Bone Densitometry	528
Opportunistic CT	528
Treatment for Osteoporosis	529
Education	529
Nutritional Supplementation	532
Secondary Causes of Osteoporosis	533
Pharmaceutical Management	533
Complications of Medical Treatment of Osteoporosis	533
Preoperative Optimization of the Spine Surgery Patient	534
Mitigation of Poor Bone Health on Spinal Outcomes	534
Preoperative Bone Health Program	534
Conclusion	537
References	537

Abstract

Poor bone mass is a common condition affecting over 50 million Americans. Consequences are fracture with increased morbidity and mortality compared to the no fracture population. The spine patients appear to be at high risk for poor health but this is often overlooked. Despite spine surgeons treating patients with

P. A. Anderson (✉)
Department of Orthopedic Surgery and Rehabilitation, University of Wisconsin, Madison, WI, USA
e-mail: Anderson@ortho.wisc.edu

© Springer Nature Switzerland AG 2021
B. C. Cheng (ed.), *Handbook of Spine Technology*,
https://doi.org/10.1007/978-3-319-44424-6_63

fragility fractures, further osteoporotic care is rarely provided, thereby increasing the risk of secondary fractures. Secondary fracture prevention programs are comprehensive programs to identify those at risk and provide counseling, nutritional recommendations, physical therapy, and medication when indicated. Programs such as *Own the Bone* are highly effective and can reduce the risk of secondary fracture by 40%. Increasingly, evidence has also linked poor bone health to poor outcomes after spine surgery. Vitamin D deficiency is almost universal, and osteoporosis, when present, is associated with increased nonunion and hardware complications. A proposal for preoperative bone health optimization using methods similar to secondary fracture prevention has been recommended. In this program, if patients are osteoporotic, then, if possible, surgery should be delayed until bone health is improved.

Keywords

Osteoporosis · Secondary fracture · Secondary fracture prevention · Fracture Liaison Service · Spine surgery · Preoperative bone health optimization · Vitamin D deficiency

Introduction

Poor bone health status from osteopenia and osteoporosis is a major health concern throughout the world. Despite accurate diagnostic tests and effective treatments, osteoporosis care is lacking for the majority of patients. For example, primary care utilizes guideline-based assessment, and treatment is less than 10% of patients for primary osteoporosis prevention (Camacho et al. 2016; Cosman et al. 2014). Secondary fracture care after an osteoporotic-related fragility fracture occurs in less than 20% of patients and has not improved in the last 20 years (Balasubramanian et al. 2014).

The purpose of this chapter is to review the epidemiology of osteoporosis and its prevalence in spinal diseases. We will review updated definitions to identify osteoporosis and the consequences of osteoporosis on morbidity and mortality of spine patients, the effects on outcomes of surgery, and the results of secondary fracture prevention programs. Finally, preoperative bone health optimization programs will be introduced with the emphasis that spine surgeons can provide such care.

Epidemiology

Osteoporosis is a complex public health condition with a prevalence in 2011 in the United States of over 10 million people and over 40 million with low bone mass (osteopenia). By 2030 this is expected to increase to 14 million and 60 million people, respectively (Wright et al. 2014). Approximately 2.1 million fragility fractures occur yearly (Burge et al. 2007). This is greater than that for stroke, heart attack, and breast cancer combined. Costs of care for fractures exceed $19 billion per year and will increase dramatically in the future. Hospitalizations for fragility fractures in 2011 occurred in 325,000 patients with hip fractures and 246,000 for spine fractures (the second most common fragility fracture requiring hospitalization). The incidence of clinical vertebral fractures is related to gender and age. For example, 3.5% of females compared to 1.6% of males require hospitalization for fragility fractures (Burge et al. 2007; Ballard et al. 2018). In patients over the age of 80, fragility fractures resulted in 6.4% of all hospital admissions compared to 1% of patients in their 60s. Burge estimated that there will be a 75% increase in fractures between 2005 and 2025 in women and a 25% increase in men. Costs will increase in the same proportions (Burge et al. 2007).

Morbidity and Mortality of Fragility Fractures

Osteoporotic fragility fractures are life-changing events resulting in reduced independence, chronic pain, loss of function, diminished health-related quality of life, and increased mortality. Tajeu found that after hip fracture, there is overall 28% mortality at 1 year which was 2.2 times higher than those without hip fracture. Disability and loss of independence were four times higher, and destitution was twice as likely after hip fracture when compared to control (Tajeu et al. 2014). Lau found the same findings in a large Medicare database study of 97,000 patients having vertebral fracture; mortality was over 2–3 times higher than expected depending on the age at time of fracture (Lau et al. 2008). Chen examined the social consequences after hospitalization for vertebral fracture (Chen et al. 2013). Forty percent were discharged to skilled nursing homes, and 15% required home health, whereas only 14% of patients were able to return home immediately.

Pain is also common after vertebral fracture. Chen found that baseline pain after fracture was of 7.8/10 and decreased to 3.4 at 6 months of follow-up (Amin et al. 2014). The baseline pain is greater than any studies for elective spine surgery. Health-related quality of life is also significantly diminished. Tosteson reviewed patients with hip and/or spine fractures using the SF-36 5 years after fracture (Tosteson et al. 2001). Only 13% of patients reported no limitations of activities of daily living (ADLs), whereas 25% of the spine fractures and over half the hip fractures have significant limitations of ADLs. Similarly, the SF-36 decreased by 1 standard deviation after spine fracture, 1.6 standard deviations for hip fracture, and 2 standard deviations for patients with combined hip and spine fractures. Svensson performed structured narrative interviews of octogenarians after vertebral fractures (Svensson et al. 2016). Patients consistently reported struggling to understand a deceiving body, breakthrough pain, and fear of isolation, dependency, and an uncertain future.

Secondary Fracture Risk

An important impetus for secondary fracture prevention is that one fragility fracture begets another. Hodsman reported a 12% increase in secondary vertebral fracture risk at 2 years, 16% by 5 years, and 25% at 10 years (Hodsman et al. 2008). Center followed patients for 16 years determining the incidence of secondary vertebral fracture. The risk per 1,000 patient years was 80 in females and 100 in males (Center et al. 2007). The relative risk of a new fracture in those with prior fracture was 2.5 in females and 6.2 in males. Kanis pooled placebo-controlled randomized controlled trials (RCT) for assessment of alendronate (Kanis et al. 2004). In the placebo group, 26% had prior fragility fracture. The relative risk of a new fracture in patients with prior fracture was 1.8 for females and 2.0 for males. Anderson reported results of a meta-analysis of nine RCTs of vertebroplasty versus control (Anderson et al. 2013). In the placebo treatment group, 18.8% of patients had a secondary fracture within 12 months of injury. Lindsay pointed out that the number of vertebral fractures significantly increased secondary fracture risk (Lindsay et al. 2001). Those patients with a single fracture had a 4% risk within 1 year, whereas those patients presenting with two or more fractures had a 25% risk.

Secondary Fracture Prevention

Osteoporosis care, even after fracture, is poorly managed throughout the world. The care is ideally suited for primary care, but unfortunately primary care physicians do not have the time to care for bone disease and are often confused by the many conflicting guidelines and complicated pathways (Binkley et al. 2017a). In 2004, the US Surgeon General reviewed bone health and osteoporosis throughout the United States noting that there was a gap between what we know and its application (American Orthopaedic Association 2005). Primary care physicians and orthopedic surgeons rarely discuss osteoporosis with patients having fractures. In response to the Surgeon General's report, the American Orthopedic Association (AOA) initiated the "Own the Bone" program in recognition of the failure of primary care (American Orthopaedic Association 2005). This is a quality improvement program to provide optimum secondary fracture prevention. In this model, fragility fractures are seen as a sentinel event, an opportunity to encourage patients to obtain management of their bone disease.

In a systematic review of secondary fracture prevention programs, Ganda identified four types (Ganda et al. 2013). In Type A, a Fracture Liaison Service (FLS) coordinator is imbedded into the fracture team and assumes the entire care of bone health. In Type B, the coordinator is also imbedded in the fracture team but refers the follow-up osteoporosis care to the primary care physician (PCP). The initial consultation is provided at the time of fracture hospitalization. In Type C programs, the patient is seen, provided educational material regarding the need for further fracture evaluation care, and communication is sent to the primary care physician. Type D simply instructs the patient to obtain further osteoporosis care through their PCP. As would be expected, treatment is directly correlated to the intensity of effort. In the Type A program, 79% of patients receive DXA and almost all have recommendation for pharmaceutical therapy. In the Type B program, 59% of patients obtain DXA and 40% are prescribed medications. In Type C, 43% receive DXA and 23.4% pharmaceutical medications, and in Type D, only 8% of patients received any secondary fracture care (Table 1).

Although secondary fracture prevention is highly effective, over the last 20 years, no improvement in the incidence of patients receiving care has occurred. Leslie reported that from 1996 to 2006, there continued to be less than 20% receiving secondary fracture prevention (Leslie et al. 2012). Balasubramanian more recently reported that the percentage of patients receiving secondary fracture prevention is actually worsening with only 19% of patients receiving fracture prevention care which was less than occurred in 2001 (Balasubramanian et al. 2014). In men, only 10% received secondary fracture care. Flais identified causes for the lack of secondary fracture prevention (Flais et al. 2017). In 35% of cases, there was a lack of awareness on the part of the PCP that further care was needed, and in 17% of cases, there was a lack of awareness of the PCP that the patient even had a fracture. These are modifiable with a comprehensive care program.

Outcomes of Secondary Fracture Prevention

Secondary fracture programs have been shown to be effective. Bawa found that only 10.6% patients were treated with anti-osteoporotic medications after fracture (Bawa et al. 2015). However, this group had a 40% reduction in secondary fractures compared to non-treated patients (Bawa et al. 2015). In England and Wales, a quality improvement program to have all hip fracture patients receive secondary fracture prevention was instituted in 2005 (Hawley et al. 2016). Hawley reported a 30% reduction in secondary fracture risk (Hawley et al. 2016). Several studies from Australia comparing hospitals with fracture liaison services to those without document a secondary fracture reduction of over 50%

Table 1 Secondary fracture prevention programs

		DXA recommendations	Pharmaceutical recommendations
Type A	FLS coordinator Imbedded in orthopedic trauma team Assumes entire care of bone health	79.4%	46.4%
Type B	FLS coordinator Imbedded in orthopedic trauma team Refers bone health care to primary care	59.5%	40.6
Type C	Inpatient consult Education Communicates with primary care	43.4%	23.4
Type D	Education Refer to primary care	–	8.0

FLS Fracture Liaison Service

(Nakayama et al. 2016; Lih et al. 2011). Curtis calculated that the number needed to treat to prevent subsequent major fracture depending on initial fracture type and patient age was between 8 and 46, thought to be well within the acceptable range for cost-effective care (Curtis et al. 2010). Secondary fracture prevention is highly effective but has proven difficult to implement across populations. It is recommended that when spine surgeons treat osteoporotic-related fractures they assure that patients receive secondary fracture prevention.

Own the Bone Quality Improvement Program

One example of a secondary fracture prevention program is "Own the Bone" (Bunta et al. 2016; Dirschl and Rustom 2018). The goal of this program is to break the fragility fracture cycle. The introduction of the Fracture Liaison Service provides assessment, education, nutrition recommendations, elimination of toxins, and, when indicated, pharmaceutical management. The *Own the Bone* program identifies fragility fracture patients older than 50, consults during the "teaching moment," and initiates multi-disciplinary care. *Own the Bone* provides documentation in a registry and many educational opportunities for patients and providers. Over 250 programs are distributed throughout the United States, and over 50,000 patients have been entered in the *Own the Bone* program. The program is adaptable to all health-care settings, not just academic medical centers; more than 50% of the programs are in community-based hospitals or physicians' practices.

Diagnosis of Osteoporosis

The World Health Organization (WHO) utilizes bone mineral density from dual-energy x-ray absorptiometry (DXA) to classify osteoporosis (Camacho et al. 2016). This was a major improvement in the understanding, diagnosis, and treatment of this disease. Because of the differences in bone mineral density obtained, depending on the type of scanner, a statistical methodology is utilized. Patient bone mineral density is compared to young healthy female between 20 and 30 age years (T-scores) or age-gender matched controls (Z-scores). The T- and Z-scores are calculated using the formula:

$$\text{T or Z Score} = \frac{\text{BMD Patient} - \text{BMD Reference}}{\text{SD Reference}}$$

The WHO criteria for osteoporosis is a T-score less than −2.5. Osteopenia (or what is now called low bone mass) is a T-score of −1.0 to −2.5. Normal bone is greater than −1.0.

The use of the World Health Organization T-scores, although an advancement, is problematic as there is poor association with the fracture in over 50% of cases and does not aid in treatment decisions. Recent guidelines by the National Osteoporosis Foundation (NOF) and the American Association of Clinical Endocrinology have combined bone mineral density with established function (Camacho et al. 2016; Cosman et al. 2014). The criteria for osteoporosis is any of the following: T-score of −2.5; a recent fragility spine or hip fracture; low bone mass (−1.0 and −2.5) and a fragility fracture; and low bone mass and high fracture risk probability (FRAX).

More important than BMD to the diagnosis of osteoporosis is the fracture risk of the patient which is based on measurable risk factors and BMD. A number of instruments have been created to assess fracture risk. The Fracture Risk Assessment Tool (FRAX) probability assesses the 10-year risk of hip and major osteoporotic fracture based on 12 known risk factors. Risk factors include demographics, known risk factors for osteoporosis, and femoral neck bone mineral density T-scores. Fracture risk can be determined without DXA, although this is more often used to screen patients for the need for DXA. The FRAX identifies the 10-year risk of hip and major osteoporotic fractures (Table 2). The criteria for recommendation for pharmaceutical treatment are when the fracture risk for hip fracture is greater than 3% and major osteoporotic fracture greater than 20% (Camacho et al. 2016).

Table 2 Fracture Risk Assessment Tool (FRAX) risk factors

Age	Smoking history	Interpretation of FRAX
Gender	Alcohol consumption >3 units/day	Without DXA >9.5% 10-year risk of major fracture indicates need for DXA
Height	Inflammatory arthritis	Osteoporosis diagnosis >3% 10-year risk hip fracture >20% 10-year risk of major fracture
Weight	Glucocorticoid use (>5 mg prednisone daily)	
Prior fracture	Secondary osteoporosis	
Parent with hip fracture	Bone mineral density (DXA), T-score, or trabecular bone score	

Dual-Energy X-Ray Absorptiometry (DXA)

The gold standard to measure BMD is DXA. It is important that spine surgeons be able to interpret their own DXA like other diagnostic tests. The BMD is taken at three regions of interest (ROI): the proximal femur, the lumbar spine, and the distal 1/3 radius. Unless unavailable, the BMD of the proximal femur T-score is used to classify bone health (Fig. 1a–d). If the proximal femur is not available due to surgery or other bony abnormalities such as osteoarthritis, then the spine and/or distal radius is used. It is common to have discordant results between ROIs and this may have important implications. For example, lower third distal radius score with relatively normal hips and spine may indicate hyperparathyroidism.

Advances in Bone Densitometry

There are several adjuncts to bone mineral density testing that may aid in classification and directing treatment. A vertebral fracture assessment (VFA) can be obtained during DXA (Fig. 1e). The patient is turned in a lateral position and a low-energy x-ray from approximately T4 to L5 is obtained. This is a true lateral, with the x-ray beam always parallel to the disc spaces avoiding parallax. Utilizing the VFA, occult fractures can be identified using morphologic criteria of Genatt (Fuerst et al. 2009). It is estimated that at least 20% of patients having DXA scans may have occult spine fractures (Jager et al. 2011) and result in a change in their diagnosis to osteoporosis. The trabecular bone score (TBS) is a new software technology that assesses the microarchitecture of bone (Harvey et al. 2015). The TBS measures the heterogeneity between pixels of bone cross sections obtained from DXA (Fig. 1e). Osteoporotic bone will have voids between pixels compared to normal bone and thus high heterogeneity. TBS is assessed as follows: <1.23 is degraded microarchitecture; 1.23–1.31 is partially degraded; and greater than 1.31 is intact. The TBS is better than BMD-based T-scores for fracture prediction and can be used in FRAX (Schousboe et al. 2016). The trabecular bone score may be of more importance to spine surgery when considering instrumentation. Further research in the use a trabecular bone score to predict surgical outcomes is needed.

Opportunistic CT

Computed tomography (CT) is a frequent diagnostic test and may be used to assess bone quality. CT is based on x-ray attenuation determined for each voxel of tissue. Attenuation is dependent upon the density and atomic mass of atoms in the tissue thus, in the case of bone, to bone mineral density. X-ray attenuation is measured by the Hounsfield unit (HU) which is normalized based on air (−1000) and water (0). The Hounsfield unit is easily obtainable from all CT using the elliptical tool available on the picture archiving and communication system (PACS) (Fig. 2). A region of interest is drawn and the mean HU in that area is displayed. Cortical bone typically will have an HU of 500 and normal trabecular bone such as vertebral body greater than 150. Using CT obtained for other reasons to estimate bone health is termed "opportunistic CT" and can aid in determining those patients that need further evaluation and predicting failure of surgical procedures (Meredith et al. 2013; Schreiber et al. 2014; Pickhardt et al. 2013).

Thresholds for opportunistic CT have been established for the lumbar spine. Pickard recommended using a threshold of >150 HU to exclude osteoporosis and HU values less than 110 rules in osteoporosis. HU values of 135 or less suggest further workup or evaluation with DXA is needed and that there is a likelihood of low bone mass or osteoporosis (Carberry et al. 2013).

Treatment for Osteoporosis

Education

The treatment of osteoporosis begins with recognizing that osteoporosis is present in the case of a fragility fracture and may potentially be present in a preoperative spine surgery candidate (Table 3) (Camacho et al. 2016). Patients with osteoporosis should receive education regarding bone health and what they themselves can do to reduce further bone loss. Education should

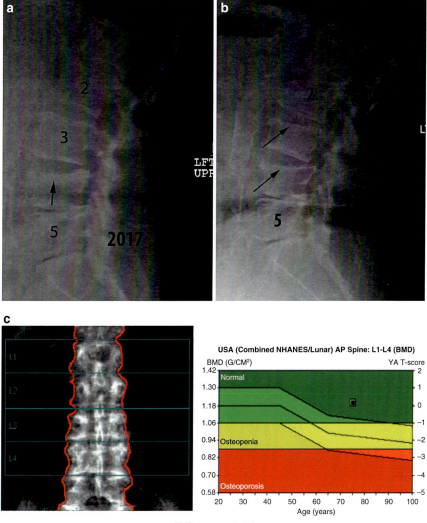

Fig. 1 (continued)

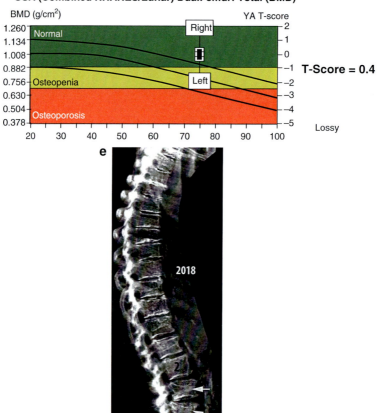

Fig. 1 (continued)

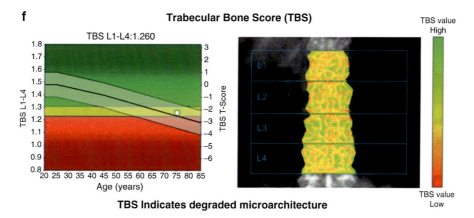

Fig. 1 (a) A 57-year-old female sustained a superior endplate fracture of L4 from a ground-level fall. No further evaluation was done despite a diagnosis of osteoporosis based on spinal fracture from ground-level fall. The patient did well with resolution of pain. (b) One year later she had a compression fracture of L3 after another ground-level fall (arrows). Laboratory evaluation revealed 25(OH) vitamin D of 12 ng/ml. (c) Spine DXA shows a T-score of 0.3. This is unreliable due to degenerative changes. (d) Hip DXA also shows normal bone mineral density. (e) Vertebral fracture assessment shows fractures at L3 and L4 (arrows). (f) Trabecular bone score (TBS) shows degraded bone microarchitecture

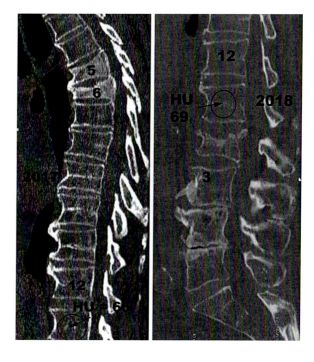

Fig. 2 Sagittal CT from 2017 and 2018. In 2017 fractures of T5 and T6 are present. An oval region of interest of L1 (ellipse) has a mean Hounsfield unit (HU) of 86 which indicated high probability of osteoporosis. In 2018 a new L2 fracture is present. The HU at L1 remained low

Table 3 General principles of osteoporotic care

1	Consider poor bone health
2	Patient education
3	Eliminate toxins
4	Provide nutrient supplements
5	Fall prevention and balance
6	Screen for secondary causes of osteoporosis
7	Assess for need for further testing (consider fracture risk)
8	Communication with primary care
9	Medications for high fracture risk

Table 4 Classification of 25(OH) vitamin D status

	25 (OH) vitamin D (ng/ml)
Normal	>30
Insufficient	20–30
Deficient	<20

include the consequences of osteoporosis, identification of risk factors, and natural history. Toxins that affect bone health should be eliminated such as smoking and drinking in excess of two alcoholic beverages per day. Nutritional needs should be assessed. Osteoporotic patients are often underweight or morbidly obese. In addition, sarcopenia (loss of muscle mass and function) is common in osteoporotic patients (Tarantino et al. 2013). Therefore, weight-bearing exercises such as walking or jogging should be discussed. Fall risk should be assessed and, if warranted, fall prevention therapy recommended. During examination, fall risk can be assessed using the "timed up and go test" in which the patient should be able to arise from a chair without using his hands and walk 3 meters, turn around, and sit again. Patients who exceed 20 s are at increased risk of fall. Another test is grip strength using a dynamometer; men should have 15 kg and women 10 kg of grip strength. Fall prevention at home should be discussed including elimination of obstructions, with special attention to the bathroom where falls frequently occur at night.

Nutritional Supplementation

Vitamin D has important effects on bone physiology. Vitamin D promotes osteoblastic differentiation and osteoblastic-mediated mineralization. Vitamin D is involved in calcium regulation, and, when under hypocalcemic conditions, it increases the differentiation of monocytes to osteoclasts. In addition, vitamin D increases collagen crosslinking leading to stronger bones. Serum vitamin D is measured as the 25(OH) vitamin D. 25(OH) vitamin D is one of the many metabolites and is in the greatest concentration. It is felt to represent long-term stores of vitamin D. 25(OH) vitamin D is bound by acute-phase reactants and therefore may be lower immediately after surgery or trauma (Binkley et al. 2017b). Although the normal vitamin D level is unknown and remains controversial, most authorities agree that normal exceeds 30 ng/ml (Table 4) (Camacho et al. 2016). An insufficient condition is between 20 and 30 ng/ml and deficient is less than 20 ng/ml.

Multiple studies have shown that the majority of patients preoperatively have vitamin D deficiencies or insufficiencies (Stoker et al. 2013; Ravindra et al. 2015a; Kim et al. 2013). Therefore, spine surgeons should consider treating all patients with vitamin D prior to surgery or after fracture. Vitamin D supplements vary in potency. Ergocalciferol (vitamin D2) is less potent and is not measured on some assays; therefore, it is not recommended. Cholecalciferol (vitamin D3) is more potent. The author recommends that all patients take 2000 U vitamin D3 daily. Preoperative patients who are insufficient should consume between 2,000 and 5,000 U daily. Deficient patients are prescribed 50,000 U vitamin D3 weekly for 6 weeks and then have levels rechecked.

In addition to vitamin D, adequate calcium intake is required. In adults, 1,200 mg of calcium intake is the recommended daily requirement (Camacho et al. 2016). This can be obtained through the diet largely through the consumption of milk products. An 8 ounce glass of milk or equivalent is approximately 250 mg. If dietary intake is insufficient, then calcium supplements should be recommended. Either calcium citrate or a calcium carbonate compound is effective. It is best if calcium can be obtained dietarily, rather than by supplements if at all possible.

Secondary Causes of Osteoporosis

Primary osteoporosis is the age-related loss of bone mass especially in females after menopause. Secondary causes from medical disease and medications are common and should be screened (Camacho et al. 2016; Cosman et al. 2014). Most important is to identify endocrine disorders such as hyperparathyroidism and hyperthyroidism, cancer especially multiple myeloma, and liver and renal disease. The author recommends screening with complete blood count, complete metabolic panel, 25(OH) vitamin D, intact parathyroid hormone, phosphate, and 24-h urine collection for calcium, sodium, and creatinine. Other tests will be dictated by results of history and physical examination.

Pharmaceutical Management

Pharmaceutical management is recommended in patients who have high fracture risk as assessed by FRAX (Camacho et al. 2016). Specifically, those patients who have a 3% or greater 10-year risk of hip fracture or 20% or greater 10-year risk of major fracture are considered candidates for pharmaceutical treatment. Two classes of medications are available: anti-resorptive and anabolic.

Anti-resorptive medications include bisphosphonates, calcitonin, denosumab, and estrogens. Bisphosphonates bind to hydroxyapatite on their resorptive surfaces preventing osteoclastic enzymatic breakdown of bone. In addition, they cause apoptosis of osteoclast cells. Denosumab is a monoclonal antibody inhibitor of the receptor activator of nuclear factor kappa-B (RANK) ligand. RANK ligand activation is required for activation of preosteoclast to osteoclast and is prevented by denosumab. Bisphosphonates and denosumab have consistently been shown to reduce fractures by 50–70% at 3 years for primary osteoporosis and after fracture (Camacho et al. 2016; Cosman et al. 2014).

Calcitonin is an older anti-resorptive medication; however, there is increased risk of cancer and this is rarely used today. Estrogens and estrogen analogs also have anti-resorptive effects but carry risk of venous thromboembolic disease and cancer so are rarely utilized for osteoporosis currently.

Two recombinant parathyroid hormone analogs are approved for osteoporosis: teriparatide and abaloparatide. The mechanism of action for anabolics is promoting osteoblastic differentiation, and they are highly effective at reducing risk of fracture in osteoporotic patients (Camacho et al. 2016; Cosman et al. 2014). The indications for which insurers will pay for the anabolic agents will vary. However, they are indicated after failure of anti-resorptive medications or in high-risk patients such as those with high FRAX risk scores or on glucocorticoids. The anabolic agents are delivered by daily injection and costs range from $1,800 to $3,200 per month for 18–24 months.

Complications of Medical Treatment of Osteoporosis

Bisphosphonates and denosumab have been linked to atypical femur fracture and osteonecrosis of the jaw. Atypical femur fractures are stress fractures located in the femoral subtrochanteric area (Shane et al. 2014). Initially there will microfracture of bone and then eventually a stress fracture will occur. This is often bilateral. Atypical femur fractures are directly related to duration of exposure (Camacho et al. 2016). In general, the increase is seen after 5 years of treatment or in cancer patients who are given large doses of the IV form of bisphosphonate. It is estimated that bisphosphonates prevent between 15 and 100 fractures for every 1 of atypical femur fractures. As a consequence, drug holidays are commonly prescribed (Camacho et al. 2016; Cosman et al. 2014). Specifically after 5–7 years, it is recommended that bisphosphonates be discontinued. Some loss of bone mineral density will occur after withdrawal of bisphosphonates, although fracture risk does not appear to be affected. However, in some patients, rapid bone loss occurs after drug withdrawal and therefore monitoring needs to continue. In very-high-risk patients, it is not recommended to use a drug holiday (Muszkat et al. 2015).

Preoperative Optimization of the Spine Surgery Patient

Spine surgeons are increasingly recognizing the linkage of poor outcomes and complications to the presence of osteoporosis and, therefore, are considering preoperative bone health optimization (Lubelski et al. 2015). Preoperative optimization of bone health requires assessment, recommendations for nutritional support, elimination of toxins, assessment of fall risk, and potentially pharmaceutical management with a delay in surgery.

Vitamin D deficiency has been examined in patients undergoing spine fusion. Stoker found that only 16% of patients had normal vitamin D levels, Ravindra only 31%, and Kim none (Stoker et al. 2013; Kim et al. 2013; Ravindra et al. 2015b). The majority of patients were actually deficient, <20 ng/ml. In addition, osteoporosis and osteopenia are also quite common in the preoperative spine patient. Bjerke found that only 31% of patients had normal BMD, 59% were osteopenic, and 10% were osteoporotic (Bjerke et al. 2017). These were findings similar to Yagi and Wagner (Yagi et al. 2011; Fujii et al. 2013). Thus, the majority of adult patients undergoing spine surgery have vitamin D deficiencies and osteopenia or osteoporosis.

Outcomes in osteoporotic patients and vitamin D-deficient patients have been examined. Ravindra found that the time to fusion was significantly lower in patients who are vitamin D-deficient (Ravindra et al. 2015a). In addition, fusion success was 3.5 times less likely in the vitamin D-deficient than normal patients. Kim found an inverse correlation between the final Oswestry Disability Index and preoperative vitamin D levels; that is, patients with high baseline vitamin D have better recovery (Kim et al. 2012).

Osteoporosis is also linked to poor outcomes of spine surgery (Fig. 3). Bjerke reviewed 140 patients who underwent spine fusion and assessed bone mineral density using the WHO criteria (Bjerke et al. 2017). In the osteoporotic patients, nonunion occurred in 46% of patients compared to 19% and 18% in normal and osteopenic patients, respectively. Complications were also significantly worse in the osteoporotic group; 50% of patients had complications compared to 33% and 22% in osteopenic and normal, respectively. Oh evaluated cage subsidence after interbody fusion in 120 osteoporotic patients finding that the severity of osteoporosis strongly correlated with the severity of cage subsidence (Oh et al. 2017). Meredith linked prediction of proximal junctional fracture after spine fusion with baseline HU. HU of <135 strongly correlated to risk of fracture in two-thirds of patients (Meredith et al. 2013).

Mitigation of Poor Bone Health on Spinal Outcomes

Multiple randomized controlled trials have demonstrated improvement in clinical results, time to fusion, fusion success, and reduction of hardware-related complications in patients with osteoporosis who were treated with bisphosphonates (Stone et al. 2017). Similar results were seen with the use of anabolic agents which also show a positive effect on bone healing and reduction of bone-related complications, although no improvement in clinical outcomes was demonstrated (Stone et al. 2017; Ebata et al. 2017). Inoue randomized osteoporotic patients to teriparatide or control and assessed insertional torque of pedicle screws (Inoue et al. 2014). He found that even with only 2 months of treatment, there was a statistical improvement of insertional torque. This is consistent with investigations that show that even at 3 months, there are decreased rates of secondary fractures with treatment with anabolic agents (Kanis et al. 2004).

Preoperative Bone Health Program

No comprehensive preoperative bone health program has established efficacy. *Own the Bone* has developed an example program, although efficacy also has not been established. It is recommended that patients older than the age of 50 years having thoracolumbar spine surgery be assessed for risk of osteoporosis as part of the preoperative workup when surgery is being scheduled. A checklist is reviewed based on current guidelines, and, if positive, the patient is scheduled for bone densitometry (DXA) (Table 5) (Camacho et al. 2016; Cosman et al. 2014). The risk factors include women older than 65 years, men greater than 70, history of inflammatory arthritis, glucocorticoid use (>5 mg of prednisone daily), diabetes mellitus, history of fracture after age of 50, and a FRAX score greater than 9.3% of major osteoporotic fracture. The FRAX is calculated without bone mineral density and is used to identify patients who need DXA. In addition, all patients are prescribed 2,000–5,000 units of vitamin D daily and 1,200 mg of calcium.

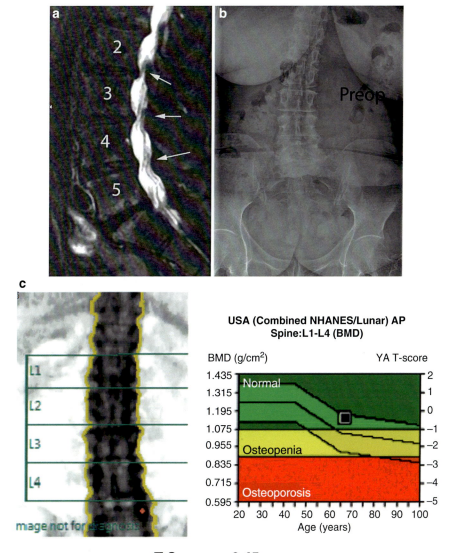

Fig. 3 (continued)

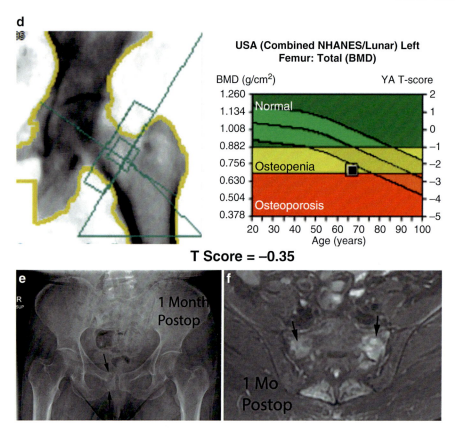

Fig. 3 (**a**) A 69-year-old female presents with neurogenic claudication from L2 to L5. Spinal stenosis is seen on T2-weighted MRI (arrows). (**b**) Her bone quality appears poor on anteroposterior radiographs. She was diagnosed 6 months earlier with osteoporosis but was untreated. (**c**) The spine DXA shows normal bone mineral density but should be ignored given her degenerative spinal disease. (**d**) The proximal hip DXA shows osteoporosis with a T-score −3.5. She was not treated for osteoporosis. (**e**) One month postoperative she presents with severe groin pain and left radicular pain and difficulty walking. Pelvic radiograph shows fracture of pubic rami (arrows). (**f**) Pelvis T2-weighted MRI shows bilateral sacral insufficiency fractures (arrows). She was subsequently started on vitamin D3, calcium replacement, and teriparatide daily injection

Table 5 Preoperative bone health assessment checklist to determine need for further evaluation and treatment

1	Females >65 years
2	Males >70 years
3	Inflammatory arthritis
4	Diabetes mellitus
5	Glucocorticoid exposure
6	Fracture after age 50
7	FRAX >9.3% (without DXA)
	Any positive criteria is evaluated by DXA

The DXA should be reviewed by the surgeon and a FRAX (with DXA) obtained. If the FRAX is greater than 3% for 10-year risk of hip or 20% risk of major osteoporotic fracture, then the patient is referred to a bone health specialist. This may be an established Fracture Liaison Service, bone health specialist, or primary care. Based on a high fracture risk, the patient should consider preoperative treatment and delay of surgery. In this case, in the author's opinion, an anabolic agent is used if insurance coverage can be obtained and the patient consents.

The surgical delay will depend upon many factors and is not evidence-based at this time. For patients with a low-risk requirement for bone healing such as laminectomy, surgical delay would only consist of 3 months. For patients who require bony fusion or who have more severe osteoporosis, a delay between 3 and 6 months should be considered. For patients who require multilevel fusion, spinal osteotomy, or coronal sagittal plane deformities, then longer treatment of 9–12 months should be considered. After surgery the medication should be restarted.

Conclusion

Poor bone health is common and increasing in prevalence. There is a need for spine surgeons to become involved in the diagnosis and assuring that patients receive adequate treatment. Any patient over the age of 50 with a fracture requires critical evaluation for osteoporosis and probably medical treatment. In addition, surgical outcomes are strongly linked to poor bone health. These can be mitigated through the use comprehensive programs such as a Fracture Liaison Service. It is critical that spine surgeons first recognize that osteoporosis is a problem and then encourage patients to delay treatment if necessary so that surgical outcomes can be optimized and complications avoided.

References

American Orthopaedic Association (2005) Leadership in orthopaedics: taking a stand to own the bone. American Orthopaedic Association position paper. J Bone Joint Surg Am 87(6):1389–1391

Amin S et al (2014) Trends in fracture incidence: a population-based study over 20 years. J Bone Miner Res 29(3):581–589

Anderson PA, Froyshteter AB, Tontz WL Jr (2013) Meta-analysis of vertebral augmentation compared with conservative treatment for osteoporotic spinal fractures. J Bone Miner Res 28(2):372–382

Balasubramanian A et al (2014) Declining rates of osteoporosis management following fragility fractures in the U.S., 2000 through 2009. J Bone Joint Surg Am 96(7):e52

Ballard DH et al (2018) Clinical applications of 3D printing: primer for radiologists. Acad Radiol 25(1):52–65

Bawa HS, Weick J, Dirschl DR (2015) Anti-osteoporotic therapy after fragility fracture lowers rate of subsequent fracture: analysis of a large population sample. J Bone Joint Surg Am 97(19):1555–1562

Binkley N et al (2017a) Osteoporosis in crisis: it's time to focus on fracture. J Bone Miner Res 32(7):1391–1394

Binkley N et al (2017b) Surgery alters parameters of vitamin D status and other laboratory results. Osteoporos Int 28(3):1013–1020

Bjerke BT, Zarrabian M, Aleem IS, Fogelson JL, Currier BL, Freedman BA, Bydon M, Nassr A (2017) Incidence of osteoporosis-related complications following posterior lumbar fusion. Glob Spine J 7(1):1–7

Bunta AD et al (2016) Own the bone, a system-based intervention, improves osteoporosis care after fragility fractures. J Bone Joint Surg Am 98(24):e109

Burge R et al (2007) Incidence and economic burden of osteoporosis-related fractures in the United States, 2005–2025. J Bone Miner Res 22(3):465–475

Camacho PM et al (2016) American Association of Clinical Endocrinologists and American College of Endocrinology Clinical Practice Guidelines for the diagnosis and treatment of postmenopausal osteoporosis – 2016. Endocr Pract 22(Suppl 4):1–42

Carberry GA et al (2013) Unreported vertebral body compression fractures at abdominal multidetector CT. Radiology 268(1):120–126

Center JR et al (2007) Risk of subsequent fracture after low-trauma fracture in men and women. JAMA 297(4):387–394

Chen AT, Cohen DB, Skolasky RL (2013) Impact of non-operative treatment, vertebroplasty, and kyphoplasty on survival and morbidity after vertebral compression fracture in the medicare population. J Bone Joint Surg Am 95(19):1729–1736

Cosman F et al (2014) Clinician's guide to prevention and treatment of osteoporosis. Osteoporos Int 25(10):2359–2381

Curtis JR et al (2010) Is withholding osteoporosis medication after fracture sometimes rational? A comparison of the risk for second fracture versus death. J Am Med Dir Assoc 11(8):584–591

Dirschl DR, Rustom H (2018) Practice patterns and performance in U.S. fracture liaison programs: an analysis of >32,000 patients from the own the bone program. J Bone Joint Surg Am 100(8):680–685

Ebata S et al (2017) Role of weekly teriparatide administration in osseous union enhancement within six months after posterior or transforaminal lumbar interbody fusion for osteoporosis-associated lumbar degenerative disorders: a multicenter, prospective randomized study. J Bone Joint Surg Am 99(5):365–372

Flais J et al (2017) Low prevalence of osteoporosis treatment in patients with recurrent major osteoporotic fracture. Arch Osteoporos 12(1):24

Fuerst T et al (2009) Evaluation of vertebral fracture assessment by dual X-ray absorptiometry in a multicenter setting. Osteoporos Int 20(7):1199–1205

Fujii T et al (2013) Dichotomizing sensory nerve fibers innervating both the lumbar vertebral body and the area surrounding the iliac crest: a possible mechanism of referred lateral back pain from lumbar vertebral body. Spine (Phila Pa 1976) 38(25):E1571–E1574

Ganda K et al (2013) Models of care for the secondary prevention of osteoporotic fractures: a systematic review and meta-analysis. Osteoporos Int 24(2):393–406

Harvey NC et al (2015) Trabecular bone score (TBS) as a new complementary approach for osteoporosis evaluation in clinical practice. Bone 78:216–224

Hawley S et al (2016) Anti-osteoporosis medication prescriptions and incidence of subsequent fracture among primary hip fracture patients in England and Wales: an interrupted time-series analysis. J Bone Miner Res 31(11):2008–2015

Hodsman AB et al (2008) 10-year probability of recurrent fractures following wrist and other osteoporotic fractures in a large clinical cohort: an analysis from the Manitoba Bone Density Program. Arch Intern Med 168(20):2261–2267

Inoue G et al (2014) Teriparatide increases the insertional torque of pedicle screws during fusion surgery in patients with postmenopausal osteoporosis. J Neurosurg Spine 21(3):425–431

Jager PL et al (2011) Combined vertebral fracture assessment and bone mineral density measurement: a new standard in the diagnosis of osteoporosis in academic populations. Osteoporos Int 22(4):1059–1068

Kanis JA et al (2004) A meta-analysis of previous fracture and subsequent fracture risk. Bone 35(2):375–382

Kim TH et al (2012) Changes in vitamin D status after surgery in female patients with lumbar spinal stenosis and its clinical significance. Spine (Phila Pa 1976) 37(21):E1326–E1330

Kim TH et al (2013) Prevalence of vitamin D deficiency in patients with lumbar spinal stenosis and its relationship with pain. Pain Physician 16(2):165–176

Lau E et al (2008) Mortality following the diagnosis of a vertebral compression fracture in the Medicare population. J Bone Joint Surg Am 90(7):1479–1486

Leslie WD et al (2012) A population-based analysis of the post-fracture care gap 1996–2008: the situation is not improving. Osteoporos Int 23(5):1623–1629

Lih A et al (2011) Targeted intervention reduces refracture rates in patients with incident non-vertebral osteoporotic fractures: a 4-year prospective controlled study. Osteoporos Int 22(3):849–858

Lindsay R et al (2001) Risk of new vertebral fracture in the year following a fracture. JAMA 285(3):320–323

Lubelski D et al (2015) Perioperative medical management of spine surgery patients with osteoporosis. Neurosurgery 77(Suppl 4):S92–S97

Meredith DS et al (2013) Lower preoperative Hounsfield unit measurements are associated with adjacent segment fracture after spinal fusion. Spine (Phila Pa 1976) 38(5):415–418

Muszkat P et al (2015) Digital vertebral morphometry performed by DXA: a valuable opportunity for identifying fractures during bone mass assessment. Arch Endocrinol Metab 59(2):98–104

Nakayama A et al (2016) Evidence of effectiveness of a fracture liaison service to reduce the re-fracture rate. Osteoporos Int 27(3):873–879

Oh KW et al (2017) The correlation between cage subsidence, bone mineral density, and clinical results in posterior lumbar interbody fusion. Clin Spine Surg 30(6):E683–e689

Pickhardt PJ et al (2013) Opportunistic screening for osteoporosis using abdominal computed tomography scans obtained for other indications. Ann Intern Med 158(8):588–595

Ravindra VM et al (2015a) Vitamin D levels and 1-year fusion outcomes in elective spine surgery: a prospective observational study. Spine (Phila Pa 1976) 40(19):1536–1541

Ravindra VM et al (2015b) Prevalence of Vitamin D deficiency in patients undergoing elective spine surgery: a cross-sectional analysis. World Neurosurg 83(6):1114–1119

Schousboe JT et al (2016) Prediction of incident major osteoporotic and hip fractures by trabecular bone score (TBS) and prevalent radiographic vertebral fracture in older men. J Bone Miner Res 31(3):690–697

Schreiber JJ, Anderson PA, Hsu WK (2014) Use of computed tomography for assessing bone mineral density. Neurosurg Focus 37(1):E4

Shane E et al (2014) Atypical subtrochanteric and diaphyseal femoral fractures: second report of a task force of the American Society for Bone and Mineral Research. J Bone Miner Res 29(1):1–23

Stoker GE et al (2013) Preoperative vitamin D status of adults undergoing surgical spinal fusion. Spine (Phila Pa 1976) 38(6):507–515

Stone MA et al (2017) Bisphosphonate's and intermittent parathyroid hormone's effect on human spinal fusion: a systematic review of the literature. Asian Spine J 11(3):484–493

Svensson HK et al (2016) A painful, never ending story: older women's experiences of living with an osteoporotic vertebral compression fracture. Osteoporos Int 27(5):1729–1736

Tajeu GS et al (2014) Death, debility, and destitution following hip fracture. J Gerontol A Biol Sci Med Sci 69(3):346–353

Tarantino U et al (2013) Osteoporosis and sarcopenia: the connections. Aging Clin Exp Res 25(Suppl 1):S93–S95

Tosteson AN et al (2001) Impact of hip and vertebral fractures on quality-adjusted life years. Osteoporos Int 12(12):1042–1049

Wright NC et al (2014) The recent prevalence of osteoporosis and low bone mass in the United States based on bone mineral density at the femoral neck or lumbar spine. J Bone Miner Res 29(11):2520–2526

Yagi M, King AB, Boachie-Adjei O (2011) Characterization of osteopenia/osteoporosis in adult scoliosis: does bone density affect surgical outcome? Spine (Phila Pa 1976) 36(20):1652–1657

Part IV
Technology: Fusion

Pedicle Screw Fixation

Nickul S. Jain and Raymond J. Hah

Contents

Introduction and History	542
Anatomy of the Pedicle	543
Design and Anatomy of the Pedicle Screw	543
Biomechanics of Pedicle Screw Fixation	546
Indications for Use	548
Insertion Techniques	548
General	548
Freehand Technique	548
Fluoroscopic-Guided Technique	552
Percutaneous Screw Placement	553
Computer-Assisted Surgery and Navigation Technique	553
Intraoperative Neuromonitoring (IONM)	554
Pedicle Screw Outcomes	555
Complications	555
Augmentation	556
Conclusions	557
References	557

Abstract

Pedicle screws and rods are a modern posterior spinal instrumentation system that has gained widespread adoption throughout the world as the gold standard for instrumentation of the spine over the last two decades. They provide significant advantages in that they provide rigid 3-column fixation of the spine from an entirely posterior approach without reliance on intact dorsal elements. However, there is a steep learning curve for their placement, and adequate training is required prior to their routine use. They are not without their own set of unique complications. Many modifications to pedicle screws exist to improve clinical

N. S. Jain · R. J. Hah (✉)
Department of Orthopaedic Surgery, University of Southern California, Los Angeles, CA, USA
e-mail: njain@scoi.com; Ray.Hah@med.usc.edu

© Springer Nature Switzerland AG 2021
B. C. Cheng (ed.), *Handbook of Spine Technology*,
https://doi.org/10.1007/978-3-319-44424-6_57

outcomes including augmentation with cement, and a variety of novel technologies can be used to help improve accuracy in their placement including fluoroscopy, computer navigation, and robotics.

Keywords

Pedicle screw · Pedicle instrumentation · Transpedicular fixation · Navigation · Screw · Dorsal instrumentation

Introduction and History

Posterior spinal instrumentation has been used for decades to allow surgeons to correct spinal deformity, stabilize fractures and instability, and promote arthrodesis. They have provided surgeons many advantages including a more stable, low-strain environment for fusion procedures and more immediate stability in unstable conditions requiring fixation (Vanichkachorn et al. 1999). This allows for early patient mobility, often times eliminating the need of external orthoses. Many indications for posterior spinal instrumentation have been described including unstable thoracolumbar fractures, metastatic tumor resulting in spine instability, spondylolisthesis, scoliosis, and pseudarthrosis (Vaccaro and Garfin 1995a).

Scoliosis was previously often treated with posterior spine fusion without instrumentation. Complications included a reported 30–40% pseudarthrosis rate with progressive loss of scoliotic correction. Harrington first described a hook-rod posterior spinal instrumentation system in 1962 which allowed for distraction and compression of the spine and marked reduction in pseudarthrosis rates (1–15%) (Harrington 1962). The system provided excellent coronal plane correction but had no rotational stability or sagittal alignment control. This predisposed patients to develop a hypolordotic "flat back" but was protective against progressive kyphosis and neurological decline. Disadvantages included loss of fixation with hook disengagement in up to 20% of cases and an inability to perform short-segment fixation (Harrington 1988).

Luque in 1980 then described the first dorsal instrumentation that allowed for segmental fixation and short constructs using sublaminar wires attached to rods. The authors demonstrated decreased pseudarthrosis rates and stable fixation; however, the system did not have the ability to resist axial load. Other complications included durotomies, neurologic injury, and wire failure (Luque 1980). Cotrel and Dubousset modified this technique to use laminae or pedicle hooks to achieve segmental fixation; however, this required intact dorsal elements including the lamina and facet joints (Cotrel et al. 1988).

Pedicle fixation allows for segmental fixation of the spine while providing the ability to control axial displacement and functions independent of the presence or absence of the dorsal elements of the spine. Additionally, they are the only posterior spinal instrumentation that allows for entire 3-column fixation of the spine which provides significant biomechanical advantage. The first posterior-based screws were described in the 1940s by King as short transfacet screws with high pseudarthrosis rates (King 1944, 1948). Boucher then described a longer screw that crossed the facet joint in 1958 (Boucher 1959).

Roy-Camille first applied screws through the entirety of the pedicle attached to plates for thoracolumbar fractures, instability after tumor resection, and lumbosacral fusion (Roy-Camille 1970). Multiple newer and improved iterations were then developed in the following decades including the AO internal fixator, the variable spinal plating (VSP) system, the Cotrel-Dubousset Universal Spinal Instrumentation (USI), the Texas Scottish Rite (TSRH), and Isola systems all providing various advantages including variable angles to ease screw-rod connection. Newer, modern designs have increased adaptability with polyaxial heads, variable diameter rods, side-to-side connectors, and modern materials including titanium and cobalt-chrome alloys.

Pedicle screws are a versatile and powerful tool for posterior spinal instrumentation. They can resist load in all planes given their 3-column fixation nature and provide a powerful fulcrum for correction of rotational, sagittal, and coronal plane deformities. Pedicle screws also allow for the surgeon to apply significant forces to the spine (including distraction, compression, and translation). They have a proven benefit in enhancing fusion rates and avoid the complications of entering the spinal canal of some of the predecessor posterior spine instrumentation systems (Lorenz et al. 1991). Additionally, they allow for earlier rehabilitation and obviate the need for postoperative external orthoses.

However, they are not without disadvantages. Pedicle screw insertion has a steep learning curve, and malpositioned screws can result in durotomies or neural injury if there is pedicle wall penetration. Their use increases operative time and cost. Additionally, they often require increased radiation exposure for both patient and surgeon, and they often obscure postoperative imaging. Additionally, the rigidity of fixation and placement of screws that violate adjacent segment facet joints may result in accelerated rates of adjacent segment degeneration. Despite these shortcomings, pedicle screws are still widely considered the gold standard for posterior spinal instrumentation today.

Anatomy of the Pedicle

The pedicle is the strongest part of the vertebra and has often been described as the "force nucleus" of the spine. The posterior elements of the vertebra converge and are linked to the anterior vertebral body and the anterior two columns of the spine by the cylindrical pedicle (Steffee et al. 1986). The pedicle is comprised of a strong shell of cortical bone with a cancellous bone core. Typically, the transverse width of the pedicle is less than the sagittal pedicle height with the exception of the low lumbar spine (Figs. 1 and 2).

Clinically, it is critical to understand the pedicle anatomy for accurate placement of screws within the pedicle. The coronal and sagittal angulation and the transverse diameter vary from level to level within the entire spinal axis. In the sagittal plane, cephalad and caudal angulation of the pedicle starts at neutral in the thoracic spine at T1 and increases to approximately 10° of cephalad angulation at T8 before decreasing back to 0° by T12 (McCormack et al. 1995). In the axial plane, beginning at T1, medial angulation decreases as one travels through the thoracic spine. In the lumbar spine, medial angulation in the axial plane increases from neutral at L1 to approximately 25–30° of medial angulation at L5. The width of the pedicle increases from L1 to S1 (Krag 1991), while the midthoracic pedicles (T4–T8) are typically considered the most narrow.

The inner diameter of the pedicle has been shown to account for 60% of the screw pullout strength and 80% of the longitudinal stiffness (Hirano et al. 1997). It has been correlated to the height of the patient. Typical screw sizes have been proposed as 4.5 mm diameter and 25–30 mm in length for T1–T3 and 4.5–5.5 mm in diameter and 30–35 mm in length from T4 to T10 (Louis 1996). Pedicles do have some plasticity and ability to undergo expansion however.

Many structures exist in close contact and surround the pedicle. Intrathecal nerve roots course along the medial aspect of the pedicle as the traversing root and have been shown to be 0.2–0.3 mm from the pedicle at T12 and touching the dura below L1. Exiting nerve roots then course beneath the pedicle and enter the neural foramen, occupying the ventral and rostral one third of the foramen (Benzel 1995a). Clinically, this is relevant as violation of the pedicle medially or caudally can injure the nerve root.

Design and Anatomy of the Pedicle Screw

The pedicle screw consists of a head, neck, body, and threads, each serving a distinct purpose (Fig. 3). The head of the pedicle screw facilitates attachment of the screw to longitudinal rods to provide fixation to adjacent segments or levels. Modern screws can have either monoaxial or polyaxial translating heads. Monoaxial screws have significant biomechanical advantages and reduce head-neck junction failure commonly seen in polyaxial screws; however some cadaveric testing has shown no differences between the two

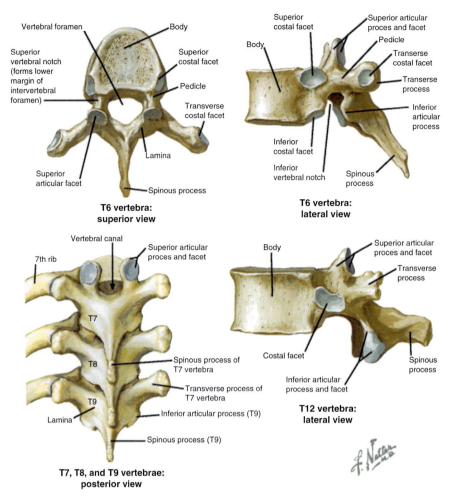

Fig. 1 Anatomy of the thoracic pedicle. (Reproduced from *Netter's Concise Orthopaedic Anatomy*, 2010 with permission from Elsevier)

in regard to construct stiffness (Fogel et al. 2003; Shepard et al. 2002). However, in exchange for this vulnerability to fatigue failure, polyaxial screws provide surgeons significant increased versatility and facilitate ease of rod to screw fixation and rod contouring across multiple levels. This helps limit implant-bone contact stress which can be increased when there is screw-plate or screw-rod mismatch. Additionally, the head-neck junction in polyaxial screws may be protective against pedicle screw breakage within a pedicle (Fogel et al. 2003).

The neck of the screw bridges the head to the body and is typically considered the weakest part of the screw. The body of a screw contains threads to obtain bony purchase. The bending or fatigue strength of a screw is proportional to the core (or inner) diameter of the screw body (Benzel et al. 1995). Liu and coauthors found fatigue strength of a screw increased 104% following a 27% increase in diameter (Liu et al. 1990).

The body of a screw can be conical or cylindrical (Fig. 4). Conical screws have been shown by some authors to have superior insertional torque with no difference in pullout strength (Kwok et al. 1996). However, other authors have advocated that conical screws, when backed out half to one full turn, lose significant purchase (Lill

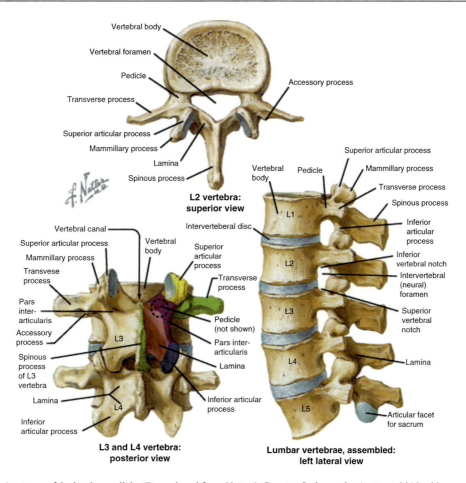

Fig. 2 Anatomy of the lumbar pedicle. (Reproduced from *Netter's Concise Orthopaedic Anatomy*, 2010 with permission from Elsevier)

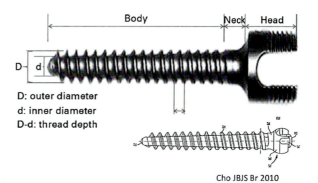

Fig. 3 Anatomy of the pedicle screw. (Reproduced from Cho et al. 2010 with permission from *J Bone Joint Surgery British*)

et al. 2006). The conical geometry of a screw may also be beneficial as 60% of the screw pullout strength is obtained from the cortical bone of the pedicle as opposed to the trabecular bone of the vertebral body (Shea et al. 2014). There has been significant debate between the two screw designs and their effectiveness with conflicting studies showing either no difference or biomechanical advantages of conical screws over cylindrical screws.

Fig. 4 Cylindrical versus conical screw design. (Reproduced from Shea et al. 2014 with permission from *Biomed Rest Int*)

The body of a screw can also be hollow to allow for screw passage over a wire in a cannulated fashion. This has been shown to be safe and effective but does decrease the bending strength of screws significantly when compared to solid bore-bodied screws. Threads are the portion of the body of the screw that allows for bony purchase. The difference in the inner and outer diameter of a screw is equal to the thread depth. The pitch is the distance between threads longitudinally across the body. Threads can be fully threaded along the entirety of the body of the screw or partially threaded across only a part and are typically cancellous type thread pattern given their fixation within the cancellous bone of the pedicle. However, some newer screw designs incorporate a dual-thread design with cortical threads dorsally along the screw to obtain cortical fixation within the pedicle and cancellous threads within the anterior column (vertebral body).

The pullout strength of screws is determined by the amount and quality of bone between the threads of a screw. Smaller thread pitches confer slightly stronger pullout strength as do deeper thread depths and more total threads (fully threaded). A general rule of thumb is that large outer diameters, small inner diameters, short pitch, and strong bone maximize pullout strength of the screw. These factors in combination with bone mineral density (BMD) help determine insertional torque of a screw which has been demonstrated to have a linear correlation with cycles to screw loosening (Zdeblick et al. 1993).

Modern pedicle screw systems typically have polyaxial heads, and diameters range from 4.5 to 8.5 mm for the thoracic and lumbar spines and lengths between 25 and 60 mm increments. They are typically made up of either stainless steel, titanium alloys, or cobalt-chrome-molybdenum alloys. Stainless steel (a nickel-chromium-iron alloy) was originally used due to its biocompatibility, low cost, and high stiffness in bending strength. However, modern screws have moved away from stainless steel as a material given their MRI incompatibility for postoperative imaging, higher corrosion rates, and the prevalence of nickel allergies. Titanium-aluminum-vanadium alloys (TiAlVa or Ti6-4) have been commonly used in bone implants given their lower modulus of elasticity than stainless steel that more closely approximates the modulus of bone. This has been hypothesized to decrease stress shielding of bone. Additionally, Ti alloys have high yield strength, are biocompatible, promote osteointegration, and are MRI safe. Cobalt-chromium alloys (CoCr) have also been more recently popularized given their superior stiffness and fatigue strength when compared to Ti alloys; however, they are often times significantly more expensive. Both titanium alloys and cobalt-chrome implants have low risk of corrosion when compared to stainless steel.

Various coatings have been added to screws in attempts to improve fixation. Hydroxyapatite coatings allow for bone ingrowth and provide thicker threads with increased initial friction and stability and have been shown to be useful in osteoporotic animal models (Sandén et al. 2001).

Biomechanics of Pedicle Screw Fixation

Spinal instrumentation functions to stabilize the spine, and its construct strength is determined by the mechanical load at which implants fail. Stiffness of a given spine construct is defined as the ability of fixation to resist axial compression as well as linear and circular moment forces.

These biomechanical characteristics of implants help define clinical success as implant failure typically leads to poor clinical outcomes.

Pedicle screws have been compared biomechanically to other dorsal spinal instrumentations. When compared to Harrington rods and Luque sublaminar wiring constructs, pedicle screw constructs have been shown to have greater torsional rigidity, overall construct stiffness and strength, and a significant reduction in the strain of flexion loading (Chang et al. 1989; Puno et al. 1987). They also have been shown to be superior in flexion-extension and lateral bending strength when compared to facet screw fixation (Panjabi et al. 1991a).

Dorsal pedicle screw systems allow for the surgeon to impart cantilever bending forces to the spine around a fixed moment arm which can provide distraction, compression, as well as tension band fixation of the spine. Since they extend past the instantaneous axis of rotation of the spine, they do allow for three-dimensional control of the spine. These constructs do become load bearing as well with adequate anterior column support for load sharing. Without additional anterior column support (i.e., corpectomy model), they can be vulnerable to construct failure (Yoganandan et al. 1990) (Fig. 5). Ensuring that maximal pedicle screw biomechanical advantage is achieved is critical to help avoid catastrophic implant failure or pullout.

Pilot holes in the dorsal pedicular cortex are used to begin cannulation of a pedicle and allow safe screw passage. Pilot hole size has been described to contribute to the insertional torque of a screw, critical for establishing both maximum pullout strength and preventing pedicle fracture. Battula et al. established the critical pilot hole size as 71.5% of the outer diameter of the pedicle screw was ideal in osteoporotic bone to optimize the balance between low insertional torque and high pullout strength (Battula et al. 2008).

Pedicle screws should be placed in a convergent fashion with medial angulation (Cho et al. 2010). This allows for a more lateral starting point resulting in longer screw lengths and reduced contact with the superior facet joint of the vertebra. Additionally, the convergence allows for an interlocking effect that increases resistance to torsional and lateral bending and up to 28.6% increase in pullout strength when compared to a straight-ahead technique (Barber et al. 1997).

They should also be placed parallel to the superior end plate to minimize screw breakage as the "straight-forward" technique paralleling the superior end plate has been shown to be biomechanically superior to the anatomic screw trajectory in the thoracic spine (Lehman et al. 2003; Youssef et al. 1999). An anatomic screw trajectory can be used as a salvage technique especially within the thoracic spine, when multiple screw attempts have been attempted and failed,

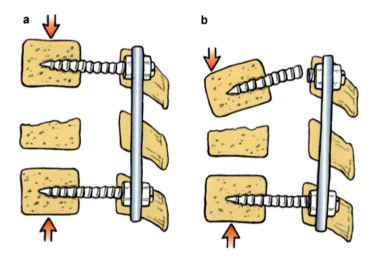

Fig. 5 Screw failure without anterior column support. (Reproduced from *Benzel's Spine Surgery*, 2017 with permission from Elsevier)

given its more cephalad starting point (Lehman et al. 2003).

Ideal screw length has been determined to be at least 60–70% across the body and total length of the pedicle. Screws placed to only 50% of anterior-posterior length of the pedicle had 30% less pullout strength than screws that spanned 80% of the width (Krag et al. 1988). Minimum engagement of at least the neurocentral junction is critical as it has been demonstrated to provide 75% of the maximum insertional torque of a screw (Lehman et al. 2003). Lateral fluoroscopy and a measured ball-tip probe can be used intraoperatively to aid in determining screw length, and care should be taken not to place screws longer than 80% of the length of the pedicle on imaging as this can penetrate the anterior cortex 10–30% of the time (Whitecloud et al. 1989a). While bicortical fixation spanning the entirety of the pedicle has been shown to improve pullout strength up to 25%, the dangers of anterior perforation to critical vascular structures are too great to advocate routine bicortical screw fixation. One exception is at the S1 level where anterior midline penetration and bicortical purchase are safe due to the capacious pedicle and absence of midline vascular structures at this level (Lonstein et al. 1999).

Ideal screw diameter should be such that the screw threads obtain purchase at the inner cortical portion of the pedicle which serves to decrease hoop stresses and cortical deformation. A screw diameter that is too large can result in risk of perforation or pedicle fracture, especially in weak or osteoporotic bone.

Indications for Use

Pedicle screws as dorsal spine instrumentation have many uses including fracture stabilization to allow early mobilization, even in the setting of posterior element injury, tumor instability, infection, spondylolisthesis, fusion assistance in degenerative conditions, and scoliotic deformity correction of the spine. Overall, they serve to provide rigid internal immobilization that allows mechanical support, early mobilization, and rehabilitation.

Contraindications to pedicle screw fixation include small pedicles, severe osteoporosis, and absence of adequate anterior column support (Orndorff and Zdeblick 2017).

Insertion Techniques

General

Placement of pedicle screws requires a thorough understanding of the anatomy of the pedicle for safe passage of a screw. In general, screws should not penetrate the pedicle and be placed away from critical neural and vascular structures as well as facet joints. An exception to this rule is the case of the "in-out-in" screw, typically reserved for severe deformity or congenital small pedicles. This method utilizes a far lateral entry point and is an extrapedicular tract through the transverse process into the pedicle (Perna et al. 2016). The dorsal cortex of the pedicle should be kept intact as much as possible to allow for maximal insertional torque and pullout strength (Daftari et al. 1994).

Freehand Technique

The pedicle screw entry point is identified by the surgeon using anatomic landmarks (described below) and careful review of preoperative imaging studies. Once a pilot hole in the dorsal cortex of the pedicle is made at the ideal starting point, typically a blunt-tipped gearshift probe can be used to cannulate the cancellous bone of the pedicle and allow for creation of a safe screw track within the cortical pedicle walls. Tactile feedback and experience are used in the freehand technique to establish this safe corridor. Typically, the pedicle probe is directed laterally for the first 15–20 mm of the pedicle before being removed and flipped 180° and then directed medially into the vertebral body once past the neurocentral junction. A sudden loss of resistance is often indicative of a cortical breach. The passing of the probe allows compaction of the cancellous bone during cannulation of the pedicle. Alternatively, a drill can be used to cannulate the pedicle without significant difference in biomechanical properties of final screw placement (George et al. 1991). A ball-tip feeler or another pedicle sounder can be used to palpate the anterior, superior, inferior, medial, and lateral margins of the pedicle to verify pedicle cortical integrity and provide a depth measurement. This however has variable accuracy even among expert surgeons.

A tap can be used to create screw threads within the pedicle prior to screw placement; however it is not required as most modern screw systems are self-tapping. Self-tapping screws do have the disadvantage of increased insertional torque and pedicle fracture risk. Tapping has demonstrated improved screw trajectory but variable effects on screw pullout strength (Erkan et al. 2010; Pfeiffer et al. 2006). Line-to-line tapping (using a tap the same size as the screw) is not recommended as it reduces screw purchase and pullout strength. However, using a tap 1 mm smaller in diameter has been shown to have the same pullout strength as untapped pilot holes (Carmouche et al. 2005; Chatzistergos et al. 2010). Tapping is typically performed just within the cortical bone of the pedicle cylinder and not extended into the cancellous bone of vertebral body as tapping cancellous bone reduces screw-bone contact and pullout strength (Chapman et al. 1996). The pedicle is then gently probed after tapping again to confirm no cortical perforations. A screw is then placed.

Freehand pedicle screw placement has a steep learning curve and requires detailed understanding of an individual patient's anatomy as it is essentially a blind technique. Accuracy rates for freehand pedicle screw placement have been reported between 59% and 91% in the lumbar spine and 45% and 97% in the thoracic spine (Perna et al. 2016).

Cervical

Traditionally, posterior instrumentation of the subaxial cervical spine has been limited to lateral mass fixation, sublaminar or interspinous wiring, and translaminar fixation. While pedicle screw fixation has been commonly described at C2 and C7 with good safety and efficacy, pedicles at C3–C6 have often been considered too dangerous to attempt screw fixation due to the proximity of the vertebral artery and the cervical nerve roots as well as the significant variability in the cervical pedicle morphology between patients.

Panjabi et al. demonstrated anatomically the ability for the cervical spine to accommodate pedicle screws (Panjabi et al. 1991b). They quantified the C2 pedicle to be the largest, the C3 to be the smallest, and the increasing pedicle size up to C7. At C4, an approximate 45° medial angulation in the coronal plane is required for insertion, and it decreases sequentially to about 30° at C7. The sagittal angle (superior-inferior) is determined by review of the preoperative imaging of the individual patient.

Cervical pedicle screw fixation has been shown to have superior biomechanical properties in regard to loosening and fatigue testing compared to other dorsal cervical spine instrumentation. Indications for cervical pedicle screw fixation have been described as trauma-induced cervical fractures and/or dislocations, multilevel instability, tumor resection, osteoporosis, or absence of dorsal spine elements (Pelton et al. 2012).

Typical cervical pedicle screw size is 3.5–4.5 mm in diameter and requires careful study of preoperative imaging for length determination and to ensure a safe passageway. For C2, the pedicle starting point has been well established as 2 mm lateral to the bisection of a horizontal line through the mid-pars of C2 and a line vertically between the midpoints of the facets. The trajectory is typically 30–45° medial angulation and 35° superior angulation. Typically at C2 cannulation of the pedicle is done with a drill as opposed to a larger gearshift probe. Laminoforaminotomy can be added to allow for palpation of the medial border of the C2 pedicle to confirm the trajectory.

For C3–C7, there is more heterogeneity in the starting point, but many authors describe it as slightly lateral to the midpoint of the lateral mass and superior (closer to the cephalad inferior articular process) (Fig. 6). Laminoforaminotomy can be added to allow for palpation of the medial border of the pedicle to confirm the trajectory.

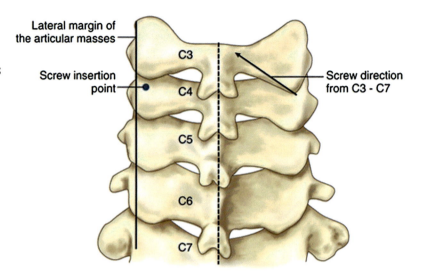

Fig. 6 Entry point for cervical pedicle screw placement. (Reproduced from Spine surgery. Operative techniques, 2008 with permission from Elsevier)

Cannulation of the pedicle can then be performed with a drill with set depth stops as the cervical pedicles are typically hard and hand-controlled instruments can slip or create too much downward pressure (Ludwig et al. 1999). At C7, some authors have described the pedicle entry point to be 1 mm inferior to the midportion of the facet joint above, with a 25–30° medial angulation and neutral sagittal plane (Ludwig et al. 1999).

Freehand technique is not usually recommended in the cervical spine, and image-guided assistance with fluoroscopy or computer-assisted stereotactic navigation is recommended as there is evidence to support improved safety (Ludwig et al. 2000).

Complications of cervical pedicle screw placement include misplacement, pedicle fracture, CSF leak, infection, nerve root injury, spinal cord injury, and vascular injury. Despite the serious consequences that can occur with cervical pedicle screw placement and previous anatomic studies suggesting vascular injury being the most likely complication of cervical pedicle screws, Kast et al. reported in their series of 26 patients with 94 total screws a 30% malposition rate with 9% being critical and 1 patient requiring revision surgery for nerve root symptoms. There were no vascular injuries. The authors described a significant learning curve for this technique that has also been reported by other authors (Kast et al. 2006; Yoshihara et al. 2013).

Thoracic
Much like the cervical spine, thoracic pedicle screws have a low margin of error due to the proximity of the spinal cord, lungs, esophagus, great vessels, and large intercostal and segmental vessels that are closely associated with the thoracic vertebrae (Vaccaro et al. 1995). Scoliosis increases the difficulty of accurate cannulation with altered trajectory from axial rotation and hypoplastic pedicles at the concavity of the curvature.

Progressing cephalad from T12, the starting points tend to be progressively more medial and cephalad up to T7, at which point they then shift to be more lateral and caudal (Parker et al. 2011; Xu et al. 1998; Chung et al. 2008). Typical medial angulation is 30° at T1–T2 and approximately 20° from T3 to T12. Sagittal angulation varies based on the level and patient, but a general rule is to cannulate the pedicle orthogonal to the dorsal spine.

Anatomic landmarks can also be used to identify the starting point and have been described as the midpoint of a triangle formed by the lower border of the superior articular facet, the medial border of the transverse process, and the pars interarticularis medially. Some authors have proposed a consistent starting point, as opposed to varying starting points, for the thoracic screws that are 3 mm caudal to the junction of the lateral aspect of the superior articular process and

Fig. 7 One method of thoracic pedicle screw entry point localization. (Reproduced from Avila and Baaj 2016 with permission from *Cureus*)

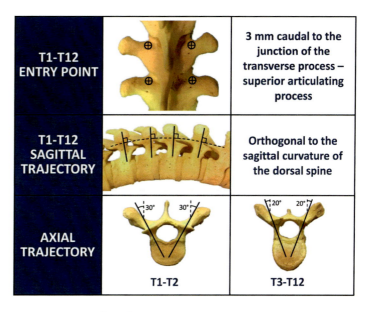

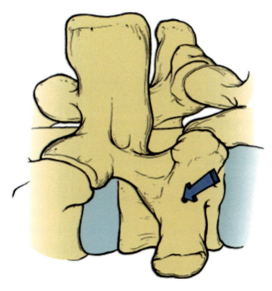

Fig. 8 Lumbar pedicle screw entry point. (Reproduced from *Benzel's Spine Surgery*, 2017 with permission from Elsevier)

transverse process (Avila and Baaj 2016) (Fig. 7). During decortication of the dorsal cortex, the surgeon can look for the pedicle blush of cancellous bleeding bone to ensure an accurate starting point.

The thoracic pedicles are most narrow between T4 and T9. Typical screw sizes are between 4.5 and 5.5 mm. Overall accuracy of freehand thoracic screws has been reported in the literature between 85% and 98% (Avila and Baaj 2016).

Lumbar

In the lumbar spine, the ideal pedicle screw starting point is the bony junction of the pars interarticularis, the transverse process, and the mammillary process or lateral facet joint. Alternatively, it can be described as the intersection of a vertical line bisecting the facet joint and a horizontal line through the midportion of the transverse process (Fig. 8). A laminoforaminotomy may also be used to palpate the medial wall of the pedicle from within the epidural space to allow guidance of the cannulation. Cannulation is performed as described above. Typical lumbar pedicle violations are lateral more commonly than medial or inferior.

In the lumbar spine, a novel pedicle screw tract known as the cortical screw has been described. It utilizes a more medial and caudal starting point and has a medial-lateral and caudal-to-cranial direction in order to increase screw-cortical bone contact to improve fixation in osteoporotic patients (Santoni et al. 2009). It does require some resection of the inferior spinous process and has been theorized to be weaker in axial rotation but does have advantages including potential increased fixation strength and less required muscle dissection (Rodriguez et al. 2014; Calvert et al. 2015). Screws are typically shorter in length and smaller in diameter but placed in a similar fashion as described above (Fig. 9).

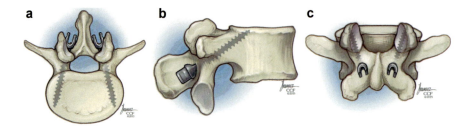

Fig. 9 Cortical screw trajectory for the lumbar spine. (Reproduced from *Benzel's Spine Surgery*, 2017 with permission from Elsevier)

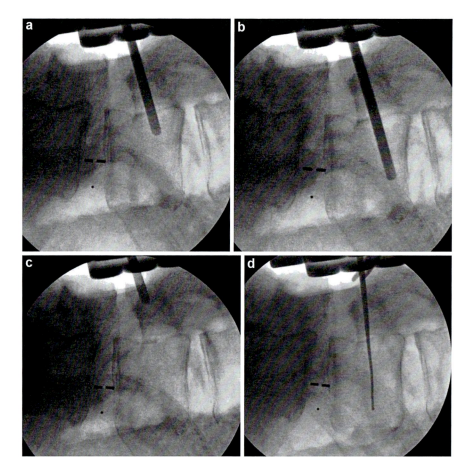

Fig. 10 Fluoroscopic-assisted screw placement. Cannulation was performed under lateral XR guidance followed by pedicle probing. Start points can be confirmed using AP and lateral fluoroscopic imaging

Fluoroscopic-Guided Technique

Intraoperative fluoroscopy can be used to aid pedicle screw placement as it provides 2D imaging of the entry point to the pedicle using radiographic markers as well as the trajectory of a pedicle to aid in cannulation. Using a combination of serial AP and lateral images with a parallel superior end plate, the pedicle cannula is started in the midpoint of the lateral most edge of the pedicle on the AP image and directed in the cranial-caudal direction of the pedicle on the lateral image (Fig. 10). While this can verify and increase accuracy rates of placement, it does not guarantee accurate trajectory.

Fluoroscopic-assisted pedicle screw placement accuracy rates have been reported to be similar to the freehand technique with one study reporting a 68.1% accuracy rate (Mason et al. 2014). 3D fluoroscopy software has more recently been implemented to allow consecutive images from different angles to create a 3D visualization to improve accuracy rates in fluoroscopic screw placement with the caveat of increased radiation exposure to the patient (Perna et al. 2016).

Percutaneous Screw Placement

Pedicle screws can also be placed via a Wiltse paraspinal approach percutaneously with the assistance of one or multiple of any of the abovementioned imaging modalities (fluoroscopy, intraoperative CT, computer-assisted navigation, or robotic-assisted systems). Purported advantages include reduced length of stay, earlier mobilization, decreased postoperative pain and blood loss, and earlier return to work. Principles for placement of pedicle screws percutaneously are no different than that of fluoroscopic or navigated screw placement and utilize imaging to guide the surgeon through the pedicle. Typically for fluoroscopic percutaneous screw placement, K-wires can be used after cannulation of the pedicle to maintain the pedicular track, and cannulated screws can be placed over these wires into the pedicle (Fig. 11).

Computer-Assisted Surgery and Navigation Technique

Computer stereotactic navigation techniques have recently been utilized to assist in pedicle screw

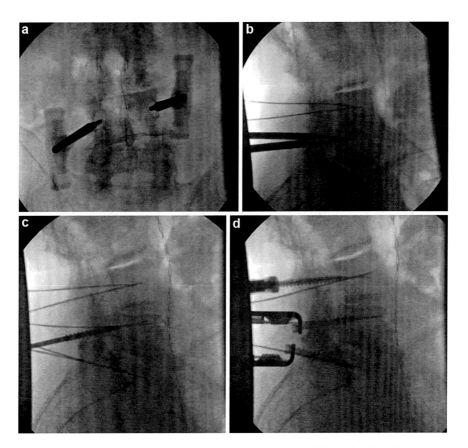

Fig. 11 Percutaneous screw placement: (**a**) Pedicle cannulation using a Jamshidi needle, (**b**) guide wire placement into cannulated pedicle, (**c**) cannulated tap using guidewire, and (**d**) cannulated screw placement over guidewires

placement by correlating a patient's preoperative or intraoperative acquired images to the patient's real-time surgical anatomy using fixed-point optical or electromagnetic markers. A computer model generation is then used to guide the surgeon in real time relative to the patient's anatomy (Fig. 12). Many authors have advocated for the safe and effective use of computer-assisted technology to make pedicle screw placement more reproducible by guiding the surgeon to the appropriate trajectory; however, effects on patient outcomes and benefit in reducing neurologic complications are unclear (Ughwanogho et al. 2012; Verma et al. 2010). The use of intraoperative cross-sectional imaging and referencing has gained popularity as it limits the inaccuracies that may develop due to patient repositioning when using computer-assisted navigation based on preoperative imaging. However, inaccuracy can still develop, and the further away one works from a reference frame, the less accurate the navigation system becomes (Scheufler et al. 2011). Disadvantages include increased radiation exposure to the patient, cost, and operative time. Overall accuracy of pedicle screw placement using navigated technology has been reported between 91.5% and 97.7%, which appear to be significantly higher than freehand or fluoroscopic placement rates, with the most benefits seen in the accuracy of thoracic pedicle screw placement (Puvanesarajah et al. 2014; Waschke et al. 2013). Additionally, repeat imaging using intraoperative CT scan can detect misplaced screws and allow the surgeon to correct them intraoperatively (reported at a rate of 1.8% in one series) (Van de Kelft et al. 2012).

Navigated optical technology has been expanded into robotic-assisted pedicle screw placement as well. Using preoperative or intraoperative imaging and appropriate patient fiducial markers, a robotic guidance arm can be used to guide pedicle cannulation trajectory and screw placement with increased reliability, reproducibility, and accuracy with potentially reduced radiation exposure. Disadvantages include significant cost, operative time, and learning curve.

Intraoperative Neuromonitoring (IONM)

Electrophysiological intraoperative testing can be useful to assess or confirm pedicle screw placement within a pedicle. Stimulation of pedicle screws or cannulation tools allows for electric currents to be transmitted into the pedicle. Cortical bone has a high resistance to electrical current resulting in minimal stimulation of nearby nerve roots if intact. Cortical breaches of the pedicle can allow for electric current to flow into soft tissues and allow for depolarization of nearby nerve roots which can be picked up on EMG recordings of specific myotomes in monitored extremities. Typically acceptable minimum thresholds of depolarization for safe screws are reported between 10 and 12 mA.

This technique of triggered EMG is useful in detecting misplaced pedicle screws as it has been shown to be highly specific; however, there is a high false-negative rate with only fair sensitivity with up to 22% of misplaced screws being

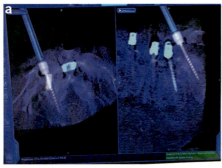 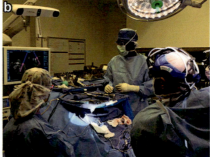

Fig. 12 Navigated screw placement and workflow

missed (Mikula et al. 2016). This technique, while primarily used for the lumbar spine given the lower extremity myotomes, has been described for monitoring thoracic nerve roots as well by selective myotome monitoring of the rectus abdominis for T6–T12 and the intercostal muscles from T3 to T6. This technique has been described for cervical screws as well as iliosacral screws.

While widely advocated for general use for safe placement of pedicle screws, there is a paucity of clinical data supporting improved clinical outcomes with routine IONM and EMG testing of screws (Reidy et al. 2001).

Pedicle Screw Outcomes

Pedicle screws first received US FDA approval as a class III device in 1995 but were frequently used prior to that throughout the world. Early transpedicular fixation screws were found by McAfee to have an approximate 80% survival rate at 10-year follow-up with 90% incidence of successful fusion in a mixed cohort of patients undergoing fusion with the early VSP device or the Cotrel-Dubousset transpedicular screw systems (McAfee et al. 1991). Yuan et al. established the safety of pedicle screw fixation in 1994 with a cohort of 303 surgeons with nearly 3,500 patients revealing very low rates (<1%) of implant failure, neurovascular injury, and dural tears in their cohort (Yuan et al. 1994). In 1998, the FDA downgraded pedicle screws to a class II device with increasing evidence of their safety.

Arthrodesis or fusion involves a surgeon-created artificial process of bone formation across a motion segment. It has a useful tool for spine surgeons to eliminate pathologic motion within the spine and provide stability to unstable segments. Fusion success is often directly proportional to construct stiffness and is dependent on a low-strain environment for primary or secondary bone healing and formation. Typically, a goal of <10% strain is desired in a construct. Wolff's law describes increased loads that result in increasing competitive strain. As bone adapts to load, bone formation occurs to add rigidity.

Pedicle screw constructs are ideal to provide a construct with adequate stiffness and provide a low-strain environment within the spine to allow for bone formation and fusion. Multiple studies have shown that pedicle fixation increases spinal arthrodesis rates. Louis in 1986 studied 266 patients in the lumbosacral spine who underwent instrumentation with pedicle screws and plates and found a 97% rate of successful fusion (Louis 1986). West et al. studied 62 patients undergoing spinal arthrodesis and found a 90% fusion rate and 2/3 of patients returned to full-time work (West et al. 1991). Zdeblick compared degenerative lumbar spine surgical patients with and without rigid pedicle screw instrumentation and found in short-term follow-up a significant difference in fusion rates (64% in uninstrumented patients and 95% in patients with pedicle screw and rigid rod instrumented fusions). However, clinical outcomes were not significantly different (87% good to excellent in uninstrumented, 95% instrumented) (Zdeblick et al. 1993; Zdeblick 1995). He also noted a significantly increased fusion rate in rigid screw-rod constructs when compared to semirigid plate and screw constructs. These findings were confirmed by Fischgrund et al. in 1997 in regard to improved fusion rates but no difference in overall clinical outcomes between instrumented and uninstrumented patients in the degenerative lumbar spine.

Complications

Pedicle screws, while consistently shown to be a safe method of posterior spinal instrumentation, are not without complications. Overall complication rates have been reported up to 25%; however, many are without significant clinical consequence, while others can be catastrophic.

Misplaced pedicle screws occur in various rates reported from 5% to 41% in the lumbar spine and from 3% to 55% in the thoracic spine and have been reported in up to 21% of posthumous cadaveric studies (Perna et al. 2016; Vaccaro and Garfin 1995b). The majority of misplaced screws are asymptomatic; however, medial-breached pedicle screws can cause nerve root injury or irritation that can be symptomatic and require screw revision (approximate incidence of 0.5%). Misplaced screws are typically classified as screws greater than 4 mm of breach (Gertzbein and Robbins 1990) or by the thoracic safe zone criteria of up to 6 mm lateral breach and 2 mm medial breach as described by Belmont et al. (2002). Most case series have shown that less than 2 mm of breach is not associated with complications (Gelalis et al. 2012; Belmont et al. 2002). Superior or rostral breach can lead to superior adjacent-level disc penetration resulting in poor screw purchase. Inferior breach can lead to nerve root or dural injury. Lateral screw placement can lead to segmental vessel injury and poor screw purchase. Nerve root injury can occur in 2.5–7.5% of cases, and removal of malpositioned screws can lead to resolution (Ohlin et al. 1994). Dural tears have been reported to be about 2–4% (Robert 2000).

Screw pullout or cutout from the pedicle is very common and dependent on not only technical surgeon-controlled factors of insertion but also implant design and host bone mineral density (Chapman et al. 1996; Zindrick and Lorenz 1997; Coe et al. 1990). Pedicle fracture can also occur resulting in loss of fixation or injury to surrounding neurovascular structures.

Implant failure or fatigue has also been reported, and early pedicle screw systems such as the VSP system reported rates as high as 17.5% screw failure (Whitecloud et al. 1989b). As technology has improved including material science and surgeon understanding of pedicle screw fixation techniques, this rate has dramatically decreased.

Posterior spinal instrumentation (and pedicle screws in particular) does increase rates of surgical site infections (SSIs) when compared to uninstrumented fusions approximately twofold from 3% to 6%.

Pedicle screw systems can also cause direct irritation symptoms to dorsal soft tissues as they are relatively raised compared to the dorsal elements of the spine. This can lead to wound breakdown or painful bursitis, especially in thin patients.

Augmentation

With a rapidly aging population, an increasing number of spine fusion procedures being performed each year with pedicle screw instrumentation, the issue of bone mineral density has become increasingly important for surgeons to be cognizant of when planning pedicle screw fixation. Many strategies have been developed to help improve pedicle screw fixation in the setting of osteoporosis or osteopenia via pedicle augmentation.

Polymethyl methacrylate (PMMA) bone cement has been described to augment pedicle screw fixation and can increase screw pullout strength in osteoporosis from 50% to 250% (Becker et al. 2008). Typically 1–1.5 mL of PMMA is placed into the vertebral body after pedicle cannulation followed by immediate screw placement to allow for hardening of the cement around the screw. Alternatively, some cannulated and fenestrated screw designs allow for cement delivery through the screw itself (Fig. 13). This technique, while effective, does pose a safety risk as cement extravasation resulting in emboli or neurovascular damage has been reported. Alternatively, biodegradable bone substitutes such as calcium sulfate or phosphate have been used in a similar fashion as a potentially safer alternative without the exothermic reaction of PMMA (Rohmiller et al. 2002; Bai et al. 2001).

Novel screw designs to aid in screw fixation in osteoporotic spines have also been described in

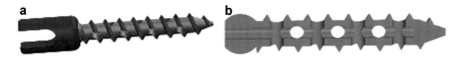

Fig. 13 Fenestrated screw design. (Reproduced from Shea et al. 2014 with permission from *Biomed Rest Int*)

an attempt to avoid PMMA use. Expandable screws that allow for finned expansion in the distal portion of a screw have been described with varying reports on the biomechanical properties of the expandable screw in different osteoporotic spine models (Cook et al. 2004; Koller et al. 2013; Gao et al. 2011; Liu et al. 2016; Lei and Wu 2006). A definitive advantage has not been shown over PMMA augmentation of traditional pedicle screws; however future research may demonstrate a clinical advantage.

Conclusions

Transpedicular fixation has been rapidly adopted among the spine surgery community in the last two to three decades due to its many advantages and ability to provide immediate three column stability to the spine and impart corrective forces all from a posterior-only approach. It is critical for surgeons to have a thorough understanding of the pedicle screw design options and flaws in order to achieve maximum fixation for a given scenario and avoid common complications of screw pullout, pedicle fracture, or misplacement. Given the potential for catastrophic neurovascular injury during pedicle screw placement, adequate training must be obtained before attempting placement of these fixation devices. New technologies such as computer-assisted and robotic navigation can aid in the safe placement of pedicle screws, but their clinical advantage and value have yet to be definitively proven. As pedicle screw technology and design continue to evolve, their widespread adoption, safety, and efficacy are likely to continue to improve.

References

Avila MJ, Baaj AA (2016) Freehand thoracic pedicle screw placement: review of existing strategies and a step-by-step guide using uniform landmarks for all levels. Cureus 8(2):e501. https://doi.org/10.7759/cureus.501

Bai B, Kummer FJ, Spivak J (2001) Augmentation of anterior vertebral body screw fixation by an injectable, biodegradable calcium phosphate bone substitute. Spine (Phila Pa 1976) 26:2679–2683

Barber JW, Boden SD, Ganey T (1997) A biomechanical study of lumbar pedicle screws: does convergence affect axial pullout strength? Paper presented at Eastern Orthopaedic annual meeting, Scottsdale, Oct 1997

Battula S, Schoenfeld AJ, Sahai V, Vrabec GA, Tank J, Njus GO (2008) Effect of pilot hole size on the insertion torque and pullout strength of self-tapping cortical bone screws in osteoporotic bone. J Trauma 64(4): 990–995

Becker S, Chavanne A, Spitaler R et al (2008) Assessment of different screw augmentation techniques and screw designs in osteoporotic spines. Eur Spine J 17: 1462–1469

Belmont PJ Jr, Klemme WR, Robinson M et al (2002) Accuracy of thoracic pedicle screws in patients with and without coronal plane spinal deformities. Spine (Phila Pa 1976) 27:1558–1566

Benzel EC (1995a) Biomechanically relevant anatomy and material properties of the spine and associated elements. In: Benzel EC (ed) Biomechanics of spine stabilization: principles and clinical practice. McGraw-Hill, New York, pp 3–16

Benzel EC (1995b) Implant-bone interfaces. In: Benzel EC (ed) Biomechanics of spine stabilization: principles and clinical practice. McGraw-Hill, New York, pp 127–134

Boucher HH (1959) A method of spinal fusion. J Bone Joint Surg Am 41-B:248–259

Calvert GC, Lawrence BD, Abtahi AM et al (2015) Cortical screws used to rescue failed lumbar pedicle screw construct: a biomechanical analysis. J Neurosurg Spine 22:166–172

Carmouche JJ, Molinari RW, Gerlinger T, Devine J, Patience T (2005) Effects of pilot hole preparation technique on pedicle screw fixation in different regions of the osteoporotic thoracic and lumbar spine. J Neurosurg Spine 3:364–370

Chang KW, Dewei Z, McAfee PC et al (1989) A comparative biomechanical study of spinal fixation using the combination spinal rod-plate and transpedicular screw fixation system. J Spinal Disord 1:257–266

Chapman JR, Harrington RM, Lee KM, Anderson PA, Tencer AF, Kowalski D (1996) Factors affecting the pullout strength of cancellous bone screws. J Biomech Eng 118:391–398

Chatzistergos PE, Sapkas G, Kourkoulis SK (2010) The influence of the insertion technique on the pullout force of pedicle screws: an experimental study. Spine (Phila Pa 1976) 35:E332–E337

Cho W, Cho S, Wu C (2010) The biomechanics of pedicle screw-based instrumentation. J Bone Joint Surg Br 92-B(8):1061–1065

Chung KJ, Suh SW, Desai S, Song HR (2008) Ideal entry point for the thoracic pedicle screw during the free hand technique. Int Orthop 32:657–662. https://doi.org/10.1007/s00264-007-0363-4. [PMID: 17437109]

Coe JD, Warden KE, Herzig MA et al (1990) Influence of bone mineral density on the fixation of the thoracolumbar implants: a comparative study of transpedicular screws, laminar hooks, and spinous process wires. Spine (Phila Pa 1976) 15:902–907

Cook SD, Salkeld SL, Stanley T, Faciane A, Miller SD (2004) Biomechanical study of pedicle screw fixation in severely osteoporotic bone. Spine J 4(4):402–408

Cotrel Y, Dubousset J, Guillaumat M (1988) New universal instrumentation in spinal surgery. Clin Orthop 227: 10–23

Daftari TK, Horton WC, Hutton WC (1994) Correlations between screw hold preparation, torque of insertion, and pullout strength for spinal screws. J Spinal Disord 7:139–145

Erkan S, Hsu B, Wu C et al (2010) Alignment of pedicle screws with pilot holes: can tap- ping improve screw trajectory in thoracic spines? Eur Spine J 19:71–77

Fischgrund JS, Mackay M et al (1997) Volvo Award winner in clinical studies: degenerative lumbar spondylolisthesis with spinal stenosis: a prospective, randomized study comparing decompressive laminectomy and arthrodesis with and without spinal instrumentation. Spine (Phila Pa 1976) 22:2807–2812

Fogel GR, Reitman CA, Liu W et al (2003) Physical characteristics of polyaxial-headed pedicle screws and biomechanical comparison of load with their failure. Spine (Phila Pa 1976) 28:470–473

Gao M, Lei W, Wu Z, Liu D, Shi L (2011) Biomechanical evaluation of fixation strength of conventional and expansive pedicle screws with or without calcium based cement augmentation. Clin Biomech 26(3): 238–244

Gelalis ID, Paschos NK, Pakos EE et al (2012) Accuracy of pedicle screw placement: a systematic review of prospective in vivo studies comparing free hand, fluoroscopy guidance and navigation techniques. Eur Spine J 21:247–255

George DC, Krag MH, Johnson CC et al (1991) Hole preparation techniques for transpedicle screws. Effect on pull-out strength from human cadaveric vertebrae. Spine (Phila Pa 1976) 16:181–184

Gertzbein SD, Robbins SE (1990) Accuracy of pedicular screw placement in vivo. Spine (Phila Pa 1976) 15:11–14

Harrington PR (1962) Treatment of scoliosis: correction and internal fixation by spine instrumentation. J Bone Joint Surg Am 44:591–610

Harrington PR (1988) The history and development of Harrington instrumentation. Clin Orthop Relat Res 227:3–5

Hirano T, Hasegawa K, Takahashi HE et al (1997) Structural characteristics of the pedicle and its role in screw stability. Spine (Phila Pa 1976) 22:2504–2509

Kast E, Mohr K, Richter H-P, Börm W (2006) Complications of transpedicular screw fixation in the cervical spine. Eur Spine J 15(3):327–334

King D (1944) Internal fixation for lumbosacral fusion. Am J Surg 66:357–361

King D (1948) Internal fixation for lumbosacral fusion. J Bone Joint Surg 30-A:560–565

Koller H, Zenner J, Hitzl W et al (2013) The impact of a distal expansion mechanism added to a standard pedicle screw on pullout resistance. A biomechanical study. Spine J 13(5):532–541

Krag MH (1991) Biomechanics of thoracolumbar spinal fixation. A review. Spine (Phila Pa 1976) 16:S84–S99

Krag MH et al (1988) Depth of insertion of transpedicular vertebral screws into human vertebrae: effect upon screw-vertebrae interface strength. J Spinal Disord 1:287–294

Kwok AW, Finkesltein JA et al (1996) Insertional torque and pull-out strength of conical and cylindrical pedicle screws in cadaveric bone. Spine (Phila Pa 1976) 21:2429–2434

Lehman RA Jr, Polly DW Jr, Kuklo TR et al (2003) Straight-forward versus anatomic trajectory technique of thoracic pedicle screw fixation: a biomechanical analysis. Spine (Phila Pa 1976) 28:2058–2065

Lei W, Wu Z (2006) Biomechanical evaluation of an expansive pedicle screw in calf vertebrae. Eur Spine J 15(3):321–326

Lill CA, Schneider E, Goldhahn J, Haslemann A, Zeifang F (2006) Mechanical performance of cylindrical and dual core pedicle screws in calf and human vertebrae. Arch Orthop Trauma Surg 126(10):686–694

Liu YK, Njus GO, Bahr PA, Geng P (1990) Fatigue life improvement of nitrogen-ion- implanted pedicle screws. Spine (Phila Pa 1976) 15:311–317

Liu D, Shi L, Lei W et al (2016) Biomechanical Comparison of Expansive Pedicle Screw and Polymethylmethacrylate-augmented Pedicle Screw in Osteoporotic Synthetic Bone in Primary Implantation: An Experimental Study. Clin Spine Surg 29(7): E351–357

Lonstein JE, Denis F, Perra JH, Pinto MR, Smith MD, Winter RB (1999) Complications associated with pedicle screws. J Bone Joint Surg Am 81(11):1519–1528

Lorenz M, Zindric M, Schwaegler P (1991) A comparison of single-level fusions with and without hardware. Spine (Phila Pa 1976) 16:S455–S458

Louis R (1986) Fusion of the lumbar and sacral spine by internal fixation with screw plates. Clin Orthop 203:18–33

Louis R (1996) Application of the Louis system for thoracolumbar and lumbosacral spine stabilization. In: Fessler RG, Haid RW (eds) Current techniques in spinal stabilization. McGraw-Hill, New York, pp 399–407

Ludwig SC, Kramer DL, Vaccaro AR, Albert TJ (1999) Transpedicle screw fixation of the cervical spine. Clin Orthop Relat Res 359:77–88

Ludwig SC, Kramer DL, Balderston RA, Vaccaro AR, Foley KF, Albert TJ (2000) Placement of pedicle screws in the human cadaveric cervical spine: comparative accuracy of three techniques. Spine (Phila Pa 1976) 25(13):1655–1667

Luque ER (1980) Segmental spinal instrumentation: a method of rigid internal fixation of the spine to induce arthrodesis. Orthop Trans 4:391

Mason A, Paulsen R, Babuska JM et al (2014) The accuracy of pedicle screw placement using intraoperative image guidance systems. J Neurosurg Spine 20:196–203

McAfee PC, Weiland DJ, Carlow JJ (1991) Survivorship analysis of pedicle spinal instrumentation. Spine (Phila Pa 1976) 16:S422–S427

McCormack BM, Benzel EC, Adams MS et al (1995) Anatomy of the thoracic pedicle. Neurosurgery 37:303–308

Mikula A, Williams SK, Anderson PA (2016) The use of intraoperative triggered electromyography to detect misplaced pedicle screws: a systematic review and meta-analysis. J Neurosurg Spine 24:624–638

Ohlin A, Karrlson M, Duppe H et al (1994) Complications after transpedicular stabilization of the spine: a survivorship analysis of 163 cases. Spine (Phila Pa 1976) 19:2774–2779

Orndorff DG, Zdeblick TA (2017) Chapter 71: Thoracolumbar instrumentation: anterior and posterior. In: Benzel spine. Elsevier, Philadelphia, PA

Panjabi MM, Yamamoto I, Oxland TR (1991a) Biomechanical stability of the pedicle screw fixation systems in a human lumbar spine instability model. Clin Biomech 6:197–205

Panjabi MM, Duranceau J, Goel V, Oxland T, Takata K (1991b) Cervical human vertebrae: quantitative three-dimensional anatomy of the middle and lower regions. Spine (Phila Pa 1976) 16:861–869

Parker SL, McGirt MJ, Farber SH, Amin AG, Rick AM, Suk I, Bydon A, Sciubba DM, Wolinsky JP, Gokaslan ZL, Witham TF (2011) Accuracy of free-hand pedicle screws in the thoracic and lumbar spine: analysis of 6816 consecutive screws. Neurosurgery 68:170–178; discussion

Pelton MA, Schwartz J, Singh K (2012) Subaxial cervical and cervicothoracic fixation techniques – indications, techniques, and outcomes. Orthop Clin North Am 43(1):19–28, vii. https://doi.org/10.1016/j.ocl.2011.08.002. Epub 2011 Oct 19

Perna F, Borghi R, Pilla F, Stefanini N, Mazzotti A, Chehrassan M (2016) Pedicle screw insertion techniques: an update and review of the literature. Musculoskelet Surg 100:165–169

Pfeiffer FM, Abernathie DL, Smith DE (2006) A comparison of pullout strength for pedicle screws of different designs: a study using tapped and untapped pilot holes. Spine (Phila Pa 1976) 31:867–870

Puno RM, Bechtold JE, Byrd JA et al (1987) Biomechanical analysis of five techniques of fixation for the lumbosacral junction. Orthop Trans 11:86

Puvanesarajah V, Liauw JA, Lo SF, Lina IA, Witham TF (2014) Techniques and accuracy of thoracolumbar pedicle screw placement. World J Orthop 5(2):112–123

Reidy DP, Houlden D, Nolan PC et al (2001) Evaluation of electromyographic monitoring during insertion of thoracic pedicle screws. J Bone Joint Surg Br 83:1009–1014

Robert WG (2000) The use of pedicle screw internal fixation for the operative treatment of spinal disorders. J Bone Joint Surg 82:1458

Rodriguez A, Neal MT, Liu A et al (2014) Novel placement of cortical bone trajectory screws in previously instrumented pedicles for adjacent-segment lumbar disease using CT image-guided navigation. Neurosurg Focus 36:E9

Rohmiller MT, Schwalm D, Glattes RC, Elalayli TG, Spengler DM (2002) Evaluation of calcium sulfate paste for augmentation of lumbar pedicle screw pullout strength. Spine J 2(4):255–260

Roy-Camille R, Roy-Camille M, Demeulenaere C (1970) Osteosynthesis of dorsal, lumbar, and lumbosacral spine with metallic plates screwed into vertebral pedicles and articular apophyses. Presse Med 78(32):1447–1448

Sandén B, Olerud C, Larsson S (2001) Hydroxyapatite coating enhances fixation of loaded pedicle screws: a mechanical in vivo study in sheep. Eur Spine J 10:334–339

Santoni BG, Hynes RA, McGilvray KC et al (2009) Cortical bone trajectory for lumbar pedicle screws. Spine J 9:366–373

Scheufler KM, Franke J, Eckardt A, Dohmen H (2011) Accuracy of image-guided pedicle screw placement using intra-operative computed tomography-based navigation with automated referencing. Part II: thoracolumbar spine. Neurosurgery 69:1307–1316

Shea T, Laun J, Gonzalez-Blohm S et al (2014) Designs and techniques that improve the pullout strength of pedicle screws in osteoporotic vertebrae: current status. Biomed Res Int 2014:748393. A) cylindrical screw B) conical screw. https://doi.org/10.1155/2014/748393

Shepard MF, Davies MR, Abayan A et al (2002) Effects of polyaxial pedicle screws on lumbar construct rigidity. J Spinal Disord Tech 15:233–236

Steffee AD, Biscup RS, Sitkowski DJ (1986) Segmental spine plates with pedicle screw fixation: a new internal fixation device for disorders of the lumbar and thoracolumbar spine. Clin Orthop 203:45–53

Ughwanogho E, Patel NM, Baldwin KD et al (2012) Computed tomography-guided navigation of thoracic pedicle screws for adolescent idiopathic scoliosis results in more accurate placement and less screw removal. Spine (Phila Pa 1976) 37:E473–E478

Vaccaro AR, Garfin SR (1995a) Pedicle-screw fixation in the lumbar spine. J Am Acad Orthop Surg 3(5):263–274

Vaccaro AR, Garfin SR (1995b) Internal fixation (pedicle screw fixation) for fusion of the lumbar spine. Spine (Phila Pa 1976) 20(Suppl 24):157S–165S

Vaccaro AR, Rizzolo SJ, Balderston RA, Allardyce TJ, Garfin SR, Dolinskas C, An HS (1995) Placement of pedicle screws in the thoracic spine. Part II: an anatomical and radiographic assessment. J Bone Joint Surg Am 77:1200–1206. [PMID: 7642665]

Van de Kelft E, Costa F, Van der Planken D, Schils F (2012) A prospective multicenter registry on the accuracy of pedicle screw placement in the thoracic, lumbar, and sacral levels with the use of the O-arm imaging system and StealthStation Navigation. Spine (Phila Pa 1976) 37(25):E1580–E1587. https://doi.org/10.1097/BRS.0b013e318271b1fa

Vanichkachorn JS, Vaccaro AR, An HS (1999) Chapter 19, Transpedicular screw instrumentation. In: An HS, Cotler JM (eds) Spinal instrumentation, 2nd edn. Lippincott Williams & Wilkins, Philadelphia

Verma R, Krishan S, Haendlmayer K, Mohsen A (2010) Functional outcome of computer-assisted spinal pedicle screw placement: a systematic review and meta-analysis of 23 studies including 5992 pedicle screws. Eur Spine J 19:370–375

Waschke A, Walter J, Duenisch P, Reichart R, Kalff R, Ewald C (2013) CT-navigation versus fluoroscopy-guided placement of pedicle screws at the thoracolumbar spine: single center experience of 4500 screws. Eur Spine J 22:654–660. https://doi.org/10.1007/s00586-012-2509-3. [PMID: 23001415]

West JL III, Bradford DS, Ogilvie JW (1991) Results of spinal arthrodesis with pedicle screw-plate fixation. J Bone Joint Surg Am 73:1179–1184

Whitecloud TS, Skalley T, Cook SD (1989a) Roentgenographic measurements of pedicle screw penetration. Clin Orthop Relat Res 245:57–68

Whitecloud TS III, Butler JC, Cohen JL et al (1989b) Complications with the variable spinal plating system. Spine 14:472–476

Xu R, Ebraheim NA, Ou Y, Yeasting RA (1998) Anatomic considerations of pedicle screw placement in the thoracic spine. Roy-Camille technique versus open-lamina technique. Spine (Phila Pa 1976) 23:1065–1068. https://doi.org/10.1097/00007632-199805010-00021. [PMID: 9589548]

Yoganandan N, Larson SJ, Pintar F et al (1990) Biomechanics of lumbar pedicle screw/plate fixation in trauma. Neurosurgery 27:873–881

Yoshihara H, Passias PG, Errico TJ (2013) Screw-related complications in the subaxial cervical spine with the use of lateral mass versus cervical pedicle screws. A systematic review. J Neurosurg Spine 19(5):614–623

Youssef JA, McKinley TO, Yerby SA, McLain RF (1999) Characteristics of pedicle screw loading: effect of sagittal insertion angle on intrapedicular bending moments. Spine (Phila Pa 1976) 24:1077–1081

Yuan HA, Garfin SR, Dickman CA, Marjetko SM (1994) A historical cohort study of pedicle screw fixation: thoracic and lumbar spine fusions. Spine (Phila Pa 1976) 19(Suppl):2279S–2296S

Zdeblick TA (1993) A prospective randomized study of lumbar fusion. Preliminary results. Spine (Phila Pa 1976) 18:983–991

Zdeblick TA (1995) The treatment of degenerative lumbar disorders. Spine (Phila Pa 1976) 20:126S–137S

Zdeblick TA, Kunz DN, Cooke ME, McCabe R (1993) Pedicle screw pullout strength. Correlation with insertional torque. Spine (Phila Pa 1976) 18:1673–1676

Zindrick MR, Lorenz MA (1997) Posterior lumbar fusion: overview of options and internal fixation devices. In: Frymoyer JW (ed) The adult spine. Lippincott-Raven, Philadelphia, pp 2175–2205

Interspinous Devices

27

Douglas G. Orndorff, Anneliese D. Heiner, and Jim A. Youssef

Contents

Introduction	562
Indications and Contraindications	564
North American Spine Society (NASS) Coverage Policy Recommendations for Interspinous Devices	564
International Society for the Advancement of Spine Surgery (ISSAS) Coverage Indications and Limitations of Coverage	564
Surgical Technique	565
Complications	568
Complications, Complication Rates, and Reoperation Rates	568
Sources of Complications	569
Outcomes of Interspinous Devices	570
Interspinous Devices	570
Interspinous Devices Versus Nonoperative Treatment	570
Interspinous Devices Versus Bony (Direct) Decompression	570
Combined Interspinous Device Plus Bony Decompression Versus Bony Decompression Alone	571
Interspinous Device Versus Laminectomy plus Instrumented Fusion	571
Conclusions	571
Cross-References	571
References	571

D. G. Orndorff (✉) · J. A. Youssef
Spine Colorado, Durango, CO, USA
e-mail: DOrndorff@spinecolorado.com; jim@youssef5.com

A. D. Heiner
Penumbra, Inc., Alameda, CA, USA
e-mail: anneliese_heiner@yahoo.com

Abstract

The purpose of an interspinous device is to distract adjacent spinous processes, thus producing flexion and limiting extension of that spine level. Interspinous devices can be designed for motion preservation or for fusion, and the device insertion is minimally invasive. The North American Spine Society (NASS) and the International Society for the

© Springer Nature Switzerland AG 2021
B. C. Cheng (ed.), *Handbook of Spine Technology*,
https://doi.org/10.1007/978-3-319-44424-6_59

Advancement of Spine Surgery (ISSAS) have provided indications and contraindications for the use of interspinous devices, with careful patient selection being paramount. The patient is typically placed in a prone position, and the interspinous device is inserted between the adjacent spinous processes to hold the spine in a slightly kyphotic position, thus increasing the canal diameter and reducing symptoms of neurogenic claudication. Complications, complication rates, and reoperation rates have been reported in a large retrospective study and in review articles. Complications can result from the design or intrinsic purpose of the device, incorrect surgical indications or patient selection, or incorrect device sizing. Outcomes of interspinous devices and how their outcomes compare with the outcomes of nonoperative treatment, bony decompression alone, and instrumented fusion have been reported.

Keywords

Clinical outcomes · Dynamic stabilization · Indications · Indirect decompression · Interspinous device · Interspinous fusion device · Lumbar spinal stenosis · Minimally invasive · Motion preservation · Surgical technique

Introduction

The purpose of an interspinous device is to distract adjacent spinous processes (Fig. 1), thus producing segmental flexion and limiting extension of that spine level (Ravindra and Ghogawala 2017; Gazzeri et al. 2015). This flexion can provide indirect decompression of the neural elements and thus alleviate the symptoms of neurogenic claudication caused by lumbar spinal stenosis (Ravindra and Ghogawala 2017; Gazzeri et al. 2014; Borg et al. 2012). Interspinous devices of appropriate design can be used alone, as a means of indirect decompression; in conjunction with bony decompression, as a means of stabilization with motion preservation; or as a fusion device. The procedure is minimally invasive, which makes it especially useful for older, higher-risk patients (Gazzeri et al. 2015), and can be used to avoid or delay more invasive procedures (Bonaldi et al. 2015).

Distracting adjacent spinous processes and inducing segmental flexion with an interspinous device have several anatomical effects (Gazzeri et al. 2014, 2015; Parchi et al. 2014; Gala et al. 2017). The spinous process distraction tightens the ligamentum flavum, thus preventing it from buckling into the spinal canal and causing stenosis. The distraction provides indirect decompression by increasing the spinal canal area and increasing the neural foramina area and height. The facet joints and posterior intervertebral disc are unloaded. Adjacent spine levels should not be affected (Parchi et al. 2014).

Interspinous devices can be designed for motion preservation or for fusion (Parchi et al. 2014). Furthermore, motion preservation devices can be static or dynamic. Static devices are rigid, thus disallowing any extension upon contact with both spinous processes (Fig. 1a). Dynamic devices are flexible and act like a compression spring between the spinous processes to prohibit excess extension (Fig. 1b).

Interspinous devices used for motion preservation allow for dynamic stabilization of the motion segment; stabilization is achieved, but with motion restricted rather than prevented as would occur with fusion (Lee et al. 2015; Bonaldi et al. 2015; Parchi et al. 2014). Motion is restricted in the directions that can cause pain, but is allowed in other directions (Lee et al. 2015; Bonaldi et al. 2015). Retaining motion at the treated level allows that segment to continue to contribute to the motion of the spine and prevents or reduces adjacent segment disease (Lee et al. 2015; Parchi et al. 2014; Serhan et al. 2011).

An interspinous fusion device (Fig. 1c) may be used as a less invasive alternative to pedicle screw-rod fixation (Bonaldi et al. 2015; Parchi et al. 2014). The interspinous fusion device holds the spinous processes in distraction and, with the addition of bone graft, facilitates development of a stable arthrodesis. Although biomechanical studies have indicated that interspinous fusion devices and pedicle screw-rod fixation

27 Interspinous Devices 563

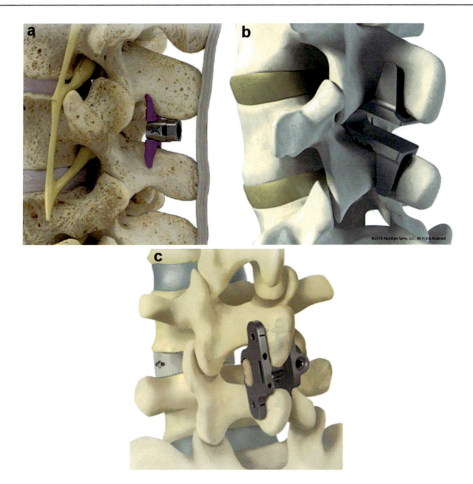

Fig. 1 Representative interspinous devices. (**a**) Static motion preservation device (Superion, Vertiflex Inc., Carlsbad, CA) (Vertiflex Inc. 2016). Note that the interspinous ligament is preserved. (Figure reprinted with permission from Vertiflex Inc.). (**b**) Dynamic motion preservation device (Coflex, Paradigm Spine, New York, NY). (Figure courtesy of Paradigm Spine). (**c**) Fusion device (Aspen MIS Fusion System, Zimmer Biomet Spine Inc., Westminster, CO) (Zimmer Biomet Spine Inc. 2016). Note the interbody cage at that level and the bone graft placed laterally through the device. (Figure reprinted with permission from Zimmer Biomet Spine Inc.)

appear to be similar in how they limit the flexion/extension range, interspinous fusion devices may be less effective in limiting axial rotation and lateral bending, as compared with bilateral pedicle screw-rod fixation (Bonaldi et al. 2015). An interspinous fusion device could be used stand-alone or in conjunction with an interbody cage (Parchi et al. 2014).

Interspinous device insertion has the advantages of a minimally invasive procedure (Pintauro et al. 2017). The device may be inserted with the patient under local anesthesia (Gazzeri et al. 2015; Borg et al. 2012; Parchi et al. 2014). Soft tissue damage, blood loss, and skin scarring are minimized, and operative time is reduced (Borg et al. 2012; Bonaldi et al. 2015). Because interspinous device insertion does not directly interfere with the spinal canal or neural foramina, dural tears or nerve root injuries are unlikely (Borg et al. 2012). Recovery time, complications, and length of stay are reduced, and the surgery could be performed on an outpatient basis (Bonaldi et al. 2015; Borg et al. 2012). Additionally, the procedure is reversible, if further decompression or fusion surgery is required later (Gazzeri et al. 2014, 2015; Borg et al. 2012).

Indications and Contraindications

North American Spine Society (NASS) Coverage Policy Recommendations for Interspinous Devices

The North American Spine Society (NASS) has provided coverage policy recommendations for each category of lumbar interspinous device: those to be used without fusion and with direct decompression (NASS Coverage Committee 2018), those to be used without fusion or direct decompression (NASS Coverage Committee 2014b), and those to be used with fusion for stabilization (NASS Coverage Committee 2014a). For lumbar interspinous devices to be used without fusion and with direct decompression (laminectomy), as an alternative to lumbar fusion, NASS provides indications and contraindications (Table 1). For lumbar interspinous devices to be used without fusion and without direct decompression, NASS also provides indications and contraindications but considers this coverage to be conditional pending further evidence (Table 2). For lumbar interspinous devices to be used with fusion for stabilization, NASS only states, "Interspinous fixation with fusion for stabilization is currently NOT indicated as an alternative to pedicle screw fixation with lumbar fusion procedures (NASS Coverage Committee 2014a)."

International Society for the Advancement of Spine Surgery (ISSAS) Coverage Indications and Limitations of Coverage

The International Society for the Advancement of Spine Surgery (ISSAS) has also provided indications and limitations of coverage for interspinous devices (referred to as interlaminar

Table 1 North American Spine Society (NASS) indications and contraindications for lumbar interspinous devices to be used without fusion and with direct decompression (NASS Coverage Committee 2018). (Reprinted with permission from NASS Coverage Recommendations, © 2014–2018, North American Spine Society)

Indications
Stabilization with an [interspinous device] without fusion in conjunction with laminectomy may be indicated as an alternative to lumbar fusion for **degenerative lumbar stenosis with or without low-grade spondylolisthesis** (less than or equal to 3 mm of anterolisthesis on a lateral radiograph) with qualifying criteria when appropriate:
1. Significant mechanical back pain is present (in addition to those symptoms associated with neural compression) that is felt unlikely to improve with decompression alone. Documentation should indicate that this type of back pain is present at rest and/or with movement while standing and does not have characteristics consistent with neurogenic claudication
2. A lumbar fusion is indicated post-decompression for a diagnosis of lumbar stenosis with a Grade 1 degenerative spondylolisthesis as recommended in the *NASS Coverage Recommendations for Lumbar Fusion*
3. A lumbar laminectomy is indicated as recommended in the *NASS Coverage Recommendations for Lumbar Laminectomy*
4. Previous lumbar fusion has not been performed at an adjacent segment
5. Previous decompression has been performed at the intended operative segment
Contraindications
[Interspinous devices] are *not* indicated in cases that do not fall within the above parameters. In particular, they are not indicated in the following scenarios and conditions:
1. Degenerative spondylolisthesis of Grade 2 or higher
2. Degenerative scoliosis or other signs of coronal instability
3. Dynamic instability as detected on flexion-extension views demonstrating at least 3 mm of change in translation
4. Iatrogenic instability or destabilization of the motion segment
5. A fusion is otherwise not indicated for a Grade 1 degenerative spondylolisthesis and stenosis as per the *NASS Coverage Recommendations for Lumbar Fusion*
6. A laminectomy for spinal stenosis is otherwise not indicated as per the *NASS Coverage Recommendations for Lumbar Laminectomy*

Table 2 North American Spine Society (NASS) indications and contraindications for lumbar interspinous devices to be used without fusion and without direct decompression (NASS Coverage Committee 2014b). (Reprinted with permission from NASS Coverage Recommendations, © 2014–2018, North American Spine Society)

Indications
Interspinous distraction devices without fusion may be indicated for the following diagnoses with qualifying criteria, when appropriate:
1. Degenerative lumbar stenosis:
(a) Associated with neurogenic claudication that is relieved by lumbar flexion
(b) Patients over 50 years old
(c) Failure of nonoperative treatment
(d) No more than 25° of degenerative scoliosis
(e) No more than a Grade 1 degenerative spondylolisthesis
(f) Open surgery (e.g., laminectomy) is not a medically safe treatment option because of comorbidities
Contraindications
Interspinous distraction devices are **NOT** indicated in cases that do not fall within the above parameters. In particular, they are not indicated in the following scenarios and conditions:
1. Degenerative spondylolisthesis of Grade 2 or higher
2. Degenerative scoliosis greater than 25° dynamic instability at the operative level
3. Symptoms are not relieved by flexion
4. Patient is medically suitable for a direct decompressive procedure (e.g., laminectomy)
5. Patient has primarily axial back pain that is unrelated to activity
6. Patients younger than 50 years old

stabilization) used with direct decompression (Table 3; Guyer et al. 2016). The ISASS awaits further data and review before providing indications and limitations of coverage for interspinous devices used without direct decompression.

Surgical Technique

Interspinous device placement can be performed from either a lateral decubitus or prone position. We prefer the prone position on a Jackson frame, as this setup seems to give better exposure and leverage with implant placement and also reduces radiation exposure to the patient and surgical team when using intraoperative fluoroscopy. The procedure described below is applicable to both motion preservation devices and fusion devices.

After the patient has undergone general anesthetic, he or she is placed in the prone position with all bony prominences well padded and checked by the surgeon. The patient's skin is typically prepped and draped in the usual fashion using chlorohexidine scrub. Fluoroscopy is then used to identify the appropriate surgical levels.

The skin is then anesthetized with 0.25% Marcaine with epinephrine, to aid with hemostasis and postoperative pain control. A 10-blade scalpel is then used to make a longitudinal incision over the length of the interspace; if a fusion procedure is to be performed, the incision can be extended over the length of the superior and inferior lamina. If multiple levels will receive an interspinous device, we recommend making a separate incision for each level, to reduce scarring and improve cosmesis. Once the skin incision has been made and hemostasis has been obtained, subperiosteal dissection is performed using Cobb elevators to expose the superior and inferior lamina. Great care is taken to avoid injury to the facet capsules and joints and to preserve the supraspinous and interspinous ligaments. Once adequate exposure is obtained, a self-retaining retractor is placed to hold the soft tissue out of the way for adequate visualization. At that point, a curved curette is placed under the superior lamina, and a lateral x-ray is taken to confirm the appropriate level.

Table 3 International Society for the Advancement of Spine Surgery (ISSAS) indications/limitations of coverage for interspinous devices (interlaminar stabilization) used with direct decompression (Guyer et al. 2016). (Reprinted with permission from the International Journal of Spine Surgery)

Indications

Patients who have all of the following criteria may be eligible for decompression with interlaminar stabilization:

1. Radiographic confirmation of at least moderate lumbar stenosis, which narrows the central spinal canal at 1 or 2 contiguous levels from L-1 to L-5 that require surgical decompression. Moderate stenosis is defined as >25% reduction of the anteroposterior dimension compared with the next adjacent normal level, with nerve root crowding compared with the normal level, as determined by the surgeon on CT scanning or MRI

2. Radiographic confirmation of the absence of gross angular or translatory instability of the spine at index or adjacent levels (instability as defined by White and Panjabi: sagittal plane translation >4.0 mm or 15% or local sagittal plane rotation >15° at L1–2, L2–3, and L3–4; >20° at L4–5 based on standing flexion-extension radiographs). Improved imaging technologies are able to better refine/detect previously undetected instability, and as these technologies become more established, surgeons should expect to refine with specificity and clear delineation of appropriate surgical candidates requiring stabilization

3. Patients who experience relief in flexion from their symptoms of leg/buttocks/groin pain, with or without back pain, and who have undergone at least 12 weeks of nonoperative treatment consisting of nonsteroidal anti-inflammatory drugs and at least one of the following: rest, restriction of activities of daily living, physical therapy, or steroid injections

Limitations (Contraindications)

Decompression with interlaminar stabilization is NOT indicated for patients with the following:

1. More than 2 vertebral levels requiring surgical decompression

2. Prior surgical procedure that resulted in gross translatory instability of the lumbar spine

3. Prior fusion, implantation of a total disc replacement, or complete laminectomy at index level

4. Radiographically compromised vertebral bodies at any lumbar level(s) caused by current or past trauma, tumor, or infection

5. Severe facet hypertrophy requiring extensive bone removal that would cause gross instability

6. Radiographic confirmation of gross angular or translatory instability of the spine at index or adjacent levels with sagittal plane translation >4.0 mm as spondylolisthesis or retrolisthesis

7. Isthmic spondylolisthesis or spondylolysis (pars fracture)

8. Degenerative lumbar scoliosis (Cobb angle >25° lumbar segmental)

9. Osteopenia and osteoporosis

10. Back or leg pain of unknown etiology

11. Axial back pain only, with no leg, buttock, or groin pain

12. Morbid obesity defined as a body mass index >40

13. Active or chronic infection – systemic or local

14. Known history of Paget disease, osteomalacia, or any other metabolic bone diseases (excluding osteopenia, which is addressed above)

15. Rheumatoid arthritis or other autoimmune diseases requiring chronic steroid use

16. Active malignancy: a patient with a history of any invasive malignancy (except nonmelanoma skin cancer), unless he/she has been treated with curative intent and there has been no clinical signs or symptoms of the malignancy for at least 5 years. Patients with a primary bony tumor are excluded as well

17. Known allergy to titanium alloys or magnetic resonance contrast agents

18. Cauda equina syndrome defined as neural compression causing neurogenic bowel or bladder dysfunction

Once the adequate level has been confirmed with fluoroscopy, the placement of the interspinous device begins.

The placement of the interspinous device depends on the implant chosen and on whether or not that implant involves preserving or resecting the supraspinous and interspinous ligaments (Fig. 2). The surgeon should have a definitive knowledge of the requirements of each device and surgical placement steps prior to the surgery. Typically, there is a step of trial sizing to choose the appropriate implant (Fig. 3). The appropriately sized implant will give good tension and distraction between the spinous processes, to

Fig. 2 Superion (Vertiflex Inc., Carlsbad, CA) interspinous device being inserted through the supraspinous ligament (Vertiflex Inc. 2016). (Figure reprinted with permission from Vertiflex Inc.)

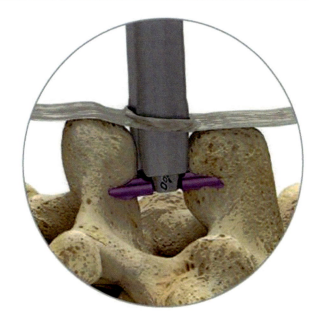

Fig. 3 Measurement of the interspinous separation distance, to select the appropriate rasp and implant size; note the preservation of the supraspinous ligament (Zimmer Biomet Spine Inc. 2016). (Figure reprinted with permission from Zimmer Biomet Spine Inc.)

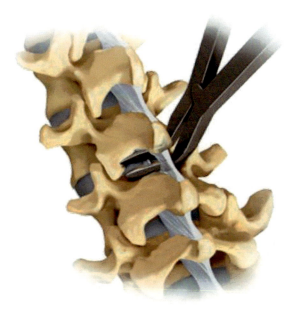

widen the interspinous spaces and indirectly decompress the spinal canal. However, an oversized implant can lead to spinous process fracture and implant subsidence, while an undersized implant can result in inadequate decompression and poorer patient outcomes. Once the interspinous device has been successfully placed and locked in position, final imaging is used to confirm the appropriate level and good placement of the implant (Fig. 4).

If a fusion follows the interspinous device placement, dissection may be carried out further around the facet joints for additional exposure. The facet joints are then removed with a Leksell rongeur and decorticated with a high-speed burr. This local bone is saved as autograft to facilitate

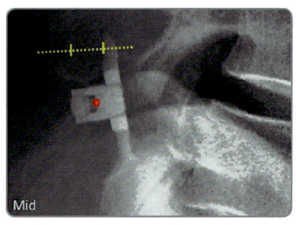

Fig. 4 Radiograph demonstrating proper placement of Superion (Vertiflex Inc., Carlsbad, CA) interspinous device (Vertiflex Inc. 2016). (Figure reprinted with permission from Vertiflex Inc.)

Table 4 Complications of interspinous device surgery (Gazzeri et al. 2015; Ravindra and Ghogawala 2017; Pintauro et al. 2017; Borg et al. 2012)

Durotomy	Device malpositioning	Instability
Hematoma	Device fracture	Nerve root distraction
Swelling	Device dislocation	New radicular deficit
Dehiscence	Spinous process fracture	Pain not improving
Infection	Spinous process erosion	Pain recurrence
		Progression of symptoms

the arthrodesis. Decortication of the superior and inferior lamina is then performed to obtain a large surface area of the bleeding bone for bony fusion. Some surgeons will elect to place pedicle screws, and some surgeons may rely only on the interspinous spacer for fusion. We recommend pedicle screw supplemental fixation, as biomechanical studies show superior fixation and increased fusion rates with pedicle screw fixation compared with interspinous spacers alone (Gonzalez-Blohm et al. 2014). Once the posterior lateral arthrodesis has been performed, the wound is copiously irrigated to obtain hemostasis. This fusion procedure does not often require a drain for postoperative management.

Closure involves approximating the fascial layer with an absorbable Vicryl suture, followed by a dermal closure, and then by skin closure, typically with an absorbable suture. For interspinous device placement, we do not feel it is indicated at this juncture to place vancomycin antibiotic powder. Once the skin is closed, a sterile dressing is applied with Steri-Strip 4 × 4's and coverall tape.

Patients are often allowed to go home the same day postoperatively. We typically recommend a lumbar corset brace for 6 weeks postoperatively to minimize patient motion and prevent implant migration. After 6 weeks, we commence a 6-week course of physical therapy involving with core strengthening and flexibility exercises. At 3 months postoperatively, patients are allowed to resume normal activity without restrictions.

Complications

Complications, Complication Rates, and Reoperation Rates

Complications (Table 4), complication rates (Table 5), and reoperation rates for interspinous devices have been reported in a large retrospective study and in review articles. In a multicenter

Table 5 Complication rates after interspinous device (ISD) surgery

	Rate	Citation and study type
Device failure (mainly spinous process fracture) or intraoperative device-related complication	4.8% mean first-generation ISDs; 2.9% mean next-generation ISDs	Pintauro et al. (2017) Review; 15 studies published 2011–2016
Postoperative complications	1.5–32.3%	Lee et al. (2015) Systematic review; 6 studies published 2004–2007
Overall complications	7%	Moojen et al. (2011) Systematic review and meta-analysis; 563 patients in 11 studies published 2004–2010
Device failures	6%	
Other (infection and postoperative leakage)	1%	
Overall complications and failures	14.81%	Gazzeri et al. (2015) Multicenter retrospective study; 1,108 patients
Recurrent pain	5.14%	
Spinous process fracture	2.44%	
Dura matter tear	2.08%	
Dislocation	1.81%	
No improvement/worsening of pain	1.81%	
Malposition (over-/underdistraction)	1.26%	
Instability	0.27%	
Infection	0%	

retrospective study of 1,108 patients (Gazzeri et al. 2015), the reoperation rate was 9.6%. The reasons for reoperation were recurrence of symptoms after an initial good outcome (3.8%), acute worsening of low-back pain secondary to spinous fracture or overdistraction of the supraspinous ligament (2.4%), implant dislocation (1.8%), and total lack of improvement (1.6%). All reoperations involved interspinous device removal. Of those reoperations, 12.15% occurred within 3 months postoperatively, and the remaining 87.85% occurred after a minimum of 24 months postoperatively. In a review of 15 studies published 2011–2016, Pintauro et al. (2017) reported interspinous device reoperation rates at a mean 24 months' follow-up as 11.1% and 3.7% for first-generation and next-generation devices, respectively. In a systematic review and meta-analysis involving 563 patients in 11 studies published 2004–2010, Moojen et al. (2011) reported a reoperation rate of 6% (follow-up period not reported).

Sources of Complications

Some complications of interspinous device implantation are related to the design or intrinsic purpose of the device. A too-stiff device can cause spinous process fracture, while a too-flexible device can lead to new radiculopathy (Serhan et al. 2011). The removal or disruption of posterior ligamentous process components during insertion of the device can cause instability (Bono and Vaccaro 2007; Kim and Albert 2007). An interspinous device with insufficient means of fixation could dislocate (Serhan et al. 2011). The V-shape of the posterior interspinous space could lead to device dislocation upon rupture of the supraspinous ligament (Gazzeri et al. 2015). (A fall can also cause device dislocation (Gazzeri et al. 2015).) The induction of segmental kyphosis with interspinous device implantation could theoretically cause adjacent segment disease (Bono and Vaccaro 2007; Kim and Albert 2007).

Incorrect surgical indications or patient selection for interspinous devices can lead to complications (Pintauro et al. 2017; Gazzeri et al. 2015). In particular, osteoporosis or osteopenia can lead to spinous process fracture (Gazzeri et al. 2015; Ravindra and Ghogawala 2017; Pintauro et al. 2017; Borg et al. 2012) and are considered as reasons for caution or as contraindications for interspinous device implantation (Bonaldi et al. 2015; Siewe et al. 2015; Pintauro et al. 2017; Gazzeri et al. 2015). The chance of spinous process fracture in patients with osteoporosis or osteopenia could be reduced by choosing a device size that results in less distraction (Gazzeri et al. 2015) or by augmenting an osteoporotic posterior vertebral arch with bone cement injection (spinoplasty) (Bonaldi et al. 2012).

Incorrect interspinous device sizing can also lead to complications. Implanting an oversized device can overdistract the supraspinous ligament, causing pain, and overcompress the spinous processes, causing spinous process fracture (Gazzeri et al. 2015). Implanting an undersized interspinous device can lessen the desired mechanism of the device to induce segmental kyphosis, leading to such results as an under-distracted ligamentum flavum being able to buckle into the spinal canal (Gazzeri et al. 2015; Gala et al. 2017).

Intraoperative durotomy is more likely at the L5–S1 level, where the dural sac is located more posteriorly than it is at higher levels. Placing the interspinous device more posteriorly (in the midportion of the level's interspinous ligament) can reduce the chance of this complication (Gazzeri et al. 2015). The spinous process of S1 is very unpredictable in size and can be small, bifid, or absent. We do not recommend interspinous device insertion at the L5–S1 level.

One study detected heterotopic ossification in 81.2% patients that had a Coflex interspinous device (Fig. 1b), at 24–57 months postoperatively (Tian et al. 2013). Although heterotopic ossification could aid in fusion and thus stabilization at the operated level (Tian et al. 2013; Gazzeri et al. 2015), there has been a case report with Coflex implantation in which heterotopic ossification caused stenosis and recurrence of neurogenic claudication (Maida et al. 2012).

Outcomes of Interspinous Devices

Interspinous Devices

Complication and reoperation rates for interspinous devices, as reported in a large retrospective study and in review articles, are discussed in section "Complications, Complication Rates, and Reoperation Rates." The large retrospective study of 1,108 patients discussed earlier (Gazzeri et al. 2015) also reported clinical outcomes for interspinous devices at a minimum 24 months' follow-up: 41.5% excellent, 34.7% good, 12.5% fair, 6.6% marginal, and 4.7% poor.

Interspinous Devices Versus Nonoperative Treatment

In their systematic review and meta-analysis, Li et al. (2017) identified 3 randomized controlled trials in 5 articles, with a total of 564 patients in the interspinous device group and 244 patients in the nonoperative group. They calculated that the interspinous device group had a lower incidence of additional surgery and a better clinical outcome than did the nonoperative group.

Interspinous Devices Versus Bony (Direct) Decompression

Two recent systematic reviews and meta-analyses have compared the outcomes of interspinous devices versus bony decompression (Zhao et al. 2017; Phan et al. 2016). Zhao et al. (2017) identified 4 randomized controlled trials in 7 articles, with a total of 200 patients in each treatment group. They concluded that both techniques were acceptable for treating lumbar spinal stenosis, but they did not have enough evidence to recommend one technique over the other. Additionally, the interspinous device group had higher reoperation rates, higher postoperative visual analog scale (VAS) back pain scores, and lower cost-effectiveness. The authors expressed a need for interspinous device studies with larger sample sizes and longer follow-up. Phan et al. (2016)

identified 7 studies with a total of 404 interspinous device patients and 424 bony decompression patients. They calculated that interspinous device implantation had significantly lower surgical complications but significantly higher long-term reoperation rates than did bony decompression. Additionally, interspinous device implantation had significantly higher postoperative VAS back pain scores than did bony decompression, but no significant difference in postoperative VAS leg pain scores or in Oswestry Disability Index (ODI) scores (for the two studies that reported ODI) was detected between the groups.

Combined Interspinous Device Plus Bony Decompression Versus Bony Decompression Alone

In their systematic review and meta-analysis, Phan et al. (2016) identified 4 articles with a total of 139 interspinous device plus bony decompression patients and 137 bony decompression alone patients. They calculated that an interspinous device plus bony decompression resulted in significantly higher surgical complications than did bony decompression alone. No significant difference was detected between the groups in reoperation rates or in postoperative VAS back or VAS leg pain scores.

Interspinous Device Versus Laminectomy plus Instrumented Fusion

In their systematic review, Li et al. (2017) identified 2 randomized controlled trials in 3 articles, with a total of 245 patients in the interspinous device group (30 with an interspinous device alone and 215 with an interspinous device plus laminectomy) and 137 patients in the laminectomy plus instrumented fusion group. One trial concluded that the interspinous device group had a lower complication rate and more improvement in VAS and in ODI than did the laminectomy plus instrumented fusion group. The other trial concluded that the two groups had comparable complication and reoperation rates and that the interspinous device group had better Zurich Claudication Questionnaire (ZCQ) scores.

Conclusions

Interspinous devices can be used in a well-selected subset of patients suffering from neurogenic claudication. The procedure can be done minimally invasive and performed in patients who have comorbidities that may preclude them from more invasive surgical procedures or in patients who want to maintain motion and avoid a fusion. Interspinous devices can also be used to supplement interbody fusion. As with any device, proper patient selection and setting realistic expectations can help achieve good results. More long-term randomized studies are necessary to help better clarify best surgical practices.

Cross-References

▶ Design Rationale for Posterior Dynamic Stabilization Relevant for Spine Surgery
▶ Lessons Learned from Positive Biomechanics and Poor Clinical Outcomes
▶ Lessons Learned from Positive Biomechanics and Positive Clinical Outcomes
▶ Posterior Dynamic Stabilization
▶ Thoracic and Lumbar Spinal Anatomy

References

Bonaldi G, Bertolini G, Marrocu A, Cianfoni A (2012) Posterior vertebral arch cement augmentation (spinoplasty) to prevent fracture of spinous processes after interspinous spacer implant. AJNR Am J Neuroradiol 33(3):522–528. https://doi.org/10.3174/ajnr.A2792

Bonaldi G, Brembilla C, Cianfoni A (2015) Minimally-invasive posterior lumbar stabilization for degenerative low back pain and sciatica. A review. Eur J Radiol 84(5):789–798. https://doi.org/10.1016/j.ejrad.2014.04.012

Bono CM, Vaccaro AR (2007) Interspinous process devices in the lumbar spine. J Spinal Disord Tech 20(3):255–261. https://doi.org/10.1097/BSD.0b013e3180331352

Borg A, Nurboja B, Timothy J, Choi D (2012) Interspinous distractor devices for the management of lumbar spinal stenosis: a miracle cure for a common problem? Br J Neurosurg 26(4):445–449. https://doi.org/10.3109/02688697.2012.680630

Gala RJ, Russo GS, Whang PG (2017) Interspinous implants to treat spinal stenosis. Curr Rev Musculoskelet Med 10(2):182–188. https://doi.org/10.1007/s12178-017-9413-8

Gazzeri R, Galarza M, Alfieri A (2014) Controversies about interspinous process devices in the treatment of degenerative lumbar spine diseases: past, present, and future. Biomed Res Int 2014:975052. https://doi.org/10.1155/2014/975052

Gazzeri R, Galarza M, Neroni M, Fiore C, Faiola A, Puzzilli F, Callovini G, Alfieri A (2015) Failure rates and complications of interspinous process decompression devices: a European multicenter study. Neurosurg Focus 39(4):E14. https://doi.org/10.3171/2015.7.Focus15244

Gonzalez-Blohm SA, Doulgeris JJ, Aghayev K, Lee WE 3rd, Volkov A, Vrionis FD (2014) Biomechanical analysis of an interspinous fusion device as a stand-alone and as supplemental fixation to posterior expandable interbody cages in the lumbar spine. J Neurosurg Spine 20(2):209–219. https://doi.org/10.3171/2013.10.Spine13612

Guyer R, Musacchio M, Cammisa FP Jr, Lorio MP (2016) ISASS recommendations/coverage criteria for decompression with interlaminar stabilization – coverage indications, limitations, and/or medical necessity. Int J Spine Surg 10:41. https://doi.org/10.14444/3041

Kim DH, Albert TJ (2007) Interspinous process spacers. J Am Acad Orthop Surg 15(4):200–207

Lee SH, Seol A, Cho TY, Kim SY, Kim DJ, Lim HM (2015) A systematic review of interspinous dynamic stabilization. Clin Orthop Surg 7(3):323–329. https://doi.org/10.4055/cios.2015.7.3.323

Li M, Yang H, Wang G (2017) Interspinous process devices for the treatment of neurogenic intermittent claudication: a systematic review of randomized controlled trials. Neurosurg Rev 40(4):529–536. https://doi.org/10.1007/s10143-016-0722-y

Maida G, Marcati E, Sarubbo S (2012) Heterotopic ossification in vertebral interlaminar/interspinous instrumentation: report of a case. Case reports in surgery 2012:970642. https://doi.org/10.1155/2012/970642

Moojen WA, Arts MP, Bartels RH, Jacobs WC, Peul WC (2011) Effectiveness of interspinous implant surgery in patients with intermittent neurogenic claudication: a systematic review and meta-analysis. Eur Spine J 20 (10):1596–1606. https://doi.org/10.1007/s00586-011-1873-8

NASS Coverage Committee (2014a) NASS coverage policy recommendations: interspinous fixation with fusion. Retrieved from https://www.spine.org/PolicyPractice/CoverageRecommendations/AboutCoverageRecommendations#currentcoverage

NASS Coverage Committee (2014b) NASS coverage policy recommendations: interspinous fixation without fusion. Retrieved from https://www.spine.org/PolicyPractice/CoverageRecommendations/AboutCoverageRecommendations#currentcoverage

NASS Coverage Committee (2018) NASS coverage policy recommendations: lumbar interspinous device without fusion and with decompression. Retrieved from: https://www.spine.org/PolicyPractice/CoverageRecommendations/AboutCoverageRecommendations#currentcoverage

Parchi PD, Evangelisti G, Vertuccio A, Piolanti N, Andreani L, Cervi V, Giannetti C, Calvosa G, Lisanti M (2014) Biomechanics of interspinous devices. Biomed Res Int 2014:839325. https://doi.org/10.1155/2014/839325

Phan K, Rao PJ, Ball JR, Mobbs RJ (2016) Interspinous process spacers versus traditional decompression for lumbar spinal stenosis: systematic review and meta-analysis. J Spine Surg 2(1):31–40. https://doi.org/10.21037/jss.2016.01.07

Pintauro M, Duffy A, Vahedi P, Rymarczuk G, Heller J (2017) Interspinous implants: are the new implants better than the last generation? A review. Curr Rev Musculoskelet Med 10(2):189–198. https://doi.org/10.1007/s12178-017-9401-z

Ravindra VM, Ghogawala Z (2017) Is there still a role for interspinous spacers in the management of neurogenic claudication? Neurosurg Clin N Am 28(3):321–330. https://doi.org/10.1016/j.nec.2017.02.002

Serhan H, Mhatre D, Defossez H, Bono CM (2011) Motion-preserving technologies for degenerative lumbar spine: the past, present, and future horizons. SAS J 5(3):75–89. https://doi.org/10.1016/j.esas.2011.05.001

Siewe J, Selbeck M, Koy T, Rollinghoff M, Eysel P, Zarghooni K, Oppermann J, Herren C, Sobottke R (2015) Indications and contraindications: interspinous process decompression devices in lumbar spine surgery. J Neurol Surg A Cent Eur Neurosurg 76(1):1–7. https://doi.org/10.1055/s-0034-1382779

Tian NF, Wu AM, Wu LJ, Wu XL, Wu YS, Zhang XL, Xu HZ, Chi YL (2013) Incidence of heterotopic ossification after implantation of interspinous process devices. Neurosurg Focus 35(2):E3. https://doi.org/10.3171/2013.3.Focus12406

Vertiflex Inc. (2016) Superion indirect decompression system – surgical technique manual. https://www.vertiflexspine.com/wp-content/uploads/2016/10/VF-LD-0001-L_Superion-Surgical-Technique-Manual.pdf

Zhao XW, Ma JX, Ma XL, Li F, He WW, Jiang X, Wang Y, Han B, Lu B (2017) Interspinous process devices (IPD) alone versus decompression surgery for lumbar spinal stenosis (LSS): a systematic review and meta-analysis of randomized controlled trials. Int J Surg 39:57–64. https://doi.org/10.1016/j.ijsu.2017.01.074

Zimmer Biomet Spine Inc. (2016) Aspen MIS fusion system – surgical technique guide. https://www.zimmerbiomet.com/content/dam/zimmer-biomet/medical-professionals/000-surgical-techniques/spine/aspen-mis-fusion-system-surgical-technique-guide.pdf

Kyphoplasty Techniques

28

Scott A. Vincent, Emmett J. Gannon, and Don K. Moore

Contents

Introduction	574
Indications	575
Contraindications	576
Initial Workup	576
History and Examination	576
Imaging	578
Preoperative Testing	579
Technique	580
Postoperative Care	582
Complications	583
Outcomes	586
Special Considerations and Topics	587
Antibiotic Prophylaxis	587
Bilateral Transpedicular Versus Unilateral Transpedicular Approach	587
Metastatic or Primary Bone Tumor Cases	588
Conclusions	589
References	590

S. A. Vincent (✉) · E. J. Gannon
Department of Orthopaedic Surgery and Rehabilitation, University of Nebraska Medical Center, Omaha, NE, USA
e-mail: scott.vincent@unmc.edu; emmett.gannon@unmc.edu

D. K. Moore
Department of Orthopaedic Surgery, University of Missouri, Columbia, OH, USA

© Springer Nature Switzerland AG 2021
B. C. Cheng (ed.), *Handbook of Spine Technology*,
https://doi.org/10.1007/978-3-319-44424-6_60

Abstract

Galibert and Deramond performed the first percutaneous vertebral cement augmentation in 1984 for the treatment of painful vertebral hemangiomas. Over the next decade, its use became more widespread and modifications to the technique led to the development of kyphoplasty. Currently, both kyphoplasty and vertebroplasty are most commonly used in the USA for the treatment of painful osteoporotic vertebral compression fractures. More than 50 million people in the USA have osteoporosis or low bone density and this number is projected to only increase with the aging population. Osteoporotic vertebral compression fractures are one of the most common manifestations of the disease, with more than 1.4 million occurring worldwide each year. These vertebral compression fractures can be a source of substantial morbidity and disability. Other common uses of vertebroplasty and kyphoplasty include the treatment of vertebral body pain or fracture secondary to metastatic disease or primary bone tumors. There have been numerous studies investigating the utility of their use. Despite the large volume of research, there is still debate on the exact role and efficacy of both vertebroplasty and kyphoplasty. Prior to recommending or performing percutaneous vertebral augmentation, physicians should weigh the potential benefits andcomplications for each individual being considered for treatment.

Keywords

Kyphoplasty · Vertebroplasty · Osteoporosis · Vertebral compression fracture · Kyphoplasty technique · Percutaneous vertebral augmentation

Introduction

Percutaneous vertebral body augmentation was first performed in France by Galibert and Deramond who percutaneously injected acrylic cement into the vertebral body for treatment of painful hemangiomas in 1984 (Galibert et al. 1987). This technique was given the name vertebroplasty and was eventually used in the USA in the early 1990s where its main use has been in the treatment of osteoporotic vertebral body compression fractures (VCFs). Kyphoplasty was later developed with the added potential of deformity correction due to the addition of an inflatable bone tamp. The bone tamp is theoretically able to improve the vertebral height and decrease the amount of kyphosis that resulted from the VCF, while also creating a cavity for the cement to be injected. Since its development, the use of kyphoplasty has had widespread use for the treatment of VCFs. In 2007, 130,000 patients with VCFs were treated with either vertebroplasty or kyphoplasty (Mauro 2014). A majority of these VCFs occur in patients with osteoporosis, however, they can also occur in other patient populations including those with hemangiomas, multiple myeloma, and metastatic lesions (Wang et al. 2015). In the USA alone, the estimated number of adults in 2010 with osteoporosis and low bone mass was greater than50 million (Wright et al. 2014). By the year 2020, it was expected that the total number of patients with severe osteoporosis will exceed 14 million (National Osteoporosis Foundation 2002). Due to the aging population, it is predicted that greater than three million osteoporotic fractures will occur in 2025, with more than one-quarter of these affecting the vertebral column (Burge et al. 2007). Osteoporotic VCFs have been shown to significantly affect a patient's quality of life, both mentally and physically. The risk of mortality is also significantly increased after an osteoporotic VCF, with a mortality risk 25% higher than after hip fracture (Cauley et al. 2000). In addition to the significant long-term effect on morbidity and mortality, VCFs can also have an immediate impact on a patient's health. In many cases, narcotics are used as a primary means to attain adequate pain control. Unfortunately, these medications carry a significant risk of serious side effects. Their use for treatment among the commonly affected elderly patient is of particular concern as the geriatric population routinely experiences more severe complications with opioid use. Bed rest is another commonly used treatment modality for patients with painful VCFs. Like narcotics, bed rest can have a substantial impact on an individual in a very short period of time. Bed rest can not only lead to extensive deconditioning, but it has been shown to have an almost immediate detrimental effect on bone quality (Kortebein et al. 2008). Therefore, the risk of short- and long-term consequences can begin to increase immediately after sustaining a 9ol compression fracture severity and complexity varies greatly, and therefore treatment decisions and management strategies should be individualized based on the clinical exam and fracture morphology. Treatment usually begins with medical and nonoperative management; however, in some cases percutaneous vertebral augmentation should be considered. Due to the significant pain some patients experience with VCFs, many physicians believe vertebral augmentation is an excellent option in the treatment pathway, as conservative treatment options fail to provide symptomatic relief. There have been numerous studies investigating the efficacy of vertebroplasty and kyphoplasty that have showed varying degrees of efficacy. Two of the more popular studies that demonstrated no benefit of vertebroplasty were published in the *New England Journal* in 2009, in which both found no difference in outcomes between vertebroplasty and a sham procedure in treatment of osteoporotic VCFs (Kallmes et al. 2009; Buchbinder et al. 2009). A significant benefit, however, was found in well-regarded articles published in the *Lancet* journal, supporting the use of vertebroplasty and kyphoplasty (Klazen et al. 2010; Wardlaw et al. 2009; Clark et al. 2016). The inconsistent findings have led to no general consensus among physicians on the role of percutaneous vertebral augmentation in the treatment of VCFs. The most recent American Academy of Orthopaedic Surgeons (AAOS) clinical guidelines for treatment of osteoporotic VCFs recommend against the use of vertebroplasty (McGuire 2011). In addition, according to the AAOS guidelines, kyphoplasty is considered an option for patients with osteoporotic VCFs with a limited strength of recommendation. Therefore, when considering using percutaneous vertebral augmentation for treatment of a VCF, a physician must consider the risks and benefits of the procedure for each individual.

Indications

The most common indication for the use of kyphoplasty is in the treatment of an unhealed vertebral compression fracture with persistent pain despite conservative therapy. Commonly accepted failures of medical therapy include inadequate relief with analgesic medications, adverse side effects with their use (namely narcotics), and hospitalization secondary to uncontrolled pain. Other medications that have been used include the initiation of osteoporotic-specific medications to prevent future fractures, namely bisphosphonates and teriparatide. Other forms of conservative care that are often used include bed rest and bracing. Much like narcotic therapy, these nonoperative methods are often poorly tolerated by the elderly population. Bed rest leads to deconditioning and has detrimental effects on bone quality, while bracing can be uncomfortable and may restrict pulmonary function. Therefore, the inherent risks and benefits of various conservative treatment modalities should be weighed based on inherent patient factors.

For those failing conservative management, kyphoplasty can be considered. The exact length of time for conservative management is still unclear. Many would consider 3–6 weeks as a reasonable time period of trialing nonoperative care and then considering cement augmentation in those that do not respond. In addition, there is advocacy for earlier utilization of vertebral augmentation in those with incapacitating pain and the inability to tolerate mobilization. The advocacy for earlier utilization of kyphoplasty in select cases is supported by the high mortality rate with VCFs and how commonly nonoperative modalities can be poorly tolerated or cause detrimental effects.

Osteoporosis is the leading cause of painful VCFs that necessitate consideration for treatment. In addition to osteoporosis, other causes of VCFs that may benefit from kyphoplasty include metastatic disease, secondary osteoporosis (i.e., steroid-induced osteoporosis), or multiple myeloma. Kyphoplasty can also be considered in patients without a vertebral fracture that exhibit a painful vertebra secondary to primary bone tumors, like a hemangioma or giant cell tumor, or in those with metastatic disease. In addition, it can be considered in patients with Kummell disease, which is the development of a vascular necrosis of the vertebral body due to a VCF nonunion. Special consideration should be taken in those with metastatic disease and primary tumors of the spine. The timing and treatment plan is very much dependent on tumor type and stage of disease. Collaboration with medical oncologists is warranted in order to determine appropriateness of treatment. It is also important to consider timing of the treatment in regards to specific chemotherapy and radiation therapy plans. There is currently no consensus on the best timing of treatment, whether before, during, or after chemotherapy or radiation treatment. There is a theoretical risk of tumor dissemination after the injection of pressurized cement, leading some physicians to recommend its use after radiation therapy in certain circumstances. The timing of cement augmentation depends largely on the tumor tissue type and planned medical or radiation treatment. For example, multiple myeloma can be treated with cement augmentation at any time as the surgical trauma is minimal and the risk of wound complication in the setting of ongoing or prior radiation therapy is extremely low.

Contraindications

There are both relative and absolute contraindications to the use of kyphoplasty. Absolute contraindications include resultant neurologic injury secondary to the fracture, active spinal or systemic infection, bleeding diatheses, and cardiopulmonary or other health compromise that would impede undergoing the necessary general anesthesia or sedation safely. Allergy to the bone filler/cement or opacification agents is also considered an absolute contraindication. Relative contraindications include instances where the risks and difficulty of performing kyphoplasty are substantially increased. These include disruption of the posterior cortex of the vertebral body, extension of a tumor into the epidural space, significant canal stenosis, and extensive loss of the vertebral height (>70%). Most of these instances result in a significantly increased risk of spinal cord or nerve root injury due to cement leakage. With advanced vertebral collapse, placement of the cannula can become significantly more challenging. Some physicians also recommend against performing kyphoplasty on more than three levels during a single procedure due to the potential risk of developing a cardiopulmonary injury secondary to cement, fat, or marrow embolization to the lungs. The presence of radiculopathy is also considered to be a relative contraindication to the use of kyphoplasty. As a result of the increased risk of complications in patients with these relative contraindications, physicians should proceed with caution and these cases should only be performed by experienced practitioners (Herkowitz and Rothman 2011; Mauro 2014) (Fig. 1).

Initial Workup

History and Examination

Obtaining a full and detailed history is essential in the initial assessment of a patient with a known or suspected VCF. Patients with an acute VCF will typically present with new onset midline back pain that is commonly worsened with standing and motion, especially flexion. Most osteoporotic vertebral compression fractures will present without a history of a fall or trauma (Savage et al. 2014). Key elements of the history include timing of symptom onset, pain severity, individual risk factors including history of previous cancer, diagnosis of osteoporosis, or signs or symptoms concerning for infection. Attempted treatments and their efficacy, including any improvement in symptoms or adverse side effects, are also very important to document. In addition, patients should be inquired on whether they have had any radicular-type symptoms or perceived neurologic changes in sensation, strength, coordination, or bowel and bladder control. Physicians should also inquire about the patient's functional status, past medical history, and use of anticoagulation therapy. Assessing the patient's overall state of health is vital in determining appropriate treatment options and strategy.

The physical examination is another essential piece in the evaluation of a patient with a VCF. Typically, patients will have tenderness to

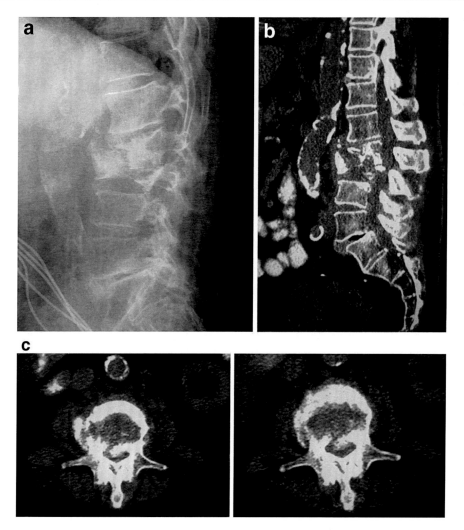

Fig. 1 Sagittal fluoroscopic (**a**) and sagittal (**b**) and axial (**c**) computed tomography images of a burst fracture. There is retropulsion of fracture fragments into the spinal canal and the posterior cortex is also noted to be compromised. These findings would be contraindications to the use of vertebral cement augmentation

palpation over the affected level's spinous process. It is critical to ascertain the level at which the patient is having symptoms, which is especially true in patients with multiple VCFs. The clinical exam and its correlation with imaging findings will then assist in determining which level(s) may benefit from intervention. It is also important to note that tenderness to palpation may not always be present in a patient with an unhealed VCF. Therefore, a lack of localizable pain with palpation should not preclude treatment. In these cases, the patient's history and imaging correlation is imperative in identifying a symptomatic VCF. In addition, a thorough neurologic assessment is the cornerstone of a complete examination and should be done in all patients. This preoperative neurologic assessment will not only identify patients that can potentially worsen with vertebral augmentation, and should be excluded from consideration, but will also aid in detecting any postoperative changes or complications.

Imaging

For every case, imaging of the spine is obtained in order to establish a diagnosis by correlating imaging results with a patient's clinical symptoms and examination findings. Imaging will not only identify potential candidates for intervention, but will also identify those in which cement augmentation would be contraindicated. Comparison to previous imaging is very beneficial in detecting new fractures and lesions or progression of those that had been previously identified. The diagnosis of a new VCF can be confirmed through either magnetic resonance imaging (MRI), serial radiographs, or bone scintigraphy.

Plain radiographs of the spine should be the first imaging modality obtained when evaluating a patient with a suspected VCF. It is an easily attainable assessment of the spine and also an excellent resource for comparison to previous or future radiographs. In addition to the wide accessibility, plain radiographs are a considerably more cost-effective source of initial evaluation when compared to more advanced imaging modalities. Furthermore, standing radiographs of the spine supply an excellent assessment of a patient's coronal and sagittal alignment and stability. This is of special importance when assessing the common kyphotic deformity that can result from a VCF, as well as any potential instability of the vertebral column.

Magnetic resonance imaging (MRI) is another useful adjunct when imaging a patient with known or suspected VCF. An important role of MRI is determining the acuity of VCFs. This can be helpful in patients without previous radiographs or in patients with a history of multiple fractures and equivocal exam findings. In these situations, having the ability to distinguish between new, symptomatic fractures and chronic fractures is essential to guide appropriate treatment when considering kyphoplasty. Findings consistent with an acute fracture include an increased signal on the short tau inversion recovery (STIR) and T2-weighted sequences, and decreased intensity on the T1-weighted sequence. Chronic fractures, which typically are not responsive to kyphoplasty, will not have an increased signal on STIR or T2-weighted sequences. For cases in which the cause of a pathologic fracture is unknown, a MRI is very useful in establishing a differential diagnosis and identifying patients that may require further diagnostic workup. Visualization of cord or nerve root compression from retropulsed fracture fragments, tumors, or other pathology is also best accomplished with a MRI.

Computed tomography (CT) can also be a beneficial resource in the preoperative evaluation of a patient with VCF or pathologic compromise of the vertebral body. A CT is most useful in assessing the integrity of the posterior cortex of the vertebral bodies. When the posterior cortex is compromised, injection of cement can lead to cement leakage or further displacement of the compromised bone posteriorly into the spinal canal. Therefore, a CT is especially valuable in patients in which the integrity of the posterior cortex is in question. It is also the imaging modality of choice for identifying other osseous injuries, and evaluation of the spine in patients involved in high-energy trauma. A CT is also useful in patients that cannot undergo a MRI safely, such as those with a pacemaker.

Bone scintigraphy is another imaging modality that can be used to differentiate between healed and unhealed fractures in patients that cannot undergo MRI. In patients with acute or unhealed fractures, a higher metabolic activity will lead to an increased uptake of technetium-99m. Although bone scintigraphy has a high sensitivity, it has a low specificity as it can continue to show increased uptake for greater than 1 year after a significant amount of healing has occurred (Savage et al. 2014). Another disadvantage of bone scintigraphy is the inability to directly visualize the spinal cord and nerve roots and the lack of spatial resolution. Single photon emission computed tomography (SPECT) is a form of bone scintigraphy that allows for improved fracture localization and characterization due to the improved spatial resolution. A MRI is still preferred over bone scintigraphy, as it is more reliable in assessing the chronicity of a fracture and provides improved visualization of the spinal cord, nerve roots, and surrounding soft tissues (Figs. 2 and 3).

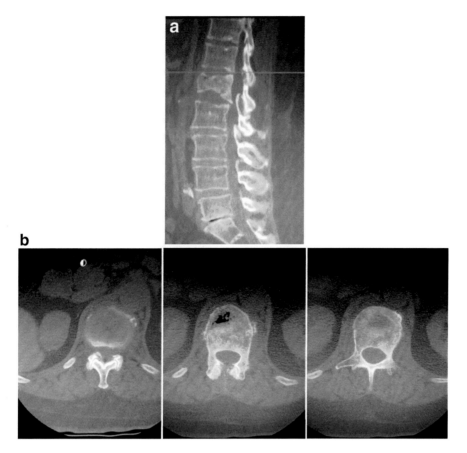

Fig. 2 Sagittal (**a**) and axial (**b**) computed tomography images of a L1 compression fracture secondary to metastatic colon cancer. No retropulsion of fracture fragments into the spinal canal noted. The integrity of the posterior cortex of the vertebral body is noted to be intact

Preoperative Testing

If a patient is considered a candidate for kyphoplasty, there are several laboratory tests that should be routinely obtained prior to proceeding. These include coagulation studies, a basic metabolic panel, and a complete blood cell count. In some instances, further testing may be warranted, such as inflammatory markers, an electrocardiogram, or a chest radiograph. Determining the necessary preoperative testing should be done on a case-by-case basis and should be based on specific patient risk factors. Ideally, this should be accomplished through a team approach that involves the physician performing the procedure, anesthesiologist, hospitalist, and in some circumstances, other medical subspecialists. It is critical to ensure a patient is medically optimized prior to the procedure in order minimize the risk of intraoperative or postoperative complications. Early involvement with referral and establishment of care with a specialist in metabolic bone disease in order to help formulate a postoperative treatment plan to prevent future osteoporotic fractures is also beneficial. Recently, the American Orthopaedic Association developed the Own the Bone program to address the need for comprehensive care of patients with metabolic bone disease. This national postfracture, system-based, multidisciplinary fragility fracture prevention initiative is designed to address physician and patient behavior in an effort to reduce the incidence of further fragility fractures.

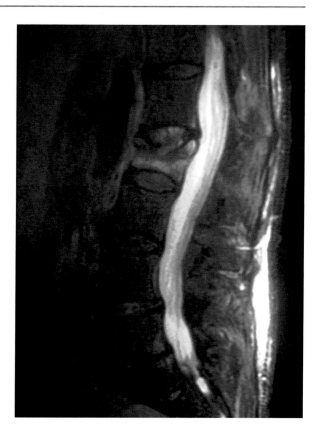

Fig. 3 Sagittal magnetic resonance image demonstrating a L2 compression fracture with accompanying increased signal within the vertebral body, indicating it is most likely an acute fracture

Technique

Before a definitive decision is made on the treatment plan and utilization of the vertebral cement augmentation, a well-informed discussion with the patient regarding the risks and benefits and alternative treatments should occur. Once a patient is determined optimized, they are brought to the operating room or radiology suite and general anesthesia or sedation is initiated. In contrast to vertebroplasty, which is generally performed under local anesthesia, kyphoplasty is usually performed under general anesthesia at most institutions. In patients with a significantly increased risk of medical complications with general anesthesia, the procedure can be performed with intravenous (IV) analgesia and sedation only, as demonstrated by Mohr et al. (2011). The decision between general anesthesia or intravenous sedation should be made in conjunction with the anesthesia provider. Adequate anesthesia should routinely be attained prior to positioning, as required movement and maneuvering can be exceedingly painful for patients with VCFs. The patient is then placed in the prone position and cushion support or chest and pelvic boosters are properly positioned to allow for spine extension. Proper positioning with adequate spine extension will facilitate reduction of the typical kyphotic deformity. The arms should also be placed toward the head of the bed to facilitate fluoroscopic visualization during the procedure. In patients with suspected limited shoulder motion, a preoperative exam testing the range of motion of both shoulders can be beneficial in anticipating lack of abduction and externals needed for positioning. In these cases, the arms may need to be placed in line with the spine. A significant portion of the patient's undergoing kyphoplasty will have underlying osteoporosis, therefore care should be taken during the transferring and positioning of the patient to prevent additional fragility fractures such as rib or sternal fractures.

After attaining adequate anesthesia and positioning of the patient, the next step is identifying the affected level(s) with fluoroscopy. Fluoroscopy is used throughout the procedure and some physicians find the use of simultaneous biplanar fluoroscopy to be beneficial. It is imperative for the correct vertebral level to be treated and close attention to preoperative and intraoperative imaging is critical in ensuring this is accomplished. For both thoracic and lumbar levels, it is helpful to count from sacrum up to the vertebral body to be addressed. Identifying transitional vertebra or anatomic variations preoperatively is very useful in order to correctly correlate with intraoperative fluoroscopic images. Obtaining both thoracic and lumbar X-rays preoperatively is imperative whenever treating thoracic level pathology in order to ensure consistency when counting from the sacrum up to the thoracic level to be treated. It can be helpful to have a discussion with the radiologist preoperatively in advance so in order to ensure the correct levels are labelled and identified prior to surgery. These steps are especially useful in cases in which the thoracic vertebral fractures reduce with positioning, which can lead to increased difficulty in identifying the correct level intraoperatively. Obtaining repetitive fluoroscopic images with a radio-opaque metallic instrument used as reference point while the counting is being done can also be extremely helpful. Placing a sterile marker such as a spinal needle adjacent to the spinous process of the vertebral body can provisionally identify the correct level. If using local anesthesia or IV sedation, a local anesthetic can be delivered via a 22-gauge needle into the skin and periosteum prior to the insertion of the larger needle and cannula. An additional benefit of this step is the ability to make adjustments to the insertion site and trajectory prior to insertion of the larger-gauge needle. A size of 11- or 13-gauge needle is sheathed in a cannula and a Jamshidi needle is then inserted. Prior to this step, a small incision can be made to allow for easier insertion and trajectory adjustments. There are two specific approaches to the vertebral body that can be utilized. These include a transpedicular approach or an extrapedicular approach. The transpedicular approach begins with needle insertion at the posterior aspect of the pedicle, followed by subsequent cannulation through the length of the pedicle and into vertebral body. The extrapedicular approach entails the needle traveling along the lateral aspect of the pedicle and then inserting into the vertebral body at the junction of the pedicle and vertebral body. One benefit of the extrapedicular approach is that it allows for a more medial tip placement of the needle in the vertebral body which may allow more centralized cement placement. This can be difficult to attain with the transpedicular approach as the path is limited by the anatomic configuration of the pedicle. An advantage of using the transpedicular approach is the utilization of an intraosseous path that protects against soft tissue structure penetration and potential neurologic injury. A general guideline for both approaches, to decrease the risk of accidental spinal canal or neural foramen penetration, is to keep the needle superior to the inferior cortex of the pedicle on the lateral fluoroscopic image and lateral to the medial cortex of the pedicle on the AP view. Advancement of the needle is done under fluoroscopic guidance, ensuring proper trajectory. A mallet or orthopedic hammer can be used to assist in needle advancement. Once the needle is advanced into the vertebral body, just anterior to the junction of the pedicle and the body, the stylet is removed and a working channel through the cannula is utilized for advancement of the balloon tamp. If necessary, biopsy needles can be used at this point to obtain samples prior to balloon tamp and cement insertion. The cannula is then brought back posteriorly to the junction of the pedicle and the vertebral body. Kyphoplasty can be performed through either a bipedicular or unipedicular approach. If the bipedicular approach is used, the Jamshidi needle or needle and cannula placement on the contralateral side is done at this time. The balloon bone tamp is then inserted and advanced within the vertebral body. The balloon tamp is then inflated under intermittent fluoroscopic visualization and pressure monitoring via a digital manometer. When inflating the balloon, inflation is stopped once the fracture has been adequately reduced; the balloon tamp reaches maximal pressure or volume, or cortical contact occurs. After one of these objectives is

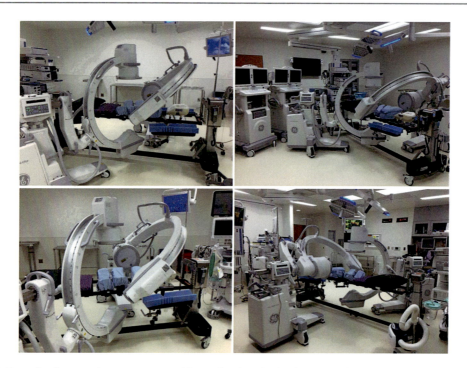

Fig. 4 Example of an operating room setup used for performing a kyphoplasty procedure. In this example, utilization of simultaneous biplanar fluoroscopy is accomplished with the use of two C-arms

attained, the balloon is then deflated and removed. The cement, most commonly polymethyl methacrylate (PMMA), is then injected through the cannula until the cavity created by the balloon tamp is filled. A radio pacifier is required to appropriately visualize cement administration fluoroscopically. Most commercially available PMMA formulations contain either barium sulfate ($BaSO_4$) or ziroconium dioxide (ZrO_2) as a radiopacifier. Radiopaque cement is necessary to monitor for extravasation and ensure adequate filling of the cavity formed by the balloon tamp. In addition to the inclusion of a radiopacifier attaining an appropriate level of viscosity prior to its injection is critical. This will assist in preventing extravasation and also facilitate cement travel through the cannula. According to Lieberman et al., cement with a low viscosity or longer liquid phase is preferred for vertebroplasty, while cement with high viscosity or longer working phase is more ideal for kyphoplasty (Lieberman et al. 2005). The patient is then left in the supine position until the cement has cured. The cement plungers are inserted into the working cannula after the delivery of the cement as close as possible to the end of the cement filler. This prevents leaving a cement column that may harden inside the cannula and thus remain in the soft tissue after the cannulas are removed. Once it has cured, the cannulas are removed, dressings are applied, and the patient is transported back to their hospital bed. (Herkowitz and Rothman 2011; Mauro 2014) (Fig. 4).

Postoperative Care

After the patient is safely transported back to the hospital bed, the patient is brought to the post-anesthesia care unit for routing postoperative monitoring. Some physicians recommend obtaining a routine postoperative chest X-ray in patients undergoing thoracic kyphoplasty to rule out iatrogenic pneumothorax. Select patients may benefit from an overnight observational stay, while most patients are safe for discharge later the same day. Most of the care postoperatively focuses on assessing for any neurologic changes and attaining adequate pain control. For the same reason kyphoplasty may be indicated, avoidance or minimized use of narcotics should be a priority when formulating a sufficient analgesic regimen in order to avoid their deleterious side effects. If the patient develops neurologic deficits, or other concerns for cement extravasation, CT imaging should be obtained urgently. Physicians must also be cognizant of the potential for pulmonary embolism, particularly if multiple levels were addressed. A chest X-ray to rule out pulmonary edema should be considered for patients with postoperative dyspnea.

Establishing appropriate follow-up is necessary for these patients as many will require treatment for their underlying cause of fracture. Most frequently, patients will require management of their underlying osteoporosis and it is important to make the appropriate referrals for necessary testing and treatment of underlying metabolic bone disease. Furthermore, many patients will be at high risk of subsequent fractures, and education regarding future risk of fracture is essential. In follow-up, if signs or symptoms of subsequent fractures occur, providers should obtain new imaging as appropriate. Routine follow-up radiographs should also be obtained and can be useful for comparison if further fracture or deformity occurred (Fig. 5).

Complications

Complications following percutaneous vertebral augmentation are generally rare; however, they can be a cause of significant morbidity. Complications that do occur are commonly a result of cement extravasation, subsequent fracture, or embolization. Other potential complications include infection, pneumothorax, nerve or spinal cord injury, pain exacerbation, hematoma formation, and intraoperative fractures (pedicle, vertebral body, and rib). The type of fracture being treated also plays an important role in risk of complications as malignancy-related fractures result in a higher complication rate compared to osteoporotic VCFs (Mathis et al. 2001; Barragan-Campos et al. 2006). When comparing kyphoplasty and vertebroplasty, the rate of procedure-related complications is significantly lower with kyphoplasty (Lee et al. 2009). Cement extravasation is a common occurrence for both kyphoplasty and vertebroplasty, but it is rarely symptomatic. In some circumstances, however, cement extravasation can lead to neurologic deficits, which may necessitate decompression and reconstruction (Savage et al. 2014). Lee et al. reported the rate of symptomatic cement extravasation is significantly lower in kyphoplasty compared to vertebroplasty (Lee et al. 2009). The study found the rate of symptomatic cement extravasation was 1.48% after vertebroplasty and 0.04% following kyphoplasty. If there is concern for complications related to cement extravasation, a CT scan is the imaging modality of choice to best visualize cement leakages. Embolization is another potential complication that is commonly asymptomatic; however, it may have severe cardiopulmonary consequences. The rate of cement embolization following percutaneous vertebral augmentation varies between 2.1% and 26% (Wang et al. 2012). The incidence appears to be lower following kyphoplasty compared to vertebroplasty. This is likely a result of the creation of a cavity that leads to the cement being injected under lower pressure. The emboli can either be from the bone marrow fat or the cement as a small fragment or as monomer that is later polymerized at a distant location. Regardless of cause, this may lead to cardiopulmonary embolism, which can be fatal in very rare cases. Clinical manifestations of cardiopulmonary embolization include patient complaints of chest pain or tightness, palpitations, and shortness of breath. Examination of the patient may reveal tachypnea, hypotension, oxygen desaturation, cyanosis, or cardiac arrhythmias with the potential development of acute respiratory distress syndrome or cardiac arrest. Physicians should be cognizant of the early signs and symptoms of cardiopulmonary embolization, as early diagnosis and treatment is critical. The development of subsequent fracture is common following vertebral augmentation. Most patients being treated will commonly have an underlying condition that already carries an increased risk of future fractures. Treatment with vertebral augmentation, however, does not appear to be an individual risk factor. A meta-analysis done by Anderson et al. demonstrated no significant difference in secondary fractures between those treated with vertebroplasty and those treated with conservative management. In this analysis, both groups had approximately 20% of patients developing a new fracture between 6 and 12 months after the procedure (Anderson et al. 2013). Because of this high rate of subsequent

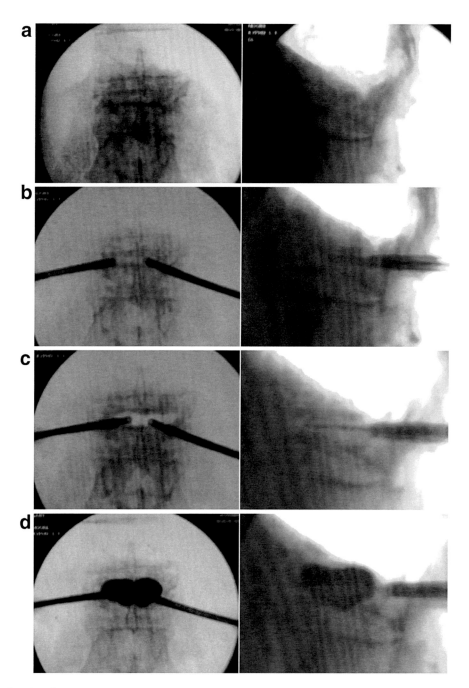

Fig. 5 (continued)

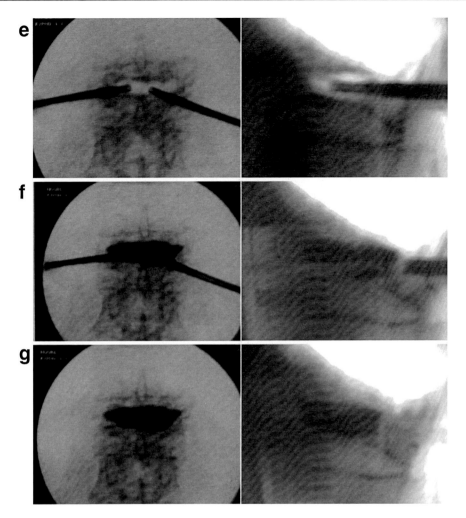

Fig. 5 Biplanar fluoroscopic images of a kyphoplasty being performed using a bilateral approach for treatment of vertebral compression fracture. (**a**) Vertebral compression fracture. (**b**), Insertion of the starting needles into the vertebral body. (**c**) Insertion of the balloon bone tamps. (**d**) Inflation of the balloon bone tamps. (**e**) Residual cavity formation noted after deflation of the balloons. (**f**), Injection of PMMA cement into the cavity. (**g**) Final AP and lateral fluoroscopic images following cement augmentation with mildly improved sagittal alignment and vertebral height

fracture, physicians should be weary of future fractures and attempt to decrease the risk by establishing appropriate treatment for underlying diseases. Although the development of an adjacent or new spinal level vertebral fracture is more common, it is also possible for patients to have a re-fracture or progression at a previously treated level. This should be of concern in patients that have no improvement, increasing pain, or worsening pain after an initial improvement period after treatment. Patients at an increased risk of re-fracture or progression include those with inadequately filled fractures or with fluid-filled vertebral fracture clefts (Jacobson et al. 2017). For these patients, a MRI or fine cut CT can assist in determining the cause for lack of improvement or early deterioration. Treatment with either observation or revision should be formulated based upon the patient's clinical status and imaging findings. Overall, complications are rare following vertebral augmentation. Treating physicians, however, should be aware of the signs and symptoms of the potential complications as late recognition may lead to significant morbidity and poor outcomes.

Outcomes

There have been numerous studies investigating the efficacy of vertebral augmentation. Despite the extensive volume of data, debate still exists regarding its effectiveness. Since the late 2000s, a number of prospective randomized controlled trials (RCTs) have been published investigating the efficacy of vertebral augmentation for treatment of osteoporotic VCFs. Wardlaw et al. published a prospective RCT in which kyphoplasty was compared to nonoperative management of VCFs (Wardlaw et al. 2009). In this study a significant improvement in the Short-Form-36 (SF-36) physical component summary scores were found in the kyphoplasty group compared to the nonoperative group at 1 month. Another prospective RCT done by Klazen et al. found beneficial results when comparing vertebroplasty to medical management of VCFs (Klazen et al. 2010). In this study there was a significant improvement found in pain scores and in secondary outcome measures in those that underwent vertebroplasty. Similar to other prospective RCTs that demonstrated beneficial results with vertebral augmentation, both studies by Klazen et al. and Wardlaw et al. did not blind the treatment and control groups. The absence of blinding has been considered a major limitation of these and similar studies as the efficacy of vertebral augmentation may be overestimated secondary to a placebo effect. In 2009, two articles, by Kallmes et al. and Buchbinder et al., were published in the *New England Journal of Medicine* in which both the treatment and control group were blinded (Kallmes et al. 2009; Buchbinder et al. 2009). In both of these prospective RCTs, vertebroplasty was found to have no beneficial effect compared to a sham procedure in the treatment of osteoporotic VCFs. Although these studies addressed a major limitation of similar RCTs, there has been concern regarding the selection criteria for patients involved in these studies. One concern entails the inclusion of patients with fractures that were up to 12 months old. The involvement of a sham procedure instead of traditional medical management has also led many to question the impact these articles should have on practice management. The RCTs by Kallmes et al., Buchbinder et al., Klazen et al., and Wardlaw et al. were subsequently utilized in a meta-analysis performed by Anderson et al. (2013). In addition to these four RCTs, two additional studies met inclusion criteria and were used to compare vertebral augmentation with conservative management in patients with osteoporotic VCFs. The study revealed a significant improvement in pain relief, functional recovery, and health-related quality of life with vertebral augmentation compared to nonoperative management or sham procedures. This significant difference was noted at early (less than 12 weeks) and long-term follow-up (6–12 months). In 2016, Clark et al. published results on a multicenter, double-blinded, prospective RCT in which 44% of patients that underwent vertebroplasty had a numeric rated pain score below 4 out of 10–14 days compared to only 21% in the control group (Clark et al. 2016). In this study, the control group underwent a process to simulate vertebroplasty in order to control for the placebo effect. Unlike the sham procedures performed in the studies done by Kallmes et al. and Buchbinder et al., there was no local anesthetic or needle infiltration of the periosteum as lidocaine use was limited to subcutaneous administration only. Following the procedure, patients were then treated by their primary physicians with standard medical care. Inclusion criteria for this study also required that the patient's painful vertebral fractures were less than 6 weeks old. This article, in addition to the meta-analysis done by Anderson et al., support the use of vertebral cement augmentation in carefully selected patients with painful VCFs (Clark et al. 2016; Anderson et al. 2013). There have also been studies comparing the results of kyphoplasty with vertebroplasty. In a systematic review done by Han et al., the authors concluded that vertebroplasty had improved short-term pain relief while kyphoplasty demonstrated better intermediate-term functional improvement (Han et al. 2011). There was found to be no difference, however, between the two in long-term pain relief or functional status. In a study done by Omidi-Kashani et al., both kyphoplasty and vertebroplasty demonstrated significant improvement in pain scores and outcome measures (Omidi-Kashani et al. 2013). Those that underwent kyphoplasty showed improved kyphosis with an average of 3.1° of correction. This study did not find a significant difference between the two in regards to pain and functional outcomes. As mentioned previously, complications have been shown to be more commonly seen with vertebroplasty, especially cement extravasation. The most recent AAOS guidelines recommend against the use of vertebroplasty and kyphoplasty carried a limited recommendation in the treatment of painful osteoporotic VCFs. Since this recommendation, there have been multiple articles published supporting the use of both vertebroplasty and kyphoplasty. With the substantial amount of data investigating the use of vertebral augmentation, physicians should make an effort to understand the strengths and limitations of the current literature in order to formulate the optimal treatment plan for each patient.

Special Considerations and Topics

Antibiotic Prophylaxis

For percutaneous vertebral augmentation, antibiotic prophylaxis can be accomplished one of two ways, via IV administration or by mixing with the PMMA during cement preparation. Although there is no data to support its use in this procedure, most practicing providers use at least one type of antibiotic prophylaxis due to the potential morbidity associated with infection (Moon et al. 2010). As with many other procedures, the most common infection-causing bacteria are *Staphylococci* and *Streptococci* species. For that reason, the most frequently used IV antibiotics include cefazolin, cefuroxime, and clindamycin. When using antibiotic impregnated cement, 1.2 g of tobramycin is ordinarily used and mixed with the PMMA cement. Both impregnated cement and IV administered antibiotics are considered appropriate as no evidence demonstrates superiority of one technique over the other. Theoretical disadvantages include increasing antibiotic resistance and the individual side effects that accompany their use. Based on the Surgical Care Improvement Project (SCIP) guidelines, the authors recommend intravenous antibiotics given within 1 h prior to surgical incision (Rosenberger et al. 2011).

Bilateral Transpedicular Versus Unilateral Transpedicular Approach

Kyphoplasty has been traditionally been performed using a bilateral transpedicular approach. This requires bilateral insertion of the balloon bone tamp and simultaneous inflation to create the cavity. Some studies, however, have shown that it can be done using a unilateral approach without negatively affecting outcomes. Chen et al. and Yılmaz et al. both demonstrated no significant difference in pain relief, kyphotic angle, and vertebral height restoration between the unilateral and bilateral approaches (Chen et al. 2014; Yılmaz et al. 2017). Both studies also found that the unilateral approach required a significantly shorter operative time and less cement. Hu et al. reported similar success with the use of a unilateral approach and, like many other authors, recommended a more medial trajectory to attain a midline position within the vertebral body (Hu et al. 2005). Yılmaz et al., however, questioned the necessity of a midline position when using the unilateral approach (Yılmaz et al. 2017). In their study, the needle trajectory was not altered from their typical trajectory and placement with the bilateral approach and, therefore, no additional effort was made to obtain a more medial start point or final midline position. This approach led to no difference in outcomes or decreased deformity correction when compared to other studies. The baseline position of needle placement in this study, however, was not reported, and therefore it is difficult to assess the significance of these findings. It does appear, however, that the unilateral approach can be used safely in kyphoplasty without negatively affecting outcomes. A recent analysis of registry data evaluated the effect of cement volume on pain relief in balloon kyphoplasty. In their analysis, they found that cement volumes greater than 4.5 ml independently predicted pain relief in patients with vertebral compression fractures (Röder et al. 2013). This data may explain why a unilateral approach may be as successful as a bilateral approach, simply by restoring the mechanical property of the cemented vertebral body. Advantages of the bilateral approach include the ability to more easily access the contralateral portion of the vertebral body for cavity formation and the facilitation of cement injection using bilateral cannulas. The shorter operative time and the avoidance of the risks associated with placing an additional needle are both benefits of the unilateral approach. Some physicians also believe attaining a more midline position when utilizing the unilateral method, which is easier to obtain using an extrapedicular approach to the vertebral body. An insertion needle with a flexible tip to allow for a modifiable curve is also currently available and may aid in obtaining a more midline position within the vertebral body. Overall, the outcomes of both the bilateral and unilateral approach appear to be similar and the decision on which approach is utilized should be based on the performing physician's experience and comfort (Fig. 6).

Metastatic or Primary Bone Tumor Cases

There are a few special considerations when pathologic fractures involve metastatic or primary bone tumors. An essential part of ensuring improvement following vertebral augmentation with these types of cases is differentiating pain related to the fracture versus the tumor. This is critical, as pain originating from the tumor is typically not improved with vertebral augmentation (Savage et al. 2014). Clinical features that would be more consistent with a painful fracture include pain that increases with load-bearing activities, such as walking, sitting, or standing. Whereas pain that is secondary to the tumor will typically be present at rest and when lying supine, patients may also experience the classic worsening of symptoms at night. If a patient is having tumor-related pain, this is most often treated more successfully with radiation therapy. Patients with pain secondary to fractures with metastatic disease or primary bone tumors, such as giant cell tumors, may benefit from vertebral augmentation. First line treatment for these types of fractures, much like that for osteoporotic VCFs, consists of medical management and appropriate analgesia. The goal for treatment of painful metastatic or primary bone tumors of the vertebral body is to attain pain control and preserve function. Radiation, chemotherapy, and bisphosphonate therapy are all options that should be discussed and considered as reasonable treatment options (Gralow and Tripathy 2007). As previously discussed with VCFs, goals of treatment and timing of intervention should be addressed utilizing a team approach and individualized based on fracture pattern and underlying pathology. If vertebral cement augmentation is indicated, special care should be taken to ensure the risk of potential complication is minimized. Careful review of pertinent imaging is important for minimizing the potential risk of complication. Important aspects of the imaging include visualization of the integrity of the posterior cortex of the vertebral body, and any potential spinal cord or nerve root compression as a result of the tumor. In addition, a biopsy may be necessary in some cases and this should be known prior to proceeding. Outcomes in the treatment of cancer-related VCFs with kyphoplasty have been promising. In a randomized-control study, Berenson et al. found a significant improvement in pain relief and overall function at 1 month postoperatively compared to the control group (Berenson et al. 2011). Dudeney et al. also showed favorable results with the use of kyphoplasty in patients with osteolytic VCFs secondary to multiple myeloma (Dudeney et al. 2002). In their study, patients experienced a significant improvement in SF-36 scores, pain, and in physical and social function compared to preoperatively. With the main goals in treatment being pain relief and maintaining function, kyphoplasty is a viable option for certain patients with primary bone tumors or metastatic disease.

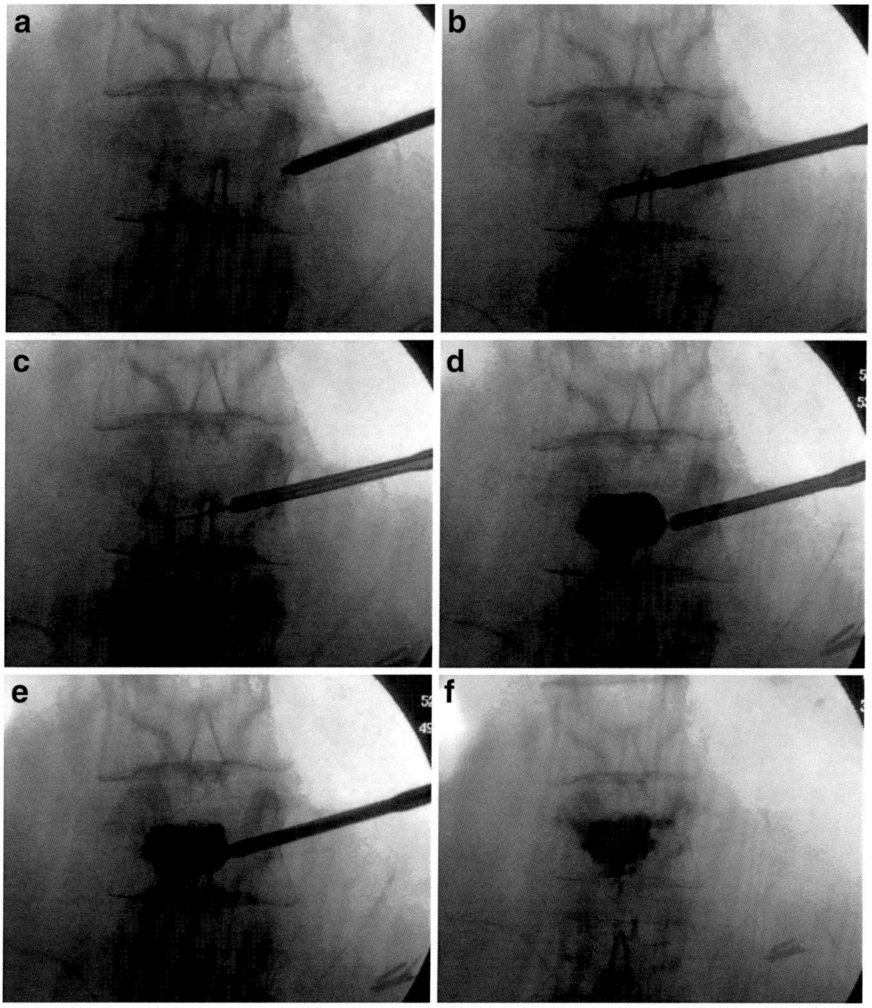

Fig. 6 Sequential intraoperative fluoroscopic images of a kyphoplasty being performed using a unilateral approach. (**a**) Initial insertion of the starting needle into the vertebral body. (**b**) Advancement of the needle utilizing a medial trajectory in order to attain a more midline final position. (**c**) Insertion of the balloon bone tamp. (**d**) Inflation of the balloon bone tamp. (**e**) Injection of PMMA cement into the cavity created by the balloon tamp. (**f**) Final AP fluoroscopic image following kyphoplasty performed via an unilateral approach

Conclusions

Despite the large volume of literature, there is still no consensus on the role of vertebral cement augmentation. Furthermore, the ideal timing of performing kyphoplasty or vertebroplasty remains controversial. Based upon the current available literature, cement augmentation should be considered for patients that meet a general set of criteria. The ideal patients being those that fail conservative management with persistent, debilitating pain, limited mobility, and an acute VCF. The length of time dedicated to conservative treatment is of debate, but generally 3–6 weeks is a commonly used time frame. Earlier consideration for patients that poorly tolerate nonoperative care, particularly narcotics and bed rest, seems to be appropriate. The use of vertebral augmentation for treatment of chronic symptomatic VCFs, metastatic disease, and primary bone tumors has also shown promise and can be considered in certain situations. Physicians should be cognizant of the potential benefits and complications of kyphoplasty and vertebroplasty when considering treatment with vertebral augmentation. In addition, it is vital for practitioners to have a solid grasp on the current literature in order to hold well-informed discussions with patients when making an individualized treatment plan.

References

Anderson PA, Froyshteter AB, Tontz WL Jr (2013) Meta-analysis of vertebral augmentation compared with conservative treatment for osteoporotic spinal fractures. J Bone Miner Res 28(2):372–382

Barragan-Campos HM, Vallee J-N, Lo D et al (2006) Percutaneous vertebroplasty for spinal metastases: complications. Radiology 238(1):354–362

Berenson J, Pflugmacher R, Jarzem P, Cancer Patient Fracture Evaluation (CAFE) Investigators et al (2011) Balloon kyphoplasty versus non-surgical fracture management for treatment of painful vertebral body compression fractures in patients with cancer: a multicentre, randomised controlled trial. Lancet Oncol 12 (3):225–235

Buchbinder R, Osborne RH, Ebeling PR et al (2009) A randomized trial of vertebroplasty for painful osteoporotic vertebral fractures. N Engl J Med 361(6):557–568

Burge R, Dawson-Hughes B, Solomon DH et al (2007) Incidence and economic burden of osteoporosis-related fractures in the United States, 2005–2025. J Bone Miner Res 22(3):465–475

Cauley JA, Thompson DE, Ensrud KC et al (2000) Risk of mortality following clinical fractures. Osteoporos Int 11(7):556–561

Chen H, Tang P, Zhao Y et al (2014) Unilateral versus bilateral balloon kyphoplasty in the treatment of osteoporotic vertebral compression fractures. Orthopedics 37(9):e828–e835

Clark W, Bird P, Gonski P et al (2016) Safety and efficacy of vertebroplasty for acute painful osteoporotic fractures (VAPOUR): a multicentre, randomised, double-blind, placebo-controlled trial. Lancet 388(10052):1408–1416

Dudeney S, Lieberman IH, Reinhardt MK, Hussein M (2002) Kyphoplasty in the treatment of osteolytic vertebral compression fractures as a result of multiple myeloma. J Clin Oncol 20(9):2382–2387

Galibert P, Deramond H, Rosat P, Le Gars D (1987) Preliminary note on the treatment of vertebral angioma by percutaneous acrylic vertebroplasty. Neurochirurgie 33 (2):166–168

Gralow J, Tripathy D (2007) Managing metastatic bone pain: the role of bisphosphonates. J Pain Symptom Manag 33(4):462–472

Han S, Wan S, Ning L et al (2011) Percutaneous vertebroplasty versus balloon kyphoplasty for treatment of osteoporotic vertebral compression fracture: a meta-analysis of randomised and non-randomised controlled trials. Int Orthop 35(9):1349–1358

Herkowitz HN, Rothman RH (2011) The spine, 6th edn. Elseviers/Saunders, Philadelphia

Hu MM, Eskey CJ, Tong SC et al (2005) Kyphoplasty for vertebral compression fracture via a uni-pedicular approach. Pain Physician 8(4):363–367

Jacobson RE, Palea O, Granville M (2017) Progression of vertebral compression fractures after previous vertebral augmentation: technical reasons for recurrent fractures in a previously treated vertebra. Cureus 9(10):e1776

Kallmes DF, Comstock BA, Heagerty PJ et al (2009) A randomized trial of vertebroplasty for osteoporotic spinal fractures. N Engl J Med 361(6):569–579

Klazen CA, Lohle PN, de Vries J, Jansen FH et al (2010) Vertebroplasty versus conservative treatment in acute osteoporotic vertebral compression fractures (VERTOS II): an open-label randomised trial. Lancet 376(9746):1085–1092

Kortebein P, Symons TB, Ferrando A et al (2008) Functional impact of 10 days of bed rest in healthy older adults. J Gerontol A Biol Sci Med Sci 63(10):1076–1081

Lee MJ, Dumonski M, Cahill P et al (2009) Percutaneous treatment of vertebral compression fractures: a meta-analysis of complications. Spine 34(11):1228–1232

Lieberman IH, Togawa D, Kayanja MM (2005) Vertebroplasty and kyphoplasty: filler materials. Spine J 5(6 Suppl):305S–316S

Mathis JM, Barr JD, Belkoff SM et al (2001) Percutaneous vertebroplasty: a developing standard of care for vertebral compression fractures. AJNR Am J Neuroradiol 22 (2):373–381

Mauro MA (2014) Image-guided interventions, 2nd edn. Saunders/Elsevier, Philadelphia

McGuire R (2011) AAOS clinical practice guideline: the treatment of symptomatic osteoporotic spinal compression fractures. J Am Acad Orthop Surg 19(3):183–184

Mohr M, Pillich D, Kirsch M et al (2011) Percutaneous balloon kyphoplasty with the patient under intravenous analgesia and sedation: a feasibility study. AJNR Am J Neuroradiol 32(4):649–653

Moon E, Tam MD, Kikano RN, Karuppasamy K (2010) Prophylactic antibiotic guidelines in modern interventional radiology practice. Semin Interv Radiol 27(4):327–337

National Osteoporosis Foundation (2002) America's bone health: the state of osteoporosis and low bone mass in our nation. National Osteoporosis Foundation, Washington, DC

Omidi-Kashani F, Samini F, Hasankhani EG et al (2013) Does percutaneous kyphoplasty have better functional outcome than vertebroplasty in single level osteoporotic compression fractures? A comparative prospective study. J Osteoporos 2013:690329

Röder C, Boszczyk B, Perler G, Aghayev E, Külling F, Maestretti G (2013) Cement volume is the most important modifiable predictor for pain relief in BKP: results from SWISSspine, a nationwide registry. Eur Spine J 22:2241–2248

Rosenberger LH, Politano AD, Sawyer RG (2011) The surgical care improvement project and prevention of post-operative infection, including surgical site infection. Surg Infect 12(3):163–168

Savage JW, Schroeder GD, Anderson PA (2014) Vertebroplasty and kyphoplasty for the treatment of osteoporotic vertebral compression fractures. J Am Acad Orthop Surg 22(10):653–664

Wang LJ, Yang HL, Shi YX et al (2012) Pulmonary cement embolism associated with percutaneous vertebroplasty or kyphoplasty: a systematic review. Orthop Surg 4(3):182–189

Wang H, Sribastav SS, Ye F et al (2015) Comparison of percutaneous vertebroplasty and balloon kyphoplasty for the treatment of single level vertebral compression fractures: a meta-analysis of the literature. Pain Physician 18(3):209–222

Wardlaw D, Cummings SR, Van Meirhaeghe J et al (2009) Efficacy and safety of balloon kyphoplasty compared with non-surgical care for vertebral compression fracture (FREE): a randomised controlled trial. Lancet 373 (9668):1016–1024

Wright NC, Looker AC, Saag KG et al (2014) The recent prevalence of osteoporosis and low bone mass in the United States based on bone mineral density at the femoral neck or lumbar spine. J Bone Miner Res 29 (11):2520–2526

Yılmaz A, Çakır M, Yücetaş CŞ et al (2017) Percutaneous kyphoplasty: is bilateral approach necessary? Spine 43 (14):977–983

Anterior Spinal Plates: Cervical

A. Karim Ahmed, Zach Pennington, Camilo A. Molina, C. Rory Goodwin, and Daniel M. Sciubba

Contents

Introduction	594
Anatomic and Biomechanical Considerations	594
Patient Evaluation	595
Diagnostic Work-Up	595
Special Populations to Consider	595
Surgery	596
Positioning and Monitoring	596
Operative Details	596
Complications	597
Case Illustration	598
References	599

Abstract

Anterior cervical diskectomy and fusion (ACDF) is one of the most common and effective spine procedures performed, indicated for the treatment of cervical degenerative disk disease. Patients frequently present with neck pain, radiculopathy, or myelopathy secondary to compression of the neural elements. A thorough patient evaluation is performed, consisting of clinical examination, imaging, and possibly nerve conduction studies. This chapter outlines the diagnostic evaluation, indications, operative details, considerations, and complications of the ACDF procedure.

Keywords

Anterior cervical diskectomy and fusion · Anterior cervical fusion · Diskectomy · Corpectomy · Subaxial cervical spine · Cervical radiculopathy · Cervical myelopathy

A. K. Ahmed · C. A. Molina
Department of Neurosurgery, The Johns Hopkins School of Medicine, Baltimore, MD, USA
e-mail: aahmed1@alumni.nd.edu; cmolina2@jhmi.edu

Z. Pennington · D. M. Sciubba (✉)
Department of Neurosurgery, Johns Hopkins Hospital, The Johns Hopkins School of Medicine, Baltimore, MD, USA
e-mail: zpennin1@jhmi.edu; dsciubba1@jhmi.edu

C. R. Goodwin
Department of Neurosurgery, Duke University Medical Center, Durham, NC, USA
e-mail: rory.goodwin@duke.edu

© Springer Nature Switzerland AG 2021
B. C. Cheng (ed.), *Handbook of Spine Technology*,
https://doi.org/10.1007/978-3-319-44424-6_61

Introduction

The anterior approach to the subaxial cervical spine was first reported by Smith and Robinson, in The Bulletin of the Johns Hopkins Hospital in 1955, where it was described for the treatment of cervical disk herniation (Robinson and Smith 1955); this reported was followed shortly thereafter by Cloward in 1958 (Cloward 1958). In contrast with posterior decompression, anterior approaches to the subaxial cervical spine provide direct decompression for compressive pathologies involving the vertebrae and/or intervertebral disks. Posterior decompressive laminectomy alone may carry a risk of progressive kyphotic deformity, especially when spanning multiple levels or when applied to patients with noted ligamentous laxity (i.e. younger patients and those with select inherited connective tissue diseases) (Song and Choi 2014; Kani and Chew 2018).

Anterior cervical diskectomy (ACDF) is one of the most common neurosurgical procedures performed, with high rates of efficacy and a relatively short recovery period. The utilization of fascial planes minimizes soft tissue disruption, and the limited range of motion in the subaxial spine does not substantially impact post-operative mobility (Robinson and Smith 1955; Cloward 1958; Song and Choi 2014; Kani and Chew 2018).

Anatomic and Biomechanical Considerations

The subaxial spine describes cervical vertebrae caudal to the axis, C2 (C3–C7). The occiput meets the cervical spine at the level of C1 (atlas), with the occipital condyles articulating with the lateral masses of the atlas. The resulting atlanto-occipital junction facilitates the majority of flexion-extension mmotion within the cervical spine. The bilateral anterior and posterior arches of C1 form the ring of the atlas and are secured to the occipital bone by the anterior and posterior atlanto-occipital membranes which help to prevent hyperextension and hyper-flexion at the O-C1 joint, respectively. The adjacent atlantoaxial junction (C1–C2) - the articulation of the atlas with the axis - allows for most of the rotational motion of the head.

The borders of the central canal are defined by the vertebral body anteriorly, the pedicles and lateral masses anterolaterally, and the laminae posteriorly. Unique to the cervical vertebrae are the foramina transversaria, which encase the V2 segment of the vertebral arteries from the level of C6 to C2, and lie within the transverse processes, located anterolateral to the lateral masses. The vertebral arteries originate from the subclavian artery and are divided into four segments: (V1) pre-foraminal, stretching from the subclavian artery to C6 foramen; (V2) foraminal, running cranially through the foramina transversaria from C6 to C2; (V3) extradural, arching laterally from the superior surface of C2, around the posterior C1 arch, through the sulcus arteriosus to the point of dural entry above C1; and (V4) intradural, running from the dural surface to the juncture of the basilar artery.

Load to the atlantoaxial junction is transmitted through the lateral masses, unlike the subaxial spine where intervertebral disks are the major shock absorbers. In the subaxial spine, the greatest degree of degeneration occurs at the C5–C6 level. In the cervical spine, each nerve root exits above the pedicle of the respectively numbered pedicle, with the exception of the C8 root which exits below the C7 vertebrae (Lang 1993; Tubbs et al. 2011; Hai-bin et al. 2012).

Motor innervation from the cervical spine is essential for upper extremity mobility and breathing. The phrenic nerve, from C3 to C5, supplies the diaphragm, with the brachial plexus supplied from C5 to T1. The musculocutaneous (C5–C7), axillary (C5, C6), radial (C5–T1), median (C6–T1), and ulnar (C8, T1) are the major terminal nerves. Patients with nerve root compression will commonly present with radiculopathy corresponding to the myotome supplied, including abduction of the arm (C5), elbow flexion (C6), elbow extension (C7), flexion of the digits (C8), and abduction of the digits (T1) (Song and Choi 2014; Kani and Chew 2018; Lang 1993; Tubbs et al. 2011; Hai-bin et al. 2012; Payne and Spillane 1957).

The combination of degenerative forces over time, and canal diameter, is important to understand the common levels of breakdown. As previously stated, disk degeneration most commonly occurs at the C5–C6 level. The canal diameter, however, is most narrow at the C4 level (13.33–17.50 mm) and widest at C1 (18.47–21.60 mm), with a mean cord diameter of 10 mm and mean overall canal diameter of 17 mm (Hai-bin et al. 2012; Payne and Spillane 1957; Gupta et al. 1982; Hashimoto and Tak 1977).

Patient Evaluation

Patients with degenerative conditions of the cervical spine frequently present with neck pain, arm pain, and/or neurologic deficits. Mechanical neck pain, exacerbated by motion, suggests instability between adjacent vertebrae. Upper extremity pain/weakness/paresthesias, however, suggest a radiculopathy due to nerve root compression. Myelopathy occurs due to spinal cord compression and may present as difficulty with fine motor skills, progressive upper and/or lower extremity weakness, gait instability, and/or bowel/bladder incontinence (Song and Choi 2014; Kani and Chew 2018; Lang 1993).

Intervertebral disk disease at C5–C6 is the most common underlying indication for patients undergoing anterior cervical diskectomy and fusion. Other indications for surgery may include fractures, neoplastic conditions, or infection (i.e., osteomyelitis) (Song and Choi 2014; Kani and Chew 2018; Angevine et al. 2003).

Diagnostic Work-Up

Plain X-ray is often the first imaging modality employed. Both dynamic radiographs (e.g. flexion/extension films) and alignment films (i.e., scoliosis imaging) are critical to assess adjacent segment motion and global balance, respectively. On dynamic films, disruption of the anterior vertebral line, posterior vertebral line, spinolaminar line, and/or posterior spinous line may suggest adjacent segment instability – manifest clinically as mechanical neck pain. Magnetic resonance imaging (MRI) is ideal for soft tissue and neural structures, critical in identifying spinal cord/nerve root compression (Kani and Chew 2018; Horne et al. 2016). Computed tomography (CT), however, is the gold standard for imaging of bony structures; the major disadvantage to this modality is the limited ability to identify compression of neural structures (Horne et al. 2016; Stanley et al. 1986). This compression, most commonly caused by disk degeneration and posterior ligamentous hypertrophy, may be readily visualized on MRI; CT is invaluable for surgical planning however, both for the aforementioned evaluation of the bony anatomy, and for the ability to identify calcification of the intervertebral disk or ossification of the posterior longitudinal ligament (OPLL) (Horne et al. 2016; Stanley et al. 1986). As such, these imaging modalities complement one another and both should be considered essential to the complete radiographic evaluation of a patient with cervical spine pain/radiculopathy/myelopathy.

Electromyogram/nerve conduction studies (EMG/NCS) are helpful diagnostic adjuvants, especially in cases of long-standing radiculopathy or where the level of symptomatic compression cannot be clearly identified by clinical exam. These may also help elucidate concomitant spinal cord, nerve root, and peripheral nerve pathology, giving a better assessment of which clinical features may be improved by surgical intervention. In addition, such studies can be beneficial in counseling patients on expected outcomes after an ACDF, offering insight into an acute or chronic radiculopathy (Alrawi et al. 2007; Carette and Fehlings 2005; Ellenberg et al. 1994).

Special Populations to Consider

Special attention should be paid to patients with connective tissue diseases and chronic inflammatory conditions known to affect the occipito-cervical junction and/or subaxial spine, as these may alter the biomechanics, range of motion, and/or ligamentous structures. Conditions known to produce these changes include Down syndrome, mucopolysaccharidoses (i.e., Morquio syndrome), Klippel-Feil syndrome, rheumatoid arthritis, skeletal dysplasias, and others like Chiari malformation type I (Frost et al. 1999; Samartzis et al. 2016; Prusick et al. 1985; Ricchetti et al. 2008; Hamidi et al. 2014).

Down syndrome, trisomy 21, is the most common chromosomal disorder and often results in ligament laxity at the craniovertebral junction. In addition to progressive atlantoaxial instability and atlanto-occipital hypermobility, patients with Down syndrome are predisposed to os odontoideum, odontoid hypoplasia, and cervical canal stenosis (Frost et al. 1999).

Patients with Morquio syndrome (mucopolysaccharidosis type IV) are predisposed to atlantoaxial instability, from dens hypoplasia, similar to a subset of patients with Down syndrome. These patients may also have aberrant retrodental soft tissue masses resulting in cervical canal stenosis and myelopathy (Samartzis et al. 2016).

Klippel-Feil syndrome involves the classic triad of short neck (Cloward 1958), low hairline (Song and Choi 2014), and limited neck mobility; the full triad of which is in 50% of patients. Klippel-Feil syndrome involves vertebral autofusion and accelerated spondylosis. Abnormal fusion of adjacent cervical vertebrae restricts motion and may lead to premature instability. This population is at high risk for atlanto-occipital and vertebral artery injury. Among patients with DiGeorge syndrome, 58% have a dysmorphic dens, and 34% have an autofusion of C2–C3 (Prusick et al. 1985; Ricchetti et al. 2008; Hamidi et al. 2014).

Recent estimates suggest that up to 80% of patients with rheumatoid arthritis, a chronic inflammatory condition, have radiographic cervical spine involvement – present within 2 years of initial diagnosis in many. Fibrovascular tissue proliferation, pannus formation, and bony erosion lead to the cascade observed: atlantoaxial instability (AAI), cranial settling (CS), and subaxial subluxation (SAS) (Nguyen et al. 2004; Wasserman et al. 2011; Krauss et al. 2010).

Surgery

Positioning and Monitoring

Patients are positioned in the supine position, with the upper extremities fully adducted and secured at the side. General anesthesia with an endotracheal tube is administered. Shoulders should be pulled down to allow for imaging, but excessive shoulder depression has been proposed as a contributing factor for post-operative C5 palsies (Alonso et al. 2018). Cervical traction (5–15 lbs) may be appropriate, especially in cases of intervertebral disk degeneration. Intraoperative monitoring, including somatosensory and transcranial motor evoked potentials, is recommended as it may help to prevent iatrogenic injury to the spinal cord or roots (Davis et al. 2013; Legatt et al. 2016).

Operative Details

Anatomic landmarks should be identified prior to skin incision, including the inferior border of the mandible, hyoid bone, thyroid cartilage, and cricoid cartilage, overlying C2/C3, C3, C4/C5, and C6, respectively (Rao et al. 2011; Moran and Bolger 2012). Due to the shorter course of the right laryngeal nerve, some surgeons opt for a left-sided approach to minimize permanent nerve injury. However, surgeon comfort and hand preference may dictate a right-sided approach which has been demonstrated to have no greater risk of permanent dysphonia (Kilburg et al. 2006).

A horizontal incision, located in the skin crease, is fashioned from the midline to the sternocleidomastoid muscle. Incisions along the skin crease are aesthetically preferred. The sternal notch can be used to approximate the midline. Dissection continues along the subcutaneous tissue, to expose the platysma. The platysma is divided with electrocautery, exposing the deep cervical fascia. The strap muscles are retracted medially, and the sternocleidomastoid laterally. The pre-tracheal fascia is bluntly dissected, with careful attention to avoid damage to the superior, middle, and inferior thyroid arteries therein. It is necessary to develop an avascular plane, medial to the carotid sheath and lateral to the esophagus and trachea, to minimize unwanted vascular injury. The carotid sheath may be palpated to confirm the position of the carotid artery within. After lateral retraction of the carotid sheath, the location of the vertebral column can be confirmed with palpation, which reveals clearly identifiable "hills" and "troughs" representing disk spaces and vertebral bodies, respectively.

A lateral radiograph is taken to determine the appropriate level. Dissection of the thin prevertebral fascia exposes the disk space and vertebral bodies. The bilateral longus colli muscles, identified at the anterior surface of the cervical spine and originating at the transverse processes, may be reflected laterally. The placement of Caspar pins facilitates distraction across the intervertebral disk space, for ease of removal.

Diskectomy is performed by an incision into the outer annulus fibrosis and complete removal with curettes and pituitary rongeurs. Disk forceps may also be employed to ensure complete resection of the disk and adequate decompression of the neural elements. Our practice is to decompression until the posterior longitudinal ligament can be visualized, but the extent of decompression is up to individual surgeon preference. Osteophytes arising from the anterior adjacent endplates can then be removed with a high-speed drill and curettes; forward-angled curettes can be used to resect osteophytes projecting into the canal and to widen the foramina, to relieve pre-surgical complaints of radiculopathy referable to nerve root compression. A Kerrison punch may also be useful to decompress posterior osteophytes extending into the spinal canal. Throughout the exposure and decompression, meticulous hemostasis should be maintained, facilitated by bipolar cautery, cottonoid, and hemostatic agents, with particular attention to epidural venous bleeding (Rao et al. 2011; Moran and Bolger 2012). After decompression of the disk space, a high speed drill is used to decorticate and prepare the superior and inferior endplates for graft placement. Allograft has been the standard graft material for years; however, titanium and polyetherether ketone (PEEK) interbodies are both widely used and acceptable options.

Cervical vertebrectomy may be indicated in cases requiring greater anterior decompression, such as metastatic spine disease, burst fracture, or multilevel degenerative conditions (i.e., posterior osteophyte and degenerative disk disease at two adjacent levels). In cases of metastatic disease, tumor invasion typically spares the intervertebral disks allowing for safe diskectomies, but diseased bone should not be used for subsequent autograft. The uncinate process limits the lateral extent of the vertebrectomy, allowing for a ~3 mm safe zone in the central third of the vertebral body which can be corpectomized without risking injury to the vertebral arteries. Leksell rongeurs may be useful for bone removal. Additional posterior instrumentation may be required for additional support and iatrogenic instability (Kilburg et al. 2006; Buckingham and Chen 2018; Xu et al. 2014; Rhee 2015).

Interbody or cage placement, following diskectomy or vertebrectomy, should maintain normal alignment and serve as anterior column structural support. Decortication and the placement of healthy autologous bone graft, or allograft (i.e., cadaveric tricortical graft), maximize the odds of fusion. An anterior plate is securing with obliquely oriented locking screws – to minimize screw pullout, screws diverge in the sagittal plane and converge in the axial plane. Watertight closure of the fascia and skin is performed in the typical fashion (Buckingham and Chen 2018; Xu et al. 2014; Rhee 2015; Chong et al. 2015; Lee et al. 2014).

Complications

Permanent complications following ACDF are exceptionally rare, and many short-term complications can be avoided with sufficient evaluation and surgical technique (Quintana 2014). The rate of post-operative complications following anterior cervical fusion increases with the number of levels operated on (Quintana 2014; Nanda et al. 2014; Bazaz et al. 2002; Park and Jho 2012; Odate et al. 2017; Zhong et al. 2013; Gaudinez et al. 2000; Hershman et al. 2017). Complications include post-operative hematoma, transient dysphagia, recurrent laryngeal nerve (RLN) palsy, Horner's syndrome (characterized by anhidrosis, miosis, and ptosis ipsilateral to the lesion), esophageal perforation, wound infection, vascular injury, and cerebrospinal fluid (CSF) leak (Quintana 2014; Nanda et al. 2014; Bazaz et al. 2002; Park and Jho 2012; Odate et al. 2017; Zhong et al. 2013; Gaudinez et al. 2000; Hershman et al. 2017). Complication rates are further eleveted in patients with ossification of the posterior longitudinal ligament (OPLL) (Odate et al. 2017). RLN palsy is one of the most feared complication as bilateraly injury can lead to airway compromise. Despite fears of injury, the rate of clinically symptomatic RLN palsy is 8.3%, decreasing to 2.5% at 3 months after surgery (Jung et al. 2005). Additionally, most lesions are unilateral. Other well-known cause of post-operative airway compromise is surgical site hematoma, which has been documented in up to 6.1% of cases. This requires immediate evacuation to prevent permanent sequelae). Other complications requiring surgical revision include CSF leak (0.2–1%) and esophageal perforation (0.25%) (Quintana 2014; Nanda et al. 2014; Bazaz et al. 2002; Park and Jho 2012; Odate et al. 2017; Zhong et al. 2013; Gaudinez et al. 2000; Hershman et al. 2017; Jung et al. 2005).

Case Illustration

This 37-year-old male presented with left-sided spastic hemiparesis that progressed after sustaining a fall 2 years prior. Imaging demonstrated a broad disk bulge at C3–C4 compressing the spinal cord, with signal change present in the parenchyma from C2 to C4 (Fig. 1). An ACDF was determined to be the most appropriate treatment.

The patient was placed supine, and an inflated bag was placed between his shoulders for neck extension, given the high location of the exposure in the cervical spine. The patient's head was then placed in a Mayfield horseshoe adaptor with roughly 15° of extension and 15° of rotation away from the surgeon to facilitate the conventional Smith-Robinson approach. Using the Mayfield adaptor, 10lbs of chin strap traction was applied to facilitate access to the C3/C4 disk space. At this time, intraoperative neuromonitoring leads were also placed for measurement of somatosensory evoked potentials (SEPs), transcranial motor evoked potentials (TcMEPs), and free-run electromyography (EMG) throughout the case. Leads for

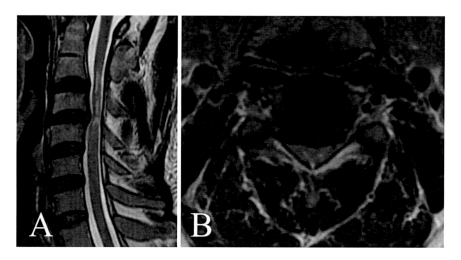

Fig. 1 Pre-operative imaging. (**a**) Sagittal T2-weighted MRI demonstrating a disk bulge at C3–C4 and cord signal change. (**b**) Axial T2-weighted MRI

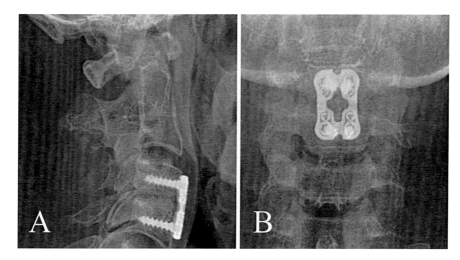

Fig. 2 Post-operative imaging. (**a**) Lateral X-ray following C3–C4 ACDF. (**b**) AP X-ray

TcMEPs were placed at the bilateral tibialis anterior, abductor hallucis, deltoid, and pollicis brevis; EMG leads were placed bilaterally at the deltoid and abductor pollicis brevis. His arms were secured and shoulders taped down; a lateral X-ray was then taken for localization and marking of the skin incision. A transverse skin incision was placed along the crease of the neck, and electrocautery was utilized to dissect the platysma. Blunt dissection proceeded medial to the carotid sheath, lateral to the esophagus and trachea. Upon encountering the spine, a repeat X-ray was taken to confirm the correct level. Caspar pins were placed in C3 and C4 for distraction. An operating microscope was brought in for the decompression. The anterior longitudinal ligament was incised with a 15-blade, and diskectomy was performed using a high-speed burr, rongeurs, and a Kerrison punch. The degenerated posterior uncovertebral joints were subsequently removed, with attention to preserve the traversing nerve. Following adequate decompression of the central canal and bilateral foramen, the endplates were decorticated. A structural allograft was sized and placed in the diskectomy defect. Distraction was released from the Caspar pins to reduce the C3 and C4 vertebrae onto the graft, and an angled curette was used to determine secure positioning. An anterior plate was placed from C3 to C4 with four locking screws. A lateral X-ray was taken showing good placement of the hardware (Fig. 2). Somatosensory and motor evoked potentials were monitored throughout the case and were stable. The wound was closed in a layered watertight fashion. The platysma was closed using 0 Vicryl, with 4–0 Biosyn sutures in the subcuticular layer. The skin surface was closed with glue and adhesive strips. The patient did well postoperatively and gradually regained function and the ability to independently ambulate over the next 2 years, working with rehabilitation.

References

Alonso F, Voin V, Iwanaga J et al (2018) Potential mechanism for some postoperative C5 palsies: an anatomical study. Spine (Phila Pa 1976) 43(4):161–166

Alrawi MF, Khalil NM, Mitchell P, Hughes SP (2007) The value of neurophysiological and imaging studies in predicting outcome in the surgical treatment of cervical radiculopathy. Eur Spine J 16(4):495–500

Angevine PD, Arons RR, McCormick PC (2003) National and regional rates and variation of cervical discectomy with and without anterior fusion, 1990–1999. Spine (Phila Pa 1976) 28:931–939

Bazaz R, Lee MJ, Yoo JU (2002) Incidence of dysphagia after anterior cervical spine surgery. Spine 27(22):2453–2458

Buckingham MJ, Chen KS (2018) Technique for performing an anterior cervical discectomy and fusion with attention to the exposure of the anterior cervical spine: 2-dimensional operative video. Oper Neurosurg (Hagerstown) 14(5):596

Carette S, Fehlings MG (2005) Clinical practice. Cervical radiculopathy. N Engl J Med 353(4):392–399

Chong E, Pelletier MH, Mobbs RJ, Walsh WR (2015) The design evolution of interbody cages in anterior cervical discectomy and fusion: a systematic review. BMC Musculoskelet Disord 25:99

Cloward RB (1958) The anterior approach for removal of ruptured cervical disks. J Neurosurg 15:602–617

Davis SF, Corenman D, Strauch E, Connor D (2013) Intraoperative monitoring may prevent neurologic injury in non-myelopathic patients undergoing ACDF. Neurodiagn J 53(2):114–120

Ellenberg MR, Honet JC, Treanor WJ (1994) Cervical radiculopathy. Arch Phys Med Rehabil 75(3):342–352

Frost M, Huffer WE, Sze CI et al (1999) Cervical spine abnormalities in down syndrome. Clin Neuropathol 18(5):250–259

Gaudinez RF, English GM, Gebhard JS, Brugman JL, Donaldson DH, Brown CW (2000) Esophageal perforations after anterior cervical surgery. J Spinal Disord 13(1):77–84

Gupta SK, Roy RC, Srivastava A (1982) Sagittal diameter of the cervical canal in normal Indian adults. Clin Radiol 33(6):681–685

Hai-bin C, Zheng-Guo W, Ling Z et al (2012) Cervical spinal canal narrowing and cervical neurological injuries. Chinese J Traum 5(1):36–41

Hamidi M, Nabi S, Husein M et al (2014) Cervical spine abnormalities in 22q11.2 deletion syndrome. Cleft Palate Craniofac J 51(2):230–233

Hashimoto I, Tak YK (1977) The true sagittal diameter of the cervical spinal canal and its diagnostic significance in cervical myelopathy. J Neurosurg 47(6):912

Hershman SH, Kunkle WA, Kelly MP et al (2017) Esophageal perforation following anterior cervical spine surgery: case report and review of the literature. Global Spine J 7(1 Suppl):28S–36S

Horne PH, Lampe P, Nguyen JT et al (2016) A novel radiographic indicator of developmental cervical stenosis. J Bone Joint Surg Am 98(14):1206–1214

Jung A, Schramm J, Lehnerdt K, Herberhold C (2005) Recurrent laryngeal nerve palsy during anterior cervical spine surgery: a prospective study. J Neurosurg Spine 2(2):123–127

Kani KK, Chew FS (2018) Anterior cervical discectomy and fusion: review and update for radiologists. Skelet Radiol 47(1):7–17

Kilburg C, Sullivan HG, Mathiason MA (2006) Effect of approach side during anterior cervical discectomy and fusion on the incidence of recurrent laryngeal nerve injury. J Neurosurg Spine 4(4):273–277

Krauss WE, Bledsoe JM, Clarke MJ, Nottmeier EW, Pichelmann MA (2010) Rheumatoid arthritis of the craniovertebral junction. Neurosurgery 66(3):A83–A95

Lang J (1993) Clinical anatomy of the cervical spine. Thieme, Stuttgart

Lee YS, Kim YB, Park SW (2014) Risk factors for postoperative subsidence of single-level anterior cervical discectomy and fusion: the significance of the preoperative cervical alignment. Spine (Phila Pa 1976) 39(16):1280–1287

Legatt AD, Laarakker AS, Nakhla JP et al (2016) Somatosensory evoked potential monitoring detection of carotid compression during ACDF surgery in a patient with a vascularly isolated hemisphere. J Neurosurg Spine 25(5):566–571

Moran C, Bolger C (2012) Operative techniques for cervical radiculopathy and myelopathy. Adv Orthop 212:916149

Nanda A, Sharma M, Sonig A et al (2014) Surgical complications of anterior cervical diskectomy and fusion for cervical degenerative disk disease: a single surgeon's experience of 1,576 patients. World Neurosurg 82(6):1380–1387

Nguyen HV, Ludwig SC, Silber J et al (2004) Rheumatoid arthritis of the cervical spine. Spine J 4(3):329–334

Odate S, Shikata J, Soeda T et al (2017) Surgical results and complications of anterior decompression and fusion as a revision surgery after initial posterior surgery for cervical myelopathy due to ossification of the posterior longitudinal ligament. J Neurosurg Spine 26(4):466–473

Park H-K, Jho H-D (2012) The management of vertebral artery injury in anterior cervical spine operation: a systematic review of published cases. Eur Spine J 21(12):2475–2485

Payne EE, Spillane JD (1957) The cervical spine: an anatomico-pathological study of 70 specimens (using a special technique) with particular reference to the problem of cervical spondylosis. Brain 80(4):571–596

Prusick VR, Samberg LC, Wesolowski DP (1985) Klippel-Feil syndrome associated with spinal stenosis. A case report. J Bone Joint Surg 67(1):161–164

Quintana LM (2014) Complications in anterior cervical discectomy and fusion for cervical degenerative disc disease. World Neurosurg 82(6):1058–1059

Rao AS, Michael ALR, Timothy J (2011) Surgical technique of anterior cervical discectomy and fusion (ACDF). In: Giannoudis P (ed) Practical procedures in elective orthopedic surgery. Springer, London

Rhee JM (2015) Anterior cervical discectomy and fusion surgery for cervical radiculopathy: is time of essence? Spine J 15(3):433–434

Ricchetti ET, Hosalkar HS, Gholve PA et al (2008) Advanced imaging of the cervical spine and spinal cord in 22q11.2 deletion syndrome: age-matched, double-cohort, controlled study. J Child Orthop 2(5):333–341

Robinson RA, Smith GW (1955) Anterolateral cervical disc removal and interbody fusion for cervical disc syndrome. Bull John Hopkins Hosp 96:223–224

Samartzis D, Kalluri P, Herman J et al (2016) "Clinical triad" findings in pediatric Klippel-Feil patients. Scoliosis Spinal Disord 11:15

Song K-J, Choi B-Y (2014) Current concepts of anterior cervical discectomy and fusion: a review of literature. Asian Spine J 8(4):531–539

Stanley JH, Schable SI, Frey GD et al (1986) Quantitative analysis of the cervical spinal canal by computed tomography. Neuroradiology 28(2):139–143

Tubbs RS, Hallock JD, Radcliff V et al (2011) Ligaments of the craniocervical junction. J Neurosurg Spine 14:697–709

Wasserman BR, Moskovich R, Razi AE (2011) Rheumatoid arthritis of the cervical spine—clinical considerations. Bull NYU Hosp Jt Dis 69(2):136–148

Xu R, Bydon M, MAcki M et al (2014) Adjacent segment disease after anterior cervical discectomy and fusion: clinical outcomes after first repeat surgery versus second repeat surgery. Spine (Phila Pa 1976) 39 (2):120–126

Zhong Z-M, Jiang J-M, Qu D-B, Wang J, Li X-P, Lu K-W et al (2013) Esophageal perforation related to anterior cervical spinal surgery. J Clin Neurosci 20 (10):1402–1405

Spinal Plates and the Anterior Lumbar Interbody Arthrodesis

30

Zach Pennington, A. Karim Ahmed, and Daniel M. Sciubba

Contents

Introduction	604
A Brief History of Lumbar Fusion	604
Anterior Versus Posterior Approaches to the Lumbar Spine	605
Relative Advantages and Disadvantages of the Anterior Approach	606
Anterior Lumbar Interbody Fusion	608
Positioning, Draping, and Mapping of the Incision Site	608
Incision and Approach	609
Addressal of the Pathology	609
Closure	610
Case Illustration	611
Lateral Lumbar Interbody Fusion	611
Description	612
Case Illustration	615
Indications for Plate Usage in Anterior and Lateral Lumbar Surgery	616
Anterior	616
Lateral	617
Anterior Plates and Other Anterior Fusion Technologies	618
Material Selection	619
Plate Design	619
Interbody Design	620
BMP and Profusion Agents	624

Z. Pennington · D. M. Sciubba (✉)
Department of Neurosurgery, Johns Hopkins Hospital, The Johns Hopkins School of Medicine, Baltimore, MD, USA
e-mail: zpennin1@jhmi.edu; dsciubba1@jhmi.edu

A. K. Ahmed
Department of Neurosurgery, The Johns Hopkins School of Medicine, Baltimore, MD, USA
e-mail: aahmed33@jhmi.edu

© Springer Nature Switzerland AG 2021
B. C. Cheng (ed.), *Handbook of Spine Technology*,
https://doi.org/10.1007/978-3-319-44424-6_125

Conclusion	625
Cross-References	625
References	626

Abstract

Each year over 200,000 lumbar spine fusions are performed. Fusion can be accomplished through an anterior, retroperitoneal approach, posterior or posterolateral approach, or more recently, through the lateral or transpsoas approach as pioneered by Pimenta and colleagues. Traditionally though, anterior interbody fusion (ALIF) has been a workhorse for discogenic pain and sagittal deformity of the lower lumbosacral spine. Most commonly this procedure involves placement of an interbody device followed by anterior tension band plating. But over the past decade and a half, new implants and new fusion devices have become available, which increase the robustness of this procedure. Here we discuss the basic indications of the ALIF procedure, provide a description of the classical ALIF procedure, and discuss current technologies.

Keywords

Lumbar fusion · Interbody fusion · Spinal fusion · Lumbar spondylosis · ALIF · Spinal implants

Introduction

The use of anterior lumbar interbody fusion (ALIF) to address degenerative spine pathologies was first described in 1933 by Burns, who described its use for the treatment of spondylolisthesis secondary to bilaminar fracture in a 14-year-old boy (Burns and Camb 1933). Since that time, it has become an increasingly popular option for addressing pathologies of the lumbosacral spine, including spondylolisthesis and intervertebral disc herniation, as well as lumbosacral kyphoscoliosis, in which it most commonly complements posterior instrumentation (Loguidice et al. 1988). At its basest level, the procedure involves an anterior retroperitoneal approach to the lower lumbar or lumbosacral spine with partial or complete discectomy followed by placement of an interbody with or without supplemental anterior plating. Over its years of use though, new technologies have been developed for both improved surgical access and spinal reconstruction, including the introduction of integrated fixation cages. Here we provide an overview of the ALIF procedure and describe the various pieces of instrumentation employed for spinal column reconstruction.

A Brief History of Lumbar Fusion

The first description of instrumented lumbar fusion in print is made by Berthold Hadra, who reported the treatment of a C6/7 dislocation using internal reduction and fixation with silver wire connecting the adjacent spinous processes (de Kunder et al. 2018). Hadra himself attributed this technique of interspinous fusion to the thoracolumbar fusions performed by Dr. W. Wilkins, though the former never published his results (Keller and Holland 1997; Peek and Wiltse 1990). Despite multiple interventions, de Kunder's wiring technique proved to have insufficient strength to maintain reduction of the patient's fracture dislocation, and the patient's symptoms recurred within weeks of surgery (de Kunder et al. 2018). As a result, it was evident that alternative methods were needed. The well-known alternative to emerge was a technique described roughly two decades later in 1909 by Fritz Lange. Lange utilized rigid rods made of celluloid – a precursor to modern plastics – along with 5-mm-thick steel and silk wire to correct scoliosis in pediatric patients, a technique far not too dissimilar to modern posterior fusion techniques (Tarpada et al. 2017). Then in 1933, Burns and Camb described the first anterior fusion in a 14-year-old boy being treated for spondylolisthesis secondary to a bilaminar fracture (Burns and

Camb 1933), using a technique first proposed by Capener 1 year earlier (Capener 1932). Over the subsequent decades, progress in anterior fusion technologies lagged behind that of posterior fusion technologies, including progress in interbody device implementation (de Kunder et al. 2018). This culminated in the development of the Harrington rod in 1962 by American orthopedist Dr. Paul R. Harrington, which was the workhorse of spinal instrumentation for several decades. Around this same time though, Melvin Watkins published his findings on far lateral fusion (M. B. Watkins 1953), paving the way for the modern oblique lateral (OLIF) and extreme lateral interbody fusion (XLIF) used today.

Part of the reason that anterior fusion advancement may have lagged behind those of posterior fusion may stem from the struggle of early spine surgeons attempting to identify a strong material capable of maintaining correction and providing long-term stabilization. The early celluloid implants of Lange were reasonably tolerated within the body; however, their flexural modulus, like that of the steel wires also employed by Lange and his predecessors, was insufficient to maintain correction over their long intended service life (Peek and Wiltse 1990). Conversely, carbon steel implants had high flexural modulus but were susceptible to degradation within the electrolytic solutions of native human tissue. This necessitated the invention of new implants with electrolytic resistance similar to the noble metals (e.g., gold) but with the relatively high flexural modulus of steel. The solution was concomitantly reached through the invention of Vitallium – a cobalt-chrome alloy – and stainless steel, both of which were developed in the late 1930s (de Kunder et al. 2018; Peek and Wiltse 1990). Despite these advances in posterior fusion, the implementation of metal instrumentation for anterior fusion did not occur until the middle of the century. Leading up to this point, anterior approaches to lumbar fusion had relied solely upon external bracing for reinforcement during the immediate postoperative period. Surgeons had also championed the use of fitted interbody devices, chief among them, Paul Harmon, who in the late 1950s and early 1960s used tibial peg grafts and iliac crest grafts as fitted interbody devices (Harmon 1960). The latter of were also employed by Freebody and Crock (Harmon 1960; Peek and Wiltse 1990). At this time though, the first description of instrumentation for anterior lumbar fusion was made by Humphries et al. in 1958, who reported the use of an intervertebral clamp in a series of 25 dogs (Humphries et al. 1958). In their abstract, Humphries et al. also described a matched control series fused without the anterior vertebral clamp; unlike the instrumented group, the non-instrumented group had a 100% pseudarthrosis rate, demonstrating the potential utility of instrumentation in anterior lumbar fusion. Humphries and colleagues further elaborated on this technique in their 1961 paper, where they reported the results in a small series of human patients (Humphries et al. 1961). This clamp served as the precursor to the modern anterior and lateral plates that maintain compression on the placed interbody to facilitate successful fusion. Finally, in the mid-1980s, Louis published his series of over 400 patients including several that had undergone anterior-only fusion using lag screw placement across the interbody device – a technique mechanically similar to the increasingly popular integrated fusion cages (Louis 1986).

Anterior Versus Posterior Approaches to the Lumbar Spine

Despite the greater cumulative experience of spine surgery with posterior instrumentation, there is no clearly superior option across all cases. As with any surgical procedure, it is imperative to first identify the indication for surgery and the goals for surgery. Based upon these goals, a surgical plan can then be developed, which includes the decision of whether they treat the pathology from an anterior-only approach (ALIF), lateral/anterolateral approach (XLIF, DLIF, OLIF), posterior-only (TLIF, PLIF), or combined anterior-posterior approach.

Determining the surgical goals begins by identifying the pathologies responsible for the patient's clinical picture, whether that is a mobile spondylolisthesis generating mechanical, low back pain or an eccentric disc producing unilateral radicular pain. For degenerative cases, this includes assessing the patient's spine for the presence of any pathologic curvature (i.e., scoliosis, flat-back syndrome, or focal kyphosis), the presence and grade of spondylolistheses within the spine, the presence of fragility (e.g., vertebral body compression fracture) or iatrogenic fractures (e.g., pars fracture), the presence and extent of cord and nerve root compression, and the overall sagittal (normal <5 cm) and coronal imbalance (normal <2 cm). Having done so, it is then necessary to identify which of these structural lesions should be addressed by surgery. Many patients have clinically appreciable spinal deformity that is asymptomatic (Kebaish et al. 2011; Schwab et al. 2005). The basis for this decision is ultimately up to both the surgeon and patient but should be based upon the patient's clinical complaints/presentation and overall health. Patients presenting with chief complaints of radicular pain or functional radiculopathy may be adequately treated with a discectomy or two-level decompression without fusion, whereas those with significant kyphoscoliotic deformity and back pain as a chief complaint will likely require a larger operation to realize clinical benefit from the procedure. Extra caution must be exercised when considering the latter option though, as not all patients have enough physical reserve to successfully recover from large, multilevel reconstructions. The International Spine Study Group has recently published on this, suggesting that frail patients may benefit most from nonsurgical management given their high risk for complications (Miller et al. 2017).

The next step after having deemed a patient to be a surgical candidate and having localized the pathology is to formulate an approach. In some cases, either posterior or anterior fusion is superior to the other, as in the cases of long-construct fusion operations for deformity, which are optimally achieved through a posterior-only approach (Geck et al. 2009; S. S. Lee et al. 2006). But in other cases, such as one- or two-level pathologies of the lumbar spine, both anterior and posterior approaches are available, and the decision should be made on the exact pathology (e.g., disc herniation), patient body habitus, and surgeon familiarity with each approach.

For the lower lumbar spine and lumbosacral spine, there are several anatomic considerations that must be made when selecting an approach. At the L5/S1 level – the most common site of lumbar intervertebral disc herniation – lateral approaches are largely precluded, as the iliac crests make access essentially impossible in most patients. This is also a concern at the L4/5 level, though less commonly so, and is based upon the actual height of the patient's ilia. At higher levels, lateral approaches become a more robust option, and they may preferable for pathologies above the bifurcation of the iliac arteries, due to the decreased risk of vascular damage relative to the anterior approach and decreased blood loss relative to the posterior approach. Furthermore, above the level of the renal vessels (L1–2), the anterior approach largely ceases to be a consideration due to its requirement for diaphragm mobilization. As a result, the main indication for an anterior approach is one- or two-level disease between the L2 and S1 vertebrae. Here the anterior fusion can be used as either a stand-alone procedure or as a supplement to a posterior procedure in cases requiring significant correction of coronal Cobb or lumbar kyphosis. Of note though, we recommend against an anterior approach in cases of prior retroperitoneal surgery or obesity due to increased complexity of access. We also recommend against its use in cases of an ankylosed level with full disc space collapse or solid arthrodesis at the surgical level. We do however feel it is appropriate to perform an ALIF with an indirect foraminal decompression, if the collapsed disc space has not undergone ankylosis. Additionally, when considering a one- or two-level fusion, it is still important to take into consideration the patient's sagittal imbalance, as failure to correct sagittal imbalance and loss of lumbar lordosis is a previously identified negative predictor of surgical outcome in short-construct fusions (B. H. Lee et al. 2017).

Relative Advantages and Disadvantages of the Anterior Approach

As with any surgery, there are advantages and disadvantages to employing an anterior as opposed to a posterior approach. Relative advantages of the anterior approach include superior visualization of the disc spaces of the inferior lumbar spine and lumbosacral junction (Giang et al. 2017) and the ability to avoid dissection of the posterior musculoligamentous structures (Mobbs et al. 2015), thereby preserving the posterior tension band and reducing in the risk of nerve root injury (Jeswani et al. 2012). Additionally, and more important to the scope of this work, the anterior approach may offer several mechanical advantages relative to a posterior (PLIF) or posterolateral (TLIF) approach. One study by Hsieh and colleagues reported a cohort of 57 patients operated for discogenic back pain (Hsieh et al. 2007). Among this cohort, 32 patients were treated via the anterior approach, and 25 patients were treated via a posterolateral approach (TLIF). The authors reported superior correction among the ALIF group in terms of restoration of foraminal height, disc angle, and lumbar lordosis. Similar findings have been observed by other groups (J. Kim et al. 2010; R. G. Watkins et al. 2014), with 10° of correction per level or upward of 20° to 30° per level with some of the more recent hyperlordotic cages (Saville et al. 2016). These larger cages can only be placed via an anterior approach due to the access restrictions posed by the exiting nerve roots in posterior and posterolateral approach, and as a result, the anterior approach may be the best option in patients requiring substantial sagittal correction. Although similar levels of sagittal correction can be achieved using pedicle subtraction osteotomies (K. Cho et al. 2005), this correction is non-physiologic in that it compresses the exiting lumbar nerve roots and may lead to postoperative radicular pain if concomitant foraminotomies are not performed. Furthermore, this approach requires compromise of the posterior tension band (Udby and Bech-Azeddine 2015), which may decrease the ability of patients to maintain large corrections postoperatively, though evidence to support this is still pending.

Despite these advantages, the anterior interbody fusion has several limitations. Chief among these is the complexity of the retroperitoneal approach (Giang et al. 2017). Multiple delicate structures must be crossed during this dissection, including the aorta and inferior vena cava in the prevertebral space above the level of L4, the hypogastric plexus within the presacral and L5 prevertebral space, and the ureters in the posterolateral retroperitoneal space. Injury to these structures can lead to significant blood loss, hydronephrosis, and retrograde ejaculation (Czerwein et al. 2011; Quraishi et al. 2013). Because of these risks and the lack of familiarity that the average surgeon may have with this approach, anterior interbody fusions often require the use of an access surgeon. This is especially true for surgically complex abdomens, such as those seen in patients with substantial abdominal obesity and in individuals who have previously undergone abdominal surgery. The latter have significant adhesions and scar tissue secondary to their prior operation(s), obscuring normal landmarks, while the former have significant abdominal soft tissue requiring retraction. This results in a narrow surgical corridor, and in the morbidly obese, extensive abdominal soft tissue may make adequate retraction impossible. Even in those cases where the anterior approach is possible, patients may also require a second, posterior approach for increased structural stability in order to prophylax against pseudarthrosis, which has been reported at a higher rate in this population (Jiménez-Avila et al. 2011).

In terms of biomechanical stability, results are mixed regarding the superiority of anterior- versus posterior-alone procedures for single-level disease. A recent cadaveric study by Liu and colleagues suggested that anterior interbody fusion with plating provides superior compressive strength compared to pedicle screw supplemented constructs (L. Liu et al. 2014). However, a previous cadaveric study by Tzermiadianos and colleagues (Tzermiadianos et al. 2008) saw superior strength with respect to lateral bending and flexion-extension in the pedicle screw fixation group. Additionally, multiple studies (Dorward et al. 2013; D. Kim et al. 2009), including recent meta-analyses by Phan et al. (2015) and Teng et al. (2017), observed no difference in fusion rates between anterior and posterior interbody techniques, suggesting that there may be no significant role for biomechanical considerations when selecting an approach for the treatment of discogenic pain. Because of this and the similar complication rates for anterior and posterior approaches (Teng et al. 2017), we recommend that the approach employed be dictated by the familiarity of the surgeon with the approach and the indication for surgery. Anterior approaches may yield superior results in patients being treated for discogenic pain conditions with only minor subluxation deformity being treated by surgeons familiar with the retroperitoneal approach. Conversely, posterior approaches are preferable for patients with high-grade spondylolisthesis, especially if the attending surgeon is uncomfortable with the retroperitoneal approach.

Anterior Lumbar Interbody Fusion

Positioning, Draping, and Mapping of the Incision Site

As with all surgical procedures, the anterior lumbar interbody fusion begins with a review of preoperative imaging. In the context of an ALIF, the goal of preoperative imaging review is to identify the angle of the target disc level (generally L4/5 or L5/S1). Patients with greater pelvic incidences will have more caudally directed anterior disc surfaces, which may require induction of additional Trendelenburg positioning. Concomitant inducement of greater lordosis is recommended to open the ventral disc space, which can be done using a roll or cushion; this may not be necessary in patients with substantial native lordosis though (Heary et al. 2017). Additionally, preoperative imaging allows for evaluation of prevertebral soft tissues and identification of potential surgical obstacles. Structures of particular concern are the iliac vessels, which generally bifurcate at the level of L3 (veins) or L4 (arteries) and consequently must be mobilized in approaches to the mid- or upper lumbar spine.

After the angle of the disc space has been identified, the surgical procedure can begin. The patient is brought to the operating room and placed under general anesthesia. The patient is then transferred to a radiolucent table, such as the Jackson table, which is placed in mild Trendelenburg position to position the target disc space perpendicular to the floor. In patients with larger body habitus, we recommend securing the feet of the patient to the table to prevent sliding during the procedure. We also recommend rotating these patients slightly, placing them in a position intermediate to supine and lateral decubitus; this helps retract intraperitoneal structures during the approach. While placing an inflatable pillow or surgical bump underneath the patient buttocks is helpful in most patients, owing to an increase in exposure of the ventral disc space, it may need to be excluded in patients with significant pelvic incidence, as it can decrease lordosis and exposure of the lower lumbar disc spaces. Similar flexibility is offered during positioning of the patient's arms. The overall goal in positioning is to remove the distal arm from the abdominal region and surgical field, which we believe can best be accomplished by abducting the right arm and securing the left arm across the patient's chest in a standard room set up. This approach allows for adequate visualization of the discs with fluoroscopy while providing access to all members of the surgical team.

After positioning, the incision line is then drawn on the patient. Several options are available including a standard midline incision (Aryan and Berta 2014), a vertical paramedian incision of the linea semilunaris (Jeswani et al. 2012), and a right-angled incision with a horizontal arm between the umbilicus and pubic symphysis (Heary et al. 2017). For L5/S1 discs, a 7-cm horizontal incision 2–3 cm above the pubic symphysis is preferred, whereas for L4/5 pathology, a vertical incision is preferred. This latter technique offers superior exposure at all levels and superior cosmetic results and can be easily extended under circumstances, such as vascular injury, that require additional exposure. Alternatively, a Pfannenstiel incision may be used, as is commonly performed in gynecological procedures (Aryan and Berta 2014). If a paramedian incision is utilized, placement on the left side of the patient is preferred as it facilitates aortic retraction and reduces the amount of retraction that the more fragile inferior vena cava must undergo (Heary et al. 2017). After the patient is positioned and the incision is marked, a rail for retractor attachment is connected to the table, the surgical field is cleared with appropriate skin preparation, and the patient is draped in the usual fashion.

Incision and Approach

Either a transperitoneal or retroperitoneal approach may be employed. Both procedures are effective means of accessing the disc space; however, previous evidence suggests that transperitoneal approaches have a 10x higher risk of hypogastric plexus injury and retrograde ejaculation (Heary et al. 2017). As such, we prefer a retroperitoneal approach, especially in male patients. It should be noted that for either the transperitoneal or retroperitoneal approaches though, it is recommended to utilize a vascular access surgeon in order to decrease the risk of injury to the genitourinary, vascular, and prevertebral nervous system structures. This is especially true for patients with risk factors for retroperitoneal fibrosis, e.g., prior abdominal surgery.

The procedure begins by making a 4–6 cm paramedian vertical incision along the planned incision line, centered roughly 2 cm to one side of the patient (left of midline for a right-handed surgeon). Incision should be through skin and the superficial fascia (Camper's, Scarpa's) to expose the anterior rectus sheath. The sheath, comprised of the aponeuroses of the external and internal oblique, as well as the transversus abdominis below the arcuate line, is divided, exposing the fibers of the rectus abdominis. To expose the posterior rectus sheath, the rectus fibers are then either dissected away from the linea alba and the released rectus muscle is retracted laterally or the fibers are dissected away from the semilunar line and the rectus is retracted medially. Both options are acceptable, and the decision is up to the individual surgeon. Note that during this dissection, splitting, as opposed to cutting of the rectus fibers, should be used to reduce the risk of postoperative hernia at the surgical site. The posterior sheath is then divided vertically with a scalpel and blunt; finger dissection is used to develop a plane between the posterior rectus sheath and the underlying peritoneum. Below the level of the arcuate line, the plane is developed between the peritoneum and more superficial transversalis fascia. Dissection continues laterally retracting the overlying soft tissues laterally and the peritoneum and intraperitoneal contents superomedially. Once the depth of the psoas muscle is achieved, the ureter should be immediately identified and protected to prevent postoperative complications; it is usually adherent to the peritoneum and can therefore be protected within this structure (Jeswani et al. 2012). The iliac vessels are then identified. For L4/5 pathologies, the iliac vessels are retracted medially along with the aorta and inferior vena cava, exposing the disc and overlying sympathetic chain medial to the psoas. The iliolumbar and segmental vessels are ligated to facilitate mobilization. For L5/S1 pathologies, the iliac vessels are retracted bilaterally away from the midline, and the middle sacral artery and vein are ligated and divided to expose the disc space. The sympathetic chain is then identified using forceps to avoid injury to it during discectomy. In some cases though, gentle retraction may have to be applied to the sympathetic chain in order to expose the pathologic level, though this is associated with an increased risk of sympathetic syndrome. Following completion of the approach, fluoroscopy is used to confirm the spinal level. A universal retractor ring is then affixed to the table centered on the level of the affected disc; this ring will be used as a fixation point for retractors during the discectomy and reconstruction.

Addressal of the Pathology

Discectomy

A #11 blade is used to outline the target disc in rectangular fashion using a templated trial selected based upon preoperative CT or MR imaging, which are used to measure the width and height of the target disc space. During incision, care must be taken to avoid damaging the sympathetic chain located immediately lateral to the disc space, a concern that is most important at the L4/5 level. Though an en bloc resection of the disc is preferred, in cases where this is not feasible, the disc can be longitudinally sectioned in several pieces to facilitate extraction. An ALIF system, straight osteotome, Cobb elevator, or laminae spreader is then used to distract the disc space (Heary et al. 2017). Additional distraction can be achieved using sequential dilator; care should be taken to avoid overdistraction, which has been associated with postoperative neurapraxia (Taher et al. 2013). The remaining disc attachments are cleared from the superior and inferior end plates using a curette, concomitantly decorticating the end plates in preparation for fusion. Loosened cartilage and semi-mobilized disc fragments can be resected using a pituitary rongeur. We recommend preserving the posterior annulus in most cases, as it protects the ventral dura and helps to stabilize the cage postoperatively. The vertebral end plates are then prepared using a forward-angled curette. Trial implants are tested in the discectomy level until an implant large enough to restore the original disc height has been identified. Preoperative imaging is useful at this point, as adjacent, normal disc spaces can be used for selecting the appropriate trial size. Note that the implant should not be too large, as this may increase the risk of implant subsidence or kickout. Similarly, the trial should not be too small, as this (1) may provide insufficient sagittal plane correction and (2) may decrease the stability of the construct by reducing the compressive force exerted on the implant by the annulus (Patwardhan et al. 2003). The trial spacer is then removed, and the permanent interbody is placed. Femoral allograft, titanium, and PEEK implants are all available, and graft material (discussed later) can include demineralized bone matrix, allograft, autograft, and osteoinductive materials such as rhBMP-2. After placement of the implant, positioning is confirmed on fluoroscopy. If positioning is satisfactory, additional graft material may be placed around the interbody device. This additional graft has the potential to facilitate fusion and decrease the risk of implant subsidence (Kumar et al. 2005); it is difficult to assure that said graft remains in place however, and many surgeons only place graft in the interbody. Screws are then placed to fix the cage to the bone (in the case of integrated fixation cages), or a plate is affixed to prevent anterior subluxation of the cage. With some implants, notably integrated fixation cages, it is necessary to trim the lateral surfaces of the vertebrae bracketing the implant to allow placement of the screws.

Closure

After final positioning is confirmed on fluoroscopy, closure of the surgical corridor begins. Retractors are removed sequentially, inspecting both the ureter and the iliac vessels during removal to confirm that they have not been injured during the procedure. If these structures have been injured, it is necessary to repair them prior to closure of the wound. The risk of these injuries in the anterior retroperitoneal approach is one reason that it may be beneficial to utilize an approach surgeon for the exposure. The corridor is closed in layers with 2–0 Vicryl in the posterior rectus sheath, a few loose 0 Vicryl sutures in the rectus abdominis to facilitate muscle reapproximation, and 2–0 Vicryl sutures in the subcutaneous tissue; the skin is closed with 4–0 Vicryl or staples.

Spondylolisthesis

An anterior stand-alone procedure is not recommended for significant spondylolisthesis (above grade I), as the ability to reduce the translation is inferior to that offered by a posterior procedure (Jeswani et al. 2012). If the indication for surgery is spondylolisthesis, the anterior procedure should be performed exactly as described above for discectomy, except that a tension band plate is not placed, unless the spondylolisthesis can be reduced sufficiently on the anterior approach. Instead, a kickout plate or cancellous screw and washer construct are attached to one of the vertebrae to retain the interbody. The wound is then closed in layers, and the patient is flipped for the posterior portion of the procedure. Pedicle screw fixation is applied as described elsewhere, reducing the spondylolisthesis as the rods are applied, if desired. Fluoroscopy is then used to confirm placement of the instrumentation. If it is satisfactory, Valsalva is performed to confirm that no dural perforation has occurred. Note that this is not necessary in cases where pedicle screw instrumentation is placed percutaneously. Lumbar drains are placed, and the posterior wound is closed in layers, with 0–0 Vicryl in the deep fascia and 2–0 Vicryl in the subcutaneous fascia. The skin may be closed with Nylon sutures, steri-strips, glue, or staples.

Case Illustration

A 32-year-old female presented to the clinic of the senior author for chronic low back pain and radicular, right lower limb pain secondary to compression of the right L4 nerve root. The patient had managed her back pain conservatively for years but underwent an L4/5 discectomy at an outside hospital for her leg pain 4 months prior to presenting to our clinic. This intervention had been minimally beneficial, relieving her leg pain for several weeks but ultimately failing to provide robust benefit for either her leg or lower back pain. Upon presenting she was neurologically intact and was advised to try conservative management prior to another surgical intervention. After failing 6 months of conservative management, including physical therapy and several epidural nerve blocks, the patient was offered an L4–5 ALIF to stabilize her pathologic level and relieve her discogenic pain.

Surgery proceeded by placing the patient supine on a Jackson table and identifying the target level. She was then draped, and vascular surgery was called in to make the approach as the patient had a history of two prior Cesarean deliveries, and it was felt that employing an access surgeon would minimize the risk posed by the potential scar tissue. A traditional left, retroperitoneal approach was adopted, beginning with an oblique incision inferolateral to the umbilicus, intermediate to McBurney and Battle incisions. Sharp dissection and monopolar cautery were used to divide the superficial tissues and expose the anterior abdominal wall, which was then divided using the monopolar cautery. The rectus muscle and anterior rectus sheath were mobilized medially, and the retroperitoneum was entered. The retroperitoneal plane was extended posteriorly until the iliac veins were exposed; these lay above the level of the L4/5 disc, and so the veins were mobilized bilaterally to facilitate the anterior approach. A #10 blade was then used to incise the annulus of the L4/5 disc, which was removed with a combination of rongeurs, high-speed drill, and curettages. After the superior L5 end plate and inferior L4 end plate were cleaned, trial spacers were placed until sufficient correction of the anterior defect was achieved. An allograft spacer with demineralized bone matrix was then placed, and an anterior plate connecting the L4 and L5 levels was placed to reduce the risk of graft kickout postoperatively. Lateral imaging demonstrated good placement of the graft, at which point vascular surgery reentered and completed the closure. The anterior abdominal wall was closed with a #1 Maxon suture, and the skin was closed in layers using 4–0 Vicryl and a subcuticular technique for the most superficial layer. No complications occurred intraoperatively, and the patient was discharged home on postoperative day 3 after an uneventful inpatient course. The patient reported significant improvement in both her back and leg pain by the 1-month follow-up appointment. The patient demonstrated solid fusion across her construct at last follow-up – 52 months following surgery – and reported minimal pain.

Lateral Lumbar Interbody Fusion

A more recently popularized anterior approach for lumbar interbody fusion is the transpsoas approach, also known as the extreme lateral (XLIF) or direct lateral approaches (DLIF) and popularized by Luiz Pimenta over the past 15 years (Ozgur et al. 2006). This technique has the chief advantage of being minimally invasive, as it requires minimal tissue disruption – an advantage over conventional anterior approaches. Similarly, it allows for more complete removal of the target disc without disrupting either the anterior or posterior tension bands and at the same time allowing the placement of a larger interbody graft as compared to posterior approaches (Winder and Gambhir 2016); it may also allow for more complete end plate preparation as compared to ALIF techniques of similar invasiveness (Tatsumi et al. 2015). Indications for this procedure are similar to those of the above-described ALIF, including low-grade spondylolisthesis, discogenic low back or radicular leg pain, and sagittal plane deformity.

The chief advantages of the XLIF and DLIF procedures relative to the more conventional ALIF procedure are that they (1) do not require the assistance of a vascular access surgeon; (2) do not require retraction of the great vessels or sympathetic chain, thereby minimizing risk of injury to these structures; (3) are more easily performed on patients with significant abdominal obesity, who are not candidates for conventional ALIF; and (4) reduce risk of injury to the presacral autonomic plexus (Laws et al. 2012). One of the chief disadvantages for the DLIF/XLIF approach is that it is relatively contraindicated for fusion at the L5/S1 level, as the iliac crest blocks the approach to this level. It also has an increased risk of intraoperative injury to the nerves of lumbar plexus, specifically the genitofemoral nerve, which lies in close proximity to the retractor as it is passed through the psoas muscle (Jahangiri et al. 2010; Uribe et al. 2010). This risk, which is reportedly greatest for surgery at the L4/5 level due to the anteroinferior trajectory of the nerve, can be reduced by utilizing neuromonitoring and a stimulating retractor (Benglis et al. 2009; Kepler et al. 2011; Regev et al. 2009; Riley et al. 2018; Tohmeh et al. 2011). Additionally, this technique increases radiation exposure to both the patient and the operating room staff, as it requires serial fluoroscopy to ensure that the correct approach is being utilized. Lastly, XLIF/DLIF may have a decreased ability to increase segmental lordosis as compared to the traditional approach (Winder and Gambhir 2016), though evidence exists that this drawback can be nullified by resecting the ALL intraoperatively, thereby allowing for inducement of greater lordosis at the treated segment (Akbarnia et al. 2014; Deukmedjian et al. 2012) (Figs. 1 and 2).

Description

Positioning is extremely important to XLIF/DLIF approach, as proper placement of the interbody relies on the surgeon being able to align the plane of the target disc with the fluoroscope and surgical working plane (Winder and Gambhir 2016). This begins by placing the patient on a bendable radiolucent table in a true lateral decubitus position with the iliac crest at the level of the table break and the patient's back flush with the edge of the table. In addition to providing the surgeon with the most direct approach to the pathology, this position also has the advantage of allowing gravity to retract the intraperitoneal contents away from the spine, which is of increased benefit in patients with substantial visceral obesity. Under most circumstances, either the left or right lateral decubitus position may be chosen; however regional vascular variability may preclude approach from one side or the other, and surgeons are encouraged to consult preoperative imaging. Similarly, in cases where the patient also has

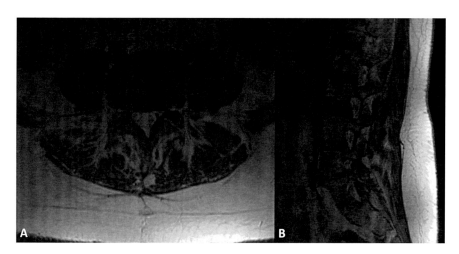

Fig. 1 Preoperative (**a**) L4/5 axial and (**b**) right parasagittal T2-weighed MR images demonstrating marked degeneration of the residual L4/5 disc leading to mild compression of the right L4 and L5 roots

30 Spinal Plates and the Anterior Lumbar Interbody Arthrodesis

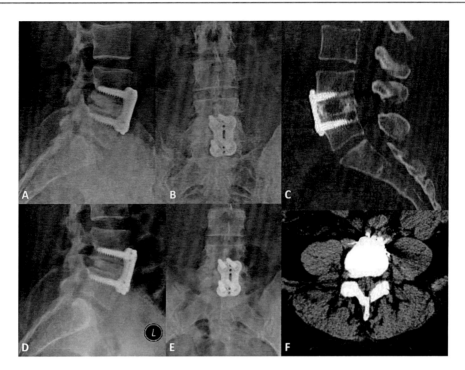

Fig. 2 (**a**) and (**b**) 3-month postoperative lateral and posterior-anterior radiographs demonstrating good alignment of the construct. (**c**) and (**f**) mid-sagittal and L4/5 axial CT views at 44-month follow-up demonstrating solid fusion with bridging of the construct by bony trabeculae. (**d**) and (**e**) Postoperative radiographs at 52-month follow-up showing successful fusion without evidence of adjacent segment disease or graft subsidence

scoliotic curvature of their lumbar spine, it is recommended that the patient be placed with the convex side facing the table (Beckman and Uribe 2017).

After the patient is placed in the lateral decubitus position on the table, their legs are flexed at the hips and knees to reduce tension in the psoas muscle, which facilitates passage of the dilators later in the procedure. Axillary and hip rolls are placed, as is a pad between the knees to reduce the risk of pressure ulcers. Positioning is then optimized using fluoroscopy to place the patient's spine in the true AP plane; the use of fluoroscopy may be most useful for patients having undergone prior lumbar surgeries, in which the native anatomy has been disturbed. It may be helpful to induce some bend at the table break to move the iliac crest out of the approach path. The patient is then secured to the table with tape at the shoulder and iliac crest, as well as at the legs, with strips parallel to the femur and tibia (Badlani and Phillips 2014). Final adjustments are then made to reestablish the true AP plane (Fig. 3a, b). Electronic monitoring leads are then placed, and preoperative signals should be acquired to establish the patient's neurologic baseline.

After skin preparation and draping, a #10 blade is used to make an incision through the dermis. As with the ALIF technique, a single 3–4 cm transverse incision can be used if only a single level is to be treated; if multiple levels are to be addressed, then a single vertical incision is planned. After passage through the dermis, monopolar cautery is used to dissect through the retroperitoneal fat until the external oblique fascia is reached. A transverse incision is formed in this fascia, and blunt dissection is used to pass through the abdominal wall musculature; examining the orientation of the fibers of the dissected muscle can be used to track that penetration of the abdominal wall has been complete. During dissection, care must be taken to remain on line with an approach to the posterior third of the disc space or middle third at the L4/5 level. Migration of the dissection

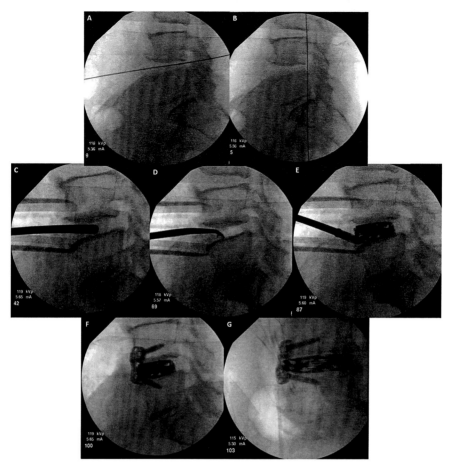

Fig. 3 (**a**) and (**b**) Fluoroscopy demonstrating effective positioning with true lateral positioning of the patient on the table. (**c**) The retractor is docked on the disc space, and a Cobb elevator is used to distract the disc space. (**d**) A forward angle curette is used to clean the superior and inferior end plates. (**e**) Titanium interbody is tamped into place, and positioning is confirmed. (**f**) and (**g**) Lateral and AP imaging demonstrating accurate positioning of the interbody and plate

path anteriorly endangers the retroperitoneal and intraperitoneal contents, whereas migration posteriorly risks damage to the nerves of the lumbar plexus.

Once the retroperitoneal space is entered, blunt finger dissection is then used to palpate the quadratus lumborum; the finger is then rotated anteriorly to palpate the psoas. The first dilator is passed along the trajectory of the finger to the top of the psoas muscle, keeping the finger anterior to the dilator to prevent injury to the retroperitoneal contents. Once the dilator has been placed, its position is confirmed on fluoroscopy, and then the dilator is passed through the psoas, docking on the intervertebral disc space; fluoroscopy is used again to confirm position. During passage of the dilator through the psoas, the target site is highly dependent upon the disc level being addressed. Previous work by Uribe et al. (2010) documenting the course of the genitofemoral nerve has established safe zones within the psoas muscle through which the retractor may be passed with minimal risk to the genitofemoral nerve. Biologic variability among patients precludes said results from eliminating nerve injury risk from the approach, but they can reduce the risk substantially. Additionally, state-of-the-art dilator models have EMG capability, which can be used during placement to prevent inadvertent nerve injury (Tohmeh et al. 2011). For such dilators, stimulation should be performed following placement in the dilator to determine the relative position of the nerves. Low thresholds on the EMG indicate closer proximity of the nerve (threshold >11 mA are safe; <5 mA suggests direct contact); current thought is that positioning of the nerve anterior to the dilator increases the risk of nerve injury upon serial dilation. Fluoroscopy should then be performed to confirm placement. If the dilator is malpositioned on either EMG or fluoroscopy, it should be adjusted prior to serial dilation. Changes in the superoinferior plane can be accomplished without removing the dilator from the psoas muscle due to the myofiber orientation, but changes in the anteroposterior plane require removal from the psoas and reinsertion. A guidewire is then placed through the dilator, and sequential dilators are passed over the guidewire. Once the largest dilator has been placed, the retractor is passed over it and docked onto the disc space; the retractor is then fixed to the surgical bed arm, and a light source is connected to the posterior blade of the retractor. The field is inspected, and suspicious structures are stimulated with triggered EMG (tEMG) to rule out the possibility that the structure is a motor root. A protector is then placed posteriorly to prevent spinal cord injury, and retractor position is confirmed on fluoroscopy. If position is adequate, the retractor is expanded, and the anterior face of the disc is defined. Note that the goal is to minimize total retraction time (<20 min, per Beckmann (Beckman and Uribe 2017)), as extensive retraction can lead to postoperative lumbar plexopathy (Bendersky et al. 2015; Winder and Gambhir 2016).

An annulotomy is then performed using a box cutter instrument, and a pituitary rongeur is used to resect the disc; the anterior and posterior annulus are left intact to hold the interbody in place unless significant sagittal plane correction is required (C. Kim et al. 2017). A Cobb elevator is then malleted vertically along the superior and inferior end plates, making sure to continue through and disrupt the contralateral annulus (Fig. 3c). The end plates are then prepared with forward-angled curettes (Fig. 3d) and a pituitary rongeur, taking care to maintain end plate integrity. Trial interbodies are then placed until one is identified that restores adequate disc height;

an equivalent interbody is then malleted into place with allograft in the central cavity (Fig. 3e). Note that some experienced centers use preoperative imaging to determine the optimal implant size and thereby spare the retraction time required for serial trial placement; this is a technique that is perhaps best left to experienced surgeons. After the interbody has been placed, hemostasis is achieved, and positioning is verified on AP and lateral fluoroscopy. It is at this point that a lateral plate can be placed to increase construct rigidity. If interbody device position is adequate (Fig. 3f, g), then the retractor is removed, and the abdominal wall musculature should return to its native position. The fascia is closed with interrupted 0 Vicryl, the subcutaneous tissue is closed with 3–0 Vicryl, and 4–0 monocryl is used to close the skin in subcuticular fashion (Beckman and Uribe 2017; Ozgur et al. 2006).

Case Illustration

A 62-year-old male presented to the neurological spine service with complaint of acute superimposed on chronic lower back pain of 7- to 8-year duration. The patient also reported minor left lower extremity radiculopathy. The patient had failed several months of conservative management, having tried epidural injections, physical therapy, and trigger point injections without robust benefit. The patient was neurologically intact on exam, and outside MR demonstrated an acute disc herniation compressing the left L5 nerve root. The patient was advised to undergo an L4–5 discectomy with pedicle screw instrumentation. The patient desired a more minimally invasive approach though. ALIF was relatively contraindicated secondary to the patient's significant abdominal obesity, and so an oblique lateral L4–5 discectomy and interbody fusion were offered.

The patient underwent a procedure as previously described in this chapter. He was placed in the left-lateral position, and sharp incision was made through the external oblique, internal oblique, and transversus abdominis in line with the muscle fibers at the level of the midaxillary line. Blunt dissection was used to navigate the retroperitoneal space and identify the psoas muscle. The psoas was mobilized with bipolar electrocautery, and neuromonitoring was used to guide a tubular retractor to the L4/5 disc space. A subtotal discectomy was performed, leaving the anterior and posterior annulus to maintain graft position. The end plates were prepared, and a size 7 expandable cage packed with allograft was then tamped into place. Placement was confirmed on fluoroscopy, and an anterolateral plate was placed. The patient had an uneventful hospital course with near-immediate improvement of his pain. He is now 6-month postoperatively and is doing well (Fig. 4).

Indications for Plate Usage in Anterior and Lateral Lumbar Surgery

Anterior

As with surgery for degenerative conditions at other levels of the spine, the prime indication for instrumentation in anterior lumbar interbody procedures is to facilitate osseous fusion across the construct – the basis for long-term construct stability. Previous studies (Song et al. 2010) have suggested that the utilization of instrumentation in the spine construct, namely, an anterior tension band plate, can help to both improve the rate of successful arthrodesis and reduce the time to radiographic fusion, which has been reported to take up to 12 months (Blumenthal et al. 1988). Biomechanical studies, such as those of Tzermiadianos and Zhang, have suggested that the reason for these superior fusion outcomes is that the instrumentation helps to maintain positioning of and reduction onto the interbody (Tzermiadianos et al. 2008; J. Zhang et al. 2012). Said opposition is necessary to allow bony ingrowth into the interbody, the first step to osseous fusion. One recent study providing evidence to this end by Liu and colleagues (L. Liu et al. 2014) compared the biomechanical stability – axial compression, flexion, extension, lateral bending, and torsion – across three techniques for L4/5 lumbar interbody arthrodesis: ALIF without instrumentation, ALIF with an anterior fusion plate, and ALIF with posterior pedicle screw instrumentation. They observed that ALIF with anterior plate instrumentation provided superior stability relative to both the ALIF-alone and ALIF with pedicle screw instrumentation. The plate-supplemented ALIF group also had superior

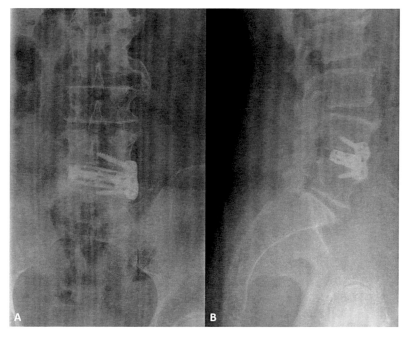

Fig. 4 (**a**) and (**b**) 6-month postoperative imaging of a patient who underwent L4–5 OLIF/XLIF with anterolateral plating for left L5 radiculopathy superimposed on chronic lower back pain

stiffness in terms of flexion, extension, lateral bending, and rotational torsion relative to the group treated with ALIF alone. Similar findings have also been reported by Gerber et al., Beaubien et al., and Tzermiadianos et al. using human cadaver models of lumbar and lumbosacral instability, though the findings of the Gerber group did not achieve the level of statistical significance (Beaubien et al. 2005; Gerber et al. 2006; Tzermiadianos et al. 2008). In all groups, supplemental stability with pedicle screw instrumentation decreased range of motion in lateral bending relative to anterior plating; Tzermiadianos also observed a significant decrease in flexion-extension range of motion relative to anterior tension band plating (Tzermiadianos et al. 2008).

In addition to facilitating long-term fusion, anterior plating can help to reduce the rate of cage migration, which is of greater concern in anterior relative to posterior procedures (Teng et al. 2017). Plating accomplishes this by having the plate function as a sort of retaining device, which traps the interbody between the plate anteriorly and the posterior longitudinal ligament posteriorly (as well as any residual portion of the annulus fibrosis). The plate also provides anterior column support, taking the role of an anterior tension band and preventing excess motion at the treated level (Yoganandan et al.). Evidence that the plate functions in this role is perhaps best supported by Bozkus et al., who compared anterolateral and lateral plating for anterior interbody fusion in a calf spine model (Bozkus et al. 2004). The authors found that while both lateral and anterolateral plates significantly decreased instability and range of motion in flexion-extension, rotation, and lateral bending relative to both sham and interbody-only constructs, the greatest decrease for each plate was in the dimension coplanar with the plate. That is to say the anterior plates provided the greatest stabilization in the flexion-extension plane, whereas lateral plates provided the greatest stabilization in the lateral bending plane, likely because the plate acts as a tension band, restricting motion that attempts to distract the anchor points of the plate.

Lateral

Indications for plating of XLIF/OLIF/DLIF constructs are similar to those of ALIF procedures. The main advantages of plating are increased construct stability (Fogel et al. 2014; Laws et al. 2012) and reduction in the risk of interbody device migration or kickout (Du et al. 2017). The latter concern is more significant in cases where the anterior longitudinal ligament has been disrupted, as such patients have lost a large portion of the annulus responsible for retaining the device. The former has received significant study in in vitro models. One series by Laws and colleagues examined stability of four constructs – stand-alone ALIF or DLIF, plated ALIF or DLIF, and DLIF with unilateral or bilateral pedicle screw fixation – under conditions of flexion-extension, lateral bending, and axial rotation (Laws et al. 2012). Plating of the DLIF construct significantly decreased range of motion in flexion-extension and axial rotation relative to unplated constructs; it also decreased range of motion in all directions relative to ALIF constructs, though these differences were not significant. These results agree with those of Heth, who also found DLIF/OLIF/XLIF to be biomechanically similar to ALIF, if not superior. Fogel et al. conducted a similar study using fresh-frozen cadaveric spines, comparing stand-alone XLIF constructs with those supplemented by either lateral plating or pedicle screw instrumentation (Fogel et al. 2014). Lateral plating provided a similar level of stability to pedicle screw fixation in lateral bending, but not flexion-extension or axial rotation (bilateral pedicle screw fixation only). All instrumentation methods provided significant reductions in axial rotation and lateral bending relative to the stand-alone construct, though the plate did not provide a significant reduction in flexion-extension range of motion. Liu et al. and Reis et al., by contrast, reported a significant reduction only in vertebral end plate stress and lateral bending with application of the plate; axial rotation and flexion-extension were not significantly decreased (X. Liu et al. 2017; Reis et al. 2016). Taken in aggregate, these results, like those of anterior plating, suggest that plates increase construct stability by functioning as tension bands. As a result, they lead to stabilization primarily in the direction of motion that directly distracts or compresses the plate; they provide only minor stabilization when shearing forces are applied, as occurs with axial rotation and flexion-extension in lateral plates and lateral bending in anterior plates.

Anterior Plates and Other Anterior Fusion Technologies

As previously described, anterior spine fusion was developed more than seven decades ago, and dedicated instrumentation for anterior fusion has been around since at least 1953, when Wenger described the use of anterior instrumentation for the treatment of scoliosis (Dwyer et al. 1969; Ghanayem and Zdeblick 1997). These systems became increasingly popular in the late 1970s though, following the increased incidence of reports on anterior fusion techniques (Bradford and McBride 1987; Dunn 1984; Kaneda et al. 1984; Kostuik 1983; Zielke et al. 1976). Some of the earliest systems to gain mainstream acceptance were the Zielke system (Zielke et al. 1976), the Kostuik-Harrington device (Kostuik 1983), and the Kaneda system (Kaneda et al. 1984). The Zielke system was developed as a less morbid alternative to posterior Harrington instrumentation in patients being treated for scoliosis (Zielke et al. 1976). The later Kaneda system, by contrast, was developed for the treatment of thoracolumbar burst fractures, which commonly involve shorter constructs (Kaneda et al. 1984). Like the anterior Harrington device and biomechanically similar Zielke device, the Kaneda relied upon conventional screw-rod architecture, with screws placed at the superior and inferior instrumented vertebrae and rigid rods bridging the two anchor points (Kaneda et al. 1984). Contemporary in vitro biomechanical studies suggested that these devices, particularly the Kaneda, provided structural stiffness equivalent or superior to posterior Harrington rod instrumentation while requiring the instrumentation of fewer levels (Gurr et al. 1988; Zdeblick et al. 1993). And clinical work suggested that they offered superior decompression of the spinal canal (Bradford and McBride 1987; Ghanayem and Zdeblick 1997). Additionally, all three systems had the advantage of allowing for both compression and distraction. To their detriment however, all three systems were hampered by their relatively bulky nature, which had the potential to irritate prevertebral and paravertebral tissues. As a result, there was an incentive to develop new, lower-profile systems. Results of these efforts including the Z-plate, the University Plate, and the Anterior Thoracolumbar Locking Plate had the benefit of being less bulky and of creating a smaller radiograph artifact due to their titanium construction. Of these different systems, the most widely used of these was the Z-plate. The Z-plate comprised a slotted plate instrumented at the superior and inferior construct levels using two bicortical vertebral body screws at each vertebra. It was designed for short anterior fusion constructs, originally developed for the treatment of thoracolumbar burst fractures (Ghanayem and Zdeblick 1997; McDonough et al. 2004). Several in vitro studies, including those of Hitchon et al. and Dick et al., demonstrated the Z-plate to be biomechanically equivalent to the earlier Kaneda device (An et al. 1995; Dick et al. 1997; Hitchon et al. 2000; Kotani et al. 1999). Of note, Hitchon et al., Kotani et al., and Dick et al. also compared the anterolateral locking plate – the direct ancestor of modern anterolateral plating systems – to these devices. Hitchon et al. noted that the anterolateral plating system produced a less stiff construct than the earlier systems upon instrumentation, but after only 5000 cycles of flexion-extension, the Z-plate system ceased to differ significantly from the anterolateral plate (Hitchon et al. 2000). By contrast, Dick et al. and Kotani et al. found the Synthes® anterior thoracolumbar locking plate to be stiffer than the Kaneda device and Z-plate in axial compression, axial rotation, and lateral flexion (Dick et al. 1997; Kotani et al. 1999). It also had a significantly longer service life as demonstrated on fatigue testing. Since that time, this anterolateral plate and other similar designs have come to dominate the anterior fusion device market.

Currently, over a half dozen different anterior and anterolateral plate systems are commercially available. The indications for the use of these devices are relatively uniform across manufacturers and include cases requiring anterior lumbar interbody fusion for any of the following reasons: trauma (fracture or dislocation), tumor, low-grade spondylolisthesis, spondylolysis, pseudarthrosis, deformity, and failed back syndrome. Similarly, contraindications to their use are relatively uniform, including active systemic or local infection, inadequate mechanical support (tumor infiltration, severe osteoporosis, metabolic bone disease), patient history of foreign body reaction or sensitivity to metal implants, risk factor for poor wound healing (e.g., insufficient tissue to cover implant), conditions that would place excessive load on implant during the healing period (e.g., significant obesity), and any general contraindications for surgery (e.g., concurrent drug use or psychiatric illness).

Material Selection

The majority of plates currently on the market are composed of titanium or titanium alloy. These materials are generally stronger for a given size than are stainless steel implants, which allow titanium plates to be thinner for any given combination of tensile and compressive strengths. Titanium instrumentation is also more amenable to postoperative follow-up with magnetic resonance imaging than is stainless steel or chrome-containing instrumentation. Unlike the latter, which are ferromagnetic, titanium and titanium alloy implants are paramagnetic and so produce substantially smaller artifact on MR (Do et al. 2018; Tahal et al. 2017). Additionally, the local tissues are generally thought to respond more favorably to the use of titanium implants as compared to either stainless steel or chromium-based implants (Gibon et al. 2017; Tahal et al. 2017). Chromium-based implants can be particularly irritating as they trigger cytotoxic reactions in the local tissue, which may weaken the surrounding osseoligamentous structures, though no evidence to date exists to suggest that this leads to significantly different clinical outcomes. Chromium and stainless steel-based implants are also more likely to generate delayed-type hypersensitivity reactions at the site of implantation due to the osteoclast-induced release of nickel and chromium ions from the implants (Gibon et al. 2017). Nickel and chromium ions are two of the most common causes of metal sensitivities among the general population, and so we prefer the utilization of titanium-based implants when available, unless other biomechanical considerations suggest the patient would benefit from the superior fatigue strength provided by chromium-based implants (Tahal et al. 2017).

In addition to the concern of local tissue toxicity, the propensity of an implant to serve as an infection nidus as well as implant fatigability is an important concern. The latter consideration is of biggest concern among young patients, given that the implant will undergo a substantial number of loading cycles over the course of its service life – estimated at two million cycles per year (Graham 2006). Previous research has failed to demonstrate substantial differences in the fatigue strength of titanium and titanium alloy plates relative to stainless steel implants (Pienkowski et al. 1998). However, several studies have demonstrated a lower incidence of biofilm formation and surgical site infection among those persons instrumented with pure titanium implants, as compared to either stainless steel or titanium alloy implants (Tahal et al. 2017).

Plate Design

As discussed, most commercially available anterior lumbar plates are composed of pure titanium or a titanium alloy (e.g., Ti-6Al-4 V/TC4, nitonol) due to the superior mechanical properties and greater tolerance by local tissues. The variability among implants is largely based upon implant shape, length, and screw fixation. Both straight and prelordosed plates are currently available, with the latter providing a better anatomic match to the native anatomy. This better approximation of the lumbar and lumbosacral lordosis may improve the rigidity of the construct and so reduce the rate of pseudarthrosis and need for reoperation, though no direct comparisons currently exist. Like the more traditional straight plates, prelordosed plates are available in both one- (25–54 mm length depending upon the manufacturer) and two-level lengths (65–89 mm depending upon the manufacturer) for instrumentation at the lumbar spine or lumbosacral junction, with most manufacturers offering their plates in 3-mm size increments. Both fixed-angle and variable-angle screw systems are available, with both systems generally utilizing two 5.5–6.0 mm screws at the inferior instrumented level and one or two 5.5–6.0 mm screws at the superior instrumented level.

Variable- Versus Fixed-Angle Devices

The relative merit of a variable-angle plate, as compared to a fixed-angle plate, is rooted in the ability of the surgeon to select the screw angle when placing instrumentation. This presupposes that certain screw angles confer superior pullout strength relative to others, a conclusion that has mixed support in the current literature. Work by DiPaola et al. on cervical spine plates in a polyurethane block model has suggested that orthogonally positioned screws have superior pullout strength relative to other orientations (DiPaola et al. 2007, 2008). This is perhaps because the orthogonal angle places each thread perpendicular to the force vector, allowing it to maximally resist pullout. Regardless of the biomechanical underpinning, the said results favor the variable screw model over fixed-angle instrumentation, as the former enables the attending surgeon to place screws along optimal trajectories even in cases of suboptimal implant-bone apposition. Rodríguez-Olaverri et al. reach a similar conclusion regarding the importance of screw trajectory (Rodríguez-Olaverri et al. 2005). In their thoracolumbar calf spine model, Rodríguez-Olaverri placed bicortical polyaxial screws linked by a single 5.5 mm cobalt-chrome and then subjected the construct to 10,000 cycles of lateral bending with a 100 N load. Unlike DiPaola, they found that superior construct stabilization was provided by angled screws; however, like the former, they concluded that screw trajectory had a significant impact on construct service life. This favors instrumentation that affords the surgeon options when placing instrumentation, as is offered by a variable-angle system.

By contrast, Rios and colleagues failed to demonstrate said advantages to variable screw angle systems (Rios et al. 2012). Using a similar polyurethane block model, they investigated the relative pullout strength of anterior ALIF plates with screws fixed in one of nine different orientations, varying in both sagittal and coronal angulation (Rios et al. 2012). Though they noticed some minor but significant differences between the strongest and weakest trajectories, they failed to observe superiority of any trajectory over the neutral or "straight-in" trajectory. This suggests that the ability to select screw trajectory confers no mechanical advantages, in turn indicating the non-inferiority of fixed-angle systems. These results were replicated by Patacxil et al. and Hadley et al., the latter of whom also found no difference between variable- and fixed-angle plates in a cadaveric model (Hadley et al. 2012; Patacxil et al. 2012). Additionally, cohort studies by Oh and Hong examining radiographic outcomes in patients having undergone anterior cervical discectomy and fusion report similar findings, with the Hong group finding that fixed-angle plates may actual reduce the risk of graft subsidence (Hong et al. 2010; Oh et al. 2013). Given these mixed results, we believe that insufficient evidence exists to recommend variable-angle designs over conventional, fixed-angle designs. Rather, the plate type used should be based upon surgeon experience.

Interbody Design

Lumbar interbody devices have been used in spine surgery for more than three-quarters of a century, having been first described in 1944 by Briggs and Milligan (de Kunder et al. 2018). Their original procedures utilized a posterior approach and made use of an allograft bone peg with external fixation to facilitate fusion. Cloward published a larger series in 1953 describing a similar technique using multiple bone grafts to reduce the procedural morbidity (de Kunder et al. 2018).

From there, interbody devices remained relatively unchanged until the invention of the Bagby basket in 1986 for the treatment of Wobbler syndrome in horses – a myelopathic condition seen secondary to cervical instability in these animals (DeBowes et al. 1984; Phan and Mobbs 2016). This device was novel in that it utilized autograft-packed titanium implants, which had significant greater compressive strength than did the earlier bone graft interbody devices utilized by Briggs and Milligan (de Kunder et al. 2018). The Bagby device, or rather its successor – the BAK device – was quickly adapted by Kuslich and colleagues (Phan and Mobbs 2016), who described its use for posterior lumbar interbody fusion in 1992 (Kuslich et al. 1998). Within the decade it was approved for anterior interbody fusion, and several biosimilar devices had been presented, including the threaded cage of Ray (1997), mesh cage of Harms and Biederman, and the tapered or trapezoid-shaped cages which have become increasingly popular (Phan and Mobbs 2016). Previous in vitro analyses have suggested gross biomechanical equivalency among these different design types, though threaded cages, such as the BAK device, may provide lower intracage pressures and thereby help to protect cancellous graft placed within the cage (Kanayama et al. 2000).

PEEK and carbon fiber interbodies were also developed during the 1990s and early 2000s in order to serve as an alternative to the titanium interbodies (Brantigan and Steffee 1993; D. Cho et al. 2002; Phan and Mobbs 2016). The former had the advantage of better approximating the elastic modulus of bone and lacked the substantial radiographic artifact produced by titanium devices (Phan and Mobbs 2016). Interbody devices made from all three materials are now available, both for the traditional retroperitoneal anterior approach (ALIF) and for the lateral approaches (OLIF, XLIF). These devices offer a range of lordosis and footplate dimensions, allowing the surgeon to optimize the device to each patient. All three of these parameters – composition, lordosis, and cross section – should be considered during surgical planning.

Material

The three main materials used for anterior lumbar interbody fusion are femoral ring allograft, titanium or titanium alloy, and polyether-ether ketone (PEEK). Each material has its relative advantages and disadvantages. Ring allograft and PEEK better approximate the elastic modulus of native bone compared to titanium interbody devices (Phan and Mobbs 2016), which may reduce the risk of pistoning and implant subsidence (Niu et al. 2010; Seaman et al. 2017). Though pistoning is concern in all patients, it is most common in patients with low bone density and should be given greater consideration in osteoporotic patients, such as postmenopausal women (Melton et al. 1992).

PEEK cages and femoral ring allograft also allow for superior radiographic follow-up, as they produce no artifact on computed tomography imaging. This allows fusion across the construct to be more easily assayed than in constructs using titanium implants. Given that 11–23% of patients treated with anterior interbody fusions may experience nonunion (C. S. Lee et al. 2011), the ability to assess osseous fusion may favor the use of these materials, especially in patients at high risk of pseudarthrosis (e.g., those with current or previous history of cigarette use) (Phan et al. 2017). Despite this, studies comparing interbody fusion with PEEK vs. Ti implants have suggested that titanium interbody devices provide equal (Niu et al. 2010) or superior rates (Spruit et al. 2005) of arthrodesis. Consequently, the superior fusion monitoring available in PEEK constructs may be negated by the inferior overall fusion rates.

The reason for inferior fusion rates in PEEK implants has not been definitively established, though current evidence suggests that PEEK devices offer less bony contact per unit end plate area. This difference is most significant when comparing to the newer plasma-treated titanium interbody devices (Pelletier et al. 2016) whose porous surfaces allow for the rapid ingrowth of cancellous bone (Chang et al. 1998). Additionally, although the greater yield strength mismatch seen in constructs containing titanium interbody devices may increase the risk of implant subsidence, the high yield strength of the titanium interbody increases the initial stability of these constructs relative to PEEK constructs (Spruit et al. 2005). As a result, such constructs are likely to better maintain graft-end plate opposition during the critical period when fusion is occurring. This may also contribute to the superior fusion rates observed in some studies. Lastly, prior in vitro studies have demonstrated PEEK to be hydrophobic and therefore biochemically inert, meaning that commercially available implants rely upon a surface coating of hydroxyapatite or biocomparable material in order to promote osseointegration (Briem et al. 2005; Dennes and Schwartz 2009; Durham et al. 2017; Noiset et al. 1999; Zhao et al. 2016).

At least some of these concerns – chemical inertness and lower contact area per end plate area – can be addressed by employing a hybrid interbody device, one comprised of a PEEK body with porous titanium face plates. The said devices have the advantage of superior osteoconduction at the titanium end plates (Han et al. 2010; X. Wu et al. 2012), while maintaining a yield modulus similar to that of natural bone. Preliminary in vivo studies by Han et al. using a rabbit model of spine fusion found that titanium-coated PEEK implants had bone-to-implant contact that was nearly twice as great as standard PEEK implants (Han et al. 2010), consistent with earlier in vitro studies demonstrating superior survival of osteoblast cell lines on titanium-coated implants (Han et al. 2010; X. Wu et al. 2012). Additionally, biomechanical analysis of an in vivo PEEK-titanium composite interbody demonstrated it to have superior strength in terms of axial rotation, flexion-extension, and lateral bending as compared to a conventional PEEK interbody device (McGilvray et al. 2017). These differences were progressively more pronounced with increasing follow-up, suggesting that their greatest advantage relative to conventional devices may be in terms of longer service life. The titanium-coated PEEK implants are not without their drawbacks though. Recently, Torstrick and colleagues reported the results of an in vitro comparison of conventional PEEK, porous PEEK, and plasma-sprayed titanium-coated PEEK interbody devices (Torstrick et al. 2018). Using a polyurethane spine model and guided weight impactor, the authors applied compressive forces to the interbody devices. While both pure PEEK devices showed minimal surface damage on SEM, the titanium-coated device showed a significant decrease in surface roughness, which may nullify the potential bone ingrowth advantages conferred by the titanium coating.

Another innovation designed to combine the advantages of PEEK and titanium interbody devices is porous titanium cages, which have only become commercially viable with the increased availability of 3D printing technologies. The first description of porous metal interbody devices for spinal fusion was by Levi et al., who described the use of porous tantalum implants in a cadaveric model of anterior cervical discectomy and fusion (Levi et al. 1998). A half-decade later, Assad and colleagues described the use of a porous titanium-nickel interbody for lumbar intervertebral fusion in a sheep model (Assad et al. 2003a, 2003b). Using standard lateral radiographs, the authors demonstrated superior bone bridging and periprosthetic radiolucency at 3 and 6 months postoperatively in animals instrumented with porous interbodies as compared to traditional titanium alloy implants (Assad et al. 2003b). Takemoto et al. reported similar findings using a porous titanium implant with 50% porosity in a canine model, which they attributed to the superior bony contact of the porous implants (Takemoto et al. 2007). Furthermore, they reported that these implants had yield compressive strengths similar to that of cortical bone – 80–120 MPa – meaning that they are theoretically less likely to suffer from implant pistoning than are traditional titanium implants. This yield strength is superior to that of porous tantalum implants, and the fact that these implants do not incorporate nickel, like those of Assad, means that they are not associated with the potential downsides of this material, namely, concerns over intraspinal metallosis and neurological deficits (del Rio et al. 2007; Takahashi et al. 2001). Fujibayashi et al. reported the first clinical series of patients treated with these implants, describing successful radiographic fusion within 6 months in five individuals who underwent TLIF with 60%

porosity implants (Fujibayashi et al. 2011). Several contemporaneous studies in animal models demonstrated that such devices had superior osseointegration as compared to PEEK implants, as well as superior mechanical stability for flexion-extension, axial rotation, and lateral bending (S. Wu et al. 2013). More recently, McGilvray and colleagues reported the result of an ovine lumbar fusion model directly comparing PEEK, porous titanium-coated PEEK, and 3D-printed porous titanium alloy interbody devices (McGilvray et al. 2018). Using micro-computed tomography and histology to evaluate fusion, and biomechanical analyses to evaluate construct stability, the authors demonstrated significant reductions in flexion-extension range of motion, increases in bony ingrowth, and increases in stiffness for the porous titanium cages relative to both the titanium-coated and convention PEEK interbody devices. The porous titanium implants are also less likely to fragment with application of compressive forces and hence may demonstrate better wear over the life of the device (Kienle et al. 2015). Further studies are required, but it appears as if the porous titanium interbodies may provide the overall best option in a cost-neutral comparison. Additionally, although some in vitro (MacBarb et al. 2017a) and in vivo animal (MacBarb et al. 2017b) evidence suggests that 3D-printed porous titanium devices are superior to conventional plasma-spray-coated implants with respect to bone infiltration, no clinical series exist to support this finding. The relative superiority of these devices is likely to depend upon the relative porosity though, with greater porosity offering greater bony ingrowth (Li et al. 2007; Taniguchi et al. 2016; de Vasconcellos et al. 2010) and lower stresses inside the implant (Z. Zhang et al. 2018) but coming at the cost of decreased construct stability (Z. Zhang et al. 2018).

Dimensions

When selecting dimensions for an anterior interbody, the goal should be to restore the native anatomy, in terms of disc height, foraminal diameter, and lumbar lordosis. To address both end points, most cage manufacturers offer their interbody devices in a variety of heights and lordosis angles, including hyperlordotic interbody devices, which may restore up to 30 degrees of lordosis per instrumented level (Saville et al. 2016). When selecting dimensions for the interbody, it is also important to consider the size of the end plates of the superior and inferior vertebral levels (Phan and Mobbs 2016). These surface areas should be determined preoperatively to select the widest interbody that can be safely placed at the discectomy level (Lowe et al. 2004). In vitro studies have shown that at bare minimum, the selected interbody should cover 30% of the end plate to prevent graft subsidence (Closkey et al. 1993). Ideally, the implant should be wide enough to engage the edges of the end plate, but not so large as to damage surrounding tissues. Doing so allows the interbody to engage cortical bone at the vertebral body in addition to cancellous bone, producing a more even distribution of stress on the adjacent vertebral levels (Kumar et al. 2005). Additionally, engaging the cortical bone, which has a significantly higher yield strength (Fyhrie and Vashishth 2000; Kutz et al. 2000), reduces the likelihood of implant subsidence – a nontrivial concern in fusions with titanium interbody devices. And there is some suggestion that the wider implants – at least for the XLIF/DLIF procedure – may provide a stiffer construct (Pimenta et al. 2012).

The risk of pistoning can be further decreased by insertion of bone graft into and around the interbody device, both of which increase the effective cross section of the implant and provide osteoconductive material to facilitate fusion (Kumar et al. 2005). Furthermore, previous research in the posterior lumbar interbody fusion literature has demonstrated fusion rates to be directly correlated to the contact area between the cage and the bony end plates. One prospective series, published by Seo et al., examined 60 patients undergoing instrumented PLIF using bilateral autograft-filled titanium cages (Seo et al. 2017). They observed that bone between the interbody devices fused in 100% of patients, whereas graft circumscribing the region bounded by the interbody devices only fused in 72.3% of patients.

Integrated Fixation Cages

Another option for interbody fusion is the use of an integrated fixation cage (IFC) or stand-alone interbody device, which first appeared on the market 15 years ago. These devices incorporate the instrumenting screws directly into the interbody and so eliminate the need for an anterior plate. This allows for a technically simpler procedure and may help to reduce operative room times.

As with any new technology, one of the questions that must be asked when considering instrumentation with an IFC is whether said device will provide similar rigidity to traditional interbody-plate constructs. A handful of publications have been presented to date directly comparing plate and IFC constructs (Beaubien et al. 2009, 2010; Kornblum et al. 2013; Palepu et al. 2017). The first of these was presented by Beaubien et al., who examined the stiffness of various interbody constructs in cadaveric spines, including traditional plate-interbody and stand-alone IFC constructs (Beaubien et al. 2009, 2010). They found that although the latter construct exhibited significantly greater motion in flexion and extension, it possessed significantly greater pullout resistances, in part, because failure required fracture of all bone surrounding the integrated screws. Similar findings were reported by Kornblum and colleagues (Kornblum et al. 2013), who noted that IFC constructs had inferior stability in flexion-extension motion relative to plate systems. More recently though, a study by Palepu and colleagues (Palepu et al. 2017) reported no difference in the flexion-extension stabilities of the two constructs. These authors compared the constructs in cadaveric human lumbar spines across 20,000 cycles of flexion-extension loading and found similar stability in flexion-extension and lateral bending motions at the time of implantation and throughout fatigue testing. Interestingly, they also found the IFC constructs to provide superior stability at implantation and throughout testing in terms of axial rotation, suggesting that newer devices may have addressed the biomechanical shortcomings of previous designs.

Of the IFC devices on the market today, the majority has similar construction – they utilize a lordosed PEEK interbody cage with an anterior titanium face plate through which 5.0 mm or 6.0 mm lag screws are placed. The interbody sizes available vary across manufacturer, but most offer interbody heights of 10–20 mm (in 2 mm increments) with anywhere between 4° and 16° degrees of lordosis and end plates between 22 × 30 mm and 28 × 40 mm. The various commercially available cages vary in the number of lag screws incorporated, using anywhere between two and four lag screws for fixation. Previous studies examining the stability of these devices (Kornblum et al. 2013) have not demonstrated significant differences between them, suggesting that cage selection should be based upon surgeon preference. Two more recent technologies include the use of variable-angle screws for fixation and the incorporation of plasma-treated titanium end plates. In a cadaveric study comparing blade-type, fixed-angle screw, and variable-angle screw devices, Freeman et al. reported that fixed-angle cages had higher pullout strengths relative to the other two technologies (Freeman et al. 2016). No confirmatory studies have been performed yet; however studies from the plate literature suggest that variable-angle and fixed-angle systems may be equivalent (Hadley et al. 2012; Patacxil et al. 2012). Similarly, no studies have been performed to compare fusion rates of the new Ti-coated PEEK IFC to those of the conventional IFC devices, but it seems likely that the former may demonstrate superior fusion rates.

BMP and Profusion Agents

As with other spine procedures, anterior interbody arthrodesis has traditionally relied on bone allograft and autograft to facilitate fusion across the construct. This bone is usually incorporated directly into the device, though as mentioned previously, it may also be packed around the implanted interbody device. Kumar et al., using a computerized model of lumbar interbody fusion, demonstrated that incorporation of cancellous graft into and around the device decreases the load placed by the implant on the adjacent end plates, presumably by increasing the effective end plate cross section of the implant itself (Kumar et al. 2005). In so doing, this supplementation may help to decrease the risk for implant subsidence.

Other osteoinductive and osteoconductive technologies are available for use in patients who are deemed to be at higher risk for pseudarthrosis. These include demineralized bone matrix, bone graft substitutes or bone putties, hyaluronic acid, and bone morphogenic protein; in fact, the ALIF procedure is the only one for which rhBMP-2 use is currently approved by the FDA. Previous evidence has suggested that demineralized bone matrix provides similar fusion rates to traditional autograft (Fu et al. 2016), and similar findings have been reported for the other bone graft extenders, including calcium phosphate crystals and ceramic-based substitutes (Gupta et al. 2015). By contrast, numerous studies have observed superior rates of fusion in patients receiving rhBMP-2 (Boden et al. 2000; Burkus et al. 2002, 2003, 2009; Glassman et al. 2008; Gupta et al. 2015; Hustedt and Blizzard 2014), including several prospective randomized trials (Boden et al. 2000; Burkus et al. 2002), which led to its approval by the FDA in 2004. BMP use has been previously associated with higher complication rates in the cervical spine despite improving fusion rates, but evidence to support this in the lumbar spine is not conclusive. As a result, it is recommended that rhBMP-2 use be considered in all patients at high risk for nonunion, including smokers (Jackson and Devine 2016), older patients (Y. J. Kim et al. 2005, 2006), obese patients, those with low osteoblast-to-osteoclast activity (Inose et al. 2018), those with hypovitaminosis D (Ravindra et al. 2015), and those with histories of nonunion (Glassman et al. 2008). Osteoporosis, while a relative contraindication to surgery due to increased risk of implant failure, may not be a risk factor for nonunion. The evidence to support this conclusion is weak however (van Wunnik et al. 2011; Zura et al. 2016) and does not disentangle low bone density from advanced age, which has been previously associated with a decreased likelihood of nonunion (Zura et al. 2017). Obesity may similarly be a risk factor for nonunion (Zura et al. 2016) and so be an indication for the use of rhBMP-2. Evidence from the area of long bone fusion supports this (Meidinger et al. 2011), but studies on the role of obesity in spinal fusion nonunion are lacking.

Conclusion

The anterior and lateral lumbar interbody arthrodeses are robust constructs that remain an effective option for multiple pathologies of the lumbosacral spine. They allow for anatomic correction of disc height loss and for placement of larger interbody devices than can be placed through more traditional posterior interbody fusions, such as the TLIF or PLIF. Conventionally, anterior fusion has been accomplished through a combination of interbody placement and anterior plating, which reduces cage kickout and increases the stability of the construct to prevent early construct failure. In recent years, integrated fusion cages (IFC) – interbody devices with incorporated lag screws to prevent pullout – have become available. These devices allow for more expedient construct completion, and recent research suggests that they may provide stability equivalent to more traditional plate-interbody systems. Additional new technologies are being brought on the market every year, including interbody and integrated fixation cage devices with plasma-treated titanium end plates, which allow for superior osseous fusion rates relative to traditional PEEK interbody devices. Because of this continued innovation, it is apparent that the ALIF remains a valued option for select patients with pathology of the lumbosacral spine.

Cross-References

▶ Anterior Lumbar Spinal Reconstruction
▶ Bone Grafts and Bone Graft Substitutes
▶ Mechanical Implant Material Selection, Durability, Strength, and Stiffness

References

Akbarnia BA, Mundis J, Gregory M, Moazzaz P, Kabirian N, Bagheri R, Eastlack RK, Pawelek JB (2014) Anterior column realignment (ACR) for focal kyphotic spinal deformity using a lateral transpsoas approach and ALL release. 27:29–39. https://doi.org/10.1097/BSD.0b013e318287bdc1

An HS, Lim T, You J, Hong JH, Eck J, McGrady L (1995) Biomechanical evaluation of anterior thoracolumbar spinal instrumentation. 20:1979–1983. https://doi.org/10.1097/00007632-199509150-00003

Aryan HE, Berta S (2014) Approach to anterior lumbar interbody fusion. In: Nader R, Berta SC, Gragnaniello C, Sabbagh AJ, Levy ML (eds) Neurosurgery tricks of the trade: spine and peripheral nerves. Thieme, New York, pp 96–99

Assad M, Jarzem P, Leroux MA, Coillard C, Chernyshov AV, Charette S, Rivard C (2003a) Porous titanium-nickel for intervertebral fusion in a sheep model: part 2. Surface analysis and nickel release assessment. J Biomed Mater Res 64B:121–129

Assad M, Jarzem P, Leroux MA, Coillard C, Chernyshov AV, Charette S, Rivard C (2003b) Porous titanium-nickel for intervertebral fusion in a sheep model: part 1. Histomorphometric and radiol anal 64:107

Badlani N, Phillips FM (2014) Lateral lumbar interbody fusion. In: Zdeblick TA, Albert TJ (eds) The spine, 3rd edn. Lippincott Williams & Wilkins, Philadelphia, pp 357–372

Beaubien BP, Derincek A, Lew WD, Wood KB (2005) In vitro, biomechanical comparison of an anterior lumbar interbody fusion with an anteriorly placed, low-profile lumbar plate and posteriorly placed pedicle screws or translaminar screws. 30:1846–1851. https://doi.org/10.1097/01.brs.0000174275.95104.12

Beaubien BP, Freeman AL, Turner JL, Castro C, Armstrong WD, Waugh LG, Dryer RF (2009) Comparative biomechanical evaluation of a lumbar spacer with integrated screws. Conference proceeding/poster abstract (Poster No. 1712) form 55th annual meeting of the Orthopaedic Research Society, Las Vegas, 22–25 February 2009

Beaubien BP, Freeman AL, Turner JL, Castro CA, Armstrong WD, Waugh LG, Dryer RF (2010) Evaluation of a lumbar intervertebral spacer with integrated screws as a stand-alone fixation device. J Spinal Disord Tech 23:351–358. https://doi.org/10.1097/BSD.0b013e3181b15d00

Beckman JM, Uribe JS (2017) MIS lateral lumbar interbody fusion. In: Steinmetz MA, Benzel EC (eds) Benzel's spine surgery. Elsevier, Philadelphia, pp 667–673

Bendersky M, Solá C, Muntadas J, Gruenberg M, Calligaris S, Mereles M, Valacco M, Bassani J, Nicolás M (2015) Monitoring lumbar plexus integrity in extreme lateral transpsoas approaches to the lumbar spine: a new protocol with anatomical bases. Eur Spine J 24:1051–1057. https://doi.org/10.1007/s00586-015-3801-9

Benglis DM, Vanni S, Levi AD (2009) An anatomical study of the lumbosacral plexus as related to the minimally invasive transpsoas approach to the lumbar spine. J Neurosurg Spine 10:139

Blumenthal SL, Baker J, Dossett A, Selby DK (1988) The role of anterior lumbar fusion for internal disc disruption. Spine 13:566–569

Boden SD, Zdeblick TA, Sandhu HS, Heim SE (2000) The use of rhBMP-2 in interbody fusion cages. Definitive evidence of osteoinduction in humans: a preliminary report. Spine 25:376–381

Bozkus H, Chamberlain RH, Perez Garza LE, Crawford NR, Dickman CA (2004) Biomechanical comparison of anterolateral plate, lateral plate, and pedicle screws-rods for enhancing anterolateral lumbar interbody cage stabilization. 29:635–641. https://doi.org/10.1097/01.BRS.0000115126.13081.7D

Bradford DS, McBride GG (1987) Surgical management of thoracolumbar spine fractures with incomplete neurologic deficits. Clin Orthop Relat Res 218:201–216

Brantigan JW, Steffee AD (1993) A carbon fiber implant to aid interbody lumbar fusion. Two-year clinical results in the first 26 patients. Spine 18:2106–2107

Briem D, Strametz S, Schröder K, Meenen NM, Lehmann W, Linhart W, Ohl A, Rueger JM (2005) Response of primary fibroblasts and osteoblasts to plasma treated polyetheretherketone (PEEK) surfaces. J Mater Sci Mater Med 16:671–677. https://doi.org/10.1007/s10856-005-2539-z

Burkus JK, Gornet MF, Dickman CA, Zdeblick TA (2002) Anterior lumbar interbody fusion using rhBMP-2 with tapered interbody cages. J Spinal Disord Tech 15:337–349

Burkus JK, Dorchak JD, Sanders DL (2003) Radiographic assessment of interbody fusion using recombinant human bone morphogenetic protein type 2. Spine 28:372

Burkus JK, Gornet MF, Schuler TC, Kleeman TJ, Zdeblick TA (2009) Six-year outcomes of anterior lumbar interbody arthrodesis with use of interbody fusion cages and recombinant human bone morphogenetic protein-2. 91:1181–1189. https://doi.org/10.2106/JBJS.G.01485

Burns BH, Camb BC (1933) An operation for spondylolisthesis. Lancet 221:1233

Capener N (1932) Spondylolisthesis. British Journal of Surgery 19(75):374–386. https://doi.org/10.1002/bjs.1800197505. https://onlinelibrary.wiley.com/doi/10.1002/bjs.1800197505

Chang YS, Gu HO, Kobayashi M, Oka M (1998) Influence of various structure treatments on histological fixation of titanium implants. J Arthroplast 13:816–825

Cho D, Liau W, Lee W, Liu J, Chiu C, Sheu P (2002) Preliminary experience using a polyetheretherketone (PEEK) cage in the treatment of cervical disc disease. 51:1343–1350. https://doi.org/10.1097/00006123-200212000-00003

Cho K, Bridwell KH, Lenke LG, Berra A, Baldus C (2005) Comparison of Smith-Petersen versus pedicle subtraction osteotomy for the correction of fixed sagittal imbalance. Spine 30:2037; discussion 2038

Closkey RF, Parsons JR, Lee CK, Blacksin MF, Zimmermant MC (1993) Mechanics of interbody spinal fusion: analysis of critical bone graft area. 18:1011–1015. https://doi.org/10.1097/00007632-199306150-00010

Czerwein JK, Thakur N, Migliori SJ, Lucas P, Palumbo M (2011) Complications of anterior lumbar surgery. J Am Acad Orthop Surg 19:251–258

de Kunder S, Rijkers K, Caelers IJMH, de Bie RA, Koehler PJ, van Santbrink H (2018) Lumbar interbody fusion, a historical overview and a future perspective:1. ePub ahead of print. https://doi.org/10.1097/BRS.0000000000002534

de Vasconcellos LMR, Leite DO, d Oliveira FN, Carvalho YR, Cairo CAA (2010) Evaluation of bone ingrowth into porous titanium implant: histomorphometric analysis in rabbits. 24:399–405. https://doi.org/10.1590/S1806-83242010000400005

DeBowes RM, Grant BD, Bagby GW, Gallina AM, Sande RD, Ratzlaff MH (1984) Cervical vertebral interbody fusion in the horse: a comparative study of bovine xenografts and autografts supported by stainless steel baskets. Am J Vet Res 45:191

del Rio J, Beguiristain J, Duart J (2007) Metal levels in corrosion of spinal implants. Eur Spine J 16:1055–1061. https://doi.org/10.1007/s00586-007-0311-4

Dennes TJ, Schwartz J (2009) A nanoscale adhesion layer to promote cell attachment on PEEK. 131:3456–3457. https://doi.org/10.1021/ja810075c

Deukmedjian AR, Dakwar E, Ahmadian A, Smith DA, Uribe JS (2012) Early outcomes of minimally invasive anterior longitudinal ligament release for correction of sagittal imbalance in patients with adult spinal deformity. Sci World J 2012:1–7. https://doi.org/10.1100/2012/789698

Dick JC, Brodke DS, Zdeblick TA, Bartel BD, Kunz DN, Rapoff AJ (1997) Anterior instrumentation of the thoracolumbar spine. A biomechanical comparison. 22:744–750. https://doi.org/10.1097/00007632-199704010-00005

DiPaola CP, Jacobson JA, Awad H, Conrad BP, Rechtine Glenn R (2007) Screw pull-out force is dependent on screw orientation in an anterior cervical plate construct. 20:369–373. https://doi.org/10.1097/BSD.0b013e31802c2a4a

Dipaola CP, Jacobson JA, Awad H, Conrad BP, Rechtine GR (2008) Screw orientation and plate type (variable- vs. fixed-angle) effect strength of fixation for in vitro biomechanical testing of the Synthes CSLP. Spine J 8:717–722

Do TD, Sutter R, Skornitzke S, Weber M (2018) CT and MRI techniques for imaging around orthopedic hardware. Rofo 190:31–41. https://doi.org/10.1055/s-0043-118127

Dorward IG, Lenke LG, Bridwell KH, O'Leary PT, Stoker GE, Pahys JM, Kang MM, Sides BA, Koester LA (2013) Transforaminal versus anterior lumbar interbody fusion in long deformity constructs: a matched cohort analysis. Spine 38:755. https://doi.org/10.1097/BRS.0b013e31828d6ca3

Du JY, Kiely PD, Bogner E, Al Maaieh M, Aichmair A, Salzmann SN, Huang RC (2017) Early clinical and radiological results of unilateral posterior pedicle instrumentation through a Wiltse approach with lateral lumbar interbody fusion. 3:338–348. https://doi.org/10.21037/jss.2017.06.16

Dunn HK (1984) Anterior stabilization of thoracolumbar injuries. Clin Orthop Relat Res 189:116–124

Durham JW, Allen MJ, Rabiei A (2017) Preparation, characterization and in vitro response of bioactive coatings on polyether ether ketone. 105:560–567. https://doi.org/10.1002/jbm.b.33578

Dwyer AF, Newton NC, Sherwood AA (1969) An anterior approach to scoliosis. A preliminary report. Clin Orthop Relat Res 62:192–202

Fogel GR, Parikh RD, Ryu SI, Turner AWL (2014) Biomechanics of lateral lumbar interbody fusion constructs with lateral and posterior plate fixation: laboratory investigation. 20:291–297. https://doi.org/10.3171/2013.11.SPINE13617

Freeman A, Walker J, Fen M, Bushelow M, Cain C, Tsantrizos A (2016) Biomechanical comparison of stand-alone anterior lumbar interbody fusion devices with secured fixation: four-screw locking plate vs. three-screw variable angle vs. blade fixation. In: Conference proceeding (ISASS16), Las Vegas

Fu T, Wang I, Lu M, Hsieh M, Chen L, Chen W (2016) The fusion rate of demineralized bone matrix compared with autogenous iliac bone graft for long multi-segment posterolateral spinal fusion. BMC Musculoskelet Disord 17. https://doi.org/10.1186/s12891-015-0861-2

Fujibayashi S, Takemoto M, Neo M, Matsushita T, Kokubo T, Doi K, Ito T, Shimizu A, Nakamura T (2011) A novel synthetic material for spinal fusion: a prospective clinical trial of porous bioactive titanium metal for lumbar interbody fusion. Eur Spine J 20:1486–1495. https://doi.org/10.1007/s00586-011-1728-3

Fyhrie DP, Vashishth D (2000) Bone stiffness predicts strength similarly for human vertebral cancellous bone in compression and for cortical bone in tension. Bone 26:169–173

Geck MJ, Rinella A, Hawthorne D, Macagno A, Koester L, Sides B, Bridwell KH, Lenke LG, Shufflebarger HL (2009) Comparison of surgical treatment in Lenke 5C adolescent idiopathic scoliosis: anterior dual rod versus posterior pedicle fixation surgery: a comparison of two practices. Spine (Phila Pa 1976) 34:1942–1951

Gerber M, Crawford NR, Chamberlain RH, Fifield MS, LeHuec J, Dickman CA (2006) Biomechanical assessment of anterior lumbar interbody fusion with an anterior lumbosacral fixation screw-plate: comparison to stand-alone anterior lumbar interbody fusion and anterior lumbar interbody fusion with pedicle screws in an unstable human cadaver model. 31:762–768. https://doi.org/10.1097/01.brs.0000206360.83728.d2

Ghanayem AJ, Zdeblick TA (1997) Anterior instrumentation in the management of thoracolumbar burst fractures. Clin Orthop Relat Res 335:89–100

Giang G, Mobbs R, Phan S, Tran TM, Phan K (2017) Evaluating outcomes of stand-alone anterior lumbar interbody fusion: a systematic review. World Neurosurg 104:259–271. https://doi.org/10.1016/j.wneu.2017.05.011

Gibon E, Amanatullah DF, Loi F, Pajarinen J, Nabeshima A, Yao Z, Hamadouche M, Goodman SB (2017) The biological response to orthopaedic implants for joint replacement: part I: metals. J Biomed Mater Res Part B Appl Biomater 105:2162–2173. https://doi.org/10.1002/jbm.b.33734

Glassman SD, Carreon LY, Djurasovic M, Campbell MJ, Puno RM, Johnson JR, Dimar JR (2008) RhBMP-2 versus iliac crest bone graft for lumbar spine fusion: a randomized, controlled trial in patients over sixty years of age. Spine 33:2843–2849. https://doi.org/10.1097/BRS.0b013e318190705d

Graham J (2006) Standard test methods for spine implants-chapter 13. In: Kurtz SM, Edidin A (eds) Spine technology handbook. Elsevier Inc, San Diego, pp 397–441

Gupta A, Kukkar N, Sharif K, Main BJ, Albers CE, El-Amin SF III (2015) Bone graft substitutes for spine fusion: a brief review. World J Orthop 6:449–456. https://doi.org/10.5312/wjo.v6.i6.449

Gurr KR, McAfee PC, Shih C (1988) Biomechanical analysis of anterior and posterior instrumentation systems after corpectomy. A calf-spine model. Journal of Bone and Joint Surgery-American Volume 70(8):1182–1191. http://www.ncbi.nlm.nih.gov/pubmed/3417703

Hadley ZS, Palmer DK, Williams PA, Cheng WK (2012) Pullout strength of anterior lumbar interbody fusion plates: fixed versus variable angle screw designs. J Spine. https://doi.org/10.4172/2165-7939.1000118

Han C, Lee E, Kim H, Koh Y, Kim KN, Ha Y, Kuh S (2010) The electron beam deposition of titanium on polyetheretherketone (PEEK) and the resulting enhanced biological properties. 31:3465–3470. https://doi.org/10.1016/j.biomaterials.2009.12.030

Harmon PH (1960) Anterior Extraperitoneal lumbar disk excision and vertebral body. Fusion 18:169–198

Heary R, Yanni DS, Halim AY, Benzel EC (2017) Anterior lumbar interbody fusion. In: Steinmetz M, Benzel EC (eds) Benzel's spine surgery. Elsevier, Philadelphia, pp 655–666

Hitchon PW, Goel VK, Rogge TN, Torner JC, Dooris AP, Drake JS, Yang SJ, Totoribe K (2000) In vitro biomechanical analysis of three anterior thoracolumbar implants. 93:252–258. https://doi.org/10.3171/spi.2000.93.2.0252

Hong S, Lee S, Khoo LT, Yoon S, Holly LT, Shamie AN, Wang JC (2010) A comparison of fixed-hole and slotted-hole dynamic plates for anterior cervical discectomy and fusion. 23:22–26. https://doi.org/10.1097/BSD.0b013e31819877e7

Hsieh PC, Koski TR, O'Shaughnessy BA, Sugrue P, Salehi S, Ondra S, Liu JC (2007) Anterior lumbar interbody fusion in comparison with transforaminal lumbar interbody fusion: implications for the restoration of foraminal height, local disc angle, lumbar lordosis, and sagittal balance. J Neurosurg Spine 7:379–386. https://doi.org/10.3171/SPI-07/10/379

Humphries AW, Hawk HA, Berndt AL (1958) Anterior fusion of the lumbar spine using an internal fixation device. Surg Forum 770–773

Humphries AW, Hawk WA, Berndt AL (1961) Anterior interbody fusion of lumbar vertebrae: a surgical technique. Surg Clin N Am 41:1685–1701

Hustedt JW, Blizzard DJ (2014) The controversy surrounding bone morphogenetic proteins in the spine: a review of current research. Yale J Biol Med 87:549–561

Inose H, Yamada T, Mulati M, Hirai T, Ushio S, Yoshii T, Kato T, Kawabata S, Okawa A (2018) Bone turnover markers as a new predicting factor for nonunion after spinal fusion surgery. Spine 43:E34. https://doi.org/10.1097/BRS.0000000000001995

Jackson KL, Devine JG (2016) The effects of smoking and smoking cessation on spine surgery: a systematic review of the literature. Global Spine J 6:695–701. https://doi.org/10.1055/s-0036-1571285

Jahangiri FR, Holmberg A, Sherman JH, Louis R, Elias J, Vega-Bermudez F (2010) Protecting the genitofemoral nerve during direct/extreme lateral interbody fusion (DLIF/XLIF) procedures. 50:321. https://doi.org/10.1080/1086508X.2010.11079786

Jeswani S, Drazin D, Liu JC, Ames C, Acosta FL (2012) Anterior lumbar interbody fusion: indications and techniques. In: Quiñones-Hinojosa A (ed) Schmidek & sweet operative neurosurgical techniques: indications, methods, and results, 6th edn. Elsevier Saunders, Philadelphia, pp 1955–1961

Jiménez-Avila JM, García-Valencia J, Bitar-Alatorre WE (2011) Risk factors affecting fusion in the treatment of lumbar spine instability. Acta Ortop Mex 25:156–160

Kanayama M, Cunningham BW, Haggerty CJ, Abumi K, Kaneda K, McAfee PC (2000) In vitro biomechanical investigation of the stability and stress-shielding effect of lumbar interbody fusion devices. J Neurosurg 93:259–265

Kaneda K, Abumi K, Fujiya M (1984) Burst fractures with neurologic deficits of the thoracolumbar-lumbar spine. Results of anterior decompression and stabilization with anterior instrumentation. Spine 9:788–795

Kebaish KM, Neubauer PR, Voros GD, Khoshnevisan MA, Skolasky RL (2011) Scoliosis in adults aged forty years and older: prevalence and relationship to age, race, and gender. Spine 36:731–736. https://doi.org/10.1097/BRS.0b013e3181e9f120

Keller T, Holland MC (1997) Some notable American spine surgeons of the 19th century. Spine 22:1413–1417

Kepler C, Bogner E, Herzog R, Huang R (2011) Anatomy of the psoas muscle and lumbar plexus with respect to the surgical approach for lateral transpsoas interbody fusion. Eur Spine J 20:550–556. https://doi.org/10.1007/s00586-010-1593-5

Kienle A, Graf ND, Wilke H (2015) Does impaction of titanium-coated interbody fusion cages into the disc space cause wear debris or delamination? 16:235–242. https://doi.org/10.1016/j.spinee.2015.09.038

Kim YJ, Bridwell KH, Lenke LG, Rinella AS, Edward CI (2005) Pseudarthrosis in primary fusions for adult idiopathic scoliosis: incidence, risk factors, and outcome analysis. 30:468. https://doi.org/10.1097/01.brs.0000153392.74639.ea

Kim YJ, Bridwell KH, Lenke LG, Rhim S, Cheh G (2006) Pseudarthrosis in long adult spinal deformity instrumentation and fusion to the sacrum: prevalence and risk factor analysis of 144 cases. 31:2329. https://doi.org/10.1097/01.brs.0000238968.82799.d9

Kim D, O'Toole JE, Ogden AT, Eichholz KM, Song J, Christie SD, Fessler RG (2009) Minimally invasive posterolateral thoracic corpectomy: cadaveric feasibility study and report of four clinical cases. Neurosurgery 64:746–752

Kim J, Lee K, Lee S, Lee H (2010) Which lumbar interbody fusion technique is better in terms of level for the treatment of unstable isthmic spondylolisthesis? J Neurosurg Spine 12:171–177. https://doi.org/10.3171/2009.9.SPINE09272

Kim C, Harris JB, Muzumdar A, Khalil S, Sclafani JA, Raiszadeh K, Bucklen BS (2017) The effect of anterior longitudinal ligament resection on lordosis correction during minimally invasive lateral lumbar interbody fusion: biomechanical and radiographic feasibility of an integrated spacer/plate interbody reconstruction device. 43:102–108. https://doi.org/10.1016/j.clinbiomech.2017.02.006

Kornblum MB, Turner AWL, Cornwall GB, Zatushevsky MA, Phillips FM (2013) Biomechanical evaluation of stand-alone lumbar polyether-ether-ketone interbody cage with integrated screws. Spine J 13:77–84. https://doi.org/10.1016/j.spinee.2012.11.013

Kostuik JP (1983) Anterior spinal cord decompression for lesions of the thoracic and lumbar spine, techniques, new methods of internal fixation results. 8:512–531. https://doi.org/10.1097/00007632-198307000-00008

Kotani Y, Cunningham BW, Parker LM, Kanayama M, McAfee PC (1999) Static and fatigue biomechanical properties of anterior thoracolumbar instrumentation systems. A synthetic testing model. 24:1406–1413. https://doi.org/10.1097/00007632-199907150-00004

Kumar N, Judith MR, Kumar A, Mishra V, Robert MC (2005) Analysis of stress distribution in lumbar interbody fusion. 30:1731–1735. https://doi.org/10.1097/01.brs.0000172160.78207.49

Kuslich SD, Ulstrom CL, Griffith SL, Ahern JW, Dowdle JD (1998) The Bagby and Kuslich method of lumbar interbody fusion. History, techniques, and 2-year follow-up results of a United States prospective, multicenter trial. 23:1267–1278. https://doi.org/10.1097/00007632-199806010-00019

Kutz M, Adrezin RS, Barr RE, Batich C, Bellamkonda RV, Brammer AJ, Buchanan TS, Cook AM, Currie JM, Dolan AM, Elad D, Einav S, Fajardo LL, Gage KL, Grimm MJ, Grotberg JB, Helmus MN, Iftekhar A, Jasti BR, Johnson AT, Joskowicz L, Keaveny TM, Kohn DH, Leamy P, Li X, Madsen MT, Manal KT, Meilander NJ, Morgan EF, Muthuswamy J, Nolan PJ, Nowak MD, O'Leary JP, Pandy M, Peterson DR, Reddy NP, Reinkensmeyer DJ, Rockett P, Rowley BA, Schaefer DJ, Shade DM, Shen SI, Silver-Thorn MB, Smith J, Snyder RW, Tackel I, Taylor R, Thomenius KE, Towe BC, Wagner WR, Wang G, Weir RF, Williams MB, Yeh OC, Zhu L (2000) Standard handbook of biomedical engineering and design. McGraw-Hill, New York

Laws CJ, Coughlin DG, Lotz JC, Serhan HA, Hu SS (2012) Direct lateral approach to lumbar fusion is a biomechanically equivalent alternative to the anterior approach: an in vitro study. 37:819–825. https://doi.org/10.1097/BRS.0b013e31823551aa

Lee SS, Lenke LG, Kuklo TR, Valenté L, Bridwell KH, Sides B, Blanke KM (2006) Comparison of Scheuermann kyphosis correction by posterior-only thoracic pedicle screw fixation versus combined anterior/posterior fusion. Spine 31:2316–2321. https://doi.org/10.1097/01.brs.0000238977.36165.b8

Lee CS, Hwang CJ, Lee D, Kim Y, Lee HS (2011) Fusion rates of instrumented lumbar spinal arthrodesis according to surgical approach: a systematic review of randomized trials. 3:39–47. https://doi.org/10.4055/cios.2011.3.1.39

Lee BH, Yang JH, Kim HS, Suk KS, Lee HM, Park JO, Moon SH (2017) Effect of sagittal balance on risk of falling after lateral lumbar interbody fusion surgery combined with posterior surgery. Yonsei Med J 58:1177–1185. https://doi.org/10.3349/ymj.2017.58.6.1177

Levi AD, Choi WG, Keller PJ, Heiserman JE, Sonntag VK, Dickman CA (1998) The radiographic and imaging characteristics of porous tantalum implants within the human cervical spine. 23:1245–1250. https://doi.org/10.1097/00007632-199806010-00014

Li JP, Habibovic P, Doel M, Wilson CE, Wijn JR, Blitterswijk CA, Groot K (2007) Bone ingrowth in porous titanium implants produced by 3D fiber deposition. Biomaterials 28:2810–2820

Liu L, Guo C, Zhou Q, Pu X, Song L, Wang H, Zhao C, Cheng S, Lan Y, Liu L (2014) Biomechanical comparison of anterior lumbar screw-plate fixation versus posterior lumbar pedicle screw fixation. J Huazhong Univ Sci Technol Med Sci 34:907–911. https://doi.org/10.1007/s11596-014-1372-3

wLiu X, Ma J, Park P, Huang X, Xie N, Ye X (2017) Biomechanical comparison of multilevel lateral interbody fusion with and without supplementary instrumentation: a three-dimensional finite element study. 18. https://doi.org/10.1186/s12891-017-1387-6

Loguidice VA, Johnson RG, Guyer RD, Stith WJ, Ohnmeiss DD, Hochschuler SH, Rashbaum RF (1988) Anterior lumbar interbody fusion. Spine (Phila Pa 1976) 13:366–369

Louis R (1986) Fusion of the lumbar and sacral spine by internal fixation with screw plates. Clin Orthop Relat Res 203:18–33

Lowe TG, Hashim S, Wilson LA, O'Brien MF, Smith DAB, Diekmann MJ, Trommeter J (2004) A biomechanical study of regional endplate strength and cage morphology as it relates to structural interbody support. 29:2389–2394. https://doi.org/10.1097/01.brs.0000143623.18098.e5

MacBarb RF, Lindsey DP, Bahney CS, Woods SA, Wolfe ML, Yerby SA (2017a) Fortifying the bone-implant Interface part 1: An in vitro evaluation of 3D-printed and TPS porous surfaces. 11:15. https://doi.org/10.14444/4015

MacBarb RF, Lindsey DP, Woods SA, Lalor PA, Gundanna MI, Yerby SA (2017b) Fortifying the bone-implant Interface part 2: An in vivo evaluation of 3D-printed and TPS-coated triangular implants. 11:16. https://doi.org/10.14444/4016

McDonough PW, Davis R, Tribus C, Zdeblick TA (2004) The management of acute thoracolumbar burst fractures with anterior corpectomy and Z-plate fixation. 29:1901–1908. https://doi.org/10.1097/01.brs.0000137059.03557.1d

McGilvray KC, Waldorff EI, Easley J, Seim HB, Zhang N, Linovitz RJ, Ryaby JT, Puttlitz CM (2017) Evaluation of a polyetheretherketone (PEEK) titanium composite interbody spacer in an ovine lumbar interbody fusion model: biomechanical, microcomputed tomographic, and histologic analyses. 17:1907–1916. https://doi.org/10.1016/j.spinee.2017.06.034

McGilvray KC, Easley J, Seim HB, Regan D, Berven SH, Hsu WK, Mroz TE, Puttlitz CM (2018) Bony ingrowth potential of 3D-printed porous titanium alloy: a direct comparison of interbody cage materials in an in vivo ovine lumbar fusion model. 18:1250–1260. https://doi.org/10.1016/j.spinee.2018.02.018

Meidinger G, Imhoff AB, Paul J, Kirchhoff C, Sauerschnig M, Hinterwimmer S (2011) May smokers and overweight patients be treated with a medial open-wedge HTO? Risk factors for non-union. Knee Surg Sports Traumatol Arthrosc 19:333–339. https://doi.org/10.1007/s00167-010-1335-6

Melton LJ, Chrischilles EA, Cooper C, Lane AW, Riggs BL (1992) Perspective. How many women have osteoporosis? 7:1005–1010. https://doi.org/10.1002/jbmr.5650070902

Miller EK, Neuman BJ, Jain A, Daniels AH, Ailon T, Sciubba DM, Kebaish KM, Lafage V, Scheer JK, Smith JS, Bess S, Shaffrey CI, Ames CP (2017) An assessment of frailty as a tool for risk stratification in adult spinal deformity surgery. Neurosurg Focus 43:E3. https://doi.org/10.3171/2017.10.FOCUS17472

Mobbs RJ, Phan K, Malham G, Seex K, Rao PJ (2015) Lumbar interbody fusion: techniques, indications and comparison of interbody fusion options including PLIF, TLIF, MI-TLIF, OLIF/ATP, LLIF and ALIF. J spine surg (Hong Kong) 1:2

Niu C, Liao J, Chen W, Chen L (2010) Outcomes of interbody fusion cages used in 1 and 2-levels anterior cervical discectomy and fusion: titanium cages versus polyetheretherketone (PEEK) cages. 23:310–316. https://doi.org/10.1097/BSD.0b013e3181af3a84

Noiset O, Schneider Y, Marchand-Brynaert J (1999) Fibronectin adsorption or/and covalent grafting on chemically modified PEEK film surfaces. 10:657–677. https://doi.org/10.1163/156856299X00865

Oh K, Lee CK, You NK, Kim SH, Cho KH (2013) Radiologic changes of anterior cervical discectomy and fusion using allograft and plate augmentation: comparison of using fixed and variable type screw. 10:160–164. https://doi.org/10.14245/kjs.2013.10.3.160

Ozgur BM, Aryan HE, Pimenta L, Taylor W (2006) Extreme lateral interbody fusion (XLIF): a novel surgical technique for anterior lumbar interbody fusion. Spine J 6:435–443

Palepu V, Peck JH, Simon DD, Helgeson MD, Nagaraja S (2017) Biomechanical evaluation of an integrated fixation cage during fatigue loading: a human cadaver study:1–8. https://doi.org/10.3171/2016.9.SPINE16650

Patacxil WM, Palmer DK, Rios D, Inceoglu S, Williams PA, Cheng WK (2012) Screw orientation and foam density interaction in pullout of anterior lumbar interbody fusion plates. Duke Orthop J 2:35–39

Patwardhan AG, Carandang G, Ghanayem AJ, Havey RM, Cunningham B, Voronov LI, Phillips FH (2003) Compressive preload improves the stability of anterior lumbar interbody fusion cage constructs. 85:1749–1756. https://doi.org/10.2106/00004623-200309000-00014

Peek RD, Wiltse LL (1990) History of spinal fusion. In: Anonymous spinal fusion. Springer, New York, pp 3–8

Pelletier MH, Cordaro N, Punjabi VM, Waites M, Lau A, Walsh WR (2016) PEEK versus Ti interbody fusion devices: resultant fusion, bone apposition, initial and 26-week biomechanics. 29:208

Phan K, Mobbs RJ (2016) Evolution of design of interbody cages for anterior lumbar interbody fusion. Orthop Surg 8:270–277. https://doi.org/10.1111/os.12259

Phan K, Thayaparan GK, Mobbs RJ (2015) Anterior lumbar interbody fusion versus transforaminal lumbar interbody fusion – systematic review and meta-analysis. Br J Neurosurg 29:705–711. https://doi.org/10.3109/02688697.2015.1036838

Phan K, Fadhil M, Chang N, Giang G, Gragnaniello C, Mobbs RJ (2017) Effect of smoking status on successful arthrodesis, clinical outcome, and complications after anterior lumbar interbody fusion (ALIF). World Neurosurg. https://doi.org/10.1016/j.wneu.2017.11.157

Pienkowski D, Stephens GC, Doers TM, Hamilton DM (1998) Multicycle mechanical performance of titanium and stainless steel transpedicular spine implants. Spine 23:782–788

Pimenta L, Turner AWL, Dooley ZA, Parikh RD, Peterson MD (2012) Biomechanics of lateral interbody spacers: going wider for going stiffer. TheScientificWorldJOURNAL 2012:381814

Quraishi NA, Konig M, Booker SJ, Shafafy M, Boszczyk BM, Grevitt MP, Mehdian H, Webb JK (2013) Access related complications in anterior lumbar surgery performed by spinal surgeons. Eur Spine J 22:16–20. https://doi.org/10.1007/s00586-012-2616-1

Ravindra V, Godzik J, Dailey A, Schmidt M, Bisson E, Hood R, Cutler A, Ray W (2015) Vitamin D levels and 1-year fusion outcomes in elective spine surgery: a prospective observational study. 40:1536–1541. https://doi.org/10.1097/BRS.0000000000001041

Ray CD (1997) Threaded titanium cages for lumbar interbody fusions. 22:667–679. https://doi.org/10.1097/00007632-199703150-00019

Regev GJ, Chen L, Dhawan M, Lee YP, Garfin SR, Kim CW (2009) Morphometric analysis of the ventral nerve roots and retroperitoneal vessels with respect to the minimally invasive lateral approach in normal and deformed spines. 34:1330–1335. https://doi.org/10.1097/BRS.0b013e3181a029e1

Reis MT, Reyes PM, BSE AI, Newcomb AGUS, Singh V, Chang SW, Kelly BP, Crawford NR (2016) Biomechanical evaluation of lateral lumbar interbody fusion with secondary augmentation. 25:720–726. https://doi.org/10.3171/2016.4.SPINE151386

Riley MR, Doan AT, Vogel RW, Aguirre AO, Pieri KS, Scheid EH (2018) Use of motor evoked potentials during lateral lumbar interbody fusion reduces postoperative deficits.. Epub ahead of print. https://doi.org/10.1016/j.spinee.2018.02.024

Rios D, Patacxil WM, Palmer DK, Williams PA, Cheng WK, Inceoğlu S (2012) Pullout analysis of a lumbar plate with varying screw orientations: experimental and computational analyses. 37:E948. https://doi.org/10.1097/BRS.0b013e318254155a

Rodríguez-Olaverri JC, Hasharoni A, DeWal H, Nuzzo RM, Kummer FJ, Errico TJ (2005) The effect of end screw orientation on the stability of anterior instrumentation in cyclic lateral bending. 5:554–557. https://doi.org/10.1016/j.spinee.2005.03.014

Saville PA, Kadam AB, Smith HE, Arlet V (2016) Anterior hyperlordotic cages: early experience and radiographic results. J Neurosurg Spine 25:713–719. https://doi.org/10.3171/2016.4.SPINE151206

Schwab F, Dubey A, Gamez L, El Fegoun AB, Hwang K, Pagala M, Farcy J (2005) Adult scoliosis: prevalence, SF-36, and nutritional parameters in an elderly volunteer population. Spine 30:1082–1085

Seaman S, Kerezoudis P, Bydon M, Torner JC, Hitchon PW (2017) Titanium vs. polyetheretherketone (PEEK) interbody fusion: meta-analysis and review of the literature. 44:23–29. https://doi.org/10.1016/j.jocn.2017.06.062

Seo DK, Kim MJ, Roh SW, Jeon SR (2017) Morphological analysis of interbody fusion following posterior lumbar interbody fusion with cages using computed tomography. 96:e7816. https://doi.org/10.1097/MD.0000000000007816

Song K, Taghavi C, Hsu M, Lee K, Kim G, Song J (2010) Plate augmentation in anterior cervical discectomy and fusion with cage for degenerative cervical spinal disorders. Eur Spine J 19:1677–1683. https://doi.org/10.1007/s00586-010-1283-3

Spruit M, Falk RG, Beckmann L, Steffen T, Castelein RM (2005) The in vitro stabilising effect of polyetheretherketone cages versus a titanium cage of similar design for anterior lumbar interbody fusion. Eur Spine J 14:752–758. https://doi.org/10.1007/s00586-005-0961-z

Tahal D, Madhavan K, Chieng LO, Ghobrial GM, Wang MY (2017) Metals in Spine. World Neurosurg 100:619–627. https://doi.org/10.1016/j.wneu.2016.12.105

Taher F, Hughes AP, Lebl DR, Sama AA, Pumberger M, Aichmair A, Huang RC, Cammisa FP, Girardi FP (2013) Contralateral motor deficits after lateral lumbar interbody fusion. Spine 38:1959–1963

Takahashi S, Delécrin J, Passuti N (2001) Intraspinal metallosis causing delayed neurologic symptoms after spinal instrumentation surgery. 26:8; discussion 1499. https://doi.org/10.1097/00007632-200107010-00024

Takemoto M, Fujibayashi S, Neo M, So K, Akiyama N, Matsushita T, Kokubo T, Nakamura T (2007) A porous bioactive titanium implant for spinal interbody fusion: an experimental study using a canine model. 7:435–443. https://doi.org/10.3171/SPI-07/10/435

Taniguchi N, Fujibayashi S, Takemoto M, Sasaki K, Otsuki B, Nakamura T, Matsushita T, Kokubo T, Matsuda S (2016) Effect of pore size on bone ingrowth into porous titanium implants fabricated by additive manufacturing: An in vivo experiment. 59:690–701. https://doi.org/10.1016/j.msec.2015.10.069

Tarpada SP, Morris MT, Burton DA (2017) Spinal fusion surgery: a historical perspective. 14:134–136. https://doi.org/10.1016/j.jor.2016.10.029

Tatsumi R, Lee Y, Khajavi K, Taylor W, Chen F, Bae H (2015) In vitro comparison of endplate preparation between four mini-open interbody fusion approaches. Eur Spine J 24:372–377. https://doi.org/10.1007/s00586-014-3708-x

Teng I, Han J, Phan K, Mobbs R (2017) A meta-analysis comparing ALIF, PLIF, TLIF and LLIF. J Clin Neurosci 44:11–17. https://doi.org/10.1016/j.jocn.2017.06.013

Tohmeh AG, Rodgers WB, Peterson MD (2011) Dynamically evoked, discrete-threshold electromyography in the extreme lateral interbody fusion approach. 14:31–37. https://doi.org/10.3171/2010.9.SPINE09871

Torstrick FB, Klosterhoff BS, Westerlund LE, Foley KT, Gochuico J, Lee CSD, Gall K, Safranski DL (2018) Impaction durability of porous polyether-ether-ketone (PEEK) and titanium-coated PEEK interbody fusion devices. 18:857–865. https://doi.org/10.1016/j.spinee.2018.01.003

Tzermiadianos MN, Mekhail A, Voronov LI, Zook J, Havey RM, Renner SM, Carandang G, Abjornson C, Patwardhan AG (2008) Enhancing the stability of anterior lumbar interbody fusion: a biomechanical comparison of anterior plate versus posterior transpedicular instrumentation. 33:E43. https://doi.org/10.1097/BRS.0b013e3181604644

Udby PM, Bech-Azeddine R (2015) Clinical outcome of stand-alone ALIF compared to posterior instrumentation for degenerative disc disease: a pilot study and a literature review. Clin Neurol Neurosurg 133:64–69. https://doi.org/10.1016/j.clineuro.2015.03.008

Uribe JS, Arredondo N, Dakwar E, Vale FL (2010) Defining the safe working zones using the minimally invasive lateral retroperitoneal transpsoas approach: an anatomical study. 13:260–266. https://doi.org/10.3171/2010.3.SPINE09766

van Wunnik BPW, Weijers PHE, van Helden SH, Brink PRG, Poeze M (2011) Osteoporosis is not a risk factor for the development of nonunion: a cohort nested case–control study. Injury 42:1491–1494. https://doi.org/10.1016/j.injury.2011.08.019

Watkins MB (1953) Posterolateral fusion of the lumbar and lumbosacral spine. 35:1014–1018. https://doi.org/10.2106/00004623-195335040-00024

Watkins RG, Hanna R, Chang D, Watkins RG (2014) Sagittal alignment after lumbar interbody fusion: comparing anterior, lateral, and transforaminal approaches. J Spinal Disord Tech 27:253–256. https://doi.org/10.1097/BSD.0b013e31828a8447

Winder MJ, Gambhir S (2016) Comparison of ALIF vs. XLIF for L4/5 interbody fusion: pros, cons, and literature review. J spine surg (Hong Kong) 2:2

Wu X, Liu X, Wei J, Ma J, Deng F, Wei S (2012) Nano-TiO2/PEEK bioactive composite as a bone substitute material: in vitro and in vivo. studies 7:1215–1225. https://doi.org/10.2147/IJN.S28101

Wu S, Li Y, Zhang Y, Li X, Yuan C, Hao Y, Zhang Z, Guo Z (2013) Porous Titanium-6 Aluminum-4 vanadium cage has better Osseointegration and less micromotion than a poly-ether-ether-ketone cage in sheep vertebral fusion. 37:E201. https://doi.org/10.1111/aor.12153

Yoganandan N, Arun MWJ, Dickman CA Benzel EC practical anatomy and fundamental biomechanics. In: Steinmetz MP, Benzel EC (eds) Benzel's spine surgery. Elsevier, Philadelphia, pp 58–82

Zdeblick TA, Warden KE, Zou D, McAfee PC, Abitbol JJ (1993) Anterior spinal fixators. A biomechanical in vitro study. 18:513–517. https://doi.org/10.1097/00007632-199318040-00016

Zhang J, Poffyn B, Sys G, Uyttendaele D (2012) Are stand-alone cages sufficient for anterior lumbar interbody fusion? Orthop Surg 4:11–14. https://doi.org/10.1111/j.1757-7861.2011.00164.x

Zhang Z, Li H, Li Y, Fogel GR, Liao Z, Liu W (2018) Biomechanical analysis of porous additive manufactured cages for lateral lumbar interbody fusion: a finite element analysis. 111:e591. https://doi.org/10.1016/j.wneu.2017.12.127

Zhao M, Li H, Liu X, Wei J, Ji J, Yang S, Hu Z, Wei S (2016) Response of human osteoblast to n-HA/PEEK – quantitative proteomic study of bio-effects of Nano-hydroxyapatite composite. 6:22832. https://doi.org/10.1038/srep22832

Zielke K, Stunkat R, Beaujean F Ventrale Derotationspondylodese. Vorläufiger Ergebnisbericht über 26 operierte Fälle. Arch orthop Unfall-Chir. 1976 January [cited Jun 26, 2018];85(3):257–77. Available from: https://link.springer.com/article/10.1007/BF00415189

Zura R, Mehta S, Della Rocca GJ, Steen RG (2016) Biological risk factors for nonunion of bone fracture. 4:1. https://doi.org/10.2106/JBJS.RVW.O.00008

Zura R, Braid-Forbes MJ, Jeray K, Mehta S, Einhorn TA, Watson JT, Della Rocca GJ, Forbes K, Steen RG (2017) Bone fracture nonunion rate decreases with increasing age: a prospective inception cohort study. Bone 95:26–32. https://doi.org/10.1016/j.bone.2016.11.006

Interbody Cages: Cervical

31

John Richards, Donald R. Fredericks Jr., Sean E. Slaven, and Scott C. Wagner

Contents

Introduction	634
Cage Design	635
Threaded Cages	636
Screw Cage Designs (Horizontal Cylinder)	636
Non-threaded Cages	637
Box-Shaped Designs	637
Vertical Ring (Cylinder) Cages	637
Other Cage Considerations/Current Trends	638
Cage Materials	639
Carbon Fiber	639
Titanium	640
Peek	640
Hybrid Cages	641
Operative Technique	641
Summary	642
References	642

Abstract

Interbody cages represent an invaluable technologic advancement in the field of spinal fusion surgery, particularly anterior cervical discectomy and fusion (ACDF). Interbody cages improve sagittal alignment, aid in fusion by allowing for containment of graft material, and restore biomechanical stability after discectomy. Various design iterations and materials have been used over the last two to three decades, and advancements in materials

---

J. Richards · D. R. Fredericks Jr. · S. E. Slaven
Department of Orthopaedics, Walter Reed National Military Medical Center, Bethesda, MD, USA
e-mail: john.richards448@gmail.com;
donaldfredericks.jr@gmail.com;
sean.e.slaven@gmail.com

S. C. Wagner (✉)
Department of Orthopaedic Surgery, Walter Reed National Military Medical Center, Bethesda, MD, USA
e-mail: scott.cameron.wagner@gmail.com

© This is a U.S. Government work and not under copyright protection in the U.S.; foreign copyright protection may apply 2021
B. C. Cheng (ed.), *Handbook of Spine Technology*,
https://doi.org/10.1007/978-3-319-44424-6_62

science and cage properties have provided improved functional utility of interbody cages in ACDF surgery. This chapter provides an overview of the history of cervical interbody cages, including improvements in design, material, and methods of manufacturing processes of cervical interbody cages that have yielded the designs most commonly utilized today.

Keywords

Cervical · Interbody · Cage · Spine · Fusion · Bone graft

Introduction

The first cervical interbody fusions were pioneered by neurosurgeons in the 1950s for early anterior cervical discectomy and fusion (ACDF). These procedures were developed as alternatives to posterior fusions to treat cervical radiculopathy and myelopathy. In 1958, Cloward described an anterior approach for cervical interbody fusions. This technique utilized bone dowels inserted horizontally between the adjacent vertebrae, from an anterior to posterior direction, following the decompression (Cloward et al. 1958). The vertebral bodies were prepared with a drill, and the bone dowel was cut slightly larger than the drilled defect and seated with distraction and impaction. These bone grafts were typically iliac crest autograft or, less commonly, allograft bone cut into cylindrical dowels. Robinson and Smith described a similar technique that utilized the same anterior approach and decompression, but instead used a rectangular graft in place of a cylindrical bone dowel. This technique required less augmentation and preparation of the adjacent vertebral bodies. Another advantage of this rectangular graft is that it provided a greater surface area compared to cylindrical grafts (30% more of similar size), which was theorized to provide better contact surface for revascularization and fusion (Robinson and Smith 1958).

In 1969 Simmons and Bhalla published a modification of this same technique. They utilized the same anterior approach and decompression, but instead used a "keystone"-shaped bone graft. This keystone shape was a beveled modification of the rectangular graft, making it thicker posteriorly than anteriorly. This required the adjacent vertebrae to be modified similarly. These "keystone" grafts performed superiorly in biomechanical studies demonstrating less extrusion in comparison to simple rectangular or dowel grafts with flexion and extension of the spine. In a clinical study, they also performed more superiorly than interbody fusions performed with dowel bone grafts with 3/17 symptomatic nonunions in the dowel group compared to none in the keystone group. Various original ACDF implant designs are shown in Fig. 1.

Donor site morbidity has been reported as high as 22% (McConnel et al. 1976), and is a known disadvantage to iliac crest autograft, and led to the development of alternative graft materials. Allograft was one of the most widely studied alternatives. It initially provided mixed results in clinical trials and had the downsides of significantly increased costs, slower fusion rates, and the theoretical risk of disease transmission (An et al. 1995; Bishop et al. 1996; Brown et al. 1976); however, recent studies have shown that a ACDF with autograft vs. allograft, particularly with anterior plate fixation, is nearly equivalent clinically and radiographically (Kaiser 2002 Neurosurgery). Poly(methyl methacrylate) (PMMA) was another alternative that was described. It offered a graft with immediate stability of the adjacent vertebrae, but it provided no ability to achieve fusion and was therefore abandoned (Bent et al. 1996). Biocompatible osteoconductive polymer (BOP) and coralline hydroxyapatite were two other promising bone graft substitutes, but both were ultimately found to be inferior to autograft in clinical trials. BOP was biomechanically similar to autograft, but demonstrated little ability to incorporate into host bone (Ibanez et al. 1998). Hydroxyapatite was structurally weak and demonstrated graft subsidence and fragmentation (McConnell et al. 2003). Despite a number of different alternatives studied, autograft continually demonstrated superior outcomes in both biomechanical studies and long-term clinical trials and therefore remained the gold standard.

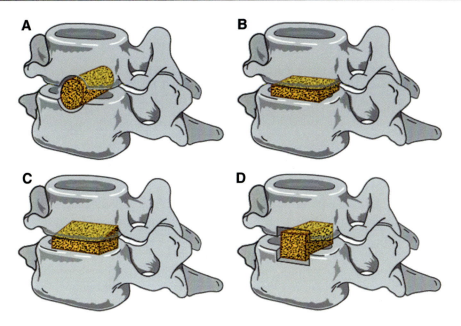

Fig. 1 Historical perspectives on ACDF implants. (**a**) Cloward dowel graft. (**b**) Smith-Robinson-based rectangular implant. (**c**) Simmons-Bhalla keystone. (**d**) Bailey-Badgley onlay strut (Reproduced from Chong et al. 2015)

The failure of graft substitutes forced a shift in the pursuit of viable alternatives. In 1988, Bagby proposed the first use of interbody cages in the human spine. He adapted a technique previously described by DeBowes to treat cervical myelopathy in horses (Bagby 1988, DeBowes et al. 1984). This device consisted of a hollow cylindrical stainless steel cage with fenestrations which was placed through an anterior approach similar to ACDF. This structure allowed bone ingrowth and provided a non-compressible scaffold (Chong et al. 2015). Interbody cage designs were developed out of necessity due to the associated morbidity of iliac crest autograft harvest. Various cage designs and materials have been since been trialed and variably adopted as the technology has advanced.

Cage Design

Cervical interbody cages may circumvent donor site morbidity associated with autograft and infection associated with allograft. For example, autologous iliac crest bone harvesting has been shown to be associated with 22% long- or short-term complications (McConnell et al. 2003). Due to their ease of use, rigid cages with internal space for local autogenous bone graft or morselized allograft with or without osteoinductive materials such as platelet-rich plasma or bone marrow aspirate fragments have become increasingly popular for cervical interbody fusions. Interbody cages have shown to be a successful means of cervical interbody body fusion with associations with less post-operative pain, shorter hospital stays, and higher rates of fusion than autologous bone graft alone (Jain et al. 2016). The ideal cage provides immediate structural supports (axial compression/anterior column distraction), adequate resistance to subsidence, and avoidance of structural allograft complications. Overall goals are to stabilize the newly operated segment to allow new boney ingrowth, maintain anatomic disc height, and avoid cage subsidence until fusion has occurred (Matgé 2002). Cage subsidence is associated with neck pain, late neurologic deterioration, and significantly lower Japanese Orthopaedic Association scores (Chen et al. 2008). Stand-alone interbody cervical cage implantation has shown to be effective with good clinical and patient-reported outcomes (93–100%) despite increased subsidence rates as compared to that with plating or cage/bone grafting hybrid techniques (Kulkarni et al. 2007).

The basic design of cervical interbody cages encompasses a small, hollow implant with upper/lower and/or lateral windows in which autogenous bone, allograft bone, and/or osteoinductive materials may be placed (Chong et al. 2015; Wilke et al. 2000). Traditionally cage designs can be categorized into threaded (screw, horizontal cylinder) and non-threaded (open box-shaped and vertical rings/cylinders) (Chong et al. 2015; Kandziora et al. 2001). Horizontal cylinders/screws are the manufactured equivalents to allograft dowel techniques, while vertical ring cages are designed to mimic cortical ring allografts. Open boxes are the equivalent of tricortical or quadricortical graft (Weiner and Fraser 1998).

Threaded Cages

Screw Cage Designs (Horizontal Cylinder)

Examples: BAK/C (Zimmer Spine), Ray Threaded Fusion Cage (Surgical Dynamics)

A variety of cervical interbody cages exist, but the most common is the threaded titanium interbody cage. Some of the earliest available cages were described as horizontal cylinder or screw design cages. These cages were based on Cloward's original technique and functioned as dowel grafts. Screw cages, also known as horizontal cylinder cages, have a circular cross section in the coronal plane. Fenestrations along the device allow for contact between prepared endplates and non-structural bone graft or osteoinductive biologic material placed into core. Reaming is usually required to allow proper fit for cages (Pisano et al. 2016). One example of this cage design is BAK-C (Zimmer Spine, Warsaw, IN) which demonstrated device stability, accelerated fusion rates, and higher stiffness when compared to more traditional non-threaded tricortical iliac crest bone graft (Matgé 2002; Chong et al. 2015; Kandziora et al. 2002) (Banco et al. 2002). Another threaded cage design, the Ray Cervical interbody cage (Surgical Dynamics, Norwalk, CT), is shown implanted at C5-6 in the representative post-operative radiograph demonstrated in Fig. 2.

Following initial success, the failures of the threaded screw design were found to be associated with decreased maximum distractive height and increased levels of cage subsidence due to adjacent vertebral endplate weakening. Additional biomechanical studies by Kandziora et al. and others showed threaded screw designs to be less stable in flexion/extension/bending in animal models (Kettler et al. 2001; Kandziora et al. 2001; Jain et al. 2016).

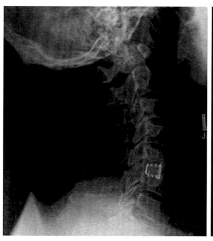 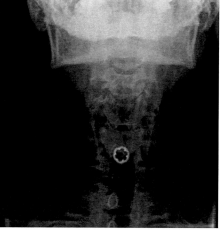

Fig. 2 (a) Anterior-posterior and (b) lateral X-rays of the cervical spine with the Ray Cervical interbody cage (Surgical Dynamics, Norwalk, CT). Used with permission from Banco et al.

By virtue of the shape of the screw cage designs, there is less space available for bone graft than obtained for either vertical ring or open box designs. The use of the BAK/C and other threaded horizontal cylinder devices has declined due to excessive endplate violation needed for insertion of device.

Non-threaded Cages

Box-Shaped Designs

Example: SynCage-C (DePuy Synthes, Raynham, MA)

Rectangular box cages initially were designed with roughened contact surfaces to improve anchorage and fusion into adjacent cervical vertebral bodies. These cages employ a design that mimics iliac crest cortical graft, with a vertically oriented box with central core allowing bone graft placement (Pisano et al. 2016). This design has demonstrated better segmental stiffness in all directions as compared to tricortical iliac crest (Kandziora et al. 2001). This design has undergone further improvements for surface fit and cage anchorage such as incorporating wedge-shaped and trapezoidal designs as well as attempting to mimic shape of vertebral endplate contours (Chong et al. 2015; Steffen et al. 2000; Kast et al. 2008). Inversely matching cervical vertebral endplate contours allows wedge-shaped box morphology to contribute to increased stability in lateral bending, axial rotation, and flexion as compared to other designs (Kast et al. 2008). Goals of wedge-like box-shaped designs are to match natural cervical lordosis by using anterior slope with 1–2 mm larger box height anteriorly than posteriorly (Gödde et al. 2003; Chong et al. 2015).

The SynCage-C (DePuy Synthes, Raynham MA) is a box form cage with both superior and inferior contact surfaces. The superior cage endplate opens to central contiguous pore (Epari et al. 2005). Within the non-threaded box-shaped design group, there is variability in manufacturer modifications. In Fig. 3, a rendition of SynCage-C is shown from the anterior view with varying superior and inferior endplate modifications.

Vertical Ring (Cylinder) Cages

Example: Harms Cage (DePuy Synthes, Raynham, MA)

Vertical ring cage design is an adaptation of cortical ring allograft interbody fusion described by Ono et al. and popularized by Kozak and O'Brien (Weiner and Fraser 1998; Kozak and O'Brien 1990; Ono et al. 1992). Vertical rings are typically cylindrical mesh cages with vertically oriented walls that have a circular cross section in the

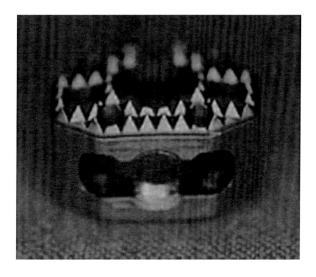

Fig. 3 SynCage-C (DePuy Synthes) is a box-shaped design-type cage. Representative image of the cage from an anterior view, used with permission from *Epari* et al.

Fig. 4 An example of a Harms cage, which is a vertical cylinder of titanium mesh. Used with permission from *Epari* et al.

axial plane. With a cylindrical shape, the central portion of this graft design accepts autograft and allograft (Pisano et al. 2016). Vertical rings or cylindrical designs commonly lack superior and inferior contact surfaces in contrast to those seen in box-type non-threaded cage designs. Proponents argue that the lack of inferior and superior contact surfaces along with decreased rigidity in meshed walls results in reduced rigidity of cage and provides ingrowth path for vessels and bone (Epari et al. 2005). Cylindrical implants have been shown to subside greater than tricortical autograft, allograft, and rectangular box constructs (Wilke et al. 2000). The Harms cage is a hollow vertical cylinder design and has the largest possible pore size for its volume (Epari et al. 2005) (Fig. 4).

Due to their ability to mimic healthy cervical anatomy and confer initial stability with improvement in initial surface contact, non-threaded cages remain superior to threaded/screw design cages biomechanically (Chong et al. 2015). When comparing subsidence between a vertical ring and box design cages, Kandziora et al. showed similar amounts of endplate penetration despite having very different endplate surface area contact (Kandziora et al. 2001; Epari et al. 2005). A reduction in stiffness of interbody fusion cages in animal models has been shown to enhance interbody fusion after 6 months and up to 3 years (Epari et al. 2005; van Dijk et al. 2002). However, the clinical literature comparing cage design and shape is limited (Chong et al. 2015).

Overall trends in surgeon usage had favored wedge-shaped/trapezoidal non-threaded cages, although the clinical outcome data comparing threaded versus non-threaded is sparse. These reported trends may be due to ease of implantation, restoration of cervical lordosis, and greater segmental stiffness with wedge-shaped/trapezoidal non-threaded cages (Chong et al. 2015).

Other Cage Considerations/Current Trends

Newer hybrid interbody cage devices incorporate plate and screw recesses. The cage functions as an anchored spacer with integrated screw construct allowing immediate fixation (Samandouras et al. 2001). The impetus for these new design concepts most likely is derived from increasing popularity of stand-alone cages and recognized issues of subsidence, anterior displacement, and varying results in fusion rates. In vitro studies have revealed that combined use of cage and plate is more stable biomechanically in flexion/extension than stand-alone cage constructs with bone graft (Shimamoto et al. 2001) (Keogh et al. 2008).

Expandable cages, originally used for larger oncologic en bloc resections, have become more popular for degenerative cervical conditions. Proponents of this technique tend to argue that the previously described methods for cage implantation ensure segmental endplate over-distraction and exceedingly tight implant-endplate contact, while distraction in situ allows more precise fit (Truumees 2011).

Cages may be classified by not only design principle or geometry but also the implant material. Interbody cages may have a variety of described material properties, but more commonly the material make-up consists of titanium, polyether ether ketone (PEEK), or carbon fiber-reinforced polymers (Pisano et al. 2016; Chong et al. 2015). Design materials and designs of interbody cages have changed considerably over the last several decades. Threaded titanium alloy cages, often filled with autogenous bone graft, became popular in the 1990s due to superior fusion rates as compared to non-threaded cages and bone grafts. Non-threaded box-shaped titanium cages as well as polyether ether ketone (PEEK) cages gained popularity due to lack of flexion/extension stability and high incidence of subsidence of threaded cages (Jain et al. 2016).

Cage Materials

Cages composed of synthetic materials were originally introduced to circumvent issues with autograft, allograft, and biocomposite grafts used in ACDF. Desirable features in an interbody cage material include an elastic modulus similar to bone, biocompatibility, achievement of bony fusion, maintenance of alignment, and the ability to avoid subsidence. Bagby's initial description of ACDF in horses was using a stainless steel implant; however, the primary materials used in modern ACDF procedures have been carbon fiber-reinforced polymers (CFRP), titanium (Ti), and polyether ether ketone (PEEK). While overall results with each material have been generally favorable, differences in material properties and outcomes have been described.

Carbon Fiber

Carbon fiber-reinforced polymer implants for interbody fusion were first described by Brantigan and Steffe (1993) for use in the lumbar spine.

Cages were later designed and adapted for use in arthrodesis of the cervical spine (Brooke et al. 1997), with excellent clinical results in small series with short follow-up. These initial implants were a composite material of long-fiber carbon in PEEK. The material properties of CFRP cages, such as compressive and tensile strength and isotropy, are dictated by the length, alignment, and volume of the carbon fibers. A biomechanical comparison to iliac crest bone graft by Shono et al. (1993) showed CFRP cages to be more rigid in flexion/extension testing and have higher stiffness in axial compression and rotation. CFRP cages have the benefit of being radiolucent to allow easier evaluation of the adjacent vertebral bodies and subsequent bony fusion on plain radiographs. Additionally, CT and MRI evaluation of the post-surgical spine is easier due to the lack of metal artifact. Early in their development, radiopaque tantalum beads were added in the corners of the cage to assist in visualizing the position of the cage.

Initial results of fusion with CFRP cages demonstrated reliable relief of neck and radicular pain, with authors reporting fusion rates as high as 87–98% in small series with short follow-up. Correction of cervical lordosis, another primary goal of ACDF surgery, was also reliably achieved with CFRP cages. Subsidence was noted frequently in CFRP cages, at a rate of 49% (Van der Haven et al. 2005); however, the occurrence of subsidence did not correlate with clinical outcomes. It should be noted that subsidence in the setting of cervical cages refers to the penetration of the cage through the vertebral endplate, rather than a collapse of the cage itself.

While CFRP had good initial success as a cage material, it has been largely replaced by PEEK in cage manufacturing and clinical use. Yoo et al. (2014) compared CFRP and PEEK cages and found lower fusion in CFRP cages (68.6%) than PEEK cages (82.6%), as well as a higher rate of subsidence in CFRP than PEEK (34.3% vs. 26.1%, respectively). The superior performance of PEEK in these respects is thought to be due to the difference in elastic modulus between CFRP (45 GPa) and PEEK (3.4 GPa), with PEEK more closely matching bone and therefore limiting potential stress shielding of the vertebrae (see Table 1).

Table 1 Elastic modulus of bone and cage materials

Material	E (GPa)
Cancellous bone	1–2
Cortical bone	12–20
PEEK	3.4
CFRP	45
Ti	106–115

Titanium

Titanium and its alloys are another material introduced early in the development of cervical cages. Like CFRP, titanium cages for vertebral arthrodesis were first described in the lumbar spine (Ray 1997) with good results and were later adapted for cervical interbody fusion, with Profeta et al. (2000) reporting favorable initial results with a threaded Ti cage compared to bone graft, noting good "immediate stability." Titanium, introduced to orthopedic practice in the 1940s, has several desirable properties for an orthopedic implant. It has a high yield strength, low density, biocompatibility, and excellent corrosion resistance due to self-passivation with titanium oxide ($TiO2$). With respect to cervical arthrodesis, modern Ti cages typically undergo surface modification to roughen the surface, via processes such as plasma spraying, which provides mechanical and biological advantages.

A roughened surface increases friction at the bone-implant interface, contributing to initial stability. The degree of micromotion at the bone-implant interface dictates the characteristics of bone and fibrous tissue formation; Jasty et al. (1997) demonstrated using porous-coated Ti alloy implants in an animal model that cyclic micromotion of under 40 micrometers reliably led to stable ingrowth of bone, whereas micromotion from 40 to 150 micrometers led to a proportionally increasing amount of fibrous tissue and fibrocartilage and less stable ingrowth. The use of roughened titanium cages thereby confers the potential advantage of improved stability and bone apposition at the bone-implant interface, and several authors have commented on the good perceived initial stability and handling characteristics of such cages.

Roughened Ti cages have also been shown to have potential biological advantages. When compared to PEEK (a typically smoother surface), roughened Ti surfaces have been shown in vivo to increase osteoblast maturation and the production of bone morphogenetic proteins, which may be of benefit in achieving osseointegration and ultimately bony fusion (Olivares-Navarrete et al. 2012). Due in part to this potential biological advantage, "empty" Ti cages (i.e., without bone graft material) have been trialed with good short-term outcomes reported by Krayenbuhl et al. (2008).

In contrast to CFRP and PEEK cages, Ti cages are radiopaque, rendering direct assessment of bony fusion through the cage difficult. As Profeta noted in his initial experience following patients with Ti cages, "signs of osseous consolidation can be detected around the cage," and fusion "may be deduced from the long-term stability and absence of bone rarefaction around the cage." A bigger concern with Ti cages has been early cage subsidence, which was originally reported in eight patients by Gercek et al. (2003). He described radiographic subsidence of stand-alone Ti cages in five of nine operated levels, with one patient developing recurrent radicular symptoms and foraminal stenosis after a period of initial improvement post-operatively. Other authors to follow redemonstrated the propensity of Ti cages to subside at rates of 13–62.5% with a meta-analysis showing a subsidence rate of 33/211 patients (15.6%) for Ti cages compared to 11/184 (6.0%) for PEEK cages (Li et al. 2016); however, it is important to note that this analysis failed to show a difference in functional outcomes between PEEK and Ti cages. The high elastic modulus relative to bone and the resultant potential for stress shielding has been cited as a possible contributing factor in Ti cage subsidence (see Table 1). With the advent of three-dimensional printing technology, modifications to cage surface architecture and implant porosity may improve bony ingrowth and biomechanical stability; however, long-term clinical outcomes for these new technologies are not yet readily available.

Peek

Polyether ethyl ketone was first investigated for use as an orthopedic implant in the 1980s and was used as a component of CFRP cages in early interbody cages such as the Brantigan cage discussed previously. After CFRP cages fell out of favor, PEEK cages emerged as the primary alternative to Ti cages. PEEK is a semi-crystalline polymer, the mechanical properties of which are dependent on its molecular weight, temperature, strain rate, and crystallinity (Kurtz and Devine 2007). PEEK used in spine cages has an elastic modulus of approximately 3.4 GPa, which is similar to adjacent bone and significantly lower than its CRFP and Ti counterparts. A lower elastic modulus may lead to less stress shielding and less subsidence, as discussed above. As with CFRP cages, PEEK cages are radiolucent (with marking beads) and allow for easier, more direct visualization of bony fusion after surgery, as well as less artifact when imaging with a CT or MRI.

The structure of PEEK is very chemically stable, and as a result PEEK implants are biologically inert. While favorable in the sense that this prevents an inflammatory response or degradation, there is a concern that inertness as well as the hydrophobic nature of PEEK, which limits cellular adhesion, can lead to limited fixation to adjacent bone. An animal study on PEEK implants (Toth et al. 2006) demonstrated the development of non-osseous tissues (cartilage, fibrous tissue) at the bone-implant interface. To attempt to improve the osseointegration of PEEK implants, efforts have been made to develop PEEK cages with an altered roughened surface, which some propose accounts for differences in osseointegration between previous PEEK and Ti implants. Torstrick et al. (2017) report on the development of a PEEK cage with a porous surface microstructure, in which they showed improved histological osseointegration compared to smooth PEEK and encouraging results in early clinical application in ACDF patients.

Clinical outcomes with PEEK cages have been excellent and have led to widespread use of PEEK implants in clinical practice. PEEK has been shown to have equivalent functional and radiographic results to autograft iliac bone graft, with the expected shorter surgical time and decreased donor site morbidity expected with use of a cage (Zhou et al. 2011). PEEK has also compared favorably to Ti in several studies. Niu et al. (2010), in a prospective comparison with 12-month follow-up, showed higher fusion rates (100% for PEEK vs. 86.5% for Ti) and lower subsidence (0% for PEEK vs. 16.2% for Ti), with equivalent rates of good to excellent clinical outcomes. With a longer follow-up period of 7 years, Chen et al. (2013) demonstrated lower subsidence (5.4% compared to 34.5%), better maintenance of angular correction, and better clinical outcomes with PEEK compared to Ti while noting that fusion was achieved in all patients in both groups.

Hybrid Cages

In an effort to combine the desirable features of both PEEK and Ti cages, hybrid cages with a PEEK body and Ti coating (via plasma-spraying) have been developed. These theoretically have an elastic modulus closer to bone with a radiolucent center due to the PEEK component while incorporating the potentially better initial fixation and subsequent osseointegration of a rough Ti surface. Pre-clinical studies using this type of cage demonstrated improved bone ingrowth to the Ti-PEEK cages as opposed to PEEK alone (Walsh et al. 2015). A concern raised for this type of cage is the delamination between the Ti and PEEK layers; however, manufacturer-published biomechanical testing demonstrated no delamination at ten million cycles (Medacta 2018). Hybrid Ti-coated PEEK cages are in the early phases of clinical usage, and their long-term performance relative to other cage materials remains to be seen, though there are concerns that the coating may be susceptible to delamination during impaction (Torstrick Spine J. 2018).

Operative Technique

After standard anterior cervical spine exposure, the author's preferred surgical technique is to utilize Caspar pins for interspace distraction. Pins are placed in the midpoint of the superior and inferior vertebral bodies. Annulotomy of the intended disc space is performed with a 15-blade scalpel knife, and the interspace is then distracted. Diskectomy is performed with a combination of pituitary rongeurs, Kerrison rongeurs, curettes, and a high-speed burr. Once a thorough diskectomy is complete, the cartilage layer on the superior and inferior endplates is removed with straight and angled curettes. Sclerotic subchondral bone is roughened with the use of a rasp. The posterior osteophytes are identified and removed with the high-speed burr, and the osseous shavings are saved as local autograft. Doing so reveals the posterior longitudinal ligament, which is then split with a nerve hook in line with the ligamentous fibers and then resected with Kerrison rongeurs. Once the posterior longitudinal ligament has been completely removed, the neural foramina are palpated and decompressed with Kerrison rongeurs. Decompression is verified using a nerve hook. Once the decompression is completed, the distracted interspace is sized for an appropriate cage. Fluoroscopic evaluation is performed to assess for alignment as well as distraction of the facet joints, to prevent "overstuffing" of the interspace. The cage is packed with local autograft, as well as osteoinductive or conductive material such as demineralized bone matrix, and distraction can be released. The cage is impacted into the interspace. The implant position is verified with intraoperative radiography. An anterior cervical locking plate is then placed in front of the cage, with screws in the cephalad and caudal vertebral bodies. Intervertebral cage devices with integrated fixation can avoid additional plating, particularly in instances such as adjacent segment degeneration above a prior fusion.

Summary

As cervical interbody fusion techniques have advanced, so too have implant designs and materials. Development of cervical interbody cages has allowed for improvements in fusion rates as well as in a reduction in the necessity of donor site morbidity for iliac crest harvest. As manufacturing processes and understanding of bone biology continue to improve, future interbody cage designs and materials will likely follow suit.

References

An HS, Simpson JM, Glover JM et al (1995) Comparison between allograft plus demineralized bone matrix *versus* autograft in anterior cervical fusion. Spine 20:2211–2216

Bagby G (1988) Arthrodesis by the distraction-compression method using a stainless steel implant. Orthopedics 11(6):931–934

Bent MJ, Oosting J, Wouda EJ et al (1996) Anterior cervical discectomy with or without fusion with acrylate. Spine 21:834–840

Bishop RC, Moore KA, Hadley MN (1996) Anterior cervical interbody fusion using autogeneic and allogeneic bone graft substrate: a prospective comparative analysis. J Neurosurg 85:206–210

Brown MD, Malinin TI, Davis PB (1976) A roentgenographic evaluation of frozen allografts *versus* autografts in anterior cervical spine fusions. Clin Orthop 119:231–236

Chong E, Pelletier MH, Mobbs RJ, Walsh WR (2015) The design evolution of interbody cages in anterior cervical discectomy and fusion: a systematic review. BMC Musculoskelet Disord 16:99. https://doi.org/10.1186/s12891-015-0546-x

DeBowes RM, Grant BD, Bagby GW et al (1984) Cervical vertebral interbody fusion in the horse: a comparative study of bovine xenografts and autografts supported by stainless steel baskets. Am J Vet Res 45:191–199

Ibanez J, Carreno A, Garcia-Amorena C et al (1998) Results of the biocompatible osteoconductive polymer (BOP) as an intersomatic graft in anterior cervical surgery. Acta Neuroshir (Wien) 140:126–133

McConnell JR, Freeman BJ, Debnath UK, Grevitt MP, Prince HG, Webb JK (2003) A prospective randomized comparison of coralline hydroxyapatite with autograft in cervical interbody fusion. Spine 28(4):317–323

Cage Design References

An HS, Simpson JM, Glover JM, Stephany J (1995) Comparison between allograft plus demineralized bone matrix versus autograft in anterior cervical fusion|a prospective multicenter study. Spine 20(Suppl):2211–2216

Banco SP, Jenis L, Tromanhauser S, Rand F, Banco RJ (2002) The use of cervical cages for treatment of cervical disc disease. Curr Opin Orthop 13(3):220–223

Cauthen JC, Kinard RE, Vogler JB, Jackson DE, DePaz OB, Hunter OL, Wasserburger LB, Williams VM (1998) Outcome analysis of noninstrumented anterior cervical discectomy and interbody fusion in 348 patients. Spine 23(2):188–192

Chen Y, Chen D, Guo Y, Wang X, Lu X, He Z, Yuan W (2008) Subsidence of titanium mesh cage: a study based on 300 cases. J Spinal Disord Tech 21(7):489–492

Chong E, Pelletier MH, Mobbs RJ, Walsh WR (2015) The design evolution of interbody cages in anterior cervical discectomy and fusion: a systematic review. BMC Musculoskelet Disord 16(1):99

Epari DR, Kandziora F, Duda GN (2005) Stress shielding in box and cylinder cervical interbody fusion cage designs. Spine 30(8):908–914

Gödde S, Fritsch E, Dienst M, Kohn D (2003) Influence of cage geometry on sagittal alignment in instrumented posterior lumbar Interbody fusion. Spine 28(15):1693–1699

Hacker RJ (2002) Threaded cages for degenerative cervical disease. Clin Orthop Relat Res 394:39–46

Jain S, Eltorai AEM, Ruttiman R, Daniels AH (2016) Advances in spinal Interbody cages. Orthop Surg 8(3):278–284

Kaiser MG, Haid RW Jr, Subach BR, Barnes B, Rodts GE Jr (2002) Anterior cervical plating enhances arthrodesis after discectomy and fusion with cortical allograft. Neurosurgery 50(2):229–236; discussion 236–8

Kandziora F, Pflugmacher R, Schäfer J, Born C, Duda G, Haas NP, Mittlmeier T (2001) Biomechanical comparison of cervical spine interbody fusion cages. Spine 26(17):1850–1857

Kandziora F, Schollmeier G, Scholz M, Schaefer J, Scholz A, Schmidmaier G, Schröder R, Bail H, Duda G, Mittlmeier T, Haas NP (2002) Influence of cage design on interbody fusion in a sheep cervical spine model. J Neurosurg Spine 96(3):321–332

Kast E, Derakhshani S, Bothmann M, Oberle J (2008) Subsidence after anterior cervical inter-body fusion. A randomized prospective clinical trial. Neurosurg Rev 32(2):207–214

Keogh A, Hardcastle P, Ali SF (2008) Anterior cervical fusion using the IntExt combined cage/plate. J Orthop Surg 16(1):3–8

Kettler A, Wilke H-J, Claes L (2001) Effects of neck movements on stability and subsidence in cervical interbody fusion: an in vitro study. J Neurosurg Spine 94(1):97–107

Kozak JA, O'Brien JP (1990) Simultaneous combined anterior and posterior fusion, an independent analysis of a treatment for the disabled low-back pain patient. Spine 15(4):322–328

Kulkarni AG, Hee HT, Wong HK (2007) Solis cage (PEEK) for anterior cervical fusion: preliminary radiological results with emphasis on fusion and subsidence. Spine J 7(2):205–209

Matgé G (2002) Cervical cage fusion with 5 different implants: 250 cases. Acta Neurochir 144(6):539–549. discussion 550

McConnell JR, Freeman BJC, Debnath UK, Grevitt MP, Prince HG, Webb JK (2003) A prospective randomized comparison of coralline hydroxyapatite with autograft in cervical interbody fusion. Spine 28(4):317–323

Ono K, Ebara S, Yonenobu K, Hosono N, Dunn EJ (1992) Prosthetic replacement surgery for spine metastasis. In: Recent advances in musculoskeletal oncology. Springer Japan, Tokyo, pp 208–218

Pisano AJ, Short TK, Formby PM, Helgeson MD (2016) Anterior cervical discectomy and fusion techniques: bone graft, biologics, interbody spacers, and plating options. Semin Spine Surg 28(2):84–89

Samandouras G, Shafafy M, Hamlyn PJ (2001) A new anterior cervical instrumentation system combining an intradiscal cage with an integrated plate: an early technical report. Spine 26(10):1188–1192

Shimamoto N, Cunningham BW, Dmitriev AE, Minami A, McAfee PC (2001) Biomechanical evaluation of stand-alone interbody fusion cages in the cervical spine. Spine 26(19):E432–E436

Smith GW, Robinson RA (1958) The treatment of certain cervical-spine disorders by anterior removal of the intervertebral disc and interbody fusion. J Bone Joint Surg 40(3):607–624

Steffen T, Tsantrizos A, Aebi M (2000) Effect of implant design and endplate preparation on the compressive strength of interbody fusion constructs. Spine 25(9):1077–1084

Truumees E (2011) Cervical instrumentation. In: Rothman Simeone the spine. Elsevier, Philadelphia, pp 1175–1218

van Dijk M, Smit TH, Sugihara S, Burger EH, Wuisman PI (2002) The effect of cage stiffness on the rate of lumbar Interbody fusion. Spine 27(7):682–688

Weiner BK, Fraser RD (1998) Spine update lumbar interbody cages. Spine 23(5):634–640

Wilke HJ, Kettler A, Claes L (2000) Primary stabilizing effect of interbody fusion devices for the cervical spine: an in vitro comparison between three different cage types and bone cement. Eur Spine J 9(5):410–416

Anterior Lumbar Interbody Fusion and Transforaminal Lumbar Interbody Fusion

32

Tristan B. Fried, Tyler M. Kreitz, and I. David Kaye

Contents

Anterior Lumbar Interbody Fusion (ALIF) 646
Introduction 646
Indications 646
Contraindications 646
Operative Setup 648
Surgical Technique 648
Complications 649
Pearls and Pitfalls 649
Postoperative Protocol 650

Transforaminal Lumbar Interbody Fusion (TLIF) 650
Introduction 650
Indications 650
Contraindications 651
Operative Setup 651
Surgical Technique 653
Complications 654
Pearls and Pitfalls 654
Postoperative Protocol 654

References 654

T. B. Fried (✉)
Sidney Kimmel Medical College, Thomas Jefferson University, Philadelphia, PA, USA
e-mail: Tristanbfried@gmail.com

T. M. Kreitz
Thomas Jefferson University Hospital, Philadelphia, PA, USA
e-mail: tyler.m.kreitz@gmail.com

I. D. Kaye
Rothman Institute, Thomas Jefferson University Hospital, Philadelphia, PA, USA
e-mail: iandavid.kaye@gmail.com; david.kaye@rothmanortho.com

Abstract

The use of interbodies as a method for lumbar fusion has been increasing over the past decades. Increased surface area and enhanced biomechanical forces for achieving fusion have made the technique more appealing than traditional posterolateral lumbar fusion. While many interbody techniques are actually less invasive than traditional open procedures, there are unique complication profiles with different approaches. The benefits of interbody fusion must be considered with the potential

© Springer Nature Switzerland AG 2021
B. C. Cheng (ed.), *Handbook of Spine Technology*,
https://doi.org/10.1007/978-3-319-44424-6_126

morbidity of these approaches. This chapter focuses on the techniques of both lumbar transforaminal and anterior interbody fusion along with potential complications with special attention toward graft/interbody selection and pearls and pitfalls of the respective technique.

Keywords

Anterior lumbar interbody fusion · Transforaminal lumbar interbody fusion · Technique · Complications · Graft

Anterior Lumbar Interbody Fusion (ALIF)

Introduction

Since its description in 1932 by Capener (1932) for the treatment of spondylolisthesis, anterior lumbar interbody fusion (ALIF) has become an accepted treatment modality for many degenerative lumbar conditions. ALIF may be used as a standalone procedure or in combination with posterior instrumentation as demonstrated in Fig. 1. Anterior lumbar fusion allows for complete discectomy, indirect decompression, maximal surface area for fusion under compression, and restoration of lumbar lordosis and sagittal alignment. The benefits of anterior fusion must be considered with the potential morbidity associated with the anterior approach or combined anterior/posterior procedure. This chapter reviews the indications, perioperative considerations, and surgical technique for anterior lumbar interbody fusion (ALIF).

Indications

- Isthmic and degenerative spondylolisthesis
- Degenerative disc disease
- Discogenic back pain
- Degenerative lumbar scoliosis
- Revision transforaminal lumbar interbody fusion (TLIF)
- Pseudarthrosis after posterior fusion (Mobbs et al. 2013a)
- Sagittal malalignment
- Osteodiscitis
- Tumor
- Trauma

A thorough history and physical with review of the appropriate imaging studies is necessary prior to considering a patient for an ALIF. In the case of degenerative disc disease with associated spondylolisthesis or deformity, the patient may present with radicular symptoms. The patient should demonstrate insufficient relief of symptoms with appropriate nonoperative management including use of NSAIDs, epidural injections, and physical therapy. Surgery may be expedited in the setting of significant motor weakness or cauda equina syndrome, a constellation of findings including saddle anesthesia, urinary retention, or loss of bowel control. Discography may be selectively used to help diagnose a discogenic source in a patient with axial low back pain.

Advantages of ALIF over posterior interbody fusion techniques (PLIF, TLIF) include the potential for more thorough discectomy, maximization of surface area for fusion, the potential for greater correction of deformity, indirect nerve root decompression, preservation of posterior structures including paraspinal musculature and facet complex, and avoidance of nerve root manipulation during graft insertion (Mobbs et al. 2013a; Richter et al. 2015).

Contraindications

Spine surgery including ALIF is not indicated when other pathology is the cause of the patient's neurologic symptoms: specifically central and peripheral neuropathy (diabetes, vitamin B deficiency, multiple sclerosis, and Guillain-Barre).

Contraindications specific to ALIF include:

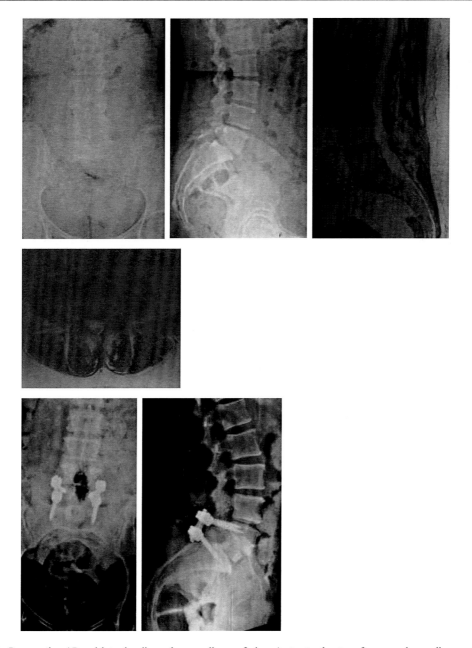

Fig. 1 Preoperative AP and lateral radiographs as well as sagittal and axial T2 MRI demonstrating stenosis and spondylolisthesis. Postoperative AP and lateral radiographs demonstrating lumbar 5 to sacral 1 anterior lumbar interbody fusion. A structural autograft was used as well as posterior instrumentation. Images courtesy of Department of Orthopaedic Surgery, Thomas Jefferson University Hospital

- Calcified great vessels (aorta, iliac vessels)
- Prior vascular reconstruction
- History of abdominal or retroperitoneal surgery
- History of pelvic inflammatory processes
- Medical comorbidities putting the patient at significant risk under general anesthesia

Operative Setup

Instruments and Materials Required
- Exposure surgeon: general or vascular
- Jackson radiolucent table
- Intraoperative radiography
- Surgical loupes or microscope
- Abdominal retractor system
- Standard spine instrument tray with curettes, Kerrison rongeurs
- Distractors and graft trials
- Interbody graft or cage

Graft Selection
Bone grafts may be used in isolation or in combination with cages, biologics (BMP-2), and anterior or posterior instrumentation. Iliac crest autograft (ICBG) is the "gold standard" for spinal fusion, but is associated with significant donor site morbidity including pain, infection, and neurovascular injury (Arrington et al. 1996). ICBG autograft may be used in combination with titanium cages or allograft material, reducing the necessary autograft quantity and associated morbidity while maintaining fusion potential (Newman and Grinstead 1992; Sasso et al. 2004). Alternatively, femoral ring allograft (FRA) or iliac crest structural allograft may be used. Allografts may lack the osteogenic and osteoinductive properties of autograft, but their versatility, availability, biologic superiority to cages, and avoidance of autograft donor morbidity make them an attractive option (Stevenson 1999; Sarwat et al. 2001; Mobbs et al. 2013a). Use of allograft in combination with posterior instrumentation demonstrates reliable rates of arthrodesis, greater than 90% among single-level procedures (Anderson et al. 2011; Mobbs et al. 2013b). Fusion rates may be enhanced with the addition of bone morphogenic protein-2 (rh-BMP2). The use of rh-BMP2 with allograft scaffold has demonstrated efficacy in ALIF procedures without the associated morbidity of ICBG autograft (Burkus et al. 2002; Anderson et al. 2011). Bone morphogenic protein-2 has been associated with ectopic bone formation, soft tissue edema, and excessive osteoclastic activity (Benglis et al. 2008; Chen et al. 2012). Titanium and composite interbody cages were developed to enhance biomechanical stability and improve fusion rates in standalone ALIF procedures, averting the risks and morbidity associated with combined anterior/posterior procedures. These interbody devices are used in combination with allo-/autograft and biologic materials (Burkus et al. 2009; Strube et al. 2012; Behrbalk et al. 2013). Titanium interbody cages in combination with BMP-2 demonstrate substantial fusion rates, greater than 90%, when used in standalone ALIF procedures without supplemental posterior instrumentation (Burkus et al. 2002, 2009; Boden et al. 2000). However, concern exists regarding titanium cage subsidence due to metal-bone modulus. Biocompatible composite material polyetheretherketone (PEEK) cages were developed as a load-bearing interbody device with similar modulus to bone to reduce rates of subsidence compared to metal cages (Galbusera et al. 2012; Behrbalk et al. 2013). Radiolucent PEEK cages also allow for easier radiographic assessment of fusion. Composite material cages carry the risk of fragmentation and extrusion over time. Ultimately, several bone graft and interbody device options have been developed for use with ALIF procedures, each with their own inherent advantages and disadvantages.

Positioning
The patient is placed supine on a flat Jackson table. Slight Trendelenburg and bump under the sacrum may remove obstructing pannus and decrease lordosis. The patient's arms are taped across the chest. Use of a Foley catheter is preferred. The patient is then prepped and draped in usual sterile fashion.

Surgical Technique

Step 1: Exposure
A left-sided retroperitoneal approach is performed with the assistance of an exposure surgeon. A low transverse incision between the umbilicus and pubic symphysis is preferred. Alternatively, a longitudinal curvilinear left-sided incision may be used for retroperitoneal approach. Dissection is carried through the dermis, and subcutaneous tissues are carried to the level of the rectus sheath. The anterior rectus sheath is divided transversely. The superior and inferior edges of the rectus fascia

are elevated off the underlying rectus abdominis muscles. The right and left rectus muscles are separated in the midline exposing the dorsal fascia and arcuate line. Underlying preperitoneal fat is separated, and the peritoneum is identified and preserved. Dissection is carried in a left lateral extraperitoneal plane. The psoas muscle with overlying genitofemoral nerve are encountered just lateral to the common iliac vessels (Spruit et al. 2005; Behrbalk et al. 2013). The left ureter may be encountered running beneath the peritoneum and should be protected. The lower lumbar spine and sacrum are identified. Overlying soft tissue is bluntly dissected exposing the spine. Middle sacral vessels may be clipped and cauterized. Exposure of the L5/S1 disc space is performed distal to the iliac vessels and cranial levels proximal to the vessels. A radiographic marker is placed in the appropriate intervertebral disc, and a plain radiograph is obtained to confirm the anatomic level.

Step 2: Discectomy

After confirmation of appropriate disc for excision, a long-handled knife is used to perform an annulotomy and define the perimeter of the identified disc. Electrocautery is avoided for risk of injury to the autonomic nervous system. A knife, high-speed burr, pituitary rongeur, and/or curette may be used sequentially to remove the identified disc posterior to the posterior longitudinal ligament (PLL). Sequential distraction can be used to allow better visualization and access to the posterior disc space. Vertebral endplates may be further defined using a cobb elevator.

Step 3: Endplate Preparation

Using a curette, the inferior endplate of the cephalic vertebrae and superior endplate of the caudal vertebrae are denuded of cartilage creating a roughened bleeding surface for fusion. Care is taken to not disrupt or compromise the integrity of the endplate resulting in stress riser and potential for implant subsidence.

Step 4: Insertion of Graft/Cage

Trial sizers may be used to determine appropriate graft size restoring intervertebral disc height. The bone graft may be asymmetrically machined to restore appropriate lordosis. The bone graft/cage complex may be augmented with a biologic to enhance bony fusion. Graft/cage positioning is confirmed using an intraoperative lateral radiograph.

Step 5: Closure

The wound is copiously irrigated. All bleeding should be controlled using hemostatic agents and electrocautery. The wound is closed in multiple (fascia, subcutaneous tissue, and skin) layers, and a sterile dressing placed.

Complications

- Sexual dysfunction, retrograde ejaculation in males
  - Due to damage to superior hypogastric (sympathetic) plexus located along the prevertebral tissues anterior to L5 vertebrae. Especially at risk when performing L5/S1 ALIF. Blunt dissection and avoidance of electrocautery are encouraged at this level.
- Iatrogenic injury to retroperitoneal structures (i.e., great vessels, ureters, bowel)
  - Care is taken during the approach and in retractor placement to identify and protect vital structures.
- Dural tear (CSF leak)
  - Rare as the epidural space is generally not encountered
- Nerve root damage
  - Rare as there is no direct nerve decompression; rather decompression occurs through distraction.
- Infection
- Pseudarthrosis
- Hardware/graft malposition or migration
- Medical complications
  - Patients are at risk of postoperative ileus and in rare instances bowel obstruction from the anterior approach to the lumbar spine.

Pearls and Pitfalls

- A bump can be used on the sacrum to retrovert the pelvis to allow for easier access to the disc space. This is particularly critical for L5–S1 isthmic spondylolistheses.

- Care should be taken during anterior exposure to identify and protect vital structures within the retroperitoneum. The use of an exposure surgeon is highly recommended to facilitate safe exposure.
- Use of electrocautery is avoided anterior to the L5 vertebrae to prevent injury to the prevertebral sympathetic plexus.
- A thorough discectomy and endplate preparation is critical for fusion.
- Bone graft/cage must be sized appropriately to restore alignment, achieve indirect decompression, and accomplish a large surface area of contact.

Postoperative Protocol

Early postoperative mobilization and physical therapy facilitate recovery and prevent the risk of medical complications associated with immobility. Due to the risk of postoperative ileus, patients are started on a clear liquid diet until they pass flatus and then advanced as tolerated. A bowel movement prior to discharge is encouraged.

Transforaminal Lumbar Interbody Fusion (TLIF)

Introduction

PLIF was first described in 1944 by Briggs and Milligan; however, it did not increase in popularity until it was described in 1953 by Cloward. Unlike the original description which used laminectomy bone chips as the interbody graft, Cloward's technique involved a graft made of iliac autograft (Briggs and Milligan 1944; Cloward 1953; Cole et al. 2009; Mura et al. 2011). A posterior approach had some advantages in that it could reduce some of the possible complications of ALIF including vascular injury and sexual dysfunction. The TLIF was then first described in 1982 by Harms and Jeszenszky as a modification of the PLIF procedure (Harms and Rolinger 1982; Cole et al. 2009; Mura et al. 2011). The TLIF approach is designed to approach the disc space unilaterally through the foraminal area (Kambin's triangle) to reduce complications associated during PLIF such as durotomy, radiculitis, and interoperative bleeding (Harms and Rolinger 1982; Craig Humphreys et al. 2001; Cole et al. 2009; Mura et al. 2011). TLIF has since evolved and can also be performed as a minimally invasive procedure that was popularized by Foley et al. in 2003 (Foley et al. 2003; Sonmez et al. 2013). This chapter reviews the indications, perioperative considerations, and surgical technique for transforaminal lumbar interbody fusion (TLIF). Figure 2 represents pre- and postoperative images from a typical TLIF procedure.

Indications

- Symptomatic spondylolisthesis
- Spinal stenosis with instability
- Degenerative lumbar scoliosis (Cloward 1953; Collis 1985; Lin 1985; Cole et al. 2009; Xiao et al. 2009)
- Discogenic low back pain
- Recurrent lumbar disc herniation (Cole et al. 2009)

Before the decision to perform a TLIF, a thorough history and physical exam must be performed. Prior to consideration of surgery, patients should have exhausted a course of nonoperative management such as NSAIDs, physical therapy, and consideration of injections. In addition, the patient's pain and symptoms should be concordant with advanced imaging.

TLIF has some distinct advantages over ALIF and PLIF procedures. In cases which require a posterior decompression, the performance of a TLIF vs. an ALIF can be significantly more time efficient. As a posteriorly based approach, performance of a TLIF can avoid complications uniquely associated with the anterior approach such as damage to the great vessels, damage to retroperitoneal structures, or sexual dysfunction. The TLIF also has advantages over the PLIF surgery in that it does not involve nerve retraction to the same degree which therefore decreased the risk of neurological damage. In addition, as TLIF can frequently be performed with unilateral facetectomy, preservation of bony structures on the contralateral side can aid in stability and ultimate fusion.

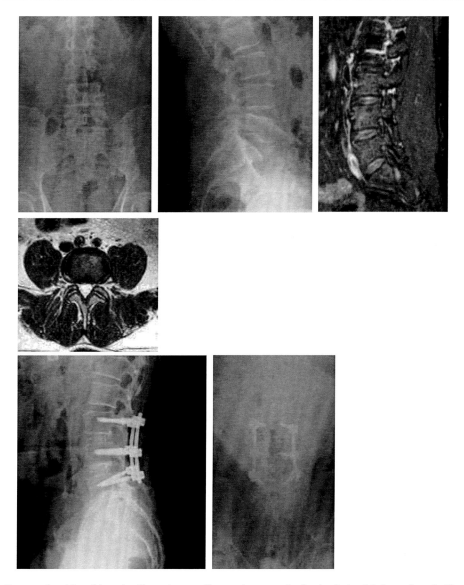

Fig. 2 Preoperative AP and lateral radiographs as well as T2 axial (L5–S1) and sagittal MRI demonstrating foraminal stenosis. Also, postoperative AP and lateral radiographs demonstrating lumbar 3–4 and 4–5 transforaminal interbody fusion. Images courtesy of Department of Orthopaedic Surgery, Thomas Jefferson University Hospital

Contraindications

Contraindications specific to TLIF include:

- Conjoined nerve root within the foramen (Holly et al. 2006)
- Previous wide posterior decompression (relative) (Ozgen et al. 1999; Lai et al. 2004; Xiao et al. 2009)
- Severe osteoporosis (Min et al. 2008; Xiao et al. 2009)
- Medical comorbidities putting the patient at significant risk under general anesthesia

Operative Setup

Instruments and Materials Required

- Exposure surgeon: general or vascular
- Jackson radiolucent table
- Intraoperative radiography
- Surgical loupes or microscope

- Abdominal retractor system
- Standard spine instrument tray with curettes, Kerrison rongeurs
- Distractors and graft trials
- Interbody graft or cage

Graft Selection

Numerous interbody and grafting options have been developed over the past decades for use in TLIF. Although generally ICBG is the gold standard for fusion, possessing osteoindicative, osteoconductive, and osteogenic potential, ICBG is rarely used for TLIF as it has been associated with significant donor site morbidity with complication rates approaching 25% (Kurz et al. 1989; Younger and Chapman 1989; Fernyhough et al. 1992; Craig Humphreys et al. 2001; Gibson et al. 2002; Lowe and Coe 2002; Coe 2004; Mummaneni et al. 2004a). An ideal substitute should be structurally capable of maintaining the reconstructed disc space's integrity (i.e., maintain disc height and neuroforaminal height) without subsidence and help create an environment conducive to lumbar fusion (Boden 2002; Mummaneni et al. 2004b; Cole et al. 2009; Xiao et al. 2009). Allograft has been used as a comparable substitute to AIBG, and while some comparative studies found increased fusion rates with AIBG, most have concluded that there is no overall difference in fusion rates between the two (Lin 1985; Rish 1989; Rompe et al. 1995; Xiao et al. 2009). Titanium interbody cages have been increasingly used over the past 5 years. Titanium cages have been reportedly more likely to subside into the endplates of the vertebral bodies due to the increased modulus of elasticity of titanium relative to bone. In addition, accurate assessment of postoperative fusion can be challenging on CT and MR imaging secondary to metallic artifact (Weiner and Fraser 1998; Lowe and Coe 2002; Cole et al. 2009). Metal artifact reducing sequencing (MARS) on MRI has made this issue less problematic. Carbon fiber and PEEK cages were developed to minimize some of the technical issues that titanium has posed. With a modulus of elasticity closer to bone, there have been theoretical reports of decreased interbody subsidence. Additionally, PEEK cages may allow more reliable assessment of fusion mass (Weiner and Fraser 1998).

There are also bone graft substitutes that have been used; however, there is limited data on these substances. The most commonly used and widely studied biological agent used as a bone graft substitute has been recombinant human bone morphogenic protein-2 (rhBMP-2) (Resnick et al. 2005; Kaiser et al. 2014). Michielsen et al. performed a level I randomized trial comparing TLIF using PEEK interbody with BMP-2 or autograft and found that for the BMP-2 cohort, there was significantly greater interbody healing on bone deniometry up to a year postoperatively ($p = 0.014$), but no difference in overall fusion rates on CT scan (all patients fused) and no clinical differences (VAS, ODI, SF-36) postoperatively (Michielsen et al. 2013). More recently, Khan et al. published their series of 191 patients undergoing TLIF with BMP-2 or autograft and demonstrated equivalent fusion rates and overall complication rates between the cohorts, but higher rates of postoperative radiculitis and postoperative seroma for the BMP-2 patients (Khan et al. 2018). Although rhBMP-2 has certainly rivaled AIBG in terms of fusion potential, the complication profile has muted more widespread enthusiasm. rhBMP-2 has been shown in multiple studies to be associated with complications including radiculitis, osteolysis, and ectopic bone formation (Wong et al. 2008; Gray and Rampersaud 2010; Helgeson et al. 2011).

Positioning

The patient is positioned prone on a Jackson table with the hip pads positioned just below the iliac crests and the chest pad positioned just below the sternal notch. The abdomen is allowed to hang freely minimizing pressure on the abdominal cavity and helping to prevent excessive epidural bleeding.

Exposure

Although the procedure can be performed via a traditional open approach or using minimally invasive techniques, this section will focus on open techniques as MIS is discussed elsewhere. An incision is made in the midline and dissection is carried through the skin and subcutaneous tissue to the level of the fascia. The fascia is then incised on either side of the spinous processes and subperiosteal elevation of the paraspinal musculature is performed.

Surgical Technique

Distraction across the disc space can facilitate discectomy, and therefore we frequently remove the interspinous ligament at the planned level (i.e., remove the ligament between L4 and 5 for an L4–5 TLIF) but keep the spinous processes intact to serve as a buttress for placement of a laminar spreader. The laminar spreader is then expanded which allows for expansion of the interlaminar space. We performed the TLIF from the side which is more symptomatic. Partial laminectomy may be performed with the use of a high-speed burr or Kerrison rongeurs. Laminectomy is begun at the midline and proceeds in a medial to lateral direction. Once the lamina has been at least partially removed (about 50% of the height in a cranial caudal direction, which is generally above the insertion of ligamentum flavum on the undersurface of the cranial lamina and at the region where the epidural fat begins), an osteotome may be used to remove the remaining lamina and inferior articular process (IAP) of the cranial vertebrae en bloc. This bone may be used as local autograft to be placed in the disc space after discectomy. Alternatively, a burr or Kerrison rongeur can be used to remove the remaining lamina and IAP piecemeal.

During the procedure, the ligamentum flavum may be left intact until after the TLIF is performed or may be removed before the discectomy. Leaving the ligament intact may afford an extra level of protection to the dura and nerve root, and decompression may be performed after the interbody has been placed.

A woodson elevator can be used to palpate the top of the pedicle of the caudal vertebra. The woodson may be left in place, and an osteotome may then be used to remove the superior articular process of the caudal vertebrae. Any remaining bone superior to the pedicle should be removed with Kerrisons so that there is no overhanging bone above the pedicle. Similarly, the medial wall of the pedicle should be skeletonized by removing some of the lamina of the caudal vertebrae. At this point, the majority of the bony work has been completed for performance of the discectomy.

A woodson elevator is used to palpate the medial and superior borders of the pedicle and identify the disc space which is just above the caudal pedicle. Next, a Penfield 4 can be used to dissect of the overlying soft tissue above the disc, and bipolar electrocautery can be used to ligate the overlying epidural plexus at the disc level. The working zone for the discectomy is classically described as a triangle (Kambin's) bordered by the traversing nerve root medially, the pedicle/superior endplate inferiorly, and the exiting nerve root superiorly. Identification of the exiting nerve root allows for a safer working zone and more aggressive lateral disc work. Assuming the nerve has been identified and is sufficiently lateral at the level of the disc, frequently, medial retraction of the traversing nerve and thecal sac is not required. If required, gently retract the nerve and thecal sac medially with the use of a nerve root retractor. Be sure to relax the traction at frequent intervals to prevent the development of a palsy or postoperative radiculitis.

Discectomy begins with creating a box annulotomy as largely as possible with the use of a #15 blade on a long handle. We often use a Kerrison rongeur to remove the posterior lip of the caudal vertebral body as well as define the extent of the annulotomy in all directions. We begin disc removal with specially designed shavers which efficiently remove a large amount of disc material. Once bony chatter is appreciated, we no longer upsize the shavers. The size of the final shaver is generally the size of the final interbody. After shavers have removed much of the disc material, we use a combination of straight and angled curettes to remove residual adherent disc and the cartilaginous endplate. Care is taken not to violate the bony endplates. Trialing of the interbody spacers may be performed. Once a thorough discectomy and endplate preparation have been performed, the disc space is packed with a combination of locally harvested autograft and allograft. After the space has been packed, the interbody is placed and impacted into final position. A more anteriorly placed graft may allow for better creation of lordosis, but a more posteriorly placed graft may allow better indirect neuroforaminal decompression. Once the interbody has been impacted into final position, a woodson elevator can be used to palpate the posterior disc space and ensure bone is palpated posteriorly to the interbody.

At this point, the laminar spreader can be removed to further compress the interbody. All remaining encroaching bone and ligamentum can more easily be removed at this point. The foramen must be reexplored to ensure no remaining stenosis or bone graft remains.

If necessary, a contralateral facetectomy may be performed (akin to a Ponte osteotomy) to allow for even greater creation of lordosis.

After performing the TLIF, pedicle screws are placed in a standard fashion according to the particular surgeon's preference. By compressing the pedicle screws, the interbody can be locked in place and restore even greater lordotic alignment.

The wounds are then thoroughly irrigated and closed over drains in a layered fashion per routine.

Complications

- Radiculitis (Yan et al. 2008)
- Screw loosening or hardware failure (Yan et al. 2008)
- Paraspinal iatrogenic injury (Craig Humphreys et al. 2001; McAfee et al. 2005; Cole et al. 2009; Sakeb and Ahsan 2013; Mobbs et al. 2015)
- Dural tear (CSF leak) (Holly et al. 2006)
- Deep infection (Hackenberg et al. 2005)
- Nerve root damage
  - Radiculopathy (Hackenberg et al. 2005)
- Pseudarthrosis (Hackenberg et al. 2005)

Pearls and Pitfalls

Removing the IAP and SAP en bloc with the use of an osteotome can be time efficient and serve as an excellent source of local autograft.

Use traction on the traversing nerve root and thecal sac judiciously to prevent inadvertent neurologic injury.

Biting the posterior inferior lip of the cranial vertebrae can allow better visualization for discectomy and easier placement of interbody graft.

Endplate violation during discectomy can lead to excessive bleeding from the disc space and can predispose to interbody subsidence.

Postoperative Protocol

The standard of treatment after spinal fusion procedures includes an early regimen of physical therapy to decrease postoperative complications.

References

Anderson DG, Sayadipour A, Shelby K, Albert TJ, Vaccaro AR, Weinstein MS (2011) Anterior interbody arthrodesis with percutaneous posterior pedicle fixation for degenerative conditions of the lumbar spine. Eur Spine J 20(8):1323–1330. https://doi.org/10.1007/s00586-011-1782-x

Arrington ED, Smith WJ, Chambers HG, Bucknell AL, Davino NA (1996) Complications of iliac crest bone graft harvesting. Clin Orthop Relat Res 329:300–309

Behrbalk E, Uri O, Parks RM, Musson R, Soh RCC, Boszczyk BM (2013) Fusion and subsidence rate of stand alone anterior lumbar interbody fusion using PEEK cage with recombinant human bone morphogenetic protein-2. Eur Spine J 22(12):2869–2875. https://doi.org/10.1007/s00586-013-2948-5

Benglis D, Wang MY, Levi AD (2008) A comprehensive review of the safety profile of bone morphogenetic protein in spine surgery. Neurosurgery 62(5 Suppl 2):ONS423–ONS431. (Discussion ONS431). https://doi.org/10.1227/01.neu.0000326030.24220.d8

Boden SD (2002) Overview of the biology of lumbar spine fusion and principles for selecting a bone graft substitute. Spine 27(16 Suppl 1):S26–S31. https://doi.org/10.1097/00007632-200208151-00007

Boden SD, Zdeblick TA, Sandhu HS, Heim SE (2000) The use of rhBMP-2 in interbody fusion cages. Definitive evidence of osteoinduction in humans: a preliminary report. Spine 25(3):376–381

Briggs H, Milligan PR (1944) Chip fusion of the low back following exploration of the spinal canal. J Bone Joint Surg Am 26(1):125–130. http://jbjs.org/content/26/1/125.abstract

Burkus JK, Gornet MF, Dickman CA, Zdeblick TA (2002) Anterior lumbar interbody fusion using rhBMP-2 with tapered interbody cages. J Spinal Disord Tech 15(5):337–349

Burkus JK, Gornet MF, Schuler TC, Kleeman TJ, Zdeblick TA (2009) Six-year outcomes of anterior lumbar interbody arthrodesis with use of interbody fusion cages and recombinant human bone morphogenetic protein-2. J Bone Joint Surg Am 91(5):1181–1189. https://doi.org/10.2106/JBJS.G.01485

Capener N (1932) Spondylolisthesis. Br J Surg 19:374–386

Chen Z, Ba G, Shen T, Fu Q (2012) Recombinant human bone morphogenetic protein-2 versus autogenous iliac crest bone graft for lumbar fusion: a meta-analysis of ten randomized controlled trials. Arch Orthop Trauma Surg 132(12):1725–1740. https://doi.org/10.1007/s00402-012-1607-3

Cloward RB (1953) The treatment of ruptured lumbar intervertebral discs by vertebral body fusion. I. Indications, operative technique, after care. J Neurosurg 10(2):154–168. https://doi.org/10.3171/jns.1953.10.2.0154

Coe JD (2004) Instrumented transforaminal lumbar interbody fusion with bioabsorbable polymer implants and iliac crest autograft. Neurosurg Focus 16(3):11. https://doi.org/10.3171/foc.2004.16.3.12

Cole CD, McCall TD, Schmidt MH, Dailey AT (2009) Comparison of low back fusion techniques: transforaminal lumbar interbody fusion (TLIF) or posterior lumbar interbody fusion (PLIF) approaches. Curr Rev Musculoskelet Med 2(2):118–126. https://doi.org/10.1007/s12178-009-9053-8

Collis JS (1985) Total disc replacement: a modified posterior lumbar interbody fusion. Report of 750 cases. Clin Orthop Relat Res 193:64–67. http://www.ncbi.nlm.nih.gov/pubmed/3882298

Craig Humphreys S, Hodges SD, Patwardhan AG, Eck JC, Murphy RB, Covington LA (2001) Comparison of posterior and transforaminal approaches to lumbar interbody fusion. Spine 26(5):567–571. https://doi.org/10.1097/00007632-200103010-00023

Fernyhough JC, Schimandle JJ, Weigel MC, Edwards CC, Levine AM (1992) Chronic donor site pain complicating bone graft harvesting from the posterior iliac crest for spinal fusion. Spine 17(12):1474–1480. https://doi.org/10.1097/00007632-199212000-00006

Foley KT, Holly LT, Schwender JD (2003) Minimally invasive lumbar fusion. Spine 28(15 Suppl):S26–S35. https://doi.org/10.1097/01.BRS.0000076895.52418.5E

Galbusera F, Schmidt H, Wilke H-J (2012) Lumbar interbody fusion: a parametric investigation of a novel cage design with and without posterior instrumentation. Eur Spine J 21(3):455–462. https://doi.org/10.1007/s00586-011-2014-0

Gibson S, McLeod I, Wardlaw D, Urbaniak S (2002) Allograft versus autograft in instrumented posterolateral lumbar spinal fusion: a randomized control trial. Spine 27(15):1599–1603. https://doi.org/10.1097/00007632-200208010-00002

Gray RJ, Rampersaud YR (2010) Comparison of the incidence of radiculitis and radiographic adverse event following minimally invasive lumbar transforaminal interbody fusions (TLIF) with and without the use of bone morphogenetic protein (BMP). Spine J 10(9):S8–S9. https://doi.org/10.1016/j.spinee.2010.07.028

Hackenberg L, Halm H, Bullmann V, Vieth V, Schneider M, Liljenqvist U (2005) Transforaminal lumbar interbody fusion: a safe technique with satisfactory three to five year results. Eur Spine J 14(6):551–558. https://doi.org/10.1007/s00586-004-0830-1

Harms J, Rolinger H (1982) A one-stager procedure in operative treatment of spondylolistheses: dorsal traction-reposition and anterior fusion. Z Orthop Ihre Grenzgeb 120(3):343–347. https://doi.org/10.1055/s-2008-1051624

Helgeson MD, Lehman RA, Patzkowski JC, Dmitriev AE, Rosner MK, Mack AW (2011) Adjacent vertebral body osteolysis with bone morphogenetic protein use in transforaminal lumbar interbody fusion. Spine J 11(6):507–510. https://doi.org/10.1016/j.spinee.2011.01.017

Holly LT, Schwender JD, Rouben DP, Foley KT (2006) Minimally invasive transforaminal lumbar interbody fusion: indications, technique, and complications. Neurosurg Focus 20(3):E6. http://ovidsp.ovid.com/ovidweb.cgi?T=JS&PAGE=reference&D=emed10&NEWS=N&AN=43713346

Kaiser MG, Eck JC, Groff MW et al (2014) Guideline update for the performance of fusion procedures for degenerative disease of the lumbar spine. Part 1: introduction and methodology. J Neurosurg Spine 21(1):2–6. https://doi.org/10.3171/2014.4.SPINE14257

Khan TR, Pearce KR, McAnany SJ, Peters CM, Gupta MC, Zebala LP (2018) Comparison of transforaminal lumbar interbody fusion outcomes in patients receiving rhBMP-2 versus autograft. Spine J 18(3):439–446

Kurz LT, Garfin SR, Booth RE (1989) Harvesting autogenous iliac bone grafts: a review of complications and techniques. Spine 14(12):1324–1331. https://doi.org/10.1097/00007632-198912000-00009

Lai PL, Chen LH, Niu CC, Fu TS, Chen WJ (2004) Relation between laminectomy and development of adjacent segment instability after lumbar fusion with pedicle fixation. Spine 29(22):2527–2532. https://doi.org/10.1097/01.brs.0000144408.02918.20

Lin PM (1985) Posterior lumbar interbody fusion technique: complications and pitfalls. Clin Orthop Relat Res 193:90–102. http://www.ncbi.nlm.nih.gov/entrez/query.fcgi?cmd=Retrieve&db=PubMed&dopt=Citation&list_uids=3882302

Lowe TG, Coe JD (2002) Resorbable polymer implants in unilateral transforaminal lumbar interbody fusion. J Neurosurg 97(4 Suppl):464–467. https://doi.org/10.3171/spi.2002.97.4.0464

McAfee PC, DeVine JG, Chaput CD et al (2005) The indications for interbody fusion cages in the treatment of spondylolisthesis. Spine 30(Suppl):S60–S65. https://doi.org/10.1097/01.brs.0000155578.62680.dd

Michielsen J, Sys J, Rigaux A, Bertrand C (2013) The effect of recombinant human bone morphogenetic protein-2 in single-level posterior lumbar interbody arthrodesis. J Bone Joint Surg Am 95(10):873–880

Min JH, Jang JS, Jung BJ et al (2008) The clinical characteristics and risk factors for the adjacent segment degeneration in instrumented lumbar fusion. J Spinal Disord Tech 21(5):305–309. https://doi.org/10.1097/BSD.0b013e318142b960

Mobbs RJ, Loganathan A, Yeung V, Rao PJ (2013a) Indications for anterior lumbar interbody fusion. Orthop Surg 5(3):153–163. https://doi.org/10.1111/os.12048

Mobbs RJ, Chung M, Rao PJ (2013b) Bone graft substitutes for anterior lumbar interbody fusion. Orthop Surg 5(2):77–85. https://doi.org/10.1111/os.12030

Mobbs RJ, Phan K, Malham G, Seex K, Rao PJ (2015) Lumbar interbody fusion: techniques, indications and comparison of interbody fusion options including PLIF, TLIF, MI-TLIF, OLIF/ATP, LLIF and ALIF. J Spine Surg 1(1):2–18. https://doi.org/10.3978/j.issn.2414-469X.2015.10.05

Mummaneni PV, Pan J, Haid RW, Rodts GE (2004a) Contribution of recombinant human bone morphogenetic protein-2 to the rapid creation of interbody fusion when used in transforaminal lumbar interbody fusion: a preliminary report. J Neurosurg Spine 1(1):19–23. https://doi.org/10.3171/spi.2004.1.1.0019

Mummaneni PV, Haid RW, Rodts GE (2004b) Lumbar interbody fusion: state-of-the-art technical advances. J Neurosurg Spine 1(1):24–30. https://doi.org/10.3171/spi.2004.1.1.0024

Mura PP, Costaglioli M, Piredda M, Caboni S, Casula S (2011) TLIF for symptomatic disc degeneration: a retrospective study of 100 patients. Eur Spine J 20(Suppl. 1):57–60. https://doi.org/10.1007/s00586-011-1761-2

Newman MH, Grinstead GL (1992) Anterior lumbar interbody fusion for internal disc disruption. Spine 17(7):831–833

Ozgen S, Naderi S, Ozek MM, Pamir MN (1999) Findings and outcome of revision lumbar disc surgery. J Spinal Disord 12(4):287–292

Resnick DK, Choudhri TF, Dailey AT et al (2005) Guidelines for the performance of fusion procedures for degenerative disease of the lumbar spine. Part 8: lumbar fusion for disc herniation and radiculopathy. J Neurosurg Spine 2(6):673–678. https://doi.org/10.3171/spi.2005.2.6.0673

Richter M, Weidenfeld M, Uckmann FP (2015) Anterior lumbar interbody fusion. Indications, technique, advantages and disadvantages. Orthopade 44(2):154–161. https://doi.org/10.1007/s00132-014-3056-x

Rish BL (1989) A critique of posterior lumbar interbody fusion: 12 years' experience with 250 patients. Surg Neurol 31(4):281–289. http://www.ncbi.nlm.nih.gov/entrez/query.fcgi?cmd=Retrieve&db=PubMed&dopt=Citation&list_uids=2928922

Rompe JD, Eysel P, Hopf C (1995) Clinical efficacy of pedicle instrumentation and posterolateral fusion in the symptomatic degenerative lumbar spine. Eur Spine J 4(4):231–237. https://doi.org/10.1007/BF00303417

Sakeb N, Ahsan K (2013) Comparison of the early results of transforaminal lumbar interbody fusion and posterior lumbar interbody fusion in symptomatic lumbar instability. Indian J Orthop 47(3):255–263. https://doi.org/10.4103/0019-5413.111484

Sarwat AM, O'Brien JP, Renton P, Sutcliffe JC (2001) The use of allograft (and avoidance of autograft) in anterior lumbar interbody fusion: a critical analysis. Eur Spine J 10(3):237–241

Sasso RC, Kitchel SH, Dawson EG (2004) A prospective, randomized controlled clinical trial of anterior lumbar interbody fusion using a titanium cylindrical threaded fusion device. Spine 29(2):113–22-122. https://doi.org/10.1097/01.BRS.0000107007.31714.77

Sonmez E, Coven I, Sahinturk F, Yilmaz C, Caner H (2013) Unilateral percutaneous pedicle screw instrumentation with minimally invasive TLIF for the treatment of recurrent lumbar disk disease: 2 years follow-up. Turk Neurosurg 23(3):372–378. https://doi.org/10.5137/1019-5149.JTN.7122-12.1

Spruit M, Falk RG, Beckmann L, Steffen T, Castelein RM (2005) The in vitro stabilising effect of polyetheretherketone cages versus a titanium cage of similar design for anterior lumbar interbody fusion. Eur Spine J 14(8):752–758. https://doi.org/10.1007/s00586-005-0961-z

Stevenson S (1999) Biology of bone grafts. Orthop Clin North Am 30(4):543–552

Strube P, Hoff E, Hartwig T, Perka CF, Gross C, Putzier M (2012) Stand-alone anterior versus anteroposterior lumbar interbody single-level fusion after a mean follow-up of 41 months. J Spinal Disord Tech 25(7):362–369. https://doi.org/10.1097/BSD.0b013e3182263d91

Weiner BK, Fraser RD (1998) Spine update lumbar interbody cages. Spine 23(5):634–640. https://doi.org/10.1097/00007632-199803010-00020

Wong DA, Kumar A, Jatana S, Ghiselli G, Wong K (2008) Neurologic impairment from ectopic bone in the lumbar canal: a potential complication of off-label PLIF/TLIF use of bone morphogenetic protein-2 (BMP-2). Spine J 8(6):1011–1018. https://doi.org/10.1016/j.spinee.2007.06.014

Xiao YX, Chen QX, Li FC (2009) Unilateral transforaminal lumbar interbody fusion: a review of the technique, indications and graft materials. J Int Med Res 37(3):908–917. https://doi.org/10.1177/147323000903700337

Yan DL, Pei FX, Li J, Soo CL (2008) Comparative study of PILF and TLIF treatment in adult degenerative spondylolisthesis. Eur Spine J 17(10):1311–1316. https://doi.org/10.1007/s00586-008-0739-1

Younger EM, Chapman MW (1989) Morbidity at bone graft donor sites. J Orthop Trauma 3(3):192–195. https://doi.org/10.1097/00005131-198909000-00002

Scoliosis Instrumentation Systems

33

Rajbir Singh Hundal, Mark Oppenlander, Ilyas Aleem, and Rakesh Patel

Contents

Introduction	658
Adult Scoliosis: Definition and Etiology	658
Clinical Evaluation	659
Imaging Evaluation	660
Spinal and Spinopelvic Parameters	660
Management	661
Early Fixation Constructs: Harrington Instrumentation, Wires, and Hooks	662
Proximal and Distal Extent of Instrumentation	663
Osteotomies	664
Smith-Petersen Osteotomy	664
Pedicle Subtraction Osteotomy	666
Vertebral Column Resection	668
Minimally Invasive Surgery	669
Conclusions	671
References	671

Abstract

Adult spinal deformity is a complex deformity that involves three-dimensional deformation in coronal, sagittal, and axial planes. Spinal and spinopelvic parameters such as SVA, pelvic tilt, pelvic incidence, and lumbar lordosis are important in understanding, characterizing, and treating adult spinal deformity. Treatment of adult spinal deformity needs to be tailored to each patient with respect to the nature of the curve and the patients' overall medical health. Operative techniques have changed substantially with time, from the early use of

---

R. Singh Hundal · M. Oppenlander · I. Aleem · R. Patel (✉)
Department of Orthopaedic Surgery, University of Michigan, Ann Arbor, MI, USA
e-mail: rhundal@med.umich.edu; moppenla@med.umich.edu; ialeem@med.umich.edu; Rockpatelmd@gmail.com

© Springer Nature Switzerland AG 2021
B. C. Cheng (ed.), *Handbook of Spine Technology*,
https://doi.org/10.1007/978-3-319-44424-6_64

Harrington rods to modern pedicle screws. Multiple osteotomies (SPO, PSO, and VCR) can be applied for the desired level of spinal correction. Operative management of adult spinal deformity is wrought with complexity and severe complications. Newer techniques involving minimally invasive surgery and interbody fusions are being increasingly used for deformity correction. In this chapter, we will discuss such operative techniques for spinal deformity correction.

Keywords

Adult deformity · Scoliosis correction · Corrective osteotomy · Minimally invasive surgery (MIS) correction

Introduction

Adult spinal deformity (ASD) is an expansive term that covers a wide variety of conditions that involve deformation or malalignment of the adult spine. Among others, terms for various forms of deformity include scoliosis, sagittal imbalance, and spondylolisthesis (regional deformity). Normal anatomic variation does exist that can account for small regional curves of the spine. Adult spinal deformity, however, exceeds this normal anatomic variation and can possibly impair horizontal gaze or the neutral center of the spine over the pelvis and femoral heads. Impairment of horizontal gaze has a dramatic impact on the quality of an individual's life and has associated morbidity. Prevalence of adult deformity does appear to vary based on multiple factors. The overall prevalence in US adults aged 25–74 is about 8.3% with women having twice the rate as men (10.7% and 5.6%, respectively) (Carter and Haynes 1987). Furthermore, the prevalence appears to increase with advancing age. In a 2005 study by Schwab et al., they suggest prevalence rates of adult scoliosis (Cobb angles >10°) may be as high as 68% among adults 60 and older (Schwab et al. 2005). While in some cases adult spinal deformity can be asymptomatic, severe spinal deformity can present in multiple ways including back pain, hip pain, functional decline, radiculopathy, neurogenic claudication, and other neurologic symptoms.

Spinal deformity was initially simplified to deformity in the coronal plane. In particular, scoliosis was described as a lateral curvature of the spine resulting in a deformity in the coronal plane. As knowledge of deformity has grown, we have learned that deformity consists of complex three-dimensional changes that can result in changes in coronal, sagittal, and axial (rotational) planes (Stokes 1994). As understanding of the adult spinal deformity has grown, operative management has also advanced. Various corrective osteotomies can be applied for deformity correction including Smith-Petersen/Ponte osteotomy, pedicle subtraction osteotomy, and vertebral column resection. Instrumentation techniques involving wires and hooks have given way to constructs using pedicle screws and cortical screws (Fig. 1).

In this chapter we briefly discuss adult scoliosis including etiology, presentation, clinical evaluation, radiographic assessment, spinopelvic parameters, and overview on treatment. The primary focus of the chapter, however, relates to operative correction of deformity. In particular, we will discuss the corrective osteotomies that can be employed to improve spinal deformity in adult patients.

Adult Scoliosis: Definition and Etiology

Scoliosis consists of a three-dimensional deformity involving the coronal, sagittal, and axial (rotational) planes. In the sagittal planes, this can manifest as kyphotic changes impacting sagittal imbalance (Stokes 1994; Aebi 2005). The three-dimensional nature of the deformity can substantially impact the position of the head, horizontal gaze, and general positioning of the spine in relation to the pelvis.

The etiology of adult spinal deformity can be multifactorial. Some cases of ASD relate to congenital abnormalities of the vertebrae or spinal cord such as Chiari malformations or myelomeningocele (spina bifida). Neuromuscular conditions that may involve spinal deformity include cerebral palsy, Friedreich's ataxia, Charcot-Marie-Tooth, spinal muscular atrophy, muscular dystrophy, and arthrogryposis (Berven and Bradford 2002). Adult deformity can also

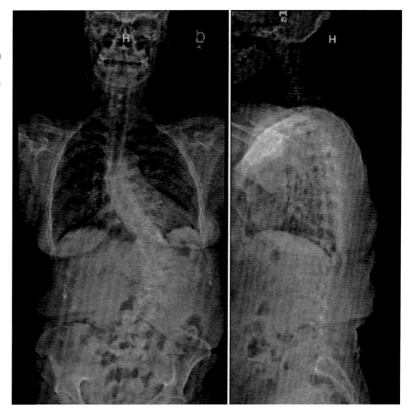

Fig. 1 Standing anteroposterior (AP) and lateral full-length spinal radiographs. AP radiograph demonstrates the deformity in the coronal plane as seen by the lateral curvature of thoracolumbar spine. Lateral radiograph demonstrates the sagittal deformity as seen by the positive sagittal imbalance

represent a progression of idiopathic scoliosis from childhood (infantile, juvenile, adolescent idiopathic scoliosis). Spinal deformity arising and developing in the adult population is often termed de novo or degenerative scoliosis. As the name suggests, this form of scoliosis is thought to relate to degenerative changes to spinal elements including the vertebral discs and zygapophyseal joints (Birknes et al. 2008). Other factors that can contribute to adult deformity include infection (poliomyelitis), spinal cord tumor, post-traumatic, and iatrogenic (post-surgical) (Berven and Bradford 2002; Birknes et al. 2008; Berven and Lowe 2007).

Clinical Evaluation

As with any complex condition of the spine, a complete history and physical examination is imperative. Pain is often a common presenting complaint that can vary from mild to severe and is often diffuse and ill-defined. The etiology of pain may be degenerative changes within the vertebral column (discs, facet joints), as well as paraspinal musculature (Birknes et al. 2008; Kostuik et al. 1973; Smith et al. 2009a, b). Given the imbalance of the spine over the pelvis and subsequently femoral heads, patients may also present with buttocks, hip, or leg pain. Patients may also present with symptoms of radiculopathy and/or stenosis (neurogenic claudication). Severe deformity may impair an individual's ability to maintain horizontal gaze. Different classification schemes, such as the Scoliosis Research Society (SRS) classifcation for adult spinal deformity, have been developed to help direct evaluation and management (Lowe et al. 2006). We will look at various spinal parameters below that can help guide evaluation and management. As part of the clinical evaluation, a full neurological exam should be performed to assess for weakness as well as additional issues such as myelopathy and cauda equina syndrome.

Imaging Evaluation

Initial imaging consists of standing full-length spinal radiographs, both PA and lateral views. Many of the spinopelvic parameters that are discussed below can be assessed on these radiographs alone. The PA view allows for evaluation of coronal alignment through measurements involving the central sacral vertical line (CSVL) and Cobb's angle. Pelvic obliquity can also be assessed on the PA radiograph. If the pelvic obliquity is related to a leg length discrepancy, repeat standing radiographs with blocks under the short leg may be needed. This is important in unmasking any perceived spinal deformity that may just relate to pelvic obliquity. The lateral radiograph allows for evaluation of the sagittal balance including any variation in lordosis and kyphosis in each spinal segment. The lateral radiographs also help to assess the sacral slope, pelvic tilt, and pelvic incidence. Additionally, the chin-brow to vertical angle, the angle formed between a line connecting the patient's chin to brow and a vertical line, can be measured in this view. Increasing chin-brow to vertebral angle suggests difficulty with maintaining horizontal gaze.

CT scans can prove useful in assessing bony morphology as part of planning for corrective osteotomies or placement of instrumentation such as pedicle screws. Being a supine study, CT scans can also be used to evaluate the flexibility of the curve in the sagittal plane when compared to upright x-rays. Given that patients may present with radicular or other neurological symptoms, an MRI can provide details regarding the location and etiology of areas of compression on the spinal cord and spinal nerves. If patient cannot undergo an MRI, a CT myelogram can be considered.

Spinal and Spinopelvic Parameters

Introduced in 1948, the Cobb method provides a quantitative measure of spinal curve on the coronal plane as seen on PA radiographs (Cobb 1948). The method consists of identifying the vertebral segment at the apex of the curve and most tilted vertebral bodies cephalad and caudal to the apex. Parallel lines are drawn along the superior end plate of the cephalad vertebral body and along the inferior end plate of the caudal vertebral. Perpendicular lines are subsequently drawn to each of the previously formed lines along the end plates. The angle formed between the intersections of the perpendicular lines is the Cobb angle. Traditionally, scoliosis is defined as a Cobb angle greater than 10° (Aebi 2005). The inter-observer error using the Cobb's method is about 5% (Mehta et al. 2009). Furthermore, studies suggest that an inherent error of up to 5° exists using the Cobb's method, meaning that only a change in the Cobb's angle of 5° or more is considered a real change (Morrissy et al. 1990). The Cobb's method can also be applied to lateral radiographs as a method of quantifying lordosis and kyphosis (Fig. 2).

While the Cobb method measures degree of curvature with respect to regional curves, the central sacral vertical line (CSVL) assesses overall coronal alignment (Lenke et al. 2001; Angevine and Kaiser 2008; O'Brien et al. 2004). A vertical line is made through the center of the sacrum. A second vertical line, C7 plumb line, is made centered on the C7 vertebral body. The difference between these two lines is the CSVL. A negative

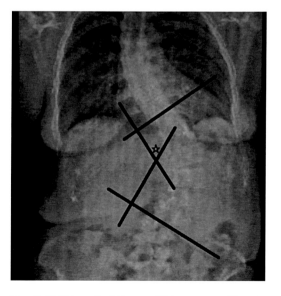

Fig. 2 Cobb's method for quantifying a curve. The star represents the Cobb's angle

value denotes that the C7 plumb line is to the left of the sacral line, while a positive value denotes that the C7 plumb line lines to the right.

The lateral radiograph provides crucial insight into the nature of the spinal deformity. Several parameters can be measured on the lateral radiographs including pelvic incidence (PI), sacral slope (SS), pelvic tilt (PT), sagittal vertical axis (SVA), and T1 pelvic angle (TPA). Pelvic incidence is the angle formed between a line perpendicular to the S1 end plate and a line between the center of the sacral end plate and the center of the femoral head (Legaye et al. 1998). Pelvic incidence also describes the sum of the sacral slope and the pelvic tilt. As a formulaic representation, PI = SS + PT. Sacral slope is the angle formed between a pure horizontal line and a line parallel to the sacral end plate. Pelvic tilt is the angle formed between a pure vertical line and a line between the center of the femoral head to the center of the sacral end plate. Of note, the pelvic tilt and sacral slope can change depending on position of the pelvis. Any movement leading to a change in pelvic inclination (i.e., increasing retroversion) will impact the pelvic tilt and sacral slope (Lafage et al. 2008; Boulay et al. 2006a; Jackson and McManus 1994; Schwab et al. 2009).

The sagittal vertical axis is also measured on the lateral radiograph. A vertical plumb line is drawn down from the center of C7 vertebral body. The distance between this plumb line and a point at the posterior-superior aspect of the sacral end plate is measured. A plumb line that lies anterior to the point on the posterior superior sacral end plate is denoted as a positive value. Normative values for the SVA are +2 to −2 cm; values outside of this range are considered positive or negative sagittal imbalance (Schwab et al. 2009; Boulay et al. 2006b; Roussouly and Nnadi 2010; Bernhardt and Bridwell 1989; Berthonnaud et al. 2005). The SVA, however, does not account for pelvic parameters and as such can be impacted by positioning and tilt of the pelvis. The T1 pelvic angle may provide more accurate insight into the overall sagittal alignment as it incorporates elements from the abovementioned pelvic parameters. The T1 pelvic angle is formed at the intersection of a line drawn from the T1 vertebral body to the center of the femoral head and a line drawn from the center of the femoral head to the center of the sacral end plate (Ryan et al. 2014). Lafage et al. introduced the TPA in 2014 as part of the International Spine Study Group. They proposed a goal/normative TPA of 10°, with a TPA greater than 20 representing a severe sagittal deformity (Ryan et al. 2014).

Scoliosis was initially viewed as a lateral curvature in the coronal plane; however, studies have not found a link between patient disability or perceived pain and degree of coronal deformity (Glassman et al. 2005a; Schwab et al. 2006a; Lazennec et al. 2009). Sagittal imbalance has been found to correlate with patient-reported pain and disability across several studies as measured by health-related quality of life measures (HRQOL). Sagittal imbalance as measured by pelvic tilt, TPA, T1 spinopelvic inclination, and SVA has been associated with worse scores on surveys such as the Oswestry Disability Index (ODI), SRS 23 Patient Questionnaire, and 12-Item Short Form Health Survey (SF-12) (Glassman et al. 2005a, b; Schwab et al. 2006a; Lazennec et al. 2009; Lafage et al. 2009). In lieu of these HRQOL studies, Schwab et al. outlined ideal thresholds with regard to key spinopelvic parameters. They found severe disability with regard to ODI with SVA exceeding 47 mm, pelvic tilt greater than 25°, and pelvic incidence minus lumbar lordosis being above 11° (Schwab et al. 2006b, 2010, 2013).

Management

Non-operative management of adult spinal deformity is usually limited to patients with mild deformity, minimal to mild pain, little disability in daily functional activities, nonprogressive symptoms, and lack of worrisome symptoms such as those of cauda equina. Non-operative management can also be applied to poor surgical candidates who have high anesthetic risks given profound comorbidities. Non-operative management modalities include massage, aqua therapy, and physical therapy which can serve to strength the surrounding paraspinal muscles and core as a whole. Additional modalities include nonsteroidal anti-inflammatory drugs, neuropathic medications (gabapentin), and epidural steroid injections (Cummins et al. 2006). The impact of non-operative modalities in improving pain and disability, however, is controversial. In 2010, Glassman et al. presented a prospective cohort study of 123 patients. Sixty-eight patients proceeded with conservative management consisting of physical therapy, bracing, bed rest, injections, and chiropractic care. Despite a mean cost of $10,815 over the course of 2 years, no significant change was found with regard to HRQOL outcomes (Glassman et al. 2010).

Indications for operative management include worsening pain, progressive deformity, declining neurological function, and failure of non-operative interventions. The spinopelvic parameters discussed earlier can help to assess the degree of deformity. Severe disability (measured with ODI) is correlated with SVA exceeding 47 mm, pelvic tilt greater than 25°, and pelvic incidence minus lumbar lordosis being above 11° (Schwab et al. 2010, 2013). In a 2009 prospective observational cohort, Bridwell et al. followed symptomatic adult scoliosis patients for 2 years. One hundred sixty patients treated either non-operatively or operatively were followed for 2 years. The non-operative cohort had no significant change in quality of life measures such as SRS and ODI. The operative cohort, however, did experience a significant improvement across all quality of life metrics (Bridwell et al. 2009). While each case of adult spinal deformity is unique, these findings do suggest that those with severe deformity and poor QOL scores may benefit from operative intervention (Bridwell et al. 2009; Smith et al. 2009c).

Operative intervention needs to be tailored to the specifics of each adult spinal deformity patient. Factors such as clinical symptoms, age, and overall medical health can help to steer direction of management. Operative modalities can include decompression, decompression with limited instrumentation, long-segment instrumentation, and corrective osteotomies. Decompression alone has a limited but important scope. Studies have shown that decompression alone may help radicular and compressive relative symptoms but risks progression of deformity (Kelleher et al. 2010). As such, decompression alone may help to address primarily compressive or radicular symptoms in an elderly individual, who may not otherwise be a candidate for extensive instrumentation or deformity correction given osteoporosis or medical comorbidities. In the following sections, we will discuss various methods of instrumentation, decision-making regarding what levels to include, and corrective osteotomies.

Early Fixation Constructs: Harrington Instrumentation, Wires, and Hooks

A key in the early development of spinal instrumentation involved the use of Harrington rods and instrumentation technique (Drummond 1988). Initially, Harrington rods were applied with the use of facet screws. However, Harrington constructs involving the use of facet screws did not prove viable in the long term as the screws were unable to accommodate the forces needed to correct spinal deformity (Harrington 1972, 1973). Subsequently, attention was directed toward new forms of spinal fixation involving sublaminar wiring and hooks. One such wiring technique, Luque wiring, was developed in Mexico. Luque wiring consisted of sublaminar wires that were twisted around rods posteriorly (Luque 1982). Since they are sublaminar, Luque wiring does place neural structures at risk during placement (Zdeblick et al. 1991). During the development of these early constructs, however, deformity was primarily understood as a problem in the coronal plane. As such, Harrington instrumentation and these early fixation models did not take into account the importance of sagittal. In the 1970s, various publications described the loss of lumbar lordosis and the development of a "flat back" resulting from Harrington distraction techniques (Doherty 1973; Grobler et al. 1978). The resultant flat back (iatrogenic fixed sagittal imbalance) made it a challenge to maintain upright posture and a horizontal gaze. To accommodate for the flat back, patient often flexes the hips and knees while extending the mobile cervical and thoracic segments (Potter et al. 2004).

Subsequent development focused on hooks as a means of providing segmental fixation that accommodated for lumbar lordosis. Examples of hooks include pedicle, laminar, supralaminar, and transverse process hooks. While adult spinal deformity is a complex malalignment involving all three vertebral columns in multiple planes, hooks primarily rely on fixation to the posterior column. Fixation through the posterior column alone may be unable to overcome the forces associated with the underlying spinal deformity required in obtaining and maintaining a correction (Rohlmann et al. 2006; Hackenberg et al. 2002). As such, these earlier techniques often involved additional anterior releases and correction to supplement the posterior fixation.

With the advent of pedicle screws, fixation could be placed across all three columns of the vertebra making it useful in deformity correction (Chang et al. 1988). Studies comparing pedicle screws versus hooks suggested that hooks had less pullout strength compared to pedicle screws (Liljenqvist et al. 2001). Clinically, pedicle screw constructs have been shown to lead to greater improvement in Cobb angles and sagittal alignment (Hamill et al. 1996). Some reports suggest increased rates of postoperative fusion with pedicle screws (Hamill et al. 1996; Gaines 2000; West et al. 1991; Thomsen et al. 1997). Multiple studies have suggested a decreased need for postoperative immobilization and bracing with the use of pedicle screws, as well as earlier process of rehabilitation (Marchesi and Aebi 1992; Suk et al. 1994, 1995). Pedicle screw placement has become safe and efficient. In particular, use of intraoperative fluoroscopy and intraoperative computed tomography and navigation has allowed for increased precision when placing pedicle screws (Miller et al. 2016; Gelalis et al. 2012).

Proximal and Distal Extent of Instrumentation

The proximal extent of the instrumentation is referred to as the upper instrumented vertebra (UIV). Generally, the UIV segment should not be at a level of segmental rotation or translation. Additionally, the UIV should not be at the apex of the curvature. Ending at a level of junctional kyphosis should be avoided. Mardjetko suggests that the UIV should be at a level within 2 cm of the coronal vertical axis and sagittal vertical axis (Shufflebarger et al. 2006). Given that spinal deformity curves may extend from the lumbar to the thoracic spine, the proximal instrumentation may need to extend to the thoracic spine. Extension to the thoracic spine, however, does raise concerns of proximal junctional kyphosis. As such, the most kyphotic range of the thoracic spine is avoided. This leaves two options for the upper instrumented vertebra: upper thoracic (T1–T6) and lower thoracic (T9–L1) (Kim et al. 2008, 2013, 2014; McCord et al. 1992). Proximal instrumentation to the upper thoracic versus the lower thoracic in adult scoliosis patients has increased operative times and blood loss but had similar levels of proximal junctional kyphosis and revision surgeries compared to UIV to the lower thoracic levels (T9–L1) (Kim et al. 2014). Mode of proximal junctional failure in the upper thoracic UIV is often ligamentous disruption compared to lower thoracic UIV in which failure is bony. While not reaching clinical significance, the total number of complications was greater with upper thoracic group, including substantial complications such as pseudoarthrosis (Kim et al. 2014).

The distal instrumented vertebra (DIV) has evolved over the years. McCord et al. defined the lumbosacral pivot point as the "intersection of the middle osteoligamentous column in the sagittal plane and the lumbosacral intervertebral disc in the transverse plane." They discuss concerns that DIV to sacrum is potentially less resistant to flexion moments and advocate for longer constructs distal to S1 and anterior to the pivot point (McCord et al. 1992). Constructs such as iliac bolts and S2-alar-iliac screws subsequently evolved to accommodate these principles. In a 2001 study, Lenke et al., they found a 95.1% fusion rate when using iliac screws for long fusions to the sacrum and severe spondylolisthesis (Kuklo et al. 2001). Another option is that of the S2-alar-iliac screws, which has a starting point at S2 with extended through the sacral ala into the ilium (Burns et al. 2016). Biomechanical studies suggest similar load to failure in comparing iliac screws to S2AI screws with the S2AI screws having the benefit of being lower profile and lining up with the lumbosacral screws obviating the need for offset connectors. Overall, such longer constructs can potentially better resist flexion moments with lower rates of failure compared to fixation ending at L5 or S1 (Kuklo et al. 2001; Burns et al. 2016; Kebaish 2010).

Osteotomies

In cases of rigid and severe spinal deformity, instrumented fusion alone may not fully correct the deformity, and additional correction through osteotomies of the vertebral column may be needed. Several osteotomies are available to aid in deformity correction including Ponte, Smith-Petersen osteotomy (SPO), pedicle subtraction osteotomy (PSO), and vertebral column resection. These osteotomies should be viewed as a spectrum with the more complex osteotomies built on the foundation of simpler osteotomies giving a greater correction. Choice in osteotomy depends on the amount of correction that is desired. Goals for deformity correction in the sagittal plane are an SVA under 5 cm, pelvic tilt less than 25°, and pelvic incidence minus lumbar lordosis being less than 11° (Schwab et al. 2010, 2013). In 2014, Schwab et al. created the comprehensive anatomical spinal osteotomy classification, which is a system to understand vertebral osteotomies. This system classifies the osteotomies into six categories based on increasing vertebral resection and destabilization; a graphic illustration can be seen in Fig. 3 (Schwab et al. 2015). While Schwab's classification system provides a systemic framework for understanding osteotomies, we will focus the discussion on the above classically described osteotomies.

Smith-Petersen Osteotomy

Developed in 1945, the Smith-Petersen osteotomy was initially developed to address flexion deformity in patients with ankylosed spines and rheumatoid arthritis (Smith-Petersen et al. 1945). As outlined in their original paper in 1945, the SPO involves removal of elements from the posterior column. The SPO does not extend into the vertebral body itself. The overall principle behind the SPO relies on an axis of rotation through the middle column. In effect, the removal of the posterior column and subsequent closing of the posterior void lead to elongation of the anterior column through osteoclasis of the anterior disc space and anterior longitudinal ligament.

Although the terms are often used interchangeably, Ponte osteotomies are distinguished from Smith-Petersen osteotomies in patient selection. A Ponte is performed in patients with an open disc space. Although it still results in a lengthening of the anterior column, it does not involve an osteoclasis of the anterior column. It is often used in conjunction with an interbody cage which serves as a fulcrum assisting to get angular correction with posterior compression. For the purposes of this chapter, we will refer to both these techniques as SPO. A SPO consists of a standard laminectomy and resection of the inferior articular facet of cranial level and superior articular facet of caudal level. They are usually performed at multiple consecutive levels in order to achieve a gradual correction. Classically these are performed for pathology such as Scheuermann's kyphosis (Ponte et al. 2018).

SPO technique differs slightly based on location: thoracic versus lumbar spine. Overall the concept is the same, resection of posterior elements allowing for compression and angular correction of approximately 5–10°. Anatomic differences particularly in facet orientation alter the sequence of steps depending on the location.

In the thoracic spine, the first step is using an osteotome to resect the inferior articular facet of the cranial level, exposing the cartilage of the superior articular facet. Next, the spinous process of the osteotomy level is removed, exposing the interlaminar space. The amount of resection of the lamina is based on the desired angular correction. More resection will potentially lead to more correction. Ideally, after the osteotomy is performed, the lamina which is resected and the

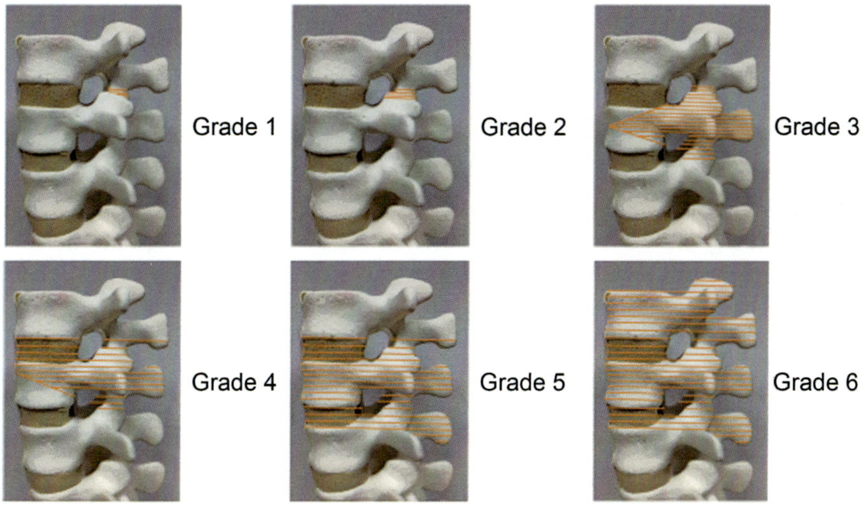

Fig. 3 Graphic illustration of Schwab et al. anatomical spinal osteotomy classification. Grade 1 involves partial resection of facet joint. Grade 2 involves complete facet joint resection. Grade 3 resects posterior elements, pedicles, and portion of vertebral body. Grade 4 resects posterior elements, pedicles, portion of vertebral body, intervertebral disc, and adjacent end plate. Grade 5 involves complete resection of vertebral segment and the adjoining intervertebral discs. Grade 6 involves complete resection of multiple vertebral segments (Schwab et al. 2015)

lamina of the caudal level should be in contact, providing a surface area for fusion. The lamina should be resected in a superiolateral direction on the midline creating a "V"-shaped bony defect. Due to the shape of the resection, these osteotomies are often referred to as chevron osteotomies. Next, the exposed ligamentum flavum is resected in the same direction exposing the spinal cord. Access to the canal can be gained through a midline defect in the ligament. After the ligament is resected, the superior articular facet of the caudal level is resected by continuing laterally with a Kerrison. Thorough excision of the ligament and superior articular facet is critical. Failure to remove these structures will lead to compression of either the spinal cord or nerve root, potentially leading to postoperative complications. If these osteotomies are performed after pedicle screws are placed, the heads of the screws may obstruct the resection of the superior articular facet. In such instances, one can either perform the osteotomy prior to placing in the screws or using a modular system where the heads are attached after the osteotomy.

In the lumbar spine, the screw heads do not interfere with the resection and therefore be inserted prior to performing the osteotomy. Additionally, due to the bony anatomy, it is not typically possible to have the resected lamina contact the caudal lamina. Therefore, a more generous laminectomy is performed. Since there is often spinal stenosis in the lumber spine which needs to be addressed as well, lamina resection at least to the origin of ligamentum flavum is recommended. Additionally, most lumbar SPOs/ Ponte are performed caudal to the conus, allowing an interbody cage to be safely placed posteriorly. An appropriately placed cage can provide a pivot point to gain more angular correction. The laminectomy is performed in the usual standard fashion. Subsequently, the pars on both sides are identified and resected with a Kerrison or a drill. Our preference is to place a Woodson in the foramen, serving to protect the exiting nerve root. Subsequently, we drill away the pars in its entirety until the Woodson is visualized. The inferior articular facet is then removed as it is no longer attached to any bony or soft tissue structures. The final step is to resect the overhanging portion of the superior articular facet. Removal with a Kerrison can be challenging given overgrowth. It is our preference to use a straight osteotome and place it in line with the superior aspect of the pedicle and remove it en bloc. There is usually venous bleeding in the foramen which can be stopped with bipolar cautery. We find this technique to be safe as the exiting nerve root typically lies in the superior third of the foramen. Therefore, even in the event of plunging with the osteotome, the exiting nerve root is safe from harm. The posterior void is subsequently closed via spinal instrumentation (Schwab et al. 2015; Bridwell 2006).

Smith-Petersen/Ponte osteotomies are best for deformities that have larger radius of curvatures as opposed to sharp curves (Cho et al. 2005). Since the anterior column is elongated, they should be avoided in cases where there is less than 5 mm of intervertebral disc present. These osteotomies allow for about 10° of correction per level performed (Cho et al. 2005).

Pedicle Subtraction Osteotomy

The pedicle subtraction osteotomy requires more resection than that for an SPO/Ponte osteotomy. While the SPO only involves the posterior column, pedicle subtraction osteotomies extend into the vertebral body at the desired level of correction, and as such is a three-column osteotomy. A PSO generally creates a triangular wedge through the vertebral body with removal of the posterior column. The osteotomy has an axis of rotation at the anterior aspect of the vertebral and shortens the posterior column without elongating the anterior column. The PSO can prove useful in patients with a rigid ALL or immobile vertebral disc, where an SPO/Ponte is usually contraindicated. Furthermore, a PSO can address sharp curves and curves that exceed 25° (Cho et al. 2005; Berjano and Aebi 2015; Chen et al. 2001). A PSO can provide 25–35° of correction (Berjano and Aebi 2015; Chen et al. 2001; Bridwell et al. 2003). Figure 4 demonstrates a case involving a PSO at L3 in conjunction with Smith-Petersen osteotomies at adjacent segments resulting in significant deformity correction.

In performing a PSO, the patient is placed prone on the operating table. When possible the PSO is below the level of the conus. There is variability as to where the conus ends; as such, it is imperative to review preoperative MRI or CT myelogram in selecting the level for the PSO. A cord-level PSO is associated with a considerably higher risk of cord injury and should be avoided when possible. The more caudal the osteotomy is performed, the greater the SVA is corrected for the same angular wedge resection. However, the more caudal the PSO is performed, the fewer fixation points will exist. For the stated reasons, L2 and L3 are commonly chosen levels. Due to significant angular correction, laminectomies are usually performed above and below the PSO site in order to prevent compression of the neural elements upon closure of the osteotomy site.

Conceptually, the building blocks of a PSO are two adjacent SPO. This will isolate a pedicle of a single level and is the first step of a PSO. The amount of angular correction is based on the angle of the wedge which is excised. This correlates to the distance between the starting points of the osteotomy along the posterior vertebral body. The limiting structures are the disc space above the pedicle and exiting nerve root below the pedicle to be excised. After two adjacent SPO are performed, these structures are identified bilaterally. The exiting nerve is followed out into the foramen. Prior to performing a PSO, all screw fixation is in place. We will routinely tap the pedicle of the osteotomy level with a large tap removing all the cancellous bone thereby making pedicle resection easier. We will also tap into the vertebral body creating a trajectory for our osteotome. The residual superior articular facet is then resected with a Leksell rongeur until flush with the transverse process. The transverse process is detached from its attachment at the lateral aspect of the pedicle. It is critical that the TP is cut flush with the lateral border of the pedicle. If it is not, when dissecting the psoas off the lateral aspect of the vertebral body, the segmental vessel is at risk. Using a large curette, the lateral wall of the pedicle and vertebral body is exposed. With the pedicle now in view circumferentially, it is removed with a rongeur. Any bony prominences need to be removed as

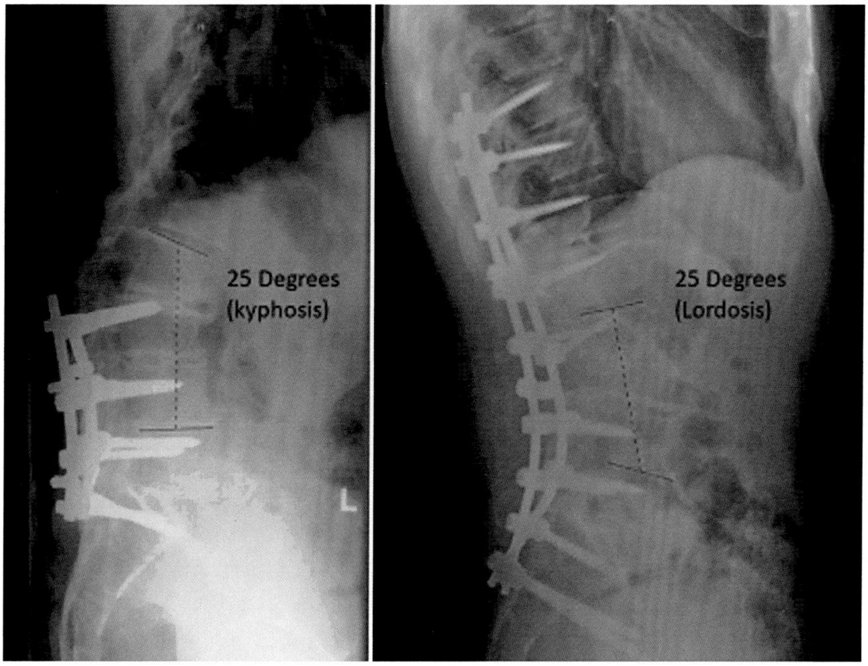

Fig. 4 A 68-year-old patient with persistent back pain status post remote L3 to S1 instrumentation and fusion. At initial evaluation, patient was found to have spinal stenosis and deformity consisting of kyphoscoliosis. Segmental Cobb angle from L2 to L4 demonstrated 25° of kyphosis. Patient underwent extension of instrumentation both proximally to T10 and distally to ilium. Additionally, a pedicle subtracting osteotomy was performed at L3 in conjunction with Smith-Petersen osteotomies at T12 to L2. Postsurgical radiographs demonstrated improvement in segmental lordosis to 25° from L2 to L4, representing an improvement in about 50°

they may cause foraminal stenosis after the osteotomy is closed. The exiting nerve root is protected with a nerve root retractor when removing the inferior wall of the pedicle. By resecting the pedicle, the two foramens have been combined making one large foramen that is housing two nerve roots. This step is performed bilaterally. A temporary stabilizing rod is now placed unilaterally, and an osteotome is used to make a wedge resection on one side. The superior cut is just caudal to the disc space above where the pedicle was, and the caudal cut is just cranial to the exiting nerve root immediately below where the pedicle was. One pass of the osteotome is directed medially and the other laterally, cutting the lateral wall of the vertebral body. The rod is moved to the opposite side, and a contralateral wedge resection is performed. The depth of the osteotome is determined by fluoroscopy or navigation. If using fluoroscopy, in the setting of a rotational deformity, the author prefers to rotate the table, so the osteotomy segment is no longer rotated. This leads to a more accurate assessment of depth of the osteotomy on fluoroscopy. After these cuts are made, a single vertical cut of the posterior vertebral body is made connecting the first two cuts. The resultant wedge is then resected and saved as autograft. Subsequently, a curette is used to remove any cancellous bone behind the remaining posterior cortex ventral to the thecal sac. A Woodson is used to develop a plane between the dura and posterior cortex. An Epstein curette or a Siefert bone tamp is used to impact the posterior wall into the defect created by removing the wedge. With the three-column osteotomy now complete, the spine should be mobile and the deformity ready to be corrected. Compression is applied on the temporary rods on either side closing the osteotomy, and wrinkling of the dura is noticed. Contact between the edges of the osteotomy marks the maximum extent of the correction obtained. If further correction is desired, the fixation is released, and further bony resection is performed. In the osteoporotic spine, if there is concern for screw loosening with compression, the patient's hips can be extended to close the osteotomy either manually or with an axis bed (Cho et al. 2005; Chen et al. 2001; Bridwell et al. 2003; Bianco et al. 2014).

If a larger correction is needed, one can perform an extended PSO. The extended PSO involves resection of the posterior aspect of the adjoining disc space and superior end plate. This creates a larger wedge and subsequently a larger correction. In order to increase the likelihood of a fusion, this procedure is often accompanied by a TLIF with the implant placed anteriorly resting on the residual superior end plate. If a patient has a multiplanar deformity, asymmetric wedges can be resected to achieve a correction in the sagittal as well as the coronal plane.

While a pedicle subtraction osteotomy allows for substantial correction of deformity, given the complex and aggressive nature of the osteotomy, it is associated with some notable complications. Several studies have reported complications rate reaching close to 50% (Bianco et al. 2014; Kelly et al. 2014). The International Spine Study Group reported 7% rate of intraoperative complications, 39% rate of postoperative complications, and 42% rate of overall complications. Additionally, they reported an average blood loss of 55% of total blood volume. Age older than 60, a thoracic three-column osteotomy, osteotomies at two or more levels, and major blood loss were all associated with increased complications (Kelly et al. 2014).

Vertebral Column Resection

Vertebral column resection builds on a PSO and allows greater segmental correction. It entails complete removal of a vertebral segment and allows for multiplanar corrections. A VCR can also prove useful in malformed vertebral segments that are not amenable to angular osteotomies such as those encountered in congenital scoliosis. Vertebral column resection was first described in the early 1980s by Bradford as a method of addressing severe and rigid spinal deformity (Bradford 1987; Lenke et al. 2010).

Setup and technique for a VCR start similar to that of a PSO. Similar to a PSO, prior to proceeding with the VCR, it is imperative to establish fixation above and below the level of correction, as the VCR will lead to destabilization of the spine. The pedicle is isolated and resected as described above. Deviating from a PSO, the authors next prepare the cranial and caudal disc spaces as one would do for a TLIF. Careful attention is paid to removing all disc material and cartilage on the inferior and superior end plates from the cranial and caudal levels, respectively. This will establish margins for resection required for a VCR and to place a cage. Subsequently, similar to a PSO, the lateral aspects of the vertebral body are accessed and protected, while the vertebral body is resected. The resection can be performed with an osteotome or a drill. Similar to a PSO, the posterior wall is resected last. Subsequently, a spacer is placed where the vertebral body was. In the lumbar spine, this can be challenging as the nerve roots block complete access to the vertebrae to be resected and to the space created during cage insertion. For this reason, we use expandable cages as they can be inserted in the interval between the nerve roots, rotated, and expanded. In the thoracic spine, the nerve roots can be resected allowing for easier access to the anterior aspect of the spine without significant neurologic repercussion. While VCRs do have the potential for significant deformity correction, they are also associated with substantial complications. In 2011 study by the Scoliosis Research Society, VCRs were associated with a complication rate of 61.1%. In contrast, they found a 28.1% complication rate in SPOs and 39.1% in PSOs (Smith et al. 2011). Suk et al. in the early to mid-2000s published several retrospective studies that detailed their preferred technique for a VCR and report outcomes. In their 2002 study, 70 patients underwent a VCR; an average correction of 61.9% in the coronal and 45.2% in the sagittal planes was achieved. Twenty-four of the 70 patients (34.2%) had a complication including 2 complete injuries to the spinal cord (Suk et al. 2002). In Suk's

2005 study, they performed 16 VCRs and achieved an average SVA correction from 4.2 to 1.6 cm. They had complications in 4 of the 16 patients (25%), including 1 involving complete paralysis (Suk et al. 2005a). These studies highlight that the potential deformity correction through a VCR comes at the cost of a technically challenging procedure with high rates of severe complications (Smith et al. 2011; Suk et al. 2002, 2005a, b).

Minimally Invasive Surgery

With technological and surgical advancements, interest has grown in minimally invasive surgery as a route to operatively address spinal deformity. Minimally invasive surgery can include use of interbody fusion through anterior and extreme lateral. A systematic review by Phan et al. in 2015 regarding direct lateral and extreme lateral interbody fusions (DLIF and XLIF) showed promise in correcting coronal deformity and regional lumbar lordosis (Phan et al. 2015). A retrospective review by Anand et al. suggests that MIS deformity correction has the potential for significant deformity correction, with less blood loss and morbidity compared to open procedures (Anand et al. 2010). Figure 5 shows correction achieved with placement of lateral retroperitoneal interbody placement in conjunction with posterior osteotomies and instrumentation.

Newer studies, however, have suggested the possibility of more substantial correction with hyperlordotic cages that can help to correct the global sagittal imbalance and improve lordosis (Gödde et al. 2003; Le et al. 2012). Additionally, the anterior longitudinal ligament resection is increasingly being appreciated as a method for additional correction. In particular, selective releases of the anterior longitudinal ligament through a minimally invasive retroperitoneal transpoas (lateral) approach can help to restore lumbar lordosis while minimizing the complex dissection and resection involved in the various posterior-based osteotomies (Deukmedjian et al. 2012a). In a 2012 cadaveric study, combination of a hyperlordotic cage and ALL releases led to an increase in 11.6° of segmental lordosis (Uribe et al. 2012). In a retrospective review of

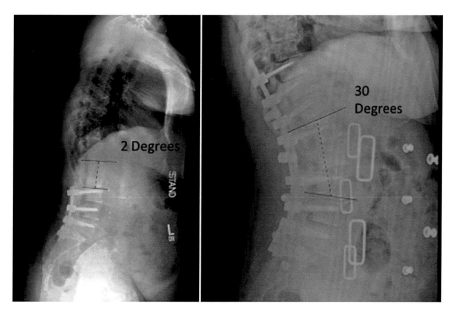

Fig. 5 A 63-year-old patient presented with global sagittal imbalance and stenosis at L2–L3. Initial radiographs on left demonstrate segmental lumbar lordosis measuring at 2° from L2 to L3. Radiographs on right demonstrate extension of fusion proximally to T10 with Smith-Petersen osteotomies at L1 and L2 with lateral retroperitoneal interbody placement at L2–L3. Segmental lumbar lordosis improved to 31°

prospectively collected data, Deukmedjian et al. assess ALL releases in patients with adult spinal deformity. In their study, they found an overall increase in lordosis of 24°, with segmental lumbar lordosis improving by 17° per level of ALL release (Deukmedjian et al. 2012b). In a cadaveric study and presentation of four clinical cases, Uribe et al. found an average increase of 10.2° per level of ALL released and 25° of overall global lumbar lordosis (Deukmedjian et al. 2012c). In a 2016 cadaveric biomechanical study by Hutton et al., they found a placement of 30° lordotic cage in addition to ALL release led to a 10.5° increase in segmental lumbar lordosis (Melikian et al. 2016). When combined with posterior facet resection and compression, one can achieve an even great degree of correction. While the individual correction values may vary in these studies, they do highlight the potential of ALL releases in deformity correction (Le et al. 2012).

Prior to performing an ALL resection, a surgeon must be comfortable performing a standard lateral interbody fusion. After prepping the disc space for the placement of an implant, soft tissue is dissected off the disc space along the anterior border of the spine. There should be a clean plane between the great vessels and the spine. If there is resistance to dissection, it is our recommendation that the ALL release should be abandoned. A retractor is then placed in front of the disc space across the anterior aspect of the spine. With a clear view of the anterior annulus and ALL, a special knife is used to cut the ALL. It is our preference to use an expanding trial to rupture any remaining fibers. We then place in a hyperlordotic implant with integrated fixation and secure it to one vertebral body in order to prevent anterior extrusion of the implant.

While MIS technology has advanced and provides a reasonable method for deformity correction in specific situations, careful patient selection and acknowledgment of MIS limitations are imperative. As in any spine case, extensive preoperative planning is critical in matching patient's diagnosis and pathology with appropriate treatment. The decision to pursue MIS, open deformity correction, or a combination of the two must match the intended degree of correction. Mummaneni et al. as part of the Minimally Invasive Section of the ISSG published an algorithm in 2014 that aimed to help in MIS and deformity decision-making (Mummaneni et al. 2014). The minimally invasive spinal deformity surgery (MISDEF) algorithm separates deformity correction into three different classes.

Class I is defined as patients with compressive symptoms relating to claudication or radiculopathy with minimal deformity. Furthermore, they use several parameters to define class I deformity: SVA less than 6 cm, PT less than 25°, LL-PI less than 10, lateral listhesis less than 6 mm, coronal Cobb angle less than 20°, and a flexible curve. They suggest that MIS techniques using decompression alone or with limited fusion are reasonable for class I deformity. Class II is defined as patients with previously mentioned compressive symptoms with a large component of back pain as well. Parameters for class II include lateral listhesis greater than 6 mm, coronal Cobb greater than 20°, and a LL-PI mismatch of 10–30°. For class II they recommend MIS surgery using decompression with multilevel interbody fusion that extends beyond just the apex of the curve (Mummaneni et al. 2014).

Class III patients are characterized by severe deformity in both coronal and sagittal imbalances. Parameters for this group include inflexible curves, SVA greater than 7 cm, LL-PI mismatch of greater than 30°, PT greater than 25°, and thoracic hyperkyphosis greater than 60. Class III patients are not readily amenable to MIS deformity correction and are better suited for traditional open deformity correction with osteotomies (as described in the previous sections). Mummaneni et al. tested the algorithm by having spine surgeons' complete surveys to classify various cases into the above classes and found MISDEF to have high intra- and interobserver reliability (Mummaneni et al. 2014).

While algorithms like the MISDEF provide a framework to understand treatment options for deformity correction, treatment must accommodate for the unique characteristics of the patient's deformity as well as the surgeon's comfort with various surgical techniques. Furthermore, MIS technology continues to advance, and patients that currently are treated with open corrective techniques may in the future be treated with MIS approaches.

Conclusions

Adult spinal deformity is complex deformity that involves three-dimensional deformation in coronal, sagittal, and axial planes. Spinal and spinopelvic parameters such as SVA, pelvic tilt, pelvic incidence, and lumbar lordosis are important in understanding, characterizing, and treating adult spinal deformity. Treatment of adult spinal deformity needs to be tailored to each patient with respect to the nature of the curve and the patients overall medical health. Operative techniques have changed substantially with time, from the early use of Harrington rods to modern pedicle screws. Multiple osteotomies (SPO, PSO, and VCR) can be applied for the desired level of spinal correction. Operative management of adult spinal deformity is wrought with complexity and severe complications. Newer techniques involving minimally invasive surgery and interbody fusions are being increasingly used for deformity correction.

References

Aebi M (2005) The adult scoliosis. Eur Spine J 14:925–948

Anand N, Rosemann R, Khalsa B, Baron EM (2010) Midterm to long-term clinical and functional outcomes of minimally invasive correction and fusion for adults with scoliosis. Neurosurg Focus 28(3):E6

Angevine PD, Kaiser MG (2008) Radiographic measurement techniques. Neurosurgery 63(suppl 3):40–45

Berjano P, Aebi M (2015) Pedicle subtraction osteotomies (PSO) in the lumbar spine for sagittal deformities. Eur Spine J 24(Suppl 1):S49–S57

Bernhardt M, Bridwell KH (1989) Segmental analysis of the sagittal plane alignment of the normal thoracic and lumbar spines and thoracolumbar junction. Spine 14(7):717–721

Berthonnaud E, Dimnet J, Roussouly P, Labelle H (2005) Analysis of the sagittal balance of the spine and pelvis using shape and orientation parameters. J Spinal Disord Tech 18(1):40–47

Berven S, Bradford DS (2002) Neuromuscular scoliosis: causes of deformity and principles for evaluation and management. Semin Neurol 22:167–178

Berven SH, Lowe T (2007) The Scoliosis Research Society classification for adult spinal deformity. Neurosurg Clin N Am 18(2):207–213

Bianco K, Norton R, Schwab F et al (2014) Complications and intercenter variability of three-column osteotomies for spinal deformity surgery: a retrospective review of 423 patients. Neurosurg Focus 36:E18

Birknes JK et al (2008) Adult degenerative scoliosis: a review. Neurosurgery 63(suppl 3):94–103

Boulay C et al (2006a) Sagittal alignment of spine and pelvis regulated by pelvic incidence: standard values and prediction of lordosis. Eur Spine J 15:415–422

Boulay C, Tardieu C, Hecquet J et al (2006b) Sagittal alignment of spine and pelvis regulated by pelvic incidence: standard values and prediction of lordosis. Eur Spine J 15(4):415–422

Bradford DS (1987) Vertebral column resection. Orthop Trans 11:502

Bridwell KH (2006) Decision making regarding Smith-Petersen vs. pedicle subtraction osteotomy vs. vertebral column resection for spinal deformity. Spine 31:S171–S178

Bridwell KH, Lewis SJ, Lenke LG et al (2003) Pedicle subtraction osteotomy for the treatment of fixed sagittal imbalance. J Bone Joint Surg Am 85-A:454–463

Bridwell KH, Glassman S, Horton W et al (2009) Does treatment (nonoperative and operative) improve the two-year quality of life in patients with adult symptomatic lumbar scoliosis: a prospective multicenter evidence-based medicine study. Spine 34(20):2171–2178

Burns CB, Dua K, Trasolini NA, Komatsu DE, Barsi JM (2016) Biomechanical comparison of spinopelvic fixation constructs: iliac screw versus S2-alar-iliac screw. Spine Deform 4(1):10–15

Carter O, Haynes S (1987) Prevalence rates for scoliosis in US adults: results from the first National Health and Nutrition Examination Survey. Int J Epidemiol 16:537–544

Chang KW, Dewei Z, McAfee PC et al (1988) A comparative biomechanical study of spinal fixation using the combination spinal rod-plate and transpedicular screw fixation system. J Spinal Disord 1(4):257–266

Chen IH, Chien JT, Yu TC (2001) Transpedicular wedge osteotomy for correction of thoracolumbar kyphosis in ankylosing spondylitis: experience with 78 patients. Spine 26:E354–E360

Cho KJ, Bridwell KH, Lenke LG et al (2005) Comparison of Smith-Petersen versus pedicle subtraction osteotomy for the correction of fixed sagittal imbalance. Spine 30:2030–2037

Cobb JR (1948) Outline for the study of scoliosis. In: Edwards JW, American Academy of Orthopaedic Surgeons (eds) Instructional course lectures. American Academy, Ann Arbor, pp 261–275

Cummins J, Lurie JD, Tosteson TD et al (2006) Descriptive epidemiology and prior healthcare utilization of patients in the Spine Patient Outcomes Research

Trial's (SPORT) three observational cohorts: disc herniation, spinal stenosis, and degenerative spondylolisthesis. Spine 31(7):806–814

Deukmedjian AR, Le TV, Baaj AA, Dakwar E, Smith DA, Uribe JS (2012a) Anterior longitudinal ligament release using the minimally invasive lateral retroperitoneal transpsoas approach: a cadaveric feasibility study and report of 4 clinical cases. J Neurosurg Spine 17(6):530–539

Deukmedjian AR, Dakwar E, Ahmadian A, Smith DA, Uribe JS (2012b) Early outcomes of minimally invasive anterior longitudinal ligament release for correction of sagittal imbalance in patients with adult spinal deformity. ScientificWorldJournal 2012:789698

Doherty J (1973) Complications of fusion in lumbar scoliosis. Proceedings of the Scoliosis Research Society. J Bone Joint Surg Am 55:438

Drummond DS (1988) Harrington instrumentation with spinous process wiring for idiopathic scoliosis. Orthop Clin North Am 19(2):281–289

Gaines RW (2000) The use of pedicle-screw internal fixation for the operative treatment of spinal disorders. J Bone Joint Surg Am 82-A(10):1458–1476

Gelalis ID, Paschos NK, Pakos EE et al (2012) Accuracy of pedicle screw placement: a systematic review of prospective in vivo studies comparing free hand, fluoroscopy guidance and navigation techniques. Eur Spine J 21(2):247–255

Glassman SD, Berven S, Bridwell K et al (2005a) Correlation of radiographic parameters and clinical symptoms in adult scoliosis. Spine 30:682–688

Glassman SD, Bridwell K, Dimar JR et al (2005b) The impact of positive sagittal balance in adult spinal deformity. Spine 30:2024–2029

Glassman SD, Carreon LY, Shaffrey CI et al (2010) The costs and benefits of nonoperative management for adult scoliosis. Spine 35(5):578–582

Gödde S, Fritsch E, Dienst M, Kohn D (2003) Influence of cage geometry on sagittal alignment in instrumented posterior lumbar interbody fusion. Spine 28(15):1693–1699

Grobler L, Moe J, Winter R et al (1978) Loss of lumbar lordosis following surgical correction of thoracolumar deformities. Orthop Trans 2:239

Hackenberg L, Link T, Liljenqvist U (2002) Axial and tangential fixation strength of pedicle screws versus hooks in the thoracic spine in relation to bone mineral density. Spine 27(9):937–942

Hamill CL, Lenke LG, Bridwell KH, Chapman MP, Blanke K, Baldus C (1996) The use of pedicle screw fixation to improve correction in the lumbar spine of patients with idiopathic scoliosis. Is it warranted? Spine 21(10):1241–1249

Harrington PR (1972) Technical details in relation to the successful use of instrumentation in scoliosis. Orthop Clin North Am 3:49–67

Harrington PR (1973) The history and development of Harrington instrumentation. Clin Orthop Relat Res 93:110–112

Jackson RP, McManus AC (1994) Radiographic analysis of sagittal plane alignment and balance in standing volunteers and patients with low back pain matched for age, sex, and size: a prospective controlled clinical study. Spine (Phila Pa 1976) 19:1611–1618

Kebaish KM (2010) Sacropelvic fixation: techniques and complications. Spine 35(25):2245–2251

Kelleher MO, Timlin M, Persaud O, Rampersaud YR (2010) Success and failure of minimally invasive decompression for focal lumbar spinal stenosis in patients with and without deformity. Spine 35: E981–E987

Kelly MP, Lenke LG, Shaffrey CI et al (2014) Evaluation of complications and neurological deficits with three-column spine reconstructions for complex spinal deformity: a retrospective Scoli-RISK-1 study. Neurosurg Focus 36:E17

Kim YJ, Bridwell KH, Lenke LG et al (2008) Proximal junctional kyphosis in adult spinal deformity after segmental posterior spinal instrumentation and fusion: minimum five-year follow-up. Spine 33:2179–2184

Kim HJ, Bridwell KH, Lenke LG et al (2013) Proximal junctional kyphosis results in inferior SRS pain subscores in adult deformity patients. Spine (Phila Pa 1976) 38:896–901

Kim HJ, Boachie-adjei O, Shaffrey CI et al (2014) Upper thoracic versus lower thoracic upper instrumented vertebrae endpoints have similar outcomes and complications in adult scoliosis. Spine 39(13):E795–E799

Kostuik JP, Israel J, Hall JE (1973) Scoliosis surgery in adults. Clin Orthop Relat Res 93:225–234

Kuklo TR, Bridwell KH, Lewis SJ et al (2001) Minimum 2-year analysis of sacropelvic fixation and L5-S1 fusion using S1 and iliac screws. Spine 26(18):1976–1983

Lafage V et al (2008) Standing balance and sagittal plane spinal deformity: analysis of spinopelvic and gravity line parameters. Spine 33:1572–1578

Lafage V, Schwab F, Patel A, Hawkinson N, Farcy J (2009) Pelvic tilt and truncal inclination: two key radiographic parameters in the setting of adults with spinal deformity. Spine 34:E599–E606

Lazennec JY, Ramare S, Arafati N et al (2009) Sagittal alignment in lumbosacral fusion: relations between radiological parameters and pain. Eur Spine J 9:47–55

Le TV, Vivas AC, Dakwar E, Baaj AA, Uribe JS (2012) The effect of the retroperitoneal transpsoas minimally invasive lateral interbody fusion on segmental and regional lumbar lordosis. Sci World J 2012:516706

Legaye J, Duval-Beaupere G, Hecquet J et al (1998) Pelvic incidence: a fundamental pelvic parameter for three-dimensional regulation of spinal sagittal curves. Eur Spine J 7:99–103

Lenke LG et al (2001) Adolescent idiopathic scoliosis: a new classification to determine extent of spinal arthrodesis. J Bone Joint Surg Am 83-A:1169–1181

Lenke LG, Sides BA, Koester LA et al (2010) Vertebral column resection for the treatment of severe spinal deformity. Clin Orthop Relat Res 468:687–699

Liljenqvist U, Hackenberg L, Link T, Halm H (2001) Pullout strength of pedicle screws versus pedicle and laminar hooks in the thoracic spine. Acta Orthop Belg 67(2):157–163

Lowe T et al (2006) The SRS classification for adult spinal deformity: building on the King/Moe and Lenke classification systems. Spine (Phila Pa 1976) 31(suppl 19): S119–S125

Luque ER (1982) Segmental spinal instrumentation for correction of scoliosis. Clin Orthop Relat Res 163:192–198

Marchesi DG, Aebi M (1992) Pedicle fixation devices in the treatment of adult lumbar scoliosis. Spine 17(8 Suppl):S304–S309

McCord DH, Cunningham BH, Shondy Y, Myers J, McAffee PC (1992) Biomechanical analysis of lumbosacral fixation. Spine 17:S235–S243

Mehta SS et al (2009) Interobserver and intraobserver reliability of Cobb angle measurement: endplate versus pedicle as bony landmarks for measurement: a statistical analysis. J Pediatr Orthop 29:749–754

Melikian R, Yoon ST, Kim JY, Park KY, Yoon C, Hutton W (2016) Sagittal plane correction using the lateral transpsoas approach: a biomechanical study on the effect of cage angle and surgical technique on segmental lordosis. Spine 41(17):E1016–E1021

Miller CA, Ledonio CG, Hunt MA, Siddiq F, Polly DW (2016) Reliability of the planned pedicle screw trajectory versus the actual pedicle screw trajectory using intra-operative 3D CT and image guidance. Int J Spine Surg 10:38

Morrissy RT et al (1990) Measurement of the Cobb angle on radiographs of patients who have scoliosis: evaluation of intrinsic error. J Bone Joint Surg Am 72:320–327

Mummaneni PV, Shaffrey CI, Lenke LG et al (2014) The minimally invasive spinal deformity surgery algorithm: a reproducible rational framework for decision making in minimally invasive spinal deformity surgery. Neurosurg Focus 36(5):E6

O'Brien MF, Kuklo TR, Blanke KM et al (2004) Radiographic measurement manual. Medtronic Sofamor Danek, Memphis, pp 47–108

Phan K, Rao PJ, Scherman DB, Dandie G, Mobbs RJ (2015) Lateral lumbar interbody fusion for sagittal balance correction and spinal deformity. J Clin Neurosci 22(11):1714–1721

Ponte A, Orlando G, Siccardi GL (2018) The true ponte osteotomy: by the one who developed it. Spine Deform 6(1):2–11

Potter BK, Lenke LG, Kuklo TR (2004) Prevention and management of iatrogenic flatback deformity. J Bone Joint Surg Am 86-A(8):1793–1808

Rohlmann A, Richter M, Zander T, et al (2006) Effect of different surgical strategies on screw forces after correction of scoliosis with a VDS implant. Eur Spine J 15(4):457–464

Roussouly P, Nnadi C (2010) Sagittal plane deformity: an overview of interpretation and management. Eur Spine J 19(11):1824–1836

Ryan D et al (2014) T1 pelvic angle (TPA) effectively evaluates sagittal deformity and assesses radiographical surgical outcomes longitudinally. Spine 39(15):1203–1210

Schwab F et al (2005) Adult scoliosis: prevalence, SF-36, and nutritional parameters in an elderly volunteer population. Spine (Phila Pa 1976) 30:1082–1085

Schwab F, Farcy JP, Bridwell K et al (2006a) A clinical impact classification of scoliosis in the adult. Spine 31:2109–2114

Schwab F et al (2006b) A clinical impact classification of scoliosis in the adult. Spine (Phila Pa 1976) 31:2109–2114

Schwab F et al (2009) Sagittal plane considerations and the pelvis in the adult patient. Spine (Phila Pa 1976) 34:828–1833

Schwab F, Patel A, Ungar B, Farcy JP, Lafage V (2010) Adult spinal deformity – postoperative standing imbalance: how much can you tolerate? An overview of key parameters in assessing alignment and planning corrective surgery. Spine 35(25):2224–2231

Schwab F et al (2013) Radiographical spinopelvic parameters and disability in the setting of adult spinal deformity. Spine 38(13):E803–E812

Schwab F, Blondel B, Chay E et al (2015) The comprehensive anatomical spinal osteotomy classification. Neurosurgery 76(Suppl 1):S33–S41

Shufflebarger H, Suk SI, Mardjetko S (2006) Debate: determining the upper instrumented vertebra in the management of adult degenerative scoliosis: stopping at T10 versus L1. Spine 31(19 Suppl):S185–S194

Smith J, Shaffrey C, Berven S et al (2009a) Operative vs. nonoperative treatment of leg pain in adults with scoliosis: a retrospective review of a prospective multicenter database with two-year follow-up. Spine 34(16):1693–1698

Smith J, Shaffrey C, Berven S et al (2009b) Improvement of back pain with operative and non-operative treatment in adults with scoliosis. Neurosurgery 65(1):86–93

Smith JS et al (2009c) Operative versus nonoperative treatment of leg pain in adults with scoliosis: a retrospective review of a prospective multicenter database with two-year follow-up. Spine (Phila Pa 1976) 34:1693–1698

Smith JS, Sansur CA, Donaldson WF et al (2011) Short-term morbidity and mortality associated with correction of thoracolumbar fixed sagittal plane deformity: a report from the Scoliosis Research Society Morbidity and Mortality Committee. Spine 36(12):958–964

Smith-Petersen MH, Larson CB, Aufranc OE (1945) Osteotomy of the spine for the correction of flexion deformity in rheumatoid arthritis. J Bone Joint Surg Am 27:1–11

Stokes IA (1994) Three-dimensional terminology of spinal deformity: a report presented to the Scoliosis Research Society by the Scoliosis Research Society Working Group on 3-D terminology of spinal deformity. Spine (Phila Pa 1976) 19:236–248

Suk SI, Lee CK, Min HJ, Cho KH, Oh JH (1994) Comparison of Cotrel-Dubousset pedicle screws and hooks in the treatment of idiopathic scoliosis. Int Orthop 18(6):341–346

Suk SI, Lee CK, Kim WJ, Chung YJ, Park YB (1995) Segmental pedicle screw fixation in the treatment of thoracic idiopathic scoliosis. Spine 20(12):1399–1405

Suk SI, Kim JH, Kim WJ, Lee SM, Chung ER, Nah KH (2002) Posterior vertebral column resection for severe spinal deformity. Spine 27:2374–2382

Suk SI, Chung ER, Kim JH, Kim SS, Lee JS, Choi WK (2005a) Posterior vertebral column resection for severe rigid scoliosis. Spine 30:1682–1687

Suk SI, Chung ER, Lee SM, Lee JH, Kim SS, Kim JH (2005b) Posterior vertebral column resection in fixed lumbosacral deformity. Spine 30:E703–E710

Thomsen K, Christensen FB, Eiskjaer SP et al (1997) The effect of pedicle screw instrumentation on functional outcome and fusion rates in posterolateral lumbar spinal fusion: a prospective, randomized clinical study. Spine 22(24):2813–2822

Uribe JS, Smith DA, Dakwar E et al (2012) Lordosis restoration after anterior longitudinal ligament release and placement of lateral hyperlordotic interbody cages during the minimally invasive lateral transpsoas approach: a radiographic study in cadavers. J Neurosurg Spine 17(5):476–485

West JL, Bradford DS, Ogilvie JW (1991) Results of spinal arthrodesis with pedicle screw-plate fixation. J Bone Joint Surg Am 73(8):1179–1184

Zdeblick TA, Becker PS, McAfee PC et al (1991) Neuropathologic changes with experimental spinal instrumentation: transpedicular versus sublaminar fixation. J Spinal Disord 4(2):221–228

SI Joint Fixation

34

J. Loewenstein, W. Northam, D. Bhowmick, and E. Hadar

Contents

Introduction	676
Anatomy	676
Biomechanics	677
Etiology of Sacroiliac Joint Pathology	678
Diagnosis and Evaluation	678
Conservative Management Strategies	679
Surgical Decision-Making	679
Instrumented Surgical Options	680
Open Surgical Approaches	680
Minimally Invasive Approaches	682
Minimally Invasive Lateral and Posterolateral Iliosacral Trans-articular Approach	682
Minimally Invasive Posterior Intra-articular Approach	682
Instrumentation Options	683
Evidence Supporting Different Surgical Techniques	683
References	685

Abstract

Surgical fixation of the sacroiliac joint (SIJ) has increased in popularity over the last few decades, especially with the recent emergence of minimally invasive techniques. The indications for this procedure are expanding and include joint dysfunction, degeneration/arthrosis, trauma, and postpartum instability, among others. With rising frequency of lumbosacral arthrodesis, interest has developed regarding the SIJ as a pain generator due to accelerated degeneration/dysfunction as an "adjacent segment" receiving more force distribution. The current body of literature suggests that a targeted history and physical examination

J. Loewenstein · W. Northam · D. Bhowmick · E. Hadar (✉)
Department of Neurosurgery, University of North Carolina, Chapel Hill, NC, USA
e-mail: Joshua.Loewenstein@unchealth.unc.edu; Weston.Northam@unchealth.unc.edu; deb_bhowmick@med.unc.edu; eldad_hadar@med.unc.edu

© Springer Nature Switzerland AG 2021
B. C. Cheng (ed.), *Handbook of Spine Technology*,
https://doi.org/10.1007/978-3-319-44424-6_65

specific to the SIJ, and provocative testing, are paramount for appropriate patient selection. A thorough understanding of the anatomy and biomechanics involving the SIJ is essential to forming a critical review of the various surgical approaches and hardware instrumentation options as they become available. The SIJ anatomy can be approached from several open corridors including ventral-ilioinguinal, posterolateral iliosacral, and posterior sacral-alar-iliac approach. Similarly, minimally invasive approaches have been developed using posterolateral iliosacral trans-articular and posterior intra-articular techniques. Multiple hardware options are available for SIJ fixation and continue to grow, including screw-plate and screw-rod constructs, trans-articular fusion rods, intra-articular cages, trans-articular threaded screws, and hollow modular anchoring screws. The epidemic nature of low back pain will likely lead to an expanding interest in SIJ fixation, and critical appraisal of the cost and efficacy of hardware and techniques will warrant greater study.

developed, with both open and minimally invasive approaches. As is the case with many spinal technologies, several different implant materials and styles have been trialed, including titanium fixation constructs and biologic materials.

As understanding of SIJ pathology grows, the clinical conditions for which surgical fixation is being utilized are also expanding. These conditions include degeneration/arthrosis, joint dysfunction, postpartum instability, trauma, pathologic fractures, and inflammatory arthropathies among others (Zaidi et al. 2015). Because imaging techniques have not demonstrated sufficient diagnostic value to determine which patients will see benefit from surgical fixation, the development of consensus over history and physical techniques, as well as provocative testing, is paramount (Elgafy et al. 2001; Dreyfuss et al. 2004). Further study of implantation techniques and materials will need to address postoperative complication profiles, rates of bony fusion, implant cost, and natural history of SIJ pathology.

Keywords

Sacroiliac joint · Fixation · Instrumentation · Hardware · Techniques

Introduction

Low back pain (LBP) is a growing global problem and a common cause of disability and lost work days (Freburger et al. 2009). It is recognized as a major driver for morbidity across the economic spectrum from low- to high-income countries (Hoy et al. 2010). Although a wide range of etiologies exists, recent literature attention has focused on the sacroiliac joint (SIJ) as an important contributor to low back pain (Zaidi et al. 2015). The SIJ may be implicated as the pain generator in as many as 30% of patients with low back pain according to recent studies (Bernard and Kirkaldy-Willis 1987; Cohen et al. 2013). As such, a variety of conservative and surgical treatment modalities are being

Anatomy

The pelvic girdle constitutes a support structure that distributes force vectors from the spine as well as the legs. As the junction between the sacrum and the remainder of the bony pelvis, the SIJs have been conceptualized as "stress relievers" between the lower extremities and the trunk (Vleeming et al. 2012). Although there is considerable anatomic variability among individuals and sexes, the SIJ usually spans the majority of S1, S2, and S3 sacral levels. The joint can be conceptualized as diarthrodial, with hyaline and fibrocartilage, and a relatively irregular articulating surface (Forst et al. 2006). Because there are synarthrotic components, the joint has also been referred to as amphiarthrodial (Vleeming et al. 2012). However, there is a relative paucity of movement across the joint under normal circumstances, with generally less than 1 mm of transverse/sagittal translation and vertical movements usually less than 2 mm (Walheim et al. 1984). The articulating surface visualized en face is roughly C-shaped, and the superior portion is predominately fibrous, whereas the inferior portion is mostly synovial (Cole et al. 1996). The overall joint orientation is arranged such that the vertical forces from gravity can be resisted (Vleeming et al. 2012). There does appear to be sexual dimorphism, with articular surface area ranging up to 18 sq. cm for females and 22.3 sq. cm for males (Sashin 1930; Miller et al. 1987).

The functional integrity of the SIJ is closely supported by several investing ligaments, including interosseous, ventral, and dorsal locations. The interosseous ligaments are also connected to the sacroiliac transverse ligaments. Ventrally, the ligamentous attachments of the SIJ form a connective tissue plane that is relatively thin and vulnerable to injury. Dorsally, the ligamentous anatomy supporting the SIJ is multilayered and more complex and includes dorsal sacroiliac ligaments categorized as long and short. These ligaments predominantly course from the crests of the sacrum to the posterior superior iliac spine. Stability across the SIJ is due in part to muscular action as well. These include the gluteus maximus, erector spinae, and multifidi, among others (Vleeming et al. 2012).

Innervation of the SIJ has been studied in both human and animal models, in an effort to help elucidate the origin of pain attributed to the joint itself. This supply appears to derive from both the ventral (L4 and L5) and dorsal rami (L5, S1, S2) and the superior gluteal nerve (Nakagawa 1966). However, subsequent analyses have determined that the majority of the supply may originate from the dorsal aspect (Forst et al. 2006). Murine neural tracer was applied to study this question in more detail and found dorsal root ganglion supply in the SIJ from the L1 to S2 levels primarily, with L1–3 innervating the ventral aspect and L4–S2 innervating the dorsal aspect. On the ventral aspect, there was innervation also emerging from the sympathetic trunk (Murata et al. 2001). Within the SIJ itself, nerve fibers were observed in human dissection that were both myelinated and unmyelinated (Grob et al. 1995), and this is consistent at least for the outer margins of the joint (Vleeming et al. 2012).

Biomechanics

An understanding of the biomechanical principles underlying the SIJ is critical to the design and implementation of instrumentation constructs and arthrodesis. Early twentieth-century literature had already established that there was a small amount of movement across the joint and that this tended to abate after approximately the fifth decade (Sashin 1930). Cadaveric analysis reveals that the adult SIJ orientation at the level of S1–2 is obliquely in the anterior-posterior direction with 20 degrees of offset from the vertical plane. Force testing showed that the bilateral natural joint construction most resisted medio-lateral vectors, with progressively more motion resulting from superior/inferior and then anterior-posterior vectors. When one joint was isolated, anterior shearing and torsion were seen to cause larger degrees of motion (Miller et al. 1987).

The largest degree of functional movement imparted by the SIJ was determined to occur with iliac rotation relative to the sacrum, on a transverse axis. This is called nutation in the forward direction and counternutation backward. When load is placed across the SIJs, when sitting or standing, the movement of nutation is seen. The degree of irregularity and surface characteristics of the articulating surfaces of the SIJ make it unique among similar human joints; it has a higher coefficient of friction than any other diarthrodial joint, which helps to resist shearing (Vleeming et al. 2012). The biomechanical properties of the SIJ differentiate it from nearby spinal segments; compared to the lumbar spine, the SIJ is more likely to fail under axial compression and axial torsion (Forst et al. 2006). In order to study motion about the SIJ, the radiostereometric analysis method has been validated and utilized (Kibsgård et al. 2012). Applied to human volunteers, this analysis has confirmed a very small degree of motion across the SIJs, approximately 0.5 degrees (Sturesson et al. 2000), which is compared to prior work showing mean rotation of 2.5 degrees, and no significant difference in motion between symptomatic and asymptomatic joints (Sturesson et al. 1989).

Etiology of Sacroiliac Joint Pathology

Over the lifespan, the human SIJ undergoes an expected degenerative process. Starting during adolescence, the joint surface is reported to become rougher with plaque formation. By the fifth decade, osteophyte formation was common along with corresponding articular surface irregularity. These osteophytes often were interdigitating across the joint by the seventh decade along with thinning of the articular cartilage (Bowen and Cassidy 1981). Aside from natural history of the joint through the aging process, a variety of pathologies can affect the SIJ to cause symptoms. These can include infection, arthritis, fracture and ligamentous injury, malignancy hypo−/hypermobility, chondromalacia, enthesitis, leg length or gait asymmetry, and scoliosis (Cohen et al. 2013).

Recent interest has surrounded SIJ pain that arises in the context of lumbosacral long-segment spinal fusion procedures. After a successful lumbosacral fusion, the distribution of motion across the SIJ increases, which can precipitate accelerated degeneration of the joint, as in the pathology of adjacent segment disease. In one prospective cohort study of lumbosacral fusion patients over 5 years of follow-up, the incidence of SIJ degeneration was 75% as determined by CT imaging. Patients were found to have degeneration regardless of the number of levels fused, and it was found that usage of iliac crest graft also had a deleterious effect on the SIJ (Ha et al. 2008). Possible causes of SIJ pathology after lumbosacral fusion is thought to be related to either adjacent segment disease as mentioned above, harvesting of bone graft in close proximity to the joint, or possibly misdiagnosis of a pre-existing SIJ syndrome. The study of the SIJ as a possible generator for LBP after long lumbosacral fusions bears considerable importance especially given the failure rate of these procedures and prevalence of LBP in this population of patients (Yoshihara 2012).

Diagnosis and Evaluation

Much of the difficulty associated with the treatment of pain originating from the SIJ derives from the variable clinical presentations that can arise as a result of this pathology. Regions of reported pain referral can include the lower back, buttocks, groin, lower extremities, and even the abdomen. However, provocative joint injections have indicated that the most common referral zone is the buttock, followed closely by lower lumbar, and patients with lower extremity pain usually localize to the posterior or lateral thigh (Slipman et al. 2000). Of the buttock region, the posterior superior iliac spine (PSIS) appears to be a common anatomic region identified by patients with SIJ pathology (Maigne et al. 1996). Further complicating the diagnosis of SIJ pain are the proximity of several other anatomic regions commonly implicated in pain from chronic degenerative disease in adults (the lumbar spine and hips, specifically) and the possibility for pain originating from these different entities concurrently.

Although many physical examination techniques have been developed for the purpose of evaluating the SIJ, none have shown sufficient sensitivity/specificity for standalone usage without more invasive testing (injection, etc.). Furthermore, the usage of multiple physical examination maneuvers did not augment the diagnostic power when compared to injection, including Gaenslen's Test, Patrick's Test, and tenderness of the sacral sulcus (just medial to the PSIS), among others. The highest sensitivity was seen with sacral sulcus tenderness (89%). Of note, there was also no consistent statistical validation for historical features such as relief when standing, sitting, walking, lying down, or aggravation with bowel movement or coughing (Dreyfuss et al. 1996). Because of the possibility of hip or lumbar spine pathology confounding diagnosis of SIJ-related pain, physical evaluation should include routine neurologic evaluation for weakness and radiculopathy, as well as provocative hip joint maneuvers. Of note, gait and leg length discrepancy are also important factors to address during the workup (Thawrani et al. 2018).

The usage of imaging (CT, MRI, bone scan, etc.) in the workup of SIJ pain has been shown to be largely unhelpful in determining whether the joint itself is likely to be the primary pain generator. However, imaging studies can be utilized to rule out other causes of pain such as the detection of fracture, neoplasm, infection, or spondyloarthropathy (Dreyfuss et al. 2004). In one retrospective review of CT imaging findings relative to patients with injection-proven SIJ provocation, the sensitivity and specificity of CT imaging were found to be only 57.5% and 69%, respectively (Elgafy et al. 2001). Plain film imaging has also been difficult to apply to SIJ evaluation given the natural history that 24.5% of patients over 50 years of age show abnormalities on these studies (Dreyfuss et al. 1995). Although radionuclide bone scanning has received some attention in the literature as it relates to SIJ pain, it is not recommended as part of the basic workup due to the relatively low sensitivity (12–46%, Thawrani et al. 2018).

After sufficient clinical suspicion for SIJ-related pain has arisen, a percutaneous SIJ block is a generally agreed-upon test to establish this diagnosis. This is best performed using contrast media and fluoroscopic guidance, with an effort to avoid over-injecting the joint space and accidentally seeing false results due to anesthetizing the nearby lumbosacral nervous anatomy (especially ventrally). Additionally, >75% pain relief is the expected standard for diagnosis and is sometimes followed by a repeat injection block later due to placebo effect (Dreyfuss et al. 2004). The false positive rate from a single SIJ block injection is approximately 20% (Hansen et al. 2007). The usage of steroid injection into and around the joint and RF ablation technologies have been investigated and may be promising options for durable pain relief but require more establishment by the literature (Cohen et al. 2013).

Conservative Management Strategies

Targeted efforts to address the particular pathology affecting the SIJ are utilized first, which may include physical therapy or orthotic options for imbalances with gait mechanics or leg length discrepancy. Strength and flexibility training would fall under this category and can be beneficial. Similarly, trials of medications can be undertaken, including non-steroidal anti-inflammatory medications, non-opioid pain medications, and others, which may be especially efficacious in the case of inflammatory SIJ disease (Dreyfuss et al. 2004).

Belt orthoses have also been trialed to relieve SIJ-related pain, as has been the case with SIJ dysfunction in the peripartum period. These belts, when worn above the greater trochanter, have shown approximately 30% reduction in joint motion, but should be weaned when able to reduce muscular weakening and dependence (Vleeming et al. 1992). Manual joint manipulation therapy has not yet been substantiated in the literature but has shown a potential benefit for certain patients (Kirkaldy-Willis and Cassidy 1985). Lastly, in addition to the aforementioned injection strategies for neurologic blockade, steroid use, and radiofrequency ablation, percutaneous viscosupplementation is also being explored (Dreyfuss et al. 2004).

Surgical Decision-Making

Surgical fixation of the sacroiliac joint was traditionally only considered in situations where joint instability was known such as fracture and/or severe ligamentous disruption from trauma and infection. However, in the mid-1980s, surgeons began considering fixation and arthrodesis techniques using a variety of instrumentation techniques for the treatment of refractory SI joint degeneration/dysfunction (Rand 1985; Smith et al. 2013), considered to be stable SI joint pathologies. Surgical treatment in these patients, however, is still largely thought of as a treatment option of last resort for patients whose symptoms have been unresponsive to all other non-surgical options. In addition to diagnosis with at least two positive SI joint injections, most surgeons require a course of non-surgical treatment lasting at least 6 months in duration. Discussions of risk in SI joint surgery should include explanation of all possible complications. These include neurovascular injury, hemorrhage requiring blood transfusion, superficial and/or deep infection, pulmonary embolism provoked by postoperative weight bearing status, refractory lower back and SI joint pain, non-union requiring surgical revision, etc.

The earliest reports supporting sacroiliac arthrodesis for non-traumatic SI joint dysfunction were published in the 1920s (Smith-Petersen and

Rogers 1926; Gaenslen 1927). However, the high level of complications, long periods of non-weight bearing, and unreliable rates of bony arthrodesis kept the technique from gaining widespread acceptance as a treatment for SI joint dysfunction. The emergence of SI joint arthrodesis garnered renewed interest over the past two decades as the creation of pain clinics sparked increasing numbers of diagnostic and therapeutic SI joint injections. Additionally, increased utilization of lumbar and lumbosacral arthrodesis procedures is thought to have brought greater attention to the SI joint as a generator of low back pain.

The modern era of SI joint arthrodesis began in the late 1980s with publication of a report by Waisbrod in which he reported a series of 21 surgical procedures performed for stable SI joint arthritis and lower back pain. He utilized an open posterior approach with intra-articular ceramic blocks and local autograft from the iliac crest, harvested during the approach. Postoperatively patients were maintained in short-leg spica casts for 8 weeks. The series achieved satisfactory results (defined by reduction of pain of greater than 50%) in 11 of 21 cases, all of which demonstrated bony arthrodesis on follow-up radiographs (Waisbrod et al. 1987). These early reports served as a demonstration that surgical instrumentation and fusion of the SI joint could be safely and effectively employed as a modality for treatment of the dysfunctional SI joint.

Although surgical fusion is now an accepted therapy for refractory lower back pain caused by the dysfunctional SI joint, many factors still need to be considered before surgery should be offered. CT imaging of the SI joints should be evaluated preoperatively to alert the surgeon to any bony sacroiliac anomalies that can affect the surgeon's ability to instrument across the joint or enter the joint space. Further, the presence of osteoporosis as a comorbid condition for patients with SI joint dysfunction may influence the decision as to whether surgical intervention is used and, if so, which approach will provide the chosen instrumentation with the greatest purchase to cortical bony surfaces. Similarly, patients with morbid obesity carry higher risk of complications when undergoing surgery for SI joint fusion. As such the surgeon will often benefit from choosing an approach that favors less traditional soft tissue dissection and favors fluoroscopy in the AP as opposed to lateral projection to avoid distortion from excess soft tissue. This will often be a posterior or posterolateral, minimally invasive approach, but surgeons should continue to evaluate these factors on a case-by-case basis. Finally, preoperative counseling on the postoperative course that can include non-weight bearing status and extensive physical therapy and rehabilitation will help establish expectations and can improve patient satisfaction and outcomes.

Instrumented Surgical Options

Open Surgical Approaches

Open Ventral-Ilioinguinal Approach

The ventral approach facilitates access to the iliac crest, entire iliac fossa, lateral sacral ala, and consequently the anterior and superior portions of the sacroiliac joint. The patient is positioned supine, ideally on a radiolucent OR table. An ilioinguinal incision is planned just inferior to the iliac crest and is taken deep through the subcutaneous tissues until the fascia overlying the external oblique muscle is encountered. This fascia is followed until the gluteus fascia is visualized. The border between these muscles is then identified and the interval is developed. The iliac crest is then identified and the external oblique muscle is elevated from the iliac crest in a subperiosteal fashion. The iliacus muscle is then identified and elevated from the iliac fossa in the same periosteal layer. Working anteromedially along the iliac fossa, the anterior sacroiliac joint capsule will be identified. Hohmann retractors can be utilized to maintain visualization of the joint capsule, with attention being paid to avoid injury of the traversing L5 nerve and superior gluteal artery and nerve inferomedially near the sciatic notch. The sacroiliac joint capsule is then incised sharply to visualize the joint. Joint cartilage is then resected using a combination of curettes and rongeurs. Morselized bone graft and other biomaterials are packed in the joint, and instrumented reconstruction with screw-plate interface is most often used.

Open Posterolateral Iliosacral Approach

The posterolateral iliosacral approach facilitates access to the posteromedial portion of the iliac crest including the posterior superior iliac spine (PSIS), posterior surface of the sacral ala, and the posterior sacroiliac joint. The patient is positioned in the prone position. A longitudinal, linear, or curvilinear incision is planned just lateral to the PSIS, following the iliac crest. The soft tissues are divided, exposing the bone of the PSIS and iliac crest. The gluteal muscles are then elevated from the posteromedial ilium in the subperiosteal plane and reflected laterally to expose an adequate portion of bone, with care being taken to avoid injury to the superior gluteal neurovascular bundle. In the same subperiosteal plane, the multifidi muscles are elevated from the sacrum and reflected medially until adequate exposure of the posterior sacral surface is exposed. Retractors are placed to maintain bony exposure. Bone of the iliac crest and PSIS that overhang the posterior sacroiliac joint are resected using osteotomes or rongeurs. Bone is morselized for use as autograft. The sacroiliac joint capsule is then incised exposing the articular portion of the joint. Cartilage in the articular faces is then resected using a combination of curettes and rongeurs. Morselized bone graft and other biomaterials and/or synthetic cages are packed in the joint, and instrumented fixation is performed using screw-plate constructs, screw-rod constructs, or trans-articular screws.

Open Posterior Sacral-Alar-Iliac Approach

The posterior sacral approach facilitates access to the bilateral posterior sacral surfaces, bilateral posterior SI joints, and potentially the most medial portion of the adjacent ilium. The patient is positioned prone. A longitudinal, linear midline incision is made from the L3 spinous process down past the S2 spinous process and taken down to the deep lumbosacral fascia. Care is taken not to violate the deep layer of the fascia in the midline. This layer can be defined from the superficial layer, as the deep layer runs in a longitudinal fashion versus the oblique fibers of the superficial layer. Splitting the layers of fascia with electrocautery or a scalpel, the deep fascial layer is followed to its lateral attachment along the medial border of the ilium. Between the fascial layers, the PSIS and approximately 8–10 cm of the dorsal iliac wing can be palpated. Care should be taken to avoid damage cluneal nerves at the cephalad and lateral portions of the dissection. The deep fascial layer is then incised over the PSIS until bone is encountered and dissection occurs medially in a subperiosteal plane, exposing the posterior transverse iliosacral ligaments. These can be removed using a combination of rongeurs. The joint can then be accessed moving anterolaterally. The joint can be prepared by remove articular cartilage and bone with a combination of angled curettes, rongeurs, and a high-speed drill. An intra-articular cage and/or bone dowel is then sized and placed, often using fluoroscopy to confirm proper placement. The paraspinal musculature is then retracted medially at the cephalad portion of the exposure, and an S1 pedicle screw is placed. A trajectory into the ilium is then chosen using fluoroscopic guidance, CT-guided navigation, or freehand technique, and the hole is created using a high-speed drill or awl. An iliac "pedicle" screw of choice is then placed. The remainder of the visible sacroiliac joint can then be decorticated and residual morselized bone graft or orthobiologic material laid within the joint. A rod is then cut to size and secured with S1 and iliac screw heads, and a compression can be performed before the set screws are tightened.

An alternative method can also be utilized using the posterior midline incision. With this approach, the incision is taken to bone in the midline avascular plane exposing the L4, L5, and possibly L3 spinous process and the posterior-most portions of the sacral spinous processes. Elevation of the lumbosacral musculature from medial to lateral is performed until the sacral foramina are visualized. Continued dissection and elevation of muscles may be needed if visualization of the posterior portion of SI joint is desired. After achieving visualization of the S1 and S2 dorsal foramina, placement of sacral-alar-iliac screws can be performed. Starting point for placement of this screw should be approximately half between the lateral edges of the S1 and S2 dorsal foramina. Trajectory of screw should be toward the anterior inferior iliac spine which can be approximated by palpating the top of the greater trochanter. Using freehand, fluoroscopy-guided, or intraoperative CT-guided techniques, fixation screws that traverse the sacral ala into the ilium can be safely placed. This technique is most commonly used in conjunction with lumbosacral arthrodesis procedures. The benefit of the sacral-alar-iliac screw is better alignment of the screw head with proximal S1 and lumbar pedicle screws for alignment of the fixation rod.

Minimally Invasive Approaches

Over the past 15 years, the popularity of minimally invasive approaches to SI joint arthrodesis has increased rapidly. These techniques aim to achieve joint fusion with decreased morbidity compared to open procedures. They emphasize smaller skin incisions, less traditional dissection using anatomic landmarks, and greater dissection using sequential dilation and tubular retractors. Additionally, these techniques have a far greater reliance on image guidance, both fluoroscopic guidance and CT-guided navigation. According to the International Society for Advancement of Spinal Surgery, by 2012, approximately 90% of all sacroiliac joint fusion were being performed using minimally invasive techniques. More than 15 distinct systems have been approved by the FDA for use in sacroiliac joint arthrodesis. The most often cited of these systems are the Rialto™ SI Joint Fusion System (Medtronic), SIJ-Fuse (SpineFrontier), iFuse® Implant System (SI-Bone), SImmetry® Sacroiliac Joint Fusion System (Zyga Technology), Silex™ Sacroiliac Joint Fusion System (Xtant Medical), SambaScrew® (Orthofix), the SI-LOK® Sacroiliac Joint Fixation System (Globus Medical), and the TriCor™ Sacroiliac Joint Fusion System (Zimmer Biomet). These systems employ delivery instruments and implants, specifically designed to fixate the ilium to the sacrum in a trans-articular fashion. To varying degrees, the technology also allows for bony decortication of the sacroiliac joint and delivery of grafting material.

Minimally Invasive Lateral and Posterolateral Iliosacral Trans-articular Approach

While the instrumentation systems each have different specifications for their technique, they follow many common themes. Although supine positioning is possible, most require prone positioning on a radiolucent table. Intraoperative fluoroscopy is used to plan the incision along the posterolateral gluteal region. Lateral imaging is the workhorse of these operations, but pelvic inlet and outlet views and to a lesser degree AP views will be utilized. Once a small skin incision is made, a guide wire is navigated to the lateral ilium until the cortical bone is encountered. Sequential dilators are then placed over the guide wire to facilitate docking of a tubular retractor. Each of the instrumentation systems then creates trajectories through the ilium into the sacral ala and body at different levels, and fixation devices are deployed across the joint. In addition, delivery of grafting material into the joint is sometimes required depending on the coating characteristics of fixation rod or screws.

Minimally Invasive Posterior Intra-articular Approach

Some minimally invasive approaches have emphasized attacking the joint along the longitudinal axis, perpendicular to the posterior joint capsule. Preoperative CT images should be utilized to measure the depth of the joint. Patients are positioned prone with either fluoroscopic guidance or CT-guided navigation. An incision is planned in the same area as the open posterolateral approach, by palpation of the PSIS or visualization of the PSIS using fluoroscopy. The incision is carried deep to the bone of the PSIS using blunt dissection or sequential dilators. Fluoroscopy is used to align the chosen instrumentation along

long axis of the joint, by shifting the axis of the C-arm cephalad and obliquely approximately 20–30° toward the contralateral side, so the images are shot posteromedial to anterolateral. From this point a variety of techniques can be utilized to implant the intra-articular graft into the joint. Hollow, threaded cages are commonly used in this technique for the ability to gain purchase on the sacral and iliac portions of the joint, while also facilitating the addition of grafting material and other orthobiologics, such as bone morphogenetic protein.

Instrumentation Options

- Screw-Plate Constructs (Fig. 1)
- Screw-Rod Constructs
- Intra-Articular (Distracting) Cage
- Trans-Articular Threaded Screw (Fig. 2)
- Trans-Articular Fusion Rod (Fig. 3)
- Hollow Modular Anchorage Screw

Evidence Supporting Different Surgical Techniques

While surgery to fixate the SI joint for patient with low back pain has been performed for approximately 100 years, until recently the literature was quite sparse regarding the safety and efficacy of SI joint fusion. Emergence of MIS techniques for these operations over the past 10 years has brought renewed attention to the problem of the dysfunctional sacroiliac joint, and the body of literature now reflects this interest. The majority of manuscripts published on the topic are derived from retrospective case series. However, during the past decade, groups have also started to study the problem and surgical interventions to treat it in a prospective, randomized fashion. Additionally, they have made efforts to measure outcomes in a more standardized, quantifiable manner using measures such as the Visual Analog Scale (VAS) for pain and Oswestry Disability Index (ODI) for functional performance.

In 2001, a single surgeon published a report of four cases in which he performed an open posterior sacroiliac joint arthrodesis using a screw-rod construct and intra-articular iliac crest autograft (Belanger and Dall 2001). The authors state that all four patients had qualitative clinical improvements with regard to lower back pain and demonstrated solid bony arthrodesis on either CT or plain radiograph. No significant complications or revisions were reported in this series, although after bony fusion was confirmed, two patient had instrumentation removed due to pain at the screw sites. This manuscript is representative of earlier reports on the topic.

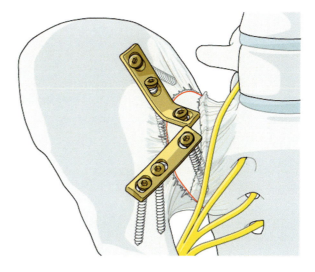

Fig. 1 Open anterior approach with use of screw-plate fixation across the SI joint (Source: AO Foundation)

Fig. 2 (**a**) Lateral approach iliosacral fixation screw, options for open or minimally invasive placement techniques. (**b**) Sacral-alar-iliac screw (Source: AO Foundation)

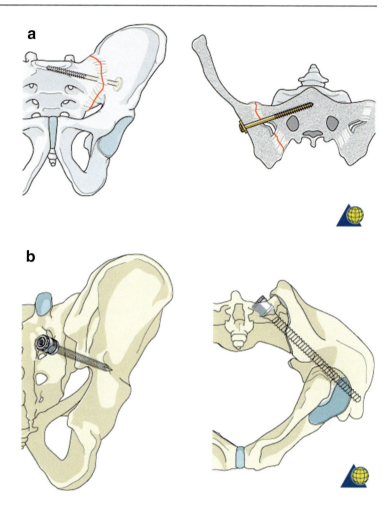

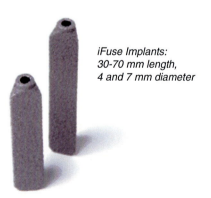

iFuse Implants: 30-70 mm length, 4 and 7 mm diameter

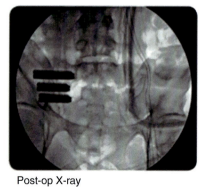

Post-op X-ray

Fig. 3 Fusion rod with titanium plasma spray (TPS) coated, placed from lateral minimally invasive approach (Source: SI-Bone)

A question of whether outcomes are different between open and MIS fusion techniques has also been postulated. In 2013, a multi-centered retrospective cohort analysis of 263 patients receiving SI joint fusion was published comparing traditional, open posterior SI joint fusion to minimally invasive SI joint fusion using a series of titanium plasma spray (TPS)-coated triangular fusion rods (iFuse® system). This study found that operating room time, estimated blood loss, and length of hospitalization were all significantly lower in minimally invasive fusions than in open surgical fusions. Furthermore, the patients who had minimally invasive fusions had significantly greater reductions in their lower back pain as measured by the VAS, even when matched for age, gender, and history of lumbar spine fusion (Smith et al. 2013).

A similar retrospective cohort study was performed at single-center, comparing 27 patients who had MIS SI joint fusion to 36 patients who had open, anterior approach SI joint fusion. MIS SI fusion was accomplished using TPS-coated triangular fusion rods and CT-guided navigation, while open fusion was accomplished using anterior screw-plate constructs. Utilizing propensity score, pairwise matching of MIS, and open SI fusion patients, the study found significantly lower length of surgery, length of hospitalization, and estimated blood loss for patients in the MIS group, but no significant difference in disability score as measured by ODI. Radiographic confirmation of bony arthrodesis could not be assessed in this study.

Additionally, clinicians treating the dysfunctional SI joint have attempted to better characterize the outcomes of surgery compared to non-surgical treatment measures. A prospective, randomized trial was performed at nine European sites to explore this question. 103 patients were randomized to either MIS SI fusion (with triangular TPS-coated fusion rods) or conservative management which included medical therapy, physical therapy, and in some cases cognitive behavioral therapy, but did not include interventional procedures such as intra-articular joint injections and radiofrequency ablation. The primary endpoint was low back pain as measured using the VAS on a scale of 0–100. Patients in the SI fusion group experienced a mean improvement of 43.3 points compared to a 6.8 point mean improvement with conservative management, a statistically significant improvement. Secondary outcome measures, such as ODI, EQ-5D-3L (a quality of life assessment), and overall satisfaction, were also statistically significantly better in the SI joint fusion group at 6-month follow-up (Sturesson et al. 2017).

The INSITE, Investigation of Sacroiliac Fusion Treatment, group also performed a multicenter, prospective, randomized trial exploring outcomes of minimally invasive SI joint fusion against non-surgical management (NSM). The triangular TPS-coated fusion rod was utilized for these interventions. The NSM group received a combination of pain medications, physical therapy, SI joint steroid injections, and radiofrequency ablation, but excluded use of cognitive behavioral therapy. Cross-over was allowed after the 6-month visit, leading 102 patients to receive SI joint fusion and 46 patients to receive NSM. The SI joint fusion group had a mean improvement in VAS of pain by 55.4 at 24-month follow-up, compared to a mean improvement of 12.2 point in the NSM group, a statistically significant improvement. Furthermore, patient in the SI fusion group demonstrated statistically significant improvements on the SF-36 disability form at 6-, 12-, and 24-month follow-ups compared to NSM (Polly et al. 2016).

Given the accelerating technological advancement in instrumentation designed to fixate the SI joint, we expect the body of literature regarding sacroiliac joint fusion to continue expanding. As such, we will be able to better characterize the effects of different instrumentation techniques on outcomes when treating patient with sacroiliac joint dysfunction.

References

Belanger TA, Dall BE (2001) Sacroiliac arthrodesis using a posterior midline fascial splitting approach and pedicle screw instrumentation: a new technique. Clin Spine Surg 14:118

Bernard JT, Kirkaldy-Willis WH (1987) Recognizing specific characteristics of nonspecific low back pain. Clin Orthop Relat Res 217:266–280

Bowen V, Cassidy JD (1981) Macroscopic and microscopic anatomy of the sacroiliac joint from embryonic life until the eighth decade. Spine 6:620–628

Cohen SP, Chen Y, Neufeld NJ (2013) Sacroiliac joint pain: a comprehensive review of epidemiology, diagnosis and treatment. Expert Rev Neurother 13:99–116. https://doi.org/10.1586/ern.12.148

Cole JD, Blum DA, Ansel LJ (1996) Outcome after fixation of unstable posterior pelvic ring injuries. Clin Orthop Relat Res 1976–2007 329:160–179

Dreyfuss P, Cole AJ, Pauza K (1995) Sacroiliac joint injection techniques. Phys Med Rehabil Clin North Am 6:785–813. https://doi.org/10.1016/S1047-9651(18)30434-0

Dreyfuss P, Michaelsen M, Pauza K et al (1996) The value of medical history and physical examination in diagnosing sacroiliac joint pain. Spine 21:2594–2602

Dreyfuss P, Dreyer SJ, Cole A, Mayo K (2004) Sacroiliac joint pain. J Am Acad Orthop Surg 12:11

Elgafy H, Semaan HB, Ebraheim NA, Coombs RJ (2001) Computed tomography findings in patients with sacroiliac pain. Clin Orthop Relat Res 1976–2007 382:112–118

Forst SL, Wheeler MT, Fortin JD, Vilensky JA (2006) The sacroiliac joint: anatomy, physiology and clinical significance. Pain Physician 9:61–67

Freburger JK, Holmes GM, Agans RP et al (2009) The rising prevalence of chronic low Back pain. Arch Intern Med 169:251–258. https://doi.org/10.1001/archinternmed.2008.543

Gaenslen FJ (1927) Sacro-iliac arthrodesis indications, author's technic and end-results. J Am Med Assoc 89:2031. https://doi.org/10.1001/jama.1927.02690240023008

Grob KR, Neuhuber WL, Kissling RO (1995) Innervation of the sacroiliac joint of the human. Z Rheumatol 54:117–122

Ha K-Y, Lee J-S, Kim K-W (2008) Degeneration of sacroiliac joint after instrumented lumbar or lumbosacral fusion: a prospective cohort study over five-year follow-up. Spine 33:1192–1198. https://doi.org/10.1097/BRS.0b013e318170fd35

Hansen HC, McKenzie-Brown AM, Cohen SP et al (2007) Sacroiliac joint interventions: a systematic review. Pain Physician 10:165–184

Hoy D, March L, Brooks P et al (2010) Measuring the global burden of low back pain. Best Pract Res Clin Rheumatol 24:155–165. https://doi.org/10.1016/j.berh.2009.11.002

Khurana A, Guha AR, Mohanty K, Ahuja S (2009) Percutaneous fusion of the sacroiliac joint with hollow modular anchorage screws. J Bone Joint Surg Br 91-B:627–631. https://doi.org/10.1302/0301-620X.91B5.21519

Kibsgård TJ, Røise O, Stuge B, Röhrl SM (2012) Precision and accuracy measurement of Radiostereometric analysis applied to movement of the sacroiliac joint. Clin Orthop 470:3187–3194. https://doi.org/10.1007/s11999-012-2413-5

Kirkaldy-Willis WH, Cassidy JD (1985) Spinal manipulation in the treatment of low-back pain. Can Fam Physician 31:535–540

Maigne J-Y, Aivaliklis A, Pfefer F (1996) Results of sacroiliac joint double block and value of sacroiliac pain provocation tests in 54 patients with low back pain. Spine 21:1889

Miller JA, Schultz AB, Andersson GBJ (1987) Load-displacement behavior of sacroiliac joints. J Orthop Res 5:92–101. https://doi.org/10.1002/jor.1100050112

Murata Y, Takahashi K, Masatsune et al (2001) Origin and pathway of sensory nerve fibers to the ventral and dorsal sides of the sacroiliac joint in rats. J Orthop Res 19:379–383. https://doi.org/10.1016/S0736-0266(00)90017-2

Nakagawa T (1966) Study on the distribution of nerve filaments over the iliosacral joint and its adjacent region in the Japanese. Nihon Seikeigeka Gakkai Zasshi 40:419–430

Polly DW, Swofford J, Whang PG et al (2016) Two-year outcomes from a randomized controlled trial of minimally invasive sacroiliac joint fusion vs. non-surgical management for sacroiliac joint dysfunction. Int J Spine Surg 10:28. https://doi.org/10.14444/3028

Rand JA (1985) Anterior sacro-iliac arthrodesis for post-traumatic sacro-iliac arthritis. A case report. J Bone Joint Surg Am 67:157–159

Sashin D (1930) A critical analysis of the anatomy and the pathologic changes of the sacro-iliac joints. J Bone Joint Surg 12:891

Slipman CW, Jackson HB, Lipetz JS et al (2000) Sacroiliac joint pain referral zones. Arch Phys Med Rehabil 81:334–338. https://doi.org/10.1016/S0003-9993(00)90080-7

Smith AG, Capobianco R, Cher D et al (2013) Open versus minimally invasive sacroiliac joint fusion: a multicenter comparison of perioperative measures and clinical outcomes. Ann Surg Innov Res 7:14. https://doi.org/10.1186/1750-1164-7-14

Smith-Petersen MN, Rogers WA (1926) End-result study of arthrodesis of the sacro-iliac joint for arthritis-traumatic and non-traumatic. J Bone Joint Surg 8:118

Sturesson B, Selvik G, Udén A (1989) Movements of the sacroiliac joints. A roentgen stereophotogrammetric analysis. Spine 14:162–165

Sturesson B, Uden A, Vleeming A (2000) A radiostereometric analysis of movements of the sacroiliac joints during the standing hip flexion test. Spine 25:364–368. https://doi.org/10.1097/00007632-200002010-00018

Sturesson B, Kools D, Pflugmacher R et al (2017) Six-month outcomes from a randomized controlled trial of minimally invasive SI joint fusion with triangular titanium implants vs conservative management. Eur Spine J 26:708–719. https://doi.org/10.1007/s00586-016-4599-9

Thawrani DP, Agabegi SS, Asghar F (2018) Diagnosing sacroiliac joint pain. J Am Acad Orthop Surg 1. https://doi.org/10.5435/JAAOS-D-17-00132

Vleeming A, Buyruk HM, Stoeckart R et al (1992) An integrated therapy for peripartum pelvic instability: a study of the biomechanical effects of pelvic belts. Am J Obstet Gynecol 166:1243–1247. https://doi.org/10.1016/S0002-9378(11)90615-2

Vleeming A, Schuenke MD, Masi AT et al (2012) The sacroiliac joint: an overview of its anatomy, function and potential clinical implications. J Anat 221:537–567. https://doi.org/10.1111/j.1469-7580.2012.01564.x

Waisbrod H, Krainick J-U, Gerbershagen HU (1987) Sacroiliac joint arthrodesis for chronic lower back pain. Arch Orthop Trauma Surg 106:238–240. https://doi.org/10.1007/BF00450461

Walheim G, Olerud S, Ribbe T (1984) Mobility of the pubic symphysis: measurements by an electromechanical method. Acta Orthop Scand 55:203–208. https://doi.org/10.3109/17453678408992338

Yoshihara H (2012) Sacroiliac joint pain after lumbar/lumbosacral fusion: current knowledge. Eur Spine J 21:1788–1796. https://doi.org/10.1007/s00586-012-2350-8

Zaidi HA, Montoure AJ, Dickman CA (2015) Surgical and clinical efficacy of sacroiliac joint fusion: a systematic review of the literature. J Neurosurg Spine 23:59–66. https://doi.org/10.3171/2014.10.SPINE14516

Lateral Lumbar Interbody Fusion

35

Paul Page, Mark Kraemer, and Nathaniel P. Brooks

Contents

Introduction	690
Indications and Contraindications	691
Relevant Anatomy	691
Lumbar Plexus	693
Vascular Anatomy Considerations	693
Preoperative Planning and Operative Window	693
Trans-psoas	694
Pre-psoas	694
Procedural Details	694
Surgical Positioning	694
Incision and Retroperitoneal Dissection	695
Retractor Positioning	695
Disc Preparation and Implant Placement	696
Posterior Instrumentation and Fusion	697
Complications and Their Management	697
Conclusions	698
References	699

Abstract

Lateral lumber interbody fusion is an important technique in the continually growing field of minimally invasive spine surgery. While it had previously been utilized in the early twentieth century for the treatment of traumatic injuries and Pott's disease, the current revolution of minimally invasive surgery has seen a recurrence of this approach and expansion of its clinical applications. Though this approach was largely abandoned in the late twentieth century for anterior and posterior approaches due to a high morbidity, a combination of improved technology and understanding of lumbar plexus anatomy has allowed for

P. Page · M. Kraemer · N. P. Brooks (✉)
Department of Neurological Surgery, University of Wisconsin, Madison, WI, USA
e-mail: page@neurosurgery.wisc.edu; kraemer@neurosurgery.wisc.edu; brooks@neurosurgery.wisc.edu

© Springer Nature Switzerland AG 2021
B. C. Cheng (ed.), *Handbook of Spine Technology*,
https://doi.org/10.1007/978-3-319-44424-6_66

its resurgence. Clinical applications of the retroperitoneal trans-psoas and pre-psoas approaches are continually expanding and frequently include scoliosis, neoplasms, traumatic injuries, and a variety of degenerative disorders. Here we describe the clinical utility of this approach, review the pertinent clinical anatomy, and describe the procedure in detail.

Keywords

Lateral lumbar interbody fusion · Trans-psoas · Pre-psoas · LLIF · OLIF · Minimally invasive spine surgery

Introduction

Interbody fusion in the lumbar spine is an established treatment for a wide variety of spinal disorders ranging from trauma, infection, degenerative disease, deformity correction, and neoplasms. The use of interbody fusion provides additional biomechanical advantages because of the ability to place a large interbody graft that provides support to the anterior and middle columns of the vertebral segment. Additionally, the ability to extend the graft across the thicker bone of the apophyseal ring of the vertebral body limits subsidence or fracture. Restoration of interbody height by interbody fusion allows for indirect decompression of the neural elements. The goals of treatment and surgical approaches to the spine vary based upon the spinal pathology. The options are circumferential, including the posterior lumbar interbody fusion (PLIF), transforaminal lumbar interbody fusion (TLIF), lateral lumbar interbody fusion (LLIF) either pre-psoas or trans-psoas, and anterior lumbar interbody fusion (ALIF). Of these techniques, the lateral interbody approach is growing in popularity due to avoidance of the vasculature anteriorly and the thecal sac posteriorly. Additionally, there is minimal disruption of the existing ligamentous structures and surrounding musculature. The LLIF has multiple trade names depending on the company. The trans-psoas approach is called the direct lumbar interbody fusions (DLIF) or extreme lateral interbody fusion (XLIF). The pre-psoas approach is called the oblique interbody fusions (OLIF).

Compared to traditional anterior and posterior approaches to the lumbar spine, the minimally invasive lateral interbody fusion is a relatively new approach as it relates to common practice. But it is important to note that variations of this approach have been described historically. Although lateral approach to the lumbar spine was originally described and utilized in the treatment of Pott's disease in the early twentieth century by Drs. Menard and Capener, it remained infrequently used due to injury to the traversing lumbar plexus and nerve roots. Despite this neurologic morbidity, the approach became more commonplace in the treatment of Pott's disease through the twentieth century. With the emergence of the minimally invasive revolution in the late twentieth century, this approach reemerged and expanded to include a wide variety of disease pathologies. This expansion is largely attributed to recent advances in minimally invasive technologies and a better understanding of the anatomic relationships of the exiting lumbar nerve roots that form the lumbar plexus. While the traditional lateral approach accessed the vertebral column using a trans-psoas corridor, Mayer in 1997 described an oblique retroperitoneal approach in which instrumentation is performed anterior to the psoas with the benefit of fewer neural injuries (Mayer 1997). Once these anatomic limitations were identified and techniques developed to limit nerve injury, lateral approaches to the lumbar spine have proven to be versatile tool in modern spine practice.

The lateral lumbar interbody fusion has demonstrated some distinct advantages compared with the anterior or posterior approaches to the lumbar spine. Compared with traditional approaches, there is minimal disruption of the posterior elements, which may provide some benefit in the stability of the construct and postoperative pain. Additionally, lateral approaches allow for a larger access to the disc space and increase the size of interbody graft compared to posterior approaches. Research has demonstrated decreased blood loss and operative time as compared with traditional approaches. One review evaluating extreme lateral lumbar interbody fusion (XLIF) demonstrated overall short operative times, 199 min, and relatively minimal blood loss of 155 ml (Youssef et al. 2010).

Compared to traditional ALIF approaches, lateral interbody fusion demonstrated similar degrees of foraminal height gain. However, there was less segmental lordosis correction than ALIF. In previous studies the amount of foraminal height following ALIF has demonstrated improvements in foraminal height of approximately 2.7 mm. Alimi et al. demonstrated similar foraminal height improvements, 2.5 mm on average, in their series of 145 patients who underwent interbody fusion from a lateral approach (Alimi et al. 2014). As compared to ALIF, however, segmental lordosis correction in lateral interbody fusion is generally less due to retention of the anterior longitudinal ligament. While specific degrees of improved lordosis vary, ALIF generally provides approximately 4.5° of lordosis, but only 2.5° following a lateral approach (Winder and Gambhir 2016).

Indications and Contraindications

The indications of the lateral lumbar interbody fusion are primarily to improve intervertebral height and to reduce deformity. Common indications are degenerative disease with loss of disc height and foraminal stenosis, spondylolisthesis, coronal imbalance, lateral vertebral subluxation, and revision of adjacent segment degeneration.

The contraindications of this approach are primarily related to anatomical considerations such as aberrant vascular anatomy, prior retroperitoneal approach, or prior abdominal infections or surgeries with associated adhesions in the retroperitoneal corridor. A high-riding iliac crest or low-lying rib is a relative contraindication.

Relevant Anatomy

The lateral lumbar approach traverses anatomy that is rarely encountered in traditional approaches. Given the narrow operative corridor, a detailed understanding of the relevant anatomy is crucial for safe, effective surgery. Traversing the retroperitoneal space involves significant risk to major vascular structures and vital organs. Evaluation of the preoperative imaging and understanding the location of these structures and their relationship to the disc space is a critical part of preoperative planning and intraoperative crisis management. Furthermore, if concern for major vessel injury arises preoperatively or intraoperatively, the surgeon should seek vascular surgery consultation.

When accessing the retroperitoneal space, care must be taken to first identify the external oblique fascia, external oblique, internal oblique, and transversus abdominal muscles at the beginning of the approach (Fig. 1). It is important to recognize the trajectory of the iliohypogastric and ilioinguinal nerve as they course through the psoas before innervating the internal and external oblique muscles. When unclear as to which muscular layer is being visualized, the surgeon should recognize the direction of the muscle fibers for reorientation. Once these muscular layers have been traversed, a layer of adipose tissue is identified in the retroperitoneal space. Deep to this, the peritoneum will be identified. Dissection is best performed bluntly with either finger dissection or use of cotton kittners. It is easier to dissect along the interior of the abdominal wall, palpate the iliac crest, and then identify the psoas than to try and dissect along the peritoneum. Further dissection leads to the lateral aspect of the psoas (Fig. 2). Care must be taken to expose the anterior psoas as it is easy to fall into a plane behind the psoas that leads to the spinal canal and foramen. The ureter typically will mobilize with the peritoneum and reflect anteriorly. However, if it is taking an unusual course, it can be identified by its visually identifying peristalsis with manipulation. Care should be taken to not overly compress or stretch the ureter. A keen awareness of the location of the great vessels anterior to the vertebral bodies is paramount. These can be palpated but care should be taken to avoid manipulation or retraction without adequate visualization. Segmental arteries will arise from the aorta in the midpoint, "valleys," of the vertebral bodies. Occasionally, the iliolumbar vein or veins will be

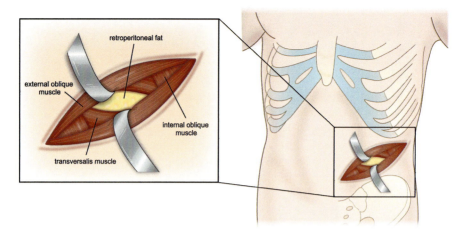

Fig. 1 This illustration shows the oblique incision, which is centered over the disc of interest and oriented along the fibers of the external oblique. Blunt dissection is performed through the external, internal, and transversalis abdominal muscles to reveal the retroperitoneal fat pad. (Reprinted with permission, University of Wisconsin © 2018. All Rights Reserved)

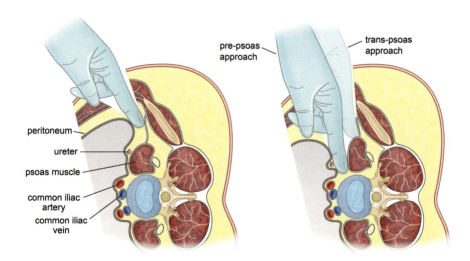

Fig. 2 Schematic demonstrating blunt dissection of the retroperitoneal fat pad from the transversalis fascia and anterior retraction of the aorta. The transverse process is first palpated before isolating the psoas muscle. If electing to perform a trans-psoas approach, instrumentation is then directed through the psoas under imaging guidance and neuromonitoring (triggered and free-running EMG). In the pre-psoas approach, the anterolateral portion of the vertebral body is identified (left) and the psoas is retracted posteriorly. Care should be taken to avoid dissection of the psoas medially, as this may irritate exiting nerve roots. (Reprinted with permission, University of Wisconsin © 2018. All Rights Reserved)

seen coursing from underneath the aorta usually at L4–L5 interval. This vein could be isolated and ligated to avoid avulsion of the vein from the inferior vena cava which can create a vascular injury that is very difficult to repair. During the course of dissection, if there is aberrant or overly large vascular anatomy, then strong consideration should be given to (1) obtaining vascular surgeon consultation, (2) aborting the procedure, and (3) operating via a posterior approach. Furthermore, a thorough review of preoperative imaging is important to understand the relation of adrenal glands, kidneys, ureters, and renal vasculature that may be encountered when utilizing this approach.

Lumbar Plexus

The lumbar plexus is deeply integrated into the psoas muscle and contains innervation from subcostal contributions from the T12 as well as the ventral rami of the first four lumbar nerve roots. The fourth lumbar nerve root additionally supplies contributions to the sacral plexus. The lumbar plexus is ultimately divided into two divisions named the anterior and posterior division. The posterior division provides innervation to the main motor component of the posterior leg via the femoral nerve with contributions from the L2 to L4, while the anterior division provides motor innervation via the obturator nerve. Sensory innervation is chiefly accomplished by the iliohypogastric, ilioinguinal, genitofemoral, lateral femoral cutaneous, and anterior femoral cutaneous nerves.

Understanding the course of the ilioinguinal and iliohypogastric nerve is vital to avoiding complications. Both nerves run posterior to the psoas major on its proximal lateral border of the vertebral bodies and then travel along the anterior border to the quadratus lumborum. After traveling anterior to quadratus lumborum, the ilioinguinal nerve pierces the lateral abdominal wall after traveling at the level of the iliac crest to supply sensory innervation to the external ring, the area over the pubic symphysis, and the lateral area of the scrotum or labia majora. Comparatively the iliohypogastric provides motor innervation to the abdominal internal oblique and transverse abdominis until it provides a terminal cutaneous branch supplies which the skin above the inguinal ligament.

The lateral cutaneous nerve consequently pierces the psoas directly through a lateral approach most frequently in the middle location of the psoas muscle. Given its location directly through the psoas muscle, this nerve is at risk during a lateral lumbar approach. Once it emerges from the psoas, it then courses across the iliacus muscle obliquely and continues to the anterior superior iliac spine. At this point it crosses under the inguinal ligament over the sartorius muscle into the thigh.

The femoral nerve is the longest and largest nerve of the entire lumbar plexus and supplies both sensory and motor innervation to the anterior compartment of the superior leg. Contributions from the lumbar plexus arise from the L2, L3, and L4 nerve roots. After arising distal to the nerves to the psoas muscles directly, it courses through the femoral triangle lateral to femoral artery.

Vascular Anatomy Considerations

When considering a lateral retroperitoneal approach, important consideration of the major vascular structures such as the inferior vena cava, abdominal aorta, and common iliac arteries and veins must be given. In order to limit the potential injury to vascular structures, a careful review of the preoperative imaging is vital. While risk to the great vessels is highest at the L4–L5 level due to their lateral migration, major vasculature injury could occur at any level. In addition to knowledge of the great vessels, care should be taken to identify and avoid avulsion of any of the segmental vessels or iliolumbar veins crossing into the disc space during removal of the annulus that could result in avulsion of the aorta or vena cava.

Preoperative Planning and Operative Window

In considering the operative corridor during interbody fusion, it is vital to understand what constitutes a safe and effective operative window. Avoiding injury to the traversing nerve roots and lumbar plexus and great vessels is paramount. It is also important to consider that surface anatomy: a high-riding iliac crest or low-riding ribs can make the approach more difficult. These surface limitations can be managed by removing rib or positioning the hip over a bump or table break. Care should be taken to not overextend the torso as this can cause thigh pain and weakness.

Trans-psoas

The trans-psoas approach avoids the great vessels but puts the lumbar plexus and peripheral nerves at greater risk of injury. The anatomy of the plexus cannot be well discerned on preoperative imaging. So, determination of an operative window is made intraoperatively via a combination of general knowledge of the lumbar plexus anatomy, visual and fluoroscopic inspection, and the use of neuromonitoring (triggered and free-running EMG).

Despite attempts to simplify the anatomical association of the lumbar plexus and the psoas muscle, the authors have demonstrated enormous variability. The plexus generally tends to migrate anteriorly as the psoas muscle enters the pelvis. Due to this relationship, the plexus is often at highest risk of injury at the L4–L5 disc space. A key development occurred in 2010 when Uribe et al. published a cadaveric study in which the zones of safest psoas disruption were identified (Uribe et al. 2010a). In this system four quartiles along the sagittal axis of the vertebral body were defined at each vertebral level. At the L1 and L2 disc space, the middle of this quartile was shown to have the lowest risk for injury to the nerve roots or lumbar plexus; however, at lower levels, the safest location migrates slightly anteriorly until the L4–L5 disc space. At the L4–L5 disc space, the safest location was the midpoint of the vertebral body. Additionally, the authors noted that the genitofemoral nerve was the nerve at most risk in the third quartile. This nerve must be a consideration to the surgeon as given its sensory function it will not be recognized by EMG and can be easily injured. This "safe entry zone" should not be considered universal, and as previously discussed, significant variation in patient anatomy may be present. Ultimately visual inspection and neuromonitoring are critical in minimizing risk to traversing nerves. In general, triggered EMG thresholds below 5 mA indicate direct contact, 5–10 mA indicate that the stimulation is in close proximity, and 11 mA indicates a farther distance from the lumbar plexus (Uribe et al. 2010b).

Pre-psoas

The key difference of the lateral pre-psoas approach is the intent of docking instrumentation between the psoas muscle and the great vessels. Due to the more anterior location on the vertebral body, there is a higher risk to the great vessels anteriorly, and it is frequently cited at a rate similar to the anterior approaches. Currently existing literature demonstrates a rate of vascular complication cited from 1.1% to 2.8% with damage to segmental arteries being the most common complications (Xu et al. 2018). Conversely, the risk of injury to the lumbar plexus is lower because the psoas is not blindly traversed. Determination of the operative window is made by preoperative planning and intraoperative visual and fluoroscopic inspection and dissection. Neuromonitoring is not necessary for this approach but can be considered.

Procedural Details

Surgical Positioning

Proper positioning is essential for successful lateral lumbar interbody fusion. The patient should be placed in the lateral decubitus position with the left side up. Right-sided approach may be considered, but this is generally discouraged due to increased risk of injuring the relatively thin-walled inferior vena cava during manipulation. If such an approach is undertaken, a trans-psoas corridor should be considered to decrease risk of IVC injury. Additionally, lateral jack-knife position may be used to improve access and visualization in certain cases, but this should be avoided if possible to avoid transient neurologic deficits (Molinares et al. 2016). An axillary roll is placed to avoid brachial plexus injury. Care should be taken to position the patient perpendicular to the floor. The fluoroscope is then brought into the field, and minor table adjustments are used to obtain a perpendicular lateral view of the target disc. Similarly, the flouroscope should be easily maneuverable to obtain a clear orthogonal AP view of the disc. Alternatively, computerized stereotactic navigation may be used, in which case the registration pin is placed into the iliac crest projecting posteriorly after prepping and draping. A small amount of hip flexion may be used to relax the psoas, which also serves to position the L4 and L5 nerve roots more posteriorly. At this time neuromonitoring (EMG) may be attached. If approaching the vertebral column using a trans-psoas approach, neuromonitoring should be used to avoid the risk of nerve injury while traversing the psoas. The patient's abdomen and flank should be prepped and draped widely despite the plan for a small incision. This will allow the laparotomy incision to be enlarged in the case of difficulty with dissection or complication.

Incision and Retroperitoneal Dissection

Under fluoroscopic or navigation guidance, the intervertebral disc of interest is identified before marking its caudocranial and anteroposterior projections on the skin. For the trans-psoas approach, a 4 cm incision is centered over the disc of interested and oriented obliquely (Fig. 1). If approaching anterior to psoas, the incision should be positioned more anteriorly from the center of the disc space (approximately 5 cm) to facilitate psoas mobilization and vertebral body visualization. The external oblique fascia is then sharply divided, and splitting of the external, internal, and transversalis abdominal musculature is performed using a Kelly clamp or bluntly. The underlying transversalis fascia is identified and divided before entering the retroperitoneal fat pad. Using a gloved index finger, the fat pad is gently dissected from the transversalis fascia before advancing more medially and posteriorly and along the anterior boarder of the quadratus lumborum to palpate the transverse process of the vertebral body (Fig. 2). After palpating the transverse process, blunt dissection is used to retract the peritoneal contents anteriorly to identify the vertebral disc space. Fluoroscopy is then used to confirm the correct intervertebral disc level.

Retractor Positioning

Trans-psoas

Fluoroscopy or navigation is used to localize the planned position to dock the retractor in the disc space. This is done by placing a k-wire in the disc space and using serial dilation to split the psoas muscle fibers. Neuromonitoring (triggered and free-running EMG) is monitored as each dilator and eventually the retractor blades are advanced. The electrical contacts are different for each company and should be studied and understood prior to surgery to evaluate the direction of stimulation. Typically, one small area of each dilator will have exposed uninsulated metal, and stimulation can proceed in quadrants to look for EMG firing. This can help the surgeon determine the direction of any at-risk nerves and reposition the retractor accordingly. If the EMGs demonstrate irritation, the retractor can be repositioned away from the direction of nerve root firing. The retractor blades can be pinned into the vertebral bodies above and below the disc space firmly into position. An ideally placed retractor will be overlying the disc at about the anterior 1/2 of the disc space, parallel to the endplates and in the coronal plane (Fig. 3). Fluoroscopy should be used to verify this position.

Pre-psoas

In the pre-psoas approach, the psoas is mobilized posteriorly for exposure of the ideal operative window. This space need only be slightly wider than the planned implant. Self-retaining retractors are then placed to retract the abdominal contents and psoas. The retractors can be rotated and slid slightly above the disc space to allow the disc prep tools and implants to be positioned in the coronal plane. A second retractor is placed medially to protect the peritoneum and great

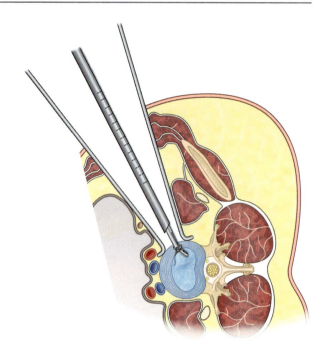

Fig. 3 In the pre-psoas approach, instrumentation is docked at the anterolateral disc space and a small annulotomy is performed followed by complete discectomy. (Reprinted with permission, University of Wisconsin © 2018. All Rights Reserved)

vessels, and if needed, a third retractor can be placed in the caudal aspect of the incision to protect peritoneal contents. Various retractor setups are available. This can also be done with handheld retractors if desired.

Disc Preparation and Implant Placement

An annulotomy is then performed followed by complete discectomy and removal of the cartilaginous endplates. This will ensure that a large surface area is available for effective fusion. Care should also be taken to bilaterally release the annulus to avoid coronal imbalances after implant placement. This may be performed by rotating a Cobb across the distal annulus of the disc space (Orita et al. 2017). Care should be taken to maintain a coronal trajectory. Failure to work in the coronal plane can lead injury to the vasculature anteriorly or neural elements posteriorly. The disc space is then sequentially distracted using spacers until the ideal height is reached. A lordotic cage filled with graft is then placed and positioned parallel to the disc space (Fig. 4) on the AP view and in line with the posterior aspect of the vertebral bodies on the lateral view. To avoid inserting the cage in a rotated alignment on the lateral view, the trials and rasps should be placed so that they are aligned with the posterior aspect of the vertebral bodies, allowing the cage to simply follow the created path.

Ideal implant placement involves adjusting the midpoint of the cage to the center of the vertebral body on AP view and between the anterior and middle-third on lateral view. Implant placement in the trans-psoas approach is directly perpendicular to the vertebral body along the planned trajectory. However, special attention needs to be used to place pre-psoas implants. The pre-psoas implant is placed obliquely from the 10 o'clock position on the disc, advanced 1/2 way into the disc space, and then the handle is rotated posteriorly perpendicular to the OR table to place it across the disc space. This is sometimes called the "orthogonal maneuver."

The surgical field is then copiously irrigated, and meticulous hemostasis is achieved before removal of the self-retaining retractor and wound closure. The surgical corridor should be inspected for any injury to the peritoneum or retroperitoneal structures.

Fig. 4 Illustration demonstrating interbody graft placement using an pre-psoas approach. Ideal graft placement is midline on the lateral view and parallel to the disc space. (Reprinted with permission, University of Wisconsin © 2018. All Rights Reserved)

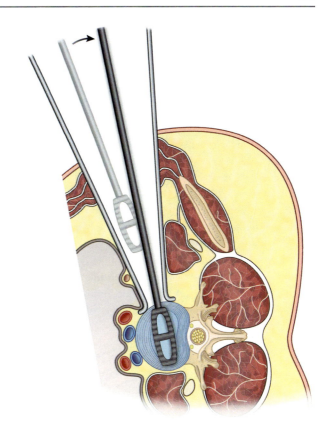

Posterior Instrumentation and Fusion

Posterior instrumentation can be considered to achieve a stable construct and increase the likelihood of fusion. This may be performed in multiple ways, but unless otherwise contraindicated, we prefer repositioning the patient in a prone position and performing bilateral percutaneous pedicle screw fixation using either fluoroscopic or stereotactic guidance.

Pre-psoas L5–S1

This is an advanced surgical technique but the pre-psoas approach does allow access to the L5–S1 level. This is performed by using a more anterior and medial incision and carefully docking the retractors between the bifurcation of the aorta and vena cava. An annulotomy is then performed, and discectomy and endplate preparation are completed. An anterior interbody cage is then implanted using a specially designed oblique introducer. Centering the implant can be challenging because of this oblique trajectory.

Complications and Their Management

Complications of the lateral approach are similar to those seen with ALIF with major complications related to damage of surrounding vascular, visceral, and neurologic structures. Yet, because there is generally no retraction of major vascular structures in the lateral approach, large series have reported no vascular or intraoperative injuries (Rodgers et al. 2011). If a vascular injury is identified, the first step is to obtain temporary control of the bleeding. This is often done using pressure from a kittner, suction, or sponge stick. If it is a large injury, then anesthesia should be notified to have blood products prepared to be administered. The second step will be to obtain improved access by making the incision larger. The third step will be to get adequate visualization of the injury. Primary repair can be attempted. Typically, prolene suture is used to suture vessels. Venous injuries must be repaired with care as the thin walls of the vessel can often tear. At any point, a vascular or general surgery consultation is encouraged to be obtained.

Other major complications include injury to the exiting nerve roots, particularly the L4 root. Permanent motor deficits have been reported between 0.7% and 3.4% (Knight et al. 2009) (Rodgers et al. 2011). Yet, when compared to other approaches, there is a high rate of transient groin and thigh pain after lateral approach which ranges from 10% up to 30%. These transient injuries, often hyperalgesia in the distribution of the iliohypogastric or ilioinguinal cutaneous nerves, are likely due to a combination of stretch and compression injury during the approach and retraction during surgery. Trans-psoas approaches are generally associated with higher complications rates (32.8%) versus oblique psoas-sparing approaches (13.5%) due to a higher likelihood of encountering the lumbar plexus during muscle dissection (Abe et al. 2017). Hip flexor weakness has also been reported and relates to manipulation of the psoas.

If the peritoneum is torn or injured, this is not typically a serious complication. Inspection should be performed to verify that no intestinal injury has occurred as this can be life-threatening if not identified. The peritoneum can be stitched closed with absorbable suture to avoid herniation of intestines. The surgery can continue.

If the intestines or abdominal organs are injured, then general surgery should be consulted for repair. Strong consideration should be given to aborting the procedure as the infection risk is very high in this situation. It can be helpful to have the patients undergo a bowel preparation prior to the surgery in order to minimize spillage of visceral contents, decrease likelihood of infection in case of incidental enterotomy, and increase the rate of repair.

Ureter injury is uncommon, but if it is encountered, then a urologist should be consulted to repair the ureter. The risks and benefits of proceeding should be weighed. Urine is typically sterile so the infection risk should be lower compared to intestinal injury.

As with ALIF, there is a risk of incisional hernia. If this occurs, then a referral to general surgery for repair is warranted.

Limitation of the parallel trajectory of the cage relative to the disc space secondary to a high-riding iliac crest is not technically a complication. However, it is often encountered at the L4–L5 level. In these cases, the senior author has proceeded with a discectomy without violating the contralateral annulus and placed a shorter cage to avoid neural compression in the canal or neural foramen.

Another key consideration when comparing anterior interbody to lateral interbody fusion is the risk of subsidence, which is defined as the potential loss of height within the neural foramen following indirect decompression with an interbody graft. While the gold standard for reducing the risk of subsidence is the anterior interbody fusion with an average of 10% risk of any subsidence without any events of neurologic consequence, both LLIF and OLIF have significant risks of subsidence and are important considerations with approach. Stand-alone LLIF has been shown to have subsidence rates of up to 30% when using standard 18 mm grafts (Marchi et al. 2013).

Conclusions

The lateral lumbar interbody fusion is a useful technique for the spine surgeon to have in his/her armamentarium. The keys to performing this technique safely and effectively are appropriate patient selection, safe lateral positioning, appropriate targeting of disc space with fluoroscopy or computerized stereotactic navigation, careful dissection and retractor placement to identify and avoid injury to intraperitoneal contents or pre-vertebral vascular structures, and aligning the tools to prepare the disc space and implant the graft in the coronal plane. If performing a trans-psoas approach, then neuromonitoring (triggered and free-running EMG) should be used to limit nerve injury. The outcomes of this procedure are similar to anterior lumbar interbody fusion but with less muscular dissection because of the lateral trajectory.

References

Abe K, Orita S, Mannoji C, Motegi H, Aramomi M, Ishikawa T et al (2017) Perioperative complications in 155 patients who underwent oblique lateral interbody fusion surgery: perspectives and indications from a retrospective, multicenter survey. Spine 42:55–62

Alimi M, Hofstetter CP, Cong G-T, Tsiouris AJ, James AR, Paulo D et al (2014) Radiological and clinical outcomes following extreme lateral interbody fusion. J Neurosurg Spine 20:623–635

Knight RQ, Schwaegler P, Hanscom D, Roh J (2009) Direct lateral lumbar interbody fusion for degenerative conditions: early complication profile. J Spinal Disord Tech 22:34–37

Marchi L, Abdala N, Oliveira L et al (2013) Radiographic and clinical evaluation of cage subsidence after stand-alone lateral interbody fusion. J Neurosurg Spine 19:110–118

Mayer HM (1997) A new microsurgical technique for minimally invasive anterior lumbar interbody fusion. Spine 22:691–699. discussion 700

Molinares DM, Davis TT, Fung DA, Liu JC-L, Clark S, Daily D et al (2016) Is the lateral jack-knife position responsible for cases of transient neurapraxia? J Neurosurg Spine 24:189–196

Orita S, Inage K, Furuya T, Koda M, Aoki Y, Kubota G et al (2017) Oblique lateral interbody fusion (OLIF): indications and techniques. Oper Tech Orthop 27:223–230

Rodgers WB, Gerber EJ, Patterson J (2011) Intraoperative and early postoperative complications in extreme lateral interbody fusion: an analysis of 600 cases. Spine 36:26–32

Uribe JS, Arredondo N, Dakwar E, Vale FL (2010a) Defining the safe working zones using the minimally invasive lateral retroperitoneal transpsoas approach: an anatomical study. J Neurosurg Spine 13:260–266

Uribe JS, Vale F, Dakwar D (2010b) Electromyographic monitoring and its anatomical implications in minimally invasive spine surgery. Spine 35:S368–S374

Winder MJ, Gambhir S (2016) Comparison of ALIF vs. XLIF for L4/5 interbody fusion: pros, cons, and literature review. J Spine Surg 2:2–8

Xu DS, Walker CT, Godzik J, Turner JD, Smith W, Uribe JS (2018) Minimally invasive anterior, lateral, and oblique lumbar interbody fusion: a literature review. Ann Transl Med 6:104

Youssef JA, McAfee PC, Patty CA, Raley E, DeBauche S, Shucosky E et al (2010) Minimally invasive surgery: lateral approach interbody fusion: results and review. Spine 35:S302–S311

Minimally Invasive Spine Surgery

36

Bilal B. Butt, Rakesh Patel, and Ilyas Aleem

Contents

What Is Minimally Invasive Spine Surgery (MIS)?	702
Advantages and Disadvantages of MIS	702
History of MIS	703
MIS Discectomy	703
MIS Laminectomy	705
MIS Transforaminal Lumbar Interbody Fusion (TLIF)	706
Lateral Interbody Fusion	707
Sacroiliac (SI) Joint Fusion	708
Application of MIS to Deformity Correction	709
Role of Lateral Interbody Fusions	712
Limitations of MIS in ADS	713
Role of Navigation in MIS	714
Robotics in MIS	714
References	714

Abstract

This chapter explores the basic principles and concepts of minimally invasive spine surgery (MIS). It provides technical insight into how these procedures are performed safely. By utilizing MIS techniques, one can largely treat the same conditions, which historically have been treated in the open fashion. Both short- and long-term advantages will be discussed including but not limited to decreased blood loss, decreased postoperative pain, and faster return to baseline. The application of these methods to deformity correction surgery and interbody fusions will also be explored. The roles of navigation and robotics in this rapidly expanding field and how they

B. B. Butt · R. Patel · I. Aleem (✉)
Department of Orthopaedic Surgery, University of Michigan, Ann Arbor, MI, USA
e-mail: bbutt@med.umich.edu; Rockpatelmd@gmail.com; ialeem@med.umich.edu

© Springer Nature Switzerland AG 2021
B. C. Cheng (ed.), *Handbook of Spine Technology*,
https://doi.org/10.1007/978-3-319-44424-6_129

can be utilized to improve accuracy are investigated. This chapter is targeted toward junior faculty members, residents, midlevel providers, and other individuals who wish to expand their knowledge base on MIS.

Keywords

Minimally invasive surgery · MIS · Pedicle screws · MIS TLIF

What Is Minimally Invasive Spine Surgery (MIS)?

Minimally invasive spine surgery (MIS) strives to correct surgical pathology, which is typically treated with larger incisions and greater tissue destruction, with the goal of better short- and long-term patient outcomes. Although long-term benefits are debatable, the short-term benefits, including decreased blood loss, decreased postoperative pain, decreased hospital stays, and faster return to baseline, have been well established (Lombardi et al. 2014; Tullberg et al. 1993; Obenchain 1991; Shamji et al. 2015; Terman et al. 2014; Parajon and Hartl 2017; Costanzo et al. 2014). Additionally, MIS techniques have been shown to decrease both the direct and indirect costs associated with certain surgical procedures (Shamji et al. 2015). By decreasing operative time, blood transfusions, and length of stay, the direct cost is significantly impacted. Earlier return to work and fewer postoperative hospital visits significantly decrease indirect costs. The goal of this chapter is to present the reader with current MIS techniques as well as a brief insight into the future of MIS.

Advantages and Disadvantages of MIS

There are several advantages and disadvantages of MIS techniques. A steep learning curve is associated with the safe implementation of MIS into one's practice, resulting in a lower than expected adaptation of this technique. Numerous studies have demonstrated that the first 20–30 cases of a surgeon's implementation of MIS may be associated with higher rates of complications (Sclafani and Kim 2014; Shamji et al. 2015; Fujibayashi et al. 2017). In addition to the steep learning curve, another barrier to adaption of MIS techniques is increased radiation exposure to the surgeon due to reliance on fluoroscopy. However, this risk may be minimized with the usage of intraoperative navigation.

Though introduced several decades ago, MIS techniques have made significant progress recently due to numerous technological advancements which have resulted in a numerous advantages of a less invasive approach. In utilizing an MIS approach, there is no need to detach the paraspinal muscles from their insertions on the spinous processes as compared to open techniques, thus minimizing muscle dissection and stripping (Pishnamaz and Schemmann 2018). Muscle injury in spinal surgery correlates with the length of time and force of the muscle retraction (Kawaguchi et al. 1996). With prolonged retraction, capillary perfusion is decreased and leads to accelerated rates of muscle fiber degeneration secondary to changes in cellular metabolism. The mechanism of this degeneration and necrosis are not yet fully elucidated, but most of these changes are believed to be associated with destruction of the sarcolemma and subsequent mitochondrial damage (Heffner and Barron 1978). Postoperative MRIs have demonstrated decreased cross-sectional area of paraspinal muscle, supporting the idea of muscle fiber atrophy following open surgery (Bresnahan et al. 2017) (Fig. 1). Stevens and colleagues used high-definition MRI to study the multifidus muscle postoperatively in patients undergoing MIS TLIF vs open TLIF. They observed significant intermuscular and intramuscular edema at the 6-month mark in those patients undergoing open TLIF. In patients who underwent MIS TLIF, no edema was present and overall the muscle appeared normal (Stevens et al. 2006). Levels of creatine kinase have also been used as a marker for muscle fiber injury. Open techniques have been shown to have a direct correlation with postoperative rises in creatine kinase levels, as compared to MIS, which show lower levels of CK

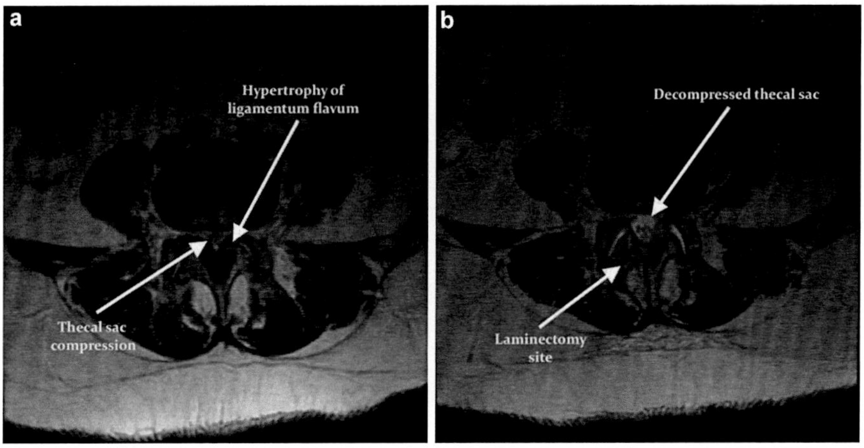

Fig. 1 (a–b) Comparison of pre-post op MIS laminectomy MRIs

(Wang et al. 2017). Cawley et al. were able to show that patients undergoing open surgery had abnormal postoperative EMG activation patterns in the lumbar multifidus as compared to those patients undergoing the same procedure via an MIS technique (Cawley et al. 2013). In evaluation of the sacrospinalis muscle using EMG, Wang et al. concluded that MIS TLIF was associated with reduced muscle damage as compared to open TLIF (Wang and FZ 2011). Newer data even suggest that with MIS techniques, the overall inflammatory state of the patient is decreased and this aids in shorter recovery periods as compared to open procedures (Lombardi et al. 2014). This is supported by lower levels of CRP, IL-6, and IL-10 following MIS procedures as compared to their conventional alternatives (Kim et al. 2006; Huang et al. 2005).

History of MIS

As a way to avoid excessive muscular retraction in spinal surgery, Wiltse et al. proposed a paraspinal sacrospinalis-splitting approach to the lumbar spine in 1968 (Wiltse 1973). The plane that Wiltse identified was an intermuscular plane between the multifidus muscle medially and the longissimus muscle laterally (Guiroy et al. 2018). Wiltse advocated that care must be taken to avoid overexposure of the vertebrae, as he had some concept of the negative consequences associated with excessive muscle stripping and damage. Because this approach utilizes an intermuscular plane, soft tissue trauma is minimized, and the posterior tension band of the spine and the supportive elements of the contralateral side are preserved (Anderson 2014). All of these taken together helped to improve patient outcomes following spinal surgery at the time.

MIS Discectomy

Disc herniations are painful and often debilitating conditions, which have a substantial impact on the function and quality of life of patients. There are also considerable social and economic impacts to society as most patients with disc herniations are of working age (Anderson et al. 2017). Given this, MIS discectomy may help mitigate some of the risks of surgery compared to open techniques and should be discussed with the patient if possible. Open surgery has been shown to be associated with longer operative times, longer incisions, increased bony resection, and increased retraction and damage to the paraspinal muscles as compared to MIS techniques (Ditsworth 1998; Rasouli et al. 2013; Alvi et al. 2018). MIS discectomy has been shown to have a shorter period of time off work, less opioid analgesia, less blood loss, and shorter hospital stays

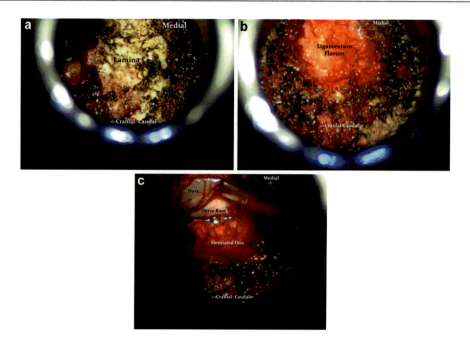

Fig. 2 (a–c) Intraoperative photos of discectomy through tubular dilator

(Tullberg et al. 1993; Kotil et al. 2007). It should be noted, however, that VAS scores both in short-term and long-term follow-up are essentially equivalent between groups of patients undergoing open procedures and those undergoing MIS procedures (Dasenbrock et al. 2012). Thus, both procedures ultimately decompress the neural elements and achieve pain relief.

As such, indications for MIS discectomy parallel those set forth for open procedures. Patients, who have failed conservative measures for a minimum of 6 weeks, have progressive motor weakness, or disabling pain can all be surgical candidates. As is standard, surgical indications should be evaluated on a case-by-case basis. The patient should always be included in the decision for surgery and appropriate informed consent should be obtained prior to surgery.

Obenchain described the first laparoscopic lumbar discectomy and was soon followed by Faubert and Caspar who published reports of lumbar percutaneous discectomy using a muscular retractor system in the early 1990s (Obenchain 1991; Faubert and Caspar 1991). This was the foundation by which Foley et al. built upon. Foley and colleagues used successive tubular dilators to achieve a desired diameter portal to which an endoscope was attached (Foley and Smith 1997). Foley's techniques were termed microendoscopic discectomy (MED) (Fig. 2).

Present day, usage of tubular retraction systems are common and are very much similar to Foley's initial description (Foley 2015). The patient should be positioned prone on a radiolucent spinal frame and prepped and draped in the usual fashion. Initially a 22-gauge spinal needle is introduced directed toward the facet joint. Careful attention is made to ensure that the needle does not aim midline, as this trajectory could puncture the dural sac and lead to a spinal fluid leak (Phillips et al. 2014). The location of the needle is confirmed using C-arm after obtaining orthogonal x-rays. Once the location is confirmed, the needle is removed, and a small, paraspinal incision is made, generally 2–2.5 cm lateral to midline. In cases where decompression of the contralateral is desired, the incision should be 3–4 cm lateral of the midline. If only an ipsilateral decompression is warranted, then the standard 2 cm from midline incision is sufficient. The incision should roughly be the same size as the

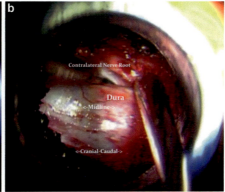

Fig. 3 (a–b) Intraoperative photos of Ipsilateral and contralateral laminectomy through tubular dilator

diameter of the intended tubular dilator (Phillips et al. 2014) (Fig. 3). Of note, in obese patients (BMIs >30), a more lateral incision may be necessary to obtain adequate visualization. There will be two distinct fascial layers present deep to skin incision. The superficial fascia represents that thoracodorsal fascial, and the deeper, thinner fascia represents that of the multifidus muscle (Schwender 2018). Both fascial incisions should extend slightly beyond that of the skin incision to allow for small adjustments by the surgeon. Once through the fascia of the multifidus, sequential tubular dilators are then used. The initial and smallest dilator is placed (docked) at the caudal edge of the lamina. Larger dilators are then placed over the initial dilator until an appropriately sized surgical window is created. Different procedures call for different diameter retractors. In the case of a microdiscectomy, 16–18 mm retractors are usually large enough for the procedure. The dilators are then removed and the retractor is placed in the muscular window. The retractor is then secured to the surgical table using a bracket mounted to the bed frame. Its location is then confirmed once again using fluoroscopy. Using a high-speed drill, a laminotomy is performed until the level of the ligamentum flavum. The flavum is then excised in a medial to lateral fashion using a Kerrison. The exposed nerve root is identified and protected and is gently retracted medially using a nerve root retractor. Using a bayonetted disc blade, an incision is made through the annulus fibrosis (Kimball and Yew 2013). Careful attention is paid to confirm the adequate decompression of the neural elements: thecal sac, nerve roots, and neural foramen. The surgical portal is then irrigated with saline, hemostasis is ensured, and the retractors are removed. The incision is closed in a layered, watertight fashion (Kulkarni et al. 2014).

MIS Laminectomy

A laminectomy in an appropriate selected patient can lead to significant reduction in neurogenic pain and its associated disability. It also has been shown to significantly improve patient-reported health-related quality of life (Shamji et al. 2015). Laminectomies are most often used to treat multi-level spinal stenosis, which is common in the aging population.

When evaluating the literature surrounding MIS laminectomy compared to open laminectomy, evidence supports that MIS procedures may be associated with less operative blood loss and shorter hospital stays (Terman et al. 2014). In a meta-analysis, Phan and Mobbs (2016) demonstrated that patients undergoing MIS laminectomies reported lower VAS scores as compared to the open approach, high rates of satisfaction, lower rates of blood loss, and thus lower rates of transfusions. They did note that reoperation rates were similar between both groups.

Much like the MIS discectomy, the MIS laminectomy utilizes the same overall approach.

The main differences that should be noted are the size of the tubular retractor is larger and there is more extensive bone and ligamentum flavum resection in order to obtain adequate neural decompression. Also it should be noted that the level of intended decompression will largely determine the necessary position for the tubular retractor. If the intended decompression is L1–L4, then the tubular retractor will be oriented more vertical and closer to the midline as compared to a decompression of L4–L5 or L5–S1 (Parajon and Hartl 2017). This is based on the anatomical bony structure of the vertebral bodies at those levels. Once the retractor is placed appropriately, a laminectomy is performed. The ligamentum flavum is then identified and removed. In cases where contralateral decompression is needed, the tube is repositioned medially; careful attention is needed as to not entrap soft tissue into the tube (Parajon and Hartl 2017). The table is then tilted away from the surgeon. The base of the spinous process is then drilled and undercut. The contralateral lamina is now removed using a high-speed drill and a Kerrison. Attention is now taken to the ligamentum flavum of the contralateral side and is removed. Some surgeons may benefit from utilization of a 90° Kerrison to aid them at this point. Once all of the flavum is removed, the table is then returned to its original position, hemostasis is ensured, and retractors are removed (Phillips et al. 2014; Watkins III and Watkins IV 2015).

MIS Transforaminal Lumbar Interbody Fusion (TLIF)

First described in early the 2000s, the TLIF provided an alternative to the standard posterior lumbar interbody fusion (PLIF) (Moskowitz 2002). A standard PLIF requires a midline incision through which exposure of the entire spinous process, bilateral lamina, and disc space is needed. This approach also places a fair amount of stress on the nerve roots as they are retracted out of the surgical field in order to garner access to the disc space. In the TILF a more lateral approach is made over the paraspinal muscles and directed toward the midline. This approach also allows for preservation of the contralateral side and requires less mobilization of the thecal sac and less risk of injury to a nerve root. There are also minimal retraction of the spinal nerves and decreased approach-related complications and morbidity as compared to the PLIF (Rosenberg and Mummaneni 2001). As a way to minimize the soft tissue trauma associated with open fusion procedures, Isaacs and colleagues described the minimally invasive transforaminal lumbar interbody fusion (MIS TLIF) (Hartl and Gelb 2017). In his original study, Issacs et al. compared their novel MIS TLIF techniques to standard single-level posterior interbody fusions at the same institutions. The authors concluded that patients undergoing the MIS TLIF had decreased hospital length of stay, decreased intraoperative blood loss, and received approximately 50% less postoperative narcotics as compared to the standard PLIF group (Isaacs et al. 2005). These outcomes were directly related the surgical approach in which normal tissue destruction was minimized. Common indications for a TLIF are foraminal stenosis, sagittal deformity, and central stenosis in patients with instability.

Following the same principles of tubular surgery described above, the MIS TLIF can be accomplished (Ozgur et al. 2006). Certain initial differences that should be highlighted are for one, the start point. The incision is initially made 4–5 cm lateral to the midline. This allows for an oblique entry into the spinal canal. As previously mentioned, the start point may have to be adjusted for larger patients. The desired visualized field for a TLIF is the inferior articulating facet joint of the level to be fused. In this, the capsule of the facet complex is entered and removed, and then the superior facet is resected down to the superior aspect of the pedicle (Hartl and Gelb 2017). The pedicle is then skeletonized. The ligamentum flavum is now exposed and can be removed in a piecemeal fashion using a Kerrison. The disc should now be visualized, and a discectomy is performed. Once the desired portion of disc is removed, the space is inspected to ensure adequate decompression. Bone graft and a structural implant are inserted to help preserve height and fuse the level. A MIS posterior fusion can sometimes be indicated. Because transverse processes are not exposed in the approach, the only surface area exposed following the decompression is the interbody space.

Lateral Interbody Fusion

First described by Pimenta at the Brazilian Spine Society Meeting in 2001 and via publication by Ozgur in the early 2000s, the lateral interbody fusion was a way to gain access into the lumbar spine via a minimally invasive far lateral approach. The procedure is performed via incisions that dissect down through the retroperitoneal fat and psoas muscle on to the vertebral body. The procedure provides good access of the anterior portion of the spine and accomplishes this without having to approach the spine via an anterior trans-peritoneal route (Ozgur et al. 2006). The use of a general surgeon is also avoided, as a spine surgeon can accomplish this minimally invasive method safely. In his report Ogzur notes that possible advantages of this procedure as compared to a standard anterior approach to the spine include no need for a general surgeon, no need to retract the aorta and IVC, simple operative technique as compared to laparoscopic methods, and avoidance complications of laparoscopic and open approaches. The entire procedure is performed under direct vision, and there is little to no impairment of the surgeon's depth perception. Serious complications of the standard anterior approach, damage to great vessels and superior hypogastric nerve plexus, are avoided because of the lateral entry. Some the most common complications associated with the lateral approach are sensory nerve injury and psoas muscle weakness (Fujibayashi et al. 2017).

Some of the main indications for patients to undergo a lateral interbody fusion are lumbar scoliosis, spondylolisthesis, foraminal or central stenosis, and according to newer reports corpectomy and stabilization in trauma patients (Isaacs et al. 2010). In this procedure, the patient is placed in the lateral decubitus position on a table that is able to flex. Attention is made to pad all bony prominences. The greater trochanter of the patient is located at the apex of the bend in the table. Of note, in choosing the entry side, a few considerations should be made. If the patient lacks a coronal plane deformity, then the preferred entry site is the left side of the patient, as the great vessels are located more anterior as compared to the right side (Pawar et al. 2015). If the patient has a coronal plane deformity, then the spine should be approached from the concavity of the lumbar curve. This allows for access to multiple levels of the spine, with a single skin incision. Once the desired side is chosen and the patient is positioned appropriately, the patient is secured to the table using tape or straps. Using fluoroscopy, true AP and lateral x-rays are taken, and the anterior and posterior borders of the vertebral body are identified. The patient is then prepped and draped in the usual sterile fashion (Fig. 4).

A skin incision is made in an oblique fashion from the anterior inferior caudal vertebral body to the posterior superior portion of the next adjacent vertebral body. The deep dissection continues through the subcutaneous fat and abdominal muscles to the retroperitoneal space. When dissecting through the abdominal muscles, attention is made to split muscles in line with the fibers. Between the internal oblique and the transverse abdominal muscle lie the iliohypogastric and ilioinguinal nerves, so care is made as to not cause excessive trauma to this region. Once at the retroperitoneal level, the surgeon can gently sweep the peritoneum anteriorly, lifting it off of the psoas muscle. Using intraoperative neuro-monitoring the fibers of the psoas muscle are splint in the anterior to middle third of the muscle (Ozgur et al. 2006). This location, coupled with neuro-monitoring, ensures that lumbar plexus nerve roots are not harmed. Once the level of disc space is reached, the location is confirmed with fluoroscopy. Now using tubular dilators, the surgical portal is enlarged until a self-retaining retractor is then introduced and secured. A discectomy is then performed in a standard fashion, and a structural implant is placed. Posteriorly, percutaneous pedicle screws can then be inserted as required.

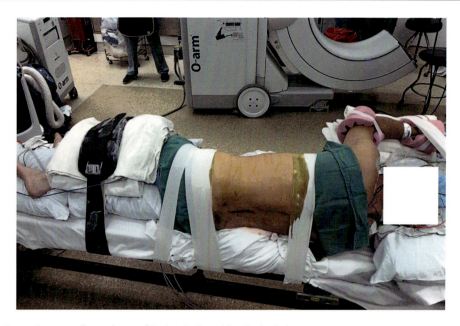

Fig. 4 Image demonstrating patient positioning for lateral interbody fusion

Sacroiliac (SI) Joint Fusion

The SI joint is a complex synovial joint that connects the spine to the pelvis via many ligamentous and muscular attachments. Imbalance between any of these can lead to altered biomechanics, which often lead to pain and disability (Hungerford et al. 2003). Often this pain is overlooked as a pain generator as patients may report many non-focal symptoms such as back, groin, or gluteal pain. Prior trauma to the pelvic region, prior lumbar fusion, and large body habitus are all risk factors for SI joint dysfunction. Once the SI joint is isolated as the source of the pain, non-operative treatments are initially recommended. Treatments such as physical therapy, exercise, steroid injections, NSAIDs, and in some cases nerve ablation are all recommended prior to surgery. If these measures fail and the patient reports persistent pain lasting greater than 6 months or a sudden worsening of nerve function, then surgery would be indicated. Historically the SI fusion initially was performed without any screws via an incision made over the posterior superior iliac spine, articular cartilage was removed, and bone graft was placed (Smith-Petersen 1921). This method called for long periods of external stabilization by either bracing or casting, to ensure that fusion occurred. Internal fixation for SI fusions began to appear in the literature in the 1980s. This eliminated postoperative bracing, but due to the morbidity of the approach, extent of bone grafting, and lengthy hospital stays, this was not favored among patients (Moore 1997). With the advent of MIS approaches to the SI joint, the open procedure fell out of favor. A 2012 survey of spine surgeons globally noted that 85% of SI joint fusions were occurring via MIS techniques (Smith et al. 2013). In comparing open fusions to MIS SI joint fusions, MIS has shown to have shorter surgical times, less blood loss, shorter duration of hospital stays, and larger decreases in postoperative VAS scores (Smith et al. 2013).

For a MIS SI fusion, the patient is positioned prone on radiolucent spine operating room table. The patient is then prepped and draped in the usual sterile fashion. Using fluoroscopy, the affected joint is localized. Using a lateral view in which the sacral slopes are super-imposed, the appropriate trajectory is identified (Miller and Block 2014). Next a 2 cm lateral incision is made. The tissue is dissected and a dilator is advanced through the incision until it contacts bone. Its location can be confirmed with fluoroscopy. Next the dilator is removed and guide pin is drilled, first into the outer cortex of the ilium. Once this location is confirmed and the pin is perpendicular to the SI joint, it is advanced until it abuts the sacral cortex. The guide pin remains in place, and a 9 mm dilator is placed over it. Attention is paid to ensure that no soft tissue becomes entrapped in the dilator. Next a cannulated drill is passed over the guidewire and only the ilium is drilled. These shavings of cortex are saved on the back table for use later in grafting. Attention is now turned to preparing the SI joint for fusion. This is accomplished via insertion of a flexible decorticator (Kube). The cartilage is removed and the joint space is partially decorticated. The joint is then irrigated with saline, and dilators are reinserted. Bone graft is inserted into the cavity. A guidewire is then replaced and passed into the sacral cortex; its location is confirmed with C-arm. A cannulated screw is then placed over the guidewire and into the sacrum. Wound closure occurs in a watertight fashion.

Application of MIS to Deformity Correction

The previous sections discussed both the origins and the applications of MIS techniques to common spinal procedures: discectomy, laminectomies, and single-level fusions. In this section we will explore the literature surrounding the usage and benefits of MIS application to the field of adult deformity surgery (ADS). Historically deformity correction surgery in adults was associated with a major complication rate around 7.6% (Glassman et al. 2007) and an overall complication rate as high as 70% (Anand et al. 2014a). Major patient risk factors for complications are a sagittal vertical axis (SVA) great than 4 cm, age greater than 60 years old, and more than three medical comorbidities (Auerbach et al. 2016). As is the case with other procedures in spinal surgery, the overall goals of deformity surgery are to decompress the neural elements that are being impinged and establish/restore the global sagittal alignment. It has been demonstrated in great detail that kyphosis is poorly tolerated in lumbar region of the spine and has a direct correlation with the severity of patient-reported symptoms (Glassman et al. 2004). In attempting to measure outcomes following major ADS, Lafage et al. (2009) noted that both SVA and pelvic tilt as a measure of pelvic position have the highest correlation with health-related quality of life. Failure to restore a SVA <50 mm and a pelvic tilt less than 20° has been show to be associated with poor clinical outcomes. These goals can now be accomplished using the MIS techniques previously described and in some instances have better patient outcomes than conventional open procedures. Each clinical scenario is unique and requires a thoughtful and methodical process in planning for surgery. While MIS techniques are often sufficient to accomplish the goal, at times, there is a mix of MIS procedures and open surgery, termed hybrid surgery.

Percutaneous pedicle screw fixation's (PPSF) role in spinal deformity and spine trauma has been shown to be a safe and efficacious alternative to open surgery. Briefly, in this application, the patient is positioned prone on a radiolucent spine table, with bony prominences padded. The type of intraoperative imaging used is at the discretion of the surgeon as both navigation and fluoroscopy have been shown to be safe for pedicle screw placement (Park et al. 2010). This overview details usage of intraoperative biplanar fluoroscopy. X-rays are taken in the AP plane prior to any incisions to ensure that the superior endplate is flat (Anderson et al. 2007) and the pedicle-spinous process interface form an imaginary inverted "V." In the lateral view, careful attention is made to ensure that a single flat superior endplate and only a single pedicle shadow are identified. In obtaining orthogonal views, the relative positions of landmarks are identified (Fig. 5) (Aleem et al. 2017). An incision is then made approximately 4.5 cm lateral to the pedicle border. The fascia is incised and blunt dissection is used to obtain access to the junction between the transverse process and facet. A Jamshidi needle is then used to violate the dorsal pedicle in a lateral to medial fashion. Using AP and lateral imaging, the Jamshidi is advanced to the posterior cortex of the

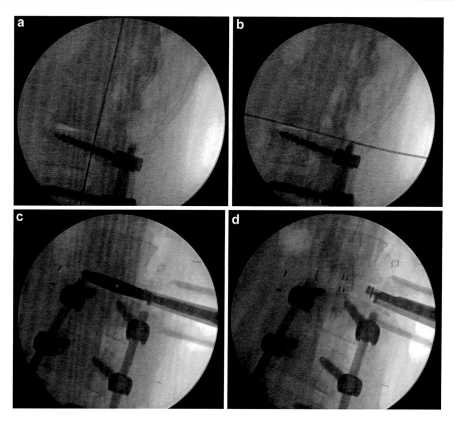

Fig. 5 (**a–d**) Intraoperative fluoroscopic images demonstrating level confirmation, endplate preparation, and implant plantation

pedicle (Figs. 6 and 7). The needle should be located in the center of the pedicle on lateral imaging, and it should never cross the medial border of the pedicle on AP imaging. Once in a satisfactory location, the needle is removed and replaced with a guidewire. A tap is used over the guidewire to expand the cortical opening. The guidewire should not be advanced beyond its initial placement, as this could potentially injure the great vessels located deep to it. Once tapping is completed, the tap is removed and replaced with a cannulated pedicle screw (Fig. 8). This process is then repeated, as indicated by the pathology. Once all screws have been placed, a rod is introduced usually from the most proximal screw's incision, and the desired reduction is performed. Aleem et al. described the technique of MIS screw fixation in detail (Fig. 9).

In his study Tinelli et al. demonstrated that using a MIS system in the setting of spinal trauma, his group was able to accurately place almost 98% of 682 pedicle screws in 131 fractures. The remaining 2% of screws were suboptimally placed, but not to the extent where revision surgery was necessary (Tinelli et al. 2014). Anand et al. were able to show that correction of adult lumbar degenerative scoliosis could be corrected with PPSF. He reported that multi-segment spinal corrections could be performed with less blood loss and less morbidity than open corrections (Anand et al. 2008).

Intraoperative fluoroscopy is a necessity in most cases when attempting PPSF in patients with deformity. For proper screw placement, it is imperative that both tilting view and wig-wag views are obtained if the case calls for it. As a technical note, one must ensure that one is orthogonal to the targeted pedicle to ensure proper location. If the operative case is not technically demanding and the surgeon is

Fig. 6 AP image showing Jamshidi needle docked at start point on lateral edge of the pedicle at roughly the 9 o'clock position

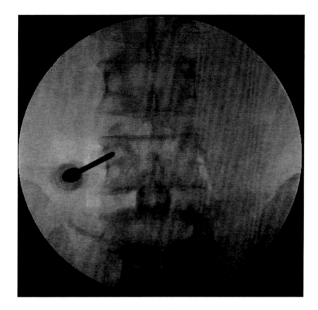

Fig. 7 Lateral fluoroscopic image showing Jamshidi needled in center of pedicle

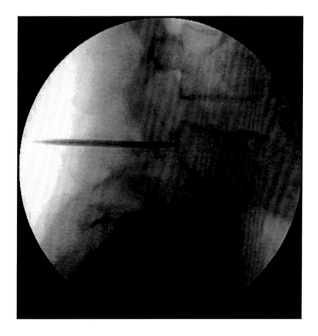

experienced enough, use of a single anteroposterior C-arm can be sufficient for proper screw placement. Ahmad and Wang (2014) demonstrated this, when 410 pedicle screws were placed in patients with at least 10° of axial rotation. He noted that he had 15 grade 1 violations, 6 grade 2 violations, and 8 grade 3 violations and only 2 screws were required to be revised. Of note the Gertzbein classification is most often used when discussing pedicle screw placement and location relative its medial or lateral wall. There are four grades in the classification ranging from 0 to 3. Grade 0 indicates that there is no breech of pedicle; grade 1, <2 mm breech; grade 2, 2–4 mm breech; and grade 3, >4 mm breech of the pedicle.

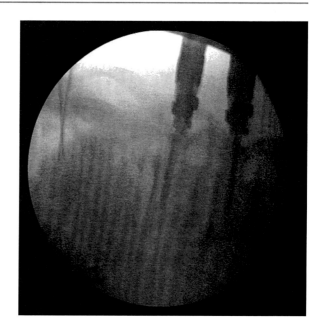

Fig. 8 Lateral fluoroscopic image showing pedicle screws with attachments

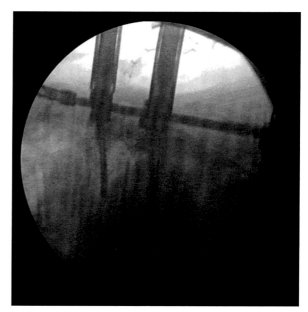

Fig. 9 Lateral fluoroscopic image showing rod capture in all screw heads

Role of Lateral Interbody Fusions

Lateral MIS approaches to the spine have numerous advantages compared to anterior approaches and however may be limited in their ability to sufficiently correct sagittal deformities in adults in isolation (Costanzo et al. 2014). In his systematic review, Costanzo et al. looked at the role of MIS lateral lumbar interbody fusions in sagittal balance and spinal deformity. He concluded that there is no clear answer with regard to how well MIS can correct sagittal balance and noted that open posterior osteotomies would continue to be the gold standard in sagittal balance correction (Costanzo et al. 2014). Acosta et al. performed a retrospective radiographic study looking at changes in coronal and sagittal plane alignments following lateral interbody fusions. Statistical improvements in the visual analog scale (VAS), the Oswestry Disability Indices (ODI), and the coronal Cobb angle were noted; however, no statistically significant change in the overall sagittal alignment was identified by a postoperative SVA measurements. They concluded that direct lateral interbody fusions alone are insufficient to correct for sagittal imbalance (Acosta et al. 2011). Deukmedjian et al. evaluated a novel technique for attempting to restore a normal SVA. In their study, they utilized a MIS lateral approach to first release the anterior longitudinal ligament and place a 30° hyperlordotic cage. Following this, percutaneous pedicle screws were placed posteriorly to help stabilize the construct. This resulted in a 17° segmental lordosis increase per level as well as an overall SVA decrease of 49 mm and a 7° pelvic tilt (Deukmedjian et al. 2012). Manwaring noted that a two-stage MIS procedure was comparable to Smith-Peterson osteotomies (SPO), because of its ability of providing disc height and correcting coronal imbalance (Manwaring et al. 2014). The first stage of the procedure involved lateral interbody fusions with or without anterior column releases (ACR). The second stage involved PPSF. A 12° improcvement in segmental lordosis and a 31 mm improvement in SVA per ACR level released were noted. Anand et al. have since adopted these principles of staged MIS procedures and proposed a protocol for MIS correction of adult spinal deformity (Anand et al. 2017). Much like Manwaring, Anand proposed that a lateral interbody fusion should occur in the first stage, with or without an ACR. He reports that avoiding an open surgery can avoid potentially serious postoperative complications.

Wang et al. (2014) described the ceiling effects for deformity correction of three different spinal surgery techniques: stand-alone (lateral MIS procedure), circumferential MIS (combined lateral with posterior), and hybrid procedures. The authors note that the ceiling effect in the coronal plane for all three procedures were as follows: 23° for stand-alone, 34 for cMIS, and 55° for the hybrid procedure. A statically significant alteration in the SVA occurred only in the hybrid procedure group, but this was overshadowed by high rates of complications in the hybrid group. Anand et al. (2014b) previously reported that the max SVA correction obtainable is 10 mm utilizing MIS techniques without osteotomies.

Limitations of MIS in ADS

As already noted, not all patients can or should undergo a MIS procedure. The decision to undergo a MIS procedure is ultimately left up to the shared decision-making of the surgeon as well as the patient. The goal should be to safely address surgical pathology and provide the best clinical outcome for the patient. In cases where the decision to proceed with a MIS procedure for ADS is made, some important patient factors should be considered, such as presenting symptoms, physical exam findings, and radiographic findings. Utilizing MIS in deformity surgery presents some unique limitations such as limited sagittal correction, decreased ability for in situ bending and compression, concern for sub-optimal correction, and pseudoarthrosis if interbody fusions are not performed. Since MIS procedures contain some level of a learning curve, inexperienced surgeons are likely to have increased operative times and increased cost of service as well as potentially increased radiation exposure to the patient and surgical team.

As a way to help surgeons select patients that can possibly benefit from MIS, the International Spine Study Group (ISSG) published a rational framework for decision-making in 2014. In this algorithm radiographic parameters are used to guide decision-making. The parameters used in the decision-making tree are SVA, PT, LL-PI mismatch, coronal Cobb angle, curve flexibility, and amount of listhesis. At its core the algorithm is based upon the idea that MIS is limited in its ability to treat sagittal plane deformities (Mummaneni et al. 2014).

Role of Navigation in MIS

As surgical technologies continue to advance, their contributions to surgical procedures are continually investigated. In the last 20 years, image guidance and navigation have come a long way in assisting the surgeon in accurate and safe positioning of hardware. Tajsic et al. (2018) evaluated and compared C-arm navigated, O-arm navigated, and conventional 2D fluoroscopy-assisted MIS techniques. Outcomes that were analyzed included operating time, radiation exposure, and the accuracy of pedicle screw placements. They concluded that pedicle screws placed with the assistance of the O-arm had the lowest rate of malpositioning (1.23%) and screws placed with 2D fluoroscopy were misplaced 5.16% of the time. However, O-arm usage was associated with highest rate of single image radiation exposure as compared to the other two modalities. Among all three modalities, operating room time was comparable. They concluded that given increased accuracy of pedicle screw placement, acceptable doses of overall radiation exposure, and comparable operating room time, the O-arm is the best form of intra-op navigation. Other studies have validated the usage of O-arm in MIS surgeries (Kleck et al. 2018; Chachan et al. 2018).

Robotics in MIS

As surgeons attempt to tackle more complex cases in the aging population, the indications for surgical fixation continue to evolve. As such, methods of attempting to reduce overall radiation to the patient and surgical team also evolve. The use of robotic-assisted pedicle screw placements has been discussed in the literature as a way to circumvent excessive intra-op radiation exposure. To our knowledge there has been only one randomized controlled trial comparing MIS robotics to open fluoroscopic-guided posterior lumbar interbody fusion (Hyun et al. 2017). The average per-screw radiation in the robotic-assisted surgeries was 37.5% of the per-screw exposure in the fluoroscopic group. Over all there was a mean reduction in radiation of 62.5% in the group undergoing robotic-assisted surgery. The results of the study are promising, but further data is needed to validate the routine use of robotics in MIS spinal surgeries.

References

Acosta FL et al (2011) Changes in coronal and sagittal plane alignment following minimally invasive direct lateral interbody fusion for the treatment of degenerative lumbar disease in adults: a radiographic study. J Neurosurg Spine 15(1):92–96. https://doi.org/10.3171/2011.3.spine10425

Ahmad FU, Wang MY (2014) Use of anteroposterior view fluoroscopy for targeting percutaneous pedicle screws in cases of spinal deformity with axial rotation. J Neurosurg Spine 21(5):826–832. https://doi.org/10.3171/2014.7.spine13846

Aleem IS, Park P, La Marca F, Patel R (2017) Minimally invasive pedicle screw placement for applications in trauma and tumor surgery. Oper Tech Orthop 27(4):217–222

Alvi MA et al (2018) Operative approaches for lumbar disc herniation: a systematic review and multiple treatment meta-analysis of conventional and minimally invasive surgeries. World Neurosurg 114:391

Anand N et al (2008) Minimally invasive multilevel percutaneous correction and fusion for adult lumbar degenerative scoliosis. J Spinal Disord Tech 21(7):459–467. https://doi.org/10.1097/bsd.0b013e318167b06b

Anand N et al (2014a) Evidence basis/outcomes in minimally invasive spinal scoliosis surgery. Neurosurg Clin N Am 25(2):361–375. https://doi.org/10.1016/j.nec.2013.12.014

Anand N et al (2014b) Limitations and ceiling effects with circumferential minimally invasive correction techniques for adult scoliosis: analysis of radiological outcomes over a 7-year experience. Neurosurg Focus 36(5). https://doi.org/10.3171/2014.3.focus13585

Anand N et al (2017) A staged protocol for circumferential minimally invasive surgical correction of adult spinal deformity. Neurosurgery 81(5):733–739. https://doi.org/10.1093/neuros/nyx353

Anderson DG (2014) Lumbar decompression using a tubular retractor system. In: Minimally invasive spine surgery: surgical techniques and disease management, by Sapan D Gandhi. Springer, New York

Anderson DG, Samartzis D, Shen FH et al (2007) Percutaneous instrumentation of the thoracic and lumbar spine. Orthop Clin North Am 38:401–408. [abstract vii]

Anderson MO et al (2017) Return to work after lumbar disc surgery is related to the length of preoperative sick leave. Dan Med J 64(7). pii:A5392

Auerbach JD et al (2016) Delayed postoperative neurologic deficits in spinal deformity surgery. Spine 41(3). https://doi.org/10.1097/brs.0000000000001194

Bresnahan LE et al (2017) Assessment of paraspinal muscle cross-sectional area after lumbar decompression. Clin Spine Surg 30(3):E162

Cawley DT et al (2013) Multifidus innervation and muscle assessment post-spinal surgery. Eur Spine J 23(2):320–327

Chachan S et al (2018) Cervical pedicle screw instrumentation is more reliable with O-arm-based 3D navigation: analysis of cervical pedicle screw placement accuracy with O-arm-based 3D navigation. Eur Spine J. https://doi.org/10.1007/s00586-018-5585-1

Costanzo G et al (2014) The role of minimally invasive lateral lumbar interbody fusion in sagittal balance correction and spinal deformity. Eur Spine J 23(S6):699–704. https://doi.org/10.1007/s00586-014-3561-y

Dasenbrock HH et al (2012) The efficacy of minimally invasive discectomy compared with open discectomy: a meta-analysis of prospective randomized controlled trials. J Neurosurg Spine 16(5):452–462

Deukmedjian AR et al (2012) Early outcomes of minimally invasive anterior longitudinal ligament release for correction of sagittal imbalance in patients with adult spinal deformity. Sci World J 2012:1–7. https://doi.org/10.1100/2012/789698

Ditsworth DA (1998) Endoscopic transforaminal lumbar discectomy and reconfiguration: a postero-lateral approach into the spinal canal. Surg Neurol 49(6):588–598

Faubert C, Caspar W (1991) Lumbar percutaneous discectomy. Neuroradiology 33(5):407–410. https://doi.org/10.1007/bf00598613

Foley KT (2015) Microendoscopic discectomy for lumbar disc herniations: paramedian and far lateral approaches. In: Surgical approaches to the spine. Springer, New York

Foley KT, Smith MM (1997) Microendoscopic discectomy. Techn Neurosurg 3:3017

Fujibayashi S et al (2017) Complications associated with lateral interbody fusion. Spine 42(19):1478–1484. https://doi.org/10.1097/brs.0000000000002139

Glassman S et al (2004) P90. The impact of positive sagittal balance in adult spinal deformity. Spine J 4(5). https://doi.org/10.1016/j.spinee.2004.05.231

Glassman SD et al (2007) The impact of perioperative complications on clinical outcome in adult deformity surgery. Spine 32(24):2764–2770. https://doi.org/10.1097/brs.0b013e31815a7644

Guiroy A et al (2018) How to perform the wiltse posterolateral spinal approach: technical note. Surg Neurol Int 9(1):38

Hartl R, Gelb D (2017) Step-by-step guide: key steps in a MIS TLIF procedure. AO Spine. https://aospine.aofoundation.org/Structure/education/online-education/mis-material/Pages/mis-material.aspx

Heffner Rr, Barron Sa (1978) The early effects of ischemia upon skeletal muscle mitochondria. J Neurol Sci 38(3):295–315

Huang T-J et al (2005) Less systemic cytokine response in patients following microendoscopic versus open lumbar discectomy. J Orthop Res 23(2):406–411

Hungerford B et al (2003) Evidence of altered lumbopelvic muscle recruitment in the presence of sacroiliac joint pain. Spine 28(14):1593–1600

Hyun S-J et al (2017) Minimally invasive robotic versus open fluoroscopic-guided spinal instrumented fusions. Spine 42(6):353–358. https://doi.org/10.1097/brs.0000000000001778

Isaacs RE et al (2005) Minimally invasive microendoscopy-assisted transforaminal lumbar interbody fusion with instrumentation. J Neurosurg Spine 3(2):98–105. https://doi.org/10.3171/spi.2005.3.2.0098

Isaacs RE et al (2010) A prospective, nonrandomized, multicenter evaluation of extreme lateral interbody fusion for the treatment of adult degenerative scoliosis. Spine 35(Supplement):S322

Kawaguchi Y et al (1996) Back muscle injury after posterior lumbar spine surgery. Spine 21(8):941–944

Kim K-T et al (2006) The quantitative analysis of tissue injury markers after mini-open lumbar fusion. Spine 31(6):712–716

Kimball J, Yew A (2013) Minimally invasive surgery for lumbar microdiscectomy. Neurosurg Focus 35(2 Suppl): Video 15

Kleck CJ et al (2018) One-step minimally invasive pedicle screw instrumentation using O-arm and stealth navigation. Clin Spine Surg:1. https://doi.org/10.1097/bsd.0000000000000616

Kotil K et al (2007) Serum creatine phosphokinase activity and histological changes in the multifidus muscle: a prospective randomized controlled comparative study of discectomy with or without retraction. J Neurosurg Spine 6(2):121–125

Kube R. et al (2016) Sacroiliac joint fusion: one year clinical and radiographic results from minimally invasive sacroiliac joint fusion surgery. Open Ortho J 10:679–689

Kulkarni A et al (2014) Microendoscopic lumbar discectomy: technique and results of 188 cases. Indian J Orthop 48(1):81

Lafage V et al (2009) Pelvic tilt and truncal inclination. Spine 34(17). https://doi.org/10.1097/brs.0b013e3181aad219

Lombardi G et al (2014) Is minimally invasive spine surgery also minimally pro-inflammatory? Muscular markers, inflammatory parameters and cytokines to quantify the operative invasiveness assessment in spine fusion. Eur J Inflamm 12(2):237–249

Manwaring JC et al (2014) Management of sagittal balance in adult spinal deformity with minimally invasive anterolateral lumbar interbody fusion: a preliminary radiographic study. J Neurosurg Spine 20(5):515–522. https://doi.org/10.3171/2014.2.spine1347

Miller LE, Block JE (2014) Minimally invasive arthrodesis for chronic sacroiliac joint dysfunction using the SImmetry SI Joint Fusion system. Med Dvices May 7;7:125–30

Moore MR (1997) Surgical treatment of chronic painful sacroiliac joint dysfunction. In: Movement, stability, and low back pain: the essential role of the pelvis. Churchill Livingstone, New York, pp 563–572

Moskowitz A (2002) Transforaminal lumbar interbody fusion. Orthop Clin N Am 33(2):359–366. https://doi.org/10.1016/s0030-5898(01)00008-6

Mummaneni PV et al (2014) The minimally invasive spinal deformity surgery algorithm: a reproducible rational framework for decision making in minimally invasive spinal deformity surgery. Neurosurg Focus 36(5). https://doi.org/10.3171/2014.3.focus1413

Obenchain TG (1991) Laparoscopic lumbar discectomy: case report. J Laparoendosc Surg. https://doi.org/10.1089/lps.1991.1.145

Ozgur BM et al (2006) Extreme lateral interbody fusion (XLIF): a novel surgical technique for anterior lumbar interbody fusion. Spine J 6(4):435–443. https://doi.org/10.1016/j.spinee.2005.08.012

Parajon A, Hartl R (2017) Step-by-step guide: minimally invasive tubular approaches to lumbar spine decompression and dural repair – surgical techniques. AO Spine. https://aospine.aofoundation.org/Structure/education/online-education/mis-material/Pages/mis-material.aspx

Park P, Foley KT, Cowan JA et al (2010) Minimally invasive pedicle screw fixation utilizing O-arm fluoroscopy with computer-assisted navigation: feasibility, technique, and preliminary results. Surg Neurol Int 1:44

Pawar A et al (2015) Lateral lumbar interbody fusion. Asian Spine J 9(6):978

Phan K, Mobbs RJ (2016) Minimally invasive versus open laminectomy for lumbar stenosis. Spine 41(2):E91

Phillips FM et al (2014) Minimally invasive spine surgery: surgical techniques and disease management. Springer, New York

Pishnamaz M, Schemmann U (2018) Muscular changes after minimally invasive versus open spinal stabilization of thoracolumbar fractures: a literature review. J Musculoskelet Nueronal Interact 18(1):62–70

Rasouli MR et al (2013) Minimally invasive discectomy versus microdiscectomy/discectomy for symptomatic lumbar disc herniation. Cochrane Database Syst Rev Sept 4;(9) CD010328

Rosenberg WS, Mummaneni PV (2001) Transforaminal lumbar interbody fusion: technique, complications, and early results. Neurosurgery 48(3):569–575. https://doi.org/10.1097/00006123-200103000-00022

Schwender JD, Shafa E (2018) Minimally invasive posterior lumbar instrumentation. In: Rothman and Simeone the spine 7e, Chapter 55, by Elsevier, Inc

Sclafani JA, Kim CW (2014) Complications associated with the initial learning curve of minimally invasive spine surgery: a systematic review. Clin Orthop Relat Res 472(6):1711–1717. https://doi.org/10.1007/s11999-014-3495-z

Shamji MF et al (2015) Minimally invasive spinal surgery in the elderly. Neurosurgery 77:S108

Smith A et al (2013) Open versus minimally invasive sacroiliac joint fusion: a multi-center comparison of perioperative measures and clinical outcomes. Ann Surg Innov Res 7(1):14

Smith-Petersen MN (1921) Arthrodesis of the sacroiliac joint. A new method of approach. J Bone Joint Surg Am 3(8):400–405

Stevens KJ et al (2006) Comparison of minimally invasive and conventional open posterolateral lumbar fusion using magnetic resonance imaging and retraction pressure studies. J Spinal Disord Tech 19(2):77–86

Tajsic T et al (2018) Spinal navigation for minimally invasive thoracic and lumbosacral spine fixation: implications for radiation exposure, operative time, and accuracy of pedicle screw placement. Eur Spine J. https://doi.org/10.1007/s00586-018-5587-z

Terman SW et al (2014) Minimally invasive versus open transforaminal lumbar interbody fusion: comparison of clinical outcomes among obese patients. J Neurosurg Spine 20(6):644–652

Tinelli M et al (2014) Correct positioning of pedicle screws with a percutaneous minimal invasive system in spine trauma. Orthop Traumatol Surg Res 100(4):389–393. https://doi.org/10.1016/j.otsr.2014.03.015

Tullberg T et al (1993) Does microscopic removal of lumbar disc herniation lead to better results than the standard procedure? Spine 18(1):24–27

Wang HL, FZ LU (2011) Minimally invasive lumbar interbody fusion via MAST Quadrant retractor versus open surgery: a prospective randomized clinical trial. Chin Med J 124:3868–3874

Wang MY et al (2014) Less invasive surgery for treating adult spinal deformities: ceiling effects for deformity correction with 3 different techniques. Neurosurg Focus 36(5). https://doi.org/10.3171/2014.3.focus1423

Wang Y-P et al (2017) Comparison of outcomes between minimally invasive transforaminal lumbar interbody fusion and traditional posterior lumbar intervertebral fusion in obese patients with lumbar disk prolapse. Ther Clin Risk Manag 13:87–94

Watkins RG III, Watkins RG IV (2015) Surgical approaches to the spine. Springer, New York

Wiltse LL (1973) The paraspinal sacrospinalis-splitting approach to the lumbar spine. Clin Orthop Relat Res 91:48–57

Cervical Spine Anatomy

37

Bobby G. Yow, Andres S. Piscoya, and Scott C. Wagner

Contents

Introduction	718
Atlantoaxial Complex	718
C1	718
C2	720
Subaxial Cervical Spine	720
Vertebral Body	722
Pedicles	722
Transverse Process	722
Facet Joint	722
Spinal Canal	723
Lateral Mass	723
Lamina and Spinous Process	723
Ligaments	723
Intervertebral Disc	724
Fascia	725
Investing	725
Pre-tracheal	725
Prevertebral	725
Muscles	725
Ventral	725
Dorsal	727

B. G. Yow (✉) · A. S. Piscoya · S. C. Wagner
Department of Orthopaedic Surgery, Walter Reed National Military Medical Center, Bethesda, MD, USA
e-mail: bobby_yow@alumni.brown.edu; apiscoya23@gmail.com; scott.cameron.wagner@gmail.com

© Springer Nature Switzerland AG 2021
B. C. Cheng (ed.), *Handbook of Spine Technology*,
https://doi.org/10.1007/978-3-319-44424-6_2

Neurovascular Structures	729
Spinal Cord	729
Meninges and Dura	730
Nerve Roots	730
Spinal Cord Blood Supply	731
Important Ventral Structures	731
Carotid Sheath	731
Vertebral Artery	731
Superior Laryngeal Nerve	731
Inferior (Recurrent) Laryngeal Nerve	733
Hypoglossal Nerve	733
Sympathetic Chain	733
Surgical Anatomy: The Cervical Triangles	734
Ventral (4 Types)	734
Dorsal (2 Types)	734
References	734

Abstract

An in depth understanding of cervical spine anatomy is essential to the diagnosis and management of cervical spine pathology. From the osseous anatomy down to the soft tissues structures that function to stabilize, maintain, and protect the spinal cord clinicians must be able to appreciate the biomechanics and the complex anatomy. For patients managed operatively, appropriate surgical planning and operative technique rely heavily upon a sound understanding of the intricate anatomy in this region. In this chapter we detail the cervical spine anatomy with particular emphasis on the osseous, muscular, ligamentous, and neurovascular tissues with the goal of providing clinicians a comprehensive review that they can depend on and refer to when treating patients with cervical spine disease.

Keywords

Anatomy · Cervical spine · Vertebrae · Review

Introduction

The cervical spine consists of seven vertebrae. Each is named according to its corresponding order from C1 cranially to C7 caudally. After completion of embryonic development first two vertebrae are unique in that they do not contain a typical vertebral body and are referred to as the "atlas" (C1) and "axis" (C2), respectively (Fig. 1). These two vertebrae form the atlantoaxial complex of the upper cervical spine. The remaining vertebra (C3–C7) is referred to as the subaxial cervical spine. The overall sagittal plane alignment is concave posteriorly, resulting in an overall lordotic curvature to a normal cervical spine. Each vertebra has an associated nerve root exiting bilaterally above the pedicle and through the foramina, which are referred to as C1–C7. The C8 nerve root exits below the pedicle of C7.

Atlantoaxial Complex

C1

The C1 vertebra (*atlas*) is a ring-shaped structure that is unique from C3 to C7 in that it lacks a robust vertebral body. It instead has an anterior tubercle, which is the attachment site for the longus colli muscle. On the posterior aspect of the ring is the posterior tubercle, which provides attachment points for the rectus capitis posterior minor muscle and the suboccipital membrane. Both the lateral aspects of the anterior and posterior tubercles create semicircular structures called the anterior and posterior arches. The junction of these arches on the lateral-most aspect of the rings

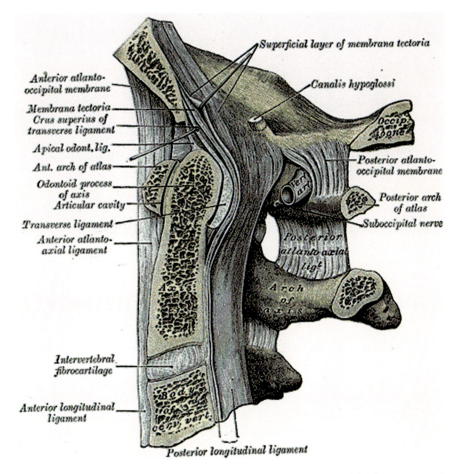

Fig. 1 Midsagittal graphical representation of the upper cervical spine. (By Henry Vandyke Carter – Henry Gray (1918) *Anatomy of the Human Body: Gray's Anatomy*, Plate 308, Public Domain)

form the lateral masses. Each lateral mass is composed of an articular process on both the superior and inferior aspects of the lateral mass. The superior articular process is concave and oriented inferiorly and medially, allowing for a congruent articulation with the occiput bilaterally. Inferiorly, the processes are oriented with a sloped angle inferiorly and laterally, allowing for an articulation with the superior articular processes of C2 (*axis*) (Daniels et al. 1983; Parke and Sherk 1989). Just posterior to the lateral mass is a subtle groove, in which the vertebral artery courses. The atlas' superior and inferior articulations allow for primarily flexion and extension, as well as lateral bending (Panjabi et al. 1991a). On the posterior aspect of the anterior tubercle, on the inner aspect of the anterior arch at the midline, is a subtle indention that allows for an articulation with the dens of the axis (Fig. 2). Additionally, there are insertion sites just anterior and medial to the lateral masses on the inner aspect of the ring for the transverse ligaments, which also attach to the posterior aspect of the dens of the axis. Extending laterally from each lateral mass is the transverse process. The transverse process at this level houses the vertebral arteries before they exit the vertebral column and enter the foramen magnum in order to become an intracranial structure. A subset of the population has an anatomic variation consisting of an osseous bridge that covers the ridge containing the vertebral artery; this variation is termed an arcuate foramen or ponticulus posticus and is estimated to have a 3–15% prevalence in the population.

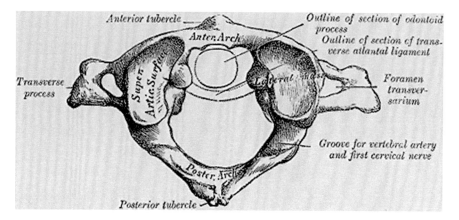

Fig. 2 Axial representation of the C1 vertebra, or atlas. (By Henry Vandyke Carter – Henry Gray (1918) *Anatomy of the Human Body: Gray's Anatomy*, Plate 86, Public Domain)

C2

Like C1, the C2 vertebra (*axis*) is unique in that has a bony prominence that projects cranially termed the odontoid process or dens. Anteriorly, the odontoid has a synovial articulation with the posterior aspect of the anterior tubercle of the atlas. This articulation allows for nearly half of the rotatory movement of the head and cervical spine. The posterior aspect of the odontoid has two subtle prominences that are the attachment sites of the alar ligaments, which span the atlas and attach medial to the posterior occipital condyles. These ligaments insert laterally at the base of the skull and provide stabilization of the atlantoaxial complex. The apical ligament attaches at the apex of the odontoid, which subsequently attaches to the anterior aspect of the foramen magnum at the midline. The primary stabilizer of the odontoid to the atlas is the transverse ligament, which is a structure that spans the anterior arch of the atlas and is a restraint to lateral displacement of the odontoid. The transverse ligament also has cephalad and caudal projections at the midline termed the cruciform ligaments which provide additional stability to the atlantoaxial complex. Posteriorly, the axis contains a bifid spinous process that serves as the attachment site for the rectus capitis posterior major and obliquus capitis inferior muscles attach. On the lateral aspects of the axis, there are both superior and inferior articular processes which articulate with the analogous structures on the adjacent vertebrae. The superior facet is sloped inferior laterally, congruent with the inferior facets of the lateral masses of the atlas. It is important to note that the sagittal diameter of the spinal canal in the upper cervical spine is greater than that of the lower cervical spine, allowing for adequate space for both the odontoid and the spinal cord. Steel's rule of thirds classically states that one-third of the canal diameter is occupied by the odontoid, one-third by the spinal cord, and the remaining one-third as free space preventing compression of the cord (Ebraheim et al. 1998). The transverse processes of the axis are directed caudally, containing the foramen transversaria that houses the vertebral arteries (Fig. 3).

Subaxial Cervical Spine

Unlike C1 and C2, the osteology of the five subaxial vertebrae share much in common. Each consists of a vertebral body anteriorly, which are separated by an intervertebral disc, bilateral transverse processes containing a foramen, pedicles, as well as two facet joints and a spinous process posteriorly. The subaxial spine, much like the entirety of the vertebral column, is to resist the compressive loads that are placed upon it. Additionally, the bony structures of the cervical spine act to protect important neurovascular structures and provide stability while allowing for functional flexion, extension, and lateral bending (Fig. 4a, b).

Fig. 3 Axial representation of the C2 vertebra, or atlas. (By Henry Vandyke Carter – Henry Gray (1918) *Anatomy of the Human Body: Gray's Anatomy*, Plate 87, Public Domain)

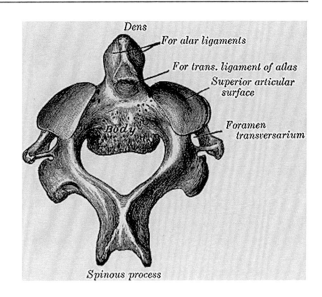

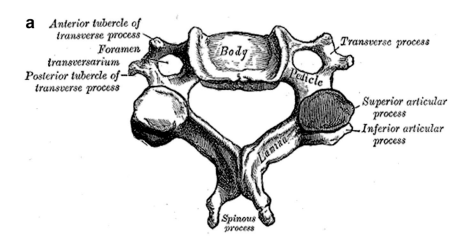

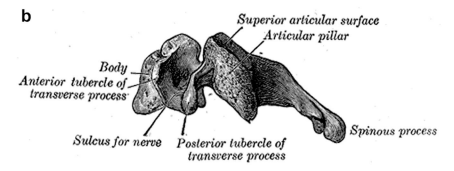

Fig. 4 (**a**, **b**) Axial and lateral representations of a subaxial cervical vertebra. (From: Henry Vandyke Carter – Henry Gray (1918) *Anatomy of the Human Body*, Plate 84–85, Public Domain)

Vertebral Body

The anterior aspect of the vertebra is a relatively cylindrical structure called the vertebral body. The body withstands the majority of the compressive loads placed on the vertebral column, and each is separated by an intervertebral disc that functions as a shock absorber. With descending levels down the spinal column, the height of the body slightly increases in height, with the occasional exception of C6, which can be shorter than C7. Each level is larger in the coronal plan than in the sagittal plane. The diameter of each body in the coronal plane is larger than that in the sagittal plane. The superior aspect is concave, and the inferior aspect is convex. Both the superior and inferior aspects contain a shell of cortical bone called the end plate, which eventually transitions into the fibrocartilaginous intervertebral disc. In the coronal plane, the lateral superior aspects of the body demonstrate a lip of bone that projects cranially is congruent with the inferior and lateral aspects of the adjacent cranial body and make up the uncovertebral joint, or joint of Luschka. The uncovertebral joint is important in resisting lateral translation of the vertebra and helps limit lateral bending. Intraoperatively, the uncovertebral joints are important anatomical landmarks that aid in identifying the lateral extent of the body and can act as a reference point for identifying the midline when placing implants during anterior-based procedures.

Pedicles

Projecting dorsolaterally each side of the body is the pedicle, which connects the body to the posterior arch. Unlike the pedicles of the thoracic and lumbar regions, those in the subaxial spine are located midway between inferior and superior end plates of the body on the coronal plane. Descending down the subaxial spine from C3 to C7, the pedicle height increases from an average 5.1 to 9.5 mm, and width increases from 3 to 7.5 mm. Average width and height increase from 5 to 7 mm, respectively (An et al. 1999; Ebraheim et al. 1997; Panjabi et al. 1991b). Additionally, the pedicle angle transitions from 45° to 30° as it descends down the subaxial spine.

Transverse Process

Projecting laterally off of the posterior aspect of the body, anterior to the pedicle, are the bilateral transverse processes. Each transverse process contains both an anterior and posterior tubercle. The anterior tubercle is the origin of the longus colli cervicalis, anterior scalene, ventral intertransverse, and longus colli muscles. The posterior tubercle is the origin of the longissimus, levator scapulae, middle scalene, posterior scalene, splenius cervicalis, and iliocostalis muscles. The anterior tubercle of the C6 transverse process is also referred to as the carotid tubercle and is an important anatomical landmark, as it marks the transverse proves that separates the carotid artery from the vertebral artery. The transverse processes of C1 through C7 each contain a transverse foramen. The vertebral artery and vein travel through the transverse foramen of C1 through C6, and in the majority of cases, not through C7. The transverse foramen is bound by the lateral aspect of the pedicle, the posterior aspect of the anterior tubercle, and the anterior aspect of the posterior tubercle. Additionally, each transverse process contains a groove on its superior surface that runs posterior to the transverse foramen. This groove carries the exiting nerve at the corresponding level after it exits the neural foramen. For example, the groove on the C3 transverse process contains the exiting C3 nerve root.

Facet Joint

Projecting from both the superior and inferior aspects of the lateral mass is an articular process that is congruent with the adjacent articular process of the neighboring vertebra, which together comprises the facet joint. The superior articular process of the vertebra articulates with the inferior articular process of the cephalad vertebra, and the inferior facet of the vertebra articulates with the facet of the superior articular process of the caudal vertebra. The facet joint is a diarthrodial joint with a relatively lax capsule that allows for appropriate motion to occur. In the sagittal plane, these joints are oriented obliquely from anterior-superior to posterior-inferior at approximately 45° (Fletcher et al. 1990). This orientation differs from the relatively vertically oriented facet joints in the coronal plane of the lumbar spine, and in conjunction with the relatively lax capsule, allows for a broad range of motion in flexion, extension, lateral bending, and rotation (Bland 1987). Just as in other diarthrodial joints throughout the body, the facet joint is susceptible to degenerative changes such as joint swelling, cartilage thinning, and osteophyte formation. Given its close proximity to the neural foramina and spinal canal, these changes can have significant clinical implications.

Spinal Canal

The cervical spinal canal is bordered ventrally by the posterior aspect of the vertebral body, ventrolaterally by the pedicles transverse near the location of the neural foramina, laterally by the lateral masses, and posteriorly by the lamina and spinous process. The lateral diameter of the spinal canal is larger than the anterior-posterior (AP) diameter at all levels of the subaxial spine. The AP diameter is approximately 17 mm at the C3 level and decreases to 15 mm at C7, which has the lowest cross-sectional area.

Lateral Mass

Located dorsolateral to the pedicle is a cylindrical piece of bone termed the lateral mass. The lateral mass is analogous to the pars interarticularis of the thoracic and lumbar spinal regions, as it is the structure that connects the superior and inferior articular surfaces that make up the cephalad and caudal facet joints. It is directly dorsal to the midportion of the transverse process, and as such it is in very close proximity to the exiting nerve root. It has a bony projection dorsomedially, which becomes confluent with the lamina. The lateral masses of the subaxial cervical spine have an average depth and width of approximately 13 mm and 12 mm, respectively, and slightly decrease at each descending level to C7 where it is thinnest (Mohamed et al. 2012). Given the relatively small pedicle size in this region of the spine, screw fixation within the lateral mass is sometimes a desired treatment option, and thus a proper understanding of the lateral mass size is crucial in avoiding complications.

Lamina and Spinous Process

The dorsomedial projection from the lateral masses is termed the lamina. As they continue posteriorly bilaterally, they merge to form the posterior-most bony prominence called the spinous process. The C2 through C6 vertebra are normally bifid, but the C7 spinous process is not. The lamina and the spinous process make up the posterior aspect of the spinal canal. An important anatomic landmark is the junction of the lamina and spinous process posteriorly.

Ligaments

The subaxial cervical spine contains an array of ligaments. The ligaments contribute significantly to the stability and alignment of the bony structures in the region and allow for motion in various planes while restricting extremes of motion that could compromise proper anatomic alignment and integrity of the local structures.

The anterior and posterior aspects of the vertebral bodies and intervertebral discs are bound by both the anterior and posterior longitudinal ligaments. The anterior longitudinal ligament (ALL) is composed of longitudinal fibers that run in a cranial and caudal direction spanning the base of the skull to the sacrum. The ALL attaches to the anterior surfaces of the vertebral bodies and intervertebral discs and acts as a restraint to hyperextension of the mobile segments of the vertebral column. The ALL is narrow and thick over the concave surface of the vertebral

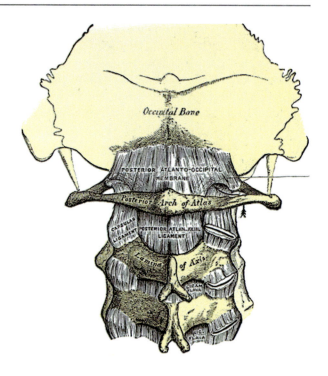

Fig. 5 Posterior representation of the upper cervical vertebra. (From: Henry Vandyke Carter – Henry Gray (1918) *Anatomy of the Human Body*, Plate 305, Public Domain)

bodies but becomes more wide and thin over the discs. The posterior longitudinal ligament (PLL) also spans the length of the vertebral column, fanning out to form the tectorial membrane at its most cranial aspect, and attaches to the sacrum caudally. Just like the ALL, the PLL is more narrow over the bodies and wide over the discs (Parke and Sherk 1989) (Fig. 1).

The ligamentum flavum is a grouping of sequential ligaments located in the posterior segment of the vertebra, with the name arising from the relatively yellow appearance (Fig. 5). Each ligament traverses adjacent lamina, attaching anteriorly near the midportion of the cephalad lamina and running obliquely to attach to the superior most margin of the caudal lamina. These ligaments have a high elastin content that have the propensity to lose their elasticity along with the aging process. In such situations, anterior buckling of the ligaments may occur during extension which may in turn produce a mass effect in the spinal canal and contribute to spinal cord compression.

The ligamentum nuchae is composed of the interspinous and supraspinous ligaments. The interspinous ligament is a relatively thin structure that connects the spinous processes of adjacent vertebra. It runs obliquely from the anteroinferior aspect of the cephalad spinous to the posterosuperior aspect of the caudal spinous process. It is bound by the ligamentum flavum anteriorly and the supraspinous ligament posteriorly. The supraspinous ligament connects the posterior tips of the spinous processes along the length of the vertebral column. However, in the subaxial spine, these two ligaments are less distinct as individual structures until the level of C7 but rather form a complex of thick ligamentous elastic tissue that is referred to the ligamentum nuchae. The ligamentum nuchae runs from the inion of the occiput to the spinous process of C7 and acts as an attachment point for the nuchal musculature in the region.

Intervertebral Disc

The cervical spine contains six intervertebral fibrocartilaginous discs which separate the vertebral bodies (Fig. 1). There is no disc between the occiput and the atlas or between the atlas and the axis. The first disc is located between C2 and the C3 body. The junction of the disc with the adjacent bodies is lined by a cartilaginous layer termed the end plates. The disc itself is composed of two primary components – the nucleus pulposus and the annulus fibrosus. The nucleus pulposus is the centrally located portion of the disc that is the remnant of the primitive notochord and is comprised with primarily type II collagen, proteoglycans, and water. This makeup of the nucleus pulposus results in a gelatinous type substance that allows for force dissipation to the annulus fibrosis and both end plates when compression is applied to the vertebral column. The annulus fibrosus is the component of the disc that surrounds the nucleus pulposus circumferentially and composed of type I collagen, proteoglycans, and water. It is characterized by multiple circumferential layers of fibers that run in an oblique pattern from the cephalad to caudal vertebral bodies. The annulus fibrosus has a high tensile strength that contributes to the stability within a pair of vertebra, which is assisted in the lateral direction by the uncovertebral joint. As the aging process progresses, the margin between the nucleus pulposus and annulus fibrosus becomes more difficult to distinguish (Bland and Boushey 1990). In the coronal view, the superior aspect of the disc is concave, and the inferior aspect is convex as to contour its respective adjacent end plates. The height of the intervertebral disc is slightly larger anteriorly than posteriorly, which contributes to the lordotic curvature of the cervical spine.

Fascia

Investing

The deep cervical layer of the neck is separated into compartments that can be used as landmarks during dissection. The investing layer is the most superficial layer and provides broad coverage to the trapezius posteriorly and wraps around anteriorly to enclose the sternocleidomastoid (SCM) as well. Superiorly, it reaches the hyoid bone and the caudal extent of the mandible and then dives inferiorly to capture both the suprasternal space and form the ceiling of both the ventral and dorsal cervical triangles (Fig. 6).

Pre-tracheal

The next layer is the pre-tracheal layer, which houses many structures and likewise is referred to by multiple names including the middle cervical fascia or the visceral layer. This multifaceted aponeurosis envelops the infrahyoid muscles as well at the omohyoid muscles which lie just superficial to the visceral space. This space is residence to important soft tissue structures such as the thyroid gland, larynx, trachea, and esophagus and deep to this layer run the thyroid vessels. Its superior attachments are the hyoid and thyroid cartilage and inferiorly to the clavicles and sternum. The carotid sheath makes up its lateral margin (Fig. 6).

Prevertebral

The prevertebral layer is a thick fascial plane that surrounds the vertebral column and its muscles. This layer includes the longus and the scalene muscles. The longus colli is a notable structure that aids in establishing midline during an anterior cervical approach. Identifying this structure also helps protect the cervical portion of the sympathetic chain during anterior dissection by retracting laterally. The alar layer is also included as part of the prevertebral layer and encloses the carotid sheath, which houses the vagus nerve, carotid artery, and internal jugular vein (Fig. 6).

Muscles

Ventral

The anterior cervical muscles can be divided into superficial and deep. The platysma is the most superficial layer and is a thin wispy muscle that begins from the mandible spreading inferiorly and laterally to the second rib and acromion process. It has neurovascular bundles that integrate into

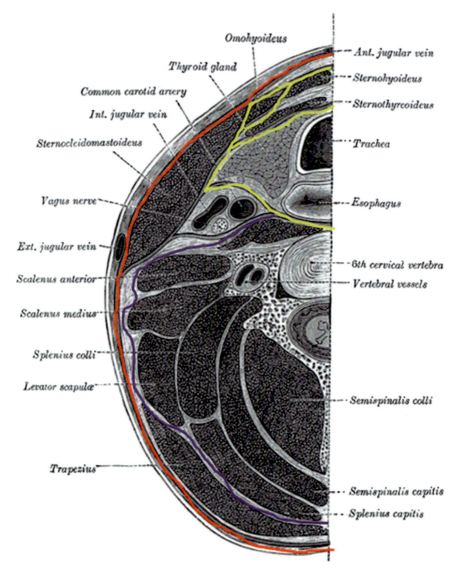

Fig. 6 Axial cuts of the neck at the level of the sixth cervical vertebrae demonstrating fascial arrangements: investing fascia (red), prevertebral fascia (purple), pre- tracheal fascia (yellow). (By Henry Vandyke Carter – Henry Gray (1918) *Anatomy of the Human Body: Gray's Anatomy*, Plate 384, Public Domain)

the skin with its main function aiding in facial expression. Just below this lies the sternocleidomastoid (SCM) which has two heads of origin: the medical clavicle and the sternum that attach to the mastoid and occipital bones. It runs obliquely and functions to turn the head the contralateral side as well as flexion to the ipsilateral side (Fig. 7). It also separates the neck into different triangles, its contents which will be discussed later.

The next layer of muscles is the infrahyoid, scalene, and longus group. The infrahyoid muscles are a group of muscles, as the name implies, that attach to the hyoid bone. These include the mylohyoid, stylohyoid, geniohyoid, digastric, and omohyoid. The strap muscles of the larynx are the sternothyroid and sternohyoid which are important structures to landmark during an anterior approach because they have no direct

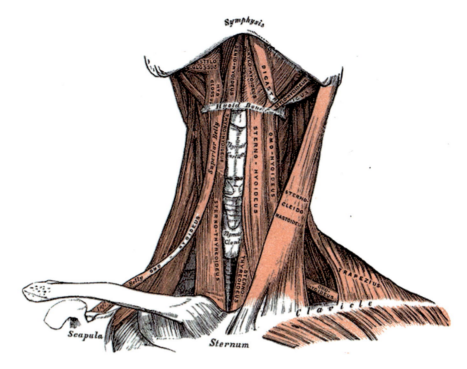

Fig. 7 Superficial ventral muscles of the neck. (By Henry Vandyke Carter – Henry Gray (1918) *Anatomy of the Human Body: Gray's Anatomy*, Plate 386, Public Domain)

involvement in cervical motion. The scalene group is made up of the anterior, medial, and posterior scalene muscles. The anterior muscle originates from the transverse process of the C3 to C6 vertebrae and inserts into the first rib. The medial arises from the posterior transverse process of C2 to C7 and also inserts on the first rib. The posterior scalene muscle has more variable course but originates from the posterior transverse process of C4 to C6 and inserts onto the second rib. This muscle group is well-known for its contribution to thoracic outlet syndrome, which results from neurovascular compression of either the subclavian artery or brachial plexus. The longus muscle group is composed of the longus colli, capitis, and rectus lateralis and as previously discussed are found within the prevertebral fascia. The longus colli originates from the anterior aspect of C3 to C6 and extends obliquely from C1 to T3 to attach onto the anterior atlas. The longus capitis arises from the anterior transverse process of C3 to C6 and attaches on the basilar aspect of the occiput. The rectus has two heads, an anterior and lateral head which originate from the lateral mass of the atlas and transverse process of the atlas, respectively. The anterior head will insert into the base of the occipital bone, while the lateral head will attach to the jugular process of the occiput (Fig. 8).

Dorsal

The dorsal muscle groups provide tension to the vertebrae to keep them in an upright position and deliver balance as well. These muscles are innervated by the dorsal rami. The erector muscles take advantage of the tension band principle to provide sagittal support and symmetric balance in an effort to preserve lordosis of the cervical region. Loss of strength, often attributed to pain, can lead to progressive loss of lordosis and a relative kyphotic deformity. In the coronal plane, the lateral tension bands provide support. Imbalance in any of these

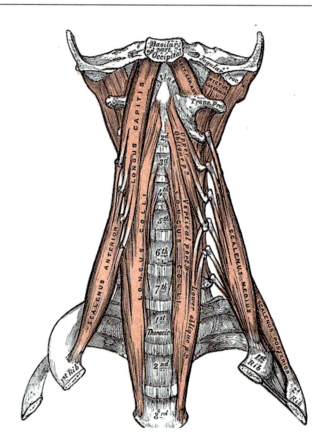

Fig. 8 Deep ventral muscles of the neck. (By Henry Vandyke Carter – Henry Gray (1918) *Anatomy of the Human Body: Gray's Anatomy*, Plate 387, Public Domain)

planes can lead to deformity seen in abnormal cervical spine curves. All the muscles in the dorsal compartment spread out into three layers discussed below.

Superficial Layer

From superficial to deep, this layer includes trapezius, splenius, and levator scapulae. These muscles work synergistically to rotate the head, extend, and laterally bend. The trapezius has a broad origin that extends along the cervical and thoracic spine. Its upper division begins at the occipital protuberance and attaches at the medial 1/3 of the clavicle. The splenius group consists of the capitis and the cervicis. The capitis begins from the ligamentum nuchae and the spinous process of C6 and inserts along the lateral 1/3 of the superior nuchal line and mastoid. The cervicis inserts along the posterior aspects of the transverse process of C1–C4. The levator scapulae originates along the same posterior tubercles of C1–C4 transverse process and insert along the medial border of the scapula between the superior medial angle and scapular spine (Fig. 9).

Intermediate Layer

This layer includes the erector spinae group which has a common original at the iliac crest, sacrum, and lumbar spinous process. These muscles consist of the iliocostalis, the longissimus, and the semispinalis from lateral to medial. They work together to extend and bend the neck in the coronal plane. The iliocostalis group inserts into the posterior tubercles of the C4–C6 transverse process; the longissimus group inserts on the mastoid process. The semispinalis group inserts along the spinous process of the cervical spine (Fig. 10).

Deep Layer

The transversospinalis group makes up the deepest layer and lies along the spinous process and lamina of the cervical spine. They consist of the multifidus and rotator muscles. They are

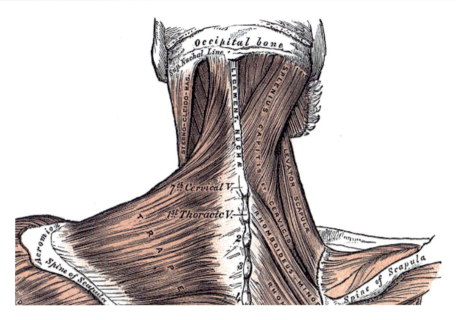

Fig. 9 Superficial dorsal muscles of the neck. (By Henry Vandyke Carter – Henry Gray (1918) *Anatomy of the Human Body: Gray's Anatomy*, Plate 409, Public Domain)

Fig. 10 Intermediate and deep dorsal muscles of the neck. (By Henry Vandyke Carter – Henry Gray (1918) *Anatomy of the Human Body: Gray's Anatomy*, Plate 389, Public Domain)

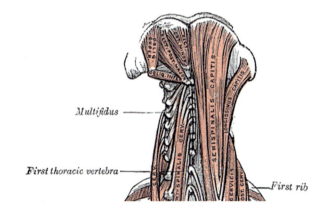

innervated by the dorsal rami of the spinal nerves of the cervical spine (Fig. 10).

Neurovascular Structures

Spinal Cord

Though a detailed description of spinal cord neuroanatomy is beyond the scope of this chapter, basic anatomic understanding is necessary. The spinal cord exits the intracranial space through the foramen magnum and terminates at approximately L2 as the conus medullaris. The spinal cord is widest at C6, measuring an average of 38 mm in circumference, which provides enough space for the increased density in neurologic structures such as the brachial plexus (Parke and Sherk 1989). The inner cord is made up gray matter which houses the nerve cell bodies and branching dendrites. It is separated into anterior, lateral, and posterior segments (horns). The anterior horn contains motor neurons controlling the skeletal muscles and is the column where the cell bodies of the alpha motor neurons are located. The posterior horn contains sensory neurons that

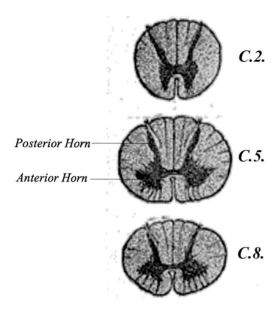

Fig. 11 Transverse section through cervical spinal cord with labeled anterior and posterior horns of the inner gray matter. (From: Henry Vandyke Carter – Henry Gray (1918) *Anatomy of the Human Body*, Plate 666, Public Domain)

transmit sensory information from the body that includes fine touch, proprioception, and vibration (Fig. 11). The lateral segment is located only within the thoracic and upper lumbar regions and contains components of the sympathetic nervous system.

The outer circumferential layer is the white matter and is composed of myelinated axons. In a similar manner to the gray matter, it is separated into posterior, lateral, and anterior columns. The lateral column houses the lateral corticospinal tract which provides efferent motor innervation control ipsilateral extremity motion. Also within the lateral column is the lateral spinothalamic tract which is sensory pathway that transmits contralateral pain and temperature. This tract decussates to the other side of the spinal cord at the anterior white commissure, usually 1–2 spinal nerve segments above the entry point. The posterior column, composed of the fasciculus gracilis and cuneatus, is the structure responsible for ascending sensory signals transmitting proprioception, vibration, and fine touch. Sensory information from this pathway is also from the contralateral extremity although its crossover is much higher, located in the brain stem. The anterior column houses both sensory and motor systems, as well as the anterior spinothalamic tract which is responsible for crude touch.

Meninges and Dura

The meninges enclose the spinal cord and are composed of three layers: the pia, arachnoid, and dura matter. The innermost layer is the pia, followed by the arachnoid and then the dura mater. The pia has lateral projections between exiting nerves that attach to the arachnoid and dura. These projections are known as the denticulate ligaments and with the aid of CSF act as a bolster for the spinal cord. The space between the pia and arachnoid is known as the subarachnoid space and contains CSF and nerve rootlets. The space between the dura and vertebral canal is the epidural space and has a rich venous plexus and adipose tissue (Fig. 12).

Nerve Roots

In the cervical spine, there are eight rootlets that exit the spinal cord that unite and form the dorsal and ventral roots. These form nerve roots at each corresponding level and pass through the dura to the intervertebral foramina. In the cervical spine, the nerve roots pass above the corresponding pedicle, except for the C8 nerve root which travels underneath the C7 pedicle. These nerves also leave the spinal cord at an angle that approximates a right angle and explains why foraminal and central herniations will affect the same nerve root. In the foramen, the nerve root takes up one-third of the space, medially it is located at the caudal portion of the superior articular process and as it travels laterally adopts a more inferior position above the pedicle (Daniels et al. 1986). When the neck is extended, the foramen size decreases in overall volume, and the nerve takes up a more superior position within the foramen; when flexed, the foramen size increases, and the nerve root assumes a position in the caudal half of the foramen (Rauschning 1991). The remaining space is filled with fat, which provides cushion to the nerve (Flannigan et al. 1987).

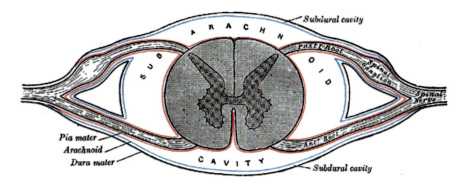

Fig. 12 Transverse section through cervical spine with labeled membranes and spinal nerve roots. (From: Henry Vandyke Carter – Henry Gray (1918) *Anatomy of the Human Body*, Plate 770, Public Domain)

Spinal Cord Blood Supply

The vertebral arteries are the primary blood supply to the cervical spine which branch of the subclavian arteries and ultimately form the basilar artery. In general, each vertebral artery enters the transverse foramen at C6 and courses rostrally until C1 (Rickenbacher et al. 1982). It is important to note, during an anterior approach that the vertebral artery is located in the middle one-third of the vertebral body, just lateral to the uncinated process. At the atlas, the vertebral arteries curve around and enter the foramen magnum to unite with the contralateral artery to become the basilar artery. Throughout their course, they give off feeding branches to the spinal cord known as the anterior and posterior spinal arteries. The anterior spinal artery supplies the anterior two-thirds of the spinal cord, while the posterior spinal artery assumes the remaining one-third.

Venous outflow of the spinal cord consists of three anterior and three posterior veins. The most prominent are the anterior venous structures and are located medial to the pedicles. The posterior venous plexus surrounds the spinal cord.

Important Ventral Structures

Carotid Sheath

The carotid sheath contains the internal jugular vein, the vagus nerve, and the common carotid artery from lateral to medial. A small branch of the hypoglossal nerve can sometimes be seen crossing anteriorly. The common carotid artery branches approximately 1 cm above the superior border of the thyroid cartilage within this triangle. The carotid sinus lies just inferior to the bifurcation and is prominent baroreceptor regulating blood pressure. The vagus nerve, lying just dorsal, gives off two important branches to the neck: the superior and inferior laryngeal nerves (Fig. 13).

Vertebral Artery

The vertebral artery is divided into four segments and has an average diameter of 4.5 mm. It travels medial to the anterior scalene muscles and enters the C6 foramen in roughly 90% of the population. After entering cranially through the foramen transversarium of C2 and C1, it then changes course and heads medially along the superior arch of C1 at which point it goes further cranially into the foramen magnum. During a posterior approach to C1, it is critical to avoid dissection greater than 1.5 cm lateral to the midline as injury to the vertebral artery is greatest in this region (Fig. 14).

Superior Laryngeal Nerve

A branch off the vagus nerve, the superior laryngeal nerve runs medial to the carotid sheath and bifurcates at the level of the hyoid to provide motor function the inferior pharyngeal

Fig. 13 Relevant ventral structures of the neck. (By Henry Vandyke Carter – Henry Gray (1918) *Anatomy of the Human Body: Gray's Anatomy*, Plate 794, Public Domain)

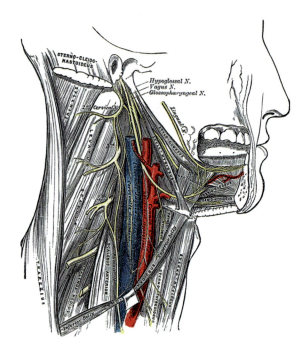

Fig. 14 Internal carotid and vertebral arteries. (By Henry Vandyke Carter – Henry Gray (1918) *Anatomy of the Human Body: Gray's Anatomy*, Plate 513, Public Domain)

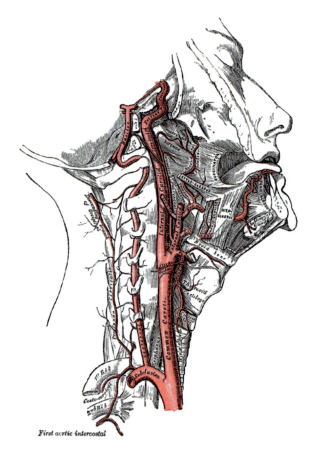

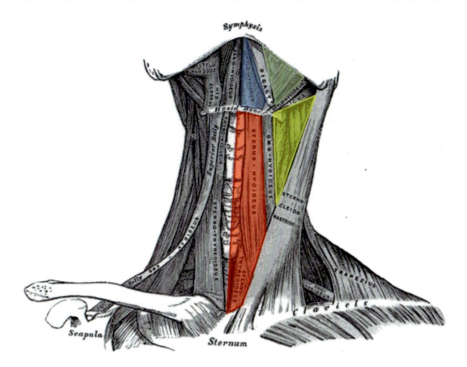

Fig. 15 Ventral cervical triangles: submental triangle (blue), muscular triangle (red), submandibular triangle (green), carotid triangle (yellow). (By Henry Vandyke Carter – Henry Gray (1918) *Anatomy of the Human Body: Gray's Anatomy*, Plate 386, Public Domain)

constrictors. It also has sensory branch that provides sensation to the base of the tongue and the larynx. Injury to this nerve can be manifested with poor gag reflex and voice control especially with high pitches. Loss of the gag reflex can be most debilitating as these patients are often at increased risk for aspiration.

Inferior (Recurrent) Laryngeal Nerve

The inferior laryngeal nerve, commonly known as the recurrent laryngeal nerve, has a U-shaped course in the thorax, specifically in the tracheoesophageal groove. As it pierces the inferior pharyngeal constrictor, it provides motor function to the intrinsic laryngeal muscles. Its course in the neck is not symmetric. On the left, it loops under the aortic arch, and on the right, it loops under the right subclavian artery.

Hypoglossal Nerve

The hypoglossal can be located in the carotid triangle, deep to the belly of the digastric muscle, and as previously discussed, in between the carotid artery and internal jugular vein. Before heading toward the oral cavity to innervate the tongue, it gives off a branch to innervate the strap muscles, which is termed the *ansa cervicalis*.

Sympathetic Chain

The sympathetic chain resides in the prevertebral fascia, just ventral to the longus colli muscles. It surrounds the vertebral artery during its ascension toward the cranial vault. Injury to this structure during an anterior approach can cause ipsilateral Horner syndrome, which is characterized by ptosis, miosis, and anhidrosis.

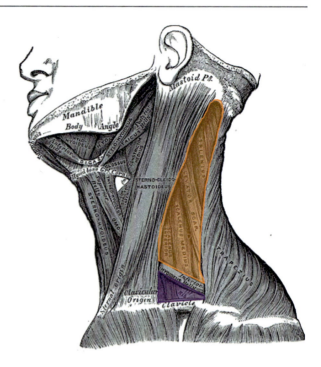

Fig. 16 Dorsal cervical triangles: occipital triangle (orange), subclavian triangle (purple). (By Henry Vandyke Carter – Henry Gray (1918) *Anatomy of the Human Body: Gray's Anatomy*, Plate 385, Public Domain)

Surgical Anatomy: The Cervical Triangles

Ventral (4 Types)

The borders of the anterior cervical triangle are the medial edge of the SCM, the inferior mandibular border, and midline of the neck. Within this triangle reside four subtriangles. The submental triangle is formed by the hyoid and the two anterior bellies of the digastric muscles; the floor of which is made up the two mylohyoid muscles. Next is the submandibular triangle, its margins being the ventral and dorsal bellies of the digastric muscle, the inferior mandibular border with the floor consisting of the hyoglossus, mylohyoid, and middle pharyngeal constrictors muscles. Important to note, is the hypoglossal nerve which passes through this triangle. The carotid triangle is bordered by the anterior margin of the SCM, the superior border of the omohyoid and inferior border of the digastric muscle. This triangle is particularly important as the common carotid artery, internal jugular vein, and the vagus nerve are found within this structure. Last, is the muscular triangle which is formed by the medial margin of the SCM, the superior belly of the omohyoid and median plane of the neck (Fig. 15).

Dorsal (2 Types)

The dorsal triangle is bordered by the lateral edge of the SCM, ventral trapezius border, and middle third of the clavicle. The floor is made up the scalene muscle group and prevertebral fascia, and the ceiling is made up of the deep cervical fascia. This triangle is divided into two smaller triangles, the occipital and subclavian triangle. The external jugular vein runs caudally through this triangle at the angle of the mandible (Fig. 16).

References

An HS, Wise JJ, Xu R (1999) Anatomy of the cervicothoracic junction: a study of cadaveric dissection, cryomicrotomy and magnetic resonance imaging. J Spinal Disord 12:519–525

Bland JH (1987) Disorders of the cervical spine. Saunders, Philadelphia

Bland JH, Boushey DR (1990) Anatomy and physiology of the cervical spine. Semin Arthritis Rheum 20:1–20

Daniels DL, Williams AL, Haughton VM (1983) Computed tomography of the articulations and ligaments at the occipito-atlantoaxial region. Radiology 146:709–716

Daniels DL, Hyde JS, Kneeland JB et al (1986) The cervical nerves and foramina: local-coil MR imaging. AJNR Am J Neuroradiol 7:129–133

Ebraheim NA, Xu R, Knight T et al (1997) Morphometric evaluation of lower cervical pedicle and its projection. Spine 22:1–6

Ebraheim NA, Lu J, Yang H (1998) The effect of translation of the C1-C2 on the spinal canal. Clin Orthop Relat Res 351:222–229

Flannigan BD, Lufkin RB, McGlade C et al (1987) MR imaging of the cervical spine: neurovascular anatomy. AJR Am J Roentgenol 148:785–790

Fletcher G, Haughton VM, Ho KC, Yu SW (1990) Age-related changes in the cervical facet joints: studies with cryomicrotomy, MR, and CT. AJNR Am J Neuroradiol 11:27–30

Mohamed E et al (2012) Lateral mass fixation in subaxial cervical spine: anatomic review. Glob Spine J 2(1): 039–045

Panjabi M, Dvorak J, Crisco J 3rd et al (1991a) Flexion, extension, and lateral bending of the upper cervical spine in response to alar ligament transections. J Spinal Disord 4:157–167

Panjabi MM, Duranceau J, Goel V et al (1991b) Cervical human vertebrae: quantitative three-dimensional anatomy of the middle and lower regions. Spine (Phila Pa 1976) 16:861–869

Parke WW, Sherk HH (1989) Normal adult anatomy. In: Sherk HH, Dunn EJ, Eismont FJ et al (eds) The cervical spine. Lippincott, Philadelphia, pp 11–32

Rauschning W (1991) Anatomy and pathology of the cervical spine. In: Frymoyer JW (ed) The adult spine. Lippincott Williams & Wilkins, Philadelphia, pp 907–929

Rickenbacher J, Landolt AM, Theiler K (1982) Applied anatomy of the back. Springer, Berlin

Thoracic and Lumbar Spinal Anatomy

38

Patricia Zadnik Sullivan, Michael Spadola, Ali K. Ozturk, and William C. Welch

Contents

Thoracic Spine	737
Bullet Points	741
Lumbar Spine	741
Bullet Points	743
Conclusion	743
References	744

Abstract

Degenerative arthropathy, trauma and congenital anomalies, as well as focal abnormalities such as facet overgrowth and disk herniations render each patient unique, and these abnormalities can impact surgical approach. The goal of this chapter is to discuss anatomic considerations that impact surgical planning and to provide a framework for thinking about patient-specific anatomy when approaching the thoracic and lumbar spines.

Keywords

Thoracic spine · Lumbar spine · Degenerative arthropathy · Intraoperative localization

Thoracic Spine

The thoracic spine is composed of 12 rib-bearing vertebrae, separated by intervertebral disks and connected posteriorly by the interspinous ligament (Fig. 1). Each thoracic vertebra has a body, a spinous process, and superior and inferior articulating facets in addition to inferior, superior, and transverse costal facets to articulate with the head of the rib. The pedicles of thoracic spinal vertebrae vary in size, with T1 pedicles being narrow and subsequent pedicles increasing in width approaching the thoracolumbar junction (Fig. 2).

The angulation of the thoracic pedicles further changes, with more caudally oriented trajectories

P. Z. Sullivan (✉) · M. Spadola · A. K. Ozturk · W. C. Welch
Department of Neurosurgery, University of Pennsylvania, Philadelphia, PA, USA
e-mail: Patricia.Zadnik@pennmedicine.upenn.edu; Michael.Spadola@pennmedicine.upenn.edu; Ali.Ozturk@pennmedicine.upenn.edu; William.welch@pennmedicine.upenn.edu

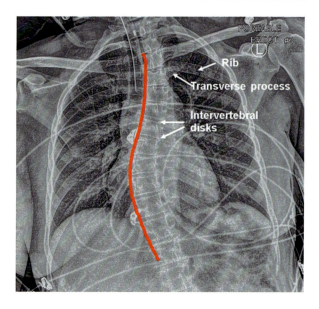

Fig. 1 Anterior-posterior x-ray of the thoracic spine demonstrating coronal scoliosis in the thoracic spine. Ribs, intervertebral disks, and bodies are shown. The red line indicates the lateral aspect of the vertebral bodies

in the upper thoracic spine. Thoracic instrumentation can be challenging due to this variability, and navigation or fluoroscopy-based techniques can assist in planning thoracic pedicle screw trajectories.

The thoracic spine is more rigid than the cervical or lumbar spine, as it is fixed to the sternum via the ribs, limiting the range of motion in the thoracic spine. The cervicothoracic junction and thoracolumbar junctions are more mobile points of transition and are thus more likely to succumb to traumatic pathology. Degenerative pathologies of thoracic spine include sagittal kyphotic deformity, typically secondary to progressive compression fractures, and coronal scoliosis (Fig. 1). Adolescent scoliosis is a common childhood disorder affecting thoracic spine alignment, and in adult patients, iatrogenic or degenerative scoliosis with coronal curvature may affect thoracic spine alignment.

Common traumatic thoracic pathologies include disk herniations and fractures. Thoracic disk herniations may occur with minimal trauma, and over time these disk herniations may become calcified (Oppenlander et al. 2016). If a thoracic disk herniation is noted on magnetic resonance imaging (MRI), computed tomography (CT) can be useful to identify the degree of calcification prior to surgical planning. If a thoracic disk is calcified and causing cord compression and neurological deficit, a transpedicular approach may be required to safely access and drill down the calcified component (Fig. 3).

In older patients with osteoporosis, falls are a common cause of compression fractures in the thoracic spine. Thoracic compression fractures may also be seen in patients with metastatic cancer involving the vertebral bodies. Pain with axial loading (i.e., standing) is a common symptom in thoracic compression fractures. Patients may develop myelopathy from cord compression following fracture or severe, radicular chest pain along the chest wall from neuroforaminal narrowing. Due to the overall stability of the thoracic spine, severe thoracic spine fractures require high velocities such as motor vehicle accidents. If the thoracic spinal column is fractured and displaced, there will likely be associated rib and sternal fractures. Spinal cord transection, although rare, can happen with these types of injuries. Patients with diffuse idiopathic skeletal hyperostosis (DISH) or ankylosing spondylitis (Rustagi et al. 2017) are more likely to experience thoracic spine fractures with low velocity accidents (Fig. 4).

Intraoperative localization in the thoracic spine can be challenging if the lesion is not readily identifiable on standard radiographic studies or

Fig. 2 Illustration showing some morphometric characteristics of the thoracic vertebrae from T1 to T12. The widths of the isthmus of the transverse pedicle are listed on the left. The pedicle entry points (+) and their relationship to the transverse process, laminae, and facets are shown in the center. The transverse pedicle angles are listed on the right. (Reprinted with permission from Hartl et al. 2004, Technique of thoracic pedicle screw fixation for trauma, *Operative Techniques in Neurosurgery*)

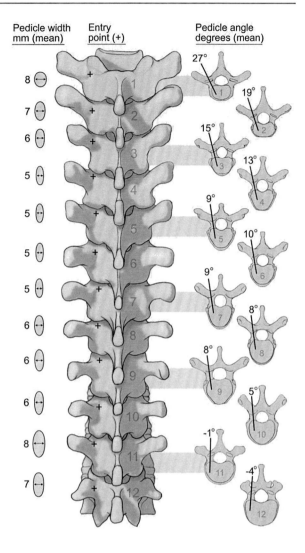

fluoroscopy. Unlike cervical and lumbar spine localization, there is no distinct body (i.e., sacral endplate or dens) available as a reference for counting. Preoperative thoracic and lumbar x-rays may be helpful to determine the true number of ribbed and non-ribbed vertebrae. This may help to correlate with the MRI if the patient has a transitional S1 that may be lumbarized or hypoplastic or an abnormal number of ribbed vertebrae.

Preoperative CT or MRI scans for localization may be obtained prior to surgery incorporating a reference body (i.e., sacral endplate, dens, or other identifiable structure). A radiologist can provide labeling of the localization scan to confirm the precise thoracic body affected. This allows congenital anomalies, such as sacralized lumbar vertebrae, or an abnormal number of rib-bearing vertebrae to be identified. These studies can be correlated with preoperative plain films to reduce the likelihood of wrong-level surgery.

Intraoperative anterior-posterior and lateral fluoroscopic images may be taken to localize the surgical level. Live intraoperative fluoroscopy may also be used for level confirmation, typically by counting up from the sacrum. Intraoperative 3-D imaging, if available, may also help to provide more definitive surgical localization. Prior instrumentation, kyphoplasty cement, or unique fractures may further help to confirm the target level.

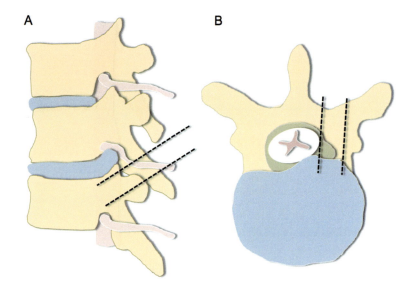

Fig. 3 Transpedicular approach for resection of calcified thoracic disk. (**a**) Lateral view of herniated thoracic disk causing deformation of exiting nerve root. Black lines indicate transpedicular approach to the calcified thoracic disk. (**b**) Axial diagram depicting approach to calcified thoracic disk

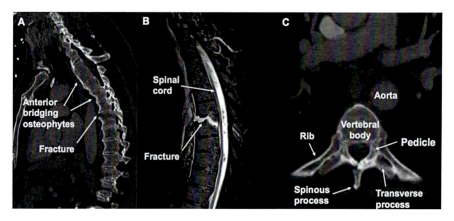

Fig. 4 T8 fracture extending though vertebral body, pedicle, and spinous process in a patient with ankylosing spondylitis. (**a**) Sagittal CT scan demonstrating anterior bridging osteophytes consistent with ankylosing spondylitis and fracture line extending through the T8 vertebral body, pedicle, and spinous process. (**b**) Sagittal T2-weighted fat-suppressed MRI demonstrating T2 hyperintensity along fracture line. The spinal cord can be seen draped ventrally along the posterior aspect of the vertebral bodies. (**c**) Axial CT scan through mid-thoracic vertebra demonstrating relationship of the rib, spinous process, transverse process, and pedicle

It is important to note that technical and patient factors (obesity, surgical position, non-radiolucent OR tables, and others) may interfere with the correct interpretation of these studies. In these instances, consultation with a radiologist should be undertaken.

Direct anterior approaches to the thoracic spine are uncommon, as the aorta and vena cava abut the ventral aspect of the thoracic vertebrae and the lungs and other key structures obstruct direct access (Fig. 5). En bloc resection of thoracic spinal tumors or metastatic pathology may merit anterior approaches (Xu et al. 2009), and these surgeries are often conducted with a cardiothoracic access surgeon. Anterolateral approaches to the thoracic spine include costotransversectomy with removal of the transverse costal facet and rib head for access to the lateral vertebral body and the lateral extracavitary approach, which removes portions of the rib head lateral to the transverse process for greater access to the vertebral body. Careful coordination is necessary when planning

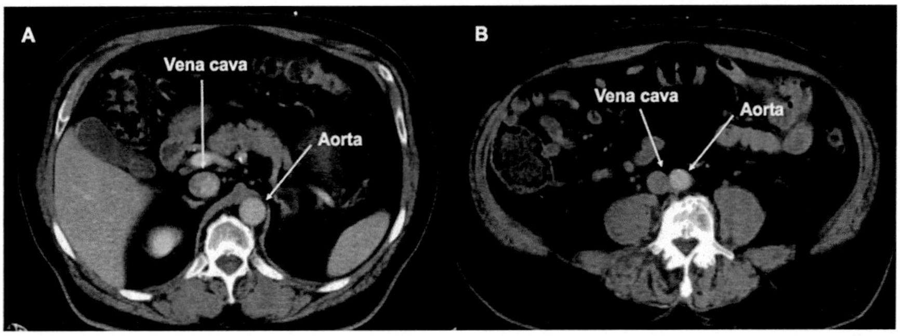

Fig. 5 Sagittal CT scan through (a) T12 and (b) L3 levels illustrating the relationship of the aorta to the thoracic and lumbar spines

an anterior thoracic approach, as the patient may be intubated with a dual-lumen endotracheal tube. This allows the anesthesiologist to selectively hold respirations in the lung adjacent to the surgical field, facilitating access to the vertebral body. Injury to the lung pleura puts patients at an increased risk of pulmonary complications in some studies, and a chest tube may be electively placed to reflate the lung and prevent pneumothorax or large pleural effusions after anterior thoracic spine approaches. Other treatments such as the application of talc powder or mechanical abrasion of the pleural surfaces to promote pleural adhesion may also be considered. Minimally invasive and endoscopic techniques have also been described to reduce complications for anterior spinal surgeries (Borm et al. 2004).

Posterior and posterolateral approaches to the thoracic spine provide limited access to the posterior vertebral body for debulking of metastatic tumors and decompression of fracture fragments. Following laminectomy, unilateral or bilateral pedicles can be resected via careful drilling to access the ventral vertebral body. In the thoracic spine below T2, nerve roots do not provide significant motor contributions, and these roots may be sacrificed lateral to the dorsal root ganglion to further expand the exposure and improve access to the ventral disk space. Nerve root avulsion or compression should be avoided as this may result in postoperative radicular pain.

For calcified disk herniations, pedicle resection can be an effective way to access the disk space. If the pedicle is sacrificed for access, unilateral or bilateral posterior fusion may be required to limit segmental motion and collapse. Posterior fusion of the thoracic spine typically involves placement of pedicle screws, and preoperative evaluation should take into account pedicle length and width. Medialized screws in the thoracic spine result in cord compression, while lateralized screw trajectories can incorporate rib or injure thoracic viscera. Intraoperative navigation is a useful tool to decide optimal screw trajectory.

Bullet Points

- Ventral disk herniations in the thoracic spine may be calcified.
- Thoracic spine localization requires careful preoperative planning.
- Unilateral nerve roots T2-12 can be sacrificed to improve surgical exposure in posterior and posterolateral approaches.

Lumbar Spine

The lumbar spine is composed of five vertebrae in lordotic alignment, joined by intervertebral disks and posteriorly via facet joints (Fig. 6). Lumbar vertebrae are composed of a body, two pedicles, two transverse processes, and a superior and inferior articulating facet. The transverse processes project laterally and may become fractured during an assault or trauma, leading to musculoskeletal discomfort. There is no load-bearing function of

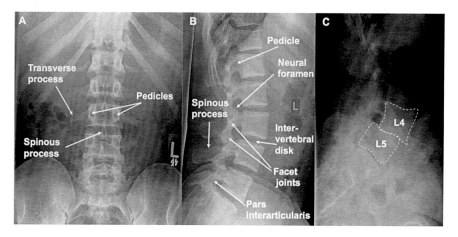

Fig. 6 (**a**) AP and (**b**) lateral x-rays of the lumbar spine illustrating the transverse process, pedicles, spinous process, facet joints, and neural foramen of the lumbar spine. (**c**) Lateral x-ray of the lumbar spine of a different patient, illustrating spondylolisthesis of L4 on L5. The vertebral bodies are outlined with dotted white lines

the transverse processes, and these do not need to be repaired in the case of fracture. The iliopsoas muscle attaches at the transverse processes along the lumbar spine and inserts on the trochanter of the femur. During posterior instrumented fusion, the transverse processes may serve as a surface to encourage fusion. The midportion of the transverse process, as determined in a superior-inferior direction, generally correlates with the midpoint of the lumbar pedicle (again as determined in the superior-inferior axial plane). This landmark can be used to help localize the starting point for pedicle probe or drill insertion.

Intraoperative localization for lumbar spine surgery is typically achieved with lateral radiographs, and the levels are identified by counting from the L5 to S1 disk space. In some patients the fifth lumbar vertebrae may be sacralized, meaning its orientation mimics a typical S1 caudal orientation. This anatomic variant should be identified prior to surgery, as it affects intraoperative localization. Spina bifida occulta may be recognized in patients undergoing evaluation for other spinal issues, and laminar defects may be identified prior to surgery. Another congenital anomaly in the lumbar spine is a pars defect. In this variant, the pars interarticularis (the bony bridge between the superior and inferior articulating facets) fails to develop. The pars interarticularis resists the vector of anterolisthesis, and when the pars is compromised, patients may be at increased risk of developing progressive spondylolisthesis (Fig. 6).

Many patients with lumbar stenosis experience worsening of symptoms with axial loading and ambulation. In contrast, MRI and CT images are traditionally acquired in a supine position. Standing, 36-in. x-rays with neutral leg position (i.e., no compensatory knee bend) provide a more accurate illustration of a patient's global alignment. Spinopelvic parameters can help surgeons to establish how much correction is needed if a lumbar fusion surgery is planned (Celestre et al. 2018). Among these measurements, pelvic incidence and lumbar lordosis are particularly relevant to lumbar spinal anatomy. Lumbar lordosis is measured by the angle between the lower T12 endplate and S1 endplates (Fig. 7).

The pelvic incidence is measured as the angle between a line drawn perpendicular to the center of the S1 endplate and a second line from the center of the S1 endplate to the center of the femoral heads. A patient's lumbar lordosis should be comparable (within 10°) to their pelvic incidence; otherwise an iatrogenic "flat-back" deformity of the lumbar spine may occur.

Common lumbar pathologies include lumbar stenosis, facet arthropathy, spondylolisthesis, and disk herniations (Issack et al. 2012). The posterior facets are prone to degenerative arthritis from

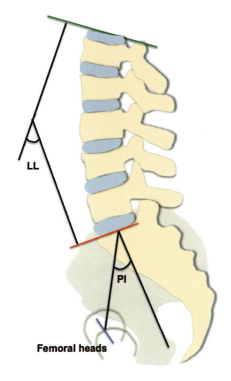

Fig. 7 Pelvic incidence (PI) and lumbar lordosis (LL). The green line indicates the interior endplate of T12 and the superior endplate of S1 (red line). Pelvic incidence is depicted as the angle subtended from the perpendicular line to the S1 endplate and the midpoint of a line drawn (blue) between the femoral heads

repeated abnormal motion, leading to facet overgrowth and synovial cyst formation (Fig. 8). These arthritic changes can compress the spinal canal and contribute to lumbar stenosis. Patients will manifest with radicular symptoms or lumbar claudication in cases of severe stenosis. In cases of spondylolisthesis (Fig. 8), subadjacent vertebral bodies may angle away from the surgeon, complicating the initial dissection. If facet overgrowth is suspected, removal of facet osteophytes may be necessary to identify the anatomic laminar edge.

Posterior approaches to the lumbar spine include laminectomy, hemilaminectomy with discectomy, and instrumented lumbar fusion. Hemilaminectomy is appropriate for patients with unilateral symptoms and a focal disk herniation; however, if the disk herniation is large or central (Fig. 9), a full laminectomy may be completed. If decompression without fusion is planned, the surgeon should avoid manipulation of the facet joint, as removal of the bone may disrupt facet integrity and lead to progressive instability. When fusion is planned, removal of facet overgrowth via rongeur or drilling can improve the surgical exposure and help to identify the laminar edge.

Anterior approaches to the lumbar spine are typically achieved with the help of an access general surgeon or vascular surgeon. The anterior lumbar interbody fusion (ALIF) involves removal of the intervertebral disk, placement of a disk replacement, and securing the disk replacement with an anterior plate and screws (Phan et al. 2017). This approach is complicated by the presence of the lumbosacral plexus (L5-S1) and the iliac bifurcation (L4-5). Anterior approaches may be utilized for patients with failed posterior fusion or in patients with severe deformity requiring anterior and posterior instrumentation to facilitate strength and reduce the risk of failure. Anterior lumbar approaches can be challenging in obese patients if the abdominal girth exceeds the length of surgical instruments. Further, retrograde ejaculation is a reported complication in men at a rate of 7.4–9.8% following manipulation of the lumbosacral plexus in ALIFs, thus compromising male fertility (Lindley et al. 2012).

Bullet Points

- Lumbar degenerative arthropathy can obscure the laminar edge on initial dissection.
- Sacralized lumbar vertebra can complicate intraoperative localization.
- Anterior lumbar fusions may be challenging in obese patients and carry the risk of retrograde ejaculation in male patients.

Conclusion

Careful patient examination and review of available preoperative imaging is crucial for success in spine surgery. CT scans provide key information regarding calcifications, and MRI scans are necessary to identify disk herniations. Prior to

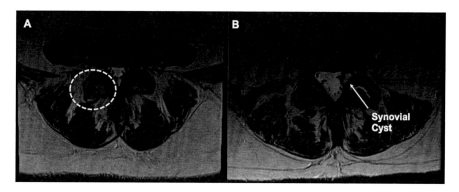

Fig. 8 Axial T2-weighted MRIs of the lumbar spine demonstrating (**a**) spine facet arthropathy with osteophytes noted at the superior and inferior articulating facets. (**b**) Synovial cyst with impingement of spinal canal

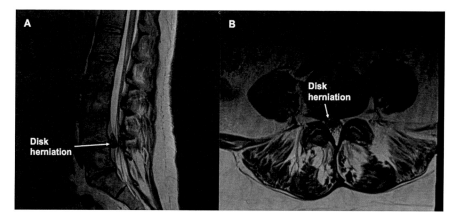

Fig. 9 Right-sided disk herniation causing compression of spinal canal and cauda equina nerve roots. (**a**) Sagittal T2-weighted MRI demonstrating compression of spinal canal and nerve roots. Swelling of nerve roots is noted caudal to the disk herniation. Heterogeneous T2 signal abnormality can be seen within the vertebral bodies of adjacent spinal levels

entering the operating suite, technical aspects of the surgery should be decided, including the type of surgical instrumentation if indication. Patient anatomical anomalies, such as sacralized vertebrae, abnormal rib-bearing vertebrae, osteophytes, and overgrown facets, should also be reviewed in detail. Upright or standing radiographs can complete the picture, as they may highlight loss of lordosis or spondylolisthesis. While knowledge of general spinal anatomy is crucial to form the foundation of spine surgery, patient-specific details must be considered to ensure the optimal outcome.

References

Borm W, Hubner F, Haffke T, Richter HP, Kast E, Rath SA (2004) Approach-related complications of transthoracic spinal reconstruction procedures. Zentralbl Neurochir 65:1–6

Celestre PC, Dimar JR 2nd, Glassman SD (2018) Spinopelvic parameters: lumbar lordosis, pelvic incidence, pelvic tilt, and sacral slope: what does a spine surgeon need to know to plan a lumbar deformity correction? Neurosurg Clin N Am 29:323–329

Issack PS, Cunningham ME, Pumberger M, Hughes AP, Cammisa FP Jr (2012) Degenerative lumbar spinal stenosis: evaluation and management. J Am Acad Orthop Surg 20:527–535

Lindley EM, McBeth ZL, Henry SE, Cooley R, Burger EL, Cain CM, Patel VV (2012) Retrograde ejaculation after anterior lumbar spine surgery. Spine (Phila Pa 1976) 37:1785–1789

Oppenlander ME, Clark JC, Kalyvas J, Dickman CA (2016) Indications and techniques for spinal instrumentation in thoracic disk surgery. Clin Spine Surg 29:E99–e106

Phan K, Lackey A, Chang N, Ho YT, Abi-Hanna D, Kerferd J, Maharaj MM, Parker RM, Malham GM, Mobbs RJ (2017) Anterior lumbar interbody fusion (ALIF) as an option for recurrent disc herniations: a systematic review and meta-analysis. J Spine Surg 3:587–595

Rustagi T, Drazin D, Oner C, York J, Schroeder GD, Vaccaro AR, Oskouian RJ, Chapman JR (2017) Fractures in spinal ankylosing disorders: a narrative review of disease and injury types, treatment techniques, and outcomes. J Orthop Trauma 31(Suppl 4):S57–S74

Xu R, Garces-Ambrossi GL, McGirt MJ, Witham TF, Wolinsky JP, Bydon A, Gokaslan ZL, Sciubba DM (2009) Thoracic vertebrectomy and spinal reconstruction via anterior, posterior, or combined approaches: clinical outcomes in 91 consecutive patients with metastatic spinal tumors. J Neurosurg Spine 11:272–284